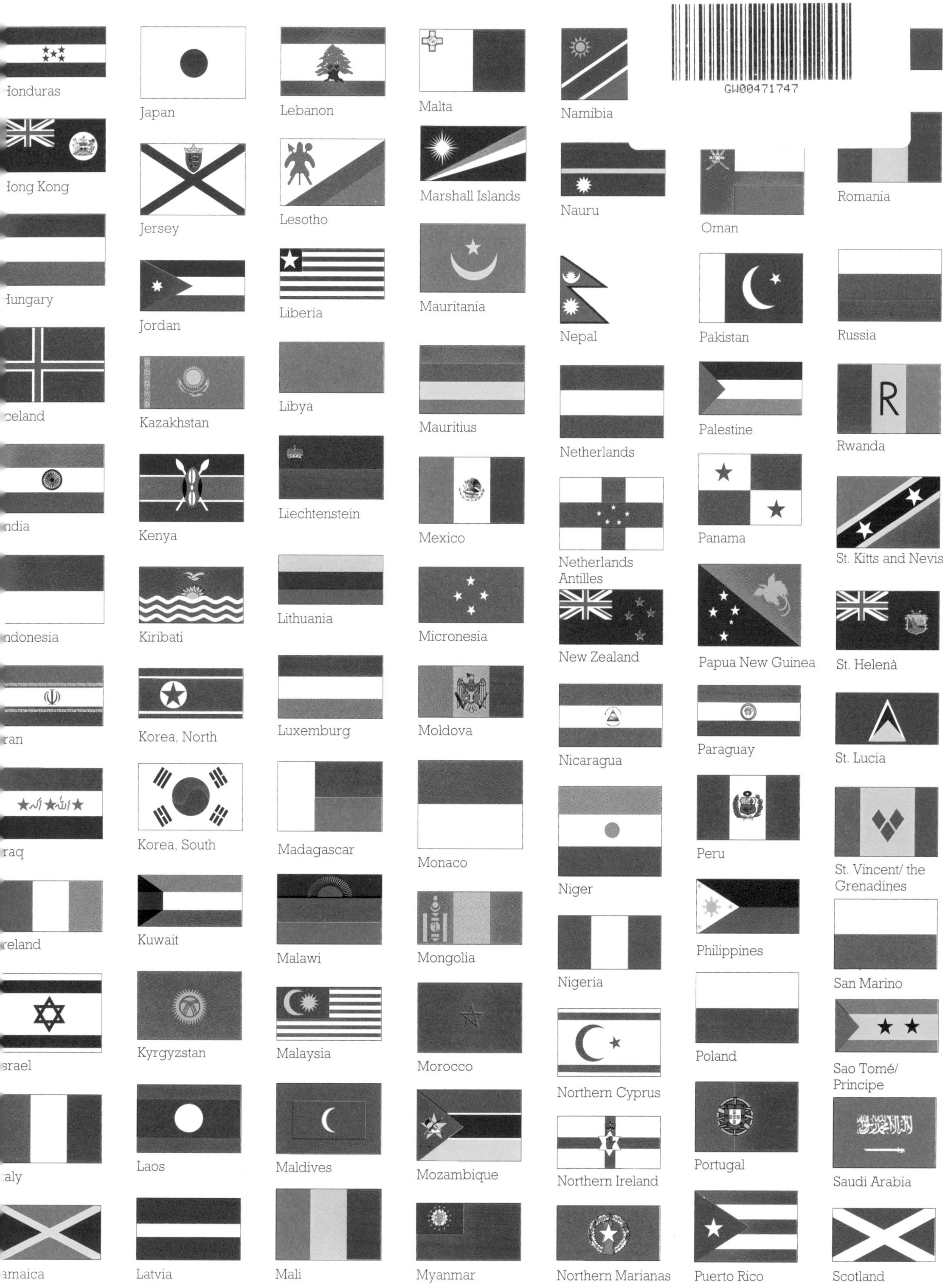
Ionduras
Iong Kong
Iungary
celand
ndia
ndonesia
ran
raq
reland
srael
aly
amaica
Japan
Jersey
Jordan
Kazakhstan
Kenya
Kiribati
Korea, North
Korea, South
Kuwait
Kyrgyzstan
Laos
Latvia
Lebanon
Lesotho
Liberia
Libya
Liechtenstein
Lithuania
Luxemburg
Madagascar
Malawi
Malaysia
Maldives
Mali
Malta
Marshall Islands
Mauritania
Mauritius
Mexico
Micronesia
Moldova
Monaco
Mongolia
Morocco
Mozambique
Myanmar
Namibia
Nauru
Nepal
Netherlands
Netherlands Antilles
New Zealand
Nicaragua
Niger
Nigeria
Northern Cyprus
Northern Ireland
Northern Marianas
Oman
Pakistan
Palestine
Panama
Papua New Guinea
Paraguay
Peru
Philippines
Poland
Portugal
Puerto Rico
Romania
Russia
Rwanda
St. Kitts and Nevis
St. Helenà
St. Lucia
St. Vincent/ the Grenadines
San Marino
Sao Tomé/ Principe
Saudi Arabia
Scotland

OXFORD
ILLUSTRATED
ENCYCLOPEDIA

Volume 9

INDEX
AND
READY REFERENCE

OXFORD ILLUSTRATED ENCYCLOPEDIA

Volume 9

INDEX AND READY REFERENCE

OXFORD
OXFORD UNIVERSITY PRESS
NEW YORK MELBOURNE
1993

Oxford University Press, Walton Street, Oxford OX2 6DP
Oxford New York Toronto
Delhi Bombay Calcutta Madras Karachi
Kuala Lumpur Singapore Hong Kong Tokyo
Nairobi Dar es Salaam Cape Town
Melbourne Auckland Madrid
and associated companies in
Berlin Ibadan

Oxford is a trade mark of Oxford University Press

First published as a set 1993

British Library Cataloguing in Publication Data
Data is available from the British Library

Library of Congress Cataloging in Publication Data
Vol. 5 has full title: Oxford illustrated encyclopedia of the arts;
v. 6 has title: Oxford illustrated encyclopedia of invention and technology.
Contents: v. 1. The physical world / volume editor, Sir Vivian Fuchs—[etc]—v. 6. Invention and technology / volume editor, Sir Monty Finniston— — v. 9. Index and ready reference.
1. Encyclopedias and dictionaries. I. Judge, Harry George. II. Toyne, Anthony. II. Title: Oxford illustrated encyclopedia of the arts. III. Title: Oxford illustrated encyclopedia of invention and technology.
AE5.094 1985 032 85-4876
ISBN 0-19-869174-2 (Volume 9)
ISBN 0-19-869223-4 (set)

Text processed by Oxuniprint, Oxford University Press
Printed in Hong Kong

General Preface

The *Oxford Illustrated Encyclopedia* is designed to be useful and to give pleasure to readers throughout the world. Particular care has been taken to ensure that it is not limited to one country or to one civilization, and that its many thousands of entries can be understood by any interested person who has no previous detailed knowledge of the subject.

Each volume has a clearly defined theme made plain in its title: no previous knowledge is required, there is no jargon, and references to other volumes are avoided. Nevertheless, taken together, the eight thematic volumes (and the Index and Ready Reference volume which completes the series) provide a complete and reliable survey of human knowledge and achievement. Within each independent volume, the material is arranged in a large number of relatively brief articles in A–Z sequence, varying in length from fifty to one thousand words. This means that each volume is simple to consult, as valuable information is not buried in long and wide-ranging articles. Cross-references are provided whenever they will be helpful to the reader.

The team allocated to each volume is headed by a volume editor eminent in the field. Over four hundred scholars and teachers drawn from around the globe have contributed a total of 2.4 million words to the Encyclopedia. They have worked closely with a team of editors at Oxford whose job it was to ensure that the coverage and content of each entry form part of a coherent whole. Specially commissioned artwork, diagrams, maps, and photographs convey information to supplement the text and add a lively and colourful dimension to the subject portrayed.

Since publication of the first of its volumes in the mid-1980s, the *Oxford Illustrated Encyclopedia* has built up a reputation for usefulness throughout the world. The number of languages into which it has been translated continues to grow. In compiling the volumes, the editors have recognized the new internationalism of its readers who seek to understand the different cultural, political, technological, religious, and commercial factors which together shape the world view of nations. Their aim has been to present a balanced picture of the forces that influence peoples in all corners of the globe.

I am grateful alike to the volume editors, contributors, consultants, editors, and illustrators whose common aim has been to provide, within the space available, an Encyclopedia that will enrich the reader's understanding of today's world.

HARRY JUDGE

Contents

Figures and places in contemporary religions, and in legend and mythology

This section contains a number of additional entries on figures and places relating to the major religions, and to those of oral tradition found around the world. The historical information given in this volume supplements the accounts of the major religions and their practices provided elsewhere, which the reader will readily find by using the Index to the Encyclopedia.

Also included in this section are entries on people and places which are considered to be mythological, but which form an integral part of our cultural heritage (such as the Greek and Germanic pantheon of gods). The genealogy of supernatural beings of myth and legend is a fluid subject since, for example in Oceanic tradition, the ancestry of a well-known figure may mutate according to its narrative source. Many tenets of belief and tradition cannot be conclusively defined, since legends have for centuries been re-composed in every generation.

A further category of entries to be found here is that of beings such as angels, djinns, fairies, and demons which are encountered in folklore and religion around the world. Many are described here, especially those familiar through literature and iconography. Linked images often co-exist side by side, for example archangels in Judaism, Christianity, and Islam. The entries in this volume illustrate the rich diversity of traditions the world over.

A

Aaron, a biblical figure, the elder brother of Moses and leader of the Israelite tribe of Levi. Remembered as the traditional founder and head of the Jewish priesthood, he was assistant and spokesman to Moses, with whom he led the Israelites out of slavery in Egypt. He wielded a magic rod which brought ten plagues upon the Egyptians, and which blossomed miraculously before the Ark of the Covenant. During Moses' long absence on Mount Sinai, Aaron erected a golden calf which was idolatrously worshipped by the people. Hārūn (Aaron) is accorded an important place in the Koran as a prophet. In Islamic legend Moussa (Moses) and Hārūn ascended Mount Horeb, not knowing which of them was to die. The coffin they found fitted Hārūn and he was taken to heaven in it.

Abel, a biblical figure, the second son of Adam and Eve. Abel, a shepherd, was killed by his brother, Cain, who was jealous that God had accepted Abel's sacrifice whereas his own had been rejected. In Christian tradition Abel is regarded as the first martyr. The story of Ḥabil (Abel) and Kabil (Cain) is also elaborated upon in Islamic legend.

Abla *'Antar.

Abraham, the first patriarch of the Hebrew nation, who is also revered by Christians and Muslims. Acting in accordance with divine inspiration, Abraham left his homeland in Mesopotamia to become a nomadic tent-dweller in Canaan. God made a covenant or compact with him, promising him a son by his wife Sarah and innumerable descendants with a land of their own. At over 90 Sarah bore Isaac. God subsequently tested Abraham's faith by asking him to sacrifice Isaac to him. When Abraham showed himself willing to do even this, God substituted a ram at the last moment. Arab Muslims call Ibrahim (Abraham) 'father', considering him their ancestor through Ishmael (Ismail), his son by Hajar (Hagar), the maidservant of Sarah. In Muslim tradition it was Ishmael who was almost sacrificed; Ibrahim was the Prophet Messenger (*nabī rasūl*) who established the primordial monotheistic faith which the Prophet Muḥammad came to restore in the form of Islam. Ibrahim is frequently mentioned in the Koran, where it is recounted that when the people tried to burn him for attacking their idols, God cooled the flames and delivered him.

Absalom, a biblical figure, the third and favourite son of David, king of Judah. Exiled for the murder of his half-brother Amnon and eventually forgiven, Absalom later led a rebellion against his father. During the battle, Absalom was found by his cousin, Joab, caught by his hair in an oak tree; despite David's command that he should not be harmed, Joab killed him.

Achilles, in Greek mythology, the greatest warrior on the Greek side in the Trojan War, during which he slew the Trojan champion, Hector, in single combat. He was the son of Peleus, king of the Myrmidons, and Thetis, a sea-nymph. His mother dipped him in the River Styx to make him immune to wounds, but the heel by which she held him remained vulnerable and he was killed by an arrow, shot by Paris, that struck him there.

Actaeon, in Greek mythology, a young hunter who inadvertently saw the goddess Artemis bathing naked and in punishment was turned by her into a stag and killed by his own dogs.

'Ad, a legendary southern Arabian people mentioned in the Koran to whom the prophet Hud was sent. Their ancestor 'Ad was allegedly descended from Nuh (Noah) through Sam (Shem), and is said to have lived more than two thousand years, and to have fathered four thousand children. One of his heirs, Shaddad, is renowned for his building of an earthly paradise, the legendary Iram of the Columns, which contained a hundred thousand palaces. His pride at this achievement was said to have been punished by God making it disappear before he and his court could occupy it.

Adam, according to Hebrew, Christian, and Islamic belief, the first man and consort of Eve. In the account given in the Book of Genesis he was created on the sixth day in the image and likeness of God, commanded to multiply, and given dominion over the earth. The word *adam* is used as a generic term for mankind in Hebrew and in several other Middle Eastern languages. In the Talmud, Lilith is said to be Adam's first wife, dispossessed by Eve. In Islam, Adam is God's vice-regent, created from potter's clay, as recorded in the Koran and greatly elaborated upon in Islamic legend. He is regarded as the first Prophet Messenger (*nabī rasūl*), and, according to one tradition, the original builder of the sacred shrine, the Ka'bah in Mecca. Some Muslims believe Adam did penance in Sri Laṅkā (where one of the peaks is named after him) following his expulsion from Paradise and before being reunited with Hawwa (Eve) in Mecca.

ad-Dajjāl (Arabic, 'the deceiver' or 'imposter'), the Muslim antichrist. According to Islamic tradition there will be a number of false prophets in history, but the last and greatest will be the red-faced and one-eyed al-Masih ad-Dajjāl, who will appear in a time of great turmoil and seek to lead people astray. He will be destroyed by Jesus or the Mahdi, and the Day of Resurrection and Judgement will follow.

Adonis, in Greek mythology, an extremely beautiful youth with whom the goddess Aphrodite fell in love. While hunting he was killed by a wild boar, but because of Aphrodite's grief he was restored to life for part of each year.

Aeneas, in Roman mythology, a Trojan prince, son of the goddess Aphrodite and Anchises. He played a prominent role in the defence of Troy against the Greeks, but when the city fell, he escaped to Carthage. There he fell in love with Dido, the widowed queen, but abandoned her in order to found the state of Rome.

Aesir and Vanir, in Norse and Germanic mythology, two tribes of gods, of whom the Vanir were generally the subordinate. Traditionally, the gods were often at war with one another, sometimes taking hostages in battle. Chief among the Aesir were Odin, served by the Valkyries, and his consort Frigga, whose name survives in 'Friday'. Other important gods were their son Balder, Tyr, god of war, Thor, god of thunder, and Loki. Loki, who was apparently hermaphrodite and could bear as well as beget children, produced three evil progeny: Hel, the goddess of death, Jörmungand, the serpent that lies coiled around the earth, and Fenir, the wolf. The Vanir were gods of fertility and riches; among their leaders were Njörd, god of the sea and of riches, and his children Freyr, protector of all living things, and his sister Freya, goddess of love and of fertility. Both tribes of gods will meet their doom at Ragnarök (the twilight of the gods).

Agamemnon, in Greek mythology, king of Mycenae and the commander-in-chief of the Greeks during the Trojan War. To get fair winds for sailing to Troy he sacrificed his daughter Iphigenia to Artemis. After his return from the war he was murdered by his wife, Clytemnestra, and avenged by his son, Orestes, who was urged on by his sister, Electra.

Agni, the Indo-Aryan god of fire and of priests, and the bearer of the oblation to the gods in the smoke of the sacrifice. Born from a lotus created by Brahma, Agni was also regarded as the protector

and friend of mankind, the giver of rain, and the bestower of immortality, who cleansed man from sin after death. Agni is now invoked by Hindu lovers and by men for virility. He is described as a red figure with two faces and seven tongues. His vehicle is a ram or goat, a symbol of creative energy.

Ahriman (Angra Mainyu), in the *Avesta*, the sacred book of the Zoroastrians, the destructive, evil spirit and opponent of the spirit of wisdom, Ahura Mazda, by whom, after 9,000 years, it is believed he will eventually be overcome. Ahriman and his followers, the malevolent *devas*, whom he created as counterparts to Ahura Mazda's good spirits, the *amesha spentas*, are considered the source of all evil, death, suffering, and destruction in the world. Ahriman's principal epithet is *Druj* (Ancient Persian, 'the lie').

Ahura Mazda (Ormazd), the creator god of ancient Iran, in particular of Zoroastrianism. Elevated to a supreme position by the Achaemenian dynasty (*c*.550–330 BC), Ahura Mazda was considered to be the creator of the universe and the cosmic order who, according to the prophet Zoroaster (7th–6th century BC), created the twin spirits of Spenta Mainyu, who chose truth, light, and life, and Angra Mainyu (see *Ahriman), who chose the lie, darkness, and death. The earth is the battleground between the two. In the sacred book of the Zoroastrians, the *Avesta*, Ahura Mazda became identified with Spenta Mainyu; later sources made him and Ahriman the twin offspring of Zurvan (Time).

Aisha bint Abi Bakr (613–678), in Islam, the favourite wife of the Prophet Muḥammad and daughter of the first caliph, Abū Bakr. Aisha is revered by Sunnite Muslims, many of whose *ḥadīths* (traditions) rely on her testimony. Shī'ite dislike of her stems from an incident where she was suspected of adultery and the caliph 'Alī advised Muḥammad to divorce her. It is believed that a revelation cleared her name and established the law whereby accusations of adultery required four witnesses. This incident is said to have caused implacable hostility between 'Alī and Aisha, culminating in her active opposition to 'Alī's caliphate in the Battle of the Camel (656), so named because Aisha was mounted on a camel.

Ajax, in Greek mythology, a warrior at the Trojan War, second in prowess on the Greek side only to Achilles. After Achilles' death, Ajax coveted his armour, and when this was awarded to Odysseus he went mad with resentment and killed himself.

Ala (Ana, Ali, Ane), the most notable goddess of the Ibo pantheon in Nigeria, West Africa. Daughter of the great god Chuku, Ala is a mother-earth deity who dispenses fertility, and receives the dead into her womb (a pocket). She is also the guardian of morality; laws and oaths are invoked in her name. Shrines to Ala often have a statue of her bearing a child.

Aladdin, in the oriental *Tales of the Thousand and One Nights*, based on the lost book of Persian folk tales, *Hazar Afsanah* (*c.* AD 800). Aladdin is depicted as the ne'er-do-well son of a poor Chinese tailor. By rubbing his Enchanted Lamp, acquired from a sorcerer, Aladdin is able to summon a genie (a marid, the most powerful of jinn), who enables him to acquire great wealth, build a palace and marry the Chinese sultan's daughter. They are transported to Africa, where the sorcerer recovers his lamp by bartering new ones for old, but eventually return to China after several adventures.

Alcestis, in Greek mythology, the wife of King Admetus of Thessaly. She was a paragon of wifely love and virtue and offered to take her husband's place when he was about to die. However, Hercules (Heracles) followed her to the Underworld and rescued her.

Alchera (also, Altjira, Altjiranga, Alcheringa, Wongar, or Djugururba) (Australian Aboriginal, 'Dream time'), in Australian Aboriginal belief, a mythological period of time during which the natural environment was shaped and mythic beings began to walk the earth. Some were responsible for creating human life, which shares a common life force with its creators and with all nature. The dream time has no foreseeable end, and the mythic beings, transsubstantiated and metamorphosed into natural features such as rocks, water holes, or ritual objects, are eternal. As a philosophy, the Alchera regards mankind and nature as one corporate whole, sometimes expressed symbolically in totem form.

Alexander the Great (Arabic, Iskandar) (356–303 BC), king of Macedonia, regarded as a god in his life-time, and an important figure in Greek, Christian, Jewish, and Islamic legend. Renowned as a great general, Alexander is a favourite figure in medieval legends and in European literature and art. As a prophet-like figure and considered a believer, he has assumed particular significance in Islam. As the *Dhu'l-Qarnayn* (he-of-the-two-horns/centuries) in the Koran he saved people from the depredations of Yajuj and Majuj. In Persian sources, his search for knowledge takes precedence over world conquest. In the *Iskandar-nāmah* (Book of Alexander) by the Persian poet Nizami, Alexander is depicted as the half-brother of the conquered King Darius, and therefore a legitimate successor to him.

Ali Baba, a hero of an oriental tale, generally regarded as one of the *Tales of the Thousand and One Nights*, a collection of stories derived from Indian, Persian, and Arabic sources. Chancing to see a band of forty robbers entering a rock face by means of a magic password, 'Open, Sesame!', Ali Baba discovers their treasure in a cave behind the rock and takes it home. The robbers find out that he has stolen their hoard, and the captain of the robbers has his men brought to the house of Ali Baba concealed in leather oil jars, intending to kill him in the night. They are defeated by the slave girl Morgiana, who destroys them with boiling oil.

'Alī ibn Abi Talib (*c*.600–661), the fourth caliph of Islam, cousin and son-in-law of the Prophet Muḥammad through his marriage to the Prophet's daughter Fatimah, and father of Ḥasan and Ḥusain. Revered in particular by Shī'ite Muslims as the rightful successor to Muḥammad, and as the first of their Imāms, 'Alī is renowned throughout the Islamic world for his piety, courage, and learning. Numerous Shī'ite and Sufi legends recount 'Alī's miraculous feats with his two-edged sword *Dhu al-faqar*, and the beauty of his eyes is celebrated in Persian poetry. By some Muslim sects, rejected by orthodox Islam, 'Alī is regarded as the incarnation of Allah.

al-Khaḍir (El-Khizr), 'the Green One', a legendary Islamic saint who found the Waters of Immortality; he is identified with the unnamed companion of Moses in the Koran, and whose actions brought good out of apparent evil. Although largely ignored by orthodox Islam, al-Khaḍir is an immensely popular figure throughout the Islamic world, appearing in various Islamic writings, including the *Arabian Nights* and the Alexander sagas, and depicted in Persian miniatures. He is believed to come to the aid of Muslims in distress, to protect wayfarers, and is associated with Sufi foundations. In India al-Khaḍir is identified with the water-god Khwadja Khiḍr, said to ride a fish, and is thus depicted appeared on the crest of the kings of Oudh.

Altjiranga *Alchera.

Amaterasu Omikami, in Shinto mythology, the daughter of Izanagi; goddess of the sun, ruler of heaven, and weaver of the garments of the gods. According to legend, Amaterasu, terrified by her brother Susan-o-o's arrogant behaviour which had caused terrible earthly destruction, fled to a cave, whereupon the universe was plunged into darkness and plagued with evil spirits. The other gods eventually enticed her out

with sounds of laughter and a mirror in which she saw her own reflection. Amaterasu, with her principal shrine at Ise in southern Honshu, Japan, is the major deity in the Shinto pantheon. Through their claim of descent from Amaterasu, the Japanese imperial family legitimized their position as rulers of Japan.

Amazons, a legendary tribe of women warriors who frequently feature in Greek mythology. They are said to have come from Scythia, near the Black Sea. The men of neighbouring tribes were used to sire their children; boys were killed or returned to their fathers, while girls were trained for war. Their right breasts were cut off so as not to impede drawing a bowstring (the name derives from the Greek *a-mazos*, 'without breast'). Among the Greek heroes involved with the Amazons was Hercules (Heracles), who killed their queen Hippolyta to obtain her girdle—one of his Twelve Labours.

Amesha spentas *Ahriman.

Amida-nyorai *Amitabha.

Amitabha (Sanskrit, 'Infinite Light'; Japanese, Amida-nyorai; Pinyin Àmítuó), in Buddhism, the great saviour deity worshipped principally by the Pure Land Sects of Japan and China. The cult emerged in China in *c.* AD 650, and spread to Japan. Amitabha is believed to have been a monk called Dharmakara, who, on obtaining Buddhahood, fulfilled the vow that he had previously made that those who believed in him would reside in his Western Paradise until they obtained nirvana. Highly esteemed in Tibet and Nepal, Amitabha is there believed to be one of the eternally existent Buddhas, who has appeared as the earthly Gautama and as the *bodhisattva* Avalokitesvara. Japanese raigo paintings of the late Heian Period (897–1185) depict Amitabha welcoming the dead. His colour is red, his posture is one of meditation, and his symbol is the begging bowl. He is depicted on a peacock mount, holding an ambrosia vase from which spill the jewels of eternal life.

Amos, a Hebrew prophet in the reign of Jeroboam II (8th century BC). A herdsman, Amos prophesied the fate of the nations surrounding Israel and of Israel itself. He denounced the social abuses of his time, strongly supporting the poor and oppressed.

Amun (Amon, Amen), in Egyptian mythology, a human-headed god represented with a ceremonial beard and two-plumed crown, personifying the breath of life and called the 'Hidden One', that is, the wind. He is sometimes depicted ram-headed, the ram being sacred to him. Amun was originally a local god at Thebes (modern Luxor), but was elevated to state god during the New Kingdom (*c.*1550–1050 BC), at which time Thebes became the capital of Egypt and its temples at Karnak and Luxor, dedicated to Amun, became religious centres. Amun assimilated with many other gods, most notably Re, to become Amun-Re, the creator god and father of all reigning Pharaohs. Amun was associated with the mother goddess Mut as his consort and with the moon god Khonsu as their son.

An or **Anu**, the ancient sky god of Sumerian origin, worshipped in Babylonian religion. The survivor of an early creation myth, An was the father of the earth god Enlil; with him and with the water god Ea, An formed the great triad of gods.

Anansi (Mr Spider), in West African and West Indian folklore, the trickster spider of countless fables. In Africa, Anansi, sometimes credited with the creation of the world, is said by some groups to be the chief official of the deity who only acquired his name after obtaining, by immense cunning, a hundred slaves for his god in return for a corn-cob. Only in his encounter with the Wax Girl, a life-sized doll to which he became stuck, was Anansi beaten. From this episode onwards he is said to have acquired his flat shape, and to frequent dark places. Spider stories are widespread in the tropical areas of Africa, while *Hare stories predominate in desert and savannah regions.

Andrew, New Testament Apostle and martyr. Brother of Simon Peter, and one of the twelve Apostles, Andrew is said to have preached the Gospel in Ethiopia, Greece (Scythia, Epirus), Russia, and Turkey. In the 8th century, St Rule is believed to have transferred Andrew's relics to Scotland, to a site that subsequently became a centre of pilgrimage and evangelization and is now named after him. For this reason St Andrew became the patron saint of Scotland in *c.*750. An apocryphal work dating probably from the 3rd century describes his death by crucifixion but makes no mention of the X-shaped cross associated with his name.

Andromeda, in Greek mythology, an Ethiopian princess who was chained to a rock as a sacrifice to a sea-monster. She was rescued by (and then married) Perseus, who turned the monster to stone with the Gorgon's head. (The story perhaps played a part in forming the myth of St George and the Dragon.)

angel (Greek, *angelos*, 'messenger'), in Judaism, Christianity, and Islam, an immortal being or spirit, the attendant or messenger of God, and usually depicted in human form, with wings. In Jewish and Christian belief angels are regarded as intermediate between God and man. In Islam *mala'ika* (sing. *malak*), angels, are considered by some to be lower than man since in the Koran they were commanded to prostrate themselves before Adam. Belief in guardian angels is widespread among Christians and Muslims. In Islamic tradition each person is said in addition to have two recording angels who note down his or her good and evil actions. (See also, *archangel, *cherubim and seraphim, *Satan, *Iblīs, *Gabriel, *Michael, *Azrael, and *Nakir.)

'Antar (fl. 6th century AD), in Arab legend, the black desert poet and warrior 'Antarah ibn Shaddād, the hero of the pre-Islamic collection of poems, *al-Mu'allaqāt*, composed anonymously between the 8th and 12th centuries. The 'Romance of 'Antar' encompasses 500 years of Arab history, court life, and warrior-chivalry. 'Antarah was the son of King Shaddad of the Banu Abs and of the negro slave-girl Zabiba. His famed love for his cousin 'Abla enabled him to overcome all the obstacles set in the way of their eventual marriage. With his fabulous horse, Abjar, and Dami, his sword, he was victorious in campaigns from Arabia to Spain, North Africa to Sudan. 'Antarah was the alleged father of two Crusaders, Godfrey de Bouillon (by a Frankish princess) and Ghadanfar Coeur de Lion (by the sister of the king of Rome).

Antigone, in Greek mythology, the daughter of King Oedipus, whom she faithfully attended when he was blind and in exile. Her brother Polynices was killed fighting against Thebes, and her uncle, Creon, king of Thebes, forbade his burial. She was condemned to death for disobeying him, but killed herself.

Anubis, the Egyptian dog or jackal god of cemeteries and of embalming, also depicted as a dog-headed man attending to a mummy or conducting the dead into the underworld Hall of Judgement to weigh their hearts in the presence of Osiris. Anubis' role in burial promoted him throughout Egypt, although his cult centre was in the 17th Upper Egyptian nome (district) near el-Qeis. In later times Anubis' role as a judge of the dead was supplanted by that of Osiris.

Aphrodite, the ancient Greek goddess of love, beauty, and fertility. She was said to be the daughter of Zeus or alternatively to have been born of the sea foam. Her husband was Hephaestos, but she had other lovers among both gods (notably Ares) and mortals (notably Adonis). (See also *Paris.)

Apis, in Egyptian mythology a god always depicted as a bull (symbolizing

strength in war and in fertility), worshipped especially at Memphis, where he was recognized as a manifestation of Ptah, the city's patron, then of Ra (the solar disc was placed between his horns), and later of Osiris. A live bull, carefully chosen, was considered to be his incarnation and kept in an enclosure. When it died it was mummified and ceremonially interred, and a young black bull with suitable markings was installed in its place.

Apollo, in Greek mythology, one of the most important of the gods, the son of Zeus and twin brother of Artemis. He was primarily the god of the sun and light, but he was also particularly associated with the forces of civilization, notably music, poetry, and medicine.

Apostle (Greek, 'one who is sent'), the official name in Christianity for the twelve disciples chosen by Jesus to be with him during his ministry, and to whom he entrusted the organization of the Church. They were James (the Great) and John, the sons of Zebedee; James (the Less), the son of Alpheus; Jude, thought to be the same as Judas Lebbaeus, surnamed Thaddeus; Simon who was re-named Peter, and his brother Andrew; Matthew; Thomas, also called Didymus; Simon the Canaanite; Philip; Bartholomew, thought to be the same as Nathanael. The twelfth, Judas Iscariot, was replaced by Matthias. Three of the Apostles, Peter, James, and John, formed an inner circle. Paul and Barnabas are also given the title of Apostle in the New Testament even though they were not among the Twelve. (See also Christian Saints, p. 40.)

Arahat, in Buddhist belief, the being who has attained nirvana (enlightenment), the freedom from re-birth. This state is the ideal of the Theravāda branch of Buddhism. For the other main branch of Buddhism, Mahāyāna, the ideal is the *bodhisattva*, the enlightened being who renounces nirvana in order to help others along the Path.

Ararat, the extinct volcano on the border of modern north-east Turkey and southern Armenia where Noah's ark is said to have come to rest when the Flood subsided. Ararat, whose Assyrian and modern name is Urartu, is also the name of a kingdom, established in 840 BC, and of an area now in eastern Armenia.

archangel, an angel of the highest rank. In Jewish and Christian belief, Gabriel, Michael, Raphael, and Uriel are the four archangels who surround the throne of God. In Islam, Jibrīl, Mikāl, and Isrāfīl correspond to the first three, but the fourth is Azrael ('Izrā'īl), the Jewish and Islamic angel of death.

Ares, the ancient Greek god of war, son of Zeus and Hera. He was widely disliked by the other gods because of his savagery, but he became the lover of Aphrodite.

Argonauts, in Greek mythology, the fifty heroes who sailed with Jason in his ship *Argo* in search of the Golden Fleece. The ship was named after the craftsman, Argos, who built it. Hercules was the most famous of the Argonauts, but he parted company from his companions when he tarried to search for his page Hylas, who, because of his beauty, had been carried off by water-nymphs.

Ariadne, in Greek mythology, the daughter of Minos, king of Crete. She fell in love with Theseus and helped him in his most famous exploit, giving him a ball of string by means of which he was able to retrace his steps from the Minotaur's maze after slaying the monster. They left Crete together, but Theseus deserted her on the island of Naxos. She was consoled by Dionysos, whom she married.

Arjuna Pandava, in Hindu mythology, the son of Indra, and boyhood friend of Krishna. Arjuna is the principal hero of the Hindu philosophical poem the *Bhagavadgītā*, where he questions the morality of a war between the Pandavas and their cousins, the Kauravas, which will involve fighting and killing his own relatives and friends. Krishna explains that it is his duty as a warrior (*kshatriya*) to fight and offer up the results to God without thought of personal reward. Arjuna's first wife was Draupadi, whom he won in an archery contest and who became the wife of all five Pandava brothers. He later married Krishna's sister, Subhadra.

ark, in the Old Testament Book of Genesis, the boat (Hebrew, *tēbāh*) made by Noah on the instruction of God in which he, his family, and one pair of every living creature survived the Flood.

Ark of the Covenant, the sacred chest in which the Israelites, after their Exodus from Egypt, carried the two tablets of the law given to Moses by God, a pot of manna, and Aaron's rod. Symbolic of the divine presence, the Ark was eventually placed in Solomon's Temple, but after the fall of Jerusalem (587 BC) it was hidden in a cave, and its fate remains unknown.

Armageddon, the Greek name for the strategically important Mount of Megiddo in biblical northern Palestine. Scene of many critical battles, its name has become a symbol of war. In the Christian Book of Revelation it is used symbolically as the scene of the final conflict between the forces of good and evil.

Artemis, the ancient Greek goddess of hunting and the moon. She was the daughter of Zeus and the twin sister of Apollo. Although she was herself a virgin, she was the helper of women in childbirth.

Arthur, legendary King of Britain and hero of Celtic and medieval Christian mythology. Son of Uther Pendragon and guided in childhood by the magician Merlin, he won the crown by drawing the magic sword Excalibur from a stone. Other characters—Launcelot, Gawain, Tristram, and Sir Galahad—gradually became associated with Arthur as his Knights of the Round Table at Camelot. Their highest quest was for the Holy Grail, the cup of the Last Supper and the symbol of perfection. Other legends give prominence to the love of Launcelot and Arthur's Queen, Guinevere, and the love of Tristram and Iseult. A rebellion led by Arthur's nephew (or son) Modred, who had seduced Guinevere, brought disaster to his kingdom and left Arthur mortally wounded. There is reason to believe that the legend may have grown around a historical chieftain or general of the 5th or 6th century.

Asclepius, in Greek mythology, the patron of healing. In some accounts he is a god—the son of Apollo. In others he is mortal and is killed by Zeus, who feared that his healing powers would save humans from death. His emblem was a snake; the sloughing of its skin symbolized the regenerative power of healing.

Asgard, in Norse and Germanic mythology, the dwelling place of the gods, the Aesir. It could only be reached by crossing the rainbow bridge, Bifrost, which connected Asgard to Midgard, the world of mankind. The realms of Asgard included Valhalla, a great hall where Odin dwelt with the Valkyries and with the souls of heroes killed in battle.

Asiya, in Islamic belief, the wife of Pharaoh, who saved the infant Moses from the Nile. She is described in the Koran as a believer and considered as one of the four 'perfect women' of Islam.

Astarte, the Greek name for Ashtoreth, the Phoenician goddess of the moon, fertility, and sexual love. In the Bible her worship is linked to that of the Canaanite Baal. The cult of Astarte, in various forms, was widespread in the eastern Mediterranean world. In Egypt she was identified with Isis, in Greece with Aphrodite, and in Babylonia with Ishtar.

Atalanta, in Greek mythology, a huntress who was a remarkably swift runner. She declared that she would marry only a man who could overtake her in a race; suitors who tried and failed were killed. Finally she was outstripped by

Hippomenes, who was helped by Aphrodite: she had given him three golden apples, which he dropped during the race, and Atalanta paused to pick them up while Hippomenes raced past her.

Aten, the (Aton), the supreme god and creative principle of Ancient Egypt worshipped during the reign of the heretic king Amenhotep, who changed his name to Akhenaten (*c.*1353–1335 BC). He is symbolized in the sun's disc, depicted with rays ending in hands which offer the ankh-sign of life to the king. Formerly an aspect of Re, during the Amarna Period (Akhenaten's reign) the Aten was worshipped to the exclusion of all other gods, whose temples were closed. As his son, Akhenaten was the sole intercessor between the Aten and mankind. He built a new city for himself and the god, naming it Akhetaten. After his death, Tutankhamen brought the court back to Thebes, and Aten-worship ceased.

Athena (Athene), the ancient Greek goddess of wisdom and of the arts and crafts. Represented as a woman of severe beauty, in armour, she was the daughter of Zeus, springing from his forehead, which Hephaestus (or Prometheus) had opened with an axe. Like Apollo, she exerted a benevolent, civilizing influence (she was a patron of war, but in its just rather than its savage aspects). Her most famous temple is the Parthenon, built on the Acropolis at Athens (447–432 BC) by Pericles to honour her as the city's patron goddess.

Atlantis, a mythical island in the western seas, said to have been a powerful kingdom before it was submerged by the sea. It is first mentioned in the writings of the Greek philosopher Plato in the 4th century BC, who describes Atlantis as an ideal state; similar legends occur in other mythological traditions.

Atlas, in Greek mythology, a giant, one of the Titans who revolted against Zeus. As a punishment for this he was condemned to support the heavens on his shoulders. Perseus turned him to stone with the Gorgon's head and he thus became the Atlas Mountains in northwest Africa.

Atum *Re.

Audumla, in ancient Germanic and Norse mythology, the primeval cow who was created from drops of melting frost in the great void. She was nourished by licking salty, frozen stones. Audumla licked the stones into the shape of a man, Buri, the grandfather of the god Odin and his brothers.

Augean Stables, in Greek mythology, the stables of Augeas, king of Elis, which housed an immense herd of cattle and had not been cleaned for thirty years. As one of his Twelve Labours, Hercules had to cleanse them in a day, which he accomplished by diverting a river through them.

Aurgelmir (Ymir), in Norse and Germanic mythology, the first being, a giant created from the ice of Niflheim, the abode of the dead, melted by the heat of Muspelheim, the land of fire. He was given milk by a cow named Audumla, who licked the salt from frosted stones for food. One such stone became Buri, grandfather of the god Odin. Odin and his brothers in turn killed Aurgelmir, whose body was broken up to create the earth, the seas, and the skies.

Avalokitesvara (Sanskrit, *avalokita*, 'looking on'; *isvara*, 'lord'; Chinese, Guānyīn; Japanese, Kannon), in India and Tibet the name of the beloved *bodhisattva* ('Buddha-to-be') of infinite mercy and compassion, venerated throughout the Buddhist world as creator of the world, and protector of man. In Buddhist belief, Avalokitesvara is a manifestation and attendant of the eternal Buddha Amitabha. In paintings, he is usually depicted in white (in Nepal, red), and is sometimes shown with eleven heads and a multitude of arms.

Avalon, in early Celtic mythology, one of the Islands of the Blessed. In later Celtic and medieval Christian mythology, it is the place to which Joseph of Arimathea traditionally brought the Holy Grail, and is linked with Glastonbury in Somerset.

Azrael ('Izrā'īl), the Islamic archangel of death. In Islam, Azrael is traditionally described as being of cosmic size with countless wings, eyes, tongues, and four faces. One of the four archangels, he is said to stand with one foot on the fourth (or seventh) heaven, the other on the razor-sharp bridge that divides paradise from hell. He is believed to keep the roll of mankind and, when Allah decides, to draw the soul from the human body within forty days. Stories, often humorous, of Azrael confronting his victim, who had thought to avoid him by moving to a different place, are widespread in the Islamic world.

B

Baal, the Phoenician god of fertility, the storm, and winter rains, whose annual struggle with Mot, the god of harvesting crops, symbolized for Phoenicians the renewal of the earth's vegetation each spring. The name Baal, as a general title meaning 'lord', came to be applied to any of the male fertility gods whose sacrificial cult, so often condemned by the Hebrew prophets, was widespread in ancient Phoenician and Canaanite lands. (See also *Tanit.)

Babel, the Hebrew name for Babylon. According to the biblical account in the Book of Genesis, the Babylonians tried to rebel against God by building a tower to reach the heavens. Their plan was frustrated by God causing linguistic confusion amongst its builders. The many languages of the world were accounted for in this way.

Bacchus, the Roman god of wine and ecstasy, identified with Dionysus in Greek mythology.

Bahrām Gūr (ruled 420–438), in Persian mythology, the legendary Sassanian king, renowned as a hunter and lover, credited with the invention of poetry, and a frequent figure in Persian poetry and iconography. Bahrām Gūr appears as a hero-king in Firdausi's epic poem *Shāhnāmah* and Nizami's *Haft Peykar* ('Seven Portraits') (1197). He marries the seven daughters of the Kings of the Seven Climates, sleeping with a different one each night of the week, and each one telling him a different tale. The tales, telling of natural and fantastic events, are seen in Persian Islamic tradition as symbolic of the purification of the mystic's soul on its way to God.

Balder, in Norse and Germanic mythology, the Aesir sun-god, the favourite of the gods, representing joy, goodness, beauty, and wisdom. The son of Odin and Frigga, Balder was killed by a branch of mistletoe, the only thing that would hurt him, thrown by his blind brother, Hoder, but guided by Loki, who alone hated him.

banshee, a supernatural being in Celtic folklore whose wail or keening outside a house is believed to portend death within.

Barabbas, a New Testament figure, the bandit whom Pilate released instead of Jesus at the request of the Jews.

Barnabas, the 1st-century saint,

accorded the rank of Apostle in Christianity although not one of the Twelve. A Jewish Cypriot and Levite, Barnabas accompanied St Paul on his first missionary journey. He evangelized Cyprus, where he is said to have been martyred in AD 61. His feast day is 11 June, and he is invoked against hailstorms and as a peacemaker. The Gospel of St Barnabas, an apocryphal account of Jesus' life in which Judas is crucified instead of him, is regarded by Christians as a medieval forgery, but as the suppressed, true gospel by many Muslims.

Barsisa, in Islamic legend, the ascetic who, tempted by the devil, continuously seduced and then killed women. Agreeing to worship the devil if his life were spared, Barsisa was then mocked by his adversary with words from the Koran. Many versions of the story exist, the 15th-century Turkish 'History of the Forty Viziers' being the most elaborate and the one on which Matthew Lewis based his novel, *The Monk* (1796).

Bastet, the Egyptian cat goddess, also depicted as a cat-headed woman. She was the daughter of Re, and her most important cult centre was at Bubastis, where a vast cemetery of mummified cats sacred to her have been found. Her feline nature permits opposing tendencies: fierce in battle (like a lioness), and fertile and caring (like a cat with kittens). As a fertility goddess she is sometimes assimilated with Hathor, and with other feline goddesses such as Sekhmet and Pakhet.

Bathsheba, a biblical figure, the mistress and then wife of David, king of Israel. Seeking to conceal his guilt at taking Bathsheba as his mistress, David sent her husband Uriah to his death in the front line of battle. This act and his subsequent marriage to Bathsheba was denounced by the prophet Nathan. Bathsheba persuaded David to appoint their sole surviving son, Solomon, as his successor.

Bāxīān *xiān.

Beelzebub, the Greek form of Baalzebub, a god of Ekron, a Philistine town in Canaan. In Hebrew, the name means 'lord of the flies', though it may originally have been a corruption of Baal-zebul, 'exalted Lord'. In New Testament times and subsequently both forms have been used as epithets for 'the prince of evil spirits', that is, the Devil.

Beersheba, an ancient oasis town in Israel, marking the southern limits of Canaan, the promised land for the Jewish people. It was an important watering place for the flocks of Abraham, Isaac, and Joseph, as well as being the scene of Hagar's trial and Abraham's covenant with Abimelech.

Belial (Hebrew, 'prince of evil'), a biblical term used to signify Satan. In the Old Testament the term 'son' or 'man of Belial' meant a worthless, ungodly, or wicked individual.

Bellerephon, a hero in Greek mythology. After being falsely accused of trying to seduce the queen of Proteus, king of Argos, Bellerephon was given a number of seemingly impossible tasks, all of which he carried out successfully. The most famous was slaying the Chimaera. He was aided by the winged horse Pegasus, but later offended the gods by trying to ride Pegasus to heaven, and ended his life a lonely outcast.

Belshazzar, co-regent of Babylon in the 6th century BC, the son of Nabonidus, the last Chaldean King of Babylon. In the biblical Book of Daniel (where he is wrongly called the son of Nebuchadnezzar), he is described as giving a great feast at which a hand is seen writing the message, '*mene, mene, tekel*' and '*parsim*' on the wall of his palace, predicting the imminent downfall of his kingdom.

Benjamin, a biblical figure, the youngest and favourite son of the Hebrew patriarch Jacob and his wife Rachel. His only other full brother, Joseph, had been sold into slavery in Egypt, where he soon rose to a position of power. Jacob's ten other sons by his first wife Leah migrated to Egypt during a famine in Israel, where they and their father, together with Benjamin, were eventually reunited with Joseph. Benjamin's descendants formed one of the twelve tribes of Israel.

Benten, in Japanese folklore, a river goddess, the guardian of eloquence, giver of wealth and of music. According to legend, she married a serpent king of the sea in the vicinity of Kamakura Island.

Bethal (Hebrew, 'house of God'), a town north of Jerusalem, of importance in biblical times. The site where Abraham first pitched his tent and built an altar, and where Jacob experienced the revelation from God, it became for a time the chief sanctuary of the Israelite tribes.

Bethany a village near Jerusalem on the eastern slopes of the Mount of Olives. It was the home of three close friends of Jesus, Martha, Mary, and Lazarus, and also of Simon the Leper. It was near Bethany that Jesus' ascension into heaven traditionally took place.

Bethesda (Hebrew, 'house of mercy'), the site of a pool, outside one of the gates of Jerusalem, where the sick came to be healed in biblical times. It was traditionally here that Jesus cured a man who had been waiting for thirty-eight years.

Bethlehem (Hebrew, 'house of bread'), a town on the West Bank of the River Jordan, south of Jerusalem, where, according to New Testament accounts, Jesus was born. Bethlehem was also the presumed birthplace and boyhood home of King David.

Bhava-chakra (Sanskrit, 'wheel of life'), an iconographical representation of the cycle of existence in Tibetan Buddhism. The Bhava-chakra is depicted as a wheel, held by the god of the underworld, Yama, and divided into six segments representing the main types of worldly existences depicting the realms of the gods, demons, human beings, animals, rapacious ghosts, and hell. The centre of the wheel contains the animals representing the causes of the cycle: the cockerel (desire), the pig (ignorance), and the serpent (aggression or hate). Forming the outer rim of the wheel are the twelve nidānas or interrelated phases in the cycle of existence: a blind woman (ignorance), a potter (power of formation), a monkey (consciousness), two men in a boat (name and form, or mind and body), a six-windowed house (the six senses), a couple embracing (contact), an arrow piercing an eye (sensation), a person drinking (craving), a man gathering fruit (grasping, attachment), copulation (becoming), a woman in labour (birth), and a man carrying a corpse (death).

Bilqis *Sheba, Queen of.

Boreas, in Greek mythology, the north wind. He is characterized as an old man with wings and flowing grey locks and is the personification of winter.

Brahma, in Hindu belief, the creator god of the triad with Vishnu, the 'preserver', and Shiva, the 'destroyer'. Brahma's consort is Sarasvati, goddess of wisdom and of learning. Of the many accounts of Brahma's origin, his emergence from the reclining Vishnu's navel, seated on a lotus, is the most common in both myth and iconography, where he is depicted with four heads and arms. His vehicle is the goose, symbol of the creative principle. Although Brahma, originally a minor Vedic god, became the personification of the impersonal Supreme Principle (the neuter brahman) in the transition of Hinduism to monotheism, his importance was soon eclipsed by that of Vishnu and Shiva, and his cult is practically non-existent in modern times.

Bran, legendary king of Britain. According to a myth recounted in the medieval Welsh *Mabinogion*, Bran was a deity of gigantic stature. After being severely wounded fighting the Irish, he requested his followers to cut off his head. It retained its independent life and contin-

ued to provide them with marvellous entertainment on their voyages, until it was laid to rest in London.

Brunhild, in the Germanic epic poem *Nibelungenlied* (*c.*1205), the beautiful, imposing Icelandic princess who became the wife of Gunther, king of the Burgundians. Brunhild was wooed and won by Siegfried on behalf of Gunther, king of the Burgundians. Later, in the guise of Gunther, Siegfried took her ring and girdle and gave it to the Burgundian princess Kriemhild. When Brunhild discovered the deception, she had Siegfried killed in revenge.

Buga, the supreme deity of the Tungus peoples of Siberia. Buga created the first two people out of iron, fire, water, and earth. Out of the earth he fashioned the flesh and bones, out of the iron the heart, out of the water, blood, and out of the fire, warmth.

Buraq̄, in Islamic legend, the miraculous steed brought to the Prophet Muhammad by the archangel Gabriel for his Night Journey (*isrā*) from Mecca to Jerusalem. Also said to have been used by Ibrahim (Abraham) to visit his son Ishmael in Mecca, Buraq is described as 'smaller than a mule, larger than an ass' and white. He is sometimes described as winged, and associated with Muḥammad's ascent to heaven (*mir'rāj*). In iconography, Buraq is often depicted, according to Indian belief, with the face of a woman and the tail of a peacock. The alleged stable of Buraq in Jerusalem has given rise to disputes between Jews and Muslims.

C

Caiaphas, a biblical figure, the Jewish high priest of the Temple (*c.* AD 18–37) who presided over the Sanhedrin when it tried Jesus, and handed him over to the Romans for crucifixion.

Cain, according to Jewish, Christian, and Muslim belief, the eldest son of Adam and Eve. Cain, a farmer, became jealous that God had accepted the sacrifice of his brother Abel, a shepherd, in preference to his own. He murdered Abel and was condemned by God to wander over the face of the earth. God placed a protective mark upon him and a curse on anyone who would lay hands on him. According to Islamic tradition, Cain and Abel (Kabil and Ḥabil) each had twin sisters, Aklima and Labuda. It was Kabil's refusal to marry Ḥabil's twin sister, Labuda, that caused Allah to reject Kabil's sacrifice and accept Ḥabil's, the latter having obeyed Adam's wish that he marry Aklima. In a fit of rage Kabil killed his brother, but not knowing what to do with the body, he wandered with it over the face of the earth for many years. Eventually, on seeing a raven digging a hole to lay a dead bird to rest, he decided to do likewise for his brother.

Calliope, in Greek mythology, the Muse of epic poetry.

Calvary (or Golgotha), the hill outside the walls of ancient Jerusalem where Christians believe Jesus was crucified. The name comes from the Latin word *calvaria*, meaning 'skull' (in Hebrew, *golgotha*). The hill may have acquired its name because of its skull-like shape or because it was regularly used for executions. It is now the site of the Church of the Holy Sepulchre.

Calypso, in Greek mythology, a nymph, a daughter of Atlas. Odysseus was shipwrecked on her island, Ogygia, and she promised to make him immortal if he would stay with her. After seven years, she let him go at the command of Zeus.

Camelot, the mythical court of the Celtic hero King Arthur. It was from here that the Knights of the Round Table set off on their many quests. Camelot is variously identified with sites in Cornwall, Somerset, Hampshire, and Caerleon in Wales.

Canaan (biblical Palestine and parts of Syria), the name given to an area in western Asia whose original, pre-Israelite inhabitants were called Canaanites. Successive Israelite tribes conquered and then occupied Canaan from the late 2nd millennium BC onwards, on the ground that it was the 'Promised Land', given to them by God. The region was to suffer further invasions, but under the leadership of King David (10th century BC) the Israelites secured control over all other groups.

Cassandra, in Greek mythology, a daughter of Priam, king of Troy, and his wife Hecuba. Apollo gave her the gift of prophecy, but because she rejected his love, he ordained that none of her predictions would be believed. Thus her warning to the Trojans about the Wooden Horse of the Greeks was disregarded. After the sack of Troy, Cassandra was taken to Greece as the prize of Agamemnon and was murdered by his wife, Clytemnestra.

Castor and Pollux, in Greek mythology, twin sons of Zeus. Their mother was Leda, queen of Sparta; Zeus came to her in the form of a swan, and from their union, the twins were hatched from an egg. Great warriors and inseparable companions, Castor and Pollux were among the Argonauts. After their deaths they became the constellation Gemini.

Centaurs, in Greek mythology, a tribe of creatures having the head, torso, and arms of a man, and the body and legs of a horse. Generally they are represented as barbarous, drunken, and lecherous, but Chiron—the best and wisest of them—was the tutor of Achilles, Castor and Pollux, Jason, and other heroes. After his death Chiron became the constellation Sagittarius.

Cerberus, in Greek mythology, a monstrous three-headed dog that guarded the entrance to Hades, preventing the dead from escaping and the living from approaching the Underworld. Hercules—as his twelfth and final Labour—brought Cerberus up to earth.

Ceres, the Roman goddess of corn, identified with the Greek Demeter.

Charon, in Greek mythology, the ferryman who carried the dead across the River Styx to Hades. The Greeks used to put a coin in the mouth of corpses as Charon's fee.

cherubim and seraphim, in Judaism, Christianity, and Islam, high-ranking angels, the attendants of God, and classified in the early Christian Church as the highest orders of the celestial hierarchy. The six-winged seraphim are described in Isaiah's vision of God. The Old Testament cherubim are depicted as griffin-like creatures, probably derived from Babylonian beliefs,

although in medieval Jewish folklore they were thought of as beautiful men. In post-medieval western European art they are usually depicted as *putti*: chubby, winged infants (cherubs).

Chimaera, in Greek mythology, a fearsome fire-breathing monster having a lion's head, a she-goat's body, and a serpent's tail. It was killed by Bellerephon.

Chuku *Ala.

Cinderella, a folktale found throughout the world but best known in Europe in Charles Perrault's version, 'Cendrillon' (1697). In this, Cinderella is mistreated by her stepmother and two stepsisters. With the help of a supernatural godmother, however, she attends a ball, dances with a prince who falls in love with her, and loses a glass slipper as she leaves in haste. After scouring his kingdom for the foot that will fit the slipper, the prince finds Cinderella and marries her.

Circe, in Greek mythology, a sorceress encountered by Odysseus and his companions on their return from the Trojan war. By means of a magic potion, she turned his companions into swine, but Odysseus himself—forewarned by Mercury—ate a herbal antidote, overcame her, and forced her to return them to their normal form.

Clio, in Greek mythology, the Muse of History.

Clytemnestra, in Greek mythology, the wife of Agamemnon and mother of Orestes and Electra. With her lover she murdered Agamemnon (and also Cassandra) and was in turn slain by Orestes.

Coatlicue, the Aztec earth goddess, one of the wives of Mixcoatl, the cloud-serpent god of hunting. All land belonged to Coatlicue, a belief which upheld Aztec law that no one could own land. Celebrated in spring grain festivals, Coatlicue, who lived on human corpses, was depicted with clawed feet and hands, wearing a skirt of writhing serpents and a necklace of human hearts and hands from which a skull was suspended. Her son was Huitzilopochtli, god of the sun, lightning, and storms, to whom a large number of human sacrifices were made.

Cockaigne (Cockayne), **Land of**, in medieval European legend, an imaginary land of plenty and idleness, where all manner of luxuries were there for the taking. A 13th-century English poem, 'The Land of Cockaygne', is a satire on monastic life.

Conchobar mac Nesse (King Conor), legendary Irish Gaelic king who held court at Emain Macha (near modern Armagh) with his Knights of the Red Branch. A group of legends and tales, the Ulaid Cycle, recorded from oral tradition between the 8th and 11th centuries, is centred around him and the heroic exploits of his people, the Ulaids.

Coyote, the trickster deity of many North American myths who assumes innumerable names and forms. It is believed that it was his desire to make the life of man more interesting, which caused the creation of sickness, sorrow, and death. Destructive natural phenomena, as well as inventions which are of benefit to mankind, are believed to be the outcome of Coyote's creative but mischievous power.

Cronus, in pre-Hellenic and Greek mythology, the youngest son of Uranus (Heaven) and Gaea (Earth), leader of the Titans. By the advice of his mother he castrated his father, thus separating Heaven from Earth. He became ruler of the Earth, but was fated to be in turn deposed by one of his own children. He swallowed Hestia, Demeter, Hera, Hades, and Poseidon at birth. However, his wife Rhea (who was also his sister) substituted a stone wrapped in swaddling-clothes for Zeus, the last-born. Zeus eventually deposed his father and the other children were vomited up.

Cuchulain (Cuchulinn, Cu Chulainn), in ancient Irish Gaelic mythology, the greatest of the Knights of the Red Branch, loyal to Conor, reputedly king of the Ulaids of north-east Ireland in the 1st century BC. The son of the god Lug of the Long Arm and a mortal woman Dechtire, he was a youth of extraordinary beauty and gentleness. In battle he took on monstrous deformities, having seven fingers on each hand, seven toes on each foot, and seven pupils in each eye. Single-handed, he defended Ulster against the forces of Medb (Maeve), queen of Connacht. By being the only man to accept the challenge of a monster, he became champion of Erin (the ancient name for Ireland), but died young, as had been prophesied.

Cupid, the Roman god of love, identified with the Greek Eros.

Cybele, in the mythology of the ancient Near East, the great mother-goddess, associated particularly with fertility. Her cult spread to Greece and Rome.

Cyclops (plural Cyclopes), in Greek mythology, one of a race of savage giants who had one eye in the middle of the forehead. They were reputedly descended from three one-eyed Titans who made thunderbolts for Zeus. Polyphemus was the most famous of the Cyclopes.

D

Daedalus, in Greek mythology, a great craftsman and inventor. He was held captive by King Minos of Crete, for whom he constructed the labyrinth that housed the monstrous Minotaur. To escape captivity he made wings for himself and his son Icarus out of wax and feathers, but Icarus flew too near the sun, the wax melted, and he fell to his death.

Dagda, the ancient Irish deity of life and death, chief of the Tuatha Dé Dannan. Also known as Aed, 'fire', he held among his sacred possessions an inexaustible cauldron and ever-laden fruit trees. His daughter was Brigid, goddess of fire, fertility, cattle, and poetry.

Dainichi-Nyorai (Japanese, the 'Great Sun Buddha'), in Buddhism, the Japanese name for Mahavairocana (Vairocana) the Great Illuminator, regarded as the supreme Buddha by many believers. Dainichi-Nyorai is considered the source of the universe by the Japanese Shingon sect, founded by the monk Kūkai (posthumously known as Kōbō Daishi) (774–835). Dainichi-Nyorai is usually depicted as a crowned figure, seated on a white lotus, and surrounded by his various emanations.

Damocles, a courtier of Dionysius I, tyrant of Syracuse in the 4th century BC. Damocles was a flatterer who extolled the position of kingship; to teach him a lesson, Dionysius seated him at a banquet with a sword hanging over his head suspended by a single hair. This symbolized the precariousness of a king's fortunes, and the phrase 'Sword of Damocles' has come to signify an impending disaster or the permanent threat of it.

Dana (also Anu, Danu), in Celtic mythology, the earth-mother and female principle. Her name was given to the legendary Tuatha Dé Danann (People of the goddess Danu), the Irish assembly of gods. In later Irish folklore these survived as the fairy folk.

Danaids, in Greek mythology, fifty daughters of Danaus, a king of Argos. He was forced to agree to his daughters marrying their fifty cousins (sons of Danaus' brother, who had driven him out of his native Egypt), but he ordered them to kill their bridegrooms on the wedding night. All but one daughter obeyed. After their death the Danaids were punished in Hades for their crimes by perpetually being made to fill leaking jars with water.

Daniel, a Hebrew prophet, hero of the Book of Daniel, traditionally active during the Babylonian Captivity of the Jews (*c.*598–538 BC). He is portrayed as an interpreter of the dreams of Nebuchadnezzar and receiver of prophetic visions, cast by King Darius into the lions' den for obeying God rather than him, but saved by divine intervention.

Deirdre, the Irish heroine whose renowned beauty brought suffering to her people of Ulster. The intended bride of King Conchobhar, Deirdre fell in love with Naoise, the son of Uisnech, and eloped with him. Eventually allowed to return from their exile in Scotland, Naoise and his brothers were killed and Deirdre died. Their story is the basis of J. M. Synge's play *Deirdre of the Sorrows* (1909).

Delphic oracle, the oracular shrine of Apollo in his temple at Delphi on Mount Parnassus. Here the priestess of the god, seated on a tripod over a fissure in the rock, uttered in a divine ecstasy incoherent words in reply to questions, which were interpreted by priests in the form of verses.

Demeter, in Greek mythology, the goddess of agriculture, fertility, and marriage. She was the sister of Zeus and also the mother of his daughter Persephone.

Devas *Div; *Ahriman.

Devi, the great goddess of Hinduism. Devi represents the different aspects of Shiva's wife under various names according to the forms she takes. As Parvati, mother of Ganésha and Karttikeya, she is the beautiful, benevolent goddess of the mountains and opponent of the demons. As Kālī, a goddess of fertility and time, she personifies the opposing forces of creation and destruction, and assumes a malevolent aspect: a black, hideous old woman, with a necklace of skulls, a belt of severed heads and a protruding tongue. As Durgā, she is the fierce goddess, often identified with Kālī, depicted with eight or ten arms, riding a tiger or lion, and slaying the buffalo demon.

Devil, the embodiment of the supreme spirit of evil in Judaism, Christianity, and Islam. Also known as Lucifer and as Satan (the Islamic Shayṭan), he is regarded as the enemy of God, contesting God's omnipotence, and tempter of mankind. He is said to be the chief of the fallen angels, and in Islam as Iblis the chief of the djinn, cast out of heaven for disobeying God. Used in the plural, devils, it denotes demons and evil spirits, traditionally important elements in all three religions.

Dhu'l Qarnayn *Alexander the Great.

Diana, in Roman mythology, the goddess of hunting, chastity, and the moon, identified with the Greek Artemis.

Dido, originally the name of a Phoenician goddess, she married her uncle Sychaens. After his murder by Pygmalion, her brother, Dido fled to Africa, where the king of Mauritania offered her as much land as might be covered by an ox-hide. By the device of cutting the hide into narrow strips, she secured space to found the city of Carthage. In Roman mythology, Dido offered refuge to the fleeing Aeneas at Carthage. Their love for each other was doomed, since Aeneas was destined to go on to found the state of Rome. On his departure, Dido took her life on a funeral pyre.

Dionysos, in Greek mythology, the god of wine and ecstasy. He was the son of Zeus and Semele, daughter of a king of Thebes. One of the most popular Greek gods, he was the subject of many legends and his worship, manifesting itself in a frenzied rout of votaries, male and female, Satyrs, Sileni, Maenads, and Bassarids, was often drunken and orgiastic. He was also regarded as a patron of the arts, inspiring music and poetry.

Div, in Persian mythology, an evil spirit or jinn. Divs are the *daevas* (*devas*) of the Zend-Avesta, malevolent spirits created as counterparts of the good spirits, the *amesha spentas* by Ahriman, who ruled over them. By contrast, the devas of Vedic Hindu belief are divine, good spirits, the gods as opposed to the demons.

Djugurba *Alchera.

Don Juan, a universal, fictitious character who is a symbol of libertinism. In the earliest, 17th-century Spanish version, as in most subsequent interpretations of the story in opera, on stage, in novels, and poems, he is cast as the 14th-century Don Juan Tenorio of Seville.

Dracula, the chief of the vampires in the novel *Dracula* (1897), written by the Irish-born civil servant Bram Stoker, and partly set in a lonely castle in Transylvania. Vlad Tepes (Vlad the Impaler), also known as Dracula, was a 15th-century Prince of Wallachia, renowned for his cruelty, and the novelist wove this name into a sinister tale of a region with which vampires and werewolves were traditionally associated.

dragon, a mythical monster like a reptile, usually with wings and able to breathe out fire. The dragon, which possesses both protective and terror-inspiring qualities, is probably the commonest emblem in oriental art, and the most ancient. A dragon form with five claws on each foot was adopted as the chief imperial emblem in China representing fertilizing power and cosmic energy. Chinese dragons resided especially in water, in rivers, lakes, and the sea; in springtime they moved in heaven among the clouds. In the art of the West the dragon appears as the principle of evil and of paganism in such contexts as St George slaying the dragon. It is used as a heraldic emblem, for example in Wales, where it was introduced as a military standard during the Roman occupation in the 1st century AD.

Dragon King(s), the *Lóngwáng.

Dream time *Alchera.

Dryads, in Greek mythology, nymphs associated with trees or forests.

Durgā *Devi.

dwarf, in Norse and Germanic mythology and folklore, a mythological being inhabiting the interior of mountains and caves. Their halls were built of gold and lined with precious stones. For the most part kindly and generous, they resembled very small, greybearded men. They were expert craftsmen, famed for their metalwork and for the forging of magical swords, rings, and necklaces. Humans who stole their treasure either met with disaster or (as with gold stolen from fairies) found the treasure turned to dead leaves in their hands. Among the treasures they forged was Thor's hammer, used by the god to bring thunder and lightning down on to the heads of mortals.

E

Ea (Enki), the water god of the Sumerians, worshipped in Babylonian religion, who evolved into the ruler of Apsu, the freshwater regions beneath the earth. A creator deity, Ea's union with the earth-mother, Ninhursaga, was believed to have made the paradise of Dilmun into a fruitful garden. According to one legend, Ea, as lord of wisdom and magic, created man from clay in response to the gods' request for servants. With the sky god An and the earth god Enlil, he was one of the divine triad. Ea was commonly represented as half-goat, half-fish creature, from which the astrological figure for Capricorn is derived.

Echo, in Greek mythology, a nymph who fell in love with the beautiful youth Narcissus and, because her love was unrequited, pined away until nothing but her voice remained.

Eden, Garden of, in the Old Testament Book of Genesis, and in earlier Sumerian records, the abode of Adam and Eve at their creation, from where they were expelled for their disobedience. Eve, tempted by Satan disguised in the form of a serpent, plucked the forbidden fruit of the tree of knowledge. On eating it, both she and Adam grew aware and ashamed of their nakedness, thus demonstrating mankind's fall from a state of innocence and bliss to a condition of the knowledge of sin and suffering.

El, the supreme deity of Phoenician and Canaanite belief, god of fertility-giving rain, rivers, and streams, and father of gods and men. He is always portrayed as a seated figure, wearing a bull's horns, the symbol of strength. El was also believed to be the father of Keret, a king of Sidon, and to have made the gift of a son, the fertility deity Aqhat, to King Dan'el.

El Dorado (Spanish, 'the gilded'), the name given by 16th-century Spanish conquistadors to a legendary gilded man, and to a land or city full of gold, hence its modern connotation of a place of abundance. With many alleged locations in North, Central, and South America, El Dorado was the goal of numerous unsuccessful expeditions. One possible source of the myth may be found in Chibcha and North Andean Indian sun and earth rituals, formalized in the Lake Guatavita ceremonies in Colombia, where rulers were ritually coated in gold dust.

Electra, in Greek mythology, the daughter of Agamemnon and Clytemnestra. She supported her brother Orestes when he killed their mother to avenge the murder of their father.

elf, in Germanic mythology, a small supernatural being, usually in male form. Elves are probably related to the *alfar* ('spirits') of Germanic mythology. They tended to be mischievous, causing milk to curdle, cattle to fall ill, and stealing human children to substitute them with their own changeling children. They have survived into modern times as the hill- and rock-dwelling *huldufolk* ('hidden people') of Icelandic folklore.

Elijah (Greek form, 'Elias'), a biblical figure, the man of the desert, fed by ravens, who is among the most revered of the Hebrew prophets. Elijah condemned both the social injustices and idolatrous cults in Ahab's northern kingdom of Israel, prophesying drought and famine because of the king's worship of Baal. He was 'translated' into heaven (without dying) in a chariot of fire. According to Christian tradition, the prophecy that he would return before the coming of the Messiah was declared by Jesus to be fulfilled in the person of John the Baptist. Ilyas (Elijah) is one of the prophets mentioned in the Koran and a popular figure in Islamic legend, which closely follows the Bible account: Ilyas is said to have been given power over the rain after Ahab rejected his teaching, and to have caused a great drought during which he was miraculously provided with food while so many Israelites died that God eventually interceded on their behalf.

Elisha, a biblical figure, the wealthy farmer's son who became the disciple, and then the anointed successor, of Elijah in Israel. Portrayed in the Bible as both the helper of all people and the king's counsellor, Elisha became renowned for his miracles far beyond Israel. His sending of a messenger to anoint Jehu, who was leading the rebellion against Ahab, profoundly affected Israel's subsequent history. Alyasa (Elisha) is mentioned in the Koran, and is described by Muslim traditions as the disciple and successor of Ilyas (Elijah).

Elysium (also called Elysian Fields), in Greek mythology, a place of perfect happiness inhabited by favoured mortals after death. Originally it was thought as being entered only by those whom the gods particularly loved, but later it was conceived of as a home for the righteous dead in general.

Endymion, in Greek mythology, a handsome youth who was loved by Selene, the moon goddess. Zeus sent him to sleep for ever so that his youth and beauty might be preserved, and Selene came to embrace him every night.

Enki *Ea.

Enlil (Bel), the ancient earth god of Sumerian origin, worshipped in Babylonian religion. He was later believed to be the controller of mankind's destiny and source of power for kings. Sometimes known as Bel in Babylonia, Enlil was responsible for order and harmony in the universe, but also could wreak havoc through terrible storms and destruction. Together with An and Ea he formed the great triad of gods.

Enoch, an early Hebrew patriarch and father of Methuselah. Enoch is said to have lived for 365 years and then, like Elijah, to have been 'translated' into heaven. According to some Jewish legends he invented writing, arithmetic, and astronomy. In Islam Enoch is usually identified with Idris.

Ephraim, the second son of Joseph and the founding ancestor of the Israelite tribe of Ephraim, which gave its name to the area in Canaan where they settled. Ephraim himself was given preference over his elder brother Manasseh by his grandfather Jacob, who foretold that he and his descendants would be greater and become a multitude of nations. The northern kingdom of Israel, established by the Ephraimite Jeroboam I in 934 BC, was often called Ephraim.

Eros, a Greek god of love, a primeval force, the son of Aphrodite. In Hellenistic times he became associated with romantic love, and was represented as a little winged archer, shooting his arrows at gods and men. The Romans identified him with Cupid.

Esau (Hebrew, 'red'), the hunter son of Isaac, older twin brother of Jacob and ancestor of the nation of Edom. Exchanging his birthright (that is, rights due to him as the eldest son) for a 'mess of pottage' (bowl of food), he was then cheated of his dying father's blessing by Jacob impersonating him. The story sets out to explain why Israel, synonymous with Jacob's kingdom, was entitled to dominate the tribes of Edom.

Eshu, the trickster figure of Yoruba mythology in Nigeria, who acts as a messenger and arbitrator between the deity Olorun and the people. Although supposed to be a guardian of humans, Eshu is known for his unpredictability, his jealous pride in his own power, and his dislike of harmony between people.

Esther, the Jewish heroine who married King Xerxes (Ahasuerus) of Persia after he had banished the Persian Queen Vashti from his court for defying his orders. Esther won legendary favour with

her husband by her great beauty and by twice saving his life; she was in time able to anticipate and prevent a massacre of the Jews, who then rounded on their enemies and destroyed them. These events, recorded in the biblical Book of Esther, are still celebrated in the popular Jewish feast of Purim.

Estsanathlehi, a deity of the Navaho Indians, daughter of Yadilyil, 'the upper darkness', who created her out of a black cloud covering the summit of a mountain. Fed on pollen brought by the sun god Tsohanoai, she grew rapidly, always changing her form. She dwelt on the great waters in the west, where the sun visited her every evening. Estsanathlehi created men and women out of small pieces of her own skin, to keep her company.

Eulenspiegel, Till, the German peasant trickster, born, according to tradition, in *c.*1300. Till represents the cunning peasant who is able to show his ultimate superiority over the townsmen, clergy, and nobility by his farcical, and often obscene, jests and practical jokes. Till, whose exploits were first recorded at Antwerp in 1515, is the subject of many folk-tales and musical and literary works.

Europa, in Greek mythology, a princess with whom Zeus fell in love. He disguised himself as a white bull and when—beguiled by the animal's docile nature—she climbed on his back, he carried her off by sea to Crete, where he resumed his normal shape and ravished her.

Eurydice, in Greek mythology, a Dryad, the wife of Orpheus. She died from a snake bite and Orpheus followed her to the Underworld. There the beauty of his musicianship persuaded Hades, the god of the Underworld, to allow him to bring her back, provided that he did not look round at her before he reached the upper world. He did look round, however, and lost her for ever.

Evangelists (the Four), traditionally the authors of the four Gospels recording the life and teaching of Jesus which are accepted as authentic in Christianity. They are Matthew, Mark, Luke, and John. On the basis of the New Testament Book of Revelation, they are symbolized by a man, a lion, an ox, and an eagle.

Eve, according to Jewish, Christian, and Islamic belief, the first woman and consort of Adam. The Book of Genesis recounts how she was created from Adam's rib and tempted by a serpent to eat of the forbidden fruit of the tree of knowledge, encouraging Adam to do likewise. Because of their disobedience, they were expelled from the Garden of Eden. Eve was further punished with the pain of childbirth. The Christian doctrines of the Fall, Original Sin, and redemption through Jesus Christ are based on these accounts. In Islamic legend, Adam and Hawwa (Eve) made the pilgrimage to Mecca, where Hawwa had her first menstruation and cleansed herself in the waters of Zamzam.

Excalibur, in Arthurian legend, the name of King Arthur's magic sword. According to one version, the young Arthur alone among many contenders could pull the sword from the stone in which it was fixed, thus confirming his kingship. In a later version, the sword was given to Arthur by the Lady of the Lake. When the king lay mortally wounded after his last battle, he ordered his knights to cast Excalibur back into the lake. The arm of the Lady of the Lake rose from the waters and, brandishing it, sank down into the deep.

Ezekiel, a Hebrew prophet deported into exile by Nebuchadnezzar in 597 BC. His visions and prophecies, set down in the biblical Book of Ezekiel, foretell the inevitability of God's judgement on Jerusalem and also its eventual restoration. Ezekiel's teaching paved the way for the religious nationalism of post-exilic Judaism. In Islam Ezekiel (Hizkil) is usually identified with Dhū-l-Kifl, a prophet mentioned in the Koran.

F

fairy, in European folklore, a supernatural being characteristically beautiful to the eye, who lives in fairyland but may manifest itself to humans to take part in their affairs. Fairies have no souls. Fairies may carry off children, leaving changelings, and female fairies may be deadly to human lovers. Traditionally, they were clothed in green and gold, and lived underground in palaces of great beauty and light.

familiar, in European mythology, a demon or spirit, often in the form of an animal, believed to serve witches, sorcerers, and magicians. Witch familiars were alleged to subsist by sucking blood from the witch's body. In the 15th to 17th centuries warts and moles found on the skins of accused persons were thought to be 'teats' and hence signs of guilt.

Fates, in Greek mythology, three daughters of Zeus and the Titaness Themis (the personification of justice) who controlled human destiny. Clotho spun the thread of life, Lachesis determined its length, and Atropos cut it off. The Fates are generally depicted as old and ugly. Some Greek writers regarded them as goddesses of destiny, to which even the Olympian gods were subject, others merely as symbols of human life.

Fātima(h) (*c.*605–633), a daughter of the Prophet Muḥammad, the wife of ʿAli, and one of the alleged four 'perfect women' of Islam. Although early Islamic writings contain little information about Fātima's relationship with Muḥammad, later traditions give her an exalted position as the person most dear to him, 'the queen of the women of Paradise'. For Shīʿites, she is the mother of the Imams Ḥasan and the martyred Ḥusain, and regarded as the embodiment of the divine in woman, whose birth was miraculous and whose marriage was divinely ordained. Fātima's hand is a symbol often seen in Islamic art.

Faunus, in Roman mythology, a god of nature and fertility, worshipped particularly by shepherds and farmers as the protector of herds and crops. He was equivalent to the Greek god Pan. Fauns were a class of minor rural deities, represented as men with pointed ears and the horns, tail, and hind legs of a goat. They were lustful but benevolent.

Faust (Faustus), the man reputed to have sold his soul to the devil in return for knowledge and power. One of the

most durable legends in Western literature, it is based on the mythical exploits of an alchemist and astrologer who lived in Germany (*c.*1490–1540). According to early versions of his life, he called up an evil spirit, Mephistopheles, with whom he made a compact to surrender his soul in return for experiencing the full potential of human delights, power, and insight. The story inspired Christopher Marlowe's *The Tragicall History of D. Faustus* (1604) and Wolfgang Goethe's *Faust* (Part I, 1808; Part II, 1832).

Finn MacCool (also MacCumaill, Fionn MacCumal, Finn, and in Scotland, Fingal), in Celtic mythology, a famed 3rd-century Irish hero-chieftain and the father of Oisín (Ossian) the poet. Only men who were skilled in poetry and had undergone a daunting physical ordeal were admitted to Finn's select band of warriors, the Fianna Éireann. Tales about Finn and his war band were first written down in the early 13th century as *The Interrogation of the Old Men*, and the Fenian cycle remains a vital part of Irish folklore.

Flora, the Roman goddess of fertility, flowers, and spring.

Flying Dutchman, the spectral ship alleged to appear in the vicinity of the Cape of Good Hope and to lure other vessels to their doom. According to one version of the legend, the ship's captain is condemned to wander until the Last Judgement because he refused to heed the warning of God not to sail round the Cape. The story is the basis of Wagner's opera *The Flying Dutchman* (1843).

Four Great Diamond Kings, the *Sì dà Tiānwáng.

Four Heavenly Kings, the *Sì dà Tiānwáng.

Frey (Yngvi, Old Norse, 'lord'), in Norse mythology, the son of the fertility god Njörd, the most handsome of the gods, ruler of peace who, together with his twin sister Freyja, brought together the two divine races, the Aesir and the Vanir. He took for his bride, Gerd, a giantess held in the clutches of the frost giants and demons of winter. Himself a god of fertility (his symbol was the boar) as well as the sun and rain, Frey owned *Skidbladnir*, a magic ship for carrying the gods, a horse named Bloodyhoof, and a victory-winning sword whose lack at the battle of Ragnarök (the doom of the gods) will lead to his defeat by Surt, the sovereign-guardian of Muspelheim, the land of fire.

Freyja (Freya) (Old Norse, 'Lady'), in Norse mythology, a goddess of wealth, fertility, love, battle, and death, from whose name Friday is derived. The sister of Frey and the most beautiful goddess of the Vanir, Freyja was granted the privilege of choosing one-half of the heroes slain in battle for her great hall in Fólkvangar, the god Odin taking the other half to Valhalla. The trickster god Loki stole her renowned necklace, Brisingamen, forged by dwarfs, and struggled with Heimdall, guardian of the gods' rainbow bridge, for possession of it.

Frigga (Frigg, Friga), in Norse and Germanic mythology, a fertility and domestic goddess, the wife of Odin, queen and mother of the gods, often confused with Freyja. Aware of impending harm to her son Balder, Frigga obtained a promise from all created things except the mistletoe to leave him untouched, and a branch of this plant in time killed him.

Fudo-myoo, in Japanese Buddhism, the guardian of wisdom and patron of ascetics. His wrath is directed against things liable to distract the pilgrim from the true path. A similar force is Aijen-myoo, who represents love transformed into desire for illumination.

Fúlù, the magical talismans, said to have been revealed by the Chinese philosopher Lǎozi, used in several schools of Daoism, and believed to ward off illness and demons. Now consisting of inscribed strips of paper, metal, or bamboo, Fúlù were originally contracts, the ones used by celestial masters being magic formulae, guaranteeing that agreements entered into with deities would be honoured.

Furies, in Greek and Roman mythology, three terrifying goddesses—Alecto, Megaesa, and Tisiphone—who pursued and punished the perpetrators of certain crimes, particularly murder of kin. They were generally represented as old women with snakes for hair.

G

Gabriel, the archangel and traditional messenger of God in the Bible whose name in Hebrew means 'mighty one of God'. He appeared to Daniel to interpret his visions, to Zacharias to foretell the birth of John the Baptist, and to Mary to announce that she would have a son called Jesus. In Islamic tradition the whole Koran was revealed to Muḥammad by the archangel Jibrīl (Gabriel), who is sometimes called the Faithful Spirit. In later legends he is alleged to have shown Adam the site of Mecca, to have helped the prophets, and to have announced the birth of Yahya (John) to Zachariya (Zacharias).

Gadarenes (Gerasenes, Gergesenes), the people of Gadara, a town south-east of the Sea of Galilee near which Jesus healed a man possessed by demons. According to the Gospel account, the demons hailed him as the Son of God and asked to be cast out into a herd of swine. This Jesus did, whereupon the herd rushed headlong into the sea and perished.

Gaea (Ge or Gaia), in Greek mythology, the goddess and personification of Earth, sprung from Chaos. She was the mother and wife of Uranus (Heaven), and their offspring included the Cyclopes, Cronus, and the Titans.

Galahad, in Celtic and medieval Christian mythology, the son of Lancelot and Elaine, daughter of the Grail King Pelles. In Malory's *Le Morte D'Arthur* (1470), Galahad alone among Arthur's knights could occupy the Siege Perilous (that seat at the Round Table reserved for the knight who should find the Holy Grail). He died in ecstasy after his vision of Christ with the Grail.

Ganesha (Ganesa), in Hinduism, the elephant-headed son of Shiva and Pārvatī. He is the patron deity of prosperity and learning, invoked at the beginning of literary works, rituals, and new undertakings as a remover of obstacles. Ganesha is usually depicted as a jolly, pot-bellied figure with four arms, a broken tusk, and riding a rat.

Ganymede, in Greek mythology, a shepherd boy who became Zeus' cup-bearer. Zeus was smitten by his beauty and turned himself into an eagle to carry him off to Olympus for the greater enjoyment of his company.

Gawain, in Celtic mythology, the son of

King Lot of Orkney and Morgan le Fay, Arthur's half-sister. An exemplary Knight of the Round Table, Gawain went in search of the Holy Grail. In his encounter with the Green Knight, however, his reputation was tested and found wanting. He died supporting Arthur in the rebellion led by Modred. He is linked with Gwalchmei, the sun-god of Welsh mythology.

Gayomart, in Persian Zoroastrian mythology, the first man, created at the same time as the Celestial Bull by Ahura Mazda. The evil Ahriman slew both, but as the seed of Gayomart fell, a rhubarb plant grew which produced the first human couple. From the Celestial Bull's seed animals were created. In Firdausi's epic, the *Shāh-nāmeh* (1010), Gayomart, called Kaiumers, is the first king.

genie *jinn.

George, St, patron saint of England. Little is known of his life, but his historical existence is now generally accepted. He may have been martyred near Lydda in Palestine some time before the reign of Constantine (d. 337), but his cult did not become popular until the 6th century, and the slaying of the dragon who threatened both a princess and the inhabitants of her city (possibly derived from the legend of Perseus) was not attributed to him until the 12th century. His rank as patron saint of England (in place of Edward the Confessor) probably dates from the reign of Edward III, who founded the Order of the Garter (*c*.1344) under the patronage of St George, who by that time was honoured as the ideal of chivalry.

ghost, the name given to the apparition of a dead person or animal, and to a person's soul or spirit. Ghost is used in the latter sense for God's spirit in the name of the Third Person of the Christian Trinity, the Holy Ghost.

ghoul (Arabic, *ghūl* (male), *ghūlah* (female)), in Islamic mythology, an evil spirit, sometimes classified as a jinn. Ghouls are believed to frequent graveyards, inhabit desert places, and lead travellers astray. In Islamic tradition Muḥammad is usually said to have denied their existence or ability to change shape.

giants, legendary beings of great size and brutish strength, shaped like humans. In Greek mythology, the giants were a race of huge creatures who attacked Olympus and were defeated by Zeus and the other gods. Giants figure in almost all mythologies and were often portrayed, as in Norse and American Indian legends, as the first race of people to inhabit the earth. Among the most familiar legends concerning giants are those of Jack the Giant Killer and David and Goliath.

Gideon, the Hebrew hero divinely appointed to help the Israelites regain Canaan after God had punished them with the Midianite invasion. Gideon, guided by an angel, first cleansed his father's house of Baal worship and then, by night, he attacked the Midianites, driving them beyond the Jordan. He refused the Israelite offer of kingship.

Gilgamesh, a legendary king of the Sumerian city-state of Uruk in southern Mesopotamia. The Gilgamesh epic, one of the best-known works of ancient literature, recounts the tale of the warrior king Gilgamesh, half divine and half human. To curb his power, the god Anu creates a savage, Enkidu, who is brought up in the wilderness. They meet in a trial of strength which Gilgamesh wins. Enkidu becomes his beloved companion, but is fated to die for having slain the divine bull of Ishtar, goddess of love. Gilgamesh, obsessed with his own mortality, wanders the earth in a quest for the plant that bestows eternal youth. He finds it, only to have it seized from him by a serpent which, by sloughing off its skin, appears to have achieved the renewal of life that has eluded Gilgamesh.

Glastonbury tor, a site in the Mendip district of Somerset, England, to which St Joseph of Arimathea reputedly brought the Holy Grail of the Last Supper; planting his staff into the ground, it grew into the Glastonbury thorn that flowers on Christmas Day. In the early Middle Ages it was claimed that the graves of King Arthur and Queen Guinevere had been discovered at Glastonbury, and that St Patrick had been laid to rest here.

goblin, a mischievous, ugly demon of European folklore. The name is probably derived from the Latin, *gobelinus*, a spirit, which is also related to *kobold*, the German demon of mines.

Gog and Magog, comprehensive terms used in Christian and later Hebrew literature to denote the powers of evil. In the Old Testament Ezekiel referred to Gog as the leader of a people and to Magog as their land. In the New Testament Book of Revelation they are the nations under the dominion of Satan. In an independent British legend, Gog and Magog are survivors of a race of giants destroyed by Brutus the Trojan, legendary founder of London.

Golden Calf, in the Bible, the idol, possibly representing the sacred bull Apis or the Canaanite fertility god Baal, that the Hebrews asked Aaron to construct while his brother Moses was absent on Mount Sinai. Moses had the Calf melted down, pulverized, and mixed with water. The people were then made to drink the mixture. Those who had not worshipped the Golden Calf and had remained faithful to their religion survived the ordeal, while the idol-worshippers died of a plague.

golem (Hebrew, 'embryo or anything incompletely developed'), in Jewish folklore, an effigy made of clay and brought to life by means of a charm. In the Middle Ages, the golem was portrayed as an automatonlike servant who carried out his master's commands. He later assumed a more powerful form as general protector of persecuted Jews. The most famous legend about a golem centres around Rabbi Juda Löw (*c*.1525–1609) of Prague, who was forced to destroy the creature which he had created after it had run amok. This legend forms the basis for Gustav Meyrink's novel *Der Golem* (1916).

Goliath, the Philistine warrior giant from Gath who challenged the Israelites in the time of Saul. He was slain by the boy David with a stone from his sling.

Gorgons, in Greek mythology, three female monsters who were sisters; they had wings, claws, and human heads with snakes for hair. Their names were Euryale, Medusa, and Sthenno. Medusa, loved by Poseidon, was the only one who was mortal. Her gaze was so terrifying that it turned anyone who beheld her to stone, but she was killed by Perseus, who looked at her reflection in a polished shield. Chrysaor and the winged horse Pegasus sprang from her blood when she died.

Graces, in Greek mythology, daughters of Zeus and of Hera, usually three in number—Aglaia (brightness), Euphrosyne (joyfulness), and Thalia (bloom).

Graiae, in Greek mythology, Pemphredo, Enyo, and Deino, goddesses represented as grey-haired women, the personification of old age. They had only one eye and one tooth between them, and were the protectresses of their sisters, the Gorgons. Perseus contrived to steal the eye, and so was able to surprise the Gorgons.

Grail, the Holy, in Arthurian legend identified as the cup of the Last Supper. The Grail was later used by Joseph of Arimathea to catch the blood of the crucified Christ; in some versions, he brought it to north Wales at the end of his lengthy wanderings. The Grail appeared uniquely to Galahad, a Knight of King Arthur distinguished by his great purity, who was overwhelmed by its beauty and died in ecstasy.

Grendel, the man-eating water mon-

ster of *Beowulf*, an Early English epic poem dating from the 8th century. Grendel plagued the mead hall of the Danish king Hrothgar. Beowulf, who later became king of the Geats in Scandinavia, grappled with him, tearing off the monster's arm and mortally wounding him. The following night Grendel's mother killed Hrothgar's trusted friend, Aschere, in revenge, but she in turn was killed by Beowulf.

griffin, in European and Near Eastern mythology, a fabulous creature with an eagle's head and wings, a lion's body, and sometimes a serpent's tail. The Old Testament cherubim are often depicted as griffin-like creatures. The chariots of the Greek gods were said to be drawn by griffins. In medieval times griffin-claws were thought to have magical properties.

Guāndì, the 3rd-century Chinese general Guānyǔ, a popular hero in Chinese literature who became the Daoist god of war. In his role as protector of the peace against internal and external foes, Guāndì was venerated by state officials. In popular Chinese belief he was Fúmodàdì, who banished demons. Depicted as a bearded giant with a scarlet face, Guāndì is frequently portrayed in armour, carrying a halberd, and standing beside his horse.

Guānyīn, the Chinese name for the *bodhisattva* ('Buddha-to-be') Avalokitesvara, often considered a great mother goddess in China, Japan, and Korea. Believed to manifest herself in diverse forms whenever help is needed, she is invoked especially by individuals threatened by water, demons, fire, or sword, and also by childless women. In China Guānyīn is sometimes depicted with a thousand arms and eyes, bearing a child, or accompanied by a maiden or by the protector of teaching, Wéituó.

Guinevere *Arthur.

H

Hades, in Greek mythology, a name applied to both the god of the underworld and to the underworld itself, where the spirits of humans went after their earthly death. Hades the god was the brother of Zeus and the husband of Persephone. Hades the place was thought of as being located underground or in the far west, separated from the world of the living by the River Styx and guarded by Cerberus. It was a place of gloominess rather than torment. Souls that the gods wished to punish were sent to Tartarus—a terrible abyss. Souls of the blessed went to Elysium.

Hahgwehdiyu, in Iroquois mythology, the benign creator deity who shaped the sky with the palm of his hand, and placed his mother's face in the sky as the sun, while from her breasts he made the moon and stars. To the earth he gave her body as a source of fertility. His evil twin brother Hagwehdaetgah placed darkness in the west, invented earthquakes and thunderstorms and, after a long struggle, was exiled to a subterranean realm.

Ḥallāj, al-Āsrar (al-Husayn ibn Mansur) (Arabic, 'the carder of hearts and consciences') (857–922), in Islam, renowned mystic and poet of Persian origin, executed for heresy in Baghdad, and the patron of cotton-carders. A controversial figure in his life-time, Ḥallāj was credited with miracles and suspected of wizardry. His painful death for claiming 'I am the Truth' (meaning God), which became a statement of Sufi mystics, is a frequent theme in Islamic literature and serves as a model for suffering lovers. His poetry with its mystical images of God's immanence has likewise remained popular.

Ham, the second son of the Hebrew figure Noah, father of Cush, Mizraim, Phut, and Canaan and supposed ancestor of the southern peoples such as the Ethiopians, Egyptians, and Babylonians. Egypt was referred to as the 'Land of Ham'.

Hanuman, in Hindu and Buddhist mythology, the agile monkey-chief son of the wind-god Vayu, whose semi-divine nature enabled him to fly and change his shape and size. As general of the army of the monkey-king Sugriva, Hanuman helped Rāma defeat Rāvaṇa's forces and save Sītā. Chinese Buddhist legend recounts how, as Sun Houzu, Hanuman helped the great Chinese pilgrim, Xuanzang, on his pilgrimage to India.

Hare, the trickster hero of countless animal stories, widespread throughout the desert and savannah lands of Africa. Clever but improvident, Hare usually manages to outwit the bigger animals. Sometimes, however, he is too clever and is outwitted in his turn, as in his race with Tortoise. The Yoruba of Nigeria explain Hare's long ears and savannah home by the tale of how, in a time of drought, Hare refused to sacrifice the tips of his ears as the other animals did to buy hoes to dig a well. When he then proceeded to drink the well-water and wash himself in it, he was driven into the grasslands by his angry compatriots. Taken by African slaves to the American continent, Hare tales became the source of the Brer Rabbit, Brer Fox, and Brer Tarrypin tales in *Uncle Remus: His Songs and his Sayings* (1880) by Joel Chandler Harris.

harpy, in Greek mythology, a type of loathsome monster with an insatiable hunger, conceived of as a repulsive bird of prey with a woman's head and trunk.

Harut and Marut, in Islam, the two fallen angels, recounted in the Koran, who taught sorcery and magic and sowed discord between husband and wife among the citizens of Babel. According to Islamic legend Harut and Marut chose suffering in this world as their punishment for succumbing to the sexual charms of a woman and for killing the man who had witnessed their crime. In a similar Jewish legend the fallen angels are 'Uzza and 'Azael.

Hātim al Ta'i (6th century), an Arab poet and exemplar of the pre-Islamic knight. Renowned for his victories and his magnanimity to those he defeated, Hatim's hospitality, inherited from his mother, was proverbial. It is alleged that he rose from his tomb after death to entertain travellers. The stories of his exploits appear in several languages throughout the Islamic world.

Haviti *Kishibojin.

Hawwa *Eve.

Hebe, in Greek mythology, the goddess of youth. She was the daughter of Zeus and Hera and the handmaiden of the gods.

Hecate, in Greek mythology, a goddess with many functions but mainly associated with black magic—the patron of witches and sorcerers. At night she wandered with the ghosts of the dead, haunting graveyards and crossroads, where offerings were put out for her.

Hector, in Greek mythology, the greatest hero on the Trojan side in the Trojan

War. He was the son of Priam and Hecuba and was killed in combat by Achilles.

Hecuba, in Greek mythology, queen of Troy, the wife of Priam and the mother of Cassandra, Hector, and Paris. After the fall of Troy she was given as a captive to Odysseus.

Heitsi-eibid, a legendary hero in the Khoisi myth of the people of Namibia and Cape Province, South Africa. The son of a cow and of the grass she had eaten, he is renowned as a magician, hunter, and fearless fighter. Heitsi-eibid rid the Khoisi tribe of the monster Ga-gorib ('thrower down'). Although allegedly killed on numerous occasions, he is believed to resurrect himself.

Hel, in early Norse mythology the world of the dead, ruled over by the goddess of death, synonymous with her kingdom. Another name for the kingdom was Niflheim, the world of mist and darkness. One of its realms was Nāströnd, the shore of corpses; here the dragon Nidhogg kept murderers, adulterers, and perjurers imprisoned in his castle, inflicting torment on them and sucking the blood from their bodies.

Helen, in Greek mythology, the most beautiful of mortal women. She married King Menelaus of Sparta, but was abducted by Paris and carried off to Troy, thus causing the Trojan War. When Troy fell ten years later, she returned to Menelaus.

Helios, in Greek mythology, the sun personified as a god.

Hephaestus, the Greek god of fire and metalworking, represented as a lame blacksmith. He was the son of Zeus and Hera and the husband of Aphrodite.

Hera, in Greek mythology, the supreme goddess, the sister and wife of Zeus. She was constantly at odds with Zeus because of his infidelities and she often plotted against his lovers and his offspring.

Heracles *Hercules.

Hercules (Heracles), in Greek mythology, the greatest of all heroes, famed for his strength and courage. He was the son of Zeus (from whom he inherited an insatiable sexual appetite) and Alcmene, a mortal. He performed a great variety of stupendous feats, triumphing over evil against all odds, but he is most famous for his Twelve Labours, undertaken as a penance for killing his wife and children in a fit of madness induced by Hera. The Labours were: killing the Nemean Lion; killing the Hydra, a seven- (or nine-) headed water serpent; capturing the Erymanthian Boar; capturing the Hind of Ceryneia; driving away the man-eating Stymphalian Birds; cleaning the Augean Stables; capturing the Cretan Bull; taming the horses of Diomedes; carrying off the girdle of Hippolyta, queen of the Amazons; capturing the Oxen of Geryon; carrying off the golden apples of the Hesperides; capturing Cerberus. After his death (brought about by treachery) he ascended to the heavens and became the only mortal to be raised to the same level as the Olympian gods. Hebe was given to him as his bride.

Hermaphroditus, in Greek mythology, the son of Hermes and Aphrodite who loved a nymph—Salmacis—and prayed to the gods to unite them in one body, which they did. Hence the term 'hermaphrodite', referring to a creature having both male and female sexual characteristics.

Hermes, in Greek mythology, the god of commerce, invention, theft, and cunning, also the messenger and herald of the other gods. He was the patron of travellers and conducted the souls of the dead to Hades, and is identified in Roman mythology with Mercury.

Hermes Trismegistus ('the thrice-great Hermes'), the Greek name for the Egyptian god Thoth, identified by the Greeks with Hermes. In Islam, Hermes Trismegistus is a thrice-incarnated figure, traces of which also exist in Egyptian legends. The first Hermes is identified with Akhnukh (the biblical Enoch and Islamic Idris), alleged in legend to have built the Egyptian pyramids and inscribed the scientific achievements of man inside them to preserve them from the Flood. The second lived after the Flood in Babylonia and revived scientific study. The third wrote in Egypt about various sciences and crafts. According to one legend Hermes hid the writings on pre-Flood science in a cave. Found by Balinus (Apollonius of Tyana), they were acquired by Aristotle, who gave them to Alexander the Great. Hidden in a monastery in 'Ammuriya at Alexander's command, they were discovered by the caliph al-Mu'tadid. Many philosophical books and collections of ethical sayings have been attributed to Hermes, including a large number of extant Arabic writings.

Hero, in Greek mythology, a beautiful priestess of Aphrodite at Sestos on the European shore of the Hellespont (the strait—now called the Dardanelles—separating Europe from Asia Minor). She was loved by Leander, a youth who lived on the opposite shore and swam across every night to see her. One stormy night he was drowned, and Hero in despair threw herself into the sea.

Hesperides, in Greek mythology, a group of nymphs, sisters (their number varies from three to seven in different accounts), who guarded the golden apples given to Hera when she married Zeus. The apples grew in a remote garden (the Garden of the Hesperides) and were guarded by a fearsome dragon. One of Hercules' Labours was to carry off the apples.

Hestia, in Greek mythology, the goddess of the hearth and symbol of home and family. She was a sister of Zeus. A hearth was consecrated to her in every home, and sacred fires were kept burning to her in every Greek city.

Hiawatha (Ojibwa, 'he who makes rivers'), the legendary North American Indian of the Onondaga tribe who, according to Iroquois tradition, lived in the late 16th century and was co-founder of the Iroquois League of Five Nations. Credited with magical powers, Hiawatha was said to have taught people agriculture, navigation, medicine, and the arts.

Hina, in Oceanic mythology, the goddess of many faces who presided over the forces of life and death. In some traditions, Hina was created by the sky god Tane from sand, and was said to be the sister of the trickster god, Maui. In time she became the consort of Tane, and bore him a daughter, Hine-Titama, who in turn became her father's wife. Accompanying her brother Ru on a voyage of exploration through the skies, Hina alighted on the moon. Seduced by its great beauty, she made it her home.

Hine-Titama (Hine-i-tau-ira, Hine-nui-te-po), in Oceanic mythology, the goddess of the underworld and of death. In some traditions, she was the daughter of the goddess Hina and the sky god Tane, who took her for his wife. Discovering that her husband was also her father, she killed herself and descended to the underworld, where she became the goddess of death.

hobgoblin, a mischievous imp or terrifying apparition of English folk belief. Also called Puck and Robin Goodfellow, the name hobgoblin is formed from Hob, the diminutive form of Robin, and goblin.

Horus (Hor, Horos), the Egyptian falcon god to the sky, magically conceived by Isis from her dead brother-husband Osiris. Whilst Osiris ruled the dead in the underworld, Horus ruled the living. Thus every Pharaoh was identified in life with Horus and in death with Osiris. Set, his uncle, plucked out Horus' eyes during a dispute over the kingship of Egypt. The act of plucking out Horus' eyes divided them into pieces, from which the Egyptians expressed fractions. Isis healed the

eyes, so that they were once again complete and restored to Horus' head. Amulets in the form of the Eye of Horus were powerful protection against illness, misfortune, and malignant influences. Depicted as a hawk-headed man, the falcon god is Re-Horakhty, an assimilation between Re and Horus. Called Harpocrates (Horus the child), he may also be depicted as a naked child wearing a royal crown and sitting on Isis' lap.

Hosea, the Hebrew prophet who preached against the sins of the northern Kingdom of Israel in the 8th century BC. The story of his marriage (which may be allegorical to show God's love for Israel, a wayward nation) recounts the prophet's forgiveness of his faithless wife, Gomer, and his apparent remarriage to her after she had deserted him.

houri (Arabic *ḥawira*, 'black-eyed'), in Islam, one of the permanent virgins who await good Muslims in Paradise, referred to as 'purified wives', and 'spotless virgins' in the Koran. Later tradition states that the number of houris allotted to each man for co-habitation will be proportionate to his fasting and good works on earth. For some Muslims the houri is a symbol of a spiritual state rather than of a physical being.

Hud, in Islamic belief, one of the four prophets sent specifically to the Arabs. Numerous legends surround Hud, who, in the Koran, preached monotheism unsuccessfully to his people, 'Ad, a tribe 'who built monuments in high places' and who are associated with Iram of the Columns, a legendary terrestrial paradise.

Huìnéng (638–713), the Chinese sage, originally a pedlar of firewood who was the legendary sixth patriarch of a dominant School of Zen Buddhism in China and in Japan. Huìnéng's radical doctrine of potentially sudden enlightenment through inner tranquillity, that is, the state of seeing one's own nature, freed of erroneous distraction, created a wide schism with the Northern School, which taught a gradual enlightenment achieved through sitting in motionless meditation over a long period of time. Huìnéng became venerated as the founder of the Southern School of Zen.

Huitzilopochtli (from the Aztec, 'humming bird' and 'left'), the Aztec war-god, god of the sun, lightning, and storms, and protector of travellers. Conceived when some hummingbird feathers, associated in Aztec belief with the soul of a dead warrior, fell from heaven, Huitzilopochtli emerged from the womb of the earth-goddess Coatlicue fully armed. He saved her from the anger of her 400 other offspring, the stars of the southern sky, and the moon goddess. His animal disguise was the eagle. Often portrayed with hummingbird feathers and snakes, Huitzilopochtli's primary weapon was a turquoise snake. He was the pre-eminent god to whom large numbers of human sacrifices were made at the time of the Spanish conquest.

Ḥusain, the second son of 'Ali and of the Prophet Muḥammad's daughter Fāṭima, and martyred third Imam of *Ithnā 'Asharī* in Shī'ite Islam. Refusing to swear allegiance to the Umaiyad caliph Yazid, Ḥusain and his small force of warriors were massacred. Under the Buyids, and later even more so under the Iranian Safavids, the martyrdom of Ḥusain and his companions became a central event in Shī'ite history, with re-enactments of the tragedy in the form of passion-plays (*ta'ziyas*) performed each anniversary, and processions where many followers cut and beat themselves with knives and chains. Ḥusain's tomb in Kerbala is the most important Shī'ite shrine. According to one legend, Ḥusain married Shahrbanu, a daughter of Yazdagird, the last Sassanian king of Persia. The Safavids, who transformed Shī'ism into the state religion of Iran, claimed descent from this union.

Hyacinthus, in Greek mythology, a beautiful youth beloved by Apollo and Zephyrus. He returned Apollo's love, inflaming Zephyrus with jealousy. In revenge, when Apollo and Hyacinthus were playing at quoits, Zephyrus blew Apollo's quoit so that it struck Hyacinthus and killed him. Apollo changed his blood into the flower—hyacinth—that bears his name.

Hydra, in Greek mythology, a many-headed monster slain by Hercules as one of his Labours. When Hercules cut off one head, two more grew in its place, but he overcame this by getting a companion to hold a burning brand to the wound as each head fell.

Hylas *Argonauts.

Hymen, in Greek and Roman mythology, the god of marriage, represented as a young man carrying a torch and a veil.

I

Iblīs, in Islam, a fallen angel also called Shayṭan, the personification of evil and the Muslim counterpart of Lucifer. In the Koran it is said that Iblīs refused God's command to bow down before Adam through pride. Instead of immediately being cast into exterior darkness for his sin, God agreed to allow him to tempt mankind until the Day of Judgement. In Islamic belief Iblīs is often considered a jinn, and chief of various demons who afflict the world.

Icarus, in Greek mythology, the son of Daedalus. Imprisoned by King Minos in Crete, his father built wings of wax and feathers for their escape. Icarus flew too close to the sun and fell to his death when the wax melted.

Idris, described in the Koran as a prophet; the Islamic patron of tailors, craftsmen, and warriors. Usually identified by Muslims with the biblical Enoch, Idris is believed to have transmitted divine revelations in several books. He is alleged to have lived for 365 years and then ascended to heaven without dying. He is considered the founder of philosophy and of the sciences, and is credited with the invention of writing, sewing, and various forms of divination.

Ifa, the Yoruba name given to a West African system of divination. Nuts, often threaded into a chain, are thrown. The patterns of successive throws are traced in sand. Associated with different signs are *ese* (verses), which contain the results of the divination, and are an important part of Yoruba oral literature. In Yoruba tradition, the Ifa system originated on earth with a diviner, also called Ifa, but its veracity stems from the deity Òrunmìlà, to whom the foundation of the city of Ifè is sometimes attributed. In the Ewe language in Togo and Ghana Ifa-type divination systems are called Afa or Fa.

Ifé, a present-day city of Nigeria, and formerly the capital of the Yoruba kingdom of Ifé, famed for its bronze and terracotta works, and allegedly founded by the diviner Ifa. According to Yoruba belief Ifé, meaning wide, is the place where creation began, and Ilé-Ifé ('the house of Ifé'), the sacred city, is the Yoruba people's place of origin.

'ifrīt (fem. 'ifritah), in Islam, a winged creature of great stature, formed of smoke, and usually identified as a powerful jinn. 'Ifrīts are generally considered malevolent, but as depicted in *The Tales of*

the Thousand and One Nights, they can be good. They are believed to live underground, to haunt ruins, to sometimes marry humans, and to be susceptible to magic but not to weapons. In Egypt, the souls of the deceased, particularly of those who had met a violent death, are sometimes called 'ifrīts.

Illuyankas, in Hittite mythology, the dragon personifying evil forces, slain by the weather god Tarhun or Teshub. Illuyankas is the initial victor, but is then overcome, tricked by the goddess Inares, aided by her lover Hupasiyas.

Imhotep, architect in ancient Egypt of the Step Pyramid of Zoser (*c*.2630–2611 BC) at Sakkarah. In later times he was the patron of craftsmen, and identified as a sage and physician. Imhotep is depicted with cropped or shaven head, seated with a roll of papyrus on his lap. The Greeks identified Imhotep with Asclepius because of his healing powers. Imhotep is important among Egyptian gods as one of the few mortals to achieve a sort of sainthood. He was thought to have been the son of Ptah of Memphis and a mortal woman called Khredu-ankh.

Indra, in Persian mythology a minor, evil god, but in Hinduism the chief of the Vedic gods with Agni and Surya. Lord of the thunderbolt and rain, Indra is the divine patron of the warrior caste who slew Vṛtra (Ahi), the serpent of drought, and rescued the sacred cows of the gods from the *asuras* (demons). Indra is often depicted holding a thunderbolt and bow, and riding either a chariot or his white elephant, Airāvata. He is sometimes referred to as the 'thousand-eyed' because his body is covered in marks resembling eyes or yonis, symbols of the female sexual organ, placed on him as a curse for seducing the wife of a sage.

Io, in Greek mythology, a beautiful princess beloved by Zeus. He changed her into a heifer to escape the jealousy of Hera, but she discovered the ruse and sent a gadfly to torment the unfortunate animal, which was forced to wander over the face of the earth. Eventually Io recovered her human shape and bore Zeus a son.

Iphigenia, in Greek mythology, a daughter of King Agamemnon and his queen Clytemnestra. To appease the wrath of Artemis, whose stag he had killed, Agamemnon offered Iphigenia as a sacrifice to obtain fair winds so that the Greek fleet could sail against Troy. Artemis was moved by Iphigenia's innocence, however, and saved her by substituting a deer for the sacrifice. Iphigenia was then entrusted with the care of Artemis' temple at Tauris.

Iram of the Columns *'Ad.

Isaac, the Hebrew patriarch, son of Abraham and Sarah, husband of Rebecca, and father of Jacob and Esau. Isaac's willingness to be sacrificed to God by Abraham has linked Isaac symbolically to Jesus in Christianity, where he is seen as a prefiguration of Jesus and his Passion. Ishaq (Isaac) is mentioned in the Koran as a gift from God to Ibrahim (Abraham). Muslims generally consider that Ibrahim intended to sacrifice Ishmael, not Ishaq.

Isaiah, the Hebrew prophet who played an important prophetic and political role during the reigns of four kings of Judah. The biblical Book of Isaiah incorporates many prophecies from a later age, concerning the Babylonian Exile, the victories of Cyrus, and the deliverance of the Jews. The fulfilment of these prophecies increased Isaiah's fame posthumously. In Christianity his prophecies concerning the Messiah are seen as fulfilled in the person of Jesus. According to one tradition Isaiah was martyred in the reign of Manasseh by being sawn in two.

Ishmael, the son of the Hebrew patriarch Abraham and Hajar (Hagar), the Egyptian serving maid of Abraham's wife Sarah. Cast out with his mother into the desert because of Sarah's jealousy, Ishmael in turn had twelve sons from whom twelve northern Arabian tribes are supposed to descend. In Islam Ishmael is considered a Prophet Messenger (*nabī rasūl*), and patriarch. In the Koran, the son whom Ibrahim (Abraham) was prepared to sacrifice is not named; he is thought by Muslims to be Ishmael, and the site to be near Mecca at Mina. The sufferings of Hajar and her son in the desert are commemorated in the pilgrimage rites at Mecca (see *Zamzam), where Ibrahim is said to have brought and then visited them, and where both are alleged to be buried. In Islamic tradition Ishmael helped Ibrahim rebuild the Ka'bah. He married a woman from the Jurhum tribe, and the Prophet Muḥammad was a descendant of this union.

Ishtar, in Babylonian mythology, a fertility and mother-goddess, the personification of Venus, sister-wife of Tammuz, and scorned lover of Gilgamesh. As daughter of the Akkadian sky-god An, Ishtar was worshipped as a goddess of love and desire. As the moon-god Sin's daughter, she was the warrior goddess, worshipped in Assyria, who sent the vanquished to the underworld.

Isis, Ancient Egyptian goddess of magical power and divine mother, sister-wife to Osiris and mother of Horus. In human form she is depicted wearing either the throne-sign of her name or the cow's horns and sun disc on her head. Hathor, goddess of love and childbirth, is often assimilated with Isis. Her magical power is illustrated by her power to revive her dismembered husband. As divine mother she is depicted suckling the king who was seen as her son. The cult of Isis was widespread, although the centre may have been at Busiris in the Delta. Isis was worshipped throughout the western world, most notably at Rome in the first century BC, and she was identified with the goddesses Astarte, Minerva, Venus, and Diana of other cultures.

Israel, the name ('he that strives with God') bestowed on the patriarch Jacob from whom the twelve biblical tribes of Israel are descended. It is also the name given to the northern kingdom, formed in 931 BC, consisting of the ten tribes who broke away from Judah and Benjamin after Solomon died. They were conquered and dispersed by the Assyrians in 721 BC. Referred to as the ten 'Lost Tribes', their fate is the subject of conjecture. Used in a religious rather than a political sense, Israel refers to the Jewish nation as God's 'chosen people'.

Israel, the Twelve Tribes of, in the Bible, the twelve tribes named after the ten sons of Jacob (whose name was changed to Israel by God) and the two sons of Jacob's son, Joseph. Jacob's first wife bore him six sons: Reuben, Simeon, Levi, Judah, Zebulon, and Issachar. Two other tribes, Gad and Asher, were named after sons born to Jacob and Zilpah, Leah's maidservant. Two further tribes, those of Dan and Naphtali, were named after sons of Jacob by Bilhah, the maidservant of his second wife, Rachel. Rachel herself bore Jacob two sons, Joseph and Benjamin. Two tribes were named after Joseph's sons, Manasseh and Ephraim. In 930 ten of the tribes formed the kingdom of Israel in the north, and the two other tribes, Judah and Benjamin, set up the kingdom of Judah in the south. Following the conquest of the northern kingdom by the Assyrians some 200 years later, the ten tribes (the 'Lost Tribes') were gradually assimilated by other peoples.

Israfil, in Islamic tradition, the counterpart of Raphael, and the archangel of awe-inspiring aspect whose trumpet clarion will announce the Day of Resurrection and Last Judgement from a holy rock in Jerusalem. Although not mentioned by name in the Koran, in Islamic tradition Israfil appears in several legends and is credited with having trained Muḥammad for prophethood for three years before Jibril (Gabriel) revealed the Koran to him.

Ixchel, in Maya mythology, a malevolent goddess who had to be appeased by human sacrifices. She assisted the sky serpent in creating the deluge, and fre-

quently caused destruction in tropical storms. The consort of Itzamna, 'lord of the heavens', she carried on her head writhing serpents, and had crossbones depicted on her robes.

Izanagi and **Izanami** (Japanese, He Who Invites and She Who Invites), in Shinto mythology, the central deities in the creation myth. They were descended from seven pairs of brothers and sisters who had appeared after heaven and earth had separated out of chaos. A mighty bridge floated between the heavens and the primeval oceans; standing on this, Izanagi and Izanami stirred the waters below with a jewelled spear to form the first land mass. Their union gave birth to the islands of Japan and to various deities. In giving birth to the fire-god Kagutsuchi (or Homusubi), however, Izanami was fatally burnt and descended to the land of darkness, Yomi. When Izanagi ventured into the underworld to seek his dead spouse, he found her alive but imprisoned in a decomposing body. Fleeing, Izanagi bathed in the sea to purify himself and in doing so gave birth to a number of deities, among them Amaterasu, the sun goddess, from his left eye, the moon-god Tsukumi from his right eye, and the storm god Susan-o-o from his nostrils. In Shinto religion, the purification practised in the *harai* ceremony commemorates Izanagi's submersion in water.

Izanami *Izanagi.

J

Jacob (Hebrew 'trickster', or 'supplanter'), Hebrew patriarch, the younger twin son of Isaac and Rebecca, whose favourite he was, and the shepherd ancestor of the Twelve Tribes of Israel. By trickery Jacob won the birthright and the father's blessing due to his older brother, Esau. Esau's anger caused Jacob to flee to their ancestral land of Haran, where he had a vision at Bethel of a ladder from heaven to earth and of his own promised blessing and prosperity. While in Haran he married his uncle Laban's two daughters, Leah and Rachel. Crossing a river on his way back to Canaan he wrestled until dawn with a mysterious, unseen presence at Peniel and for his perseverance received from it the name of 'Israel' ('he that strives with God'). Ya'qub (Jacob), in early Meccan *shūras* of the Koran, appears as the brother of Ishaq (Isaac), and the son of Ibrahim (Abraham). He is regarded as a prophet, and was temporarily blinded by the grief he experienced at the loss of Yusuf (Joseph). In post-Koranic legend, Ya'qub and Esau are said to have fought in their mother's womb over who should be born first. Ya'qub allowed his brother preference to spare his mother pain.

Jade Emperor, the *Yùhuáng.

Jael, Hebrew heroine, the wife of Heber the Kenite (believed to be descendants of Cain), who, when the Israelites were being attacked by the Canaanites, gave refuge and hospitality in her tent to an enemy of her people, the Canaanite commander, Sisera. Lulling him to sleep, she killed him by driving a tent peg through his skull.

Jagannatha (Sanskrit, 'lord of the world'), in Hinduism, the form under which the god Krishna is worshipped in Puri, Orissa (India). His image is placed on a chariot and drawn through the streets by his followers during the annual festival of Asadha, held in June or July. Hundreds of devotees are required to pull it, the journey taking several days. The English word juggernaut is derived from this name.

James (the Great), Christian saint, one of the Twelve Apostles, the fisherman son of Zebedee and Salome, and brother of John. Put to the sword by Herod Agrippa in AD 44, he was the first Apostle to be martyred. Later Christian documents allege James preached and was buried in Spain, where he is greatly venerated.

James (the Just), Christian saint, referred to as 'the Lord's brother' in the Gospels, and regarded as the Apostle James the Less by Roman Catholics. James, with Peter, became leader of the Christian Church in Jerusalem. He was reputedly put to death by the Sanhedrin. The Epistle of St James, ascribed to him, stresses that faith without action is valueless.

Jamshid, the legendary Persian king whose pride in his own power caused God to plunge his empire into chaos. In the epic poem by Firdawsi, *Shāhnāmah* (*c.*1000), Jamshid is credited with the invention of iron weaponry, the art of medicine, the making of clothing, jewellery, and perfume. He is eventually sawn in two by the evil king Zahhak. Much of Firdawsi's account of Jamshid is drawn from the mythology that accumulated around the figure of Yima, progenitor of the human race in ancient Iranian lore.

Janus, in Roman mythology, the god of gates, doorways, and bridges. He was represented with two heads facing opposite ways, suggesting vigilance—looking both fore and aft. He was regarded as the guardian of the Roman state during war, when the doors of his temple were left open; in times of peace they were closed.

Jason, in Greek mythology, one of the most famous of heroes. In order to recover his father's kingdom, usurped by his uncle Pelias, Jason undertook to capture the Golden Fleece (the magic fleece of a winged ram) owned by King Aetes of Colchis. Accompanied by the Argonauts and helped by Medea (Aetes' daughter) he accomplished this. He recovered his kingdom and married Medea, but later deserted her for Creon's daughter. In revenge, the passionately jealous Medea killed her own children and Jason's new wife.

Jataka Tales, the collection of stories dealing with the former lives of the Buddha in which he appears in a variety of beings, for example as king, outcast, god, or elephant, to exhibit some important virtue. Although contained in the Canon of the *Theravāda*, the Jataka are found throughout the Buddhist world and provide the themes for many Buddhist works of art. They have been absorbed into the folklore of many countries.

Jehovah (from the Hebrew name for God, YHWH (IAO), traditionally meaning 'I am that I am'), a name used by Christians to describe God. Known as the Tetragrammaton (Greek, 'four letters'), it was replaced from *c.*300 BC by the word *adonai*, Hebrew for 'lord'.

Jehu, the army commander and then king of Israel who was called upon by the prophet Elisha to destroy the house of Ahab and the Baal-worship it had encouraged. Famous for his swift action and for driving his chariot at great speed, Jehu has become a synonym for a fast driver.

Jephthah, the mighty warrior and ninth judge of Israel who led the Israelites against the Ammonites. Vowing that, if God granted him victory in battle, he would sacrifice the first living creature that met him on his return, he felt obliged to order the death of his daughter in fulfilment of the vow.

Jeroboam II, Hebrew leader, the powerful fourteenth king of Israel who delivered his country from the Syrian yoke. During his reign, however, idolatrous cults flourished and the wealthy lived a life of licentious immorality at the expense of the poor. The prophets Amos and Hosea denounced Jeroboam and the oppressive corruption prevailing in the nation.

Jezebel, the Phoenician princess, married to Ahab of Israel, whose name has become synonymous with female wickedness. She encouraged the cult of Baal and persecuted the Hebrew prophets. Denounced by Elijah for this and for her role in Naboth's death, Jezebel was flung from a palace window and eaten by dogs when Jehu led the rebellion which terminated Ahab's dynasty.

Jīāngxī *Mǎzŭ Dàoyī.

Jibrīl *Gabriel.

Jimmu Tenno (*c*.7th century BC), the legendary first emperor of Japan, founder of the current imperial dynasty, and alleged descendant of the sun-goddess Amaterasu and of the storm god Susan-o-o. It is widely considered that the subjugation, by Jimmu, of Japan in the 7th century BC probably refers to Japanese imperial expansion eastward at a later date. The story is reported in the *Kojiki* (Japanese, 'Records of Ancient Matters') and *Nihon Shoki* ('Chronicles of Japan'), the earliest official history of Japan, compiled from oral tradition in the 8th century AD.

jinn (djinn, or genie), in Islam, an order of spirits, lower than angels, sometimes referred to as *jann*, who, being of flame or air, are believed to have the power of assuming human or animal form and exercising supernatural influence over men and women. Said in the Koran to have been created from smokeless fire, jinn are given legal status in Islam. Muḥammad is alleged to have converted some, and the Mosque of the Jinn in Mecca commemorates the place where he came to an agreement (*bay'ah*) with them. In Islamic belief there are various orders of jinn, some beneficent, others malevolent. Among the most evil are the five sons of Iblīs, who, between them, are thought to cause much of the disaster and sin in the world. Favourite figures in stories throughout the Islamic world, the jinn are alleged to have helped build Solomon's Temple after he gained mastery over them. In Persian mythology jinn are sometimes referred to as divs when evil (male and female), and paris when good (female). (See also *ghoul, *ʿifrīt, *Satan.)

Jizo, the Japanese name for the Mahāyāna Buddhist *bodhisattva* ('Buddha-to-be'), Kshitigarbha, known as Dìzàng in China. Numbered among the eight great bodhisattvas, Jizo is believed to be concerned with the welfare of children, travellers, and lost souls, and to be the guardian of highways and of mountain passes. He is sometimes shown in a series of six images, which include his depiction as the monks Jizo Bosatsu and Raku Jizo, and as the war-god Shogun Jizo on horseback.

Job, the Hebrew hero of the biblical Book of Job, which deals with the fundamental problem of undeserved suffering. Afflicted through Satan with the loss of family and possessions, and then with disease, the upright Job accepted all as the will of God. Only after friends had argued with him that suffering was the result of sin, did Job, sure of his faithfulness, lose patience and question God's omnipotence. In the epilogue, probably a later addition, he is restored to his former fortunes when he submits again to the will of God, which, however, remains mysterious and inscrutable. Ayyub (Job) and his sufferings are mentioned in the Koran.

Jok, creator god of the Alur tribe of Uganda and Zaire. Also known as 'Jok Odudu', 'god of birth', he is creator of the *djok*, spirits who manifest themselves in snake forms or in large rocks, and are part of their ancestor world.

John (also called John the Evangelist or John the Divine), the Galilean fisherman son of Zebedee and one of the twelve Apostles. According to Christian tradition John, referred to as 'the disciple whom Jesus loved', is credited (probably erroneously) with the authorship of the fourth Gospel, the Book of Revelation, and three epistles of the New Testament. Present at many of the major events in Jesus' ministry, he was the only disciple known to have witnessed the crucifixion and it was there that Jesus entrusted his mother, Mary, to his care. John accompanied Peter in the early years of the Church. It is traditionally thought that he was exiled to Patmos under the Roman Emperor Domitian, who, according to one legend, had him thrown into a cauldron of boiling oil from which he emerged unharmed.

John the Baptist, Jewish prophet, son of Elizabeth, a relative of Mary the mother of Jesus, and of the temple priest Zechariah. Living an ascetic life in the Judaean desert until he was about 30, John then started preaching of the imminence of God's Final Judgement, baptizing those who repented, in the River Jordan. Among those who came to him was Jesus, whom he baptized, hailing him as the Messiah. He was imprisoned for denouncing Herod Antipas' marriage to Herodias, formerly the wife of Herod's half brother. He was beheaded at the request of Herodias' daughter, Salome, in return for dancing at Herod's birthday feast. As the prophet Yahya, John has a prominent place in the Koran, esteemed for his gentleness and chastity. In Islamic legend, the blood from his decapitated head is said to have boiled. A tomb, alleged to be that of Yahya, is in the Great Mosque of Damascus.

Jonah, the recalcitrant Hebrew prophet who sought to avoid God's call for him to prophesy disaster to the Assyrian city of Nineveh, notorious for its excessive wickedness. Taking passage in a ship that would carry him away from Nineveh, he was caught in a great storm and swallowed by a fish, to be regurgitated alive three days later on dry land. Jonah then went to Nineveh, and his prophecies caused its inhabitants to repent. He pitied a plant that perished, but abhorred the idea that God's compassion should extend to Gentiles. God chastised him, affirming that his mercy encompassed Jews and Gentiles alike. As Yunus, Jonah is mentioned in the Koran as a prophet, and is frequently depicted in Islamic painting with a fish.

Jonathan, the warrior son of the Hebrew king Saul and beloved friend of the future King David. After defeating the Philistines many times, Jonathan was killed at the battle of Mount Gilboa. David lamented the deaths of Saul and Jonathan in a moving elegy.

Jörmungand *Loki; *Ragnarök; *Thor.

Joseph (New Testament), a carpenter of Nazareth, the earthly father of Jesus and husband to the Virgin Mary. A descendant of the House of King David, he is venerated in Roman Catholicism as the patron of the universal church.

Joseph (Old Testament), the favoured son of the Hebrew patriarch Jacob and his wife Rachel. Sold into slavery by his

half-brothers, who were jealous of their father's gift to Joseph (a 'coat of many colours'), Joseph eventually attained high office in Egypt because of his power to interpret the Pharaoh's dreams. His foresight saved Egypt from a famine, during which his brothers came from Israel for food. Joseph, forgiving them, sent for Jacob and for their families to settle in Egypt under his care. His name eventually became equated with all the tribes that made up the northern Kingdom. As Yusuf, Joseph is a favourite figure in Islam, renowned for his beauty and regarded as a prophet. The biblical account of Joseph's life, with minor differences, is told in the Koran and embellished in Islamic legend, where the Egyptian who buys him is called Kitfir (the biblical Potiphar), who has Zulaikha for his wife. The story of Yusuf and Zulaikha, immortalized in verse by the Persian Sufi poet Jami (1414–92), is one of the most popular love stories in the Islamic world.

Joseph of Arimathaea (New Testament), a wealthy member of the Jewish Sanhedrin who was a secret disciple of Jesus. With Nicodemus, Joseph obtained the body of Jesus and laid it in his own tomb. According to the apocryphal Gospel of Nicodemus, Joseph helped found one of the first Christian communities at Lydda. A medieval legend associates him with Glastonbury in England, where he is alleged to have come with the Holy Grail, the cup used by Jesus in the Last Supper, and to have planted a thorn tree that flowers on Christmas Day.

Joshua, the Hebrew leader who succeeded Moses. After the Exodus from Egypt, Joshua led the Israelites across the miraculously dried-up River Jordan into the Promised Land of Canaan. An inspired fighter, his campaigns, including the razing of the walled city of Jericho, helped to consolidate Israel's hold on Canaan.

Judah, the fourth son born to Jacob and his first wife, Leah. He gave his name to the Israelite tribe of Judah and to the kingdom founded in the southern part of ancient Palestine in 936 BC. The kingdom of Judah fell to Nebuchadnezzar of Babylon in 586 BC. The Jews returned from their Exile in 537 BC, and the land they occupied became known as Judaea, the Greek form of the name, from the 4th century onwards.

Judas Iscariot, the Apostle destined to betray Christ to the Jewish authorities. After the Last Supper he led an armed band to the Garden of Gethsemane, where he identified Christ to the soldiers by kissing him. For this he was given 30 pieces of silver, the price commonly paid for a slave. He bought a potter's field, and there hanged himself.

Judith, the Jewish heroine of the apocryphal Book of Judith who saved her city of Bathulia when it was besieged by Nebuchadnezzar's general, Holofernes. Judith, a widow of great beauty, went to the enemy's camp, beguiled Holofernes into a drunken stupor with her charms, and then beheaded him with a sword.

Juno, the greatest of the Roman goddesses, sister and wife of Jupiter, identified with Hera in Greek mythology. She was particularly concerned with marriage and the well-being of women.

Juok, the supreme god of the Sudanese peoples of Shilluk, who allegedly divided the earth with the River Nile. The Shilluk believe that human beings were expelled from the land of Juok for eating forbidden fruit which made them sick, and that the founder of their royal dynasty was a man from heaven called Nyikang. Nyikang is alleged to have married a crocodile-woman, the patron-protector of birth and infants, to whom offerings are still made on the banks of the Nile.

Jupiter, in Roman mythology, the supreme god, identified with the Greek god Zeus.

K

Kaang (Cagn, Kang, Kho, Thora), the creator deity whose exploits form the basic cycle of African Bushman mythology. Generally regarded as the invisible spirit within natural phenomena, Kaang is particularly associated with the *kaggen* (mantis), the *ngo* (caterpillar), and through his daughter's marriage with *one*, the serpent. After creating the world and withdrawing from it because of human stupidity—a common theme in African myth—his two sons, Cogaz and Gewi, are alleged to have introduced humans to tools with sharp stone points, for digging the earth.

Kaʿba, in Islam, the cube-shaped shrine containing the sacred Black Stone in the Great Mosque at Mecca, and considered the most sacred place on earth by Muslims. According to the Koran, Abraham and Ishmael built the Kaʿba, while in popular tradition it was Adam, helped by the angel Gabriel, with the Black Stone being originally the tent given by God for Adam to live in after the Fall.

kachina, a sacred cult of Hopi and other Pueblo American Indian religious belief systems, representing supernatural beings or ancestral spirits. Each tribe has its own particular kachinas, who are thought to reside with them for half the year and to become visible through the correct performance of traditional ceremonies, where men wear kachina masks depicting these spirits. Kachinas are also depicted in the form of elaborately decorated, carved wooden dolls which serve as toys for teaching children about the tribal spirits.

Kālī *Devi.

Kalila and Dimna, the central characters in the ancient Sanskrit animal fables collected in the *Panchatantra* (Sanskrit, 'five chapters') between 100 BC and AD 500. They were translated into Arabic in the 8th century and since then have become part of Islamic folklore. A Hebrew version, written by Rabbi Joel in the 12th century, became the source of the earliest European version, known as *The Fables of Bidpai*. Slavic, Turkish, Latin, and Javanese versions followed, some through oral transmission. In Europe they became a source for Jean de La Fontaine's *Fables* (1668). The scheming jackals, Kalila and Dimna, with their band of comrades, appear in stories which mirror a world of human intrigue, immorality, and shrewd statecraft. The

aphorisms which round off each fable glorify worldly wisdom.

kami, in Shinto, the sacred being(s) who are worshipped as gods. These include the sun goddess Amaterasu and other spirits, illustrious ancestors, and forces of nature, both good and evil. Animate and inanimate phenomena such as rocks, plants, birds, animals, and fish may all be treated as kami. There are believed to be an infinite number of kami. Humans, especially those with special talents, may be considered to be kami in their own right. Kami are a source of power and benefit that enrich human life. They may be manifested in, or take residence in, a symbolic object such as a mirror, in which form they are usually worshipped in shrines.

Kannon, in Buddhism, the Japanese name for the *bodhisattva* (Buddha-to-be') Avalokitesvara, a manifestation of Amitabha. In Japan, Kannon, usually represented as female, is the most popular Buddhist figure, widely venerated for her mercy and compassion. She is believed to assume diverse forms, one of which is a four-handed figure with eleven heads.

Khadījah (*c.*555–619), the Prophet Muḥammad's first wife, some fifteen years his senior, and a wealthy widow with her own caravan trade. Revered by Muslims as one of the four 'perfect women' of Islam, Khadījah's proposal and marriage to Muḥammad are surrounded by legendary features. Tradition always depicts her favourably for her support of Muḥammad and for his revelations, and attests to his deep grief at her death.

Khizr *al-Khaḍir.

Khwadja Khidr *al-Khaḍir.

Kintu and Nambi, in the mythology of some Bantu-speaking peoples in Uganda, the first man and his consort, the daughter of Gulu, the king of heaven. Kintu is said to have undergone many ordeals before being allowed to marry Nambi because her relatives despised his solitary life as herdsman. Kintu is the alleged ancestor of other legendary figures, including Kato Kimera, the first king of the Nyoro State. He is also identified with the first king of Ganda (late 14th century).

Kishibojin (Kishimojin), in Buddhism, the Japanese name for Hariti, a child-eating ogress, converted by the Buddha into a protectress of children and of women in childbirth. Sometimes regarded as a feminine form of Kannon, Hariti is usually depicted either surrounded by children or carrying a child, and holding a pomegranate or cornucopia.

Kobo Daishi, the posthumous name given to the monk, scholar, artist, and calligrapher Kukai (774–835), founder of the esoteric Shingon Buddhist sect in Japan. According to legend Kukai did not die but entered into eternal meditation at Mount Koya. Legend also attributes to him the founding of the pilgrimage to the eighty-eight temples of Shikoku, as well as countless miracles. Throughout Japan the appearance of fresh water and hot springs is attributed to him, and he remains one of the most popular figures of folkloric worship in Japan today.

Krishna, the eighth incarnation or *avatar* of the Hindu god Vishnu. In Hinduism, Krishna the Dark One is the preserver and Almighty Prince of Wisdom. A mass of legends and fables have collected about him, relating to three phases of his life: as a mischievous child-god, which includes his slaying of Kaṃsa of Mathura, his wicked uncle; as the cowherd in love with the milkmaid, Rādhā, herself a reincarnation of Lakshmi, in which role he embodies the perfect deification of life; and finally as Arjuna, Pāṇḍava's divine charioteer. Krishna was killed by an arrow fired by the hunter Jara ('old age').

Kshitigarbha *Jizo.

Kurma, in Hinduism, the second incarnation or *avatar* of Vishnu, where he assumed the form of a giant tortoise in response to the gods' weakened strength because of a curse laid on Indra. By supporting Mount Mandara, which served as a churning stick, Kurma enabled the gods and demons to churn the ocean and retrieve the treasures of the Vedic tribes, lost in the Deluge. The gods were eventually restored to their former strength and authority.

Kvasir, in Norse and Germanic mythology, the god whose death brought poetry into the world. Famed for his wisdom, Kvasir was murdered by dwarfs, but Odin caught the blood and transformed it into a sacred mead which, in the form of poetry, he brought to the dwelling place of the gods. A giant stole the mead, but Odin, disguised as an eagle, rescued it and carried the liquid back in his crop.

Kwoth, the great spirit deity of the Nuer tribes in southern Sudan. According to Nuer belief, the universe came into existence through Kwoth's will, and continues because of his activity. He is believed to uphold justice, punish wickedness, and show compassion to the weak and unfortunate.

L

Lakshmi (Lakṣmi), the Hindu goddess of fortune and prosperity, renowned for her beauty, wife of Vishnu. Born from the churning of the ocean by the gods and demons, Lakshmi assumed various forms to accompany her husband in his incarnations, becoming Sītā when he was Rāma and Rādhā when he was Krishna. Lakshmi is usually depicted seated or standing on a lotus, and holding a lotus and a pot of ambrosia (the nectar of immortality).

Lares and Penates, in Roman mythology, household gods. The Lares were regarded as protectors of particular localities or houses. The Penates were guardians of the store-cupboard. Eventually they became barely distinguishable. The term 'lares and penates' has come to refer to treasured household possessions.

Lazarus (from Hebrew, 'God has helped'), a friend of Jesus, the brother of Mary and Martha in Bethany, who, after four days in his tomb, was brought back to life by Jesus. Another Lazarus, in a parable told by Jesus, is cast as the beggar who lay suffering at the gates of a rich man, traditionally referred to as Dives. In life, Dives ignored the plight of Lazarus, but after death the rich man, parching in hell, pleaded in vain for water from Lazarus, now in heaven.

Leander *Hero.

Leda, in Greek mythology, a queen of Sparta who was beloved by Zeus. He took the form of a swan to seduce her, and the children he fathered by her—Castor and Pollux—were hatched from an egg. Her other children were Clytemnestra and Helen, but accounts vary as to whether they were fathered by Zeus.

Legba, the trickster spirit of Fon mythology in Benin (formerly Dahomey) who is closely associated with the oracle Fa of the Ifa tradition. According to Fon belief, Legba originally acted only on the orders of the omnipotent deity, but he tired of being blamed for the sufferings visited on innocent people. His ruse to make mankind blame his god instead, caused the latter's withdrawal from the world. Legba subsequently assumed the role of reporter of human affairs to and intermediary between God and man. Legba is also a god of the Haitian religion, Voodoo.

Leila *Majnūn.

Lêr, or Lir, in Gaelic mythology, the sea-god, one of the Tuatha Dé Danann (Gaelic, 'People of the Goddess Danu'), perhaps to be identified with the British sea-god Lyr. The three children of Lêr were changed into swans by their jealous stepmother Aoife, and condemned to spend 900 years on the seas and lakes of Ireland.

Lethe, in Greek mythology, one of the rivers of Hades; it caused forgetfulness in those who drank its waters.

Levi, the third son of Leah and of the Hebrew patriarch Jacob, whose descendants, the Levites, were given by God the sole right to the Jewish priesthood in the time of Aaron. Referred to as priests before the Exile in Babylon, they were not assigned a specific territory. Many of them became dispersed along with the other northern tribes of Israel, and in time became Temple servants with subordinate duties.

Leviathan, the reptile, mentioned in the Old Testament, variously identified as a crocodile, whale, or dolphin. Possibly related to the Phoenician snake-god Lotan, in the Book of Isaiah, the name Leviathan also symbolizes the mighty Assyrian and Babylonian empires, hence its use to denote anything very large or powerful.

Lilith, in Hebrew folklore, the female demon who attempts to kill new-born children. In the Talmud, Lilith is said to be Adam's first wife, dispossessed by Eve. In another version of the legend, Lilith deserts Adam to become a fiend.

limbo, in medieval Christian theology, a region on the border of hell, the supposed abode of pre-Christian righteous persons and of unbaptized infants and adults.

Lohengrin, the Knight of the Swan, a hero of the German version of a widely spread legend. Lohengrin is summoned from the temple of the Grail to help a lady in distress, Elsa of Brabant. He is borne on a swan-boat to Antwerp and saves Elsa from an unwanted suitor. Lohengrin will marry Elsa if she does not ask his origin; but she does, and the swan-boat carries him back to the castle of the Holy Grail. The story is the subject of Wagner's opera *Lohengrin* (1850).

lokapalas, in classical Hinduism, the protective deities who act as guardians of the eight quarters of the world. They include Agni (south-east), Indra (east), Surya (south-west), and Yama (south). The lokapalas play an important part in the design of temples and cities: Indian architecture is largely based on the square with fixed cardinal points, each marked by a guardian deity.

Loki, the Norse and Germanic god of the Aesir. A cunning and mischievous trickster, he had the ability to change his sex and age at will. His three offspring were Fenrir, his wolf-son, Hel, the goddess of death, and Jörmungand, the evil serpent that lies coiled around the earth. Loki stole Freyja's necklace, fashioned by dwarfs. For causing Balder's death, the gods chained him to a rock, but he was saved from the venomous bite of a serpent by his faithful wife, Sigyn. At Ragnarök, Loki will side with the giants and monsters who confront the Aesirs in battle.

long, the dragon of Chinese folklore. The long is essentially a benevolent divinity and held in high regard. He is the rainbringer, the lord of the waters in clouds, rivers, marshes, lakes, and the seas. His appearance may be composite, with the horns of a stag, the head of a camel, the eyes of a demon, the neck of a snake, the scales of a fish, the claws of an eagle, the pads of a tiger, the ears of a bull, and the long whiskers of a cat. The five-clawed long once served as the imperial symbol.

Lóngwáng (English, 'the Dragon King(s)', in Daoist mythology, beneficent dragons answerable to the Yuánshǐ Tiānzūn (English, 'Celestial Venerable of the Primordial Beginning'), to whom they submit yearly reports. Believed to have power over rain and funerals, Lóngwáng are invoked in times of drought and in funerary rites, to ward off misfortune to the dead person's descendants. Lóngwáng include the Celestial Dragon Kings, the Dragon Kings of the Four Oceans who live in sumptuous palaces on the ocean floor, and those of the Five Cardinal Points.

Lorelei, an echoing rock in the River Rhine, in German legend the haunt of the siren, Lorelei, who lures fishermen to their destruction with her song. According to legend the maiden Lorelei became a siren when she threw herself into the Rhine on being abandoned by a faithless lover. The German writer Clemens Brentano claimed to have invented the story in his novel *Godwi* (1800–2).

Lot, the nephew of the Hebrew patriarch Abraham, who accompanied him to Canaan. When their flocks became too large, Lot settled on the outskirts of Sodom near the southern end of the Dead Sea in Israel. A messenger from God came to warn the God-fearing Lot of the imminent destruction of the city because of the wickedness of its inhabitants, and urged him and his family to flee. Lot's wife was turned to a pillar of salt for disobeying God's order not to look back. In Islam, Lut (Lot) is mentioned several times in the Koran for having warned his people of God's impending punishment for their indecency and perverse sexual acts. In Islamic legend, Lut's wife is said to have been turned into a pillar of salt because she used ostentatiously to borrow salt from neighbours to show that her husband was entertaining forbidden guests.

Lucifer, in early Christian belief, the angel who rebelled against God and was cast out of heaven with his followers to become the devil (Satan), who tempts people to evil. He is also conceived of as the personification of evil in Judaism and Islam.

Lugh, in early Irish Celtic mythology, the sun god and chieftain who killed the leader of the Fomors, the giant predecessors of the Tuatha Dé Danann in Ireland. Lugh had a spear of deadly accuracy and was skilled in the arts of war and peace. The name of Lugh was assumed by the heroes of much of Irish mythology.

Luke, an Apostle and physician, possibly the son of a Greek freedman of Rome, closely associated with St Paul and traditionally the author of the third Gospel and the Acts of the Apostles.

Luqman (Lokman), in Islamic belief, a wise man of various legendary origins, and alleged author of Arabic proverbs and fables. He is mentioned in the Koran as a wise father, admonishing his son. Sometimes identified by Western commentators with Aesop, Luqman, according to one tradition, was a deformed slave who, when offered the gift of either wisdom or prophecy by God, chose wisdom. The Prophet Muḥammad is said to have esteemed him for his proverbs. The fables attributed to him first appeared in the 13th century.

Lyr *Lêr.

M

Ma'at, the Egyptian goddess of cosmic order, truth, and harmony. Daughter of Re, she is depicted as wearing an ostrich feather on her head. It was the duty of every Pharaoh to uphold *ma'at* (cosmic order) during his reign; failure to do so would wreak chaos and disorder. The feather of Ma'at was used as a counterbalance against the hearts of the dead during the Weighing of the Heart in front of Osiris.

Maenads, in Greek mythology, 'mad women', votaries of Dionysus who, intoxicated or possessed, danced about him, tearing animals to pieces.

Maeve *Medbh.

Magi, the, in Christian tradition, the men from the East who, possessing astronomical and astrological wisdom, followed a star that led them to the birthplace of the infant Jesus, in Bethlehem. There they paid homage to him with gold, the gift bestowed on kings, frankincense, used to worship at the altar of God, and myrrh, an embalming agent for the dead. In later tradition the magi became three kings (Caspar, a moor, Melchior, and Balthasar).

Majnūn, the legendary Arabian lover of Leila (Laylā). In the legend Qays, later known as Majnūn, 'the demented one', falls in love with Leila, whose father marries her to another man. Crazed with grief, Majnūn spends the rest of his life in the hills of Najd, composing poetry about his unhappy love. The story of Leila and Majnūn, celebrated throughout the Islamic world, provides the theme for many Turkish, Persian, and Indian miniaturists and writers, notably the 12th-century Persian poet Nizami. For Sufis, Majnūn symbolizes the quest of the soul for unity with God.

Manco Capac (Mama Occlo), in Inca mythology, the brother-sister husband and wife who allegedly founded the sacred city of Cuzco and became the first rulers of the Inca empire. Sometimes said to be one pair of four brothers and sisters, in one version of the legend they are the children of the sun-god.

mandala (Sanskrit, 'circle'), a symbolic diagram representing the universe and in itself a focus of universal forces. As a tantric or sacred text it is used in the performance of sacred rites and as an instrument of meditation in the mystical practices of Hinduism and Buddhism. In meditation, it is used to guide the devotee through a cosmic process of disintegration towards reintegration. Mandalas are basically of two types, those in which the movement is from the one to the many, and those in which it is from the many to the one. Mandalas can be painted on cloth or paper, drawn with coloured threads or rice powders, or fashioned out of bronze or stone.

Manjusri (Chinese, Wénshū; Japanese, Monju), in Mahayānā Buddhist belief, the *bodhisattva* ('Buddha-to-be'), personifying supreme wisdom. Although usually considered a celestial bodhisattva, Manjusri is alleged to have taken various human, historical forms, including that of the monk Vairocana. He is usually depicted with princely ornaments, holding a palm-leaf manuscript and sword, and seated on a lion or lotus.

Manu, the name of several sages in Hindu cosmology, the best known being Manu Vaivasvata, a 'Noah' figure. He survived the Deluge by building a ship and was rescued by Matsya. He thus became the father of the present race of men and women and first king of Ayodhya. Manu Vaivasvata was also the legendary author of the *Laws of Manu*, the most famous code of Hindu religious law.

Mara, in Buddhist belief, the demon 'Lord of the Senses' who tempted the Buddha on several occasions. In the Buddha's bodhisattva form of Gautama, Mara attempted unsuccessfully to distract him as he meditated under the Bo tree. After his Enlightenment, Mara tried to persuade him not to preach the law. In Hindu mythology, Mara is sometimes an aspect of Kama, god of desire and love.

Marduk, the lord and ruler of the Assyro-Babylonian pantheon of gods. Originally a fertility and agricultural deity, Marduk was alleged to have gained his divine supremacy either by slaying Tiamat, from whose body he created the heavens and earth, or by recovering the Tablets of Fate from the storm-god Zu.

Mars, the Roman god of war. He was identified with the Greek Ares, but was a much more important figure in Roman mythology than Ares in Greek mythology, ranking second only to Jupiter among the gods.

Martha, a friend of Jesus Christ, the sister of Mary and Lazarus of Bethany, who chose to express her devotion to Jesus by preparing a meal for him while her sister Mary sat in his presence. In Christian allegory she symbolizes the active life united to God, and her sister the contemplative.

Mary Magdalene, a follower of Jesus. He is said to have cast seven demons out of her before she became his devout disciple. Traditionally identified with both Mary of Bethany and the sinner who anointed Jesus' feet, drying them with her hair, Mary Magdalene witnessed his crucifixion and was the first to see him after his resurrection.

Mary the Virgin, the mother of Jesus. Traditionally thought to be the daughter of Anne and Joachim, she is the foremost saint in Roman Catholic and Orthodox Christianity. She is also revered by Muslims. The events of Mary's life mentioned in the Gospels include the annunciation by the angel Gabriel of Jesus' birth; her visitation to her kinswoman Elizabeth, mother of John the Baptist; the nativity or birth of Jesus; the purification (or presentation) when Jesus was blessed by Simeon in the Temple; her presence at Jesus' crucifixion, when he committed her to the care of the apostle. Referred to variously as Our Lady, the Madonna, the Blessed Virgin Mary, she was given the title of Mother of God at the Council of Ephesus (AD 431). Surrounded by an ancient tradition of pious devotion and legend, Mary is also the subject of Christian dogma: the major beliefs being that Jesus had no human father, but was conceived by the power of the Holy Spirit (the doctrine of the Virgin Birth, which because of its biblical origin is generally accepted in Roman Catholic, Orthodox, and most Protestant theologies); that she was born without original sin (much disputed, and accepted only as dogma by the Roman Catholic Church, it is commemorated on the Feast of the Immaculate Conception); that she was taken body and soul directly into heaven on her death (commemorated on the Feast of the Assumption by the Roman Catholic Church, and on the Feast of the Dormition or 'Falling Asleep' by the Orthodox Churches). Her apparitions, which have become pilgrimage centres, include those at Guadalupe (Mexico), Lourdes (France), and Fatima (Portugal). She is also associated with Ephesus (Turkey), Loreto (Italy), and Walsingham (England). Maryam (Mary) is greatly revered in Islam. In the Koran it is said that she was the daughter of Imran, raised by the temple priest Zachariya (see *Zacharias), and that she gave birth to 'Isa (Jesus) under a palm tree. As a child she used to retreat to a prayer niche in the temple, where she was miraculously fed. According to Islamic tradition, as in Catholic doctrine, she was born sinless and remained a virgin. Muḥammad is alleged to have ordered the destruction of all icons in the Ka'bah except that of Maryam and the Infant 'Isa when he conquered Mecca.

Matsya, in Hinduism, the first incarnation of Vishnu as a horned fish to save the

world from destruction. When Matsya was a very small fish, Manu saved its life by taking it from the ocean and placing it in even larger bowls as it grew. In return for this Matsya saved Manu from the Deluge by acting as moorage for his ship.

Maui-tiki-tiki, a Melanesian and Polynesian god and hero-trickster. He seized the life-giving force of fire from his ancestress Mahuike in the bowels of the earth and placed it in a tree. Since that time mankind has been able to make fire from the wood of trees, for example through fire boring. Maui is revered both as a general symbol and as a local divinity. In various forms and sometimes aided by an enchanted jawbone, he is renowned from Hawaii to New Zealand. Maui is credited with raising the sky, slowing down the sun, and fishing islands from the depths of the ocean. In Hawaiian mythology, Maui obtained fire from the Alae, a wading bird, sacred to the goddess Hina. His attempt to gain immortality for man, however, by killing Hina failed. In the Tuamotu Archipelago, Maui is said to have vanquished the monster eel-husband of Hina with his phallus. In Hawaii, Maui is thought to have been killed by the inhabitants, tired of his tricks. It is his blood that coloured the rainbow and gave shrimps their red hue.

Mawu-Lisa, the female-male supreme creator deity of the Fon people of Benin and Togo, West Africa. Mawu is associated with the moon, fertility, motherhood, gentleness, and wisdom; Lisa with the sun, work, war, power, and strength. A 17th-century Capuchin catechism from Dahomey (Benin) identifies Mawu as God the Father, and Lisa as Jesus.

Măzŭ Dàoyī (Jīāngxī), a great master of Zen Buddhism in China. Măzŭ is renowned for his unorthodox training methods (the sudden shout, unexpected blow, paradoxical answer) to liberate his students from habits of conceptual thinking and help them experience enlightenment. He was said to have a tiger's glance, a buffalo's gait, a tongue that could cover his nose, and the Buddhist wheel (bhava-chakra) on the soles of his feet.

Măzŭpó, in Chinese Daoist mythology, the familiar name for Tiānfēi, the Heavenly Concubine or Holy Mother of Heaven. Măzŭpó is the goddess of sailors and navigators, and is invoked by barren women.

Medbh (Maeve) (Gaelic, 'drunken woman'), the legendary queen of Connacht in Ireland. A fierce goddess, she led her people in battle against the forces of Ulster. Among her many lovers were king Ailil and the hero Fergus. A bird and a squirrel, sitting on her shoulder, are associated with her.

Medusa *Gorgons.

Meleager, in Greek mythology, a prince of the town of Calydon. At his birth the Fates said he would live as long as a piece of wood on the fire was unconsumed; his mother snatched it off and kept it safely. Meleager grew into a great warrior and hunter and when Artemis sent a huge boar to terrorize Calydon, he led the band of heroes who killed it (the chase is known as the Calydonian boar hunt). When he gave the head to Atalanta (who had been the first to wound it), his mother's brothers tried to take it and he killed them. His outraged mother burned the carefully hidden piece of wood and Meleager died.

Menelaus, in Greek mythology, a king of Sparta, the brother of Agamemnon and the husband of Helen. When Helen was abducted by Paris, Menelaus appealed to the other Greek kings to join him in waging war on Troy. After Troy's defeat, Helen returned with him to Sparta.

Mercury, in Roman mythology, the god of commerce, travel, and theft, and the messenger of the other gods—associated with Hermes in Greek mythology.

Merlin, in Celtic mythology, the enchanter and wise man associated with the Arthurian legends, whose alleged prophecies were a potent influence in medieval Europe. Linked with personages in earlier Celtic mythology, in particular with Myrddin in Wales, Merlin used his magic arts to bring about the birth of Arthur and acted as counsellor for his future kingship. In some versions, his love for Nineve, the Lady of the Lake, led to his death.

mermaid, in world mythology, a sea-creature with the head and body of a woman, and a fish's tail in place of legs. Thought of as beautiful but treacherous, mermaids are said to lure sailors to their death. Mermaids and their male counterparts, mermen, have been a part of maritime mythology since ancient Babylonian, Semitic, and Greek civilizations. It is suggested that the manatee (sea-cow), related to the dugong, seen from a distance, may have added credence in such creatures.

Methuselah, the Hebrew patriarch who was the son of Enoch and grandfather of Noah. According to the Bible he fathered Lamech at 187 and died at the age of 969 years.

Micah (Micheas), the Hebrew prophet who foretold the doom and eventual redemption of Judah and Israel. Teaching that true religion must be based on justice, mercy, and humble communion with God, he is traditionally seen as having prophesied the coming of the Messiah (redeemer) and of his birth in Bethlehem.

Michael, in Judaism, Christianity, and Islam (as Mikāl), the archangel, and guardian of mankind against the devil. Mentioned in Christian apocryphal writings as fighting the serpent Satan, Michael is depicted in armour standing over the devil in the shape of a dragon. In Islamic legend Mikāl and Jibrīl (Gabriel) were the first to obey God's command to bow down before Adam.

Midas, in Greek mythology, a king who was granted a wish by Dionysus and asked that everything he touched be turned to gold. When even his food turned to gold and became inedible, he realized how foolish he had been and he was relieved of his unwanted power by bathing in the River Pactolus. In another story, Midas offended Apollo by declaring that Pan was a better musician, and was punished by the god by having his ears turned into those of an ass.

Midgard (also called Manna-heim, the Home of Man), in Norse mythology, the middle universe and domain of mankind, situated between Muspelheim, the world of the fire giants, and Niflheim, the misty underworld. Midgard was connected to Asgard by the rainbow bridge of Bifrost.

Minerva, in Roman mythology, the goddess of wisdom and the arts, identified with Athena in Greek mythology.

Minotaur, in Greek mythology, a monster with a man's body and a bull's head. It was kept by Minos, king of Crete, in a labyrinth built by Daedalus. The citizens of Athens had to send a yearly tribute of seven youths and seven maidens to be eaten by the Minotaur, until Theseus—with Ariadne's help—killed it.

Mithra (Greek and Latin, Mithras), an ancient Indo-Iranian sun-god, and the Zoroastrian god of light and truth, champion of humanity against the evil spirit Ahriman. As fearless antagonist of the powers of darkness, Mithra captured and sacrificed a sacred bull from whose body all earthly blessings sprang. From Persia the cult of Mithra spread throughout Asia Minor and Greece to become, as Mithraism, one of the principal religions of the Roman Empire.

Moloch, the Canaanite god of fire to whose image children were sacrificed as burnt offerings. It has been suggested that Moloch was not a god but the term used to describe this Canaanite rite.

Moses, Hebrew leader and prophet, who delivered his people from slavery and founded the religious community

called Israel. According to biblical accounts, he was a Hebrew foundling adopted and reared at the Egyptian court. In Midian in north-west Arabia he saw a burning yet unconsumed bush and experienced the voice of God (Yaweh) commanding him to lead his people from Egypt. In the Exodus that followed, the pursuing Egyptians were engulfed by the sea. On Mount Sinai (Horeb) Yaweh revealed the Covenant, including the Ten Commandments, between himself and the people of Moses. Moses died within sight of Canaan, the Promised Land allegedly at Moab.

Mr Spider *Anansi.

Mujaji, the Rain Queen of the Lovedu, a Bantu tribe of Transvaal in South Africa, and source of Rider Haggard's novel *She* (1887). According to Lovedu myth, Mujaji was the descendant of Mambo, a 17th-century king of Zimbabwe, whose daughter had fled south with his rain charm and sacred beads to found the Lovedu tribe. The Lovedu were ruled by a succession of queens called Mujaji, all believed to have power over the rain.

Mulungu (also Murungu, Mungu, Mlungu, Mluku, Mugu), the name given to the creator god and the impersonal spiritual force that are central to the religious beliefs of many East African peoples. Mulungu is sometimes associated with storms and with tribal ancestors.

Munkar *Nakir.

Muses, in Greek mythology, nine daughters of Zeus and Mnemosyne (Memory), each of whom presided over a particular branch of art, literature, or science. Their names and responsibilities vary in different ancient accounts, but are generally given as follows: Calliope (epic poetry), Clio (history), Erato (love poetry and the lyre), Euterpe (lyric poetry and the flute), Melpomene (tragedy), Polyhymnia (hymns), Terpsichore (dance and choral song), Thalia (comedy), and Urania (astronomy).

Muspelheim, in Norse mythology, the land of the south, hot and bright, guarded by the fire giant Surt. Its geographical opposite is Niflheim, the cold, misty, and dark world of the dead, ruled by the goddess Hel. Between these regions lies Midgard, or Middle Earth, the home of men and women, surrounded by a fence made from the eyebrows of the giant Aurgelmir. Sparks from Muspelheim have become our sun, moon, and stars. At the great battle of the gods, Ragnarök, the giants of Muspelheim will swarm over the earth and destroy it by fire.

N

Na Atibu *Nareau.

nabī rasūl *Prophet Messenger.

Naboth, the owner of the fertile vineyard, coveted by King Ahab of Israel. Jezebel, Ahab's wife, obtained the land for him by having Naboth stoned to death. The curse of the prophet Elijah for this deed forecast the downfall of Ahab's dynasty.

naga, a serpent-geni figure in the mythologies of Hinduism, Jainism, and Buddhism. As water gods, nagas inhabit the bottoms of rivers, lakes, and seas, in splendid, jewel-studded palaces ever alive with dancing and song. In temple architecture, nagas stand guard at the portals of shrines. In South India naga kals, stones decorated with a single serpent or an entwined serpent-pair, are set up as votive offerings by women desiring offspring.

Naiad, in Greek mythology, a water nymph living in fresh water such as brooks, fountains, and springs. Sea water nymphs were called Oceanids.

Nakir and **Munkar**, in Islamic belief, the two angels who interrogate newly buried corpses about the Prophet Muḥammad. Those who answer correctly are said to receive air from Paradise, while incorrect responses are met with beatings and ever-increasing pressure by the angels on the grave of the corpse.

Narasinha (Narasimha), in Hinduism, the fourth incarnation or *avatar* of Vishnu as half-man (*nara*, 'man', his lower half) and half-lion (*simha*, 'lion', his upper half) in response to the demon-king Hiranyakashipu's attempt to usurp the place of the gods. According to the myth, Hiranyakashipu had obtained a boon from Brahma granting him invincibility from man and beast, by day and night, indoors and outdoors. When his son Prahlada persisted in his devotion to Vishnu, the demon-king mockingly asked him if Vishnu, who Prahlada claimed was everywhere, was in a pillar which he then kicked, not realizing that his invincibility was inoperative: it was dusk, the pillar was on the threshold, and Vishnu as half-man and half-lion was able to tear him to bits.

Narcissus, in Greek mythology, a beautiful youth who spurned the love of the nymph Echo and in punishment was made to fall in love with his own reflection; he pined away gazing at himself in a pool and at his death was changed into the flower bearing his name.

Nareau, the creator deity of Gilbertese traditions. Entering the primordial darkness, he created out of sand and water two beings, Na Atibu and Nei Teukez. From their union sprang the gods. The world of men and women was created from the dismembered parts of Na Atibu's body.

Nasreddin (Nasrudin) **Hoja**, known in Persia as Mullah Nasreddin, the sage and folk hero of countless humorous tales told throughout eastern Europe as well as in the Islamic world. The humble, unassuming Nasreddin, a Turkish teacher and preacher (*hoca*), apparently naïve and stupid, often shows up the absurdities of society and of the powers-that-be in tales which always contain a moral. Nasreddin's historical existence is questionable, although a tomb bearing his name, dated 1284–5, exists in Akshehir in Turkey. Another Nasreddin, Nasreddin Mahmud, a hero of Turkish folk-songs, allegedly was a warrior in the early 15th century. Recent versions of the tales have been given in English by Idries Shah in *The Pleasantries of the Incredible Mulla Nasrudin* (1968).

Naströnd *Hel.

Nathan, the Hebrew prophet who condemned King David's seduction of Bathsheba, likening it to the theft of the poor man's sole ewe lamb. Nathan's advice ensured that Solomon, son of David and Bathsheba, rather than Adonijah, succeeded to the kingdom on David's death.

Nemesis, in Greek mythology, the goddess of retribution or vengeance. According to Hesiod, Nemesis was a child of Night. She was a personification of the gods' resentment at, and consequent punishment of, defiance (*hubris*) towards themselves.

Neptune, in Roman mythology, the god of the sea, associated with Poseidon in Greek mythology. He is often represented carrying a trident.

Nestor, in Homeric legend, the eldest of the Greek kings at the Trojan War. He was an elder statesman figure, garrulous but wise, counselling moderation in the quarrel of the leaders.

Ngewo, the sky god of the Mende tribe of Sierra Leone. He manifests himself in natural phenomena, for example by sending rain to fall on his wife, the earth. Between Ngewo and mankind are the spirits, ancestral spirits and *dyinyinga*, genii which are associated with rivers, forests, and rocks.

Nibelung, in Germanic mythology, a member of a Scandinavian race of dwarfs, owners of a hoard of gold and magic treasures, who were ruled by Nibelung, king of Nibelheim (land of mist). In the epic poem, *The Nibelungenlied* (*c.*1205), a Nibelung is any supporter of Siegfried, the subsequent possessor of the hoard, as well as any of the Burgundians who stole it from him.

Nicodemus, the Pharisee and member of the Sanhedrin who was a secret disciple of Jesus. He protested at Jesus' unjust trial before the Sanhedrin and, after the crucifixion, brought spices for his enbalming, helping Joseph of Arimathea to lay his body in the tomb.

Nidhogg *Hel; *Yggdrasil.

Niflheim *Hel; *Midgard.

Nimrod, a legendary biblical figure, the 'mighty hunter', and descendant of Cush. As King of Shinar and founder of the Babylonian Empire, Nimrod's dominion of the world was alleged, in Hebrew tradition, to have been acquired through his possession of the garments worn by Adam and Eve. To him is attributed the building of the Tower of Babel, caused by his desire to dominate heaven, too. Assyria was sometimes referred to as the land of Nimrod.

Niobe, in Greek mythology, a queen of Thebes whose children were killed by Apollo and Artemis to punish her for her boastfulness. Niobe was turned into a stone figure that wept tears.

Noah (Noe), a biblical patriarch who, forewarned by God because of his blameless piety, built a great ark or ship and, together with his family and with the creatures he took on board, survived the Flood sent to punish the wickedness of the human race. According to the Book of Genesis the whole of mankind descends from Noah's sons, Shem, Ham, and Japheth. The rainbow, which manifested itself after the Flood, was the visible sign of God's future protection of mankind against catastrophe. In Islam Nūh (Noah) is regarded as a Prophet Messenger (*nabi rasūl*). The story of Nūh is told in the Koran in various *Suras*, and greatly embellished in Islamic legend. The origin of Āshūrā (the tenth day of Muḥarram) is said to commemorate the day the ark came to rest on Mount Judi (Ararat), as well as the day on which God created Adam and Eve, heaven, hell, life, death, and the pen.

Nommo, in Dogon mythology in Mali, West Africa, the two pairs of twin offspring of the creator god, Amma, who are central figures in Dogon life and culture. The Dogon believe that humans, like Nommo, are composed of both male and female principles although bodily they may be either one or the other. Male and female circumcision is thought of as a Nommo invention to 'fix' sexual identity. Nommo include Yorugu or Ogo (see *Pale Fox), who polluted the earth and caused death; the creator Nommo, who saved the earth from his depradations; the blacksmiths of the sky to whom Dogon blacksmiths owe their high status; as well as Nommo descendants who became the ancestors of humankind.

Num, the sky god of the Nenet or Yurak people of Arctic Russia, who at the beginning of time sent an eagle over the limitless waters. The eagle laid its egg on the knee of Vainemoinen, the sorcerer slumbering in the seas. When the egg fell into the water, it broke; the yoke became the sun and moon, while the pieces of shell formed the earth and the stars.

Nun, in Egyptian mythology, the primeval waters out of which the creator-god emerged. He can sometimes be referred to as the Nile and also the sacred lakes at temple sites. Nun was the 'father of the gods', one of the Ogdoad of the Hermopolitan creation myth worshipped at Khemnu (Eighth town). The Ogdoad comprised four pairs of elements present at creation: primordial abyss, darkness, infinity, and hidden power. Nun and his female counterpart, Naunet, were the abyss in the form of water.

Nut, the sky goddess of ancient Egypt. Usually portrayed as a naked, giant woman, her arched back, supported by Shu, the air, contained the heavens. Day and night were accounted for in terms of solar birth. The Sun entered the mouth of Nut each evening, passing during the night through her body. In the morning, it was born again from her womb.

nymphs, in Greek and Roman mythology, minor goddesses conceived as beautiful maidens associated with particular places or types of place; Dryads lived in woods, Naiads in fresh water, and Oceanids in the sea.

O

Oberon (Alberon), king of the elves in French medieval legend. As Alberich, he is king of the dwarfs in medieval German tales. A member of the Nibelung, he steals the magic gold from the Rhine maidens. In English Tudor literature, Oberon's queen is Titania, the name also given by Ovid to Diana and Circe as descendants of the Titans.

Oceanus, in Greek mythology, the name of a river that was supposed to encircle the earth and also of a Titan who was god of this river and the father of all water nymphs and river gods.

Odin, the chief deity of Norse mythology, husband of Frigg and father of seven sons, including Balder and Thor. To acquire wisdom, he entered a near-death state by suspending himself on the World Tree, Yggdrasill, for nine days and nights. Identified with the Anglo-Saxon Woden (hence Wednesday from Woden's day) and the Germanic Wotan, Odin was god of the wind, war, magic, and poetry, leader of souls and king of the Aesir. With his brothers, he created the Earth from the carcass of the giant Aurgelmir. He gave life to Ask and Embla, the first man and woman. He lived in Asgard, where he was kept informed by two ravens. He had an eight-legged horse, Sleipnir; a spear called Gungir; and Draupnir, a magic ring. Odin is doomed to be consumed by the wolf Fenrir at Ragnarök.

Odysseus (English, Ulysses), in Greek mythology, one of the heroes of the Trojan War, the son of Laertes, king of Ithaca. After the fall of Troy, he spent ten years trying to reach home, frustrated by Poseidon (whose son Polyphemus the Cyclops was blinded by Odysseus) and other hazards (see *Calypso, *Circe, *Sirens). These episodes, and Odysseus' final return from Troy and subsequent vengeance on the suitors of his faithful wife, Penelope, are recounted in the *Odyssey*, an epic poem by Homer.

Oedipus, in Greek mythology, the son of Laius, king of Thebes, and his wife Jocasta. He was abandoned as a child because of a warning by Apollo that he would kill his father and marry his mother. He was rescued and later unwittingly fulfilled both predictions; when he found out the truth of his birth, he blinded himself and went into exile.

Ogo *Pale Fox.

Oisín (Ossian), legendary Irish warrior-poet. A member of Fianna Éireann, an élite corps of huntsmen skilled in poetry, he was lured by a fairy princess to Tír na n Óg, the Land of the Young. There he lived happily for three hundred years. Wishing to return to Ireland, the fairy woman warned him to remain mounted on his white horse and not let his foot touch Irish soil. The horse slipped, Oisín fell to the ground, and his body instantly shrivelled into that of a blind old man.

Olorun, the chief deity of the Yoruba tribe in Nigeria, West Africa. He is also called Oba-Orun (king of the sky), Ólódùmarè (owner of endless space), Eleda (creator), Oluwa (lord), and Òrìshà-Oke (sky god). According to Yoruba belief, Olorun is the creator of the universe who sees into people's hearts and determines their destiny.

Olympus, a mountain in Thessaly, the highest peak in Greece, regarded as the home of the gods in Greek mythology.

Om, in Buddhism, Hinduism, and Jainism, a mystic syllable, a universal affirmation, considered the most sacred mantra (Sanskrit, 'instrument of thought'). It represents several important triads: the three spheres of heaven, air, and earth; the three major Hindu gods, Vishnu, Shiva, and Brahma; the three sacred Vedic scriptures, *Sāmaveda*, *Yaruveda*, and *Rigveda*; and the triad composed of the gods Vishnu and Lakshmi, and the worshipper him- or herself.

Ometecuhtli, in Aztec mythology, the 'dual lord' of all gods and source of all existence. Outside space and time, he was the unity of opposing forces, male and female, light and darkness, order and chaos; the supreme being beyond the stars, far above the events of the world.

oni, in Buddhist and Daoist folklore in Japan, demons associated with disease and misfortune, often depicted as cruel and malicious. Their chief exorcists are Buddhist priests. Human in form, *oni* are depicted as having horns, three eyes, wide mouths, and three sharp talons on their feet and hands. An annual ceremony of banishing *oni*, and of calling in good fortune, takes place at *setsubun*, a festival in early February that traditionally heralds the first day of spring.

Orestes, in Greek mythology, the son of Agamemnon and Clytemnestra. With his sister Electra, Orestes avenged his father's murder by killing his mother and her lover.

Orion, in Greek mythology, a giant and hunter of Boeotia, the subject of various legends, according to which he was deprived of sight by Dionysus, or killed by Artemis (either from jealousy because he was loved by Eos, the Dawn, or because he challenged her to throw the discus against him), or stung to death by a scorpion, by the same goddess's design, while ridding the earth of wild beasts. Another story is that he pursued the Pleiades and both he and they were turned into constellations.

Orpheus, in Greek mythology, a musician of sublime artistry; he played the lyre so beautifully that he could charm wild animals and even trees and rocks. He is said to have taken part in the expedition of the Argonauts, and by his song helped them to resist the lure of the Sirens. He married Eurydice, a Dryad. When Eurydice died, Orpheus went down to Hades to recover her. By his music he induced Persephone to let Eurydice go, on condition that he should not look back at her. He failed, and lost her. Later, he was torn to pieces by the Maenads and his head, still singing, reached the Isle of Lesbos.

Osiris, one of the most important gods of ancient Egypt, originally perhaps connected with fertility. In mythology he was king of Egypt, killed by his jealous brother Set who dismembered his body. Osiris' sister-wife, Isis, gathered the pieces and reassembled them as the first mummy with the aid of Anubis, breathing new life into his corpse as goddess of magical powers, and conceiving from Osiris a son, Horus, through her function as fertility goddess. Osiris, magically revived, ruled the dead in the underworld as king of the dead, symbolizing resurrection and regeneration, whilst his son Horus ruled the living. During the Old Kingdom, deceased pharaohs were identified with Osiris, but later all the dead could be termed an 'Osiris'.

P

Pale Fox (Yorugu, Ogo), in Dogon mythology in Mali, West Africa, one of the original twin creatures called Nommo, conceived by the creator god, Amma. Seeking to possess the universe, Pale Fox, in the form of Ogo, forced his own premature birth and created the earth from his own placenta. Without his twin sister, however, his creation was imperfect and impure. In his attempt to find her he violated the earth, incestuously procreating in his own placenta. He was eventually punished by being turned into a fox and deprived of the power of speech. Foxes are believed by the Dogon to predict the future and to reveal their incipient ability to speak in the language of their pawmarks; this forms the basis for Dogon divination.

Pan, in Greek mythology, the god of flocks, herds, woods, and fields, represented as an ugly but merry man with the legs and usually the horns and ears of a goat. He played the pan-pipes, which he invented, and his musical skills helped him to seduce the nymphs he constantly pursued (he was regarded as a personification of Lust). Pan was thought to be responsible for the sudden alarm felt by people, especially travellers in remote and desolate places—hence the term panic.

Pandora, in Greek mythology, the first woman on earth, sent by Zeus as a punishment for Prometheus' crime of stealing fire from the gods. Zeus gave Pandora a box which, when opened, let out all the ills that have since beset mankind; Hope, however, remained inside it as a comfort.

Pan gu, in Chinese mythology, the primeval man, born of the cosmic egg. Like the egg, Pan gu split into a number of parts: his head formed the sun and moon, his blood the rivers and seas, his hair the forests, his sweat the rain, his breath the wind, his voice thunder and, last of all, his fleas became the ancestors of mankind.

Papa *Rangi.

Paris, in Greek mythology, a Trojan prince, son of Priam, of whom it was predicted at his birth that he would bring destruction to Troy. He was left to die, but was brought up by the shepherds who rescued him. He grew into the most handsome of mortal men and was appointed judge in the dispute among the goddesses Aphrodite, Athena, and

Hera as to who was the most beautiful. All three tried to bribe Paris and he succumbed to Aphrodite, who promised him the most beautiful woman in the world. This was Helen, and Paris' abduction of her caused the Trojan War. In the war, Paris—a skilful archer—killed Achilles, but he was himself killed by an arrow shot by Philoctetes.

Parnassus, a mountain in Greece, regarded in ancient times as sacred to Apollo and the Muses and thus the home of poetry and music.

Parvati *Devi.

Pegasus, in Greek mythology, a winged horse that sprang from the blood of Medusa the Gorgon when Perseus killed her. Pegasus was ridden by Perseus when he rescued Andromeda, and by Bellerephon when he killed the Chimaera.

Pelops, in Greek mythology, the son of Tantalus. When he was a child, his father killed him and served his flesh to the gods at a banquet to see if they could tell it from a wild animal's. Demeter ate part of the shoulder, but the other gods rejected the food. Pelops was restored to life, the missing shoulder was replaced by one of ivory, and Tantalus was punished.

Penelope, in Greek mythology, the faithful wife of Odysseus. During his absence she was wooed by many suitors; unwilling to repulse them outright, she prevaricated by saying that before marrying she had first to finish weaving the robe on her loom, which she worked every day and then unpicked each night until Odysseus returned.

peri (pari), in Persian mythology, a benign female spirit, endowed with grace and beauty. Peris, as a race of superhuman beings, were thought in earlier mythology to be malevolent creatures, attractive in appearance but demonic in action.

Persephone, in Greek mythology, a beautiful goddess, the daughter of Zeus and Demeter, goddess of agriculture. She was carried off by the god Hades and made queen of the Underworld. Demeter sought her everywhere, lighting her torches at the fires of Mount Etna, while the earth became barren at her neglect. Though Zeus yielded at length to Demeter's lamentations, Persephone could not be entirely released from the lower world because she had eaten some pomegranate seeds there. She was allowed to spend part of each year on earth and the remainder in Hades.

Perseus, in Greek mythology, a hero, the son of Zeus and Danae (a mortal). He killed Medusa the Gorgon and rescued and married Andromeda. A constellation is named after him.

Phaethon, in Greek mythology, the son of Helios the sun god. His father allowed him to drive the sun-chariot across the sky for one day, but Phaethon, unable to control the horses, came too low and nearly set the earth on fire. Zeus saved the situation by killing him with a thunderbolt.

Philemon and **Baucis**, in Greek mythology, a poor, old couple who entertained Zeus and Hermes hospitably when the gods visited the earth in disguise and were repulsed by the rich. For this, Philemon and Baucis were saved from a deluge and their hut was transformed into a temple. They were also granted their request to die at the same time, and were changed into trees whose boughs intertwined.

Philoctetes, in Greek mythology, a warrior, famous for his prowess with bow and arrows (which he had inherited from Hercules). On his way to the Trojan War he was bitten by a serpent and abandoned on the island of Lemnos because of his terrible cries and the stench from his fetid wound. After many years the Greeks were told in a prophecy that Troy would not fall without Philoctetes' help, so he was rescued from Lemnos and healed by Machaon. Philoctetes killed Paris with his arrow, thus helping to bring about the downfall of Troy.

phoenix, in Egyptian and oriental mythology, a sacred bird, said to renew itself every few hundred years from the ashes of the pyre of flames on which it places itself. The phoenix, used in literature as a symbol of death and resurrection, is alleged to resemble an eagle, but with red and gold plumage. As a sacred symbol of Egypt, the phoenix represented the sun, which sinks each night and rises again each morning.

Pontius Pilate, the Roman Prefect of Judaea by whom Jesus was tried. Depicted in the Gospels as wanting to release Jesus, Pilate, before handing him over to the high priests, attempted to evade responsibility for his death by washing his hands in a symbolic gesture in front of the crowd who demanded his crucifixion.

Poseidon, in Greek mythology, the god of the sea, of earthquakes, and of horses. He was a brother of Zeus. The Romans identified him with the water-god Neptune.

Priam, in Greek mythology, the king of Troy at the time of the Trojan War, the husband of Hecuba and the father of Hector and Paris. Priam was killed when Troy fell to the Greeks.

Prometheus, in Greek mythology, a Titan who made the first man from clay and stole fire from the gods to give to mankind. In revenge for the theft, Zeus chained Prometheus to a rock, where his liver was eaten every day by an eagle, only to grow again every night. Hercules eventually rescued him. Prometheus has been seen as a symbol of freedom, rebellion against tyranny, and creative imagination.

Prophet Messenger (Arabic, *nabī* 'prophet', *rasūl* 'messenger' or 'envoy'), in Islam, the title given to those prophets believed to have received a major new revelation from God to transmit to their people. In one Islamic tradition God has sent 124,000 prophets (sometimes interpreted as a symbolic number to indicate that there is no people to whom a prophet has not been sent) but only twelve Prophet Messengers. These are Adam, Shith (Seth), Nuh (Noah), Ibrahim (Abraham), Lut (Lot), Ishmael (Ismail), Moussa (Moses), Salih, Hud, Shu'ayb, 'Isa (Jesus), and Muḥammad.

Proteus, in Greek mythology, a sea-god who had the power of assuming different shapes to escape capture. If he was captured, however, he would use his gift of prophecy to answer questions about the future.

Psyche (Greek, 'soul'), in Roman fable, a princess of outstanding beauty loved by Cupid. Their love was initially thwarted by Cupid's mother, Venus, who was jealous of Psyche's beauty. Cupid placed her in a remote palace, but only visited her in total darkness and forbade her to see him. One night she lit a lamp and discovered that the figure sleeping at her side was the god of love himself. A drop of oil from her lamp fell on him and Cupid awoke in anger at her disobedience. He fled away, and Psyche wandered the earth in search of him. Finally Jupiter made her immortal and gave her in marriage to Cupid.

Ptah, a very ancient Egyptian god, patron of craftsmen and god of creation in the Memphite cosmogony. He is depicted standing on a wedge-shaped plinth, wearing a skull cap and close-fitting garment, like a mummy. He holds the sceptre of dominion which incorporates in its head the *ankh*-sign (life) and *djed*-sign (stability). The animal sacred to Ptah was the Apis bull. The great temple at Memphis (now destroyed) was dedicated to Ptah. In the creation myth of Ptah, he is self-created, existing before Nun, the primeval waters, and is sometimes called Ptah-Nun. Ptah then created the other gods, men, and towns through his thought and tongue. At Memphis, Ptah was associated with the lioness goddess Sekhmet as his consort, and with the

personification of the primeval lotus Nefertem as their son.

Puck (Irish, *púca*; Welsh, *pwcca*), the 16th-century English name for a mischievous demon or goblin, also known as Robin Goodfellow and Hobgoblin, and commonly believed to haunt, in a variety of shapes, the countryside. As Puck he appears in Shakespeare's *A Midsummer Night's Dream* (1595). In earlier superstition he was considered an evil demon, luring travellers off their path and young girls to disaster.

purgatory, in Roman Catholic and Orthodox Christian doctrine, the state of existence or condition of a soul which still needs purification from venial (that is, pardonable) sin before being admitted to heaven. It is believed that souls in purgatory can be helped by the prayer and good works of the faithful. Those who die unrepentant, having totally and deliberately rejected God, are said to go direct to hell.

Pygmalion, in Greek mythology, a king of Cyprus who fell in love with the statue of a beautiful woman (in some accounts he carved it himself). He prayed to Aphrodite to give him a wife as beautiful as the statue, and the goddess brought the statue to life; Pygmalion married the woman so created (sometimes called Galatea).

Pyramus and Thisbe, the tragic Babylonian lovers of classical legend, immortalized in Shakespeare's *Midsummer Night's Dream* (1595). As related by the poet Ovid, the lovers, forbidden to marry, planned to elope. Pyramus killed himself when he mistakenly thought that Thisbe had been devoured by a lion, and Thisbe killed herself on finding Pyramus dead. Their blood was said to have turned the mulberries on a nearby tree, previously white, to red.

Python, in Greek mythology, a dragon or serpent who guarded a shrine of Mother Earth at Pytho (later Delphi) on Mount Parnassus. Apollo slew the Python, ousted the deity, and established his famous oracle.

Q

Quat, the spirit hero of Banks Island, Melanesia, who created night. One myth of the origin of life describes Qat as carving the bodies of men and women from a tree and, by dancing and beating a drum around them for three days, bringing them to life.

Quetzalcóatl, in American-Indian Toltec and Aztec belief, the plumed serpent-god to whom human sacrifices were made. Quetzalcóatl was the god of the morning star, wind, life, fertility, wisdom, and practical knowledge, and invented agriculture, the calendar, and various arts and crafts. He is also identified with a legendary priest-king who sailed away, promising to return. When Montezuma II, the last Aztec king, heard of Hernán Cortés and his men coming in 1519, he believed it was Quetzalcóatl, and he ordered his men to welcome them. This event led to the destruction of Aztec civilization. In addition to his guise as a plumed serpent, Quetzalcóatl was represented as the wind god Ehécatl, shown with a mask and two protruding tubes through which the wind blew.

Quirius, a major Roman deity, war god of the Sabines and later adopted by Rome among the state gods of the city, ranking close to Jupiter and Mars. His worship was identified with that of Romulus, the founder of Rome.

R

Rabi'a al-'Adawiya of Basra (AD 714–801), Islamic mystic, ascetic, and poet, credited with many miracles; one of the first Sufis to teach the doctrine of Pure Love. Said to have originally been a poor girl, sold into slavery as a singer, Rabi'a is renowned for her recitals of mystical love and for her witty responses. In popular Islamic belief her tomb is confused with that of Saint Pelagia of Jerusalem. Another Rabi'a (9th century), Rabi'a of Syria, was the devout wife of a leading Sufi, with whom she lived in an unconsummated love.

Rachel, one of the four Jewish matriarchs; the beloved second wife of Jacob. Tricked by his avaricious maternal uncle, Laban, into first marrying her elder sister, Leah, Jacob served him fourteen years before marrying the younger Rachel. Childless for many years, Rachel eventually bore Joseph. Both sisters fled with Jacob after his dispute with Laban, Rachel taking her father's holy images with them. She died giving birth to Benjamin.

Ragnarök (old Icelandic, 'the fate of the gods' or 'twilight of the gods'). In Norse and Germanic mythology, an apocalyptic disaster that will engulf the world and bring to an end the present order of life. The struggle of the gods to keep the demons of destruction at bay will founder when the god Loki and his son, the savage hound Fenir, break their chains, and giants and monsters will attack the world from all directions. The slain warriors of Valhalla will arise to do battle at Odin's side, and Thor will engage in mortal combat with the serpent Jömungand, in which both will be destroyed. The sun will turn black and the earth will sink beneath the sea, only to rise again, cleansed and fertile. The cosmic ash tree Yggdrasil will be attacked from root to branch, but will survive and grow in strength. On the death of Odin, his son Vidar will ascend to power and Balder will return to inhabit the dwelling of the gods.

Rāma, the seventh incarnation or *avatar* of the Hindu god Vishnu as the son of King Dasharatha of Oudh, descendant of an ancient solar dynasty. His incarnation was Vishnu's response to the threat of the demon-king Rāvaṇa's power in Sri Lanka. When his wife Sītā, an incarnation of Lakshmi, was abducted by Rāvaṇa, Rāma and his half-brother Lakshmana, aided by Hanuman and an

eagle, overcame the demon-king. Rāma and Sītā represent the stereotypical ideal Hindu man and woman.

Rangi and **Papa**, in Oceanian mythology, the sky and the earth, progenitors of all gods and all creatures. According to Maori legend their separation by their son, the sky god Tane, gave rise to mankind's affliction with storms and floods. Papa and Rangi's sorrow at their separation is the origin of summer mists, dew, and ice.

Rāvaṇa, in Hindu mythology, the thrice-incarnated demon who was Vishnu's implacable foe. As Hiranyakashipu he was overcome by Narasinha. As Shishupala, born with four arms and three eyes, he was restored to normality on the knees of Krishna as had been prophesied, and was eventually slain by him. Rāvaṇa's increased power as the ten-headed Rakshasa, king of Sri Lanka, caused the fear of the gods and Vishnu's incarnation as Rāma. Rāvaṇa abducted Rāma's wife Sītā, an incarnation of Lakshmi, but was finally killed by a mighty arrow from Rāma's bow.

Re (Ra), in Egyptian mythology, god of the sun and of creation. He is depicted most frequently as a falcon-headed man with a sun disc on his head, but can also be seen as a ram-headed man. His chief cult centre was On (Heliopolis), from whence emerged the Heliopolitan creation myth. In this the lotus emerged from the primordial waters of chaos (Nun) containing the newborn sun as Re. Self-existent and alone, he brought forth the divine part of Shu, air, and Tefnut, moisture. From their union sprang the earth god Geb and the sky god Nut. The arch-enemy of Re was the serpent Apophis, the equivalent of the Babylonian she-dragon Tiamat, slain by the sun-god Marduk. During creation, which is repeated in the daily cycle of the sun, Re assimilates with other solar gods such as Khepri, the scarab beetle god (symbol of the rising sun), and with Atum, the human-headed ancient sun-god (symbol of the setting sun). Re travels daily in his bark through the sky, and during the hours of darkness traverses the underworld. During the Amarna Period (*c.*1353–1335 BC), the heretic Pharaoh Akhenaten chose to worship the visible sun's disc which is called the Aten, a form of the god Re. By the 5th dynasty (*c.*2495–*c.*2345 BC) every Pharaoh claimed to be both the son of Re, and even Re himself.

Rebecca, one of the four Jewish matriarchs, wife of Isaac, and sister of Laban, mother of the twin brothers Esau and Jacob. Rebecca devised the deception of the blind and aged Isaac which enabled Jacob, her favourite, to obtain the father's blessing and become the traditional ancestor of the people of Israel.

Rehoboam, the son of Solomon by an Ammonite princess, and his successor. Faced with a revolt on his accession, caused by his father's harsh policies, Rehoboam antagonized the rebels with threats. This led to the break-up of the unified Hebrew kingdom into Israel under Jereboam I, and Judah under Rehoboam.

Reuben, the eldest son of Jacob and Leah, and ancestor of the tribe of Reuben, one of the ten Israelite tribes later taken captive by the Assyrians. In the Bible he is depicted as being more merciful than the other brothers to his half-brother Joseph.

Rhea, in Greek pre-Hellenic mythology, a Titan. A daughter of Uranus (Heaven) and Gaea (Earth), she was the sister and wife of Cronus and the mother of Zeus and of other Olympian gods.

Robin Hood, the legendary English outlaw who stole from the rich to give to the poor. The earliest extant ballads, dating from the 14th century, place his exploits in Yorkshire, though this was later revised to Nottingham. Post-medieval ballads give Robin Hood companions in Sherwood Forest, among them Maid Marian, Friar Tuck, Little John, Will Scarlett, and Allan-a-Dale.

Roland, the Emperor Charlemagne's leading knight-errant or paladin, and hero of the French epic poem, 'The Song of Roland' (*c.*1100). Roland, trapped against overwhelming odds by the treachery of his stepfather Ganelon, who had sided with the Saracen king, was the paragon of the ideal warrior, triumphant even in defeat.

Romulus, the legendary founder of Rome. He and his twin brother Remus were the children of a Vestal Virgin who had been ravished by Mars; they were abandoned to die, but were suckled by a she-wolf and brought up by a herdsman. In 753 BC Romulus is said to have founded Rome; he killed Remus, who had ridiculed him by jumping over the beginnings of the city wall. After his death, Romulus was regarded as a god and was identified with the god Quirius.

Ruth, the Moabite great-grandmother of King David, ancestress through Joseph of Jesus and central figure in the biblical Book of Ruth. The Gentile Ruth, although widowed, remained faithful to her Hebrew mother-in-law Naomi, and came with her to Bethlehem. There she gleaned the fields of her wealthy kinsman, Boaz, who, struck by her beauty and devotion, married her.

S

Saints for Christian Saints, see p. 40.

Salmān al-Fārisī (Salman the Persian) (7th century), in Islam, a Companion of the Prophet, the patron of workmen's corporations, and a popular legendary figure. Converted from Zoroastrianism to Christianity as a boy, Salman is said to have travelled from his native Iran to central Arabia, where he was told that he would find a prophet who would revive the religion of Abraham. On his way he was sold into slavery, but after recognizing Muḥammad as the True Prophet from a birthmark on his back, he, with Muḥammad's help, bought his freedom, was converted to Islam, and went on to become a notable figure in early Muslim religious thought. Salman is credited with introducing the *khandak* (trench) in siege warfare, first used in the Battle of the Ditch (627), in which the city of Medina was successfully defended. He is regarded as a national hero in Shīʿa Iran, a link in the mystical chain (*silsila*) of many Sufi orders and, by the Alawites, the chief figure in their Trinity.

Salome, the step-daughter of one of the sons of Herod the Great and daughter of Herodias. Struck by the gracefulness of her dancing, Herod Antipas offered to reward her with whatever gift she desired. Prompted by her mother, she asked for the head of John the Baptist on a platter.

Samson, judge of Israel, the Hebrew hero whose legendary strength came from his long hair, which, as a Nazirite, he had vowed to God never to cut. Bribed by the enemies of his people, the Philistines, Delilah seduced him and cut it off, whereupon he was blinded and imprisoned in Gaza. He regained his strength as his hair grew long again, and he exacted revenge by pulling down the pillars of the temple of the Philistines, destroying himself and those who had mocked him.

Sarah (Sarai), one of the four Hebrew matriarchs, the wife of Abraham. Sarah accompanied Abraham in his desert wanderings from Ur to Canaan. Presented there by Abraham as his sister rather than his wife, she was taken into the harems of the Egyptian pharaoh and the Philistine king, Abimelech. After giving birth to Isaac at over 90, Sarah became jealous of Hagar, her Egyptian handmaid, for having given birth to

Abraham's son Ishmael, and had both cast out into the desert to die.

Sarasvati, originally an early Vedic river-goddess, but in later Hinduism, the beautiful wife of Brahma, born from his body. Goddess of wisdom, learning, the arts, and music, Sarasvati is often shown with four arms, playing the stringed instrument, the vina, holding a book or rosary, and seated on a lotus. Like Brahma, her usual mount is a goose; she is also associated with the peacock, parrot, and ram. Her main temple is at Dilwara.

Satan (Hebrew, 'adversary'), one of the many names given to the devil, the prince of evil spirits and personification of evil in Judaism, Zoroastrianism (see *Ahriman), Christianity (where his name, in the Gospels, is Beelzebub), and Islam (shayṭān), where his other name is Iblīs. Believed to be Lucifer, a fallen angel who was originally created good but chose of his own free will to become evil, he rebelled against God and was cast out of heaven. Satan may also appear in the form of a woman, snake, serpent, or dragon. In the folklore of the European Middle Ages he was given the attributes of cloven hoofs, a forked tail, horns, and the strong smell of sulphur. Shayṭān is, in Islamic mythology, one of the evil and unbelieving jinn who has numerous offspring called shayṭāns. In Islamic folklore shayṭāns are said to be ugly, hooved creatures who prefer shade to light. They live on dirt and excrement, and use disease as a weapon.

Saturn, in Roman mythology, a god of agricultural plenty, later identified with the Greek god Cronus. His festival was the Saturnalia, celebrated from the 17th to 19th December, during which presents were given, candles were lit, and slaves were given licence to fool.

Satyr, in Greek mythology, a type of creature of the woods and hills, mainly human in form, but with some bestial aspect, usually the legs of a goat. Lustful and fond of revelry, satyrs often attended Dionysus. The Roman equivalent were fauns (see *Faunus).

Scylla, in Greek mythology, a female sea-monster who lived in a cave opposite the whirlpool Charybdis (traditionally they were situated in the Straits of Messina between Italy and Sicily). In seeking to avoid one of the dangers, sailors ran the risk of being killed by the other—hence the phrase 'to be between Scylla and Charybdis' means to be in a situation where two possible courses of action are equally dangerous.

Selene, in Greek mythology, the Moon, the daughter of Hyperion and the Titaness Theia. She is sometimes identified with Artemis.

Semiramis, in Greek legend, the daughter of a Syrian goddess, and second wife of Ninus, king of Assyria, with whom she founded Nineveh. After Ninus' death, Semiramis is alleged to have ruled for many years, conquering many lands and founding the city of Babylon. At her death she vanished from the earth in the shape of a dove and was thereafter worshipped as a deity. The historical figures behind the legend are thought to be two powerful Assyrian queen-mothers, Sammuramat (9th century BC) and Naqi'a (7th century BC).

Serapis, the supreme god of the Graeco-Roman period, worshipped during the reign of Ptolemy I Soter (*c.*304 BC). He combined elements of Egyptian gods with those of the Greeks, providing a recognizable deity for the ruling Greeks to worship in Egypt. Serapis is the assimilation of Osiris and Apis (the sacred bull of Ptah at Memphis), depicted in Greek style as a man with long curly hair and beard, also wearing a ram's horns. Serapis' functions were more related to Greek than Egyptian gods. Like Helios and Zeus, he became the supreme solar deity; like Dionysos he distributed fertility; like Hades he was a link with the afterlife; and like Asclepius he possessed the power of healing.

Set (Seth, Sutekh), in Egyptian mythology, the god of the forces of chaos and of the hostile desert lands. His cult centres were in the delta and at Ombos in Upper Egypt. He is depicted as a donkey or monstrous typhon-headed man, and is also represented in the form of hippopotami, serpents, the desert oryx, pigs, crocodiles, and some types of birds. Set was violent from birth, tearing himself from his mother Nut. He was brother to Osiris, Isis, and Neith, the latter being also his consort. Set, jealous of Osiris' kingship of Egypt, killed him and cut the corpse into pieces. He then contended for the kingship with Horus, son of Osiris, and lost. For these deeds Set became the traditional enemy of *ma'at* (cosmic order) and is depicted in animal guise on temple walls being killed by the pharaoh in order to uphold everlasting order over chaos.

Seth, a biblical figure, the third son of Adam and Eve who replaced the slain Abel. In Islamic belief Seth (Shith, Shath) is a Prophet Messenger (*nabī rasūl*) and the executor of Adam's will, who was instructed by Adam how to worship God. Said to have lived in Mecca and performed the rites of pilgrimage, Seth is believed to have collected the 'leaves' revealed to Adam and himself, and to be the founder of craft guilds from whom craft initiation, closely related to Hermeticism, originated.

Seven Sleepers of Ephesus, in 6th-century Christian legend, and in Islam, the Sleepers are the youths walled up in a cave after taking refuge there from the persecution of Decius (*c.*250). In the Christian version God put them to sleep and they woke up some two centuries later to find their city Christian. Shūra 18 of the Koran ('The Cave') contains the Islamic version of the story, which differs in detail and includes a dog, later called Kitmir and cited as one of the ten animals allowed into Paradise.

Shaddad *'Ad.

Shahrazad (Scheherazade), the wife of the Sultan Shahriyar, legendary king of Samarkand. As narrator of the tales of *The Thousand and One Nights*, Shahrazad's own story provides the framework for the tales believed to have been taken from the lost book of Persian folk tales, *Hazar Afsānah* ('Thousand Tales'). From the first night of their marriage onwards, Shahrazad sets out to break the practice of the king, of having his brides executed after the consummation of their marriage. She entertains him with tales each night for 1,001 nights, firing his curiosity by interrupting each tale at a crucial moment in the narrative, and postponing the continuation until the next night.

Shakti (Śakti), in Hinduism, a feminine noun meaning 'energy' or 'power' which represents the potency of the male gods, personified as a goddess. This notion of female divine power is particularly associated with the mythology of Shiva and the worship of the great goddess Devi.

Shamba (also called Shamba of the Bonnet), the legendary inventor-king of the Bushongo and related tribes of the Congo. Shamba is credited with introducing embroidery, tobacco, weaving, and the cassava plant, as well as the game of Mankala to prevent gambling. He is renowned for his wisdom, allegedly refusing to accept hearsay in court, and for abolishing the use of bows, arrows, and the shongo (a four-bladed knife).

Shango, the storm-god of the West African Yoruba tribe in Nigeria, identified in myth with the magician-king of Oyn (the ancient capital of Yorubaland), who is alleged to have destroyed his own family with lightning. His symbols are the axe and ram. He is associated with Voodoo cults in the Caribbean and South America, where he is variously regarded as the thunder, a great general, and a protector against sickness. In Haiti, Shango is identified with St Barbara because of her association with thunder. In the Afro-Brazilian Macumba sects, Candomblé and Umbanda, Shango (as

Xangô) is identified with St Jerome, one of whose symbols is a ram, and with John the Baptist because of his association with fires. There is also a Brazilian cult named Xangô, and Shango is the Afro-Caribbean religion of Trinidad and Tobago.

Shayṭān *Satan.

Sheba, Queen of, the wise ruler of the Saba', the biblical name of a wealthy area and people (also known as the Sabaeans) of south-west Arabia (present-day Yemen). In the biblical account, the queen visited King Solomon to test for herself his famed wisdom. Her visit is also mentioned in the Koran, and greatly embellished in Islamic legend, where she is called Bilqis (Balkis). According to one account, the jinn mischievously informed Solomon that she had hooves and hairy legs. When the latter proved true, Solomon commanded the jinn to have the hair from her legs removed before their union. In a Hebrew version, their union produced Nebuchadnezzar; in Ethiopia, where she is known as Makeda, their son Menelik I is said to have founded the Ethiopian royal dynasty. In Ethiopian iconography, their descendants are depicted with skin that is mottled black and white.

Scheherazade *Shahrazad.

Shén, in Daoism, the gods and spirits at places such as temples and graves. Originally a term in Chinese mythology referring to 'sacred power', Shén has come to denote all that is thought to manifest such power including divinities, spirits, temples, talismans, and holy men who have attained immortality, Shenién ('spirit-men'). (See also *xiān.)

Shénrén *Shén.

Shiva (Śiva), in Hinduism, one of the most important gods who, with Brahma and Vishnu, forms the Hindu trinity (*trimurti*). Sometimes called Rudra (Sanskrit, 'howler'), Shiva may have evolved from an Aryan deity, who was the dreaded god of the storm and of the dead. Shiva is worshipped in various destructive and creative aspects: as the terrifying destroyer, the naked ascetic, the lord of beasts, as Naṭarāja (lord of rhythm and dance), and as the phallus. In benevolent aspect he lives with Pārvatī (see *Devi) in the Himalayas and rides his bull, Nandi. In human form he is often shown with three eyes, wearing a skull necklace entwined with writhing snakes, and carrying a trident.

Shu'ayb, in Islamic belief, the third of the four prophets sent to the Arabs, and sometimes identified with the father-in-law of Moses, Jethro. In the Koran Shu'ayb preached monotheism to the Midianites and also exhorted them to honesty. Their disbelief and corrupt practices were said to have caused their destruction by God in an earthquake.

Sì dà Tiānwáng (Japanese, Shitennō, 'the Four Heavenly Kings'), also known as Sì dà Jīngāng, the Four Great Diamond Kings, in Buddhism, the guardians of the universe and controllers of the four elements (fire, air, earth, water). Often seen as immense images bearing symbols (which vary according to the country) and in full armour, at the entrance to temples, the four deities are said to have helped Gautama Buddha at important stages of his life. According to Buddhist mythology they were originally demon-kings who, after conversion, were assigned to guard the centre of the universe, Xūmí Shan. Duō Wén, or Vaisravana, is guardian of the North and associated with autumn. He is associated with black, and his symbol is a pearl and snake. He rules the Yakshas (good spirits). Zēng Zhǎng, or Virudhaka, is guardian of the South and associated with spring. His colour is red, his symbol a circular canopy of cloth, and his attendants the Kubhandas (deformed demons). Chí Guó, or Dhritarashta, guards the East and is associated with summer. His colour is blue and his symbol a stringed instrument. He commands the Gandharas (musicians) and Pisatchas (powerful vampires). Guǎng Mù, or Virupaksa, guards the West, and is associated with winter. His colour is white, his symbol is a sword, and he leads an army of nagas (serpent-gods). The Four Great Kings of Daoism, corresponding to the Buddhist Heavenly Kings, are the crab-featured Mó Lǐqīng with a magic lance, Mó Lǐhóng with his magic canopy, Mó Lǐhǎi with his stringed instrument, and Mó Lǐshòu with his magic whip and 'rat' that can become a winged elephant.

Sibyl, a general name given by the ancient Greeks and Romans to various prophetesses. They sometimes had individual names, for example Herophile (also known as the Erythraean Sibyl, from her birthplace Erythrae), who prophesied to Hecuba before the Trojan War.

Siegfried (Sigurd), in Germanic mythology, a hero of outstanding strength, beauty, and courage, who killed the dragon Fafnir with a magic sword. Sprayed with the dragon's blood, Siegfried became invulnerable except where a leaf settled between his shoulders. He learned the language of the birds and overcame the dwarf Nibelungs, taking their treasure and the magic ring that he gave to Brunhild after rescuing her. He was killed at Brunhild's instigation after he married Kriemhild, sister of the Burgundian king, Gunther. The 13th-century Icelandic *Volsunga Saga* tells a similar story with Sigurd, Brynhild, Gudrun, and Gunnar corresponding to the Germanic characters. (See also *Nibelung.)

Silenus, in Greek mythology, a satyr, the foster-father and attendant of Dionysus. He was a fat, jolly drunkard, who was, however, wise and could foretell the future. Sometimes the term silenus (plural sileni) is applied to any elderly satyr.

Silvanus, in Roman mythology, the god of wild land, beyond the cultivated fields. He was roughly identified with Pan in Greek mythology, although he is sometimes considered closer to the satyrs.

Simurgh, the fabulous bird of Persian mythology, and a recurring image in Persian literature. In Firdawsī's epic *Shāhnāmah*, Simurgh is the rescuer and guardian of Zal, who helps at his son Rustam's birth and later extracts arrows from Rustam and his horse Rakhsh. Another Simurgh, monstrous and evil, is slain by Isfandiyar. In Sufism, Simurgh became the symbol of God, as exemplified in the Persian mystic Aṭṭar's 'Conference of the Birds' (*Mantiq-al-Tayr*), where the quest of the thirty birds (*sī murgh* in Persian) reveals their reflection, the ultimate reality of their Oneness with God.

Sinai, Mount (Arabic, Jabal Musa), a place of pilgrimage in north-east Egypt for Jews, Christians, and Muslims because of its associations with Moses. Jebel Musa (Mount of Moses), part of Mount Sinai, is where Moses is said to have received the Ten Commandments. At its foot is the Greek Orthodox monastery of St Catherine, built on the alleged site of the burning bush, and housing the remains of St Catherine of Alexandria, miraculously transported to Sinai after her execution. It also holds a document sent by the prophet Muḥammad, who allegedly visited the monastery as a merchant, giving protection to the monks.

Sindbad the Sailor, Persian hero of one of the *Tales of a Thousand and One Nights*, who narrates his adventures on seven voyages to Sindbad the Porter. Sindbad escapes death miraculously on several occasions, including strangulation by the Old Man of the Sea. He acquires wealth from the Valley of Diamonds, the king of Serendib, and the burial ground of the elephants. Sindbad's tales figure world-wide in children's books.

Sirens, in Greek mythology, two (or in some accounts three) sea-nymphs who—by the irresistible beauty of their

singing—lured sailors to destruction on a rocky coast. When his ship was about to pass their island, Odysseus ordered his men to stop their ears with wax but left his own unblocked and had himself lashed to the mast so that he could be the only person to experience the beauty of their singing and live to tell the tale. According to one account, the Sirens killed themselves in vexation at his escape.

Sisyphus, in Greek mythology, a cunning king of Corinth who, for his various crimes, was, after his death, sentenced to eternal punishment in Hades. He was made to roll a large stone to the top of a hill, only for it to roll down again, so that he had to toil uphill endlessly.

Sītā (Sanskrit, 'furrow'), in Hindu belief, the incarnation of Lakshmi as the foster-child of Janaka, king of Videha. Won by Rāma in an archery contest, Sītā is the model of the ideal Hindu wife, following her husband into exile in the forest. There she was abducted by the demon Rāvaṇa which led to war between India and Sri Lāṇka, and Rāvaṇa's death. Sītā subsequently proved her chastity by surviving the fire-ordeal unharmed. As her name suggests, she may originally have been an agricultural goddess.

Síva *Shiva.

Sodom and **Gomorrah**, two of the five cities of the plain, possibly now covered by the shallow waters near the southern end of the Dead Sea. Notorious for the corrupt lives led by their citizens and for the perverted sexual acts in which they indulged, the inhabitants of both cities, with the exception of Lot and his family, were destroyed by 'brimstone and fire' rained on them by God.

Sohrab and **Rustam** (Rustum), in Persian mythology, two heroes, son and father, of several episodes in the epic poem by the Persian poet Firdawsī, the *Shāhnāmah*. Rustam, son of Zal, displayed miraculous strength and courage from childhood. As general of the Persian king's army, and aided by his faithful horse Rakhsh, he survived numerous ordeals in order to free his captured king. In the course of his quests, Rustam married the king of Samangan's daughter. Fearing that Rustam would take their son Sohrab when he was born, the princess pretended that she had given birth to a daughter. From his early youth onwards Sohrab longed to find his father and joined the Turanian army to do so. In a battle between the Turanians (nomads from the central Asian steppes) and Iranians, Sohrab and Rustam were fated to meet in single combat; they only learned each other's identity when Sohrab had been mortally wounded by his father. In Persian iconography Rustam is depicted in armour with a leopardskin cuirass.

Solomon (10th century BC), king of the ancient Hebrews, the youngest son of David and Bathsheba and traditionally regarded as the greatest king of Israel. His wisdom was proverbial, his reign was peaceful, and trade and commerce flourished. He built the first Hebrew Temple in his new capital, Jerusalem. His fame as sage and poet prompted the queen of Sheba to visit him. The Bible account, however, also tells of his apostasy and ruthlessness towards his opponents, and how heavy taxation eventually led to the break-up of the kingdom under his son Rehoboam. Sulayman (Solomon), a popular figure in Islamic legend, has a prominent place in the Koran, where he is considered an apostle and prophet. It is said he knew the language of the birds and ants, that he was visited by Bilqis, queen of Sheba, and that he had mastery over the jinn, with whose help he built shrines and statues. The Koranic account also says God placed a phantom on his throne to try him but that he repented, which probably refers to Solomon's brief lapse into idolatry.

sphinx, a mythological creature with a lion's body and human head, an important figure in ancient Egyptian and Greek culture. To the ancient Egyptians the sphinx was a representation of the Pharaoh with a lion's body (the Pharaoh was endowed with the vigour of the bull and the lion) and with a man's head, wearing the traditional striped headcloth. The best known of these statues is the monumental sphinx at Giza (*c*.2613–2494 BC), bearing the face of Khephren and probably positioned by his pyramid as a guardian. Later tradition identified this sphinx as Haurun, a Canaanite god, and Harmachis, a form of Horus, probably because the statue faces eastwards towards the rising sun.

In Greek mythology, the sphinx was a monster with a woman's bust and the body of a winged lion. She devoured all who could not solve the riddle disclosed to her by the Muses, namely: 'What is it that has one voice, walks on four legs in the morning, on two at midday, and on three in the evening?' Oedipus alone gave the answer, 'man', who crawls on all fours as an infant, walks upright in his prime, and in old age leans on a stick. The sphinx thereupon destroyed herself.

Styx, in Greek mythology, the main river of Hades, across which the dead were ferried by Charon.

Surya, in early Indian religion, a sun-god who, with Indra and Agni, formed a trinity of Vedic gods. One of several solar deities in Vedic religion, Surya became the sun-god of later Hinduism, superseded by the former minor solar deity, Vishnu.

Susan-o-o, in Shinto belief, the god of the sea, storm, and fertility, born from Izanagi's nose. It is alleged that Susan-o-o, challenging his sister Amaterasu's ascendancy, performed various insulting actions which disturbed the natural order of the world. As a result he was banished from his sister's realm to Izumo on the island of Honshu, from where he went to Korea and then returned to Izumo, where he remains an important deity. At Izumo he gave battle to a serpent with eight monstrous heads. When he cut the slain creature to pieces, a sword fell from its tail, and the god sent the marvellous weapon to his sister Amaterasu as a token of his submission. The sword, along with a mirror and a jewel, now form the insignia of the ruling family. In Shinto tradition, all the gods gather at Izumo in the month of October, which is designated Kamiari Juki (month with gods). Elsewhere in Japan, October is known as Kanna Juki (month without gods).

T

Tammuz (Sumerian, Dumuzi), the Assyro-Babylonian god of crops, livestock, and vegetation, believed to die each winter to be reborn each spring. According to mythology, Tammuz was rescued annually from the nether world, either by his sister-wife Inanna, goddess of heaven, or by the fertility goddess Ishtar, the Phoenician Astarte. His festival, commemorating the yearly destruction and rebirth of vegetation, corresponded to the festivals of the Phoenician and Greek god Adonis and of the Phrygian god Attis.

Tane, the Polynesian god known as Kane in Hawaii, where he is particularly revered as the god of regeneration. Depicted with a dazzling phallus, he is the Hawaiian ancestor of chiefs and commoners, creator of the heavens and earth. Tane, as Tane-mahuta, is also the sky-god son of Rangi and Papa, protector of birds, forests, and wood-workers. In some Maori mythology he is the creator-husband of Hina, whom he fashioned out of sand, breathing life into her mouth. Their first offspring was a bird's egg, but their second was a daughter, Hine-Titama, who was in turn to become his wife.

Tanit (Tinith, Tiunit), Phoenician fertility goddess, chief goddess of the city-state of Carthage, she was probably the consort of Baal Hammon, and is associated with Astarte. Although she seems to have been a mother goddess, children, probably firstborn, were sometimes sacrificed to her. The worship of Tanit, often in the form of household effigies of the goddess, was also widespread along northern Africa and in Malta, Sardinia, and Spain.

Tantalus, in Greek mythology, a king of Lydia, son of Zeus and a nymph. At first the intimate friend of the gods, he offended them in serving his son's flesh at their table, in stealing nectar and ambrosia, the food of the gods, to give to men, and in revealing their secrets. He was banished to Tartarus, the prison beneath the underworld and set in a pool of water which always receded when he tried to drink from it, and under trees whose branches the wind tossed aside when he tried to pick their fruit (hence the word 'tantalize').

Tartarus, in Greek mythology, a region below Hades where the wicked were punished for their misdeeds on earth.

taʿziya *Ḥusain.

Tell, William, a legendary hero of the liberation of Switzerland from Austrian oppression, who was required to hit with an arrow an apple placed on the head of his son; this he successfully did. The events are placed in the 14th century, but there is no evidence for a historical person of this name. Similar legends of a marksman shooting at an object placed on the head of a man or child are of widespread occurrence.

Tezcatlipoca (Nahuatl, 'smoking mirror'), the all-powerful chief deity of Aztec mythology, but probably of Toltec origin. Credited with several aspects, Tezcatlipoca was the sun-god turned into a tiger by Quetzalcóatl, his constant opponent; as patron of warriors he was identified with Huitzilopochtli; in his association with witches and evil-doers he was a trickster god. He is often depicted as one-footed, the other having been bitten off by the monster, Earth, as Tezcatlipoca raised him from the primeval waters.

Themis, in Greek mythology, a goddess and the personification of justice. By Zeus she was the mother of the Fates.

Theseus, in Greek mythology, one of the greatest of heroes, especially venerated in Athens, of which city he became king. His most famous feat was, with the aid of Ariadne, killing the Minotaur. He was a friend of Hercules and like him one of the Argonauts.

Thor, the Norse god of the sun and of thunder, the son of Odin, an enemy to the race of giants but benevolent to mankind. The attribute most commonly associated with him is thunder and lightning, which he created by casting his hammer Mjollnir to earth. He possessed iron gloves and a belt which greatly increased his strength. Thor's perpetual enemy was the serpent Jörmungand, symbol of evil, which is coiled around the world of man. Thor's day (Thursday) was considered propitious for marriages.

Thoth (Djehuty), in Egyptian mythology originally a moon-god, depicted mostly as an ibis-headed man, but also as an ibis or squatting baboon, and in all cases wearing a sun disc and moon crescent on his head. Thoth was the scribe of the gods, recording their dictates, transmitting wisdom through writing, science, and mathematics, acting as impartial arbitrator in disputes, and recording the result of the Weighing of the Hearts ceremony performed before Osiris in the afterlife. At Hermopolis (Khemun) Thoth was revered as a creator god, bringing forth the lotus flower that arose from the primeval waters of Nun, to reveal the beautiful child creator of the world, the infant sun. As god of wisdom, he uttered words which sprang into life, thus animating the first beings.

Tiamat, the ancient sky divinity of Sumerian origin, worshipped in Babylonian religion. With Apsu, the abyss of fresh water beneath the earth, Tiamat became the parent of the first gods. Angered at the commotion made by their descendants, however, Tiamat, in serpent form, representing the forces of chaos and evil, made ready to destroy them. The gods were eventually saved by Marduk, who slew the serpent. In another version of the myth, Tiamat was a great she-dragon, reigning over chaos. Challenged by the younger god Marduk, she opened her jaws to swallow him, but he killed her with an armoury of winds. He split her carcass into two parts: one he pushed upwards to form the heavens, the other he used to make a floor above the deep. In the world between, he created man and woman.

Timothy in the New Testament, Paul's companion and 'son in the faith'. The son of Eunice, a Judean Christian, Timothy was found in Lystra by Paul, circumcized to placate the Jews, and left in charge of the church in Ephesus. Said to have been killed in a riot after denouncing the worship of Artemis, Timothy became a Christian saint and martyr.

Tiniran, in Oceanic mythology, the deity of the oceans and of all creatures that dwell in it. At once destructive and creative, he is sometimes referred to as the consort of the goddess Hina.

Tír na nÓg *Oisín.

Titans, in Greek mythology, six sons and six daughters of Uranus (Heaven) and Gaea (Earth); they were the older generation of gods who were overthrown by the Olympian gods, led by Zeus. Some of their children (notably Atlas and Prometheus) were also regarded as Titans.

tjurunga, in Australian Aboriginal mythology, a ritual object, usually made of stone or wood, that is a representation of a sacred thing. It may take the form of a stone, an earth mound, ritual pole, ground painting, headgear, body decoration, or sacred song. Tjurungas may be made of wood carved with designs of intricate mythological significance, or they may be seen as the mythical being itself. In essence, they are the symbol and expression of communion between the living and their spiritual heritage of *Alchera or the Dream time.

Tlaloc, the rain-god of ancient (possibly Olmec) Mexico, worshipped by the Toltecs and their successors, the Aztecs. The serpent-masked Tlaloc was ruler of the sun of the third universe and con-

troller of the clouds, rain, lightning, springs, and mountains. Chalchevhtlicue, a rain-goddess, was one of his wives. His kingdom, the earthly paradise 'Tlalocan', received the spirits of those killed by thunderbolts, water, leprosy, and contagious diseases.

Tobias *Tobit.

Tobit and **Tobias**, the pious father and son heroes of the apocryphal Book of Tobit in the Bible. The blind Tobit, exiled to Nineveh, sent Tobias, accompanied by his dog, to the house of Sarah, equally afflicted by a demon. Led by the angel Raphael through many difficulties, Tobias married Sarah, from whom he had exorcised a demon, and Tobit regained his sight. The Book of Tobit ends with a prophecy by Tobit of the restoration of Jerusalem.

Tom Thumb, the diminutive character of an old English nursery tale of the same name. In an early version of the legend Tom Thumb was the son of a ploughman in the time of King Arthur, and he was the same size as his father's thumb. Similar characters appear under a variety of names in other European and West Asian folklore.

Tonatiuk, the Aztec sun god, who received the hearts and blood of men daily to quench his thirst. The Aztecs conceived of this god as a power being constantly burnt up, threatened by the colossal task of its daytime birth and journey as well as its struggle and death at nightfall. Only through continuous sacrifice and moral virtue would its daily cycle be sustained.

Tristram, in Celtic mythology, a Cornish knight, the prototype of the courtly lover. Injured in a battle which freed Cornwall of the levy of men exacted by the Irish, Tristram, disguised as the minstrel Tantris, was healed by the Irish Queen Iseult. Later he was by mistake given a love-potion intended for her daughter, Iseult, and her betrothed, the king of Cornwall. Tristram and Iseult were bound thereafter in endless passion; they were nevertheless forced to part and Tristram eventually married a Breton princess, Iseult of the White Hands. On being fatally wounded, however, it was to his great love, the second Iseult, that he appealed for help. Tristram died when he was wrongly told that his beloved Iseult was not on board the ship approaching Brittany.

Triton, in Greek mythology, a sea-god, son of Poseidon. More loosely, the term is applied to a type of merman—half-man, half-fish. Tritons are usually shown blowing wind instruments made of conch-shells.

troll, in Norse mythology, originally a member of the supernatural race of giants associated with Thor's battles. In later folklore trolls came to be seen as dwarfs who live in caves and mountains, credited with skilful craftsmanship but limited intelligence. The probably related *trows* of Orkney (Scotland) were believed to be evil, witch-like beings.

Tuatha Dé Danann (Gaelic, 'People of the goddess Danu'), in Celtic mythology, a company of gods who destroyed the sea giants, the Fomors, at the battle of Moytura, and made Ireland their home. Early accounts relate that they had been banished from heaven on account of their knowledge of magic, and had descended on the island in a cloud of mist. Among their gods were the earth-goddess Dana, the sun-god Lugh, and Lêr, the god of the sea. They were in turn overcome by the Milesians (ancestors of the modern Irish), and took refuge in the hills, continuing to live in their palaces and halls underground. In popular legend they are linked with the fair folk that inhabit the Irish landscape.

Tyr (Anglo-Saxon, Tiw), in Norse and Germanic mythology, a war-god and deity of athletes. Identified with Mars, 'Tiw's day' became Tuesday, corresponding to the Roman *martis dies* (day of Mars). Cited variously as the son of Frigg, Odin, and Hymir, Tyr chained Fenrir, the wolf-son of Loki, and his hand was bitten off. He is doomed to die at Ragnarök, slaying Garm, the watchdog at the entry to Hel, the world of the dead.

U

Ujigami, originally the common ancestor and tutelary deity of a Japanese clan (*uji*), and worshipped at a clan shrine. Now, as the Japanese clan system has disappeared, ujigami has come to refer to the tutelary deity at the local Shinto shrine, that protects all in the local area. Ujigami are thus now regional rather than ancestral in nature.

Ulaids *Conchobar mac Nesse; *Cuchulain.

Ulysses *Odysseus.

unicorn, a mythical animal resembling a horse with a single horn on its forehead. An enigmatic symbol, the first known mention of it comes in ancient Sanskrit texts. Its first known depiction is on Assyrian reliefs. As *ch'i-lin*, it was considered an animal of good omen in China as early as the 27th century BC. According to Islamic and Western medieval legend it was difficult to capture and could be tamed only by a virgin girl. As the biblical animal of the Old Testament, the unicorn was taken as an allegorized interpretation of the Christian Church, often likened to Christ.

Unkulunkulu (Nguni, 'chief'), the creator sky-god of the Zulu, a nation of Nguni-speaking people of southern Africa. According to Zulu myth, human beings became mortal when the lizard, sent by Unkulunkulu with the message of death, reached them first because the chameleon with the message of life loitered on the way. Unkulunkulu then instituted marriage in compensation, to enable people to continue their lives through those of their children.

Uranus, in Greek mythology, the personification of the heavens and the first ruler of the universe. The husband of Gaea (Earth) and father of the Titans and Cyclopes, he was overthrown by his son Cronus.

Utnapishtim, in Mesopotamian mythology, the hero of the story of the Flood. Like the biblical Noah, he survived cosmic destruction by heeding the divine command to build an ark. But unlike Noah, who, when the waters subsided, was committed by God to build a better world, Utnapishtim and his wife were taken from the earth and made immortal.

V

Vainamoinen, in Finnish mythology, the archetypal sorcerer of the far North, son of Ilmatar the air goddess. He played music on the *kantele*, an instrument he created out of the jaw of a horse and strung with the hairs from its tail. When he played the instrument wild animals grew tame and the turmoil of the elements ceased.

Valhalla *Asgard.

Valkyrie (Old Norse, 'chooser of the Slain'), in Norse and Germanic mythology, the warrior maidens, chief of whom was Brynhild. They lived with Odin as his attendants and messengers in Valhalla, Odin's hall in Asgard. The Valkyries rode out on horses through the air and over the seas to fetch the heroes slain in battle whom Odin had selected for their outstanding valour. The Valkyrie, who were the foreboders of bloodshed and death on the battlefield, were nevertheless associated with fairness and with a bright beauty.

Vāmana, in Hindu belief, the fifth incarnation or *avatar* of Vishnu as a dwarf. Assuming this form he freed heaven and earth from the usurper demon Bali by modestly requesting a patch of earth he could cover in three strides. When Bali consented, Vāmana, transformed into a cosmic giant, covered the universe in two strides and drove the demon into the underworld with the third. Vāmana is depicted either as a small, youthful, Vedic student or as a hunchbacked, big-bellied dwarf.

vampire, a creature supposed to suck the blood of a sleeping person. A vampire may be seen as a reanimated corpse which at night takes on the form of a bat-like demon; once bitten, the victim in turn becomes a vampire. To protect the living, a stake could be driven through the buried corpse, or it was exhumed and burned. Vampires, which are said to cast neither shadow nor mirror-reflection, appear in mythologies as far apart as Europe (where Count Dracula is the archetypal example), Asia, and Africa. In Hindu mythology, a species of female vampire is commonly held to debilitate a sleeping man by sucking blood from his toes.

Vanir *Aesir.

Varāha, in Hinduism, the third incarnation of Vishnu as a wild boar. When the earth was ruled by powerful demons during the Deluge, Varāha saved it by diving into the water, retrieving the *Vedas* (the Hindu sacred scriptures), and slaying the demon ruler, Hiranyaksha, who was holding the earth down. Varāha is depicted either as a powerful man with a boar's head, or as a massive boar. The earth, personified as a goddess, is often shown as a diminutive female figure holding on to one of his tusks.

Varuna, an Indo-Aryan deity, and one of the oldest of the Vedic gods of the *Rig Veda*, the collection of sacred songs that form part of the ancient Vedic literature of India. Originally considered the sovereign lord of the universe (superseded first by Indra, then by Shiva and Vishnu), and god of law and justice, Varuna became associated in Hinduism with the moon, where, with Yama the first human, he was lord of the dead. In this capacity he was also guardian of the sacred plant, soma, which was alleged to give strength, wisdom, and immortality. Varuna was a god of justice with many informers who reported any infringements of the law which he sought to uphold.

Vashti, in the biblical book of Esther, the Persian wife of King Ahasuerus (Xerxes), who refused his summons to display her beauty at a banquet before the drunken governors and people of his kingdom. Accepting his counsellors' advice that she should be punished so as to deter other women from disobeying their husbands, Ahasuerus banished Vashti, and replaced her as queen with Esther.

Venus, the Roman goddess of love and beauty, identified with Aphrodite in Greek mythology. Venus Verticordia was the goddess, worshipped by Roman matrons, who turned married women's hearts to chastity.

Vesta, the Roman goddess of the blazing hearth, identified with Hestia in Greek mythology. In Vesta's temple in Rome, the sacred fire on the state hearth was never allowed to die out, except on 1 March, the start of the new year, when it was ceremonially renewed. The fire was tended day and night by Vestal Virgins—priestesses who were sworn to chastity.

Viracocha, an ancient Peruvian god and chief deity of the Inca empire, believed to inhabit the depths of Lake Titicaca with his sister-wife Mama Cocha. God of water and all life, creator of the universe and man, Viracocha, according to one of many variations of the myth, was displeased with his first creation, and destroyed it in a deluge. In his subsequent creation he wandered as a beggar, teaching man the rudiments of civilization. He is depicted with the sun as a crown, rain as tears, and holding thunderbolts.

Vishnu, originally a minor solar deity (see *Surya) but now one of the most important gods of modern Hinduism. The 'preserver' in the triad (trimurti) with Brahma and Shiva, Vishnu, whose consort is Lakshmi, he is believed to go through ten principal animal and human incarnations (*avatars*), which take place when the world needs them. He has already been a fish, Matsya, a tortoise, Kurma, a boar, Varaha, a man-lion, Narasinha, a dwarf, Vamana, as well as Parashurama, Ramachandra (Rāma), Krishna, and the Buddha. The tenth *avatar*, still to come, will be Kalki, a white horse or a man riding a white horse, signalling the end of the current cycle of time. Often depicted as a blue-skinned youth clad in yellow regal robes, Vishnu holds a mace, conch-shell, disc, and lotus in his four hands.

Vodun (voudoun), the general name for a god in the Ewe and Fon languages of Togo and Benin (Dahomey) in West Africa, from which the name for the Haitian possession-cult religion, Voodoo, is probably derived. The Vodun of Voodoo are deified spirits of the dead, believed to take possession of the believer through the medium of a *hungan* (priest) or *mambo* (priestess). Celestial marriages of a woman or man with a spirit of the opposite sex, by which the former acquires permanent possession and protection by the spirit, are a central feature of Voodoo. Ezila, or Erzuli, the patron goddess of lovers, is a frequently cited Vodun in such marriages.

Voodoo *Vodun.

Vulcan, an early Roman god of fire and metalworking, identified with Hephaestus in Greek mythology.

W X Y

Wax Girl *Anansi.

Wayland (Weland) **the Smith**, a smith of outstanding skill; according to some Norse, Germanic, and Early English mythology, a lord of the elves. Captured by the Swedish king, Nidud, Wayland is lamed and forced to work in the king's smithy. In revenge he kills the king's sons, forging their skulls into drinking bowls, and rapes their sister. Among his alleged burial places is Wayland's Smithy in Berkshire in England, Sisebeck in Sweden, and a site in Vellerby in Jutland.

Weighing of the Hearts ceremony (hesbet), in ancient Egyptian mythology, the moment when the dead stood in judgement before Osiris and his forty-two assessors. Placed by the jackal god Anubis on a pair of scales, the soul was weighed against the feather of truth. The result of the weighing was inscribed on a tablet by Thoth. Those whose souls were not on a level with truth were fed to the monster Ammut, part crocodile, part lion, and part hippopotamus. The others departed for the 'other land', west of the setting sun.

Wénshū *Manjusri.

werewolf, a mythical being that at times changes from a person to a wolf. Belief in werewolves is found throughout the world; in countries where wolves are uncommon, the creature is said to assume the form of another dangerous animal such as a tiger or a bear. Traditionally, werewolves are held to be most active at full moon, devouring animals, humans, or corpses. A person who is bitten by a werewolf (or a vampire) is turned into one her- or himself.

witch, in European mythology, a sorceress, a woman said to have dealings with the devil or evil spirits, and popularly depicted with a black cat (her familiar) and broomstick. The witch's male counterpart is named a wizard, sorcerer, or warlock. The most notable sorceress of ancient Greece was the legendary Medea. By the late Middle Ages witches were considered as Christian heretics who had entered a pact with the devil. They were accused of demonic possession, midnight orgies (sabbaths) with the devil, and night-flying, singly or in covens (groups).

Woden *Odin.

Wongar *Alchera.

Xangô *Shango.

xiān, in Daoism, humans who are thought to have gained immortality through spiritual practice. Unlike shénién ('spirit men') xiān have no attachment to particular places and are not associated with cults, sacrificial rituals, or relics. Although described in innumerable ways, xiān possess certain basic characteristics: they are thought of as capricious creatures who frequent the heavens, mountains, and grottoes; they possess superhuman powers; their fantastical appearance is not visible to mortals. Of the various classes of xiān, the best-known are the Bāxīān or Eight Immortals, a heterogeneous group of sages each of whom earned the right to immortality, and have become popular figures on stage, in art, and in folklore.

xiāng shēng xiāng kè, in Daoism, the concept that the five elements (wood, fire, earth, metal, and water), which may give rise to one another, may also destroy one another. In this theory, of central relevance to Chinese traditional medicine, the elements are viewed as abstract and counterbalancing forces rather than material substances.

Xīwángmǔ (Queen of the West), in Chinese Daoist mythology, the queen of the immortals (see *xiān) and female genies (of the West Flower paradise) whose realm is Mount Kūnlún. Xīwángmǔ is said to have visited the Emperor Wǔ Dì in person in 110 BC and to have presented him with the elixir of immortality in the form of the pántáo, a flat peach. Sometimes described as having been a semi-human mountain spirit with a panther's tail and tiger's teeth who was transformed into a beautiful woman, Xīwángmǔ is a popular figure in traditional Chinese belief, especially revered by women.

Xiuhtecuhtli, the Aztec fire-god, also known as Huehueteotl (Nahuati, 'old god'), ruler of the sun of the present universe, to whom human sacrifices were made. As the leader of souls, Xiuhtecuhtli helped the spirits of the dead to be absorbed into the earth. His mysterious spindle extended throughout the universe from 'Mictlan', land of the dead, through the household fires on earth to the heavens. He is sometimes depicted bearing a brazier on his head.

Yājūj and Mājūj, in Islamic belief, the two hostile forces which will ravage the earth at the end of time. The Muslim counterpart of the biblical Gog and Magog, they are mentioned in the Koran in connection with Dhu'l-Qarnayn (Alexander the Great), who, in response to a request made by people being terrorized by these forces, is said to have built a wall of iron and brass between two mountains to keep them out.

Yama, in Hindu mythology, the first man to die, becoming lord of the dead. Later considered as god of death, Yama is usually depicted carrying a club representing authority and punishment, and a noose with which he seizes the souls of his victims. His mount is a buffalo. In Chinese Buddhism, Yama is usually called Yánluó, and is believed to have been a king of Vaishali. As a punishment he was re-born as king of demons and hells, over which he and his consort Yamina rule. Depicted by the Chinese in his capacity as judge of future punishments, Yánluó is said to send his messengers (old age and sickness) to human beings to keep them from leading immoral, excessive lives.

Yanauluha, the great medicine-man of Pueblo Zuni mythology. At the beginning of time the first men and women were born as misshapen creatures from the earth: black and scaly, with short tails, owl's eyes, huge ears, and webbed feet. Yanauluha brought a vessel from the primeval ocean, as well as seeds, and a staff which had power to give life. He taught the Zuni the arts of civilization: agriculture, husbandry, and the regulation of social life. His potent medicine staff, painted in bright colours and decorated with feathers, shells, and stones, is the emblem of the first chief priest.

Yánluó *Yama.

Yggdrasil (Old Norse, Mimameidr), the cosmic ash tree of Norse and Germanic mythology. The tree supports the universe: its roots reach down to Jötunheim, land of the giants; to Niflheim, abode of the dragon monster Nidhogg; and to Asgard, home of the gods. At its base are three wells: Urdarbrunnr, the Well of Fate from which the tree is watered by the Norns or Fates; Hvergelmir, the Roaring Water Pot, in which the Nidhogg dwells, gnawing at the tree's roots; and Mímisbrunnr, the Well of Mimir, source of Wisdom. Its branches overhang the world, and reach

up beyond the heavens. The god Odin discovered the secret of runic wisdom by sacrificing himself to the tree, remaining suspended in it for nine days and nine nights. After the terrible battle of Ragnarök, the cosmic tree, though badly harmed, will be the source of new life.

Yima, in ancient Iranian belief, the son of the Sun, the first man, king and founder of civilization, replaced by Gayomart in Zoroastrianism. According to one legend, Yima refused Ahura Mazda's request to act as his prophet and became instead a king under whose rule man and beast prospered and multiplied so much that Yima had to enlarge the world. According to another, warned of destruction by a great winter, he and the best of men, women, and beasts survived in an underground cavern. Under the name of Jamshid, Yima is the subject of many tales known throughout the Islamic and Hindu world.

yīn and yáng (in Japanese, *in-yō*), two opposing forces whose complementary interaction forms and sustains the universe according to Eastern thought. Originally 'yin' described the cold, northern side of a mountain, and 'yang' the hot, southern slope. By the 3rd century BC, yin, the passive force, was taken in China to represent the earth, darkness, and all that is feminine and receptive, while yang, the active force, represents light, the sky, and all that is masculine and penetrating. The common symbol of yin and yang resembles the light and dark halves of a circle, curving one into the other. Each holds a small particle of the other in it. This balance between yin and yang applies to all human affairs as well as to all physical processes.

Ymir *Aurgelmir.

Yùhuáng (the Jade Emperor, also known as Dōng Wánggōng), the mythological figure of later Daoist China to whom annual offerings of silk and jade were made. Also known as the August Personage of Jade, he was believed to be the creator of man and lord of heaven, meting out rewards and punishments to the gods according to their performance. He is usually depicted in green robes embroidered with dragons, and seated on a throne.

Z

Zacchaeus, the tax-collector of Jericho who, being small, had to climb a sycamore tree to see Jesus pass by. Noticing him, Jesus invited himself to his house, thus incurring the criticism of the crowd. His visit prompted Zacchaeus to give half his wealth to the poor, and restore fourfold anything he had fraudulently acquired.

Zacharias (Zachariah), in the Bible and Koran (as Zakariya), a temple priest and father of John the Baptist. In the New Testament Zacharias is struck dumb until his son's circumcision because he disbelieves the angel Gabriel's message that his aged and barren wife Elizabeth will bear a child. In the Koranic account, Zakariya is also the guardian of the Virgin Mary.

Zal, in the Persian epic *Shāhnāmeh*, the father of the great Rustam, and beloved of Rudabeh. The infant Zal is exposed on a mountain to die by his father Sam because he is born white-haired. He is rescued and brought up by the fabulous bird Simurgh. Later Sam repents, eventually finding Zal when he is a young man. The white-haired Zal is a frequent figure in Persian iconography.

Zamzam, in Islam, the sacred well in the city of Mecca, sometimes referred to as Ishmael's Well. According to legend, the archangel Gabriel opened the well for Hajar (Hagar), desperate for water for her son Ishmael after their expulsion from Abraham's land. The site of the well was later forgotten but then allegedly re-discovered by ʿAbd al-Muttalib, the Prophet Muḥammad's grandfather. Hajar's search for water is commemorated in the pilgrimage (*ḥajj*) rites, and pilgrims take water from Zamzam back to their countries, often dipping their shrouds in it before leaving.

Zanahary, the supreme deity of the Madagascan pantheon. A multiple deity with female and male aspects, Zanahary made, according to one creation myth, the earth, and gave life to the images of human beings and animals that a lesser divinity, Ratovoantany, had shpaed out of clay. Thereafter, Zanahary always insisted that he should take their lives back when they had run their course.

Zàojūn (Chinese, 'Lord of the Stove or Hearth'), a Daoist household deity, protector of the family, and an important figure in Chinese folk belief. The picture of Zàojūn surrounded by children is placed above the hearth and venerated by the family on days of the full and new moon. Zàojūn is believed to deliver reports on the household each New Year to the Jade Emperor Dòng Wánggōng and, for this reason, his mouth is sometimes smeared with honey on the eve of the festival.

Zeus, in Greek mythology, the son of Cronus, whom he overthrew and succeeded as the supreme god. The greatest of the gods and ruler of the universe, he was married to his sister Hera, and was consort of a number of goddesses and lover of mortal women. The latter legends may be accounted for in some cases by the claim of royal houses to be descended from him. Zeus was the giver of laws; he saw that justice was done and liberty maintained. Supreme among gods, his power was limited only by the mysterious dictates of the Fates.

Zion, the Jebusite hilltop fortress captured by David, identified with Ophel in modern Jerusalem. In the Bible Zion is used to mean Jerusalem and by Christians to mean the Christian Church or the Kingdom of Heaven. For Jews Zion also symbolizes their return to the Promised Land.

zombi, in the Voodoo cult of Haiti, a soulless body and the slave of a magician. A zombi can be a living person whose soul has been removed by magic, or a revived corpse whose soul has been separated from it by death. It is believed that the Ghede, the top-hatted death spirits, have the power to re-animate corpses as zombis.

Zurvan, in Persian mythology, the god of time and fate, father in later Zoroastrianism of Ahura Mazda and Ahriman. Zurvanism appeared in Persia during the Sassanian period (3rd–7th century AD) from older roots. Opposed to the orthodox dualistic belief of Zoroastrianism of that time, Zurvanism propagated the belief of the universe as an evolutionary development of primeval matter, represented as Zurvan, rather than as a creative act of God. It was this form of Zoroastrianism which influenced both Mithraism, in which Zurvan was a god, and Manichaeism.

Christian saints

1. The dates given are those generally accepted by Anglican and Roman Catholic communions. Many were different in the past and some may vary now in certain localities, as do many saints' feast-days in the Orthodox and Eastern Churches.
2. Unofficial title given to ecclesiastical authors of the early centuries of Christianity whose authority on doctrinal matters carried special weight.
3. Title formally conferred on certain Christian theologians of outstanding merit and saintliness.
4. Roman Catholic.
5. In the reform of the Roman calendar initiated by Pope John XXIII in 1958.
6. Anglican Book of Common Prayer.
7. Alternative Service Book, the supplement to the Book of Common Prayer authorized in 1980.
8. A group of fourteen saints, also known as Auxiliary Saints, venerated for the supposed efficacy of their prayers on behalf of human necessities.

Name and feast-day[1]	Identity and associations	Emblem	Where active	Date of death
Agnes 21 January	Virgin martyr; patron of betrothed couples, maidens, and gardeners	lamb on book;sword in her throat	Rome (Italy)	c.350
Ambrose 7 December	Bishop, Father[2] and Doctor[3] of Latin Church; great opposer of heresy; associated with Ambrosian chant; patron of bee-keepers and domestic animals	episcopal vestments; scourge, beehive, dove	Milan (Italy)	397
Andrew 30 November	Apostle and martyr; brother of St Peter; patron saint of Scotland and Russia; patron of fishermen and sailors	saltire (X) cross (flag: white cross on blue field); fishing net	Palestine Greece Constantinople	c.60
Anselm 21 April	Benedictine monk; Doctor[3] of the Church; archbishop; influential theologian; founder of scholasticism	episcopal vestments; ship	Rome (Italy) France Canterbury (England)	1109
Anthony of Egypt 17 January	Abbot; patriarch and patron of monks; healer of men and animals	tau (T) cross/T-shaped staff; pigs, bells, crow	Egypt	356
Anthony of Padua 13 June	Franciscan friar and priest; Doctor[3] of the Church; preacher and teacher; patron of lost things	book and lily/Infant Jesus; fish	Italy France	1231
Augustine of Canterbury 26 May; 27 May (RC[4] and outside Britain) 13 May at Canterbury	Monk, prior, and archbishop; evangelizer of the Anglo-Saxons	episcopal vestments; lily, book	Rome (Italy) England	c.604
Augustine of Hippo 28 August	Bishop and Doctor[3] of Latin Church; son of St Monica; author of the influential *Confessions, On the Trinity,* and *City of God*; patron of theologians	episcopal vestments with pastoral staff; book; heart on fire or bleeding, pierced by arrows	North Africa Italy	430
Barbara (4 December suppressed 1969[5])	Virgin martyr; patron of those in danger of sudden death by fire, explosion, or lightning i.e. miners, gunners, firemen; patron of builders; one of the Fourteen Holy Helpers[8]	tower; bolt of lightning; sword, chalice, host	Nicomedia (Turkey) Tuscany (Italy) Heliopolis (Egypt) Rome (Italy)	235
Barnabas 11 June	Apostle (not one of the Twelve) and martyr; accompanied St Paul on first missionary journey; according to legend Milan's first bishop	roses	Cyprus Antioch (S. Turkey) Milan	61
Bartholomew 24 August	Apostle and martyr; patron saint of tanners and all who work with skins	flaying-knife	Armenia Persia India	1st century
Benedict of Nursia 11 July	Abbot; founder of the Benedictine order; patriarch of western monasticism; Patron Saint of Europe; patron of schoolboys and coppersmiths	monk's habit, holding his Rule or a rod; a broken chalice; raven; sword and scroll	Italy	c.550
Bernard of Clairvaux 20 August	Abbot and founder of 163 Cistercian monasteries; Doctor[3] of the Church; patron of bee-keepers	three mitres on book; in white monk's habit adoring Virgin	France	1153
Boniface 5 June	Archbishop and martyr; Apostle of Frisia and Germany; patron of brewers and tailors	mitre and staff; book pierced by sword; axe in root of tree	Crediton (England) Frisia (Germany, The Netherlands, Denmark)	c.754
Bridget of Kildare (Brigid/Bride) 1 February	Abbess; second patron saint of Ireland; also venerated in Wales, Alsace, Flanders, and Portugal; patron of poets, blacksmiths, dairymaids or cattle, fugitives, healers	oil lamp and oak wreath; goose, cheese, cow at her feet	Ireland	c.523
Catherine of Alexandria (25 November, suppressed 1969[5])	Virgin martyr; patron of young women, students, scholars, the clergy (especially philosophers and apologists), attorneys, nurses, wheelwrights, spinners, millers, and the dying; one of the Fourteenth Holy Helpers[8]	spiked wheel; ring; book	Egypt	c.310

Christian saints

Name and feast-day[1]	Identity and associations	Emblem	Where active	Date of death
Cecilia (Cecily/Celia) 22 November	Virgin martyr; patron of musicians	organ, lute, viol, harp	Rome	2nd or 3rd century
Christopher 25 July (reduced to a local cult 1969[5])	Martyr; patron of sailors, travellers, motorists; invoked against water, tempest, plague, and sudden death; one of the Fourteen Holy Helpers[8]	giant carrying Christ-child on his back; hermit with river, boats, fishes, flowering staff, arrow, mermaid, lantern	Turkey	3rd century
Columba of Iona 9 June	Abbot and missionary; Apostle of Scotland	Iona cross; basket of bread; orb of world in ray of light	Ireland Iona (Scotland) Scotland	597
Cornelius 16 September (RC)[4]; September 26 (BCP),[6] 13 (ASB)[7]	Pope and martyr; associated with St Cyprian in Western Church	papal vestments, holding triple cross and bull's horn	Rome	253
Cuthbert 20 March	Monk, prior, and bishop; most popular saint of Northern England; patron of shepherds and seafarers	cross with lions; crowned head of St Oswald	Lindisfarne, Ripon, Melrose, Durham (England)	687
David of Wales 1 March	Monk and bishop; patron saint of Wales and only canonized Welsh saint; nicknamed Aquaticus because he drank no wine or beer	episcopal vestments; standing on mound with dove on shoulder	Wales	*c.*601
Dominic 8 August, 4 August (ASB)[7]	Founder of the Order of Friars Preachers (Dominicans); sent to preach against Albigensian heresy in Languedoc	black and white monk's habit; holding rosary; black and white dog with torch in mouth; lily, star	Spain France Italy	1221
Dorothy (Dorothea) 6 February	Virgin martyr; associated in legend with the lawyer St Theophilus; patron of brides	basket of apples and roses; with the Christ-child	Cappadocia (Turkey)	*c.*313
Dunstan 19 May	Benedictine monk, abbot, and reformer; Archbishop of Canterbury; musician, illuminator and metal-worker; patron of goldsmiths, jewellers, locksmiths, blacksmiths, and of the blind	gold cup; pincers; holding devil by nose with goldsmith's tongs; host of angels	Glastonbury, Canterbury (England)	988
Eligius (Loi, Eloi) 1 December	Bishop; evangelizer of Flanders; goldsmith; patron of all smiths and metal-workers	depicted holding devil by nose with pincers, shoeing a horse; anvil, hammer, horse's leg	Flanders (now part of France, The Netherlands, Belgium)	660
Elizabeth of Hungary/ Thuringia 17 November	Queen; Franciscan tertiary; benefactor of orphans and hospitals which she founded; patron of beggars and bakers	double crown or three crowns; roses, bread; beggars	Hungary Germany	1231
Francis of Assissi 4 October	Friar, founder of the Franciscan Order and Franciscan Tertiaries; preacher; lover of poverty, lepers, and nature; bearer of the stigmata; patron of ecologists, animals	brown habit with cord; stigmata (marks of crucifixion) on hands, feet, side; birds, wolf	Italy Eastern Europe Egypt Palestine	1226
Gabriel 29 September (with St Michael and All Angels)	One of seven archangels of the Bible; the divine messenger; patron of post office, telephone and telegraph workers	winged figure with lily	—	—
George 23 April (reduced to local cult, 1969[5])	Martyr; soldier; patron saint of England, also of Venice, Genoa, Portugal, Catalonia; among Fourteen Holy Helpers[8] in Germany; invoked against plague, leprosy, syphilis	knight with dragon; red cross on shield or banner	Palestine	*c.*303
Giles 1 September	Hermit and abbot; one of Fourteen Holy Helpers[8] in Germany; patron of cripples, lepers, beggars, nursing-mothers, blacksmiths	abbot's robe with staff and hind; doe pierced by arrow	France Greece	*c.*710
Gregory the Great 12 March; 3 September (RC[4] and ASB[7])	Pope; Doctor[3] and Father[2] of Latin Church; Apostle of the English; prolific writer and benefactor of poor; patron of musicians, associated with Gregorian chant	papal vestments with Holy Spirit in form of dove; depicted writing or saying Mass with suffering Christ above	Sicily Rome (Italy) Constantinople (Turkey)	604
James the Great 25 July	First Apostle-Martyr; son of Zebedee and elder brother of St John, both of whom were nicknamed 'Boanerges', 'sons of thunder', indicating impetuous, fiery character; patron saint of Spain; patron of pilgrims	pilgrim's hat; scallop shells	Palestine Spain	44

Christian saints

Name and feast-day[1]	Identity and associations	Emblem	Where active	Date of death
Jerome (Hieronymus) 30 September	Monk; Doctor[3] of the Latin Church; biblical scholar and controversialist of immense learning; patron of students, archaeologists, librarians	cardinal's dress and hat; in cave with lion at feet; book	Italy Antioch (Turkey) Chalcis (Syria) Egypt Palestine	420
John the Baptist 24 June	Hermit then preacher; son of the temple priest Zacharias and his wife Elizabeth, cousin of the Virgin Mary; the forerunner of Jesus Christ patron of farriers and tailors	wearing camel hair, carrying staff and scroll saying 'Ecce Agnus Dei'; bearing book or dish with lamb on it; with wings; axe	Palestine	c.30
John the Evangelist 27 December	Apostle and evangelist; son of Zebedee, brother of St James (the Great); author of the Fourth Gospel; 'the disciple whom Jesus loved', entrusted with care of Mary; patron of theologians, booksellers, printers, publishers, writers, painters	chalice with serpent; book and eagle; cauldron of oil	Palestine Ephesus (Turkey) Patmos (Aegean island)	1st century
Joseph of Nazareth 19 March; 1 May, St Joseph the Worker	Foster-father of Jesus and husband of Virgin Mary; carpenter; patron of fathers of families, bursars, procurators, manual workers (especially carpenters), engineers, and all who desire a holy death	depicted with Mary and the Infant Jesus, carrying him or leading him by hand; carpenter's square; madonna lily	Palestine	1st century
Jude (Thaddaeus/Lebbaeus) 28 October (with St Simon)	Apostle and martyr; author of the Epistle of St Jude; patron of hopeless cases, invoked in desperate situations	holding a ship or lance; club, axe	Palestine Persia	1st century
Lawrence (Laurence) 10 August	Deacon and martyr; prominent almsgiver; patron of cooks, cutlers, armourers, schoolboys	in deacon's robes; bearing long cross on shoulder; a gridiron; money-bag	Spain Rome (Italy)	258
Luke 18 October	Evangelist; author of both the Third Gospel and Acts; physician and possibly painter; disciple of St Paul; patron of artists, sculptors, doctors, surgeons	winged ox or bull; depicted writing or painting	Greece Antioch (Turkey) Rome (Italy)	1st century
Margaret of Antioch (Marina) (20 July, suppressed 1969)[5]	Virgin and martyr; became a shepherdess when turned out of home; one of Fourteen Holy Helpers[8]; patron of women (especially when pregnant and in childbirth), nurses, peasants; invoked for divine protection on death-bed	depicted with dragon, transfixing it with spear	Rome (Italy) Turkey	304
Mark 25 April	Evangelist; author of the Second Gospel; patron saint of Venice; patron of secretaries, notaries, glaziers	book and winged lion	Palestine Cyprus Rome, Venice (Italy) Alexandria (Egypt)	c.74
Martin of Tours 11 November	Monk and bishop; pioneer of western monasticism; destroyer of heathen temples; healer of lepers; patron of beggars, innkeepers, and tailors; a patron saint of France	escarbuncle; globe of fire; goose; depicted dividing cloak to clothe beggar	France Hungary Italy Balkans	397
Mary Magdalene 22 July	Follower of Christ, 'out of whom he had cast seven devils'; often identified with Mary of Bethany, sister of Martha and Lazarus; the sinner who anointed Jesus' feet; present at the crucifixion and first witness of his resurrection; patron of repentant sinners and the contemplative life	long hair and jar of ointment; in scenes of Passion and resurrection; tear-drops	Palestine Ephesus (Turkey) Provence (France)	1st century
Matthew 21 September	Apostle, evangelist and martyr; author of the First Gospel; publican (tax-collector) patron of bankers and tax-collectors	winged man; seated at desk, writing book with angel guiding his hand or holding inkwell; wearing spectacles; spear, sword, halberd; money-bag or box	Palestine Ethiopia Persia Tarsuana (east of Persian Gulf)	1st century
Michael 29 September (with All Angels)	One of seven archangels of the Bible; messenger of God; principal fighter against the devil or dragon; the great captain of celestial armies; rescuer of souls from hell; invoked for care of sick; patron of radiologists, artists, soldiers	winged angel; depicted weighing souls on scales/ slaying devil or dragon; pommée cross	—	—

Christian saints

Name and feast-day[1]	Identity and associations	Emblem	Where active	Date of death
Nicholas 6 December	Bishop; Santa Claus of the Christmas legend; miracle-worker; patron saint of Russia and many towns in West; patron of children, unmarried girls, sailors, merchants, apothecaries, perfumiers, pawnbrokers	three balls; three children in a tub; ship	Myra (Turkey)	4th century
Patrick 17 March	Bishop; Apostle and patron saint of Ireland; wonder-worker	in bishop's vestments, treading on snakes; shamrock; saltire (X) cross	England Scotland Ireland	*c.*461
Paul (Saul of Tarsus) 29 June (with St Peter) 25 January (his conversion)	Apostle to the Gentiles and martyr; a former Pharisee and persecutor of Christians; tent-maker; tireless missionary; author of the Epistles; the source of much Christian doctrine; patron of missionaries	book on sword, with inscription 'Spiritus gladius' (The Sword of the Spirit)	Palestine Damascus (Syria) Arabia Antioch (Turkey) Ephesus (Turkey) Cyprus Macedonia Malta Rome (Italy) Spain	*c.*65
Peter (Simon) 29 June (with St Paul)	Prince or leader of the Apostles and martyr; given the name Peter, meaning 'rock' by Jesus, who entrusted him with his Church; fisherman; patron saint of the Church and papacy, who opened Church to Gentiles; invoked as universal saint and heavenly door-keeper	dressed in toga or papal vestments and tiara; short, curly beard; crucified upside down; ship, fish, cock, keys; inverted cross	Palestine Rome (Italy)	*c.*64
Raphael 29 September (with St Michael and All Angels)	One of the seven archangels of the Bible; venerated especially in the East as a healer and guide of souls; patron saint of guardian angels; patron of the blind	winged figure, pictured with Michael and Gabriel in company of Tobit with a fish; jar of ointment; fishes' gall; staff and wallet	—	—
Sebastian 20 January (with Fabian)	Martyr; soldier and captain of Roman pretorian guards; one of Fourteen Holy Helpers; invoked against the plague; patron of archers, soldiers, armourers, ironmongers, potters	pierced with arrows or carrying them; elderly bearded man with crown	Rome Milan (Italy)	*c.*288
Stephen 26 December	The first deacon of the Christian Church and its protomartyr; stoned for blasphemy with St Paul present; patron of deacons; invoked against headaches	book and stone; palm	Jerusalem (Palestine)	*c.*35
Teresa of Ávila (Theresa) 15 October	Virgin and mystic; Carmelite nun; foundress of reformed Carmelites and, with St John of the Cross, the Carmelite friars; spiritual writer and contemplative; first woman Doctor[3] of the Church (1970); invoked by those in need of grace	dressed in Carmelite habit with fiery arrow or dove above head and heart pierced by Christ or angel with arrow; heart with rays, inscribed with IHS (monogram for Jesus)	Spain Portugal	1582
Thomas (Didymus) 3 July (RC[4] and ASB[7]); 21 December (BCP[6])	Apostle and martyr; the disciple who refused to believe Jesus' resurrection until he touched His wounds; alleged evangelizer of Parthians and Syrian Christians of Malabar; patron of architects, builders, carpenters, geometricians, the blind	holding spear or lance, a builder's T-square or carpenter's rule	Parthia (Persia) India	1st century
Ursula 21 October (reduced to local cult 1969)[5]	Virgin martyr; associated in legend with 11,000 virgins; teaching order of Ursulines named after her; patron of schoolgirls, teachers, drapers	depicted with crown and sceptre, sometimes with her female companions under her cloak; princess holding arrow; cross	Cologne (Germany) Britain Basle (Switzerland) Brittany (France)	4th century
Wenceslas 28 September	Martyr; Duke of Bohemia; patron saint of Bohemia whose crown is regarded as a symbol of Czech nationalism and independence. (Content of 'Good King Wenceslas' Christmas carol wholly imaginary)	wearing crown; dagger; eagle and staff	Bohemia	*c.*929
Zita (Sitha, Citha) 27 April	Serving-maid; patron of domestic servants, housewives (especially when they lose their keys), those in danger from rivers or crossing bridges	serving-maid with key and rosary	Lucca (Italy)	1272

Books of the Christian Bible

Old Testament

Genesis, Exodus, Leviticus, Numbers, Deuteronomy — Pentateuch

Genesis	Second Book of Chronicles	Hosea
Exodus	Ezra	Joel
Leviticus	Nehemiah	Amos
Numbers	Esther	Obadiah
Deuteronomy	Job	Jonah
Joshua	Psalms	Micah
Judges	Proverbs	Nahum
Ruth	Ecclesiastes	Habakkuk
First Book of Samuel	Song of Songs, Song of Solomon, Canticles	Zephaniah
Second Book of Samuel	Isaiah	Haggai
First Book of Kings	Lamentations of Jeremiah	Zechariah
Second Book of Kings	Ezekiel	Malachi
First Book of Chronicles	Daniel	

Apocrypha

First Book of Esdras	Wisdom of Solomon	Susanna
Second Book of Esdras	Ecclesiasticus, Wisdom of Jesus the Son of Sirach	Bel and the Dragon
Tobit	Baruch	Prayer of Manasses
Judith	Song of the Three Children	First Book of Maccabees
Rest of Esther		Second Book of Maccabees

New Testament

Gospel according to St Matthew	Epistle to the Ephesians	Epistle to the Hebrews
Gospel according to St Mark	Epistle to the Philippians	Epistle of James
Gospel according to St Luke	Epistle to the Colossians	First Epistle of Peter
Gospel according to St John	First Epistle to the Thessalonians	Second Epistle of Peter
Acts of the Apostles	Second Epistle to the Thessalonians	First Epistle of John
Epistle to the Romans	First Epistle to Timothy	Second Epistle of John
First Epistle to the Corinthians	Second Epistle to Timothy	Third Epistle of John
Second Epistle to the Corinthians	Epistle to Titus	Epistle of Jude
Epistle to the Galatians	Epistle to Philemon	Revelation, Apocalypse

Books of the Tanach (Hebrew Scriptures)

Torah

Genesis
Exodus
Leviticus
Numbers
Deuteronomy

Nevi'im

Joshua
Judges
Samuel
Kings
Isaiah
Jeremiah
Ezekiel
Twelve Minor Prophets

Ketuvim

Psalms
Proverbs
Job
Song of Songs
Ruth
Lamentations
Ecclesiastes
Esther
Daniel
Ezra-Nehemiah
Chronicles

Books of the Islamic Faith

The Holy Koran (Qur'an) (the literal words of God)

The Hadiths (teachings of the Prophet Muhammad)

(*The most important books from the Hadiths*)

Al-Jamí al-Sahieh	Imam Muhammad bin Ismail al-Bukhari
Al-Jamí al-Sahieh	Imam Muslim bin al-Hajaj al-Qushairi
Kitab al-Sunan	Imam Abo Dawood al-Sijistani
Kitab al-Sunan	Imam Abo Isa al-Tirmizi
Kitab al-Sunan (al-Mujtaba)	Imam Abo Abdul Rahman al-Nasaie
Kitan al-Sunan	Imam bin Maja al-Qazwini
Al-Muwata	Imam Malik bin Anas
Al-Musnad	Imam Ahmad bin Hanbal

Ready Reference

The following ready reference tables are organized by theme, beginning with alphabets; then through weights, measures, mathematics, and scientific tables; to ready reference tables on subjects as diverse as signs of the zodiac and punctuation marks. There are major groups of tables which deal with dynasties and ruling houses; heads of state and popes; and the areas and regions of different countries of the world.

Further tables are found throughout Volumes 1–8 of the Encyclopedia, under their relevant headwords. In addition, there is a major section of tables on the countries of the world to be found at the back of Volume 7: *Peoples and Cultures*. Tables dealing with data on the stars and planets are situated at the end of Volume 8: *The Universe*.

Alphabets

Roman

Roman
A a
B b
C c
D d
E e
F f
G g
H h
I i
J j
K k
L l
M m
N n
O o
P p
Q q
R r
S s
T t
U u
V v
W w
X x
Y y
Z z

Greek

Letter	Name	Transliteration
Α α	alpha	a
Β β	beta	b
Γ γ	gamma	g
Δ δ	delta	d
Ε ϵ	epsilon	e
Ζ ζ	zeta	z
Η η	eta	ē
Θ θ	theta	th
Ι ι	iota	i
Κ κ	kappa	k
Λ λ	lambda	l
Μ μ	mu	m
Ν ν	nu	n
Ξ ξ	xi	x
Ο ο	omicron	o
Π π	pi	p
Ρ ρ	rho	r, rh
Σ σ ς	sigma	s
Τ τ	tau	t
Υ υ	upsilon	u
Φ ϕ	phi	ph
Χ χ	chi	kh
Ψ ψ	psi	ps
Ω ω	omega	ō

Russian

Letter	Transliteration
А а	a
Б б	b
В в	v
Г г	g
Д д	d
Е е	e
Ё ё	ë
Ж ж	zh
З з	z
И и	i
Й й	ĭ
К к	k
Л л	l
М м	m
Н н	n
О о	o
П п	p
Р р	r
С с	s
Т т	t
У у	u
Ф ф	f
Х х	kh
Ц ц	ts
Ч ч	ch
Ш ш	sh
Щ щ	shch
Ъ ъ ('hard sign')	"
Ы ы	y
Ь ь ('soft sign')	'
Э э	é
Ю ю	yu
Я я	ya

Hebrew

Letter	Name	Transliteration
א	aleph	'
ב	beth	b, bh
ג	gimel	g, gh
ד	daleth	d, dh
ה	he	h
ו	waw	w
ז	zayin	z
ח	ḥeth	ḥ
ט	ṭeth	ṭ
י	yodh	y
כ ך	kaph	k, kh
ל	lamedh	l
מ ם	mem	m
נ ן	nun	n
ס	samekh	s
ע	'ayin	'
פ ף	pe	p, ph
צ ץ	ṣadhe	ṣ
ק	qoph	q
ר	resh	r
שׂ	śin	ś
שׁ	shin	sh
ת	taw	t, th

Arabic

				Name	Transliteration
ا	ـا			'alif	'
ب	ـب	ـبـ	بـ	bā'	b
ت	ـت	ـتـ	تـ	tā'	t
ث	ـث	ـثـ	ثـ	thā'	th
ج	ـج	ـجـ	جـ	jīm	j
ح	ـح	ـحـ	حـ	ḥā'	ḥ
خ	ـخ	ـخـ	خـ	khā'	kh
د	ـد			dāl	d
ذ	ـذ			dhāl	dh
ر	ـر			rā'	r
ز	ـز			zay	z
س	ـس	ـسـ	سـ	sīn	s
ش	ـش	ـشـ	شـ	shīn	sh
ص	ـص	ـصـ	صـ	ṣād	ṣ
ض	ـض	ـضـ	ضـ	ḍād	ḍ
ط	ـط	ـطـ	طـ	ṭā'	ṭ
ظ	ـظ	ـظـ	ظـ	ẓā'	ẓ
ع	ـع	ـعـ	عـ	'ayn	'
غ	ـغ	ـغـ	غـ	ghayn	gh
ف	ـف	ـفـ	فـ	fā'	f
ق	ـق	ـقـ	قـ	qāf	q
ك	ـك	ـكـ	كـ	kāf	k
ل	ـل	ـلـ	لـ	lām	l
م	ـم	ـمـ	مـ	mīm	m
ن	ـن	ـنـ	نـ	nūn	n
ه	ـه	ـهـ	هـ	hā'	h
و	ـو			wāw	w
ى	ـى	ـيـ	يـ	yā'	y

Japanese writing system

Japanese writing comprises three types of symbol *kanji, hiragana*, and *katakana*. Roman letters (*rōmaji*) are also sometimes used for the convenience of foreigners.

Kanji are characters of Chinese origin, which are often referred to as ideographs but are properly speaking logographs, that is, symbols used to write a word or part of a word. They usually have at least two different readings, reflecting native Japanese words and those derived from Chinese, and may have more than one meaning.

Hiragana and *katakana* are two different varieties of *kana*, which are phonetic signs used to write a complete syllable consisting of a consonant followed by a vowel. (Nasal N, the only final consonant in Japanese, is also considered a distinct syllable.)

Although it is possible to write Japanese entirely in *kana* (as, for instance, in books for very young children), a mixture of *kanji* and *kana* is normally used; *kanji* for nouns and the stems of verbs and adjectives, with *kana* for adding inflections and other grammatical and syntactic elements.

Hiragana is the form most commonly used, while *katakana* is kept for foreign names and loan-words, and also for highlighting words, rather as italics are used in English.

a	あ	ア	i	い	イ	u	う	ウ	e	え	エ	o	お	オ									
ka	か	カ	ki	き	キ	ku	く	ク	ke	け	ケ	ko	こ	コ	kya	きゃ	キャ	kyu	きゅ	キュ	kyo	きょ	キョ
sa	さ	サ	shi	し	シ	su	す	ス	se	せ	セ	so	そ	ソ	sha	しゃ	シャ	shu	しゅ	シュ	sho	しょ	ショ
ta	た	タ	chi	ち	チ	tsu	つ	ツ	te	て	テ	to	と	ト	cha	ちゃ	チャ	chu	ちゅ	チュ	cho	ちょ	チョ
na	な	ナ	ni	に	ニ	nu	ぬ	ヌ	ne	ね	ネ	no	の	ノ	nya	にゃ	ニャ	nyu	にゅ	ニュ	nyo	にょ	ニョ
ha	は	ハ	hi	ひ	ヒ	fu	ふ	フ	he	へ	ヘ	ho	ほ	ホ	hya	ひゃ	ヒャ	hyu	ひゅ	ヒュ	hyo	ひょ	ヒョ
ma	ま	マ	mi	み	ミ	mu	む	ム	me	め	メ	mo	も	モ	mya	みゃ	ミャ	myu	みゅ	ミュ	myo	みょ	ミョ
ya	や	ヤ				yu	ゆ	ユ				yo	よ	ヨ									
ra	ら	ラ	ri	り	リ	ru	る	ル	re	れ	レ	ro	ろ	ロ	rya	りゃ	リャ	ryu	りゅ	リュ	ryo	りょ	リョ
wa	わ	ワ										o	を	ヲ									
n	ん	ン																					
ga	が	ガ	gi	ぎ	ギ	gu	ぐ	グ	ge	げ	ゲ	go	ご	ゴ	gya	ぎゃ	ギャ	gyu	ぎゅ	ギュ	gyo	ぎょ	ギョ
za	ざ	ザ	ji	じ	ジ	zu	ず	ズ	ze	ぜ	ゼ	zo	ぞ	ゾ	ja	じゃ	ジャ	ju	じゅ	ジュ	jo	じょ	ジョ
da	だ	ダ	ji	ぢ	ヂ	zu	づ	ヅ	de	で	デ	do	ど	ド									
ba	ば	バ	bi	び	ビ	bu	ぶ	ブ	be	べ	ベ	bo	ぼ	ボ	bya	びゃ	ビャ	byu	びゅ	ビュ	byo	びょ	ビョ
pa	ぱ	パ	pi	ぴ	ピ	pu	ぷ	プ	pe	ぺ	ペ	po	ぽ	ポ	pya	ぴゃ	ピャ	pyu	ぴゅ	ピュ	pyo	ぴょ	ピョ

Kana double consonants

kka	っか	kki	っき	kku	っく	kke	っけ	kko	っこ
ssa	っさ	sshi	っし	ssu	っす	sse	っせ	sso	っそ
tta	った	tchi	っち	ttsu	っつ	tte	って	tto	っと
ppa	っぱ	ppi	っぴ	ppu	っぷ	ppe	っぺ	ppo	っぽ

Examples

食ーべーる	ta-be-ru	(I) eat
事	koto	thing
食ー事	shoku-ji	meal

NATO alphabet

A	Alpha	J	Juliet	S	Sierra
B	Bravo	K	Kilo	T	Tango
C	Charlie	L	Lima	U	Uniform
D	Delta	M	Mike	V	Victor
E	Echo	N	November	W	Whisky
F	Foxtrot	O	Oscar	X	X-ray
G	Golf	P	Papa	Y	Yankee
H	Hotel	Q	Quebec	Z	Zulu
I	India	R	Romeo		

Alphabets for the deaf, and for the blind

Finger spelling

Braille

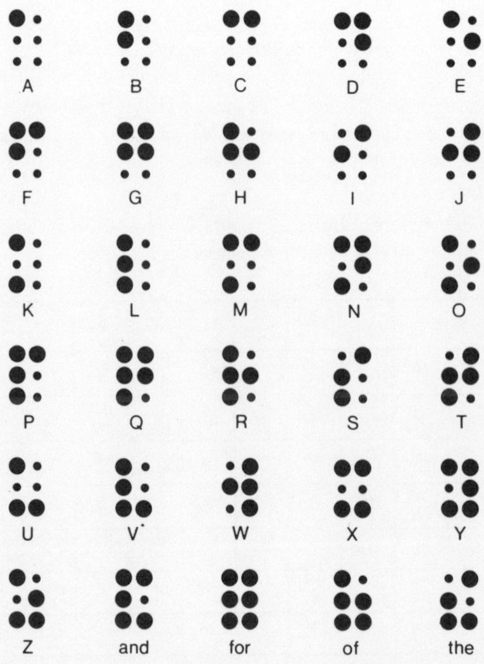

Alphabets for signalling

Semaphore

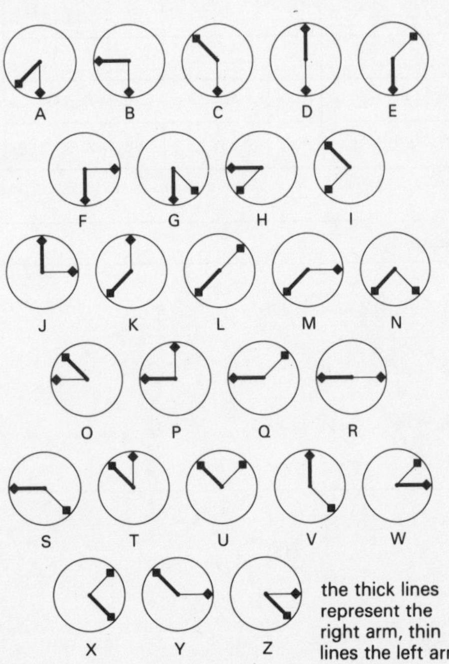

the thick lines represent the right arm, thin lines the left arm

Morse code

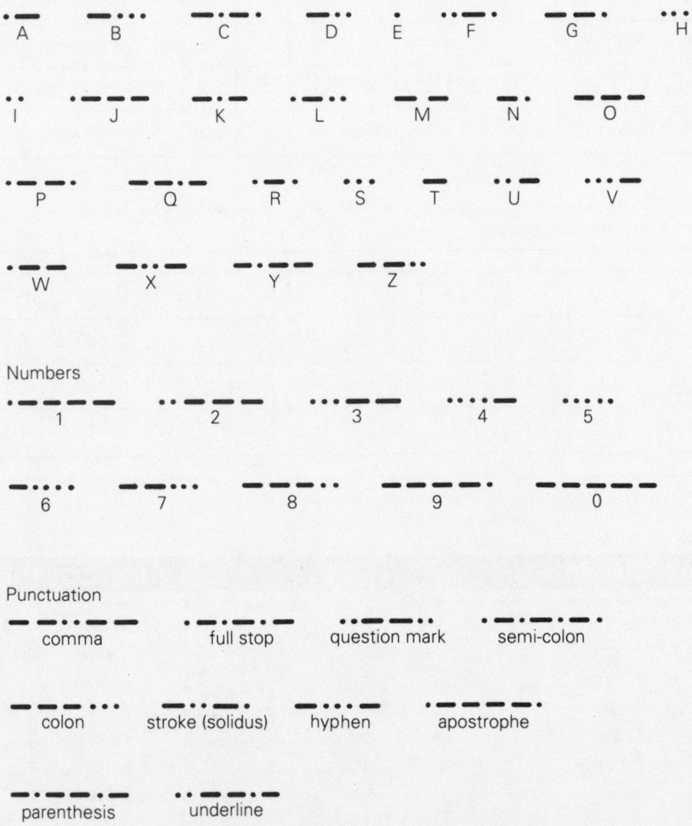

Weights, measures, and notations

1. British and American, with metric equivalents

Linear measure

1 inch	= 25.4 millimetres exactly
1 foot = 12 in.	= 0.3048 metre exactly
1 yard = 3 ft.	= 0.9144 metre exactly
1 (statute) mile = 1,760 yd.	= 1.609 kilometres

Square measure

1 square inch	= 6.45 sq. centimetres
1 square foot = 144 sq. in.	= 9.29 sq. decimetres
1 square yard = 9 sq. ft.	= 0.836 sq. metre
1 acre = 4,840 sq. yd.	= 0.405 hectare
1 square mile = 640 acres	= 259 hectares

Cubic measure

1 cubic inch	= 16.4 cu. centimetres
1 cubic foot = 1,728 cu. in.	= 0.0283 cu. metre
1 cubic yard = 27 cu. ft.	= 0.765 cu. metre

Capacity measure

British	
1 pint = 20 fluid oz.	= 0.568 litre
= 34.68 cu. in.	
1 quart = 2 pints	= 1.136 litres
1 gallon = 4 quarts	= 4.546 litres
1 peck = 2 gallons	= 9.092 litres
1 bushel = 4 pecks	= 36.4 litres
1 quarter = 8 bushels	= 2.91 hectolitres
American dry	
1 pint = 33.60 cu. in.	= 0.550 litre
1 quart = 2 pints	= 1.101 litres
1 peck = 8 quarts	= 8.81 litres
1 bushel = 4 pecks	= 35.3 litres
American liquid	
1 pint = 16 fluid oz.	= 0.473 litre
= 28.88 cu. in.	
1 quart = 2 pints	= 0.946 litre
1 gallon = 4 quarts	= 3.785 litres

Avoirdupois weight

1 grain	= 0.065 gram
1 dram	= 1.772 grams
1 ounce = 16 drams	= 28.35 grams
1 pound = 16 ounces	= 0.4536 kilogram
= 7,000 grains	(0.45359237 exactly)
1 stone = 14 pounds	= 6.35 kilograms
1 quarter = 2 stones	= 12.70 kilograms
1 hundredweight = 4 quarters	= 50.80 kilograms
1 (long) ton = 20 hundredweight	= 1.016 tonnes
1 (short) ton = 2,000 pounds	= 0.907 tonne

2. Metric, with British equivalents

Linear measure

1 millimetre	= 0.039 inch
1 centimetre = 10 mm	= 0.394 inch
1 decimetre = 10 cm	= 3.94 inches
1 metre = 10 dm	= 1.094 yards
1 decametre = 10 m	= 10.94 yards
1 hectometre = 100 m	= 109.4 yards
1 kilometre = 1,000 m	= 0.6214 mile

Square measure

1 square centimetre	= 0.155 sq. inch
1 square metre = 10,000 sq. cm	= 1.196 sq. yards
1 are = 100 sq. m	= 119.6 sq. yards
1 hectare = 100 ares	= 2.471 acres
1 square kilometre = 100 hectares	= 0.386 sq. mile

Cubic measure

1 cubic centimetre	= 0.061 cu. inch
1 cubic metre = 1,000,000 cu. cm	= 1.308 cu. yards

Capacity measure

1 millilitre	= 0.002 pint (British)
1 centilitre = 10 ml	= 0.018 pint
1 decilitre = 10 cl	= 0.176 pint
1 litre = 10 dl	= 1.76 pints
1 decalitre = 10 l	= 2.20 gallons
1 hectolitre = 100 l	= 2.75 bushels
1 kilolitre = 1,000 l	= 3.44 quarters

Weight

1 milligram	= 0.015 grain
1 centigram = 10 mg	= 0.154 grain
1 decigram = 10 cg	= 1.543 grains
1 gram = 10 dg	= 15.43 grains
1 decagram = 10 g	= 5.64 drams
1 hectogram = 100 g	= 3.527 ounces
1 kilogram = 1,000 g	= 2.205 pounds
1 tonne (metric ton) = 1,000 kg	= 0.984 (long) ton

The conversion factors are not exact unless sc marked. They are given only to the accuracy likely to be needed in everyday calculations.

SI units and equivalents in common use

Quantity	Unit name	Abbreviation	Definition or equivalent basic units
Basic SI units			
Length	metre	m	One metre equals the distance travelled by light in a vacuum in 1/299,792,458 of a second
Mass	kilogram	kg	One kilogram equals the weight of a substance equal to the weight of the International Prototype Kilogram, held at Sèvres in France
Time	second	s	One second equals 9,192,631,770 periods of the radiation corresponding to the transition between the two hyperfine levels of the ground state of the caesium-133 atom
Electric current	ampere	A	One ampere in each of two infinitely long parallel conductors of negligible cross-section 1 m apart in a vacuum will produce on each a force of 2×10^{-7} N/m
Thermodynamic	kelvin	K	One kelvin is 1/273.16 of the thermodynamic temperature of the triple point of water
Luminous intensity	candela	cd	The luminous intensity of a black-body radiator at the temperature of freezing plantinum at a pressure of 1 standard atmosphere viewed normal to the surface is 6×10^5 cd/m^2
Amount of substance	mole	mol	One mole is the amount of a substance containing as many elementary units as there are carbon atoms in 0.012 kg of carbon-12. The elementary unit (atom, molecule, ion, etc.) must be specified
Derived SI units			
Acceleration	metre/second squared	m/s^2	
Area	square metre	m^2	
Capacitance	farad	F	A s/v (C/v) (ampere-seconds per volt or coulombs per volt)
Charge	coulomb	C	A s (ampere-seconds)
Density	kilogram/cubic metre	kg/m^3	
Energy (including heat)	joule	J	N m (newton-metres)
Force	newton	N	kg m/s^2 (kilogram-metres per square second)
Frequency	hertz	Hz	l/s^{-1} (seconds^{-1})
Inductance	henry	H	V s/m^2 (volt-seconds per square metre)
Magnetic flux density	tesla	T	V s (volt-seconds)
Power	watt	W	J/s (joules per second)
Pressure	pascal	Pa	N/m^2 (newtons per square metre)
	bar	10^5 Pa	
Resistance	ohm	Ω	V/A (volts per ampere)
Velocity	metre/second	m/s	
Voltage	volt	V	J/s/A (W/A) (joules per second per ampere, or watts per ampere)
Volume	cubic metre	m^3	

Mathematical expressions

Below are some of the more common symbols and expressions used in mathematics, geometry, and statistics.

- $+$ plus
- $-$ minus
- $\pm$ plus or minus
- $\times$ (is) multiplied by
- $\div$ (is) divided by
- $=$ is equal to
- $\neq$ is not equal to
- $\simeq$ approximately equal to
- $\equiv$ is equivalent to
- $<$ is less than
- $\nless$ is not less than
- $\leqslant$ is less than or equal to
- $>$ is more than
- $\ngtr$ is not more than
- $\geqslant$ is more than or equal to

- % per cent
- ∞ infinity
- $\propto$ varies as
- 3:9::4:12 three is to nine, as four is to twelve
- $\in$ is an element of (a set)
- $\notin$ is not an element of (a set)
- $\emptyset$ *or* {} is an empty set
- $\cap$ intersection
- $\cup$ union
- $\subset$ is a subset of
- $\Rightarrow$ implies
- $\log_e$ natural logarithm *or* logarithm to the base e
- $\sqrt{\ }$ square root
- $\sqrt[3]{\ }$ cube root
- x^2 x squared

- x^3 x cubed
- x^4 x to the power four
- π pi
- r = radius of circle
- πr^2 formula for area of circle
- n! n factorial
- $\int$ the integral of
- $\angle$ angle
- ∟ right angle
- $\triangle$ triangle
- $\parallel$ is parallel to
- ∪ is perpendicular to
- ° degree
- ′ minute (of an arc)
- ″ second (of an arc)

Shapes and forms in mathematics

Circles and cones

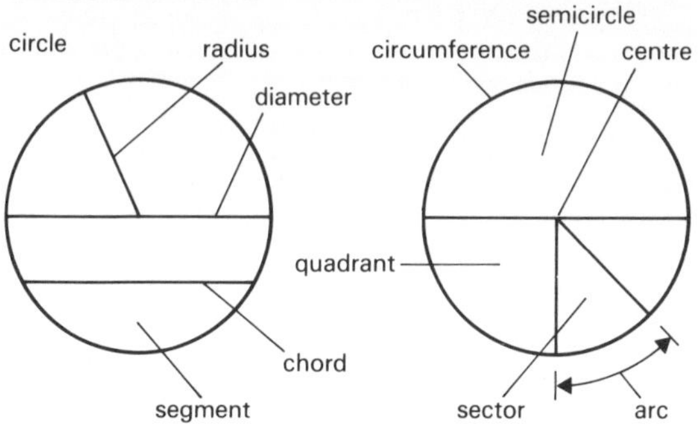

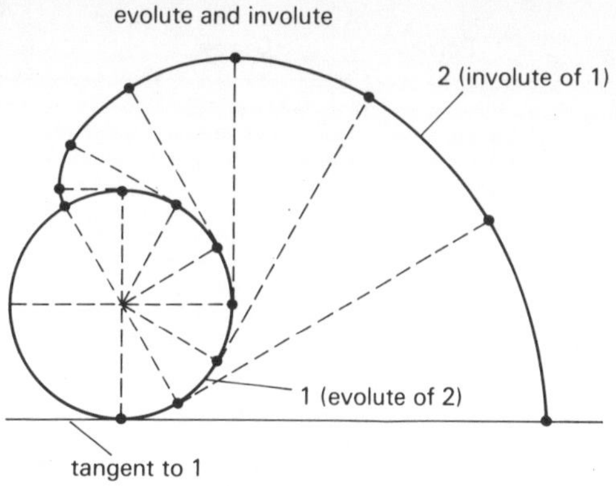

conic sections

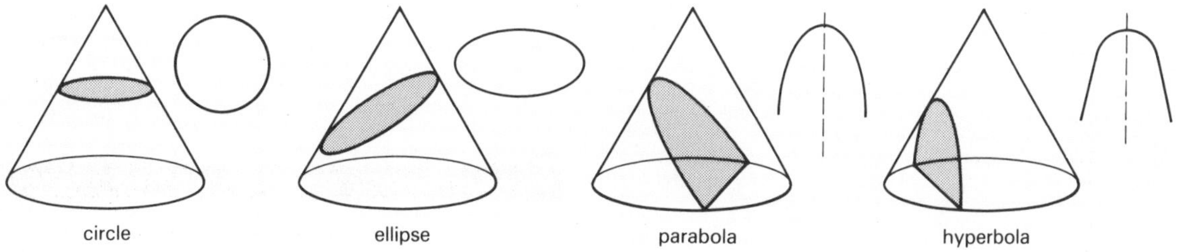

Plane figures

triangles

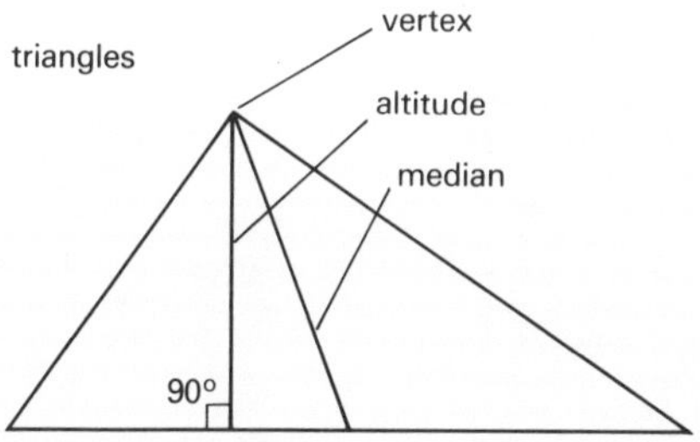

trigonometrical ratios

a/b = sine x	b/c = secant x
c/b = cosine x	b/a = cosecant x
a/c = tangent x	c/a = cotangent x

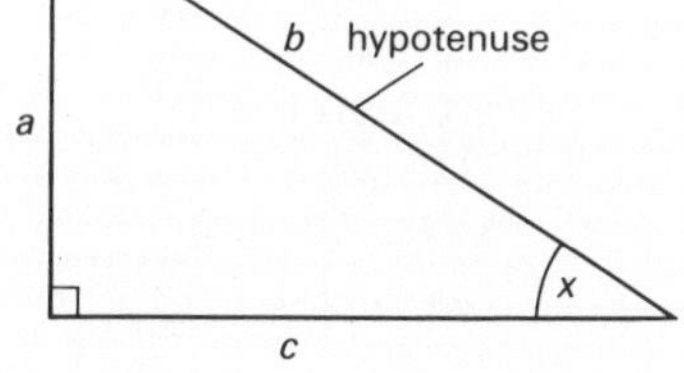

quadrilaterals

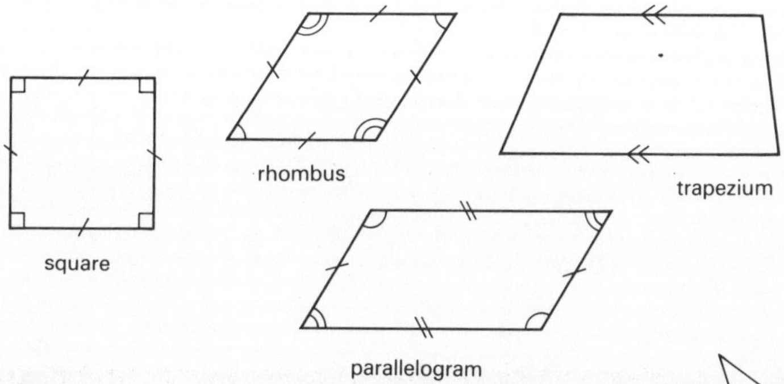

pentagons

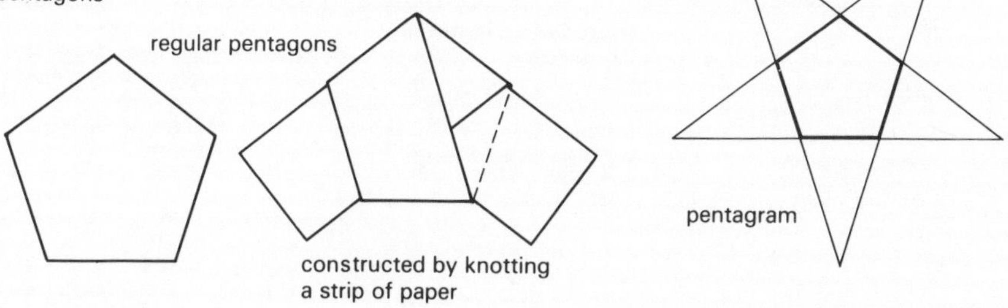

Solids

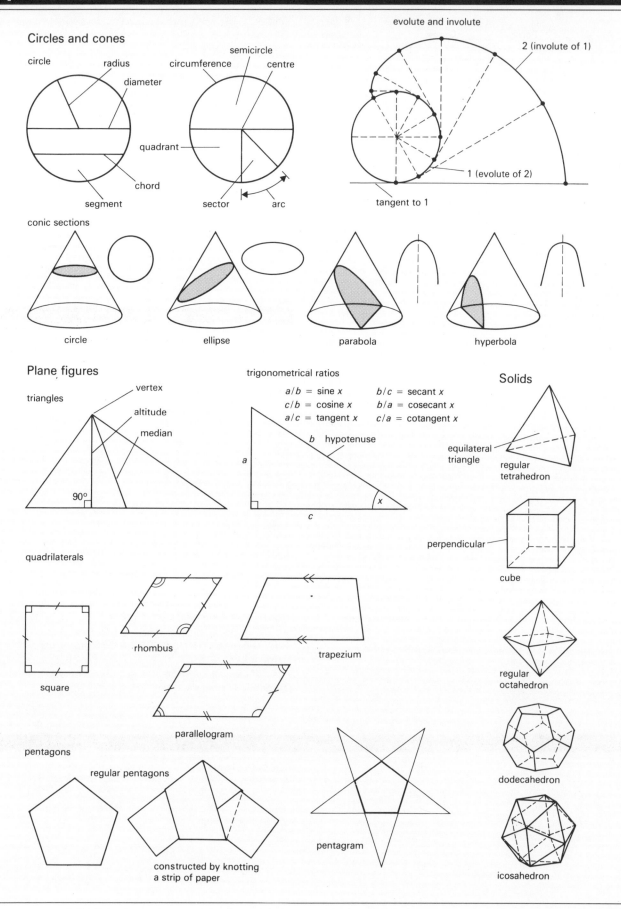

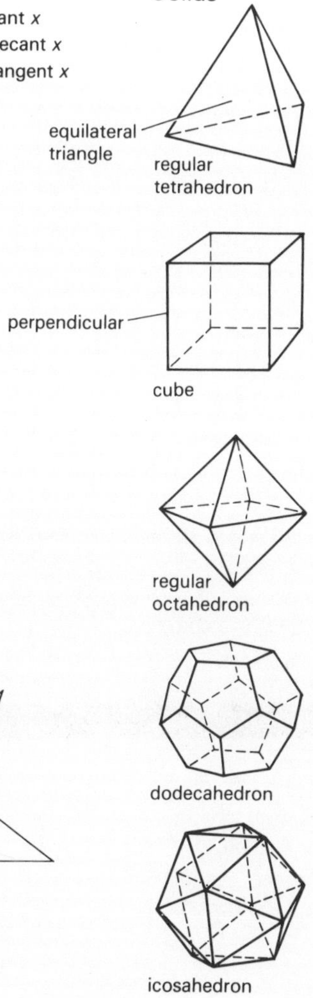

Numerical equivalents

Arabic	Roman		Greek	Binary numbers
1	I	i	α'	1
2	II	ii	β'	10
3	III	iii	χ'	11
4	IV	iv	δ'	100
5	V	v	ε'	101
6	VI	vi	ς'	110
7	VII	vii	ζ'	111
8	VIII	viii	η'	1000
9	IX	ix	θ'	1001
10	X	x	ι'	1010
11	XI	xi	ια'	1011
12	XII	xii	ιβ'	1100
13	XIII	xiii	ιγ'	1101
14	XIV	xiv	ιδ'	1110
15	XV	xv	ιε'	1111
16	XVI	xvi	ιξ'	10000
17	XVII	xvii	ιζ'	10001
18	XVIII	xviii	ιη'	10010
19	XIX	xix	ιθ'	10011
20	XX	xx	κ'	10100
30	XXX	xxx	λ'	11110
40	XL		μ'	101000
50	L		ν'	110010
60	LX		ξ'	111100
70	LXX		ο'	1000110
80	LXXX		π'	1010000
90	XC		,ο'	1011010
100	C		ρ'	1100100
200	CC		σ'	11001000
300	CCC		τ'	100101100
400	CD		υ'	110010000
500	D		φ'	111110100
1 000	M		,α	1111101000
5 000	$\overline{V}$		,ε	1001110001000
10 000	$\overline{X}$		,ι	10011100010000
100 000	$\overline{C}$		,ρ	11000011010100000

In Roman numerals, a letter placed after another letter of greater value adds, e.g. VI = 5 + 1 = 6.

A letter placed before a letter of greater value subtracts, e.g. IV = 5 − 1 = 4.

A dash placed over a letter multiplies the value by 1,000; thus $\overline{X}$ = 10,000, and $\overline{M}$ = 1,000,000.

Binary system

Only two units (0 and 1) are used, and the position of each unit indicates a power of two.
One to ten written in binary form:

	eights (2^3)	fours (2^2)	twos (2^1)	one
1				1
2			1	0
3			1	1
4		1	0	0
5		1	0	1
6		1	1	0
7		1	1	1
8	1	0	0	0
9	1	0	0	1
10	1	0	1	0

i.e. ten is written as 1010 (2^3+0+2^1+0); one hundred is written as 1100100 (2^6+2^5+0+0+2^2+0+0).

Metric prefixes

	Abbreviation or symbol	Factor
deca-	da	10
hecto-	h	10^2
kilo-	k	10^3
mega-	M	10^6
giga-	G	10^9
tera-	T	10^{12}
peta-	P	10^{15}
exa-	E	10^{18}
deci-	d	10^{-1}
centi-	c	10^{-2}
milli-	m	10^{-3}
micro-	μ	10^{-6}
nano-	n	10^{-9}
pico-	p	10^{-12}
femto–	f	10^{-15}
atto–	a	10^{-18}

These prefixes may be applied to any units of the metric system: hectogram (abbr. hg) = 100 grams; kilowatt (abbr. kW) = 1,000 watts; megahertz (MHz) = 1 million hertz; centimetre (cm) = 1/100 metre; microvolt (μV) = one millionth of a volt; picofarad (pF) = 10^{-12} farad, and are sometimes applied to other units (megabit, microinch).

Large numbers

		US	European countries
1 000 000 000	10^9	one billion	one thousand million
1 000 000 000 000	10^{12}	one trillion	one billion
1 000 000 000 000 000	10^{15}	one quadrillion	one thousand billion
1 000 000 000 000 000 000	10^{18}	one quintillion	one trillion

Electronics: some symbols and components

Circuits and wiring

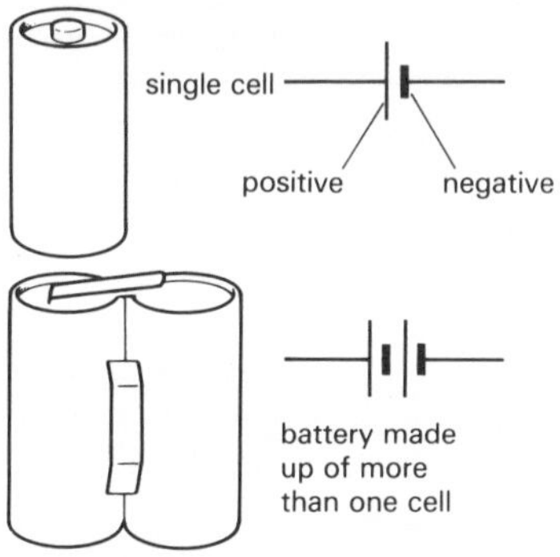

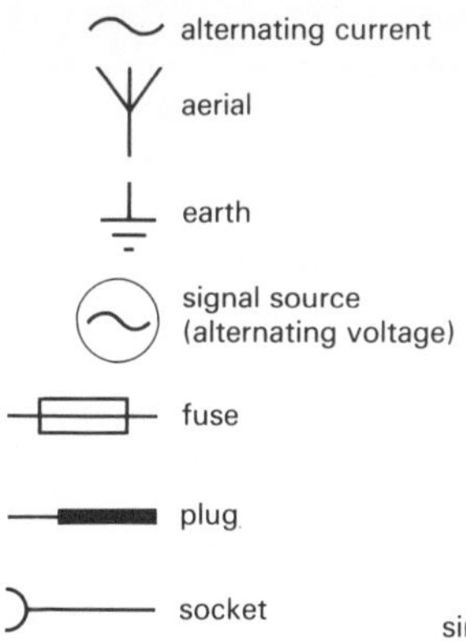

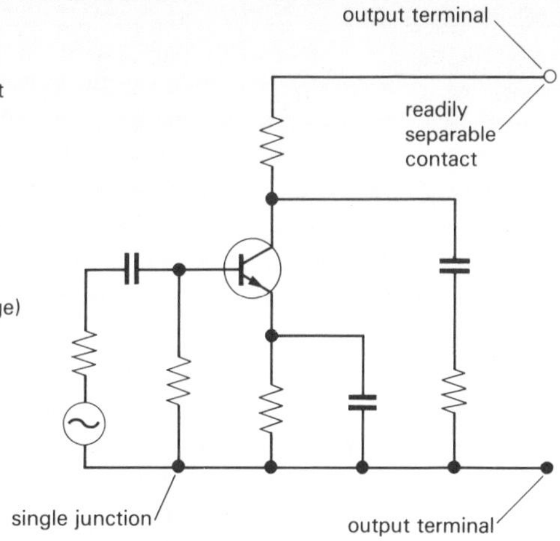

Switches
(making and breaking electrical contact)

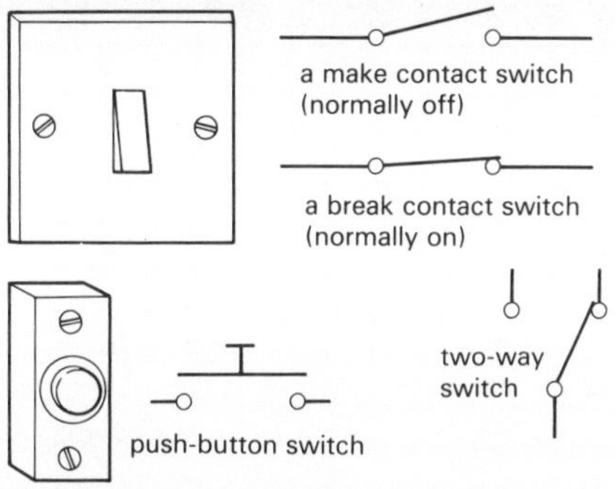

Composite symbols
general symbols can be combined, e.g.

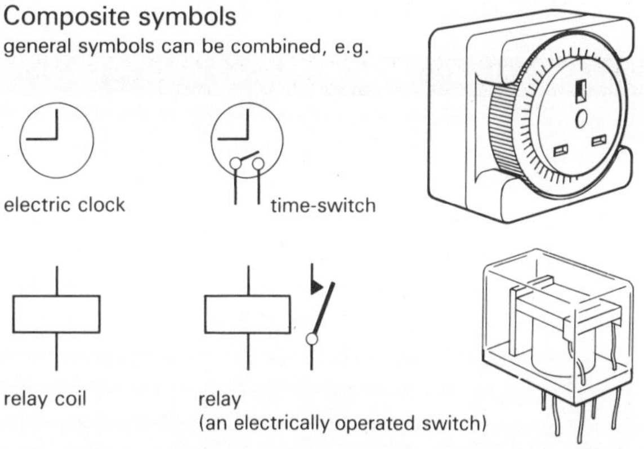

Resistors
(restricting the flow of current)

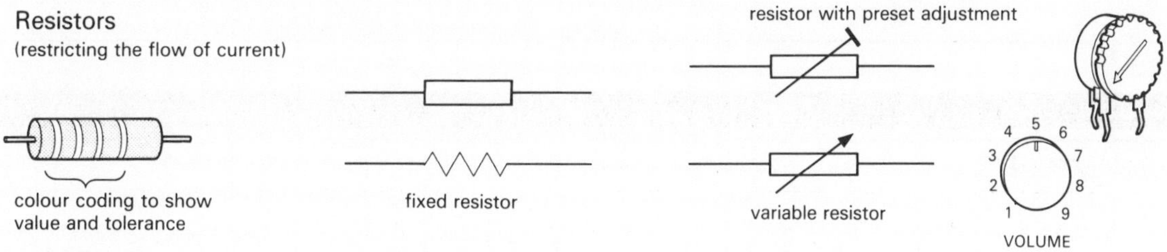

Transistors
(allowing current to be controlled or amplified)

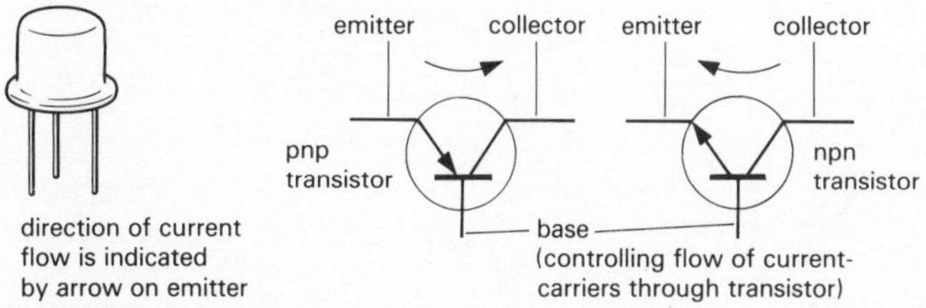

Capacitors
(retaining electric charge)

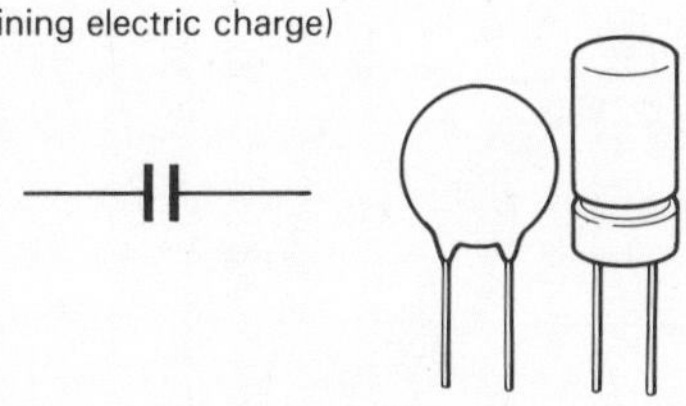

Integrated circuit
(a complete circuit compressed into one component)

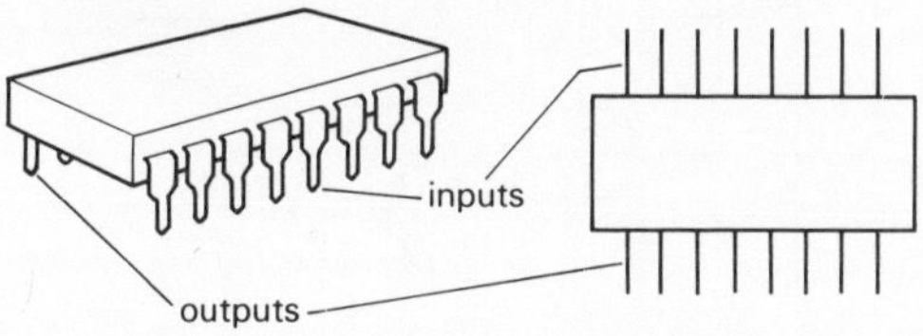

Transformer
(increases or decreases a varying voltage)

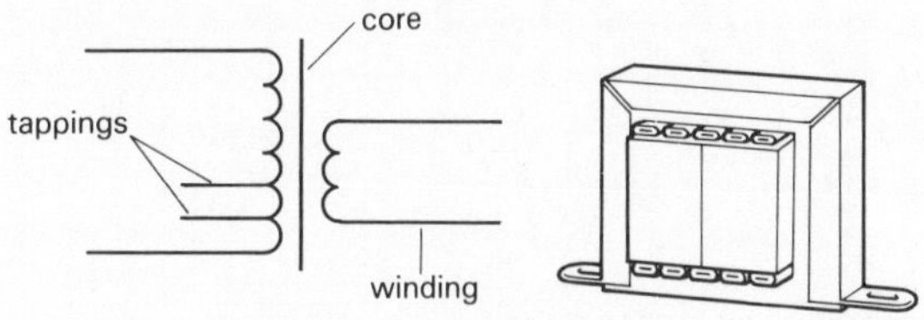

Fundamental constants

Constant	Symbol	Value in SI units
acceleration of free fall	g	$9.806\,65\ \mathrm{m\,s^{-2}}$
Avogadro constant	L, N_A	$6.022\,52 \times 10^{23}\ \mathrm{mol^{-1}}$
Boltzmann constant	$k = R/N_A$	$1.380\,622 \times 10^{-23}\ \mathrm{J\,K^{-1}}$
electric constant	ε_0	$8.854 \times 10^{-12}\ \mathrm{F\,m^{-1}}$
electronic charge	e	$1.602\,192 \times 10^{-19}\ \mathrm{C}$
electronic rest mass	m_e	$9.109\,558 \times 10^{-31}\ \mathrm{kg}$
Faraday constant	F	$9.648\,670 \times 10^{4}\ \mathrm{C\,mol^{-1}}$
gas constant	R	$8.314\,34\ \mathrm{J\,K^{-1}\,mol^{-1}}$
gravitational constant	G	$6.664 \times 10^{11}\ \mathrm{N\,m^2\,kg^{-2}}$
Loschmidt's constant	N_L	$2.687\,19 \times 10^{25}\ \mathrm{m^{-3}}$
magnetic constant	μ_0	$4\pi \times 10^{-7}\ \mathrm{H\,m^{-1}}$
neutron rest mass	m_n	$1.674\,92 \times 10^{-27}\ \mathrm{kg}$
Planck constant	h	$6.626\,196 \times 10^{-34}\ \mathrm{J\,s}$
proton rest mass	m_p	$1.672\,614 \times 10^{-27}\ \mathrm{kg}$
speed of light	c	$2.997\,924\,58 \times 10^{8}\ \mathrm{m\,s^{-1}}$
Stefan–Boltzmann constant	σ	$5.6697 \times 10^{-8}\ \mathrm{W\,m^{-2}\,K^{-4}}$

Temperature

Fahrenheit: Water boils (under standard conditions) at 212° and freezes at 32°.
Celsius or Centigrade: Water boils at 100° and freezes at 0°.
Kelvin: Water boils at 373.15 K and freezes at 273.15 K.

Celsius	Fahrenheit	Celsius	Fahrenheit
−17.8°	0°	50°	122°
−10°	14°	60°	140°
0°	32°	70°	158°
10°	50°	80°	176°
20°	68°	90°	194°
30°	86°	100°	212°
40°	104°		

To convert Celsius into Fahrenheit: multiply by 9, divide by 5, and add 32.
To convert Fahrenheit into Celsius: subtract 32, multiply by 5, and divide by 9.

Properties and uses of radio-frequency wavebands

Waveband	Frequency	Wavelength	Uses
Low frequency (long wave)	Less than 300 kHz	Greater than 1 km	AM radio broadcasts, ship-to-shore communications, navigation
Medium frequency (medium wave)	300–3000 kHz (3 MHz)	100 m–1 km	AM radio, marine communications
High frequency (short wave)	3–30 MHz	10–100 m	Intercontinental telephony, AM radio, marine and aeronautical communications, amateur radio
Very high frequency (VHF)	30–300 MHz	1–10 m	FM radio, short-range communications (e.g. police, taxis)
Ultra-high frequency (UHF)	300–3000 MHz (3 GHz)	100 cm–1 m	Television broadcasts, meteorological and space communications
Microwave	3–30GHz	10–100 cm	Satellite communications, telephony

International time zones

World times at 12 noon GMT

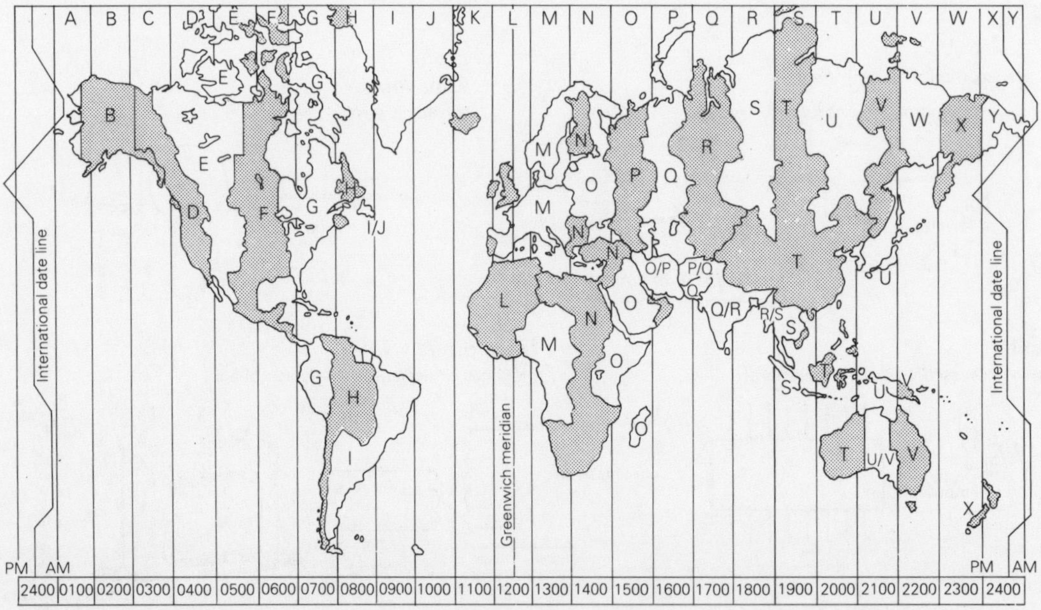

Those countries which have adopted half-hour time zones are indicated on the map as a combination of two coded zones. For example, it is 1730 hours in India at 1200 GMT. The standard times shown are subject to variation in certain countries where Daylight Saving/Summer Time operates for part of the year.

Time

Seconds

1 ephemeris second	= 1/31 556 925.975 of the length of the tropical year at 1900.0
1 mean solar second	= 1.002 737 909 mean sidereal seconds

Days

1 mean solar day	= 1.002 737 909 sidereal days
	= 24 hours 3 minutes 56.55 seconds mean sidereal time
	= 86 636.55 mean sidereal seconds
1 sidereal day	= 0.997 269 566 mean solar days
	= 23 hours 56 minutes 4.09 seconds mean solar time
	= 86 164.09 mean solar seconds

Months	days	hours	minutes	seconds
synodic	29	12	44	03
sidereal	27	07	43	12
anomalistic	27	13	18	33
tropical	27	07	43	05
nodical	27	05	05	36

Years	days	hours	minutes	seconds
tropical	365	5	48	46
sidereal	365	6	9	10
anomalistic	365	6	13	53
Julian	365	6	0	0
eclipse	346	14	52	57

Geological timescale: major divisions

Era	Period	Epoch	Millions of years ago
Cainozoic (Cenozoic)	Quaternary	Holocene	0.01
		Pleistocene	1.6
	Tertiary	Pliocene	5.3
		Miocene	23.7
		Oligocene	36.6
		Eocene	57.8
		Palaeocene	66.4
Mesozoic	Cretaceous		144
	Jurassic		208
	Triassic		245
Palaeozoic	Permian		286
	Carboniferous		360
	Devonian		408
	Silurian		438
	Ordovician		505
	Cambrian		570
	Precambrian		4,600

Chemical elements

Element	Symbol	Atomic number	Element	Symbol	Atomic number	Element	Symbol	Atomic number	Element	Symbol	Atomic number
actinium	Ac	89	europium	Eu	63	mercury	Hg	80	samarium	Sm	62
aluminium	Al	13	fermium	Fm	100	molybdenum	Mo	42	scandium	Sc	21
americium	Am	95	fluorine	F	9	neodymium	Nd	60	selenium	Se	34
antimony	Sb	51	francium	Fr	87	neon	Ne	10	silicon	Si	14
argon	Ar	18	gadolinium	Gd	64	neptunium	Np	93	silver	Ag	47
arsenic	As	33	gallium	Ga	31	nickel	Ni	28	sodium	Na	11
astatine	At	85	germanium	Ge	32	niobium	Nb	41	strontium	Sr	38
barium	Ba	56	gold	Au	79	nitrogen	N	7	sulphur	S	16
berkelium	Bk	97	hafnium	Hf	72	nobelium	No	102	tantalum	Ta	73
beryllium	Be	4	hahnium*	Ha	105	osmium	Os	76	technetium	Tc	43
bismuth	Bi	83	helium	He	2	oxygen	O	8	tellurium	Te	52
boron	B	5	holmium	Ho	67	palladium	Pd	46	terbium	Tb	65
bromine	Br	35	hydrogen	H	1	phosphorus	P	15	thallium	Tl	81
cadmium	Cd	48	indium	In	49	platinum	Pt	78	thorium	Th	90
caesium	Cs	55	iodine	I	53	plutonium	Pu	94	thulium	Tm	69
calcium	Ca	20	iridium	Ir	77	polonium	Po	84	tin	Sn	50
californium	Cf	98	iron	Fe	26	potassium	K	19	titanium	Ti	22
carbon	C	6	krypton	Kr	36	praseodymium	Pr	59	tungsten	W	74
cerium	Ce	58	kurchatovium*	Ku	104	promethium	Pm	61	uranium	U	92
chlorine	Cl	17	lanthanum	La	57	protactinium	Pa	91	vanadium	V	23
chromium	Cr	24	lawrencium	Lr	103	radium	Ra	88	xenon	Xe	54
cobalt	Co	27	lead	Pb	82	radon	Rn	86	ytterbium	Yb	70
copper	Cu	29	lithium	Li	3	rhenium	Re	75	yttrium	Y	39
curium	Cm	96	lutetium	Lu	71	rhodium	Rh	45	zinc	Zn	30
dysprosium	Dy	66	magnesium	Mg	12	rubidium	Rb	37	zirconium	Zr	40
einsteinium	Es	99	manganese	Mn	25	ruthenium	Ru	44			
erbium	Er	68	mendelivium	Md	101	rutherfordium*	Rf	104			

* Names formed systematically and without attribution are preferred by IUPAC for numbers from 104 onward, and are used exclusively for numbers from 106 onward. Names based on the atomic number are formed on the numerical roots *nil* (= 0), *un* (= 1), *bi* (= 2), etc. (e.g. *unnilquadium* = 104, *ununbium* = 112, etc.).

Meteorological symbols

Cloud amount

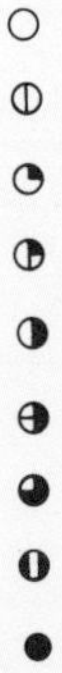

- 0
- 1 or less
- 2
- 3
- 4
- 5
- 6
- 7 or more
- 8 (oktas)

Weather

- = mist
- ≡ fog
- ; drizzle
- , rain and drizzle
- · rain
- ⁑ rain and snow
- * snow

Air pressure

isobars at 4 mb intervals

——— 1024 ———

- **H** high pressure cell
- **L** low pressure cell

Temperature

05 in degrees Celsius

Wind speed (knots)

- ◎ calm
-  1–2
-  3–7
-  8–12
-  13–17

for each additional half-feather add 5 knots

Prevailing winds

Arrows fly with the wind: the heavier the arrow, the more regular ('constant') the direction of the wind

Fronts

warm

cold

occluded

The Beaufort scale of wind force

Beaufort number	Equivalent speed at 10 m above ground: knots	miles per hour	metres per second	Description of wind	Specification for use at sea	Specification for use on land
0	<1	<1	0.0–0.2	Calm	Sea like a mirror	Calm; smoke rises vertically
1	1–3	1–3	0.3–1.5	Light air	Ripples with the appearance of scales formed but without foam crests	Direction of wind shown by smoke drift but not by wind vanes
2	4–6	4–7	1.6–3.3	Light breeze	Small wavelets, still short but more pronounced; crests have a glassy appearance and do not break	Wind felt on face; leaves rustle; ordinary vanes moved by wind
3	7–10	8–12	3.4–5.4	Gentle breeze	Large wavelets; crests begin to break; foam of glassy appearance; perhaps scattered white horses	Leaves and small twigs in constant motion; wind extends light flag
4	11–16	13–18	5.5–7.9	Moderate breeze	Small waves, becoming longer; fairly frequent white horses	Dust and loose paper raised; small branches are moved
5	17–21	19–24	8.0–10.7	Fresh breeze	Moderate waves, taking a more pronounced long form; many white horses are formed; chance of some spray	Small trees in leaf begin to sway; crested wavelets form on inland waters
6	22–27	25–31	10.8–13.8	Strong breeze	Large waves begin to form: the white foam crests are more extensive everywhere; probably some spray	Large branches in motion; umbrellas used with difficulty
7	28–33	32–38	13.9–17.1	Near gale	Sea heaps up and white foam from breaking waves begins to be blown in streaks along the direction of the wind	Whole trees in motion: inconvenience felt when walking against wind
8	34–40	39–46	17.2–20.7	Gale	Moderately high waves of greater length; edges of crests begin to break into spindrift; the foam is blown in well-marked streaks along the direction of the wind	Twigs broken off trees; progress generally impeded
9	41–47	47–54	20.8–24.4	Strong gale	High waves; dense streaks of foam along the direction of the wind; crests of waves begin to topple, tumble, and roll over; spray may affect visibility	Slight structural damage occurs (chimney-pots and slates removed)
10	48-55	55–63	24.5–28.4	Storm	Very high waves with long overhanging crests; the resulting foam, in great patches, is blown in dense white streaks along the direction of the wind; on the whole, the surface of the sea takes a white appearance; the tumbling of the sea becomes heavy and shock-like; visibility affected	Seldom experienced inland; trees uprooted; considerable damage occurs
11	56–63	64–72	28.5–32.6	Violent storm	Exceptionally high waves (small and medium-sized ships might be for a time lost to view behind the waves); the sea is completely covered with long white patches of foam lying along the direction of the wind; everywhere the edges of the wave crests are blown into froth; visibility affected	Very rarely experienced; accompanied by widespread damage
12	⩾64	⩾73	⩾32.7	Hurricane	The air is filled with foam and spray; sea completely white with driving spray; visibility very seriously affected	—

Geology: fossil history

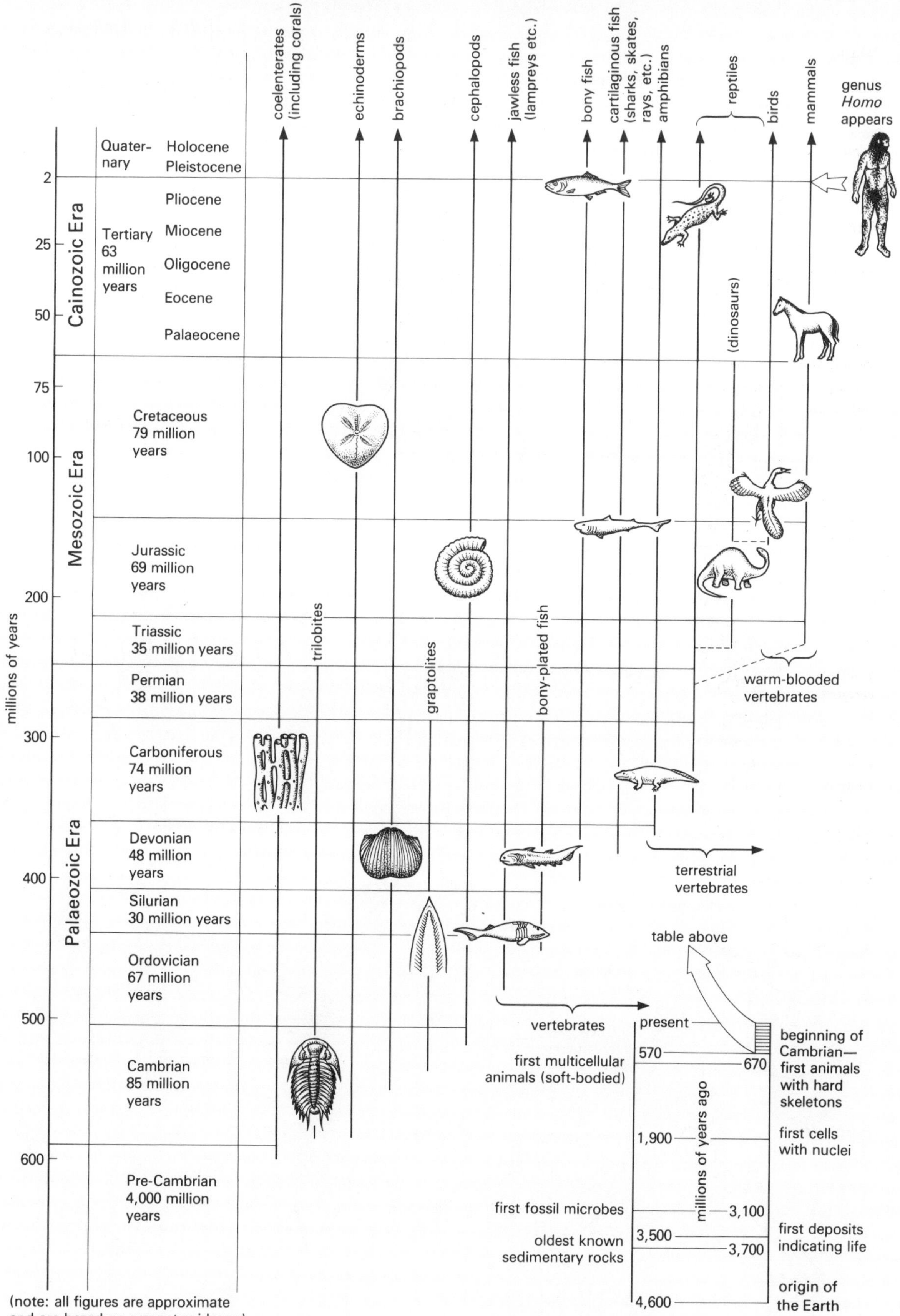

(note: all figures are approximate and are based on present evidence)

Time-scale of events in the Earth's history

The human body

Teeth

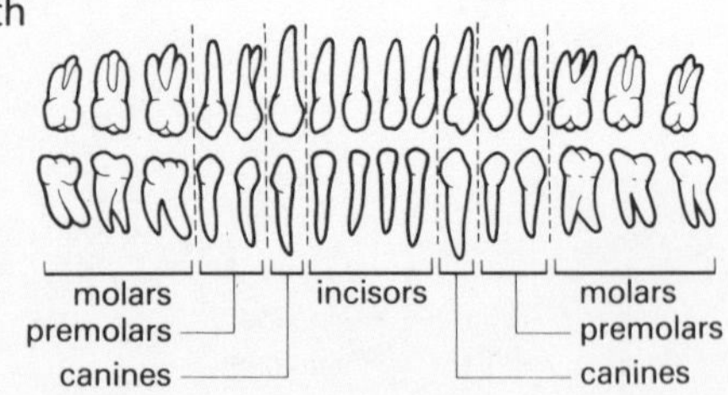

Skull

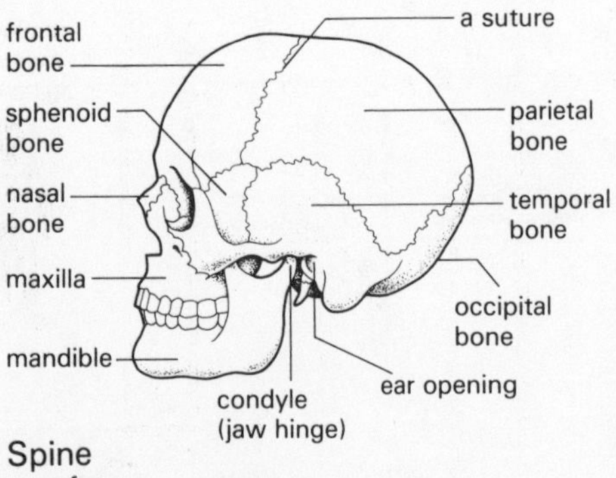

Skeleton

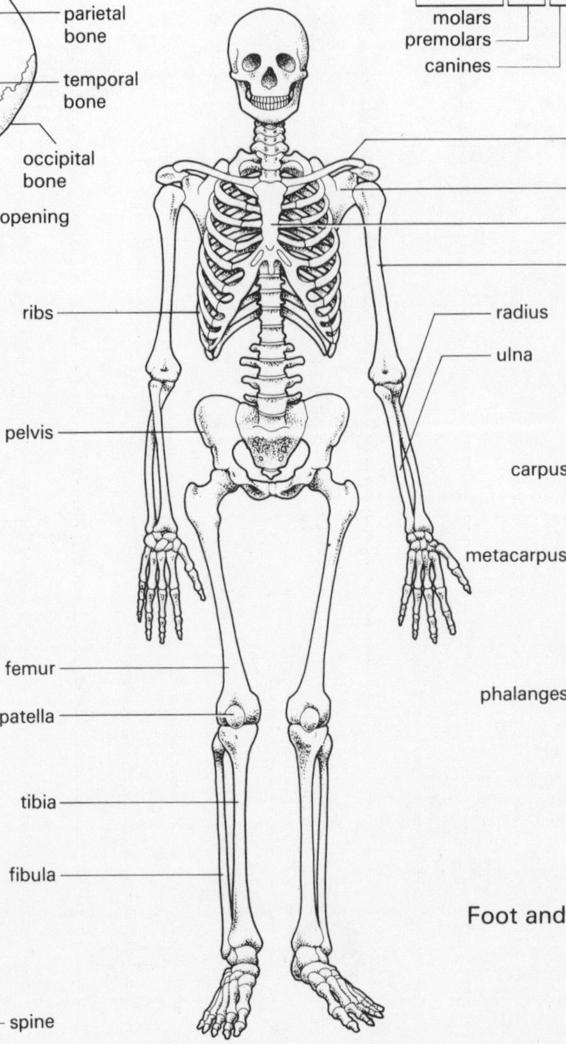

Spine
seen from the left side

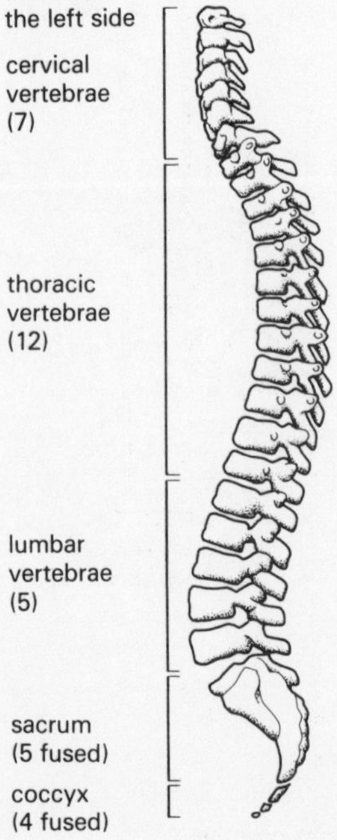

Hand and wrist
(palm upwards)

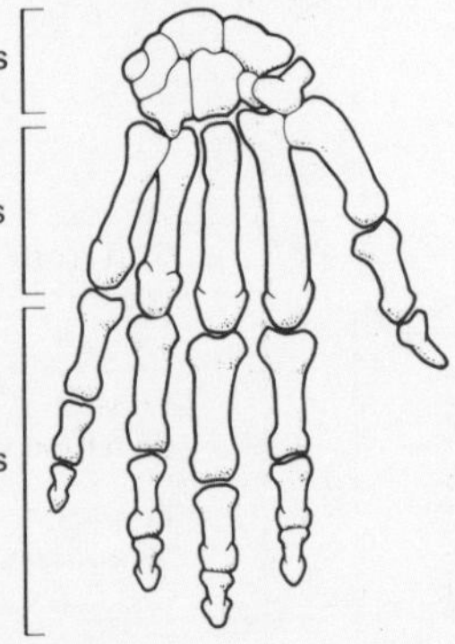

Foot and ankle

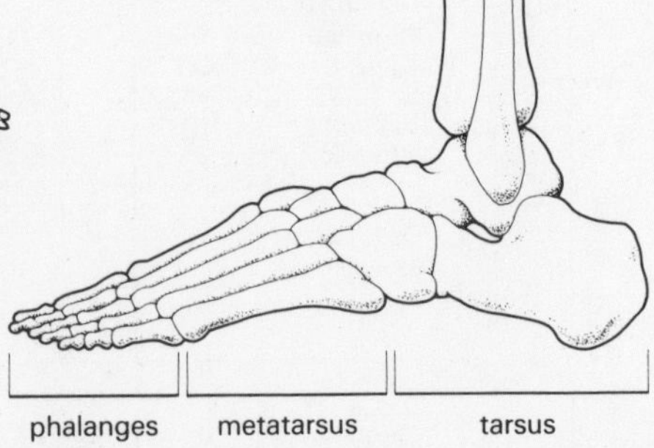

a thoracic vertebra

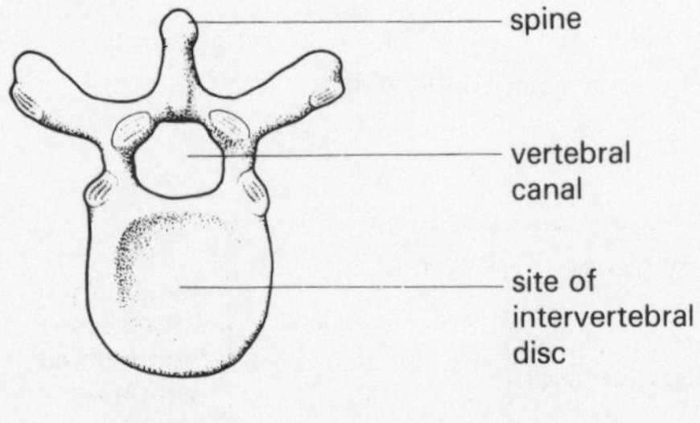

Fractures

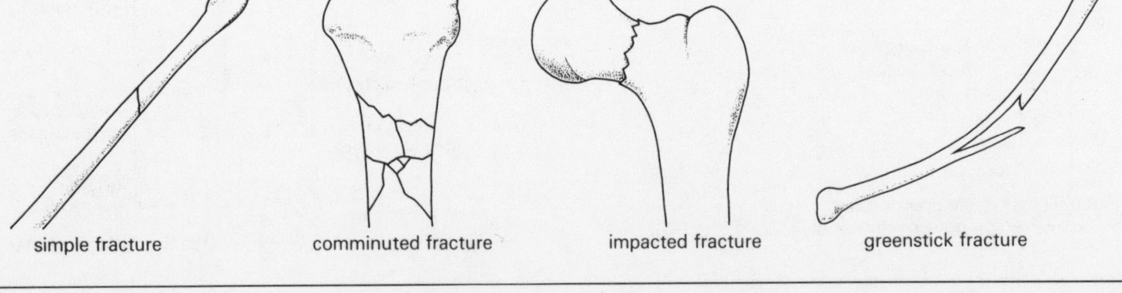

The human body

Ball and socket joint

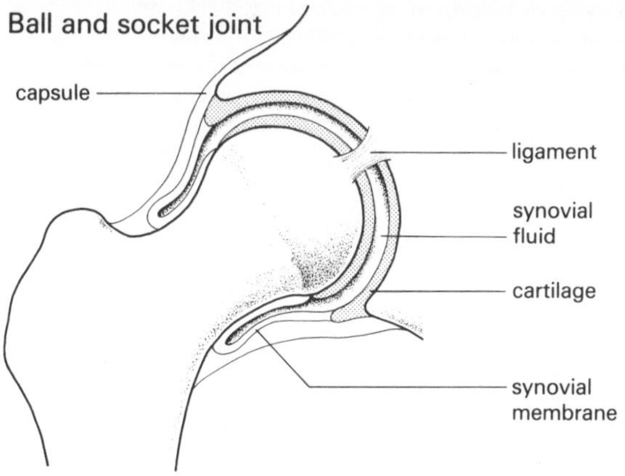

Lower abdomen

male

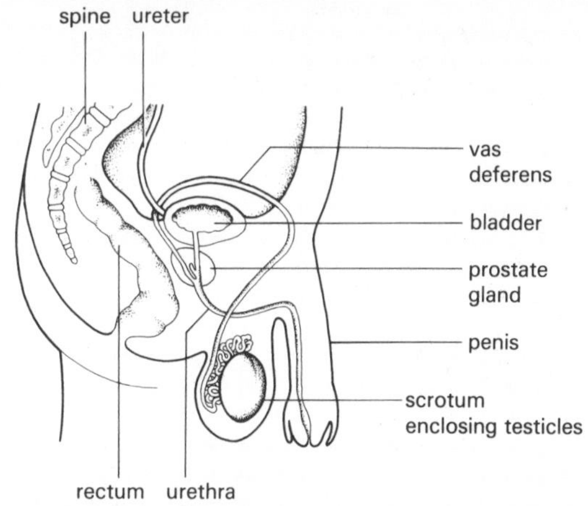

Parts of a muscle

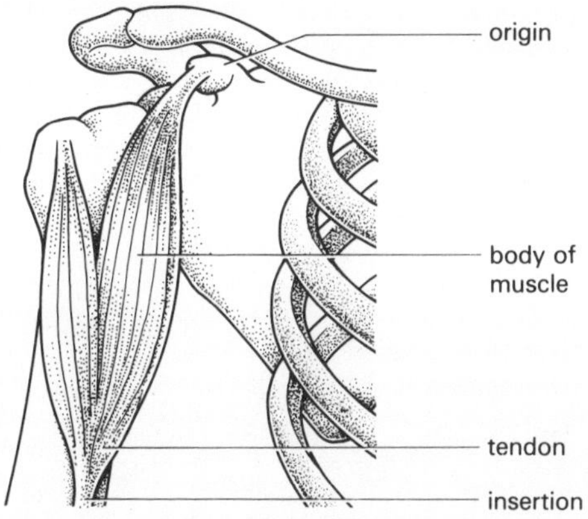

female

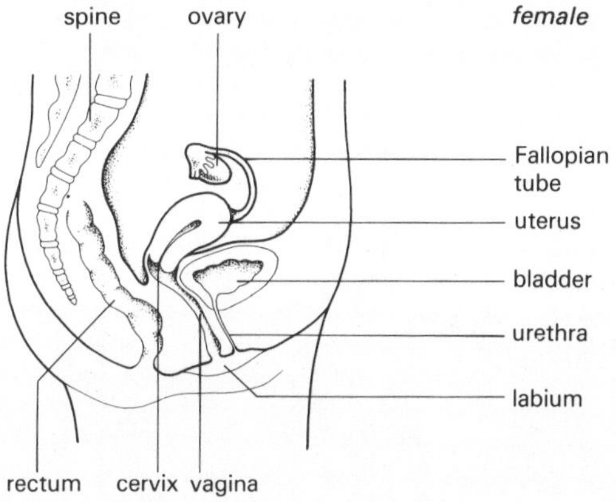

The alimentary canal

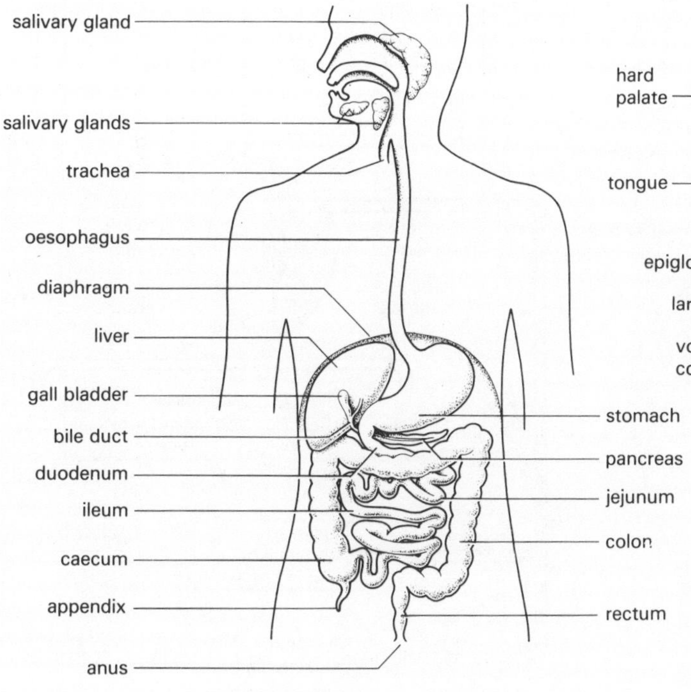

Nose, mouth, and throat

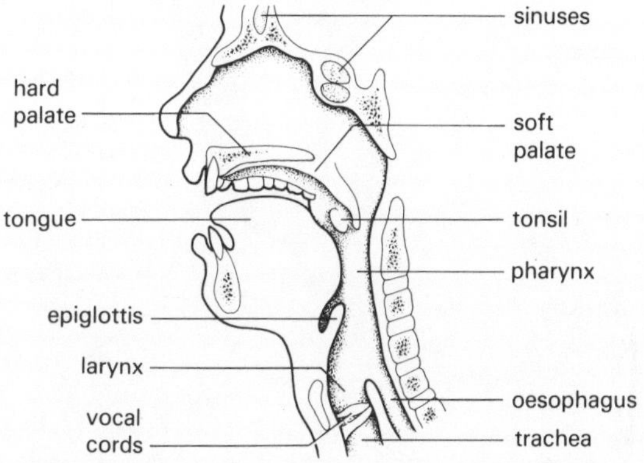

The human body

Respiration

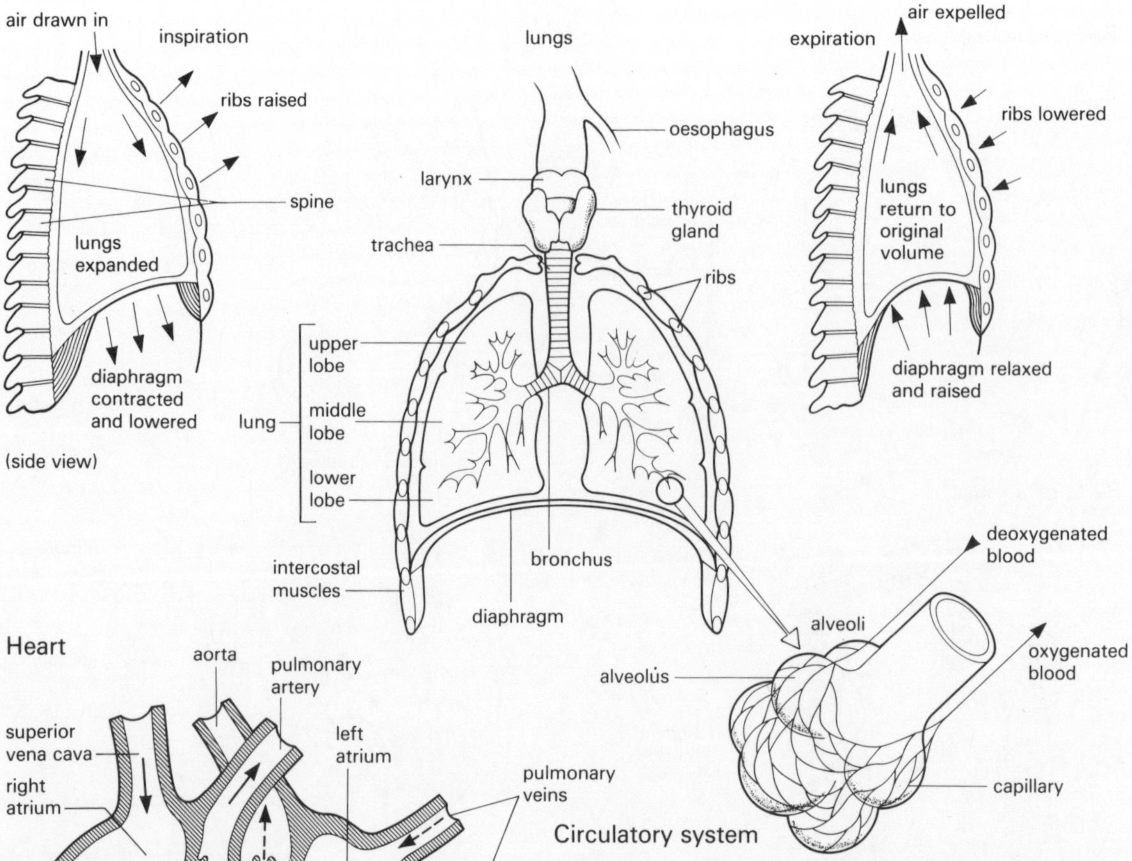
air drawn in
inspiration
ribs raised
spine
lungs expanded
diaphragm contracted and lowered
(side view)
lungs
oesophagus
larynx
thyroid gland
trachea
ribs
lung
upper lobe
middle lobe
lower lobe
intercostal muscles
bronchus
diaphragm
expiration
air expelled
ribs lowered
lungs return to original volume
diaphragm relaxed and raised
deoxygenated blood
alveoli
oxygenated blood
alveolus
capillary

Heart

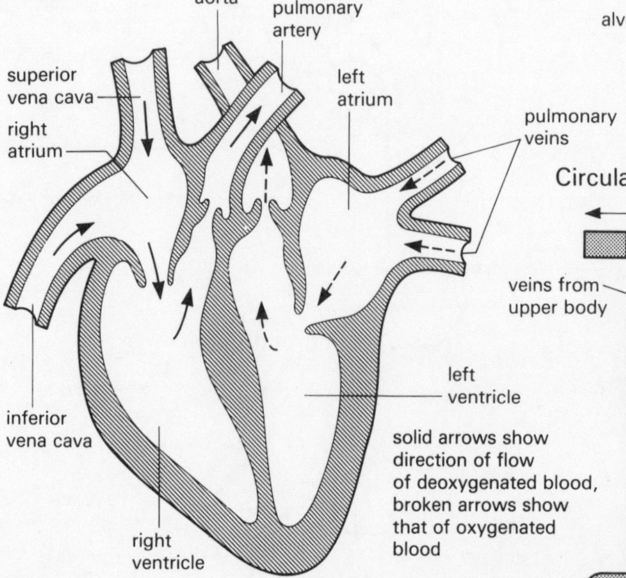
aorta
pulmonary artery
superior vena cava
left atrium
right atrium
pulmonary veins
inferior vena cava
left ventricle
right ventricle
solid arrows show direction of flow of deoxygenated blood, broken arrows show that of oxygenated blood

Circulatory system

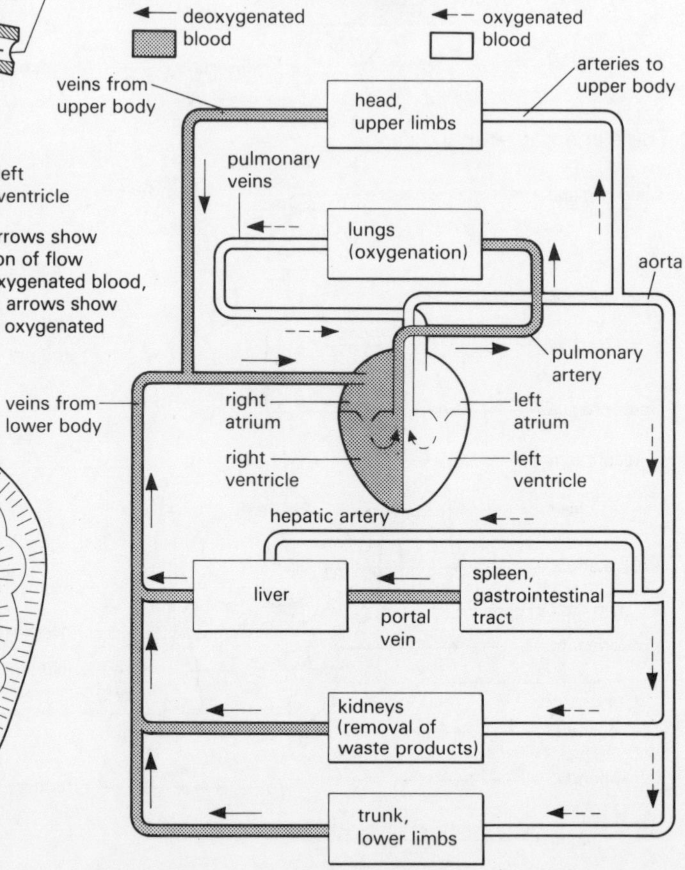
deoxygenated blood
oxygenated blood
veins from upper body
head, upper limbs
arteries to upper body
pulmonary veins
lungs (oxygenation)
aorta
pulmonary artery
veins from lower body
right atrium
left atrium
right ventricle
left ventricle
hepatic artery
liver
spleen, gastrointestinal tract
portal vein
kidneys (removal of waste products)
trunk, lower limbs

Kidney

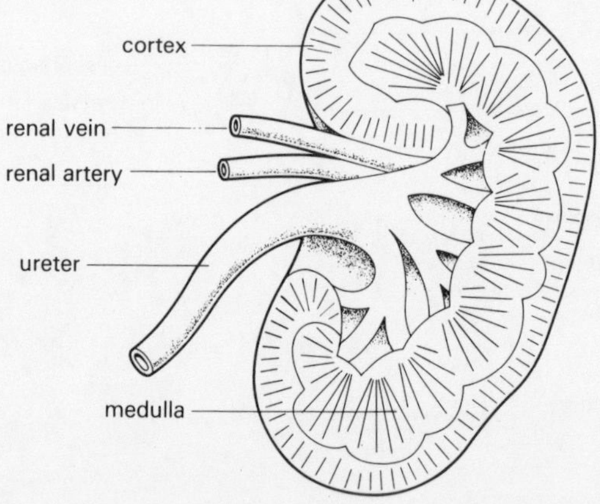
cortex
renal vein
renal artery
ureter
medulla

The human body

Brain

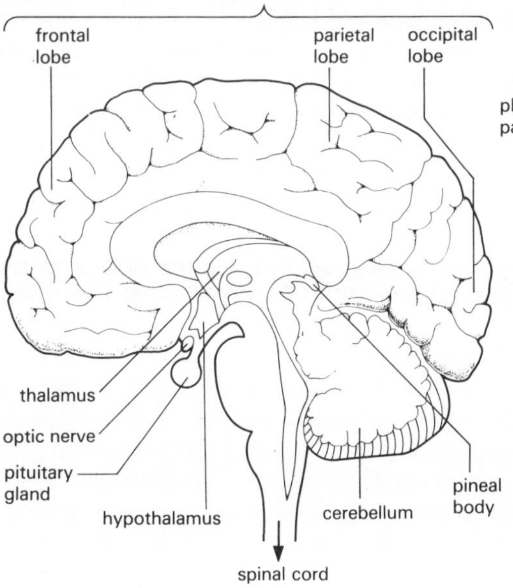

Tongue

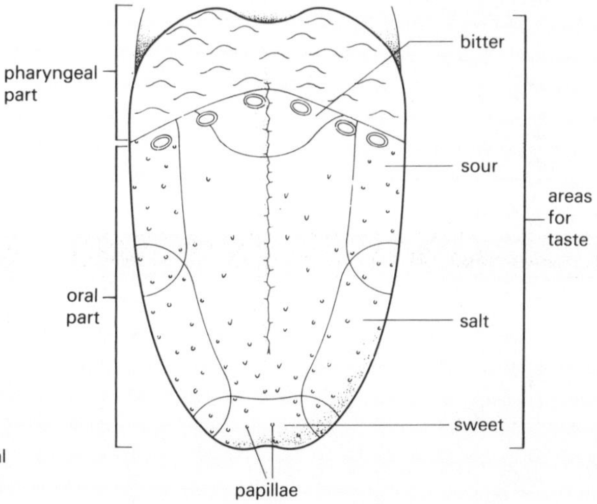

Skin
(section)

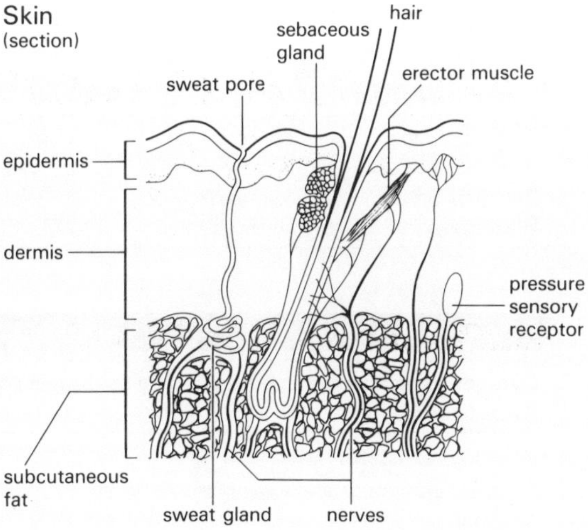

Fingerprints
(unique to each individual)

the six basic formations

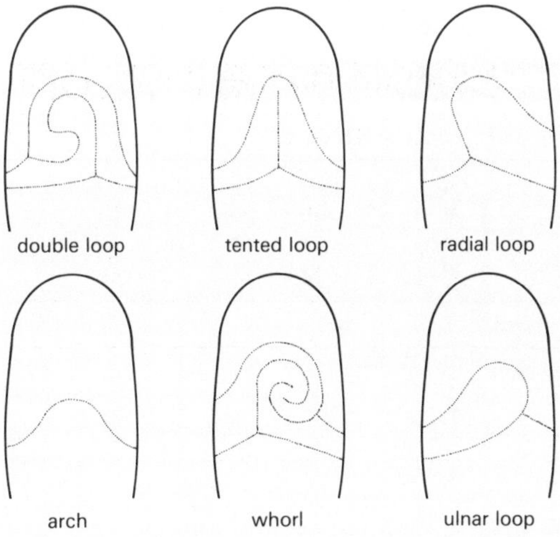

Ear

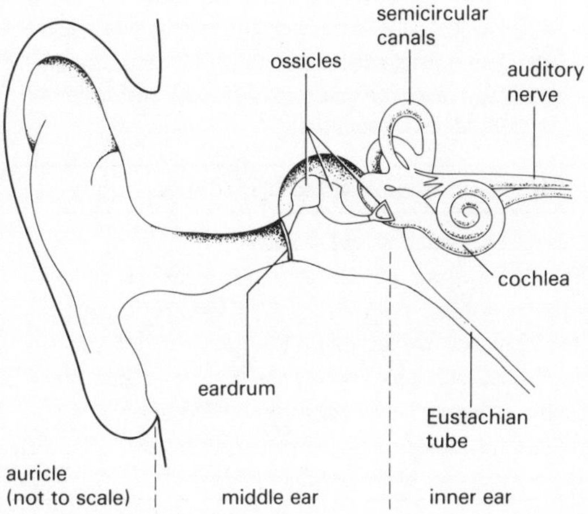

Eye

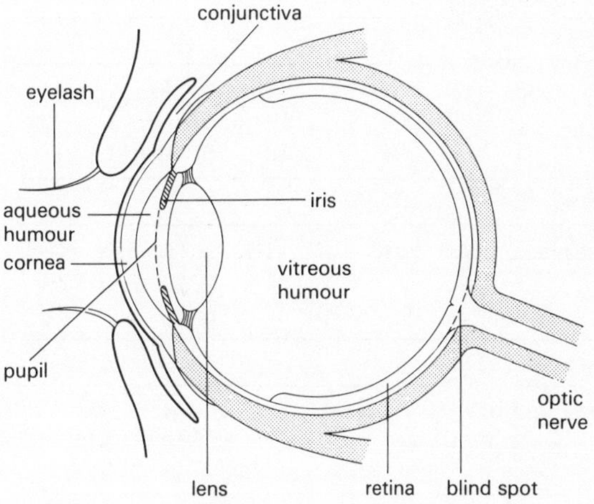

Wine bottle sizes

(in terms of standard bottles 75 cl)

Litre	1·5 bottles
Magnum	2 bottles
Jeroboam/Double Magnum (Champagne)	4 bottles
Jeroboam/(Bordeaux)	5 bottles
Rehoboam	6 bottles
Methuselah	8 bottles
Salmanazar	12 bottles
Balthazar	16 bottles
Nebuchadnezzar	20 bottles
Imperiale (Clarets)	8 bottles

Oven temperatures

Gas mark ¼	225 °F	110 °C
Gas mark ½	250 °F	120 °C
Gas mark 1	275 °F	140 °C
Gas mark 2	300 °F	150 °C
Gas mark 3	325 °F	160 °C
Gas mark 4	350 °F	180 °C
Gas mark 5	375 °F	190 °C
Gas mark 6	400 °F	200 °C
Gas mark 7	425 °F	220 °C
Gas mark 8	450 °F	230 °C
Gas mark 9	475 °F	240 °C

Clothing sizes

Women's dresses, coats, and blouses

France	—	40	42	44	46	48	50
Italy	—	44	46	48	50	52	54
UK	8	10	12	14	16	18	20
USA	—	8	10	12	14	16	18

Women's shoes

Europe (mainland)	36	37	38	39	40	41	42
UK	3	4	5	6	7	8	9
USA	4½	5½	6½	7½	8½	9½	10½

Men's suits

Europe (mainland)	44	46	48	50	53	54	56	58
UK	34	36	38	40	42	44	46	48
USA	34	36	38	40	42	44	46	48

Men's shirts

Europe (mainland)	36	37	38	39/40	41	42	43	44
UK	14	14½	15	15½	16	16½	17	17½
USA	14	14½	15	15½	16	16½	17	17½

Men's socks

Europe (mainland)	38/39	39/40	40/41	41/42	42/43	43/44	44/45
UK	9½	10	10½	11	11½	12	12½
USA	9½	10	10½	11	11½	12	12½

Men's shoes

Europe (mainland)	40	41	42	43	44	45
UK	6	7	8	9	10	11
USA	7	8	9	10	11	12

The signs of the zodiac

Aries (Ram)	♈	Libra (Scales)	♎
Taurus (Bull)	♉	Scorpio (Scorpion)	♏
Gemini (Twins)	♊	Sagittarius (Archer)	♐
Cancer (Crab)	♋	Capricorn (Goat)	♑
Leo (Lion)	♌	Aquarius (Waterbearer)	♒
Virgo (Virgin)	♍	Pisces (Fishes)	♓

Symbols of the Sun and planets of the solar system

Sun	☉	Jupiter	♃
Mercury	☿	Saturn	♄
Venus	♀	Uranus	♅
Earth	⊕	Neptune	♆
Mars	♂	Pluto	♇

The seasons

Vernal Equinox	Spring begins	20 March
Summer Solstice	Summer begins	21 June
Autumnal Equinox	Autumn begins	23 September
Winter Solstice	Winter begins	22 December

Film and video certificates (UK)

Categories of film certificates issued by the British Board of Film Classification

U	Universal: suitable for all.
PG	Parental guidance: some scenes may be unsuitable for young children.
12	Passed only for persons of twelve years and over.
15	Passed only for persons of fifteen years and over.
18	Passed only for persons of eighteen years and over.
R18	For restricted distribution only (through specially licensed cinemas or sex shops to which no one under the age of eighteen is admitted).

The classification of video tapes differs slightly

U	Universal: suitable for all.
Uc	Universal: particularly suitable for children.
PG	Parental guidance: general viewing, but some scenes may be unsuitable for young children.
15	Suitable only for persons of 15 years and over.
18	Suitable only for persons of 18 years and over.
R18	Restricted: to be supplied only in licensed sex shops to persons of not less than 18 years.

British hallmarks

All marks shown relate to silver except where otherwise indicated

A hallmark

maker's mark standard mark Assay Office mark date letter

Maker's mark (from 1363)

originally symbols, now initials

symbol

symbol and initials

initials

Assay Office mark (from 1300)

now only London, Birmingham, Sheffield, and Edinburgh

London

gold and silver (leopard's head uncrowned from 1821; mark includes platinum from 1975)

Britannia silver (prior to 1975)

Edinburgh

gold and silver (also platinum from 1975)

Birmingham

gold (also platinum from 1975)

silver

Sheffield

silver (prior to 1975)

gold (also silver and platinum from 1975)

Some earlier Assay Office marks (with dates of closure)

Norwich (1702)

York (1856)

Exeter (1883)

Newcastle (1884)

Chester (1962)

Glasgow (1964)

Standard mark (from 1544)

Marks guaranteeing pure metal content of the percentage shown

sterling silver 92.5%

marked in England

marked in Scotland (from 1975)

marked in Scotland (prior to 1975)

Britannia standard silver (1697–1720, also occasional use since) 95.8%

gold (crown followed by millesimal figure of the standard)

750 i.e. 18 carat 75%

916 22 carat 91.6%

585 14 carat 58.5%

375 9 carat 37.5%

(prior to 1975 marks incorporated the carat figure, and Scottish 18 and 22 carat gold bore a thistle mark instead of the crown)

Date letter (from 1478)

one letter per year before changing to next style of letter and/or shield

cycles vary between Assay Offices

London date letters (A-U used, excluding J) showing style of first letter and years of cycle

a	1498–1518[1]	A	1598–1618	A	1697[3]–1716	A	1796–1816	a	1896–1916
A	1518–1538	a	1618–1638	A	1716–1736	a	1816–1836	a	1916–1936
A	1538–1558	A	1638–1658	a	1736–1756	A	1836–1856	A	1936–1956
a	1558–1578	A	1658–1678	A	1756–1776	A	1856–1876	a	1956–1974[2]
A	1578–1598	a	1678–1697[2]	a	1776–1796	A	1876–1896	A	1975[4]–

Notes
1. Letter changed on 19 May until 1697
2. No U used in these cycles
3. A from 27 March–28 May 1697; year letters then changed on 29 May until 1975
4. Year letter changed with each calendar year; from 1975 all UK Offices use the same date letters and shield shape

Proof-correction marks

Instruction to printer	Textual mark	Marginal mark
Correction made in error. Leave unchanged	- - - - - under character(s) to remain	⊘
Remove extraneous marks or replace damaged character(s)	Encircle marks to be removed or character(s) to be changed	×
(Wrong fount) Replace by character(s) of correct fount	Encircle character(s) to be changed	⊗
Insert in text the matter indicated in the margin	⋏	New matter followed by ⋏
Delete	Stroke through character(s) to be deleted	ծ7
Substitute character or substitute part of one or more word(s)	/ through character or ⊢—⊣ through word(s)	New character or New word(s)
Set in or change to italic type	—— under character(s) to be set or changed	⊔⊔
Change italic to roman type	Encircle character(s) to be changed	⊔/⊔
Set in or change to capital letter(s)	≡ under character(s) to be set or changed	≡
Change capital letter(s) to lower-case letter(s)	Encircle character(s) to be changed	≢
Set in or change to small capital letter(s)	= under character(s) to be set or changed	=
Change small capital letter(s) to lower-case letter(s)	Encircle character(s) to be changed	≠
Set in or change to bold type	∿∿ under character(s) to be changed	∿∿
Set in or change to bold italic type	∿∿ under character(s) to be changed	⊔⊔ ∿∿
Invert type	Encircle character to be inverted	∩
Substitute or insert character in 'superior' position	/ through character or ⋏ where required	⅂ under character (e.g. $\frac{e}{⅂}$)
Substitute or insert character in 'inferior' position	/ through character or ⋏ where required	∠ over character (e.g. $\frac{∠}{8}$)
Insert full point or decimal point	⋏ where required	⊙
Insert colon, semi-colon, comma, etc.	⋏ where required	⊙/ ;/ ,/ (/)/ [/]/ ⅂ ⅂ ⅂ ⅂
Rearrange to make a new paragraph here	_⌐ before first word of new paragraph	_⌐
Run on (no new paragraph)	⊂⁀ between paragraphs	⊂⁀
Transpose characters or words	⊔⊓ between characters or words to be transposed, numbered where necessary	⊔⊓
Insert hyphen	⋏ where required	⊢⊣
Insert rule	⋏ where required	⊢1em⊣ ⊢4mm⊣ (i.e. give the size of the rule in the marginal mark
Insert oblique	⋏ where required	⊘
Insert space between characters	| between characters	Y
Insert space between words	Y between words	Y
Reduce space between characters	| between characters	⊤
Reduce space between words	⊤ between words	⊤
Equalize space between characters or words	| between characters or words	Ӿ

Punctuation marks

full stop	.	Used at the end of a sentence or abbreviation
comma	,	Indicates a slight pause or break between parts of a sentence, or separates words or figures in a list
semicolon	;	Used where there is a more distinct break than that indicated by a comma but less than that indicated by a full stop.
colon	:	Used (i) to show that what follows is an example, list, or summary of what precedes it, or a contrasting idea; (ii) between numbers that are in proportion
question mark	?	Placed after and indicating a question
exclamation mark	!	Placed after and indicating an exclamation
apostrophe	'	Shows the possessive case, or the omission of letters, or the plural of letters
quotation marks	' ' or " "	Used at the beginning and end of quoted passages or words
brackets	( ), [ ], or { }	Used in pairs to enclose words or figures
dash.	—	a horizontal stroke to mark a break in the sense, omitted words, etc.
hyphen	-	The sign used to join words together; to mark the division of a word at the end of a line; or to divide a word into parts.

The Roman Empire

The Julio-Claudian Emperors

27 BC–AD 14	Augustus
14–37	Tiberius
37–41	Caligula
41–54	Claudius
54–68	Nero
68–69	Galba
69	Otho
69	Vitellius

The Flavian Emperors

69–79	Vespasian
79–81	Titus
81–96	Domitian

The Five Good Emperors

96–98	Nerva
98–117	Trajan
117–138	Hadrian
138–161	Antoninus Pius
161–169	Lucius Verus
161–180	Marcus Aurelius
180–192	Commodus
193	Pertinax
193	Didius Julianus

The Severi

193–211	Septimius Severus
211	Geta
211–217	Caracalla
217–218	Macrinus
218	Diadumenian
218–222	Elagabalus
222–235	Severus Alexander

The Soldier-Emperors

235–238	Maximinus the Thracian
238	Gordian I; Gordian II
238	Balbinus; Pupienus Maximus
238–244	Gordian III
244–249	Philip I, the Arabian
247–249	Philip II
249–251	Decius
251	Herennius Etruscus
251	Hostilian
251–253	Trebonianus Gallus
251–253	Volusian
253	Aemilian
253–260	Valerian
253–268	Gallienus
260	Saloninus
268–270	Claudius II, Gothicus
270	Quintillus
270–275	Aurelian
275–276	Tacitus
276	Florian
276–282	Probus
282–283	Carus
283–284	Numerian
283–285	Carinus

The 'Gallic Empire'

260–269	Postumus
269	Laelian
269	Marius
269–271	Victorinus
271–274	Tetricus

Diocletian and the Tetrarchy

284–305	Diocletian
286–305	Maximian
305–306	Constantius I, Chlorus
305–311	Galerius
306–307	Severus
307–312	Maxentius

Dynasty of Constantine

307–337	Constantine I, the Great
308–324	Licinius
310–313	Maximinus II
316–317	Valerius Valens
324	Martinian
337–340	Constantine II
337–350	Constans
337–361	Constantius
350–353	Magnentius
360–363	Julian the Apostate
363–364	Jovian

Dynasty of Valentinian

364–375	Valentinian I
364–378	Valens
375–383	Gratian
375–392	Valentinian II

Dynasty of Theodosius

379–395	Theodosius I, the Great
383–388	Maximus
387–388	Victor
392–394	Eugenius

Western Roman Emperors

395–423	Honorius
421	Constantius III
423–425	John
425–455	Valentinian III
455	Petronius Maximus
455–456	Avitus
457–461	Majorian
461–465	Libius Severus
467–472	Anthemius
472	Olybrius
473–474	Glycerius
474–480	Julius Nepos
475–476	Romulus Augustus
476/80	End of direct imperial rule in the West

The Holy Roman Empire

Carolingian House

800–814	Charles I, the Great (Charlemagne)
814–840	Louis I, the Pious
840–855	Lothair I
855–875	Louis II
875–877	Charles II, the Bald
881–887	Charles II, the Fat
887–899	Arnulf of Carinthia
900–911	Louis III, the Child

House of Franconia

911–918	Conrad I

House of Saxony

919–936	Henry I, the Fowler
936–973	Otto I, the Great
973–983	Otto II
983–1002	Otto III
1002–24	St Henry II

Salian House

1024–39	Conrad II
1039–56	Henry III
1056–1105	Henry IV
1077–80	[Rudolf of Swabia]
1081–8	[Herman of Salm]
1087–98	Conrad
1105–25	Henry V

House of Supplinburg

1125–37	Lothair II of Saxony

House of Hohenstaufen

1138–52	Conrad III
1147–50	Henry
1152–90	Frederick I, Barbarossa
1190–7	Henry VI
1198–1208	Philip of Swabia

House of Welf

1198–1218	Otto IV of Brunswick

House of Hohenstaufen

1212–50	Frederick II
1220–35	Henry
1246–7	[Henry Raspe of Thuringia]
1247–56	[William of Holland]
1250–4	Conrad IV
1257–72	[Richard of Cornwall]

House of Habsburg

1273–91	Rudolf I

House of Nassau

1292–8	Adolf

House of Habsburg

1298–1308	Albert I of Austria

House of Luxemburg

1308–13	Henry VII

House of Wittelsbach

1314–47	Louis IV of Bavaria
1314–30	[Frederick of Austria]

House of Luxemburg

1346–78	Charles IV
1349	[Günther of Schwarzburg]
1378–1400	Wenceslas

House of Wittelsbach

1400–10	Rupert of the Palatinate

House of Luxemburg

1410–37	Sigismund
1410–11	[Jobst of Moravia]

House of Habsburg

1438–9	Albert II of Austria
1440–93	Frederick III
1493–1519	Maximilian I
1519–58	Charles V
1558–64	Ferdinand I
1564–76	Maximilian II
1576–1612	Rudolf II
1612–19	Matthias
1619–37	Ferdinand II
1637–57	Ferdinand III
1658–1705	Leopold I
1705–11	Joseph I
1711–40	Charles VI

House of Wittelsbach

1742–5	Charles VII of Bavaria

House of Habsburg-Lorraine

1745–65	Francis I of Lorraine
1765–90	Joseph II
1790–2	Leopold II
1792–1806	Francis II Renunciation of the title of Holy Roman Emperor

Ancient Egypt

Dynastic period	Dynasty	Dates	Important rulers
Early Dynastic period	First Dynasty	c.3100–2905 BC	Menes unites Upper and Lower Egypt
	Second Dynasty	c.2905–2755 BC	Mastaba tombs
Old Kingdom	Third Dynasty	c.2755–2680 BC	Zozer Step pyramid at Saqqara
	Fourth Dynasty	c.2680–2544 BC	Cheops pyramid at Giza
	Fifth Dynasty	c.2544–2407 BC	
	Sixth Dynasty	c.2407–2255 BC	
First Intermediate Period	Seventh–Tenth Dynasties	c.2255–c.2035 BC	Collapse of central rule from Memphis
Middle Kingdom	Eleventh Dynasty	c.2134–1991 BC	Thebes emerges as capital under Mentuhotep II
	Twelfth Dynasty	c.1991–1786 BC	Sesostis III conquers Nubia
Second Intermediate Period	Thirteenth–Seventeenth Dynasties	c.1786–1570 BC	Hyksos kings rule
New Kingdom	Eighteenth Dynasty	c.1570–1293 BC	Amenhotep III, Tuthmosis III, civilization at height; Akhenaten heresy Tutankhamun tomb
	Nineteenth Dynasty	c.1293–1185 BC	Thebes at height of power under Ramesses II
	Twentieth Dynasty	c.1185–1070 BC	Sea Peoples invade. Defeated by Ramesses III
Third Intermediate period	Twenty-first–Twenty-sixth Dynasties	c.1070–525 BC	Nubians invade, followed by Assyrians who sack Thebes c.655 BC
Late Dynastic period	Twenty-seventh–Thirty-first Dynasties	525–332 BC	Persian domination Alexander the Great arrives 332 BC

Dynasties of China

2550–2140 BC	Five Emperors
2140–1711 BC	Xia
1711–1066 BC	Shang
1066–256 BC	Zhou
[1066–771 BC	Western Zhou]
[770–256 BC	Eastern Zhou]
475–221 BC	Warring States
221–206 BC	Qin
206 BC–AD 220	Han
[206 BC–AD 25	Western Han]
[AD 25–220	Eastern Han]
220–280	Three Kingdoms
265–420	Jin
[265–317	Western Jin]
[317–420	Eastern Jin]
420–589	South & North Dynasties
581–618	Sui
618–907	Tang
907–960	Five Dynasties
960–1279	Song
[960–1127	Northern Song]
[1127–1279	Southern Song]
1279–1368	Yuan
1368–1644	Ming
1644–1911	Qing
1912–	Republic
1949–	People's Republic

Dynasties of Islam

The Caliphates

632–661	Orthodox Caliphate
661–750	Umayyad Dynasty
750–1258	Abbāsid Dynasty

The Ottomans

1280–1924	Osmanli Dynasty

Mogul Emperors (India)

1526–40	Mogul Dynasty
1540–55	Sūrī Dynasty
1555–1858	Mogul Dynasty

Japan

Eras

1600–1868	Tokugawa (or Edo) Era
1868–1912	Meiji Era
1912–16	Taishō Era
1926–89	Shōwa Era
1989–	Heisei Era

Shoguns

1603–5	Tokugawa Ieyasu
1605–23	Hidetada
1623–51	Iemitsu
1651–80	Ietsuna
1680–1709	Tsunayoshi
1709–12	Ienobu
1713–16	Ietsugu
1716–45	Yoshimune
1745–60	Ieshige
1760–86	Ieharu
1787–1837	Ienari
1837–53	Ieyoshi
1853–8	Iesada
1858–66	Iemochi
1867–8	Yoshinobu

Emperors

1586–1611	Go-Yōzei
1611–29	Go-Mizunoo
1629–43	Meishō
1643–54	Go-Kōmyō
1655–63	Go-Sai
1663–87	Reigen
1687–1709	Higashiyama
1709–35	Nakamikado
1735–47	Sakuramachi
1747–62	Momozono
1762–71	Go-Sakuramachi
1771–9	Go-Momozono
1780–1817	Kōkaku
1817–46	Ninkō
1846–67	Kōmei
1868–1912	Mutsuhito
1912–26	Yoshihito
1926–89	Hirohito
1989–	Akihito

Mogul (Mughal) Empire of India

Mogul Dynasty

1526–30	Babur
1530–40	Humayun

Sūrī Dynasty

1540–5	Shir Shah Sur
1545–53	Islam Shah
1553–5	Muhammad Adil
1555	Ibrahim III
1555	Sikander III

Mogul Dynasty

1555–6	Humayun
1556–1605	Akbar I, the Great
1605–27	Jahangir
1628–58	Shah Jahan I
1658–1707	Aurangzib Alamgir I
1707–12	Bahadur Shah I
1712–13	Jahandar Shah
1713–19	Farrukhsiyar
1719	Rafi al-Darajat
1719	Shah Jahan II
1719–48	Muhammad Shah
1748–54	Ahmad Shah
1754–9	Alamgir II
1759–1806	Shah Alam II
1806–37	Akbar II
1837–58	Bahadur Shah II

Rulers of England and of the United Kingdom

Saxon Line

955–959	Edwy
959–975	Edgar
975–978	Edward the Martyr
978–1016	Ethelred the Unready
1016	Edmund Ironside

Danish Line

1017–35	Canute (Cnut)
1035–40	Harold I
1040–2	Hardicanute (Harthacnut)

Saxon Line

1042–66	Edward the Confessor
1066	Harold II (Godwinson)

House of Normandy

1066–87	William I (the Conqueror)
1087–1100	William II
1100–35	Henry I
1135–54	Stephen

House of Plantagenet

1054–89	Henry II
1189–99	Richard I
1199–1216	John
1216–72	Henry III
1272–1307	Edward I
1307–27	Edward II
1327–77	Edward III
1377–99	Richard II

House of Lancaster

1399–1413	Henry IV
1413–22	Henry V
1422–61	Henry VI

House of York

1461–83	Edward IV
1483	Edward V
1783–5	Richard III

House of Tudor

1485–1509	Henry VII
1509–47	Henry VIII
1547–53	Edward VI
1553–8	Mary I
1558–1603	Elizabeth I

House of Stuart

1603–25	James I of England and VI of Scotland
1625–49	Charles I

Commonwealth (declared 1649)

1653–8	Oliver Cromwell, Lord Protector
1658–9	Richard Cromwell

House of Stuart

1660–85	Charles II
1685–8	James II
1689–1702	William III and Mary II (Mary d. 1694)
1702–14	Anne

House of Hanover

1714–27	George I
1727–60	George II
1760–1820	George III
1820–30	George IV
1830–7	William IV
1837–1901	Victoria

House of Saxe-Coburg-Gotha

1901–10	Edward VII

House of Windsor

1910–36	George V
1936	Edward VIII
1936–52	George VI
1952–	Elizabeth II

Rulers of Sweden

House of Vasa

1523–60	Gustavus I Vasa
1560–8	Eric XIV
1568–92	John III
1592–1604	Sigismund
1604–11	Charles IX
1611–32	Gustavus II Adolphus
1632–54	Christina
1654–60	Charles X
1660–97	Charles XI
1697–1718	Charles XII
1718–20	Ulrica Eleanora (m. Frederick of Hesse-Cassell, King of Sweden)
1720–51	Frederick of Hesse-Cassell
1751–71	Adolphus Frederick
1771–92	Gustavus III
1792–1809	Gustavus IV Adolphus
1809–18	Charles XIII

House of Bernadotte

1818–44	Charles XIV
1844–59	Oscar I
1859–72	Charles XV
1872–1907	Oscar II
1907–50	Gustavus V
1950–73	Gustavus VI Adolphus
1973–	Charles Gustavus XVI

The Austrian Empire

House of Habsburg-Lorraine

1804–35	Francis I
1835–48	Ferdinand I
1848–1916	Francis Joseph I
1916–18	Charles I
	Proclamation of the republic

Rulers of Scotland

1034–40	Duncan I
1040–57	Macbeth (usurper)
1057–93	Malcolm III
1093–8	Donalbane
1098–1107	Edgar
1107–24	Alexander I
1124–53	David I
1153–65	Malcolm IV
1165–1214	William the Lion
1214–49	Alexander II
1249–86	Alexander III
1286–90	Margaret, Maid of Norway
1292–6	John Balliol
1306–29	Robert I
1329–70	David II

House of Stuart

1371–90	Robert II
1390–1406	Robert III
1406–37	James I
1437–60	James II
1460–88	James III
1488–1513	James IV
1513–42	James V
1542–67	Mary Stuart
1567–1625	James VI (1603–25, as James I, King of England)

Kings of Prussia

House of Hohenzollern

1688–1713	Frederick III(I)
1713–40	Frederick William I
1740–86	Frederick II, the Great
1786–97	Frederick William II
1797–1840	Frederick William III
1840–61	Frederick William IV

German Emperors

1861–88	William I
1888	Frederick III
1888–1918	William II
1918	Proclamation of the republic

Rulers of Serbia

Obrenović and Karageorgević Families

1804–17	Kara George Petrović (as ruler)
1817–39	Milos Obrenović (as Prince)
1839	Milan I Obrenović
1839–42	Michael III Obrenović
1842–58	Alexander Karageorgević
1858–60	Milos Obrenović I
1860–8	Michael III Obrenović
1868–82	Milan II Obrenović (as Prince)
1882–9	Milan II Obrenović (as King)
1889–1903	Alexander Obrenović
1903–21	Peter I Karageorgević
1921–34	Alexander I Karageorgević
1934–45	Peter II Karageorgević

The Kingdom of France

Carolingian House

751–768	Pepin the Short
768–771	Carloman
768–814	Charles the Great (Charlemagne)
814–840	Louis I, the Pious
840–877	Charles I, the Bald
877–879	Louis II, the Stammerer
879–882	Louis III
882–884	Carloman
885–888	Charles II, the Fat

Robertian House

888–898	Eudes

Carolingian House

893–923	Charles III, the Simple

Robertian House

922–923	Robert I
923–936	Rudolf

Carolingian House

936–954	Louis IV of Outremer
954–986	Lothair
986–987	Louis V, the Sluggard

Capetian House

987–996	Hugh Capet
996–1031	Robert II, the Pious
1017–25	Hugh
1031–60	Henry I
1060–1108	Philip I
1108–37	Louis VI, the Fat
1129–31	Philip
1137–80	Louis VII, the Younger
1180–1223	Philip II, Augustus
1223–6	Louis VIII, the Lion
1226–70	St Louis IX
1270–85	Philip III, the Bold
1285–1314	Philip IV, the Fair
1314–16	Louis X, the Stubborn
1316	John I
1316–22	Philip V, the Tall
1322–8	Charles IV, the Fair

House of Valois

1328–50	Philip VI
1350–64	John II, the Good
1364–80	Charles V, the Wise
1380–1422	Charles VI, the Mad
1422–61	Charles VII, the Victorious
1461–83	Louis XI
1483–98	Charles VIII

Line of Orléans

1498–1515	Louis XII

Line of Angoulême

1515–47	Francis I
1547–59	Henry II
1559–60	Francis II
1560–74	Charles IX
1574–89	Henry III

House of Bourbon

1589–1610	Henry IV
1610–43	Louis XIII
1643–1715	Louis XIV
1715–74	Louis XV
1774–92	Louis XVI
1793–95	Louis XVII

First Republic

1792–5	National Convention
1795–9	Directory
1799–1804	Consulate: Napoleon Bonaparte, First Consul

House of Bonaparte – First Empire

1804–14, 1815	Napoleon I
1815	Napoleon II (as Duke of Reichstadt)

House of Bourbon

1814–24	Louis XVIII
1824–30	Charles X

Line of Orléans

1830–48	Louis Philippe I

Second Republic

1848–52	Louis Napoleon Bonaparte, President

House of Bonaparte – Second Empire

1852–70	Napoleon III Proclamation of the Third Republic

The Kingdom of Spain

House of Habsburg

1516–56	Charles I
1556–98	Philip II
1598–1621	Philip III
1621–65	Philip IV
1665–1700	Charles II

House of Bourbon

1700–24	Philip V
1724	Louis I
1724–46	Philip V (again)
1746–59	Ferdinand VI
1759–88	Charles III
1788–1808	Charles IV
1808	Ferdinand VII

House of Bonaparte

1808–13	Joseph Napoleon

House of Bourbon

1813–33	Ferdinand VII
1833–68	Isabel II
1868–70	Provisional Government

House of Savoy

1870–3	Amadeus I
1873–4	First Republic

House of Bourbon

1874–85	Alfonso XII
1886–1931	Alfonso XIII
1931–9	Second Republic
1939–75	Spanish State: Francisco Franco Bahamonde, chief of state

House of Bourbon

1975–	Juan Carlos I

Kings of Italy

House of Savoy

1849–78	Victor Emmanuel II
1878–1900	Umberto I
1900–46	Victor Emmanuel III
1946	Umberto II Proclamation of the Republic

Tsars of Russia

1533–84	Ivan IV, the Terrible
1584–98	Theodore I

House of Godunov

1598–1605	Boris Godunov
1605	Theodore II
1605–6	Dimitri

House of Shuiskii

1606–10	Basil IV Shuiskii

House of Romanov

1613–45	Michael Romanov
1645–76	Alexis
1676–82	Theodore III
1682–96	Ivan V
1682–1725	Peter I, the Great
1725–7	Catherine I (Martha)
1727–30	Peter II
1730–40	Anne
1740–1	Ivan VI
1741–62	Elizabeth

House of Holstein-Gottorp-Romanov

1762	Peter III
1762–96	Catherine II, the Great (Sophia of Anhalt)
1796–1801	Paul I
1801–25	Alexander I
1825–55	Nicholas I
1855–81	Alexander II
1881–94	Alexander III
1894–1917	Nicholas II Provisional government, then Soviet rule

The Kingdom of The Netherlands

House of Orange-Nassau

1813–40	William I
1840–9	William II
1849–90	William III
1890–1948	Wilhelmina
1948–80	Juliana
1980–	Beatrix

Prime Ministers of Great Britain and of the United Kingdom

[1721]–1742	Sir Robert Walpole	Whig
1742–3	Earl of Wilmington	Whig
1743–54	Henry Pelham	Whig
1754–6	Duke of Newcastle	Whig
1756–7	Duke of Devonshire	Whig
1757–62	Duke of Newcastle	Whig
1762–3	Earl of Bute	Tory
1763–5	George Grenville	Whig
1765–6	Marquis of Rockingham	Whig
1766–8	Earl of Chatham	Whig
1768–70	Duke of Grafton	Whig
1770–82	Lord North	Tory
1782	Marquis of Rockingham	Whig
1782–3	Earl of Shelburne	Whig
1783	Duke of Portland	coalition
1783–1801	William Pitt	Tory
1801–4	Henry Addington	Tory
1804–6	William Pitt	Tory
1806–7	Lord William Grenville	Whig
1807–9	Duke of Portland	Tory
1809–12	Spencer Perceval	Tory
1812–27	Earl of Liverpool	Tory
1827	George Canning	Tory
1827–8	Viscount Goderich	Tory
1828–30	Duke of Wellington	Tory
1830–4	Earl Grey	Whig
1834	Viscount Melbourne	Whig
1834	Duke of Wellington	Tory
1834–5	Sir Robert Peel	Conservative
1835–41	Viscount Melbourne	Whig
1841–6	Sir Robert Peel	Conservative
1846–52	Lord John Russell	Whig
1852	Earl of Derby	Conservative
1852–5	Earl of Aberdeen	coalition
1855–8	Viscount Palmerston	Liberal
1858–9	Earl of Derby	Conservative
1859–65	Viscount Palmerston	Liberal
1865–6	Earl Russell	Liberal
1866–8	Earl of Derby	Conservative
1868	Benjamin Disraeli	Conservative
1868–74	William Ewart Gladstone	Liberal
1874–80	Benjamin Disraeli	Conservative
1880–5	William Ewart Gladstone	Liberal
1885–6	Marquis of Salisbury	Conservative
1886	William Ewart Gladstone	Liberal
1886–92	Marquis of Salisbury	Conservative
1892–4	William Ewart Gladstone	Liberal
1894–5	Earl of Rosebery	Liberal
1895–1902	Marquis of Salisbury	Conservative
1902–5	Arthur James Balfour	Conservative
1905–8	Sir Henry Campbell-Bannerman	Liberal
1908–16	Herbert Henry Asquith	Liberal
1916–22	David Lloyd George	coalition
1922–3	Andrew Bonar Law	Conservative
1923–4	Stanley Baldwin	Conservative
1924	James Ramsay MacDonald	Labour
1924–9	Stanley Baldwin	Conservative
1929–35	James Ramsay MacDonald	coalition
1935–7	Stanley Baldwin	coalition
1937–40	Neville Chamberlain	coalition
1940–5	Winston Spencer Churchill	coalition
1945–51	Clement Richard Attlee	Labour
1951–5	Sir Winston Spencer Churchill	Conservative
1955–7	Sir Anthony Eden	Conservative
1957–63	Harold Macmillan	Conservative
1963–4	Sir Alexander Douglas-Home	Conservative
1964–70	Harold Wilson	Labour
1970–4	Edward Heath	Conservative
1974–6	Harold Wilson	Labour
1976–9	James Callaghan	Labour
1979–90	Margaret Thatcher	Conservative
1990–	John Major	Conservative

Presidents of the United States of America

1	1789–97	George Washington	Federalist
2	1797–1801	John Adams	Federalist
3	1801–9	Thomas Jefferson	Democratic-Republican
4	1809–17	James Madison	Democratic-Republican
5	1817–25	James Monroe	Democratic-Republican
6	1825–9	John Quincy Adams	Independent
7	1829–37	Andrew Jackson	Democrat
8	1837–41	Martin Van Buren	Democrat
9	1841	William H. Harrison	Whig
10	1841–5	John Tyler	Whig, then Democrat
11	1845–9	James K. Polk	Democrat
12	1849–50	Zachary Taylor	Whig
13	1850–3	Millard Fillmore	Whig
14	1853–7	Franklin Pierce	Democrat
15	1857–61	James Buchanan	Democrat
16	1861–5	Abraham Lincoln	Republican
17	1865–9	Andrew Johnson	Democrat
18	1869–77	Ulysses S. Grant	Republican
19	1877–81	Rutherford B. Hayes	Republican
20	1881	James A. Garfield	Republican
21	1881–5	Chester A. Arthur	Republican
22	1885–9	Grover Cleveland	Democrat
23	1889–93	Benjamin Harrison	Republican
24	1893–7	Grover Cleveland	Democrat
25	1897–1901	William McKinley	Republican
26	1901–9	Theodore Roosevelt	Republican
27	1909–13	William H. Taft	Republican
28	1913–21	Woodrow Wilson	Democrat
29	1921–3	Warren G. Harding	Republican
30	1923–9	Calvin Coolidge	Republican
31	1929–33	Herbert Hoover	Republican
32	1933–45	Franklin D. Roosevelt	Democrat
33	1945–53	Harry S. Truman	Democrat
34	1953–61	Dwight D. Eisenhower	Republican
35	1961–3	John F. Kennedy	Democrat
36	1963–9	Lyndon B. Johnson	Democrat
37	1969–74	Richard M. Nixon	Republican
38	1974–7	Gerald R. Ford	Republican
39	1977–81	James Earl Carter	Democrat
40	1981–9	Ronald W. Reagan	Republican
41	1989–93	George H. W. Bush	Republican
42	1993–	William J. Clinton	Democrat

Prime Ministers of Australia

1901–3	Edmund Barton
1903–4	Alfred Deakin
1904	John C. Watson
1904–5	George Houstoun Reid
1905–8	Alfred Deakin
1908–9	Andrew Fisher
1909–10	Alfred Deakin
1910–13	Andrew Fisher
1913–14	Joseph Cook
1914–15	Andrew Fisher
1915–23	William M. Hughes
1923–9	Stanley M. Bruce
1929–31	James H. Scullin
1932–9	Joseph A. Lyons
1939–41	Robert Gordon Menzies
1941	Arthur William Fadden
1941–5	John Curtin
1945–9	Joseph Benedict Chifley
1949–66	Robert Gordon Menzies
1966–7	Harold Edward Holt
1968–71	John Grey Gorton
1971–2	William McMahon
1972–5	Gough Whitlam
1975–83	J. Malcolm Fraser
1983–91	Robert J. L. Hawke
1991–	Paul Keating

Presidents of France

Third Republic

1899–1906	Emile Loubet
1906–13	Armand Fallières
1913–20	Raymond Poincaré
1920	Paul Deschanel
1920–4	Alexandre Millerand
1924–31	Gaston Doumergue
1931–2	Paul Doumer
1932–40	Albert Lebrun

Fourth Republic

1947–54	Vincent Auriol
1954–8	René Coty

Fifth Republic

1958–69	Charles de Gaulle
1969–74	Georges Pompidou
1974–81	Valéry Giscard d'Estaing
1981–	François Mitterand

Ireland

Governors General

1922–7	Timothy Michael Healy
1927–32	James McNeill
1932–6	Donald Buckley

Presidents

1938–45	Douglas Hyde
1945–59	Sean Thomas O'Kelly
1959–73	Eamon de Valera
1973–4	Erskine H. Childers
1974–6	Caroll Daly
1976–90	Patrick J. Hillery
1990–	Mary Robinson

Prime Ministers (Taoiseach)

1919–21	Eamon de Valera
1922	Arthur Griffiths
1922–32	William Cosgrave
1932–48	Eamon de Valera
1948–51	John Aloysius Costello
1951–4	Eamon de Valera
1954–7	John Aloysius Costello
1957–9	Eamon de Valera
1959–66	Sean Lemass
1966–73	John Lynch
1973–7	Liam Cosgrave
1977–9	John Lynch
1979–82	Charles Haughey
1982–7	Garret Fitzgerald
1987–92	Charles Haughey
1992–	Albert Reynolds

German Weimar Republic

Presidents

1919–25	Friedrich Ebert
1925–34	Paul von Hindenburg
(1933	Hitler appointed Chancellor)

Third Reich

President and Führer

1934–45	Adolf Hitler

German Federal Republic

Presidents

1949–59	Theodor Heuss
1959–69	Heinrich Lübke
1969–74	Gustav Heinemann
1974–9	Walter Scheel
1979–84	Karl Carstens
1984–	Richard von Weizsäcker

Chancellors

1949–63	Konrad Adenauer
1963–6	Ludwig Erhard
1966–9	Kurt Georg Keisinger
1969–74	Willy Brandt
1974–82	Helmut Schmidt
1982–	Helmut Kohl
1990–	(after unification with German Democratic Republic) Helmut Kohl

German Democratic Republic

President

1949–60	Wilhelm Pieck

Chairmen of the Council of State

1960–73	Walter Ernst Karl Ulbricht
1973–6	Willi Stoph
1976–89	Erich Honecker
1989	Egon Krenz
1989–	Gregor Gysi

Premiers

1949–64	Otto Grotewohl
1964–73	Willi Stoph
1973–6	Horst Sindermann
1976–89	Willi Stoph

Chairmen of the Council of Ministers

1989	Hans Modrow
1989–90	Lothar de Maizière

The Soviet Union

Presidents

1917	Leo Borisovitch Kamenev
1917–19	Yakov Mikhailovitch Sverlov
1919–46	Mikhail Ivanovitch Kalinin
1946–53	Nikolai Shvernik
1953–60	Klimentiy Voroshilov
1960–4	Leonid Brezhnev
1964–5	Anastas Mikoyan
1965–77	Nikolai Podgorny
1977–82	Leonid Brezhnev
1982–3	Vasily Kuznetsov (*Acting President*)
1983–4	Yuri Andropov
1984	Vasily Kuznetsov (*Acting President*)
1984–5	Konstantin Chernenko
1985	Vasily Kuznetsov (*Acting President*)
1985–8	Andrei Gromyko
1988–90	Mikhail Gorbachev

Executive President

1990–1	Mikhail Gorbachev

General Secretaries

1922–53	Josef Stalin
1953	Georgiy Malenkov
1953–64	Nikita Krushchev
1964–82	Leonid Brezhnev
1982–4	Yuri Andropov
1984–5	Konstantin Chernenko
1985–91	Mikhail Gorbachev

Chairmen (Prime Ministers)

Council of Ministers

1917	Georgy Evgenyevich Lvov
1917	Aleksandr Fyodorovich Kerensky

Council of People's Commissars

1917–24	Vladimir Ilyich Lenin
1924–30	Aleksei Ivanoch Rykov
1930–41	Vyacheslav Mikailovich Molotov
1941–53	Josef Stalin

Council of Ministers

1953–5	Georgiy Malenkov
1955–8	Nikolai Bulganin
1958–64	Nikita Krushchev
1964–80	Alexei Kosygin
1980–5	Nikolai Tikhonov
1985–91	Nikolai Ryzhkov

Prime Ministers of Canada

1867–73	John A. Macdonald
1873–8	Alexander Mackenzie
1878–91	John A. Macdonald
1891–2	John J. C. Abbott
1892–4	John S. D. Thompson
1894–6	Mackenzie Bowell
1896	Charles Tupper
1896–1911	Wilfrid Laurier
1911–20	Robert L. Borden
1920–1	Arthur Meighen
1921–6	W. L. Mackenzie King
1926	Arthur Meighen
1926–30	W. L. Mackenzie King
1930–5	Richard B. Bennett
1935–48	W. L. Mackenzie King
1948–57	Louis Stephen St Laurent
1957–63	John George Diefenbaker
1963–8	Lester B. Pearson
1968–79	Pierre Elliott Trudeau
1979–80	Joseph Clark
1980–4	Pierre Elliott Trudeau
1984	John Turner
1984–	Brian Mulroney

The Russian Federation

President

1991–	Boris Yeltzin

Prime Ministers of New Zealand

1856	Henry Sewell
1856	William Fox
1856–61	Edward William Stafford
1861–2	William Fox
1862–3	Alfred Domett
1863–4	Frederick Whitaker
1864–5	Frederick Aloysius Weld
1865–9	Edward William Stafford
1869–72	William Fox
1872	Edward William Stafford
1872–3	George Marsden Waterhouse
1873	William Fox
1873–5	Julius Vogel
1875–6	Daniel Pollen
1876	Julius Vogel
1876–7	Harry Albert Atkinson
1877–9	George Grey
1879–82	John Hall
1882–3	Frederick Whitaker
1883–4	Harry Albert Atkinson
1884	Robert Stout
1884	Harry Albert Atkinson
1884–7	Robert Stout
1887–91	Harry Albert Atkinson
1891–3	John Ballance
1893–1906	Richard John Seddon
1906	William Hall-Jones
1906–12	Joseph George Ward
1912	Thomas Mackenzie
1912–25	William Ferguson Massey
1925	Francis Henry Dillon Bell
1925–8	Joseph Gordon Coates
1928–30	Joseph George Ward
1930–5	George William Forbes
1935–40	Michael J. Savage
1940–9	Peter Fraser
1949–57	Sidney G. Holland
1957 (Aug.–Nov.)	Keith J. Holyoake
1957–60	Walter Nash
1960–72	Keith J. Holyoake
1972	John R. Marshall
1972–4	Norman Kirk
1974–5	Wallace Rowling
1975–84	Robert D. Muldoon
1984–9	David Lange
1989–90	Geoffrey Palmer
1990	Michael Moore
1990–	James Bolger

Poets Laureate of England

1617	Ben Jonson
1638	Sir William Davenant
1668	John Dryden
1689	Thomas Shadwell
1692	Nahum Tate
1715	Nicholas Rowe
1718	Laurence Eusden
1730	Colley Cibber
1757	William Whitehead
1785	Thomas Wharton
1790	Henry Pye
1813	Robert Southey
1843	William Wordsworth
1850	Alfred, Lord Tennyson
1896	Alfred Austin
1913	Robert Bridges
1930	John Masefield
1968	Cecil Day Lewis
1972	Sir John Betjeman
1984	Ted Hughes

The post of Poet Laureate was only officially established in 1668

Secretaries General of the United Nations

Secretary General	Country of origin	Dates in office
Trygve Lie	Norway	1946–52
Dag Hammarskjöld	Sweden	1953–61
U Thant	Burma	1962–72
Kurt Waldheim	Austria	1972–81
Javier Pérez de Cuéllar	Peru	1982–91
Boutros Boutros-Ghali	Egypt	1992–

Popes and antipopes

(antipopes are in bold)

–c.64	Peter
c.67–76/79	Linus
76–88 or 79–91	Anacletus
88–97 or 92–101	Clement I
c.97–c.107	Evaristus
105–115 or 109–119	Alexander I
c.115–c.125	Sixtus I
c.125–c.136	Telesphorus
c.136–c.140	Hyginus
c.140–155	Pius I
c.155–c.166	Anicetus
c.166–c.175	Soter
c.175–189	Eleutherius
c.189–199	Victor I
c.199–217	Zephyrinus
217–222	Calixtus I (Callistus)
222–230	Urban I
230–235	Pontian
235–236	Anterus
236–250	Fabian
251–253	Cornelius
251	**Novatian**
253–254	Lucius I
254–257	Stephen I
257–258	Sixtus II
259–268	Dionysius
269–274	Felix I
275–283	Eutychian
283–296	Gaius
291/296–304	Marcellinus
308–309	Marcellus I
309/310	Eusebius
311–314	Miltiades (Melchiades)
314–335	Sylvester I
336	Mark
337–352	Julius I
352–366	Liberius
355–358	**Felix (II)**
366–384	Damasus I
366–367	**Ursinus**
384–399	Siricius
399–401	Anastasius I
401–417	Innocent I
417–418	Zosimus
418–422	Boniface I
418–419	**Eulalius**
422–432	Celestine I
432–440	Sixtus III
440–461	Leo I
461–468	Hilary
468–483	Simplicius
483–492	Felix III (or II)
492–496	Gelasius I
496–498	Anastasius II
498–514	Symmachus
498, 501–c.505/507	**Laurentius**
514–523	Hormisdas
523–526	John I
526–530	Felix IV (or III)
530	**Dioscorus**
530–532	Boniface II
533–535	John II
535–536	Agapetus I
536–537	Silverius
537–555	Vigilius
556–561	Pelagius I
561–574	John III
575–579	Benedict I
579–590	Pelagius II
590–604	Gregory I
604–606	Sabinian
604	Boniface III
608–615	Boniface IV
615–618	Deusdedit (also called Adeodatus I)
619–625	Boniface V
625–638	Honorius I
640	Severinus
640–642	John IV
642–649	Theodore I
649–655	Martin I
654–657	Eugenius I
657–672	Vitalian
672–676	Adeodatus II
676–678	Donus
678–681	Agatho

Popes and antipopes

682–683	Leo II
684–685	Benedict II
685–686	John V
686–687	Conon
687–701	Sergius I
687	**Theodore**
687	**Paschal**
701–705	John VI
705–707	John VII
708	Sisinnius
708–715	Constantine
715–731	Gregory II
731–741	Gregory III
741–752	Zacharias (Zachary)
752	Stephen (II)
752–757	Stephen II (or III)
757–767	Paul I
767–768	**Constantine (II)**
768	**Philip**
768–772	Stephen III (or IV)
772–795	Adrian I
795–816	Leo III
816–817	Stephen IV (or V)
817–824	Paschal I
824–827	Eugenius II
827	Valentine
827–844	Gregory IV
844	**John**
844–847	Sergius II
847–855	Leo IV
855–858	Benedict III
855	**Anastasius (Anastasius the Librarian)**
858–867	Nicholas I
867–872	Adrian II
872–882	John VIII
882–884	Marinus I
884–885	Adrian III
885–891	Stephen V (or VI)
891–896	Formosus
896	Boniface VI
896	Stephen VI (or VII)
897	Romanus
897	Theodore II
898–900	John IX
900	Benedict IV
903	Leo V
903–904	**Christopher**
904–911	Sergius III
911–913	Anastasius III
913–914	Lando
914–928	John X
928	Leo VI
929–931	Stephen VII (or VIII)
931–935	John XI
936–939	Leo VII
939–942	Stephen VIII (or IX)
942–946	Marinus II
946–955	Agapetus II
955–964	John XII
963–965	Leo VIII
964–966?	Benedict V
965–972	John XIII
973–974	Benedict VI
974	**Boniface VII (1st time)**
974–983	Benedict VII
983–984	John XIV
984–985	**Boniface VII (2nd time)**
985–996	John XV (or XVI)
996–999	Gregory V
997–998	**John XVI (or XVII)**
999–1003	Sylvester II
1003	John XVII (or XVIII)
1004–09	John XVIII (or XIX)
1009–12	Sergius IV
1012	**Gregory (VI)**
1012–24	Benedict VIII
1024–32	John XIX (or XX)
1032–44	Benedict IX (1st time)
1045	Sylvester III
1045	Benedict IX (2nd time)
1045–46	Gregory VI
1046–47	Clement II
1047–48	Benedict IX (3rd time)
1048	Damasus II
1049–54	Leo IX
1055–57	Victor II
1057–58	Stephen IX (or X)
1058–59	**Benedict X**
1059–61	Nicholas II
1061–73	Alexander II
1061–72	**Honorius (II)**
1073–85	Gregory VII
1080–1100	**Clement (III)**
1086–87	Victor III
1088–99	Urban II
1099–1118	Paschal II
1100–02	**Theodoric**
1102	**Albert (also called Aleric)**
1105–11	**Sylvester (IV)**
1118–19	Gelasius II
1118–21	**Gregory (VIII)**
1119–24	Calixtus II (Callistus)
1124–30	Honorius II
1124	**Celestine (II)**
1130–43	Innocent II
1130–38	**Anacletus (II)**
1138	**Victor (IV)**
1143–44	Celestine II
1144–45	Lucius II
1145–53	Eugenius III
1153–54	Anastasius IV
1154–59	Adrian IV
1159–81	Alexander III
1159–64	**Victor (IV)**
1164–68	**Paschal (III)**
1168–78	**Calixtus (III)**
1179–80	**Innocent (III)**
1181–85	Lucius III
1185–87	Urban III
1187	Gregory VIII
1187–91	Clement III
1191–98	Celestine III
1198–1216	Innocent III
1216–27	Honorius III
1227–41	Gregory IX
1241	Celestine IV
1243–54	Innocent IV
1254–61	Alexander IV
1261–64	Urban IV
1265–68	Clement IV
1271–76	Gregory X
1276	Innocent V
1276	Adrian V
1276–77	John XXI
1277–80	Nicholas III
1281–85	Martin IV
1285–87	Honorius IV
1288–92	Nicholas IV
1294	Celestine V
1294–1303	Boniface VIII
1303–04	Benedict XI
1305–14	Clement V (at Avignon, from 1309)
1316–34	John XXII (at Avignon)
1328–30	**Nicholas (V) (at Rome)**
1334–42	Benedict XII (at Avignon)
1342–52	Clement VI (at Avignon)
1352–62	Innocent VI (at Avignon)
1362–70	Urban V (at Avignon)
1370–78	Gregory XI (at Avignon, then Rome from 1377)
1378–89	Urban VI
1378–94	**Clement (VII) (at Avignon)**
1389–1404	Boniface IX
1394–1423	**Benedict (XIII) (at Avignon)**
1404–06	Innocent VII
1406–15	Gregory XII
1409–10	**Alexander (V) (at Bologna)**
1410–15	**John (XXIII) (at Bologna)**
1417–31	Martin V
1431–47	Eugenius IV
1439–49	**Felix (V) (also called Amadeus VIII of Savoy)**
1447–55	Nicholas V
1455–58	Calixtus III (Callistus)
1458–64	Pius II
1464–71	Paul II
1471–84	Sixtus IV
1484–92	Innocent VIII
1492–1503	Alexander VI
1503	Pius III
1503–13	Julius II
1513–21	Leo X
1522–23	Adrian VI
1523–34	Clement VII
1534–49	Paul III
1550–55	Julius III
1555	Marcellus II
1555–59	Paul IV
1559–65	Pius IV
1566–72	Pius V
1572–85	Gregory XIII
1585–90	Sixtus V
1590	Urban VII
1590–91	Gregory XIV
1591	Innocent IX
1592–1605	Clement VIII
1605	Leo XI
1605–21	Paul V
1621–23	Gregory XV
1623–44	Urban VIII
1644–55	Innocent X
1655–67	Alexander VII
1667–69	Clement IX
1670–76	Clement X
1676–89	Innocent XI
1689–91	Alexander VIII
1691–1700	Innocent XII
1700–21	Clement XI
1721–24	Innocent XIII
1724–30	Benedict XIII
1730–40	Clement XII
1740–58	Benedict XIV
1758–69	Clement XIII
1769–74	Clement XIV
1775–99	Pius VI
1800–23	Pius VII
1823–29	Leo XII
1829–30	Pius VIII
1831–46	Gregory XVI
1846–78	Pius IX
1878–1903	Leo XIII
1903–14	Pius X
1914–22	Benedict XV
1922–39	Pius XI
1939–58	Pius XII
1958–63	John XXIII
1963–78	Paul VI
1978	John Paul I
1978–	John Paul II

The United Kingdom

England

Counties

Avon
Bedfordshire
Berkshire
Buckinghamshire
Cambridgeshire
Cheshire
Cleveland
Cornwall
Cumbria
Derbyshire
Devon
Dorset
Durham
East Sussex
Essex
Gloucestershire
Greater London
Greater Manchester
Hampshire
Hereford & Worcester
Hertfordshire
Humberside
Isle of Wight
Kent
Lancashire
Leicestershire
Lincolnshire
Merseyside
Norfolk
Northamptonshire
Northumberland
North Yorkshire
Nottinghamshire
Oxfordshire
Shropshire
Somerset
South Yorkshire
Staffordshire
Suffolk
Surrey
Tyne and Wear
Warwickshire
West Midlands
West Sussex
West Yorkshire
Wiltshire

Scotland

Regions

Borders
Central
Dumfries & Galloway
Fife
Grampian
Highland
Lothian
Strathclyde
Tayside

Island Areas

Orkney
Shetland
Western Isles and Inner Hebrides

Northern Ireland

Counties

Antrim
Armagh
Down
Fermanagh
Londonderry
Tyrone

Wales

Counties

Clwyd
Dyfed
Gwent
Gwynedd
Mid Glamorgan
Powys
South Glamorgan
West Glamorgan

Other UK islands

Channel Islands

Alderney
Guernsey
Jersey
Sark

Isle of Man

Isle of Wight

New Zealand

Local Government Regions

Aorangi
Auckland
Bay of Plenty
Canterbury
Clutha
Coastal – North Otago
East Cape
Hawke's Bay
Horowhenua
Marlborough
Manawatu
Nelson Bays
Northland
Southland
Taranaki
Tongariro
Thames Valley
Waikato
Wairarapa
Wanganui
Wellington
West Coast

Canada

(with official abbreviations)

Provinces	Capitals
Alberta (Alta.)	Edmonton
British Columbia (BC)	Victoria
Manitoba (Man.)	Winnipeg
New Brunswick (NB)	Fredericton
Newfoundland and Labrador (Nfld.)	St John's
Nova Scotia (NS)	Halifax
Ontario (Ont.)	Toronto
Prince Edward Island (PEI)	Charlottetown
Quebec (Que.)	Quebec
Saskatchewan (Sask.)	Regina
Northwest Territories (NWT)	Yellowknife (seat of government)
Yukon Territory (YT)	Whitehorse (seat of government)
Federal capital	Ottawa

The Commonwealth of Australia

States	Capitals
New South Wales	Sydney
Queensland	Brisbane
South Australia	Adelaide
Tasmania	Hobart
Victoria	Melbourne
Western Australia	Perth
Australian Capital Territory	Canberra (federal capital)
Northern Territory	Darwin

India

States	Capitals
Andhra Pradesh	Hyderabad
Arunachal Pradesh	Itanagar
Assam	Dispur (temporary)
Bihar	Patna
Goa	Panaji
Gujarat	Gandhinagar
Haryana	Chandigarh
Himachal Pradesh	Shimla
Jammu and Kashmir	Srinagar (summer), Jammu (winter)
Karnataka	Bangalore
Kerala	Trivandrum
Madhya Pradesh	Bhopal
Maharashtra	Bombay
Manipur	Imphal
Meghalaya	Shillong
Mizoram	Aizawl
Nagaland	Kohima
Orissa	Bhubaneshwar
Punjab	Chandigarh
Rajasthan	Jaipur
Sikkim	Gangtok
Tamil Nadu	Madras
Tripura	Agartala
Uttar Pradesh	Lucknow
West Bengal	Calcutta

Union Territories	
Andaman and Nicobar Islands	Port Blair
Chandigarh	Chandigarh
Dadra and Nagar Haveli	Silvassa
Daman and Diu	Daman
Delhi	Delhi
Lakshadweep	Kavaratti
Pondicherry	Pondicherry

States of the United States of America

(with official and postal abbreviations)

States	Capitals	Popular names
Alabama (Ala., AL)	Montgomery	Yellowhammer State, Heart of Dixie, Cotton State
Alaska (Alas., AK)	Juneau	Great Land
Arizona (Ariz., AZ)	Phoenix	Grand Canyon State
Arkansas (Ark., AR)	Little Rock	Land of Opportunity
California (Calif., CA)	Sacramento	Golden State
Colorado (Col., CO)	Denver	Centennial State
Connecticut (Conn., CT)	Hartford	Constitution State, Nutmeg State
Delaware (Del., DE)	Dover	First State, Diamond State
Florida (Fla., FL)	Tallahassee	Sunshine State
Georgia (Ga., GA)	Atlanta	Empire State of the South, Peach State
Hawaii (HI)	Honolulu	The Aloha State
Idaho (ID)	Boise	Gem State
Illinois (Ill., IL)	Springfield	Prairie State
Indiana (Ind., IN)	Indianapolis	Hoosier State
Iowa (Ia., IA)	Des Moines	Hawkeye State
Kansas (Kan., KS)	Topeka	Sunflower State
Kentucky (Ky., KY)	Frankfort	Bluegrass State
Louisiana (La., LA)	Baton Rouge	Pelican State
Maine (Me., ME)	Augusta	Pine Tree State
Maryland (Md., MD)	Annapolis	Old Line State, Free State
Massachusetts (Mass., MA)	Boston	Bay State, Old Colony
Michigan (Mich., MI)	Lansing	Great Lake State, Wolvering State
Minnesota (Minn., MN)	St Paul	North Star State, Gopher State
Mississippi (Miss., MS)	Jackson	Magnolia State
Missouri (Mo., MO)	Jefferson City	Show Me State
Montana (Mont., MT)	Helena	Treasure State
Nebraska (Nebr., NB)	Lincoln	Cornhusker State
Nevada (Nev., NV)	Carson City	Sagebrush State, Battleborn State, Silver State
New Hampshire (NH)	Concord	Granite State
New Jersey (NJ)	Trenton	Garden State
New Mexico (N. Mex., NM)	Santa Fe	Land of Enchantment
New York (NY)	Albany	Empire State
North Carolina (NC)	Raleigh	Tar Heel State, Old North State
North Dakota (N. Dak., ND)	Bismarck	Peace Garden State
Ohio (OH)	Columbus	Buckeye State
Oklahoma (Okla., OK)	Oklahoma City	Sooner State
Oregon (Oreg., OR)	Salem	Beaver State
Pennsylvania (Pa., PA)	Harrisburg	Keystone State
Rhode Island (RI)	Providence	Little Rhody, Ocean State
South Carolina (SC)	Columbia	Palmetto State
South Dakota (S. Dak., SD)	Pierre	Coyote State, Sunshine State
Tennessee (Tenn., TN)	Nashville	Volunteer State
Texas (Tex., TX)	Austin	Lone Star State
Utah (UT)	Salt Lake City	Beehive State
Vermont (Vt., VT)	Montpelier	Green Mountain State
Virginia (Va., VA)	Richmond	Old Dominion
Washington (Wash., WA)	Olympia	Evergreen State
West Virginia (W. Va., WV)	Charleston	Mountain State
Wisconsin (Wis., WI)	Madison	Badger State
Wyoming (Wyo., WY)	Cheyenne	Equality State

The Russian Federation

Republics	Population (millions)	Republics	Population (millions)
Adygei	0.43	Karelia	0.79
Bashkiria	3.94	Khakassia	0.57
Buryatia	1.04	Komi	1.25
Checheno-Ingushetia	1.27	Mari	0.75
Chuvashia	1.34	Mordovia	0.96
Dagestan	1.80	North Ossetia	0.63
Gorno-Altai	0.19	Tatarstan	3.64
Kabardino-Balkaria	0.75	Tuva	0.31
Kalmykia	0.32	Udmurtia	1.61
Karachai-Circassia	0.42	Yakutia	1.09

Source: 1989 census

The Commonwealth

The Commonwealth is a free association of the 49 sovereign independent states listed below, together with their associated states and dependencies.

Antigua and Barbuda
Australia
Bahamas
Bangladesh
Barbados
Belize
Botswana
Brunei
Canada
Cyprus
Dominica
Gambia, the
Ghana
Grenada
Guyana
India
Jamaica
Kenya
Kiribati
Lesotho
Malawi
Malaysia
Maldives
Malta
Mauritius
Nauru
New Zealand
Nigeria
Pakistan
Papua New Guinea
St Christopher and Nevis
St Lucia
St Vincent and the Grenadines
Seychelles
Sierra Leone
Singapore
Solomon Islands
Sri Lanka
Swaziland
Tanzania
Tonga
Trinidad and Tobago
Tuvalu
Uganda
United Kingdom
Vanuatu
Western Samoa
Zambia
Zimbabwe

The Soviet Union

Former republics	Capitals
Armenia	Yerevan
Azerbaijan	Baku
Belorussia	Minsk
Estonia	Tallinn
Georgia	Tbilisi (formerly Tiflis)
Kazakhstan	Alma-Ata
Kirghizia	Frunze
Latvia	Riga
Lithuania	Vilnius (formerly Vilna)
Moldavia	Kishinev (formerly Chişinău)
Russia	Moscow
Tadjikistan	Dushanbe
Turkmenistan	Ashkhabad
Ukraine	Kiev
Uzbekistan	Tashkent

Members of the European Community (EC)

Belgium
Denmark
France
Germany
Greece
Ireland
Italy
Luxemburg
The Netherlands
Portugal
Spain
United Kingdom

Italy

Regions

Abruzzi
Basilicata
Calabria
Campania
Emilia Romagna
Friuli-Venezia-Giulia
Lazio
Liguria
Lombardia
Marche
Molise
Piedmonte
Puglia
Sardegna
Sicilia
Toscana
Trentino-Alto-Adige
Umbria
Valle d'Aosta
Veneto

Greece

Prefectures

Aegean Islands
Central Greece and Euboea
Crete
Epirus
Greater Athens
Ionian Islands
Macedonia
Peloponnessos
Thessaly
Thrace

Spain

Regions

Andalusia (Andalucía)
Aragon
Asturias
Balearic Islands (Islas Baleares)
Basque Provinces (Provincias Vascongadas)
Canary Islands (Islas Canarias)
Catalonia (Cataluña; Catalunya)
Extremadura (Estremadura)
Galicia
León
Murcia
Navarre (Navarra)
Valencia

Republic of Ireland

Counties

Carlow
Cavan
Clare
Cork
Donegal
Dublin
Galway
Kerry
Kildare
Kilkenny
Laoighis
Leitrim
Limerick
Longford
Louth
Mayo
Meath
Monaghan
Offaly
Roscommon
Sligo
Tipperary, North Riding
Tipperary, South Riding
Waterford
Westmeath
Wexford
Wicklow

Germany

Länder

Baden-Württemberg
Bavaria (Bayern)
Berlin
Brandenburg
Bremen
Hamburg
Hesse (Hessen)
Lower Saxony (Niedersachsen)
Mecklenberg-West Pomerania (Mecklenberg-Vorpommern)
North Rhine-Westphalia (Nordrhein-Westfalen)
Rhineland Palatinate (Rheinland-Pfalz)
Saarland
Saxony (Sachsen)
Saxony-Antalt (Sachsen-Anhalt)
Thuringia (Thuringen)

Switzerland

Cantons

Aargau (Argovie)
Appenzell – Outer- and Inner-Rhoden
Basel (Bâle)
Bern (Berne)
Fribourg (Freiburg)
Genève (Genf)
Glarus (Glaris)
Graubünden (Grisons)
Jura
Neuchâtel (Neuenberg)
St Gallen (St Gall)
Schaffhausen (Schaffhouse)
Schwyz
Solothurn (Soleure)
Thurgau (Thurgovie)
Ticino (Tessin)
Unterwalden – Upper and Lower
Uri
Valais (Wallis)
Vaud (Waadt)
Zug (Zoug)
Zürich (Zurich)

France

Departments

Alsace
Aquitaine
Auvergne
Basse-Normandie
Bretagne
Bourgogne
Centre
Champagne
Corse
Franche-Comté
Haute-Normandie
Languedoc-Rousillon
Limousin
Lorraine
Midi-Pyrénées
Nord-Pas-de-Calais
Pays de la Loire
Picardie
Poitou-Charentes
Provence-Côte d'Azur
Region Parisienne
Rhône Alpes

Belgium

Provinces

Antwerp (Anvers)
Brabant
East and West Flanders
Hainaut
Liège
Limbourg
Luxembourg
Namur

The Netherlands

Provinces

Groningen
Friesland
Drenthe
Overijssel
Flevoland
Gelderland
Utrecht
Noord- and Zuid-Holland
Zeeland
Noord-Brabant
Limburg

World population

Continent	Area km²	Area square miles	Estimated population mid-1992
Africa	30,313	11,704	654,000,000
North America (including Greenland and Hawaii)	21,525	8,311	283,000,000
Latin America (Mexico and Americas south of USA)	20,547	7,933	453,000,000
Asia (including European Turkey but none of former Soviet Union)	27,549	10,637	3,207,000,000
Europe (excluding European Turkey and former Soviet Union)	4,961	1,915	511,000,000
Former Soviet Union	22,402	8,649	284,000,000
Oceania (including Australia, New Zealand, and Micronesian, Melanesian, and Polynesian islands)	8,510	3,286	28,000,000
Total	135,807	52,435	5,420,000,000

Source: *1992 World Population Data Sheet*, Population Concern, London, UK

Index

A User's Guide to the Index

This index is designed for easy use, but the following notes may be helpful to the reader. The page references to Volumes 1, 2, and 4 in the index apply to the revised and updated volumes in the boxed-set edition. Earlier editions of these volumes follow a slightly different pagination.

Emboldening Each entry is prefixed by an emboldened volume number, followed by a page number. For example, **2.**36 indicates page 36 in Volume **2.** The general subject of an entry can be deduced from these volume prefixes. Thus **1** is the physical and earth sciences; **2** is natural history; **3** and **4** are history; **5** is the arts; **6** is technology and applied science; **7** is human society; **8** is astronomy; and **9** is religion, mythology, and tables.

Main entries The page references for major encyclopedia entries are shown in *italics*. For example, in **natural selection 2.**194, ***2.**225*, **7.**11, the reference in italics identifies the main entry. Every detail could not be included, but further facts can be obtained by turning to the pages where the main entries appear in the Encyclopedia. Additional cross-references to related topics in the main entries may be given, indicated by asterisks (*). For example, **terrorism** in Volume 4 contains eight cross-references within the text.

Sub-entries group longer entries into their constituent topics within an entry, thus enabling the user to find specific information more quickly.

Strings of page references have been avoided wherever possible, though they are occasionally necessary. Sub-entries are generally arranged alphabetically, not chronologically.

Alphabetization, ordering, and abbreviations Index entries are in word-by-word order, and hyphenated words are regarded as one word for alphabetization. For example, **leaf**; **leaf insect**; **leaf-nosed bat**.

Prepositions, articles, and conjunctions are ignored in alphabetization. For example, **Field of the Cloth of Gold** precedes **field equations**.

Names prefixed with ad-, al-, an-, el-, ul-, are alphabetized according to the significant name. For example, **al-Kabir** is alphabetized under **Kabir**.

Titles (for example, Lord, Sir, Dame, Bishop, Saint) are ignored for alphabetization purposes. Thus, **Beaufort, Bishop Henry** comes after **Beaufort, Francis**.

Names with numerals precede those without. For example, **James I** precedes **James, Henry**.

Dates and other qualifications are omitted except where there are entries which are identical or very similar. For example**, Congreve, William (1670–1729)**; **Congreve, Sir William (1772–1828)**; **file (computer)**; **file (tool)**.

Abbreviations for place- and proper-names beginning with 'Saint' or their foreign equivalents (St, Ste, Sta) are alphabetized as if they were spelt out. For example, **St Albans**; **Saint Elias Mountains**; **St Ives School**; **Sainte-Beuve, Charles-Augustin**.

Proper names having the prefix 'Mac' or its contracted form 'Mc' are all alphabetized as 'Mac'. For example, **Macintosh, Charles**; **Mackenzie river**; **McLuhan, Marshall**.

Countries Many of the entries for countries are very long. For this reason, sub-entries for natural history (Volume 2), where many of the species are ubiquitous, have been omitted.

Numerals are alphabetized as though the numbers were spelt out. For example, **fifth-generation computer**; **tenth planet**; **twentieth-century music**.

Running heads and continuation lines In order to help the reader find information more quickly, the first and last entry on each page of the index is given in the running heads. Thus the key word of the first entry is given on the top left-hand side of the page and that of the last entry is given on the top right-hand side of the page. Continuation lines at the top of a column indicate where an entry runs on from one column to the next.

ANN HALL

A

a priori 7.15, 7.178, 7.259, 7.312
  see also nativism
Aalto, Alvar 5.1
AAMs *see* air-to-air missiles
aardvark 2.1
aardwolf 2.1
Aaron 9.2, 9.5, 9.14
abacus 6.1
Abahai of Manchus 3.218
abalone 2.1
abattoir 6.1
Abbas I (the Great), king of Persia 3.1, 5.344, 5.397
Abbasid dynasty 3.1, 3.157–8
  arts 5.1, 5.358
    see also arts *under* Islam
Abbe, Ernst 6.1, 6.310, 6.391
Abbott, Berenice 5.1, 5.347
'Abd al-Muttalib 9.39
Abd al-Aziz ibn Saud 4.303
Abd el-Krim 4.1, 4.224
Abd al-Rahman ibn Mauwiya 3.240, 3.365
Abdal, Pir Sultan 5.439
Abdallah, Khalifa 4.207
abdication crisis 4.1
abdomen 2.1
Abdorra 7.338
Abduh, Muhammad 7.168
Abdul Aziz (Morocco) 4.55
Abdul Hamid II, Ottoman sultan 4.1, 4.25
Abdulaziz, Ottoman sultan 4.253, 4.338
Abdullah, Sheikh Muhammad 4.1
Abdulmecid I, Ottoman sultan 4.253, 4.338
Abdurrahman Khan 4.3
Abel 9.2, 9.8
Abel, Carl Friedrich 5.28
Abel, Niels Henrik 1.1
Abélard, Peter 3.1, 3.40, 5.257, 7.1
Abelian group 1.1
Aberdeen, George Hamilton Gordon, 4th earl of 4.1, 4.251
Aberhart, William 4.317
aberration 8.1
el-Abidine Ben Ali, Zine 4.354
Abla 9.4
ABMs *see* anti-ballistic missiles
abolitionists 4.1
  see also abolition and emancipation *under* slavery
Aborigines, Australian 3.29, 4.1–2, 7.1
  Aranda 7.15
  circumcision 7.51
  languages 7.22
  music 5.1
  Myall Creek massacre 4.229
  myths and legends 9.3, 9.35
  poetry 5.1
  technology 6.42
abortion 2.1, 7.1, 7.115
Aboukir Bay (Nile battle) 3.254
Abraham 3.1, 3.58, 3.192, 7.73, 7.175, 7.272, 9.2
  as Ibrahim 9.2, 9.8, 9.18, 9.19
Abraham, E. P. 6.69
Abramovitsh, Sh. Y. 5.497
abrasion (geomorphology) 1.1
abrasives 6.1, 6.12, 6.68, 6.129, 6.345, 6.365
  see also silicon carbide
ABS plastic 6.273
Absalom 9.2
abscess 2.1
absentee landlord 3.1
absolute government 7.144
absolute magnitude 8.90
absolute zero 1.1
absorption:
  interstellar 8.74
  line 8.157
abstract art 5.1–2, 5.243–4, 5.299, 5.301, 5.318, 5.350, 5.391
  Britain 5.205, 5.303, 5.318, 5.350
  colour field 5.107
  France 5.23, 5.128, 5.244, 5.260, 5.286
  New York School 5.318
  Op art 5.328, 5.383, 5.469
  Suprematism 5.440
  see also Abstract Expressionism; modern art
Abstract Expressionism 5.2, 5.27, 5.441
  United States 5.2, 5.14, 5.127, 5.184, 5.306, 5.355, 5.375, 5.391
  see also abstract art; Expressionism; modern art
abstract music 5.367, 5.496
Absurd, Theatre of 5.2, 5.10, 5.40–1, 5.201, 5.225, 5.349, 5.484, 5.491
'Abū Bakr, first caliph of Islam 3.1, 3.11, 3.58, 7.39
Abu Dhabi 7.70, 7.136
Abū Nuwās, al-Ḥasan ibn Hani' 5.2
Abū'l-'Atāhiya 5.19
Abushiri Revolt 4.2
Abydos 3.2
abyssal plain 1.1, 1.236
Abyssinia *see* Ethiopia
Abyssinian Campaigns 4.2
a.c. *see* alternating current
acacia 2.2
academic freedom 7.1
Académie des Sciences 3.2, 6.1
Académie française 5.2
Academy (Athens) 3.2
Academy of Motion Picture Arts and Sciences 5.2
Academy of Sciences of the USSR 6.1
Acadia *see* Nova Scotia
acanthocephalan 2.322
acanthus 2.2
ACAS (Advisory, Conciliation, and Arbitration Service) 7.158
Accademia Nationale dei Lincei 6.1
acceleration 1.1, 1.355, 6.1, 6.149, 6.368
accelerator:
  linear 6.196, 6.258
  particle *see* accelerator *under* particle(s)
  proton 6.258
accelerometer 6.1, 6.178
accent 7.1
accentors 2.2, 2.154
acceptance house 7.1
accidents and disasters 7.91–2
  airship 6.10
  building collapse 6.344
  chemical industry 1.91–2, 6.35, 6.275, 6.314
  disaster relief 8.92
  first railway death 4.159
  mythical 9.30
  nuclear reactors 6.9, 6.72, 6.243, 6.245, 6.275, 6.281, 6.355
  prevention *see* safety
  Tacoma Narrows 1.281
  volcanic eruptions 1.221
accordion 5.2
accounting systems 7.1–2
  management 7.190–1
accretion disc 8.1–2
accumulator 6.1–2, 6.32
ACE 6.85
acetate 6.273, 6.295
acetic (ethanoic) acid 1.114, 6.66, 6.253
acetone (propanone) 1.268, 6.284, 6.295
acetylcholine 6.240
acetylene *see* ethyne
Achaean League 3.2
Achaemenid dynasty and empire 3.2, 3.95, 3.97, 3.271, 3.276, 3.316–17, 3.387
  arts 5.344
Achebe, Chinua 5.2, 5.402
Acheh and Acheh War 3.2, 4.2
Achernar 8.2
Acheson, Dean 4.2
Acheulian culture 3.2, 3.50
Achilles (asteroids) 8.2
Achilles (Greek hero) 9.2, 9.3
achondrite 8.2
achromatic lens 6.2, 6.310
acid anhydride 1.2
acid igneous rocks 1.165
  see also granite; rhyolite
acid rain 1.2, 6.2, 7.2, 7.240
acid-base titration 1.2
acids 1.1–2, 6.2
  acetic (ethanoic) 1.114, 6.66, 6.253
  acid-base titration 1.2
  amphoteric compound 1.12
  ascorbic 2.357, 6.141
  aspartic 6.342
  and bases *see* salts
  in battery 6.32
  bile 2.29
  buffer solution 1.45
  carbolic *see* phenol
  carboxylic 1.52, 1.213, 6.252
  chemical industry 6.70–1
  citric 2.181, 2.187
    see also citrus
  ethanoic (acetic) 1.114, 6.66, 6.253
  fatty *see* fats, oils, and fatty acids
  formic (methanoic) 1.213, 6.222
  hydrochloric *see* hydrochloric acid
  hydrofluoric 6.173, 6.177
  indicator of 1.197
  lactic 2.115, 6.127
  methanoic (formic) 1.213, 6.222
  neutralization 1.59, 1.226
  nicotinic 2.357
  nitric *see* nitric acid
  pH 1.161, 6.2, 6.267
  phenol *see* phenol
  phosphoric 6.2, 6.71, 6.267
  prussic 2.56
  pyruvic 2.138
  sulphuric *see* sulphuric acid
  see also amino acids; DNA; nucleic acids; RNA
Ackermann, Rudolph: Ackermann principle 6.333–4
acne 2.2
aconite 2.2
acorn worm 2.2
acoustics 6.2, 6.223
ACP (African, Caribbean, and Pacific countries) 7.103
acquired immune deficiency syndrome *see* Aids
acropolis 3.2
acrostic 5.2
Acrux 8.2
acrylic fibre 6.2–3, 6.58, 6.61
acrylic paint 5.2–3
act 7.2, 7.319
  Helsinki Final 7.2, 7.64
  Single European 7.99, 7.103, 7.111
  of Union 3.365, 4.2
  see also law and justice *and* laws/legal system *under individual countries*
Actaeon 9.2
acting 5.3
  see also under cinema *and* drama
actinide metals 1.2
  see also plutonium; thorium; uranium
actinium 1.106
Action Française 4.2
action painting 5.3
action and reaction (physics) 1.2
Actium 3.3
activation energy 1.2–3
active layer (geomorphology) 1.3
active and semi-active radar homing 6.162
activity series 1.3, 1.102
actuary 7.2
acupuncture 2.2, 2.3, 6.3, 7.2
'Ad 9.2
ADA 6.86
Adam 9.2, 9.11, 9.12, 9.23, 9.32
Adam, Robert 5.3, 5.5, 5.105, 5.351
Adam de la Halle 5.3, 5.5
Adams, Abigail 4.117
Adams, Ansel 5.5, 5.488
Adams, Henry 5.5
Adams, John (1735–1826, US president) 3.3, 3.124, 3.386
Adams, John Couch (1819–92, astronomer) 8.2, 8.11, 8.84, 8.106–7
Adams, John Quincy (1767–1848, US president) 4.2, 4.95, 4.117
Adams, Samuel 3.2–3
Adamson, Robert 5.206
Adanson, Michel 2.2–3
adaptation (evolution) 2.3
Addams, Jane 4.2
adder 2.3
adder's tongue 2.11
adding machine *see* calculator
Addington, Henry, 1st viscount Sidmouth 4.2–3
Addison, Joseph 5.5
Addison, Thomas: Addison's disease 2.3
addition reaction 1.59–60
additive:
  food 6.16, 6.27, 6.46, 6.140, 6.176
  fuel 6.147, 6.265
  see also lead tetraethyl
Addled Parliament 3.3
address (computer) 6.3
Aden 4.3
Adenauer, Konrad 4.3, 4.135
adenine 2.95
adenoids 2.3
adenosine triphosphate 2.18, 2.138, 2.222
adhesive 6.3, 6.162, 6.170
Adi Granth 7.2, 7.169
adiabatic change 1.3
Adirondack Mountains 1.3
adit mining 6.3, 6.79
adjective 7.229
adjustment 7.2–3
Adler, Alfred 2.4, 7.214
Adler, Dankmar 5.439
administrative law 7.3, 7.213, 7.319
adobe 1.3, 1.4, 6.235
Adolf of Nassau 3.168
Adonis (asteroid) 8.2
Adonis (mythology) 9.2
adoption 7.3
Adorno, Theodor 7.124, 7.245, 7.292
Adowa, battle of 4.3
Adrastea 8.2
adrenal glands 2.4
adrenalin 6.344
Adrian IV, pope 3.3, 3.121, 3.132
Adrianople, battle of (378) 3.3
Adrianople, Treaty of (1829) 4.3
Adriatic Sea 1.3
adult education 7.68–9
Advaita Hinduism 7.3, 7.21, 7.155, 7.156, 7.203, 7.263

advanced gas-cooled reactor 6.3–4, 6.243–4
advection 1.3
Advent 7.3
Adventists 7.3, 7.202
see also eschatology; millenarianism
adventitious food additive 6.140
adverb 7.229
adversarial procedure 7.3
advertising 7.3–4, 7.152–3, 7.216
Advisory, Conciliation, and Arbitration Service 7.158
advocate 7.3, 7.4
Ady, Endre 5.5
adze 6.4, 6.27, 6.357–8
Aegean civilization *see* Mycenaean
Aegean Sea 1.3, 1.84, 1.282, 1.317, 3.160
Aegospotami, battle of 3.3
Aemilius Paullus 3.59
Aeneas 9.2, 9.10
Aeneid 5.477
aeolian harp 5.5
Aeolian Islands 1.196
aeons, geological 1.3–4, 1.133, 1.134
Archaean 1.18
Proterozoic 1.133, 1.134
see also Phanerozoic
aerial *see* antenna
aerial perspective 5.5
aerial photography/surveying 6.215, 6.268
aerodynamics 6.4–5, 6.65, 6.135, 6.386, 7.188
see also flight
aerofoil 6.5, 6.135
aeronautics 6.134–6
aeroplane *see* aircraft
aerosol 1.4, 6.5, 6.80
Aeschylus 5.5–6, 5.191, 5.459
Aesir and Vanir 9.2
Aesop 5.6, 9.23
Aesthetic Movement 5.6
aesthetics 7.4
aestivation *see* dormancy; hibernation
Aetius 3.64
Aetolia 3.3
Aezanas of Axum 3.259
Afa 9.17
Afars and Issas 4.101
affirmative action 7.243
'affluent society' 7.4–5
al-Afghani, Jamal al-Din 4.3, 4.168, 7.168
Afghanistan 1.4, 3.3, 3.4, 4.3, 7.337
ethnic groups 4.262, 7.231
exploration 1.164
infant mortality 7.159
language 7.157
medicine and health care 7.140
Muslim Brotherhood 7.166
physical geography and geology 1.4, 1.156, 1.183, 1.246
politics and government 7.255
refugees 7.261–2
regions 3.33, 4.247–8, 4.354
religion/theology 7.168
rulers 3.209, 3.216, 4.102, 4.311
and UN 7.321
wars and defence:
Anglo-Afghan 4.13
invaded by USSR 7.6, 7.34, 7.255
Aflaq, Michel 4.27
AFL-CIO 7.314
Africa 1.4, 3.3–4, 4.3–4, 4.116–17
African Charter on Human and People's Rights 7.5, 7.149
agriculture 6.7, 6.46, 6.68, 6.351, 7.6, 7.232, 7.237
see also under pastoralism; subsistence
aid to 7.8, 7.121
anthropology 7.112
arts 3.24, 5.6–9
architecture 5.6
Africa (*cont.*):
arts (*cont.*):
body 5.6
costume and textiles 5.7
masks and masquerades 5.7
pottery 5.8
rock art 5.8
sculpture and carving 3.40, 3.174, 3.254, 5.8–9
textiles 5.7
see also literature; music *below*
Berlin Conference on 4.307
colonies 4.306–7, 7.191
decolonization 4.4, 4.83
see also under empire *or* colonialism *under* Britain; France; Germany; Portugal
dance 5.9
demography:
birth rate 7.32
contraception 7.115–16
dependency ratio 7.85
fertility rate 7.119
infant mortality 7.159
infertility 7.159
life expectancy 7.184
migration 7.200
mortality 7.207
old age 7.221
population density 7.241
population growth 7.226, 7.241–2
desertification 7.86
diseases and disorders 7.9, 7.59
dowry 7.95
economy and trade 3.311, 7.88
Conventions 4.198, 4.386
debt 7.2
migrant labour 7.200
pay 7.203, 7.328
Royal African Company 3.311
in SADCC 7.296
slaves 3.330
education 7.93, 7.101, 7.102, 7.103, 7.141, 7.155, 7.185, 7.277
environment 7.82
ethnic groups 7.137, 7.176, 7.185, 7.268, 7.315
hunters and hunter-gatherers 7.178, 7.256, 7.285
see also Azande; Maasai; Nuer
evolution 3.120
exploration *see under* exploration
famine 7.116, 7.117–18, 7.292
finance 7.124, 7.201
forests 6.143, 7.349
government *see* politics *below*
languages 7.106
Afro-Asiatic 7.30, 7.43, 7.78, 7.109, 7.278–9
Berber 7.30
Chadic 7.43
Cushitic 7.78
Khoisan 7.176
Niger–Congo 7.30, 7.217
Nilo-Saharan 7.217
Semitic 7.109, 7.278–9
laws/legal system 7.40, 7.51, 7.54, 7.79
literature 5.2, 5.7, 5.184, 5.334, 5.402, 5.406
drama 5.9, 5.165–6, 5.310, 5.427
poetry 5.7, 5.86, 5.427
see also under novels
maps 1.4, 3.4, 3.120, 3.220, 4.4, 4.306
matrilineality 7.18, 7.19, 7.30
migration 7.200
monarchy 7.97, 7.139
mortality 7.207
music 5.7–9, 5.117, 5.321
pastoralists *see under* pastoralism
patrilineality *see under* patrilineality
Africa (*cont.*):
physical geography and geology 1.4
cape 1.5
climate 1.65
continental drift 1.135
deserts 1.31, 1.179, 1.271, 1.289
lakes 1.57, 1.203, 1.331, 1.356
map 1.4
mountains 1.23–4, 1.95, 1.135, 1.182, 1.183
plate tectonics 1.260
Red Sea 1.277–8
rivers *see under* rivers
volcano 1.99
politics, government, and foreign policy 4.194
ACP 7.103
African Charter 7.5, 7.149
Bandung Conference 4.30
Berlin Conference 4.307
Brussels Conference 4.307
corruption 7.71
coup d'état 7.73
East African Community 4.106
ECOWAS 4.108
electoral systems 7.124
military 7.201
Non-Aligned Movement 7.218
OAU 4.251, 7.5, 7.224–5
Pan-Africanism 4.257
parliamentary 7.229
see also under individual countries in particular South Africa
population *see* demography *above*
poverty 7.244
prehistory *see under* prehistory
refugees 7.260–1
religion/theology:
Christianity 7.5, 7.203
fetishism 7.119
Islam 3.4, 7.178, 7.256, 7.285, 7.302
myths 5.7, 9.21
gods 7.5, 9.26, 9.28
tricksters 9.11, 9.15, 9.22
see also in particular under Egypt; South Africa; West Africa
new 7.5
reincarnation 7.263
shamanism 7.282
Sikhism 7.286
spiritualism 7.297
traditional 7.5
slavery 3.330, 7.287
states 3.193
technology and science *see entries above and below prefixed* 1. *and* 6.
urban areas 7.298, 7.323
vehicles and transport 6.292, 6.300
wars and defence:
civil 7.52
communal conflict 7.59
expenditure 7.201
mercenaries 7.198
weapons 6.42
water supply 7.328
women's movement 7.332
see also East Africa; North Africa; South Africa; West Africa *and also individual countries*
African, Caribbean, and Pacific (ACP) countries 7.103
African Charter on Human and People's Rights 7.5, 7.149
African lily 2.190
African National Congress 4.4–5, 4.201, 4.209, 7.238
Afrikaans language 7.131, 7.156
Afrikaner 4.5
see also Boers
Afro-American music 5.9
Afro-American religion 7.32
Afro-Asiatic languages 7.30, 7.43, 7.78, 7.109, 7.278–9
AFTA (ASEAN free trade area) 7.20
afterlife 7.5–6, 7.140, 7.141, 7.217
Aga Khan 3.25, 4.5, 7.168
Agade 3.6
Agadja of Dahomey 3.96
Agamemnon 3.361, 9.2, 9.18
agamid lizards 2.4, 2.124
agar 6.141
agaric 2.4, 2.88, 2.160, 2.222
Agassiz, Louis 1.4
agate 1.4–5, 1.76
agave 2.4
age (astronomy):
of Earth 8.2
of Galaxy 8.2
of solar system 8.2
of stars 8.2
of stellar clusters 8.2
of universe 8.2
age (social sciences) 7.6
see also ageing; old age *and also* age *under* population
ageing 2.4, 7.6
disorders 2.16, 2.90, 2.242, 7.83
senescence 2.303
see also age; old age *and also* age *under* population
Agena 8.2
agenda-setting 7.6
Agent Orange 6.98
agent provocateur 7.6
Agesilaus of Sparta 3.387
agglomerate (geology) 1.5
agglomeration (megacity) 7.197
aggradation 1.5
aggregate 6.5
aggression (animals) 2.4–5
aggression (international relations) 7.6
see also wars
aggression (psychology) 7.7
Agha Muhammad 4.279
Agincourt, battle of 3.5
Agnes, Saint 9.40
Agni 9.2–3
Agnon, Shmuel Yosef 5.9–10
agnosticism 7.7
agonists and drugs 6.266
agora 3.5
agouti 2.5
AGR (advanced gas-cooled reactor) 6.3–4, 6.243–4
Agricola, Georgius 6.5, 6.227, 6.385
Agricola, Gnaeus Julius 3.5
agriculture 3.5, 4.5, 4.7, 6.7, 6.124–5, 6.126, 7.7–8
Agricultural Revolution 3.5
Anglo-Saxon 3.67
animals *see* livestock
arable *see* agrochemicals; crops
cereals *see* cereals and grains
chemicals *see* agrochemicals
collective 4.78, 7.56, 7.179–80, 7.191
communes 4.80
crop rotation 3.91–2
dairy *see* dairy farming
dry farming 6.109
economics 7.6, 7.7, 7.8
employment percentage 7.106
factory farming 6.125, 6.205, 6.279
field systems 3.126
fruit 6.76, 6.147, 6.271
green revolution 6.7, 6.161, 6.364
harvest *see* harvesting
haymaking *see* crop *under* grass
horticulture 6.170, 6.371
improvements 4.5, 4.7
intensification 6.142
Irish Land League 4.190
irrigation 6.7, 6.56, 6.95, 6.161, 6.188, 6.193, 6.378, 7.6
land *see* land
lime used in 6.202

agriculture (*cont.*):
machinery ***6**.5–6*, **6**.7
baler **6**.6, **6**.28, **6**.166
combine harvester **6**.6, **6**.81–2
cultivator **6**.93
forage harvester **6**.143
gantry harvesting **6**.150
harrow **6**.6, **6**.166
hoe **6**.169
hydraulic power **6**.173
mowing **6**.6, **6**.196, **6**.235
plough **6**.6–7, **6**.231, **6**.273–4
reaper **6**.5, **6**.172, **6**.210, **6**.295
seed drill **6**.6, **6**.311, **6**.364
threshing machine **6**.355
tractor **6**.51, **6**.103, **6**.359–60
minimum tillage **6**.19, **6**.227
Norfolk four-course system **6**.359
organic farming **6**.18, **6**.183, **6**.253, **7**.224
plantations **3**.284–5
Rebecca riots **4**.284
research and development ***6**.6–7*, **6**.80
rotation **3**.91–2, **6**.7, **6**.253, **6**.303
seeding **6**.6, **6**.227, **6**.311, **6**.364
sharecropping **4**.310, **7**.179, **7**.232, **7**.268, **7**.282
shifting *see* shifting cultivation
strips **3**.342–3
subsistence *see* peasants; shifting cultivation; subsistence farming
threshing **6**.355
tillage **6**.227, **6**.273–4
tropical **6**.363–4
see also plantation
vegetables *see* vegetables
see also agrochemicals; technology *under* food *and under individual countries*
Agrippa, Marcus Vipsanius **3**.3, ***3**.6*
Agrippina **3**.79
agrochemicals ***6**.7–8*
crop-spraying **6**.6, **6**.92
pioneers **6**.43, **6**.196
pollution **6**.7–9, **6**.377
see also fertilizers; herbicides; insecticides; pesticides
agroforestry **6**.143, **6**.364
Aguinaldo, Emilio ***4**.5*, **4**.268
Agulhas, Cape ***1**.5*
Agung of Mataram **3**.186
Agustín I, emperor of Mexico *see* Iturbide, Agustín
Ah Cheng **5**.95
Ahab, king of Israel ***3**.6*, **3**.114, **9**.26
Ahasuerus *see* Xerxes
ahimsā ***7**.8*, **7**.169
Ahlberg, Allan and Janet **5**.91
Ahmad, Muhammad *see* Muhammad Ahmad
Ahmad, Nazir **5**.467
Ahmad ibn Said of Oman **3**.262
Ahmad Shah of Afghanistan **3**.3
Ahmad Shah of Persia **4**.279
Ahmed, Khan **3**.182
Ahriman ***9**.3*, **9**.10, **9**.14
Ahura Mazda ***9**.3*, **9**.39
AI *see* artificial insemination; artificial intelligence
aid (development aid) ***7**.8–9*
Colombo Plan **7**.57
disaster relief **7**.92
food **7**.116–18, **7**.121
in South America **4**.320
from United States **4**.74, **4**.212, **4**.271, **7**.8, **7**.9, **7**.57, **7**.121
Aidan, Saint ***3**.6*
Aids (acquired immune deficiency syndrome) ***2**.5*, ***6**.8*, ***7**.9*
and contraception **6**.87
HIV virus **6**.373
and homosexuality **7**.147
incidence **7**.59
and leucocytes **2**.188
Aids (*cont.*):
transmission **7**.95, **7**.281
Aiken, Howard **6**.84
aikido **7**.194
aileron **6**.135–6
Ailey, Alvin ***5**.10*
Ainu ***7**.9*
air and atmosphere ***1**.5*, ***1**.24*, ***1**.25*, ***8**.74–5*
air masses ***1**.5*, **1**.76, **1**.216, **1**.367
air-jet weaving machine **6**.208
aurora **1**.26, **8**.12–13
bearing **6**.33
boundary layer **1**.42
circulation ***1**.63–4*, **1**.120–1
convection **1**.74, **1**.334
convergence **1**.74
Coriolis force **1**.76
divergence **1**.92–3
jet stream **1**.177–8
subsidence **1**.325
thermal **1**.334
waves **1**.362–3
see also jet stream; whirlwinds; winds
compression *see* compressor
conditioning ***6**.8*, **6**.167
dust *see under* dust
halo **1**.150, **1**.151
haze **1**.152
humidity **1**.159
insolation **1**.169, **8**.73
ionizer **6**.186
ionosphere *see* ionosphere
mesosphere **1**.211
mixing **1**.216
planetary ***8**.12*
pollution *see* pollution
pressure *see* atmospheric *under* pressure
purification **6**.120
resistance ***1**.5–6*
sky **1**.306–7
speed measurement **6**.14
stability **1**.318–19
stratosphere and stratopause **1**.109, **1**.322
streamlines **1**.323
temperature **1**.25
thermosphere **1**.334
transport ***4**.7*
air traffic control ***6**.9–10*, **6**.330
see also aircraft
tropopause **1**.345
troposphere *see* troposphere
and weather *see* meteorology
see also aerodynamics
air brush ***5**.10*
air traffic control ***6**.9–10*, **6**.330
Airborne Warning and Control System ***6**.26–7*, **6**.375
air-brake ***6**.8*, **6**.383
aircraft ***4**.6–7*, **4**.350, ***6**.5*
aerial photography/surveying **6**.215, **6**.268
aeroplane **6**.4–5
air traffic control ***6**.9–10*, **6**.330
airbus **6**.4
airforce *see* airforce
airliner ***6**.8–9*, **6**.138–9, **6**.176, **6**.385
airport ***6**.9–10*, **6**.330
airship *see* airships
altimeter **6**.12, **6**.290
autogyro **6**.25
automatic landing system **6**.25, **6**.290
automatic pilot **6**.25
aviation fuel **6**.26, **6**.91, **6**.179, **6**.257, **6**.264
aviators **6**.40
see also Wright, Wilbur and Orville
avionics **6**.26, **6**.139–40, **6**.244
AWACS **6**.26–7, **6**.375
balloons *see* balloons
aircraft (*cont.*):
biplane **6**.38, **6**.139, **6**.202
black box **6**.39, **6**.136
bomber **6**.42, **6**.129, **6**.139, **6**.176, **6**.224
bombs **6**.41–2
brakes **6**.8, **6**.44
carrier ***6**.8*, **6**.64
cellulose acetate **6**.66
compass **6**.213
computers in **6**.5
delta-wing **6**.98
designer **6**.98, **6**.176
ejector seat **6**.113
electronics **6**.4, **6**.26, **6**.136
fighters and fighter-bombers **6**.129, **6**.139, **6**.220, **6**.224
first powered flight **6**.389
flying *see* flight
flying boat **6**.8, **6**.140
fuel *see* aviation fuel *above*
gas-turbine *see* jet
glider *see* glider
helicopter **6**.12, **6**.138, **6**.167, **6**.190, **6**.225, **6**.319
hi-jacking **7**.142
human-powered **6**.138, **6**.172
and hydraulic power **6**.173
hypersonic flight **6**.175, **6**.386
jet *see* jet engines
kite **6**.191, **6**.221
landing **6**.9–10, **6**.25, **6**.290, **6**.330
lighter-than-air *see* balloons
manufacture **6**.3, **6**.98
microlight **6**.222
military *see under* military technology
navigation ***6**.237*, ***6**.239*
altimeter **6**.12
artificial horizon **6**.21
gyro-compass **6**.163
inertial **6**.178–9
pioneer **6**.65
radio *see under* radio
satellite **6**.239
and nuclear weapons **6**.246
parachute **6**.8, **6**.113, **6**.138, **6**.257
propeller-driven **6**.25, **6**.136, **6**.284
radar **6**.9–10, **6**.26–7, **6**.139, **6**.224, **6**.379–80
reconnaissance **6**.296
refuelling, in-flight **6**.179
rockets *see* rocketry
safety-belt **6**.304
seaplane **6**.138–40, **6**.311
space-plane **6**.326
sports **7**.26, **7**.132
stacking system **6**.330
STOL **6**.25, **6**.225, **6**.335
supersonic **6**.5, **6**.138–40, **6**.339–40
swing-wing **6**.342
tyre **6**.5
vertical take-off **6**.138, **6**.225, ***6**.374*
wheel **6**.5
wide-bodied **6**.385
see also flight; space-flight
airforce ***4**.6–7*
see also under World War *and under* wars *under individual countries*
airgun ***6**.8*
airliner ***6**.8–9*, **6**.138–9, **6**.176, **6**.385
airlock ***6**.9*, **6**.98, **6**.193
airport ***6**.9–10*, **6**.330
air-raid shelter ***6**.10*
airships **1**.154, **1**.161, ***6**.10*, **6**.134–5
Zeppelins **6**.10, **6**.138, **6**.139, **6**.224, **6**.391
air-to-air missiles **6**.228–9
air-to-surface missiles **6**.228–9
Airy, Sir George Biddel ***8**.2–3*, **8**.193
Aisha bint Abi Bakr ***9**.3*
Aitmatov, Chingiz **5**.462
Aix-la-Chapelle, Treaty of ***3**.6*
Aiyar, V. V. S. **5**.446
Ajaṇṭā caves ***5**.10*, **5**.196
Ajax (mythology) ***9**.3*
Akali ***7**.9–10*
Akbar, Mogul emperor ***3**.6*, **3**.196, **3**.235, **3**.268, **3**.298, **3**.328, **5**.307
Akhenaten, Pharaoh ***3**.6*, **3**.249, **9**.6, **9**.31
Akhmatova, Anna ***5**.10*, **5**.60
Akhnukh **9**.16
al-Akhṭal **5**.19
Akkad ***3**.6*, **3**.345, **3**.367
Akkadian language **7**.278
Akkamahadevi **5**.134
Aksyonov, Vasily ***5**.10*
Akutagawa Ryūnosuke **5**.235
Ala ***9**.3*
alabaster ***1**.6*, ***5**.10*, **6**.163
Aladdin ***9**.3*
Alain-Fournier, Henri ***5**.10*
Alamans **3**.98
Alamein, El, battle of ***4**.7*
Alamo ***4**.7*
Alan, Sir Hugh **4**.255
Alanbrooke, Alan, 1st viscount ***4**.7*
Åland Islands ***4**.8*
Alans people ***3**.6–7*
Alaric I, king of Visigoths ***3**.7*, **3**.144
Alas y Ureña, Leopoldo ***5**.10*
Alaska ***1**.6*
ethnic groups **7**.219
exploration **1**.184
languages **7**.108, **7**.219
map **1**.231
physical geography and geology ***1**.6*
map **1**.231
mountains **1**.201, **1**.285, **1**.289
patterned ground **1**.248
permafrost **1**.254
rock flour **1**.285
tsunami **1**.347
Purchase ***4**.8*
Ala-ud-Din, king of Delhi (1296–1316) **3**.150, **3**.195, **3**.268
Ala-ud-din Bahman Shah of India (1347) **3**.33
Alaungpaya of Burma **3**.54
Alba, Fernando, duke of ***3**.7*, **3**.108
albacore **2**.350
Alban, St **3**.223
Albania ***1**.6*, ***4**.8*, ***7**.337*
education **7**.265
king **4**.8, **4**.391
language ***7**.10*, **7**.156, **7**.157
laws/legal system **7**.282
physical geography and geology **1**.6, **1**.301
religion/theology **7**.282
in Warsaw Pact **4**.371, **7**.328
see also Balkan States
Albany Congress ***3**.7*
albatross ***2**.5*
albedo ***1**.6*, **8**.3
Albee, Edward ***5**.10*, **5**.273
Albemarle, George, 1st duke of **3**.237
Albéniz, Isaac ***5**.10*
Alberich **9**.27
Alberoni, Giulio ***3**.7*
Albert I, king of Belgium **4**.33
Albert I, king of Romans, duke of Austria (1248–1308) **3**.153, **3**.168
Albert II, king of Romans, Bohemia, and Hungary (1397–1439) **3**.153, **3**.168
Albert of Brandenburg **3**.352
Albert of Jerusalem, St **3**.61
Albert Francis Charles Augustus Emmanuel, prince consort of England ***4**.8*, **4**.140, **4**.352, **4**.365
Alberta ***1**.6–7*
Alberti, Leon Battista **3**.301, ***5**.10–11*
Alberti, Rafael ***5**.11*

Albertus Magnus **1**.7, **3**.18
Albigensians ***3**.7*, **3**.64, **3**.105
Albinoni, Tomaso Giovanni ***5**.11*
Albion ***3**.7*
Albrecht of Brandenburg-Ansbach **3**.294
albumen **6**.3
albumin **6**.40
Albuquerque, Alfonso de ***3**.7*, **3**.143, **3**.217
Alcaeus **5**.191
Alcázar ***3**.7–8*
Alcazarquivir ***3**.8*
Alcestis ***9**.3*
alchemy ***1**.7*, ***6**.10*
Alchera ***9**.3*
Alcibiades ***3**.8*, **3**.275
Alcock, John **4**.350, **6**.138
alcohol ***1**.7*, ***6**.10*
abuse ***2**.5–6*, **2**.66, ***7**.10*
agave **2**.4
breathalyser **6**.46
drink ***6**.10–11*, **6**.37, **6**.46, **6**.68, **6**.386, **7**.10
fermentation **2**.115
gin **2**.33
glycerol **6**.157
methyl (methanol) **1**.213, **6**.222, **6**.261
molasses **6**.338
in perfume **6**.260
production **6**.37, **6**.127, **6**.253
Prohibition era **4**.277, **4**.369, **7**.10
from sap **2**.294
in thermometer **6**.354
whiskey **2**.198
see also brewing
alcoholism ***2**.5–6*
Alcott, Louisa May ***5**.11*, **5**.91
Alcuin ***3**.8*
Aldebaran ***8**.3*
aldehyde ***1**.7*, **1**.212–13, **6**.252
alder ***2**.6*
alder fly ***2**.6*
alderman ***3**.8*
Aldington, Richard **5**.217
Aldrin, Edwin 'Buzz' **4**.322–3, **6**.17, **8**.172
ale *see* brewing
aleatory music ***5**.12*
Aleichem, Sholem **5**.497
Alemán, Mateo **5**.428
Alemán, Miguel **4**.220
ALEPH **6**.117
Alessandri, Arturo ***4**.8*
Aleutian Islands ***1**.7*, **7**.108
Aleut language **7**.108
Alexadrou, Aris **5**.191
Alexander I, emperor of Russia (1777–1825) ***4**.8*
advisers **4**.91, **4**.326
and Austerlitz **3**.23
and Congress of Vienna **4**.366
and Finland **4**.120
and Georgia **3**.140
and Poland **4**.271
and Tilsit treaties **4**.345
Alexander I, king of Scotland (1078–1124) ***3**.8*
Alexander I, king of Yugoslavia (1888–1934) ***4**.8*, **4**.178, **4**.387
Alexander II, emperor of Russia (1818–81) ***4**.8*, **4**.88, **4**.243, **4**.271, **4**.275
Alexander II, king of Scotland (1198–1249) ***3**.8*
Alexander III (the Great), king of Macedonia (356–323 BC) ***3**.8*
army **3**.17, **3**.22, **3**.23, **3**.278
conquests **3**.3, ***3**.9*, **3**.33, **3**.60, **3**.76, **3**.117, **3**.177, **3**.215, **3**.276–7, **3**.302, **3**.316–17, **3**.350, **3**.364
as god **9**.3
Hellenistic civilization **3**.160
Alexander III (the Great) *(cont.)*:
and Islamic myth ***9**.3*, **9**.38
and science **9**.16
tutor **3**.21
Alexander III, emperor of Russia (1845–94) ***4**.9*
Alexander III, pope **3**.132, **3**.161
Alexander III, king of Scotland (1241–86) **3**.8
Alexander VI, pope **3**.317, **3**.359
Alexander, prince of Battenburg **4**.54
Alexander, king of Serbia **4**.249, **4**.309
Alexander, William **3**.297
Alexander Nevsky ***3**.8*
Alexander of Tunis, Harold Rupert Leofric George Alexander, 1st earl ***4**.9*, ***4**.55*
Alexandria ***3**.8–9*, **3**.26
alexandrine ***5**.12*
Alexis Romanov of Russia **3**.309
alfalfa ***2**.6*
Alfieri, Vittorio, count ***5**.12*
alfisol ***1**.7*, **1**.311
Alfonsin, Raul **4**.18
Alfonso I, king of Portugal **3**.289
Alfonso II, king of León **3**.10, **3**.113
Alfonso V (the Magnanimous), king of Aragon ***3**.9*
Alfonso VI, VII, and X, kings of León **3**.62
Alfonso X (the Wise), king of Aragon ***3**.9–10*
Alfonso XII, king of Spain **3**.46, **4**.62
Alfonso XIII, king of Spain **3**.46, **4**.62, **4**.124
Alfred (the Great), king of Wessex ***3**.10*, **3**.15, **3**.118, **3**.380
and fleet **3**.247
and fyrd **3**.136
and learning **3**.43, **3**.97, **3**.231
and shires **3**.327
Alfvén, Hannes Olaf Gösta: theory ***8**.3*, **8**.90
algae ***2**.6*
algal bloom **2**.6
blue-green *see* cyanobacteria
and chlorophyll **2**.64
colony or mat **2**.327–8
Euglena **2**.111
fucus **2**.128, **2**.129
kelp **2**.178
in lichen **2**.188, **2**.189
plankton **2**.15, **2**.257, **2**.258
seaweeds **2**.184, **2**.302
Spirogyra **2**.321, **2**.322
Algardi, Alessandro ***5**.12*
algebra ***1**.7–8*
binomial expansion **1**.39
Boolean **1**.41, **1**.98, **6**.207
differential equations **1**.90
equations **1**.90, **1**.110
hyperbola **1**.72, **1**.73, **1**.163
polynomial **1**.264
Algeciras Conference ***4**.9*
Algeria ***1**.8*, ***3**.10*, ***4**.9*, ***7**.338*
in Arab League **7**.15
colonialism **4**.306
economy **4**.252, **7**.315
independence **4**.4, **4**.114, **4**.127–8
industrial relations **7**.315
languages **7**.30
laws/legal system **7**.221, **7**.284
leaders **4**.34, **4**.47
old age **7**.221
physical geography and geology **1**.8, **1**.65, **1**.89
religion/theology **7**.284
technology **6**.236
alginate ***6**.11*
ALGOL (computers) **6**.86
Algol (star) ***8**.3*
Algonkian language **7**.219
Algonquin Indians ***3**.10*, **3**.289–90
Algonquin Provincial Park ***1**.8*
algorithm ***1**.8*, ***6**.11*
Alhazen **6**.55
'Alī, caliph of Islam **3**.58, **3**.326, **3**.365, ***7**.10*, **7**.39, **7**.138, **7**.153, **7**.166, **7**.284, **7**.302
Ali Baba ***9**.3*
'Alī ibn Abi Talib ***9**.3*
Ali, Muhammad and Shaukat (India) **4**.182
Ali Muhammad, Sayyid of Shira **4**.27
Ali Pasha, Mehmed Emin **4**.8, ***4**.10*, **4**.140
'Alī Shīr Nevā'ī **5**.462
alicyclic compound ***1**.8*
alienation ***7**.10*, **7**.194
see also anomie
alimentary canal ***2**.6*, ***9**.59*
see also gastro-intestinal tract
alimony **7**.189
aliphatic compounds ***1**.8*, **1**.9
alizarin **6**.260
alkali/alkalinity ***1**.8*, ***6**.11*, **6**.96–7
hydroxide ion **1**.163
indicator **1**.197
soil ***1**.9*
see also alkali metals; alkaline earth metals; lime; sodium carbonate; sodium hydroxide
alkali metals ***1**.8*, ***6**.11*
caesium **1**.47, **1**.106
francium **1**.106, **1**.126–7
lithium **1**.106, **1**.196
rubidium **1**.107, **1**.287, **6**.96
see also potassium; sodium
alkaline earth metals ***1**.8–9*
strontium **1**.107, **1**.324
see also barium; beryllium; calcium; magnesium; radium
alkaloid ***2**.6*
alkanes ***1**.9*, **6**.252
see also butane; propane
alkanol *see* alcohol
alkene ***1**.9*, **6**.252, **6**.261
alkyl group ***1**.9*, **6**.252
alkylation **6**.265
alkyne ***1**.9*, **6**.252
Allah **7**.10, **7**.133, **7**.166, **7**.209
Allam II, Shah **3**.80
Allegheny Mountains ***1**.9*
allegory ***5**.12*, **5**.46, **5**.339, **5**.359, **5**.457
music **5**.138
poetry **5**.25, **5**.187, **5**.255, **5**.387–8, **5**.391, **5**.428
sculpture **5**.483
see also morality plays
alleles ***2**.6*, **2**.223–4, **2**.252
allelopathy ***2**.6*
allemande ***5**.12*
Allen, Bryan **6**.138, **6**.172
Allen, Ethan, Ira, and Levi ***3**.10*, **3**.357, **3**.372
Allen, Woody ***5**.12*
Allenby, Edmund Henry Hynmam, 1st viscount ***4**.10*, **4**.218
Allende, Salvador ***4**.10*, **4**.72
allergies ***2**.6–7*, **2**.153, **2**.353, **6**.16
alliances ***7**.10*, **7**.315
see also international organizations and alliances
alligator ***2**.7*
alligator bug **2**.182, **2**.183
alliteration ***5**.12–13*, **5**.294, **5**.326, **5**.486
allopathy ***2**.7*
allotropes ***1**.9*
alloys ***6**.11–12*, **6**.20
aluminium **1**.11
antimony **6**.265
arsenic **6**.20
boron **6**.43
brass *see* brass
bronze *see* bronze
carbon **6**.11, **6**.185
ceramic **6**.318
chrome **6**.73, **6**.354
chromium *see under* chromium
alloys *(cont.)*:
cobalt **6**.212, **6**.241
and composites **6**.82
copper *see under* copper
invar **6**.18, **6**.185, **6**.241, **6**.355
iridium and platinum **6**.186
iron *see under* iron
lead **6**.265
list of major **6**.11
lithium **6**.12
magnesium **6**.11–12, **6**.212–13, **6**.213
magnetic **6**.170, **6**.212
manganese *see under* manganese
memory metal **6**.219
molybdenum **6**.11, **6**.232, **6**.241, **6**.333
nickel *see under* nickel
permalloy **6**.11, **6**.241
pewter **5**.346, **6**.265
platinum **6**.186, **6**.272
rhenium **6**.241
silicon **6**.11, **6**.185, **6**.212, **6**.333
silver **6**.12, **6**.320
and soldering **6**.324–5
steel *see under* iron
superplastic **6**.339
tantalum **6**.241, **6**.345
tempering **6**.351
tin **6**.11–12, **6**.49, **6**.265, **6**.356
titanium *see under* titanium
tungsten **6**.11, **6**.235, **6**.333, **6**.365
vanadium **6**.370
zinc **6**.11, **6**.45, **6**.49, **6**.391
zirconium **6**.11–12
allspice **2**.254
Allston, Washington ***5**.13*
all-terrain vehicle **6**.209–10
alluvium ***1**.9–10*
alluvial fan ***1**.9*
floodplain **1**.123–4
soils (meadow) **1**.208–9, **1**.311
allyl plastics **6**.272
Alma Ata agreement **7**.59
Almagest, see under Ptolemy
Almagro, Diego de **3**.283
almanac ***8**.4*
Alma-Tadema, Sir Lawrence ***5**.13*
Almeida, Francisco d' **3**.236
Almohad dynasty **3**.4, ***3**.10*
almond ***2**.7*
Almoravid dynasty **3**.4, ***3**.10–11*
Almqvist, Carl Jonas Love **5**.404
almshouse ***3**.11*
aloe ***2**.7*
Alonso, Alicia **5**.30
Alonso, Dámaso **5**.173
Alonso de Orozco **5**.311
Alp Arslan **3**.321
alpaca ***2**.7*
Alpha Centauri ***8**.4*
alpha particle ***1**.10*
alpha receptor **6**.266
alpha wave **6**.116
alphabets ***7**.10–11*, ***9**.46–8*
Braille **6**.44, **7**.35, **9**.48
cipher **7**.50
Cyrillic *see* Cyrillic
phonetic **7**.235
see also writing and scripts
alpha-radiation **6**.153, **6**.288
Alpher, Ralph **8**.98
alphorn ***5**.13*
alpine plants ***2**.7*
edelweiss **2**.104
gentians **2**.59, **2**.134
saxifrage **2**.294, **2**.295
Alps ***1**.10*, **1**.79, **1**.135, **1**.207, **1**.220
Alsace-Lorraine ***4**.10*
Alston, Richard ***5**.13*, **5**.299
Altaic languages ***7**.11*
Altair ***8**.4*
Altan Khan **7**.79
alt-azimuth system **8**.22, **8**.104
Altdorfer, Albrecht ***5**.13*, **5**.254
Alter, David **8**.147

alternating-current (a.c.) 1.102, 6.114–15, 6.153–4, 6.253
induction motor 6.351
radio-frequency heating 6.290
reactance 6.295
rectified 6.323
transformer 6.361
see also circuits *and* generation *under* electricity
alternation of generations *see* cycle *under* life
alternative energy 1.109
see also biofuel; geothermal energy; solar power *and also* energy and power *under* tides/tidal *and* waves (ocean)
alternative medicine *see* complementary medicine
alternative technology *see* appropriate technology
alternator 6.153
Altgeld, John P. 4.148
Althusser, Louis 7.11
altimeter 6.12, 6.290
altitude:
of celestial object 8.22
and climate (orographic effect) 1.241, 1.274–5
contour 1.74
datum-plane 1.84
measurement 6.12, 6.290
artificial horizon 6.21
astrolabe 6.23
cross-staff 6.92
quadrant 6.287
sextant 6.294, 6.315
snow-line 1.309
and temperature (lapse rate) 1.188
Altjiranga 9.3
altocumulus and altostratus clouds 1.66, 1.128
altruism 7.11
alum 1.10, 6.12, 6.211, 6.321
alumina 1.10–11, 6.1, 6.12, 6.82, 6.318
aluminium 1.10, 1.106, 6.12
alloy 1.11
anodized 6.15
cable 6.53
canning 6.57
dust 6.270
in glass ceramic 6.156
–lithium alloys 6.11, 6.12
ore *see* bauxite
as pollutant 6.377
production *see under* electrolysis
scaffolding 6.310
in thermite 6.383
in thermocouple 6.354
wheel 6.384
aluminium chloride 6.12
aluminium oxides 6.32, 6.90, 6.297
corundum 1.77
see also alumina; bauxite
aluminium phosphate 1.348
aluminium silicates and aluminosilicates
beryl 1.38, 1.108
china clay 1.61, 6.276, 6.382
jade 1.177
staurolite 1.320
topaz 1.339
zeolite 1.373
see also feldspar
aluminium sulphate 1.10
Alva, duke of 3.7
Alvarado, Pedro de 3.114
Alwyn, Patricio 4.73
Alyasa 9.11
Alzheimer's disease 6.377, 7.83
AM (amplitude modulation) 6.65, 6.231, 6.289–90
AM Herculis stars 8.4
Amadeus I, king of Spain 4.168
Amalfi, doge of 3.104
amalgam 1.11, 6.12
Amalthea 8.4
Amar Das, Guru 5.418, 7.137, 7.285
Amara Dunkas of Sudan 3.136
Amāravatī sculpture 5.13
Amaru 5.401
amaryllis 2.7–8
Amaterasu Omikami 9.3–4, 9.19, 9.20, 9.22, 9.34
Amati, Andrea, Girolamo, and Nicolò 5.477
Amazon 1.11
deforestation 7.83
Indians 7.11–12, 7.335
Amazons (women) 9.4
Ambartzumyan, Viktor Amazaspovich 8.4
Ambedkar, Bhimrao Ramji 4.10
amber 1.11
ornaments 5.13
ambergris 6.328
Amboina 3.11
Ambrose, Saint 3.11, 3.28, 3.231, 9.40
ambrosia beetle 2.8
ambulance 6.12
Amenhotep III, pharaoh 3.352
America (before 1783) 3.44, 3.314
after 1783 *see* United States of America
arts 5.13–14
colonies 3.258, 3.344, 8.213
Pilgrim Fathers and Plymouth 3.48, 3.62, 3.282, 3.286
Proclamation Line 3.293, 3.354
southern 3.331
Stamp Act 3.339
Thirteen 3.100–1, 3.140, 3.224, 3.252–3, 3.256, 3.275, 3.334–5, 3.344, 3.352, 3.354, 3.355, 3.374
first 3.305
promoter 3.331
see also New England
discovery and exploration of 3.82, 3.122, 3.185, 3.203, 3.372, 3.381
fiscal innovator 3.154–5
immigration *see* colonies; discovery *above*
laws and legal system 3.80, 3.87, 3.264
maps 3.252, 3.257, 3.323, 3.354
plantations 3.284–5
politics and government:
Continental Congress 3.87
Culpeper's Rebellion 3.93
Mason–Dixon line 3.224
patriots 3.10, 3.79, 3.163, 3.302
political party 3.124
statesmen and political leaders 3.40, 3.186
United Empire Loyalists 3.365
prehistory 3.80, 3.102, 3.170, 3.335
religion/theology 3.112, 3.145, 3.148, 3.225, 3.275
states *see* colonies *above and also under* United States
wars and defence 3.306
American Revolution *see* Independence, American War of
army leaders 3.23, 3.75, 3.137–8, 3.148, 3.199, 3.295
see also leaders *under* American Civil War; Independence, American War of
Boston Tea Party and Massacre 3.46–7, 3.156
Green Mountain Boys 3.148
navy leader 3.191
Paxton Boys rebellion 3.272–3
Queen Anne's 3.134, 3.297
Seven Years 3.323
America (before 1783) (*cont.*):
wars and defence (*cont.*):
uprisings and rebellions 3.33, 3.205, 3.300, 3.325–6, 3.333, 3.359
American Association for the Advancement of Science 6.12–13
American Ballet Theatre 5.14, 5.274, 5.461
American chameleon 2.11
American Civil War 4.10–11
Appomattox surrender 4.16
battles and campaigns:
land 4.15, 4.71, 4.122, 4.135–6, 4.364, 4.368, 7.52
naval 4.222–3
casualties 4.101
Confederacy 4.82, 4.331
Hampton Roads Conference 4.146
leaders 4.139, 4.171, 4.192, 4.203, 4.311
prisoners-of-war 4.12
Reconstruction Acts 4.284
Trent affair 4.352
American Colonization Society 4.10
American Convention on Human Rights 7.12
American elk 2.360
American Federation of Labor 4.10–11
–Congress of Industrial Organizations 7.314
American Indians *see* Indians, North American; Indians, South American (and Central American)
American Philosophical Society 3.11
American Scene Painting 5.14
American Standard Code for Information Interchange 6.22
American System 4.11
American War of Independence *see* Independence, American War of
Americas *see* America; Canada; Central America; Pan-Americanism; South America; United States
americium 1.106
Amerindians *see* Indians, North American; Indians, South American (and Central American)
Amesha Spentas 9.3, 9.10
amethyst 1.11
Amharic language 7.279
Amherst, Jeffery, baron 3.11, 3.357
Amichai, Yehuda 5.203
Amida-nyorai 9.4
Amin, Hafijullah 4.3
Amin, Idi 4.11, 4.145, 4.356
amine 1.11, 6.252
amino acids 2.8, 6.140, 6.252
see also proteins
amino adhesives 6.3
aminobenzene *see* aniline
Amis, Kingsley 5.15
Amish Church 7.198
Amitabha 9.4, 9.22
Amma 9.28
ammeter 6.13, 6.149
ammonia 1.11, 6.13
and chemical industry 6.71, 6.262
and fertilizer 6.128, 6.338
liquefaction 6.150
and nitric acid 6.241
production 6.70
see also Haber–Bosch *under* Haber
refrigerant 6.297
salts from 6.2
ammonite 2.8
ammonium nitrate 6.13, 6.124, 6.241
ammonium phosphates 6.2, 6.141
ammonium sulphate 1.326, 6.2, 6.13, 6.338
ammunition:
bullet 6.52, 6.131, 6.272
shell 6.315–16
shrapnel 6.318
see also weapons
amnesia 7.197–8
amnesty 7.12
Amnesty International 7.12, 7.218, 7.239, 7.313
amniocentesis 2.8, 6.13, 6.218
amoeba 2.8
amoebic dysentery 7.59
Amor (asteroid) 8.4
Amor group of asteroids 8.4, 8.48
amorphous carbon 6.270
amortization 7.12
Amos 9.4, 9.20
Amour, Salanin 4.388
Ampère, André-Marie 1.11, 1.12, 6.115, 6.117
ampere 1.11, 6.75, 6.321, 6.368
amphetamines 7.95
amphibians 2.8
brain 2.41
endangered species 2.108
limbless 2.49
tailed:
newt 2.227
siren 2.310
see also salamander
tail-less *see* frogs; toads
amphibious vehicle 6.13
amphiboles 1.11–12
hornblende 1.158
see also asbestos
amphioxus 2.182
amphipods 2.8–9, 2.293
amphisbaenid 2.9
amphitheatre 5.15
amphora 3.11–12
amphoteric compound 1.12
amplifier 6.13–14, 6.215, 6.289
amplitude 1.12
modulation *see* AM
Amritsar massacre 4.11
Amu Darya 1.12
Amubi Singh 5.279
amulets 5.15–16, 7.5, 7.14, 7.112
Amun 9.4
Amundsen, Roald 1.12, 1.231, 1.295, 4.15, 4.17
Amur 1.12
amyl nitrate 6.147
amylase 6.122
An 9.4, 9.11, 9.14
Anabaptism 3.12
anabatic wind 1.12
anabolic reactions 6.140
anabolic steroids *see* steroids
Anacletus II, antipope 3.204
anaconda 2.9
Anacreon 5.191
anaemia 2.9, 2.161, 6.164, 6.186
anaerobic adhesives 6.3
anaesthesia 2.9, 6.13, 6.14, 6.266
acupuncture 6.3
barbiturate 6.30
chloroform 6.14, 6.72–3, 6.218, 6.320
curare 6.94
in dentistry 6.99
ether 6.14, 6.218, 6.233, 6.252, 6.320
local 6.192
in ophthalmology 6.250
pioneers 6.218, 6.233, 6.320
respirator 6.298
see also surgery
analgesia 2.235, 6.14, 6.236, 6.248, 6.265
analog 6.104
computers 6.14, 6.320
-to-digital converter 6.319
analysis (chemistry) 1.12–13

analytic philosophy 7.12, 7.68, 7.173, 7.185
see also Russell, Bertrand; Wittgenstein
analytical engine 6.27, 6.84, 6.189
analytical psychology 7.173
anamnesis 7.237
anamorphosis 5.16
Anand, Mulk Raj 5.16
ānandalahārī 5.353
Anansi 9.4
Anantavarma Codagangadeva 3.263
anarchism 4.11–12, 7.12–13
see also terrorism and anarchism
Anasazi Indians 3.12, 3.79
Anatolia 3.23
languages 7.13, 7.156, 7.157
see also Asia Minor; Turkey
anatomy 2.9–10
anatomists *see* physiologists *under* physiology
cytology 2.86
histology 2.29, 2.158
morphology 2.9, 2.219
see also physiology
Anawrahta, King 5.67
Anaximander 8.61
ancestor worship 7.5, 7.9, 7.13, 7.30, 7.47, 7.197, 7.199, 7.295
see also spirits
anchor 6.14
anchovy 2.10
ancien régime 3.12, 7.213
al-Andalus 3.335
Andalusia 1.13, 3.12
Andean Common Market 4.12
Andean culture 5.425–6
Andean Group 7.13
Andersch, Alfred 5.175
Andersen, Hans Christian 5.16
Anderson, Carl 1.155
Anderson, Elizabeth Garrett 4.12
Anderson, Lindsay 5.163
Anderson, Maxwell 5.215
Anderson, Robert 4.122
Andersonville Prison 4.12
Andes 1.13
erosion 1.310
formation 1.135
lake 1.338
Tierra del Fuego 1.337
volcano 1.78
andesite 1.13
Andorra 1.14, 7.269
andosol 1.14
Andover scandal 4.12
Andrada e Silva, José Bonifacio 4.12
Andrassy, Julius, count 4.12
Andre, Carl 5.297
André, John 3.23
Andrea del Sarto 5.16
Andreossy, François 6.299
Andrew, Saint 9.4, 9.5, 9.40
Andrewes, Lancelot 3.12, 5.411
Andrianampoinimerina, king of Madagascar 3.215
Andrić, Ivo 5.16
Andriscus 3.215
Andromeda (constellation) 8.4, 8.5
Andromeda Galaxy 8.5
Andromeda (mythology) 9.4
Andromedids 8.5, 8.194
Andropov, Yuri 4.181
Andrzejewski, Jerzy 5.16
anemometer 6.14
anemone (flower) 2.10
anemone, sea 2.297, 2.298
aneurysm 2.10
ANFOs (explosives) 6.124
Angad Das 5.418
Angel Falls 1.14
angelfish 2.10
angelica 2.10
Angelico, Fra 5.16
Angelou, Maya 5.16–17
angels 9.4, 9.6, 9.8–9, 9.18, 9.26
archangels 9.5, 9.6, 9.13, 9.18, 9.25
fallen 9.10, 9.14, 9.15, 9.17, 9.23, 9.32
angelshark 2.217–18
Angevins 3.12–13, 3.89, 3.161, 3.284, 3.302–3
see also Henry II; John; Richard II
angina pectoris 2.10
angiosperms 2.10
flower 2.122
latex from 2.183–4
see also annual plants; deciduous trees and shrubs; Dicotyledons; evergreen plants; grass; Monocotyledons; perennial plants; shrubs
Angkor and Angkor Wat 3.13, 5.247–8
angle 1.14
in astronomy:
angular diameter 8.5
angular measure 8.5, 8.192
angular momentum 8.5–6
angular velocity 8.6
of elongation 8.46
of dip 1.92
of incidence and reflection 1.278
measurement 6.157
see also trigonometry
anglerfish 2.10
Angles (people) 3.13, 3.98
Anglicans 3.13–14, 4.12–13, 7.13
canon law 7.39
church building 7.50
and homosexuality 7.146
leaders 7.48, 7.49
monasticism 7.205
parson 3.271
printing 3.66
recusants 3.300
sacraments 7.271
and Scots, attempt to impose on 3.41–2, 3.89–90
Thirty-Nine Articles 3.355
Uniformity, Acts of 3.365
angling *see* fishing
Anglo-Afghan Wars 4.13
Anglo-Australian telescope 8.150
Anglo-Burmese Wars 4.13
Anglo-Dutch Wars 3.14, 6.375
Anglo-Egyptian Sudan *see* Sudan
Anglo-Frisian languages 7.156
Anglo-Japanese Alliance 4.13
Anglo-Maori Wars 4.13–14, 4.294, 4.338
Anglo-Saxons and Saxons 3.15
agriculture 3.67
Anglo-Saxon Chronicle 3.14–15
arts and literature 5.17, 5.326
British resistance to 3.242
Hengist and Horsa 3.160
invasion of Roman empire 3.98
kingdoms *see* Mercia; Wessex
military force 3.136
nobleman 3.352
politics and government 3.8, 3.384
resistance to Normans 3.164
site 3.346
wergild 3.379
see also Saxon *under* rulers *under* Britain
Angola 1.14, 4.14, 7.338
colonized and decolonized 4.4, 4.306
economy and trade 7.296
languages 7.176, 7.269
physical geography and geology 1.14, 1.237, 1.372
refugees 7.261
wars and defence 7.52, 7.59, 7.232
Angoulême line 3.81
listed 9.68
Angoulême line 3.81
Angoulême line (*cont.*):
see also Francis I; Henry II; Henry III
Angra Mainyu 9.2
Angry Young Men 5.15, 5.17, 5.333
angst 7.14
Angström, Anders Jonas: angström 1.14
Anguilla 1.14, 4.14
angular diameter 8.5
angular measure 8.5
angular momentum 1.14
angular velocity 1.14
anhinga 2.87
anhydride, acid 1.2
anhydrite 1.14
anhydrous compound 1.14
aniline and aniline dye 6.14, 6.27, 6.112
animals 2.10–11
as heterotrophs 2.157
aggression 2.4–5
altruism 7.11
aquatic *see* fish; whales *and under* water
behaviour 7.287, 7.294
biogeography 2.30
breeding *see* breeding
camouflage and disguise *see* protective coloration
carbon cycle 2.53
carnivorous *see* carnivores
cell 6.65
commensalism 2.73
conservation 2.75, 2.254, 7.212
diseases *see* diseases and disorders of animals
domestic *see* domesticated animals
ecology *see* ecology
endangered species 2.107, 2.108, 2.241, 7.106
ethology 7.109
evolution *see* evolution; natural selection
extinct *see under* extinct species
farm *see* livestock
fibre 6.128, 6.231
see also wool
food *see* feedstuffs, animal
fossil *see under* fossils
glue from 6.3
herbivorous 2.155
husbandry *see* livestock
imprinting 7.154
learning 7.110
leather 6.197, 6.281, 6.345
mammals *see* mammals
migration 2.210
motivation 7.207
mythical 9.21–2, 9.37
beetle 9.31
birds 9.29, 9.33
bulls and cow 9.4–5, 9.6, 9.14, 9.25
cat 9.7
dogs and wolf 9.4, 9.8, 9.38
dragons 9.10, 9.23, 9.30
goats 9.12, 9.32
griffin 9.15
horses 9.8, 9.29, 9.34
mermaid 9.25
monkeys 9.15
reptile 9.23
serpents 9.30
spider 9.4
unicorn 9.36
vampire 9.37
see also monsters
oxygen cycle 2.238
pest control by 6.37
pheromones 2.252
physiology *see* physiology of living organisms
plankton 1.258, 2.15, 2.257, 2.258
protected species 2.270
reproduction *see* reproduction
animals (*cont.*):
rights 7.13
senses *see* senses
skin 6.258
social 2.316–17
and sport 7.37–8, 7.54, 7.129, 7.136, 7.150
see also under horses
temperature 2.337
totem 7.313
veterinary medicine 6.371
welfare groups 6.279
zoology *see* zoology
zoos 2.375, 7.336
see also individual animals
animated model 6.25, 6.84, 6.168
animation 5.17, 5.131
animism 7.11, 7.14, 7.16, 7.25
see also spirits
anion 1.14
anise 2.11
Anjou 3.15–16
annals 3.16
Annam 3.16, 4.14
Annandale, Robert Bruce, earl of 3.34
Annapurna 1.14
annatto 2.101
Anne, queen of Great Britain and Ireland 3.16, 3.56, 3.143, 3.322, 3.343, 5.371
Queen Anne style 5.371–2, 5.415
Queen Anne's War 3.134, 3.297, 3.338
Anne of Austria 3.16
Anne of Cleves 3.16
annealing 6.15, 6.272, 6.304, 6.388
annelids 2.11
bristleworms 2.42
earthworm 2.102–3
fanworm 2.113–14
leech 2.186
scale worm 2.295
sea mouse 2.300–1
see also worms
annihilation of matter 1.14
annual plants 2.11
anise 2.11
basil 2.24
brassica *see* brassica
broomrape 2.43
bryony 2.43
candytuft 2.52
cannabis 2.52, 2.154
chick-pea 2.63
chrysanthemum 2.65
clover 2.69
corn salad 2.77
cranesbill 2.80
delphinium 2.90
fenugreek 2.115
forget-me-not 2.125
fumitory 2.129
godetia 2.140
groundsel 2.145
kenaf 2.178
knotweed 2.180
lobelia 2.192
lupin 2.196
mallow 2.200
melon 2.207
mesembryanthemum 2.208
mustard 2.223, 2.299
nasturtium 2.225
nemesia 2.226
Nicotiana 2.227–8
okra 2.233–4
pea *see* pea family
peanut 2.245, 2.246
petunia 2.250
phlox 2.251
pimpernel 2.254
plantain *see* plantains
rose of Jericho 2.288–9
safflower 2.291
salpiglossis 2.292

annual plants (*cont.*):
salvia **2.**292–3
scabious **2.**295
schizanthus **2.**297
sunflowers *see* sunflower family
sweet pea **2.**332
teasel **2.**337
thistle **2.**339
tomato **2.**343
watermelon **2.**362
see also cereals; grass
annuity ***7.14***
annular solar eclipse **8.**42
anoa **2.**44
anode ***1.15***, ***6.15***, **6.**197
in photomultiplier **6.**269
in polarography **6.**274
in thermionic converter **6.**354
see also battery; Castner–Kellner; cathode-ray tube; circuits; electrode; thermionic valve
anodizing ***6.15***
anole ***2.11***
anomalistic month **8.**100
anomalistic year **8.**189
anomaly ***8.6***
anomie ***7.14***, **7.**97
see also alienation
anorexia nervosa ***2.11***, **7.**99
Anouilh, Jean ***5.17–18***, **5.**61
Anschluss ***4.14***, **4.**24
Anschütz-Kaempfe, Hermann **6.**163, **6.**328
Anse aux Meadows, L' ***3.203***, **3.**373
Anselm, Saint ***3.16–17***, **7.**223, **7.**309, ***9.40***
Ansky **5.**197
Anson, George, Baron Anson ***3.17***
answering machine ***6.15***
ant ***2.2***, **2.**2, **2.**11, **2.**12–13
antagonists and drugs **6.**266
Antall, Jozsef **7.**104
antapex **8.**6
Antar ***9.4***
'Antara **5.**19
Antarctic Ocean **1.**314
Antarctica ***1.15***, ***4.14–15***, ***7.338***
Antarctic Treaty **7.**91
exploration *see under* exploration
map **1.**15
ozone layer **6.**254
physical geography and geology ***1.15***, **1.**314
climate **1.**65
ice **1.**164, **1.**244, **1.**286, **1.**287, **1.**296
map **1.**15
mountains **1.**111, **1.**112
volcanoes **1.**112
pollution **6.**9
Antares ***8.6***
antbird ***2.11***
ant-eaters ***2.12***, **2.**103, **2.**277
antelopes ***2.11–12***
duiker **2.**100
eland **2.**104
gazelles **2.**133–4, **2.**323
gnu (wildebeest) **2.**138–9
goral **2.**141
hartebeest **2.**151
klipspringer **2.**180
kudu **2.**181
nilgai **2.**228
oryx **2.**236–7
sable **2.**291
sitatunga **2.**311
springbok **2.**323
takin **2.**333
see also goat family
antenna ***6.15***, **6.**45, **6.**290
radio and television **6.**291, **6.**375, **6.**390
satellite dish **6.**308
telescope **6.**347
anthem ***5.18***
anthology ***5.18***
Anthony of Egypt, Saint ***9.40***
Anthony of Padua, Saint ***9.40***
anthracene oil **6.**79
anthracite ***1.15***, **1.**67, **6.**78
strike ***4.15***
anthrax ***2.12***, **2.**180, **2.**243
anthropology ***7.14***
'armchair' **7.**124
descent **7.**86
ethnography **7.**109, **7.**112, **7.**190
feud **7.**119–20
fieldwork **7.**14, **7.**120, **7.**180
gifts *see* gifts
magic *see* magic
meaning **7.**196
myth *see* myths and legends
and psychology **7.**77
relativism **7.**263
structuralist **7.**183
symbolism **7.**303
taboo *see* taboos
totems *see* totemism
see also in particular Malinowski; Mead *and also* ethnic groups *under individual countries*
anthroposophy **7.**299
anti-ballistic missiles ***6.15***, **6.**231, **6.**335–6
antibiotics ***2.12***, ***6.16***, **6.**27, **6.**72, **6.**254, **6.**340, **6.**375
see also penicillin
antibodies ***2.12***, **2.**169–70, **2.**188, **2.**212, **6.**57, **6.**153, **6.**177, **6.**232, **6.**290–1
see also immune system
anticlericalism ***3.17***
anticline ***1.15***, **1.**125
anticoagulant **2.**186, ***6.16***
Anti-Comintern Pact ***4.15***
Anti-Corn Law League ***4.15***
anticyclone ***1.15***
anti-depressants **6.**285, **7.**96
Antietam (Sharpsburg), battle of ***4.15***
antifreeze ***1.15***, ***6.16***, **6.**222
antigens **2.**169, **6.**177
Antigone ***9.4***
Antigonus I, king of Macedonia ***3.17***
Antigonus II, king of Macedonia **3.**215
Antigua and Barbuda ***1.15***, ***7.338–9***
see also Caribbean; CARICOM
antiknock ***1.15***
Antilles **1.**52
Anti-Masonic Party ***4.15***
antimatter and antiparticles **1.**14, ***1.15–16***
antimony ***1.16***, **1.**106, ***6.16***, **6.**265
antimony sulphate **6.**100
antimony trisulphide **6.**16
antinode ***1.16***
Antioch, prince of ***3.17***, **3.**44
Antiochus III (the Great), king of Mesopotamia **3.**33, **3.**321
Antiochus III, king of Syria **3.**214
Antiochus Epiphanes **7.**138
antioxidants ***1.16***, **6.**16, **6.**140–1, **6.**147
antiparticle *see* antimatter
antipodes ***1.16***
Antipodes Islands ***1.16***
antipopes ***3.17***, **3.**131, **3.**133, **3.**204, **4.**175, **4.**291
list of ***9.71***
see also Innocent III
antirrhinum ***2.12***
anti-semitism ***4.15***
see also under Jews and Judaism
antisepsis ***6.16***, **6.**72, **6.**218, **6.**267, **6.**340
pioneers **2.**291, **6.**204, **6.**259
Antisthenes of Athens **3.**94, **3.**104
anti-tank gun ***6.16***, **6.**131
Anti-Trust laws ***4.15–16***
antlers ***2.12***
Antlia ***8.6***
ant-lion ***2.12***
Antoine de Bourbon **3.**46–7
Antoku, emperor of Japan **3.**231
Antonello da Messina ***5.18***
Antonescu, Ion ***4.16***
Antonine Wall ***3.17***
Antonines ***3.17***
Antonioni, Michelangelo ***5.18***
Antony, Mark **3.**222–3
Anu ***9.4***, **9.**11, **9.**14
Anubis ***9.4***, **9.**28
anvil ***6.16***, **6.**39
anxiety *see* depression and anxiety; phobias; stress
Anyang ***3.17***
ANZAC ***4.16***
ANZUS ***4.16***, ***7.14***
Aoife **9.**23
Aotearoa, Mana Motuhake **4.**283
Aoun, General **4.**191
Apache Indians ***3.17–18***, ***7.14***, **7.**131, **7.**219
apartheid ***4.16***, **7.**278
see also black people *under* South Africa
apatite ***1.16***
Apelles ***5.18***
Apennines ***1.16***
aperture (camera) ***6.16***
aperture synthesis ***8.6***
apes ***2.13***
chimpanzee **2.**63, **2.**64
gibbon **2.**135–6
gorilla **2.**141–2
orang-utan **2.**235–6
see also monkeys
apex ***8.6***
of Sun's way **8.**6
aphasia ***7.15***
aphelion ***8.6***
aphid ***2.13***
aphorism ***5.18***, **5.**368
Aphrodite **9.**2, ***9.4***, **9.**6, **9.**28–9, **9.**30
apiculture ***6.16–17***
Apis ***9.4–5***, **9.**32
apoapse **8.**7
apoastron ***8.6***
apochromatic lens **6.**1
apogalacticon ***8.6***
apogee ***8.6***
Apollinaire, Guillaume ***5.18***
Apollo (asteroid) ***8.7***
Apollo (god) ***9.5***, **9.**17, **9.**25, **9.**27
Apollo and *Apollo–Soyuz* space projects **1.**273, ***6.17***, **6.**196, **6.**214, **6.**301, **8.**128, **8.**196, **8.**197
Apollo–Amor objects **8.**2, ***8.7***, **8.**65
Apollonius of Perga **8.**58
Apollonius of Rhodes **5.**190
Apollonius of Tyre **9.**16
Apostles **3.**104, ***9.5***, **9.**12
Andrew **9.**4, **9.**5
Bartholemew **9.**5
James (the Great) **9.**5, **9.**19
James (the Just/the Less) **9.**5, **9.**19
John **3.**41, **3.**104, **9.**5, **9.**20
Judas Iscariot **9.**5, **9.**21
Jude **9.**5
Peter (Simon) *see* Peter, Saint
Philip **9.**5
secret and later **9.**6–7, **9.**21, **9.**26
see also Paul, Saint
Simon the Canaanite **9.**5
Thomas (Didymus) **9.**5, **9.**43
Appalachian Mountains ***1.16***, **1.**41, **1.**135
apparent magnitude **8.**90
appeal ***7.15***, **7.**318
appeasement ***4.16***
appendicitis ***2.13***
appendix ***2.13***
Appert, Nicolas François ***6.17***, **6.**57
Appian Way ***3.18***
apple ***2.13***, **2.**129
Appleton, Edward: Appleton layer ***1.16***
appliances *see* domestic appliances
applications program ***6.17***, **6.**179, **6.**329–30
applied arts *see* decorative arts
applied mathematics ***6.17***, **6.**391
modelling **6.**17, **6.**110, **6.**126–7, **6.**320
simulation **1.**220, **6.**17, ***6.320***
see also quantum mechanics
Appomattox ***4.16***
appropriate technology ***6.18***
see also alternative energy; insulation; organic farming; recycling
apricot ***2.13***
apse/apsis/apsides ***8.7***
line of **8.**86
Apsu **9.**35
Aptidon, Hassan Gouled **4.**102
Apuleius **5.**257
Apus ***8.7***
Aqhat **9.**11
aqua regia ***1.17***
aquaculture ***6.18***, **6.**133, **6.**316
aqualung ***6.18***, **6.**97
aquamarine ***1.16–17***
Aquarids ***8.7***, **8.**195
aquarium and ornamental fish:
golden orfe **2.**140
goldfish **2.**141
guppy **2.**148
platy **2.**259
swordtail **2.**332
tetra **2.**339
Aquarius ***8.7***
aquatic organisms *see* living organisms *under* water
aquatint ***5.18***
aqueduct ***6.18–19***, **6.**379
aquiclude ***1.17***
aquifer ***1.17***, **1.**20
Aquila ***8.7***
aquilegia ***2.13***
Aquilera, Jaimé Roldos **4.**108
Aquinas, Saint Thomas ***3.18***, **5.**257, ***7.15***
and Aristotle **3.**21
and Dominicans **3.**105
and natural law **7.**213
and scholasticism **3.**318, **7.**309
Aquino, Benigno **4.**211, **4.**268
Aquino, Corazon **4.**268–9, **7.**72
Aquitaine ***4.18***
Ara ***8.7***
arabesque ***5.18***, **5.**344
Arabi Pasha ***4.16–17***, **4.**340
Arabia ***1.17***
exploration **1.**94
geographers **1.**8, **1.**10, **1.**192
historian **1.**10
holy cities **3.**227–8
peoples **3.**316, **3.**321
physical geography and geology **1.**17, **1.**277–8
religion **9.**43
see also Islam
see also Arabs; Saudi Arabia
Arabian Sea *see* Gulf (Persian Gulf)
Arabic language **7.**217, **7.**278, **7.**279
alphabet and script **7.**10, **7.**11, **7.**23, **7.**30, **7.**43, **7.**157, ***9.46***
diglossia **7.**90
literature ***7.18–19***
see also literature *under* Islam
Arabic numerals ***9.52***
arable farming ***6.19***, **6.**109
see also agrochemicals; crops
Arabs:
Arab League ***4.17***, ***7.15***, **7.**227
arts **5.**18–19, **5.**486–7
literature 5, **5.**2, **5.**18–19, **5.**198–9, **5.**218, **5.**311, **5.**454, 454
music **5.**486–7
Umayyad **5.**465–6
see also arts *under* Islam

Arabs (*cont.*):
conquests ***3**.18–19*
explorer **1**.164
language *see* Arabic language
literature ***7**.18–19*
see also under Islam
mathematics **6**.53
medicine and health care **6**.342
Orientalism **7**.225
politics **4**.27, **7**.24
technology and science **6**.10, ***6**.19*, **6**.72, **6**.148, **6**.215, **6**.346
see also technology and science *under* Islam
vehicles and transport **6**.195, **6**.215
wars and defence:
Abushiri revolt **4**.2
Arab Revolt **4**.191
Arab–Israeli **4**.315–16, **4**.386
conquests **3**.18, ***3**.19*
conventional **7**.69
weapons **6**.162
see also Islam; Middle East *and individual countries*
arachnids ***2**.13*
harvestman **2**.151
mite *see* mites
pedipalp **2**.247
pseudoscorpion **2**.270
scorpion **2**.297
sea scorpion **2**.301
sea spider **2**.16, **2**.301
spiders *see* spiders
sun spider **2**.330
ticks **2**.287, **2**.306, **2**.341
Arafat, Yasir ***4**.17*, **4**.117, **4**.335, **7**.227, **8**.227
Arafura Sea ***1**.17*, **1**.53
Arago, François (Francis) ***1**.17*, **8**.119
Aragon **3**.9, ***3**.19*, **3**.124–5, **3**.153
aragonite ***1**.17*
Arahat ***9**.5*
Arakan ***3**.19*
Aral Sea ***1**.17*
Aramaic language **7**.279, **7**.304
aramid fibre **6**.204, **6**.247
see also nylon
Aranda ***7**.15*
Arany, János **5**.214
Arapaho Indians ***3**.19*
arapaima ***2**.13–14*, **2**.23
Ararat ***1**.17*, ***9**.5*
Aratus of Sicyon **3**.2
Araujo, Juan de **5**.257
Arawaks **3**.60
Arbeau, Thoineau **5**.350
Arbenz Guzmán, Jacobo ***4**.17*, **4**.143
arbitrage ***7**.15*
arbitration **7**.92, **7**.158
Arbuckle Mountains ***1**.17*
Arbus, Diane ***5**.19*, **5**.347
Arbuthnot, John **5**.357
arc lamp ***6**.19–20*
arc welding **6**.383
Arcadia ***3**.19*
arch (architecture) ***5**.19–20*, **6**.47, **6**.49, **6**.336–7
see also vault
arch (geomorphology) ***1**.18*, **1**.224
Archaean aeon ***1**.18*, **1**.133, **1**.134
archaeology ***3**.19–20*
industrial **6**.178
see also prehistory
Archaeopteryx **2**.14
archangels ***9**.5*, **9**.6, **9**.13, **9**.18, **9**.25
Archelaus of Judaea **3**.164
Archer (Sagittarius) **8**.143
Archer, Frederick **6**.268
Archer, Thomas ***5**.21*
archerfish ***2**.14*
archery ***3**.20–1*, ***7**.15–16*
see also bow (archery)
Arches National Park ***1**.18*, **1**.225
Archilochus **5**.190
Archimedes ***1**.19*, **1**.161, **1**.358, ***6**.19*
Archimedes screw ***6**.19*, **6**.134, **6**.385
archipelagic apron ***1**.18*
archipelago ***1**.18*
Aleutians **1**.7, **7**.108
Azores **1**.29, **3**.30
Balearics **1**.30
Caymans **1**.55, **7**.346
Cyclades **1**.82
island arc **1**.174–5
Kurile Islands **1**.185, **7**.9
Seychelles **1**.300, **4**.310, **7**.380
Solomons **1**.312, **4**.318, **7**.296
Tierra del Fuego **1**.337
Tuscan **1**.101
Tuvalu **1**.348, **4**.355, **7**.296
see also Bahamas; Cape Verde; Caribbean; islands; Samoan Islands
Archipenko, Alexander ***5**.21*
architecture:
Abbasid **5**.1
acoustics **6**.2
Africa **5**.6
America *see* United States *below*
arch *see* arch
Art Deco **5**.23
art nouveau **5**.24
Assyria **5**.24–5
Australia **5**.26–7
Austria **5**.156, **5**.205–6, **5**.208, **5**.268, **5**.480
Aztec **5**.27
Babylon **5**.28
Baroque *see under* Baroque
basilica **5**.37
Belgium **5**.212, **5**.470
Brazil **5**.320
brickwork ***5**.59–60*
Britain:
15th to 17th centuries **3**.34, **5**.88, **5**.176, **5**.210
Baroque **5**.21, **5**.201, **5**.445
Palladian *see under* Palladio
Tudor **5**.423, **5**.461–2
see also Wren
18th and 19th centuries **5**.342, **5**.480
Baroque **5**.21, **5**.201, **5**.445, **5**.468
Gothic Revival **5**.369, **5**.407–8, **5**.415, **5**.482–3, **5**.495
Greek Revival **5**.353, **5**.422, **5**.437
Neoclassical **5**.3–4, **5**.351, **5**.377, **5**.423
Palladian **5**.241, **5**.245–6, **5**.365, **5**.447–8
Queen Anne style **5**.371–2, **5**.415
Regency **5**.312–13, **5**.377
see also Georgian architecture; Victorian architecture
20th century **5**.408, **5**.428, **5**.484
Arts and Crafts and Vernacular **5**.271, **5**.480, **5**.484
Expressionist **5**.290
'high-tech' **5**.162, **5**.386
Modern Movement **5**.448, **5**.480
Post-Modernist **5**.433
Tecton **5**.448
Elizabethan **5**.462
Anglo-Saxon **5**.17
Baroque **5**.21, **5**.201, **5**.468
Foster Associates **5**.162
Georgian *see* Georgian architecture
Gothic Revival **5**.66, **5**.68, **5**.369, **5**.407–8, **5**.415, **5**.482–3, **5**.495
Greek Revival **5**.353, **5**.422, **5**.437
architecture (*cont.*):
Britain (*cont.*):
Ireland **5**.169
Neoclassical *see under* Neoclassicism
Palladianism *see under* Palladio
Regency **5**.312–13, **5**.377
Scotland **5**.73, **5**.353
Brutalism **5**.64, **5**.394, **5**.446
Buddhist **5**.64
Burma **5**.66–7
buttress **5**.68, **6**.337
Byzantine **5**.69–70
capital **5**.76
Carolingian **5**.79
Central America *see* South and Central America *below*
China **5**.91–2
church **7**.48, **7**.50
classicism **5**.128, **5**.280, **5**.330
community **5**.108
Denmark **5**.232, **5**.467
dome *see* dome
Egypt **5**.142, **5**.278
Expressionist **5**.290
Fatimid **5**.152
Finland **5**.1
folly **5**.161
France:
12th and 13th centuries **5**.375, **5**.476
16th to 18th centuries **5**.56, **5**.128, **5**.168, **5**.198, **5**.261–2, **5**.280, **5**.425, **5**.469
19th and 20th centuries **5**.170–1, **5**.195–6, **5**.252–3, **5**.477
see also Le Corbusier
Baroque and neo-Baroque **5**.170–1, **5**.198, **5**.262
Fontainebleau School **5**.161
Gothic **5**.375, **5**.475, **5**.477
Mannerist **5**.161
Neoclassicism **5**.56, **5**.260, **5**.425
gargoyle **5**.170
Georgian *see* Georgian architecture
Germany **5**.38–9, **5**.42, **5**.210, **5**.317, **5**.405, **5**.498
see also Bauhaus
Gothic *see* Gothic architecture; Gothic Revival
Greece, ancient **5**.190, **5**.203, **5**.311, **5**.330
see also Greek Revival
Hittite **5**.208
India **3**.315, **5**.64, **5**.184, ***5**.219*, **5**.247, **5**.417, **5**.456
Gupta **5**.196
Hindu **5**.143, **5**.145, **5**.196, **5**.206–7, **5**.221, **5**.247, **5**.275, **7**.143, **7**.144
Mauryan **5**.286–7
Mogul **5**.307, **5**.456
rock cut temples **3**.268, **5**.143, **5**.145, **5**.275
Indonesia **5**.221
industrial *see under* industrialization
Iran **5**.344
Ireland **5**.169
ironwork *see* iron *under* buildings
Islam *see under* Islam
Italy:
14th to 16th centuries *see under* Renaissance
17th century **5**.45–6, **5**.53–4, **5**.195, **5**.267–8
18th century **5**.350
20th century **5**.317, **5**.356
Jain **5**.232
Japan **5**.233–4, **5**.446
Khmer **5**.247–8, **5**.276
Korea **5**.250–1
Mamluk **5**.278
Manueline **5**.281
masonry *see* masonry
mausoleum **5**.287
architecture (*cont.*):
Mayan **5**.287–8
Mexico **5**.74, **5**.84–5, **5**.101, **5**.449, **5**.458
Minoan **5**.297
modern *see* Modern Movement
Mogul **5**.307, **5**.456
Moorish **5**.303–4
Neoclassical *see under* Neoclassicism
Nepal **5**.316
Netherlands **5**.468
Norman **5**.388–9
Orders **5**.330
ornamental motif **5**.332
Ottoman **5**.333–4
Ottonian **5**.334
Palladian *see* Palladio
pediment **5**.343
Persia **5**.344
plasterwork **5**.351
Plateresque **5**.352
portico **5**.359
Portugal **5**.281
post-modern *see under* Post-Modernism
proportion **5**.368
Regency **5**.312–13, **5**.377
Rococo **5**.100–1, **5**.168, **5**.317, **5**.498
Roman **5**.330, **5**.386–7, **5**.478, **6**.373
Romanesque **5**.388–9
Russia and USSR **5**.371, **5**.375, **5**.394–5
Safavid **5**.397, **5**.399
Scotland **5**.73, **5**.353
screen **5**.409
Seljuq **5**.344, **5**.410–11
Sikh temple **7**.137
South and Central America **5**.84–5, **5**.100–1, **5**.320, **5**.425–6, **5**.449, **5**.458
see also Mexico *above*
Spain **5**.74, **5**.100–1, **5**.171, **5**.189, **5**.352
Sri Lanka **5**.429
Sumeria **5**.439–40
Sweden **5**.24
synagogue **7**.304
Thai **5**.450
timber *see* wood *under* buildings
Timurid **5**.456
tracery **5**.459
Tudor **5**.423, **5**.461–2
Turkey **5**.333–4, **5**.410–11, **5**.420
Umayyad **5**.465–6
United States ***5**.13–14*
18th and 19th centuries **5**.208, **5**.258, **5**.261, **5**.383
Chicago School **5**.14, **5**.91, **5**.439
20th century **5**.166, **5**.194, **5**.241, **5**.243, **5**.295, **5**.317–18, **5**.394, **5**.397, **5**.471
firm **5**.14, **5**.229, **5**.421–2
Bauhaus **5**.194, **5**.295, **5**.394
Brutalist **5**.394
Federal style **5**.153
Greek Revival **5**.258
Minimalist **5**.397
Neoclassicism **5**.258
Post-Modernism **5**.241, **5**.471
see also Modern Movement
vault *see* vault
vernacular **5**.472
Victorian **5**.66, **5**.68, **5**.342, **5**.369, **5**.474
see also Gothic Revival
Vietnam **5**.474–5
Viking **5**.476
see also buildings; functionalism
architecture (computer) **6**.85
Archytas of Tarentum **6**.25
Arcimboldo, Giuseppe ***5**.21*

Arctic *4.17*
exploration *see under* exploration
physical geography and geology **1**.30, *1.33*, **1**.232
see also Arctic Ocean
societies *7.16*
see also Inuit; Lapps
Arctic Ocean: physical geography and geology *1.18*, **1**.33
current **1**.18, **1**.235
Hudson Bay **1**.159
ice **1**.296
islands **1**.328
map **1**.18
Arcturus *8.7*
Ardashir of Persia **3**.271, **3**.316
Ardennes Campaign *4.17*
area measurement **6**.368
Arecibo Radio Observatory **8**.7, *8.8*
arenaceous rock *1.18*, **1**.20
Arend-Roland Comet *8.8*, **8**.194
Arendt, Hannah *7.16*, **7**.313
Areopagus *3.21*
Ares *9.5*
arête *1.18*, **1**.19, **1**.140
Aretino, Pietro *5.21*
Arezzo, Guido d' **5**.424
argali *2.14*
Argand, Aimé: Argand lamp *6.20*, **6**.201
Argand, Jean Robert *1.18*
Argelander, Friedrich Wilhelm **8**.22, **8**.159
Argentina *1.18–19*, *3.21*, *4.17–18*, *7.339*
economy and trade **4**.31, **4**.320, **7**.248
independence **4**.43
languages **7**.269
literature **5**.53, **5**.114, **7**.216
Patagonia *1.248–9*, **4**.262
physical geography and geology *1.18–19*
pampas **1**.246, *1.246*, **2**.240, **2**.241
plain **1**.142
plateau **1**.248–9
rivers **1**.19, **1**.258
politics and government **4**.103, **4**.166, **7**.173, **7**.242
population **4**.320
religion/theology **7**.171
technology **6**.250
urban areas **7**.197
wars and defence **7**.220
military/revolutionary leaders **4**.177, **4**.224, **4**.266–7, **4**.293, **4**.302
Paraguayan War **4**.259–60
argentite *1.19*
argillaceous rock *1.19–20*
argon *1.20*, **1**.106, *6.20*, **6**.96, **6**.240
argonaut (invertebrate) **2**.241
Argonauts (mythology) *9.5*
Argos *3.21*
argument, dialectic **7**.89
argument of perihelion **8**.114
Argyll, Archibald Campbell, 8th earl of **3**.239
Argyll, John Campbell, 2nd duke of **3**.326
Argyre Planitia *8.8*
arhat *7.16*
aria *5.21*, **5**.442
Ariadne *9.5*
Ariane **6**.196
Arianism *3.21*
Ariarthes IV, king of Cappadocia **3**.60
Arias, Oscar **4**.87
arid areas *see* deserts and arid areas
aridisol *1.20*, **1**.311
Ariel (astronomy) *8.8*
Aries (astronomy) *8.8*
Ariosto, Ludovico *5.21*, **5**.148, **5**.387
Aristarchus of Samos *8.8*, **8**.11, **8**.33, **8**.62, **8**.150
Aristide, Jean-Bertrand **4**.145, **7**.184
Aristides *3.21*
aristocracy *7.16*
Aristophanes **5**.6, *5.21*, **5**.100
Aristotle *3.21*, **3**.51, *7.16*, **7**.234
and aesthetics **7**.4
and astronomy **8**.61
and axioms **1**.28, **1**.264
commentaries and translations of **3**.43, **3**.318, **7**.23, **7**.33
and diving bell **6**.105
and ethics **7**.16, **7**.108–9
and Hermes Trismegistus **9**.16
and justice **7**.174
and literature **5**.190, **5**.212
tragedy **5**.82, **5**.425, **5**.459
and natural law **7**.213
and postulate **1**.264
and rhetoric **5**.382
and substance **7**.301
and syllogism **7**.303
and teleology **7**.307
arithmetic:
arithmetic mean **1**.209
calculator **6**.53–4
fixed-point **1**.123
floating-point **1**.123
and logic unit *see* central processing unit
see also mathematics
Arius **3**.21, **3**.26
Arjan Dev, Guru (fifth) **3**.328, **5**.418, **7**.2, **7**.137, **7**.194, **7**.285
Arjuna Pāṇḍava *9.5*, **9**.22
ark *9.5*
Ark of the Covenant *9.5*
arkose *1.20*
Arkwright, Sir Richard *6.20*, **6**.62, **6**.92, **6**.190, **6**.329
Arlen, Harold **5**.309
arm (limb) *2.14*, **2**.29
armadillo *2.14*, **2**.15, **2**.138, **2**.277
Armageddon *9.5*
Armagnac *3.23*
Armah, Ayi Kwei **5**.7
armature *6.20*
armed forces *7.16*
conscientious objection to joining **7**.65–6, **7**.227
conscription **7**.66
mercenaries **7**.198
militia **7**.202
special armed services **7**.296–7
see also airforce; armies; navies
Armenia *1.20*, *3.23*, *4.18*, *7.339*
astronomy **8**.4
in CIS **7**.59
ethnic groups **4**.188, *7.17*, **7**.178
language *7.17*, **7**.156, **7**.157
physical geography and geology **1**.20
religion **9**.40
Apostolic Church *7.16–17*
armies and soldiers *3.22–3*, *4.18–19*
condottiere **3**.84–5
fusiliers **3**.136
fyrd **3**.136
gladiators **3**.141–2, **3**.338–9
hussars **3**.173
Janissaries **3**.185
military religious orders **3**.198
militia **3**.231, **3**.360
minutemen **3**.233
musketeers **3**.22, **3**.243
New Model Army **3**.253
praetorian **3**.290
see also military technology *and* wars *under individual countries*
armillary sphere *6.20*, **6**.340, *8.8*
Arminius, Jacobus *3.23*
armonica **5**.179
Armory Show **5**.14
armour *see* weapons *and also* armour *under* body
armoured vehicle **6**.20
see also tank
arms (weapons) *see* weapons
Armstrong, Edwin Howard *6.20*
Armstrong, Louis *5.22*, **5**.238
Armstrong, Neil **4**.322–3, **6**.17, **8**.172
army *see* armies
army worm *2.14*
Arne, Thomas Augustine **5**.5, *5.22*, **5**.453
Arnhem, battle of **4**.18
Arnhem Land *1.20*, **1**.21
Arnold, Benedict **3**.10, *3.23*, **3**.357
Arnold, Malcolm *5.22*
Arnold, Matthew *5.22*, **5**.105, **5**.143, **5**.156, **7**.78
Arnold of Brescia **3**.132
Arnolfo di Cambio *5.22*
aromatic compounds *1.20*, **1**.224, **6**.71, **6**.252, **6**.261–2, **6**.265
see also benzene; hydrocarbons
aromatic and scented plants:
artemisia **2**.15
basil **2**.24
chamomile **2**.60
citronella **2**.66
dill **2**.92–3
frangipani **2**.127
gardenia **2**.133
jasmine **2**.175
lavender **2**.184
lilac **2**.190
see also mint; spices, herbs, and flavourings
Arp, Jean (Hans) *5.22–3*, **5**.298
Arpád, prince of Magyars **3**.216
arpeggione *5.23*
Arrau turtle **2**.144
Arrest, Heinrich d' **8**.2, **8**.106
arrest *7.18–19*
Arrhenius, Svante August *1.20*
Arrian **3**.117
arrow *see* bow (archery)
arrow worm *2.15*
arrow-poison frog *2.14–15*
arrowroot *2.15*
arroyo *1.20*
Ars Antiqua, Ars Nova *5.23*
Arsaces I, emperor of Parthia **3**.271
arsenic *1.20*, **1**.106, *6.20*, **6**.72, **6**.312
Arslan, Alp **3**.321
arson **7**.250
art:
aesthetics **7**.4
copyright law **7**.70
direction *5.23*
history *5.23*, **5**.217
museum **7**.209
performing rights **7**.233
symbolism **7**.303
see also art nouveau; Arts and Crafts; literature; music; performing arts; visual arts *and entries prefixed* **5**.
Art Deco *5.23*
art nouveau *5.23–4*
Belgium **5**.212, **5**.470
Britain **5**.39, **5**.177, **5**.273–4
Czech **5**.307
France **5**.168, **5**.195–6
United States **5**.455
Artabanus, emperor of Parthia **3**.271, **3**.316
Artaud, Antonin *5.23*
Artaxerxes, emperor of Sassanian empire **3**.93
Artaxerxes II, king of Persia *3.23*, **3**.95, **3**.122, **3**.316
Artaxerxes III, king of Persia **3**.23
Artemis **9**.2, *9.5*, **9**.18, **9**.25
artemisia *2.15*
arteries *2.15*
disorders **2**.10, **2**.15, **2**.77, **6**.60
see also blood
arteriosclerosis and atherosclerosis *2.15*, **2**.77
artesian spring *1.20*
arthritis *2.15–16*, **6**.101
arthropods *2.16*
chitin **2**.64
exoskeleton **2**.84
horseshoe crab **2**.16, **2**.163
myriapods **2**.16, **2**.59, **2**.212, **2**.224
shells (exoskeleton) **2**.306, **2**.307
ticks **2**.287, **2**.306, **2**.341
trilobite **2**.16, **2**.348
woodlouse **2**.370
see also arachnids; crustaceans; insects
arthroscope **6**.253
Arthur, king *3.23–4*, **3**.139, **3**.242, *9.5*
myths and legends **5**.278, *9.5*, **9**.8, **9**.12, **9**.14, **9**.25
Arthur, Chester Alan **4**.18, **4**.148
ārtī 7.19
artichoke *2.16*
article (grammar) **7**.229
Articles of Association *7.19*
articulated locomotive *6.20–1*
artificial elements **6**.197
artificial horizon *6.21*
artificial insemination **6**.15, *6.21*
artificial intelligence *6.21*, **6**.86, **6**.129, **6**.180, *7.19*
expert systems **6**.124, **6**.179
machine translation **6**.211
neural network **6**.21, **6**.180, **6**.240
optical character recognition **6**.251
pattern recognition **6**.259
speech recognition **6**.240, **6**.328
see also cybernetics
artificial limb *6.21*, **6**.36
artificial products *see entries beginning* synthetic
artificial respiration **6**.298
artificial satellites *see* satellites, artificial
Artigas, José Gervasio *4.20*
artillery *6.21–2*
ammunition **6**.21
anti-tank gun **6**.16, **6**.131
armoured vehicle **6**.20
see also tanks
artillerymen **3**.22
ballista **6**.22, **6**.29
cannon **6**.22, **6**.57, **6**.130, **6**.318
catapult **6**.22, **6**.64, **6**.225, **6**.318
howitzer **6**.22, **6**.172
manufacture **6**.385
mortar **6**.20, **6**.232–3
rockets *see* rocketry
self-propelled gun **6**.312
shell **6**.315–16
on ships *see* warships
shrapnel **6**.318
see also firearms; weapons
Artois, Charles, Comte d' *see* Charles X, king of France
Arts and Crafts Movement *5.23*, **5**.24, **5**.128–9, **5**.305, **5**.415, **5**.484
arts *see* art
Aruba *7.339*
arum *2.16*, **2**.17
Arundel, Richard Fitzalan, earl of **3**.210
Arundel, Thomas **3**.209
Arusha Declaration *4.20*
Arya Samaj *4.20*
Aryans *3.24*
aryl group *1.20*
Asad, president of Syria **4**.17
Asan, Kumaran **5**.136
Asante *3.24*, *4.20*, **4**.276, **7**.18, *7.19*
asbestos *1.20–1*, **1**.22, *6.21–2*, **6**.44, **6**.128, **6**.130, **7**.240
asbestosis **6**.22
ascarid *2.16–17*, **2**.242

ASCC (computer) 6.84
ascending node 8.108
Ascension 7.19
asceticism 7.9, 7.19, 7.86, 7.210
Asch, Solomon 7.64, 7.292
Ascham, Roger 3.20, 3.24
ASCII 6.22
Asclepius 9.5, 9.18
ascorbic acid 2.357, 6.141
ASDIC (sonar) 6.325
ASEAN (Association of South-East Asian Nations) 4.21, 4.321, 7.20
asepsis 6.22, 6.218, 6.250
asexual reproduction 2.68, 2.282
Asgard 9.5
ash (tree) 2.17
ash, volcanic 1.21
al-Aʿshā 5.19
Ashanti see Asante
Ashbee, C. R. 5.24
Ashburton, Lord 4.373
Ashcan School 5.14, 5.24
Ashcroft, Dame Peggy 5.24
ashen light 8.8
Ashikaga shogunate 3.24
Ashkenazi 7.19, 7.335
Ashley, William Henry 4.20–1
Ashley Cooper see Shaftesbury
Ashoka, emperor 3.24, 3.25, 3.226
and art 5.219, 5.286–7, 5.336, 5.400
and Buddhism 3.53, 3.315, 7.36
ashram 7.9, 7.19, 7.205
āshrama 7.19
Ashton, Sir Frederick 5.24, 5.160, 5.341, 5.393
Ashtoreth 9.5
ʿāshūrā' 7.19–20
Ashurbanipal of Assyria 3.24–5, 3.26, 3.364
Ashur-uballit of Assyria 3.26
Asia 1.21–2
adoption 7.3
agriculture 7.237, 7.268
sharecropping 7.179, 7.232, 7.282
shifting cultivation 7.175, 7.197, 7.211
see also under subsistence and under technology and science below
arts see music below and also under individual countries
astrology 7.20
colonialism 3.7, 3.11, 3.106–7, 3.216, 4.281
decolonization 4.320, 4.321
see under empire under France; Portugal
demography:
birth rate 7.32
contraception 7.115–16
fertility rate 7.119
infant mortality 7.159
life expectancy 7.6, 7.184
population density 7.241
population growth 7.226, 7.241–2
dowry 7.95
education 7.102, 7.141, 7.154–5, 7.185, 7.277
energy, renewable 7.106
ethnic groups 7.9, 7.197, 7.202, 7.315
see also Trobriand Islanders and under hunter-gatherers; patrilineality; South Asia; South-East Asia
evolution 3.120
exploration 1.75, 1.83, 1.84, 1.164, 1.205
famine 7.116
foreign policy and trade:
aid to 7.8
ASEAN 4.21, 4.321, 7.20
Bandung Conference 4.30
Asia (cont.):
foreign policy and trade (cont.):
Colombo Plan 4.57, 7.57
Greater East Asia Co-Prosperity Sphere 4.140
Non-Aligned Movement 7.218
SEATO 4.320
land reform 7.179–80
languages 7.11, 7.95, 7.106, 7.227, 7.286
Afro-Asiatic 7.30, 7.43, 7.78, 7.109, 7.278–9
Austro-Asiatic 7.22–3
laws/legal system 7.40
maps 1.22, 3.42, 3.238
Buddhism 3.52
trade 3.107, 3.329
medicine and health care 7.159
monarchy 7.97, 7.139
Mongol empire 3.238
mortality 7.207
music 5.86, 5.426–7, 5.486–7
Muslim Brotherhood 7.166
nomadic invasion from 3.30
Orientalism 7.225
Pacific see Pacific Ocean area
physical geography and geology 1.21–2
deserts 1.141, 1.330, 1.334, 1.348
earthquakes and volcanoes 1.99
lakes 1.30, 1.31, 1.54, 1.185, 1.339
map 1.22
mountains see under mountains
plate tectonics 1.260
rivers see under rivers
seas 1.17, 1.29
volcanoes 1.99, 1.129, 1.184
politics and government 7.60, 7.124, 7.332
poverty 7.244
prehistory see under prehistory
refugees 7.261
region 3.363
religion/theology:
Buddhism 3.52
see also under individual countries
Islam 3.2, 7.166
shamanism 7.282
see also Confucianism; Hinduism; Jainism; Shinto; Sikhism
science see technology and science below
slavery 7.287
sport: martial arts 7.194
Tartars 3.350
technology and science 6.53, 6.132, 6.178, 6.307
agriculture 6.7, 6.64, 6.68, 6.204, 6.364
buildings 6.308
printing 6.281, 6.387
textiles 6.208, 6.353
vehicles and transport 6.101, 6.166, 6.190, 6.215, 6.298–300, 6.363
wood 6.387
see also under individual countries
trade see foreign policy above
urban areas 7.288, 7.323
wars and defence see under individual countries
see also East Asia; South Asia; South-East Asia and also individual countries
Asia Minor 1.22
maps 3.9, 3.55, 3.75, 3.125, 3.266, 3.307
peoples 3.166, 3.320
physical geography and geology 1.22
regions and kingdoms 3.60, 3.136, 3.214, 3.280, 3.288
ruler 3.234
urban areas 3.117, 3.231, 3.361
asiento de negros 3.25
Asiya 9.5
Aske, Robert 3.281–2
Askia Muhammad I, king of Songhay 3.25, 3.333
ASMs see air-to-surface missiles
Asoka see Ashoka
asparagus 2.17
aspartame 6.141, 6.342
aspartic acid 6.342
Aspdin, Joseph 6.276
aspect ratio 6.135
Asper, Hans 3.391, 5.168
asphalt 1.22, 6.39
asphodel 2.17
aspidistra 2.17
aspirin 6.14
Asplund, Gunnar 5.24
Asquith, Herbert Henry, 1st earl of Oxford and Asquith 4.21
ass see asses
Assam 7.305
assassin and assassination 3.25, 4.161, 4.183, 7.20
assassin bug 2.17
assaying 1.341, 6.5, 6.23, 6.357, 6.359
assemblage 5.24
assembly language 6.23
assembly line 6.23, 6.143, 6.216, 6.278, 7.249
see also automation
asses 2.17
donkey 2.97
hybrid (mule) 2.221
kiang 2.178
assessment 7.20
assets 7.20
inventory 7.165
wealth 7.330
see also capital; resources
Assis, Machado de 5.256
assistance see social security
assizes 3.25–6
association, stellar 8.159
Association of South-East Asian Nations see ASEAN
associativity 1.22
assurance 7.20
see also insurance
Assylmouratova, Attynai 5.248
Assyria 3.26
arts 3.26, 3.280, 5.24–5
religion and myths 9.35, 9.36
rulers 3.24–5, 3.321
technology 6.322
Astaire, Fred 5.25, 5.309
Astarte 9.5
astatine 1.22, 1.106
asterism 8.8
asteroids 8.9
Amor and Amor group 8.4, 8.48
Apollo 8.7
Apollo–Amor objects 8.2, 8.7, 8.65
Ceres 8.24
Chiron 8.26
Hidalgo 8.67
Icarus 8.72
Juno 8.78
list of biggest 8.193
in multiple star 8.21
Trojan group 8.2, 8.64, 8.174–5
asthenosphere 1.98–9
asthma 2.17
Astley, Philip 5.101
Astor, John Jacob 3.136, 4.21
astrobiology 8.49
astrograph 8.9
astrolabe 6.23, 8.9
astrology 7.20–1, 8.9–10
signs of the Zodiac 8.190, 9.62
astrometry 8.10
astronautics 8.10
see also space-flight
astronomers 8.10
see also astronomy
Astronomers Royal 8.10
see also under astronomy under Britain
astronomical clock 8.10
astronomical units 8.10–11, 8.192
astronomical year 8.189
astronomy 8.11–12
armillary sphere 6.20, 6.340
artificial horizon 6.21
astrolabe 6.23
astronomical clock 6.76
gamma-ray 8.57
Greek 6.285
horizon 1.158
infra-red 8.73
interferometer 6.184
observatory see observatories
pendulum 6.260
photography 6.268
prehistoric 8.127
quadrant 6.287
sextant 6.315
spherical 8.158
telescope see telescopes
ultraviolet 8.177
universal joint 6.369
XUV 8.188
see also space-flight and under individual countries and entries prefixed 8.
astrophysics 8.12
Asturias, Miguel Angel 5.25
Astyages of Media 3.95
asymptote 1.23
Atacama Desert 1.23
Atahualpa, king of Incas 1.88, 3.26, 3.175–6, 3.277, 3.283, 3.298
Atalanta 9.5–6
Atatürk, Mustafa Kemal 4.21, 4.310, 4.354, 7.11
Aten 9.6, 9.31
Atget, Eugène 5.25
Athabasca, Lake 1.23
Athanasius, St: and Athanasian Creed 3.26, 7.74
atheism 7.21
Athelstan, king of Wessex 3.15, 3.26
Athena 9.6, 9.28–9
Athens 3.26–7
acropolis 3.2
Areopagus 3.21
Athenian empire 3.26–7
democracy 3.26
Draconian laws 3.105
empire 3.26–7
orator 3.101
ostracism 3.264
philosophy 3.2, 3.332
statesmen 3.21, 3.276, 3.332, 3.353
wars and defence:
leaders 3.8, 3.21, 3.353
Peloponnesian War 3.3, 3.8, 3.274–5
against Persians 3.221
see also Greece, ancient
atherosclerosis 2.15, 2.77, 6.60
athlete's foot 2.285, 2.342
athletics 7.21, 7.232
Atiba, Alafin 4.386
Atish 5.466
Atkinson, Sir Harry Albert 4.21
Atlantic, battle of 4.21–2
Atlantic Charter 4.22
Atlantic Ocean 1.23
exploration 1.154
islands 7.339
Canaries 1.50, 8.82, 8.83
Cape Verde see Cape Verde
Madeira 1.201, 3.215
Saint Helena 1.289
São Tomé and Principe 1.293, 4.4, 7.379
South Georgia 1.314–15
South Orkney 1.315
South Sandwich 1.315

Atlantic Ocean (cont.):
islands (cont.):
South Shetland 1.315
Tristan da Cunha 1.344
see also Bahamas; Bermuda; Cape Verde; Falkland Islands; Greenland; Iceland
maps 1.24, 3.330
physical geography and geology 1.23, 1.24
bay 1.129
Bermuda Triangle 1.38
currents 1.23, 1.36, 1.148, 1.185, 1.230, 1.235
islands 1.24
see also islands above
map 1.24
Mid-Atlantic Ridge 1.213–14
plate tectonics 1.260
ridges 1.213–14, 1.295
Sargasso Sea 1.293
trenches 1.24
Atlantis 3.27, 9.6
Atlas (giant) 9.6
Atlas (satellite) 8.12
atlas moth 2.17–18
Atlas Mountains 1.23–4
ATM (automatic teller machine) 6.118
ātman 7.3, 7.6, 7.21, 7.295
atmosphere see air and atmosphere; ozone; pressure
atoll 1.24, 1.26
atom see atomic; atoms
Atomba Singh 5.279
atomic bombs (fission bombs) 6.246
see also nuclear weapons
atomic clock 6.77
atomic energy see nuclear energy; nuclear reactors
atomic number 1.26
atomic physics:
colour used metaphorically 1.69
electromagnetic see electromagnetic radiation
elementary particles see elementary particles
energy levels 1.108–9
fermion 1.120
orbitals 1.239, 1.240, 8.131
pion 1.222–3, 1.258
string theory 1.323–4
strong force 1.324
tunnelling 1.346
see also atoms; nuclear physics; physics
atomic pile 6.23, 6.243, 6.245
atomic power see nuclear energy and power
atomic time 1.338, 6.356
International 8.172
atomic-hydrogen welding torch 6.194
atomism 7.21
atoms 1.24, 1.26, 6.23–5, 6.113
'atom trap' 6.194
atomic orbitals 1.239
band theory 1.31–2
in crystals 6.389
diatomic molecules 1.90
electrically charged see ions
electrons see electrons
heteroatoms 1.155
hydrogen bond 1.59, 1.161
relative atomic mass 1.280
valency 1.352
weight 1.280
see also electrons; ions; molecules; neutrons; nucleus (physics); protons; quantum mechanics and quantum theory and entries beginning atomic
atonality 5.25, 5.439, 5.484–5
atonement 7.21
ATP see adenosine triphosphate
attachment 7.21–2, 7.110, 7.153–4, 7.291
attainder 3.27
Attalus I and III, kings of Pergamum 3.136, 3.276
'Aṭṭār, Farid al-Din 5.25
al-Aṭṭār, Murshid 3.348
Attendolo, Muzio and Francesco 3.324
attention 7.22
Attica 3.27
Attila the Hun 3.23, 3.27, 3.97, 3.173, 3.264
Attlee, Clement Richard, 1st earl Attlee 4.22, 4.115, 4.276, 4.374
Atum 9.31
Atwater, Wilbur Olin 2.18
Atwood, Margaret 5.25–6
aubergine 2.18
aubretia 2.18
Aubrey, John 5.26
Aubriet, Claude 2.18
Auchinleck, Sir Claude John Eyre 4.22, 4.246, 4.346
auction 7.22
Auden, Wystan Hugh 4.164, 5.26, 5.143, 5.325, 5.436, 5.449
audio see sound
audiometer 6.25, 6.254
audion 6.291
audiotape see tape recorder
audit 7.22
Audubon, John James 2.18, 5.26
Audumla 9.6
Aue, Hartmann von 5.174
Auer, Carl Freiherr von Welsbach 6.25, 6.150
Augean Stables 9.6
auger 6.107
augite 1.26
Augsburg 3.27–8
Confession of 3.27
Interim of 3.27
League of 3.27–8
Peace of 3.27
augur 3.28
Augustan Age 5.26
Britain 5.5, 5.26, 5.117, 5.183, 5.357, 5.384, 5.417, 5.431, 5.441–2
France 5.26, 5.113, 5.301, 5.372
Augustine of Canterbury, Saint 3.15, 3.28, 3.193, 3.380, 9.40
Augustine of Hippo, Saint 3.28, 3.99, 7.22, 7.202, 7.225, 9.40
Augustinians 3.28, 3.194
Augustus, Roman emperor (Octavian) 3.3, 3.6, 3.28, 3.51, 3.57, 3.307–9, 3.365
conquests 3.19, 3.23, 3.38, 3.62, 3.79
Augustus II (the Strong), king of Poland 3.29, 3.257
Augustus III, king of Poland 3.287
auks 2.18, 2.271, 2.279
Aung San 4.22–3, 4.229
Aung San Suu Kyi, Daw 4.229
Aurangzeb, Mogul emperor 3.29, 3.41, 3.235, 3.325, 3.329, 7.132
Aurelian, Roman emperor 3.29, 3.95, 3.374
Aurgelmir 9.6, 9.25
Auric, Georges 5.421
auricula 2.269
Auriga 8.12
Aurignacian 3.367
Auriol, Vincent 4.23
auroch 2.18
aurora 1.26, 1.173, 8.12–13
Ausdrucktanz 5.300
Ausgleich 4.23
Austen, Godwin 1.179
Austen, Jane 5.26, 5.348
Austerlitz, battle of 4.23
Austin, G. 7.35
Austin, General (Grenada, 1983) 4.142
Austin, J. L. 7.185
Australia 1.26–7, 4.23–4, 7.340
agriculture and land 4.308, 4.326, 6.7, 6.351
aid from 7.8
anthropology 7.77
ANZUS 4.16, 7.14
arts 5.26–7
painting 5.1, 5.26–7, 5.319–20
see also literature; music below
astronomy 8.62
Anglo-Australian telescope 8.150
Australia Telescope National Facility 8.13
as colony:
British administration 4.13–14, 4.93, 4.142, 4.190, 4.294, 4.338, 4.370
see also penal colony below
Commonwealth of 4.23
domestic appliances 6.297
economy and trade:
mining 1.75, 1.174, 4.113–14, 6.101
in OECD 7.221
pay 7.203
education 7.275, 7.276
energy and fuel 6.385
environment 7.212, 7.296
ethnic groups see Aborigines
exploration of 1.45, 1.83, 1.191, 4.121–2
by British 1.118, 1.123, 1.201, 1.216, 1.324, 3.96, 4.46
explorer from 1.366
finance 7.155, 7.189
gold diggers' rebellion 4.113–14
government see politics and government below
immigrants 4.238, 4.254, 4.377, 7.105, 7.153
see also penal colony below
industrial relations 7.159
languages 7.22, 7.106
laws/legal system:
family 7.115
habeas corpus 7.138
industrial relations 7.159
International Court of Justice 7.164
judicial review 7.173
punishment 7.255
surveillance 7.303
literature 5.57, 5.103, 5.383, 5.431, 5.488
map 4.24
medicine and health care 7.196, 7.340
music 5.1, 5.188, 5.204, 5.289, 5.313, 5.409, 5.441
national parks 7.212
outlaws and bushrangers 4.56, 4.102
penal colony:
Bigge Inquiry 4.38
emancipists 4.111
exclusionists 4.114
ticket-of-leave 4.344
transportation and settlements 4.46, 4.85, 4.264
uprising 4.65
see also colony above
physical geography and geology 1.26–7
Ayers Rock 1.28, 1.170
bays 1.144–5
caves 1.177
coasts 1.53
continental drift 1.135
coral island 1.76
deserts 1.139, 1.146, 1.306
Great Barrier Reef 1.145
islands 1.180, 1.230, 1.331–2
lakes 1.118, 1.130, 1.340
Australia (cont.):
physical geography and geology (cont.):
mountains 1.27, 1.123, 1.145, 1.151, 1.184, 1.200, 1.211, 1.309
regions see states below
rivers 1.75, 1.84, 1.123, 1.223
salt-flats 1.291
spalling 1.315
Tasman Sea 1.332
politics and government:
bicameral 7.31
cabinet 7.38
Commonwealth of 4.23
as Dominion 7.58
elections and electoral system 7.104, 7.190, 7.260
Federal movement 4.23, 4.31, 4.94, 4.121–2, 4.260, 7.119
Free Trader 4.286
governors 4.205
Labor 4.72, 4.90, 4.114, 4.120, 4.158, 4.189, 4.202, 4.307, 4.326, 7.179, 7.246
Liberals 4.94, 4.195
Nationalists 4.53, 4.158, 4.217, 4.234
ombudsman 7.223
prime ministers see under prime ministers
Protectionism 4.94
United Australia Party 4.53, 4.202, 4.217
Whitlam crisis 4.378
women's suffrage 7.332
prehistory 3.29
see also australopithecines
radio 6.250
region see states below
religion/theology 7.187
Rum Rebellion 4.294–5
science see entries above and below prefixed 1. and 6.
social services and security 4.318, 7.330
in South Pacific Forum 7.296
sport and recreation:
cricket 7.75
football 7.122, 7.123
lacrosse 7.179
softball 7.295
speedway racing 7.297
surfing 7.303
tennis 7.308
states, regions, and territories 1.20, 1.21, 1.328
listed 9.73
New South Wales 1.227, 4.239–40
Northern territory 1.231, 4.247
Queensland 1.273, 4.280
South Australia 1.314, 4.320
Tasmania 1.331–2, 2.236, 3.29, 4.339
Victoria 1.356, 4.365
Western Australia 1.365, 4.374–5
technology and science see entries above and below prefixed 1. and 6.
telecommunications 7.307
unions and strikes 4.211, 4.311, 4.326
vehicles and transport 6.47, 6.64, 6.76, 6.202
weapons 6.42
World War I 4.129
Australian Alps 1.27
Australian copperhead 2.75
Australian federal movement 4.23
see also under politics under Australia
Australoids 3.29
see also Aborigines

australopithecines/*Australopithecus* 2.126, *3.29*, 3.120, 3.152, 3.201, 3.261, 3.341
- *see also* prehistory

Australasia 3.51
Austria *1.27*, *3.29*, *4.24–5*, *7.340*
- 1867–1919 *see* Austro-Hungarian empire
- aid from 7.8
- arts:
  - architecture 5.156, 5.205–6, 5.208, 5.268, 5.480
  - graphic art and design 5.249–50
  - painting 5.249–50, 5.405
  - *see also* literature; music *below*
- astronomy 8.15
- Ausgleich 4.23
- chemistry 1.266, 6.25, 6.132, 6.310
- economics 7.139, 7.276
  - Hayek 7.139, 7.157, 7.215, 7.291
- economy and trade 6.188, 7.139, 7.221
  - and EC 7.99
  - in EFTA 7.111
- empire *4.25*
- foreign policy 4.23
  - Concert of Europe 7.135
  - Council of Europe 7.71
  - Three Emperors' League 4.344
  - Triple Alliance 4.352
  - Villafranca di Verona Treaty 4.368
  - *see also* wars and defence *below*
- languages 7.131
- literature 5.60, 5.208–9, 5.243, 5.310
  - *see also under* drama; novels; poetry
- mathematician 1.210
- medicine and health care 6.137, 6.257
- music 5.51, 5.63, 5.214, 5.313, 5.406, 5.435, 5.491
  - conductors 5.260, 5.276
  - opera 5.44, 5.306
  - serialism 5.405, 5.484
  - *see also under* composers; opera
- neutrality 7.214
- philosophy 7.12, 7.197, 7.241
  - Vienna Circle 7.187, 7.188
  - *see also* Wittgenstein
- physical geography and geology *1.27*
  - lake 1.73
  - mountains 1.10, 1.16
  - rivers 1.84, 1.282
- physicists 1.41, 1.210, 1.294–5, 1.320
- politics and government 3.194, 4.102, 4.214, 4.218, 4.285, 4.305, 4.317, 7.67, 7.332
- prehistory 3.154
- provinces 4.356
- psychology 7.113, 7.252, 7.293
- radar 6.288
- rulers:
  - before 1800 3.153, 3.168, 3.190, 3.206
  - after 1800 4.23, 4.25, 4.26, 4.123–4, 4.144, 4.157
  - list of *9.67*
  - *see also* Francis Joseph II; Maria Theresa
- social security 4.357
- technology and science 6.211
  - *see also entries above and below prefixed* 1. *and* 6.
- urban areas 7.288
- vehicles and transport 6.300
- wars and defence:
  - annexed by Hitler 4.14, 4.24
  - Austro-Prussian War 4.26, 4.299
  - Bavarian Succession 3.37
  - Crimean 4.88, 4.260
  - with France and Sardinia 4.206
  - leaders 4.280–1, 4.305
  - Napoleonic 4.369
  - Quadruple Alliance 4.279
  - Revolutionary Wars 4.288
  - Revolutions of 1848 4.288–9
  - World War II 4.14, 4.24, 4.385
  - *see also* Austrian Succession; Seven Years; Spanish Succession *and also* foreign policy *above*

Austrian Succession, War of 3.6, *3.29–30*, 3.102, 3.128, 3.134, 3.197
Austro-Asiatic languages 7.22–3
Austro-Hungarian empire (1867–1919) *4.25–6*, 4.49, 4.169, 4.369
Austronesian languages *7.23*, 7.192, 7.197, 7.228, 7.240
Austro-Prussian War *4.26*, 4.299
authoritarian state 7.23, 7.223, 7.296
authority 7.23, 7.221
- revolt against *see* independence; mutinies; secession; separatism

autism 2.18, 7.23
autobahn 6.300
autobiography and memoirs *5.27*, 5.250
- Britain 5.66, 5.188, 5.202, 5.270, 5.286, 5.382, 5.490
- China 5.416
- Denmark 5.49
- France 5.89, 5.368
- Ireland 5.242
- Italy 5.124
- Japan 5.236, 5.304
- Russia and USSR 5.65
- United States 5.5, 5.16–17, 5.133, 5.295–6, 5.431, 5.463, 5.491
- *see also* biography

AUTOCODE 6.86
autofocus *6.25*
autogyro *6.25*
automata *6.25*, 6.84, 6.168
automatic exposure *6.25*
automatic landing *6.25*, 6.290
automatic pilot *6.25*
automatic teller machine 6.118
automatic train control *6.25–6*
automatic transmission *6.26*, 6.152
automation *4.26*, *6.26*, 6.216, 7.248–9
- coal-mining 6.79
- and computers 6.26, 6.85, 6.179
- flexible manufacturing system 6.134, 6.211
- navigation 6.237
- *see also* assembly line; mass production

automatism *5.27*
automobile *see* motor car
autonomic nervous system *2.18*, 2.222
- *see also* nervous system

autotrophic organism *2.18–19*
autumn 1.297
- crocus *2.19*
- equinox 8.47

autunite 1.239
auxins 6.271
avalanche *1.27*, 1.28
Avalokitesvara 5.401, *9.6*, 9.15, 9.22
Avalon *9.6*
Avars *3.30*
avatars *7.23*, 7.23
- of Lakshmi 9.22, 9.34
- of Vishnu 9.22, 9.24–5, 9.26, 9.30–1, 9.37

Avedon, Richard *5.27*
Aveling, Thomas 6.359
averages *1.27*
- mean 1.209
- median 1.209–10
- mode 1.217

Avercamp, Barent 5.27
Avercamp, Hendrick *5.27*
Averroës (Ibn Rushd) 3.11, *3.30*, *7.23*
Avesta 7.23, 9.3
aviation fuel *6.26*, 6.91, 6.179, 6.257, 6.264
- *see also* aircraft; flight

Avicenna (Ibn Sina) *2.19*, 5.344, *7.23–4*, 7.310
Avidan, David 5.203
avionics *6.26*, 6.139–40, 6.224
Avis dynasty 3.289
avocado pear *2.19*
avocets *2.19*, 2.326
Avogadro, Amadeo, Conte di Quaregna *1.28*
avoidance, zone of 8.190
avoirdupois weight *1.28*, *9.49*
Avril, General Prosper 4.145
AWACS (Airborne Warning And Control System) *6.26–7*, 6.375
Awami League *4.26*
awl 6.107
axe *6.27*, 6.165, 6.357–8
axiom (mathematics) *1.28*
axis (astronomy):
- celestial 8.23
- of Earth *1.28*
- polar 8.125
- semi-major 8.148

Axis Powers *4.22*, *4.26*
- *see also* wars *under individual countries*

axle *6.27*, 6.300, 6.384
axolotl *2.19*, 2.226
Axum *3.30*
Ayala, Ramón Pérez de 5.173
āyātollāh *7.24*
- *see also* Khomeini

aye-aye *2.19*
Ayer, Sir Alfred Jules *7.24*, 7.105, 7.109, 7.187
Ayers Rock *1.28*, 1.170
Ayub Khan, Muhammad *4.26*, 4.256
Ayurvedic medicine *2.19*, *7.24*
Ayuthia 3.327
Ayyubid dynasty 3.314
azalea 2.284
Azande *7.24*, 7.331
- and witchcraft 7.24, 7.92, 7.112, 7.263

Azcona, José 4.156
azeotrope *1.28*
Azerbaijan *1.28–9*, *4.26*, 7.17, 7.59, *7.340–1*
Azikiwe, (Benjamin) Nnamdi *4.26*
azimuth *1.29*, 8.22
azo dye 6.14, *6.27*, 6.161
azonal soil *1.29*, 1.311
Azores *1.29*, 1.84, 3.30
Azov, Sea of *1.29*
Azreal *9.6*
AZT (drug) 6.8
Aztecs *3.30–1*
- art and music 5.15, *5.27*
- city 3.351
- languages 7.199–200
- map 3.31
- religion and myths: deities 3.297–8, 9.9, 9.17, 9.28, 9.30, 9.35–6, 9.38
- ruler 3.239
- technology 6.73, 6.219–20, 6.287, 6.346

B

Ba Jin 5.95, 5.96
Ba Maw 4.229
Baade, Walter *8.14*, 8.72
Baader-Meinhof group *4.27*, 7.308, 8.308
Baal *9.6*
Ba'al Shem Tov 7.138
Ba'athism *4.27*, *7.24*, 7.223
Babangida, General Ibrahim 4.243
Babbage, Charles 4.81, *6.27*, 6.84, 6.189
babblers *2.20*, 2.56, 2.61
Babcock, Alpheus 5.347
Babel *9.6*
Babel, Isaak *5.28*
Babeuf, François Noël *3.32*, 3.104
Babington, Sir Anthony: Plot *3.32*, 3.151
babirusa 2.253
Bábism *4.27*
Babits, Mihály *5.28*
baboons *2.20*, 2.201
Bābur, Mogul emperor *3.32*, 3.209, 3.235, 3.268, 5.462
baby 6.178
Babylon *3.32*
- arts 3.155, *5.28*, 5.170, 5.177
- astrology 7.20
- kings of 3.39, 3.155, 3.247
- religion and myths 9.6, 9.7, 9.30
  - Babylonian Talmud 7.305
  - deities 9.4, 9.5, 9.11, 9.18, 9.24, 9.35
- technology 6.7, 6.54, 6.214, 6.368

Bacchus *9.6*
Bacchylides 5.191
Bach, Alexander 4.25
Bach, Carl Philipp Emanuel *5.28*
Bach, Johann Christian *5.28*
Bach, Johann Sebastian *5.28*, 5.47, 5.48, 5.68, 5.285, 5.291, 5.341
backbone *see* spine
backgammon *7.24*
background radiation *1.29*
back-staff 6.287
Backus, John 6.85
backwash *1.29*
Bacon, Francis (b.1909) *5.29*
Bacon, Francis, 1st Baron Verulam, viscount St Albans (1561–1626) *3.32–3*, 3.319, *5.28–9*, 5.149
Bacon, Nathaniel: Rebellion *3.33*, 3.40, 3.93
Bacon, Roger *1.7*, 1.29
bacteria *2.20*, *6.27*, *6.65*
- bacteriology 2.121, 2.142, 2.281, 6.112, 6.218
  - *see also* Pasteur *and also* pathogenic *below*
- cheese-making 6.70
- control *see* preservation *under* food
- cyanobacteria 2.6, 2.36, 2.84, 2.229–30
- eradication *see* antisepsis
- food poisoning 2.39
- genetic engineering 6.153, 6.154
- growth inhibited 6.338
- and infections *see* infections
- nitrogen fixation 2.186
- pathogenic 6.27, 6.37, 6.134, 6.164, 6.191, 6.192, 6.259
- resistant 6.16
- and sewage 6.314
- *see also* fermentation

Bactria *3.33*

bactrian camel 2.50
Baden-Powell, Robert, 1st Baron *4.27*
badgers *2.20*, 2.278
badlands *1.29*
Badlands National Monument *1.29*, 1.30
badminton 7.25
Badoglio, Pietro *4.27*
Badon, Mount 3.242
al-Badr, Muhammad 4.386
Baekeland, Leo Hendrik *6.27*
Baer, Karl Ernst von *2.20*
Báez, Buenaventura 4.102
Baez, Joan 5.29, 5.385
Baffin, William *1.29–30*, 1.231
Baffin Island *1.30*
Baganda *7.25*
bagatelle *5.29*
Bagehot, Walter *4.27*
Baghdad Pact *see* Central Treaty Organization
bagpipes 5.13, *5.29*, 5.138, 5.349, 5.376
bagworm moth *2.20*
Bahadur Khan 4.162
Bahāʿī/Bahāʿīsm *4.27*, *7.25*
Bahamas *1.30*, *3.33*, 7.307, *7.341*
see also Caribbean; CARICOM
Bahaʿuʿllāh 4.27, 7.25, 7.250
Bahmani dynasty *3.33*
Bahrain *1.30*, *4.27*, *7.341*
economy and trade 7.15, 7.136, 7.309
Bahram Gur *9.6*
Bai Juyi *5.29*
Baikal, Lake *1.30*
bail *7.25*
Bailey, Donald: Bailey bridge 6.47
bailiff *3.33*
Bailly, Jean 3.351
Baird, John Logie 4.81, *6.28*, 6.349, 7.307
Baird, Spencer Fullerton *2.20*
Bajazet I, sultan of Turkey 3.253
bakelite 6.27, 6.271
Baker, H. B. 1.73
Baker, James 4.56
Baker, Dame Janet *5.29*
Baker, Sir Samuel White *4.27*
bakery *6.28*
Bakewell, Robert 3.5, *6.28*
Baki 5.462
Bakst, Léon 5.29, 5.30
Bakunin, Mikhail 4.11, *4.27*, 7.13
Balaguer, Joaquin 4.45, 4.102
Balakirev, Mily 5.156
Balaklava, battle of *4.27–8*
balalaika *5.29*
balance 6.383
balance of payments *7.25*
balance of power *7.25*
balance of trade *7.25*
balance sheet *7.25*
Balanchine, George *5.29*
Balassi, Bálint 5.214
Balban, Ghiyas ud-Din, Mameluke sultan 3.218
Balboa, Vasco Núñez de *3.33–4*, 3.268
Balder 9.2, *9.6*, 9.13, 9.23
Baldwin I, king of Jerusalem *3.34*, 3.265
Baldwin, James Arthur *5.29*
Baldwin, Robert *4.28*, 4.259
Baldwin, Stanley, 1st earl *4.28*, 4.234, 4.339
Balearic Islands *1.30*
baleen (whalebone) 6.384
see also whalebone whales
baler 6.6, *6.28*, 6.166
Balewa, Alhaji Sir Abubakar Tafawa *4.28*, 4.34, 4.242
Balfour, Arthur James, 1st earl *4.28–9*
Balfour Declaration *4.29*
Bali (demon) 9.37
Bali (island) *1.30–1*, *3.34*, *4.29*, 7.54
Balinese *7.25–6*
Balinus *see* Apollonius of Tyre
Balkan States *3.34*, 3.355–6, *4.29*
Balkan Peninsula *1.31*
Balkan Wars *4.29*
religion 9.42
see also individual countries
Balkhash, Lake *1.31*
Ball, John *3.34*, 3.364
ball and socket joint *9.59*
Balla, Giacomo *5.29*
ballade *5.30*, 5.100, 5.283, 5.476
ballads *5.29–30*, 5.181, 5.225, 5.422, 5.428, 5.476
see also poetry
Ballala II, king of India 3.170
Ballance, John *4.29*
Ballard, J. G. 5.429
ball-bearing 6.32, 6.33, 6.301
ballet *5.30–4*
ballerina *5.30*
ballet de cour *5.33–4*
choreography *see* choreographers
companies *see* ballet and dance companies
France 5.106, 5.128, 5.270, 5.373–4, 5.392, 5.402
India 5.220
Italy 5.121
positions 5.32
Russia and USSR 5.29, 5.130, 5.180, 5.247, 5.436, 5.448
United States 5.14
see also ballet and dance companies; dance
ballet and dance companies *5.34*, *5.122*
Britain 5.13, 5.24, 5.62, 5.68, 5.126, 5.129, 5.267, 5.274, 5.299, *5.373*, 5.373, 5.393, 5.461
see also Rambert
choreography *see* choreographers
Denmark 5.393
France 5.33–4, 5.106, 5.122, 5.345
Germany 5.39, 5.117, 5.122, 5.489
modern *see under* modern dance
Postmodern *see* Post-Modernism
Russia and USSR 5.29, 5.33, 5.34, 5.51, 5.130, 5.248–9, 5.285, 5.320, 5.364
United States 5.14, 5.29, 5.62, 5.120, 5.187–8, 5.214, 5.274, 5.318, 5.397, 5.415–16, 5.447, 5.452, 5.461
see also ballet; dance
Balliol family (Edward and John) *3.34*, 3.154, 3.319
ballista 6.22, *6.29*
ballistic missiles *6.29*, 6.224, 6.228, 6.235, 6.246, 7.220
anti-ballistic missiles 6.15, 6.231, 6.335–6
early-warning system 6.111
guided 6.162
intercontinental 6.29, 6.228, 6.231
opposed 7.332
pioneer 6.45
warheads 6.375
ballistite 6.242
balloons *6.29*, *6.30*, 6.134–5
airships *see* airships
ballooning 1.58, *7.26*
meteorological 6.221, 6.291–2
military 6.224
pioneers 1.58, 1.59, 6.138, 6.139, 6.232
ballpoint pen 6.259
ballroom dancing 5.89, 5.394, 5.398, 5.423–4, 5.435, 5.446, 5.482
Balmaceda, José Manuel *4.29*
Balmer, Johann 1.31
Baltic Entente *4.29*
Baltic States 3.204, 3.287
arts 5.34
Baltic Sea *1.31*
Baltic Shield 1.120
independence 7.91
languages *7.26*, 7.156, 7.157
literatures *5.34*
see also Estonia; Finland; Latvia; Lithuania
Baltimore, Lord George Calvert 3.224, 3.251
Balzac, Honoré de *5.34*
bamboo 2.21, 6.308, 6.387
bananas *2.21*, 6.271
Bancroft, Edward *3.34*
band (music) 5.47. 5.431, 5.58
band theory (atomic physics) *1.31–2*
Banda, Hastings Kanzu *4.30*, 4.207
Bandaranaike, Solomon and Sirimavo *4.30*, 4.327
Bandello, Matteo Maria *5.34*
bandicoot *2.21*, 2.274
bandkeramik *3.35*
bandora *5.34*
Bandung Conference *4.30*, 7.218
Bandura, Albert 7.7, 7.30, 7.233, 7.291
bandwidth *6.29*, 6.346
Bandyopadhyay, Bibhutibhushan 5.43
Bangladesh *1.32*, *4.30*, *7.341*
creation of 7.277
aid to 7.8
arts 5.43
diseases 7.80
economy and trade 7.87
famine 7.116, 7.117
flooding 7.136
independence 4.321
languages 7.157, 7.286
literacy 7.185
physical geography and geology 1.32, 1.36
politics and government 4.30, 4.227
population density and growth 7.241, 7.242
poverty 7.244
refugees 7.261
see also Pakistan
banjo *5.34*
bank swallow 2.293
bank vole *2.21–2*
Banks, Sir Joseph *2.21–2*, 2.43, 2.176, 4.46
banks *7.26*
Bank of England *3.34–5*, 7.43
bank rate *7.26*
central *see* central bank
clearing house 7.53
commercial 7.26
credit 7.74
letter of credit 7.183
merchant 7.1, 7.168
nationalized 7.212
savings 7.37
and technology 6.85, 6.93, 6.118, 6.179
bankruptcy 7.161
Banksia 2.270
al-Banna, Hasan 4.168
Bannockburn, battle of *3.35*
banshee *9.6*
Bantam (port) *3.35*
Banting, Sir Frederick Grant *2.22*, *6.29*
Bantus 3.213
Bantustan *4.30–1*
Bantus (*cont.*):
language 7.30, 7.217, 7.285
see also Nguni
banyan *2.22*
Bao Dai, emperor of Japan *4.31*, 4.367
baobab *2.22*
Baptism 7.185, 7.271
Baptists *3.35*, *4.31*, 4.183, *7.26*, 7.50
bar (geomorphology):
marine *1.32–3*, 1.292, 1.339
river 1.33, 1.261
bar (unit) 6.368
bar chart 1.144
Bar Cochba *3.35–6*, 3.188
bar code *6.30*, 6.179
bar mitzvah *7.26–7*
Barabbas *9.6*
Baranauskas, Antanas 5.34
Barbados *1.33*, *3.35*, 7.292, *7.341*
see also Caribbean; CARICOM
Barbara, Saint *9.40*
Barbari, Jacopo de' 5.433
Barbarossa (corsair) *3.35*, 3.265
Barbarossa (emperor) *see* Frederick I, Holy Roman Emperor
barbary ape *2.22*
Barbary Wars 4.352
Barbegal water-mill 6.30, 6.377–8
barbel *2.22*
Barber, Samuel *5.34*
Barberini, Maffeo 5.46
barberry *2.22*
barbet *2.22–3*
barbiturate *6.30*
Barbizon School *5.34–5*, 5.296, 5.392
Barbuda *see* Antigua and Barbuda
barcarole *5.35*
barchan *1.31*, 1.33
Barclay, James 4.190
Barcoo River 1.75
bard *5.35*
Bardāī, Cand 5.206
Bardeen, John 6.317–18
Barebones Parliament *3.36*
Barents, Willem *1.33*
Barents Sea *1.33*, 1.365–6
barge *6.30*, 6.56, 6.306
Baring, Evelyn (Cromer) 4.89
Baring crisis *4.31*
barite *1.33*
barium *1.33*, 1.106, *6.30*, 6.96
barium oxide 6.43
barium sulphate 1.33, 6.30, 6.255
barium titanate 6.116
bark of trees *2.23*
baobab 2.22
bark beetles 2.8, *2.23*
cinchona 2.66
cork 2.77
paper from 2.87
barley *2.23*
see also cereals and grains
barn owl *2.23*
Barnabas, Saint *9.6–7*, *9.40*
barnacle goose *2.23*
barnacles *2.23*
Barnard, Christiaan Neethling *6.30*
Barnard, Edward 8.4
Barnard's Star *8.14*
Barnardo, Thomas John *4.31*
Barnet, battle of *3.36*
Barnum, Phineas T. 5.102
Barocci, Federico *5.35*
baroclinicity *1.33*
Baroja, Pío 5.172–3
barometer *6.30*, 6.294
baron *3.36*
Barons' Wars *3.36*, 3.120, 3.194, 3.239–40
leader *see* Montfort
Baroque arts *5.35–6*
architecture 5.21, 5.35–6, *5.53–4*, 5.201, 5.471
Britain 5.21, 5.201, 5.445, 5.468

Baroque arts (*cont.*):
architecture (*cont.*):
Churrigueresque 5.100–1
France 5.170–1, 5.198, 5.262
Italy 5.45–6, 5.53–4, 5.161, 5.267–8
music 5.36, 5.175, 5.197
painting and painters 5.53, 5.74, 5.393, 5.454
France 5.88, 5.259, 5.480
Italy 5.45–6, 5.114, 5.195, 5.255, 5.282
sculpture and carving 5.45–6
barque 6.30, 6.298, 6.306
barracouta 2.315–16
barracuda 2.23, 2.24
barramundi 2.23
Barrault, Jean-Louis 5.23, 5.36, 5.296
barred spiral galaxy 8.14
barrel organ 5.288
barrel-fish 2.23
Barrie, Sir James Matthew 5.36
barrier island 1.33
barrier reef 1.33
barrister 7.27
barrow 3.36
Barrow, Sir John 4.17
Barry, Sir Charles 5.36
Barry, James 5.36
Barry, Sir John Wolfe 5.36
Barsisa 9.7
barter 7.27
Barth, Heinrich 1.33, 4.31
Bartholemew, Saint 9.5, 9.40
Bartók, Béla 5.37, 5.321
Bartolommeo, Fra 5.37
Barton, Clara 4.31
Barton, Sir Edmund 4.23, 4.31
Barton, Otis 6.31, 6.33
Bartram, John and William 2.23–4
Baruch, Bernard 4.31–2
baryons 1.33
see also neutrons; protons
Baryshnikov, Mikhail 5.37
barytes (barite) 1.33
baryton 5.37
bas relief 5.377
basalt 1.32, 1.33, 1.70, 1.138, 1.301
Basava 5.134
bascule bridge 6.30–1
base (chemistry) *see* bases
base level (rivers) 1.34
baseball 7.27, 7.294
baseflow (rivers) 1.34
baseline (surveying) 1.34
basement complex (geology) 1.34
bases (chemistry) 1.33–4, 6.2
acid-base titration 1.2
and acids *see* salts
amphoteric compound 1.12
neutralization 1.59, 1.226
see also alkali/alkalinity; hydroxide
al-Bashir, General Omar Hassan 4.332
Bashō 5.37
BASIC (computers) 6.86
basic igneous rocks *see* basalt; gabbro
basic oxygen process 6.31, 6.185
Basie, William 'Count' 5.47, 5.239
basil 2.24
Basil the Great, St 3.164, 7.205
basilica 5.37
basilisk 2.24
'basin' irrigation 6.188
basins, lunar 8.140
Baskerville, John 5.464
basket star 2.42
basketball 7.27–8
basket-makers 3.36
basket-work 6.31, 6.281
basking shark 2.24
Basov, N. G. 6.214
Basques 7.28, 7.109
language 7.28
Basques (*cont.*):
separatism 7.28, 7.212, 7.279, 7.308
sport and recreation 7.232
bass (fish) 2.24, 2.146
Bassano, Jacopo 5.37
basset horn 5.37
bassoon 5.36, 5.38, 5.352, 5.376
Bastet 9.7
Bastidas, Rodrigo de 3.35
Bastille 3.36
Basu, Ramram 5.43
Basutoland *see* Lesotho
bat *see* bats
Batavia *see* Dutch East Indies
batch production 7.248
Bate, John: Bate's case 3.94
bateleur 2.24
Bates, Henry Walter 2.24
Batesian mimicry 2.212
Bateson, Gregory 7.116
Bateson, William 2.24
bathawk 2.152
batholith 1.34, 1.166
bathos 5.38
Bathsheba 9.7, 9.26
Bathyany, Count 4.94
bathysphere and bathyscaphe 6.31, 6.33, 6.97
batik 5.38
Batista y Zaldívar, Fulgencio 4.32, 4.65, 4.89
Batlle y Ordóñez, José 4.32, 4.361
Batoni, Pompeo 5.38
bats 2.24, 2.25
bulldog 2.45
disc-winged 2.94
false vampire 2.113
flight 2.122
flying fox (fruitbat) 2.123
free-tailed 2.127
hibernation 2.157
hispid 2.158
horseshoe 2.164
leaf-chinned 2.184
leaf-nosed 2.185
mouse-tailed 2.220
pipistrelle 2.255
sheath-tailed 2.306
short-tailed 2.308
smoky 2.314
sucker-footed 2.329
vampire 2.354
al-Battānī 8.75
Battenberg, prince Louis 4.32
battering ram 6.318
battery 6.31–2
accumulator 6.1–2
in electric motor 6.115
pioneers 6.197, 6.374
see also cell (electrical)
battery farming 6.32, 6.125
battleships 6.32, 6.188
see also warships
Battuta 3.217
Batu Khan 3.143
Baucis 9.29
baud rate 6.32
Baudelaire, Charles 5.38, 5.424, 5.442
Baudin, Nicolas 4.247
Baudot, J.-M.-E. 6.32, 6.235
Baudouin, king of Belgium 4.33
Bauhaus 5.38–9, 5.194, 5.295, 5.301, 5.347, 5.394
Bausch, Pina 5.39
bauxite 1.34, 6.12, 6.66
see also aluminium *under* electrolysis
Bavaria 1.34–5, 3.37, 3.163, 4.32
Bavarian Succession, War of 3.37
Bax, Sir Arnold 5.39, 5.442
Baxian *see* xiān
Baxter, Richard 3.37
bay 1.35, 1.144–5
bay ducks 2.24–5, 2.260, 2.297, 2.298, 2.349
bay laurel 2.25, 2.184
Bay of Pigs 4.32
bay rum tree 2.25
Bayard, Pierre Terrail, seigneur de 3.37
Bayer, Johann 8.21, 8.159
Bayer letters 8.14
Bayeux tapestry 3.158, 3.382, 5.39
Bayezid I, Ottoman sultan 3.37, 3.185, 3.201, 3.349
Bayezid II, Ottoman sultan 3.37
Bayinnaung of Burma 3.359
Bayle, Pierre 3.37
Baylis, Lilian 5.39, 5.327, 5.393
bayonet 6.226
bayou 1.35
Bazaine, Achille François 4.32, 4.124
Bazalgette, Sir Joseph William 6.32
bazooka 6.130
BBC *see* British Broadcasting Corporation
beach flea 2.293
beaches 1.35
raised 1.275
ridge 1.60
shingle 1.301
beaded lizard 2.136
beaked whale 2.25
Beaker cultures 3.37
Beale, Dorothea 4.32
beam (structural) 6.336
beam-engines 6.32, 6.241, 6.330
beans 2.25–6, 2.42
bear *see* bears
bearded vulture 2.181
beard-fish 2.26
Beardsley, Aubrey 5.39
bearings 6.32, 6.33, 6.301, 6.384
bears 2.26
cave 2.58
grizzly 2.26
hibernation 2.157–8
polar 2.261, 2.262
bear's breeches 2.2
Beat Movement 5.39–40, 5.68
Beatles 5.39
Beaton, Sir Cecil 5.40
Beaton, David 3.37
Beatrix, queen of Netherlands 4.238
Beatty, David, 1st earl 4.32, 4.177
Beauchamp, John de 3.36
Beaufort, Edmund, duke of Somerset 3.310
Beaufort, Francis: Beaufort scale 1.35
Beaufort, Bishop Henry 3.38, 3.142, 3.164–5, 3.290
Beaufort, Henry, earl of Somerset 3.38
Beaufort, John, earl of Somerset 3.37–8
Beaufort, Margaret 3.38
Beaufort, Thomas, duke of Exeter 3.37
Beaumarchais, Pierre-Augustin Caron de 5.40, 5.306
Beaumont, Sir Francis 5.40
Beaumont, William 2.26
Beauregard, Pierre Gustave Toutant 4.32–3, 4.122
Beauvoir, Simone de 4.117, 5.40, 7.332
Beaux-Arts, École des 5.40, 5.304
beaver 2.26, 2.220
Beaverbrook, William Maxwell Aitken, Baron 4.33
Bebel, August 4.317
bebop 5.239
Beche, Sir Henry Thomas de la 1.35
Bechuanaland *see* Botswana
Beck, David 4.339
Becker, Gary 7.149
Becker, Howard 7.179
Becket, Thomas à, Saint 3.38, 3.39, 3.78, 3.161
Beckett, Samuel 5.2, 5.40–1, 5.197, 5.302
Beckford, William 5.41, 5.495
Beckmann, Max 5.41
Becquerel, Antoine César 1.35
Becquerel, Henry 1.35, 1.274
becquerel 1.35
Becrux 8.14
bed load 1.35
bedbug 2.26
bedding (geology) 1.34, 1.35
strata 1.322
strike 1.323
Bede 3.15, 3.38, 3.39, 3.108, 3.160, 3.193, 3.258, 3.375
Bedford, John, duke of 3.38
Bedil (Bidil) 5.345
Bedou (Bedouin) 7.28, 7.29
bedrock 1.35
bee *see* bees
Beebe, Charles William 2.27, 6.31, 6.33
beech 2.27
Beecham, Thomas 5.110
Beecher, Lyman 4.33
bee-eater 2.27
beef cattle 6.64
see also livestock
bee-fly 2.27
Beelzebub 9.7
beer *see* brewing
Beerbohm, Max 5.78, 5.340
Beersheba 9.7
bees 2.26–7
bee-keeping 6.16–17
bumblebee 2.45
carpenter 2.55
cuckoo 2.83
honeybee 2.159, 2.358
leaf-cutter 2.184
mason 2.204
mimicry 2.213
beeswax 6.33
beet:
beetroot 2.28, 2.235
spinach 2.321
sugar-beet 2.329
Beethoven, Ludwig van 5.29, 5.41, 5.270, 5.285, 5.298, 5.405, 5.442
beetles 2.27
ambrosia 2.8
bark 2.8, 2.23
blister 2.35
burying 2.46
cardinal 2.53
carrion 2.46, 2.55
chafer 2.60, 2.288
click 2.67–8, 2.83
Colorado 2.72
cucuju 2.83
death-watch 2.88
dor 2.97–8
dung 2.100
firefly 2.119
furniture 2.130
ground 2.145, 2.341
Hercules 2.155–6
ladybird 2.30, 2.181
leaf 2.184
longhorn 2.193
mythical 9.31
oil 2.232
pill 2.254
rove 2.289
scarab 2.296
Spanish fly 2.318
stag 2.324
tiger 2.341
water 2.361
weevil 2.364
wireworm 2.67–8
beetroot 2.28, 2.235
BEF *see* British Expeditionary Force

Begin, Menachem *4.33*
Camp David Accord **4**.59, **4**.299, **7**.15
and Irgun **4**.167
begonia *2.28*
Behan, Brendan **5**.226, **5**.452
behaviour *2.28*
aggression **2**.4
animals **7**.287, **7**.294
cognitive dissonance **7**.54–5
genetics *7.28–9*, **7**.252
instinct *see* instinct
intelligence *see* intelligence
learned *see* learning
organizational theory **7**.224
personality *see* personality
rational choice theory **7**.259
ritualization **2**.286
therapy *7.30*
see also behaviourism (psychology); ethology
behaviourism (philosophy) *7.29*, **7**.203
behaviourism (psychology) *7.29–30*, **7**.113, **7**.207, **7**.294, **7**.330
see also behaviour; conditioning
Behn, Afra *5.41*, **5**.323
Behrens, Peter *5.42*
Behring, Emil von **6**.177, **6**.191
Bei Dao **5**.98
Beilstein, Friedrich Konrad *1.35*
Beit, Sir Alfred *4.33*
Béjart, Maurice *5.42*
Bel **9**.11
bel canto *5.42*
Bel Geddes, Norman **5**.413
Belarus (Belorussia) *4.33*, **7**.59, **7**.220
language **7**.156, **7**.288
physical geography and geology *1.35–6*, **1**.268
Belaúnde, Terry Fernando *4.33*, **4**.148, **4**.267
belemnite *2.28*
Belgae *3.38*
Belgian Congo **4**.306, **4**.353
see also Burundi; Rwanda; Zaïre
Belgium *1.36*, *3.38*, *4.33–4*, *7.342*
aid from **7**.8
arts **5**.147, **5**.212, **5**.470
literature **5**.275, **5**.420
painting **5**.147, **5**.275, **5**.470
astronomy **8**.4, **8**.28
colonies **4**.4, **4**.83, **4**.306–7, **4**.354
economy and trade *see* EC; EEC
education **7**.155
foreign policy **4**.372, **4**.375
Council of Europe **7**.71
languages **7**.31, **7**.59, **7**.131, **7**.269
monarchy *see* rulers *below*
in OECD **7**.221
physical geography and geology **1**.36, **1**.294
politics and government:
Christian Democracy **7**.47
coalition **7**.54
communism **4**.165
elections and electoral system **7**.104, **7**.250
leader **4**.322
parliamentary **7**.229
provinces **3**.154, *9.75*
religion **9**.41
rulers **4**.33, **4**.193, **4**.388, **7**.97, **7**.139
social services and security **7**.330
sport and recreation **7**.236
technology and science **6**.115, **6**.158, **6**.198, **6**.219
chemistry **6**.70, **6**.232, **6**.325
telecommunications **7**.39
Belgium (*cont.*):
wars and defence:
World War II **4**.262
see also NATO; WEU
see also Flanders
Belgrade Conference on Security and Co-operation in Europe **7**.64
Belgrade Non-Aligned Conference **7**.218
Belial *9.7*
Belinda (Uranian satellite) **8**.178
Belisarius *3.38–9*, **3**.62, **3**.264, **3**.327
Belize *1.36*, *4.34*, **4**.67, **7**.82, *7.342*
see also Caribbean; CARICOM
Bell, Alexander Graham *6.33*, **6**.347
Bell, Sir Charles *2.28*
Bell, Clive **5**.50
Bell, Currer, Ellis, and Acton *see* Brontë
Bell, Daniel **7**.152, **7**.243
Bell, Johannes **4**.364
Bell, Patrick **6**.295
Bell, Vanessa **5**.50, **5**.494
bell magpie *2.28*
bell moth **2**.344–5
bell founding *6.33*
belladonna lily **2**.7–8
bell-bird *2.28*, **2**.77
Bellerephon *9.7*, **9**.9
bellflower **2**.50–1
Bellingshausen, Fabian Gottlieb von *1.36*, **4**.14
Bellini, Gentile **3**.371, *5.42*
Bellini, Giovanni *5.42*, **5**.71, **5**.265
Bellini, Jacopo *5.42*
Bellini, Vincenzo *5.43*, **5**.342
Bellman, Carl Michael **5**.404
Bello, Alhaji Sir Ahmadu *4.34*
Bello, Muhammed **4**.128
Belloc, Hilaire *5.43*
Bellotto, Bernardo *5.43*
Bellow, Saul *5.43*
Bellows, George Wesley **5**.24
bells *5.43*
carillon **5**.78
change-ringing **5**.88
tubular **5**.461
see also shawms
belly-dancing **5**.229
Belorussia *see* Belarus
Belshazzar, king of Babylon *3.39*, *9.7*
belt drive and chain drive *6.33–4*
belts (astronomy) **8**.78, **8**.145
beluga **2**.328
Bely, Andrey *5.43*
Bemba *7.30*
Bembo, Pietro **5**.231
Ben Bella, Ahmed *4.34*, **4**.127
Ben Nevis *1.37*
Benares (Varanasi) **3**.370
Benavente, Jacinto **5**.173
Benbow, John *3.39*
bench-mark *1.36*
'bends' **6**.97
Benedek, Ludwig von **4**.299
Benedict of Aniane, Saint **3**.39
Benedict of Nursia, Saint *3.39*, **3**.77, **3**.237, **7**.205, **8**.205, *9.40*
Benedictines *3.39*, **3**.80, **7**.205
benefit of clergy *3.39*
benefits *see* social security
Beneš, Edvard *4.34*, **4**.92, **4**.197, **4**.213, **4**.227
Bengal *1.36*, *3.39*
arts **5**.445
battle **3**.285
Bengali language **7**.157
governors **3**.79, **3**.158–9
literature *5.43*
Pala dynasty **3**.267
see also Bangladesh
Benguela Current *1.36*
Ben-Gurion, David *4.34*
Benin (Dahomey) *1.36–7*, *3.96*, *4.4*, *4.34*, **7**.124, *7.342*
in Franc Zone **7**.124
Benin (Dahomey) (*cont.*):
physical geography and geology **1**.36–7
religion and myths **9**.22, **9**.25, **9**.37
Benin (Nigeria) **3**.4, *3.39–40*, *4.34–5*
Benjamin *9.7*
Bennett, Arnold *5.43–4*
Bennett, Richard B. **4**.85
Bennington, battle of *3.40*
Benson, Arthur **5**.143
Bent, Jeffrey Hart **4**.205
Benten **9**.7
Bentham, Jeremy *4.35*, **7**.324
benthic **1**.42
Bentinck, Lord William Cavendish (1774–1839) *4.35*, **4**.344
Bentinck, William Henry Cavendish, 3rd duke of Portland (1738–1809) *4.35*
Bentley, Edmund Clerihew **5**.104
Benz, Karl Friedrich **4**.225, **4**.350, *6.34*, **6**.52, **6**.234
benzene *6.34*, **6**.71, **6**.79, **6**.125, **6**.173, *6.261*, **6**.262, **6**.267
ring *1.36*, **1**.37, **1**.181, **1**.339
benzodiazepine **6**.285
Beowulf *5.44*, **5**.148, **9**.15
Berbers *3.40*, *7.30*
dynasties **3**.4, **3**.10–11
languages *7.30*
Tuareg **7**.217, **7**.317
Berceo, Gonzalo de **5**.428
bereavement *7.30*
Berenson, Bernard **5**.6
Berent, Waclaw *5.44*
Berg, Alban *5.44*
Berger, Hans *6.34*, **6**.116
Berger, Victor **4**.94
Bergeron, Tor *1.37–8*
Bergius, Friedrich **6**.43
Bergman, Ingmar *5.44*
Bergman, Ingrid **5**.391
bergschrund *1.38*, **1**.140
Bergson, Henri **5**.343, **5**.368
Beria, Lavrenti Pavlovich *4.35*, **4**.229, **4**.244, **4**.277
beri-beri **2**.357
Bering, Vitus Jonassen *1.38*
Bering Sea *1.37*, **1**.38
Berio, Luciano *5.44–5*
Berkeley, Busby *5.45*, **5**.309
Berkeley, George *3.40*, *7.30–1*, **7**.151
Berkeley, Sir Lennox *5.45*
Berkeley, Sir William **3**.33, *3.40*
berkelium **1**.106
Berkman, Alexander **4**.156
Berlin, Irving *5.45*, **5**.373
Berlin, Isaiah **7**.124
Berlin:
Airlift *4.36*
Conference on Africa **4**.307
Congress of *4.35–6*
Wall *4.36*
Berliner, Emil **6**.158, **6**.296
Berliner Ensemble *5.45*
Berlioz, Hector *5.45*, **5**.321, **5**.442
berm *see* beaches
Bermuda *1.38*, *3.40*, *4.36*, *7.342*
Bermuda rig **6**.143, **6**.305, **6**.306
Bermuda Triangle *1.38*
Bermundez, Juan de **3**.40
Bernadotte, House of:
Bernadotte, Folke, count *4.36*, **4**.329
Bernadotte, Jean-Baptiste *see* Charles XIV, king of Sweden
rulers listed *9.67*
Bernanos, Georges *5.45*
Bernard, Claude *2.28*
Bernard, Duke of Saxe-Weimar **3**.214
Bernard of Anhalt **3**.318
Bernard of Clairvaux, Saint **3**.1, **3**.7, *3.40*, *9.40*
Bernardino de Laredo **5**.311
Bernays, Edward L. **7**.254
Berne Convention **7**.70
Bernhardt, Sarah *5.45*, **5**.73, **5**.307, **5**.362
Bernini, Gianlorenzo **5**.35, *5.45–6*, **5**.53, **5**.78
Bernini, Pietro **5**.45
Bernoulli, Jacob (Jacques) and Johannes (Jean) and Daniel *1.38*
Bernstein, Edward **7**.267, **7**.291
Bernstein, Leonard *5.46*, **5**.110, **5**.309, **5**.384, **5**.424
berries *2.28*, **2**.29, **2**.33, **2**.35–6, **2**.278, **2**.336
Berry, Chuck **5**.382
Berthelot, (Pierre Eugène) Marcelin *1.38*
Berthollet, Claude Louis, Comte *1.38*, **1**.60
Berthoud, Ferdinand **5**.105
Bertolucci, Bernardo *5.46*
Berwick, Treaty of *3.40*
beryl *1.38*, **1**.108
beryllium *1.38*, **1**.106, **6**.34, *6.34*
Berzelius, Jöns Jakob, Baron *1.38*
Besant, Annie *4.36*, **4**.48, **4**.115, **4**.213, **7**.310
Bessel, Friedrich Wilhelm *8.14*, **8**.35, **8**.116
Bessemer, Sir Henry: Bessemer converter *6.34*, **6**.185, **6**.186, **6**.188, **6**.190
Best, Charles Herbert **6**.29
bestiary *5.46*
Bestuzhev-Riumin, Count Alexis **3**.114
Beta Centauri *8.14*
Beta Lyrae stars *8.14*, **8**.15
beta particle *1.38*
beta receptor **6**.266
beta wave **6**.116
beta-blocker *6.34*, **6**.266
Betancourt, Rómulo *4.36–7*, **4**.353, **4**.363
beta-radiation **6**.153, **6**.245, **6**.288
betatron **6**.258
betel nut **2**.240
Betelgeuse *8.14*
Bethal *9.7*
Bethany *9.7*
Bethe, Hans **8**.11, **8**.98
Bethe cycle **8**.20
Bethesda *9.7*
Bethlehem *9.7*
Bethmann-Hollweg, Theobald von *4.37*
Bethune, David *3.37*
Betjeman, Sir John *5.46*
Betti, Ugo **5**.231
betting *see* gambling
Beuys, Joseph *5.46*
Bevan, Aneurin *4.37*
Bevan, Edward John **6**.66, **6**.92, **6**.295
Bevan, Robert **5**.72
beverage crops **6**.7, *6.34*, **6**.271, **6**.364
see also cacao; coffee; tea
Beveridge, William Henry, 1st Baron: Beveridge Report *4.37*, **4**.181, **4**.234, **4**.318, **4**.348, **4**.357, **4**.374
Bevin, Ernest *4.37*
Bewick, Thomas **5**.46, **5**.493, **6**.387
Bewick's swan *2.28*
Beza, Theodore **3**.199
Bhagavadgītā *3.40–1*, *5.46*, **5**.276, *7.31*, **9**.5
Bhakti literature *5.46*
Bhaktivedanta Swami, A.C. **7**.138
Bharata-nātyam *5.46–7*
Bharati, Subrahmanya **5**.446
Bhāravi **5**.401
Bhāsa **5**.401

Bhashani, Maulana Abdul Hamid **4**.26
Bhatṛhari **5**.401
Bhattacharya, K. C. **7**.156
Bhatti **5**.401
Bhavabhūti **5**.401
Bhava-chakra *9.7*
Bhave, Vinoba *4.37*
Bhoodan movement *4.37*
Bhopal *6.35*, **6**.275
Bhumibol, king of Thailand **4**.342
Bhutan *1.38–9*, *7.342–3*
Bhutto, Benazir **4**.37–8, **4**.256
Bhutto, Zulfikar Ali *4.37–8*, **4**.256
Biafra *4.38*, **7**.130, **7**.151, **7**.277
Bialik, Chaim Nahman *5.47*
Bianca (Uranian satellite) **8**.178
bias *6.35*
Biber, Heinrich Ignatz Franz von *5.47*
Bibiera, Guiseppe **5**.33
Bible *3.41*, *7.31*, **7**.48
  books listed *9.44*
  translators **3**.41, **3**.187, **3**.364, **3**.386
bicameral assembly *7.31*, **7**.182
bicarbonates **1**.39, **1**.47–8, **1**.310
biceps *2.29*
Bichat, Marie François Xavier *2.29*
bichir *2.29*
bicycle **4**.350, *6.35*
  brakes **6**.35, **6**.44
  dynamo **6**.153
  gear *6.35–6*, **6**.152
  ratchet and pawl **6**.295
  sport **7**.79
  tricycle **6**.363
  tyre **6**.35, **6**.366
  wheel **6**.384
Bidault, Georges *4.38*
Bidil (Bedil) **5**.345
Biedermeier *5.47*
Biela, Wilhelm von: Biela's Comet **8**.5, *8.15*, **8**.28, **8**.194
biennial plants *2.29*
  angelica **2**.10
  brassica **2**.41
  campion **2**.51
  delphinium **2**.90
  foxglove *see* foxglove family
  mallow **2**.200
  parsley **2**.243
  parsnip **2**.243
  salvia **2**.292–3
  teasel **2**.337
  wallflower **2**.359
  woad **2**.369
Biffen, Sir Rowland Henry *6.36*
bifocal lens **6**.327
Bifrost **9**.5, **9**.25
big band *5.47*
Big Bang theory *8.15*, **8**.33
Big Bend National Park *1.39*
Big Crunch *8.15*
Big Dipper **8**.15
Bigge, John: Bigge Inquiry *4.38*
bighorn sheep *2.29*
Bihar *3.41*, **3**.267
Bihzād, Kamāl al-Dīn *5.47*, **5**.230, **5**.344
Bijapur *3.41*
Bikini Atoll *4.38*
Biko, Steve *4.38*, **4**.39
bilberry *2.29*
bilby **2**.274
Bildt, Carl **4**.335
Bildungsroman *5.47–8*
bile and bile duct *2.29*, **2**.131–2, **2**.175, **2**.192
Bilhaṇa **5**.401
bilharzia *see* schistosomiasis
bilingualism *7.31*
bilirubin **2**.175
bill (finance):
  of exchange **7**.1, *7.31*, **7**.198
  of lading *7.31*
bill (legislation) *7.31*
  Bills of Rights *3.41*, *7.32*
bill (weapon) **6**.165
billiards *7.31*, **7**.289
Billingham, John **8**.148
Billings, William **5**.75
Billy the Kid *4.38*, **4**.39, **4**.378
Bilqis **9**.33
bimetallic strip *6.36*, **6**.350, **6**.355
binary code **6**.22, *6.36*, **6**.84, **6**.210
binary form *5.48*
binary stars *8.15–16*, **8**.20, **8**.35
  binary pulsar *8.15*
  eclipsing **8**.3, **8**.14, **8**.43
  first-magnitude **8**.2, **8**.6, **8**.14
  tenth-magnitude **8**.14
  third-magnitude **8**.29
  X-ray **8**.36, **8**.188
binary system and numbers (mathematics) *1.39*, **9**.52, *9.52*
Binchois, Gilles **5**.66
Bindusara, Mauryan emperor **3**.226
bindweed *2.29*
Binet, Alfred **2**.29, *7.32*, **7**.162, **7**.235
Bini, Lucio **6**.116
binnacle *see* compass *under* magnet
binoculars *6.36*
binomial expansion *1.39*
Bintley, David **5**.103
binturong **2**.66–7
biochemistry *2.29*, **6**.141
  see also biogeochemical cycles
biodiversity *2.29*
bioengineering *6.36*
biofuel *6.36–7*, **6**.121
biogas **6**.36, **7**.106
biogeochemical cycles *2.30*
  carbon **2**.53
  nitrogen **2**.228
  oxygen **2**.235
biogeography *2.30*
biography and biographers *5.48*
  Bengali **5**.43
  Britain **5**.26, **5**.55, **5**.184, **5**.408, **5**.435
  Greek **5**.353
  Italy **5**.469
  United States **5**.400
  see also autobiography
biology *2.30*
  biological pest control *2.30*, **2**.250, *6.37*, **6**.267
  biological warfare *7.44*
  exobiology **8**.49
  microbiology **2**.209
  sociobiology **7**.294
  see also botany; genetics; natural selection; zoology *and entries prefixed* **2.**
biomass *2.30*
  power **7**.106
biome *2.30*
biophysics *2.30*
biopsy *6.37*
bioreactor *6.37*
bioremediation **6**.27
biosensor *6.37*
Biot, Jean-Baptiste: Biot–Savart law *1.39*
biotechnology *6.37–8*, **6**.170, **6**.193, **6**.268
  see also antibiotics; fermentation *and* genetic engineering *under* genetics
biotite *1.39*, **1**.143
biplane *6.38*, **6**.139, **6**.202
bipolar transistor **6**.361
Birch, John: Society *4.38–9*
birch (tree) *2.30*
bird of paradise *2.31*
bird of paradise flower *2.31*
birds *2.30–1*
  accentors **2**.2, **2**.154
  auks **2**.18, **2**.271, **2**.279
birds (*cont.*):
  babblers **2**.20, **2**.56, **2**.61
  barbet **2**.22–3
  bee-eater **2**.27
  birdsong transcribed **5**.291
  blackbirds **2**.33, **2**.94
  bower **2**.39, **2**.40, **2**.57
  bower-bird **2**.40
  broadbill **2**.42
  bulbul **2**.45
  buntings **2**.45, **2**.144, **2**.236, **2**.373
  bustard **2**.46
  butcher **2**.47
  cardinal **2**.53
  catbird **2**.56–7
  chats **2**.61, **2**.280–1
  cotinga **2**.28, **2**.70, **2**.77, **2**.352
  courser **2**.78
  cowbird **2**.78–9
  crane **2**.80, **2**.90, **2**.324
  creeper **2**.80
  crow *see* crow family
  cuckoos **2**.78, **2**.83, **2**.286
  curlews **2**.84, **2**.327
  domesticated *see* cage-birds; poultry
  doves *see* pigeons *below*
  drongo **2**.99
  egret **2**.104
  endangered species **2**.108
  extinct **2**.96, **2**.243, **2**.317, **2**.362
  feathers **2**.114, **2**.121–2
  finch *see* finch family
  firecrest **2**.119
  flightless *see* flightless birds
  flowerpecker **2**.122
  flycatcher *see* flycatcher family
  fossil (Archaeopteryx) **2**.14
  frog-mouth **2**.127, **2**.128
  game *see* game birds
  goldcrest **2**.140
  guano from **2**.146
  honey-eater **2**.160
  honeyguide **2**.160
  hoopoe **2**.161, **2**.370
  hornbill **2**.161, **2**.162
  hummingbirds **2**.158, **2**.165–6, **2**.277
  ibis **2**.168
  jacamar **2**.174
  kagu **2**.177
  larks **2**.183, **2**.198, **2**.313
  lyre-bird **2**.196
  martins **2**.164, **2**.204, **2**.293
  merganser **2**.314
  mesite **2**.208
  migration **2**.210–11, **2**.236
  motmot **2**.220
  mousebird **2**.220
  mythical **9**.29, **9**.33
  nest **2**.226–7
  nightjar **2**.228
  nuthatch **2**.230
  ocean *see* birds *under* water
  oilbird **2**.232
  oriole *see* oriole family
  ornithology and ornithologists **2**.18, **2**.20, **2**.142, *2.236*, **2**.236, **2**.368
  ovenbird **2**.238
  owls *see* owls
  palm-chat **2**.240
  parrot *see* parrot family
  pets *see* cage-birds
  pigeons **2**.98, **2**.243, **2**.253, **2**.371
  pitta **2**.256
  plantcutter **2**.258
  plovers **2**.182–3, **2**.259
  pochard **2**.297
  potoo **2**.266
  of prey *see* eagles; falcon family; hawk family
  puffbird **2**.271
  quetzal **2**.274
  rail *see* rail
  roller **2**.287–8
  sand grouse **2**.293
birds (*cont.*):
  sandpiper *see* sandpiper family
  screamer **2**.297
  sea *see* birds *under* water
  secretary bird **2**.302
  seriema **2**.303
  sharpbill **2**.306
  sheathbill **2**.306
  shrikes **2**.249, **2**.309
  snakebirds **2**.87, **2**.371, **2**.372
  song *2.31–2*
  sparrows **2**.263, **2**.318–19, **2**.347
  starling family **2**.215, **2**.224, **2**.324
  stork family **2**.150, **2**.202, **2**.326, **2**.365
  sugarbird **2**.329
  sun bittern **2**.329
  sunbird **2**.329
  swallows *see* swallow family
  swan *see* swans
  swifts **2**.80, **2**.332
  tanager **2**.334
  tapaculo **2**.334
  thrush *see* thrush family
  tinamou **2**.341
  tody **2**.343
  toucan **2**.277, **2**.345
  treecreeper **2**.346
  trogon **2**.348
  turaco **2**.350
  vireo **2**.356
  wading *see* wading birds
  warbler *see* warblers
  water *see* birds *under* water
  wattlebird **2**.362
  waxbill **2**.363
  waxwing **2**.363
  whistler **2**.366
  white-eye **2**.366
  wings **2**.114, **2**.121–2
  wood hoopoe **2**.370
  wood swallow **2**.371
  woodcreeper **2**.370
  woodpecker **2**.370, **2**.371
  wren **2**.371
  wryneck **2**.372
bird's foot delta *1.39*
bird's nest fern *2.31*
Birdseye, Clarence *6.38*
birdwing butterfly *2.31*, **2**.32
Birendra Bir Bikram of Nepal **4**.237
Birkeland, Kristian Olaf Bernhard *6.38*
Birla family *4.39*
Biro, Lazlo **6**.259
Birr Castle *8.16*
birth *7.45*
  defect *see* inherited disorders *under* genetics
  fertility rate **7**.119
  illegitimate **7**.152
  infertility **7**.159–60
  midwifery **7**.200
  premature **6**.178
  rate *7.32*
  see also contraception; obstetrics; pregnancy
Birtwistle, Sir Harrison *5.48*
Bishan Das **3**.32
Bishop, Maurice **4**.142
bishops **7**.48–9
Bishops' wars *3.41–2*, **3**.90
Bismarck, Otto von *4.39–40*, **4**.112, **4**.278
  and Austro-Prussian War **4**.26
  and Congress of Berlin **4**.35
  and Franco-Prussian War **4**.124
  and German Second empire **4**.134
  and Kiel Canal **4**.183
  and Reichstag **4**.286
  and social welfare **4**.234, **4**.357, **4**.374
  and Three Emperors' League **4**.344
  and Triple Alliance **4**.88, **4**.352
bismuth *1.39*, **1**.106, **6**.12, *6.38*

bison 2.32
Bissaker, Robert 6.322
Bissell, Melville 6.61
bistre 5.48
bit (computer) 6.38
al-Bitar, Salah al-Din 4.27
biting housefly 2.324
biting lice 2.32
bitonality 5.48
bittern 2.32, 2.329
bitumen 1.22, 1.39, 6.38–9, 6.53, 6.79, 6.145, 6.263, 6.315
bituminous 6.78
bivalves 2.32
  clam 2.67
  cockle 2.70
  mussels 2.223, 2.246, 2.331
  oysters 2.238, 2.246
  piddock 2.253
  razor-shell 2.279
  scallop 2.295, 2.296
  shipworm 2.307
  swan mussel 2.331
biwa 5.349
Biya, Paul 4.59
Bizet, Georges 5.48, 5.329
Bjerknes, Vilhelm Friman Koren and Jakob Aall Bonnevie 1.39–40
Björling, Jussi 5.48
Björnson, Björnstjerne 5.48
Black, Sir James Whyte 2.32–3
Black, Joseph 1.40
black beetle 2.70–1
black body 8.16–17
  radiation 1.40
'black box' flight recorder 6.39, 6.136
Black Churches 7.32, 7.44, 7.50
black currant 2.33
Black Death 3.42, 6.191, 7.74
black earth see chernozem
black fly 2.33, 2.234
Black Forest 1.40
Black Hand 4.40
Black Hawk, Chief: Black Hawk War 4.40
Black Hills 1.40
black hole (astronomy) 8.17
  event horizon 8.48
Black Hole of Calcutta 3.42
Black Mountains 1.40
Black Muslim Movement 4.40
black people:
  and religion 7.32, 7.44, 7.50, 7.258
  see also civil rights; race and also in particular under South Africa; United States
Black Power Movement 4.40–1
Black Prince see Edward the Black prince
Black Sea 1.40, 1.42
Black September 4.41
black widow spider 2.33, 2.34
Black-and-Tans 4.40
Blackbeard, Teach 3.282
blackberry 2.33
blackbirds 2.33, 2.94
blackcap 2.360
blackcock 2.146
black-eyed Susan 2.2, 2.74
Blackfoot Indians 3.42, 7.219
blackmail 7.33
Blackmore, Richard Doddridge 5.48
Blackshirts 4.41
blacksmithing 6.39
  see also forging
Blackstone, Sir William 7.58
blackthorn 2.33–4
Blackwell, Elizabeth 4.41
bladder 2.34
  disorders 6.205, 6.340, 6.342
bladderwort 2.33, 2.34
Blagg, Mary 8.17
Blaine, James Gillespie 4.41, 4.131, 4.328
Blaise, Herbert 4.142
Blake, Peter 5.357
Blake, Robert 3.14, 3.42, 3.361
Blake, William 5.48–9, 5.338, 5.340, 5.368, 5.491
Blanc, Louis 4.41, 4.288
Blanc, Mont 1.79, 1.220
Blanchard, Jean-Pierre 6.138
Blanche, Jacques-Emile 5.421
Blanco, Antonio Guzmán 4.144, 4.363
Bland, Hubert 4.115
blank verse 5.49, 5.136, 5.165, 5.262, 5.269, 5.296, 5.441, 5.453
Blanqui, Louis Auguste 4.41
Blasis, Carlo 5.49
blasphemy 7.32, 7.33
blast-furnace 6.39–40, 6.186, 6.297
Blaue Reiter, Der 5.23, 5.49, 5.243–4, 5.249, 5.282
Blaumanis, Rudolfs 5.34
Blavatsky, Helena 7.310
Blazhko effect 8.142
bleach 6.40, 6.174, 6.351
Blenheim, battle of 3.43
Blenkinsop, John 6.293
blenny 2.34
Blériot, Louis 4.350, 6.40, 6.138
blewit 2.4
Bley, Carla 5.239
Bligh, William 4.41–2, 4.47, 4.294
blind fish 2.34
blind snake 2.34–5
blindness see disorders under vision
Bliss, Sir Arthur 5.49
Bliss, Nathaniel 8.193
blister beetle 2.35, 2.318
Blitzkrieg 4.42
Blixen, Karen 5.49
blizzard 1.40
Bloch, Ernest 5.49
Bloch, Marc 4.153
block mountains 1.40
block and tackle 6.285
Bloemaert, Abraham 5.49
Blok, Aleksandr 5.49–50
Blondel 3.303
Blood, Thomas 3.43
blood 2.34, 2.35, 9.60
  albumen glue 6.3
  alcohol in 6.46
  anticoagulant 6.16
  artery see arteries
  cells 2.34
  circulatory system 9.60
  clotting 6.16, 6.60
  disorders:
    anaemia 2.9, 2.161, 6.164, 6.186
    aneurysm 2.10
    arteriosclerosis and atherosclerosis 2.15, 2.77
    blood fluke 2.35
    coronary thrombosis 2.77
    drugs for 6.266
    haematology 6.164, 6.218
    haemodialysis 6.101, 6.164
    haemophilia 2.149, 6.40
    haemorrhage 2.149
    leukaemias 2.189
    septicaemia 2.303
    thalassaemia 2.339
    thromboses 2.77, 2.340
  erythrocytes 2.110
  fibrinogen 2.116
  flow 6.367
  gas analysis 6.40
  groups 2.35, 6.40, 6.194
  haemoglobin see haemoglobin
  leucocyte 2.188–9, 2.272
  and muscles 2.222
  pioneers 2.113, 2.150, 2.152
  plasma 2.116, 2.258–9
blood (cont.):
  plasmapheresis 6.271
  pressure 2.35
  pus 2.1, 2.272
  serum 6.194
  sphygmomanometer 6.218, 6.328
  study of 6.164
  transfusion 6.40, 6.194, 6.340
  vein 2.340, 2.355
  see also antibodies
blood fluke 2.35
Blood River, battle of 4.42
bloodworm 2.210
Bloody Assizes 3.43
Bloom, Benjamin S. 7.33
Bloomsbury Group 5.50
Blow, John 5.18, 5.50
blow fly 2.35
blowfish 2.271, 2.272
blowlamp 6.40
Blücher, Gebhard Leberecht von 4.42, 4.192, 4.372
blue (butterfly) 2.35, 2.36
blue chip 7.33
Blue Ridge 1.41
blue whale 2.36, 2.200
bluebell 2.50–1, 2.166
blueberry 2.35–6
bluebird 2.36
bluebottle 2.35
bluebuck 2.228
blue-green algae see cyanobacteria
blues (music) 5.50
  rhythm and 5.358, 5.382, 5.425
bluff, river 1.350
Blum, Léon 4.42–3, 4.273
Blume, Judy 5.91
Blumlein, Alan 6.40, 6.334, 6.349
Blunden, Edmond 5.174
Boadicea see Boudicca
boar 2.37
board-games see board and table under games
boas 2.36–7, 2.277, 2.293, 2.346
Boas, Franz 7.14
boat fly 2.361
boat-billed heron 2.37
boats see ships and boats
boatswain bird 2.348
Boaz 9.32
bobolink 2.37
Boccaccio, Giovanni 5.50–1, 5.141
Boccherini, Luigi 5.51
Boccioni, Umberto 5.51, 5.480
Bodawpaya, king of Burma 3.19
Bode, Johann 8.159
  Bode's Law 8.17
Bodensee 1.73
Bodenwieser, Gertrud 5.51
Bodhisattva 7.16, 7.33, 9.6, 9.15, 9.20
  Avalokiteshvara 7.79, 7.189
Bodin, Jean 3.43
Bodley, George 6.89
Bodoni, Giambattista 5.464
body:
  armour 5.22, 6.39, 6.40–1, 6.69, 6.225, 6.226
  arts 5.6, 5.289, 5.447
    adornment 7.33
    cosmetics 6.90–1, 6.260
    theatrical makeup 5.276–7
  fluid 2.37
  language see non-verbal communication
  parts of see physiology
  scanner see tomography
Boehm, Theobald 5.159
Boeotia 3.43, 3.67, 3.285, 3.352–3
Boerhaave, Hermann 2.37–8
Boers:
  Afrikaners 4.5
  Boer Wars 4.42–3, 4.82, 4.363, 7.63
  Great Trek 4.140
  leader 4.276
Boethius, Aricius Mantius Severinus 3.43, 3.318, 5.288, 7.33
bogie 6.41
bogs 1.41
  soils 1.311
  see also swamps and marshes; wetlands
Bohemia 1.41, 3.43, 4.43
  arts 5.47
  philosophy 7.34
  physical geography and geology 1.41, 1.101, 1.325
  religion 3.173
    saint 3.378–9, 9.43
  rulers 3.70, 3.124, 3.133, 3.153, 3.168, 3.328, 3.378–9
  wars 3.290
Bohemia–Moravia 4.43
  see also Bohemia; Moravia
Bohemond I, prince of Antioch 3.44
Bohr, Niels Henrik David 1.24, 1.40, 1.41, 1.70, 1.272
Boiardo, Matteo Maria 5.51
Boileau, Nicolas 5.51, 5.371, 5.402
boiler 6.41, 6.280, 6.311
  beam-engines 6.32, 6.241, 6.330
  fuel-oil 6.147
  safety valve 6.304
boiling-point 1.41
boiling-water reactor 6.281
Boilly, Louis Léopold 3.305
Bokassa, Jean Bedel 4.67
bolas 6.41
Boleslav I and II, kings of Bohemia 3.379
Boleyn, Anne 3.44
Bolger, Jim 4.240
bolide 8.97
Bolingbroke, Henry of see Henry IV
Bolingbroke, Henry St John, 1st viscount 3.44, 3.210
Bolívar, Simón 4.43–4, 4.78, 4.267, 4.319, 4.324, 4.363
Bolivia 1.41, 3.44, 4.44, 7.343
  economy and trade 4.12, 4.263, 4.320, 7.13
  independence 4.43
  leaders 4.263, 4.332
    see also Bolívar
  mining 7.343
  physical geography and geology 1.41, 1.142, 1.338, 1.339
    mountains see Andes
  population 4.320
  religious centre 3.356
  war against Chile 4.254
Böll, Heinrich 5.51
Bologna, Giovanni 5.176
Bologna, Jacopo de 5.379
bolometer 6.291, 8.17
bolometric magnitude scale 8.17
Bolshevik 4.44
Bolshoi Ballet Company 5.51, 5.285
bolts 6.247
Boltzmann, Ludwig Eduard 1.41, 1.208
  Maxwell–Boltzmann distribution 1.208
Bolzano, Bernhard 7.34
Bombay duck (fish) 2.38
bomb/bombing 6.41–2, 6.225
  depth-charge 6.99
  disposal 6.42
  Molotov cocktail 6.232
  mortar 6.232
  napalm 6.236
  offensive 4.44
  see also bomber; nuclear weapons; terrorism; weapons
bomber 6.42, 6.129, 6.139, 6.176, 6.224
Bonald, Louis de 7.16
Bonaparte, House of:
  Bonaparte, Charles J. 4.117
  Bonaparte, Elisa 4.355

Bonaparte, House of (*cont.*):
Bonaparte, Joseph 3.327, 4.265, 4.312
rulers listed *9.68*
see also Napoleon; Napoleon III
Bonar Law, Andrew 4.191
Bond, George 8.71, 8.120
Bondi, Hermann 8.33
bonding (chemistry) 1.59
bonding (psychology) 7.21–2
bonds (finance) *7.34*, 7.80, 7.132, 7.155
bone 2.37, *2.38*
backbone *see* spine
carpals 2.150
cartilage 2.55–6, 2.93
carving *5.51*
china *5.51*, 6.276
clavicle (collar bone) 2.67
coccyx 2.70, 2.355
disorders 2.236, 2.321, 6.253
endoskeleton 2.109
femur 2.186
fibula 2.186
fractures 2.126, 2.127, 9.58
humerus 2.14
jaw 2.175
joint *see* joints
metacarpals 2.150
metatarsals 2.124
patella 2.244
pelvis 2.247–8
phalange 2.124
radius 2.14
ribs 2.284
scapula 2.14
skull 2.54, 2.312
spine *see* spine; vertebrae
sternum 2.284, 2.325
tarsals 2.124
tibia 2.186
ulna 2.14
see also cartilage; endoskeleton
Bongo, Omar 4.129
bongos *5.51*
Bonhoeffer, Dietrich *4.44–5*, 4.177
Boniface, Saint *9.40*
Boniface VIII, pope *3.44*, 3.211
Boniface IX, pope 3.269
Bonington, Richard Parkes *5.51*
bonito *2.38*
Bonnard, Pierre *5.51–2*
Bonnet, Charles *2.38*
Bonneville, Lake *1.41*
Bonnie Prince Charlie *see* Stuart, Charles Edward
Bonus Army *4.45*
bony fish *2.38*
anglerfish 2.10
bichir 2.29
Bombay duck 2.38
heart 2.153
lungfish 2.195–6
swim bladder 2.332
swordfish 2.332
wrasse 2.371
booby *2.38*
book louse *2.38*, 2.88
bookkeeping *7.34*
books:
bookbinding *5.52–3*, *6.42*, 6.282
illustration *5.52–3*, 5.67, 5.87, 5.119, 5.133, 5.208, 5.209, 5.295
children 5.90, 5.120
see also illuminated manuscripts
incunabula 5.219
library 7.184
publishing 7.254
see also literature
Boole, George: Boolean algebra *1.41*, 1.198, 6.207
boomerang *6.42*
boomslang 2.73
Boone, Daniel *1.41*, *3.44*
Boötes *8.17–18*
Booth, Ballington 4.301
Booth, Catherine 4.45
Booth, Charles *4.45*
Booth, Edwin Thomas *5.53*
Booth, Felix 1.286
Booth, Hubert 6.369
Booth, John Wilkes *4.45*, 4.196
Booth, Joseph 4.73
Booth, William *4.45*, 4.301, 7.272
booth *5.53*
bootlace fungus 2.160
bootlace worm 2.284
bootstrap *6.42*
bora *1.41*
borage 2.125
borate 6.245
borax *1.41*, *6.42*, 6.43, 6.156
Borchert, Wolfgang 5.175
bord-and-pillar mining *6.42*, 6.79, 6.228
Borden, Sir Robert Laird *4.45*, 4.85
Borduas, Paul-Émile 5.73
bore, tidal 1.336
boreal climate 1.65, 1.66
Boreas *9.7*
Borges, Jorge Luis *5.53*
Borghese, Pauline 5.74
Borgia, Cesare *3.44*, 6.199
Borgia, Lucrezia *3.44–5*
boring machine and tools *6.42–3*, *6.101*, 6.212
see also drilling
Boris III, king of Bulgaria 4.55
Boris Godunov, Tsar of Russia *3.45*, 3.312
Bormann, Martin *4.45*
Born, Max *1.41–2*
Born, Nicholas 5.175
Borneo *1.42*, *3.45*
Kalimantan 4.321
Sabah 4.299, 4.321
Sarawak 4.52, 4.302, 4.321
see also Brunei
bornhardt *see* inselbergs
Bornu *see* Kanem-Bornu
Borobudor 5.221
Borodin, Alexander *5.53*
Borodino, battle of *4.45*
boron *1.42*, 1.106, *6.43*, 6.212, 6.245, 6.312
borough *3.45*
Borovsky, Karel Havlíček 5.120
Borromeo, Charles 3.89
Borromini, Francesco *5.53–4*
Borrow, George *5.54*
Boscán, Juan 5.428
Boscawen, Edward *3.45–6*, 3.324
Bosch, Carl *6.43*
see also Haber–Bosch *under* Haber
Bosch, Hieronymus *5.54–5*
Bosch, Juan *4.45*, 4.102
Bose, Satyendra Nath 1.42
Bose, Subhas Chandra *4.46*
Bosnia-Herzogovina *1.42*, *3.46*, *4.46*, *7.343–4*
bosons *1.42*
alpha particles 1.10
see also photon
Bosporus *1.42*
boss (igneous rock) 1.166
Bossuet 3.299
Boston 3.45, *3.46–7*, 3.80, 3.156
Boston fern *2.38*
Boswell, James *5.55*, 5.241, 5.460
Bosworth Field, battle of *3.47*
bot fly *2.39*
botany *2.38–9*
botanical gardens 2.21, *2.38*
botanists:
Adanson 2.2–3
Banks 2.21–2, 2.43, 2.176, 4.46
Bartrams 2.23–4
Bonnet 2.38
Brown 2.43
Garden 2.133
Jussieu 2.177
botany (*cont.*):
botanists (*cont.*):
Logan 2.193
Smith 2.314
Theophrastus 2.38, 2.339
Torrey 2.344
Withering 2.369
Woodville 2.371
see also naturalists
classification *see* nomenclature
see also plants
Botany Bay *4.46*
Both, Jan *5.55*
Botha, Louis *4.46*, 4.350
Botha, P.W. 4.234, 4.319
Bothe, Walther 8.31
Bothwell, James Hepburn, 4th earl of *3.47*, 3.97
Botswana (*earlier* Bechuanaland) *4.46*
colonized 4.306
famine 7.118
language 7.176
leader 4.181–2
physical geography and geology *1.42*, 1.179, 1.195, 1.237
in SADCC 7.296
Böttger, Johann Friedrich *6.43*, 6.276
Botticelli, Sandro *5.55*, 5.448
bottle sizes *9.62*
bottlebrush *2.39*
Botto, Ján 5.422
botulism *2.39*
Boucher, François 3.287, *5.55–6*
Bouchon, Basile 6.84
Boucicault, Dion 5.290
Boudicca, queen of Iceni 3.5, *3.47*, 3.174
Boudin, Eugène *5.56*
Bougainville, Louis Antoine de *1.42*
bougainvillea *2.39–40*
Boulanger, Georges Ernest *4.46–7*
Boulanger, Lili *5.56*
Boulanger, Nadia 5.45, 5.56
boulder *1.42*
boulder clay *1.42*, 1.337
boule (pétanque) 7.234, 7.287
Boulez, Pierre 5.12, *5.56*
Boulle, André-Charles *5.56*, 5.269
Boullée, Etienne-Louis *5.56*
Boulogne, Jen 5.176
Boulsover, Thomas 5.416
Boult, Adrian 5.110
Boulton, Matthew 5.105, *6.43*, 6.380, 6.385
Boumédienne, Colonel Houari 4.9, 4.34, *4.47*
boundary layer *1.42*
Bounty mutiny *4.47*
'Bourbaki, Nicolas' *1.42*
Bourbons *3.46*, *3.47*
Condé 3.84
rulers listed *9.68*
in Spain *see under* rulers *under* Spain
see also Charles X; Henry IV; Louis XIII; Louis XIV; Louis XV; Louis XVI; Louis XVIII
Bourdon gauge 6.181, 6.281
Bourguiba, Habib Ali *4.47*, 4.354
Bournonville, August *5.56*
Bourse 7.299
Boussingault, Jean-Baptiste Joseph Dieudonné *6.43*
Boutros-Ghali, Boutros 4.307
bouzouki *5.56*
Bovidae *see* cattle family
bow (archery) 5.56, 6.29, *6.44*
crossbow 6.92, 6.225, 6.226
see also archery
bow (music) *5.56*
bow, musical 5.15, *5.56*
bow harp *5.56*
Bow Street Runners *3.47*
Bowdich, Edward 4.20
bow-drill 6.107
bowel *see* gastro-intestinal tract
Bowen, Edward 8.119
bower-bird 2.39, *2.40*, 2.57
Bowers, H. R. 1.295
Bowes-Lyon, Elizabeth 4.133
bowfin *2.40*
Bowie, James 4.7
bowl cultures *see* Western Neolithic
Bowlby, John 7.21–2, 7.110, 7.253, 7.291
bowling *7.34*
bowls *7.34*
bowstring arch bridge 6.48
box (plant) *2.40*
Boxer rising *4.47–8*
box-fish 2.348
box-frame *see* frame, building
box-girder *6.44*, 6.293, 6.336, 6.341
boxing *7.34*, 7.305
boy actors' companies *5.56–7*
boyar *3.47–8*
Boyce, William *5.57*
Boycott, Charles Cunningham: boycott *4.48*, *7.34*
Boyd, Arthur 5.27
Boyd, Martin A'Beckett *5.57*
Boyle, Charles (4th earl of Orrery) 8.115
Boyle, Richard, 3rd earl of Burlington 5.66
Boyle, Robert *1.42*, 1.60, 1.157, 3.319, 5.150, 7.186–7, 8.6
Boylston, Zabdiel *2.40*
Boyne, battle of the *3.48*
Boys, Sir Charles Vernon *1.43*
Bozeman, John 4.251
brachiopods 2.182
bracken *2.40*
bracken bug 2.151
bracket fungi *2.40*
Braddock, Edward *3.48*, 3.134, 3.323
Bradford, William *3.48*, 3.62, 3.282, 3.286
Bradlaugh, Charles *4.48*
Bradley, James 8.1, *8.18*, 8.181, 8.193
Bradley, Omar Nelson *4.48*, 4.245
Brady, Jim 4.38
Brady, Mathew 5.482
Braganza dynasty *3.48*
Bragg, General 4.71
Bragg, Sir William Henry and Sir Lawrence *1.43*, 1.189, 2.158, 6.303
Brahe *see* Tycho Brahe
Brahmā *7.34*, *9.7*
Brahman 7.3, *7.34–5*, 7.133, 7.203, 7.263, 7.274, 7.295
see also *ātman*
Brahmaputra *1.43*
Brahmo Samaj movement *4.48*, 7.14
Brahms, Johannes *5.57*
braiding (rivers) *1.43*, 1.140
Braille, Louis: braille *6.44*, *7.35*, *9.48*
Brain, Dennis *5.57*
brain *2.40–1*, *9.61*
disorders:
'brain death' 6.183
cerebral palsy 2.60
cognitive damage 7.55
dementia 2.90
electroencephalography 6.34, 6.116
epilepsy 2.109–10
magnetic resonance imaging 6.213
meningitis 2.12, 2.207, 6.260
neurosurgery 6.240–1
and pollution 6.377
scanner *see* tomography
stereotaxic surgery 6.334
stroke 2.242, 2.327
see also mental handicap; mental illness
hippocampus 2.185

brain (*cont.*):
hypothalamus **2**.167
intelligence *see* intelligence
pineal gland **2**.254
see also learning; memory; mind
Braine, John **5**.17
brake-van ***6**.44*
braking system **6**.35, ***6**.44–5*
air-brakes **6**.8, **6**.383
hydraulic power **6**.35, **6**.44, **6**.173
Bramah, Joseph ***6**.45*, **6**.173, **6**.217
Bramante, Donato **3**.301, ***5**.57–8*
bramble **2**.68
Bran ***9**.7–8*
Branca, Giovanni **6**.333
Brancusi, Constantin **5**.2, ***5**.58*
Brandenburg ***3**.48–9*, **3**.132, **3**.133, **3**.167
Brando, Marlon **5**.430
Brandreth, Jeremiah **4**.266
Brandt, Bill ***5**.58*, **5**.347
Brandt, Willy ***4**.48*, **4**.135
Brandt Report ***4**.48*, **7**.218
Ostpolitik **4**.97, **4**.252
Brandywine, battle of ***3**.49*
branle ***5**.58*
Brant, Isabella **5**.393
Brant, Joseph **3**.180
Braque, Georges ***5**.58*, **5**.119
brass **6**.11, ***6**.45*, **6**.89, **6**.221, **6**.391
memorial ***5**.58*
brass instruments ***5**.59*
brass band ***5**.58*
cornet **5**.113
euphonium **5**.150
flugelhorn **5**.158–9
ophicléïde **5**.329
pitch **5**.352
posthorn **5**.362
sackbut **5**.397
saxhorn **5**.403
tuba **5**.59, **5**.352, **5**.461
see also bugle; horn; trombone; trumpet; woodwind
'Brassaï' ***5**.58*
brassica ***2**.41*
kohlrabi **2**.41, **2**.180
mustard **2**.223, **2**.299
rape **2**.276, **2**.278
swede **2**.331
turnip **2**.351
see also cabbage
Brathwaite, Edward Kamau ***5**.59*
Brattain, Walter **6**.317–18
Brauchitsch, Walter von ***4**.48*
Braudel, Fernand **4**.153
Braun, Eva **4**.247
Braun, Karl Ferdinand ***6**.45*, **6**.215
Braun, Wernher Magnus Maximilian von **4**.323, ***6**.45*, **6**.179, **6**.301
Braverman, H. **7**.86–7
Bravo, Manuel Alvarez ***5**.59*
Brazil ***3**.49*, ***4**.48–9*, ***7**.344*
anthropology **7**.183
architecture **5**.320
demography **4**.320, **7**.6, **7**.241, **7**.242
diseases and disorders **7**.9, **7**.140
economy and trade **4**.320, **7**.215
energy and fuel **6**.37, **6**.95, **7**.106
ethnic groups **7**.335
food and drink **2**.41, **6**.324
forests **6**.111, **7**.83
independence **4**.43
literacy **7**.185
mining **6**.101
music **5**.476
physical geography and geology ***1**.43*
climate **1**.65
plain **1**.142
plateau **1**.207
rivers **1**.166, **1**.201, **1**.248
see also Amazon
waterfall **1**.166
Brazil (*cont.*):
political leaders **4**.12, **4**.105, **4**.187, **4**.362–3
population **4**.320, **7**.6, **7**241, **7**.242
region *see* Amazon
religion/theology **7**.171, **7**.327, **9**.33
rulers *see* John VI; Pedro
technology and science **6**.37, **6**.95, **6**.101, **6**.111, **6**.234
see also entries above and below prefixed **1.** *and* **6.**
urban areas **7**.197, **7**.216, **7**.283
vehicles and transport **6**.234
wars and defence **4**.259–60, **7**.18, **7**.220
brazil nut tree ***2**.41*
brazing ***6**.45*
Brazza, Pierre Savorgnan de **4**.83
Brazzaville Conference ***4**.49*
see also Congo People's Republic
bread **2**.365, **6**.28, ***6**.45–6*
breadfruit ***2**.41*
break-even analysis ***7**.35*
Breakspear, Nicholas *see* Adrian IV
breakwater **6**.276
bream ***2**.41*, **2**.265, **2**.280, **2**.298
Brearley, Harry ***6**.46*, **6**.330
breasts *see* mammary glands
breathalyser ***6**.46*
breathing *see* respiration
breccia ***1**.43*
Brecht, Bertolt ***5**.59*, **5**.255, **5**.485
Breckinridge, John Cabell ***4**.49*
Breda ***3**.49*
breeding:
animals **6**.7, ***6**.14–15*, **6**.28, **6**.125, **6**.155, **6**.283
eugenics **7**.110
plants **6**.36, **6**.155, **6**.271
see also genetics; reproduction
breeze ***1**.43–4*, **9**.56
Breitenfeld, battles of ***3**.49*
bremsstrahlung ***1**.44*, ***8**.18*
Brendan, Saint ***3**.49*
brent geese **2**.141
Brentano, Clemens ***9**.23*
Brentano, Franz **7**.35, **7**.162
Brest-Litovsk, Treaty of ***4**.49*
Brétigny, Treaty of ***3**.49*
Breton, André ***5**.59*, **5**.440
Breton language **7**.42, **7**.156
Bretton Woods Conference ***4**.49*, **7**.134, **7**.176
Breuer, Josef **7**.126
brewing **2**.135, **2**.317, **6**.10–11, **6**.37, ***6**.46*, **6**.68
hops **2**.68, **2**.161
Brewster, David **8**.53
Brezhnev, Leonid Ilyich ***4**.49*, **5**.396, **7**.6
Brian Boru, High King of Ireland ***3**.49*, **3**.179
Briand, Aristide ***4**.49*, **4**.180
bricks and bricklaying ***6**.46–7*, **6**.51, **6**.82, **6**.144, **6**.145–6, **6**.235, **6**.372
adobe **1**.3
and arts ***5**.59–60*, **5**.285
bridewealth **7**.95
bridge (card-game) **7**.40
Bridges, Robert ***5**.60*
bridges ***6**.47*, ***6**.48*
aqueduct **6**.18–19, **6**.379
bascule **6**.30–1
bowstring arch **6**.48
box-girder **6**.44, **6**.293, **6**.341
cable-staged **6**.48
cantilever **6**.47, **6**.48, **6**.57
clapper **6**.48
disaster **1**.281
drawbridge **6**.106
and hydraulic power **6**.173
iron and steel **6**.47, **6**.51, **6**.111, **6**.112, **6**.187, **6**.188, **6**.280, **6**.297, **6**.334, **6**.385
bridges (*cont.*):
masonry **6**.47, **6**.48, **6**.372
military **6**.260–1
plate girder **6**.272
pontoon **6**.47
railway **6**.49, **6**.111, **6**.293
reinforced concrete **6**.213–14
steel **6**.44, **6**.111, **6**.293
suspension **6**.47, **6**.48, **6**.311, **6**.341, **6**.350
swing **6**.342
transporter **6**.361, **6**.362
truss **6**.48, **6**.337
tubular-girder **6**.334
viaduct **6**.293, **6**.371–2
Bridget of Kildare, Saint ***9**.40*
Bridgewater, Francis Egerton, 3rd duke of ***3**.49*
brig ***6**.47*, **6**.76, **6**.298, **6**.405, **6**.408
Brigantes ***3**.49–50*
brigantine **6**.47
Brighouse, Harold **5**.212, **5**.259
Bright, John **4**.15, ***4**.49*, **4**.77, **4**.209
Bright, Richard **2**.41
Bright, Timothy **6**.318
brightening, limb **8**.86
brightening agent ***6**.47*, **6**.376
brightness of star *see* luminosity
brill ***2**.41*
brimstone butterfly ***2**.41*
Brindley, James **3**.49, ***6**.47*, **6**.56
brine ***1**.44*
briquetting ***6**.47*
Brisingamen **9**.13
Brissot de Warville, Jacques Pierre ***3**.50*, **3**.141
bristle fly ***2**.41*
bristletail **2**.31, **2**.310
bristleworm ***2**.42*
Britain (and United Kingdom) ***3**.50*, ***3**.115*, ***4**.50*, ***7**.387–8*
agriculture:
Agricultural Revolution **3**.5, **3**.360
agrochemicals **6**.128, **6**.180, **6**.196
crops **6**.127, **6**.359
rotation **3**.91–2
land **3**.83, **3**.342–3
see also land *below*
livestock **6**.28, **6**.315
machinery **6**.80, **6**.196, **6**.227, **6**.273, **6**.295, **6**.355, **6**.364
organic **7**.224
research and development **6**.6–7, **6**.36, **6**.359
uprising **4**.335
aircraft:
aerodynamics **6**.65
airport **6**.9
designer **6**.98
ejection seat **6**.113
flying boat **6**.140
glider **6**.157
history **6**.138, **6**.139
human-powered **6**.172
jets **6**.189, **6**.384–5
military **6**.129, **6**.374
STOL **6**.335
supersonic **6**.339–40
see also space *below*
as Albion **3**.7
almshouses for poor **3**.11
Anglo-Saxon *see* Anglo-Saxons
anthropology **7**.14
arts:
abstract **5**.205, **5**.303, **5**.318, **5**.350
architecture *see under* architecture
art nouveau **5**.39, **5**.177, **5**.273–4
Augustan *see* Augustan Age
caricature **5**.119, **5**.177, **5**.392–3
ceramics and pottery *see under* ceramics
cinema *see under* cinema
Britain (and United Kingdom) (*cont.*):
arts (*cont.*):
copyright **7**.70, **7**.233
decorative **5**.140, **5**.327
drawing **5**.39
engraving *see under* engraving
etching **5**.114, **5**.118, **5**.338
furniture *see under* furniture
garden *see under* art *under* garden
graphic *see under* graphic art
illuminated manuscripts **3**.111, **5**.17
interior design **4**.191, **5**.245–6, **5**.304–5, **5**.489
literature *see* literature *below*
metalwork **5**.79, **5**.176, **5**.177
museums **7**.209
music *see* music *below*
ornaments **3**.10, **3**.180, **3**.346
painting and drawing **5**.119, **5**.177, **5**.392–3
see also under landscape painting; painting; portraiture
photography *see under* photography
print-makers **5**.208
sculpture *see under* sculpture and carving
set and stage design **5**.117, **5**.208, **5**.241, **5**.269
woodwork **5**.46, **5**.176
see also under furniture
see also Arts and Crafts Movement
astronomy **8**.11, **8**.36, **8**.69
Astronomers Royal **8**.1, **8**.2–3, **8**.10, **8**.18, **8**.35, **8**.51, **8**.142, **8**.193
see also Flamsteed
black holes **8**.64
comets **8**.46, **8**.63
by Cook **8**.13, **8**.173
ionosphere **8**.74
laboratory **8**.76
Moon **8**.87, **8**.147
observatories **6**.291, **6**.303, **8**.76, **8**.105, **8**.142
photography **8**.120
planets and their satellites **8**.8, **8**.17, **8**.61, **8**.71, **8**.106–7, **8**.145, **8**.174, **8**.177
radio **8**.33, **8**.35
SETI **8**.148
stars **8**.43, **8**.53, **8**.59, **8**.84, **8**.90, **8**.129, **8**.159, **8**.175
catalogue **8**.21–2, **8**.51
see also Boyle, Robert; Halley; Herschel; Hooke; Newton
bailiffs **3**.33
Bounty mutiny **4**.47
bridges **6**.47
cantilever **6**.57
Iron Bridge **6**.188, **6**.385
railway **6**.49, **6**.293
suspension **6**.341, **6**.350
swing **6**.342
Tower Bridge **6**.31
transporter **6**.361, **6**.362
buildings **6**.50, **6**.51, **6**.66, **6**.276, **6**.280, **6**.297, **6**.322, **6**.344, **6**.387
see also under architecture
canals **6**.47, **6**.56, **6**.350, **6**.385
cars *see* roads *below*
ceramics *see under* ceramics and pottery
chemistry and chemists:
elements **1**.80, **1**.227
gases **1**.40, **1**.84–5, **1**.267, **1**.275–6
industry and technology **6**.69, **6**.70, **6**.79, **6**.96–7, **6**.125, **6**.191, **6**.316, **6**.318
sulphur and sulphuric acid **6**.197, **6**.232
textiles **6**.92, **6**.211, **6**.260

Britain (and United Kingdom) (*cont.*):
chemistry and chemists (*cont.*):
physical **1**.142
see also Boyle *and* iron and steel; metallurgy *below*
children and young people **4**.27, **4**.31, **4**.249
see also education *below and* literature *under* children
clocks **6**.77, **6**.166
colonial *see* empire *below*
computers **6**.27, **6**.84, **6**.85, **6**.86, **6**.361, **6**.366
contraception **4**.48, **4**.330
co-operative movement **4**.86, **4**.254
country house **3**.89
crime **3**.165, **7**.120, **7**.124, **7**.146, **7**.174, **7**.250, **7**.301
see also laws *below*
dandy **3**.246
demography **4**.208, **7**.105, **7**.153, **7**.160
immigration **7**.105, **7**.153
see also Malthus
divorce **7**.94
domestic appliances **6**.89, **6**.172, **6**.355, **6**.369
economy/economics and trade and finance:
with Africa **3**.311
agreements **4**.252, **4**.338–9
see also EC; EEC
agriculture *see* agriculture *above*
aid from **7**.8
alimony **7**.189
banks **7**.26, **7**.43
Bank of England **3**.34, **3**.35
broker **7**.35
commodity market **7**.57
companies **7**.19, **7**.62
Continental System **4**.85
co-operatives **7**.69
depression of 19th century **4**.159
economists **4**.27, **4**.62–3, **4**.77, **4**.102
see also classical economics; Keynesianism; Manchester School; Marshall, Alfred; Malthus; Mill, John Stuart; Ricardo; Smith, Adam
and EFTA **7**.111
exchange rate **7**.112
financier **3**.149
free trade **4**.49, **4**.77, **4**.159
friendly society **7**.127
Giffen goods **7**.131–2
Gold Standard **4**.138
in Group of Seven **7**.136
index **7**.155
industry *see* industry *below*
livery companies **3**.208
Luddites **4**.201
Merchant Adventurers and Staplers **3**.229
money supply **7**.206
motorcar manufacture **4**.248
multinationals **7**.209
nationalization **7**.212
Navigation Acts **3**.247
in OECD **7**.221
pay **7**.107
'People's Budget' **4**.266
pipe-roll **3**.282
poll tax **3**.287
regulation **7**.120
retail trade **7**.302
RPM **7**.266
with Russia **3**.243
scarcity **7**.275
scutage **3**.320
securities **7**.132, **7**.155, **7**.299
ship-money **3**.327
slaves **3**.25
South Sea Bubble **3**.334, **3**.335
stock exchange **7**.299

Britain (and United Kingdom) (*cont.*):
economy/economics and trade and finance (*cont.*):
tariff reform **4**.338–9
taxation **3**.287, **3**.320, **3**.339, **3**.347, **3**.357, **3**.359, **3**.363, **3**.383, **7**.306
tithe **3**.357
tunnage and poundage **3**.363
unemployment **7**.320
Wardrobe **3**.377
wool **3**.229, **3**.385
see also East India Company *and also* agriculture *above and* industry *below and individual industries above and below*
education **7**.276
assessment **7**.20
comprehensive **7**.63, **7**.276
curriculum **7**.78
distance **7**.92–3
dyslexia **7**.97
higher **7**.92–3, **7**.141
history of **3**.252, **3**.298, **4**.32, **4**.36, **4**.122, **4**.340
independent **7**.155
intelligence tests **7**.162
language **7**.180
primary **7**.246
scientific **6**.177, **6**.248
teaching **7**.307
electricity **6**.53, **6**.125, **6**.134, **6**.173, **6**.280, **6**.355–6, **6**.387
generation and supply **6**.114, **6**.127
lamps **6**.177, **6**.342
see also energy *below and under* physicists *below*
electronics **6**.118, **6**.183
empire/colonialism ***4**.50–2*
Africa **4**.171–2, **4**.201, **4**.220, **4**.289, **4**.306–7, **7**.151
America *see* colonies *under* America
Australia and New Zealand **4**.13–14, **4**.93, **4**.142, **4**.190, **4**.294, **4**.338, **4**.370
China **4**.110
colony defined **7**.57
Commonwealth *see* Commonwealth of Nations
decolonization **4**.4, **4**.321
dominions **4**.102
Europe **3**.141
exploration *see* explorers *below*
Hong Kong **7**.57
India *see* British Raj
legislation **4**.162, **4**.376
mandates **7**.191
Middle East **4**.89, **4**.138, **4**.169, **4**.220
North America **4**.110
South-East Asia **4**.281, **4**.321
trade routes **3**.107
see also British Raj; Commonwealth of Nations *and also* explorers; foreign policy *below and individual countries*
energy and fuel **6**.20, **6**.378
coal **6**.78, **6**.79, **6**.80, **6**.275
gas **6**.150, **6**.359, **6**.386
from water **6**.173, **6**.355–6, **6**.378, **6**.381
see also steam *and also* electricity *above and* nuclear power *below*
engines **6**.32, **6**.322, **6**.335, **6**.359, **6**.385
steam **6**.41, **6**.43, **6**.241, **6**.330–1, **6**.333, **6**.363, **6**.380
environment **7**.107, **7**.135
pollution **6**.275, **7**.2
eugenics **7**.110

Britain (and United Kingdom) (*cont.*):
evolutionism/Darwinism **4**.317, **4**.326
explored by Greek **1**.271
explorers **3**.122
Africa *see under* exploration
Antarctic **1**.75, **1**.286, **1**.295, **1**.300
Arctic **1**.84, **1**.127, **1**.248, **1**.286
Asia **1**.84
Australia and Pacific **1**.75, **1**.83, **1**.95, **1**.118, **1**.123, **1**.201, **1**.216, **1**.324, **3**.96, **3**.106
Canada *see under* Canada *under* exploration
Central America and West Indies **3**.159, **3**.298–9
circumnavigators **1**.95
Middle East **1**.94
Russia **1**.58
South and Central America and South Atlantic **1**.47, **1**.70, **1**.84, **1**.88, **1**.275, **3**.106, **3**.159, **3**.298–9
United States *see under* exploration
see also empire/colonialism *above*
finance *see* economy/economics *above*
food and drink **6**.73, **6**.105, **6**.134, **6**.170, **6**.342
foreign policy:
appeasement **4**.16
Concert of Europe **7**.135
Council of Europe **7**.71
entente cordiale **4**.112
Munich Pact **4**.227
Stresa Conference **4**.330–1
Sykes–Picot Agreement **4**.335
treaty with United States **4**.373
Washington Conference **4**.372
Western European Union **4**.375
see also empire/colonialism *above*; wars and defence *below*
genetics **6**.105
government *see* politics and government *below*
Great Exhibition **6**.160–1
Great Plague **3**.145–6
Greenwich meridian *see* Greenwich Mean Time
health *see* medicine *below*
Hellfire Club **3**.160
historiography **3**.141, **3**.383, **4**.153, **5**.43, **5**.78, **5**.176, **5**.408
housing **4**.58
immigration **7**.105, **7**.153
industry **7**.277
industrial relations *see* unions and labour *below*
Industrial Revolution **3**.178, ***6**.178*, **7**.158
see also chemistry; economy/economics *above and* technology and science *below and also individual industries above and below*
inventor **4**.151
Ireland *see* Ireland
iron and steel industry ***6**.186*, ***6**.188*
alloys **6**.235
Bessemer **6**.34
cast iron **6**.385
coke first used **6**.96
crucible process **6**.172
open-hearth **6**.250, **6**.319
puddling **6**.90
rolling mill **6**.302
sherardizing **6**.316
stainless **6**.46
islands **1**.153, **1**.241, **1**.295, **1**.301, **1**.318, ***9**.73*
kingdoms, early *see in particular* Mercia; Northumbria; Wessex

Britain (and United Kingdom) (*cont.*):
knighthood **3**.198
labour movement *see* unions *below*
land:
common **3**.83, **3**.115, **7**.57–8
demesne **3**.101
Domesday Book **3**.104–5
enclosures **3**.115, **7**.179
franklins **3**.132
glebe **3**.142
marcher lords **3**.221, **3**.241, **3**.276
reform **3**.339, **7**.179
socage **3**.331
strips **3**.342–3
yeomen **3**.388
see also agriculture *above*
languages **7**.59, **7**.106–7, **7**.287
accent **7**.1
writing **6**.259, **6**.318
see also English language
laws/legal system ***7**.318–19*
acts **7**.2
anti-papacy **3**.290
anti-Radical **4**.315
assizes **3**.25–6
attainder **3**.27
barrister **7**.27
blasphemy **7**.33
Bloody Assizes **3**.43
broadcasting **7**.35
case **7**.41
censorship **7**.42
chancery **3**.68
charters **3**.71
Coercive Acts **3**.80, **3**.87
colonial **4**.162, **4**.376
Combination Acts **4**.79
common *see* common law
company **7**.62, **7**.197
compensation **7**.62
consumer **7**.68
Corn Laws **4**.86
Anti-Corn Law League **4**.15
court-martial **7**.73
courts **7**.15, **7**.73, **7**.250, **7**.258
crime *see* crime
criminal responsibility **7**.76
customary **7**.78
defamation **7**.82
Defence of the Realm **4**.94
detention without trial **7**.87
discrimination **7**.92
Disestablishment Acts **4**.100
ecclesiastical courts **3**.345
equal pay **7**.107
equity **7**.108
extradition **7**.114
family **7**.94
frankpledge **3**.132
habeas corpus **7**.138
homicide **7**.146
homosexuality **7**.146, **7**.147
hue and cry **3**.170
illegitimacy **7**.152
industrial relations **4**.79, **4**.252, **4**.270, **7**.54, **7**.159
injunction **7**.94
Inns of Court **7**.161
judges **3**.186, **7**.173
judicial review **7**.173
jury **7**.174
labour **4**.79, **4**.252, **4**.270, **7**.54, **7**.159
Labourers Statute **3**.201
lawyers **3**.81, **3**.321, **3**.332–3
legal aid **7**.182
magistrate **7**.189
mutiny **3**.243–4
Navigation Acts **3**.247
outlaw **3**.306
petty sessions **3**.278
pillory and stocks **3**.282
poaching **3**.286
police **3**.47

Britain (and United Kingdom) (cont.):
laws/legal system (cont.):
Poor Laws 3.288–9, 3.344, 3.385, 4.12, 4.273
precedent 7.245
prisons 3.170, 3.298, 4.128, 6.362
property 7.250
Race Relations 4.280
Reform Acts 4.285
and religion see law under religion/theology below
Riot Act 3.304
sanctuary 3.315–16
security services 7.278
Settlement Acts 3.322
Six Acts 4.315
solicitor 7.295
Star Chamber 3.340
suicide 7.301
Supremacy Acts 3.346
Taff Vale case 4.336
Test Acts 3.352
Toleration Acts 3.358
Townshend Acts 3.359
tribunal 7.315
trust 7.316
Uniformity Acts 3.365
unions and labour 4.79, 4.252, 4.270, 7.54, 7.159
wages 4.353
watch and ward 3.378
wergild 3.379
see also under religion below
Levellers 3.207–8
literature 4.191, 5.28–9, 5.146, 5.294, 5.326
diarists 3.275–6, 5.150, 5.343
journalism 7.216
libraries 7.184
myths see myths and legends below
tragedy 5.162, 5.282, 5.295; 5.459
see also under autobiography; biography; children; comedy; critics; drama; epics; essays; historiography; journalism; naturalism; poetry; Romanticism; Renaissance; satire; stories and travel-writing under travel
machine-tools 6.45, 6.217, 6.236, 6.385
manor houses 3.218–20
maps 6.341
Masonic Order 7.195
mathematics and mathematicians:
applied 1.146
calculator 6.53
calculus 1.332
geomorphological 1.260
logarithms 1.224
logic 1.87, 1.365
pure 1.55
slide-rule 6.322
see also Boole; Newton and also computers above
medicine and health care 7.293
alternative (complementary) 7.62
cohort study 7.56
finance 7.140
medical profession 7.196
psychiatric 7.116, 7.198
public health 4.69
technology 6.85, 6.113, 6.251, 6.254, 6.387
disease prevention 6.176–7, 6.189, 6.218
drugs 6.69, 6.134, 6.188, 6.260
surgery and anaesthesia 6.13, 6.16, 6.73, 6.204, 6.320, 6.340
women in 4.12, 4.57, 4.66, 4.243

Britain (and United Kingdom) (cont.):
metallurgy and mining 6.277
alloys 6.11
arts 5.177
coal-mining 6.79
hallmarks 6.63
plating 6.120, 6.315
powder metallurgy 6.387
see also iron and steel industry above
mountaineer 1.366
museums 6.310
music 5.22, 5.60, 5.116, 5.291, 5.486
consorts 5.69, 5.176, 5.234, 5.304
polyphony 5.69, 5.139, 5.176, 5.393
symphonies 5.340, 5.393, 5.457
see also under ballet and dance; choreographers; composers; counterpoint; opera; organ; pianoforte; sacred; songs
myths and legends 9.13, 9.27, 9.36
deities and monsters 9.10, 9.23
Gaelic 9.9, 9.23, 9.25
heroes 9.7–8, 9.9, 9.13, 9.14, 9.25, 9.28, 9.31, 9.36, 9.38
legendary rulers 9.7–8
see also Arthur
Middle English 5.294
see also Knights of the Round Table
national parks 7.212
naturalists 7.294
see also Darwin
nuclear disarmament 4.58, 4.248
nuclear power 6.3, 6.54, 6.57, 6.123, 6.213, 6.243, 6.245
optics and glass 6.136, 6.156, 6.282, 6.294
paper 6.105, 6.145
parental leave 7.228
Parliament 3.270, 4.260–1
Addled 3.3
Barebones 3.36
Cavalier 3.65
five members 3.127
Houses of see House of Commons; House of Lords
impeachment 3.175
Long 3.144, 3.209–10
Model 3.234
Parliamentary government 7.52, 7.229, 7.247
Petition of Right 3.278
plot to destroy 3.150–1
Pride's Purge 3.292
Rump 3.312
Self-Denying Ordinance 3.321
see also politics below
philosophy and philosophers 1.365
analytic 7.12, 7.185
empiricists 7.30–1
see also Locke; Russell
idealism 7.324
linguistic 7.185
logical positivism 7.187
morality and religion 3.54, 3.190, 3.383
natural 3.32–3
nominalist 3.383
political 3.157, 3.167, 3.208–9, 7.144
utilitarianism 4.35, 7.324
see also Mill
physical geography and geology 1.44, 1.109–10
cave 1.318
cliffs 1.32, 1.34, 1.57
climate and meteorology 1.65, 1.140, 1.149, 1.300
coasts 1.32, 1.34, 1.57, 1.199, 1.324

Britain (and United Kingdom) (cont.):
physical geography and geology (cont.):
English Channel 1.110
geographer 1.150
geologists 1.35, 1.133, 1.160, 1.200, 1.223, 1.275, 1.298, 1.308
glaciation 1.323, 1.353
gorge 1.59
islands 1.153, 1.241, 1.295, 1.301, 1.318
lakes 1.186, 1.331
limestone 1.238
meteorology see climate above
mountains see under mountains
North Sea 1.231
rivers 1.67, 1.211, 1.300, 1.334, 1.336, 1.342
schist 1.294
screes 1.296
slate quarry 1.306
tor 1.340
turbidite 1.347
unconformity 1.349–50
physicists:
atomic and nuclear 1.42, 1.57, 1.67, 1.83, 1.221, 1.309, 1.335, 1.366, 6.258
see also Rutherford and nuclear above
electricity/electromagnetics 1.43, 1.80, 1.84, 1.139, 1.153, 1.208, 1.365
see also electricity above
gas 1.324
light 1.321, 1.371–2
quantum theory 1.92
thermodynamics 1.335
see also Boyle; Cavendish; Faraday; Newton
plastics 6.276
politics and government 7.318
administration 3.171, 3.300, 3.326, 3.327
alderman 3.8
anti-slavery 4.378
assassination plot 4.65–6
Bill of Rights 3.41
bills 7.31
cabal 3.56
cabinet 3.56, 7.38
Chartism 4.70–1
civil service 7.52
Commonwealth (Republican) 3.83
Conservatives/Tories
New Right 7.69, 7.86, 7.215
see also Conservative Party (Britain)
constitution unwritten 7.67
devolution 7.88
electoral system 7.124, 7.190
suffrage and electoral reform 4.70–1, 4.159, 4.258, 4.333, 7.332
English Protectorate 3.293
Exchequer 3.121
executive 7.113
fascist 4.225
Field of Cloth of Gold 3.126
Glorious Revolution 3.142
Houses of Parliament see House of Commons; House of Lords and Parliament above
intelligence services 7.162
'Junius' letters 3.192
Labour 4.88–9, 4.129, 4.146, 4.189, 4.190, 4.203, 4.225, 4.344, 7.315, 7.317
see also Socialists below
leaders and statesmen see statesmen and political leaders below
legislature 7.148, 7.182
Levellers 3.207–8

Britain (and United Kingdom) (cont.):
politics and government (cont.):
Liberals 4.145, 4.195, 4.290
Magna Carta 3.216
National Government 4.233–4
ombudsman 7.223
parish 3.270
parliament see Parliament above
peerage 3.36, 3.37–8
philosophy of 3.157, 3.167, 3.208–9, 7.144
pocket borough 4.271
prime ministerial 7.247
prime ministers see under prime ministers: listed 9.69
Primrose League 4.276
Radicals 3.358, 4.49, 4.77, 4.158, 4.159, 4.200, 4.270
and religion 7.265
rotten borough 4.293
security services 7.278
separation of powers 7.279
Settlement Acts 3.322
Social Democrats 4.317
Socialists 4.115, 4.317–18
see also Labour above
spy 4.128
suffrage see electoral system above
Tamworth manifesto 4.337
theorist 3.54, 3.267
Tories see Conservative Party
two-party system 7.317
unitary state 7.320
and United Nations 4.104
Whigs see under Whigs
Young England 4.386–7
Zionism 4.373
see also foreign policy above and rulers; statesmen below
pollution 6.275, 7.2
population see demography above
pottery see under ceramics and pottery
poverty 4.12, 4.31, 4.37, 4.45, 4.69, 4.77, 4.294, 4.372
prehistory 3.345, 3.346, 6.218, 6.335
printing 6.105, 6.387
professions 7.249
psychology 7.21–2, 7.129, 7.290
radar 6.288, 6.379
railways:
Automatic Warning System 6.26
Channel Tunnel 6.69, 6.365, 7.67
construction 6.292–3
gauge 6.151
high-speed 6.169
history 6.293
hover-train 6.161
locomotives 6.102, 6.115
see also Trevithick
underground 6.367
wagon-way 6.375
Regency see Regency
regions 3.108, 3.258, 3.342
listed 9.73
see also kingdoms above
religion/theology:
archbishops 3.16–17, 3.28, 3.38, 3.106, 3.202–3, 3.204, 3.241, 3.286, 4.340
bishops 3.12, 3.54, 3.188, 3.204, 3.310, 3.350, 3.383
'Brownists' 3.51
Catholics 3.58, 3.385, 4.210
clergy 3.130
Congregationalism see Congregationalists
Cranmer 3.90
Jews and Judaism 7.139, 7.171, 7.335
and laws 3.352, 3.358, 3.365, 4.100, 4.315–16, 4.345
leaders 7.50
Methodists 7.199

Britain (and United Kingdom) (*cont.*):
religion/theology (*cont.*):
monasteries, dissolution of **3**.236
mystic **3**.128
myths *see* myths and legends *above*
Oxford Movement **4**.238
palmers **3**.268
parish **3**.270
Pelagianism **3**.274
pentecostal **7**.232
Plymouth Brethren **7**.237
and politics **7**.265
preacher **3**.313
Puritans **3**.37, **3**.103–4, **3**.127, **3**.294, **3**.383
Quakers **3**.130, **7**.256
Rastafarianism **7**.258
rebel **3**.34
reformers **3**.386
saints **9**.40, **9**.41, **9**.43
see also individual names
Salvation Army *see* Salvation Army
scholar **3**.8
Sikhism **7**.286
spiritualism **7**.298
and state **3**.38, **3**.78, **7**.265
Sunday School **3**.298, **4**.333
theologians **3**.169, **3**.383
tithe **3**.357
Toleration Act **3**.358
Unitarians **7**.320
United Reform Church **7**.65, **7**.322
Western Rising **3**.380
Whitby synod **3**.380
see also Anglicans; Methodists
retail trade **7**.43
riots and uprisings:
leader **3**.56
Newport **4**.239
peasant **3**.195, **3**.273, **3**.364
Pentrich **4**.266
perloo **4**.268
Pilgrimage of Grace **3**.281–2
Rebecca riots **4**.284
strikes *see under* unions *below*
'Swing, Captain' **4**.335
Western Rising **3**.380
roads and vehicles **6**.234
bicycles **6**.35, **6**.384
breathalyser **6**.46
building **6**.210, **6**.299–300, **6**.350
carriages and coaches **6**.61
Cat's eye ***6**.64*
engines, car **6**.102, **6**.369
motor cycles **6**.235
Roman **3**.118, **3**.129, **3**.378
steering **6**.333–4
traffic control **6**.360
tyres **6**.109, **6**.366
Roman *see* Roman Britain
Royal Society **3**.311–12
rubber **6**.158, **6**.165
rulers (chronologically) **7**.97, **7**.139
list of ***9**.67*
Saxon line (955–1016, 1042–66) **3**.110, **3**.111–12
see also Anglo-Saxons; Edgar; Ethelred; Harold II
Danish line (1017–42) **3**.157
see also Canute
Normandy, House of (1066–1154) **3**.160–1, ***3**.255*, **3**.341
see also William I; William II
Plantagenet, House of (1054–1399) ***3**.255*, ***3**.284*
see also Angevins; Edward I; Edward II; Edward III; John, king of England; Henry II; Henry III; Lancastrians; Richard II; Yorkists
Lancaster, House of (1399–1461) **3**.138, ***3**.202*, **3**.284, **3**.310

Britain (and United Kingdom) (*cont.*):
rulers (chronologically) (*cont.*):
see also Henry IV; Henry V; Henry VI
York, House of (1461–85) **3**.284, **3**.310, ***3**.388*
see also Edward IV; Edward V; Richard III
Tudor, House of (1485–1603) **3**.224, ***3**.361–2*
see also Edward VI; Elizabeth I; Henry VII; Henry VIII
Stuart, House of **3**.224, ***3**.343*
see also Anne, queen; Charles I; Charles II; James I; James II; William III
Commonwealth (1649–59) *see* Cromwell
Restoration *see* Restoration
Hanover, House of (1714–1901) ***4**.146*, **4**.379
see also George I; George II; George III; George IV; Regency; Victoria
Saxe-Coburg-Gotha, House of (1901–10) **4**.365
Bill of Rights **3**.41
Windsor, House of (since 1910) **4**.1, **4**.80, **4**.105, **4**.108, **4**.110–11, **4**.133, ***4**.380*
Queens and consorts **3**.61, **3**.65, **3**.113, **3**.114, **3**.149, **3**.160, **3**.161, **3**.181, **3**.222, **3**.224
see also under Henry VIII
Settlement Acts **3**.322
see also pretenders
sanitation **6**.32, **6**.37, **6**.314, **6**.377
scholars **3**.38, **3**.90, **3**.364
science:
scientific societies, academies, and institutes **6**.47, **6**.217, **6**.236, **6**.303
see also chemistry; physicists *above and* technology and science *below*
Scotland *see* Scotland
self-help **4**.316
shipping:
building and design **6**.237, **6**.316–17
coracle **6**.89
dinghy **6**.104
dredgers **6**.106, **6**.277
hovercraft **6**.79–80, **6**.171
lifeboat **6**.200
liner **4**.201
navigation **6**.73, **6**.201, **6**.287, **6**.303, **6**.315
Plimsoll line **6**.273
propellor **6**.284
sailing **6**.76, **6**.90, **6**.306, **6**.322, **6**.390
steam **6**.49, **6**.106, **6**.147, **6**.160, **6**.236, **6**.277, **6**.332–3
warships **6**.32, **6**.99, **6**.188, **6**.322, **6**.357, **6**.375
slavery **4**.76
social services and security **4**.37, **4**.318, **4**.357, **7**.289, **7**.293
sociobiology **7**.294
sociology **7**.78
sport and recreation:
billiards **7**.31
bowls **7**.34
boxing **7**.34
cock-fighting and dog-fighting **7**.54
cricket **7**.75
cross-country running **7**.76
darts **7**.80
dominoes **7**.94
fives **7**.121
football and rugby **7**.122, **7**.123
game-shooting **7**.129
hockey **7**.144

Britain (and United Kingdom) (*cont.*):
sport and recreation (*cont.*):
horse-riding and horse-racing **6**.166, **7**.147
lacrosse **7**.179
motor-racing **7**.208
netball **7**.214
pigeon-racing **7**.236
rounders **7**.270
rowing **7**.270
snooker **7**.289
speedway racing **7**.297
squash **7**.298
tennis **7**.308
statesmen and political leaders:
16th century **3**.32–3, **3**.66, **3**.91, **3**.240–1, **3**.385
17th century **3**.32–3, **3**.51–2, **3**.66, **3**.78, **3**.81, **3**.96, **3**.103, **3**.143, **3**.251, **3**.274, **3**.295–6, **3**.325, **3**.342, **3**.345, **3**.370
18th century **3**.54, **3**.143, **3**.360, **3**.376–7, **4**.65, **4**.145, **4**.377
19th century **4**.49, **4**.62, **4**.65, **4**.77, **4**.122, **4**.142, **4**.158, **4**.290, **4**.295, **4**.310
20th century **4**.88, **4**.89, **4**.129, **4**.145, **4**.146, **4**.190
Whigs *see under* Whigs
see also prime ministers *and* politics *above*
steam *see* steam power
steel industry *see* iron and steel *above*
technology and science **6**.76, **6**.79
appropriate **6**.18
automata **6**.25
measurement **6**.36, **6**.351
operational research **6**.250
see also science *above and references above and below prefixed* **1.** *and* **6.**
telecommunications **6**.285, **6**.373, **7**.35
news agency **7**.216
radio and BBC **4**.286, **6**.48, **6**.168, **6**.208, **6**.215, **6**.290, **7**.28, **7**.254, **7**.307
telegraphy **6**.88, **6**.112, **6**.232, **6**.346, **6**.384
telephony **6**.65, **6**.347
television **6**.28, **6**.40, **6**.349, **7**.307
textiles and fibre **6**.58, **6**.178, **6**.351, **6**.384
machinery **6**.20, **6**.62, **6**.92, **6**.140, **6**.165, **6**.190, **6**.191, **6**.208, **6**.329
towns **3**.45, **3**.77
Toynbee Hall **4**.348
trade *see* economy/economics *above*
trains *see* railways *above*
transport *see* vehicles *below*
tribes **3**.13, **3**.47, **3**.49–50, **3**.60, **3**.94, **3**.174
and UN **7**.326
unions and labour **7**.314
ACAS **7**.158
child labour **4**.249
co-operatives **4**.86, **4**.254
and law **4**.79, **4**.252, **4**.270, **7**.54, **7**.159
leaders **4**.36, **4**.37, **4**.55, **4**.344
strikes **4**.132–3, **4**.198–9, **4**.213, **4**.271
Taff Vale case **4**.336
Tolpuddle Martyrs **4**.346–7
tribunals **7**.315
TUC **4**.348–9
working hours **4**.310
uprisings *see* riots and uprisings *above*
urban areas **7**.216, **7**.323
utilitarianism **4**.362
vehicles and transport:
aircraft *see* aircraft *above*

Britain (and United Kingdom) (*cont.*):
vehicles and transport (*cont.*):
bridges *see* bridges *above*
canals **6**.47, **6**.56, **6**.350, **6**.385
diving **6**.97, **6**.105
military **6**.129, **6**.226, **6**.301, **6**.344, **6**.374
railways *see* railways *above*
road *see* roads *above*
ships *see* shipping *above*
space **6**.301, **6**.326
tunnels **6**.69, **6**.365
Wales *see* Wales
wars and defence:
airforce **4**.6, **4**.103, **4**.275, **4**.351
air-raid shelter **6**.10
with America **2**.251
1812 **4**.136, **4**.189–90, **4**.238, **4**.342, **4**.371
see also Independence, American War of
Anglo-Dutch **3**.14
Anglo-Maori **4**.13–14, **4**.294, **4**.338
arms *see* weapons *below*
army **3**.297, **4**.19, **4**.51–2, **4**.155
army leaders:
in Africa *see* Kitchener
in China **4**.138, **4**.381
in Crimea **4**.281, **4**.381
in Europe **3**.144, **3**.223, **3**.339–40, **4**.223, **4**.281
in India **4**.147, **4**.185, **4**.230–1, **4**.291, **4**.374
in Near East **4**.138, **4**.185, **4**.191, **4**.316, **4**.380, **4**.381
in North America **3**.11, **3**.48, **3**.54, **3**.79, **3**.88, **3**.136, **4**.381
in South Africa **4**.126, **4**.185, **4**.291
in South-East Asia **4**.316, **4**.340–1, **4**.381
in World War I **4**.7, **4**.9, **4**.10, **4**.22, **4**.126, **4**.144–5, **4**.185, **4**.372
in World War II **4**.7, **4**.9, **4**.22, **4**.316, **4**.340–1, **4**.372
see also knighthood *and under* English Civil War
Austrian Succession **3**.29–30
Boer **4**.42, **4**.43, **4**.82, **4**.363
Bosworth Field **3**.47
castle **3**.63
CENTO **4**.68
with China **4**.32, **4**.138, **4**.230, **4**.250, **4**.381
civil *see* Barons' Wars; English Civil War
Cod War **4**.78
conscription **7**.66
Crimean **4**.27–8, **4**.88, **4**.164, **4**.260
with Danes **3**.96–7, **3**.217, **4**.86, **4**.87
with Egypt **4**.32, **4**.340
1812 *see under* America *above*
Etaples, Treaty of **3**.118
with Ethiopia **4**.206
Falklands **4**.116, **7**.69
with France **3**.189, **3**.313
in Canada **3**.284, **3**.313
in Caribbean **3**.313
Napoleonic **4**.279, **4**.349–50, **4**.372, **4**.374
Nine Years War **3**.254
see also Hundred Years War
fyrd **3**.136
invaded *see* Anglo-Saxons; Conquest *under* Normans
Lord Lieutenants **3**.210–11
militia **3**.360
mutiny **3**.243–4, **3**.339
national security state **7**.212

Britain (and United Kingdom) (cont.):
wars and defence (cont.):
navy 3.248, 3.275–6, 4.103
mutiny 3.339
pressgang 3.291
see also mutinies
navy leaders:
in Americas 3.39, 3.45–6, 3.106, 3.149, 3.169, 3.170, 3.187, 3.372, 4.77
in China 4.32
in Egypt and Sudan 4.32
in Europe 3.17, 3.42, 3.45–6, 3.55, 3.106, 3.149, 3.159, 3.169, 3.170, 3.187, 3.237, 3.249–50, 3.306, 3.372, 4.77, 4.237
in Pacific 4.41–2
in World War I 4.32, 4.120, 4.173, 4.225
in World War II 4.225
with Netherlands 3.14, 3.58, 6.375
Norman invasion see Conquest under Normans
Northern Rising 3.256
Nuclear Test-Ban Treaty 4.248
Oregon Boundary dispute 4.251
Peninsular 4.368–9, 4.374
Revolutionary 4.288
with Scotland see under Scotland
with Spain 3.32, 3.186, 3.335–6
special services 7.297
Suez 4.332
terrorism 7.308
Trafalgar 4.349–50
Trent affair 4.352
with United States see America above
Western European Union 4.375
in WEU 7.269
World War I 4.51–2, 4.94
see also under army leaders; navy leaders above
World War II 4.193, 4.217–18, 4.345–6
Battle of Britain 4.49–50
Destroyer Bases Deal 4.97
aftermath 4.22, 4.85
see also under army leaders; navy leaders above
see also English Civil War; Hundred Years War; NATO; Seven Years War; Spanish Succession, War of
water 6.136, 6.139, 6.377, 6.378
weapons 6.42, 6.111, 6.318, 6.357
arms control 7.64
guns 6.210, 6.217, 6.226
missiles 6.228–30, 6.301
nuclear 7.220
disarmament 4.59, 4.248
trade 7.18
see also wars and defence above
welfare see social services above
women's movement 7.332
writing 6.259, 6.318
Britain, battle of 4.49–50
British Association for the Advancement of Science 6.47
British Broadcasting Corporation 4.286, 7.28, 7.28, 7.254, 7.307
British Central Africa see Malawi
British Columbia 1.44
British East Africa see Kenya; Uganda
British empire see empire under Britain
British Expeditionary Force 4.51–2
British Film Institute 5.60
British Guiana 4.43
British Honduras see Belize
British Isles 1.44
see also Britain
British North America Act 4.52
British Raj 4.52
British Raj (cont.):
administration and government 3.79, 3.158–9, 4.61, 4.89, 4.90, 4.92, 4.106, 4.191, 4.225, 4.372, 4.374
boundaries 4.161, 4.204
constitutional changes 4.223, 4.224
East India Company 4.107–8, 4.313
law 4.162, 4.294, 7.143
princely States 4.276–7
province 4.278
resistance to 4.344
Amritsar massacre 4.11
Indian Mutiny 4.162, 4.200–1
Maratha Wars 4.211
non-co-operation and passive resistance see Gandhi, Mohandas
Pindaris 4.269
Red Shirts 4.285
Salt March 4.301
Sikh Wars 4.313
Round Table Conferences 4.294
British Somaliland see Somalia
British Standard 6.385
British Thermal Unit (BTU) 6.166
Brittany 3.50
Breton language 7.42, 7.156
Britten, Benjamin 5.57, 5.60, 5.75, 5.117, 5.279, 5.320, 5.391
brittle star 2.42
broaching machine 6.212
broad bean 2.42
Broadbent, Ed 4.238
broadbills 2.42
broadcasting 6.47–8, 7.35
broadcast database 7.35
live 7.186
public service 7.254
see also radio; telecommunications; television
Broadway 5.60
Broadwood, John 5.348
Broca, Paul 7.15
broccoli 2.49
Broch, Hermann 5.60
Brock, Isaac 4.371
Brod, Max 5.243
Brodsky, Iosif 5.60, 5.396
Broken Hill Man 3.50
broker 7.35
bromeliad 2.42–3
bromine 1.44, 1.106, 6.48–9, 6.71
Brongniart, Alexandre-Théodore 5.315
Brontë, Anne 5.60
Brontë, Charlotte 5.27, 5.48, 5.60
Brontë, Emily 5.60
brontosaurus 2.43
bronze 6.11, 6.49, 6.89, 6.221, 6.356
art (bronzework) 5.61
Africa 5.9
Assyrian 5.24–5
casting 5.82
Celtic 5.84
China 5.92–3, 5.96–7
Coptic 5.112
Etruscan 5.149–50
Greece, ancient 5.190
India 5.99
ormolu 5.332
Sri Lanka 5.429
Tibet 5.455
tools and weapons 6.4, 6.72, 6.225, 6.280
Bronze Age 3.50, 3.228, 3.362
art 5.61
grave-mounds 3.36
technology 6.343, 6.357
Urnfield cultures 3.368
Bronzino, Agnolo 5.61
Brook, Peter Stephen Paul 5.23, 5.61
Brooke, Alan (Alanbrooke) 4.7
Brooke, Sir James 4.52, 4.302
Brooke, Rupert 5.61–2
broom 2.43
broomrape 2.43
Brougham, Henry Peter, 1st Baron 4.52
brougham 6.61
Brouwer, Adriaen 5.62
Browder, earl 4.122
Brown, Arthur 4.350, 6.138
Brown, Ernest 8.87, 8.88, 8.101
Brown, Ford Madox 5.62
Brown, Jacob 4.371
Brown, John 4.10, 4.52–3
Brown, Kenneth 5.267
Brown, Lancelot 'Capability' 5.62, 5.348
Brown, Robert 1.44, 2.43
Brown, Trisha 5.62
Brown v. Board of Education of Topeka 4.53
brown (butterfly) 2.43, 2.154, 2.285
brown dwarfs 8.161
brown earth (brown forest soil) 1.44
brown soil 1.44, 1.311
Browne, Gore (New Zealand Governor) 4.338
Browne, Hablot K. ('Phiz') 4.201, 5.130
Browne, Robert 3.51, 3.85, 3.254
Browne, Sir Thomas 5.62
Brownian motion 1.44
Browning, Elizabeth Barrett 5.62
Browning, Robert 3.73, 5.62, 5.217, 5.424
Brownover, Sylvester 3.209
Brownshirt 4.53
Bruce, Christopher 5.62
Bruce, James 1.44–5, 5.321
Bruce, Robert see Robert I
Bruce, Stanley Melbourne, viscount 4.53
Bruch, Max 5.62
Brücke, Die 5.63, 5.248, 5.320
Bruckner, Anton 5.63
Bruegel, Pieter 5.63
Brueghel, Jan 5.63
Brueghel, Pieter (the Younger) 5.63, 5.157
bruising 2.43
Brundtland, Gro Harlem 4.248
Brunei 3.45, 4.53, 4.321, 7.20, 7.344
Brunel, Isambard Kingdom 6.49, 6.53, 6.123, 6.160, 6.293, 6.316, 6.331
Brunel, Sir Marc Isambard 6.49, 6.217, 6.316
Brunelleschi, Filippo 5.63–4, 5.132
Bruner, Jerome 7.35, 7.54, 7.246
Brunhild 9.8, 9.33, 9.37
Brunhilda, Visigothic queen 3.51
Brüning, Heinrich 4.53
Brunner, J. T. 6.232
Bruno, Giordano 3.51, 5.231, 8.18
Bruno of Cologne, St 3.51, 3.62
Brunswick, Friedrich Wilhelm 3.369
brush-footed butterflies 2.43
browns 2.43, 2.285
fritillary 2.127
morpho 2.219
painted lady 2.239
peacock 2.246
purple emperor 2.272
red admiral 2.279
white admiral 2.366
Brusilov, Aleksky 4.53
Brussels Conference on Africa 4.307
Brussels sprout 2.43
Brussels tapestries 5.64
Brutalism 5.64, 5.394, 5.446
Brutus, Dennis 5.7
Brutus, Marcus Junius 3.28, 3.51, 3.57, 3.307, 3.342, 5.51
Brutus the Trojan 9.14
Bruyas, Alfred 5.376
Bryan, William Jennings 4.53–4
Bryce Canyon 1.45
bryony 2.43
bryophytes 2.43
liverwort 2.192
mosses 2.219, 2.282, 2.319–20
bryozoan 2.219
BSE (bovine spongiform encephalography) 6.127
bubble chamber 1.104, 6.258
bubonic plague see Black Death
buccaneers 3.51, 3.241
Buchanan, George 3.184
Buchanan, James 4.54, 7.253
Buckingham, George Villiers, 1st duke of 3.51–2
Buckminster Fuller, R. 6.147, 6.155
buckthorns 2.44, 2.176
buckwheat 2.44
Buddha 3.41, 3.52–3, 3.236, 7.36, 9.6, 9.9, 9.24, 9.33
rebirth awaited 7.190
and Theravāda 7.310
see also Buddhism
Buddhism 3.53, 4.52, 4.54, 7.36–7
ahimsā 7.8
and arts 5.64, 5.65, 9.169, 9.236
see also under arts under India and entries below prefixed 5.
Bodhisattva see Bodhisattva
caste system 7.41
China 3.73, 5.64, 7.36–7
monasticism 7.205, 7.273
myths and characters 9.4, 9.6, 9.15, 9.17, 9.20, 9.24, 9.25, 9.38
White Lotus Society 3.380–1
Zen 7.336
cremation 7.75
deities 9.4, 9.15, 9.22, 9.24, 9.26, 9.33
demons 9.28
dependent origination 7.85, 7.133
dharma 7.89
dukkha 7.97, 7.311
education 7.265
festivals 5.37
Four Noble Truths 7.97, 7.123, 7.217
heaven 7.140
hell, multiple 7.141
heroes 9.5, 9.19
and Hindu law 7.143
India 3.157, 3.315, 5.10, 5.64, 5.286, 5.336, 5.400, 5.401, 7.36–7, 7.85, 9.6
Indochina (Khmers) 3.196
Indonesia 5.221
Japan 3.389–90, 5.236, 7.36–7, 7.284, 7.336
myths and characters 9.4, 9.6, 9.9, 9.15, 9.17, 9.20, 9.22
Kampuchea 7.143
karma 7.175
Korea 5.251, 9.15
Laos 7.36
law 7.89
literature 5.336
'birth-stories' 5.238, 9.19
see also Sanskrit
Lotus Sutra 7.187
Maitreya 7.190
mantra 7.191, 9.28
meditation 7.196
miracles 7.203
monasticism 5.65, 7.37, 7.205, 7.273
monastic dance 5.65
music 5.65
Myanmar 7.36, 7.143
mysticism 7.191
Nepal 5.316
nirvana 7.217, 9.5
Noble Eightfold Path 7.97, 7.123, 7.217
philosophy 7.155, 7.211
pilgrimages 3.281

Buddhism (*cont.*):
Pure Land **7**.140, **7**.255
reincarnation **7**.6, **7**.263
sacred buildings ***7**.37*
samādhi **7**.272
samsara **7**.169, **7**.263, **7**.273
sangha **7**.205, **7**.273
Shingon **7**.284, **9**.9, **9**.22
skandha **7**.287
Sri Lanka **3**.339, **5**.429, **7**.36–7
stories **5**.238, **9**.19
stupa **5**.438
tantra **7**.306
Thai **5**.450–2, **7**.36, **7**.143
Three Marks of Existence **7**.97, **7**.263, **7**.311
Threefold Refuge (Three Jewels) **7**.311–12
Tibetan *see* religion *under* Tibet
vegetarianism **7**.325
Vietnam **7**.36
see also Buddha; Mahāyāna; Theravāda; Zen
Budding, Edwin **6**.196
buddleia ***2**.44*
budgerigar ***2**.44*
budget ***7**.37*
state **7**.201
Buero Vallejo, Antonio **5**.428
buffalo ***2**.44*
Buffalo Bill **4**.315, **4**.378
buffalo gnat **2**.33
buffalo weaver ***2**.44*
buffer (railway) ***6**.49*
buffer solution ***1**.45*
Buffon, Georges Louis Leclerc, Comte de ***2**.44*, **7**.336
Buga **9**.8
Buganda *see* Uganda
Bugie, E. **6**.375
Bugis ***3**.53*
bugle (musical instrument) ***5**.65*
flugelhorn **5**.158–9
ophicléïde **5**.329
see also tuba
bugle (plant) ***2**.45*
bugs ***2**.44–5*
aphid **2**.13
assassin **2**.17
bedbugs **2**.26
cicada **2**.65
froghopper **2**.127, **2**.128
giant water bug **2**.135
jumping plant louse **2**.176
lacebug **2**.181
lantern fly **2**.182
leafhopper **2**.185
mirid **2**.213
pond skater **2**.263, **2**.264
red **2**.279
scale-insects **2**.30, **2**.142, **2**.295
shieldbug **2**.307
treehopper **2**.346
water boatman **2**.361
water scorpion **2**.362
whitefly **2**.366–7
see also insects
building society ***7**.37*
buildings and building materials ***6**.49–51*
aggregate **6**.5
air conditioning **6**.8
arch *see* arch
architecture *see* architecture
box-girder **6**.44, **6**.293, **6**.336, **6**.341
brick *see* bricks
bridge *see* bridges
caisson **6**.53, **6**.112
cement *see* cement
concrete *see* concrete
damp-proof course **6**.96
demolition **6**.98
Eiffel Tower **6**.112
employment in industry **7**.106
escalator **6**.123
buildings and building materials (*cont.*):
expansion joint **6**.124
fire-resistance **6**.132
floors *see* floor
foundation *see* foundations
frame *see* frame
girders **6**.2, **6**.44, **6**.272, **6**.293, **6**.336, **6**.341
heating *see* heating, domestic
hull-and-core **6**.321
insulation *see* insulation, thermal
iron **5**.226–7, **6**.50, **6**.51, **6**.90, **6**.112
lath and plaster **6**.146, **6**.195
lifts **6**.201, **6**.322
lime *see* lime
limestone **6**.202
manufactured board **6**.214, **6**.387
mortar **6**.5, **6**.66, **6**.232–3
moulding **6**.235
mud **6**.49, **6**.235
piling **6**.270
plaster **6**.195, **6**.271
plumbing **6**.274
prefabrication **6**.44, **6**.146, **6**.280, **6**.317, **6**.343
pyramid **6**.287
roof **6**.51, **6**.145, **6**.146, **6**.274, **6**.301–2, **6**.337
sandblasting **6**.308
scaffolding **6**.308–10
settlement **6**.314
shell **6**.51, **6**.316
skyscraper **6**.51, **6**.201, **6**.321, **6**.322
space frame **6**.51, **6**.326
steel **6**.44, **6**.50, **6**.51, **6**.82, **6**.144–5, **6**.188, **6**.316, **6**.322
stone *see* masonry
suspended structure **6**.340, **6**.351
system **6**.280, **6**.343
tiles **6**.51, **6**.76, **6**.146, **6**.302, **6**.356
vault **6**.49, **6**.51, **6**.302, **6**.337, **6**.371
walls *see* walls
wattle and daub **6**.146, **6**.235, **6**.380
windows **6**.106
wiring **6**.387
wood **5**.455, **6**.51, **6**.145–6, **6**.301–2, **6**.387
see also environmental engineering
Bukharin, Nikolai Ivanovich ***4**.54*
bulbs, plants with ***2**.45*
daffodil **2**.86, **2**.316
garlic **2**.133
grape hyacinth **2**.143
hyacinth **2**.166
iris **2**.119, **2**.136, **2**.173, **2**.218
narcissus **2**.225
bulbul ***2**.45*
Bulgakov, Mikhail ***5**.65*
Bulganin, Nikolai Alekandrovich ***4**.54*
Bulgaria ***1**.45*, ***3**.53*, ***4**.54–5*, ***7**.344–5*
arts **5**.74
in COMECON **7**.57
language **7**.156, **7**.288
leader **4**.99
physical geography and geology **1**.45
religion/theology **7**.98
and war **4**.26
in Warsaw Pact **4**.371, **7**.328
see also Balkan States; Eastern Europe
Bulge, battle of **4**.17
bulk carrier ***6**.51*, **6**.219
bulkhead ***6**.51*
Bull (Taurus constellation) **8**.169
Bull, Hedley **7**.331
Bull, John ***5**.65*
Bull, Ole **5**.193
bull:
bull-fighting ***7**.37–8*
mythical **9**.4–5, **9**.14, **9**.25
see also cattle
bulldog bat ***2**.45*
bulldozer ***6**.51*
bullet **6**.52, **6**.131, **6**.272
bullfinch ***2**.45*
bullfrog ***2**.45*
bullhead ***2**.45*
bullion **7**.133–4
bullock-drawn vehicle **6**.165
bull-roarer ***5**.65*
Bülow, Bernard, prince von ***4**.55*, **4**.135
Bülow, Hans von **5**.110
bumbass ***5**.65*
bumblebee ***2**.45*
Bunche, Ralph ***4**.55*
Bunin, Ivan ***5**.65*
Bunker Hill, battle of ***3**.53–4*
bunraku **5**.66, **5**.236, **5**.369
Bunsen, Robert: and Bunsen burner **1**.183, **6**.25, ***6**.52*, **6**.193, **8**.157
buntings ***2**.45*, **2**.144, **2**.236, **2**.373
Buñuel, Luis ***5**.65–6*
Bunyan, John **5**.27, ***5**.66*
Buontalenti, Bernardo **5**.114, **5**.412
buoy ***6**.52*, **6**.193
buoyancy **1**.124, **6**.200–1
Buraq ***9**.8*
Burbage, Richard ***5**.66*, **5**.414
burbot ***2**.45*
bureaucracy ***7**.38*, **7**.52, **7**.330
complaints against (ombudsman) **7**.223
burette ***6**.52*
Burger, Warren E. ***4**.55*
Burges, William ***5**.66*
Burgess, Anthony ***5**.66*
Burgh, Hubert de **3**.161
Burghley, Lord *see* Cecil, William
Bürgi, Joost **1**.224
burglary **7**.250
Burgoyne, John **3**.40, ***3**.54*, **3**.176, **3**.316, **3**.357
Burgundy ***3**.54*, **3**.71, **3**.98, **3**.280
Burgundian school ***5**.66*
burial **7**.75, **7**.128
Burke, Edmund ***3**.54*, **3**.130, **3**.158, **5**.438, **7**.16, **7**.66
Burke, Robert O'Hara ***1**.45*, **1**.201
Burke, T. H. **4**.269
Burkina Faso ***1**.45–6*, ***4**.55*, ***7**.345*
colonized and decolonized **4**.4, **4**.306
in Franc Zone **7**.124
leader **4**.301–2
physical geography and geology **1**.45–6
reafforestation **7**.82
burlesque ***5**.66*, **5**.154
Burlington, Lord **5**.66
Burma *see* Myanmar (Burma)
Burma Campaigns ***4**.55–6*
burn (injury) ***2**.45–6*
Burnap, Danile **5**.105
Burne-Jones, Sir Edward ***5**.67*, **5**.367
Burnet, Gilbert ***3**.54*
burnet moth ***2**.46*
Burney, Fanny ***5**.67*
Burnham, Forbes **4**.144
burning *see* combustion
Burns, George (1839) **4**.90
Burns, John Elliott ***4**.55*
Burns, Robert ***5**.67–8*
Burnside, Ambrose **4**.368
Burr, Aaron ***4**.56*, **4**.337
Burrel, Charles **6**.359
Burroughs, William Seward ***5**.68*
Burrows, Larry **5**.482
bursa ***2**.46*
burster ***8**.18*
Burton, Sir Richard Francis ***1**.46*, **1**.317, ***4**.56*, **4**.324, **5**.460
Burton, Robert ***5**.68*
Burundi ***1**.46*, **4**.4, ***4**.56*, **7**.261, ***7**.345*
Buryat language **7**.11
burying beetle ***2**.46*
bus (computer) ***6**.52*
Buschmann, Christian **5**.199
buses and coaches ***6**.52*, **6**.292, **6**.363
busing **4**.57
Bush, George ***4**.56*, **4**.287, **4**.330
Bush, Jack **5**.73
bush cricket **2**.193
bush dog ***2**.46*
bush pig ***2**.46*
bushbaby ***2**.46*, **2**.47
bushes *see* shrubs
bushido ***7**.38*
Bushmen **7**.178, **9**.21
Bushnell, David **6**.337
bushranger ***4**.56*
Busi **9**.6
business *see* companies and business
busing ***4**.57*
Buson **5**.197
Busoni, Ferruccio Dante Michelangiolo Benvenuto ***5**.68*
Buss, Frances Mary **4**.32
bust ***5**.68*
Bustamante, Anastasio ***4**.57*
Bustamante, José Luis **4**.148
Bustamante, Sir William Alexander ***4**.57*, **4**.171
bustard ***2**.46–7*
butadiene **6**.261, **6**.303–4
butane ***1**.46*, ***6**.52*, **6**.150, **6**.204, **6**.237, **6**.264
Butcher, Rosemary ***5**.68*
butcher birds ***2**.47*, **2**.309
butcher's broom ***2**.47*, **2**.48
Bute, John Stuart, 3rd earl of ***3**.54*
butene **6**.261
Buthelezi, Gatsha **4**.391
Butler, Joseph ***3**.54*
Butler, Josephine Elizabeth ***4**.57*
Butler, Samuel ***5**.68*
Butler of Saffron Walden, Lord Richard Austen ***4**.57*, **4**.103, **4**.340
Buto dance **5**.141
Butor, Michel **5**.323
butte ***1**.46*
butter ***6**.52*, **6**.215
buttercup family ***2**.47*
aconite **2**.2
anemone **2**.10
clematis **2**.67
delphinium **2**.90
hellebore **2**.154
kingcup **2**.179
Butterfield, William ***5**.68*
butterflies ***2**.47*
birdwing **2**.31, **2**.32
blue **2**.35, **2**.36
brimstone **2**.41
brown **2**.43, **2**.154, **2**.285
brush-footed *see* brush-footed
cabbage white **2**.49
caterpillar **2**.57
chrysalis **2**.65
copper **2**.75, **2**.76
fritillary **2**.127
hair-streak **2**.149
heath **2**.154
metal mark **2**.127, **2**.209
monarch **2**.209, **2**.216
morpho **2**.218, **2**.277
painted lady **2**.239
peacock **2**.246
purple emperor **2**.272
red admiral **2**.279
ringlet **2**.285
skipper **2**.312
starfish **2**.324
swallowtail **2**.331
white **2**.366

butterflies (cont.):
  white admiral 2.366
  wood-white 2.371
  yellow 2.373
butterfly diagram 8.18, 8.164
butterfly effect 8.25
butterfly fish 2.48
butterfly flower 2.297
butternut 2.48
butterwort 2.48
button quail 2.48
buttons 6.52, 6.77
buttress 5.68, 6.337
butyl rubber 6.304
butylated hydroxyonisole 6.141
Buxar, battle of 3.54
Buxtehude, Dietrich 5.68
buy-out, management 7.191
Buyoya, Pierre 4.56
Buys Ballot, Christoph Hendrik Didericus 1.46
buzzard 2.48
BWR (boiling-water reactor) 6.281
Byng, George, viscount Torrington 3.54
Byng, John 3.55
Byng, Julian H .G., 1st viscount 4.368
Byrd, Richard Evelyn 1.46
Byrd, William 5.18, 5.69
Byron, George Gordon, 6th Baron 5.69–70, 5.333, 5.402, 5.427
byte 6.52
Byzantine empire 3.55
  arts 5.69, 5.70, 5.216–17, 5.305
  rulers 3.83–4, 3.164, 3.180, 3.192–3, 3.231, 3.353–4
  wars with Turks 3.221
Byzantium see Constantinople

C

cabal 3.56
cabaret 5.70
cabbage family 2.49
  brussels sprout 2.43
  cauliflower 2.58
  kohlrabi 2.180
  see also brassica
cabbage white 2.49
Cabeza de Vaca, Alvar Núñez 3.56
Cabinda (Angola) 4.4
cabinet (government) 3.56, 4.184, 7.38, 7.113, 7.318
cabinet-making see furniture
cable, electrical and telecommunications 6.53, 6.79, 6.114, 6.183, 6.319, 6.347
  cable television 6.38–9, 6.53, 7.38–9
  cable-ship 6.53
cable, structural 6.48, 6.53
  cable-car 6.53
Cabot, John 1.47, 3.56, 3.59, 3.122, 3.251
Cabot, Sebastian 1.47, 3.56, 3.122, 3.217
Cabral, Amilcar 4.58, 4.143
Cabral, Pedro Alvares 3.334
Cabrillo, Juan Rodriguez 1.47
cacao (cocoa) 2.49, 6.34, 6.271, 6.364
Cacciatore, Niccolo 8.37
Caccini, Guilio 5.328
cache memory 6.219
CACM see Central American Common Market
cacomistle 2.285
cacti 2.49, 2.250
CAD (computer-aided design) 6.85, 6.237, 7.63
cadastral survey 6.340
Cadbury, George 4.58, 7.216
Cadbury, Richard and Elizabeth 4.58
caddis fly 2.49
Cade, Jack 3.56
cadence 5.70
cadmium 1.47, 1.106, 6.53
  in battery 6.32
  electroplating 6.120
  in nuclear reactor 6.245
  pollution 6.377, 7.2
Cadwalader of Gwynedd 3.56, 3.275
CAE (computer-aided engineering) 7.63
caecilian 2.49
Caelum 8.18
Caesar, Gaius Julius 3.57, 3.307–8
  and Belgae 3.38
  and Britain 3.106, 3.115, 3.306
  calendar 8.18, 8.76
  and Carthage 3.61
  and Cleopatra 3.78, 3.79
  and Gallic Wars 3.38, 3.130, 3.137, 3.138
  and Germany 3.140
  and Pompey 3.278
  resistance to 3.65
  and Rubicon 3.278, 3.312
  and Senate 3.309
Caesars 3.57
caesium 1.47, 1.106, 6.72, 6.77, 6.290
Caetano, Marcello José das Neves Alves 4.58
Cafferi, Jacques 5.105
Cage, John 5.70, 5.219
cage-birds:
  budgerigar 2.44
  canary 2.51
  cockatoo 2.70
  mynah 2.224, 2.324
  parakeet 2.241–2
  parrot 2.242
Caiaphas 9.8
Cain 9.2, 9.8
Cain, James 5.478
Cainozoic (Cenozoic) era 1.47, 1.133, 1.134
  periods see Quaternary; Tertiary
Cairngorms 1.47
Cairo Conference 4.58
caisson 6.53, 6.112
calabrese 2.49
calamine 1.308
calamite 2.49–50
calc tufa 1.346
calceolaria 2.50
calciferol 2.357
calcite 1.47
  crystal 1.80, 1.81
  in marble 1.205
calcium 1.47, 6.96
calcium bicarbonate 1.47–8
calcium carbonate 1.47, 1.93–4, 1.194, 6.202, 6.378
  aragonite 1.17
  calcrete 1.48
  hard water 6.378
  marl 1.206
  stalactites and stalagmites 1.318, 1.319
  travertine 1.342
  tufa 1.346
  see also calcite; lime; limestone
calcium chloride 6.323
calcium fluoride 1.124
calcium hydrogencarbonate 1.47–8
calcium hydroxide 1.48, 1.195, 6.40, 6.202
calcium magnesium carbonate 1.93–4
calcium nitrate 6.241
calcium oxide see lime
calcium phosphate 1.16
calcium sulphate 1.48, 1.326, 6.163, 6.255
  anhydrous (anhydrite) 1.14
  see also gypsum
calcrete 1.48
calculator 6.1, 6.53–4, 6.84, 6.259, 6.322
  computer as development of 6.54
calculus 1.48–9
  differentiation 1.90, 1.91
  integration 1.169–70
  logarithm 1.198
  see also under Leibniz; Newton
Calcutta, Black Hole of 3.42
Calder, Alexander 5.70–1, 5.298
Calder Hall reactor 6.54
caldera 1.48, 1.49, 1.79, 1.166
Calderón de la Barca, Pedro 5.71
Caldwell, Erskine 5.71
Caledonia 3.57
calendar 3.57, 6.54, 8.18–19
  Gregorian 8.19, 8.62
  Islamic 7.166
  Jewish 7.171
  Julian 8.76
  lunar 8.87
  year 8.189
calendering 6.145
Calhoun, John Caldwell 4.58, 4.95, 4.184, 4.248
caliche 1.48, 1.49, 6.186
California, Gulf of 1.49
California Current 1.49
Californian poppy 2.110
californium 1.106
Caligula, Roman emperor 3.57, 3.94, 3.290
caliph 7.39
  see also ʻAbū Bakr; ʻAlī; ʻUmar
caliphates 3.58, 3.365, 3.369
Calixtus II, pope 3.179
Callaghan, James 4.58
Callaghan, Morley 5.269
Callas, Maria 5.71
Callejas, Rafael 4.156
Calles, Plutarco Elías 4.58
calligraphy 5.71, 5.91, 5.344
  see also writing
Callimachus 5.71–2
Calliope 9.8
calliper see Vernier
Callisto 8.19
calories 2.18, 2.50, 2.92, 6.166, 6.368
calorimeter 6.54–5
Caloris Basin 8.19, 8.20
calotype 6.344
Calvary 9.8
Calvin, John: and Calvinism 3.58, 3.293, 7.260, 7.310
  and Huguenots 3.170–1
  and predestination 7.245
  and Presbyterians 3.291
  and Puritans 3.295
  and Reformation 3.300, 3.301
Calvino, Italo 5.72
Calypso (nymph) 9.8
Calypso (satellite) 8.175
calypso (song) 5.72
Calzabigi, Raniero 5.181
CAM (computer-aided manufacture) 6.85, 7.63
cam (wheel) 6.55
Camacho, Avila 4.219, 4.220
Camargo, Marie 5.30
Camargue 1.49
Cambacérès, Jean-Jacques 4.78, 6.57
cambering (geomorphology) 1.49
Camberwell beauty 2.43
Cambodia (Kampuchea) 1.49, 4.58, 7.345
  art and architecture 3.13
  and ASEAN 7.20
  decolonization 4.321
  famine 7.116
  genocide 7.130
  Khmer Rouge 4.182, 7.20
  languages 7.22
  law 7.143
  literature 5.72
  physical geography and geology 1.49, 1.210, 1.339
  politics, government and leaders 4.273, 4.312–13, 7.94, 7.255
  refugees 7.261
  religion/theology 7.143
  wars and defence 7.52
Cambrai, League of 3.58
Cambrian Mountains 1.49
Cambrian period 1.49, 1.133, 1.134
Cambyses II, king of Persia 3.2
camcorder 6.372
Camden, battle of 3.58
Camden Town Group 5.72, 5.318, 5.417
camelids 2.50
  alpaca 2.7
  camel 2.50
  guanaco 2.146
  llamas 2.146, 2.192, 2.356
  vicuña 2.356
camellia 2.50
Camelopardalis 8.19
Camelot 9.8
cameo 5.72
camera 6.55
  aperture 6.16
  autofocus 6.25
  automatic exposure 6.25
  camerawork 5.72
  CCD 8.22
  cine-camera 6.73, 6.75, 6.210

camera (*cont.*):
cinematography **6**.73–4, **6**.210
DX coding **6**.109
exposure meter **6**.124
film *see* film, photographic
flash **6**.134
gamma **6**.243
iconoscope **6**.118, **6**.175, **6**.349, **6**.391
lens **6**.199, **6**.268
motor-drive **6**.235
Polaroid Land **6**.274–5
shutter **6**.318
television **6**.175, **6**.241, **6**.348, **6**.349
video **6**.372
see also cinema; photography
camera obscura ***6.55–6***, **6**.94
Cameron, Charles ***5.72–3***
Cameron, Julia Margaret ***5.73***, **5**.347, **5**.449
Cameroon ***1.49***, ***4.58***, ***7.345–6***
in Franc Zone **7**.124
history **4**.4, ***4.59***, **4**.306
language **7**.43
physical geography and geology **1**.49, **1**.57
Camisards ***3.58***
Camoës, Luís de **3**.96, ***5.73***, **5**.148
camomile **2**.60
cam-operated contact-breaker **6**.175
camouflage ***2.50***, ***6.56***
evolutional changes **2**.225, **2**.249
see also protective coloration
Camp David Accord ***4.59***, **4**.299, **7**.15
Campaign for Nuclear Disarmament ***4.59***
Campanella, Tomaso **5**.231
campanula ***2.50–1***
Campbell, Colen ***5.73***
Campbell, Sir Colin **4**.201
Campbell, Glen **5**.115
Campbell, Mrs Patrick ***5.73***
Campbell-Bannerman, Sir Henry ***4.59***
Camperdown, battle of ***3.58***
camphor **2**.66, **6**.66
Campin, Robert ***5.73***
Campion, Edmund, St ***3.58***
campion ***2.51***
Camus, Albert ***5.73***
Canaan ***3.58–9***
religion and myths **9**.7, ***9.8***, **9**.11, **9**.14, **9**.25
Canada ***1.49–50***, ***3.59***, ***4.59–60***, ***7.346***
agriculture **6**.5, **6**.7, **6**.68, **6**.76, **6**.355
aid from **7**.8
arts ***5.73***, **5**.244
see also literature *below*
astronomy **6**.166
Canada–France–Hawaii telescope ***8.19***
computers **6**.84, **6**.211
demography **7**.6, **7**.85, **7**.105, **7**.153
discovery and colonization *see under* exploration
economy and trade **4**.90
'affluent society' **7**.4
employment **7**.106
in Group of Seven **7**.136
in OECD **7**.221
education **7**.277
electoral system **7**.190
energy and fuel **6**.81, **6**.95, **6**.145, **6**.236, **6**.263, **6**.315
ethnic groups *see* Inuit *and also* Indians *below*
exploration *see under* exploration
explorers from **1**.237
fishing **6**.133
government *see* politics and government *below*
Canada (*cont.*):
immigration **7**.105, **7**.153
Indians:
Algonquin **3**.10, **3**.289–90
arts ***5.321***, ***5.322***
Cree **3**.90, **7**.74, **7**.219
Ottawa **3**.264, **3**.288
uprising **3**.288
see also Iroquois; North-west Indians
languages **7**.59, **7**.106, **7**.108, **7**.219, **7**.257, **7**.269, **7**.335
laws/legal system:
compensation **7**.62, **7**.75
habeas corpus **7**.138
judicial review **7**.173
police **7**.238
privacy **7**.247
literature **5**.25–6, **5**.126, **5**.259, **5**.303, **5**.383, **5**.422
maps **1**.231, **3**.252, **3**.257, **4**.60
medicine and health care **6**.29
nuclear technology **6**.57, **6**.243, **6**.245
physical geography and geology ***1.49–50***
bay **1**.129
Canadian Shield **1**.50
geologist **1**.198
glacier **1**.221
Grand Banks **1**.142
Hudson Bay **1**.159
islands **1**.30, **1**.105, **1**.335, **1**.353, **1**.357
Labrador **1**.185
lakes **1**.23, **1**.57–8, **1**.145, **1**.146, **1**.186, **1**.367–8
map **1**.231
mountains **1**.16, **1**.54, **1**.67, **1**.183, **1**.198, **1**.221, **1**.285, **1**.289
muskeg **1**.223
national and provincial parks *see under* national parks
peninsula **1**.350
plateaux **1**.145–6
provinces *see* provinces and regions *below*
rivers *see under* rivers
waterfalls **1**.63, **1**.228
politics and government:
Bill of Rights **7**.32
constitutional reform **4**.105
as Dominion **7**.58
established as Dominion **4**.52
Mackenzie's Rebellion **4**.204
ombudsman **7**.223
Pacific scandal **4**.254–5
Papineau's Rebellion **4**.258–9
parties **4**.85, **4**.86, **4**.195, **4**.238
prime ministers: listed **9**.70
regional policy **7**.262
United Empire Loyalists **3**.365
pollution **6**.2
prehistory **3**.203
provinces and regions:
eastern and Atlantic **1**.53, **1**.226, **1**.232, **1**.238, **1**.268, **3**.251, **3**.259
see also Quebec
listed ***9.73***
northern **1**.231–2, **1**.372
prairie **1**.6–7, **1**.204, **1**.293
western **1**.44
psychology *see* Bandura
religion/theology **7**.171, **7**.187, **7**.286, **7**.335
science *see entries above and below prefixed* **1**. *and* **6**.
social services and security **4**.317, **7**.91, **7**.330
sport and recreation:
ice hockey **7**.144–5
ice-skating **7**.151
lacrosse **7**.179
Olympic Games **7**.223
Canada (*cont.*):
sport and recreation (*cont.*):
softball **7**.295
technology and science *see entries above and below prefixed* **1**. *and* **6**.
telecommunications **6**.128, **7**.39
see also McLuhan
vehicles and transport **6**.57, **6**.161, **6**.206, **6**.209, **6**.296, **6**.311
railways **4**.460, **6**.169, **6**.216, **6**.292–4
wars and defence:
1812 war **4**.342
and Louisburg **3**.212
Plains of Abraham **3**.284
Riel Rebellions **4**.290
Seven Years War **3**.323
see also NATO
Canada geese ***2.51***
Canadian Pacific Railway ***4.60***, **4**.255
Canadian Shield ***1.50***
Canaletto ***5.74***
canals ***4.60–1***, **4**.350, ***6.56***, **6**.322
barge **6**.30, **6**.56
Britain **3**.385, **6**.47, **6**.56, **6**.350
builders **6**.47, **6**.199, **6**.350
construction **3**.49
Grand Canal of China **6**.159
Kiel **4**.183
land reclamation **6**.193
Languedoc **6**.298–9
lock **6**.206
of Mars ***8.19–20***
Panama **6**.256
see also Canal *under* Panama *and* Suez
Canaris, Wilhelm **4**.287
canary ***2.51***
canary creeper **2**.225
Canary Islands ***1.50***, **8**.82, **8**.83
Cancer (constellation) ***8.20***
cancer (disease) ***2.51***, **2**.52, ***6.56–7***, ***7.39***
biopsy **6**.37
blood **6**.164
breast **6**.214
causes **6**.104, **6**.373, **6**.377, **7**.226, **7**.288–9
cervical **6**.68, **6**.248
leukaemias **2**.189
oncologist **2**.266
screening **6**.68
treatment ***6.56–7***, **6**.72, **6**.112, **6**.184, **6**.243, **6**.290, **6**.292
tumours **2**.51, **2**.350
Cancer, Tropic of **1**.344
Candela, Félix ***5.74***
candela ***1.50***, **6**.321, **6**.368
candle ***6.57***, **6**.382
CANDU reactor ***6.57***, **6**.167, **6**.243, ***6.244***, **6**.369
candytuft ***2.52***
cane rat ***2.52***
Canes Venatici ***8.20***
Canetti, Elias ***5.74***
canids *see* dogs
Canis Major and Canis Minor ***8.20***
Canmore, Malcolm *see* Malcolm III
cannabis (marijuana) ***2.52***, **2**.154, **7**.95, **7**.96
Cannae, battle of ***3.59***
Canning, Charles John, 1st earl ***4.61***, **4**.162
Canning, George ***4.61***
canning **6**.17, ***6.57***, **6**.105, **6**.142, **6**.147
Cannizzaro, Stanislao ***1.50***
cannon **6**.22, ***6.57***, **6**.130, **6**.318, **6**.377
on ships *see* warships
Cano, Alonso ***5.74***
canoe ***6.57***, **6**.375
canoeing ***7.39***
canon (music) ***5.75***, **5**.223
canon law ***7.39***
canonist ***3.59***
Canopus ***8.20***
Canossa ***3.59***
Canova, Antonio ***5.75***
cantaloupe **2**.207
cantata **5**.21, ***5.75***
canticle ***5.75***
cantilever bridge **6**.47, **6**.48, ***6.57***
Canton *see* Guangzhou
Cantor, Georg **1**.13, ***1.50***, **1**.116, **1**.247, **1**.341
cantor ***7.39***
cantus firmus **5**.351
Canute, king of England, Denmark, Norway, and Sweden **3**.15, ***3.59–60***, **3**.101, **3**.108
canvas ***5.75***, ***6.57***, **6**.382
canyon ***1.50***
submarine **1**.325
terrestrial *see* gorges
canzona ***5.75***
Cao Cao ***3.60***
Cao Xueqin **5**.136
Cao Yu ***5.75***
Capa, Robert ***5.76***
capacitance ***1.50–1***, ***1.119***, **6**.368
capacitor ***6.57–8***, **9**.53
measurement (farad) **1**.119, **6**.368
capacity: measurement ***9.49***
Cape Breton **1**.232
Cape gooseberry **2**.141
Cape hunting dog **2**.166
Cape Province ***1.51***, ***4.61***
Colony: history **3**.304, **4**.306, **4**.384
physical geography and geology ***1.51***, **1**.142, **1**.341
mountains **1**.95, **1**.330
Cape St Vincent, battle of ***3.60***
Cape Town **3**.304
Cape Verde Islands ***1.51***, **4**.4, **7**.118, ***7.346***
Čapek, Karel ***5.76***, **6**.301
Capella ***8.20***
caper ***2.52***
capercaillie ***2.52***
Capet, Hugh, king of France **3**.60, **3**.174
Capetian dynasty **3**.18, ***3.60***, **3**.69, **3**.113, **3**.161, **3**.174, **3**.278–9
listed ***9.68***
see also Louis IX; Philip IV
capillary action ***1.51***
see also surface tension
capital (architecture) ***5.76***
capital (economics) ***7.39***
defined **7**.266
human **7**.149
infrastructure **7**.160
joint-stock company **7**.172
see also assets; capitalism; investment; resources
capital punishment **3**.150, ***7.40***, **7**.146
capitalism ***4.61***, ***7.39–40***
critique of *see* Marx
Frankfurt School **7**.124, **7**.138, **7**.192
see also capital; competition
Capone, Al ***4.61–2***, **4**.277
Caporetto, battle of ***4.62***
Capote, Truman ***5.76***
Cappadocia ***3.60***
cappella ***5.76***
Capra, Frank ***5.77***
capriccio ***5.77***
Capricorn, Tropic of **1**.344
Capricornus ***8.20***
Caprivi, Leo, Graf von ***4.62***
capsicum ***2.52***, **2**.63
capsid **2**.213–14
capstan ***6.58***
capuchin monkey ***2.52***, **2**.277
capybara ***2.52***

car *see* motor car
car ferry 6.60
see also ferry
caracara 2.52
Caratacus, king of Catuvellauni 3.38, 3.50, 3.60, 3.94
Caravaggio, Michelangelo Merisi da 5.77, 5.90, 5.173, 5.382
caravan (vehicle) 6.58
caravanserai 3.60
caravel 3.60, 6.58, 6.306
carbamate 6.180
carbine 6.131
carbohydrates 2.52–3, 2.92, 6.140, 6.141
alginate 6.11
cellulose 6.66, 6.387
glycogen 2.138
in plants *see* photosynthesis
starch 6.330
see also cereals; monosaccharides; polysaccharides; starch; sugars
carbolic acid *see* phenol
carbon 1.51, 1.106, 6.58, 6.125
alloy 6.11, 6.185
steel *see* iron and steel
amorphous 6.270
in battery 6.32
carbon-14 *see* radio-carbon *below*
carbonaceous chondrites 8.26
carbon-nitrogen cycle 8.20
charcoal 5.89, 6.69–70, 6.162
cycle 1.51, 2.53
diamond *see* diamond
fibre *see* carbon fibre
graphite *see* graphite
microphone 6.347
radio-carbon dating 6.96, 6.199–200, 6.290
see also cracking; hydrocarbons; organic compounds; petroleum; wax
carbon dioxide 1.51–2, 6.58
in aerosol 6.5
in blood 6.40
dry ice 1.96
in nuclear reactor 6.3, 6.213
pollution 6.78, 6.161, 6.280, 7.135, 7.239
see also greenhouse effect
and trees 7.83
carbon disulphide 6.338
carbon fibre 6.21, 6.58–9, 6.129
see also fibre-reinforced plastics
carbon monoxide 1.52, 6.39, 6.58
carbon tetrachloride 1.52
Carbonari 4.62
carbonates 1.51, 6.197, 6.202, 6.378
bicarbonates 1.39, 1.47–8, 1.310
magnesium 1.93–4, 1.202
see also calcium carbonate; sodium carbonate
carbon-black 6.58
Carboniferous period 1.52, 1.133, 1.134
Mississippian 1.134, 1.216
Pennsylvanian 1.134, 1.250
carbonless transfer paper 6.59
carbon-paper 6.59
carbonyl group 1.52
carborundum *see* silicon carbide
carboxylic acid 1.52, 1.213, 6.252
carbuncle 2.53
carburettor 6.59, 6.73, 6.94, 6.370, 6.371
cardamom 2.53
cardboard 6.59–60, 6.255
Cárdenas, Lázaro 4.62
Cardew, Michael 5.398
card-games 7.40
Cardigan, Lord 4.27
cardinal (bird) 2.53
cardinal beetle 2.53
cardinal fish 2.53, 2.54
carding 6.60, 6.165, 6.329, 6.390
cardiology and cardiac surgery *see under* disorders *under* heart
Cardoso, F. H. 7.85
cards, playing 5.353
Carducci, Giosuè 5.77–8
Cardwell, Edward, viscount 4.62
Carew, Thomas 5.82, 5.108
cargo cults 7.40
cargo ship *see* merchant and trading *under* ships
Caribbean Community and Common Market 4.62, 7.40
Caribbean Sea 1.52–3, 1.55
Caribbean and West Indies 1.52, 3.60–1, 4.376
ACP countries 7.103
agriculture 3.284–5, 7.232, 7.237
arts 5.78, 5.487–8
see also literature; music *below*
astronomy 8.7, 8.8
CARICOM see CARICOM
Common Market *see* CARICOM
demography 7.115, 7.119
discovery and exploration 3.122
education 7.93, 7.101, 7.102
ethnic groups 3.61, 7.40, 7.295
exploration 1.69, 1.70, 1.127, 1.355
independence 4.376
infant mortality 7.159
islands and regions:
Anguilla 1.14, 4.14
Antigua and Barbuda 1.15, 7.338–9
Hispaniola 1.156, 3.166
Monserrat 1.221
Saint Kitts and Nevis 1.290, 7.378–9
Saint Lucia 1.290, 7.379
Saint Vincent and Grenadines 1.290, 7.379
Virgin Islands 1.357, 4.468
see also Barbados; Cuba; Grenada; Jamaica; Trinidad and Tobago *and also* physical geography *below*
languages 7.295
leader 4.379
literature 5.59, 5.254, 5.312, 5.481
Lomé Convention 4.198
music 5.72, 5.78, 5.358, 5.431
physical geography and geology 1.52
Anguilla 1.14
Antigua 1.15
Barbados 1.33
Caribbean Sea 1.52–3, 1.55
Cuba 1.81
Grenada 1.147
Hispaniola 1.156
Jamaica 1.177
Montserrat 1.221
Saint Kitts and Nevis 1.290
Saint Lucia 1.290
Saint Vincent and Grenadines 1.290
Trinidad and Tobago 1.344
Virgin Islands 1.357
volcano 1.221, 1.233
waterspout 1.361
see also islands and regions *above*
poverty 7.244
religion/theology 7.202, 7.327, 9.4
carnival 7.40, 7.41
social services and security 7.292
societies 7.40
Spanish Main 3.336–7
sport and recreation 7.75
technology 6.90, 6.110, 6.143
wars and defence 3.313
see also Central America *and individual countries*
caribou 2.282
Caribs 3.61, 7.295
Carib language 7.295
caricature and caricaturists 5.78, 5.80, 5.125
Britain 5.119, 5.177, 5.392–3
see also comedy
CARICOM 4.62, 7.40
carillon 5.78
Carina 8.20–1
Carlist 4.62
Carloman, king of Franks 3.68
Carlos, Don 3.287
Carlsbad Caverns 1.53
Carlson, Chester Floyd 6.60
Carlyle, Thomas 4.62, 4.63, 5.78
Carmelites 3.61
Carmichael, Frank 5.73
Carmichael, Stokeley 4.40, 4.41
Carmina Burana 5.78
Carnac 3.61
Carnap, Rudolf 7.187
carnation 2.255
Carné, Marcel 5.79
Carnegie, Andrew 4.62–3, 7.184
carnival 5.78, 7.40, 7.41, 7.268
carnivores 2.53–4
teeth 2.54
see also bears; cats; crocodiles; dogs; genet; hyena; mongoose; racoon; weasel
carnivorous plants 2.54
bladderwort 2.33, 2.34
butterwort 2.48
pitcher plant 2.256
sundew 2.329
Venus's fly-trap 2.355
Carnot, Lazare Nicolas Marguerite 3.61, 4.63
Caro, Sir Anthony 5.79
Caro, Joseph 3.61, 3.348
carob 2.54–5
Carol I, king of Romania 4.63, 4.155, 4.292
Carol II, king of Romania 4.16, 4.63, 4.197, 4.292
carole or carol 5.79
Caroline islands 3.231, 4.63–4
Caroline of Ansbach, queen 3.61
Carolingian empire 3.61, 3.275
art 5.79
rulers:
Angevin dynasty 3.12–13
in France 3.61, 3.68, 3.69, 3.70, 3.137, 3.211, 3.275
see also Charlemagne
listed 9.68
see also under Holy Roman empire
Carothers, Wallace Hume 6.60, 6.247, 6.275, 6.384
carp family 2.55
barbel 2.22
bream 2.41, 2.265, 2.280, 2.298
chub(b) 2.65
crucian 2.82
golden orfe 2.140
goldfish 2.141
gudgeon 2.146
roach 2.286
tench 2.337
Carpaccio, Vittore 5.79
carpal 2.150
Carpathian Mountains 1.53
Carpentaria, Gulf of 1.53
carpenter bee 2.55
carpenter worm 2.139
carpentry and joinery 6.60
see also furniture; hand tools
carpet moth 2.55
carpet shark 2.55
carpet sweeper 6.61
carpetbaggers 4.64
carpets and rugs 5.79–80, 5.456, 6.60–1, 6.125
Carr, Edwin 5.80
Carr, Emily 5.73
Carr, Henry 4.45
Carr, John 5.80
Carr, Robert 3.265–6
Carrà, Carlo 5.80, 5.292
Carracci, Agostino and Ludovico 5.80–1
Carracci, Annibale 5.80, 5.254
carrack 6.307
Carranza, Venustiano 4.64, 4.219
Carrera, Rafael 4.142
carriages and coaches 6.61–2
railway 6.292
road 6.30, 6.61–2, 6.341
carrier wave 6.231
carrion beetles 2.46, 2.55
carrion crows 2.55, 2.82
Carroll, Lewis 5.81
carrot family 2.55
celery 2.59
fennel 2.115
parsnip 2.243
sea holly 2.299
carrying capacity 2.55
cars *see* motor car; roads
Carson, Edward Henry, Baron 4.64, 4.357
Carson, Kit 1.53, 4.64, 4.225
cartels 7.40–1, 7.222, 7.266
quota 7.257
see also in particular OPEC
Carter, Elliott 5.81
Carter, Howard 3.364
Carter, James (Jimmy) 4.64, 4.95
and Camp David Accord 4.59, 4.299
and Iran hostage crisis 4.165
and Panama 4.258
Carter, Robert 'King' 3.284
Cartesian coordinates 1.52, 1.53
see also Descartes
Carthage 3.61–2
generals 3.154, 3.158
wars against Rome 3.59, 3.294–5, 3.319
writer 3.352
Carthusians 3.51, 3.62
Cartier, Sir George-Etienne 4.64
Cartier, Jacques 1.53, 3.59, 3.62, 3.68, 3.122
Cartier-Bresson, Henri 5.81
cartilage 2.55–6, 2.93
cartilaginous fish 2.56
chimaera 2.63
monkfish 2.217–18
ratfish 2.63
rays 2.202, 2.278–9, 2.294, 2.326, 2.344
sawfish 2.294
skate 2.311
see also sharks
Cartimandua, queen of Brigantes 3.50
cartography 1.53–4, 1.204, 1.210–11, 1.241–2
cartoons (animation) 5.17, 5.18, 5.131
cartoons (drawings) 5.81
cartridge 6.62, 6.131
carts and wagons 6.61–2, 6.165, 6.298, 6.300, 6.375, 6.384
Cartwright, Edmund 6.62, 6.208
Caruso, Enrico 5.81
Carver, George Washington 2.56
Carver, John 3.62
carving *see* sculpture and carving
caryatid 5.81
Casablanca Conference 4.64
Casals, Pablo 5.82
Casanova de Seingalt, Giovanni Giacomo 3.62
Cascade Range 1.54
case law 7.41, 7.245
see also common law
case-bearer moth 2.56
case-hardening 6.333
casein 6.3, 6.95
Casement, Roger David 4.64, 4.107

Casey, Richard Gardiner, Baron *4.65*
Cash, Johnny 5.115
cash:
  cards 6.93
  crop 7.284
    see also plantation
  flow *7.41*
  register *6.62*
cashew nut *2.56*
Casimir III (the Great), king of Poland *3.62*, 3.286
Casimir IV, king of Poland 3.183
Casimir, John 3.286
Caslon, William 5.464
Caspian Sea *1.54*
Cassandra *9.8*
Cassatt, Mary *5.82*
cassava *2.56*, 6.142, 6.302
Cassegrain telescope 8.135
cassette, tape 6.213
Cassini, César-François (Cassini III) *8.21*, 8.117
Cassini, Giovanni Domenico (Cassini I) *8.21*
  Cassini's division 8.145
  and Paris Observatory 8.116
  and Saturn's satellites and rings 8.37, 8.72, 8.139, 8.140, 8.145, 8.171
  and zodiacal light 8.190
Cassini, Jacques (Cassini II) *8.21*, 8.117
Cassini, Jacques-Dominique (Cassini IV) *8.21*, 8.117
Cassiopeia *8.21*
cassiterite *1.54*, 6.356
Cassius 3.28, 3.51, 3.57, 3.307
cassowary *2.56*, 2.57
castanets *5.82*
caste system 3.298–9, *7.41*, 8.41
Castiglione, Baldassare *5.82*
Castile *3.62*
  rulers 3.9, 3.124–5, 3.153, 3.179–80, 3.188–9, 3.278
Castilla, Ramón *4.65*, 4.267
casting and founding *5.82*, 5.101, 6.33, *6.62*, 6.145
  cast iron 6.11, 6.186, 6.188, 6.297
  die-casting 6.102
  investment 6.184, 6.185
  see also bronze
castle *3.62–4*
Castle Hill Uprising *4.65*
Castlereagh, Robert Stewart, viscount *4.65*, 4.66, 4.279, 4.366
Castner–Kellner cell *6.62*, 6.63, 6.71, 6.72, 6.174, 6.323
Castor (mythology) *9.8*
Castor (star) *8.21*
castor bean tick 2.306
castration 3.119
castrato *5.82*
Castro, Cipriano 4.363
Castro, Fidel 4.32, *4.65*, 4.89, 4.111, 4.143
casuarina *2.56*
casuistry *7.41*
cat *see* cats
CAT (computer-aided testing) 7.63
CAT scan *see* computerized axial tomography
catabolic reactions 6.140
cataclysm *see* accidents and disasters
cataclysmic variable *see under* variable stars
catacomb *3.64*
Çatal Hüyük 3.64
Catalan language 7.156
Catalaunian Fields *3.64*
catalogue (astronomy) *8.21–2*
Catalonia *3.64*
catalpa *2.56*
catalyst *1.54*, *6.62*
  clay 6.91
  cobalt 6.79
  copper 6.14
catalyst (*cont.*):
  enzymes 6.122
  hydrofluoric acid 6.173
  nickel 6.164, 6.241
  palladium 6.256
  platinum 6.273, 6.296
  polypropylene 6.276
  vanadium 6.87
catalytic converter 6.9, *6.62*, *6.64*, 6.124
catalytic cracking 6.91, 6.171, 6.175
catalytic reforming 6.34, 6.262
catamaran *6.64*
catapult 6.22, *6.64*, 6.225, 6.318
cataract (of eye) *2.56*
cataract (water) *see* rapids; waterfalls
catarrh 2.57
catastrophe: theory *1.54*, *8.22*
  see also accidents and disasters
catbird *2.56–7*
catch 5.75
categorical imperative 7.109
catena *1.54*
catenary *1.54*
catering employment 7.106
caterpillar track *6.64*, 6.65
caterpillars *2.57*, 2.220
  army worm 2.14
  carpenter worm 2.139
  hornworm 2.152
  leaf miner 2.253
  looper 2.194, 2.198
  puss moth 2.273
  silk worm 2.310
  tent 2.337–8
  webworm 2.144
  witchetty grub 2.139
  see also larvae
Catesby, Robert 3.150–1
catfish *2.57*, 2.105
Cathar heresy 3.7, *3.64*
catharsis *5.82*, 5.459
Cather, Willa (Sibert) *5.82*
Catherine I, Empress of Russia 3.64, 3.300
Catherine II (the Great), Empress of Russia *3.64–5*, 3.309, 3.312
  and arts and literature 3.103, 5.73, 5.420, 5.444
  and Jewish pale 3.267
Catherine of Alexandria, Saint 9.33, *9.40*
Catherine of Aragon *3.64*
Catherine of Braganza *3.65*, 3.70
catheter *6.64*
cathode *1.54*, *6.64*, 6.197
  in photomultiplier 6.269
  in polarography 6.274
  ray 1.104
    see also cathode-ray tube
  in thermionic converter 6.353–4
  see also battery; Castner–Kellner; circuits; thermionic valve
cathode-ray tube *6.64*, 6.119
  amplifier 6.13
  oscilloscope 6.45, 6.253–4
  television 6.349–50
  visual display unit 6.64, 6.373
Catholic Emancipation *3.65*, *4.65*
  see also Roman Catholicism
Catiline 3.76
cation *1.55*
catkin *2.57*
Cato, Marcus Porcius (Elder and Younger) *3.65*, 3.342, 3.354, 5.257
Cato Street Conspiracy *4.65–6*
cats *2.56*, 2.58
  cave lion 2.58
  cheetah 2.54, 2.61–2
  clouded leopard 2.69
  as deities 9.7
  jaguar 2.174
  leopard 2.69, 2.187–8, 2.316
  lion 2.58, 2.191
cats (*cont.*):
  lynx 2.196
  ocelot 2.277
  puma 2.271–2
  sabretoothed tiger 2.291
  snow leopard 2.316
  tiger 2.291, 2.341, 2.342
  wild 2.367
Cat's-eye *6.64*
Catskill Mountains *1.55*
Cattell, James McKeen 2.58
cattle family (Bovidae) *2.57–8*
  auroch 2.18
  bison 2.32
  buffalo 2.44
  bull-fighting *7.37–8*
  cattle farming 4.66, *6.64*
  cattle trails *4.66*
  diseases 2.306, 2.349
  eland 2.104
  gaur 2.133
  gazelles 2.133–4, 2.323
  impala 2.20
  mythical 9.4–5, 9.6, 9.14, 9.25
  ox 2.238
  stomach 2.290
  yak 2.373
  see also antelopes; goat family; sheep
Catullus, Gaius Valerius *5.82*
Catuvellauni tribe, leaders of 3.60, 3.94
Caucasian languages *7.41–2*
Caucasoids *3.65*
Caucasus Mountains *1.55*, 1.135
Cauchy, Augustin-Louis, Baron 1.13, *1.55*
caudillo *4.66*
cauliflower *2.58*
caulking *6.64*, 6.316
causation *7.42*
caustic soda *see* sodium hydroxide
caustic solutions *see* alkali
Cavaço Silva, Anibal 4.275
Cavafy, Constantine *5.82*
Cavagnari, Sir Louis 4.13
Cavalcanti, Guido *5.82*
Cavaliers *3.65*, 3.155
  Cavalier Parliament *3.65*
  Cavalier poets *5.82*, 5.148, 5.205, 5.269, 5.284
Cavalli, Francesco *5.83*
cavalry 3.173, 6.225
cave *see* caves
Cave, Peter la 3.335
cave bear *2.58*
cave fish 2.34
cave lion *2.58*
cave-dwellers *3.66*
Cavell, Edith *4.66*
Cavendish, Lord Frederick 4.269
Cavendish, Henry (1731–1810) *1.55*, 1.84, 1.275
Cavendish, Lord William (1774–1839) *4.35*, 4.344
Cavendish, Lord William Henry (1738–1809) *4.35*
caves *1.55*
  animals 2.34, 2.58
  Australia 1.177
  cave-dwellers *3.66*
  caving *7.42*
  France 1.188
  glacial 1.247
  limestone 1.53, 1.180, 1.181, 1.318, 1.319
  paintings *5.83*
  stalactites and stalagmites 1.180, 1.181, 1.318, 1.319
  United States 1.53, 1.247, 1.367, 1.368
  see also under rock art
cavitation *6.64*, 6.367
cavity wall *6.64*
Cavour, Camillo Benso, count *4.66*, *4.67*, 4.131, 4.206, 4.290, 4.318
cavy and cavy-like animals *2.58*
  agouti 2.5
  capybara 2.52
  coypu 2.80, 2.362
  guinea pig 2.147–8
  gundi 2.148
  hutia 2.166
  spiny rat 2.322
  tuco-tuco 2.349
  viscacha 2.356
Caxton, William *3.66*, 3.269
Cayley, Arthur *1.55*
Cayley, Sir George *6.65*, 6.138, 6.139, 6.157, 6.384
cayman *2.58*
Cayman Islands *1.55*, 7.346
Cazalis, Henri 5.124
CB (citizen's band) radio *6.65*
CCD 6.70, 6.372
  camera *8.22*
CD *see* compact disc
cD-type galaxies 8.55
Ceausescu, Nicolae 4.292, 7.116, 7.239, 7.313
Ceawlin of Wessex 3.379
Cecchetti, Enrico *5.83*
Cecil, Robert, 1st earl of Salisbury and 1st viscount Cranborne *3.66*, 3.151
Cecil, William, 1st Baron Burghley *3.66*, 3.114, 3.149
Cecilia, Saint *9.41*
cedar *2.58–9*
Ceiriog *see* Hughes, John
Cela Trulock, Camilo José 5.428
celadon *5.83*
Celebes *see* Sulawesi
celery *2.59*
celesta *5.83*
celestial axis 8.23
celestial equator 8.22
celestial latitude 8.23
celestial longitude 8.23
celestial mechanics *8.22*
celestial meridian 8.97
celestial poles 8.22
celestial sphere 6.238, *8.22–3*
celestite 6.336
cell (electrical) 6.115, *6.147*
  Daniell 6.32, 6.96
  Leclanché 6.31–2, 6.197
  see also battery
cell (organic) *see* cells of living organisms
Cellini, Benvenuto 3.115, *5.83*, 5.183
cello and cellists 5.51, 5.82, *5.83–4*, 5.347, 5.352, 5.391
Cellophane *6.65*
cells of living organisms *2.59*, 6.176
  ATP in 2.18, 2.138, 2.222
  biologist 2.240
  cytology 2.86
  cytoplasm 2.86
  differentiation 2.92
  division 2.207, 2.214–15
  eukaryotic *see* eukaryotes
  gametes *see* gametes
  genes *see* chromosomes; genes
  histology 2.158
  mitochondrion 2.214
  plants 2.59
  plasma membrane 2.207
  prokaryotic *see* prokaryotes
  ribosome 2.284
  structure *6.65*
  tissue 2.342
  see also nucleic acids; protozoa
cellular telephone *6.65*
celluloid *6.66*, 6.172, 6.271
cellulose *2.59*, *6.66*, 6.387
cellulose acetate *6.66*, 6.92, 6.129, 6.153, 6.272
  see also rayon
cellulose nitrate and trinitrate *see* nitro-cellulose

Celman, Juarez **4**.31
Celsius, Anders: Celsius scale ***1**.55–6*, **1**.332, **6**.166, **6**.351
Celsus, Aulus Cornelius **6**.14
Celtiberians **3**.174
Celtic Church ***3**.66*, **9**.8, **9**.23, **9**.30, **9**.33, **9**.37, **9**.38–9
Celtic peoples ***3**.66–7*
arts **5**.11, **5**.35, ***5**.84*, **5**.84
literature **5**.35, **5**.225–6, **5**.408–9, **5**.485–6
Belgae **3**.38
Brigantes **3**.49–50
Celtiberians **3**.174
Druids **3**.106
Gaul **3**.138, **3**.371–2
Hallstatt Iron Age ***3**.154*
Helvetii **3**.160
La Tène Culture ***3**.203*
languages ***7**.42*, **7**.156, **7**.157
religion and myths (including Germanic) **9**.30, **9**.98–9
deities **9**.2, **9**.6, **9**.13, **9**.22, **9**.23, **9**.36
dwarfs and elves **9**.10, **9**.11, **9**.27, **9**.28, **9**.38
giants **9**.6
human and supernatural characters **9**.5, **9**.6, **9**.8, **9**.13–14, **9**.23, **9**.25, **9**.33, **9**.36, **9**.37
places **9**.5
tricksters **9**.12, **9**.23
see also Arthur, king; Celtic Church
Scots **3**.319
technology **6**.69
Urnfield cultures **3**.368
see also Iron Age
cement **6**.51, ***6**.66*, **6**.202
composite **6**.82
hydraulic **6**.66, **6**.86, **6**.322
Portland **6**.66, **6**.86, **6**.271, **6**.276
see also concrete; mortar
cementation ***6**.66–7*
Cenozoic era *see* Cainozoic
censorship ***7**.42*, **7**.43
lack of *see* pornography *and also* expression *and* information *under* freedom
census ***7**.42–3*
Centaurs ***9**.8*
Centaurus (constellation) ***8**.23*
Centaurus A (radio galaxy) ***8**.23*, **8**.24
centaury ***2**.59*
centigrade *see* Celsius
centipede ***2**.59*
CENTO *see* Central Treaty Organization
Central Africa **3**.199
Central African Federation ***4**.67*
see also individual countries
Central African Republic ***1**.56*, ***4**.67*, ***7**.346*
decolonization **4**.4
ethnic groups **7**.24
finance **7**.124
languages **7**.43
physical geography and geology **1**.56
Central America ***1**.56*, ***3**.67*, ***4**.67–8*
adoption **7**.3
agriculture **7**.232, **7**.237, **7**.306
arts **3**.296, **5**.59, ***5**.84–6*, **5**.384
architecture ***5**.84–5*, **5**.100–1, **5**.449, **5**.458
sculpture and carving **5**.85, **5**.327, **5**.457, **5**.498
see also literature; music *below*
colonization *see* exploration *below*
demography:
contraception **7**.115
fertility rate **7**.119
infant mortality **7**.159
Central America (*cont.*):
demography (*cont.*):
infertility **7**.159
population growth **7**.226, **7**.241–2
economy and trade **7**.88
Central American Common Market **4**.68
COMECON **7**.57
debt **7**.2
dependency theory **7**.85
pay **7**.203, **7**.328
unemployment **7**.320
education **7**.93, **7**.101, **7**.102, **7**.141, **7**.185, **7**.265, **7**.277
ethnic groups *see* Aztecs; Indians, South American (and Central American); Mayans; Mixtecs; Toltecs
exploration **1**.47, **1**.70, **1**.88, **1**.159, **3**.159, **3**.298–9
government *see* politics and government *below*
land reform **7**.179–80
languages **7**.199–200, **7**.219, **7**.269, **7**.295, **7**.335
laws/legal system **7**.164, **7**.174
literature ***5**.85–6*, **5**.256
maps **3**.31, **3**.226, **4**.67
medicine and health care **7**.159
music **5**.14–15, **5**.257, **5**.282
in Organization of American States **7**.225
in Organization of Central American States **4**.251
Pan-Americanism **4**.258
physical geography and geology ***1**.213*, **1**.214, **1**.248, **1**.264, **1**.372
mountains **1**.285, **1**.303
politics and government:
Christian Democracy **7**.47
communist **7**.60
corruption **7**.71
domino theory **7**.94
military **7**.173, **7**.201
population *see* demography *above*
poverty **7**.244
refugees **7**.261–2
religion/theology **3**.297–8, **7**.49
Jews and Judaism **7**.335
liberation theology **7**.184
myths and characters **9**.22, **9**.32, **9**.37, **9**.39
pentecostal **7**.232
shamanism **7**.282
social services and security **7**.293
sport and recreation **7**.54, **7**.232, **7**.295, **7**.303
technology and science **6**.54, **6**.110, ***6**.219–20*, **6**.287, **6**.346
agriculture **6**.7, **6**.34, **6**.76
see also entries above and below prefixed **1.** *and* **6.**
urban areas **7**.288, **7**.298
wars and defence **7**.52, **7**.82
see also Caribbean and West Indies *and individual countries in particular* Mexico
Central American Common Market ***4**.68*
Central Asian music ***5**.86*
central bank **7**.26, ***7**.43*, **7**.53, **7**.133
Bank of England **3**.34–5, **7**.43
Central Europe *see* Eastern Europe
central heating **6**.66, ***6**.67*, **6**.104, **6**.167
Central Intelligence Agency ***4**.68*, **7**.72, **7**.162
Central Pacific Railroad ***4**.68*
central processing unit ***6**.67*, **6**.83, **6**.218–19, **6**.222–3, **6**.240
Central Treaty Organization ***4**.68*, **7**.315
centralization ***7**.43*
Centre (politics) ***7**.43*
centre of gravity ***1**.56*
centre of mass **8**.61
Centre National de la Recherche Scientifique ***6**.67*
centreboard ***6**.67*
Centre-Right parties ***7**.47*
centrifugal force ***1**.56–7*
centrifugal pump **6**.286
centrifuge ***6**.67*
centripetal force **1**.5, ***1**.56*
centurion ***3**.67*
Century Guild **5**.274
century plant **2**.4
ceorl ***3**.67*
cephalochordate **2**.182
cephalopods ***2**.60*
ammonite **2**.8
belemnite **2**.28
cuttlefish **2**.84
octopus **2**.232, **2**.241, **2**.308
paper nautilus **2**.241
pearly nautilus **2**.245, **2**.246
squid **2**.323
vision **2**.308
cephalosporins **6**.16, **6**.26
Cepheid variable stars ***8**.23–4*
Cepheus ***8**.24*
ceramics and pottery ***5**.358–61*, ***5**.363–4*, ***6**.67–8*, ***6**.278–9*
Abbasid **5**.1
Africa **5**.8
bone china **5**.51
Britain **3**.37, **5**.105, **5**.258–9, **5**.485, **6**.276, **6**.382
see also Arts and Crafts; Wedgwood
Burma **5**.67
Byzantine **5**.70
China **3**.387, **5**.83, **5**.91, **5**.92, ***5**.93–4*, **5**.152, **6**.276, **6**.278
Coptic **5**.112
creamware **5**.118
earthenware **5**.141, **5**.276, **5**.288, **5**.384, **5**.450
Etruscan **5**.149
folk art **5**.159
France **5**.337
glaze *see* glaze/glazing
Greece, ancient **3**.11–12, **5**.311
Hittite **5**.208
Islam **5**.207, **5**.228, **5**.229
Iznik ware **5**.231
Japan ***5**.234–5*, **5**.237, **5**.246, **5**.258–9, **5**.373, **6**.67
Khmer **5**.248
Korea **5**.251
lustreware **5**.207, **5**.229, **5**.271
maiolica **5**.276
Minoan **5**.297
Moorish **5**.303
Ottoman **5**.231, **5**.333–4
Persia **5**.344
Polynesia **5**.356
porcelain *see* porcelain
potter's wheel **6**.278
Rome **3**.11–12
slip **5**.422
South American Indians **5**.426
Spain **5**.348
Stone Age **5**.434
stoneware **5**.434
studio potter **5**.258, **5**.438
Sumerian **5**.305
technology of ***6**.67–8*
as abrasive **6**.12
alloy **6**.318
alumina **6**.1, **6**.12, **6**.82, **6**.318
brick *see* bricks
carborundum **6**.1, **6**.68, **6**.82, **6**.297, **6**.319–20
cement *see* cement
cermet **6**.68, **6**.82
composite **6**.82
concrete *see* concrete
ceramics and pottery (*cont.*):
technology (*cont.*):
electroceramics **6**.116, **6**.127, **6**.339
extrusion **6**.124
glass ceramic **6**.156, **6**.373
glazing *see* glaze/glazing
as insulator **6**.12
kiln *see* kiln
plasticizer **6**.272
porcelain *see* porcelain
silicon nitride **6**.318, **6**.320
tile *see* tiles
vitrification **6**.373
zirconia **6**.391
terracotta **5**.288, **5**.450
Thai **5**.450
tiles **6**.51, **6**.76, **6**.146, **6**.302, **6**.356
transfer printing **5**.459
Vietnam **5**.474–5
Zapotec **5**.498
Cerberus ***9**.8*
cereals and grains ***2**.60*, **6**.7, **6**.19, ***6**.68*
barley **2**.23
bread **6**.28, **6**.45–6
corn **2**.77, **2**.198
diseases **2**.110, **2**.211
flour **6**.142
green revolution **6**.161
harvesting *see* harvesting
hybridization **6**.36
increased production **7**.8
irrigation **6**.188
maize **2**.198, **2**.332
millet **2**.211
milling **6**.142, **6**.225–6
oats **2**.231–2
processing **6**.142
propagation **6**.271
rice *see* rice
rye **2**.110, **2**.291
sorghum **2**.317
storage ***6**.158*
sweet corn **2**.332
wheat **2**.365
winnowing **6**.386
see also grass
cerebral palsy **2**.60
cerebrospinal fluid **6**.210
Cerenkov radiation ***8**.24*
Ceres (asteroid) ***8**.24*, **8**.193
Ceres (goddess) ***9**.8*
Cerezo, Vinico **4**.142
cerium **1**.106, **6**.11, **6**.25
Cerletti, Ugo **6**.116
cermet ***6**.68*, **6**.82
CERN *see* European Laboratory for Particle Physics
Cerro, Sánchez **4**.147
Cerro Tololo Inter-American Observatory ***8**.24*
Cervantes Saavedra, Miguel de ***5**.86*, **5**.323, **5**.348, **5**.387
cervical cancer **6**.68
screening ***6**.248*
Césaire, Aimé **4**.309, ***5**.86*
cesium *see* caesium
cesspit **6**.312
cetane number **6**.103
Cetshwayo, king of Zulus ***4**.68*, **4**.168, **4**.391
Cetus ***8**.24*
Cevallos, Rodrigo Borja **4**.108
Ceylon *see* Sri Lanka
Cézanne, Paul ***5**.87*
CFCs *see* chloro-fluorocarbons
CFE *see* Conventional Forces in Europe
Chabrier, Emmanuel ***5**.87*
Chabrol, Claude **5**.323
Chaco **1**.142
War ***4**.68*
chaconne ***5**.87*

Chad ***1.**57*, ***4.**68–9*, ***7.**346–7*
finance **7.**124
languages **7.**43, **7.**217
physical geography and geology **1.**57
refugees **7.**261
Chad, Lake ***1.**57*
Chadic languages ***7.**43*
Chadli, Benjedid **4.**9
Chadwick, Sir Edwin ***4.**69*, **4.**99
Chadwick, Sir James ***1.**57*, **1.**153, **1.**178
Chaeronea ***3.**67*
chaetognathan **2.**15
chafers ***2.**60*, **2.**288
chaffinch ***2.**60*
Chagall, Marc ***5.**87*
Chagas' disease **2.**348
Chain, Sir Ernest Boris **2.**122, ***6.**69*, **6.**134, **6.**137, **6.**260
chain drive ***6.**33–4*
chain reaction ***6.**69*, **6.**242, **6.**245
chain store ***7.**43*
chain-mail ***6.**40*
chain-saw ***6.**69*, **6.**387
chairs **5.**167
Chaitanya, Guru **5.**43, **7.**138
Chalcedon, Council of ***3.**67*, **7.**309
chalcedony ***1.**57*
agate **1.**4–5, **1.**76
cornelian **1.**76–7
onyx **1.**238
Chalchithuitlicue/Chalchevhtlicue **3.**31, **9.**35–6
chalcid ***2.**60*
Chalcis ***3.**67*
chalcopyrites **6.**89
Chaldea **3.**39, ***3.**67*
Chalgrin, Jean-François **5.**315
Chaliapin, Fyodov ***5.**87*
chalk ***1.**57*
in cob **6.**79
for drawing ***5.**87*
calcium carbonate **1.**47
erosion **1.**75, **1.**96
wold **1.**368
Chamaeleon ***8.**25*
chamber music ***5.**88*
chamber of commerce ***7.**43*
Chamberlain, Sir Austen ***4.**69*
Chamberlain, Joseph **4.**43, ***4.**69*, **4.**79, **4.**84, **4.**125, **4.**338
Chamberlain, Neville **4.**16, ***4.**69*, **4.**227, **4.**234
Chamberlin, Thomas Chrowder ***1.**57*
Chambers, Whittaker **4.**152
Chambers, Sir William ***5.**88*, **5.**98
chameleon **2.**11, **2.**50, ***2.**60*, **2.**61
chamois ***2.**60*
chamomile ***2.**60*
Chamorro, Violeta **4.**241
Chamoun, Camile **4.**211
Champa ***3.**67–8*
Champagne (province) ***3.**68*
Champaigne, Jean-Baptiste de ***5.**88*
Champaigne, Philippe de **3.**304, ***5.**88*
Champlain, Samuel de ***1.**58*, **3.**62, ***3.**68*
and Indians **3.**173, **3.**180, **3.**264
map by **3.**252
Quebec founded **3.**59, **3.**297
Champlain, Lake ***1.**57–8*
Champollion, Jean-François **3.**310
Chancellor, Richard ***1.**58*, **3.**243
chancery ***3.**68*
Chandidas **5.**43, **5.**46
Chandler, Raymond **5.**129
Chandler, Seth: Chandler wobble ***8.**25*
Chandraguptā I, emperor of India **3.**24, **3.**151, **3.**225–6, **5.**401
Chandraguptā II, emperor of India **3.**151
Chandrasekhar, Subrahmanyan ***8.**25*
limit ***8.**25*
Chandrasekhar, Subrahmanyan (*cont.*):
Schönberg–Chandrasekhar limit ***8.**146*
Chang Heng *see* Zhang Heng
Changamire of Rozvi ***3.**68*, **3.**244, **3.**312
change-ringing ***5.**88*
channel (telecommunications) ***6.**69*
channel (structural) *see* tunnel
channel (geomorphology) ***1.**58*
Channel Islands ***1.**58*, **7.**347
Channel Tunnel ***6.**69*, **6.**365, **7.**67
chanson **5.**88
de geste ***5.**88*, **5.**368
chantarelle ***2.**61*
Chantrey, Sir Francis ***5.**88*
Chanute, Octave **6.**139
chaos ***1.**58*, ***8.**25*
chaparral ***2.**61*
Chapel, Cornaro **5.**35
Chapel Royal ***5.**88*
members of **5.**50, **5.**65, **5.**69, **5.**103, **5.**176, **5.**258, **5.**304, **5.**370
Chaplin, Sir Charles Spencer ***5.**88–9*
Chapman, Tracy **5.**358
Chappe, Claude ***6.**69*, **6.**312
Chapultepec Conference ***4.**69*
characin ***2.**61*, **2.**341
characteristic (mathematics) **1.**198
charango ***5.**89*
charcoal ***5.**89*, ***6.**69–70*, **6.**162
Chard, Lieutenant **4.**168
Chardin, Jean-Baptiste-Simeon ***5.**89*, **5.**433
charge (electrical) *see under* electricity
Charge of the Light Brigade **4.**27–8
charge-coupled device ***6.**70*, **6.**372
chariot ***3.**68*, ***6.**70*, **6.**165, **6.**225, **6.**300
charismatic movement ***7.**44*, **7.**232, **7.**250
charities **7.**218, **7.**256, **7.**327
Charlemagne ***3.**68*, **3.**231
and arts **5.**79, **5.**88
conquests **3.**141, **3.**172, **3.**318
and learning **3.**8, **3.**97
and pope **3.**205
see also Carolingian empire
Charles I, king of Anjou, Naples, and Sicily (1226–85) ***3.**69*, **3.**327
Charles I, king of Austria (Charles IV of Hungary) **4.**26, **4.**144, **4.**157
Charles I, king of Great Britain and Ireland (1600–49) ***3.**69*, **3.**343
and arts **5.**286, **5.**311, **5.**393
see also Cavalier poets *under* Cavaliers
and Commons **3.**83
and Divine Right **3.**104
execution **3.**69
and Ireland **3.**85, **3.**180
and law **3.**340
and parliament **3.**127, **3.**144, **3.**209–10, **3.**231, **3.**270, **3.**292, **3.**363
queen **3.**160, **3.**161
and religion **3.**89, **3.**295
and Scots **3.**40, **3.**41–2
and taxes **3.**94, **3.**327
and war **3.**278
see also English Civil War
Charles I, king of Spain *see* Charles V, Holy Roman Emperor
Charles II (the Bald), king of West Franks (823–77) ***3.**69*, **3.**211
Charles II, king of England, Scotland, and Ireland (1630–85) **3.**40, **3.**43, **3.**51–2, ***3.**69–70*
and Africa **3.**311
and America **3.**256, **3.**275, **3.**334
and arts **5.**384
see also under Restoration
Charles II, king of England, Scotland, and Ireland (*cont.*):
and astronomy **8.**142
Breda Declaration **3.**49
Declarations of Indulgence **3.**100
finances **3.**94, **3.**96
and Hudson's Bay Company **3.**170
and learning **3.**311, **3.**319
ministers (cabal) **3.**56
mistress **3.**152
and parliament **3.**56, **3.**65, **3.**83, **3.**270, **3.**380
plots to kill **3.**289, **3.**312
and religion **3.**90
and science **6.**303
and war **3.**14
yacht **6.**390
see also Restoration
Charles II, king of Spain, Naples, and Sicily (1661–1700) **3.**153, **3.**337
Charles III, king of France **3.**255
Charles III (the Fat), king of Franks (832–88) and emperor of Germany ***3.**70*
Charles III, king of Spain, Naples, and Sicily (1716–88) **3.**46, ***3.**70*, **3.**74
Charles IV, Holy Roman Emperor (1711–40) **3.**168
Charles IV, Holy Roman Emperor and King of Bohemia (1316–78) ***3.**70*, **3.**168
Charles IV, king of Hungary *see* Charles I, king of Austria
Charles IV, king of Sweden **4.**69–70
Charles V (the Wise), king of France (1337–80) ***3.**70*
Charles V, Holy Roman Emperor and Charles I of Spain (1500–58) ***3.**70*, **3.**168, **3.**335, **4.**62
and Augsburg **3.**27
and colonies **3.**336
family tree **3.**152, **3.**153
finances **3.**135
and Golden Bull **3.**48
Holy League against **3.**167
and Malta **3.**198, **3.**217
and painting **5.**457
and religion **3.**103, **3.**385
royal progress **3.**293
and war **3.**37, **3.**126, **3.**272, **3.**318
Charles V, king of Spain **3.**335
Charles VI, king of France **3.**89, **3.**171–2
Charles VI, Holy Roman Emperor (1685–1740) **3.**153, **3.**290, **3.**338, **3.**369
Charles VII, king of France (1403–61) ***3.**70–1*, **3.**137, **3.**189, **5.**325
Charles VII, Holy Roman Emperor (1742–5) **3.**168, **3.**384
Charles VIII, king of France **3.**50, **3.**317, **3.**369
Charles IX, king of France **3.**81
Charles X, king of France **3.**46, **3.**47, ***4.**69*, **4.**177
Charles X, king of Sweden **3.**311, **3.**318, **3.**347
Charles XI, king of Sweden **3.**318
Charles XII, king of Sweden **3.**28, ***3.**71*, **3.**246, **3.**256–7, **3.**287, **3.**347
Charles XIII, king of Sweden **4.**335
Charles XIV, king of Sweden and Norway ***4.**69–70*, **4.**288, **4.**335
Charles Albert, king of Bavaria **3.**385
Charles Albert, king of Piedmont **4.**269, **4.**281, **4.**290
Charles the Bold, duke of Burgundy (1433–77) **3.**54, ***3.**71*, **3.**211
Charles of Hohenzollern-Sigmaringen, prince **4.**155
Charles the Mad **3.**70
Charles Martel of Franks ***3.**71*, **3.**230
Charles, Jacques **1.**58, ***1.**59*, **6.**138, **6.**139
Charles, Ray **5.**382
Charleston ***5.**89*
Charlie, Bonnie Prince *see* Stuart, Charles Edward
charm (physics) ***1.**59*
Charon ***9.**8*
Charon (Pluto's satellite) **8.**25, **8.**198
Charpentier, Marc-Antoine ***5.**89*
charr ***2.**61*
charter ***3.**71*
Charter 77 ***4.**70*
chartered company ***3.**71*
Chartism ***4.**70–1*
charts *see* maps and charts
Chase, Lucia **5.**14
Chassey culture **3.**380
Chateaubriand, Vicomte François-René de ***5.**89*
chats ***2.**61*, **2.**280–1
Chattanooga Campaign ***4.**71*
chatterer ***2.**61*
Chatterji, Bankim Chandra **7.**143
Chatterton, Thomas ***5.**89*
Chattopadhyay, Bankimschandra **5.**43
Chattopadhyay, Saratchandra **5.**43
Chatwin, Bruce **5.**460
Chatwood, Samuel **6.**304
Chaucer, Geoffrey **3.**43, **3.**132, **3.**281, ***5.**89*, **5.**204, **5.**367, **5.**368, **5.**402
Chávez, Carlos **5.**257
Chavez, Cesar **4.**72
Chavez, Frederico **4.**331
Chavín culture ***3.**71*, **5.**426
Chayessteh, Narguesse **7.**239
Che Guavara **7.**136
Cheddar Gorge ***1.**59*
Chedworth Roman Villa **3.**309
cheese manufacture ***6.**70*
cheese-mite ***2.**61*
cheese-skipper ***2.**61*
cheetah **2.**54, ***2.**61–2*
Cheever, John ***5.**90*
Cheiffou, Amadou **4.**242
CHEKA ***4.**71*
Cheke, Sir John **3.**252
Chekhov, Anton Pavlovich ***5.**90*, **5.**250, **5.**383, **5.**430, **5.**458
Chekhov, Olga Knipper **5.**90
chelation ***1.**59*
Chelčický, Peter **5.**120
Chelmsford, Frederic Augustus, 2nd Baron **4.**168
Chelmsford, Frederick John Thesiger, 3rd Baron **4.**223
chemicals/chemical:
and agriculture *see* agrochemicals
energy stored **6.**121
engineering ***6.**70*
food additive **6.**140
industry ***6.**70–1*, **6.**123
accidents **6.**35, **6.**275, **6.**314
alkali industry **6.**96–7
and chlorine **6.**60, **6.**70, **6.**71, **6.**308
chloro-fluorocarbon production **6.**123, **6.**139
hydrocarbons **6.**173
hydrochloric acid **6.**173
lime used in **6.**202
natural gas **6.**237
sodium carbonate and hydroxide **6.**323–4
sulphuric acid **6.**197
see also petrochemicals *and* refining *under* petroleum
machining **6.**211
and physiology *see* drugs
vapour deposition **6.**370

chemicals (*cont.*):
warfare **1**.62, ***4**.71*, ***6**.71–2*, **6**.98, **6**.164, **6**.235, **6**.239–40, **6**.267, ***7**.44*, ***8**.44*
see also acids; bases; chemistry; petrochemicals; salts *and also individual chemicals*
chemistry ***1**.60*
activation energy **1**.2–3
activity series **1**.3, **1**.102
alchemy **1**.7, **6**.10
amalgam **1**.11, **6**.12
analysis **1**.12–13
azeotrope **1**.28
chemical bond ***1**.59*
chemical compounds *see* compounds
chemical equation ***1**.59*
chemical reactions ***1**.59–60*
endothermic **1**.3, **1**.108
enthalpy **1**.110
equilibrium **1**.111
exothermic *see* exothermic
explosions **1**.117–18
hydrolysis **1**.163
latent heat **1**.188–9
mechanisms of **1**.209
photochemical **1**.256–7
polymerization **1**.263
stoichiometry **1**.320–1
synthesis **1**.59, **1**.329
see also catalyst; corrosion
chemical symbol **1**.38, ***1**.60*
chemists *see under individual countries*
chirality **1**.61
chromatography **1**.61, **1**.62, **6**.73, **6**.217
compounds *see* compounds; organic compounds
concentration **1**.71
configuration **1**.71
conformation **1**.71, **1**.72
cosmological ***8**.25*
critical point **1**.80
diffusion **1**.91
dissociation **1**.20, **1**.92
drying agent **1**.96
of Earth *see* geochemistry
equilibrium **1**.111
free radicals **1**.127
functional groups **1**.129
gas *see* gas
heteroatoms **1**.155
homologous series **1**.157
indicators **1**.197
industry *see under* chemicals
inhibitor **2**.170
isomerism **1**.175
ligand **1**.193
and living organisms:
biochemistry **2**.29
biogeochemical cycle **2**.30, **2**.53, **2**.228, **2**.235
inhibitor **2**.170
see also metabolism
mixture *see* mixtures
molecules *see* molecules
neutralization **1**.59, **1**.226
nucleation **1**.233
organic *see* organic chemistry
oxidation *see* oxidation and reduction
periods **1**.254
precipitation **1**.59, **1**.116, **1**.266, **1**.342
reduction *see* oxidation and reduction
resonance effects **1**.281
solutions *see* solutes; solvent and solutions
stereochemistry **1**.320
sublimation **1**.325, **6**.186
synthesis **1**.59, **1**.329
titrations **1**.2, **1**.338
see also acids; bases; elements
chemofossil ***2**.62*
chemotherapy **2**.104, **6**.56, ***6**.72*, **6**.112, **6**.184
see also drugs, therapeutic
Chen Duxiu **4**.74
Chen Yi **4**.337
Cheng Ho *see* Zheng He
chenier ***1**.60*
Chénier, André ***5**.90*
Chernobyl **6**.9, ***6**.72*, **6**.243, **6**.245, **6**.275, **7**.91
chernozem ***1**.60*, **1**.269, **1**.311
Chernyshevsky, Nicolai **4**.275
Cherokee Nation ***3**.73*, ***4**.71*, ***7**.44*, **7**.219
Cherry, G. F. **3**.357
cherry ***2**.62*
chert ***1**.60*, **1**.123
see also flint
cherubim and seraphim ***9**.8–9*
Cherubini, Luigi ***5**.90*
Chesapeake Bay ***1**.60*
chess ***7**.44*
Chesterfield, earl of *see* Stanhope, James
Chesterton, Gilbert Keith ***5**.90*
chestnut soil ***1**.60*, **1**.311
chestnut tree ***2**.62*
Chesworth, John **5**.373
chevrotain ***2**.62–3*
chewing **2**.205
see also ruminant
chewing-gum **2**.294
Cheyenne **7**.219
Chiabrera, Gabriello **5**.231
Chiang Ching-kuo **4**.337
Chiang Kai-shek ***4**.71–2*, **4**.73, **4**.188, **4**.210, **4**.384
and Cairo Conference **4**.58
and Japan **4**.315
and Northern Expedition **4**.246
and Taiwan **4**.337
and warlords **4**.371
chiaroscuro ***5**.90*, **5**.382
Chibcha Indians **9**.11
Chibchan–Paezan languages **7**.295
Chicago School (architecture) **5**.14, ***5**.91*, **5**.439
Chicago School (sociology) ***7**.44*
see also symbolic interactionism
Chicanos ***4**.72*
chickadee **2**.342
Chickasaw Indians ***3**.73*
chickenpox ***2**.63*
chickens ***2**.63*, **2**.243
chick-pea ***2**.63*
chicory ***2**.63*
chief ***7**.44*
Chien-lung *see* Qianlong
chiffchaff **2**.360
Chifley, Joseph Benedict ***4**.72*
Chikamatsu Monzaemon **5**.235
Chikuyo of Rozvi empire **3**.68
Chilana, Chakufwa **4**.205
child *see* children
Childe, V. G. **7**.78
Childebert II, king of Visigoths **3**.51
Childeric III, king of Franks **3**.275
children:
abuse and neglect **4**.31, ***7**.45*, ***7**.46*
adoption **7**.3
attachment and imprinting **7**.21–2, **7**.110, **7**.153–4, **7**.291
autism **7**.23
benefit ***7**.45*
books for *see* literature *below*
childbirth *see* birth; pregnancy
childcare services ***7**.45*
childhood ***7**.45–6*
Children's Crusade ***3**.73*
cognitive development **7**.54
see also Piaget
and divorce **7**.189
and domestic violence **7**.94
education *see* education
guardian **7**.136
in institutions **7**.313
children (*cont.*):
juvenile court and delinquency **7**.174
labour ***4**.72*, **4**.249, **7**.45, ***7**.46*
language development **7**.180
literature ***5**.91*
Britain **5**.128, **5**.188, **5**.259, **5**.296, **5**.324, **5**.432
Canada **5**.303
Denmark **5**.16
first **5**.90, **5**.120
Germany **5**.194
illustrated **5**.90, **5**.120
United States **5**.11, **5**.200
malnutrition **7**.190
medicine and health care **7**.9, **7**.56
mortality **7**.119, **7**.140, **7**.153, **7**.159
play and games **7**.237, **7**.286, **7**.300
psychology **7**.35, **7**.55
psychosexual development **7**.253
puberty rites **7**.253
rights **7**.45, **7**.46, **7**.149, **7**.319
sexual abuse ***7**.46*
sexual exploitation ***7**.46*
sexuality **7**.126
skills development **7**.208
socialization **7**.291
toys **5**.459, **6**.25, **6**.73–4
and UN **7**.45, **7**.46, **7**.149, **7**.319
see also UNICEF
Chile ***1**.60*, ***3**.73*, ***4**.72–3*, ***7**.347*
astronomy **8**.24
foreign investment **4**.320
independence **4**.43
infant mortality **7**.159
leaders **4**.8, **4**.10, **4**.29, **4**.125–6, **4**.250, **4**.269, **4**.275
literature **5**.316–17
malnutrition **7**.190
medicine and health care **7**.140, **7**.159
physical geography and geology **1**.13, **1**.23, ***1**.60*, **1**.65
politics and government **7**.23, **7**.47
refugees **7**.262
social services and security **7**.293
technology **6**.133
wars **4**.254, **4**.336
Chilembwe, John ***4**.73*
chilli ***2**.63*
Chiluba, Frederick **4**.180, **4**.388
chimaera (fish) ***2**.63*
Chimaera (mythology) ***9**.9*
Chimonas, G. **5**.191
chimpanzee ***2**.63*, **2**.64
Chimú ***3**.73*, **5**.426
Chin dynasty *see* Jin
Ch'in dynasty *see* Qin
Chin Peng **4**.208
China ***1**.61*, ***3**.72*, ***4**.73–4*, ***7**.347*
agriculture **6**.7, **6**.34, **6**.204, **6**.273, **6**.351, **7**.56, **7**.179, **7**.232
aid to **7**.8
arts ***5**.91–8*
architecture ***5**.91–2*
bronzes ***5**.92–3*, **5**.96–7
and Buddhism **5**.64
ceramics *see under* ceramics
decorative ***5**.94*
metallurgy **5**.92–3, **5**.96–7
music **5**.29, ***5**.95–6*
painting *see under* painting
sculpture and carving **3**.155, **5**.91–2, **5**.94
see also literature *below*
astrology **7**.20
astronomy ***8**.25–6*, **8**.88
Boxer Rising **4**.47–8
buildings **6**.50, **6**.51, **6**.308
chemistry **6**.10, **6**.16, **6**.71
Civil War ***4**.73–4*
clocks **6**.76, **6**.340
colonialism in **3**.214, **4**.110
communes **4**.80
China (*cont.*):
Communists ***4**.74*, **7**.60
Gang of Four **4**.130–1
Hua Guofeng **4**.96, **4**.157
Liu Shaoqi **4**.96, **4**.197
Peng Pai **4**.74, **4**.264–5
see also Deng Xiaoping; Mao Zedong *and also* politics *below*
cosmetics **6**.91
crime **7**.225
Cultural Revolution **4**.90, **4**.285, **7**.191, **7**.255, **7**.267
deforestation **7**.83
demography:
contraception **7**.116
infant mortality **7**.159
population growth **7**.242
zero population growth **7**.336
dynasties and rulers ***3**.72*, ***9**.66*
Five Dynasties **3**.72
Han **3**.72, **3**.155–6, **5**.95
Jin **3**.188, **3**.218, **3**.380
Liao **3**.72, **3**.207
Manchu **3**.72, **3**.218, **4**.76, **7**.11
mandarins **3**.281
Ming **3**.72, **3**.232, **3**.349, **3**.350, **3**.388
Northern Wei **3**.258
Qin **3**.72, **3**.296, **3**.326
see also Shi Huangdi
Qing *see* Qing dynasty
Shang **3**.72, **3**.325
Song (Sung) **3**.333
Sui **3**.344
Tang **3**.72, **3**.232, **3**.347, **3**.348, **3**.349, **3**.387
Three Kingdoms **3**.72, **3**.356
Western Jin **3**.72, **3**.380
Xia **3**.72, **3**.387
Yuan **3**.72, **3**.201, **3**.388–9
Zhou **3**.72, **3**.390
see also politics and government *below*
economy and trade **4**.80, **4**.140, **4**.197, **4**.250–1, **4**.350, **4**.351
planned/market socialism **7**.193, **7**.236
education **7**.154–5, **7**.185, **7**.245, **7**.277
religious (banned) **7**.265
emperors *see* dynasties *above*
energy and fuel **6**.37, **6**.78, **6**.79, **6**.150, **6**.236, **6**.263, **6**.355, **6**.379, **7**.106
exploration **1**.164, **1**.205, **3**.122, **3**.390
explosives **6**.132, **6**.148
famine **7**.116
fishing **6**.133, **6**.146
food and drink **6**.133
foreign policy and treaties **3**.196, **3**.250, **4**.73, **4**.356, **4**.357, **4**.372
and UN **7**.277, **7**.326
funerals **7**.128
Great Leap Forward **4**.140
Great Wall **3**.146, **6**.160, **6**.161
historian **3**.328
Industrial Revolution **6**.178
-Japan Peace and Friendship Treaty **4**.73
land reform **7**.179
languages and writing **7**.11, **7**.151, **7**.169, **7**.177, **7**.286–7, **7**.305, **7**.334
laws/legal system **7**.40, **7**.239
literature ***5**.98*, **5**.110, **5**.198, **5**.270, **5**.370, **5**.416, **7**.216
modern ***5**.95*
traditional ***5**.94–5*
see also under drama; novels; poetry
Long March *see under* wars and defence *below*

China (*cont.*):
- maps 3.72, 3.349, 4.73, 4.351, 4.385
- mathematics 6.1, 6.17, 6.53, 6.391
- medicine and health care 6.3, 6.14, 6.99, 6.340, 7.24, 7.46, 7.140, 7.159, 7.196
- metallurgy 5.92–3, 5.96–7
- music 5.29, 5.95–6
- office machinery 6.366
- opium trade 4.197, 4.250–1
- paper 6.72, 6.256
- philosophy *see* religion and philosophy *below*
- physical geography and geology 1.61
  - deserts 1.141, 1.330, 1.348
  - mountains 1.246, 1.335, 1.369
  - peninsula 1.200
  - regions *see* regions *below*
  - rivers *see under* rivers
- politics and government:
  - communism *see* Communists *above*
  - Gang of Four 4.130–1
  - as great power 7.135
  - Hundred Days 4.158
  - Hundred Flowers 4.158
  - imperial leader 4.196
  - intelligence services 7.162
  - Jiangxi Soviet 4.173
  - nationalist 4.214–15
  - officials 3.218
  - parties 4.74, 4.187–8
  - protest 7.239
  - purges 7.255
  - revolutionaries 4.334, 4.389
  - rulers *see* dynasties *above*
  - United Nations 4.104
  - women's suffrage 7.332
  - *see also* wars and defence *below*
- pottery *see under* ceramics
- poverty 7.244
- prehistory 3.120, 3.210, 3.274, 3.387, 3.390
- printing 6.199, 6.281
- refugees 7.261
- regions 1.204, 1.219, 1.303, 1.369, 3.387, 3.389, 4.344, 4.354, 4.384
- religion and philosophy 7.46–7
  - ancestor worship 7.13
  - Buddhism *see under* Buddhism
  - Christianity 7.203
  - *li* 7.183
  - myths and deities 7.47, 9.10, 9.23, 9.26, 9.28
  - Taiji 7.305
  - White Lotus Society 3.380–1
  - yin and yang 7.335
  - *see also* Confucianism; Daoism; Mencius
- revolutions *see under* wars and defence *below*
- rulers *see* dynasties *above*
- science *see* technology and science *below*
- seismology 6.311
- Self-Strengthening Movement 4.308
- space 6.162
- sport and recreation:
  - boxing 7.305
  - mahjong 7.189
  - martial arts 7.194
  - street 7.300
  - table tennis 7.305
- technology and science 6.16, 6.72, 6.92, 6.346, 6.367–8
  - scientific institution 6.72
  - *see also entries above and below prefixed* 1. *and* 6.
- treaty ports 4.350, 4.351
- Triad Societies 4.352
- urban areas 3.17, 3.149, 7.197

China (*cont.*):
- vehicles and transport:
  - aircraft 6.191
  - bridges 6.47, 6.293, 6.341
  - canal 6.56, 6.159
  - railways 6.294, 6.332
  - road 6.70, 6.165, 6.299, 6.300, 6.384
  - ships 6.190, 6.219, 6.255
    - navigation 6.181, 6.212, 6.214–15
    - sailing 6.51, 6.76, 6.190, 6.219, 6.304, 6.306–7
  - space and rocketry 6.162, 6.196, 6.301
- wars and defence:
  - with Britain 4.32, 4.138, 4.230, 4.250, 4.381
  - armed forces 3.60, 7.66
  - armour 6.40
  - arms trade 7.18
  - Civil War and Long March 4.73–4, 4.199
  - guerrilla 7.136
  - Indo-Chinese War 4.163
  - leaders 3.390, 4.196–7, 4.370, 4.387, 4.389, 4.391
  - Northern Expedition 246
  - nuclear weapons 7.220
  - Opium Wars 4.230, 4.241–2, 4.250–1
  - revolutions 4.74, 4.384, 7.53, 7.136, 7.267
  - Sino-French 4.314
  - Sino-Japanese 4.312, 4.314–15
  - Sino-Soviet border dispute 4.315
  - Taiping Rebellion 4.337
  - Tibet invaded 7.265
  - treaties with Russia 3.196, 3.250
  - Twenty-One Demands 4.355–6
  - warlords 4.371
  - weapons 6.92, 6.162
  - World War II 4.385
  - Xiongnu attacks 3.387
  - *see also* foreign policy and treaties *above*
- writing and scripts *see* languages *above*

china (pottery) *see* ceramics and pottery; porcelain
china clay 1.61, 6.276, 6.382
China Sea 1.61
china-mark moth 2.63–4
chinchilla 2.64
chinchilla rat 2.64
Chinese Academy of Sciences 6.72
Chinese language *see* languages *under* China
Ch'ing dynasty *see* Qing
chinoiserie 5.98
chinook 1.61
chipboard 6.214
chipmunk 2.64
Chippendale, Thomas (Elder and Younger) 5.98
chirality 1.61
Chirico, Giorgio de 5.98, 5.292
Chiron 8.26, 9.8
chiropractic 6.72
chisel 6.72, 6.357, 6.358
Chissano, Joaquim 4.225
chitin 2.64
chivalry 3.73–4, 5.294
chive 2.64
chlorides 1.61
- aluminium 6.12
- calcium 6.323
- cobalt 1.67
- mercury 6.147
- methyl 6.201
- polyvinyl *see* PVC
- potassium 6.100
- zinc 6.391
- *see also* sodium chloride

chlorine 1.61–2, 1.106, 6.72, 6.125
- acid from *see* hydrochloric acid

chlorine (*cont.*):
- as bleach 6.40
- from Castner–Kellner cell 6.62
- chemical industry and production 6.62, 6.70, 6.71, 6.308
- in chemical warfare 6.72
- chlorinated hydrocarbons 6.325
- water 1.62, 6.379
- *see also* chlorides

chlorite 1.62
chloro-fluorocarbons 1.62, 6.72
- in aerosols 1.4, 6.5
- developed 6.224
- in dry-cleaning 6.109
- in fire-extinguisher 6.131
- and ozone layer 6.72, 6.254, 7.226
- production of 6.123, 6.139
- in refrigerator 6.297
- *see also* greenhouse effect

chloroform 6.14, 6.72–3, 6.218, 6.320
chlorophyll 2.64, 2.252, 6.270
- plants without *see* fungi

chlorpromazine 6.285, 6.360
Chlothar of Neustria 3.18
chocolate 2.46, 6.34, 6.73, 6.87
Choctaw Indians 3.74
choirs and choral music 5.8, 5.62, 5.66, 5.76, 5.78, 5.99, 5.110, 5.340, 5.430
- *see also* oratorio; songs

Choiseul, Étienne François, duc de 3.74
choke 6.73
Chōla dynasty 3.74
- sculpture 5.99

cholera 2.64, 6.32, 6.192, 6.377, 7.59
- chicken 2.243

cholesterol 2.64, 2.131–2, 6.126
Chomsky, Noam 7.29, 7.47, 7.180
chondrite 8.26
chondrules 8.26
Chong Son 5.251
Chopin, Fryderyk Franciszek 5.99, 5.288, 5.366, 5.405
Chopin, Kate 5.99
chopper 6.358
choral music *see* choirs and choral music
chorale 5.99
chordate 2.64–5
- cephalochordate 2.182
- urochordate 2.301–2
- *see also* vertebrates

chords 5.70, 5.99
- *see also* harmony

choreographers and dance teachers and arrangers 5.99–100
- Australia 5.204
- Austria 5.51
- Britain 5.13, 5.24, 5.62, 5.68, 5.126, 5.129, 5.461, 5.484
- cinema 5.245
- companies *see* ballet and dance companies
- dance notation 5.123–4
- Denmark 5.56
- France 5.42, 5.324, 5.345
- Germany 5.39, 5.242, 5.489
- Hungary 5.252
- Italy 5.49, 5.83
- Russia and USSR 5.29, 5.37, 5.159, 5.285, 5.320, 5.324
- South Africa 5.117
- Switzerland 5.238
- United States 5.10, 5.166, 5.452
  - cinema 5.25, 5.45, 5.128
  - modern 5.120, 5.139, 5.187–8, 5.214, 5.397, 5.415–16, 5.447, 5.450
  - musicals 5.245, 5.384
  - postmodern 5.62, 5.342, 5.373, 5.397

choreographers and dance teachers and arrangers (*cont.*):
- *see also* ballet; ballet and dance companies; dance

chorionic gonadotrophin 6.280
chorus 5.100
Chou Chuang Wang 8.88
Chou dynasty *see* Zhou
Chou En-lai *see* Zhou Enlai
chough 2.65
CHP *see* combined heat and power generation
Chrétien, Henri: Ritchey–Chrétien optics 8.140
Chrétien de Troyes 5.100, 5.368
Christ *see* Jesus Christ
Christadelphians 7.47
Christian I, king of Denmark, Norway, and Sweden 3.74
Christian IV, king of Denmark 3.102, 3.355
Christian V, king of Denmark 3.318
Christian, Fletcher 4.47
Christian Democrats 4.75, 7.47, 7.66
Christian Science 4.75, 7.48
Christian Socialism 4.75
Christianity 3.74–5, 4.74–5, 7.47–8
- Advent 7.3
- in Africa 7.5, 7.203
- afterlife 7.6
- and alphabet 7.11, 7.17
- Ascension 7.19
- asceticism 7.19
- atonement 7.21
- Baptism 7.185, 7.271
- Bible *see* Bible
- carnival 7.268
- caste system 7.41
- catacombs 3.64
- Catholics *see* Roman Catholicism
- Celtic Church 3.66
- Christadelphians 7.47
- Christmas 7.48
- Churches 7.47
  - *see also in particular* Eastern Orthodox; Monophysite; Nestorians; Oriental Orthodox; Protestants; Roman Catholicism
- Churches of Christ 7.48
- covenant 7.73
- creed 7.74
- cremation 7.74–5
- Dead Sea Scrolls 3.99
- Easter 7.98, 7.134
- Epiphany 7.16, 7.48, 7.107
- eschatology 7.108
- Eucharist *see* Mass (Eucharist)
- fundamentalism 7.127
- God 7.125, 7.133, 7.263, 7.316
- heaven 7.140
- hell 7.141
- and homosexuality 7.146
- Jehovah's Witnesses 7.171
- Jerusalem 3.187
- law 3.59
- Lent 7.183
- liberation theology 7.184
- liturgy 7.185–6
  - *see also* Baptism; Mass
- map 3.75
- martyrs *see* martyrs
- meditation 7.196
- miracles 7.203
- missionaries *see* missionaries
- monasteries *see* monasticism
- Monophysite *see* Monophysite Church
- mythical figures and places 9.2–39 *passim*
- Nestorian Church *see* Nestorians
- Orthodox *see* Eastern Orthodox; Oriental Orthodox
- Pentecost 7.232
- pilgrimage 7.236
- predestination 7.245

Christianity (cont.):
prophets 7.250
see also John the Baptist
Protestants see Protestants
Quakers see Quakers
resurrection 7.266
sacraments 7.123, 7.185–6, 7.271, 7.272
saints see under saints
sin 7.225, 7.286
soul 7.295
and state 3.38, 3.78, 7.265
and suicide 7.301
text see Bible
theology 7.309–10
Trinity 7.316
World Council of Churches 4.382–3, 7.13, 7.48, 7.269, 7.333
year 7.48
see also religion
Christie, Dame Agatha 5.100
Christina, queen of Sweden 3.75, 3.76, 3.102
Christine de Pisan 5.100
Christmas 7.48
Christopher, Saint 7.272, 9.41, 9.41
Christopher II, king of Denmark 3.102
Christus, Petrus 5.100
Christy, James 8.25
chromaticism (music) 5.200
chromatic scale 5.403–4
chromatography (physics) 1.61, 1.62, 6.73, 6.217
chromatic aberration 6.2, 8.1
chrome 6.73, 6.354
chromite 6.73
chromium 1.62–3, 1.106, 6.73
alloys and in steel 6.11, 6.46, 6.235, 6.330, 6.333
bearings 6.32
electroplating 6.90, 6.120
chromium dioxide 6.213
chromophore 6.109
chromosomes 2.65
disorders see inherited disorders under genetics
genome 2.134
meiosis 2.207, 2.214
mitosis 2.214–15, 2.215
mutation 2.223–4
in nucleus 2.230
see also genes
chromosphere 8.26, 8.51
chronometer 6.73, 6.77, 6.166, 6.303
see also clocks
chronophotography 6.74
chrysalis 2.65
chrysanthemum 2.65
Chrysaor 9.14
Chryse Planitia 8.26
Chrysostom 5.411
Chu Hsi see Zhu Xi
Chu Teh see Zhu De
Chuang-tzu see Zhuangzi
chub 2.65
chuck-Will's-widow 2.228
Chuku 9.3
Chulalongkorn, king of Siam 4.75, 4.341
Chun Doo Hwan 4.320
Church, Benjamin 3.75–6
Church, William 4.81
church 7.48, 7.50
leaders 7.48, 7.50
see also religion
Church of England 7.13
see also Anglicans
Church of Scotland 3.90, 7.50, 7.245
Churches of Christ 7.48
Churchill, John see Marlborough, John
Churchill, Lord Randolph Henry Spencer 4.75, 4.276
Churchill, Sarah 3.223
Churchill, Sir Winston Leonard Spencer 4.32, 4.75–6, 4.385, 5.441
advisor 4.7–8
and Atlantic Charter 4.22
and Cairo Conference 4.58
and Casablanca Conference 4.64
and Destroyer-Bases Deal 4.97
and Gallipoli 4.120, 4.129
on Iron Curtain 4.167
as manic depressive 7.191
and Potsdam Conference 4.276
and Second Front 4.307
and Teheran Conference 4.340
and Tonypandy 4.347
and Yalta Conference 4.385
Churchill Falls 1.63
churl (ceorl) 3.67
Churriguera, José de: and Churrigueresque 5.100–1
CIA see Central Intelligence Agency
Ciano, Count Galeazzo 4.76
Cibber, Colley 5.101
cicada 2.65
Cicero, Marcus Tullius 3.76, 3.77, 3.342, 5.101, 5.382
cichlids 2.66, 2.341
Ciconia, Johannes 5.379
Cid Campeador see El Cid
Cierva, Juan de la 6.25
cigarette lighter 6.73, 6.132
see also smoking
Cilicia 3.76
CIM see computer-integrated manufacturing under computers
Cimabue 5.101
Cimarosa, Domenico 5.101
Cimarron 1.63
Cimon 3.264
cinchona 2.66
Cincinnatus, Lucius Quinctius 3.76–7
cinder cone 1.63
Cinderella 9.9
cinema 5.154–5
acting 5.3, 5.447
see also under United States below
animation 5.17, 5.131
art direction 5.23
Britain 5.60, 5.132
certificates 9.62
directors and producers 5.207–8, 5.249, 5.259, 5.365, 5.376
music 5.22, 5.481
Shakespeare 5.327
camera and camerawork 5.72, 6.73, 6.75, 6.210
certificates 9.62
cinematography 6.73, 6.74–5, 6.210
cinéma-vérité see cinéma-vérité
comedy:
France 5.101–2, 5.447
United States 5.12, 5.88–9, 5.270, 5.411
copyright 7.70
directors and producers 5.44, 5.205, 5.309
see also under Britain above and France and United States below
documentary 5.101, 5.131, 5.132, 5.163
France 5.106, 5.447
comedy 5.101–2, 5.447
directors and producers 5.79, 5.101–2, 5.329–30, 5.335, 5.380, 5.447, 5.460–1
innovative 5.169, 5.270–1, 5.290, 5.323
Nouvelle Vague 5.323, 5.460–1
Free 5.163
cinema (cont.):
Germany 5.132, 5.205, 5.255, 5.309, 5.329–30
India 5.240, 5.375
Italy 5.18, 5.46, 5.153, 5.341, 5.391, 5.478
Japan 5.252
montage 5.302
music and dance 5.22, 5.25, 5.45, 5.128, 5.155, 5.175, 5.245, 5.251, 5.481
see also musicals
Nouvelle Vague, see under France above
Russia and USSR 5.132, 5.142, 5.447
Senegal 5.334
sexual behaviour 7.281
Spain 5.65–6
Sweden 5.44
and technology 6.73, 6.74–5, 6.210
cine-camera 6.73, 6.75, 6.210
cine-projector 6.284
optical sound reproduction 6.325–6
United States:
Academy Awards 5.2
actors and actresses 5.12, 5.88–9, 5.245, 5.437, 5.485
animation 5.131
directors and producers 5.77, 5.214–15, 5.244, 5.255, 5.429, 5.432, 5.437, 5.485
comedy 5.12, 5.88–9, 5.270, 5.411
innovative 5.193–4, 5.252
Westerns 5.128, 5.162
documentary 5.132
Hollywood 5.210
music and dance 5.45, 5.128, 5.175, 5.245, 5.251
voyeuristic 5.482
Westerns 5.128, 5.162, 5.487
writers 5.12, 5.88–9, 5.245
wide-screen 5.488–9
see also photography
cinematography 6.73, 6.74–5, 6.210
see also cinema
cinéma-vérité 5.101
cinnabar 1.63, 6.219
cinnamon 2.66
Cinq, Les see Five, the
Cinque Ports 3.77
cipher 6.93, 7.50
see also codes
Circe 9.9
Circinus 8.26
circle 9.51
as ellipse 1.72, 1.73
radian 1.273
on surface of sphere see great circles
circuits, electrical and electronic 6.75–6, 6.113–14
amplifier 6.13–14
circuit-breaker 6.75
current see current, electrical
demodulation 6.98, 6.289, 6.346
earthing 6.111
filter 6.130
impedance 6.76, 6.177
inductor 6.178
integrated see integrated circuit
oscillators 6.215, 6.253
oscilloscope 6.45, 6.64, 6.253–4
printed 6.281
smoothing 6.323
track 6.359
tuner 6.364–5
wiring 6.387, 9.53
see also electricity; electron tube
circulation (climatology) see under air and atmosphere
circulatory system (body) see blood
circumcision 7.50, 7.51
circumnavigators of globe 1.42, 1.95, 1.97, 1.184
circumpolar star 8.26
circumpolar vortex 1.64
circus 3.77, 5.101, 5.102
cire perdue (investment casting) 5.101, 6.184, 6.185
cirque 1.62, 1.64, 1.140
glacier 1.38
cirrhosis 2.66
cirrus and cirrostratus clouds 1.64, 1.66, 1.128
CIS see Commonwealth of Independent States
Cistercians 3.40, 3.77, 3.237, 7.205
citadel 3.77
CITES (Convention on International Trade in Endangered Species) 7.106
Citha, Saint 9.43
cities see urban areas
citizens' band radio 6.65
citizenship 7.50
citric acid 2.187
cycle 2.181
citronella 2.66
citrus fruit 2.66, 6.76, 6.271
clementine 2.67
grapefruit 2.143
lemon 2.187
mandarin 2.201
orange 2.235
shaddock 2.66, 2.305
cittern 5.101
civet 2.66–7, 2.240
Civiale, Jean 6.205
civil defence 7.50
civil disobedience 4.262, 4.303, 7.91
civil engineering 6.76, 6.122
coastal defences 6.79, 6.193
dams and flood control 6.95, 6.96, 6.136, 6.294–5
land drainage 6.79, 6.193, 6.376
pioneers 6.49
water supply 6.378–9
see also bridges; buildings; canals; mechanical engineering; ports; railways; roads; sewage; tunnel
civil law 7.41, 7.50–1, 7.318
arrest 7.18–19
bail 7.25
class action 7.52
company law 7.62
compensation 7.62
contract law 7.69
homicide 7.146
inquisitorial procedure 7.161
judge 7.172–3
mercantile 7.198
Napoleonic, see Code Napoléon
obligations 7.221
plaintiff 7.236
Scots 7.276
tort 7.313
United States 7.52, 7.323
see also Roman law
civil rights 7.52–3
Acts 4.76
Movement 7.32, 7.124
see also Bills of Rights; rights, human and under United States
civil service 7.38, 7.52
see also bureaucracy
civil society 7.52
civil wars 7.52
Lebanon 7.52, 7.96, 7.193, 7.364
Yemen 7.52
Yugoslavia 4.46, 4.387
see also American Civil War; English Civil War; revolution (political) and under China; Nigeria; Roman empire; Russia; Spain
civilization 3.77
Cixi, empress dowager of China 4.48, 4.76, 4.178, 4.279

cladding ***6.76***
Clair, René ***5.101–2***
Clairaut, Alexis-Claude **8.**87
Clairmont, Claire **5.**69
Clairvaux, abbot of **3.**40
clairvoyance **7.**228
clam ***2.67***
clamp ***6.76***
clan ***3.78***, ***7.52***, **7.**86, **7.**336
clapper bridge **6.**48
clappers (music) ***5.102***
Clapperton, Hugh ***1.64***, **1.**186
Clare, John ***5.102***
Clare election ***4.76***
Clarence, George Plantagenet, duke of ***3.78***
Clarence, Lionel, 1st duke of ***3.78***
Clarendon, Edward Hyde, 1st earl ***3.78***, **3.**302
Clarendon, Constitutions of ***3.78***
clarinets **5.**37, ***5.103***, **5.**352
Clark, Alvan **8.**14
Clark, Joe **4.**85
Clark, J. P. **5.**7
Clark, Kenneth **5.**378
Clark, William *see* Lewis, Merriwether
Clarke, Jeremiah ***5.103***
Clarke, Marcus ***5.103***
Clarkson, Thomas ***4.76***, **4.**316
class action (law) ***7.52***
class (biology) ***2.67***
class (social) *see* social class
classical economics ***7.52–3***, **7.**100–1
see also in particular *laissez-faire*; Mill, John Stuart; Malthus; Marshall, Alfred; neo-classical economics; Ricardo; Smith, Adam
classical music ***5.103***
see also under classicism
classicism:
in architecture **5.**128, **5.**280, **5.**330
in economics *see* classical economics
in literature **5.**90, **5.**241–2, **5.**249, **5.**254, **5.**267, **5.**277, **5.**371, **5.**372
in music ***5.103***
dance **5.**29, **5.**33–4, ***5.103***, **5.**139, **5.**324, **5.**345
see also composers
in painting **5.**364–5
revived *see* Renaissance
see also Neoclassicism
classification of animals and plants *see* nomenclature and classification
clastic sediments **1.**43, ***1.64***, **1.**72
Claude, Georges ***6.76***
Claude de Lorraine (1496–1550) **3.**150
Claude Gellée (Le Lorrain, 1600–82) ***5.103***, **5.**104
Claudel, Paul ***5.103***
Claudet, Antoine François Jean ***5.103***
Claudius, Roman emperor (10 BC – AD 54) **3.**49–50, **3.**60, ***3.78–9***, **3.**306, **3.**308
Claudius II, Roman emperor (d.270) **3.**29, **3.**374
Clausewitz, Karl von ***4.76***, **7.**328
Clausius, Rudolf Julius Emanuel ***1.64***
clavichord ***5.103–4***
clavicle ***2.67***
claws ***2.67***, **2.**225
Clay, Henry **4.**11, ***4.77***, **4.**80, **4.**95, **4.**377
clay ***1.64***, **6.**67–8, **6.**79, **6.**91, **6.**276
boulder **1.**42, **1.**337
china *see* china clay
illite **1.**166
marl **1.**206
mineral **1.**220–1
shale **1.**300
cleaning:
dry **6.**109, **6.**253
see also detergent; soap
clearing house ***7.53***
clearwing moth ***2.67***
cleavage (geology) ***1.64***
clef **5.**322
cleft palate ***2.67***
cleg **2.**167
Cleisthenes **3.**26, **3.**27
clematis ***2.67***, **2.**68
Clemençeau, Georges ***4.77***, **4.**268, **4.**304
Clement V, pope **1.**29, **3.**146
Clement VI, pope **3.**146
Clement VII, pope **3.**131
Clement VIII, pope **3.**70
Clement XIV, pope **3.**187
Clement of Alexandria **7.**107
Clementi, Muzio ***5.104***
clementine ***2.67***
Cleopatra VII, queen of Egypt **3.**3, **3.**78, ***3.79***
Clephane, James **6.**219
clepsydra **6.**93
clergy:
benefit of **3.**39
see also religion *under individual countries*
clerihew ***5.104***
Cleveland, Grover ***4.77***, **4.**278
click beetles ***2.67–8***, **2.**83
clientelism ***7.53***
cliffs **1.**32, **1.**34, **1.**57, ***1.64***, **1.**322
cliff-dwellers ***3.79***
climate and climatology ***1.65–6***
and altitude **1.**241, **1.**274–5
change ***1.65–6***, **7.**135–6
nuclear winter **6.**246
see also greenhouse effect; ozone depletion
circulation *see under* air and atmosphere
climatic types **1.**65, ***1.66***
climatologists *see under individual countries*
microclimate **1.**214
pluvials **1.**260
water balance **1.**360
see also meteorology
climbing mountains **1.**366, **7.**208
climbing perch ***2.68***
climbing plants ***2.68***
acacia **2.**2
begonia **2.**28
bryony **2.**43
clematis **2.**67
Dutchman's pipe **2.**100–1
fern ***2.68***
gourd **2.**142
grape **2.**142–3
honeysuckle **2.**160–1
hop **2.**161
hydrangea **2.**167
ivy **2.**173
jasmine **2.**174–5
kiwi fruit **2.**180
lianas **2.**39–40, **2.**173, **2.**189
nasturtium **2.**225
passion vine **2.**243
pea *see* pea family
pepper **2.**249
periwinkle **2.**250
plumbago **2.**259–60
poison ivy **2.**260
pumpkin **2.**272
rattan **2.**278
vetch **2.**355
Virginia creeper **2.**356
wisteria **2.**368
Cline, Patsy **5.**115
clingfish ***2.68***
clinical linguistics ***7.53***
clinical psychology ***7.53***
clinker-built ships **6.**316, **6.**317
Clinton, Bill, US President ***4.77***, **4.**95
Clinton, George ***3.79***, **3.**87, **3.**176
Clinton, Sir Henry ***3.79***
Clio ***9.9***
clipper ***6.76***, **6.**219
clitoris ***2.68***
Clive of Plassey, Robert, baron **3.**39, ***3.79***, ***3.80***, **3.**106, **3.**109, **3.**284, **3.**324
clock (computer) ***6.76***, **6.**83
clock star ***8.26***
clocks ***5.104–5***, ***6.76–7***, **6.**192
astronomical **8.**10
chronometer **6.**73, **6.**77, **6.**166, **6.**303
hydro-mechanical **6.**340
pendulum **6.**149, **6.**166, **6.**260
Clodius, Publius Claudius Pulcher ***3.79***
cloisonné enamel **5.**70, **5.**94, ***5.105***
clones and cloning ***2.68***, **2.**354, ***6.77***, **6.**153, **6.**271
closed shop ***7.53–4***, **7.**266
closed-circuit television ***6.77***
Closterman, John **5.**370
closure (mathematics) ***1.66***
cloth *see* textiles
clothes moth ***2.68–9***
clothing **6.**125
armour *see* armour *under* body
button **6.**52
dry-cleaning **6.**109, **6.**253
dryer **6.**364
manufacture ***6.77***
sizes ***9.62***
washing **6.**376
waterproof **6.**211
zip **6.**391
see also sewing; textiles
cloud chamber **6.**258
clouded leopard ***2.69***
clouds **1.**37–8, ***1.66–7***
cirrus and cirrostratus **1.**64, **1.**66, **1.**128
cloud cover ***1.67***
coalescence **1.**67
cumulus and cumulonimbus **1.**66, **1.**81, **1.**128
in depression **1.**88
and fronts **1.**128
lightning **1.**194
nimbus and nimbostratus **1.**128, **1.**229
stratus and stratocumulus **1.**66, **1.**128, **1.**323
symbols **9.**56
Clouet, Jean and François ***5.105***
Clough, Arthur Hugh ***5.105***
clove ***2.69***
clover ***2.69***, **2.**244
Clovis, king of Franks ***3.79***, **3.**130, **3.**132, **3.**314
Clovis culture ***3.80***
clown ***5.105***
club foot ***2.69***
clubmoss ***2.69***
Cluny, Order of ***3.80***
cluster-fly ***2.69***
clusters of galaxies **8.**52, **8.**55, **8.**86, **8.**118
clusters of stars ***8.27***, **8.**113
age of **8.**2
double **8.**38–9
Hyades **8.**71
Jewel Box **8.**76
Pleiades **8.**123–5
see also globular clusters
clutch ***6.77–8***
Clyde ***1.67***
Clytemnestra ***9.9***
CNC (computer numerical control) **6.**247
CND *see* Campaign for Nuclear Disarmament
cnidarian ***2.69***
jellyfish **2.**175
Portuguese man-of-war **2.**265–6
sea anemone **2.**297, **2.**298
see also coral; polyp
Cnut *see* Canute
coaches:
buses **6.**292, **6.**363
carriages **6.**30, **6.**61–2, **6.**341
railway **6.**292
coal ***1.67***, ***6.78–9***
anthracite **1.**15, **1.**67, **4.**15, **6.**78
briquetting **6.**47
coal-cutter ***6.79***
coal-fired power station **6.**275, **6.**279
combustion **6.**2, **6.**138
gas from *see* coal-gas
for heating **6.**167
lignite *see* lignite
liquefied **6.**343
mining **6.**78, ***6.79***, **6.**173, **6.**207, **6.**228, **6.**270
peat *see* peat
pollution **6.**78, **6.**275
products *see* plastics
resources **6.**121
Schuman Plan **4.**305
in stove **6.**89
see also coke
Coalbrookdale Iron Bridge **6.**187, **6.**188
coalescence ***1.67***
coalfish ***2.69–70***
coal-gas **1.**341, **6.**71, **6.**78, ***6.79***, **6.**236, **6.**386
see also lighting and heating *under* gas
coalitions:
government ***7.54***, **7.**67
in warfare *see* Axis Powers; Gulf War (1991)
Coalsack, Southern **8.**151
coal-tar ***6.79***, **6.**261
Coast Mountains ***1.67***
Coast Ranges ***1.67***
coasts ***1.67***
bays **1.**35, **1.**144–5
coastal defences ***6.79***, **6.**193
currents **1.**199, **1.**283–4
deposition:
barrier island **1.**33
bars **1.**32–3, **1.**292, **1.**339
beach *see* beaches
chenier **1.**60
conglomerates **1.**72
gravel **1.**144
longshore drift **1.**199
sand-bar **1.**292
sand-spit **1.**292–3, **1.**339
shingle **1.**301
tombolo **1.**339
erosion **1.**55, **1.**363
arch **1.**224
cliffs **1.**32, **1.**34, **1.**57, **1.**64
fiord **1.**122
lagoon **1.**185–6
rias **1.**282, **1.**283
shore platform **1.**301
stack **1.**318, **1.**319
strandline **1.**322
gulf **1.**148
salt-marshes **1.**291
strata **1.**322, **1.**324
tides *see* tides/tidal
waves *see* waves (ocean)
see also oceans and seas
Coates, Joseph Gordon ***4.77***, **4.**120
coati ***2.70***
Coatlicue ***9.9***
coaxial cable **6.**53, ***6.79***
cob (building material) ***6.79***, **6.**235
cobalamins **2.**357
cobalt ***1.67***, **1.**106, ***6.79***
alloys **6.**212, **6.**241
in cermet **6.**68
cobalt-60 **6.**292
magnets **6.**212
cobalt chloride ***1.67***
cobalt oxide **6.**156
Cobbett, Samuel **5.**460
Cobbett, William ***4.77***

Cobden, Richard **4**.15, ***4**.77*, **4**.209
COBOL **6**.86
cobras ***2**.70*, **2**.179, **2**.285–6
Coburn, Alvin Langdon ***5**.105–6*
cocaine **6**.192, **7**.95, **7**.96
Cocconi, Giuseppe **8**.115
coccyx ***2**.70*, **2**.355
Cochin (India) **3**.195
Cochin China ***3**.80*
Cochise ***4**.77*
cochlea ***2**.70*
Cochran, Sir Charles Blake ***5**.106*
Cochrane, Thomas, 10th earl of Dundonald ***4**.77*
Cockaigne (Cockayne), Land of ***9**.9*
cockatoo ***2**.70*
cockchafer **2**.60
Cockcroft, Sir John Douglas ***1**.67*, **6**.258
Cockerell, Charles Robert ***5**.106*
Cockerell, Sir Christopher ***6**.79*, **6**.171
cock-fighting ***7**.54*
cockle ***2**.70*
cock-of-the-rock ***2**.70*
cockroach ***2**.70–1*
cockscomb (plant) **2**.195
cocoa *see* cacao; chocolate
coconut ***2**.71*, **2**.75–6
coconut crab ***2**.71*
Cocteau, Jean ***5**.106*, **5**.421
cod family ***2**.71*
  brain **2**.41
  coalfish **2**.69–70
  pollack **2**.262
  rockling **2**.287
  whiting **2**.367
Cod War ***4**.78*
coda ***5**.106*
Code Napoléon ***4**.78*, **7**.51, ***7**.54*, **7**.269
codes ***6**.93*, ***7**.54*
  binary **6**.22, **6**.36, **6**.84, **6**.210
  cipher **7**.50
  Enigma machine **6**.84, **6**.366
  genetic **2**.134
  and meaning **7**.196
  Morse *see* Morse
  semaphore **6**.312, **9**.48
  signalling **9**.48
  see also semiotics
codlin moth ***2**.71*
Cody, William Frederick **4**.315, **4**.378
coelacanth ***2**.71*
coelenterate *see* cnidarian
Coello, A. Sanchez **3**.279
coelom ***2**.72*
coelostat ***8**.27*
coenzymes ***2**.72*
  see also adenosine triphosphate
Coercive Acts ***3**.80*, **4**.87
coffee ***2**.72*, **6**.34
  coffee-house ***3**.80*, **3**.81
cog (ship) **6**.80, **6**.307
cognitive psychology **7**.30, **7**.35, ***7**.55*
  attention **7**.22
  and computers **7**.252
  cognitive development ***7**.54*
    see also Piaget
  cognitive dissonance ***7**.54*
  memory *see* memory
  mind *see* mind
  perception **7**.232–3, **7**.234, **7**.279
  see also brain
cohabitation ***7**.55–6*
Cohan, Robert **5**.267
coherent radiation ***1**.68*
cohort ***7**.56*
coil (contraception) **6**.87
coil (electrical) ***6**.80*
  electric motor **6**.115
  ignition system **6**.175–6
  as inductor **6**.178
  microphone **6**.222
  relay **6**.297
  solenoid **6**.325
coil (electrical) (*cont.*):
  transformer **6**.361
  see also generator
coins **3**.260, **3**.278, ***5**.106*, **5**.196, **5**.228
  collecting ***7**.56*
coir **2**.71, **6**.128
Coke, Sir Edward ***3**.81*
Coke, Dr Thomas (1747–1814) **3**.230
Coke, Thomas William, earl of Leicester (1752–1842) **2**.72, **3**.5, **3**.271, **3**.278, ***6**.80*
coke **6**.58, **6**.78, ***6**.80*, **6**.89
  coke-oven ***6**.80*
  for smelting iron *see* blast-furnace
Cokroaminoto, H. Q. S. **4**.302
col ***1**.68*
Colbert, Jean-Baptiste **3**.2, ***3**.81*, **3**.130, **3**.212, **5**.419, **6**.299, **8**.116
cold (disease) ***2**.72*
cold (temperature) *see* cooling; freezing; ice; temperature
Cold War ***4**.78*, **7**.99, **7**.212, **7**.326
Cole, Thomas **5**.213
Coleman, Ornette **5**.239
Coleridge, Samuel Taylor **5**.13, **5**.30, ***5**.106*, **5**.325
Colette ***5**.106*
Coligny, Gaspard de, seigneur de Châtillon ***3**.81*, **3**.313
colitis ***2**.72*
collage ***5**.106*, **5**.119
collar-bone **2**.67
collateral ***7**.56*
collecting **7**.55, ***7**.56*, **7**.298
collective agriculture and collectivization ***4**.78*, **7**.56, **7**.179–80, **7**.191
collective bargaining ***7**.56*
collective ownership ***7**.56*
collective security ***7**.56*
collectivism ***7**.56–7*
Collet, Henri **5**.421
colligative properties ***1**.68*
collinear points **8**.82
Collins, Michael (astronaut) **6**.17
Collins, Michael (Irish patriot) ***4**.78*, **4**.167
Collins, Wilkie ***5**.106–7*
Collins, William ***5**.107*
Collins Foster, Stephen **5**.357
Collip, James Bertram **6**.29, **6**.80
collodion **6**.268
colloid ***1**.68*, **6**.80
  see also aerosol; emulsion
Collor de Mello, Fernando **4**.49
collotype printing ***6**.80*
colluvium ***1**.68*
Colman, Saint **3**.380
Colombia ***1**.68*, ***3**.81*, ***4**.78–9*, ***7**.347–8*
  Andean Group **7**.13
  economy and trade **4**.12, **4**.320, **8**.13
    drug trafficking **7**.96
  education **7**.93
  independence and War **4**.43, **4**.324
  Leticia dispute with Peru **4**.193–4
  literature **5**.169–70
  physical geography and geology **1**.68, **1**.202, **1**.241
    mountains *see* Andes
  population **4**.320
  religion and myths **9**.11
  sport and recreation **7**.217
  technology **6**.95
Colombo Plan ***4**.79*, ***7**.57*
colon ***2**.72*
colonialism ***7**.57*
  see also decolonization; imperialism *and under* empire *or* colonialism *under individual countries*
Colonna, Vittoria **5**.230
colony ***7**.57*
Colorado beetle ***2**.72*
Colorado River ***1**.68*
  Grand Canyon **1**.50, **1**.142–3
colorimetry *see* spectroscopy
Colosseum ***3**.82*
Colossus **6**.84
colour ***1**.68–9*, ***6**.80*, ***6**.81*
  and astronomy
    chromatic aberration **8**.1
    false **8**.50
    index ***8**.27*
    luminosity (Hertzsprung–Russell) **8**.64, **8**.66–7, **8**.160
  blindness ***2**.72*
  chromatography **1**.62
  coloration in living organisms *see* protective coloration
  dye *see* dyes and dyeing
  and ergonomics **7**.108
  field painting ***5**.107*, **5**.269
  food colouring **6**.140
  and language **7**.273
  and lens **6**.199
  metaphoric use in atomic physics **1**.69
  photography **6**.75, **6**.81, **6**.109, **6**.130, **6**.210, **6**.268
  printing **6**.81, **6**.164
  refraction **6**.2
  television **6**.28, **6**.81, **6**.349, **6**.350
  see also painting; pigment
colposcope **6**.248
Colt, Samuel: Colt revolver ***6**.81*, **6**.130, **6**.270
Coltrane, John **5**.239
colubrid snakes ***2**.72–3*, **2**.188
colugo ***2**.73*
Colum, Padraic **5**.226
Columba (constellation) ***8**.27*
Columba of Iona, Saint **3**.66, ***3**.82*, **3**.319, **9**.41, ***9**.41*
Columbanus, Saint **3**.66, ***3**.82*
Columbia Plateau ***1**.69*
Columbia River **1**.63, ***1**.69–70*
columbine **2**.13
columbite **6**.241
columbium *see* niobium
Columbus, Christopher **1**.69, ***1**.70*, ***3**.82*, **3**.121, **3**.336, **6**.215
  and Caribbean **3**.33, **3**.61, **3**.184, **3**.294, **3**.334, **3**.360
  map of travels **3**.122
  ships **3**.60, **6**.58
columnar jointing ***1**.70*
Colville, Alex **5**.73
coma (astronomy) ***8**.27*
coma (unconsciousness) ***2**.73*
Coma Berenices ***8**.27*
Comanche Indians ***3**.82*
comb jelly **2**.299
combe **1**.75
Combination Acts ***4**.79*
combine harvester **6**.6, ***6**.81*, **6**.82
combined heat and power generation **6**.79, ***6**.81*, **6**.138, **6**.150
combined-cycle technology ***6**.81*, **6**.280
combustion ***1**.70*, **6**.254
  see also fire; internal combustion engines
COMECON (Council for Mutual Economic Assistance) ***4**.79*, ***7**.57*
comedy and humour in arts ***5**.107*
  cinema **5**.12, **5**.88–9, **5**.101–2, **5**.245, **5**.411, **5**.447
  clown **5**.105
  drama **5**.40, **5**.249, **5**.285–6, **5**.301, **5**.317, **5**.491
    Britain *see under* drama
    France **5**.3, **5**.40, **5**.113, **5**.154, **5**.301
    Greece, ancient **5**.21, **5**.191, **5**.290
    Ireland **5**.183, **5**.415, **5**.417, **5**.431, **5**.443, **5**.489
comedy and humour in arts (*cont.*):
  drama (*cont.*):
    Italy **5**.183, **5**.211
    music-hall **5**.310
    Roman **5**.387
    Russia **5**.182–3
    slapstick **5**.310, **5**.422
  essays and sketches **5**.259, **5**.339, **5**.454
  farce **5**.152, **5**.154
  fiction **5**.302, **5**.339, **5**.423, **5**.432
  fool **5**.116, **5**.162, **5**.307
  novels *see under* novels
  poetry **5**.21, **5**.43, **5**.104, **5**.259, **5**.313, **5**.339
  Yiddish **5**.497
  see also caricature; farce; musicals; satire
Comenius, John Amos (Komenský) **5**.52, **5**.120
comets ***8**.27–8*, ***8**.194*
  Arend–Roland ***8**.8*, **8**.194
  Biela's *see* Biela
  coma **8**.27
  Donati's **8**.38, **8**.194
  Encke's **8**.28, **8**.46, **8**.194
  Faye's **8**.50, **8**.194
  Kohoutek's **8**.80–1, **8**.196
  Lexell's **8**.84–5, **8**.194
  Oort–öpik Cloud ***8**.112*
  structure of **8**.28
  tail **8**.169
  Tempel's **8**.170, **8**.194
  Tunguska Event **8**.175
  Whipple's **8**.186
  see also Halley's comet
comfrey ***2**.73*
Cominform (Communist Information Bureau) ***4**.79*
Comintern (Communist International) **4**.15, ***4**.79*, **4**.164–5
comitia ***3**.82*
command economy **7**.236–7
commedia dell'arte ***5**.107–8*, **5**.159, **5**.256
commensalism ***2**.73*
commerce, chamber of **7**.43
  see also companies and business; retail trade
commercial banks **7**.26
commercial law ***7**.57*, **7**.198, **7**.282
commercials *see* advertising
Commines, Philippe de ***3**.83*
Committee of Permanent Representatives **7**.111
Committee of Public Safety ***3**.83*
command module **6**.17
commodity ***7**.57*, **7**.58
  market ***7**.57*, **7**.297
  see also goods and services
Common Agricultural Policy **7**.103, **7**.124, **7**.314
common cold **2**.74
common land ***3**.83*, **3**.115, ***7**.57–8*
common law **7**.41, ***7**.58*
  adversarial procedure **7**.3
  arrest **7**.18
  bail **7**.25
  commercial law **7**.57
  company law *see under* companies
  compensation **7**.62, **7**.75, **7**.79
  conspiracy **7**.67
  contempt of court **7**.68
  contract **7**.69
  customary **7**.78–9
  damages **7**.79
  equity **7**.108
  evidence **7**.112
  habeas corpus **7**.138
  homicide **7**.146
  inquest **7**.161
  inquisitorial procedure **7**.161
  judge **7**.172
  marriage **7**.55
  mercantile **7**.198

common law (cont.):
negligence 7.214
nuisance 7.220
perjury 7.233
tort 7.312
trust 7.316
United States 7.323
violent crime 7.326
common markets 7.58
see also in particular CARICOM; European Community
common security 7.58
Commons, House of see House of Commons
Commonwealth (English republican government) 3.83
see also Cromwell
Commonwealth, British see Commonwealth of Nations
Commonwealth of Independent States 7.59
before 1991 see Russia; Union of Soviet Socialist Republics
created 7.132
in CSCE 7.64
ethnic groups 7.181
languages 7.11
religion 7.98
secession 7.277
see also Armenia; Azerbaijan; Belarus; Kazakhstan; Kyrgyzstan; Moldova; Russia; Tadhzikistan; Turkmenistan; Ukraine; Uzbekistan
Commonwealth of Nations (British Commonwealth) 4.79–80, 7.58–9
and CARICOM 7.40
and law see common law
members listed 9.74
parliamentary 7.229
religion 7.40
sport and recreation 7.144, 7.214
see also empire under Britain
communal conflict 7.59
see also civil war; separatism
commune 3.83, 4.80
Commune of Paris 4.260
communicable diseases see infections
communication(s) 4.80, 4.81, 7.59–60
in animals 2.78, 2.94–5
see also protective coloration
code see codes
cybernetics 6.94
employment percentage 7.106
mass see advertising; books; media; printing
non-verbal 7.114, 7.218, 7.219, 7.281
rational 7.138
revolution 4.80, 4.81
satellite 6.81, 6.308, 6.347, 6.349
symbolism 7.268, 7.303
see also language; linguistics; meaning; speech; telecommunications; writing
communism/Communist Parties 4.80, 7.60–1
Chinese see under China
Cominform 4.79
Comintern 4.15, 4.79, 4.164–5
Communist Manifesto 4.80, 7.53
Eurocommunism 7.110
France 4.343
Greece 4.110
India 4.80
Indonesia 4.270
Internationals 4.164–5
Italy 4.139, 4.346
Kampuchea 4.182
Laos 4.262–3
leaders 4.280
see also politics under individual socialist countries
communism/Communist Parties (cont.):
Malaya 4.208
nomenklatura 7.218
theories on 7.192
United States 4.122
attitudes to 4.38–9, 4.202, 4.203
see also Eastern Europe; Marx; socialism; Union of Soviet Socialist Republics and under politics under individual countries
community (ecology) 2.74
ecosystem 2.30, 2.103–4, 2.328
food chain and web 2.124
isolated 2.30
succession 2.328
community (society) 7.61
architecture 5.108
care 7.61
penalty (law) 7.61
theatre 5.108
commutativity 1.70
commutator 6.153
Comnena, Anna 3.83
Comnenus, Alexius 3.44, 3.55, 3.83–4
Como, Perry 5.118
Comoro Islands 1.70, 4.4, 7.124, 7.348
compact disc 1.91, 6.81–2, 6.179, 6.180, 6.372
companies and business 7.62
audit 7.22
balance sheet 7.25
bookkeeping 7.34
budget 7.37
cartel 7.40–1
cash flow 7.41
conglomerate 7.65, 7.316, 7.336
consortium 7.67
corporatism 7.70
cost accounting 7.71
critical path method 7.76
dividend 7.93
entrepreneur 7.107
firm, theory of 7.120
holding company 7.145
integration 7.162
inventory 7.165
investment trust 7.165–6
joint-stock 7.172
law 7.19, 7.62, 7.197, 7.319
limited liability 7.184
location 7.186
loss 7.187
management see management
marketing 7.192
multinational see multinational corporations
oligopoly 7.222
operations research 7.224
organic growth 7.224
partnership 7.229
paternalism 7.230–1
personnel management 7.234
public limited 7.172
public relations 7.254
shares see shares
takeovers and mergers 7.65, 7.199, 7.224, 7.305
company painting 5.108
Compaore, Blaise 4.55
comparative linguistics 7.144
compass:
gyro-compass 6.163, 6.328
inductor 6.178
magnetic 6.72, 6.212–13, 6.215
compatibilists 7.125–6
compensation 3.379, 7.62, 7.75, 7.79
competition 7.52, 7.101
law 7.62
perfect and imperfect 7.222, 7.233
see also capitalism
compiler 6.82
complaints 7.223
complementarity (physics) 1.70
uncertainty principle 1.70, 1.349
wave–particle duality 1.70, 1.363
complementary medicine 2.74, 7.62–3
acupuncture 2.2, 7.46
anthroposophy 7.299
aromatherapy 6.123
chiropractic 6.72
essential oil 6.123
herbal 6.168, 6.387, 7.46
homoeopathy 2.7, 2.159, 6.169–70, 6.257
hypnosis 7.126, 7.150
complex numbers 1.18, 1.70
components of binary star 8.15
components of vector 1.70–1
composers:
Australia 5.188, 5.409
Austria 5.63
Expressionism and atonality 5.44, 5.465–6, 5.484
operetta and waltz 5.260, 5.435
organist 5.63
pianist 5.214
Romanticism 5.44, 5.276
serialism 5.405, 5.484
songs 5.276, 5.491
see also in particular Haydn; Mozart
Bohemia 5.47
Brazil 5.476
Britain 5.66
Chapel Royal 5.50, 5.65, 5.69, 5.88, 5.103, 5.176, 5.258, 5.304, 5.370
cinema 5.22, 5.481
consorts 5.69, 5.134, 5.176, 5.304
counterpoint 5.65, 5.69, 5.153–4, 5.176, 5.485
dances 5.65, 5.134
folk 5.210
Impressionistic 5.39, 5.128, 5.225
Ireland 5.154
masques 5.50, 5.258
musicals 5.267
opera and operetta 5.48, 5.210, 5.258, 5.370, 5.430, 5.439, 5.456–7
oratorios and choral 5.143, 5.340, 5.430, 5.456
organists 5.59, 5.65, 5.69, 5.103, 5.176, 5.485, 5.486
pastoral and folk 5.212, 5.469
piano and harpsichord 5.104, 5.225, 5.393
polyphony 5.69, 5.139, 5.176, 5.393
Romanticism 5.143
sacred music 5.50, 5.139, 5.176, 5.212, 5.370, 5.430, 5.486
songs see Britain under songs
symphonies 5.340, 5.393, 5.457
theatre 5.370, 5.439
viol 5.154
vocal works 5.66, 5.143, 5.155, 5.212, 5.340, 5.430, 5.456, 5.482
Czechoslovak 5.140, 5.233, 5.284, 5.422, 5.439
Denmark 5.318, 5.320
Finland 5.417
France:
abstract 5.496
ballet 5.128, 5.270, 5.373–4, 5.392, 5.402
counterpoint 5.325
Impressionist 5.126–7, 5.216, 5.375
Les Six 5.211, 5.295, 5.364, 5.421
masses 5.89, 5.137, 5.256, 5.325
opera and operetta see under opera
oratorio 5.291
composers (cont.):
France (cont.):
organists 5.115, 5.153, 5.163, 5.187
pianists and harpsichordists 5.87, 5.115, 5.153, 5.373, 5.398
polyphony 5.89, 5.256
Romanticism 5.45, 5.153
sacred music 5.89, 5.137, 5.256, 5.325
serialism 5.291
songs 5.137, 5.153, 5.187, 5.242, 5.256, 5.270
symphonies 5.138, 5.392, 5.398
theatre 5.89, 5.128, 5.270
violinists 5.254, 5.270
Germany:
atonality 5.485
Baroque 5.197
counterpoint 5.68, 5.377, 5.406
educationalists 5.330, 5.365
electronic 5.434
Expressionism 5.206, 5.435, 5.485
galant 5.448
Mannheim School 5.280, 5.430
Neoclassicism 5.204, 5.206
opera 5.181, 5.197, 5.204, 5.206, 5.214, 5.429, 5.435–6, 5.480–1, 5.484
oratorio 5.197–8, 5.291
organists 5.68, 5.377
polyphony 5.365
sacred music 5.365, 5.406–7
serialism 5.204, 5.434
songs and choral 5.62, 5.68
symphonies 5.204, 5.291, 5.429, 5.430, 5.435
theatre 5.214, 5.291, 5.330
violinists 5.62, 5.291, 5.429, 5.430
see also Bach; Handel; Strauss, Richard
Hungary 5.36–7, 5.240, 5.250, 5.260, 5.264, 5.266
Italy 5.11
ballet 5.121
Baroque 5.175
cellist 5.51
concertos 5.113, 5.168, 5.478
counterpoint 5.164, 5.336
electronic 5.44–5, 5.320
lutenist 5.175
opera see Italy under opera
oratorio 5.405
organists 5.83, 5.164, 5.168, 5.336
pianists and harpsichordists 5.367, 5.373, 5.393–4, 5.405, 5.409
polyphony 5.336
Romanticism 5.43, 5.491
sacred music 5.90, 5.303, 5.336, 5.343–4, 5.398
serialism 5.44–5, 5.121, 5.320
songs 5.168, 5.175, 5.283, 5.302
theatre 5.172, 5.344
violinist 5.113, 5.172, 5.335, 5.447, 5.478
Japan 5.445
musicals 5.45, 5.174, 5.175, 5.359, 5.385–6
Netherlands 5.242, 5.256, 5.441
New Zealand 5.80
Norway 5.192–3
Poland 5.99, 5.271, 5.335, 5.343, 5.443
Russia and USSR:
ballet 5.180, 5.247, 5.436, 5.448
concertos 5.247, 5.373, 5.436, 5.448
Neoclassicism 5.436
opera 5.53, 5.310, 5.367, 5.383–4, 5.417, 5.436
Romanticism 5.247, 5.373

composers (*cont.*):
  Russia and USSR (*cont.*):
    serialism 5.436
    songs 5.310, 5.448
    symphonies 5.53, 5.180, 5.367, 5.373, 5.384, 5.409, 5.417, 5.448
  South and Central America 5.257
  Spain 5.10, 5.82, 5.151, 5.188, 5.386, 5.474
  Sweden 5.49
  Switzerland 5.211
  United States 5.188, 5.200, 5.242, 5.252, 5.291, 5.412, 5.453, 5.468
    atonal and polyphonic 5.231, 5.394
    cinema 5.45, 5.175, 5.251
    jazz 5.46, 5.112, 5.144, 5.175
    minimal 5.179, 5.377
    Romantic 5.34, 5.251
    songs and musicals 5.175, 5.246, 5.359, 5.385–6, 5.424
  see also classical music
composite *6.82*, 6.156, 6.193
  see also bricks; fibre-reinforced; laminate; reinforced concrete
compost and manure *2.74*, *2.202*, *6.82–3*, 6.127–8, 6.253, 6.377
  dung as fuel 6.36
compound-engines *see* steam-engines *under* steam power
compounds *1.71*, 6.23, 6.24
  amphoteric compound 1.12
  anhydrous 1.14
  covalent 1.78–9
  decomposition *see* electrolysis
  enantiomer 1.108
  ester 1.113
  hydrate 1.160
  hydrocarbons *see* hydrocarbons
  interstitial 1.171
  ionic 1.172
    see also salts
  and mixtures, difference between 1.216–17
  organic *see* organic compounds
  valency 1.352
  see also acids; alkali; bases; oxides; salts; sulphates; sulphides *and also names of individual compounds*
comprehensive education *7.63*
compression ratio *6.83*
compression-ignition engines *see* diesel
compressor 6.8, *6.83*, 6.274, 6.339
Compromise of 1850 *4.80*, 4.82
Compromise of Breda 3.49
Compton, Arthur Holly *1.71*
Compton, Sir William 5.462
Compton-Burnett, Dame Ivy *5.108*
computational linguistics *7.63*
computerized axial tomography *6.85*, 6.213, 6.218, 6.292, 6.356, 6.357
computers and information technology *6.83–6*, *6.179–80*, *7.160*
  address 6.3
  AI *see* artificial intelligence
  aircraft 6.5
  algorithm 6.11
  analog 6.14, 6.320
  applications program 6.17, 6.179, 6.329–30
  architecture *6.85*
  ASCII 6.22
  associativity 1.22
  Babbage's analytical engines 6.27, 6.84, 6.189
  bandwidth 6.29
  bar code 6.30, 6.179
  baud rate 6.32
  binary code 6.22, 6.36, 6.84, 6.210
computers and information technology (*cont.*):
  bit 6.38
  and building 6.50
  bus 6.52
  byte 6.52
  calculator 6.54
  central processing unit 6.67, 6.83, 6.218–19, 6.222–3, 6.240
  clock 6.76, 6.83
  and cognitive psychology 7.252
  compiler 6.82
  computational linguistics 7.63
  computer numerical control 6.246–7
  computer-aided design and drafting *6.85*, 6.237, 7.63
  computer-aided engineering 7.63
  computer-aided manufacture *6.85*, 7.63
  computer-aided testing 7.63
  computer-integrated manufacturing 6.26, *6.85*, 6.179, 7.63
  see also automation
  computerized axial tomography *6.85*, 6.213, 6.218, 6.292, 6.356, 6.357
  control system 6.87
  crime *6.85*, *7.63*
  data *see* data
  desktop publishing 6.85, 6.99, 6.118
  disks 6.137, 6.165
  electronic funds transfer 6.118, 6.179
  electronic mail 6.118, 6.180
  electronic scanning 6.118, 6.164
  expert system 6.124, 6.179
  fifth-generation 6.21, 6.129
  file 6.129
  flexible manufacturing system 6.134, 6.211
  flip-flop circuit 6.136
  floppy disk 6.137
  flow chart 6.137
  graphics 6.159, 6.184
  ground station 6.161–2
  hard disk 6.165
  hardware *see* hardware
  human–computer interface 6.184
  hypertext 6.175
  image processing 6.176, 6.240
  inertial navigation 6.1, 6.162, 6.178–9
  information retrieval 6.179
  input device 6.180
  interactive system 6.183
  interface *see* interface
  language 6.21, 6.23, *6.85–6*, 6.185, 6.283
  logic gate 6.206–7
  machine code 6.210
  machine translation 6.211
  mainframe 6.214
  maps 6.215
  memory 6.218–19
  menu 6.219
  meteorology 6.221
  microcomputer 6.222, 6.223
  minicomputer 6.227
  modem 6.179, 6.180, 6.231
  mouse 6.180, 6.235
  music 5.12, *5.108*, 5.443
    see also music *under* electronics
  network *see* network
  operating system 6.219, 6.250
  optical character recognition 6.251
  output device 6.254, 6.351
    printer 6.180, 6.195, 6.281
    see also modem; visual display unit
  parallel processing 6.257
  and particle detector 6.259
  peripheral 6.180, 6.254, 6.260, 6.351
    see also output device *above*
computers and information technology (*cont.*):
  PET *see* positron emission tomography
  pioneers 6.27, 6.84, 6.189, 6.209, 6.217, 6.240, 6.366
  point-of-sale terminal 6.179
  and printing industry 6.282
  program *see* program
  protocol 6.285
  RAM and ROM 6.83, 6.218–19
  real time 6.295
  RISC 6.299
  serial processing 6.312–14
  shipbuilding 6.317
  signal processing 6.319
  simulations 6.17, 6.320
  software *see* software
  speech recognition 6.240, 6.328
  spreadsheet 6.329–30
  supercomputer 6.85, 6.339, 7.19
  switch 6.342
  systems analysis 6.344
  systems engineering 6.253, 6.344
  terminal 6.351
  transputer 6.361
  typesetting 6.366
  videotex 6.180, 6.373
  visual display unit 6.64, 6.373
  window 6.385–6
  word processor 6.179, 6.388
  workstation 6.179, 6.389
Comte, Karl Auguste 4.153, *7.63*
conceit (literature) *5.108*
concentration camps *4.82*, *7.63–4*
  map of 4.154
concentration (chemistry) *1.71*
conceptual art *5.108*, 5.299
Concert of Europe 7.135
concertina *5.108–9*
concerto *5.109*
  grosso *5.109*
  Italy 5.109, 5.113, 5.168, 5.478
  Russia and USSR 5.247, 5.373, 5.436, 5.448
conch *5.109*
Conchobar mac Nesse *9.9*, 9.10
Conciliar Movement 3.84
conciliation 7.92, 7.158
Concord, battle of Lexington and *3.207*
Concordat *4.82*
Concorde (aircraft) 6.138, 6.340
concrete 6.50, 6.51, 6.66, *6.86*
  aggregate 6.5
  formwork 6.144
  foundations 6.144–5
  mixer *6.86*
  pre-cast 6.44, 6.76, 6.280, 6.343
  pre-stressed 6.146, 6.281, 6.372
  road 6.300
  skyscrapers 6.322
  see also reinforced concrete
concrete painting *5.109*
concrete (pattern) poetry 5.342
concretion (geology) 1.70, *1.71*
concussion *2.74*
Condé family *3.84*
  the Great Condé 3.84, 3.135, 3.144
condensation *1.71*
  hydrological cycle 1.162
  nuclei *1.71*
  polymer 6.275
  reaction *see* chemical reactions *under* chemistry
condenser (electrical) 6.57–8
condiments *see* spices, herbs and flavourings
'conditioned arising' 7.85, 7.133
conditioning (learning) 2.185, 2.245, 7.29, 7.30, *7.64*, 7.231, 7.287, 7.330
condom 2.75, 6.87
condor *2.74*, 2.358
Condor Legion *4.82*
Condorcet, Antoine Nicolas, marquis de *3.84*, 3.115, 4.117
condottiere *3.84–5*, 3.324–5
conduction and conductors (electrical) *1.71*
  copper 6.89
  electrolyte 1.102–3
  graphite 6.160
  lightning 6.202
  non-conductor *see* insulator *under* electricity
  silver 6.320
  superconductivity 1.327, 6.114, 6.241, 6.339
  see also current, electrical
conduction and conductors (thermal) *1.71*
conduction and conductors (atomic physics): band theory 1.32
conductors (musical) *5.110*
  Austria 5.260, 5.276
  Britain 5.60
  France 5.56
  Germany 5.280, 5.291, 5.435–6
  Hungary 5.260
  Russia and USSR 5.373
  Spain 5.82
  United States 5.46, 5.391
cone sheets 1.166
coneflower *2.74*
confectionery *6.86–7*
Confederacy *4.82*, 4.331
confederation *7.64*
  see also federalism
Confederation of the Rhine *4.82–3*
Conference on Disarmament 7.44, *7.64*
Conference on Security and Co-operation in Europe *7.64*, 7.87, 7.272
Confession of Augsburg 3.27
'confessional' poetry 5.270, 5.352
configuration (chemistry) *1.71*
Confirmation 7.271
Conflans, admiral 3.298
conflicts:
  communal 7.59
  dispute resolution 7.92
  see also wars
conformation (chemistry) 1.71, *1.72*
conformity (geology) *1.72*
conformity (psychology) *7.64*
Confucianism and Confucius 3.73, *3.84–5*, 3.390, 7.47, *7.64–5*, 7.265, 7.335
  and arts 5.251, 5.417
    literature *5.110*
  Neo-Confucianism 7.46, 7.65, 7.305
conger eel *2.74–5*
conglomerate (companies) *7.65*, 7.316, 7.336
conglomerate (geology) *1.72*
Congo *1.72*, *7.348*
  colonization and decolonization 4.4, 4.306
  crisis *4.83*
  physical geography and geology 1.72
  political leaders 4.179, 4.201
  religion and myths 9.32
  wars and defence 7.231
  see also Zaïre
congo eel *2.75*
Congo People's Republic (Brazzaville) *4.83*
Congo River *1.72*
Congregationalists *3.85*, *7.65*, 7.245
  see also United Reform
Congress, Indian National *4.83–4*
Congress of Industrial Organizations 4.10–11
Congress of United States *4.84*, 4.287, 4.308–9, 7.324
Congreve, William (1670–1729) *5.110*, 5.382, 5.418

Congreve, Sir William (1772–1828) **6**.228, **6**.301
conic section **1**.72, ***1**.73*, **1**.246–7, **9**.51
conies *see* cony
conifers ***2**.75*
cedar **2**.58–9
cypress **2**.85–6
Douglas fir **2**.98
firs **2**.98, **2**.119
as gymnosperms **2**.148
hemlock spruce **2**.154
larch **2**.183
metasequoia **2**.209
monkey puzzle tree **2**.217
pines *see* pines
redwood **2**.281
Scots pine **2**.297
spruce **2**.154, **2**.323
taiga **2**.333
yew **2**.374
see also evergreen; pines
conjunction (grammar) **7**.229
conjunction of planets **8**.122
conjunctivitis ***2**.75*
Conkling, Roscoe ***4**.84*, **4**.148, **4**.328
Conn, High king of Ireland **3**.349–50
Connaught ***3**.85*
Connecticut ***3**.86*
Connecticut River ***1**.73*
Connolly, James **4**.107
Conor, king **9**.9
conquistadors, see under Spain
Conrad II, Holy Roman Emperor **3**.168, **3**.294, **3**.341
Conrad III, Holy Roman Emperor **3**.167, **3**.168
Conrad IV, Holy Roman Emperor **3**.168
Conrad, count of Palatinate **3**.267
Conrad, Joseph ***5**.110*, **5**.162, **5**.324, **5**.376
conscientious objection ***7**.65–6*, **7**.227
consciousness **7**.169
hypnosis **7**.150
loss of **2**.73, **2**.74
see also anaesthesia
-raising **7**.332
conscription ***7**.66*
consent ***7**.66*
consequentialism ***7**.66*, **7**.324
conservation ***2**.75*, **2**.254, ***7**.66*, **7**.107
animals **7**.212
Conventions on **7**.66, **7**.106
energy **1**.73, **6**.217
national park **7**.212
organic farming **7**.224
recycling **7**.260
see also ecology; environment *and also* environment *under individual countries*
conservation laws (physics) ***1**.73*
conservatism (politics) **4**.85, ***7**.66*
see also Conservative Party (Britain); New Right; Right
Conservative Judaism ***7**.66–7*, **7**.304, **7**.305
Conservative Party (Britain) ***4**.84–5*, **4**.276, **4**.386, **7**.309, **7**.317
leaders **3**.44, **3**.145, **4**.57, **4**.100–1, **4**.103, **4**.149, **4**.191, **4**.204–5
Conservative Party (Canada) ***4**.85*
Conservatoire National des Arts et Métiers ***6**.87*
consistence ***1**.73*
consociationalism ***7**.67*
consortium ***7**.67*
consorts (musical) **5**.69, ***5**.110*, **5**.134, **5**.176, **5**.304
conspiracy ***7**.67*
Constable, John ***5**.110–11*
Constance, Council of ***3**.86*
Constance, Lake ***1**.73*
Constans, Byzantine emperor **3**.105
Constant, Benjamin ***5**.111*
constantan **6**.354
Constantine I (the Great), Roman emperor **3**.55, ***3**.86*
and armed forces **3**.142, **3**.290, **3**.308
and Christianity **3**.74, **3**.92, **3**.187, **3**.253
Constantine II, king of Greece **4**.140, **4**.363
Constantine II, Roman emperor **5**.72
Constantine Moshoeshoe II, king of Lesotho **4**.193
Constantine V, Byzantine emperor **3**.174
Constantine VI, Byzantine emperor **3**.180
Constantinople (Byzantium, Istanbul) ***3**.86*, **3**.280, **6**.149
Council of **7**.309
Constantius Chlorus **3**.104
constants ***9**.54*
cosmological **8**.33
gas **1**.132
gravitational **8**.61
constellations ***8**.28–9*
asterism **8**.8
list of names **8**.195
see also individual names
maps of **8**.30, **8**.31, **8**.123
constipation **2**.75
constitutions ***7**.67*
constitutional government ***7**.67*
and human rights **7**.32, **7**.125
Japan **4**.172, **4**.215–16
judicial review **7**.173
United States *see under* politics *under* United States
constructional engineering *see* bridges; buildings; civil engineering
Constructivism ***5**.111*, **5**.168, **5**.266, **5**.277, **5**.346, **5**.447
consulate **7**.105
consumer ***7**.67–8*
credit **7**.74
law ***7**.68*
consumption (disease) *see* tuberculosis
consumption (eating) *see* food
consumption (economics) **7**.68
contact adhesives **6**.3
contact improvisation **5**.342
contact lens ***6**.87*
contact process **6**.70, ***6**.87*, **6**.338, **6**.339
contagion ***2**.75*
see also infections
container ships ***6**.87*, **6**.219, **6**.277
containment ***4**.85*
Conté, Lansana **4**.143
Conté, Nicolas-Jacques **5**.343
contempt of court ***7**.68*
Continental Congress ***3**.87*
continental drift ***1**.73*
Gondwana **1**.135, **1**.142
Laurasia **1**.135, **1**.189
maps **1**.73, **1**.135
Pangaea **1**.135, **1**.246
Tethys Sea **1**.135, **1**.334
continental philosophy ***7**.68*
analytic philosophy **7**.12, **7**.68, **7**.173, **7**.185
phenomenology **7**.68, **7**.199, **7**.234
see also existentialism
Continental System ***4**.85*
continents ***1**.74*
continental margin **1**.236
continental rise ***1**.73*, **1**.236
continental shelf ***1**.73–4*, **1**.236
continental slope ***1**.74*, **1**.236
drift *see* continental drift
see also Africa; Antarctica; Asia; Australia; Europe; North America; South America
contingency fee ***7**.68*
continuing education ***7**.68–9*
see also higher education
continuo ***5**.111*
continuous representation ***5**.111*
continuously variable transmission **6**.26
contour ***1**.74*
contraception ***2**.75*, ***6**.87*
abortion **7**.1, **7**.115
family planning ***7**.115–16*
pill **6**.127, **6**.170, **6**.334
pioneers **4**.36, **4**.48, **4**.330
sterilization **6**.87, **6**.334
contract, law of ***7**.69*
contracting-out ***7**.69*
contraction of matter ***1**.74*
contrapposto ***5**.111–12*
control, locus of **7**.187
see also power (political and social)
Control Commissions ***4**.85*
control in technology:
control system ***6**.87–8*
cybernetics **6**.94, **6**.385
fluidics **6**.137
servo-mechanism **6**.312, **6**.314
convection ***1**.74*, **1**.334
Conventional Forces in Europe, Treaty on **7**.17, **7**.64
conventional warfare **7**.69
see also wars
Conventions ***7**.69*
on Biological Weapons **7**.91
on International Trade in Endangered Species of Wild Flora and Fauna **7**.106
on Laundering, Search, Seizure, and Confiscation of Proceeds of Crime **7**.206
Lomé **7**.103
on Prevention and Punishment of Crime of Genocide **7**.130
Relating to Status of Refugees **7**.238, **7**.261
on Rights of Child (UN) **7**.45, **7**.46, **7**.149, ***7**.319*
for Suppression of Unlawful Acts against Safety of Civil Aircraft **7**.142
for Suppression of Unlawful Seizure of Aircraft **7**.142
on Wetlands of International Importance **7**.66
see also European Conventions; international organizations and alliances *and* Conventions *under* Geneva
convents *see* monasticism
convergence (atmosphere) ***1**.74*
convergent series (numbers) ***1**.74*
conversation ***7**.69*, **7**.109
pieces ***5**.112*, **5**.129, **5**.168, **5**.499
conversions, religious *see* missionaries
convertible issues **7**.282
conveyancing ***7**.69*
conveyor ***6**.88*
convicts:
pardoned **4**.111
transportation ***4**.85*
see also penal *under* Australia
see also prisons
convolvulus **2**.29
convoy system ***4**.85*
cony **2**.253
hyrax **2**.168
pika **2**.253
Cook, Frederick Albert ***1**.75*
Cook, Captain James **1**.74, ***1**.75*, ***3**.87*, **3**.122
and astronomy **8**.13, **8**.173
and Australasia **3**.122, **4**.23, **4**.46
and Bligh **4**.41
and Canada **3**.254
and natural history **2**.21
and Pacific **4**.147, **4**.238, **4**.347
Cook, Mount ***1**.75*
Cook Islands **7**.296
Cook Strait ***1**.75*
Cooke, Sir William Fothergill ***6**.88*, **6**.232, **6**.346, **6**.384
cookers and stoves ***6**.88–9*, **6**.167
Franklin **6**.146
microwave **6**.224
pressure-cooker **6**.281
thermostat **6**.355
coolabah **2**.110
Coolidge, Calvin ***4**.85–6*
Coolidge, William David ***6**.89*, **6**.177
cooling:
contraction of matter **1**.74
system (motor) ***6**.89*
see also freezing; ice; temperature
Coomaraswamy, A. K. **7**.156
coomb ***1**.75*
Cooper, Alexander **5**.112
Cooper, Anthony Ashley *see* Shaftesbury
Cooper, Sir Astley Paston **2**.75
Cooper, James Fenimore ***5**.112*
Cooper, Samuel ***5**.112*
co-operative behaviour **7**.11
Co-operative Commonwealth Federation ***4**.86*
co-operatives **7**.56–7, ***7**.69*, **7**.70, **7**.127, **7**.180
Co-operative Movement ***4**.86*, **4**.354
Cooper's Creek ***1**.75*
co-orbital satellites ***8**.29*, **8**.47, **8**.76
coordinates ***1**.75*
Cartesian **1**.52, **1**.53
coordinate system in astronomy ***8**.29*, **8**.58
polar **1**.262, **1**.263
coot ***2**.75*
Coote, Eyre **3**.377
Cope, Sir John **3**.291
Copeau, Jacques ***5**.112*
Copenhagen, first battle of ***4**.86*
Copenhagen, second battle of ***4**.86*
Coper, Hans **5**.438
Copernicus, Nicolas **3**.51, **3**.301, **3**.318, **7**.276, **8**.11, ***8**.29*
Copernican system ***8**.29*, **8**.32
Copland, Aaron ***5**.112*
Copley, John Singleton **5**.13, ***5**.112*
co-polymer **6**.275–6
copper (butterfly) ***2**.75*, **2**.76
copper (metal) ***1**.75*, **1**.106, ***6**.89*
alloys **6**.11, **6**.12, **6**.45, **6**.49, **6**.212, **6**.265, **6**.320, **6**.370
arts ***5**.112*, **5**.346, **5**.416
see also engraving
in battery **6**.374
catalyst **6**.14
electroplating **6**.73
extraction **6**.221, **6**.253
ore **1**.239
pipe **6**.274
prehistoric use **3**.87, **6**.280
Sheffield plate **6**.315
in telephony **6**.347
in thermocouple **6**.354
Copper Age ***3**.87*
see also prehistory
copper aluminium phosphate **1**.348
copper oxide ***1**.75–6*, **1**.81, **6**.89, **6**.90, **6**.156
copper sulphate **1**.59, ***1**.76*, ***6**.89*, **6**.147
copper sulphide **6**.253
copperhead ***2**.75*
copra **2**.71, ***2**.75–6*
Coptic art ***5**.112*
Coptic Church ***3**.87*, ***7**.69–70*, **7**.217
copyright ***7**.70*, **7**.184
cor anglais ***5**.113*, **5**.352
Cor Caroli ***8**.29*
coracle ***6**.89*
coral ***2**.76*
and art (coral work) ***5**.112–13*
atoll **1**.24
barrier reef **1**.33
fossil ***2**.76*

coral (*cont.*):
- gorgonian **2**.141, **2**.279, **2**.298–9
- islands ***1.76***, **1**.203, **1**.245, **1**.309, **1**.339, **1**.348
- organ-pipe **2**.236
- red **2**.279
- reefs *see* reefs
- sea fan **2**.298–9
- soft **2**.317

Coral Sea ***1.76***
coral snake ***2.76–7***
Coram, Thomas **3**.130
Corday d'Armont, Charlotte ***3.87***
Cordelia (Uranian satellite) **8**.178
Cordero, Febres **4**.108
cordillera ***1.76***
cordite *see* nitro-cellulose
cordless telephone **6**.65, **6**.292, **6**.347
Córdoba, Fernández **3**.369
core of Earth **1**.98–9
core of Sun **8**.163
Corelli, Arcangelo ***5.113***
COREPER (Committee of Permanent Representatives) **7**.111
Corfu ***1.76***
- incident ***4.86***

Corinth ***3.87***
Coriolis, Caspar Gustave de: Coriolis force ***1.76***
cork ***2.77***
- cork oak **2**.231

cormorants ***2.77***, **2**.305
corms, plants with ***2.77***
- crocus **2**.19, **2**.81, **2**.291–2
- gladiolus **2**.136
- iris **2**.119, **2**.136, **2**.173, **2**.218
- montbretia **2**.218
- taro **2**.335
- *see also* rhizome; tubers

corn ***2.77***, **2**.198
corn bunting **2**.45
Corn Laws **4**.15, ***4.86***
corn salad ***2.77***
corncrake **2**.275
Corneille, Pierre **3**.299, **5**.2, ***5.113***
cornelian ***1.76–7***
Cornelius, Saint ***9.41***
cornet ***5.113***
cornett **5**.59, ***5.113***
cornflower ***2.77***
Cornu, Paul **6**.167
Cornwall, Piers Gaveston, earl of **3**.139
Cornwallis, Charles, 1st Marquis **3**.58, **3**.79, ***3.88***, **3**.176, **3**.357, **3**.388, **4**.388
Coromandel ***3.88***
corona ***1.77***, ***8.29***, **8**.42
- Corona Australis ***8.29–30***
- Corona Borealis ***8.30***

Coronado, Francisco Vasquez de **3**.12, **3**.88
coronary care *see* disorders *under* heart
coronary thrombosis ***2.77***
Corot, Jean-Baptiste Camille ***5.113***
corporatism ***7.70***
corrasion ***1.77***
Corrêa, Gaspar **5**.359
Correggio ***5.113–14***
correlation (statistics) ***1.77***, ***7.70***
correspondence courses **7**.68, **7**.92–3
corrie *see* cirque
corrosion ***1.77***, **1**.288, ***6.90***, **6**.221
- resistance **6**.15, **6**.73, **6**.316, **6**.333, **6**.357
  - chromium plating **6**.90, **6**.120
  - galvanizing **6**.149
  - vanadium **6**.370
  - *see also* stainless steel

corrugated iron ***6.90***
corruption ***7.70–1***
corsair **3**.35, ***3.88***
- *see also* pirates

Corsica ***1.77***, ***3.88***
Corso, Gregory **5**.39
Cort, Henry ***6.90***, **6**.186
Cortázar, Julio ***5.114***
Cortés, Hernan, Marqués del Valle de Oaxaca **3**.31, **3**.86, ***3.88***, **3**.114, **3**.239, **3**.253, **3**.351, **9**.30
corticosteroids **6**.177, **6**.334
Cortona, Pietro da **5**.53, ***5.114***
Cortot, Alfred **5**.82
corundum ***1.77***
- ruby **1**.287
- sapphire **1**.293

corvette ***6.90***
Corvus ***8.30***
Coryat, Thomas **5**.460
Cosgrave, Liam **4**.87
Cosgrave, William Thomas ***4.86–7***, **4**.119
cosine ***1.77***
Cosmati work ***5.114***, ***6.90***
cosmetics *see* arts *under* body
cosmic composition ***8.30–1***
cosmic rays ***1.77–8***, ***8.31–2***
- background radiation **8**.98–9

cosmogony ***8.32–3***
cosmology ***8.33–4***
- cosmological constant ***8.33***
- cosmological principle **8**.33

cosmonauts *see* space-flight
Cossacks ***3.88***, **3**.294
Cossigo, Francesco **4**.170
Cossío, José María de **5**.173
cost (economics) ***7.71***
- *see also* capital; interest; rent

cost accounting ***7.71***
Costa Rica ***3.88***, ***4.87***, ***7.348***
- infant mortality **7**.159
- medicine and health care **7**.159
- in OCAS **4**.251
- physical geography and geology **1**.78
- social services and security **7**.293

cost-benefit analysis ***7.71***
Costello, John **4**.119
cost-of-living index **7**.266
costume ***5.114***
Côte d'Ivoire (Ivory Coast) ***1.78***, ***4.87***, **4**.170, ***7.348–9***
- colonization and decolonization **4**.4, **4**.157, **4**.306
- economy and trade **7**.113, **7**.124
- leaders **4**.157, **4**.302
- physical geography and geology **1**.78
- religion **7**.5

cotinga ***2.77***
- bell bird **2**.28
- cock-of-the-rock **2**.70
- umbrella bird **2**.352, **2**.353

Cotman, John Sell ***5.114***
cotoneaster ***2.77–8***
Cotopaxi ***1.78***
cotton ***2.78***, ***5.115***, ***6.91***, **6**.128, **6**.364, **6**.384
- bleaching **6**.351
- canvas **6**.57
- carding **6**.165
- dye **6**.161
- gin ***6.91***
- insecticide and **6**.180
- pests on **6**.267
- printing **4**.200, ***5.115***, **6**.353
- spinning **6**.20, **6**.92, **6**.328–9
- waste for biofuel **6**.36
- yarn **6**.390
- *see also* textiles

cotton spinner (echinoderm) **2**.298
cotton stainer (bug) **2**.279
Cottrell, F. G. **6**.120
cotyledon ***2.78***
Coubertin, Baron de **7**.222
coucal ***2.78***
couch grass ***2.78***
Coudé telescope **8**.135, **8**.136
cougar **2**.272
Coughlin, Charles Edward ***4.87***
coulee **1**.20
Coulomb, Charles-Augustine de: Coulomb's Law **1**.78
coulomb ***1.78***, **6**.368
Council for Mutual Economic Assistance (COMECON) ***4.79***, ***7.57***
Council of Europe ***4.87***, ***7.71***
counselling (psychology) **7**.30, ***7.71–2***, **7**.116
counterglow **8**.57–8
counter-insurgency ***7.72***
counter-intelligence services ***7.72***
counterpoint ***5.115***, **5**.325
- Britain **5**.65, **5**.69, **5**.153–4, **5**.176, **5**.485
- canon **5**.75
- fugue **5**.166
- Germany **5**.68, **5**.377, **5**.406
- Italy **5**.164, **5**.336

Counter-Reformation ***3.89***
counter-revolution ***7.73***
country house ***3.89***, ***4.87***
country music ***5.115***
country rock (geology) ***1.78***
coup d'état **7**.72, ***7.73***, **7**.201
Couperin, Charles and Louis **5**.115
Couperin, François **5**.29, ***5.115***
couplet ***5.115***
coupling, railway ***6.91***
courante ***5.115***
Courbet, Gustave **5**.38, ***5.116***, **5**.299, **5**.376
courgette **2**.204
courser ***2.78***
court (law) *see* courts of law
court (royal):
- *ballet de cour* **5**.33–4
- courtly love and tradition (particularly poetry) ***5.116***, **5**.297, **5**.368, **5**.428, **5**.466, **5**.491
- drama and dance **5**.236, **5**.427
- fool ***5.116***
- orchestra **5**.222

court-martial ***7.73***
Courtrai, battle of ***3.89***
courts of law ***7.73***, **7**.318, **7**.324
- appeal **7**.15, **7**.318
- barrister **7**.27
- community penalty **7**.61
- contempt of court **7**.68
- court-martial **7**.73
- damages *see* compensation
- detention without trial **7**.87
- European Court of Human Rights **7**.110–11
- European Court of Justice **7**.111, **7**.173
- evidence **7**.112
- family court **7**.115
- injunction **7**.161
- Inns of Court **7**.161
- inquest **7**.161
- international *see* International Court of Justice
- judges **7**.172–3
- jury **7**.112, **7**.174
- juvenile **7**.174
- legal aid **7**.181–2
- perjury **7**.233
- plaintiff **7**.236
- political trial **7**.239
- precedent **7**.245
- prosecution **7**.250
- Scots **7**.276
- sentences *see* sentences (court)
- solicitor **7**.295
- *see also* law and justice

courtship ***2.78***, **2**.79
Cousteau, Jacques-Yves **6**.18
covalent bond **1**.59
covalent compound ***1.78–9***
covellite **1**.239
covenant (law) ***7.73***
covenant (theology) ***7.73***
Covenant (international relations) *see* Conventions; International Covenants
Covenanter ***3.89–90***
Covent Garden Theatre ***5.116***
Coverdale, Miles ***3.90***, **5**.368
Coverdale, William **3**.41
cow *see* cattle
Coward, Sir Noël ***5.116***, **5**.259
cowbird ***2.78–9***
cowboy ***4.87–8***
Cowley, Abraham ***5.116***, **5**.325
Cowper, A. E. **4**.151
Cowper, William ***5.116***, **5**.325
Cowper-Coles, Sherard **6**.316
cowry ***2.79***
cowslip **2**.269
Cox, David ***5.116***
Coyote (god) ***9.9***
coyote (animal) ***2.79***
Coypel, Noël, Antoine, and Noël-Nicholas ***5.116***
coypu ***2.80***, **2**.362
Cozens, Alexander ***5.116–17***
Cozens, John Robert ***5.117***
CPS (Crown Prosecution Service) **7**.250
CPU *see* central processing unit
Crab (constellation) **8**.20
crab apple **2**.13
Crab Nebula ***8.34***
Crabbe, George **5**.117
crabs ***2.80***
- coconut **2**.71
- hermit **2**.73–4, **2**.156
- horseshoe **2**.16, **2**.163
- king **2**.179
- spider **2**.32
- stone **2**.327

cracking ***6.91***, **6**.171, **6**.175, **6**.261, **6**.265
crafts **5**.159, **6**.31, **6**.281
- *see also* Arts and Crafts

Craig, Edward Gordon ***5.117***, **5**.413
Craig, Sir James **4**.357
Cranach, Lucas (the Elder) **3**.213, ***5.117***
Cranach, Lucas (the Younger) ***5.117***
cranberry ***2.80***
Crane, Stephen ***5.117***
crane (bird) ***2.80***, **2**.90, **2**.324
crane (mechanical) ***6.91–2***, **6**.158
crane fly ***2.80***
cranesbill ***2.80***
Cranko, John ***5.117***
crankshaft ***6.92***
Cranmer, Thomas **3**.44, **3**.64, ***3.90***, **3**.110, **3**.204, **3**.355
Cranston, Catherine **5**.274
Cranston, Samuel **3**.302
craquelure ***5.117***
Crashaw, Richard ***5.117–18***
Crassus, Marcus Licinius **3**.57, ***3.90***, **3**.271
Crater (constellation) ***8.34–5***
Crater Lake **1**.48, ***1.79***
craters ***1.79***, ***8.34***
- caldera **1**.48, **1**.49, **1**.79, **1**.177
- ejecta **8**.45
- meteorite **1**.213
- Moon **1**.79, **8**.57, **8**.114–15

Crates of Thebes **3**.94
craton *see* shield (geology)
crawfish ***2.80***
crayfish ***2.80***
crayon ***5.118***
Crazy Horse ***4.88***, **4**.197
creamware ***5.118***
creation myths ***7.73***, **7**.127
- creator gods **9**.3, **9**.4, **9**.6, **9**.7, **9**.19, **9**.26, **9**.28

creativity ***7.73–4***
Crécy, battle of ***3.90***
credit ***7.74***
- bill of exchange **7**.31
- card ***6.92***, **7**.68

credit (*cont.*):
union *7.74*
see also debt
Crédit Mobilier of America *4.88*
Cree Indians *3.90*, *7.74*, **7**.219
creed *7.74*
Creek Indians *3.90*
creep (metallurgy) **6**.221
creep (soil) **1**.49, *1.79*
creeper *2.80*
cremation *7.74–5*
Creole culture **7**.40
languages *7.235–6*
creosote **6**.79
crescendo *5.118*
cress *2.80*
Cressida (Uranian satellite) **8**.178
crested swift *2.80*
Cretaceous period *1.79*, **1**.133, **1**.134, **1**.135
Crete *1.79*, *3.90–1*, **3**.232–3, *4.88*, **5**.189
see also Minoan civilization
crevasses *1.79–80*, **1**.140
crewel work *5.118*
Crick, Francis Harry Compton *2.80–1*, **2**.95, **6**.105
cricket (insect) *2.81*, **2**.215
cricket (sport) **4**.326, *7.75*
crime *7.75–6*
blackmail **7**.33
compensation for victims **7**.62, **7**.75
computer **6**.85, **7**.63
conspiracy **7**.67
Convention on **7**.206
criminal law **7**.76, **7**.318, **7**.319
criminal responsibility *7.76*, **7**.146
criminology *7.76*
deviance as **7**.88
drug offences **7**.96
extradition **7**.113–14
forgery **7**.123
fraud **7**.104, **7**.123, **7**.124, **7**.314
genocide **7**.130
hi-jacking **7**.142
hostage-taking **4**.165, **7**.147–8
juvenile delinquency **7**.174
Mafia **7**.188
money laundering **7**.205–6
organized **7**.225
poaching **3**.286
property **7**.250
prosecution **7**.250
punishment *see* punishment; sentences
rape **7**.258
sexual offences **7**.40, **7**.255, **7**.258, **7**.281
suicide as **7**.301
treason **7**.315
violent **7**.326
war **7**.328
white-collar **7**.331
see also homicide; murder; police
Crimea *1.80*, *3.91*
Crimean War *4.88*
battles and campaigns **4**.27–8, **4**.164, **4**.309
Paris peace conference **4**.260
technology **6**.12, **6**.49, **6**.226, **6**.316
crinoid **2**.299
Cripps, Sir (Richard) Stafford *4.88*
Crispi, Francesco *4.88*, **4**.170
Cristiani, Felix **4**.111
Cristofori, Bartolemeo **5**.347
critical path method *7.76*
critical point *1.80*
critical theory *7.76*, **7**.124, **7**.138
critics, art **5**.50, **5**.341–2, **5**.394
critics, literary *5.266*
Britain:
16th to 18th centuries **5**.136, **5**.241, **5**.418
critics, literary (*cont.*):
Britain (*cont.*):
19th century **5**.22, **5**.43–4, **5**.129, **5**.184, **5**.202, **5**.254
20th century **5**.43–4, **5**.66, **5**.162, **5**.184, **5**.258, **5**.259, **5**.262, **5**.366, **5**.428, **5**.493–4
France **5**.163, **5**.368, **5**.397, **5**.429, **5**.479
Germany **5**.262, **5**.405
Greece, ancient **5**.268
Italy **5**.77
Peru **5**.468
Poland **5**.227
Russia and USSR **5**.278
Spain **5**.10
United States **5**.212–13, **5**.269, **5**.290, **5**.353–4, **5**.364, **5**.490
critics, music **5**.208, **5**.482, **5**.491
critics, photography **5**.145, **5**.432–3
Critz, Juan de **3**.184
Croatia *1.80*, *3.91*, *4.88–9*, *7.349*
language and literature *see* Serbo-Croat
Crockett, David **4**.7, *4.89*
Crocodile River **1**.195
crocodiles *2.81*
alligator **2**.7
cayman **2**.58
gharial **2**.135
crocus **2**.19, *2.81*, **2**.291–2
Croesus, king of Lydia *3.91*, **3**.95, **3**.117, **3**.214, **3**.231, **3**.322
Cro-Magnons *3.91*, **3**.120, **3**.367
Crome, John Bernay *5.118*
Crome, John ('Old Crome') *5.118*
Cromer, Evelyn Baring, 1st earl of *4.89*
Crompton, Samuel *6.92*, **6**.329
Cromwell, Oliver *3.91*
and Barebones Parliament **3**.36
and Commonwealth **3**.83
and Fifth Monarchy Men **3**.127
and Ireland **3**.180
New Model Army **3**.246, **3**.253, **3**.321
and religion **3**.85, **3**.295
and Rump Parliament **3**.210, **3**.312
and Scots **3**.50, **3**.70, **3**.106, **3**.291
death **3**.302
Cromwell, Richard *3.91*, **3**.293, **3**.302
Cromwell, Thomas, earl of Essex **3**.90, *3.91*, **3**.163, **3**.236, **3**.282
Cronus *9.9*
Crook, George **4**.88
Crookes, Sir William *1.80*
Crookes radiometer **6**.291
crookesite **6**.353
crooning *5.118*, **5**.358
crops **6**.7, *6.19*, **6**.109
baler **6**.6, **6**.28, **6**.166
beverage **6**.7, **6**.34, **6**.271, **6**.364
breeding and hybridization **6**.36, **6**.155, **6**.371
cash *see* plantation
cereals *see* cereals and grains
and chemicals *see* agrochemicals
fibre **6**.128
see also cotton; flax
forage *see* feedstuffs
fruit *see* fruit plants
genetic resource **6**.155
grass *see* grass
green revolution **6**.7, **6**.161, **6**.364
greenhouse **6**.161, **6**.170, **6**.175
horticulture **6**.170, **6**.371
hydroponics **6**.175
intercropping **6**.183, **6**.364
irrigation *see* irrigation
organic farming **6**.18, **6**.183, **6**.253, **7**.224
crops (*cont.*):
processing **6**.142
propagation **6**.271
root **6**.7, **6**.18, **6**.271, **6**.302, **6**.303, **6**.371
rotation *3.91–2*, **6**.7, **6**.253, **6**.303
row cultivation **6**.170, **6**.303
silage *see* feedstuffs
spraying **6**.6, *6.92*
terrace cultivation **6**.351, **6**.352
vegetables *see* vegetables
see also agriculture
croquet *7.76*
Crosby, Bing **5**.118, **5**.358
Crosland, Anthony *4.89*
Cross, Charles Frederick **6**.66, *6.92*, **6**.295
cross *7.76*
see also under Jesus Christ
crossbill *2.82*
crossbow *6.92*, **6**.225, **6**.226
cross-country running *7.76*
cross-cultural psychology *7.77*
cross-fertilization *2.82*
Crossman, Richard Howard Stafford *4.89*
cross-staff *6.92*
crow family *2.82*
carrion **2**.55, **2**.82
chough **2**.65
fantail **2**.113
jackdaw **2**.174
jay **2**.175
magpies **2**.28, **2**.198
raven **2**.278
rook **2**.288
Crow Indians *3.92*, **7**.219
crow-bar *6.92*
crowd psychology *7.77*
Crown *7.77*
jewels *5.118–19*
prerogative *7.78*
Prosecution Service **7**.250
see also monarchy
Crown Film Unit **5**.132
crown of thorns starfish *2.82*
CRT *see* cathode-ray tube
crucian carp *2.82*
crucible steel **6**.172, **6**.186
crucifixion *3.92*, **7**.76
see also cross *under* Jesus Christ
Cruelty, Theatre of **5**.23
Cruikshank, F. **2**.18
Cruikshank, George **4**.286, *5.119*
Cruikshank, Isaac **5**.119
cruise control *6.92–3*
Cruise missiles *6.93*, **6**.162, **6**.228, **6**.246, **7**.220
cruiser *6.93*
crumhorn *5.119*
crumple zones **6**.233
Crusades **3**.73, *3.92–3*, **3**.143, **3**.179, **3**.253
crust of Earth **1**.98–9
discontinuities **1**.98, **1**.217
sial **1**.302
sima **1**.305
see also plates of Earth's crust
crustaceans **2**.16, *2.82*
amphipod **2**.8–9, **2**.293
barnacles **2**.23
crab *see* crabs
crawfish **2**.80
crayfish **2**.80
daphnia **2**.87
Dublin Bay prawn **2**.100
lobster **2**.192
mantis shrimp **2**.202
prawn **2**.267–8
shellfish **2**.307
shrimp **2**.110, **2**.111, **2**.202, **2**.309
spider crab **2**.320
woodlouse **2**.370
Crux *8.35*
crwth *5.119*
cryolite **6**.164
crypt *5.119*
cryptic coloration **2**.50
cryptogram plants *2.82*
clubmoss **2**.69
horsetails **2**.49–50, **2**.164
spores **2**.322–3
see also bryophytes
cryptography *6.93*
see also codes
Crystal Palace **4**.140, **6**.51, **6**.160–1
crystal set *see* receiver *under* radio
crystals and crystalline materials *1.80–1*
cleavage **1**.64
crystallography, X-ray **6**.389
decrepitation **1**.86
deliquescence **1**.86
diamond **1**.90
efflorescence **1**.100
galena **1**.130
hydrate **1**.160
Iceland spar **1**.47
interstitial compounds **1**.171
isomorphism **1**.175
liquid *see* liquid-crystal
in oil refining **6**.265
optical activity **1**.239
phenocryst **1**.256
potassium permanganate **1**.265
recrystallization **1**.277
spar **1**.316
water of crystallization **1**.100, **1**.160, **1**.361
zone refining **6**.155, **6**.391
see also minerals
CS gas **6**.346
CSCE *see* Conference on Security and Co-operation in Europe
ctenophoran **2**.298, **2**.299
Ctesibius **6**.76, *6.93*, **6**.168
Ctesiphon *3.93*
Cuauhtémoc, king of Aztecs **3**.31, **3**.88
Cuba *1.81*, *3.93*, *4.89*, *7.349*
agriculture **7**.232
in COMECON **7**.57
education **7**.185, **7**.265, **7**.277
infant mortality **7**.159
land reform **7**.179
leaders **4**.32, **4**.65, **4**.131, **4**.199
medicine and health care **7**.159
physical geography and geology **1**.81
politics and government **7**.60
wars and defence **4**.32, **4**.324, **4**.368, **7**.136
Cuban Missile crisis *4.90*
cube **9**.51
cubic measurement *9.49*
Cubism *5.119*, **5**.299, **5**.410
France **5**.58, **5**.87, **5**.128, **5**.129, **5**.260, **5**.265, **5**.286, **5**.339, **5**.403, **5**.498
Spain **5**.194, **5**.348
cuchia *2.82*
Cuchulain *9.9*
cuckoo **2**.78, *2.83*, **2**.286
cuckoo bee *2.83*
cuckoo wasp *2.83*
cuckoo-shrike *2.83*
cuckoo-spit **2**.127
cucujo *2.83*
cucumber *2.83*
cud chewing *see* ruminant
cueca *5.119*
cuesta **1**.113
Cugnot, Nicolas-Joseph **4**.225, *6.93*, **6**.234, **6**.332
Cui, César **5**.156
Cullberg, Birgit **5**.14
Culloden, battle of *3.93*
culmination *8.35*
Culpeper, John: Rebellion *3.93*
Culpeper, Nicholas *2.83*, **2**.84
cultivar *6.93*
cultivator *6.93*

Cultural Revolution *4.90*, **4.**285, **7.**191, **7.**255, **7.**267
culture *7.78*
Marxist views of **7.**11
see also art; sociology and society; UNESCO
Cumberland, William Augustus, duke of *3.93–4*, **3.**128, **3.**129
Cumbria **3.**15, **3.**342
cummings, e. e. *5.119–20*, **5.**371
cumulus and cumulonimbus clouds **1.**66, *1.81*, **1.**128
Cunard, Sir Samuel *4.90*, **4.**350
Cundall, Charles **4.**105
cuneiform script *3.94*, **6.**280, **7.**13, **7.**278
Cunningham, Andrew Browne, viscount Cunningham of Hyndhope *4.90*
Cunningham, Merce *5.120*, **5.**300
Cunobelinus, king of Catuvellauni *3.94*, **3.**174
cupellation *6.93*
Cupid *9.9*, **9.**29
cuprite *1.81*
cupro-nickel **1.**11
curare **6.**14, *6.94*
curassow *2.83–4*
Curie, Marie and Pierre *1.81–2*, **1.**106, **1.**178, **1.**258, **1.**263, **1.**274, **6.**218, **6.**292
curing fish **6.**132–3
curium **1.**106
curlew *2.84*, **2.**327
Curragh incident *4.90*
currants **2.**33, *2.84*, **2.**143, **2.**279
currawong **2.**28
currency *7.78*
devaluation **7.**87
exchange rate **7.**112–13
Franc Zone **7.**124
futures markets **7.**128
and IMF *see* Special Drawing Rights
money *see* money
speculator **7.**297
current, electrical *1.102*, **6.**75–6, *6.94*, **6.**114, **6.**374
AC *see* alternating-current
accumulator **6.**1–2, **6.**32
ampere *see* ampere
conduction **1.**71
DC *see* direct-current
electrometer **6.**118, **6.**267
fuse **6.**148, **6.**316
measurement **6.**13, **6.**149, **6.**368
rectifier *see* rectifier
resistor *see* resistance (electrical)
signal **6.**319
see also circuits; electricity
currents, ocean **1.**63, *1.235*
Arctic **1.**18, **1.**235
Atlantic **1.**23, **1.**36, **1.**148, **1.**185, **1.**230, **1.**235
Benguela **1.**36
California **1.**49
Gulf Stream **1.**148
gyre **1.**149
Indian **1.**167, **1.**235
Kamchatka **1.**180
Kuroshio **1.**185
Labrador **1.**185
longshore drift **1.**199
maps of **1.**208, **1.**235
North Atlantic Drift **1.**230
Pacific **1.**49, **1.**180, **1.**185, **1.**235, **1.**255
Peru **1.**255
rip current **1.**283–4
turbidity **1.**347
curriculum **7.**33, **7.**78
Currier, Nathaniel *5.120*
Cursus Honorum *3.94*
curtain walls **6.**8, **6.**50, **6.**51, *6.94*, **6.**146, **6.**322
Curtin, John Joseph *4.90*
Curtis, Edward S. **4.**176
Curtis, Lionel **4.**106
Curtiss, Glenn **6.**138, **6.**139, **6.**140, **6.**311
curvature *8.35*
curve:
catenary **1.**54
ellipse *see* ellipse
hyperbola **1.**72, **1.**73, **1.**163
least squares method **1.**191
parabola **1.**72, **1.**73, **1.**246–7
Curwen, John **5.**424
Curzon, George Nathaniel, 1st marquis of Kedleston **4.**52, *4.90–1*
cuscus **2.**251
cushion star **2.**324
Cushitic languages *7.78*
custard apple **2.**317–18
Custer, George Armstrong **4.**88, *4.91*, **4.**197, **4.**315, **4.**378
customary law *7.78–9*
customs and excise *3.94*
customs union **7.**58
duties *7.79*
Cuthbert, Saint **3.**258, *9.41*, **9.**41
cuticle *2.84*
cutlery **6.**46
cut-off grade *1.82*
cutter (ship) *6.94*
cutting tools **6.**46, **6.**191, **6.**357
cuttlefish *2.84*, **2.**85
cutworm *2.84*
Cuvier, G. L. C. F. Dagobert, Baron *2.84*
Cuyp, Aelbert, Benjamin Gerritsz., and Jacob Gerritsz. *5.120*
Cuzco *3.94*
CVD (chemical vapour deposition) **6.**370
CVT (continuously variable transmission) **6.**26
cwm *see* cirque
cyanide *1.82*
cyanobacteria (blue-green algae) **2.**6, *2.36*, *2.84*, **2.**229–30, **2.**361
fossil **2.**240
Cyaxares, king of Medes **3.**227
Cybele *9.9*
cybernetics *6.94*, **6.**385
see also artificial intelligence
cycad *2.84*
Cyclades *1.82*
cyclamen *2.85*
cycle of erosion *1.82–3*, **1.**84
peneplain **1.**250
see also deposition; transportation
cycle (vehicle):
motor **6.**77, **6.**94, **6.**234–5
pedal *see* bicycle
cyclohexane **1.**71
cyclone *1.83*, **1.**344, **1.**345
Cyclops (mythology) *9.9*
Cyclops Project (astronomy) *8.35*
cyclotron *see* accelerator *under* particle(s)
61 Cygni (star) *8.35*
Cygnus (constellation) *8.35*
Cygnus A (radio galaxy) *8.35*, **8.**36
Cygnus Loop (supernova remnant) *8.35–6*
Cygnus X-1 (binary star) *8.36*
cylinder snake *2.85*
cymbals *5.120*
Cymbeline *see* Cunobelinus
Cynics *3.94–5*, **3.**104
cypress *2.85–6*
Cyprus *1.83*, *3.95*, *4.91*, *7.349*
in Council of Europe **7.**71
EOKA **4.**112–13
leaders **4.**113, **4.**207
physical geography and geology **1.**83
religion/theology **7.**98, **7.**193, **9.**40, **9.**42, **9.**43
trade **3.**50
Cyprus (*cont.*):
wars and defence **7.**231
Cypselus **3.**87
Cyrano de Bergerac, Savinien *5.120*
Cyrenaica **4.**306
Cyril, St *3.95*, **8.**288
Cyrillic alphabet **7.**11, **7.**108, **7.**269, **7.**288, **7.**323, *9.46*
Cyrus II (the Great), king of Persia **3.**2, **3.**32, **3.**39, **3.**67, *3.95*, **3.**230, **3.**386–7
and Jews **3.**121, **3.**267
Cyrus the Younger (king of Persia) *3.95*
cyst *2.86*
cystic fibrosis *2.86*
cytology *2.86*
cytoplasm *2.86*
cytosine **2.**95
cytotoxic drugs **6.**57, **6.**292
Czartoryski, Adam Jerzy, Prince *4.91*, **4.**271
Czech Republic *1.83*, *7.349–50*
arts:
art nouveau **5.**307
engraving and engravers **5.**210
graphic art and design **5.**250, **5.**307
painting **5.**210, **5.**250, **5.**307, **5.**410
see also literature; music *below*
astronomy **8.**80–1
in COMECON **7.**57
foreign affairs **4.**197, **4.**227, **4.**288–9, **4.**371
languages **7.**156, **7.**288
literature *5.120*, **5.**210, **5.**422
see also under drama; novels; poetry; stories
music **5.**140, **5.**233, **5.**284, **5.**422, **5.**439
physical geography and geology **1.**83, **1.**101, **1.**236, **1.**325
politics and government:
consociationalism **7.**67
dissidents **7.**99
federalism **7.**119
leaders **4.**34, **4.**70, **4.**104, **4.**139, **4.**213, **4.**256
legislature **7.**182
regions **3.**43, **3.**240, **4.**43, **4.**332
religion/theology **7.**98
sport and recreation **7.**145, **7.**297
technology and science **6.**272, **6.**301
wars and defence **7.**202
invaded by USSR **7.**6
revolution **7.**267
in Warsaw Pact **7.**328
see also Bohemia; Czechoslovakia; Eastern Europe; Moravia; Slovakia
Czechoslovakia *4.92 see also* Czech Republic; Slovakia
Czolgosz, Leon **4.**204

D

Da Gama, Vasco *1.83*, **1.**90, **3.**95–6
da Vinci *see* Leonardo
dab *2.86*
dabchick *2.86*, **2.**144
dace *2.86*
Dacian wars *3.95*
Dacron **6.**384
dactyl *5.121*
Dada *5.121*, **5.**137, **5.**148, **5.**298, **5.**299, **5.**318, **5.**410, **5.**440
Germany **5.**195, **5.**407
see also collage; montage; Surrealism
Daddah, Moktar Ould **4.**214
daddy-long-legs **2.**80, **2.**151
Daedalus *9.9*
daffodil *2.86*, **2.**316
Dagda **9.**9
dagger **6.**226
Daguerre, Louis Jacques Mandé **4.**81, *5.121*, *6.94*, **6.**241, **6.**268, **8.**119
daguerrotype **6.**209
Dahl, Roald **5.**91
Dahl, Robert **7.**244
dahlia *2.86*
Dahomey *see* Benin
daily life **7.**109, **7.**133
Daimbert, Patriarch of Jerusalem **3.**34
Daimler, Gottlieb **4.**225, **4.**350, *6.94*, **6.**209, **6.**234
daimyo *3.96*, *4.92*
Dainichi-Nyorai *9.9*
dairy farming and produce *6.94–5*
butter **6.**52, **6.**215
cheese **6.**70
cream-separator **6.**196
see also milk
daisy **2.**86
ād-Dajjāl *9.2*
Dakota (Sioux) *3.96*, **3.**180
Daladier, Édouard **4.**16, *4.92*, **4.**227
Dalai Lama **3.**202, **3.**356, **4.**344, **5.**455, *7.79*, **7.**311, **7.**312
Dalen, Gustav **6.**89
Dalhousie, James Ramsay, 1st Marquess of *4.92*, **4.**162, **4.**313
Dalí, Salvador *5.121*, **5.**440
Dall, An Clarsair **5.**409
Dallapiccola, Luigi *5.121*
Dalton, John **1.**60, *1.83*
Dalwigk, Baron **4.**150
Daly, Mary **7.**332
dam *see* dams
damages *7.79*
Damas, Léon **4.**309
damascening *5.121*
Damdinsürün, Ts. **5.**302
Damocles *9.9*
Dampier, William *1.83*, *3.96*, **4.**23
damp-proof course *6.96*
dams and reservoirs *6.95*, **6.**96, **6.**294–5, *6.297*, **6.**379
damselfly *2.86–7*
damson *2.87*
Dana, James Dwight *1.83*
Dana (goddess) *9.9*
Danaids *9.9*
Danaus, king of Argos **9.**9
Danby, Thomas Osborne, 1st earl of *3.96*
Dance, George *5.121–2*
dance *5.122–4*
Africa **5.**7, **5.**9
allemande **5.**12
arabesque **5.**18

dance (cont.):
  ballet see ballet; ballet and dance companies
  ballroom see ballroom dancing
  big band 5.47
  branle 5.58
  Broadway 5.60
  Buddhist monastic 5.65
  cabaret 5.70
  carole or carol 5.79
  chaconne 5.87
  choreography see choreographers
  classical 5.29, 5.33–4, 5.103, 5.139, 5.324, 5.345
  community 5.108
  companies see ballet and dance companies
  courante 5.115
  cueca 5.119
  dancers see choreographers
  of death 5.124
  direction see choreographers
  East Asian 5.141
  farandole 5.152
  film see music and dance under cinema
  flamenco 5.157
  folk see dance under folk
  galliard 5.168
  gavotte 5.172
  ghost 7.131
  gigue 5.177
  Greece, ancient 5.82
  hornpipe 5.212
  India 5.219–20, 5.244, 5.279, 5.307
  Islam 5.229–30
  Italy 5.46–7
  Japan 5.236
  jazz 5.239
  mazurka 5.288
  medieval 5.122–3
  minuet 5.297–8
  modern see modern dance
  mudrā 5.307
  music-hall 5.310
  notation 5.123–4
  opéra-ballet 5.329
  Pacific Ocean islands 5.325
  pas de deux 5.341
  pavan 5.342
  polka 5.355
  polonaise 5.355–6
  Post-Modern see under Post-Modernism
  Renaissance 5.379
  rumba 5.394
  Russia and USSR 5.285
  samba 5.398
  South and Central America 5.257
    see also ballroom dancing
  South-East Asia 5.426–7
  tango 5.446
  teaching see choreographers
  theorists 5.128, 5.129
  waltz 5.435, 5.482
  West Indies 5.78
  see also ballet
Dance-Holland, Nathaniel 1.74
Danckerts, Thomas 3.302
dandelion 2.87
Daṇḍin 5.401
Dandolo, Enrico 3.96
Danegeld 3.96–7
Danelaw 3.15, 3.97
Daniel 9.10
Daniell, John Frederic: and Daniell cell 1.84, 6.32, 6.96
Danilova, Alexandra 5.30
D'Annunzio, Gabriele 4.92, 5.124
Dante Alighieri 3.301, 5.12, 5.124–5, 5.141, 5.266
Danton, Georges Jacques 3.83, 3.97
Danube River 1.84
Danube School 5.125
  see also Altdorfer; Cranach
danzón 5.394
Daodejing 7.79
Daoism (Taoism) 3.73, 3.349, 7.79–80
  Dao 7.79
  funerals 7.128
  monasticism 3.236, 7.205
  myths and legends 9.13, 9.28, 9.38, 9.39
    deities 9.15, 9.25, 9.33, 9.38, 9.39
    Dragon Kings 9.23
  xīān 9.38
  see also Lǎozi; Zhuangzi
Daoud Khan, Mohammad 4.3
Daphne 2.87
Daphnia 2.87
Darazi, Ismail 7.96
D'Arblay, General 5.67
Darby, Abraham I 3.178, 6.80, 6.96, 6.186
Darby, Abraham III 6.188, 6.385
Darby, John Nelson 7.237
Dard 5.466
Dardanelles 1.84, 4.92–3
Dardic language 7.156
D'Areo, Guittone 5.230
Darío, Rubén 5.125
Darius I (the Great), king of Persia 3.97, 3.276, 3.316, 3.320, 9.9
Darius III, king of Persia 3.277
Dark Ages 3.97, 3.98
dark matter 8.54
dark nebula 8.106
darkening, limb 8.86
darkness (night) 1.229
Darlan, François 4.93, 4.246, 4.364
Darling, Sir Ralph 4.93
Darling River 1.84
Darnley, Henry Stuart, Lord 3.47, 3.97
Darrow, Clarence Seward 4.93
darter (bird) 2.87
darter (fish) 2.87
darts 7.80
Darwin, Charles Robert 2.87–8, 4.326, 7.80, 7.129
  and altruism 7.11
  and emotion 7.77, 7.80, 7.105
  and ethology 7.109
  and facial expression 7.114
  and sexual selection 2.304
  and teleology 7.307
  as travel writer 5.460
  see also evolution; natural selection
Das, Kashiram 5.43
Das, Tulsi 5.46
Dasam Granth 5.418, 7.80
Dashwood, Sir Francis 3.160
DAT (digital audio tape) 6.103
data (computer) 6.96
  baud rate 6.32
  database 6.96, 6.179
  information retrieval 6.179
  processing 6.84, 6.96
  protection 7.80
  scrambler 6.310
  storage 6.137, 6.183
date palm 2.88
date (time):
  date-line, international 1.37, 1.171
  see also calendar; dating
dating 3.97–8, 6.96
  dendrochronology 6.96, 6.290
  radioactive 8.133
  radio-carbon 6.96, 6.199–200, 6.290
  rocks 1.133, 1.136, 1.274
datum plane 1.84
datura 2.340
Daubigny, Charles-François 5.34
Daudet, Alphonse 5.125
Daughters of the American Revolution 4.93
Daulat Khan 3.209
Daumier, Honoré 5.125
Dauphiné 3.98
D'Avenant, Sir William 5.354
Davenport, James 3.145
David I, king of Scotland 3.98–9, 3.319
David II, king of Scotland 3.40, 3.99, 3.251
David, king of Israel 3.99, 3.187, 3.314, 3.317, 3.364
  and myths 9.2, 9.7, 9.8, 9.20, 9.26
David of Wales, Saint 3.99, 3.376, 9.41, 9.41
David, Antonio 3.291
David, Jacques-Louis 4.231, 5.125–6
Davies, Sir Peter Maxwell 5.126
Davies, Robertson 5.126
Davies, Samuel 3.145
Davies, Siobhan 5.122, 5.126
Davies, Thomas 3.88
Davis, Jefferson 4.10, 4.33, 4.82, 4.93, 4.146
Davis, John 1.84, 3.123
Davis, Miles 5.126, 5.239
Davis, William Morris 1.82, 1.84
Davisson, Clinton Joseph 1.84
Davitt, Michael 4.93–4, 4.189
Davout, Louis Nicolas, duke of Auerstedt 4.94, 4.173
Davy, Sir Humphry 1.84–5, 1.119, 6.19, 6.96, 6.303, 6.374
Dawes, Charles G.: Dawes Plan 4.94
Dawes, William: Dawes limit 8.36
Dawson, Charles 3.282
Dayan, General 4.316
day/diurnal 8.36, 9.55
  aberration 8.1
  libration 8.85
  motion 8.38
  sidereal 8.36, 9.55
  solar 8.36, 9.55
Day-Lewis, Cecil 5.354
day-lily 2.190
d.c. see direct current
ddl (drug) 6.8
DDT 1.85, 6.96, 6.180, 6.275
de Broglie, Louis: de Broglie waves 1.85, 1.295
de Ferranti, Sebastian Ziani 6.114, 6.127
De Forest, Lee 6.20, 6.98, 6.134, 6.291, 6.316, 6.354
de Gasperi, Alcide 4.94
de Gaulle, Charles André Joseph Marie 4.94–5, 4.123, 4.273
  and Algeria 4.9, 4.114, 4.127, 4.251
  and Canada 4.280
  and EEC 4.114
  and Free French 4.125
  and student revolts 4.331
De Havilland, Sir Geoffrey: and De Havilland aircraft 6.3, 6.38, 6.42, 6.98
de Klerk, Fredrik Willem 4.5, 4.16, 4.95, 4.234, 4.319
de Kooning, Willem 5.127
de la Halle, Adam 5.5
de la Mare, Walter 5.128
De Moivre, Abraham: De Moivre's theorem 1.86
De Morgan, Augustus 1.87
De Morgan, William 5.128–9
de Pisan, Christine 5.100
De Quincey, Thomas 5.129
de Sitter, Willem 8.37
  Einstein–de Sitter universe 8.34
De Soto, Hernando 1.88, 3.73, 3.74, 3.90, 3.102
de Valera, Eamon 4.97–8, 4.166
de Valois, Dame Ninette 5.129, 5.393
De Wint, Peter 5.130
de Witt see Witt, Jan de
Deacon, Henry 6.96–7
dead reckoning see navigation
Dead Sea 1.85
  Scrolls 3.99
deafness see disorders under hearing
Deák, Francis 4.12, 4.23, 4.94
Deakin, Alfred 4.94, 4.377
dealfish 2.284
Dean, John and Charles 6.97
Dean, Silas 3.34
Deane, Emmeline 4.239
death 2.88
  afterlife 7.5–6, 7.140, 7.141, 7.217
  ancestors see ancester worship
  bereavement 7.30
  capital punishment 3.150, 7.40, 7.146
  causes 7.39, 7.59, 7.80, 7.140
  in childbirth 7.45
  of children 7.119, 7.140, 7.153, 7.159
  cremation 7.74–5
  euthanasia 7.111–12, 7.301
  forensic medicine 6.143
  funerals see funeral customs
  intensive-care unit 6.183
  martyr 7.194
  penalty 7.40, 7.146
  from radiation 6.125
  rate see mortality
  reincarnation see reincarnation
  resurrection from 7.266
  squads 7.114
  suicide 7.97, 7.138, 7.275, 7.301–2
  will 7.248, 7.331
  see also mortality; murder
death cap 2.88
Death Valley 1.85
death watch beetle 2.88
death's-head hawk moth 2.152
debentures 7.34, 7.80
Debs, Eugene V. 4.93, 4.94, 4.278, 4.317
debt 7.80–1
  amortization 7.12
  crisis 7.81
  insolvency 7.161
  national 7.211
  see also credit
Deburau, Jean Gaspard 5.296, 5.338
Debussy, Claude 5.10, 5.126–7, 5.218, 5.277, 5.320, 5.366
Déby, Idriss 4.69
Debye, Peter 1.272
decathlon 7.232
Decatur, Stephen 4.94, 4.352
decay see decomposition
Decca navigator 6.97
Deccan 1.85, 3.33, 3.41, 3.100, 3.143, 3.254
Decembrists 4.94
decentralization 7.81
decibel 1.85–6, 6.97, 6.319
deciduous trees and shrubs 2.88, 2.89
  alder 2.6
  ash 2.17
  baobab 2.22
  barberry 2.22
  beech 2.27
  birch 2.30
  blackthorn 2.33–4
  buckthorn 2.44
  buddleia 2.44
  catalpa 2.56
  catkins 2.57
  chestnut 2.62
  dogwood 2.97
  Dutchman's pipe 2.100–1
  elder 2.104–5
  elm 2.106–7
  fig 2.117
  forsythia 2.125
  fruit see fruit plants
  fuschia 2.128
  guelder rose 2.146–7
  hawthorn 2.152–3

deciduous trees and shrubs (*cont.*):
hazel **2**.152, **2**.153
hibiscus **2**.158
hydrangea **2**.167
japonica **2**.174
laburnum **2**.181
larch **2**.183
lilac **2**.190
lime tree **2**.190
magnolias **2**.198, **2**.350
maple **2**.202
mesquite **2**.208
metasequoia **2**.209
mock orange **2**.215
oak **2**.231
plane **2**.257–8
poplar **2**.264
rhododendron **2**.284
Robinia **2**.287
rowan **2**.289
sassafras **2**.294
spindle tree **2**.321
Spiraea **2**.322
sycamore **2**.332
tree of heaven **2**.347
tulip tree **2**.350
tupelo **2**.350
Virginia creeper **2**.356
walnut **2**.359
willow **2**.367–8
wisteria **2**.368
witch-hazel **2**.369
decimal system **6**.1, **6**.178
Decius, Roman emperor **3**.74, **9**.32
Declaration of Breda **3**.49
Declaration of Independence *3.100*
Declarations of Indulgence *3.100*
declination **8**.36
decolonization:
of Africa **4**.4, **4**.83
mandates **7**.191
of South America **4**.43, **4**.324
of South-East Asia **4**.320–1
see also colonialism
decomposition (chemistry):
reaction **1**.59
see also electrolysis
decomposition (decay) *2.88–9*
compost *see* compost and manure
detritivore **2**.91
humus **2**.166
pus **2**.272
putrefaction **2**.272
decompression of diver **6**.97–8
deconstruction **7**.81–2, **7**.243
Decorated style *5.127*
decorative arts:
Art Deco **5**.23
Britain **5**.140, **5**.327
Burma **5**.67
Celtic **5**.84
China **5**.94
chinoiserie **5**.98
Cosmati work **5**.114
damascening **5**.121
Directoire style **5**.131
Egypt **5**.142
fan **5**.152
featherwork **5**.153
Federal style **5**.153
folk art **5**.159
Greece, ancient **5**.190, **5**.192, **5**.193
hornwork **5**.212
Islam **5**.230
Italy **5**.366
ivorywork **5**.231
Japan **5**.317
mother-of-pearl **5**.306
Nepal **5**.316
North American Indians **5**.321
Omega Workshops **5**.327
ormolu **5**.332
ornamental motif **5**.332
Rococo **5**.385
Romanesque **5**.389
decorative arts (*cont.*):
shellwork **5**.416
Stone Age **5**.434
straw-work **5**.436
Sumerian **5**.305
Tibet **5**.455
veneer **5**.470
see also art nouveau; Arts and Crafts; furniture; interior design; ornaments *and* arts *under* glass; metals *and* illustration *under* books
decrepitation *1.86*
Dedekind, Richard **1**.13
deduction *7.82*
deed **7**.73
deep-freezing food **6**.38, *6.97*
see also freezing
deep-sea diving *see under* diving
deep-sea fish *2.89–90*
see also fish *under* oceans and seas
deer *2.90*
antlers **2**.12
caribou **2**.282
display **2**.94
elk *see* elks
fallow **2**.113
marsh **2**.204
moose **2**.218
muntjac **2**.222
musk **2**.222–3
red **2**.279–80
reindeer **2**.282
roe **2**.287
wapiti **2**.360
deer mice **2**.370
deerfly **2**.163
defamation *7.82*
defence (in law court) **7**.3
defence (military) *7.82*
see also armed forces; fortifications; military; wars; weapons *and under* wars and defence *and* weapons *under individual countries*
defence (of organisms):
of animals and plants *see* poisonous; protective coloration
defence mechanisms:
of body *see* immune system
of mind *7.82*
Defence of the Realm Acts *4.94*
deferent **8**.129
deficiency (dietary) diseases *2.90*, **2**.140, **2**.181, **2**.357, **7**.190, **7**.198
deficit financing *7.83*
definite integral **1**.170
deflation (geomorphology) *1.86*
deflation (monetary) **7**.160
Defoe, Daniel **1**.83, **3**.96, **3**.131, *5.127*, **5**.323, **5**.348, **5**.460
defoliant *6.98*
deforestation **6**.143, **6**.161, **7**.82, *7.83*, **7**.349
Degas, Edgar **5**.18, **5**.126, *5.127*
degenerate matter *8.36*
degenerate stars **8**.36
degenerative disorders *7.83*
see also disorders *under* ageing
degrees (latitude and longitude) *1.86*
dehumidifier *6.98*
dehydration *see* drying
Deimos *8.36–7*
Deirdre *9.10*
deities *7.132–3*
Africa **7**.5, **9**.26, **9**.28
agnostics and **7**.7
biographical details 2–39 *passim*
blasphemy towards **7**.33
China **7**.47
Christian **7**.125, **7**.133, **7**.263, **7**.316
covenant with **7**.73
creator gods **9**.3, **9**.4, **9**.6, **9**.7, **9**.19, **9**.26, **9**.28
deities (*cont.*):
existence denied **7**.21
and free will **7**.125
and heaven **7**.140
Hindu **7**.133, **9**.2–3, **9**.13, **9**.15, **9**.23
avatars see *avatars*
major *see* Brahmā; Brahman; Kālī; Shakti; Shiva; Vishnu
Vedic **9**.18, **9**.32, **9**.34, **9**.37
Islam (Allah) **7**.10, **7**.133, **7**.166, **7**.209
Judaism **7**.172, **9**.19
meditation **7**.196
ontological argument **7**.223
polytheism **7**.133
predestination **7**.245
and sin **7**.286
theocracy **7**.309, **7**.310
will revealed *see* prophets
see also under individual religions and countries
Dekker, Thomas *5.127*
Delacroix, Eugène *5.127–8*
Delaunay, Charles **8**.87, **8**.101
Delaunay, Robert *5.128*
Delaunay-Terk, Sonia **5**.128
Delaware Indians *3.101*
Delaware River *1.86*
Delaware (state) **3**.100–1
Delcassé, Théophile *4.95*
delegation **7**.88
delftware **5**.276
Delhi *3.101*
Delian League *3.101*
Delibes, Léo *5.128*
Delilah **9**.31
delinquency **7**.10
deliquescence *1.86*
Delius, Frederick *5.128*
della Casa, Giovanni **5**.231
delocalized electrons *1.86*
Delorme, Philibert *5.128*
Delors, Jacques **4**.114, **7**.111
Delphi and Delphic oracle *3.101*, *9.10*
delphinium *2.90*
Delphinus *8.37*
Delporte, Eugene **8**.4, **8**.28
Delsarte, François *5.128*
deltas **1**.39, *1.86*
delta-wing *6.98*
Delvig, Baron Anton **5**.371
Demachy, Robert *5.128*
demand (economics) **7**.67, *7.83*, **7**.193
supply and **7**.302–3
trade cycle **7**.314
dementia *2.90*, *7.83*, **7**.222
demesne *3.101*
Demeter **9**.9, *9.10*, **9**.29
demilitarized zones **4**.169
deMille, Agnes *5.128*
deMille, Cecil Blount *5.128*, **5**.210
democracy **3**.26, *7.83–4*
consociationalism **7**.67
and economics **7**.276
industrial **7**.158
social *see* social democracy
Democratic Party *4.95*
see also under politics *under* United States
Democritus *1.86*, **7**.21
demodulation **6**.289, **6**.346
demodulator circuit *6.98*
demography *7.84*
demographic transition *7.84*
see also population *and under individual countries*
demoiselle crane *2.90*
demolition *6.98*
demons **9**.14, **9**.23, **9**.31
Demosthenes **3**.3, *3.101*, **5**.190
denaturation *2.90*
dendritic drainage **1**.95
dendrochronology **6**.96, **6**.290
Deneb *8.37*
Denebola *8.37*
Deng Xiaoping *4.96*, **4**.122, **4**.157, **4**.338
denial (psychology) **7**.82
Denikin, Anton Ivanovich *4.96*, **4**.296–7
Denis, Maurice *5.129*
Denmark *1.87*, *3.101–2*, *4.96*, *7.350*
aid from **7**.8
arts **5**.232, **5**.454, **5**.467
see also literature; music *below*
astronomy **8**.11, **8**.87, **8**.162, **8**.178
see also Roemer; Tycho Brahe
in Council of Europe **7**.71
economy and trade **7**.107, **7**.111, **7**.221
see also EC; EEC
explorers **1**.38, **1**.276
languages **7**.131, **7**.156
laws/legal system **7**.255
literature **5**.16, **5**.49, **5**.209–10, **5**.239, **5**.326, **5**.404
medicine and health care **7**.196
music **5**.56, **5**.318, **5**.320, **5**.393
philosophy *see* Kierkegaard
physical geography and geology **1**.87, **1**.127, **1**.179, **1**.306
physicists **1**.41, **1**.236
politics and government **4**.317, **7**.223, **7**.260
religion **9**.40
rulers **3**.193, **7**.97
10th to 15th centuries **3**.49–50, **3**.59–60, **3**.74, **3**.102, **3**.347
17th and 18th centuries **3**.102, **3**.311, **3**.318, **3**.324, **3**.355
social security **4**.357
technology and science **6**.18, **6**.123, **6**.200, **6**.277, **6**.345, **6**.386
see also entries above prefixed **1.** *and* **6.**
wars and defence:
with Britain **3**.96–7, **3**.217, **4**.86, **4**.87
with Sweden **3**.311, **3**.318, **3**.324
Northern War **3**.246, **3**.256–8, **3**.260, **3**.287
see also NATO
Denner, Johann Christoph **5**.103
density *1.87*, *8.37*
measurement **6**.174
dentistry **6**.12, *6.98–9*, **6**.99, **6**.233, *7.84–5*
drill **6**.32, **6**.33, *6.98*, **6**.99, **6**.274
see also teeth
denudation (geology andgeomorphology) *1.87*
deontology **7**.66, **7**.109
deoxyribonucleic acid *see* DNA
dependency (colonial) **7**.57
dependency ratio (population) *7.85*, **7**.222
dependency theory (economics) *7.85*
dependent origination *7.85*, **7**.133
depopulation *7.85*
deportation *7.85*
deposition (geology/geomorphology) **1**.82, *1.87–8*
agglomerate **1**.5
aggradation **1**.5
clastic sediments **1**.43, **1**.64, **1**.72
coastal *see under* coasts
colluvium **1**.68
conglomerate **1**.72
diagenesis **1**.90
drift **1**.95–6
evaporites **1**.80, **1**.81, **1**.116, **1**.149, **1**.150–1
by glaciers *see under* glaciation
and gravity **1**.43, **1**.64, **1**.72
ocean floor **1**.236, **1**.238
phosphates **1**.256

deposition (geology and geomorphology) (*cont.*):
playas **1**.259–60
regolith **1**.280
by rivers *see under* rivers
rocks *see* sedimentary rocks
sand *see* sand
saprolite **1**.293
scree **1**.295–6
silt **1**.3, **1**.197–8, **1**.304–5
by wind *see under* winds
see also cycle of erosion; sedimentary rocks; sediments; soils
depreciation **7**.12, ***7**.85–6*
depression (economics) ***7**.86*, **7**.260
the Great ***4**.96–7*, ***7**.86*, **7**.260
depression (meteorology) **1**.39–40, ***1**.88*, **1**.234, **1**.262
depression and anxiety ***7**.14*, ***7**.86*
learned helplessness **7**.181
manic **7**.191
treatment for **6**.116, **6**.285, **6**.360, **7**.96
see also phobias
Deprit, André **8**.101
depth, measurement of **6**.111, **6**.197
depth-charge ***6**.99*
derailleur gear **6**.35
Derain, André ***5**.129*
Derby, Edward George Geoffrey Smith Stanley, 14th earl of **4**.84, ***4**.97*
deregulation ***7**.86*
derivative ***1**.88*
dermatitis ***2**.90*
dermis **2**.311–12
Dermot McMurrough of Leinster ***3**.102*
Derrida, Jacques **7**.81–2
Derry (Londonderry) **3**.209
dervish ***7**.86*
dance **5**.229–30, **5**.394
Déry, Tibor **5**.214
Deryaguin, B. V. **6**.101
Derzhavin, Gavriil ***5**.129*
Desai, Anita ***5**.129*
Desai, Morarji **4**.83, ***4**.97*, **4**.130, **4**.172
desalination ***6**.99*
Desani, Gavindas Vishnoodas ***5**.129*
Descartes, René ***3**.102*, ***7**.86*
in Académie des Sciences **6**.1
and Cartesian coordinates **1**.52, **1**.53
and dualism **7**.96–7, **7**.202–3, **7**.233, **7**.301
and ideas **7**.186
and knowledge and justification **7**.234
and ontology **7**.223
and Pascal **6**.259
and rainbows **1**.275
and rationalism **7**.259
and reason **7**.259–60
and refraction **1**.308
and Spinoza **7**.297
descending node **8**.108
descent ***7**.86*
see also lineage
Desdega, Peter **6**.52
Desdemona (Uranian satellite) **8**.178
desegregation ***4**.97*
opposed **4**.101
see also under civil rights *under* United States
deserts and arid areas ***1**.88*, **1**.89, ***2**.90*
Africa **1**.31, **1**.179, **1**.271, **1**.289
animals ***2**.90*
Asia **1**.141, **1**.330, **1**.334, **1**.348
Australia **1**.139, **1**.146, **1**.306
China **1**.141, **1**.330, **1**.348
Desert cultures ***3**.102*
desertification **1**.88, **1**.96, ***7**.86*
Egypt **1**.271
India **1**.334
deserts and arid areas (*cont.*):
Iran **1**.146
Israel **1**.225
Middle East **1**.146, **1**.225
Pakistan **1**.334
physical geography and geology:
badlands **1**.29
climate **1**.65, **1**.66
desertification **1**.88, **1**.96, ***7**.86*
dunes **1**.33, **1**.97
erg **1**.112
mesa **1**.211
oasis **1**.234
pediment **1**.249
playa **1**.259–60
reg **1**.280
salt-flats **1**.291
salt-pan **1**.291
sand sea **1**.292
wadi **1**.359
plants ***2**.91*
see also succulent plants
South America **1**.23
United States **1**.85, **1**.145, **1**.313
see also sand
desiccant *see* drying
design ***5**.129*
see also functionalism; graphic art; interior design; set design; typography *and* design *under* industrialization
deskilling ***7**.86–7*
desk-top publishing **6**.85, ***6**.99*, **6**.118
desman **2**.215
Desmarest, Nicholas ***1**.88*
Desmoulins, Camille ***3**.102*
Despenser, Hugh le, earl of Winchester ***3**.102*
Dessalines, Jean Jacques ***4**.97*, **4**.145
destroyer ***6**.99*, **6**.357
Destroyer-Bases Deal ***4**.97*
detective story **5**.100, ***5**.129*, **5**.134, **5**.353–4, **5**.420
détente ***4**.97*, ***7**.87*, **7**.99
detention without trial ***7**.87*
detergent **6**.47, ***6**.99–100*, **6**.147, **6**.261
washing powder **6**.40, **6**.267, **6**.376
determinant ***1**.88*
determinism **7**.125
deterrence **7**.87, **7**.220
detonator ***6**.100*, **6**.131, **6**.246
detritivore ***2**.91*
Dettingen, battle of ***3**.102*
deus ex machina ***5**.129*
deuterium ***1**.88–9*, **6**.167, **6**.245, **6**.246, **6**.369
devaluation ***7**.87*
Devanagari script **7**.11, **7**.157, **7**.305
Devas **9**.3, **9**.10
developed countries *see* industrialized countries
developing, film **6**.100, ***6**.101*, **6**.130, **6**.268
developing countries (Third World) ***7**.87*, ***7**.310–11*
adjustment **7**.2
adoption **7**.3
agriculture **7**.7
demography:
birth rate **7**.32
contraception **7**.115–16
fertility rate **7**.119
life expectancy **7**.6
population growth **7**.6, **7**.226, **7**.241–2
dependency theory **7**.85
diseases and disorders **7**.9
economy and trade:
debt **7**.2
employment **7**.106
see also New International Economic Order; newly industrializing countries
education **7**.101–2, **7**.246, **7**.297
developing countries (Third World) (*cont.*):
emigration **7**.105, **7**.153
imperialism *see* colonialism *or* empire *under individual countries*
industrial relations **7**.53–4
malnutrition **7**.190
Non-Aligned Movement **7**.218, **7**.311
North–South relations **7**.219
politics and government **7**.332
poverty **7**.244
religion/theology **7**.48
technology **6**.7, **6**.18
and UNCTAD **7**.319
urban areas **7**.197, **7**.216, **7**.282, **7**.283, **7**.288, **7**.298, **7**.323, **7**.336
westernization **7**.331
women's movement **7**.332
see also Africa; Asia; Central America; South America *and individual countries*
development ***7**.87–8*
aid *see* aid
economics ***7**.87–8*, **7**.210
see also developing countries
Devereux, Robert *see* Essex, Robert
Devi ***9**.10*
Devi, Sitara **5**.244
deviance ***7**.88*, **7**.179
Devil ***9**.10*
Beelzebub **9**.7
Belial **9**.7
and Faust **9**.12–13
Lucifer **9**.17, ***9**.23*, **9**.32
Satan (and Shaitan/Shayṭān) **7**.141, **9**.17, ***9**.32*
devil ray **2**.202
devil's apple **2**.200
devil's fiddle **5**.65
Devils Tower National Monument ***1**.89*
Devis, Arthur ***5**.129–30*
devolution ***7**.88*
Devolution, War of ***3**.102–3*
Devonian period ***1**.89*, **1**.133, **1**.134
Devoy, John **4**.118
dew ***1**.89–90*
frozen *see* frost
Dewar, James **6**.303
Dewey, George ***4**.98*, **4**.209
Dewey, John ***7**.88–9*, **7**.105, **7**.244, **7**.246
Dewey decimal classification **7**.184
dharma ***7**.89*
Dhlakama, Alfonso **4**.225
dhole ***2**.91*
dhow ***6**.101*
Dhū'l Qarnayn *see* Alexander the Great
Dhū-l-Kifl **9**.12
diabase **1**.93
diabetes mellitus **2**.22, ***2**.91*, **2**.283, **6**.29, **6**.170
diagenesis ***1**.90*
Diaghilev, Serge **5**.33, ***5**.130*, **5**.320, **5**.364
dialect **7**.1, ***7**.89*, **7**.90
accent **7**.1
diglossia **7**.90
see also language
dialectic ***7**.89*, **7**.141, **7**.294
dialectical materialism ***7**.89*, **7**.194, **7**.195
dialysis ***6**.101*
diameter, angular **8**.5
diamond **1**.51, ***1**.90*, ***6**.101*
as abrasive **6**.1, **6**.58, **6**.68
mining **1**.90, **1**.182, **1**.183
as polymer **6**.275
diamond ring effect ***8**.37*
Diana ***9**.10*
Diana monkey ***2**.91–2*
Diane de Poitiers, duchesse de Valentinois ***3**.103*
dianthus **2**.255
diaphragm (body) **2**.1, **2**.195
diaphragm (contraception) **2**.75, **6**.87
diaphragm (photography) **6**.16
diaphragm pump **6**.286
diarists **5**.150, **5**.341, **5**.343, **5**.349
diarrhoea **2**.101
Dias, Diego **3**.215
diaspora ***3**.103*
diastase **6**.344
diathermy ***6**.101*, **6**.251
diatomic molecule ***1**.90*
diatoms **2**.232–3
diatonic scale **5**.404
Diaz, Narcisse-Virgile **5**.34
Díaz, Porfirio **4**.64, ***4**.98*, **4**.205, **4**.219, **4**.220, **4**.369
Diaz de Novaes, Bartholomew **1**.83, ***1**.90*, **3**.60, **3**.95, ***3**.103*, **3**.122, **3**.301, **6**.58
diazepam **6**.360
diazo printing ***6**.101–2*
dibber **6**.6
dice ***7**.89*
Dickens, Charles **5**.48, ***5**.130*, **5**.259
Dickinson, Emily ***5**.130–1*
Dickinson, Lowes **4**.230
Dickson, J. T. **6**.384
Dicotyledons ***2**.92*, **2**.123
see also deciduous trees and shrubs; forests and trees; perennial plants; shrubs
dictatorship ***7**.89–90*
dictionaries **5**.2, **5**.241, **7**.107, **7**.196
didactic literature **5**.90
Diderot, Denis ***3**.103*, **3**.115, **3**.300, ***5**.131*, **5**.359, ***6**.102*, **6**.346
didjeridu ***5**.131*
Dido (goddess) ***9**.10*
die-casting ***6**.102*
see also extrusion; forging *and* wire-drawing *under* wire
Diefenbaker, John George **4**.85, ***4**.98*
Diego, Gerardo **5**.173
dielectric heating **6**.101, **6**.224, **6**.251, **6**.290
Dienbienphu ***4**.98*, **4**.99
Dieppe raid ***4**.98*
Diesel, Rudolf ***6**.102*
diesel:
engines **6**.33, ***6**.102*, **6**.263, **6**.339
locomotives ***6**.102*, **6**.150, **6**.294
oil (gas oil) **6**.91, ***6**.102–3*, **6**.145, **6**.249, **6**.264
diet ***2**.92*
see also food
Diet (meeting) ***3**.103*, ***4**.99*
Dietrich, Marlene **5**.432
differential equation (algebra) ***1**.90*
differential gear (motor) ***6**.103*
differentiation (cells) ***2**.92*
differentiation (mathematics) ***1**.90*, **1**.91
derivative **1**.88
diffraction ***1**.90–1*, **6**.381, ***8**.37*, **8**.37
corona **1**.77
grating ***6**.103*, **6**.328
pattern **8**.37
X-rays **1**.43, **1**.90, **2**.158–9
diffuse nebulae **8**.106
diffusion ***1**.91*
osmosis **2**.237
Digby, Sir Everard **3**.151
Digby, Sir Kenelm ***3**.103*
digestion ***2**.92*, **6**.140
see also gastro-intestinal tract
Digger (Australia) ***3**.103–4*
digging machines and tools ***6**.103*, **6**.123–4
digital audio tape (DAT) ***6**.103*
digital computer *see* computers
digital proofing ***6**.104*
digitalis **2**.369, **6**.387
digital-to-analog converter **6**.319
digitization **6**.93, ***6**.104*, **6**.285, **6**.326, **6**.346, **6**.347

diglossia *7.90*
diglyceride 6.141
dike (embankment) 6.79
dike (geology) 1.97, 1.166
dikkop 2.327
dill *2.92–3*
Dilmun 9.11
dimension *1.91*
dimethyl butadiene 6.303
diminishing returns, law of *7.90*
Dimitrov, Georgi *4.99*
Dimna *9.21–2*
ad-Din, Rashad 3.139
ad-Din Yazdi, Shaval 3.348
Dingaan, king of Zulus 4.42, *4.99*
dinghy 6.67, *6.104*, 6.308, 6.317
Dingiswayo, king of Zulus *4.99*, 4.310
dingo *2.93*
Dinizulu, chief 4.391
dinosaurs *2.93*
  brontosaurus 2.43
  iguanodon 2.169
  megalosaurus 2.206
  triceratops 2.347
  tyrannosaurus 2.352
Diocletian, Roman emperor 3.55, 3.86, *3.104*, 3.214
diode *6.104*, *6.313*, 6.354
  light-emitting 6.201, 6.204, 6.314
  in radio receiver 6.291
  see also semiconductor; thermionic valve
Diodotus I, king of Bactria 3.33
Diogenes *3.104*
Dione *8.37–8*
Dionysius Exiguus 8.19
Dionysius I of Syracuse 9.9
Dionysus 9.5, *9.10*, 9.24, 9.25
diorite *1.91*
Dioscorides, Pedanius *2.93*, 2.129
Diouf, Abdou 4.309
dioxides 6.1, 6.156
  see also carbon dioxide
dioxin *1.91–2*, 6.98, *6.104*, 6.314
dip (geological) *1.92*
  dip slope on escarpment 1.113
dip-circle *6.104*
diphtheria *2.93*, 7.153
  vaccine 6.177, 6.191
diplegia 2.242
diplomacy *7.90–1*, 7.105
  diplomatic immunity *7.91*
  gunboat 4.144
  see also foreign policy *under individual countries*
dipole *1.92*, 6.15, *6.104*
Diponegoro *4.99*, 4.173
dipper *2.93*
diptych *5.131*
Dirac, Paul Adrien Maurice *1.92*, 1.120
direct action 7.90, *7.91*
Direct Cinema 5.101
direct-current (d.c.) 1.102, 6.114, 6.115, 6.153, 6.253, 6.323
Directoire style *5.131*
Directory, French *3.104*
dirges 5.7, *5.131*
disability and handicap *7.91*
  institutional care 7.161
  mental 7.198
  services for *7.91*
  special education 7.297
  wheelchair 6.384
disarmament *4.99*, *7.91*
  Conference on 7.64
  Strategic Arms Limitation Talks *4.330*, 6.15, 7.17, 7.87
  see also weapons
disasters *see* accidents and disasters
disc (in spine) 2.56, *2.93–4*
discharge (rivers) *1.92*
disciples of Jesus *3.104*
  see also Apostles
discontinuities in Earth's crust 1.98, 1.217
discount *7.92*
discovery, voyages of *see under* exploration
discrimination and prejudice 7.91, *7.92*, *7.245*
  positive 7.243
  racial *see* racism
  sexism 7.280
  stereotype 7.299
disc-winged bat *2.94*
diseases and disorders of animals *6.217*
  anthrax 2.12, 2.180, 2.243
  BSE 6.127
  chicken cholera 2.243
  lungworm 2.196
  rabies 2.243, 2.274, 2.354
  redwater fever 2.306
  and tsetse fly 2.349
  veterinary medicine 6.371
diseases and disorders of humans *2.94*, *4.99–100*
  abscess 2.1
  Aids *see* Aids
  allergies 2.6–7, 2.153, 2.353
  Alzheimer's 6.377, 7.11
  amoebic dysentery 7.59
  anaemia 2.9, 2.161, 6.164, 6.186
  aneurysm 2.10
  anorexia nervosa 2.11, 7.99
  antisepsis *see* antisepsis
  anxiety *see* depression and anxiety
  appendicitis 2.13
  arteriosclerosis 2.15, 2.77
  arthritis 2.15–16, 6.101
  asbestosis 6.22
  asthma 2.17
  atherosclerosis 2.15, 2.77
  athlete's foot 2.285, 2.342
  bilharzia (schistosomiasis) 2.29
  birth defects *see* inherited disorders *under* genetics
  bladder 6.205, 6.340, 6.342
  blood *see under* blood
  brain *see under* brain
  breast cancer 6.214
  bursitis 2.46
  cancer *see* cancer
  cataract 2.56
  causes *see* pathogens; vectors
  chickenpox 2.63
  cholera *see* cholera
  cold 2.72
  colitis 2.72
  coma 2.73
  communicable *see* infections
  conjunctivitis 2.75
  contagion 2.75
  deafness *see under* hearing
  deficiency *2.90*, 2.140, 2.181, 2.357, 7.190, 7.198
  degenerative *see under* ageing
  dental caries *see* dentistry
  depression *see* depression and anxiety
  dermatitis 2.90
  diabetes 2.22, 2.91, 6.29, 6.170
  diagnosis with antibodies 6.177
  dioxin-caused 1.92, 6.104
  diphtheria 2.93, 6.177, 6.191, 7.153
  disability *see* disability and handicap
  diverticular 2.95–6
  Down's syndrome 6.13
  drugs for *see* drugs, therapeutic
  dysentery 2.101, 7.59
  ear *see under* hearing
  eating 2.11, 7.99
  elephantiasis 2.106
  endocrine system 6.177, 6.266
  epidemics *see* epidemics and epidemiology
  epilepsy 2.109–10, 6.30, 6.116, 6.285, 6.334
diseases and disorders of humans (*cont.*):
  eye *see under* vision and eyes
  fever *see* fevers
  fistula 2.119
  fluid retention 6.104–5
  from fungi 2.129
  gallstones 2.131–2, 2.326
  gastro-intestinal *see under* gastro-intestinal system
  genetic *see* inherited disorders *under* genetics
  genito-urinary system 6.266
  germs 2.126
  haemophilia 2.149, 6.40
  hearing *see under* hearing
  heart *see under* heart
  hepatitis 2.155, 6.87, 6.176
  hernia 2.156–7
  herpes 2.63, 2.157, 2.307
  history of control 4.100
  hormone deficiency 6.29, 6.170
  immunity and immunology *see* immune system
  immunization *see* immunization
  industrial 2.241, 2.276
  infection *see* infections
  infertility 6.127, 6.170, 7.159–60
  inherited *see under* genetics
  injuries *see* injuries
  intestinal *see under* gastro-intestinal tract
  kidneys 2.326, 6.101, 6.164
  kwashiorkor 7.190
  language 7.53
  legionnaire's disease 2.186
  leprosy 2.188
  leptospirosis 2.364
  leukaemia 1.82
  liver *see under* liver
  malaria *see* malaria
  measles 2.206, 7.153
  meningitis 2.121, 2.207, 6.260
  mental *see* mental illness
  migraine 2.210
  morbidity rate 7.207
  multiple sclerosis 2.221–2, 6.241
  musculo-skeletal system 6.104, 6.266
  nausea 2.225
  nervous system *see under* nervous system
  nose 6.254, 6.342
  occupational 6.22, 6.248–9
  onchocerciasis 2.234, 2.286, 6.377
  operations *see* surgery
  osteoarthritis 2.16
  pain 2.239
  Parkinson's 2.242, 6.241, 6.334
  pathogen 2.244
  pathology and pathologists 2.122, 2.244, 2.354, 2.364
  peritonitis 2.250
  plagues 3.42, 3.145–6, 3.284, 6.191
  pneumonia 2.260
  polio 2.262, 6.253, 6.298, 7.153
  psychiatric *see* mental illness
  psychosomatic 7.253
  rabies 2.243, 2.274, 2.354, 6.177, 6.259
  radiation 6.125, 6.249
  resistance to *see* immune system; immunization
  respiratory *see under* respiration
  rheumatism 6.101
  ringworm 2.285
  river blindness 2.234, 2.286
  Rocky Mountain spotted fever 2.287, 2.352
  St Anthony's fire 2.110
  scabies 2.294–5
  scarlet fever 2.296–7
  schistosomiasis 2.29, 6.377, 7.59
  sciatica 2.94
  septicaemia 2.303
diseases and disorders of humans (*cont.*):
  sexually-transmitted *see* sexually-transmitted diseases
  shingles 2.63, 2.307
  shock 2.307–8
  sight *see* vision and eyes
  silicosis 6.248
  skin 2.90, 2.352, 6.104
  smallpox *see* smallpox
  and smoking 7.288–9
  speech 7.15
  spina bifida 6.13
  stones 2.326
  stress 7.300–1
  stroke 2.242, 2.327
  sunstroke 2.330
  surgery *see* surgery
  syphilis *see* syphilis
  tetanus 2.339, 7.153
  throat 6.254
  thrombosis 2.77, 2.340
  thyroid 2.140
  tinea 2.341–2
  trypanosomiasis 2.17
  tuberculosis 2.349, 6.253, 6.375, 7.59, 7.153
  tumours 2.51, 2.350
  typhoid and paratyphoid 2.281, 2.352, 6.192, 6.377
  typhus 2.352
  vaccination *see* immunization
  vectors *see* vectors
  venereal *see* sexually-transmitted diseases
  vision *see under* vision and eyes
  whooping cough 2.367, 7.153
  yellow fever 2.373
  see also drugs, therapeutic; medicine and health care
diseases and disorders of plants:
  Dutch elm disease 2.106–7
  ergot 2.110
  fungi 6.147
  grapes 2.142
  mildew 2.210
  tobacco mosaic virus 2.343
  see also under pests
Disestablishment Acts *4.100*
disguise: camouflage 6.56
dish (astronomy) *8.38*
dishwasher *6.104*
disinfectant *see* antisepsis
disk (computer) 6.137, 6.165
Disney, Walt *5.131*
disorder (medical) *see* diseases and disorders
disorder (thermodynamics): entropy 1.110
display *2.94–5*, 2.119–20
dispute resolution *7.92*
Disraeli, Benjamin, 1st earl of Beaconsfield 4.35, 4.84, 4.97, *4.100–1*, 4.387, *5.131*
dissociation (chemistry) 1.20, *1.92*
Dissolution of the Monasteries 3.236
dissolved load *1.92*
distaff 6.329
distance education 7.68, *7.92–3*
distance in astronomy *8.192*
  light-year 8.86
  modulus *8.38*
  zenith 8.22
distance measurement 6.249, 6.259
distillation 6.11, *6.104*
  mercury produced by 6.219
  molecular 6.231
  steam 6.330
  vacuum 6.210, 6.264, 6.370
  see also fractional distillation
distribution (economics) *7.93*
district heating *6.104*, 6.167, 6.377, 7.106
diuretic *6.104–5*
diurnal *see* day/diurnal
Div *9.10*

Dīvālī 7.93, 7.142
diver (bird) 2.95
diver (human) *see* diving
divergence (air) 1.92–3
divergent series (mathematics) 1.93
diversification 7.93
diverticular disease 2.95–6
divertimento 5.131
Dives 9.22
divide (geomorphology) 1.93, 1.361
divided amplitude interferometer 8.98
dividends 7.93, 7.282
divination 9.17
Divine, Father 4.101
Divine Right of Kings 3.104
diving 7.93
  deep-sea 6.97–8, 6.105, 6.167
    airlock 6.9
    aqualung 6.18, 6.97
    diving-bell 6.97, 6.105, 6.322
    and helium 6.167
    lead weight 1.191
    snorkel 6.323
diving duck *see* bay ducks
divinities *see* deities
division of labour 7.93
  see also alienation; assembly line
divorce 7.93–4, 7.189
Dix, Dorothea Lynde 4.101
Dix, Otto 5.131
Dixiecrat 4.101
Dixieland 5.131
Dixon, Jeremiah 3.224
Djibouti 1.93, 4.4, 4.101–2, 7.15, 7.117, 7.350
Djugurba 9.3
DNA 2.95, 2.96, 6.105
  chromosomes *see* chromosomes
  cloning *see* clones
  denaturation 2.90
  double helix discovered 2.81
  fingerprinting 6.105
  and fingerprints 7.120
  genetics *see* genes; genetics
  genome 2.134
  and heredity 2.156
  mutation 2.223–4
  transgenic organism 2.345
  and viruses 6.8, 6.373
Dnieper River 1.93
Döbereiner, Johann Wolfgang 1.93
Dobrovský Josef 5.120
Dobson fly 2.6
dock (plant) 2.96
dock (port) 6.105
  see also under ports and harbours
Dockstader, Lew 5.469
doctors *see* physicians
documentary 5.132
  film 5.101, 5.131, 5.132, 5.163
  theatre 5.132, 5.351, 5.352
dodder 2.96
dodecahedron 9.51
Dodecanese Islands 1.282
dodo 2.96
Doe, Samuel 4.195
dog *see* dogs
Dog Star (Sirius) 8.150
doge 3.104
dogfish 2.96–7
Dogger Bank 1.93
doggerel 5.132
Dogon 7.94, 9.27
dog-rose 2.288
dogs (canids) 2.96
  bush dog 2.46
  coyote 2.79
  dhole 2.91
  dingo 2.93
  foxes 2.114–15, 2.126
  hunting 2.166
  jackal 2.174
  mythical 9.4, .9.8
  wolf 2.369
dog's-tooth violet 2.97
dogwood 2.97
Dokuchaiev, Vasili Vasiliievich 1.93, 1.205, 1.303, 1.311
Dolby, Ray: Dolby system 6.105, 6.345
doldrums 1.93, 1.171
dole *see* assistance *under* unemployment
dolerite 1.93
Dolin, Anton 5.122
doline 1.93, 1.180, 1.181
Dolland, John 8.137
dollar diplomacy 4.102
Dollard, John 7.7, 7.291
Dollfus, Audouin 8.76
Dollfuss, Engelbert 4.24, 4.102
dolmen *see* megaliths
dolomite 1.93–4, 6.297
  metamorphosed (marble) 1.205
Dolphin (constellation) 8.38
dolphins 2.97, 2.286
Domagk, Gerhard 6.105, 6.338
dome (building) 5.132, 5.166, 6.51, 6.147, 6.155
dome (geology) 1.93, 1.94
  anticline 1.15, 1.125
  batholith 1.34, 1.166
  exfoliation 1.28, 1.94
  salt 1.291
Domenichino 5.132
Domenico Veneziano 5.132
Domesday Book 3.104–5
domestic appliances:
  carpet sweeper 6.61
  cooking *see* cookers and stoves
  dishwasher 6.104
  food processors and mixers 6.142
  heating *see* heating, domestic
  ionizer 6.186
  lighting *see* lighting
  microwave cooker 6.224
  tumble-drier 6.364
  vacuum cleaner 6.369
  washing-machine 6.376
domestic life *see* family; home
domestic violence 7.94, 7.115
domesticated animals 6.15
  birds 2.63, 2.266–7, 2.351
    see also cage-birds
  cane rat 2.52
  cat 2.56
  cattle *see* cattle
  dog 2.96
  donkey and ass 2.17, 2.97, 2.178, 2.221
  ferret 2.115
  genetic resources 6.155
  goat 2.139
  horse *see* horses
  reindeer 2.282
  sheep *see* sheep
  treadmill 6.362
  see also draught and pack; horse-drawn; livestock; pets
Domingo, Placido 5.132
Dominic, Saint 3.7, 3.28, 3.105, 7.205, 9.41, 9.41
Dominica 1.94, 7.350
  see also Caribbean; CARICOM
Dominican friars 3.105, 3.352, 3.359, 7.15, 7.205
Dominican Republic 1.156, 4.45, 4.102, 4.353, 7.350–1
dominion 4.102, 7.58
domino theory 7.94
dominoes 7.94
Domitian, Roman emperor 3.5, 3.95, 3.117
Don Carlos VII, king of Spain 4.62
Don Juan 9.10
Don Pacifico affair 4.102
Donahoe, John 4.102
Donatello 5.61, 5.132–3, 5.377
Donati, Giambattista: Donati's Comet 8.38, 8.194
Donatism 3.105
Dondi, Giovanni de' 6.77
Dongbei 1.204
  see also Manchuria
Doniphan, Colonel 4.219
Donizetti, Gaetano 5.133, 5.441
donkey 2.97
Donkin, Bryan 6.57, 6.105, 6.145
Donne, John 5.108, 5.133, 5.411, 5.424, 5.481
Doolittle, Hilda 5.217
dopamine 6.241
doping 6.312, 6.313
Doppler, Christian: Doppler effect 1.94, 6.105–6, 6.367, 8.38
dor beetle 2.97–8
DORA (Defence of the Realm Acts) 4.94
Dorado 8.38
Doré, Gustave 5.133
Doria, Andrea 3.105, 3.265
Dorians 3.105
dormancy 2.90, 2.135
  hibernation 2.157–8
dormouse 2.98, 2.157
Dorothy, Saint 9.41
Dorst, Tancred 5.175
Dos Passos, John 5.133
dosimeter 6.106
Dost Muhammad of Afghanistan 4.3, 4.13, 4.102
Dostoyevsky, Feyodor 5.80, 5.133–4, 5.177, 5.233, 5.252
Dosunmu, king of Nigeria 4.242
double, optical 8.113
double bass 5.134, 5.352
Double Cluster in Perseus 8.38–9
double star *see* binary stars
double-glazing 6.106
Doughty, Charles Montagu 1.94
Douglas, Clifford 4.102, 4.317
Douglas, Gavin 5.294
Douglas, T. C. 4.238
Douglas fir 2.98
Douglas-Home, Sir Alec 4.103
Douglass, Frederick 4.1, 4.103
Douglass, William 2.98
douroucouli 2.228
dover sole 2.317
doves 2.98, 2.253
  see also pigeon family
Dowding, Hugh Caswall Tremenheere, 1st Baron Dowding 4.103
Dowell, Anthony 5.393
Dowland, John 5.65, 5.134
Downs, Anthony 7.253
Down's syndrome 2.98–9, 6.13
downthrow 1.94
dowry 7.95
Doyle, Sir Arthur Conan 5.134
D'Oyly Carte, Richard 5.439
Drabble, Margaret 5.134
Draco 8.39
Draconian laws 3.105
draconitic month 8.100
Dracula 9.10
drag-line 6.106
Drago, Luis Maria 4.103
Dragon Kings 9.23
dragonet 2.99
dragonfish 2.99
dragon-fly 2.99, 2.206
dragons 9.10, 9.23, 9.30
dragoon 3.105
drain 6.106
drainage:
  basin 1.94–5
  patterns 1.95, 1.342
  storm-water 6.314, 6.335
  watershed 1.93, 1.361
  see also rivers; sewage *and* reclamation *under* land
Drake, Sir Francis 1.95, 1.127, 3.106, 3.122, 3.252, 3.282, 3.337
Drake, Frank 8.115
Drakensberg Mountains 1.95
drama and dramatists 5.452
  Absurd, Theatre of *see* Absurd
  acting 5.3
    see also under country entries below
  Africa 5.7, 5.9, 5.165–6, 5.318, 5.427
  Arabic 5.19
  Australia 5.103
  Austria 5.194, 5.208–9, 5.317, 5.377, 5.405
  Baltic 5.34
  Belgium 5.275
  Bengali 5.43, 5.445
  Britain 5.41, 5.43–4, 5.171, 5.192, 5.286, 5.296, 5.349, 5.434–5
    actors and actresses 3.152, 5.56–7
      comedy 5.101, 5.171
      as directors 5.177, 5.327
      Shakespearean 5.24, 5.66, 5.177, 5.227, 5.245, 5.327, 5.383, 5.450, 5.454
      tragedy 5.171, 5.418
    Angry Young Men 5.17, 5.333
    ballad opera 5.172
    booths 5.53
    comedy and satire 5.36, 5.40, 5.101, 5.127, 5.154, 5.162, 5.171, 5.183, 5.285, 5.295, 5.496
      see also comedy *under* Restoration
    Covent Garden 5.116
    directors and producers 5.61, 5.116, 5.177, 5.197, 5.227, 5.250, 5.327
    Drury Lane 5.136, 5.245
    farce and burlesque 5.154
    fringe 5.165
    Globe 5.144
    historical 5.162
    impresario 5.106
    Irish *see* Ireland *below*
    kitchen-sink 5.249, 5.486
    managers 5.101, 5.116, 5.197, 5.212, 5.227, 5.245, 5.327, 5.452
    miracle 5.294, 5.311
    moralistic 5.366, 5.428
    morality plays 5.294, 5.303–4, 5.421
    mummers *see* mummers' plays
    music-hall 5.309–10
    neoclassical 5.5
    Old Vic 5.327
    pantomime 5.338
    pastoral 5.40
    psychological 5.366
    repertory 5.380
    set design 5.117
    Theatre Workshop 5.452
    tragedy 5.162, 5.285, 5.295
    tragi-comedy 5.285, 5.435
    Wales 5.486
    women 5.41, 5.428
    see also Elizabethan theatre; Restoration comedy; Shakespeare
  Bulgaria 5.74
  Burma 5.67
  China 5.75, 5.94, 5.95, 5.195, 5.256, 5.446
  community 5.108
  Czechoslovak 5.76, 5.120, 5.201, 5.318, 5.468
  Denmark 5.209–10, 5.326, 5.404
  directors *see under country entries above and below*
  documentary 5.132, 5.351, 5.352
  Dravidian 5.136
  epic 5.59, 5.148, 5.351
  epilogue 5.148
  Expressionism 5.151
  Finland 5.165

drama and dramatists (*cont.*):
Formalism 5.162
France 3.103, 5.17–18, 5.73
Absurd, Theatre of 5.225
actors and actresses 5.23, 5.36, 5.45, 5.112, 5.282, 5.372
booths 5.53
classical 5.372
comedy 5.3, 5.40, 5.113, 5.154, 5.301
directors and producers 5.23, 5.45, 5.112
drame bourgeois see *drame bourgeois*
Existentialism 5.173, 5.402
Grand Guignol 5.188
historical 5.303
on love 5.283
music 5.89, 5.128, 5.270
mystery and miracle 5.3, 5.5
négritude 5.86
Romanticism 5.139, 5.214, 5.310–11, 5.475
Surrealism 5.106
Symbolism 5.103, 5.178, 5.238
tragedy 5.113, 5.372
unanimisme 5.386
Germany 5.175
Absurd, Theatre of 5.484
alienation effect 5.59
Berliner Ensemble 5.45, 5.59
comedy 5.249
company 5.45
documentary 5.351
Enlightenment 5.262
epic 5.59, 5.351
Expressionist 5.243, 5.457
Naturalism 5.201
Sturm und Drang 5.181–2, 5.405, 5.438
tragedy 5.202, 5.249, 5.262
Greece, ancient *5.190*, *5.191–2*
chorus 5.100
comedy 5.21, 5.191, 5.290
deus ex machina 5.129
masks 5.284–5
pantomimus 5.296, 5.338
tragedy 5.5–6, 5.150, 5.425, 5.459
happenings 5.46, 5.198, 5.375
harlequinade 5.199
Hungary 5.214
India 5.244, 5.279, 5.314
intermedio 5.224
Ireland 5.40–1, 5.324
comedy 5.183, 5.415, 5.417, 5.431, 5.443, 5.489
Irish Literary Revival 5.442–3, 5.497
Islam 9.17
Italy 4.92, 5.21, 5.140, 5.231, 5.350, 5.447
comedy 5.183, 5.211
commedia dell'arte 5.107–8, 5.159
Japan 5.235, 5.236
Kenya 5.318
Latin 5.15, 5.147, 5.257, 5.296, 5.387, 5.411
lighting, stage 5.264–5
liturgical 5.267
make-up 5.276–7
masks 5.284–5
masque 5.285
melodrama 5.290
method acting *see* Stanislavsky
mime *see* mime
miracle plays 5.294, 5.311
morality plays 5.294, 5.303–4, 5.421
mummers' plays *5.307*
music *see* musicals *and also* theatre music *below*
mystery 5.230, 5.294, 5.311, 5.341
national theatre 5.313–14
Nigeria 5.427
No 5.235, 5.236
Norway 5.48, 5.209–10, 5.216

drama and dramatists (*cont.*):
opéra-ballet 5.329
pageant 5.335, 5.336
pantomime 5.338
passion play 5.341, 9.17
Peru 5.468
Pierrot 5.349
Poland 5.183–4, 5.251–2, 5.320, 5.355, 5.377, 5.491, 5.496
political 5.355
Portugal 5.359, 5.474
producers *see* directors and producers *under country entries above and below*
prologue 5.367–8
Provençal 5.368
puppets *see* puppetry
and religion 5.267, 5.303–4, 5.311, 5.341
repertory 5.380
revue 5.309, 5.381, 5.498
Romanticism 5.139, 5.214, 5.310–11, 5.390, 5.475
Russia and USSR 5.396, 5.462, 5.498
acting 5.430
Bio-Mechanics 5.293
comedy and satire 5.182–3
directors and producers 5.250, 5.293, 5.430–1
Formalism 5.293
Futurism 5.287
Jewish 5.197
post-revolutionary 5.65
psychological 5.90
realistic 5.250, 5.333
Shakespearean 5.370
Sanskrit 5.243
soliloquy 5.424
South Africa 5.165–6
South and Central America 5.257
South-East Asia 5.427
Spain 5.71, 5.86, 5.169, 5.343, 5.428, 5.467, 5.470
stage *see* set design
Sweden 5.44, 5.254, 5.436
Switzerland 5.139–40, 5.165
Thai 5.452
theatre building 5.452, 5.453
theatre music 5.89, 5.128, 5.172, 5.214, 5.270, 5.291, 5.330, 5.344, 5.370, 5.439
Tibet 5.455
United States 5.10, 5.43, 5.262, 5.295, 5.327–8, 5.489
actors and actresses 5.53
directors and producers 5.244
film 5.245
Follies 5.498
Living Theatre 5.267
method acting 5.3, 5.430–1
repertory ('stock') 5.380
showboat 5.417
Southern 5.489–90
'variety' *see* music-hall
Wales 5.486
West Indies 5.481
Yiddish 5.497
see also cinema; comedy; set design; tragedy
dramatic monologue 5.134
drame bourgeois 5.40, 5.131, 5.134
Draper, John William 8.120
Draper classification of spectral classes *8.39*
draught and pack animals *6.106*, 6.165–6, 6.205, 6.255, 6.300
Dravidian:
languages *7.95*
literature *5.134*, *5.136*, 5.232, 5.445–6
see also Mahābhārata and literature *under* Sanskrit
drawbridge *6.106*
drawdown *1.95*

drawing 6.122, 6.259
bistre 5.48
Britain 5.39
cartoon 5.81
chalk 5.87
charcoal 5.89
chiaroscuro 5.90
Egypt 5.141
golden mean 5.183
hatching 5.201
Italy 5.261
Netherlands and Flemish 5.140
pencil 5.343
see also painting and painters *and* illustration *under* books
Dreadnoughts *4.103*, 6.32
Dream of the Red Chamber 5.136
'Dream time' 9.3
dreams *7.95*, 7.126
Dred Scott decision *4.103*
dredger *6.106*, 6.277
Dreikaiserbund 4.344
Dreiser, Theodore *5.136*
Dresden raid *4.103*
Drexler, Anton 4.236
Dreyfus, Alfred *4.103–4*, 4.173
drift chamber (physics) 6.259
drift (geology) *1.95–6*
see also continental drift
Driftless Area *1.96*
drift-mining *6.106–7*
drill (baboon) 2.201, 2.202
drill (invertebrate) *2.99*
drill (vertebrate) *see* mandrill
drilling *6.107*
dental drill 6.32, 6.33, 6.98, 6.99, 6.274
rig (oil) *6.107–9*, 6.263
tools 6.107, 6.211, 6.212, 6.357
ultrasound 6.211
drinker moth 2.104
drinks:
beverage crops 6.7, 6.34, 6.271, 6.364
see also cacao; coffee; tea
fluoridated water 6.139
herbal 2.206
soft 6.324
tonic water 2.66
see also alcohol
Drinkwater, John 5.174
driving *see* motor vehicles; roads
drizzle *1.96*
Drnovsek, Janez 4.316
Drogheda, Siege of 3.91
Drogheda, Statutes of 3.290
dromedary 2.50
drone (music) 5.119, *5.136*, 5.427, 5.490
drongo *2.99*
drop-forging 6.144
drought *1.96*, 7.136
desertification 1.88, 1.96, 7.86
dust bowl 1.97
see also deserts and arid areas
drowning prevention 6.52, 6.200–1
drugs, addictive *2.99*, 2.225
abuse *7.95–6*
and Aids 7.9
alkaloid 2.6
barbiturate 6.30
betel nut 2.240
cannabis 2.52, 2.154
heroin 2.235
hypodermic syringe 6.175
mescalin 2.251, 7.306
offences *7.96*
opium *see* opium and derivatives
from plants *see under* plants
trafficking 7.205–6
capital punishment 7.40
see also coffee; tea; tobacco
drugs, therapeutic *2.99*, 6.72, *6.109*, 6.257
Aids treatment 6.8
alkaloid 2.6

drugs, therapeutic (*cont.*):
anaemia 6.186
anaesthetic *see* anaesthesia
analgesic *see* analgesia
antibiotics *see* antibiotics
antidepressants 6.285
barbiturate 6.30
beta-blocker 6.34, 6.266
cannabis 2.52
chemotherapy 2.104, 6.56, 6.72, 6.112, 6.184
contraceptive pill 6.127, 6.170, 6.334
corticosteroids 6.177, 6.334
cytotoxic 6.57, 6.292
digitalis 6.387
diuretic 6.104–5
Factor IX 6.40
fertility 6.127
heart 6.60, 6.266, 6.387
herbal 6.168, 6.387, 7.46
homoeopathy 6.169–70, 6.257
from hormones 6.127, 6.170, 6.334
and immune system 6.177, 6.184, 6.334
interferon 6.184, 6.188
lumbar puncture 6.210
pain-killers *see* analgesia
penicillin *see* penicillin
pharmaceutical text 2.93
pharmaceutics and pharmacology 2.32–3, 6.109, 6.232, *6.265–6*
from plants *see under* plants
psychotropic 6.285, 6.360, 7.14, 7.86, *7.96*, 7.233, 7.275
quinine 2.66
resistance to 6.260, 6.338
steroids 6.126, 6.170, 6.177, 6.334
sulphonamide 6.105, 6.338
thalidomide 6.265
tranquillizer 6.285, 6.360
see also diseases; medicine and health care
Druid *3.106*
drumlin *1.96*, 1.140
Drummond, General 4.371
drums *5.136*
bongos 5.51
chime *5.136*
frame 5.162–3
friction 5.164–5
kettledrum 5.246
mṛdaṃga 5.306
slit 5.422
tabla 5.444
tabor (tambourin) 5.444
talking 5.445
tambourine 5.445
timpani 5.455
waisted 5.481
West Indies 5.78
drupe *2.99*, 2.129
Drury Lane Theatre *5.136*, 5.245
Druze sect *4.104*, *7.96*, 8.96
dry battery 6.31–2
dry dock *6.109*
dry farming *6.109*
dry ice *1.96*
dry rot fungus *2.99*
dry valleys *1.96–7*
coomb 1.75
nullah 1.233
wind-gap 1.367
Dryads 9.10, 9.12, 9.27
dry-cleaning *6.109*, 6.253
Dryden, John 5.12, 5.103, *5.136*
drying:
agent *1.96*
clothes 6.364
food 6.132, 6.141, 6.146, 6.147
oil *6.109*
drypoint *5.136*
DTP *see* desk-top publishing

du Barry, Marie Jeanne, comtesse 3.106
Du Bellay, Joachim 5.136, 5.391
Du Bois, William Edward Burghardt 4.104, 4.257, 4.372
Du Fu 5.137
Dual Monarchy see Austro-Hungarian empire
dualism 7.96–7, 7.202–3, 7.233, 7.301
Dubček, Alexander 4.92, 4.104
Dube, J.W. 4.4
Dublin Bay prawn 2.100
Dubuffet, Jean 5.137
Duccio di Buoninsegna 5.134–5, 5.137, 5.275
Duchamp, Marcel 5.108, 5.137, 5.298, 5.324
ducks 2.100
bay 2.24–5, 2.260, 2.297, 2.298, 2.349
eider 2.104
garganey 2.133
grebes 2.86, 2.144
long-tailed 2.193–4
mallard 2.199
mergansers 2.208, 2.314
pintail 2.255
pochard 2.260, 2.297, 2.349
scoter 2.297
sea 2.104, 2.193, 2.297, 2.298
shelduck 2.306
shoveller 2.307, 2.308
smew 2.314
stifftail 2.325–6
teal 2.337
tufted 2.349
wigeon 2.36
duckweed 2.100
duct flute 5.137, 5.376
ductility 6.220
Dudley, John, duke of Northumberland 3.110, 3.195, 3.258
Dudley, Robert, earl of Leicester 3.114, 3.205
due process in law 7.97
Dufay, Guillaume 5.137
Dufay, Louis 6.109, 6.268
Duff, Michael Madhusudan 5.43
Dufour, Henri 6.311
Dufy, Raoul 5.138
dugong 2.100
dug-out 6.57
duiker 2.100
Dukas, Paul 5.138, 5.181
dukkha 7.97, 7.311
dulcimer 5.138, 5.402, 5.497
Dulles, John Foster 4.104, 4.110, 4.332
Duma 4.104
Dumas, Alexandre (fils) 5.139, 5.471
Dumas, Alexandre (père) 5.139, 5.207
Dumas, Jean Baptiste André 1.97, 6.43
Dumbarton Oaks Conference 4.104
Dumb-bell Nebula 8.39
Dummer, Geoffrey 6.183
Dumont d'Urville, Jules Sébastien César 1.97
Dumouriez, Charles-François 3.141, 3.369
dumper truck 6.209
dumping 7.97
Dunama, king of Kanem-Bornu 3.193
Dunant, Henri 4.284, 7.130
Dunant, Jean-Henri 4.284
Dunbar, William 5.139
Dunbar, battle of 3.106
Duncan I, king of Scotland 3.106, 3.342
Duncan II, king of Scotland 3.106
Duncan, E. 4.250
Duncan, Isadora 5.139
Dundonald, Thomas Cochrane, earl of 4.77
dunes 1.33, 1.97, 1.298
dung beetle 2.100
Dunkirk evacuation 4.104–5
dunlin 2.100
Dunlop, John Boyd 6.109
Dunn, Robert 5.362
dunnock 2.154
Duns Scotus, John 3.106, 3.318, 7.97
Dunstable, John 5.139, 5.332
Dunstan, Saint 3.106, 3.109, 3.110, 9.41
duodenal ulcer 2.352
Dupleix, Joseph-François, marquis de 3.106, 3.109, 3.134
duplicating machine 6.109
Durbar 4.105
Dürer, Albrecht 5.139, 5.368, 5.482, 8.158
Durey, Louis 5.421
Durga 7.142, 7.281, 7.285, 9.10
Durgā-pūjā 7.87, 7.97, 7.142
Durham Report 4.105
durian 2.100
duricrust see hardpan
Durkheim, Emile 7.97
and anomie 7.14
and ritual 7.268
and social psychology 7.292, 7.294
and substance 7.301
and totem 7.313
durmast oak 2.231
Dürrenmatt, Friedrich 5.139–40
Durych, Jaroslav 5.140
Duse, Eleonora 5.140
dust:
astronomy 1.212, 8.39
in atmosphere 1.97
condensation nuclei 1.71
haze 1.152
smog 1.307, 1.308
bowl 1.97
meteoric 1.212
see also greenhouse effect
Dutch see Netherlands
Dutch East India Company 3.106–7
Dutch East Indies 3.107
see also Indonesia
Dutch empire see empire under Netherlands
Dutch Guiana 4.43
Dutch Reformed Church 7.260
Dutch Revolts 3.108, 3.141
Dutchman's pipe 2.100–1
Dutra, Eurico Gaspar 4.105
Dutton, Clarence Edward 1.97
duty 7.109
of care breached (negligence) 7.214
Duvalier, François 4.105–6, 4.145
Duvalier, Jean Claude 4.145
Duwez, Pol 6.220
Dvořák, Antonín 5.140
dwarf fan palm 2.240
dwarf novae 8.21
dwarf stars 8.2, 8.4, 8.39, 8.51
in binary stars 8.2, 8.3, 8.29
brown 8.161
Chandrasekhar limit 8.25
in multiple star 8.21
subdwarf 8.163
variable 8.37
see also red dwarfs; Sun; white dwarfs
dwarfism 6.170
dwarfs 9.10, 9.26, 9.36
DX coding 6.109
dyarchy 4.106
Dyce, William 5.140
Dyck, Sir Anthony Van 3.204, 5.140, 5.359, 5.393, 5.482
Dyer, R. H. 4.11
dyes and dyeing 6.109–10, 6.388
aniline 6.14, 6.112
azo 6.14, 6.27, 6.161
dyes and dyeing (cont.):
batik 5.38
chromatography 1.62
dye plants 2.100, 2.101
henna 2.154
pokeweed 2.261
safflower 2.291
saffron 2.291–2
sumac 2.329
turmeric 2.351
woad 2.369
illustrated 3.128
industry 6.70
mordants 6.12, 6.89, 6.110, 6.232
pioneers 6.161, 6.260
ruthenium red 6.304
selenium 6.312
synthetic 6.14, 6.79, 6.109–10, 6.112, 6.260
Dyke, Willard van 5.488
dyke (geology) 1.97, 1.166
Dylan, Bob 5.160, 5.384
Dymaxion 6.147
dynamic equilibrium 1.97, 8.47
dynamic modelling 6.110, 6.126–7, 6.320
dynamic random-access memory chips 6.183
dynamic time 1.338, 6.356
dynamical parallax 8.116
dynamics 6.138
stellar 8.159
dynamite 6.110, 6.228, 6.242
dynamo 6.153, 6.158
dynamometer 6.110
dynasty 7.97
dynode 6.269
dysentery 2.101, 7.59
dyslexia 7.97
Dyson, Frank Watson 8.193
dysprosium 1.106
Dzerzhinski 4.71

E

Ea 9.11
Eads, James Buchanan 6.111, 6.293
eagle owl 2.101–2
eagles 2.101
bateleur 2.24
golden 2.140
harpy 2.151, 2.277
sea 2.298
Eakins, Thomas 5.14, 5.141
Eanes, Antonio 4.275
earl 3.108
early English style 5.141
early-type star 8.40
early-warning system 6.111
earnings 7.98
see also income; pay
ears 2.70, 2.102, 9.61
see also hearing; sound
Earth 1.98–9, 8.40
age of 8.2
atmosphere see air and atmosphere
axis of 1.28
crust see crust of Earth
Gaia hypothesis 1.130, 7.129
geodesy 1.133
hydrosphere 1.163
interior 8.40
magnetic polarity of 1.202
oblateness of 1.234
orbit 8.17, 8.196
precession of 8.25, 8.127
revolution 1.281–2, 8.139
rotation 1.229, 1.286, 8.141, 8.142
satellite see Moon
from Space 8.41
statistical details 8.196
terrestrial radiation 1.108, 1.109
see also earth sciences
Earth resources satellite 6.111, 6.215, 6.285
earth sciences 1.99
glaciology see glaciation
oceanography 1.208, 1.235–6
see also physical geography under oceans
see also Earth; geochemistry; geology; geomorphology; geophysics; meteorology
earth star (fungus) 2.102
earthenware 5.141, 5.276
terracotta 5.288, 5.384, 5.450
earthing 6.111
earthquakes 1.99
catastrophes 6.343, 7.92
Richter scale 1.282, 1.299
seismograph 6.311–12, 6.382, 6.391
seismology 1.298, 1.299, 8.147
and tectonic plates 1.261
tsunamis 1.123, 1.345–6
earthshine 8.40
earthworm 2.102–3
earwig 2.103
East Africa:
Abushiri Revolt 4.2
aid to 7.8, 7.121
in Arab League 7.15
East African Community 4.106
empire 3.312
evolution 3.120
exploration 1.44–5, 1.46, 1.335, 4.27, 4.56
languages 7.78, 7.217, 7.279
physical geography and geology 1.229
see also Great Rift Valley

East Africa (*cont.*):
  religion/theology **7**.168, **7**.284, **7**.286, **7**.302, **9**.22, **9**.26
  ruler **3**.68
  see also individual countries
East Anglia *3.108*
East Asia:
  Co-Prosperity Sphere **4**.140
  dance *5.141*
  see also China; Japan; Korea
East China Sea **1**.61
East Germany *see* German Democratic Republic
East India Company, Dutch *3.107*
East India Company, English **3**.54, *3.109*, **3**.245, **3**.285, *4.107–8*, **4**.313
East India Company, French **3**.106, **3**.134
East Indies:
  Dutch East Indies **3**.107
  see also Indonesia; Malaysia
East Pakistan **4**.26
  see also Bangladesh
East Timor *1.99*, *7.351*
Easter *7.98*, **7**.134
Easter Island *3.108–9*
Easter Rising *4.106–7*
easterlies *1.99*
eastern elongation **8**.46
Eastern and Central Europe:
  aid from **7**.8
  economy and trade:
    adjustment **7**.3
    COMECON **4**.79, **7**.57
    market socialism **7**.193
    planned **7**.236
    privatization **7**.248
    see also under individual countries
  ethnic groups **7**.109, **7**.137
  family policy **7**.116
  finance **7**.201
  Iron Curtain **4**.167
  land reform **7**.179
  languages **7**.156, **7**.269, **7**.288, **7**.335
  literature **7**.272
    see also under individual countries
  physical geography and geology **1**.53
    see also under individual countries
  politics and government:
    communist **7**.60
    consociationalism **7**.67
    dissidents **7**.99
    élite **7**.104
    federalism **7**.119
    legislature **7**.182
    Ostpolitik **4**.252, **7**.87
    pluralism **7**.223
    police state **7**.238
    protest **7**.239
    sedition against **7**.278
    state **7**.238
    women's suffrage **7**.332
    see also under individual countries
  pollution **7**.2
  religion/theology **3**.103, **7**.98, **7**.138, **7**.335
  social services and security **7**.45
  socialism **7**.11
  sport and recreation **7**.145, **7**.297, **7**.305
  and USSR **4**.159, **7**.6, **7**.99
    see also Warsaw Pact
  wars and defence:
    expenditure **7**.201
    militia **7**.202
    revolution **7**.267
    and USSR **7**.6
    World War II **4**.106, **4**.107
    see also Warsaw Pact *and under individual countries*
  see also East–West relations *and also individual countries*
Eastern Front Campaigns *4.106*, **4**.107
Eastern Orthodox Church *3.263–4*, *4.252*, *7.98–9*
  and arts **5**.216–17
  asceticism **7**.19
  Coptic **3**.87
  creed **7**.74
  cremation **7**.75
  East–West Schism *see* East–West Schism
  Epiphany **7**.48
  iconoclastic controversy **3**.174
  leaders **7**.48
  monasticism **7**.205
  Nestorians **3**.251
  pope **7**.240–1
  sacraments **7**.271
  theology **7**.309
  see also Oriental Orthodox
eastern prime vertical **8**.128
eastings *1.99*
Eastman, George **6**.27, *6.111*, **6**.268
East–West relations **7**.87, *7.99*, **7**.326
East–West Schism *3.109*, **3**.146
eating *see* food
ebb, tidal *1.99–100*
EBCDIC 22
Ebert, Friedrich *4.108*, **4**.373
Eboué, Félix *4.108*
EC *see* European Community
Eça de Queiroz, José Maria **5**.359
eccentric anomaly **8**.6
eccentricity of ellipse **8**.45–6
ECG *see* electrocardiography
echidna *2.103*
echinoderms *2.103*
  brittle star **2**.42
  crown of thorns starfish **2**.82
  feather-star **2**.114
  sanddollar **2**.293
  sea cucumber **2**.298
  sea lily **2**.299
  sea urchin **2**.293, **2**.302
  starfish **2**.82, **2**.324, **2**.325
Echo (mythology) *9.11*, **9**.26
echo (physics) *1.100*
  echocardiography **6**.60
  echo-sounding *6.111–12*, **6**.325
Eckert, J. Presper **6**.85, **6**.217
eclipse *8.40*, *8.42–3*
  corona **8**.29, **8**.42
  year **8**.189
eclipsing binary stars **8**.3, **8**.14, *8.43*
ecliptic **8**.43, *8.44*
  latitude and longitude **8**.23
  obliquity of **8**.43, **8**.127
eclogue *5.141*, **5**.418
Eco, Umberto **5**.363, **7**.243
ecology *2.103*, *7.99–100*
  adaptation **2**.3
  biogeography **2**.30
  biomass **2**.30
  biome **2**.30
  community *see* community
  ecosystem **2**.30, *2.103–4*, **2**.328
  endangered species **7**.106
  environment **2**.109, **7**.107
  forests *see* deforestation; forests and trees
  Green politics **7**.135
  habitat *see* habitat
  inventor of term **2**.149
  niche **2**.3, **2**.227
  population **2**.264–5
  South Pacific Forum **7**.296
  territory **2**.338
  see also animals; conservation; plants; pollution *and also* environment *under individual countries*
econometrics *7.100*
Economic Community of West African States *4.108*
Economic and Monetary Union **7**.103, **7**.111
economics:
  adjustment **7**.2–3
  'affluent society' **7**.4–5
  agricultural **7**.6, **7**.7–8
  aid *see* aid
  balance of payments **7**.25
  barter **7**.27
  business *see* companies and business
  capital *see* capital
  capitalism *see* capitalism
  cartel *see* cartels
  classical *see* classical economics
  collectivism **7**.56–7
  competition *see* competition
  consumption **7**.68
  co-operatives **7**.56–7, **7**.69, **7**.70, **7**.127, **7**.180
  cost *see* cost
  debt *see* debt
  deficit financing **7**.83
  demand *see* demand
  depreciation **7**.85–6
  depression *see* depression (economics)
  devaluation **7**.87
  development **7**.87–8, **7**.210
  diminishing returns, law of **7**.90
  distribution **7**.93
  diversification **7**.93
  earnings **7**.98
  econometrics *see* econometrics
  economic growth **7**.100
  economies of scale **7**.101
  elasticity **7**.103–4
  enterprises *see* companies
  equilibrium **7**.107–8
  externality **7**.113
  factors of production **7**.114
  finance *see* finance
  firms *see* companies
  flotation **7**.121
  forecasting *7.100*
  franchise **7**.123
  free market **7**.125
  freedom **7**.124
  Giffen goods **7**.131–2
  goods *see* goods
  Gresham's Law **7**.136
  gross domestic product **7**.129, **7**.211
  gross national product *see* GNP
  Group of Seven **7**.136
  growth *7.100*
  human capital **7**.149
  income *see* income
  indexation **7**.155
  individualism **7**.157
  industry *see* industrialization
  inflation **7**.160
  informal sector **7**.160
  infrastructure **7**.160
  integration **7**.162
  interdependence **7**.163
  interest *see* interest and interest rate
  intervention **7**.165
  investment *see* investment
  labour *see* labour and employment
  labour theory of value **7**.179
  laissez-faire **3**.202, **7**.139, **7**.179
  land **7**.179
  liberal *see* liberalization *and under* liberalism
  liquidity **7**.185
  long run **7**.187
  marginal analysis **7**.192
  market *see* market
  mercantilism **7**.198
  mixed economy **7**.203
  money and monetary systems *see* currency; finance; monetarism; money
  monopoly **7**.206, **7**.212, **7**.222, **7**.266
  monopsony **7**.206
  multiplier **7**.209
economics (*cont.*):
  neo-classical *see* neo-classical economics
  oligopoly **7**.222
  organic growth **7**.224
  Pareto utility and optimum **7**.228–9
  planned economy **7**.236–7
  positive **7**.243
  post-industrial society **7**.243
  prices *see* prices
  private sector **7**.203, **7**.247
  privatization **7**.247–8
  production *see* production
  profit **7**.249
  Protestant ethic **7**.251, **7**.330
  public sector **7**.203
  quota **7**.257
  recession **7**.260
  reflation **7**.260
  rent **7**.265
  resale price maintenance **7**.266
  resources *see* resources
  retail price index **7**.266
  scarcity **7**.275
  schools of *7.100–1*
    see also classical economics; Keynesianism; mercantilism; monetarism; physiocrats
  speculator **7**.297
  stabilization policies **7**.298
  supply *see* supply
  tax *see* taxation
  Third World *see* developing countries
  trade *see* trade
  utility **7**.324
  value **7**.325
  wealth **7**.330
  see also economy; micro-economics; macro-economics; New International Economic Order
economies of scale *7.101*
economy:
  centralized **7**.43
  and United Nations **4**.49
  see also economics; industrialization; trade *and under individual countries*
ecosystem **2**.103–4
  biome **2**.30
  succession **2**.328
ECOWAS *see* Economic Community of West African States
ECSB (European System of Central Banks) **7**.43
ECSC (European Coal and Steel Community) **7**.99, **7**.103, **7**.110
ECT (electroconvulsive therapy) **6**.116
ectoparasites *see* parasites
ECU (European Currency Unit) **7**.78, **7**.111
Ecuador *1.100*, *3.109*, *4.108*, *7.351*
  Andean Group **7**.13
  economy and trade **4**.12, **4**.320
    see also OPEC
  independence **4**.43
  in OPEC **4**.252
  physical geography and geology **1**.100
    mountains *see* Andes
  political leaders **4**.131
  population **4**.320
Edda *5.141*
Eddington, Sir Arthur Stanley *8.43*, **8**.93, **8**.159
  Eddington limit **8**.71
Eddy, Mary Baker **4**.75, **7**.48
edelweiss *2.104*
Eden, Anthony, 1st Earl of Avon *4.108*
Eden, Fanny **4**.344
Eden, Garden of *9.11*

edentates:
armadillo **2**.14, **2**.138
pangolin **2**.241
sloth **2**.313
Edgar, king of Northumbria, Mercia and England (the Peaceable) (944–75) **3**.15, **3**.97, ***3**.109*, **3**.380
Edgar the Aetheling ***3**.109*
edge tool **6**.191
Edgehill, battle of ***3**.109–10*
Edgeworth, Francis Ysidro **7**.101
Edgeworth, Maria ***5**.141*
Edison, Thomas Alva ***6**.112*, **6**.383
and camera **6**.75
and electricity generation **6**.114, **6**.351
and incandescent lamp **6**.118, **6**.177, **6**.342
and microphone **6**.347
and phonograph **6**.158
and thermionic emission **6**.134, **6**.354
Edmund I, king of England ***3**.110*
Edmund II (Ironside), king of Wessex **3**.60, ***3**.110*
Edo, Treaty of ***4**.108*
EDSAC **6**.84, **6**.85
education ***4**.304*, ***7**.101–3*
access to ***7**.101*
assessment **7**.20
and communal conflict **7**.59
comprehensive **7**.63
continuing (for adults) **7**.68–9
curriculum **7**.33, **7**.78
and development ***7**.101–2*
distance **7**.92–3
failures **7**.113
financing ***7**.102–3*
higher *see* higher; tertiary; universities
independent **7**.154–5
kindergarten **7**.127
literacy **7**.185
medical **2**.9, **2**.37–8, **2**.190, **2**.237, **2**.241, **2**.364
New Learning **3**.252
pre-school **7**.206, **7**.220, **7**.245
primary **7**.102, **7**.180, **7**.206, ***7**.246*
progressive **7**.88
psychology **7**.35
religious **7**.265
scholasticism *see* scholasticism
school system ***4**.304–5*, ***7**.275–6*
secondary *see* secondary education
special **7**.297
student revolts **4**.331–2
Sunday School **3**.298, **4**.333
teaching ***7**.307*
Bloom **7**.33
Bruner **7**.35, **7**.54, **7**.246
Dewey **7**.88–9, **7**.105, **7**.244, **7**.246
Froebel **7**.127
Montessori **7**.206, **7**.246
Steiner **5**.43, ***7**.299*
see also Piaget
and technology **6**.177
tertiary **7**.102
theory **7**.88–9
universities *see* universities
vocational **7**.326
see also UNESCO; universities *and also* knowledge *under* science *and under individual countries*
EDVAC **6**.85
Edward I, king of England ***3**.110*
and Balliol, John **3**.34
and Barons War **3**.120
and castles **3**.62, **3**.142
and Falkirk, battle of **3**.123
and fyrd **3**.136
and law **3**.170
and parliament **3**.234, **3**.270
Edward I, king of England (*cont.*):
and Scots **3**.319
and taxes **3**.94, **3**.377
and Wales **3**.376
wife **3**.113
Edward II, king of England ***3**.110*, **3**.255
and Bannockburn **3**.35
as Prince of Wales **3**.376
and tunnage and poundage **3**.94
wife **3**.181
Edward III, king of England, Ireland, and France ***3**.110*, **3**.255, **3**.310
and archers **3**.20, **3**.90, **3**.154
and Balliol, Edward **3**.34, **3**.154
and code of chivalry **5**.294
and Crécy, battle of **3**.90
and David II **3**.40, **3**.251
and Gascony **3**.137
and Hundred Years War **3**.49, **3**.171
and Order of Garter **9**.14
and Salic Law **3**.314
and wool trade **3**.385
Edward IV, king of England ***3**.110*, **3**.255, **3**.284
and Roses, War of **3**.241, **3**.310, **3**.313, **3**.352, **3**.360
Edward V, king of England ***3**.110*, **3**.255, **3**.284, **3**.303, **3**.310
Edward VI, king of England **3**.13, **3**.36, ***3**.110*, **3**.362
Edward VII, king of Great Britain and Ireland ***4**.108*, **4**.365
Edward VIII, king of Great Britain and Ireland **4**.1, **4**.80, ***4**.108*
Edward the Black prince **3**.90, **3**.110, ***3**.111*, **3**.171–2, **3**.286
Edward the Confessor, Saint, king of England **3**.15, ***3**.111*
Edward the Elder, king of Wessex **3**.15, ***3**.111*, **3**.342
Edward the Martyr, St ***3**.111–12*
Edwards, Jonathan ***3**.112*, **3**.145, **5**.412
Edwin, earl of Mercia **3**.339
Edwin, king of Northumbria **3**.15, **3**.56, **3**.275
EEC *see* European Economic Community
EEG *see* electroencephalography
eel ***2**.104*
conger **2**.74–5
elver **2**.107
Moray **2**.218
sand **2**.293
snipe-eel **2**.315
eelgrass **2**.104
eelworm ***2**.104*
effective temperature ***8**.43*
Effelsberg ***8**.43*
efficiency, technological **6**.81, ***6**.112*, **6**.135, **6**.166, **6**.332
Effingham, Lord Howard of **3**.106
Effiong, General **4**.38
efflorescence ***1**.100*
EFT (electronic funds transfer) ***6**.118*, **6**.179
EFTA *see* European Free Trade Association
egalitarianism ***7**.103*
Egan, John **5**.200
Egbert, king of Wessex ***3**.112*, **3**.380
egg ***2**.104*, **2**.238
eggar moth ***2**.104*
Egmont, Lamoral, count of, prince of Gavre ***3**.112*
Egmont, Mount ***1**.100*
egoism ***7**.103*
egret ***2**.104*, **2**.105
Egypt ***1**.101*, ***3**.112–13*, ***4**.108–9*, ***7**.351*
agriculture **6**.7
aid to **7**.8, **7**.9
in Arab League **7**.15
Egypt (*cont.*):
arts **3**.78, **3**.243, **3**.249, **3**.299, ***5**.141–2*, **5**.278
astronomy **8**.18, ***8**.43–4*
buildings **6**.50, **6**.51, **6**.82, **6**.287
calendar **6**.54
canal *see* Canal *under* Suez
cities *see* urban areas *below*
colonialism **4**.306
cosmetics **6**.91
decolonization **4**.4
dynasties **3**.112, **3**.124, **3**.173, **3**.217–18, **3**.259, ***9**.66*
see also Ptolemy dynasty *and also* rulers *below*
food **6**.86
and Gulf Co-operation Council **7**.137
hieroglyphs **3**.165, **3**.310, **3**.311
Jews captive in **3**.121
land reform **7**.180
languages **7**.30, **7**.78, ***7**.103*
laws/legal system **7**.284
medicine **6**.340
mummy **3**.243
Muslim Brotherhood **7**.166
Nile *see* Nile
Non-Aligned Movement **7**.218
Nubians **3**.259–60
papyrus **3**.269
Pharaohs *see* Pharaohs
physical geography and geology **1**.101, **1**.206, **1**.271
river *see* Nile
politics and government **4**.369, **7**.182
see also under rulers *below*
religion/theology:
Copts **3**.87, **7**.69–70
Islam **7**.284, **7**.302
Maronites **7**.193
myths and ancient religion **3**.194, **9**.27, **9**.29, **9**.34, **9**.40
goddesses **9**.7, **9**.18, **9**.24, **9**.27
gods **9**.4–5, **9**.6, **9**.16–17, **9**.28, **9**.29–30, **9**.31, **9**.32, **9**.35
saints **9**.40, **9**.41, **9**.42
rulers **3**.314
modern kings and political leaders **4**.16–17, **4**.27, **4**.109, **4**.138, **4**.233, **4**.299, **4**.332
see also Farouk; Mehemet Ali; Nasser; Sadat
see also Pharaohs *and also* dynasties *above*
technology and science **6**.37, **6**.57, ***6**.112*, **6**.158, **6**.346, **6**.368
see also entries above and below prefixed **1**. *and* **6**.
tools **6**.107, **6**.357, **6**.358
in United Arab Republic **4**.358–9
urban areas **3**.2, **3**.160, **3**.228–9, **7**.197
Thebes **3**.352, **3**.353, **3**.369
vehicles and transport **6**.322
navigation **6**.201
ships **6**.56, **6**.302, **6**.306, **6**.307, **6**.316, **6**.330, **6**.333, **6**.375
see also Canal *under* Suez
wars and defence:
armour **6**.40
and Britain **4**.32, **4**.340
with Israel **4**.169
leader **4**.215
World War II **4**.7
see also War *under* Suez
water **6**.19
writing and scripts **6**.257, **6**.259
see also hieroglyphs
Ehrlich, Paul ***2**.104*, **4**.99, **6**.72, ***6**.112*, **6**.177
Eichendorff, Joseph Freiherr von ***5**.142*
Eichmann, Adolf ***4**.109*
eider ***2**.104*
eidophor ***6**.112*, **6**.113
Eiffel, Alexandre Gustave ***6**.112*, **6**.201, **6**.293
1812 War *see* War of 1812
Einstein, Albert ***1**.101*, ***8**.44–5*
and astronomy **8**.44–5, **8**.131
see also relativity
and bosons **1**.42
and complementarity **1**.70
and conservation laws **1**.73
Einstein–de Sitter universe **8**.34
einsteinium **1**.106
and Fitzgerald–Lorentz contraction **1**.199
and Hume **7**.234
and Mach **7**.188
and nuclear energy **6**.242
and quantum mechanics and quantum theory **1**.258, **1**.271, **8**.45, **8**.131
and space–time *see* space–time
see also relativity
einsteinium **1**.106
Einthoven, Willem ***6**.112–13*, **6**.218
Eire *see* Ireland
Eisenhower, Dwight David ***4**.109–10*, **4**.287
and arms race **7**.202
and desegregation **4**.197
Doctrine **4**.104, ***4**.109–110*, **4**.332
and domino theory **7**.94
and Dresden raid **4**.103
and McCarthyism **4**.203
and Normandy landings **4**.245
Eisenstein, Sergey ***5**.142*, **5**.143
Eisner, Kurt **4**.32
ejecta ***8**.45*, **8**.67
ejection seat ***6**.113*
Ekelöf, Gunnar **5**.404
Ekwensi, Cyprian **5**.7
El ***9**.11*
El Alamein **4**.7
El Cid ***3**.113*
El Dorado ***3**.113*, ***9**.11*
El Niño–Southern Oscillation event (ENSO) ***1**.108*
El Salvador ***3**.114*, ***4**.111*, ***7**.351*
civil war **7**.52
exploration **1**.47
independence **4**.67
martyr **7**.194
in OCAS **4**.251
physical geography and geology **1**.108
refugees from **7**.261
eland ***2**.104*
Elara ***8**.45*
ELAS ***4**.110*
elastic (textiles) ***6**.113*
elastic collision ***1**.101*
elasticity (economics) ***7**.103–4*, **7**.193
elasticity (physics) ***1**.101*, **6**.113, **6**.330
Hooke's law **1**.157
elastomer ***6**.113*, **6**.271
see also rubber
Elba (island) ***1**.101*
Elbe (river) ***1**.101*
elder ***2**.104–5*
Elder, John Adams **4**.171
Eleanor of Aquitaine **3**.18, ***3**.113*, **3**.137, **3**.161, **5**.368
Eleanor of Castile ***3**.113*
elecampane ***2**.105*
elections *see* electoral system; voting
Elector ***3**.113*
Electoral College ***4**.110*
electoral fraud ***7**.104*
electoral system ***7**.104*
Electoral College *see* Electoral College
fraud **7**.104
majoritarian representation **7**.190
primaries **4**.276, **7**.246

electoral system (*cont.*):
proportional representation **7.**250
referendum **7.**260
reform **4.**70–1, **4.**159
see also suffrage; voting
Electra **9.**2, ***9.**11*
Electra complex **7.**253
electricity/electric ***1.**102*, ***6.**113–14*
accumulator **6.**1–2, **6.**32
alternative sources *see* alternative energy
ammeter **6.**13, **6.**149
aurora **1.**26, **1.**173, **8.**12–13
batteries *see* battery
bus (trolleybus) **6.**363
cable **6.**53
capacitance *see* capacitance
cell *see* cell (electrical)
charge ***1.**101–2*, **6.**368
capacitance **1.**50–1, **1.**119
coulomb **1.**78
dipole **1.**92
ionization **1.**172
see also ions
see also negative charge; positive charge
and chemical industry **6.**70–1
circuits *see* circuits, electrical
coil *see* coil
combined heat and power generation **6.**79, **6.**81, **6.**138, **6.**150
cooker **6.**89
current *see* current, electrical
Daniell cell **6.**32, **6.**96
electric-arc furnace ***6.**113*, **6.**185, **6.**188, **6.**214
engineering ***6.**113*
field ***1.**102*
and lines of force **1.**195, **1.**196
fire *see* heating, domestic
fish ***2.**105*, **2.**106
furnace *see* electric-arc furnace *above*
gain **6.**13, **6.**149
galvanometer **6.**13, **6.**113, **6.**149
generation and supply ***6.**114*, **6.**121
combined-cycle technology **6.**81, **6.**280
gas-turbine **6.**150
generator *see* generator
geothermal *see* geothermal
natural gas **6.**237
nuclear *see* nuclear reactors and power stations
pioneer **6.**127
solar *see* solar power
steam-turbine **6.**333
thermionic converter **6.**353–4
transmission **6.**114
from water *see* hydroelectricity *and also under* energy *under* tides; waves (ocean)
see also power stations
gramophone **6.**158–9
insulator **6.**12, **6.**183
lighting *see* lamps; lighting
in living organisms **2.**105, **2.**106
locomotives ***6.**114–15*, **6.**293–4
and magnetism ***6.**113–14*
see also electromagnetic
measurement **6.**13, **6.**114, **6.**120, **6.**149, **6.**368
meter ***6.**114*
mixer **6.**142
motor ***6.**115*, **6.**153, **6.**334
armature **6.**20
car **6.**147, **6.**234
see also linear motor
piezo-electric effect **1.**258, **6.**1, **6.**77, **6.**116, **6.**175, **6.**269, **6.**383
pioneers **6.**89, **6.**127, **6.**146
see also Edison; Faraday
potential *see* potential (electrical)
pressure transducer **6.**181
refrigeration **6.**297

electricity/electric (*cont.*):
resistance *see* resistance (electrical)
series connection **6.**74, **6.**76
solar *see* solar
static **1.**102, **1.**194
storage **6.**121
symbols **6.**75
telegraphy *see* telegraphy
vehicles ***6.**115*, **6.**116, **6.**147, **6.**363
voltage *see* voltage
wiring **6.**387
see also in particular circuits; current, electrical *and also entries beginning* electro
electrocardiography **6.**112–13, ***6.**116*, **6.**218, **6.**240–1
electroceramics ***6.**116*, **6.**127, **6.**339
electrochemical cell *see* battery
electrochemical series ***1.**102*
electroconvulsive therapy ***6.**116*
electrode **6.**20, **6.**32, ***6.**116*, **6.**147, **6.**152–3
see also anode; cathode; diode; electrolysis; triode
electroencephalography **6.**34, ***6.**116*, **6.**218, **6.**240–1
electrolysis ***1.**102*, **6.**70–1, ***6.**116*
aluminium **6.**12, **6.**139, **6.**164, **6.**168, **6.**221
anodizing **6.**15
barium **6.**30
calcium **6.**53
chlorine **6.**71, **6.**72
cobalt **6.**53
copper **6.**89
electroplating **6.**73, **6.**89, **6.**119–20, **6.**319
fluorine **6.**139
gold **6.**157
lithium **6.**204
magnesium **6.**212
nickel **6.**241
potassium **6.**278
precious metal **6.**157, **6.**253
sodium **6.**323
tin **6.**356
electrolyte ***1.**102–3*, **6.**32, **6.**211, **6.**374
see also battery
electromagnetic radiation and electromagnetism **1.**39, ***1.**103–4*, ***6.**117–18*
absorbed by ozone **6.**254
and astronomy **8.**45
in bicycle dynamo **6.**153
bremsstrahlung **1.**44, **8.**18
coherent **1.**68
diffraction **1.**77, **1.**90–1
Doppler effect **6.**105–6
electromagnet **6.**20, **6.**114, ***6.**116–17*, **6.**158
electromagnetic force ***1.**103*, **1.**103
electromagnetic induction **6.**114, ***6.**117*
and fusion reactor **6.**148
interference **6.**184
see also generation *under* electricity
electroscope **6.**120, **6.**192
field of view **8.**50
gamma *see* gamma-radiation
heat **6.**166
infra-red *see* infra-red
ionizing **1.**172
and ions **6.**186
light *see* light
measurement **6.**291
microwaves **6.**117, **6.**215, **6.**224, **6.**380
photoelectric effects **1.**257
photons **1.**257, **8.**121
pioneer *see* Faraday
radar *see* radar
radio *see* radio
short wave **1.**103, **1.**301, **6.**380, **9.**54
spectra **1.**103, **8.**45

electromagnetic radiation and electromagnetism (*cont.*):
synchotron **8.**168
thermal **1.**273
transmission or reception **6.**15
ultraviolet *see* ultraviolet
vibrations **1.**356
waves *see* wavelengths
X-ray *see* X-rays
see also Doppler; electricity; gamma-radiation; infra-red; interferometry; magnet/magnetism; polarization; Sun/solar; spectroscopy
electromechanical device ***6.**118*
loudspeaker **6.**13, **6.**209, **6.**289
solenoid in **6.**325
switch **6.**75, **6.**136, **6.**296, **6.**297, **6.**342
telephone exchange **6.**347
see also computers *and* motor *under* electricity
electrometer ***6.**118*, **6.**267
electromotive force *see* battery
electron *see* electrons
electronegativity ***1.**104*
electronics/electronic ***6.**118*, ***9.**53*
aircraft **6.**4, **6.**26, **6.**136
amplifier **6.**13–14, **6.**215, **6.**218
banking **6.**85, **6.**93, **6.**118, **6.**179
capacitor **6.**57–8, **9.**53
circuits *see* circuits; integrated circuit
configuration **1.**104
flight recorder **6.**136
flight simulator **6.**136
Fourier analysis **1.**125
funds transfer ***6.**118*, **6.**179
image intensifier **6.**176, **6.**375
mail ***6.**118*, **6.**180
measurement of signal **6.**97
music **5.**12, **5.**44–5, **5.**108, ***5.**142*, **5.**310, **5.**320, **5.**434, **5.**443
page make-up ***6.**118*
scanning ***6.**118*, **6.**164
switch **6.**75, **6.**136, **6.**297, **6.**342, ***9.**53*
symbols and components ***9.**53*
telephone exchange **6.**347
thermography **6.**354, **6.**355
transformer **6.**114, **6.**361, **9.**53
warfare ***6.**118–19*, **6.**375
weighing machine **6.**192
see also computers; electricity; electrons; radio; resistance (electrical); semiconductor; television; thermionic valve; transistor
electrons ***1.**104*, ***6.**23–4*, **6.**25, ***6.**118*
antiparticle **1.**264
band theory **1.**31–2
beam and ions **6.**186
see also electron tube *below*
delocalized **1.**86
electron particle accelerator **6.**258
electron particle detector **6.**259
electron spin **1.**317
electron spin resonance **6.**96, ***6.**119*
electron transport chain ***2.**105*
electron tube ***6.**119*, **6.**269
see also cathode-ray tube; thermionic valve; visual display unit *and also* microscope *below*
electronic configuration **1.**104
electron-volt ***1.**104–5*
electrophile **1.**105
electropositivity **1.**105
energy levels **1.**108–9
exclusion principle **1.**116
gun *see* electron tube *above*
lone pair **1.**199
loss *see* ions
maser **6.**215
in metal **6.**220

electrons (*cont.*):
microscope **6.**65, ***6.**119*, **6.**192, **6.**223, **6.**268
nucleophile **1.**233
see also atoms; electronics
electrophile ***1.**105*
electrophoresis **6.**73
electroplating **6.**73, **6.**89, ***6.**119–20*, **6.**319
electropositivity ***1.**105*
electrorheological fluid ***6.**120*
electroscope ***6.**120*, **6.**192
electrostatics ***1.**105*
generator **6.**153, **6.**258, **6.**370, **6.**385
precipitator ***6.**120*
see also electricity
electrum **6.**12
elegy ***5.**142–3*
element *see* elements
elementary education *see* primary education
elementary particles ***1.**105*
flavour **1.**123, **1.**272
neutrinos **1.**123, **1.**226
spin **1.**317
see also bosons; electrons; hadrons; leptons; neutrons; protons
elements ***1.**105*, ***1.**106–7*, ***6.**23–5*, ***8.**45*
abundance of **8.**45
allotrope **1.**9
artificial **6.**197
atomic number **1.**26
lists of **1.**106–7, **1.**252–3, **9.**55
neutron differences *see* radioactive elements and isotopes
periodic groups **1.**251
periodic law **1.**251
periodic table *see* periodic table
radioactive *see* radioactive elements and isotopes
symbols **1.**38
trace **2.**345
valency **1.**352
see also artificial elements; atoms; electrons; metalloids; metals; non-metallic elements
elephant ***2.**105–6*, **2.**200, **2.**205
elephant grass ***2.**106*
Elephant River **1.**195
elephant shrew ***2.**106*
Elephanta caves ***5.**143*
elephantiasis ***2.**106*
elephant-snout fish ***2.**106*
elf ***9.**11*, **9.**27, **9.**38
Elgar, Sir Edward **5.**58, ***5.**143*, **5.**154, **5.**468
Elgin, James Bruce, earl **3.**2, **4.**28, ***4.**110*, **4.**361
'Elgin Marbles' **5.**190
Eliade, Mircea **7.**210
Eligius, Saint ***9.**41*
Elijah **3.**59, ***3.**114*, ***9.**11*, **9.**20
elimination reaction **1.**60
Eliot, George ***5.**143*
Eliot, Thomas Stearns **5.**100, **5.**134, ***5.**143–4*, **5.**284, **7.**78
Elis, Islwyn Ffowc **5.**486
Elisha **3.**59, **3.**114, ***9.**11*, **9.**20
élite ***7.**104*, **7.**228
aristocracy **7.**16
meritocracy **7.**199
special armed services **7.**296–7
technocracy **7.**307
see also monarchy
Elizabeth I, queen of England and Ireland ***3.**114*, **3.**205, **3.**292, **3.**362
and exploration **3.**106
and Ireland **3.**179, **3.**365
plots against **3.**32, **3.**356
and religion **3.**199
royal progress **3.**89, **3.**293
tutor **3.**24, 24
see also Elizabethan

Elizabeth II, queen of Great Britain and Ireland 4.110–11
Elizabeth, queen mother 4.133
Elizabeth of Hungary, Saint 9.41
Elizabeth Petrovna, Empress of Russia 3.64, 3.114, 3.312, 3.324, 3.360
Elizabethan architecture 5.462
Elizabethan theatre 5.40, 5.66, 5.100, 5.127, 5.144, 5.162, 5.241–2, 5.283, 5.285–6, 5.295, 5.430, 5.484
  see also Shakespeare
Elkington, George Richards and Henry 6.120
elks:
  American 2.360
  giant 2.173
  Irish 2.173
  moose 2.218
  red deer 2.279–80
Ellesmere Island 1.105
Ellias, Jorge Serrano 4.143
Ellice, Edward 4.355
Ellice Islands see Tuvalu
Ellicott, Andrew 8.84
Ellington, Duke 5.144
Elliot, J. W. 6.293
ellipse 1.72, 1.73, 1.105, 8.45–6
  axes 8.45
  eccentricity 8.45–6
ellipsoid 8.46
elliptical galaxy 8.46
Ellis, Havelock 7.146
Ellis Island 4.111
Ellmann, Richard 5.48
Ellora caves 5.144
Ellsworth, Lincoln 1.105
elm 2.106–7
elongation 8.46
Elsa of Brabant 9.23
Elssler, Fanny 5.30
Elton, Geoffrey 4.153
Éluard, Paul 5.145
eluviation 1.108
elver 2.107
elves 9.11, 9.27, 9.38
Elysium 9.11
Elysium Planitia 8.46
Elytis, Odysseus 5.191
emancipation:
  Catholic 3.65, 4.65
  slaves see abolition and emancipation under slavery
  see also civil rights under United States
emancipists (pardoned convicts) 4.111
embargo see sanctions
embassy 7.105
embroidery 5.145
  Africa 5.7
  Byzantine 5.70
  China 5.94
  crewel work 5.118
  sampler 5.398, 5.400
  see also tapestries
embryo 2.107, 2.268
embryology and embryologists 2.20, 2.107, 2.107, 2.113, 2.149, 2.369
Emecheta, Buchi 5.7
emerald 1.108
emergency powers 7.105
Emerson, Peter Henry 5.145
Emerson, Ralph Waldo 5.145
emery 6.1
emigration 4.274–5, 7.105, 7.171
  see also migration (people)
Emin Pasha 4.112
emirates 3.115, 4.128
emission:
  line 8.157
  nebulae 8.106
  spectrum 8.157
Emma, Duchess of Bohemia 3.379
Emma, queen of Northampton 3.59
Emmet, Dan 5.357
Emmet, Robert 4.112
emotion 7.105, 7.169
  conditioned 7.330
  and facial expression 7.114
emotivism 7.105, 7.109
empathy 7.11, 7.268
emperor moths 2.107, 2.310
emperors of Rome (in chronological order)
  listed 9.65
  Julio-Claudian 3.57, 3.94, 3.175, 3.290, 3.307, 3.356
    see also Augustus; Claudius; Nero
  Flavian 3.5, 3.95, 3.117, 3.188
    see also Vespasian
  five Good see Hadrian; Marcus Aurelius; Trajan
  Severi 3.17, 3.308, 3.324
  soldier-emperors 3.29, 3.74, 3.95, 3.374, 9.32
  'Gallic empire' 3.29
  Diocletian and Tetrarchy 3.104
    see also Diocletian; Severus
  Constantine and dynasty 3.86, 3.104, 3.105, 3.192, 5.72
    see also Constantine I
  Valentinian dynasty 3.3, 3.374
  Theodosius and dynasty see Theodosius I
  western 3.7, 3.308, 5.72
Empire style 5.146
empiricism 7.105
  empirical formula 1.218
  logic see logic
  primary and secondary qualities 7.30–1
  substance 7.301
  see also epistemology; Hume; Locke
employment 7.105–6
  see also labour and employment
Empson, Sir William 5.146
Emre, Yunus 5.439, 5.462
EMS see European Monetary System
Ems telegram 4.112
emu (bird) 2.107
EMU see Economic and Monetary Union
emulsion 6.80, 6.120
  adhesives 6.3
  alginate 6.11
  in food 6.140
  paint 6.120, 6.255
  photographic 6.111, 6.129–30, 6.268
  of sulphur 6.146
  and ultrasound 6.367
enamel 5.146, 6.120
  cloisonné 5.70, 5.94, 5.105
  famille rose, famille verte 5.152
  transfer printing 5.459
enanga 5.146
enantiomer 1.108
encaustic painting 5.146, 5.147
Enceladus 8.46
Encke, Johann Franz: Encke's Comet 8.28, 8.46, 8.194
enclosures 3.115, 7.179
Encyclopedists and encylopedias 3.115, 5.131, 5.249, 5.353, 5.370
end flute 5.146
endangered species 2.107, 2.108, 2.241, 7.106
  see also conservation; extinct species
Endara, Guillerino 4.257
endive 2.107–8
Endō Shūsaku 5.146
endocrine glands and system 2.108–9, 6.266
  adrenal 2.4
  hormones 2.162, 2.257, 2.340–1, 6.170
  and pancreas 2.240
endocrine glands and system (cont.):
  pituitary 2.256–7
  thyroid 2.140, 2.340
endometriosis 6.127
endoparasites see parasites
endoscopy 6.37, 6.120, 6.248, 6.251, 6.340
endoskeleton 2.64, 2.109
  see also bone; vertebrates
endothermic reaction 1.3, 1.108
Endymion 9.11
energy 1.108, 6.120–1, 8.46–7
  activation 1.2–3
  alternative see alternative energy
  budget (atmosphere) 1.108, 1.109
  conservation 1.73, 6.217
  electrical see electricity
  electron-volt see under electrons
  equipartition of 1.111–12
  excitation 1.116
  fermentation 2.115
  heat see heat
  internal 1.110, 1.171
  kinetic see kinetic energy
  levels (atomic physics) 1.108–9
  measurement 1.104–5, 1.178
    see also calories
  non-renewable see coal; fossil fuels; natural gas; petroleum
  potential 1.133, 1.265
  renewable 7.106
    see also biofuel; geothermal; solar power and also under tides; waves (ocean); winds
  resources 1.109, 6.121
  storage 6.121, 6.140, 6.286
  transducer 6.13, 6.181, 6.267, 6.360–1
  transmission see transmission
  turbine see turbine
  wind see under winds
  zero-point 1.373
  see also fuel; heat; kinetic energy; mechanical energy; nuclear energy; steam and energy under chemicals
Enescu, George 5.291
Engels, Friedrich 4.80, 4.112, 4.212, 7.53, 7.194, 7.195, 7.291
Enghien, Louis Antoine Henri de Bourbon-Condé, duc d' 3.115, 4.112
engine see engines
engineering 6.19, 6.122, 6.346
  bioengineering 6.36
  ceramics 6.68
  chemical see chemicals
  computer-aided 6.85, 7.63
  drawing 6.122
  electrical see electricity
  Fourier analysis 1.125
  genetic 6.37, 6.77, 6.153–5, 6.218
  human-factors 6.122
  mechatronics 6.179, 6.217
  military see military technology
  systems 6.253, 6.344
  see also civil engineering; engines; environmental engineering; mechanical engineering; motor; structural engineering
engines 6.121–2
  beam-engines 6.32, 6.241, 6.330
  diesel 6.33, 6.102, 6.263, 6.339
  distinct from motors 6.121–2
  gas-turbine see jet engines
  jet see jet engines
  prime mover 6.281
  reciprocating see reciprocating engines
  rotary 6.303
  silencer 6.319
  starter motor 6.330
  steam see steam-engine and steam-turbine under steam power
engines (cont.):
  Stirling 6.335
  triple-expansion 6.263, 6.330, 6.333, 6.363
  see also internal-combustion; locomotives; machinery; motor
England 1.109–10, 3.115
  maps 3.15, 3.116
  see also Britain
England, Church of see Anglican
English Channel 1.110
English Civil War 3.115–16, 7.144, 8.50
  army 3.253
  battles and campaigns 3.109–10, 3.223, 3.246, 3.291
  Cavaliers 3.65
  leaders 3.118, 3.123, 3.180, 3.202, 3.237, 3.260–1, 3.312
  Roundheads 3.118, 3.180, 3.253, 3.311
  Solemn League and Covenant 3.332
English language 7.106–7, 7.156
  accent 7.1
  and BBC 7.28
  as lingua franca 7.185
  literatures 5.146–7
    see also under individual countries
  pidgin 7.235–6
engraving and engravers 5.147, 6.122
  Aboriginal 5.1
  aquatint 5.18
  Britain 5.46, 5.48–9, 5.177, 5.209, 5.255, 5.284, 5.437–8
  Czech 5.210
  drypoint 5.136
  France 5.55
  hatching 5.201
  intaglio see intaglio
  laser 6.195
  mezzotint 5.293
  Netherlands and Flemish 5.49, 5.270
  niello 5.318
  process 6.283
  Sumerian 5.305
  wood-engraving 5.367, 5.492, 5.493
  see also etching; print
ENIAC 6.85
Enigma machine 6.84, 6.366
Enki 9.11
Enkidu 9.14
enlarger 6.100, 6.101
Enlightenment 3.116–17, 3.375, 5.147
  France 5.27, 5.148, 5.302, 5.359, 5.392
    see also Voltaire
  Germany 5.181–2, 5.262, 5.405, 5.489
  Russia 5.396
  Scotland 5.147
  and women's movement 7.332
  Yiddish 5.497
  see also naturalism and realism
Enlil 9.11
Ennius, Quintus 5.147
Enoch 9.11, 9.16, 9.17
enosis 4.112
ENSO (El Niño–Southern Oscillation event) 1.108
Ensor, James 5.147
enstatites 8.26
entente cordiale 4.112
enthalpy 1.110
  decrease see exothermic reactions
entisol 1.110, 1.311
entomology and entomologists 2.24, 2.38, 2.109, 2.113
  see also insects
entrepôt (free port) 7.125
entrepreneur 7.107

entropy 1.110
Enver Pasha 4.112, 4.387
environment 2.109, 7.107
 Conventions on 7.66, 7.106
 and living organisms 2.109
 see also ecology
 UN Programme 7.66, 7.86, 7.321
 see also conservation; ecology; pollution and also environment under individual countries
environmental engineering 6.122
 air-conditioning 6.8, 6.167
 heating see heating, domestic
 lighting see lighting
 sewage see sewage
 water 6.139, 6.378–9
enzymes 2.72, 2.109, 2.292, 6.70, 6.122, 6.140, 6.153, 6.376
Eocene epoch 1.110, 1.133, 1.134
Eohippus 2.109
EOKA 4.112–13
eon see aeons
Epaminondas 3.19, 3.353
EPC (European Political Co-operation) 7.99, 7.111
ephemeral see annual plants
ephemeris and Ephemeris Time 8.47, 8.172
Ephesus 3.117
 Seven Sleepers of 9.32
Ephialtes 3.26, 3.354
Ephraim 9.11
epics (mainly poetry) 5.147–8
 Babylon 5.177
 Britain 5.44, 5.254, 5.294, 5.296, 5.357, 5.428
 Cambodia 5.72
 China 5.483
 Dravidian 5.136
 Germany 5.59, 5.181, 5.249, 5.318, 5.351
 Greece, ancient 5.210
 India 5.134, 5.276, 5.373, 5.401, 5.447, 7.258
 Italy 5.21, 5.51, 5.369, 5.447
 Latin 5.121, 5.147, 5.270, 5.335, 5.477
 Mongolian 5.301–2
 Norse sagas 5.397
 Persia 5.155, 5.345
 Russia and USSR 5.10, 5.396, 5.461, 5.463
 Sanskrit 5.276, 5.373
 Slovak 5.422
 Spain 5.427
 theatre 5.59, 5.148, 5.351, 5.447
 Tibet 5.455
 Turkey 5.462
 Yiddish 5.497
 see also poetry
Epictetus 3.117
Epicurus 3.117
epicycle (astronomy) 8.47
epicyclic gear 6.151
Epidaurus 3.117
epidemics and epidemiology 2.109, 6.32, 6.191, 7.59, 7.107
 epidemiologists 2.121, 2.289, 2.332–3
 see also cholera; plague
epidermis 2.311–12
epidural analgesia 6.248
epigenesis 2.369
epigrams 5.148, 5.278, 5.283
epilepsy 2.109–10, 6.30, 6.116, 6.285, 6.334
epilogue 5.148
Epimenedes 7.228
Epimetheus 8.29, 8.47
Epiphany 7.16, 7.48, 7.107
epiphytes 2.42–3, 2.110, 2.236
Epirus 3.117, 3.296
Episcopal Churches 7.13
epistemology 7.107
 ontology 7.223
 perception 7.232
 a priori 7.15, 7.178, 7.259, 7.312
 rationalism 7.86, 7.259
 scepticism 7.275
 substance 7.301
 see also empiricism
epistolary novel 5.148
epitaph 5.148
epithalamion 5.148
epithet 5.148
epochs, geological 1.110
 Eocene 1.110, 1.133, 1.134
 Holocene 1.133, 1.134, 1.157
 Miocene 1.133, 1.134, 1.216
 Oligocene 1.133, 1.134, 1.237
 Palaeocene 1.134, 1.135, 1.244
 Pleistocene 1.133, 1.134, 1.170
 Pliocene 1.133, 1.134, 1.260
epoxy resins 6.3, 6.122, 6.129, 6.272
Epsom salts 6.122
Epstein, Brian 5.39
Epstein, Sir Jacob 5.148, 5.359
equal pay 7.107
equations:
 algebra 1.90, 1.110
 chemical 1.59
 dimensions 1.91
 of motion (physics) 1.110
 of time 8.47
equator 1.110, 8.47
 celestial 8.22
 equatorial climate 1.65, 1.66
 galactic 8.54
Equatorial Africa 4.49
equatorial coordinate system 8.22
Equatorial Guinea 1.111, 4.4, 4.113, 4.306, 7.124, 7.352
Equatorial language 7.295
equatorial mounting 8.104
Equiano, Olaudah 5.7
equilibrium (economics) 7.107–8, 7.328
equilibrium (geomorphology) 1.97
equilibrium (physics) 1.111, 8.47
equinoctial year 8.189
equinoxes 1.111, 1.297, 8.47
 precession of 1.266
equipartition of energy 1.111–12
equity (finance) 7.33
 see also shares
equity (law) 7.58, 7.108
Equuleus 8.47
ER (electrorheological fluid) 6.120
Érard, Sébastien 5.200, 5.348
eras, geological 1.112, 1.133, 1.134
 see also Cainozoic; Mesozoic; Palaeozoic; Pre-Cambrian
Erasmus, Desiderius 3.117–18, 3.301, 5.148
 as humanist 7.149
 and Aesop 5.6
 and Greek 2.190
 and Latin 5.258
 and printing 3.293
Erastus, Thomas 3.118
Eratosthenes 1.112, 1.133, 6.214, 8.11, 8.47–8, 8.62
erbium 1.106
Erebus, Mount 1.111, 1.112
erg 1.112
ergodic theory 1.112
ergonomics 5.223, 6.122, 7.108
ergot 2.110
Erhard, Ludwig 4.113
Eric X, king of Sweden 3.127
Eric XVI, king of Sweden 3.324
Eric the Red 1.112, 3.118, 3.121, 3.148
Erichsen, Vigilius 3.6
Ericsson, John 6.123, 6.284
Ericsson, Leif 3.118, 3.374
Eridanus 8.48
Erie, Lake, battle of 4.189–90
Erik see Eric
Erikson, Erik 7.214, 7.253
Eritrea 4.113, 4.306–7, 7.117, 7.279
 languages 7.78, 7.279
ERM see Exchange Rate Mechanism under exchange rate
ermine 2.326
Ermine Street 3.118
Ernest Augustus of Hanover 3.157, 4.146
Ernst, Max 5.148, 5.440
Eros (asteroid) 8.48
Eros (mythology) 9.11
erosion 1.112–13
 abrasion 1.1
 badlands 1.29, 1.30
 chalk 1.75, 1.96
 corrasion 1.77
 corrosion 1.77
 creep 1.79
 cycle of see cycle of erosion
 deflation 1.86
 denudation 1.87
 folding and 125
 gully 1.148–9
 by ice see under glaciation
 karren 1.180
 limestone see under limestone
 pediment 1.249
 peneplain 1.250
 residual features:
  butte 1.46
  inselbergs 1.28, 1.169, 1.170
  mesa 1.211
  monadnock 1.219
  spurs 1.317, 1.318, 1.345
  stack 1.318, 1.319
  tor 1.340
 by sea see under coasts
 sheet 1.300
 soil see under soils
 stripped land surface 1.324
 by water see under coasts; rain; rivers
 by wind see under winds
 see also weathering
eroticism see pornography
Ershad, Hosain Mohammad 4.30
eruptive variable see under variable stars
erythrocyte 2.110
ESA see European Space Agency
Esarhaddon, king of Babylon 3.32
Esau 9.11, 9.19
escalator 6.123
escape velocity 8.48
escarpment 1.113, 1.156
eschatology 7.108
Escher, Maurits Corneille 5.148
eschscholzia 2.110
Escrivá de Balaguer, Josemaria 7.224
Eshu 9.11
eskers 1.113, 1.140
Eskimo see Inuit
Eskimo–Aleut languages 7.108
Esnault-Pelterie, Robert 8.10, 8.144
ESO see European Southern Observatory
esparto grass 2.110
Esperanto 7.108
espionage see spying
Espronceda, José de 5.428
Espy, James Pollard 1.113
essays and essayists 5.149
 Argentina 5.53
 Austria 5.60, 5.208–9, 5.310
 Baltic 5.34
 Britain:
  17th century 5.68, 5.116, 5.127, 5.296, 5.481
  18th century 5.5
  19th century 4.203, 5.129, 5.188, 5.202, 5.239, 5.254, 5.341–2
  20th century 5.43, 5.90, 5.162, 5.215, 5.333
 Canada 5.259
 China 5.198, 5.270
essays and essayists (cont.):
 Czechoslovak 5.210
 Denmark 5.239
 France 5.40, 5.73, 5.176–7, 5.256, 5.302, 5.340, 5.343, 5.368, 5.467
 Germany 5.203, 5.279
 Greece, ancient 5.496
 Hungary 5.28
 Ireland 5.431
 Italy 5.82
 Poland 5.296
 Russia and USSR 5.60
 Spain 5.333
 United States:
  19th century 5.145, 5.269, 5.435, 5.453–4, 5.463–4
  20th century 5.29, 5.276, 5.290, 5.295–6, 5.454, 5.490
essence (philosophy) 7.178
essential fatty acids 6.126
essential oil 6.123, 6.260
Essex, Arthur Capel, 1st earl 3.312
Essex, Robert Devereux, 2nd earl (1567–1601) 3.110, 3.118, 3.179
Essex, Robert Devereux, 3rd earl (1591–1646) 3.118
Essex, Thomas Cromwell, earl of see Cromwell, Thomas
Estates-General 3.340
ester 1.113, 6.252
 see also fats, oils, and fatty acids
Esterhazy, Major (spy) 4.104
Esterházy, prince Nikolaus Joseph (music patron) 5.37
Esther 7.256, 9.11–12
Estigarribia, José Félix 4.68, 4.224
Estonia 1.113, 3.208, 4.29–30, 4.113, 7.352
 astronomy 8.112, 8.135
 in Baltic Entente 4.29
 language 7.26, 7.322
 physical geography and geology 1.113
Estsanathlehi 9.12
estuary 1.114
ETA (Basque separatists) 7.28, 7.308
Etaples, Treaty of 3.118
etching 5.149, 6.123
 Britain 5.114, 5.118, 5.338
 chiaroscuro 5.90
 France 5.457
 Italy 5.339–40, 5.350, 5.391
 Netherlands and Flemish 5.378, 5.410
 see also engraving
ethane 1.114, 6.91, 6.237, 6.253, 6.264
ethanoic (acetic) acid 1.114, 6.66, 6.253
ethanol 1.114, 1.213, 6.37, 6.66, 6.123, 6.253, 7.106
 see also alcohol
Ethelbald, king of Mercia 3.15, 3.229
Ethelbert, king of Kent 3.15, 3.28
Ethelred I, king of Wessex 3.118
Ethelred II (the Unready), king of England 3.60, 3.112, 3.118–19
Ethelwold, St 3.109
ethene see ethylene
ether 1.114
 anaesthesia 6.14, 6.218, 6.233, 6.252, 6.320
 hypothetical substance in space 8.48
Etherege, Sir George 5.381
ethics 7.108–9
 Aristotelian 7.16, 7.108–9
 consequentialism 7.66
 hedonism 7.141
 medical see ethics under medicine
 Plato 7.237
 in psychiatry 7.251
 Spinoza 7.297

ethics (*cont.*):
  stoic *see* Stoics
  utilitarianism *see* utilitarianism
  see also morality; virtues
Ethiopia *1.114*, **3**.4, *3.119*, *4.113*, *7.352*
  aid to **7**.8, **7**.121
  decolonization **4**.4
  emperors **4**.145, **4**.216–17
  ethnic groups **7**.114
  famine and drought **7**.117–18, **7**.136
  independence during colonial period **4**.306
  land reform **7**.180
  languages **7**.78, **7**.217, **7**.279
  physical geography and geology **1**.114, **1**.229, **1**.237, **1**.331
  prehistory **3**.152
  refugees **7**.261
  religion/theology **3**.87, **7**.109, **7**.171, **7**.258, **9**.33, **9**.42
  technology **6**.379
  telecommunications **7**.308
  torture **7**.313
  town **3**.30
  wars and defence **4**.206, **7**.52
    Abyssinian Campaign and Italian occupation **4**.2, **4**.3, **4**.4
Ethiopian Orthodox Church *7.109*
ethnic groups *7.109*, **7**.315
  conflict **7**.59
  see also anthropology; race *and under individual countries*
ethnography *7.109*, **7**.112, **7**.190
  see also anthropology
ethnolinguistics *7.109*
ethnomethodology *7.109*
ethology and ethologists **2**.28, *2.110*, **2**.194, **2**.341, **2**.358, *7.109–10*, **7**.154
ethyl alcohol *see* ethanol
ethylene (ethene) *1.114*, **6**.71, *6.123*, **6**.261
  ethylene glycol **1**.141, **6**.16, **6**.123, **6**.261
  ethylene oxide **6**.334
  see also polythene
ethylene (plant hormone) **6**.271
ethyne (acetylene) *1.114*, **6**.76, *6.123*, **6**.297, **6**.383
Etna *1.114*
Etruscans/Etruria **3**.119, **3**.350
  arts *5.149–50*
etymology **7**.110
eucalyptus **2**.110, **2**.148
Eucharist *see* Mass (Eucharist)
Euclid *1.114*, **1**.136
  and axiom **1**.28
  and golden ratio **1**.142
  postulates **1**.247, **1**.264
  and triangle **1**.343
Euctemon **8**.98
Eugène, king of Savoy **3**.43, **3**.119, **3**.338
eugenics **2**.132, *7.110*, **7**.198
Eugénie, Marie de Montijo, Empress of France *4.113*
Euglena *2.111*
eukaryotes *2.111*, **2**.313, **6**.65, **6**.127, **6**.147–8
  see also algae; fungi; protozoa
Eulenspiegel, Till *9.12*
Euler, Leonard *1.114*, **1**.257–8, **8**.25, **8**.87
Eumenes II, king of Pergamum **3**.276
eunuch **3**.119
euphausid shrimp **2**.110, *2.111*
euphonium *5.150*
euphorbia **2**.323
Euphrates *1.114–15*
  see also Mesopotamia
Eurasia *1.115*
  plate tectonics **1**.260
  see also Asia; Europe
Euratom *6.123*, **7**.99. **7**.110, **7**.111
Eureka Rebellion *4.113–14*
Euripides **5**.100, *5.150*, **5**.454, **5**.459
Eurocommunism **7**.61
Eurodollar *7.110*
Europa (asteroid) *8.48*, **8**.193
Europa (mythology) *9.12*
Europe *1.115*, *3.119–20*
  aid from **7**.8
  arts *see under individual countries*
  astrology **7**.20
  astronomy **8**.11, **8**.48
    see also under individual countries
  Concert of **7**.135
  Congress of Berlin **4**.35–6
  Continental System **4**.85
  Council of **4**.87, **7**.71
  Dark Ages **3**.97, *3.98*
  demography:
    birth rate **7**.32
    divorce **7**.94
    fertility rate **7**.119
    infant mortality **7**.159
    life expectancy **7**.184
  Divine Right of Kings **3**.104
  Eastern *see* Eastern Europe
  economy and trade:
    'affluent society' **7**.4
    co-operatives **7**.69
    industrial relations **7**.158, **7**.159, **7**.304
    migrant labour **7**.200
    in OECD **7**.221
    post-industrial **7**.309
    privatization **7**.248
    unemployment **7**.320
    see also EC; EEC *and under individual countries*
  education **7**.155, **7**.245, **7**.277
  emigration **7**.105, **7**.261–2
    see also immigration *under* United States
  energy, renewable **7**.106
  environment **7**.2, **7**.83, **7**.140
  ethnic groups **7**.137
  eugenics **7**.110
  feudalism **3**.125–6, **3**.370
  Grand Tour **3**.144–5
  Holy Leagues **3**.167
  humanism *see* humanism
  immigration **7**.153, **7**.200
  land reform **7**.179
  languages:
    Baltic **7**.26, **7**.156, **7**.157
    Germanic **7**.131, **7**.156, **7**.157
    Romance **7**.108, **7**.269
    Slavic **7**.156, **7**.157, **7**.287–8
    Uralic **7**.322–3
    see also Indo-European *and under individual countries*
  laws/legal system:
    capital punishment **7**.40
    civil **7**.51
    courts **7**.174
    homosexuality **7**.147
    industrial relations **7**.159
    police **7**.237
    see also under European Community *and under individual countries*
  literature *see under individual countries*
  manorialism **3**.220
  maps:
    Black Death **3**.42
    Byzantine empire **3**.55
    Christianity **3**.75
    Dark Ages **3**.98
    Hannibal's conquests **3**.156
    Napoleonic Wars **4**.232
    Ottoman empire **3**.266
    physical geography and geology **1**.115
    population migration **4**.274
    Reformation **3**.301
    Roman empire **3**.307
    Seven Years War **3**.323
Europe (*cont.*):
  maps (*cont.*):
    Utrecht, Peace of **3**.368
    Vienna, Congress of **4**.366
    Viking invasion **3**.373
    Westphalia, Treaty of **3**.381
    World Wars **4**.154, **4**.245, **4**.343, **4**.382, **4**.385
  Marshall Plan **4**.212
  Middle Ages **3**.231
  monarchy **7**.97, **7**.139
    see also individual names
  mortality **7**.207
  parental leave **7**.228
  philosophy *see* continental philosophy *and under individual countries*
  physical geography and geology *1.115*
    climate **1**.65
    continental drift **1**.135
    islands *see under individual countries*
    lakes **1**.54, **1**.73, **1**.132, **1**.176, **1**.185, **1**.301
    map **1**.115
    mountains *see under individual countries*
    North Sea **1**.231
    plate tectonics **1**.260
    rivers *see under individual countries*
    seas **1**.17, **1**.29, **1**.31, **1**.40, **1**.42
      see also Mediterranean
  politics and government:
    cabinet **7**.38
    communist **7**.61
    military **7**.201
    parliamentary **7**.229
    protest **7**.239
    and religion **7**.265
    women's suffrage **7**.332
    see also under individual countries
  pollution **7**.2, **7**.140
  population *see* demography *above*
  prehistory *see under* prehistory
  Reformation **3**.273–4, **3**.300, **3**.301
  refugees **7**.261–2
  religion/theology:
    Jews and Judaism **7**.19
    monasteries **7**.205
    and politics **7**.265
    witchcraft **7**.331
    see also under individual countries
  Renaissance *see* Renaissance
  sport and recreation **7**.34, **7**.76
    hockey **7**.144
    ice-skating **7**.151
    see also under individual countries
  technology and science:
    European Atomic Energy Community **6**.123
    European Laboratory for Particle Physics **6**.117, **6**.123, **6**.258
    telecommunications **7**.258, **7**.307
    see also under individual countries and also entries above and below prefixed **1**. *and* **6**.
  tourism **7**.313
  unions **7**.158, **7**.159, **7**.304
  urban areas **7**.288
    see also under individual countries
  wars and defence:
    Conference on Security and Co-operation **7**.64, **7**.87, **7**.272
    Conventional Forces Treaty **7**.17, **7**.64
    major *see* Hundred Years War; Napoleonic Wars; Seven Years War; Thirty Years War; World War I; World War II
    Paris, Belgrade, and Ottawa conferences **7**.64
    Western European Union **4**.375, **7**.330–1
Europe (*cont.*):
  wars and defence (*cont.*):
    see also NATO *and under individual countries*
  water supply **7**.328
  see also Eastern Europe; European Community *and also individual countries*
European Atomic Energy Community *6.123*, **7**.99, **7**.110, **7**.111
European Coal and Steel Community **4**.114, **4**.305, **7**.99, **7**.103, **7**.110
European Commission **7**.99, *7.110*
European Community (*earlier* European Economic Community) *4.114*, *7.99*, **7**.318
  aid from **7**.9, **7**.121
  Commission **7**.99, **7**.110
  Common Agricultural Policy **7**.103, **7**.124, **7**.314
  Council of Ministers **7**.99, **7**.111, **7**.326
  finance and currency **7**.43, **7**.78, **7**.103, **7**.111
  and Group of Seven **7**.136
  labour **7**.191
  laws/legal system **7**.92, **7**.319
    commercial **7**.57
    competition **7**.62
    consumer **7**.68
    Courts **7**.15, **7**.99, **7**.111, **7**.173
    discrimination *7.92*
    European Parliament **4**.114, **7**.99, **7**.111
    politics and government **7**.99, **7**.111
  list of members *9.75*
  multinationals **7**.209
  political co-operation **7**.99, **7**.111
  Single European Act **7**.99, **7**.103, **7**.111
  technology **6**.64, **6**.123, **6**.139
  and UN Security Council **7**.277
  wars and defence **7**.308, **7**.331
  workings of **7**.100
  and Yaoundé Convention **4**.386
  see also European Economic Community
European Conventions:
  on Human Rights *7.110–11*, **7**.125, **7**.149
  on Status of Children Born out of Wedlock **7**.152
  on Suppression of Terrorism **7**.309
  see also Conventions
European Council of Ministers **7**.99, *7.111*, **7**.326
European Court of Human Rights **7**.15, **7**.110–11
European Court of Justice **7**.15, **7**.99, *7.111*, **7**.173
European Currency Unit **7**.78, **7**.111
European Economic Area **7**.111
European Economic Community *4.114*, *7.103*
  Lomé Convention **4**.198
  Schuman Plan as precursor **4**.305
  Treaties of Rome **4**.292, **7**.62, **7**.99, **7**.103, **7**.315
  see also European Community
European Economic and Monetary Union **7**.103, **7**.111
European Exchange Rate Mechanism **7**.111
European Free Trade Association *4.114*, **7**.103, *7.111*
European Laboratory for Particle Physics *6.123*
European Monetary Co-operation Fund **7**.111
European Monetary System **7**.103, *7.111*

European Parliament 4.114, 7.99, 7.111
European Patent Convention 7.230
European Political Co-operation System (EPC) 7.99, 7.111
European Recovery Program see Marshall, George
European Southern Observatory 8.48
European Space Agency (ESA) 8.48, 8.68
European System of Central Banks 7.43
europium 1.106, 6.295
Eurydice 9.12, 9.28
eurypteria 2.301
eustatic change 1.115–16
euthanasia 7.111–12, 7.301
Euthydemus I, king of Bactria 3.33
eutrophication 2.6, 2.111
evangelicalism 7.112
Evangelical Church 7.65
Evangelists (the Four) 9.12
see also John, Saint; Luke; Mark; Matthew
Evans, Sir Arthur 3.198
Evans, Edgar 1.295
Evans, Oliver 6.41, 6.123, 6.293, 6.330
Evans-Pritchard, Sir Edward Evan 7.24, 7.112, 7.220, 7.331
evaporation 1.116
hydrological cycle 1.162
transpiration 1.162, 1.341, 2.345
see also vapour
evaporites 1.116
halite 1.80, 1.81, 1.150–1
see also gypsum
evapotranspiration 1.116
see also evaporation; transpiration
Evatt, Herbert Vere 4.114
Eve 9.2, 9.11, 9.12
evection 8.48
Evelyn, John 3.127, 5.150
evening primrose 2.111
evening star see Venus
event horizon 8.48
Everest, Sir George 1.116
Everest, Mount 1.116, 1.117, 7.208
Everglades 1.116
evergreen plants 2.111
aspidistra 2.17
aubretia 2.18
avocado pear 2.19
bamboo 2.21
barberry 2.22
bay laurel 2.25
beech 2.27
bottlebrush 2.39
box 2.40
buckthorn 2.44
buddleia 2.44
cacao 2.49
camellia 2.50
cinchona 2.66
cinnamon 2.66
coniferous see conifers
cotoneaster 2.77–8
Daphne 2.87
flamboyant tree 2.119
heather 2.29, 2.80, 2.154, 2.327
holly 2.159
hyssop 2.168
ivy 2.173
laurel 2.25, 2.184
lignum vitae 2.190
litchi 2.191
loquat 2.194
magnolia 2.198
mango 2.201–2
myrtle 2.224
oak 2.231
olive 2.234
papaya 2.241
pistachio 2.256
protea 2.269–70
evergreen plants (cont.):
rain forest 2.276
rosemary 2.288
sapodilla 2.294
strawberry tree 2.327
tea 2.336
wintergreen 2.368
yucca 2.374
Evershed effect 8.48
everyday life 7.109, 7.133
Evesham, battle of 3.120
Évian Agreements 4.114
evidence 7.112
evil eye 7.112
evolution 2.111, 3.120–1
adaptation 2.3
astronomical 8.49
Galápagos finches 2.130
natural selection 2.194, 2.225, 2.304–5, 7.11
peppered moth 2.225, 2.249
social 7.112
stellar 8.159–60
theorists:
Gray 2.144, 2.344
Huxley, T. H. 2.166
Lamarck 2.111, 2.181
Marsh 2.204
Mayr 2.205
Wallace 2.30, 2.111, 2.359
see also Darwin
see also prehistory
Evren, Kenan 4.354
Ewan, George 8.176
examinations (education) 7.20
Excalibur 9.12
excavator 6.49, 6.123–4, 6.316, 6.365, 6.367
exchange rate 7.112–13, 7.136
Exchange Rate Mechanism 7.111
Exchequer 3.121
excise see customs and excise
excitation (physics) 1.116
Exclusion crisis (James II) 3.121
exclusion principle (physics) 1.116
exclusionist (Australia) 4.114
excommunication 3.121
excretion 2.111
execution:
capital punishment 3.150, 7.40, 7.146
extra-judicial 7.114
executive (government) 7.113, 7.279, 7.318
exegesis 5.150
Exeter, Thomas Beaufort, duke of 3.37
exfoliation 1.28, 1.94, 1.116
exhaust system (motor) 6.62, 6.64, 6.124
exhibition, trade 6.160–1
Exile of Jews 3.121
existence see ontology
existence theorems 1.116
existentialism 5.188, 7.113
and drama 5.173, 5.402
and Heidegger 7.141
and novels 5.40, 5.73, 5.173, 5.278, 5.402
see also Sartre
exobiology 8.49
exocrine gland 2.111–12
Exodus of Israelites 3.121
exoparasites see parasites
exorcism 7.139, 7.297
exoskeleton 2.112
cuticle 2.84
shell 2.306, 2.307
exothermic reactions 1.3, 1.117
see also combustion; neutralization
expanding universe 8.33
expansion 1.117
expansion joint 6.124
expectation (psychology) 7.113
expenditure, public-sector 7.120
experimental psychology 7.334
expert systems 6.124, 6.179
exploration 3.121–2
Afghanistan 1.164
Africa (mainly by British)
central 1.197, 1.317, 1.320, 4.56, 4.198, 4.324, 4.326
east 1.44–5, 1.46, 1.335, 4.27, 4.56
west 1.33, 1.64, 1.90, 1.154, 1.183, 1.186–7, 1.248, 4.260
Alaska 1.184
Antarctica 1.12, 1.36, 1.97, 1.366
American 1.46, 1.105
British 1.75, 1.286, 1.295, 1.300
Arabia 1.94
Arctic:
American 1.46, 1.75, 1.105, 1.146, 1.249, 1.366
Australian 1.366
British 1.84, 1.127, 1.248, 1.286
Danish 1.38, 1.276
Dutch 1.33
Norwegian 1.12, 1.224
Asia 1.75, 1.83, 1.164, 1.205
Atlantic Ocean 1.154
Australia see under Australia
Canada:
British 1.29–30, 1.75, 1.84, 1.95, 1.127, 1.139, 1.159, 1.200–1, 1.286, 1.353, 3.141, 3.197–8, 3.254
French 1.58, 1.188, 1.190, 3.68, 3.203
see also Cartier, Jacques
Genoese 1.47
Hudson's Bay Company 3.170
New France 3.251–2
Spanish 3.254
Vikings 1.112, 3.374
see also discovery and colonization under Canada
Caribbean and West Indies 1.69, 1.70, 1.127, 1.355
Central America 1.47, 1.70, 1.88, 1.159, 3.159, 3.298–9
China 1.164, 1.205, 3.122, 3.390
East Africa 1.44–5, 1.46
explorers see under individual countries
Greenland 1.112, 1.146, 1.159, 1.224
India 1.83, 1.164
Italy 1.47, 1.70, 1.355
Middle East 1.30
New Guinea 1.75
New Zealand 1.75, 1.331
Pacific Ocean 1.42, 1.95, 1.97, 1.366, 3.96, 3.106
South America 1.88, 1.159, 1.275, 1.355–6, 3.106, 3.159
of space see space-flight
Swedish 1.230, 1.231
United States:
American 1.41, 1.53, 1.258, 4.64, 4.89, 4.126, 4.225, 4.316–17
Lewis and Clark 1.63, 1.64, 1.193, 3.122, 4.194
British 1.75, 1.95, 1.159, 1.275, 1.286, 1.353, 3.106, 3.185, 3.197–8
Canadian 1.237
French 1.206, 1.228
Spanish 1.88
West Africa 1.33, 1.64, 1.154, 1.248
see also navigation and routes under trade and under individual countries
explosions 1.117–18
explosives 6.124
bombs see bomb
detonator 6.100, 6.131, 6.246
dynamite 6.110, 6.228, 6.242
fuse 6.148, 6.316
gelignite 6.153
explosives (cont.):
grenade 6.64, 6.161, 6.226
gunpowder see gunpowder
mining 6.226–7, 6.228
Molotov cocktail 6.232
nitrates in 6.147, 6.162, 6.241
nitro-cellulose and cellulose trinitrate 6.66, 6.153, 6.241–2, 6.268, 6.310, 6.387
nitro-glycerine 6.110, 6.153, 6.242
in oil exploration 6.263
plastic 6.272
RDX 6.124, 6.272
'Semtex' 6.272
shell 6.315–16, 6.318
TNT 6.124, 6.357
see also weapons
exponential function 1.118
exports 7.113
voluntary export restraint 7.314
see also trade
exposure (photography) 6.16, 6.25, 6.124, 6.318
meter 6.124
Expressionism 5.150–1
in architecture 5.290
in dance 5.123, 5.242, 5.252, 5.300, 5.489
in drama and novels 5.243, 5.324, 5.327–8, 5.457, 5.485
see also Dostoyevsky
in literature 5.262
in music 5.25, 5.44, 5.150–1, 5.405–6, 5.484
Germany 5.206, 5.435, 5.485
in painting 5.49, 5.147, 5.182–3, 5.189, 5.195, 5.307–8, 5.320, 5.427, 5.497
Germany 5.41, 5.63, 5.248, 5.282
see also Abstract Expressionism
external-combustion engines 6.122
see also steam-engine under steam power
externality 7.113
extinct species 2.108
animals:
auroch 2.18
cave bear 2.58
cave lion 2.58
dragon-fly (Meganeura) 2.206
Eohippus 2.208
giant sloth 2.135
glyptodont 2.138
ichthyosaur 2.168–9
iguanodon 2.169
Irish elk 2.173
mammoth 2.200
mastodon 2.205
Mesohippus 2.208
placodem 2.257
plesiosaurus 2.259
pteraspis 2.271
pterodactyl 2.271
reptiles 2.168–9, 2.259, 2.271, 2.333
see also dinosaurs
sabretoothed tiger 2.291
sea urchin 2.302
Steller's sea cow 2.298
trilobite 2.348
woolly rhinoceros 2.371
birds:
dodo 2.96
huia 2.362
passenger pigeon 2.243
solitaire bird 2.317
plants: lepidodendron 2.188
see also endangered species; fossils
extinction (astronomy) 8.49
extradition 7.113–14
extragalactic nebula 8.49
extra-judicial execution 7.114
extraterrestrial life 8.49, 8.148
Extreme Unction 7.271

extrinsic variable *see under* variable stars
extroversion/introversion **7**.*114*, **7**.173
extrusion (industry) **6**.*124*, **6**.142, **6**.188, **6**.221, **6**.272, **6**.295
extrusive (volcanic) igneous rocks **1**.*118*
  andesite **1**.13
  archipelagic apron **1**.18
  basalt **1**.32, **1**.33, **1**.70, **1**.138, **1**.301
  ignimbrite **1**.165–6
  obsidian **1**.234
  pumice **1**.269
  pyroclasts **1**.270, **1**.271, **1**.295
  scoria **1**.295
  see also igneous rocks
Eyadema, Gnassingbe **4**.346
Eyck, Hubert van **5**.151
Eyck, Jan van **5**.18, **5**.*151*, **5**.326, **5**.346
Eyde, Samuel **6**.38
eyes *see* vision and eyes
eyeworm **2**.289
Eyre, Edward John **1**.*118*, **4**.171
Eyre, Lake **1**.118
Eysenck, Hans **7**.114
Ezana of Axum **3**.30
Ezekiel **3**.121, **3**.*122*, **9**.*12*
Ezra **3**.*122*

F

Fa **9**.17
Fabergé, Peter Carl **5**.*151*
Fabians **4**.*115*
  see also in particular Besant; Shaw, George Bernard; Webb, Beatrice; Webb, Sidney
Fabius, Quintus **3**.*123*, **3**.294
fables **5**.6, **5**.*151*, **5**.254
  see also myths and legends; stories
Fabre, Henri (pilot) **6**.311
Fabre, Jean Henri (entomologist) **2**.*113*
fabric (cloth) *see* textiles
fabric (geology) **1**.*119*
Fabricius, David **8**.100
Fabricius, Hieronymus **2**.*113*
Fabritius, Barent **5**.151
Fabritius, Carel **5**.*151*
face:
  facial expression **7**.*114*
  identikit picture of **6**.175
facet-cutting **5**.239
Fachang (Muqi) **5**.308
facsimile transmission *see* fax
Factor IX **6**.40
factorial **1**.*119*
factors of production **7**.*114*
factory:
  farming **6**.*125*, **6**.205, **6**.279
  ship **6**.384
  see also industrialization
Factory Acts **4**.*115*
faculae **8**.*50*
Fadden, Arthur William **4**.234
faeces **2**.*113*
Faeroe Islands **1**.*119*, **7**.156
Fahd, king of Saudi Arabia **4**.303
Fahlberg, Constantine **6**.342
Fahrenheit, Gabriel: Fahrenheit scale **1**.*119*, **1**.332, **6**.350
Faidherbe, L. L. **4**.308
faïence **5**.276
failures:
  education **7**.113
  rate in technology **6**.297
    metal fatigue **6**.220, **6**.221
fair **3**.*123*
  world **4**.383
Fair Deal **4**.*115*
Fairfax, Thomas, 3rd Baron **3**.*123*, **3**.246, **3**.253
fairy **9**.*12*
fairy moss **2**.361
Faisal I, king of Iraq **4**.*115*, **4**.166, **4**.175
Faisal II, Ibn Abd al-Aziz, king of Saudi Arabia **4**.*115*, **4**.303
Faisal II, king of Iraq **4**.283
Faiz **5**.467
Falange **4**.*115–16*
Falasha **7**.*114*, **7**.171
falcon family **2**.*113*
  gyrfalcon **2**.148
  hobby **2**.158
  kestrel **2**.178
  lanner **2**.182
  merlin **2**.208
  peregrine **2**.250
  saker **2**.292
Falconbridge, Alexander **3**.328
Falconet **5**.385
Faleto, E. **7**.85
Falkirk, battles of **3**.*123*
Falkland Islands (Malvinas) **1**.*119*, **3**.*123*, **4**.*116*, **7**.57, **7**.*352*
  physical geography and geology **1**.84, **1**.119
Falkland Islands (Malvinas) (*cont.*):
  War **4**.*116*, **7**.18, **7**.69
Fall, Albert B. **4**.339
Falla, Manuel de **5**.*151*
Falloppio, Gabriello: fallopian tubes **2**.*113*
fall-out **6**.*72*, **6**.125, **6**.245
fallow deer **2**.*113*
false colour **8**.*50*
false consciousness **7**.152
false scorpion **2**.270
false vampire bat **2**.*113*
falsificationism **7**.241
Faludi, Susan **7**.332
familiar **9**.*12*
famille rose, famille verte **5**.94, **5**.*152*
family **7**.*114–16*
  adoption **7**.3
  cohabitation **7**.55–6
  court **7**.*115*
  divorce **7**.93–4, **7**.189
  domestic violence **7**.94, **7**.115
  household **7**.148
  incest **7**.46, **7**.154
  law **7**.55, **7**.*115*
  and old age **7**.222
  parental leave **7**.228
  planning *see* contraception
  policy **7**.*116*
  socialization **7**.291
  therapy **7**.*116*
  see also children; marriage; pregnancy
family (nomenclature) **2**.*113*
famine **4**.167, **7**.*116–17*, **7**.292
  relief **7**.*116*, **7**.*118*, **7**.121
fan, alluvial **1**.9
Fan Chengda **5**.98
fan (cooling) **5**.*152*
fan palm **2**.240
fantail **2**.*113*
fantasia **5**.*152*
fantasy **7**.*118*
  painting **5**.54–5, **5**.187
  see also science fiction; horror
Fanti Confederation **4**.*116*
Fantin-Latour, Henri **5**.*152*
fanworm **2**.*113*, **2**.114
FAO *see* Food and Agriculture Organization
al-Farabī **7**.*10*, **7**.23, **7**.24
farad **1**.*119*, **6**.368
Faraday, Michael **1**.85, **1**.*119*, **6**.*125*
  and benzene **6**.34
  and Bunsen burner **6**.52
  and electric motor **6**.115
  and electrolysis **1**.102
  and electromagnetic induction **6**.117
  and Royal Institution **6**.303
farandole **5**.*152*
farce **5**.*152*, **5**.154
  see also comedy
al-Farghānī **8**.75
Farīḍ, Ibn **5**.19, **5**.439
farmers-general **3**.*123–4*
farming *see* agriculture
Farnaby, Giles **5**.*152*
Farnese, Alessandro, duke of Parma **3**.124, **3**.335
Farnese family **3**.*124*, **3**.368
Farouk, king of Egypt **4**.109, **4**.*116*, **4**.233, **4**.332, **4**.369
Farquhar, George **5**.*152*
Farrell, Suzanne, **5**.30
Farsi **7**.157
fasces **3**.*124*
fascism **4**.41, **4**.76, **4**.78, **4**.*116*, **4**.228, **4**.280, **7**.*118–19*
  see also Nazis
Fashoda incident **4**.*116–17*
Fasilidas, emperor of Ethiopia **3**.119
Fassbinder, Rainer Werner **5**.155
fast breeder reactor **6**.*126*, **6**.243, **6**.244, **6**.245, **6**.274
fast nova **8**.110
fasting **7**.258
Fata Morgana **1**.216
al-Fatah **4**.*117*
fatalism **7**.245
Fates **9**.*12*
Fath Ali Shah **4**.279
Fatima(h) **9**.*12*
Fatimid dynasty **3**.4, **3**.*124*
  art **5**.*152*, **5**.153, **5**.229
fats, oils, and fatty acids **2**.*114*, **6**.*126*
  candle **6**.57
  in food **2**.92, **2**.114, **6**.52, **6**.126, **6**.140, **6**.141, **6**.215, **6**.384
  soap from **6**.323
fatwa **7**.282
Faubus, Orval **4**.197
faucet (tap) **6**.345
Faulkner, William **5**.*152–3*, **5**.436
Faulkner of Downpatrick, Brian Deane, Baron **4**.*117*
fault-block mountains **1**.40
faults, geological **1**.*119–20*
  downthrow **1**.94
  overthrust **1**.243
  reverse **1**.120, **1**.281
  San Andreas **1**.292
  strike-slip (tear) **1**.120, **1**.323
  upthrow **1**.351
Faunus **9**.*12*
Fauré, Gabriel **5**.*153*
Faust **9**.*12–13*
Fauvism **5**.129, **5**.*153*, **5**.238, **5**.286, **5**.299, **5**.339, **5**.478
Fawcett, Elizabeth **4**.82
Fawcett, Dame Millicent Garrett **4**.*117*
Fawkes, Guy **3**.151
fax **6**.*126*, **6**.278, **6**.347
Faye's Comet **8**.*50*, **8**.194
al-Fazari **3**.141
FBI *see* Federal Bureau of Investigation
fear **2**.167
  see also depression and anxiety
feathers **2**.114
  featherwork **5**.*153*, **5**.356
feather-star **2**.*114*
Febvre, Lucien **4**.153
Federal Bureau of Investigation **4**.*117*, **7**.238
Federal Republic of Germany *see* German Federal Republic
Federal style **5**.*153*
federalism **4**.*117*, **7**.31, **7**.81, **7**.*119*
  see also under politics *under* Australia; United States
Federalist Party (USA) **3**.*124*, **4**.*117*
  see also under politics *under* United States
Federated States of Micronesia *see* Micronesia
Federation of Rhodesia and Nyasaland **4**.67
feedback **6**.88, **6**.*126–7*
feedforward system **6**.88
feedstuffs, animal **6**.*127*, **6**.*142–3*, **6**.205
  alfalfa **2**.6
  barley **2**.23
  beetroot **2**.28
  date seeds **2**.88
  forage **6**.19, **6**.142–3, **6**.302
  peanuts **2**.246
  rye **2**.291
  silage **6**.142, **6**.143, **6**.166, **6**.235, **6**.319, **6**.320
  single-cell protein **6**.321
  sorghum **2**.317
  soya bean **2**.318
  thistle **2**.339
feelings *see* emotion
feet (body) **2**.*124*, **2**.342, **9**.*58*
  footwear **6**.142
Feisal *see* Faisal
feldspar **1**.*120*
  crystal **1**.80

feldspar (*cont.*):
  moonstone **1.**221
  orthoclase **1.**143
  plagioclase **1.**258, **1.**259
  in porcelain **6.**276
Felix V, antipope **3.**17
Fell, Dr John **5.**465
Fellini, Federico ***5.****153*
felsenmeer ***1.****120*
felt **6.**125, ***6.****127*, **6.**242
female mammals:
  clitoris **2.**68
  fallopian tubes **2.**113
  menstruation and menopause **2.**208
  uterus **2.**353
  vagina **2.**354
  see also mammary glands; pregnancy; women
feminism *see* women's movement
femur **2.**186
fen ***1.****120*
fencing ***7.****119*
Fénelon, François de Salignac de la Mothe- ***5.****153*
Feng Guifen **4.**308
Fenians ***4.****118*
Fenir **9.**2, **9.**23, **9.**30
fennec fox ***2.****114*
fennel ***2.****115*
Fennoscandian Shield ***1.****120*
fens *see* wetlands
Fenton, Roger **5.**482
fenugreek ***2.****115*
Ferdinand I, king of Austria (1835–48) **4.**25
Ferdinand I, king of Castile (10th century) **3.**206
Ferdinand I, Holy Roman Emperor (1503–64) **3.**27, **3.**152, **3.**153, **3.**168
Ferdinand I, king of Romania **4.**292, **4.**365
Ferdinand I, king of Two Sicilies *see* Ferdinand IV, king of Naples
Ferdinand II, king of Aragon and Castile *see* Ferdinand V
Ferdinand II, Holy Roman Emperor, king of Bohemia and Hungary (1578–1637) ***3.****124*, **3.**153, **3.**168
Ferdinand II, king of Tuscany (1644) **1.**332, **6.**350
Ferdinand III, grand-duke of Tuscany **4.**355
Ferdinand III, Holy Roman Emperor, king of Hungary, Bohemia, Romans (1608–57) **3.**153, **3.**168
Ferdinand III, king of León and Castile (13th century) **3.**62
Ferdinand IV, king of Bohemia and Romans (1633–54) **3.**153
Ferdinand IV, king of Naples (*also* Ferdinand I, king of Two Sicilies, 1751–1825) **3.**46, **4.**228, **4.**312
Ferdinand V (*also* Ferdinand II of Aragon and Castile), king of León , Sicily, and Naples (1452–1516) ***3.****124–5*, **3.**327
  and Columbus **3.**82
  and Inquisition **3.**336
  and war **3.**12, **3.**86, **3.**335, **3.**355
  wife (Isabella of Castile) **3.**19, **3.**62, **3.**125, **3.**335
Ferdinand VI, king of Spain **3.**46
Ferdinand VII, king of Spain **3.**46, **4.**62, ***4.****118*, **4.**160, **4.**324
Ferdinand, prince of Brunswick (1759) **3.**232
Ferdinand, prince of Saxe-Coburg **4.**54
Ferguson, Sir Samuel **5.**226
Ferlinghetti, Lawrence **5.**39
Fermat, Pierre de ***1.****120*
fermentation ***2.****115*, ***6.****127*, **6.**142, **6.**259
  bioreactor and biotechnology **6.**37
  brewing **6.**10–11, **6.**37, **6.**46, **6.**68
  food **6.**27, **6.**46, **6.**68
  sewage **6.**37
  in silo **6.**320
  wine **6.**386
Fermi, Enrico **1.**106, ***1.****120*, **6.**23, ***6.****127*, **6.**242, **6.**245
fermions ***1.****120*
  muons **1.**222–3
  see also electrons; neutrons; positron; protons
fermium **1.**106
Fernandez de Moratín, Leonardo **5.**428
Fernández Hidalgo, Gutierre **5.**257
ferns ***2.****115*, **2.**116
  bird's nest **2.**31
  Boston **2.**38
  bracken **2.**40
  climbing **2.**68
  clubmoss **2.**69
  hart's tongue **2.**151
  horsetails **2.**49–50, **2.**164
  lepidodendron **2.**188
  maidenhair **2.**198
  male **2.**199
  polypody **2.**263
  pteridophyte *see* pteridophyte
  Rhynea **2.**284
  stag's horn **2.**324
  tree **2.**346
  walking **2.**358
  water **2.**361
Ferrabosco, Alfonso (and son) ***5.****153–4*
Ferranti, Sebastian Ziani de **6.**114, ***6.****127*
Ferreira de Castro, J. M. **5.**359
Ferrel, William ***1.****120–1*
ferret ***2.****115*
ferric oxide *see* iron oxide
Ferrier, Kathleen ***5.****154*
ferrisol ***1.****121*
ferrite ***6.****127*
ferromagnesian minerals ***1.****121*
  augite **1.**26
  biotite **1.**39, **1.**143
  hornblende **1.**158
  olivine **1.**237, **1.**250–1, **1.**256
ferromagnetism *see* magnet/magnetism
ferrous oxide *see* iron oxide
Ferry, Jules François Camille ***4.****118*
ferry **6.**60, ***6.****127*, **6.**172, **6.**277
Fertile Crescent ***3.****125*
fertility:
  drug ***6.****127*
  rate ***7.****119*
    infertility **7.**159–60
  soil *see* fertilizers
  see also reproduction
fertilization ***2.****115–16*
  cross-fertilization **2.**82
  seed **2.**303
  see also reproduction
fertilizers ***6.****127*, ***6.****129*
  natural **6.**127, **6.**129, **6.**170
    guano **2.**146
    from horn **6.**170
    humus **2.**166
    see also compost and manure
  synthetic **6.**7–8, **6.**129, **6.**196, **6.**267
    green revolution **6.**161
    nitrates **6.**13, **6.**124, **6.**128, **6.**241
    pioneer **6.**196
    pollution **6.**377
    potassium **6.**127–8, **6.**278
    production **6.**37, **6.**71
    sulphates **6.**2, **6.**13, **6.**338
fertilizers (*cont.*):
  synthetic (*cont.*):
    superphosphates **6.**344
    see also Haber–Bosch *under* Haber
Fessenden, Reginald Aubrey **6.**20, **6.**48, **6.**111, ***6.****128*, **6.**290
Festinger, Leon **7.**54
FET (field-effect transistor) **6.**361
fetch of waves ***1.****121*
fetish ***7.****119*
fetus *see* embryo
feud ***7.****119–20*
feudalism **3.**96, ***3.****125–6*, **3.**370, **4.**92
  see also serfs
Feuillet, Raoul **5.**123
feverfew **2.**65
fevers ***2.****116*
  Rocky Mountain spotted **2.**287
  scarlet **2.**296–7
  tick **2.**287
  typhoid and paratyphoid **2.**281, **2.**352
  typhus **2.**352
  yellow **2.**281, **2.**373
  see also diseases and disorders
Feydeau, Georges ***5.****154*
Feynman, Richard ***1.****121*
FG Sagittae ***8.****50*
Fíanna Éireann **9.**27
Fíanna Fáil ***4.****119*
Fibonacci, Leonardo: Fibonacci numbers ***1.****121*
fibre ***6.****128*, **6.**295
  from animals **6.**128, **6.**231
    mohair **6.**128, **6.**231
    reindeer **2.**282
    silk *see* silk
    wool *see* wool
  asbestos **6.**21–2
  carbon *see* carbon fibre
  optics *see under* optics/optical
  from plants ***2.****116*, **6.**128, **6.**364
    agave **2.**4
    baobab **2.**22
    cotton *see* cotton
    flax *see* flax
    hemp (cannabis) **2.**52, **2.**154
    hemp (sisal) **2.**310
    jute **2.**177
    kapok **2.**178
    kenaf **2.**178
    lime tree **2.**190
    palms **2.**71, **2.**88
    sisal **2.**4, **2.**310
  see also spinning; synthetic fibres; textiles
fibreboard **6.**338
fibreglass **6.**86, **6.**129, **6.**156
fibre-reinforced plastics **6.**5, **6.**82, ***6.****129*, **6.**156, **6.**157, **6.**204, **6.**247
fibrinogen ***2.****116*
fibula **2.**186
fiction *see* novels; stories
fiddle ***5.****154*
  crwth **5.**119
  hardingfele **5.**198
  nyckelharpa **5.**324
  sāraṅgī **5.**402
  see also violin
fief ***3.****126*
Field, John ***5.****154*, **5.**320
Field of the Cloth of Gold ***3.****126*
field equations **8.**137
field mouse **2.**370–1
field (physics) ***1.****121*
  electric **1.**102, **1.**195, **1.**196
  magnetic **1.**195, **1.**196, **1.**202
field systems (land) ***3.****126*
field of view ***8.****50*
field vole ***2.****117*
field-effect transistor **6.**361
fieldfare ***2.****116*
Fielding, Henry **3.**47, ***5.****154*, **5.**348
Fields, W. C. **5.**469
fieldwork (anthropology) **7.**14, ***7.****120*, **7.**183
Fifteen Rebellion ***3.****126–7*, **3.**326
fifth-generation computer **6.**21, ***6.****129*
Fifth-Monarchy Men ***3.****127*
fig **2.**22, ***2.****117*
fig wasp ***2.****118*
fighters and fighter-bombers ***6.****129*, **6.**139, **6.**220, **6.**224
fighting fish ***2.****117*
Figueiredo, president of Brazil **4.**49
Figueres, José **4.**87
figures of speech ***5.****154*
  conceit **5.**108
  hyperbole **5.**215
  irony **5.**227
  metaphor **5.**108, **5.**292, **5.**339
  onomatopoeia **5.**328
  oxymoron **5.**335
  simile **5.**420
  see also alliteration
Fiji ***1.****121*, ***4.****119*, **7.**221, **7.**296, ***7.****352–3*
filament ***8.****50*
filar micrometer ***8.****50*
file (computer) ***6.****129*
file (tool) ***6.****129*, **6.**358
filibusters ***4.****119*
filigree ***5.****154*
Filipino language **7.**23
Fillmore, Millard ***4.****119*, **4.**185
film, photographic **6.**66, ***6.****129–30*, **6.**268
  developing and printing **6.**100, **6.**101, **6.**130
  gel **6.**153
  see also camera; cinema
films *see* cinema
filter circuit ***6.****130*
filtration ***6.****130*, **6.**379
fin ***2.****118*
fin whale **2.**118, ***2.****119*
finance:
  accounting **7.**1–2, **7.**190–1
  arbitrage **7.**15
  assets **7.**20, **7.**165, **7.**330
  audit **7.**22
  balance sheet **7.**25
  bank rate **7.**26
  bank *see* banks
  bill of exchange **7.**31
  bookkeeping **7.**34
  break-even analysis **7.**35
  broker **7.**35
  budget **7.**37, **7.**201
  building societies **7.**37
  capital *see* capital
  cash flow **7.**41
  clearing house **7.**53
  collateral **7.**56
  commodity market **7.**57
  company ***7.****120*
  cost accounting **7.**71
  credit **7.**31, **7.**74
  credit union **7.**74
  currency *see* currency
  debt *see* debt
  deficit financing **7.**83
  discount **7.**92
  dividend **7.**93
  economics *see* economics
  education **7.**102–3
  employment percentage **7.**106
  equity **7.**33
  Eurodollar **7.**110
  financial regulation ***7.****120*
  financial sector **7.**309
  fiscal policy **7.**120
  Franc Zone **7.**124
  friendly societies **7.**127
  futures markets **7.**128
  gilt-edged securities **7.**132
  hedging **7.**141
  index **7.**155
  insider trading **7.**161

finance (*cont.*):
insolvency **7**.161
insurance *see* assurance; insurance
interest *see* interest and interest rate
investment *see* investment
liquidity **7**.185
loans *see* loans
loss **7**.187
management accounting **7**.190–1
medicine and health care **7**.139–40
military expenditure **7**.201
money *see* currency
money *see* money
public **7**.254
public sector borrowing requirement **7**.254
savings **7**.37, **7**.127, **7**.207
securities *see* securities
shares *see* shares
subsidy **7**.301
take-over **7**.65, **7**.224, **7**.305
trust **7**.316
underwriting **7**.319
see also economics; economy *and under individual countries*
finch family ***2**.118*
bullfinch **2**.45
buntings **2**.45, **2**.144, **2**.236, **2**.373
chaffinch **2**.60
crossbill **2**.82
Galápagos **2**.3, **2**.130
goldfinch **2**.140–1
greenfinch **2**.144
grosbeak **2**.144
hawfinch **2**.144, **2**.152
linnet **2**.191
rosefinch **2**.288
siskin **2**.310–11
twite **2**.352
wren thrush **2**.372
fin-de-siècle ***5**.155*
see also Beardsley; Symbolism; Wilde
Fine Gael ***4**.119*
finfoot ***2**.118*
Fingal **9**.13
finger lake ***1**.121*
fingers **2**.118, **2**.225, ***9**.58*
fingerprints **2**.132, ***7**.120*, ***9**.61*
genetic (DNA) fingerprinting **6**.105, **7**.120
Finland ***1**.121*, ***3**.127*, ***4**.120*, ***7**.353*
aid from **7**.8
arts **5**.1, **5**.417
in Council of Europe **7**.71
economy and trade **7**.99, **7**.111, **7**.221, **7**.315
ethnic groups **7**.181
languages **7**.322–3
literature ***5**.155*, **5**.165, **5**.404
myths **9**.37
physical geography and geology **1**.120, **1**.121, **1**.188
politics and government **7**.223, **7**.250
regions **1**.188, **4**.8
ruler **4**.120
sport and recreation **7**.145
war:
Finnish–Russian ***4**.120*
leader **4**.209–10
see also Baltic States
Finn MacCool ***9**.13*
Finno-Ugric languages **7**.322–3
fins **2**.119
Finston, Michael L. **8**.8
Finzi, Gerald ***5**.155*
Fiordland National Park ***1**.122*
fiords ***1**.122*, **1**.310
Fiorentino, Rosso **5**.366
fir **2**.98, ***2**.119*
Firdausī, Abū al-Qāsim ***5**.155*, **9**.6, **9**.14, **9**.19, **9**.34
and myths **9**.6, **9**.14, **9**.19, **9**.34
fire ***1**.122*
fire-extinguisher ***6**.131*
fire (*cont.*):
fire-fighting and prevention ***6**.131–2*, **6**.172, **6**.322–3, **6**.330
fire-making device ***6**.132*
fireproofing ***6**.132*
fire-resistance ***6**.132*
flame-thrower **6**.134
Fire of London ***3**.127*
firearms ***6**.130–1*, **6**.225
airgun **6**.8
anti-tank gun **6**.16, **6**.131
bullet **6**.52, **6**.131, **6**.272
cannon **6**.22, **6**.57, **6**.130, **6**.318, **6**.377
cartridge **6**.62, **6**.131
gun-sight **6**.162–3
machine-gun **6**.131, **6**.210, **6**.217, **6**.226, **6**.337
pistol and revolver **6**.81, **6**.130, **6**.131, **6**.270
propellants **6**.62, **6**.242
rifle **6**.130, **6**.131, **6**.162–3, **6**.226, **6**.298
self-propelled gun **6**.312
on ships *see* warships
shotgun **6**.318
silencer **6**.319
sub-machine-gun **6**.130, **6**.131, **6**.337
see also artillery; weapons
fireball **8**.34, ***8**.50*
firebrats ***2**.119*, **2**.310
firecrest ***2**.119*
fire-fighting *see under* fire
firefly ***2**.119*
fireless locomotive ***6**.132*
fireweed **2**.368
fireworks **1**.265, ***6**.132*, **6**.301
firm, theory of ***7**.120*
see also companies and business
firn ***1**.122*
First Point of Aries ***8**.50*
First Point of Libra ***8**.51*
first quarter of Moon **8**.119
First World War *see* World War I
first-past-the-post electoral system **7**.190
fiscal policy ***7**.120*
Fischer von Erlach, Johann Bernhard ***5**.156*
Fischer-Dieskau, Dietrich ***5**.155–6*
Fischer–Tropsch process **6**.343
fish ***2**.119*
aquarium *see* aquarium and ornamental fish
bony *see* bony fish
brain **2**.41
camouflage **2**.50
cartilaginous *see* cartilaginous fish
eels *see* eel
electric **2**.105
endangered species **2**.108
fighting **2**.117
fin **2**.118
flat *see* flatfish
fossil (*Jaymotius*) **2**.175
freshwater *see* freshwater fish
gill **2**.136
heart **2**.153
ichthyology **2**.168
jawless *see* jawless fish
lateral line **2**.183
mail cheek **2**.198
ocean *see under* oceans and seas
ornamental *see* aquarium and ornamental fish
perch *see* perch
river *see* freshwater fish
scales **2**.295
sea-horse **2**.299
shark *see* sharks
see also fishing
Fisher, Andrew ***4**.120*
Fisher, Irving ***7**.120–1*, **7**.256
Fisher, John Arbuthnot, 1st Baron ***4**.120*
Fisher, Sir Ronald Aylmer ***2**.119*
fish-hawk **2**.237
fishing as pastime ***7**.121*
fishing industry ***6**.133*
child labour in **7**.45
and conservation **7**.296
controlled **6**.215
curing ***6**.132–3*
fish farming **6**.18, ***6**.133*
freshwater **6**.133, **6**.146
long-line **6**.133, **6**.207
marine resource management **6**.215
nets **6**.133, **6**.239, **6**.240
and pollution **6**.377
products ***6**.133–4*
shellfish **6**.18, **6**.133, **6**.316
traps **6**.133, **6**.362
trawler **6**.239, **6**.362
whaling *see* whaling
see also fish; hunter-gatherers
fission, nuclear ***1**.232–3*, ***6**.134*, **6**.218, ***8**.110*
chain reaction **6**.69, **6**.242, **6**.245
fission bombs **6**.246
see also nuclear energy and power; nuclear reactors
fistula ***2**.119*
Fitzalan, Walter **3**.343
Fitzgerald, Edward ***5**.156*
Fitzgerald, Francis Scott ***5**.156*, **5**.244
Fitzgerald, George:
Fitzgerald–Lorentz contraction ***1**.122–3*, **8**.86
Fitzgerald, Zelda **5**.156
Fitzherbert, Maria **4**.133
Fitzpatrick, Thomas **4**.225
Five, The **5**.53, **5**.87, ***5**.156*, **5**.310, **5**.383–4
Five Dynasties (China) **3**.72
five members of parliament ***3**.127*
Five Nations *see* Iroquois
Five Pillars of Islam ***7**.121*
five positions **5**.32, ***5**.156*
fives ***7**.121*
fixed-point arithmetic ***1**.123*
Fizeau, Armand Hippolyte Louis ***1**.123*
flag iris ***2**.119*
flagellant ***3**.127*
flagellate ***2**.119–20*
flageolet ***5**.156*, **5**.376
Flagstad, Kirsten ***5**.156*
Flaherty, Robert J. **5**.132
flail tank ***6**.134*
Flamboyant style ***5**.156–7*
flamboyant tree ***2**.119*
flamenco ***5**.157*
flame-thrower ***6**.134*
flamingo ***2**.119*
Flamsteed, John **8**.10, ***8**.51*
atlas **5**.159
Catalogue **8**.21–2, **8**.51
Flamsteed numbers ***8**.51*
and Royal Observatory **6**.303, **8**.141
Flanders/Flemish ***3**.127–8*
arts ***5**.157–8*
see also under Netherlands
cartographers **1**.204, **1**.210–11, **1**.241–2
language **7**.131
poet and historian **3**.135
religion **9**.41
statesmen/soldiers **3**.112, **3**.170, **3**.357
wars **3**.89
flare ***8**.51*
flare star ***8**.51*
flash (camera) ***6**.134*
flash coloration ***2**.119–20*
flash spectrum ***8**.51*
Flaška, Smil **5**.120
flatfish ***2**.120*, ***2**.121*
bream **2**.41, **2**.265, **2**.280, **2**.298
dab **2**.86
flounder **2**.120, **2**.122
halibut **2**.150
lemon sole **2**.187
plaice **2**.121, **2**.257
sole **2**.317
turbot **2**.350–1
flatworms ***2**.120*, **2**.259
see also flukes
Flaubert, Gustave ***5**.157*, **5**.207
flavour (particle physics) ***1**.123*, **1**.272
flavourings *see* spices, herbs, and flavourings
flax ***2**.120*
fabric from **6**.16, **6**.57, **6**.203
linseed oil from **6**.109, **6**.255
Flaxman, John ***5**.157*
fleabane ***2**.121*
fleas ***2**.120–1*, **2**.293, **2**.352
fleet *see* navies
Fleischmann, Michael **6**.148
Fleming, Sir Alexander ***2**.121*, **2**.122, **6**.16, **6**.69, ***6**.134*, **6**.137, **6**.260
Fleming, Sir John Ambrose **6**.98, **6**.118, ***6**.134*, **6**.291, **6**.354
Flemish *see* Flanders
flesh fly ***2**.121*
Fletcher, John ***5**.40*, **5**.286, **5**.414
Fleury, André Hercule de ***3**.128*
flexible manufacturing system ***6**.134*, **6**.211
flexible response ***7**.121*
Flexner, Simon ***2**.121*
flexography ***6**.134*
flies ***2**.122*
bee-fly **2**.27
black **2**.33, **2**.234
blow **2**.35
bot **2**.39
bristle fly **2**.41
cheese-skipper **2**.61
cluster-fly **2**.69
crane **2**.80
flesh **2**.121
fruit **2**.128, **2**.130–1
fungus gnat **2**.129
gall midge **2**.131
gallfly **2**.130–1
horse **2**.163
housefly **2**.164
hoverfly **2**.164
larva: maggot **2**.198
louse **2**.194
midges **2**.131, **2**.209
robber **2**.286
sand **2**.293
stable **2**.324
tsetse **2**.348, **2**.349
see also mosquitoes
flight ***6**.134–6*, **6**.138
flying boat **6**.8, **6**.140
history ***6**.137*, ***6**.139–40*
human-powered **6**.138, **6**.172
hypersonic **6**.175, **6**.386
and living organisms ***2**.121–2*
see also bats; birds; gliding creatures; insects
recorder ***6**.136*
simulator ***6**.136*
see also aircraft; space-flight
flightless birds:
cassowary **2**.56
emu **2**.107
kiwi **2**.179
moa **2**.215
notornis **2**.230
ostrich **2**.237
penguin **2**.247, **2**.248
rhea **2**.283–4
solitaire bird **2**.317
Flinders, Matthew ***1**.123*, **1**.127, **2**.43, **4**.247
Flinders Range ***1**.123*

flint 1.60, 1.123, 6.1, 6.136, 6.227
  tools and weapons 6.4, 6.72, 6.280–1
flintlock 6.130, 6.131
flip-flop circuit 6.136
FLN see Front de Libération nationale
float glass 6.136, 6.137
float switch 6.136
floating dock 6.109
Floating World 5.37, 5.158, 5.217
floating-point arithmetic 1.123
flocculi 8.51
Flodden Field, battle of 3.128
floodplains 1.123–4
floods 1.123, 7.92, 7.136
  control 6.96, 6.136
  sheet 1,301
  and storm surge 1.321
  see also dams
floor 6.203, 6.213, 6.337
  see also carpets and rugs
floppy disk 6.137
Flora 9.13
Florence 3.128, 3.227–8, 3.372
  arts see Florentine art
  rulers see Medici
Florentine art 5.158
  architecture and sculpture 5.114, 5.158, 5.178, 5.330, 5.378, 5.472
    Brunelleschi 5.63–4, 5.132
    Michelozzo 5.294
    Pisano 5.350
    see also Michelangelo; Raphael
  painting 5.114, 5.158, 5.175, 5.177–8, 5.265, 5.378
    Angelico 5.16
    Bartolommeo 5.37
    Botticelli 5.55, 5.448
    Bronzino 5.61
    Cimabue 5.101
    Domenico 5.132
    Masaccio 5.284
    Masolino 5.285
    Orcagna 5.330
    Pollaiuolo 5.355
    see also Giotto; Leonardo; Michelangelo; Raphael
Florey, Howard Walter, Baron 2.122, 6.16, 6.69, 6.134, 6.137, 6.260
Florida 3.128
  Keys 1.124
flotation 1.124
  ore 6.89, 6.241, 6.253
flotation (economics) 7.121
flounder 2.120, 2.122
flour 6.142
Flourens, J. P. M. 6.320
flow:
  in geomorphology see mass flow
  turbulence 1.347
  see also fluid; gas; liquid; water and under rivers
flow chart 6.137
flow production 7.248–9
flower 2.122
flowering onion 2.234
flowering plants/flowers 2.122, 2.123
  see also angiosperms; Dicotyledons; Monocotyledons
flowerpecker 2.122
Fludd, Robert 3.128, 3.311
flugelhorn 5.158–9
fluid:
  body 2.37
  critical point 1.80
  drive 6.26, 6.152
  fluidics 6.137
  hydrodynamics 1.161, 6.138
  mechanics 6.138
  viscosity 1.357, 6.138
  vorticity 1.359
  see also flow; flotation; gas; liquids
fluidized-bed furnace 6.78, 6.138
flukes 2.122
flukes (cont.):
  blood 2.35
  liver-fluke 2.192
  see also flatworms
fluorescence 1.124, 6.47, 6.267
  fluorescent lamp 6.138–9, 6.177, 6.202
fluoridation 6.139
fluorine 1.106, 1.124, 6.71, 6.139
fluorite 1.124
fluorocarbons 6.72, 6.139
  see also chloro-fluorocarbons; teflon
fluoroscopy see radiology
fluorspar see fluorite
flute 5.159
  American Indian 5.15
  duct 5.137
  harmonic 5.199
  kaval 5.244
  nose 5.325
  notch 5.323, 5.414
  panpipes 5.338
  piccolo 5.348
  pitch 5.352
  shakuhachi 5.414
  side-blown 5.15
  vessel 5.324, 5.472, 5.473, 5.474
  whistle 5.156
flux 6.139, 6.325
  magnetic 1.202
fly see flies
fly agaric 2.88
fly-by-wire system 6.139, 6.140
flycatcher family 2.123–3
  kingbird 2.178
  manakin 2.200–1
  silky 2.310
  umbrella bird 2.352, 2.353
flying see flight
flying boat 6.8, 6.140
Flying Dutchman 9.13
flying fish 2.123, 2.124
flying fox 2.123
flying frog 2.123–4
flying gurnard 2.124
flying lemur 2.73
flying lizard 2.124
flying phalanger 2.124
flying shuttle 6.140, 6.190, 6.208
flywheel 6.140, 6.163
FM (frequency modulation) 6.6, 6.20, 6.231
FMS (flexible manufacturing system) 6.134, 6.211
f-number 6.16
Fo, Dario 5.159
foam-nesting frog 2.124, 2.125
focal length 6.198–9, 8.51
focal ratio 8.51
Foch, Ferdinand 4.120, 4.376
focus 8.51
  of ellipse 8.45
fodder see feedstuffs, animal
foetus see embryo
fog 1.124
  smog 1.307, 1.308
Fogazzaro, Antonio 5.159
Fogel, Robert W. 4.153
Föhn winds 1.61, 1.124
Fokine, Michel 5.159, 5.285
folds, geological 1.124, 1.125
  anticline 1.15, 1.125
  monocline 1.219
  mountains 1.23–4, 1.285
    and tectonic plates 1.261
    see also Alps
  nappe 1.224
  syncline 1.125, 1.137, 1.329
foliation 1.124
folk:
  arts and crafts 5.159
    see also Arts and Crafts
  dance 5.159–60, 5.219–20, 5.310, 5.427
  festivals 5.160
folk (cont.):
  lore and literature 5.307, 5.396, 5.417, 7.121
    see also myths and legends
  music 5.160–1
    collectors see Bartók; Dvořák; Grainger; Kodály; Sharp, Cecil
    instruments see bagpipes; banjo; bull-roarer; drone; fiddle; hardingfele; bow, musical; piobaireachd; rattles; shawms; xylophone
    integrated in nationalist work see Bartók; Grieg; Vaughan Williams
    Korea 5.251
    North Africa 5.321
    pop 5.358
    songs 5.29–30, 5.160–1
    see also dance above
folly (building) 5.161
Fomalhaut 8.51
Fomors 9.23
Fon people: religion and myths 9.22, 9.25, 9.37
Fontaine, Hippolyte 6.158
Fontaine, Pierre-François 5.146
Fontainebleau, School of 5.161
Fontana, Carlo 5.161
Fontane, Theodor 5.161–2
Fontenoy, battle of 3.30, 3.128
Fonteyn, Margot 5.30, 5.324, 5.393
food 2.231, 6.140–1, 7.121, 7.123
  additives 6.16, 6.27, 6.46, 6.140, 6.141, 6.176
    see also spices, herbs, and flavourings
  aid 7.116–18, 7.121
  animal see feedstuffs
  aquaculture 6.18, 6.133, 6.316
  bread 6.28, 6.45–6
  breadfruit 2.41
  buckwheat 2.44
  butter 6.52, 6.215
  calories 2.18, 2.50
  carbohydrates see carbohydrates
  cereals see cereals and grains
  chain 2.124, 6.96, 6.125
  chocolate and cocoa 2.46, 2.49, 6.34, 6.73, 6.87
  cocoa 2.49
  confectionery 6.86–7
  consumption see gastro-intestinal tract
  crops see crops
  customs 7.121, 7.123
  deficiency see deficiency diseases
  diet 2.92
  digestion see gastro-intestinal tract
  drink see drinks
  eating disorder see anorexia nervosa
  fats and oils see fats, oils, and fatty acids
  fish see fishing and also in particular cod; herring; mackerel; pilchard; shellfish; sprats; whiting
  flavouring see spices, herbs, and flavourings
  fruit see fruit
  fungi as 6.147–8
  honey 6.16–17, 6.86
  hunger see hunger
  ice-cream 6.95
  kosher 7.178
  lack of see famine; hunger
  legumes see legumes
  malnutrition 7.190
  margarine 6.215, 6.384
  meat see meat
  milk 2.211
  nutrients 2.231
  nutrition 2.231
  poisoning 2.39
  preparation see cookers and stoves
food (cont.):
  preservation and storage 6.141–2
    additives see additives above
    canning 6.17, 6.57, 6.105, 6.142, 6.147
    curing fish 6.132–3
    drying 6.132, 6.141, 6.146, 6.147
    extrusion 6.142
    fermentation 6.27, 6.46, 6.68
    freeze-drying and freezing 6.38, 6.97, 6.142, 6.146
    heat-sterilization see canning
    irradiation 6.188
    pasteurization 6.259
    pickling 6.141
    refrigeration see refrigeration
    smoking 6.132–3
  processors and mixers 6.142
  prohibitions see totemism
  and religion 7.281
  starch see starch
  sweetener 6.140, 6.342
    see also sugars
  technology 6.142
    see also preservation above
  vegetables see vegetables
  vegetarianism 7.8, 7.325
  vitamins see vitamins
  web 2.124
  see also agriculture
Food and Agriculture Organization (UN) 4.120, 7.180, 7.321
fool 5.116, 5.162, 5.307
fool's gold 1.270
foot see feet
football 4.326, 7.122, 7.123
footman moth 2.124–5
footwear manufacture 6.142
forage 6.142–3, 6.302
  harvester 6.143
  see also feedstuffs, animal
foraminiferan 2.125
Forbes, George William 4.77, 4.120
forbidden band (band theory) 1.32
forbidden line (astronomy) 8.51
Force Acts 4.121
force (aggression) see violence; wars
force (physics) 1.124, 6.143, 8.51–2
  interactions 1.170
  lever 6.92, 6.199, 6.200, 6.217
  measurement 1.124, 1.227
  mechanical advantage 6.151–2, 6.199, 6.217, 6.285
  moment 1.218
  strong 1.324
  torque 6.103, 6.110, 6.152, 6.357
  weak 1.364
Ford, Ford Madox 4.63, 5.162
Ford, Gerald R. 4.121, 4.287
Ford, Henry 4.121, 4.213, 4.225, 4.350, 6.23, 6.26, 6.143, 6.216, 6.234
Ford, John (1586–1639, dramatist) 5.162
Ford, John (1895–1973, film director) 5.72, 5.162
Ford, Robert 4.171
fore-and-aft rig 6.143
  barque 6.30
  brigantine 6.47
  cutter 6.94
  ketch 6.190, 6.306
  sails 6.304
  schooner 6.76, 6.298, 6.308, 6.310
  see lateen rig
forecasting:
  economic 7.100
  weather see meteorology
foreign aid and investment see aid
foreign policy see under individual countries
forensic medicine 6.143, 6.217
forester moth 2.46

forests and trees *2.125*, *2.345–6*, *2.347*
bark *see* bark
coniferous *see* conifers
deciduous *see* deciduous trees and shrubs
defoliant **6.**98
deforestation and forestry **6.**69, ***6.****143*, **6.**143, **6.**161, **6.**364, **6.**387, **7.**82, ***7.****83*, **7.**349, **8.**82, ***8.****83*
forester **2.**254
evergreen *see* evergreen plants
fire **6.**132
forestry *see* deforestation *above*
fossil **2.**137
legumes **2.**54–5, **2.**56, **2.**208
and pollution **6.**2
rain-forest *see* tropical rain-forests
in succession **2.**328
taiga **2.**333
tree rings, dating from **6.**96, **6.**290
turpentine from **6.**366
see also wood
forgery ***7.****123*
forget-me-not ***2.****125*
forging and blacksmithing ***6.****39*, ***6.****143–4*, **6.**221, **6.**383
anvil **6.**16, **6.**39
steam-hammer **6.**236, **6.**331
steel **6.**144, **6.**188
trip-hammer **6.**363
welding **6.**383
Forlanini, Enrico **6.**173
formaldehyde **1.**212–13, **6.**353
see also under phenol
Formalism ***5.****162*, **5.**293
formic (methanoic) acid **1.**213, **6.**222
Formigny, battle of ***3.****128*
Formosa *see* Taiwan
Forms, Theory of **7.**237, **7.**322
formwork **6.**144
Fornax (constellation) ***8.****52*
Fornax System (dwarf galaxy) ***8.****52*
Forrest, Sir John, 1st Baron ***4.****121–2*, **4.**195
Forrest, Nathan Bedford ***4.****121*
Forster, Edward Morgan ***5.****162*, **5.**259
Forster, Thomas **3.**126
Forster, William Edward ***4.****122*, **4.**333
forsythia ***2.****125*
Fort Stanwix ***3.****128*
Fort Sumter ***4.****122*
Fort William (India) **3.**109
fortification (wine) **6.**11
fortifications (defensive) **6.**161, **6.**225
FORTRAN **6.**84, **6.**85, **6.**86
Forty-Five Rebellion **3.**123, ***3.****128–9*, **3.**291
forum ***3.****129*
forward markets **7.**128
Foscolo, Ugo ***5.****162*
Fosse Way ***3.****129*
fossil fuels ***1.****125*, **2.**126, ***6.****144*
burning and pollution **6.**2, **6.**9, **6.**144, **6.**161, **7.**135
see also coal; lignite; natural gas; peat; petroleum
fossils ***1.****125*, ***2.****125–6*
amber **1.**11
animals:
ammonite **2.**8
Archaeopteryx **2.**14
belemnite **2.**28
coelacanth **2.**71
coral **2.**76
giant elk **2.**173
graptolites **2.**143
hominids ***2.****126*
Irish elk **2.**173
Jaymotius **2.**175
lungfish **2.**195
synapsid **2.**333
chemofossil **2.**62
fossils (*cont.*):
chronological table **9.**57
fuel *see* fossil fuels
oil exploration **2.**232–3
plants ***2.****258*
calamite **2.**49–50
clubmoss **2.**69
glossopteris **2.**137
horsetails **2.**164
Rhynea **2.**284
stromatolite **2.**327–8
time scale ***9.****57*
trace **2.**345
see also extinct species; palaeontology
Foster, Norman (Foster Associates) ***5.****162*
Foster, William Zebulon ***4.****122*
Foucault, Jean Bernard Léon ***1.****125*, **8.**52–3
Foucault's knife-edge test ***8.****52*
Foucault's pendulum **1.**125, ***8.****52–3*
gyro-compass **6.**163
Foucault, Michel ***7.****123*
Fouché, Joseph, duc d'Otranto ***3.****129*
foundations (buildings) **6.**66, **6.**112, ***6.****144–5*, **6.**270, **6.**314, **6.**367
Founding Fathers ***3.****129–30*
see also in particular Franklin, Benjamin; Madison, James; Washington, George
foundling hospital ***3.****130*
foundry *see* casting and founding
fountain pen **6.**259
Fouquet, Jean ***5.****162*
Fouquet, Nicolas, vicomte ***3.****130*
Four Days War **3.**14
Four Evangelists **9.**12
Four Great Diamond Kings **9.**33
Four Heavenly Kings **9.**33
Four Modernizations ***4.****122*
Four Noble Truths **7.**97, ***7.****123*, **7.**217
Fourcroy, A. F. **6.**197
Fourdrinier, Henry and Sealy: machine ***6.****145*, **6.**256
Fourier, François Marie Charles **4.**317
Fourier, Joseph: Fourier analysis ***1.****125*
Fourneyron, Benoît ***6.****145*
Fournier, Rafael Calderón **4.**87
four-stroke engines **6.**254, **6.**262, **6.**263, **6.**303
see also engines *under* petrol
Fourteen Points ***4.****122*
Fourteenth Amendment ***4.****122*
four-wheel drive **6.**145
Fowler, John **6.**231
Fowles, John ***5.****162*
Fox, Charles James **3.**54, ***3.****130*, **3.**158, **3.**380, ***4.****122*, **4.**142
Fox, George ***3.****130*, **3.**297, **7.**256
Fox, William **7.**302
Fox family (spiritualists) **7.**297
Fox Talbot, William Henry **5.**445, **6.**268, **6.**344
Foxe, John **3.**90, ***3.****130*, **3.**364
foxes ***2.****114*, **2.**126
foxglove family **2.**29, ***2.****126*
calceolaria **2.**50
digitalis from **2.**369
lousewort **2.**195
speedwell **2.**319
Fra Mauro formation ***8.****53*
Fracastoro, Girolamo ***2.****126–7*
fractal ***1.****125–6*
fractional distillation ***6.****145*, **6.**369
argon **6.**20
chloro-fluorocarbons **6.**72
coal-tar **6.**79
fractionation column **6.**264
heavy water **6.**167
lubricating oil **6.**210
fractional distillation (*cont.*):
neon **6.**239
nitrogen **6.**242
oxygen **6.**254
fractures, bone **2.**126, ***2.****127*, ***9.****58*
Fragonard, Jean-Honoré ***5.****162*
FRAM (ferroelectric random-access memory) **6.**21
frame, building ***6.****145–6*
cast iron **6.**297
curtain wall **6.**94, **6.**146
portal **6.**276
skyscraper **6.**322
space **6.**51, **6.**326
system building **6.**343
frame drum ***5.****162–3*
Franc Zone ***7.****124*
France, Anatole ***5.****163*
France ***1.****126*, ***3.****130*, ***4.****122–3*, ***7.****353*
agriculture **6.**43, **6.**295, **7.**163
aircraft **6.**138, **6.**139
airships **6.**10, **6.**156
balloons **6.**29, **6.**224, **6.**232, **6.**257
Concorde **6.**339, **6.**340
military **6.**224
monoplane **6.**40
navigation **6.**163
parachute **6.**257
saddles **6.**166
seaplane **6.**311
anthropology *see* Lévi-Strauss
arts **3.**203, **5.**254, **5.**371
abstract **5.**23, **5.**128, **5.**244, **5.**260, **5.**286
Académie française **5.**2
architecture *see under* architecture
art nouveau **5.**168, **5.**195–6
Augustan Age **5.**26, **5.**113, **5.**301, **5.**372
caricature and caricaturists **5.**125
ceramics and pottery **5.**337
cinema *see under* cinema
criticism *see under* critics
engraving and engravers **5.**55
Enlightenment **5.**27, **5.**148, **5.**302, **5.**359, **5.**392, **5.**402, **5.**479
etching **5.**457
furniture and cabinet-making **5.**56, **5.**168
garden art and landscape gardening **5.**103, **5.**261
glass **5.**168, **5.**254
graphic art and design *see under* graphic art
illuminated manuscripts **5.**162, **5.**339
interior design **5.**56, **5.**146, **5.**259, **5.**261, **5.**269
Neoclassical *see under* Neoclassicism
ornaments **3.**203, **5.**254
painting *see under* painting
photography **5.**25, **5.**81, **5.**103, **5.**121, **5.**128, **5.**312, **5.**375
print-makers **5.**133, **5.**171–2
Renaissance **5.**162, **5.**366, **5.**411
sculpture *see under* sculpture and carving
set and stage design **5.**58, **5.**87, **5.**348
see also literature; music *below*
astrologer **3.**258–9
astronomy **8.**11, **8.**19
astronautics **8.**10
celestial mechanics **8.**125
comets **8.**15, **8.**46
constellations **8.**178
galaxies **8.**161
Moon **8.**87
observatory **8.**116–17
optics **8.**140
photography **8.**119–20
planets **8.**19, **8.**76, **8.**84
France (*cont.*):
astronomy (*cont.*):
stars **8.**87, **8.**187
see also Bessel; Cassini; Lacaille; Lagrange; Lalande; Laplace
buildings **6.**51, **6.**112, **6.**146
see also under architecture
chemistry and chemists **1.**191
chemical industry **6.**70
compounds **1.**38, **6.**76, **6.**197
dry-cleaning **6.**109
electrolysis **6.**168
gases **1.**97, **1.**132, **1.**190, **6.**76
halides **1.**147
nitro-cellulose **6.**242
organic **1.**368
physical **1.**276
colonies *see* empire and colonialism *below*
computers **6.**32, **6.**86
domestic appliances **6.**281
economy, trade, and finance **7.**155, **7.**299
co-operatives **7.**69
economists **3.**297, **4.**222, **7.**100, **7.**328
financiers **3.**123–4, **3.**160, **3.**205, **3.**249
Franc Zone **7.**124
Gold Standard **4.**138
in Group of Seven **7.**136
industrial relations **7.**304, **7.**315
laissez-faire **3.**202
nationalization **7.**212
in OECD **7.**221
taille tax **3.**347
trade routes **3.**107
see also European Community; European Economic Community *and also* finance *below*
education **7.**20, **7.**78, **7.**245, **7.**293
scientific institutions **6.**1, **6.**67, **6.**87
student revolts **4.**331
electoral system **7.**190
empire and colonialism ***4.****126–7*
in Africa **4.**49, **4.**55, **4.**108, **4.**116–17, **4.**127, **4.**306, **4.**307
see also French Equatorial Africa; French West Africa
in America **1.**206, **1.**228, **3.**213, **4.**199
see also wars *below*
in Asia **3.**106, **3.**288, **4.**127
see also French Indo-China
in Canada **1.**58, **1.**188, **1.**190, **3.**68, **3.**135, **3.**203
see also Cartier, Jacques
decolonization **4.**4, **4.**321
mandates **7.**191
in South America **4.**43, **4.**127
wars:
in America **3.**7, **3.**134, **3.**197–8, **3.**202, **3.**284
in French Indo-China **4.**98, **4.**99, **4.**127, **4.**133
see also East India Company, French *and also* explorers *below*
energy and fuel **6.**89, **6.**117, **6.**145, **6.**178, **6.**197, **6.**294, **6.**324, **6.**355, **6.**377, **7.**106
engines **6.**363
environment **7.**260
ethnic groups *see* Basques
explorers and navigators:
Africa **4.**211
Canada and North America **1.**53, **1.**58, **1.**188, **1.**190, **1.**206, **1.**228, **3.**62, **3.**68
see also Cartier, Jacques
map **3.**122
Pacific and circumnavigators **1.**42, **1.**97

France (*cont.*):
explorers and navigators (*cont.*):
see also empire *above*
finance *see* economy, trade, and finance *above*
food and drink **6.**17, **6.**46, **6.**142, **6.**215, **6.**386
foreign policy:
aid from **7.**8
Concert of Europe **7.**135
Council of Europe **7.**71
entente cordiale **4.**112
Évian Agreements **4.**114
Munich Pact **4.**227
Stresa Conference **4.**331–2
Sykes–Picot Agreement **4.**335
Tilsit Treaties **4.**345
Villafranca di Verona Treaty **4.**368
Washington Conference **4.**372
Western European Union **4.**375
see also wars and defence *below*
government *see* politics and government *below*
historiography **3.**83, **3.**190, **4.**153, **5.**184, **5.**397
immigration **7.**105
intellectuals **3.**299
land reform **7.**179
language **7.**156, **7.**269
as lingua franca **7.**185
pidgin **7.**235
in Québec **7.**257
laws/legal system **3.**239, **4.**78
advocate **7.**4
case **7.**41
civil, see *Code Napoléon*
commercial **7.**282
courts **7.**15, **7.**73, **7.**250
International Court of Justice **7.**164
judges **7.**173
judicial review **7.**173
legal aid **7.**182
literature ***5.**163*
autobiography and memoirs **5.**89, **5.**368
encyclopedist *see* Diderot
journalism **3.**50, **3.**102, **5.**172, **5.**400, **7.**216
Renaissance **5.**302, **5.**372
stories **5.**34, **5.**286, **5.**291, **5.**372
tragedy **5.**113, **5.**372, **5.**459
see also under critics; drama; essays; poetry; naturalism; novels; Romanticism *and also* philosophy *below*
map **3.**137
Masonic Order **7.**195
mathematics and mathematicians **1.**59, **1.**86, **1.**191, **3.**102
'Bourbaki' group **1.**42
calculator **6.**53–4, **6.**259
catastrophe theory **1.**334–5
engineering **6.**122
Galois theory **1.**131, **1.**186
mathematical physics **1.**120, **1.**261
'universal' **1.**55, **1.**261
see also Pascal
measurement, scientific **6.**368–9, **6.**371
medicine and health care **3.**297, **6.**16, **6.**105, **6.**177, **6.**193, **6.**205, **6.**218, **6.**253, **6.**259, **6.**340
drugs **6.**105
immunization **6.**177
physician **3.**297
stethoscope **6.**193, **6.**218
surgery and antisepsis **6.**16, **6.**205, **6.**253, **6.**257, **6.**340
see also Pasteur
metallurgy **6.**168, **6.**188
museum **6.**255
music **5.**56, **5.**291

France (*cont.*):
music (*cont.*):
keyboard **5.**87, **5.**115, **5.**153, **5.**163, **5.**187, **5.**373, **5.**398
polyphony **5.**89, **5.**256
Romantic **5.**45, **5.**153, **5.**266
sacred **5.**89, **5.**137, **5.**256, **5.**325
symphonies **5.**138, **5.**392, **5.**398
troubador **3.**361
violin **5.**254, **5.**270
see also under ballet; ballet and dance companies; choreographers; composers; opera; songs
nuclear technology **6.**57, **6.**245, **6.**292
old age **7.**221
optics and glass **6.**156
paper **6.**257
philosophy:
economics **3.**160
encyclopedist *see* Diderot
Enlightenment **3.**116–17
phenomenology **7.**199
political **3.**43, **3.**84, **5.**302, **7.**279
religion and theology **3.**1, **3.**37, **3.**43, **3.**84, **3.**271, **7.**279
scientific and mathematical **3.**37, **3.**102, **3.**271
social **4.**278, **4.**300, **7.**123
see also continental philosophy; Descartes; existentialism; Rousseau; Voltaire
photography **6.**74, **6.**75, **6.**94, **6.**109, **6.**210, **6.**241, **6.**268
physical geography and geology ***1.**126*
caves **1.**188
Channel (La Manche) **1.**110
geologists **1.**88, **1.**148
islands **1.**352
mountains *see under* mountains
plateau **1.**206–7
regions **1.**49, **1.**248
see also regions *below*
rivers **1.**198, **1.**282, **1.**294, **1.**298, **4.**299
physicists **1.**17, **1.**35, **1.**39, **1.**81–2, **1.**123, **1.**127, **1.**132, **1.**178, **1.**255, **6.**169
see also Foucault, Jean
politics and government:
Action Française **4.**2
ancien régime **3.**12, **7.**213
Bill of Rights **7.**32
cardinals **3.**302, **3.**304
Christian Democracy **7.**47
civil service **7.**52
coalition **7.**54
communism **4.**343
conservative **4.**192
democratic **4.**221
diplomacy **3.**74, **3.**348, **4.**337, **4.**342
Directory **3.**104
electoral system **7.**260
Fascist **4.**191
Field of Cloth of Gold **3.**126
financier **3.**249
Girondin **3.**141
Huguenot ***3.**170–1*, **3.**345
'Hundred Days' **4.**158
intelligence services **7.**162
intendant **3.**179
interest group **7.**163
Jacobins **3.**129, **3.**183
Jacquerie **3.**183
journalist **3.**56
legislature **7.**182
Liberals **4.**118, **4.**144
Marxist *see* Althusser
National Assembly **3.**246
papal envoy **3.**233
Parlement **3.**270
philosophers **3.**43, **3.**84, **5.**302, **7.**279

France (*cont.*):
politics and government (*cont.*):
political analyst **4.**346
populism **7.**242
presidents: listed **9.**70
republican **4.**77, **4.**129–30
resistance **4.**38
revolutionary **4.**41, **4.**63
Socialists **4.**23, **4.**42–3, **4.**92, **4.**173, **4.**216, **4.**221, **4.**300, **4.**317
States-General **3.**340
terrorist organization (OAS) **4.**251
women's suffrage **7.**332
XYZ affair **3.**386
see also Revolution; rulers *below*
prehistory **3.**61, **3.**342, **6.**218
psychology **7.**32, **7.**77
racism **4.**137
regions and provinces:
east **3.**68, **3.**98, **3.**130, **3.**211, **3.**293–4, **4.**10
list of ***9.**75*
north **1.**248, **3.**50, **3.**68, **3.**211
physical geography **1.**49. **1.**248
south **1.**49, **3.**18, **3.**23, **3.**54, **3.**98, **3.**137, **3.**203, **3.**293–4
west **3.**15–16, **3.**18, **3.**23, **3.**50, **3.**137, **3.**203, **3.**371
religion/theology **7.**171, **7.**205, **7.**335
Albigenisans **3.**7
Huguenots ***3.**170–1*
Jews **3.**103
megalithic **3.**61
monk **3.**277–78
philosophers **3.**1, **3.**37, **3.**271
Protestants **3.**58, **3.**170–1
saints **3.**16–17, **3.**40, **3.**51, **3.**189, **9.**40–3
see also individual names
see also Calvin
Revolutions:
of 1789 ***3.**134*, **7.**66
Bastille stormed **3.**36
battles **3.**369
and class struggle **7.**53
Committee of Public Safety **3.**83
and conservatism **7.**66
guillotine **3.**150
Jacobins **3.**129, **3.**183
leaders **3.**32, **3.**97, **3.**102, **3.**159, **3.**221, **3.**305, **3.**306, **3.**313–14, **3.**328
Rights of Man and Citizen **3.**304
sans-culottes **3.**316
Tennis Court Oath **3.**351
Terror, reign of **3.**351–2
of 1848 **4.**288
July **4.**177
Paris Commune **4.**260
rivers *see under* physical geography *above*
rulers (chronologically):
Carolingians (751–888, 893–923, 936–87) *see under* Carolingian empire
Capetians (987–1328) *see* Capetian dynasty
Valois, House of (1328–1498) *see* Valois
Orléans, line of (1498–1515) **3.**50, **3.**263, **3.**369
Angoulême line (1515–89) *see* Angoulême line
Bourbon, House of (1589–1795, 1814–30) *see* Bourbons
Bonapartes and First and Second empires **9.**68
see also Napoleon
listed ***9.**68*
Queens **3.**16, **3.**113, **3.**222, **3.**227

France (*cont.*):
safety **6.**248–9
ships **6.**237
lifeboat **6.**200
navigation **6.**163, **6.**201
port **6.**277
sailing **6.**90, **6.**146
steamship **6.**332
submarines **6.**147, **6.**337
warships **6.**106, **6.**188
social services and security **7.**289, **7.**293
sociology **7.**63, **7.**307
see also Durkheim
sport and recreation **7.**76
bull-fighting **7.**37–8
cycling **7.**79
horse-racing **7.**147
pétanque (boule) **7.**234, **7.**287
tennis **7.**308
technology and science **6.**57, **6.**132
see also entries below and above prefixed **1.** *and* **6.**
telecommunications **6.**69, **6.**235, **6.**349, **7.**216
textiles **6.**189, **6.**208
and UN **7.**277, **7.**326
urban areas **6.**277
see also Paris
vehicles and transport **6.**166, **6.**344
bridges **6.**18, **6.**112, **6.**146, **6.**260–1, **6.**293, **6.**311, **6.**341, **6.**361
canals **6.**18, **6.**56, **6.**199, **6.**298–9
railways **6.**161, **6.**168–9, **6.**294, **6.**367
road **6.**52, **6.**93, **6.**234
space and rocketry **6.**111, **6.**301, **6.**307
tunnels **6.**69, **6.**280
see also aircraft; ships *above*
wars and defence:
in America *see under* empire *above*
arms *see* weapons *below*
army **4.**19, **4.**127
dragoons **3.**105
leaders **4.**63, **4.**211
in America **3.**201–2, **3.**239
Austrian Succession **3.**318
Crusades **3.**374
Hundred Years War **3.**149
and Louis XIV **3.**370
Napoleonic Wars **4.**94, **4.**227–8, **4.**240
Revolutionary Wars **3.**61, **4.**319
Spanish Succession **3.**371
World War I **4.**120, **4.**130, **4.**174, **4.**268, **4.**377
World War II **4.**94–5, **4.**268, **4.**273, **4.**289
see also under Revolutions *above*
with Austria **4.**206
Austrian Succession **3.**29–30
with Britain *see under* Britain
Cambrai, League of **3.**58
castle **3.**63
with China **4.**314
colonial *see under* empire *above*
conscription **7.**66
Courtrai, battle of **3.**89
Crimean **4.**27–8, **4.**88, **4.**164
Devolution (with Spain) **3.**102–3
with Dutch **3.**254
Etaples, Treaty of **3.**118
French Foreign Legion **4.**127
French and Indian wars **3.**134, **3.**197–8, **3.**202, **3.**284
Fronde street fighters **3.**135
and Ireland **3.**366
knight **3.**37
national security state **7.**212
navy leaders **4.**93
Nine Years **3.**254, **3.**312
nuclear weapons **7.**220

France (*cont.*):
wars and defence (*cont.*):
Peninsular War **4**.368–9
Polish Succession **3**.287
with Prussia *see* Franco-Prussian War
Revolution *see* Revolutions *above*
Revolutionary Wars **3**.61, **4**.288, **4**.319
siege warfare **6**.371
spy scandal **4**.103–4
Suez **4**.332
Trafalgar **4**.349–50
Viking invasion **3**.306
Wars of Religion *3.134*, **3**.245, **3**.313, **3**.356
in WEU **7**.330
World War I **4**.211
World War II **4**.94–5, **4**.97, **4**.125, **4**.211, **4**.268, **4**.273, **4**.289, **4**.364
see also Napoleonic Wars; NATO; Seven Years; Spanish Succession *and also* foreign policy *above and* weapons *below*
water **6**.18
weapons:
arms control **7**.64
bullet **6**.52
guns **6**.210, **6**.312
missiles **6**.228, **6**.229–30
tanks **6**.344
trade **7**.18
see also wars *above*
welfare **7**.289, **7**.293
Franche-Comté *3.130*
franchise (economics) *7.123*
franchise (suffrage) *6.123–4*
see also electoral system; voting
Francia, José Gaspar Rodriguez de *4.123*, **4**.259
Francis I, king of Austria (*previously* Francis II, Holy Roman Emperor, 1768–1835) **3**.168, **4**.23, **4**.25, *4.123*
Francis I, king of France (1494–1547) **3**.70, **3**.126, *3.130–1*, **3**.167, **3**.263, **3**.272, **3**.369, **6**.199
Francis I, Holy Roman Emperor (1745–65) **3**.168
Francis II, Holy Roman Emperor *see* Francis I, king of Austria
Francis, Sir Philip **3**.158, **3**.192
Francis of Assisi, Saint **3**.28, *3.131*, **5**.230, **7**.205, *9.41*
Francis Ferdinand, Archduke of Austria *4.123*
assassinated *see* Sarajevo
Francis Joseph II, emperor of Austria and king of Hungary **4**.23, **4**.25, **4**.26, *4.123–4*, **4**.144, **4**.318, **4**.368
Francis turbine **6**.173
Francis Xavier, Saint *3.131*
Franciscans *3.131*, **7**.97, **7**.205
francium **1**.106, *1.126*
Franck, César *5.163*
Franco, Francisco *4.124*, **4**.322, **4**.324–5, **7**.28, **7**.224
François I, king of France **1**.53
Franco-Prussian War **4**.112, *4.124*, **4**.173, **4**.307–8
generals **4**.32, **4**.46, **4**.151
frangipani *2.127*
Frank, Anne *4.124*
Frankel, Zachariah **4**.176, **7**.66, **7**.260
Frankfurt, Treaty of *4.124*
Frankfurt School *7.124*
Adorno **7**.124
and capitalism **7**.124, **7**.138, **7**.192
and communism **7**.192
critical theory **7**.76, **7**.138
Habermas **7**.124, **7**.138
Marcuse **7**.124, **7**.192, **7**.215
frankincense pine **2**.192
Franklin, Benjamin **3**.34, *3.131–2*, *6.146*
and American Philosophical Society **3**.11
autobiography **5**.27
and cooker design **6**.89
as Founding father **3**.129
and Indians **3**.7
and lightning conductor **6**.202
and musical glasses **5**.179
and Peace of Paris **3**.269
and 'Plan of Union' **3**.7
and spectacles **6**.328
Franklin, Sir John *1.127*, **4**.17
franklin *3.132*
frankpledge *3.132*
Franks *3.98*, *3.132*
rulers **3**.68, **3**.69, **3**.70, **3**.71, **3**.79, **3**.211, **3**.275
Merovingians **3**.51, **3**.229–30
Verdun Treaty **3**.372
Salic law **3**.314
Franz Josef *see* Francis Joseph
Frasch, Herman: Frasch process *6.146*, **6**.338
Fraser, Malcolm **4**.195, **4**.378
Fraser, Peter *4.124*
Fraser River *1.127*
fraud *7.124*
electoral **7**.104
faked goods **7**.314
forgery **7**.123
Fraunhofer, Joseph: Fraunhofer line *8.53*
Frazer, Sir James George **7**.14, **7**.112, *7.124*, **7**.188, **7**.210, **7**.313
Frederick I (Barbarossa), Holy Roman Emperor (1123–90) *3.132*, **3**.167–8, **3**.209
Frederick I, duke of Austria and king of Romans (1286–1330) **3**.153
Frederick I, king of Prussia (1657–1713) **3**.49, *3.132*, **3**.133, **3**.267
Frederick II (the Great), king of Prussia (1740–86) **3**.132, *3.133*, **3**.294
and army **3**.23
and Austrian Succession, War of **3**.6, **3**.29
and Bavarian Succession, War of **3**.37
and Enlightenment **3**.300
and Seven Years War **3**.323–4
Frederick II, king of Denmark **3**.311, **3**.324
and astronomy **8**.176
Frederick II, Holy Roman Emperor (1194–1250) *3.132–3*, **3**.167–8, **3**.179, **3**.269, **3**.352
Frederick III, Holy Roman Emperor (1415–93) *3.133*, **3**.153, **3**.168
Frederick III, king of Sicily **3**.327
Frederick V (the Winter King), Elector Palatine and King of Bohemia (1596–1632) *3.133*, **3**.267, **3**.343
Frederick V, Holy Roman Emperor (1415–93) **3**.153
Frederick Augustus III, king of Saxony **4**.303
Frederick Louis, prince *3.133*
Frederick the Warrior, king of Saxony (1423) **3**.318
Frederick William I, king of Prussia *3.133*
Frederick William II, king of Prussia *3.133*, **3**.294
Frederick William III, king of Prussia *4.125*, **4**.146, **4**.278, **4**.345, **4**.366
Frederick William IV, king of Prussia *4.125*, **4**.134, **4**.278
Frederick William (the Great Elector) *3.133*
Fredro, Aleksander *5.163*
Free Cinema **5**.132, *5.163*
Free French *4.125*
free market *7.125*, **7**.126, **7**.139
free oscillation *1.127*
free path **8**.113
free port *7.125*, **7**.126
see also Singapore
free radicals *1.127*, **6**.188
Free Soil Party *4.125*
free speech **7**.82, **7**.124–5
free trade *see under* trade
free verse *5.163*, **5**.244, **5**.400, **5**.486, **5**.488
free will **7**.109, *7.125–6*, **7**.199
freedom *7.124*
of expression *7.124–5*
of information *7.125*
of press *7.125*
see also independence *and entries beginning* free
Freedom Riders *4.125*
Freemasons *3.133–4*, **4**.15, *7.195*
free-reed instruments *5.163*, **5**.376
accordion **5**.2
concertina **5**.108–9
harmonica **5**.199
harmonium **5**.200
Jew's harp **5**.240
mouth organ **5**.306
see also reed instruments
Free-Silver Movement *4.125*
free-tailed bat *2.127*
freezing:
and freeze-drying food **6**.38, *6.97*, **6**.142, *6.146*
freezing-point *1.127*
see also ice; refrigeration
Frege, Gottlob *1.127*, *7.126*, **7**.187, **7**.270
Frei (Montalva), Eduardo *4.125–6*
Frelimo War *4.126*
Frémont, John Charles *4.126*, **4**.218
Frémont (explorer) **1**.53
French, Sir John **4**.52
French, John Denton Pinkstone, 1st earl of Ypres *4.126*
French *see* France
French and Indian Wars *3.134*
French Community *4.126*
French East India Company **3**.106, *3.134*
French Equatorial Africa *4.127*, **4**.306
decolonization *see* Cameroon; Central African Republic; Chad; Congo; Gabon
French Foreign Legion *4.127*
French Guiana **4**.43, *4.127*
French Guinea **4**.306
French horn **5**.57, *5.163*, **5**.352
French Indo-China *4.127*
War **4**.98, **4**.99, *4.127*, **4**.133, *4.163*
French Revolution *see* Revolutions *under* France
French Somaliland **4**.306
French Wars of Religion *3.134*, **3**.245, **3**.313, **3**.356
French West Africa **4**.49, *4.127*
decolonization *see* Algeria; Benin; Burkina Faso; Côte d'Ivoire; French Guinea; Guinea; Mali; Mauritania; Niger; Senegal
freons *see* chloro-fluorocarbons
frequency *1.127*, **6**.117, *6.146*
and antenna **6**.15
bandwidth **6**.29
measurement **1**.154
modulation *see* FM
radio *9.54*
frequency (*cont.*):
sound **6**.325
and waveband **6**.380
Frescobaldi, Girolamo *5.164*
frescoes and murals *5.163–4*, *5.308*
France **5**.371
Germany **5**.498
grotesque **5**.195
Hittite **5**.208
Italy **5**.16, **5**.175, **5**.178, **5**.281, **5**.293, **5**.455
Korea **5**.251
Minoan **5**.297
plaster **5**.351
Sri Lanka **5**.429
Umayyad **5**.466
see also painting
freshwater fish **6**.133
air breathing **2**.181
angel-fish **2**.10
arapaima **2**.13–14, **2**.23
archerfish **2**.14
barramundi **2**.23
bichir **2**.29
bowfin **2**.40
bullhead **2**.45
burbot **2**.45
carp *see* carp
catfish **2**.57
characin **2**.61, **2**.341
charr **2**.61
cichlid **2**.66, **2**.341
climbing perch **2**.68
clingfish **2**.68
cuchia **2**.82
dace **2**.86
darter **2**.87
eel *see* eel
electric fish **2**.105, **2**.106
elephant-snout fish **2**.106
fishing *6.146*
garfish **2**.133
gizzard shad **2**.136
golden orfe **2**.140
goldfish **2**.141
gourami **2**.142
grayling **2**.144
gudgeon **2**.146
guppy **2**.148
killifish **2**.178
labyrinth fish **2**.181
lamprey **2**.182
loach **2**.192
milkfish **2**.211
mullet **2**.221
muskellunge **2**.223
paddlefish **2**.239
perch **2**.68, **2**.87, **2**.249–50, **2**.348
pickerel **2**.253
pike **2**.223, **2**.253
pike-perch **2**.253
piranha **2**.255
platy **2**.259
pufferfish **2**.271, **2**.272
rudd **2**.290
salmon **2**.61, **2**.292
shad **2**.136, **2**.305
snakehead **2**.314
stickleback **2**.325
sturgeon **2**.328
sunfish **2**.329–30
thread-fin **2**.34
tilapia **2**.341
trout perch **2**.348
Fresnel, Augustin Jean: and Fresnel lens **1**.17, *1.127*, *6.146*, **6**.201
fret *5.164*
Freud, Anna **7**.82, **7**.127
Freud, Lucian *5.164*, **5**.359
Freud, Sigmund *7.126*, *7.127*
and anxiety **7**.82
and arts **5**.368, **5**.440, **5**.499
and cocaine **6**.192
and dreams **7**.95
'Freudian slips' **7**.319

Freud, Sigmund (*cont.*):
and hypnosis **7.**150
and hysteria **7.**150
and instincts **7.**207
and learning **7.**291
neo-Freudians **7.**214
see also Jung
and personality **7.**233, **7.**253
and prejudice (motivation) **7.**245
and psychoanalysis **7.**252
and totem **7.**313
and unconscious **7.**319
Frey ***9.**13*
Freya **9.**2
Freyberg, Bernard Cyril, 1st Baron Freyberg ***4.**127*
Freyja ***9.**13*
Freyr **9.**12
Freyssinet, Eugène ***6.**146*, **6.**281
friars ***3.**134–5*
Dominican **3.**105, **3.**352, **3.**359, **7.**15, **7.**205
Franciscan **3.**131, **7.**97, **7.**205
see also monasticism
Frick, Henry Clay **4.**156
friction ***1.**127*, ***6.**146*
air resistance **1.**5–6
tidal **8.**171
friction drum ***5.**164–5*
Friedan, Betty **7.**332
Friedland, duke of *see* Wallenstein
Friedman, Milton ***7.**126–7*
Friedmann, Aleksandr Aleksandrovich ***8.**53*
Friedrich, Caspar David ***5.**165*, **5.**389
Friendly Islands *see* Tonga
friendly society ***7.**127*
Friends of the Earth **7.**135
frigate ***6.**146*, **6.**188
frigate bird ***2.**127*
Frigga **9.**2, ***9.**13*
frilled lizard ***2.**127*
fringe benefits ***7.**127*
fringe theatre ***5.**165*
Frink, Dame Elisabeth ***5.**165*
Frisch, Max ***5.**165*
Frisch, Otto **1.**150, **1.**210
Frisian Islands ***1.**127*
Frisians ***3.**135*
language **7.**131, **7.**156
Frisius, Gemma **6.**219
Frith, William Powell ***5.**165*
fritillary (butterfly) ***2.**127*
fritillary (plant) ***2.**127*
Froberger, Johann **5.**68
Frobisher, Sir Martin **1.**84, ***1.**127*, **1.**139, **1.**231
Fröding, Gustaf ***5.**165*
Froebel, Friedrich ***7.**127*
froghopper (bug) ***2.**127*, **2.**128
frogmouth (bird) ***2.**127*, **2.**128
frogs ***2.**127*
arrow-poison **2.**14–15
brain **2.**41
bullfrog **2.**45
flying **2.**123–4
foam-nesting **2.**124, **2.**125
rain **2.**276
Seychelles **2.**305
tadpoles **2.**333
tailed **2.**333
tree **2.**123–4, **2.**277, **2.**346
Xenopus **2.**372
Froissart, Jean ***3.**135*, **3.**273, ***5.**165*
Fromentin, Eugène ***5.**165*
Fronde ***3.**135*
front (meteorology) ***1.**128*, **1.**262, **9.**56
see also depression (meteorology)
Front de Libération nationale ***4.**127–8*
Frontenac, Louis de Buade, comte de ***3.**135*, **3.**197
Frontinus, Sextus Julius ***6.**146–7*
Frost, John **4.**239
Frost, Robert ***5.**165*
frost **1.**128, ***1.**129*, **1.**254
Froude, William **6.**237
fructose **6.**337
fruit bat **2.**123
fruit fly ***2.**128*, **2.**130–1
fruit plants (*mainly trees and shrubs*) ***2.**127–8*, ***2.**129*
apple **2.**13, **2.**129
apricot **2.**13
avocado pear **2.**19
berry *see* berries
black currant **2.**33
blackberry **2.**33
blackthorn **2.**33–4
blueberry **2.**35–6
cherry **2.**62
citrus *see* citrus
cranberry **2.**50
cultivation **6.**76, ***6.**147*, **6.**271
currants **2.**33, **2.**84, **2.**279
damson **2.**87
date palm **2.**88
drupe **2.**99, **2.**129
durian **2.**100
elder **2.**104–5
figs **2.**22, **2.**117
gooseberry **2.**141
granadilla **2.**142
grape **2.**142–3
greengage **2.**144
guava **2.**146
hickory **2.**158
huckleberry **2.**165
japonica **2.**174
jujube **2.**176
kiwi fruit **2.**180
litchi **2.**191
loquat **2.**194
mango **2.**201–2
mangosteen **2.**202
medlar **2.**206
melon **2.**207
mulberry **2.**221
nectarine **2.**225
nuts *see* nuts
papaya **2.**241
passion fruit **2.**142
passion vine **2.**243
peach **2.**245
pear **2.**246
persimmon **2.**250
pineapple **2.**254–5
plums **2.**87, **2.**129, **2.**144, **2.**259
pome **2.**263
pomegranate **2.**263
pumpkin family **2.**194, **2.**272, **2.**323
quince **2.**274
raspberry **2.**278
red currant **2.**279
rhubarb **2.**284
snowberry **2.**316
soursop **2.**317–18
squash **2.**323
strawberry **2.**327
sweet pepper **2.**332
tamarind **2.**334
tayberry **2.**336
tomato **2.**129, **2.**343
tree tomato **2.**347
watermelon **2.**362
see also nuts; vegetables
frustration **7.**7
Fry, Captain **4.**368
Fry, Elizabeth ***4.**128*
Fry, Roger **5.**6, **5.**50, **5.**327, **5.**362
Fry, W. T. **1.**227
Fuad I, king of Egypt **4.**108
Fuad II, king of Egypt **4.**116
Fuchs, Klaus ***4.**128*
fuchsia ***2.**128*
fucus ***2.**128*, **2.**129
Fudo-myoo **9.**13
fuel:
additive ***6.**147*
cell *see* cell (electrical)
injection *see* carburettor
smokeless **6.**323
synthetic **6.**343
wood **6.**36, **6.**89, **6.**167
see also energy; fossil fuels
fuel-oil ***6.**147*, **6.**265
Fuentes, Carlos **5.**165–6
Fugard, Athol ***5.**165–6*
Fugger, Sigismund **6.**257
Fugger family ***5.**135*
Fugitive Slave Acts ***4.**128*
fugue ***5.**166*
Fujimori, Alberto Keinya **4.**267
Fujiwara family ***3.**135*, **3.**185, **3.**232
Fujiyama ***1.**129*
Fulani empire ***3.**136*, **4.**128
language **7.**217
Fulbright, James William ***4.**128*
Fulk I, count of Anjou **3.**12
Fulkerson, Mary **5.**363
full moon **8.**118, **8.**119
Fuller, Loie ***5.**166*
Fuller, Richard Buckminster **5.**14, **5.**162, ***5.**166*, ***6.**147*, **6.**155
fuller's earth **1.**220–1
fulling ***6.**147*
fulmar ***2.**129*
Fulton, Robert ***6.**147*, **6.**332, **6.**337
Fúlù **9.**13
fumarole ***1.**129*
fumitory ***2.**129*
Fúmodàdì **9.**15
functional groups (chemistry) ***1.**129*
functionalism (arts) ***5.**166*, **5.**259–60, **5.**295, **5.**347, **5.**494–5
see also Bauhaus
functionalism (philosophy) ***7.**127*, **7.**203
functionalism (social science) **7.**14, **7.**112, ***7.**127*, **7.**190, **7.**229
functions (mathematics) ***1.**129*
fundamentalism ***4.**128*, ***7.**127*, **7.**166, **7.**168, **7.**331
Fundy, Bay of ***1.**129*
funeral customs **7.**74–5, ***7.**127–8*
see also tombs
fungi ***2.**129*, **6.**65, **6.**127, ***6.**147–8*
agaric **2.**4, **2.**88, **2.**160, **2.**222
bracket **2.**40
chantarelle **2.**61
and disease **2.**129, **2.**211, **2.**260, **2.**285, **2.**291, **2.**314, **2.**341–2
dry rot **2.**99
earth star **2.**102
fungicide ***6.**147*, **6.**311
honeyfungus **2.**160
in lichen **2.**188, **2.**189, **2.**282
mildew **2.**210
morel **2.**218
moulds **2.**220
mushrooms **2.**222, **6.**148
mycorrhiza **2.**224
puffball **2.**271
ringworm and **2.**285
rust **2.**291
shaggy ink cap **2.**304, **2.**305
smut **2.**314
stinkhorn **2.**326
and termites **2.**338
toadstool **2.**342
truffle **2.**348
yeast **2.**373
fungus gnat ***2.**129*
funicular railway ***6.**148*
Funj ***3.**136*
fur **2.**149
-farming **2.**64, **2.**80, **2.**213, **2.**223, ***6.**148*, **6.**205
trade ***3.**136*
Furies **9.**13
furnace:
solar **6.**324
see also blast-furnace; boiler; fluidized-bed furnace
furniture beetle ***2.**130*
furniture and cabinet-making ***5.**70*, ***5.**166*
Assyrian **5.**24–5
Bauhaus **5.**38
Britain **5.**3, **5.**5, **5.**98, **5.**105, **5.**204, **5.**304–5, **5.**315, **5.**351, **5.**416–17
chair, development of **5.**167
Empire style **5.**146
France **5.**56, **5.**168
Germany **5.**295
Knoss **3.**199
marquetry **5.**168, **5.**283, **5.**436
ormolu **5.**332
panelling **5.**338
Shaker **5.**413
Sumerian **5.**440
United States **5.**153, **5.**295, **5.**413
veneer **5.**470
see also carpets and rugs; clocks; interior design; tapestries
further education *see* higher education
Furtwängler, Wilhelm **5.**110
fuse ***6.**148*, **6.**316
Fuseli, Henry ***5.**166*
fusilier ***3.**136*
fusion, nuclear ***1.**232–3*, ***6.**148*
and astronomy ***8.**110*
carbon-nitrogen cycle **8.**20
proton–proton reaction ***8.**128–9*
bombs *see* hydrogen bombs
see also nuclear energy and power; nuclear reactors
Fust, Johann **6.**163
Futabatei Shimei **5.**235
futures markets ***7.**128*
Futurism **5.**29, **5.**51, **5.**80, ***5.**166*, **5.**287, **5.**347, **5.**396
Fuẓūlī **5.**462
Fyodor, king of Russia **3.**45, **3.**309
fyrd ***3.**136*

G

gabbro ***1**.130*
Gabo, Naum ***5**.168*
Gabon ***1**.130*, ***4**.129*, ***7**.353*
  decolonization **4**.4
  economy and trade **4**.252, **7**.124
  exploration **1**.183
  physical geography and geology **1**.130
gaboon viper ***2**.130*, **2**.131
Gabor, Dennis ***6**.149*, **6**.169
Gaboriau, Émile **5**.129
Gabriel, Angel **7**.166, **7**.175, **7**.209, **7**.250, ***9**.13*
Gabriel, Saint ***9**.41*
Gabriel, Ange-Jacques ***5**.168*, **5**.315
Gabriel, Jacques **5**.168
Gabrieli, Andrea **5**.168
Gabrieli, Giovanni ***5**.168*
Gadarenes ***9**.13*
Gadda, Carlo Emilio **5**.231
gadfly **2**.39, **2**.163
gadolinium **1**.106
Gadsden Purchase ***4**.129*
Gaea *see* Gaia
Gaelic:
  football **7**.123
  language **7**.42, **7**.156
  literature **5**.225–6, **5**.409, **7**.42, **7**.156
    Irish **5**.225–6
    Scottish **5**.408–9
  religion and myths **9**.9, **9**.23, **9**.25
gaff rig **6**.143, **6**.306
Gagarin, Yuri **4**.322, **4**.323, **6**.214, **8**.10, **8**.154
Gage, Thomas **3**.54, ***3**.136*, **3**.207
Gagnon, Emile **6**.18
Gaia (Gaea) ***9**.13*
  hypothesis ***1**.130*, ***7**.129*
Gaidar, Yegor **4**.386
gain, electrical **6**.13, ***6**.149*
Gainsborough, Thomas ***5**.168*, **5**.359
Gairdner, Lake ***1**.130*
Gairy, Matthew **4**.141
Gaiseric *see* Genseric
Gaitskell, Hugh Todd Naylor ***4**.129*
galactic *see* galaxies
galago **2**.46, **2**.47
Galahad, Sir **9**.5, ***9**.13*, **9**.14
Galápagos finch **2**.3, ***2**.130*
Galápagos Islands ***1**.130*, **1**.190
Galatia ***3**.136*
galaxies ***8**.54–6*
  of our solar system **8**.56
  age of **8**.2
  Andromeda **8**.5
  cD-type **8**.55
  elliptical **8**.55
  galactic clusters **8**.52, **8**.55, **8**.113, **8**.118
  galactic equator ***8**.54*
  galactic fountain **8**.54
  galactic halo ***8**.54*
  galactic latitude ***8**.54*
  galactic longitude ***8**.54*
  galactic pole ***8**.54*
  galactic rotation ***8**.54*
  galactic spur **8**.109
  lenticular **8**.84
  magnitude of **8**.38
  Markarian **8**.91–2
  N-type **8**.55–6
  recession of **8**.134
  Sculptor Group **8**.146
  Seyfert **8**.148–9
  Sombrero **8**.151
galaxies (*cont.*):
  spiral **8**.55, ***8**.158*
    barred **8**.14
  spiral arms of **8**.158
  Stéphan's Quintet **8**.161
  Whirlpool **8**.186
  see also Local Group; radio galaxies
Galaxy of our solar system ***8**.56*
Galba, Roman emperor **3**.307
Galbraith, John Kenneth **7**.4–5, **7**.125
Galczyński, Konstanty Ildefons ***5**.168*
gale ***1**.130*, **9**.56
Galen, Claudius **1**.332, ***2**.130*, **2**.190, **6**.350
galena ***1**.130*, **6**.197
Galerius **3**.104
Galicia ***3**.136*
Galilean satellites of Jupiter **8**.19, **8**.43, **8**.48, ***8**.56–7*, **8**.74, **8**.144
Galilee, Sea of ***1**.130*
Galileo Galilei ***1**.130–1*, ***6**.149*, **7**.276, ***8**.56–7*
  astronomical discoveries by **8**.19, **8**.48, **8**.57, **8**.74, **8**.140, **8**.145
  and catenary **1**.54
  as heretic **7**.1
  and moon **8**.147
  and parabola **1**.247
  and pendulum **6**.260
  and philosophy **7**.234
  and Scientific Revolution **3**.318
  and telescope **8**.11, **8**.137
  and temperature **1**.332
  and thermoscope **6**.350
galingale ***2**.130*
gall, oak apple **2**.132, **2**.231
gall bladder ***2**.130*
  gallstones **2**.131–2, **2**.326
gall midge ***2**.131*
gall wasp ***2**.132*, **2**.231
Galla (Oromo) language **7**.78
Gallager, John **7**.153
Gallatin, Abraham Alfonse Albert ***4**.129*
Gallé, Émile ***5**.168*
Galle, Johann **8**.2, **8**.106
galleon ***3**.136*, ***6**.149*, **6**.306, **6**.307
Galleotti **5**.393
galley and trireme ***3**.136*, **3**.248, ***3**.361*, **3**.361, ***6**.149*, **6**.276, **6**.306, **6**.307, **6**.375
gallfly ***2**.130–1*
Galli, Boutros Boutros **7**.322
galliard ***5**.168*
Gallic wars ***3**.137*
Gallicanism ***3**.136–7*
gallinule ***2**.131*
Gallipoli Campaign ***4**.129*
gallium **1**.106, ***1**.131*, ***6**.149*
gallium arsenide **6**.20, **6**.149, **6**.312, **6**.324
gallstones ***2**.131–2*, **2**.326
Galois, Évariste ***1**.131*
Galsworthy, John ***5**.168–9*
Galtieri, Leopoldo **4**.116, **7**.173
Galton, Sir Francis **1**.28, ***2**.132*, **7**.11, **7**.80, ***7**.129*, **7**.252
Galvani, Luigi **6**.374
galvanizing ***6**.149*
galvanometer **6**.13, **6**.113, ***6**.149*
Gambetta, Leon ***4**.129–30*
Gambia ***1**.131*, **4**.4, ***4**.130*, **4**.306, **7**.8, ***7**.353–4*
gambling **7**.40, ***7**.129*
game birds:
  curassow **2**.83–4
  grouse *see* grouse
  guineafowl **2**.147
  junglefowl **2**.176
  mallee fowl **2**.199–200
  partridge **2**.243
  pheasants *see* pheasant family
game birds (*cont.*):
  quail **2**.48, **2**.273
  shooting **7**.129
game theory ***1**.131*, **6**.240, **7**.259
gamelan ***5**.169*, **5**.222
Gamelin, Maurice Gustave ***4**.130*
games:
  board and table ***7**.33*
    backgammon **7**.24
    billiards **7**.31
    card-games **7**.40
    chess **7**.44
    dice **7**.89
    dominoes **7**.94
    mahjong **7**.189
    table tennis **7**.305
  singing **7**.286
  street **7**.300
  see also hobbies and pastimes; leisure; Olympic Games; play; sport and recreation
game-shooting ***7**.129*
gametes ***2**.132*
  cross-fertilization **2**.82
  fertilization **2**.115–16
  production **2**.207
  see also sexual reproduction
gamma camera **6**.243
gamma globulin **6**.40
gamma-radiation ***1**.131*, **6**.288
  diffraction **1**.90
  frequency **6**.117
  gamma-ray astronomy ***8**.57*
  measurement **6**.153
  in nuclear reactor **6**.245
  in particle accelerator **6**.258
  and radiotherapy **6**.292
Gamov, George ***8**.33*, **8**.57, **8**.98
Gamsakhurdia, Zviad **4**.133
Gance, Abel ***5**.169*, **5**.488
Gandhara sculpture ***5**.169*
Gandhi, Indira **4**.83, ***4**.130*, **4**.131, **4**.161, **4**.381, **7**.20, **7**.74
Gandhi, Mohandas Karamchand **4**.10, ***4**.130*, **4**.131, **4**.161, **4**.301, **7**.156
  and caste system **7**.41
  followers **4**.37
  and Hinduism **4**.152
  hunger-strike **7**.150
  and Indian National Congress **4**.83
  non-co-operation and passive resistance **4**.244–5, **4**.262, **4**.294, **4**.303, **7**.8, **7**.91, **7**.169, **7**.227
  and Round Table Conferences **4**.294
  and Salt March **4**.145
Gandhi, Rahul **7**.74
Gandhi, Rajiv **4**.84, **4**.130, **4**.161, **7**.20, **7**.74
Gandon, James ***5**.169*
Ganesha ***9**.13*
Gang of Four ***4**.130–1*
Ganges **1**.132
gangrene **2**.364
gangue **1**.132
Ganivet, Angel **5**.172
gannet ***2**.132–3*
gantry harvesting ***6**.150*
Ganymede (mythology) ***9**.13*
Ganymede (satellite of Jupiter) ***8**.57*
Gao *see* Songhay
gaper **2**.67
Garcia, Alan **4**.267
García Lorca, Federico ***5**.169*, **5**.170
García Márquez, Gabriel ***5**.169–70*
Garcia Moreno, Gabriel ***4**.131*
Garcia of León **3**.62
Garcia y Iñigues, Calixto ***4**.131*
Garcilaso de la Vega ***5**.170*
Garda, Lago di ***1**.132*
Garden, Alexander **2**.133
garden:
  art and landscape gardening ***5**.170*, **5**.171
    Britain **5**.62, **5**.103, **5**.245–6, **5**.348, **5**.380
    France **5**.103, **5**.261
  botanical **2**.21, **2**.38
  gardening technology **6**.196, **6**.294
  zoological **2**.375
  see also national parks
gardenia ***2**.133*
Gardiner, Stephen ***3**.137*
Gardner, Alexander **5**.482
Garfield, James Abram **4**.18, ***4**.131*, **4**.271, **4**.328
Garfield, Leon **5**.91
Garfinkel, Harold **7**.109
garfish ***2**.133*
garganey ***2**.133*
Gargantuan Basin ***8**.57*
gargoyle ***5**.170*
Garibaldi, Giuseppe **4**.66, **4**.88, ***4**.131*, **4**.291, **4**.312
Garibaldi Provincial Park **1**.132
Garland, Hamlin **5**.213
garlic ***2**.133*
Garnerin, Jacques **6**.138, **6**.257
garnet **1**.131
Garnett, David **5**.50
Garnier, Charles ***5**.170–1*
Garrett, Almeida **5**.359
Garrett, Pat **4**.38
Garrick, David ***5**.171*, **5**.499
Garrison, William Lloyd **4**.1
Garrod, Sir Archibald Edward ***2**.133*
Garvey, Marcus Moziah ***4**.131–2*, **4**.257, **7**.258
gas ***1**.132*, ***6**.23–4*, **6**.150
  acetylene (ethyne) **1**.114, **6**.76, **6**.123, **6**.297, **6**.383
  aerosol **1**.4, **6**.5
  air *see* air and atmosphere
  ammonia *see* ammonia
  argon **1**.20, **1**.106, **6**.20, **6**.96, **6**.240
  Avogadro's hypothesis **1**.28
  in biofuel **6**.36–7
  blood gas analysis **6**.40
  Bunsen burner **6**.25, **6**.52, **6**.193
  butane **1**.46, **6**.52, **6**.150, **6**.204, **6**.237, **6**.264
  carbon dioxide *see* carbon dioxide
  centrifuge **6**.67
  chlorine *see* chlorine
  chromatography **6**.73, **6**.217
  coal-gas *see* coal-gas
  compressor **6**.83
  constant **1**.132
  diffusion **1**.91
  electricity conducted through *see* electronics
  engines ***6**.150*, **6**.198
  ethane **1**.114
  ethene *see* ethylene
  ethyne *see* ethyne
  expansion **1**.117
  fluid mechanics **6**.137
  fluorine **1**.106, **1**.124, **6**.71, **6**.139
  grid ***6**.150*
  helium *see* helium
  Henry's law **1**.154
  hydrocarbons **6**.5, **6**.52, **6**.123, **6**.222
  hydrogen *see* hydrogen
  ideal **1**.132, **1**.165
  in incandescent lamp **6**.177
  kinetic theory **1**.183
  laser **6**.194–5
  lighting and heating **6**.79, ***6**.150*, **6**.197, **6**.201, **6**.236, **6**.386
    gas-discharge **6**.76, **6**.150, **6**.202
    gas-mantle **6**.25
  liquefaction/liquefied **1**.203, **6**.76, ***6**.150*, **6**.150, **6**.237
  liquid petroleum *see* liquefied petroleum gas

gas (*cont.*):
Maxwell–Boltzmann distribution ***1.208***
methanal **1.**212–13
methane *see* methane
mustard **6.**72, **6.**235
natural *see* natural gas
neon **6.**195, **6.**239
nerve **6.**72, **6.**239–40
nitrogen *see* nitrogen
oil *see* oil *under* diesel
oxygen *see* oxygen
'phlogiston' **1.**267
phosgene **6.**72, **6.**267
pioneers **1.**84–5, **6.**96
pipe **6.**150, **6.**236, **6.**270
plasma **1.**258, **1.**259
plumbing **6.**274
poison *see* poison and poison gases *and also* warfare *under* chemicals
standard temperature and pressure **1.**320
stove **6.**89
sulphur dioxide *see* sulphur dioxide
tear-gas **6.**345–6
town **1.**341, **6.**359
transportation **6.**76
turbine ***6.150***
ultrasound **6.**367
van der Waal's forces **1.**354
vapour *see* vapour
warfare *see under* chemicals
welding **6.**383
see also anaesthesia
gas gangrene **2.**364
Gascony ***3.137***
gas-cooled reactor **6.**3–4, **6.**150, **6.**243, **6.**244
gas-discharge lighting **6.**76, ***6.150***, **6.**202
Gaskell, Elizabeth Cleghorn ***5.171***
'gasohol' **6.**37
gasoline *see* petrol
Gass, Patrick **4.**194
Gassendi, Pierre **8.**12
gastraphetes **6.**92
gastric disorders *see under* gastro-intestinal tract
gastro-intestinal tract and intestines ***2.6***, ***2.92***, ***2.172–3***, ***9.59***
appendix **2.**13
colon **2.**72
digestion **2.**92
disorders **6.**266
cholera **2.**64
colitis **2.**72
diarrhoea **2.**101
diverticular disease **2.**95–6
dysentery **2.**101, **7.**59
peritonitis **2.**250
ulcer **2.**352
enzymes **2.**72, **2.**109, **2.**292
gastric physiologist **2.**26
mastication **2.**205
oesophagus **2.**232, **2.**331
rectum **2.**279
stomach **2.**326
swallowing **2.**331
see also kidney; liver; pancreas
gastropods:
cowry **2.**79
limpet **2.**189, **2.**190
see also shellfish; snails
gastroscope *see* endoscopy
gas-turbine ***6.150***
see also jet engines
Gates, Horatio **3.**58, ***3.137–8***, **3.**316
GATT *see* General Agreement on Tariffs and Trade
gaucho ***4.132***
Gaudí y Cornet, Antoni ***5.171***
Gaudier-Brzeska, Henri ***5.171***, **5.**327
gauge, railway ***6.150–1***
Gauguin, Paul ***5.171–2***, **5.**356
Gaukler **5.**297
Gaul ***3.138***, **3.**371–2
Gaulier, Philippe **5.**296
Gaunt, John of ***3.138***, **3.**202
gaur ***2.133***
Gauss, Karl Friedrich ***1.132***, **1.**191, **8.**9, **8.**46, **8.**114
Gautama Siddhartha *see* Buddha
Gautier, Théophile ***5.172***
Gavelis, Ričardas **5.**34
Gaveston, Piers ***3.139***
gavial **2.**135
gavotte ***5.172***
Gawain **9.**5, ***9.13–14***
'Gawain and the Green Knight, Sir' ***5.172***
Gawan, Mahmud **3.**33
Gay, John ***5.151***, **5.**172, **5.**485
'gay' people *see* homosexuality
Gaye, Marvin **5.**358
Gay-Lussac, Joseph Louis **1.**132
Gayomart ***9.14***
Gayoon, Maumon Abdul **4.**208
gazelles ***2.133–4***, **2.**323
GCC (Gulf Co-operation Council) ***7.136–7***
G-clamp **6.**358
GDP (gross domestic product) ***7.129***, **7.**211
Ge language **7.**295
gean **2.**62
gear ***6.151–2***
bicycle **6.**35–6, **6.**152
differential **6.**103
gearbox **6.**26, ***6.152***
gear-cutting machine **6.**212
oil **6.**210
pump **6.**286
see also transmission in vehicles
Gebel, Matthes **5.**289
gecko ***2.134***
geese ***2.141***
barnacle **2.**23
Canada **2.**51
greylag **2.**144
Hawaiian **2.**152
pink-footed **2.**255
snow **2.**316
gegenschein ***8.57–8***
Geiger, Hans: Geiger–Müller counter ***6.152–3***, **6.**186
Geikie, Sir Archibald ***1.133***
gelatine/gelatinous substance **6.**256, **6.**268, **6.**321
gel ***6.153***
Gelber, Jack **5.**267
gelignite ***6.153***
Gemayel, Pierre **4.**211
Gemini ***8.58***
Geminiani, Francesco ***5.172***
Geminids ***8.58***, **8.**195
gemstones (precious and semi-precious) ***1.133***
emerald **1.**108
garnet **1.**131
jade **1.**177
lapis lazuli **1.**187–8
moonstone **1.**221
opal **1.**238, **1.**239
peridot **1.**250–1
ruby **1.**287
sapphire **1.**293
topaz **1.**339, **1.**341
turquoise **1.**348
see also diamond; ornaments and jewellery
gender ***7.129–30***, **7.**278, **7.**280
see also men; women
gene *see* genes
genealogy *see* descent
genera **2.**134
General Agreement on Tariffs and Trade ***4.132***, ***7.129***, **7.**251, **7.**321
General Assembly of United Nations ***4.132***
general practitioner **7.**196
General Strike ***4.132–3***
Generation of '27 **5.**11, **5.**169, **5.**170, ***5.173***, **5.**240
Generation of '98 ***5.172–3***, **5.**333
generator, electrical **6.**2, **6.**113, ***6.153***
armature **6.**20
magnetohydrodynamic **6.**213
pioneer **6.**158, **6.**258
solar **6.**115, **6.**324
van de Graaff **6.**258, **6.**370
wind power **6.**386
see also battery; cell (electrical)
genes ***2.134***
alleles **2.**6, **2.**223–4, **2.**252
genome **2.**134, **2.**345
mutation **2.**223–4
pool **6.**155
see also chromosomes; genetics
Genet, Jean ***5.173***
genet (animal) **2.**134
genetics ***2.134***
abnormalities *see* inherited disorders *below*
behaviour **7.**28–9, **7.**252
cloning **6.**153, **6.**271
code ***2.134***
DNA and DNA fingerprinting **6.**105
engineering **2.**354, **6.**37, **6.**77, ***6.153–5***, **6.**218
see also clones
fingerprints **6.**105, **7.**120
genes *see* genes
geneticists **2.**24, **2.**119, **2.**205, **2.**207
genomes **2.**134, **2.**345
genotype **2.**134, **2.**252
inherited disorders **2.**65, **2.**133
cleft palate **2.**67
club foot **2.**69
cystic fibrosis **2.**86
Down's syndrome **2.**98–9
haemophilia **2.**149
spina bifida **2.**321
thalassaemia **2.**339
see also amniocentesis
mutations **2.**223–4, **6.**104
and psychology **7.**129
resources ***6.155***
sex determination **2.**304
transgenic organism **2.**345
translocation **2.**345
see also breeding; chromosomes; heredity; natural selection
Geneva ***3.139***
Conference ***4.133***
Conventions ***4.133***, **7.**2, **7.**44, **7.**69, ***7.130***, **7.**212, **7.**328
Genghis Khan ***3.139***, **3.**188, **3.**195, **3.**238
genie **9.**20
genito-urinary system, disorders of **6.**266
see also sexually-transmitted diseases
Gennep, Arnold van **7.**268
Genoa **3.**82, **3.**104, ***3.139***
genocide ***7.130***
see also Jews *under* Nazis
genomes ***2.134***, **2.**345
genotype ***2.134***, **2.**252
genre ***5.173***
painting and painters **5.**89, **5.**141, **5.**173, **5.**192, **5.**268, **5.**296, **5.**309, **5.**489
Netherlands and Flemish **5.**13, **5.**62, **5.**211, **5.**262, **5.**292, **5.**431, **5.**449
see also Intimisme
Genscher, Hans-Dietrich **4.**185
Genseric **3.**62, ***3.139***, **3.**369
gentians **2.**59, ***2.134***
Gentile da Fabriano ***5.173***
Gentileschi, Artemisia and Orazio ***5.173***
genus ***2.134***
geocentric coordinates ***8.58***
geocentric parallax **8.**116
geocentric system ***8.58***
geochemistry ***1.133***
see also chemistry
geochronology ***1.133***, **1.**136, **1.**274
geodesic (astronomy) **8.**35
geodesic dome **6.**147, ***6.155***
geodesy ***1.133***
Geoffrey, Count of Anjou **3.**13, **3.**284
Geoffrey of Monmouth **3.**24, ***3.139***
geography **6.**285
biogeography **2.**30
human ***7.130–1***
see also geology; geomorphology; maps and charts; *and* physical geography *under individual countries*
geoid ***1.133***
geology and rocks ***1.133–6***, ***1.284***
agglomerate **1.**5
anthracite *see* anthracite
aquiclude **1.**17
arenaceous rock **1.**18, **1.**20
argillaceous rock **1.**19–20
basalt **1.**32, **1.**33, **1.**70, **1.**138, **1.**301
basement complex **1.**34
batholith **1.**34, **1.**166
bedding *see* bedding
bedrock **1.**35
boulder **1.**42
breccia **1.**43
caliche **1.**48, **1.**49
chalk *see* chalk
clay *see* clay
cleavage **1.**64
coal *see* coal
columnar jointing **1.**70
concretion **1.**70, **1.**71
conformity **1.**72
continents *see* continents
country **1.**78
crystals *see* crystals and crystalline materials
cut-off grade **1.**82
dating **1.**133, **1.**136, **1.**274
deposition *see* deposition; sediments
diorite **1.**91
dip **1.**92
dolerite **1.**93
dome *see* dome (geology)
dyke **1.**97, **1.**166
erosion *see* erosion
extrusions *see* extrusive (volcanic) igneous rocks
fabric **1.**119
faults *see* faults
flint *see* flint
folds *see* folds
foliation **1.**124
fossils *see* fossils
gabbro **1.**130
geochemistry *see* geochemistry
geochronology **1.**133, **1.**136, **1.**274
geologists *see under individual countries*
geomagnetic dating **1.**136
geomorphology *see* geomorphology
geophysics *see* geophysics
gneiss **1.**141
grade **1.**82, **1.**142
granite **1.**143, **1.**178, **1.**184, **1.**315
granulites **1.**143
greensand **1.**147
grits **1.**147, **1.**215
ignimbrite **1.**165–6
inlier **1.**169
intrusions *see* intrusive (plutonic) igneous rocks
isostasy **1.**175–6
joint **1.**178, **1.**301
laccolith **1.**166, **1.**185
landslide **1.**187
lignite *see* lignite
limestone *see* limestone

geology and rocks (*cont.*):
limonite **1**.195
magnetic anomalies **1**.202
marble **1**.47, **1**.205, **5**.282
meerschaum **1**.210
millstone grit **1**.215
mineralogy *see* minerals *and under* igneous rocks; metamorphic rocks; sedimentary rocks
Mohorovičić discontinuity **1**.98, **1**.217
molten *see* lava; magma
mudstone **1**.222
mylonite **1**.223
nodules (concretion) **1**.71
obsidian **1**.234
outcrop **1**.242
outlier **1**.242
overburden **1**.242
palaeontology *see* fossils
pebble **1**.249
pedestal **1**.249
pegmatite **1**.250
peridotite **1**.183, **1**.251
permeability **1**.254
petrology *see* petrology
phyllite **1**.257
of planets (planetology) **8**.123
plutonic rocks *see* intrusive (plutonic) igneous rocks
porosity **1**.264
porphyry **1**.264
profile **1**.268, **1**.269
pumice **1**.269
pyroclasts **1**.270, **1**.271, **1**.295
quartzite **1**.272
rhyolite **1**.282
rock flour **1**.284, **1**.285
rock glacier **1**.284–5
rockfall **1**.284
sandstone **1**.146, **1**.147, **1**.215, **1**.293
schist **1**.124, **1**.294
sediment *see* deposition; sediments
shale **1**.300
sheeting **1**.301
shields **1**.120, **1**.301
sills **1**.166, **1**.304
siltstone **1**.222, **1**.305
slate **1**.306, **1**.307, **6**.287
soil *see* soils
strata **1**.322
stratigraphy **1**.136, **1**.322
strike **1**.323
stripped land surface **1**.324
structure **1**.324
subduction **1**.324–5
subsidence **1**.325
survey **1**.35
symmetry **1**.329
syneite **1**.329
tectonics *see* tectonics
tillite **1**.337
timescale ***1**.133*, **1**.134, **1**.135, ***9**.55*
see also aeons; epochs; eras; periods (geology and rocks)
travertine **1**.180, **1**.342
turbidite **1**.347
unconformity **1**.349–50
unloading **1**.351
volcanic *see* extrusive (volcanic) igneous rocks
xenolith **1**.369
see also deposition; erosion; geomorphology; sediments; stone
geomagnetism **1**.136
geomagnetic dating ***1**.136*
geometric isomerism (chemistry) **1**.175
geometric mean (average) **1**.209
geometric series (numbers) ***1**.136*
geometrical scale (maps) **1**.294
geometry ***1**.136*, **6**.285
angle *see* angle
geometry (*cont*):
asymptote **1**.23
fractals **1**.125–6
golden ratio **1**.142
parallel postulate **1**.247
pi **1**.257–8, **5**.347
pole **1**.263
polygon **1**.263
polyhedron **1**.263
Pythagorean *see* Pythagoras
radian **1**.273
shapes illustrated ***9**.51*
symmetry **1**.329
tangent **1**.331
topology **1**.339–40
trapezium **1**.342
volume **1**.87, **1**.358
see also trigonometry
geomorphology ***1**.136*
active layer **1**.3
coastal *see* coasts
col **1**.68
deposition *see* deposition
divide **1**.93, **1**.361
dynamic equilibrium **1**.97
erosion *see* erosion
geomorphologists *see under individual countries*
ice *see* glaciation
mass movements *see* mass flow
rivers *see under* rivers
weathering *see* weathering
see also geology and rocks; topography
geophysics ***1**.136–7*, ***8**.58*
geomagnetism **1**.136
palaeomagnetism **1**.244
see also meteorology; seismology
geopolitics ***7**.131*
see also international relations
George, Saint ***9**.14*, ***9**.41*
George I, king of Great Britain and Ireland ***3**.139–40*, **3**.157, **3**.322, **3**.343, **4**.146
and cabinet **3**.56
rebellions against *see* Jacobites
and Whigs **3**.359
George II, king of Great Britain and Ireland **3**.61, **3**.94, **3**.128, ***3**.140*, **3**.359
Austrian Succession war **3**.29, **3**.102
George II, king of Greece ***4**.133*, **4**.140
George III, king of Great Britain and Ireland ***3**.140*, ***4**.133*, **4**.285
and America **3**.100, **3**.355
and art **5**.499
and Catholic emancipation **4**.65
Junius' letter to **3**.192
and parliament **3**.54, **3**.56, **3**.130, **4**.122
son as regent *see* Regency
and spa towns **3**.335
George IV, king of Great Britain and Ireland **3**.335, ***4**.133*, **4**.285, **5**.377
Regency for father *see* Regency
George V, king of Great Britain and Ireland **4**.105, ***4**.133*
see also Georgian poetry
George VI, king of Great Britain and Ireland ***4**.133*
George, Lake, battle of **3**.202
George, prince of Greece (1897) **4**.88
George, Stefan ***5**.173*, **5**.484
George Tupou I, king of Tonga **4**.347
Georgia (*earlier* Soviet republic) ***1**.137*, ***3**.140*, ***4**.133*, ***7**.354*
Armenians in **7**.17
language **7**.41–2
physical geography and geology **1**.137
religion/theology **7**.98
Georgia (USA) ***3**.140*
Georgian architecture (George I-IV) ***5**.174*, **5**.491
see also Chinoiserie; Gothic Revival; Neoclassicism; Palladio; Queen Anne; Regency
Georgian poetry (George V) **5**.61–2, **5**.128, ***5**.174*, **5**.188–9, **5**.258, **5**.284, **5**.402
Georgias of Leontini **3**.334
Georgius Agricola **1**.215
geostationary orbit **6**.308, ***8**.58*
geosynchronous orbit ***8**.58*
geosyncline ***1**.137*
geothermal activity **1**.138, **1**.159, **1**.326
geothermal energy ***1**.137*, **6**.104, ***6**.155*, **6**.167, **7**.106
geranium ***2**.134–5*, **2**.247
gerbil ***2**.135*
Gerd **9**.13
Géricault, Théodore ***5**.173*, **5**.174
Gerlach, Walther: Stern–Gerlach experiment ***1**.320*
Germain (Belgian communist) **4**.165
German Confederation ***4**.134*
German Democratic Republic (1945–89) ***4**.134*, **4**.250
aid from **7**.8
arts *see under* Germany
Berlin Wall **4**.36
in COMECON **7**.57
map **4**.385
Oder–Neisse line **4**.250
politics and government **4**.317, **4**.356–7, **4**.371, **7**.238
leaders **4**.134, **4**.357, **9**.70
souvenirs of **7**.55
technology and science *see under* Germany
wars and defence:
militia **7**.202
in Warsaw Pact **7**.328
see also Germany; Ostpolitik
German East Africa **4**.207, **4**.306
German Federal Republic (1945–89) ***4**.135*, **4**.385
aid from **7**.8
arts *see under* Germany
in Group of Seven **7**.136
languages **7**.31
map **4**.385
in NATO **7**.328
politics and government **4**.48, **4**.141, **4**.317
Ostpolitik **4**.252, **7**.87
presidents and chancellors **4**.3, **4**.113, **4**.135, **4**.304, **9**.70
religion **4**.242
states **4**.32, **4**.150
technology and science *see under* Germany
terrorism **4**.27
Western European Union **4**.375
see also Chancellors of West Germany; Berlin; Germany; 'Ostpolitik'
German South-West Africa **4**.306, **4**.352
Germanic languages ***7**.131*, **7**.156, **7**.157
Germanic peoples:
Friesians **3**.135
myths *see* religion *under* Celtic peoples
Teutons **3**.352
see also Anglo-Saxons and Saxons; Franks; Goths; Huns; Jutes; Ostrogoths; Vandals; Visigoths
germanium **1**.106, ***1**.137*, ***6**.155–6*, **6**.312, **6**.391
Germantown, battle of ***3**.140*
Germany ***1**.137–8*, ***3**.140–1*, ***4**.134–5*, ***7**.354*
Germany (*cont.*):
1945–89 *see* German Democratic Republic; German Federal Republic
agriculture **6**.68, **6**.128
aid from **7**.8
aircraft:
airport **6**.9
airships *see* Zeppelin
fighters **4**.6
history and pioneers **6**.138, **6**.139, **6**.202
jet **6**.189, **6**.249
military **6**.129, **6**.220, **6**.224
navigation **6**.163
arts:
architecture *see under* architecture
cinema **5**.132, **5**.205, **5**.255, **5**.309, **5**.329–30
critics **5**.262, **5**.405
Enlightenment **5**.181–2, **5**.262, **5**.405, **5**.489
furniture and cabinet-making **5**.295
graphic art and design *see under* graphic art
historians **5**.405, **5**.490
Neoclassicism **5**.204, **5**.206, **5**.291, **5**.490
painting *see under* painting
print-makers **5**.131
Renaissance **5**.139
sculpture and carving **5**.46, **5**.248, **5**.250
set and stage design **5**.351, **5**.405
see also literature; music *below*
astronomy **1**.132, **8**.11, **8**.194
asteroids **8**.7, **8**.48, **8**.78, **8**.183
black holes and black bodies **8**.17, **8**.146, **8**.187
comets **8**.46, **8**.170
constellations **8**.14
cosmic rays **8**.31
measurement **8**.9, **8**.35, **8**.53
orbits **8**.114
planet **8**.106
see also Copernicus; Kepler
buildings **6**.50
ceramics and pottery **6**.43, **6**.276
chemistry and chemists **1**.93, **1**.149
chromatography **1**.366
compounds **1**.35
dyes **6**.110, **6**.161
gases **6**.72–3, **6**.235, **6**.239–40, **6**.267
organic **1**.156, **1**.181, **1**.193, **1**.368, **6**.297, **6**.391
physical **1**.64, **1**.226, **1**.242
refrigeration **6**.202
synthetic rubber **6**.304
see also Bunsen; Haber–Bosch *under* Haber
colonies **4**.150, **4**.207, **4**.306–7, **4**.352, **7**.191
computers **6**.84, **6**.240
economy and trade:
bankers and merchants **3**.135
chambers of commerce **7**.43
currency and monetary policy **4**.303
Gold Standard **4**.138
in Group of Seven **7**.136
industrial relations **7**.158, **7**.159
Industrial Revolution **7**.158
in OECD **7**.221
post-industrial **7**.309
printer **3**.152
unemployment **7**.320
Zollverein **4**.391
see also European Community; European Economic Community
education **4**.187, **7**.92, **7**.127, **7**.326
energy and fuel **6**.47, **6**.114

Germany (*cont.*):
environment **7**.140, **7**.260
explorers **1**.33, **1**.159, **1**.191, **1**.224, **4**.31, **4**.112
finance **7**.43, **7**.155, **7**.299
foreign policy:
Algeciras Conference **4**.9
Anschluss **4**.14
Anti-Comintern Pact **4**.15
Berlin Congress **4**.35–6
Confederation of Rhine **4**.82–3
Council of Europe **7**.71
German Confederation **4**.133–4
Munich Pact **4**.227
government *see* politics and government *below*
historiography **4**.153, **5**.405, **5**.490
immigration **7**.153
Kulturkampf **4**.188
languages **7**.31, **7**.90, ***7**.131*, **7**.156
laws/legal system:
civil **7**.51
company **7**.62
consumer **7**.68
industrial relations **7**.159
judges **7**.173
judicial review **7**.173
legal aid **7**.182
punishment **7**.255
surveillance **7**.303
literature ***5**.174–5*, **5**.194, **5**.201, **5**.243, **7**.216
epics **5**.59, **5**.181, **5**.249, **5**.318, **5**.351
essays **5**.203, **5**.279
tragedy **5**.202, **5**.249, **5**.262
see also under drama; novels; poetry; Romanticism; stories
mathematics and mathematicians:
calculators **6**.53, **6**.54
geometry **1**.183–4, **1**.215, **1**.217, **1**.282–3
logician **1**.127, **7**.126
number theory **1**.132, **1**.215
set theory **1**.50
universal **1**.155, **1**.226, **1**.282–3
see also Leibniz
medicine and health care:
bacteriology **6**.192
drugs **6**.72, **6**.105, **6**.112, **6**.257
homoeopathy **6**.169, **6**.257
immunology **6**.177
psychiatry **6**.34
technology **6**.251, **6**.254, **6**.389
X-rays **6**.389
metallurgy and mining **6**.5, **6**.12, **6**.214, **6**.250, **6**.319
migrant labour in **7**.200
music **5**.28, **5**.68, **5**.377
conductors **5**.280, **5**.291, **5**.435–6
counterpoint **5**.68, **5**.377, **5**.406
dance **5**.39, **5**.117, **5**.122, **5**.242, **5**.489
Expressionism **5**.206, **5**.435, **5**.485
organ **5**.28, **5**.68, **5**.377
piano **5**.57, **5**.68, **5**.291
sacred **5**.365, **5**.406–7
violin **5**.62, **5**.280, **5**.291, **5**.429, **5**.430
see also under composers; opera; Romanticism; songs; symphonies
national parks **7**.212
nuclear technology **6**.127, **6**.152, **6**.218, **6**.242
optics and glass **6**.1, **6**.310, **6**.391
philosophy:
dialectical materialism **7**.89
existentialism **7**.141
idealism *see* Kant; Schopenhauer
Nietzsche **7**.216–17
phenomenology **7**.234

Germany (*cont.*):
philosophy (*cont.*):
see also continental philosophy; Frankfurt School; Goethe; Hegel; Kant; Leibniz; Marx *and under* Enlightenment
physical geography and geology ***1**.137–8*
climatologist **1**.184
geologists **1**.250, **1**.364
see also Wegener
geomorphologist **1**.250
islands **1**.127, **1**.153
lake **1**.73
mountains **1**.152
regions **1**.34–5, **1**.40, **1**.289
see also regions *below*
rivers **1**.84, **1**.282, **1**.364
soil scientist **1**.185
physicists:
atomic and nuclear **1**.149, **1**.153, **1**.226, **1**.320
electromagnetics **1**.132, **1**.154, **1**.183, **1**.189
quantum theory **1**.41–2
see also Einstein; Planck
thermodynamics **1**.64, **1**.226, **1**.242
see also Mach; Röntgen
politics and government:
Bill of Rights **7**.32
Chancellors and presidents **4**.37, **4**.39–40, **4**.53, **4**.55, **4**.108, **4**.258, **4**.304, **9**.70
Christian Democracy **7**.47
civil service **7**.52
coalition **7**.54
fascism *see* Nazism
Guelph faction **3**.149
intelligence services **7**.162
nationalism **7**.212
Ostpolitik **4**.252, **7**.87
presidential **7**.246
Reichstag **4**.286
reunification **7**.320
revisionism **7**.267
Second empire **4**.134–5
Social Democrats **4**.195–6, **4**.304, **4**.317
Socialists **4**.324
Third Reich *see* Third Reich
Weimar Republic **4**.373, **9**.70
women's suffrage **7**.332
see also Nazis; Third Reich
printing **6**.163, **6**.205, **6**.281
psychology **7**.334
see also Weber
regions, kingdoms, and states:
Black Forest **1**.40
Hesse **4**.150
listed ***9**.75*
Palatinates **3**.133, **3**.267
Saar **1**.289, **4**.299
Swabia **3**.346
see also Bavaria; Brandenburg; Hanover; Prussia; Saxony
religion/theology **3**.300, **3**.352, **4**.188, **4**.305
Jews and Judaism **3**.103, **7**.138, **7**.183–4, **7**.260
Protestants **3**.213, **3**.228, **3**.279, **4**.44–5, **4**.242, **7**.50, **7**.187
saints **9**.40, **9**.41, **9**.43
see also individual names
roads **6**.34, **6**.200
bus and trolleybus **6**.52, **6**.363
cars and engines **6**.94, **6**.102, **6**.234, **6**.254, **6**.303
motorway **6**.360
and Roman empire **3**.23
rulers:
listed ***9**.67*
see also Holy Roman empire *and under* Prussia
science *see* technology and science *below*

Germany (*cont.*):
social security **4**.318, **4**.357
societies **6**.217
sociology **7**.61
see also Frankfurt School; Weber
sport and recreation **7**.223, **7**.236
states *see* regions *above*
technology and science **6**.25, **6**.213, **6**.217
see also entries above and below prefixed **1**. *and* **6**.
telecommunications **6**.45, **6**.88, **6**.126, **6**.168, **6**.241, **6**.319, **6**.349, **7**.307
and UN **7**.277
urban areas **3**.27–8, **3**.157
vehicles and transport **6**.13
bridges **6**.44
diving **6**.97
railways **6**.114, **6**.161, **6**.293, **6**.294
rivers **4**.294, **4**.299
ships **6**.97, **6**.163, **6**.345
space and rocketry **6**.301, **6**.326
submarines **6**.323
see also aircraft; roads *above*
wars and defence **3**.205
airforce **4**.6
army **4**.19
Franco-Prussian War **4**.151
World War I **4**.151, **4**.201, **4**.304, **4**.308
World War II **4**.48, **4**.142, **4**.180, **4**.181, **4**.263, **4**.292, **4**.295
Kapp putsch ***4**.179*
navy **4**.303, **4**.304, **4**.345
Peasants' War **3**.273–4
Revolutions of 1848 **4**.288
terrorism **7**.308
Thirty Years War **3**.49, **3**.214, **3**.355
war crimes **7**.328
in WEU **7**.330
reparations **4**.94
see also NATO; Nazis; World War I; World War II *and* weapons *below*
weapons **6**.64, **6**.134
chemical **6**.72, **6**.235, **6**.239–40, **6**.267
explosives **6**.310
guns **6**.8, **6**.131
missiles **6**.29, **6**.93, **6**.228–30
see also wars and defence *above*
welfare **4**.318, **4**.357
see also Prussia
Germer, Lester **1**.84
germination ***2**.135*
germs **2**.126
see also infections
Generalic, Ivan **5**.312
Gernsback, Hugo **5**.407
Gerö, Erno **4**.158
Geronimo ***4**.135*
Gerry, Elbridge **3**.129
gerrymandering **7**.104
Gershwin, George **5**.174, ***5**.175*
Gershwin, Ira **5**.175
gesso ***5**.175*
Gestalt psychology ***7**.131*
Gestalt therapy ***7**.131*
Gestapo ***4**.135*
gestation **2**.268
gesture *see* non-verbal communication
Gesualdo, Carlo ***5**.175*, **5**.275
Gettysburg, battle of ***4**.135–6*
geum ***2**.135*
geyser ***1**.138*
Ghadanfar Coeur de Lion **9**.4
Ghaffar Khan, Abdul **4**.285
Ghalib **5**.466
Ghana (ancient kingdom) ***3**.141*

Ghana (*earlier* Gold Coast) ***1**.138*, **3**.4, ***3**.141*, ***4**.136*, ***7**.354*
chiefdoms **4**.20
decolonization **4**.4
ethnic groups *see* Asante
Fanti Confederation **4**.116
physical geography and geology **1**.138, **1**.358
political leader **4**.244
port **6**.277
gharial ***2**.135*
Ghaznavid dynasty ***3**.141*, **3**.216
al-Ghazzālī **3**.10, **3**.344, **5**.344, ***7**.10*
Ghent, Pacification of ***3**.141*
Ghent, Treaty of ***4**.136*
ghetto ***7**.131*
Ghiberti, Lorenzo **5**.23, **5**.63, ***5**.175*
Ghirlandaio, Benedetto, Davide, and Ridolfo **5**.175
Ghirlandaio, Domenico **3**.18, ***5**.175*
Ghosh, Sri Aurobindo **7**.156
ghost moth **2**.332
ghost pipefish ***2**.135*
ghosts ***9**.14*
ghost dance ***7**.131*
see also spirits
ghoul ***9**.14*
Giacometti, Alberto ***5**.175–6*
Gia-Long, emperor of Annam **3**.80, **4**.14, **4**.367
Giambattista, Giandomenico **5**.455
Giambattista, Lorenz **5**.455
Giambologna ***5**.176*
giant *see* giants
giant horsetail **2**.49–50
giant otter shrew **2**.171
giant planets ***8**.58*
see also Jupiter; Neptune; Saturn; Uranus
giant salamander ***2**.135*
giant sloth ***2**.135*
giant stars ***8**.58–9*
in binary stars **8**.14, **8**.20
evolution of **8**.160
see also red giants; supergiants
giant tortoise **2**.344
giant water bug ***2**.135*
giants and Titans ***9**.14*, ***9**.35*
in Greek myths **9**.9, **9**.27, **9**.28, **9**.29, **9**.31, **9**.35
in Norse and Germanic myths **9**.6, **9**.13, **9**.14, **9**.36
Giant's Causeway ***1**.138*
Giap, General **4**.98
Giazotto, Remo **5**.11
gibberellins **6**.271
Gibbon, Edward ***3**.141*, **3**.308, ***5**.176*
Gibbon, John Heynsham ***6**.156*, **6**.166
gibbon ***2**.135–6*
Gibbons, Grinling ***5**.176*
Gibbons, Orlando **5**.18, **5**.176
gibbous moon **8**.118, **8**.119
Gibbs, James ***5**.176*
Gibbs, Josiah Willard ***1**.138–9*
Gibraltar ***1**.139*, ***3**.141*, ***4**.136*, **7**.57, ***7**.354*
Gibson Desert ***1**.139*
Gide, André ***5**.176–7*
Gideon ***9**.14*
Gielgud, Sir John **5**.24, ***5**.177*
Giffard, Henri **6**.10, **6**.138, ***6**.156*
Giffen, Sir Robert: Giffen goods ***7**.131–2*
gifts ***7**.132*, **7**.215
kula **7**.178, **7**.197, **7**.316
potlatch **7**.178, **7**.219, **7**.243
sacrifice **7**.271–2
gig (carriage) **6**.61
gig mill ***6**.156*
gigue ***5**.177*
gila monster ***2**.136*
Gilbert, Sir Alfred ***5**.177*
Gilbert, Grove Karl ***1**.139*
Gilbert, Sir Humphrey **1**.127, ***3**.139*, ***3**.141*, **3**.251, **3**.282

Gilbert, J. H. 6.196
Gilbert, William (1544–1603, physicist) 1.139, 3.319
Gilbert, Sir William Schwenk (1836–1911, librettist) 5.6, 5.439
Gilbert Islands *see* Kiribati
gild *see* guilds
Gildas 3.23, 3.141
'Gilded Age' 4.136
gilding 5.177
Giles, Saint 9.41
Gilgamesh 3.141, 3.368, 9.14
  epic of 5.148, 5.177
Gill, Eric 5.177, 5.464
gill 2.136
Gillespie, John 'Dizzy' 5.239
Gillray, James 3.158, 3.190, 3.291, 5.177
Gilman, Harold 5.72
Gilpin, Robert 7.141
gilt-edged securities 7.132
gin 2.33
Ginastera, Alberto 5.257
ginger 2.136
ginkgo 2.136
Ginsberg, Allen 5.39
Ginzburg, Natalia 5.231
Giolitti, Giovanni 4.136
Giordano, Luca 5.177–8
Giorgione 5.178, 5.457
Giotto di Bondone 5.178, 5.179, 5.194, 5.284, 5.460
Giotto probe of Halley's Comet 8.28, 8.59, 8.60
giraffe 2.136, 2.137, 2.233
Giraudoux, Jean 5.178
girder:
  box-girder 6.44, 6.293, 6.336, 6.341
  plate 6.272, 6.336
Girondin 3.141
Girtin, Thomas 5.179
Gislebertus 5.179
Gissing, George Robert 5.179
Giulio Romano 5.179
gizzard shad 2.136
glacial mills 1.221
glaciation 1.139, 1.140
  deposition 1.140
    boulder clay 1.42
    drumlins 1.96, 1.140
    esker 1.113, 1.140
    gravel 1.144
    kame 1.140, 1.180
    kettle-hole 1.182
    moraines 1.140, 1.221
    outwash 1.140, 1.242
    till 1.42, 1.337
    valley train 1.353
  erosion 1.140
    arête 1.18, 1.19, 1.140
    channel 1.58
    cirque 1.38, 1.62, 1.64, 1.140
    corrasion 1.77
    drumlins 1.96, 1.140
    finger lake 1.121
    fiord 1.122, 1.310
    hanging valleys 1.140, 1.151
    lunette 1.199
    moulin 1.221–2
    pyramidal peak 1.140
    roches moutonnées 1.284
    rock flour 1.284, 1.285
    striations 1.323
    tarn 1.331
    truncated spur 1.318
  eustatic change 1.116
  glaciers 1.139–40
    bergschrund 1.38, 1.140
    caves 1.247
    crevass 1.79–80
    divide 1.93
    firn 1.122
    hydrological cycle 1.162
    ice-fall 1.140, 1.165
glaciation (*cont.*):
  glaciers (*cont.*):
    rock 1.285
    snout 1.140
    Sperry 1.362
    valley 1.140
    see also cirque *under* erosion *above*
  interglacial 1.170
  isostasy 1.175–6
  unloading 1.351
  see also ice; snow
Glacier National Park 1.140
gladiators 3.141–2, 3.338–9
gladiolus 2.136
Gladstone, William Ewart 4.79, 4.136, 4.195, 4.220
  and Boer Wars 4.43
  and income tax 4.160
  and Ireland 4.100, 4.118, 4.155, 4.344
Glaisher, James 1.140
glaive 6.165
glands 2.136
  adrenal 2.4
  exocrine 2.111–12
  lymph 2.4, 2.106, 2.196, 2.343
  and pancreas 2.240
  pineal 2.254
  pituitary 2.256–7
  thymus 2.340
  thyroid 2.140, 2.340
  see also endocrine glands; hormones
Glasgow School 5.179, 5.273–4
glasnost 7.132
Glass, Philip 5.179, 5.297
glass 6.156
  annealing 6.15
  arts (glassware) 5.168, 5.179–80, 5.254, 5.318
    see also stained glass
  -blowing 6.156
  building *see* Crystal Palace; greenhouse
  ceramic 6.156, 6.373
  fibre 6.86, 6.129, 6.156
  float 6.136, 6.137
  harmonica 5.179
  as insulator 6.183
  laboratory equipment 6.193
  laminate 6.193
  light bulbs 6.177
  lithium in 6.156, 6.310
  metallic 6.220
  moulding 6.235
  open hearth furnace 6.250
  optics *see* lenses; optics/optical
  plate 6.136, 6.272
  powdered *see* enamel
  recycling 6.296
  safety-glass 6.156, 6.304
  silica 6.68, 6.156, 6.319
  vitrification 6.373
glass, volcanic 1.141
glass snake 2.137
glass-fibre-reinforced concrete 6.86, 6.156
glass-fibre-reinforced plastics 6.129
  see also fibre-reinforced plastics
glassware *see* arts *under* glass
glasswort 2.137
Glastonbury tor 9.14
glaucoma 2.137
Glaucon 7.103
glauconite 1.141
glaze/glazing 5.180, 5.234, 6.156–7, 6.278
Glazunov, Aleksandr 5.180
glebe 3.142
glee 5.340
gleemen 5.297
Glencoe, Massacre of 3.142
Glendower, Owen 3.142, 3.376
Glenn, John G. 4.322, 4.323, 8.10
gleying 1.141
gliders and gliding 6.38, 6.65, 6.138, 6.157, 6.202, 6.389, 7.132
  hang-glider 6.165, 6.222
gliding creatures:
  flying fish 2.123
  flying frog 2.123–4
  flying lemur 2.73
  flying lizard 2.124
  flying phalanger 2.124
  scaly-tailed squirrel 2.295–6
Glinka, Konstantin Dimitrievich 1.205
Glinka, Mikhail 5.156, 5.180
global *see entries beginning* international; world
Global Positioning System 6.239
Global Protection Against Limited Strikes 6.336, 7.300
global warming *see* greenhouse effect and global warming
globe 1.141
globe flower 2.137, 2.138
Globe Theatre 5.144
globular clusters 8.26, 8.27, 8.117
glockenspiel 5.180–1
Glorious Revolution 3.142
glory lily 2.190
glossopteris 2.137
Gloucester, Humphrey, duke of 3.38, 3.142–3
Gloucester, Richard, duke of *see* Richard III
Gloucester, Thomas, duke of *see* Thomas of Woodstock
glow-worm 2.118, 2.119
gloxinia 2.138
Gluck, Christoph Willibrand von 5.154, 5.181, 5.329, 5.334
glucose 2.138, 6.66, 6.337
glue 6.3, 6.170
Glume, J. G. 3.132
gluons 1.69, 1.141
glutton 2.369
glycerol 6.126, 6.157, 6.241, 6.323
glycogen 2.138
glycol 1.141, 6.16, 6.123, 6.261
glycolysis 2.138
glyphs 5.181
glyptodont 2.138
GMT *see* Greenwich Mean Time
gnats 2.129, 2.219
gneiss 1.141
gnomon (on sundial) 8.165
Gnosticism 7.19, 7.310
GNP (gross national product) 7.132, 7.133, 7.211
  aid as a percentage of 7.8
  and economic growth 7.100
  and medicine and health care 7.140
  standard of living 7.299
gnu 2.138–9
Goa 3.143, 4.136–7
goanna 2.217
goat and goat-antelope family 2.139
  chamois 2.60
  goral 2.141
  ibex 2.168
  markhor 2.203
  musk ox 2.223
  mythical 9.12, 9.32
  serow 2.303–4
  see also sheep
goat moth 2.139
goatsucker 2.228
Gobelin, Jean and Philibert 5.181
Gobi Desert 1.141
Gobind Singh, Guru 3.328, 7.132
  and *Adi Granth* 7.2, 7.137
  birthday 7.285
  Dasam Granth 5.418, 7.80
  and *Khalsa* 7.137, 7.176, 7.286
  and poems 7.80
Gobineau, Joseph Arthur, comte de 4.137, 4.236
goblin 9.14
goby 2.139, 2.221
God and gods *see* deities
Go-daigo, emperor of Japan 3.24
Godard, Jean-Luc 5.323
Goddard, Robert Hutchings 4.323, 6.157, 6.301, 8.10, 8.144
goddesses *see* deities
Gödel, Kurt 1.141, 1.247
Goderich, Frederick J. R., viscount 4.137
godetia 2.140
Godfrey, Thomas 6.315
Godfrey of Bouillon 3.92, 3.143, 9.4
Godolphin, Sydney, 1st earl of 3.143
gods and goddesses *see* deities
Godse, Nathuram 4.130
Godunov, Boris, Tsar of Russia 3.45, 3.312
  House of 9.68
Godwin, earl of Wessex 3.143
Godwin, Mary 5.416
Godwin, William 5.491
Godwin Austen 1.179
godwit 2.140
Goebbels, Joseph 4.82, 4.137, 4.286
Goering, Hermann Wilhelm 4.135, 4.137, 4.248, 4.286
Goes, Hugo van der 5.181
Goethe, Johann Wolfgang von 5.48, 5.181–2, 5.197, 5.219, 5.266, 5.309, 5.460, 9.13
Goffman, Erving 7.133, 7.278, 7.292, 7.299, 7.303, 7.313
Gog and Magog 9.14
Gogarty, Oliver St John 5.226
Gogh, Vincent van 5.182–3
Gogol, Nikolay 5.182
goitre 2.140
Gokhale, Gopal Krishna 4.137
Gold, Thomas 8.33
Golan Heights 1.141–2
Golconda 3.143
gold 1.106, 1.142, 6.157
  and alchemy 6.10
  in amalgam 6.12
  arts (goldwork) 5.183
    gilding 5.177
    India 5.196
    Italy 5.472
    Phoenician 5.346
    Russia 5.151
    Scythian 5.410
    South America 5.425–6
    Vietnam 5.475
    Zapotec 5.498
  assaying 1.341, 6.357, 6.359
  bullion 7.133–4
  diggers' rebellion 4.113–14
  El Dorado 9.11
  electroplating 6.120
  gold-lead electroscope 6.120
  hallmarks 9.63
  mining 6.157, 6.173
  rushes 4.137–8
  standard 4.138, 7.134
Gold Coast *see* Ghana
gold moth 2.141
goldcrest 2.140
Golden Calf 9.14
golden eagle 2.140
Golden Horde 3.143
golden mole 2.140
golden orfe 2.140
golden ratio 1.142
golden section 5.183, 5.368
Golden Temple at Amritsar 7.285, 7.286
Golden Triangle 7.225
goldfinch 2.140–1
goldfish 2.141
Golding, William 5.183
Goldoni, Carlo 5.183
Goldsmith, Oliver 5.183
gold-tail moth 2.251–2
goldwork *see* arts *under* gold
golem 9.14

golf 7.134
Goliath 3.99, 9.14
Goliath beetle 2.155–6
Gomarus, Franciscus 3.23
Gombrowicz, Witold 5.183–4
Gomer 9.17
Gómez, Juan Vicente 4.138, 4.363
Gómez de Quevedo, Francesco 5.428
Gomulka, Wladyslaw 4.138, 4.272
gonad 2.141
gonadotrophin 6.127, 6.280
Goncalves, Nuno 1.154
Goncharov, Ivan 5.184
Goncourt, Edmond and Jules de 5.184
Gondomar, Diego Sarmiento de Acuña, Conde de 3.143
Gondwana 1.135, 1.142
Gong, prince 4.308
Góngora y Argote, Luis de 5.184
gongs 5.184, 5.427
goniometer 6.157
gonorrhoea 2.141
González, Felipe 4.322
Gooch, Daniel 6.53
Good Friday 7.134
Good Hope, Cape of 1.142
Good Neighbor Policy 4.138
Goodall, T. F. 5.145
Goodman, Benny 5.47, 5.239
Goodnow, J. 7.35
Goodricke, John 8.3, 8.14, 8.23, 8.59
goods (and services) 7.134
  commodity 7.57, 7.58, 7.297
  consumer 7.67–8
  demand see demand
  distribution 7.93
  diversification 7.93
  dumping 7.97
  futures markets 7.128
  Giffen 7.131–2
  marketing 7.192
  for production of other goods see capital
  recycling 7.260
  scarcity 7.275
  speculator 7.297
  supply 7.192, 7.193, 7.302–3
  trade mark 7.314
  utility 7.324
  see also commodity; manufacturing; retail trade; trade
goods wagon 6.157
Goodyear, Charles 6.157–8
goose see geese
gooseberry 2.141
gopher 2.141
gopura 5.184
goral 2.141
Gorbachev, Mikhail 4.138, 4.358
  and free market 7.136
  glasnost and perestroika 5.396, 7.59, 7.132, 7.218
  and political prisoners 4.277
Gordimer, Nadine 5.184
Gordon, Charles George (1833–85) 4.138–9, 4.207, 4.332, 4.337
Gordon, David 5.362
Gordon, Lord George (1751–93): Gordon riots 3.143
Gordy, Benny, Jr 5.358
Gore, Spencer 5.72
Gorges, Sir Ferdinando 3.217
gorges 1.142
  Grand Canyon 1.50, 1.142–3
  limestone 1.181
  Tanzania see Olduvai
Gorgias 7.295
gorgonian coral 2.141, 2.279, 2.298–9
Gorgons 9.14
gorilla 2.141–2
Gorky, Arshile 5.184
Gorky, Maksim 5.184, 5.185, 5.252, 5.380, 5.430
Górnicki, Lukasz 5.355
gorse 2.142
Goschen, G. J. 4.220
Goscinny 5.91
Goss, Professor 6.293
Gossaert, Jan 5.184
Gosse, Sir Edmund William 5.184
Gothic architecture 5.185–6
  buttress 5.68
  Decorated 5.127
  early English 5.141
  Flamboyant 5.156–7
  France 5.375, 5.475
  gargoyle 5.170
  international see International Gothic
  Perpendicular 5.344
  Rayonnant style 5.375
  revived see Gothic Revival
  tracery 5.459
  vault 5.344, 5.469–70
  see also Gothic Revival; International Gothic; stained glass
Gothic language 7.131, 7.156
Gothic novels 5.186, 5.207, 5.348, 5.372–3, 5.416, 5.481
Gothic Revival in architecture 5.186
  Britain 5.66, 5.68, 5.369, 5.407–8, 5.415, 5.482–3, 5.495
  France 5.477
  Germany 5.405
  India 5.219
  see also Gothic architecture; International Gothic
Goths 3.64, 3.143–4, 3.260
  see also Ostrogoths; Visigoths
Gotland 3.144
Gottfried von Strassburg 5.186
Gottwald, Klement 4.34, 4.92, 4.139
gouache 5.186–7
Goudsmidt, Samuel Abraham 1.142
gouges 6.72
Goulart, Joao 4.49
Gould, Benjamin Apthorp 8.28, 8.60, 8.120
  Gould's Belt 8.59–60
Gould, Jay 4.291
Gould, John 2.142
Gounod, Charles 5.187, 5.298
Goupy, Joseph 3.156
gourami 2.142
gourd 2.142
gout 2.16
government 7.134
  absolute 7.144
  cabinet see cabinet
  coalition 7.54, 7.67
  constitutional 7.67
  counter-insurgency 7.72
  coup d'état 7.72, 7.73, 7.201
  democratic see democracy
  direct action against 7.91
  emergency powers 7.105
  federal see federalism
  financial regulation 7.37, 7.120
  functionalism 7.127
  housing 7.134
  intelligence services 7.162
  junta 7.173
  and legislation see legislature
  local 7.186
  mandate 7.191
  military see military government under military
  national debt 7.211
  nationalism 7.211–12
  parliament see parliamentary government
  paternalism 7.230–1
  public sector 7.254
  puppet 7.255
  regional policy 7.262
  regulation see regulation
government (cont.):
  security services 7.277–8
  sedition against 7.278
  separation of powers 7.279
  separatism 7.279
  see also politics; state and politics and government under individual countries and also entries beginning public
governor (in machine) 6.158
Gower, John 5.187
Gowon, Yakubu 4.108, 4.139, 4.242
Goya y Lucientes, Francesco José de 4.118, 5.18, 5.187
Gozzoli, Benozzo 3.227
GPALS (Global Protection Against Limited Strikes) 6.336, 7.300
GPS (Global Positioning System) 6.239
Gqozo, Brigadier 4.31
Graaf, Regnier de 2.142
grab 6.158
graben 1.142, 1.143
Gracchus, Gaius Sempronius 3.144, 5.257
Gracchus, Tiberius Sempronius 3.144
Graces 9.14
grackle 2.142
grade (geology) 1.82, 1.142
grader 6.158
Graebe, Carl 6.260
Graeco-Roman technology 6.158
  see also technology and science under Greece, ancient; Roman empire
Graf Spee 4.271
grafting 6.147, 6.158
Grafton, Augustus Henry Fitzroy, 3rd duke of 3.144
Graham, Billy 4.288
Graham, John, of Claverhouse, Viscount Dundee 3.197
Graham, Martha 5.187–8, 5.300
Graham, Sir Robert 3.184
Graham, Thomas 1.142
Grahame, Kenneth 5.188
Graham-Smith, Francis 8.35, 8.193
Graiae 9.14
Grail, the Holy 9.14
Grainger, Percy 5.188
grains see cereal
Gram, Hans Christian Joachim 2.142
grammar 7.16, 7.47, 7.134–5, 7.181
  English 7.107
  Esperanto 7.108
  parts of speech 7.229
  see also language
Gramme, Zénobe-Théophile 6.158
gramophone 6.33, 6.112, 6.158–9
Grampian Mountains 1.142
Gramsci, Antonio 4.139, 7.135, 7.141
Gran Chaco 1.142
Gran Colombia 4.43, 4.67
Granada 3.144
granadilla 2.142
Granados, Federico Tinoco 4.87
Granados y Campiña, Enrique 5.188
Granby, John Manners, marquis of 3.144
Grand Alliance, War of see Nine Years War
Grand Banks 1.142
Grand Canal, China 6.159
Grand Canyon 1.50, 1.142–3
Grand Guignol, Théâtre du 5.188
Grand Remonstrance 3.144
Grand Siècle 3.144
Grand Tour 3.144–5
grand unified theory 1.143, 8.60
Granger Movement 4.139
granite 1.143, 1.178, 1.184, 1.315, 6.287
Grant, Cary 5.207
Grant, Duncan 5.50
Grant, James 1.317, 4.326
Grant, Ulysses Simpson 4.10, 4.71, 4.139, 4.192, 4.364, 4.368
granulation 8.60
granulites 1.143
Granvelle, Cardinal Antoine 3.112
Granville, John Carteret, 1st earl 3.145
grape 2.142–3
grape hyacinth 2.143
grapefruit 2.143
graph see graphs
graphic art and design 5.188
  Austria 5.249–50
  Britain 5.165, 5.240, 5.303, 5.304–5, 5.313, 5.350, 5.484
  Czechoslovak 5.250, 5.307
  France 5.51–2, 5.127, 5.138, 5.286, 5.296, 5.351, 5.376, 5.391–2
  Germany 5.41, 5.139, 5.248, 5.250, 5.264, 5.320
  Netherlands and Flemish 5.148, 5.264
  Russia 5.266, 5.298
  Spain 5.187, 5.348
  Switzerland 5.249
  United States 5.264, 5.488
  see also engraving; illustration; print; typography
graphic image manipulation 6.159
graphics (computers) 6.159, 6.184
graphite 1.51, 1.144, 6.58, 6.159–60
  in Castner–Kellner cell 6.62
  in nuclear reactor 6.3, 6.160, 6.213, 6.245
  as polymer 2.275
graphs 1.143–4
  histogram 1.156
  hydrograph 1.162
  least squares method 1.191
  saddle points 1.289
graptolite 2.143
Grass, Günter 5.188
grass 2.143–4
  bamboo 2.21
  citronella 2.66
  couch 2.78
  crop 6.160, 6.166, 6.235
    mowers 6.6, 6.196, 6.235
    see also harvesting
  disease (ergot) 2.110
  elephant 2.106
  esparto 2.110
  marram 2.204
  pampas 2.240
  poa 2.260
  reed 2.281
  rye-grass 2.244
  sugar cane 2.329
  timothy 2.341
  tussock-grass 2.351
  see also cereals and grains; grasslands
grass moth 2.144
grass snake 2.144
grasshoppers 2.143–4
  cricket 2.81
  locust 2.192–3
  longhorn 2.193, 2.194
  shorthorn 2.192–3, 2.308
grasslands:
  pampas 2.240, 2.241
  pasture 2.243–4
  prairie 2.266
  savannah 1.294, 2.294
  steppe 2.325
  in succession 2.328
  see also grass
Grattan, Henry 3.145, 3.180, 3.290, 4.139–40
gravel 1.144
Graves, Michael 5.363
Graves, Robert van Ranke 5.188–9

Gravettian culture 3.367
gravitation/gravity *1.144*, *8.60–1*
  centre of gravity 1.56
  deposition 1.43, 1.64, 1.72
  gravitational constant 8.61
  gravitational potential energy 6.121
  gravitational waves *8.60*
  potential energy 1.133
  surface 8.166
  and velocity 1.355
  weight 1.364
  see also under Newton
graviton *1.144*
gravure printing 6.123, *6.160*, 6.282
Gray, Asa *2.144*, 2.344
Gray, Elisha 6.235
Gray, Thomas 5.143, *5.189*
grayling *2.144*
graywacke 1.147
Graziani, Rodolfo *4.140*
grease, lubricating 6.145, 6.210
Great Australian Bight *1.144–5*
Great Awakening *3.145*
Great Barrier Reef *1.145*
Great Basin *1.145*
Great Bear 8.178
Great Bear Lake *1.145*
Great Britain *1.145*
  see also Britain
Great Britain, SS *6.160*, 6.236
great circle *1.145*, 8.22
  see also ecliptic; equator; meridians
Great Dividing Range *1.145*
Great Exhibition *4.140*, *6.160–1*
Great Lakes *1.145*
Great Leap Forward *4.140*, 7.191
Great Leopard Nebula 8.35
Great Looped Nebula 8.35–6
Great Nebula in Orion *8.61*
Great Plague *3.145–6*, 3.284
Great Plains *1.145–6*
great power *7.135*, 7.302
  see also Union of Soviet Socialist Republics; United States
Great Red Spot *8.61*
Great Rift Valley 1.143, *1.146*, 1.203, 1.331, 1.356
Great Salt Desert *1.146*
Great Salt Lake *1.146*
Great Sandy Desert *1.146*
Great Schism *3.146*
Great Slave Lake *1.146*
Great Smoky Mountains *1.146*
Great Trek *4.140*
Great Victoria Desert *1.146*
Great Wall of China *3.146*, 6.160, *6.161*
Greater East Asia Co-Prosperity Sphere *4.140*
Greathead, Henry 6.200
greaved turtle *2.144*
grebes 2.86, *2.144*
Greco, El *5.189*
Greece, ancient *3.146–7*
  arts *5.189–92*, *5.311*
    architecture *5.190*, 5.203, 5.311, 5.330
      see also Greek Revival
    ceramics and pottery 5.311
    decorative 5.190, 5.192, 5.193
    metallurgy 5.190
    music 5.82
    painting 5.18, 5.190, 5.192, 5.498
    vases *5.192*, 5.193
    see also under sculpture *and* literature *below*
  astrology 7.20
  astronomy 1.271, 6.285, 8.8, 8.11, 8.33, *8.61–2*
    see also Hipparchus; Ptolemy
  buildings 3.2, 6.50, 6.51, 6.225
  cities and city-states *see* regions *below*
  energy and fuel 6.78
  engines 6.330
Greece, ancient (*cont.*):
  explorer 1.271
  Hellenistic civilization 3.160
  helots 3.160
  hippodrome 3.166
  historians *see under* literature *below*
  Indo-Greek dynasties 3.177–8
  islands 3.90–1, 3.181–2
  language 7.10–11, *7.135*
    alphabet *8.192*, *9.46*
  laws, Draconian 3.105
  literature *5.190–1*, 5.210, 5.268, 5.353, 5.496
    historians 3.164, 3.287, 3.356, 3.386–7, 5.190, 5.205, 5.454, 5.496
    novel *5.189*
    tragedy 5.5–6, 5.150, 5.425, 5.459
    see also under drama; poetry *and also* philosophy *below*
  maps 3.147, 3.274
  mathematics 1.18, 6.19, 6.168, 6.285
    numerals 9.52
    see also Euclid; Pythagoras
  medicine and health care 6.21, 6.99, 6.340, 6.350, 7.24
  Mycenaean civilization 3.244, 5.311
  myths *see under* religion *below*
  orator 3.101
  philosophy 3.285–6, 3.386, 7.237
    Academy 3.2
    Cynics 3.94–5, 3.104
    stoicism 7.300
    see also Aristotle; Neoplatonists; Plato; Socrates; Stoics
  physical geography and geology 1.238, 1.271
    see also Ptolemy *and under* Greece, modern
  politics and government 7.21
    Achaean League 3.2
    city-states *see* regions and city-states
    democracy 3.26
    ostracism 3.264
    statesmen 3.21, 3.276, 3.332, 3.353, 4.363, 4.387
  pottery 6.278
  regions and city-states:
    central 3.3, 3.27, 3.43, 3.67, 3.285, 3.352–3, 3.354
    citadel 3.77
    Delian League 3.101
    north 3.77, 3.174–5
      see also Macedonia (Greece)
    Peloponnese *see* Peloponnesian area
    polis 3.286–7
    ports 3.87, 3.382
    see also Athens; Sparta
  religion/theology 3.26, 3.95, *3.147–8*, 3.261–2, 3.263
    mythical figures and places 5.6, 9.2–39 *passim*
    saints 9.40, 9.41
  rhetoric teachers 3.334
  technology and science 6.10, 6.25, 6.93, *6.158*, 6.346, 6.368
    scientists 1.18, 1.86, 1.112
    see also entries above and below prefixed 1. *and* 6.
  tools 6.191, 6.310
  urban areas *see* regions *above*
  vehicles and transport 6.105, 6.172, 6.365, 6.375
    navigation 6.201, 6.214
    ships 6.149, 6.219, 6.306, 6.307, 6.375
  wars and defence:
    Dorian invasion 3.105
    Greek–Persian wars 3.101, *3.147*, 3.221, 3.314
    hoplites 3.170, 3.278
Greece, ancient (*cont.*):
  wars and defence (*cont.*):
    military leaders 3.8, 3.221, 3.353
    weapons 6.29, 6.327
    see also Peloponnesian area
Greece, modern *1.146*, *4.140*, *7.354–5*
  astronomy 8.22
  in Council of Europe 7.71
  economy and trade 7.221
    see also European Community; European Economic Community
  languages 7.10–11, *7.135*, 7.269
    alphabet *8.192*, *9.46*
  literature 5.82, *5.191*, 5.244–5, 5.410
  physical geography and geology *1.146*
    Corfu 1.76
    islands 1.82, 1.172, 1.282
    mountains 1.238
    regions 1.200
  politics and government 4.110, 4.363, 4.387, 7.71, 7.73, 7.201
  regions *9.75*
    see also under Greece, ancient
  religion/theology 7.98, 7.205
  rulers 4.88, 4.133, 4.140, 4.141
  technology and science 6.46, 6.56, 6.123, 6.138
  wars and defence 4.140–1, 4.218
    Greek War of Independence *4.140–1*
    see also NATO
Greek fire *6.161*, 6.225
Greek Revival *5.191*, 5.258, 5.260, 5.353, 5.405, 5.422, 5.437
Greely, Adolphus Washington *1.146*
Green, George (1875, inventor) 6.98
Green, George (1793–1841, mathematician) *1.146*
Green, Henry *5.192*
Green Mountain Boys *3.148*
Green Movement *4.141*
Green Paper 7.31
Green (politics) 7.66, 7.135, 7.260
  see also environment
green revolution 6.7, *6.161*, 6.364
Greenaway, Kate 5.53
Greenback Party *4.141*
Greene, Graham 5.162, *5.192*
Greene, Nathanael 3.58, *3.148*, 3.388
greenfinch *2.144*
greenfly 2.13
greengage *2.144*
greenhouse *6.161*, 6.170, 6.175
greenhouse effect and global warming *1.141*, *1.146–7*, *6.161*, *7.135–6*, *8.62*
  and combustion 6.78, 6.280, 6.301
  and deforestation 7.83
  and desertification 7.86
  and eustatic change 1.116
  and pollution 7.240
Greenland *1.147*, *3.148*
  early-warning system 6.111
  exploration 1.112, 1.146, 1.159, 1.224
  languages 7.108
  physical geography and geology 1.135, 1.147
Greenpeace 7.66, 7.135
greensand *1.147*
Greensboro incident *4.141*
greenshank *2.144*
Greenwich Mean Time and Meridian *1.147*, 1.267, 1.338, 6.237, 6.238, *8.62*, 8.62, *8.128*, 8.172, 8.177–8
Greer, Germaine 7.332
Gregg, John 6.318
Gregorian calendar 6.54, 8.19, *8.62*
Gregory, Lady Augusta 5.226
Gregory I (the Great), pope and Saint 3.28, 3.82, *3.148*, 5.351, *9.41*
Gregory VII, St, pope 3.59, 3.80, 3.121, *3.148*, 3.179, 3.268, 3.367
Gregory IX, pope 3.59, 3.132–3, 3.179
Gregory XIII, pope 3.57, 3.250
  see also Gregorian calendar
Gregory of Tours, St *3.148*
Grenada *1.147*, *3.148–9*, *4.141–2*, *7.355*
  see also Caribbean; CARICOM
grenade 6.64, *6.161*, 6.226
Grenadines *see* Saint Vincent and Grenadines
Grendel *9.14–15*
Grenville, George *3.149*, 3.339
Grenville, Sir Richard 3.30, *3.149*
Grenville, William Wyndham, Baron *4.142*
Gresham, Sir Thomas: Gresham's Law *3.149*, *7.136*
Greuze, Jean-Baptiste *5.192*
Grew, Nehemiah 6.122
Grey, Charles, 2nd earl Grey *4.142*
Grey, Sir Edward, viscount Grey of Fallodon *4.142*
Grey, Sir George (1812–98, colonial administrator) 4.14, *4.142*, 4.183
Grey, George (1860s, explorer) 1.45
Grey, Lady Jane *3.149*
grey whale *2.144*
greyhound racing *7.136*
greylag goose *2.144*
greywacke *1.147*
Griboyedov, Aleksandr 5.396
grid system (energy) 6.114, 6.150
grid system (maps) 6.391
Grieg, Edvard *5.192–3*
Griess, Johann Peter 6.27, *6.161*
grieving 7.30
griffin *9.15*
Griffith, Arthur *4.142*, 4.314
Griffith, David Wark *5.193–4*
griffon vulture *2.144*, 2.145, 2.358
Grignard, François Auguste Victor *1.147*
Grigorovich, Yuri 5.51
Grillparzer, Franz *5.194*
Grimaldi, Francesco 8.147
Grimaldi, Joseph 5.105, 5.199
Grimm, Jacob and Wilhelm *5.194*
Grimm, S. H. 3.289
Grimmelshausen, Johann Jakob Christoffel von *5.194*, 5.348
grinding machine *6.161*, 6.211, 6.212
Gris, Juan *5.194*
grisaille *5.194*
grit (rock) *1.147*, 1.215
Grivas, Colonel Georgios 4.112–13
grizzly bear 2.26
Grock (Charles Wettach) 5.105
Grolier, Jean 5.52
Gropius, Walter 5.38, *5.194–5*
grosbeak *2.144*
Gross, Samuel David *2.144*
gross domestic product *see* GDP
gross national product *see* GNP
Grosz, Georg *5.195*
grotesque *5.195*
Grotius, Hugo *3.149*, 7.164, 7.165
ground beetle *2.145*, 2.341
ground cuckoo 2.78
ground elder *2.145*
ground squirrels *2.145–6*
  chipmunk 2.64
  gopher 2.141
  marmots 2.203–4, 2.270
  prairie dog 2.267–8
ground station *6.161–2*, 6.222, 6.308

ground-guidance system 6.161, 6.162, 6.367
groundhog 2.370
groundnut 2.246
groundsel 2.145
groundwater 1.147
  perched water-table 1.250
Group of 7 7.136
Group f.64 5.5, 5.195, 5.488
Group of 77 7.319
group theory (mathematics) 1.147
grouper 2.146
groups:
  clan 7.52, 7.86, 7.336
  co-operative see co-operatives
  crowd psychology 7.77
  ethnic see ethnic groups
  feud 7.119–20
  interest see interest groups
  social identity 7.290
  therapy 7.136
  tribes 7.315
  see also community; family; kinship; social class; sociology and society
grouse 2.146
  capercaillie 2.52
  prairie chicken 2.266–7
  ptarmigan 2.270, 2.280
  red 2.280
  turkey 2.351
growth (economic) see development
growth of living organisms 2.146
  germination 2.135
  growth hormone 6.170
  inhibition (allelopathy) 2.6
  life cycle 2.189
  meiosis 2.207, 2.214
  mitosis 2.214–15
  reproduction see reproduction
  stages see caterpillars; embryo; larvae; metamorphosis; puberty; pupae
Grub Street 5.195
Gruffydd, Rhys ap 3.221
Grün, Max von der 5.175
Grünewald, Mathis 5.195, 5.356
grand unified theory (GUT) 8.60
Grus 8.62
Gu Cheng 5.98
Gu Kaizhi 5.196
Guadeloupe 7.355
Guadet, Marguerite 3.141
guaiacum 2.190
Guam 1.147, 4.142
guan 2.84
Guan Hanqing 5.195
guanaco 2.146
Guandao, Lui 3.200
Guāndì 9.15
Guangxu, emperor of China 4.76
Guangzhou (port) 3.149
guanine 2.95
guano 2.146, 6.128
Guānyīn 9.15
Guanzhong, Luo 5.387
Guardi, Francesco 5.195
Guardi, Gianantonio 5.195
guardian 7.136
Guarini, Guarino 5.195
Guatemala 1.148, 3.149, 4.142, 7.355
  arts 5.25
  economic organization 4.251
  exploration 1.47
  independence 4.67
  leaders 4.17
  physical geography and geology 1.148
  refugees 7.261
  religion/theology 7.49
  technology 6.54
  wars and defence 7.82
guava 2.146
Guderian, Heinz 4.142
gudgeon 2.146
guelder rose 2.146–7
Guelph 3.149
Guercino, Il 5.195
Guericke, Otto von 1.24
Guernsey 1.58
Guerrero, Vincent 4.160
guerrilla warfare 4.142–3, 4.366–7, 7.136
Guesclin, Bertrand du 3.70, 3.149
Guest, Lady Charlotte 5.273
Guettard, Jean Étienne 1.148
Guevara, Ernesto 'Ché' 4.44, 4.65, 4.143
Guiana see French Guiana; Guyana; Surinam
Guicciardini, Francesco 3.149, 5.231
Guiccioli, Countess Teresa 5.69
guidance system see ground-guidance; guided missiles; warning and detection system
guided missiles 6.162, 6.228, 6.231, 6.235, 6.375
guilds 3.149–50, 3.191–2, 3.208
guillemot 2.147
guillotine 3.150
Guimard, Hector 5.195–6
Guinea 1.148, 4.4, 4.143, 4.347, 7.355
Guinea-Bissau 1.148, 4.4, 4.58, 4.143, 4.306, 7.356
guineafowl 2.147
guinea-pig 2.147–8
Guinevere, queen 9.5, 9.14
Guiraud, Ernst 5.326
güiro 5.196
Guiscard, Robert 3.150, 3.255
Guiscard, Roger 3.306
Guise (ducal house), Claude de Lorraine, Francis, Charles, Mary, and Henry 3.150, 3.356
guitarfish 2.148
guitars and guitarists 5.89, 5.196, 5.410, 5.465, 5.497
Guizot, François Pierre Guillaume 4.143
Gujarat 3.150
Gulab Singh 4.179
Gulags 7.64
gulf 1.148
Gulf (Persian Gulf) 1.148
  Chaldea 3.39, 3.67
  exploration 1.30
  Gulf Co-operation Council 7.136–7
  Gulf States in Arab League 7.15
  islands 1.30
  Trucial States 4.352–3
  war see Gulf War
  see also individual countries
Gulf Stream 1.148
Gulf War (1980s) 4.165–6
Gulf War (1991) 4.143, 4.159, 4.166, 4.188
  American troops 7.328, 7.329
  and Arab League 7.15
  and arms trade 7.18
  and Ba'athism 7.24
  collective response to 7.56
  and Gulf Co-operation Council 7.137
  hostage-taking 7.148, 7.328
  and journalism 7.172
  and Kurds 7.178
  and PLO 7.227
  propaganda 7.249
  sanctions 7.273
  and UN 7.164, 7.321, 7.326
  war crimes 7.148, 7.328
  weapons and technology 4.143, 6.15, 6.167, 7.44, 7.69, 7.220
gulls 2.148, 2.157, 2.179
gully 1.148–9
gum 6.162
Gum, Colin: Gum Nebula 8.62
gum-trees 2.2, 2.148, 2.350
gunboat diplomacy 4.144
gun-cotton see nitro-cellulose
gundi 2.148
Gunnbjörn, Ulf-Krakuson 3.148
gunnel 2.34
gunnera 2.148
gunpowder 1.118, 6.162, 6.225, 6.228
  and cannon 6.57
  China 6.16
  fireworks 6.132
  production 6.338
Gunpowder Plot 3.150–1
guns:
  gun-barrel detonation 6.246
  gun-sight 6.162–3
  see also artillery; firearms; weapons
Gunther, king 9.8, 9.33
Guo Moruo 5.98
guppy 2.148
Guptas 3.151–2, 3.152
  art 5.196
Gurdwara 7.137
gurnard 2.148
Gurney, Sir Goldworth 4.225
gurus 3.245, 4.282–3, 7.137
  Sikh see Amar Das; Arjan Dev; Gobind Singh; Hargobind; Nanak; Tegh Bagadur
gusher see geyser
Gustave IV, king of Sweden 4.335
Gustavus I Vasa, king of Sweden 3.127, 3.152, 3.318, 3.347
Gustavus II Adolphus, king of Sweden 3.49, 3.152, 3.214, 3.318, 3.347, 3.355
Gustavus III, king of Sweden 3.347
GUT see grand unified theory
Gutenberg, Johann 3.152, 3.293, 6.163, 6.199, 6.281
Gutenberg discontinuity 1.98
Guthrum, king of Danes 3.10, 3.97
Gutierrez, Gustavo 7.184
Gutiérrez de Padilla, Juan 5.257
Guy de Lusignan 3.95, 3.314
Guyana 1.149, 1.275, 4.144, 7.356
  see also Caribbean; CARICOM
Guzmán, Abimael 4.267
Guzmán Blanco, Antonio 4.144
Gwalchmei 9.14
Gwalior see Sindhia
Gwilym, Dafydd ap 5.486
Gwyn, Nell 3.152
Gwynedd, Owain 3.221
Gyaw, James Hla 5.67
gymnastics 7.137
gymnosperms 2.148
  cycad 2.84
  ginkgo 2.136
  see also conifers
gynaecology see obstetrics and gynaecology
gypsies 7.137
gypsophila 2.255
gypsum 1.6, 1.149, 6.53, 6.163, 6.255, 6.271, 6.378
gypsy moth 2.148
gyre 1.149
gyrfalcon 2.148
gyro-compass 6.163, 6.328
gyromagnetic compass see compass under magnet/magnetism
gyroscope 6.163, 6.178, 6.213, 6.330

H

Haakon IV Haakonson (the Old), king of Norway 3.8, 3.152, 3.258
Haakon VII, king of Norway 4.144, 4.247
Haavikko, Paavo 5.155
habeas corpus 7.138
Haber, Fritz 6.72, 6.164
  Haber–Bosch process 6.13, 6.43, 6.62, 6.71, 6.128, 6.164, 6.174
Habermas, Jürgen 7.124, 7.138
Habil (Abel) 9.2, 9.8
Habimah 5.197, 5.314
habitat (ecology) 2.149
  carrying capacity 2.55
  community 2.74
  niche 2.227
Habitat (UN) 7.197, 7.282, 7.321
Habré, Hissène 4.69
Habsburg dynasty 3.152, 3.272, 4.123, 4.144
  family tree 3.153, 3.338
  rulers listed 3.153, 9.67
  see also under Holy Roman empire; Spain
Habsburg-Lorraine, House of see under Holy Roman empire
Habyarimana, Juvénal 4.298
hacienda 4.144
hackling machine 6.164
Hadar 3.152
haddock 2.149
Hades 9.9, 9.12, 9.15
ḥadīth 7.138, 9.44
Hadley, George 1.149
Hadley, John 6.315
Hadley Rille 8.63
Hadrian I, pope 3.205
Hadrian, Roman emperor 3.152, 3.154
  Hadrian's Wall 3.154
hadrons 1.105, 1.149
  mesons 1.211
  see also baryons; quarks
Haeckel, Ernst Heinrich 2.107, 2.149
haematite 1.149, 1.239, 6.39, 6.186
haematology 6.112, 6.164, 6.218
  see also blood
haemodialysis 6.101, 6.164
haemoglobin 2.149, 6.164
  breakdown to form bile 2.29
  deficiency (anaemia) 2.9, 2.161
haemophilia 2.149, 6.40
haemorrhage 2.149
Ḥāfiẓ, Shams al-Dīn Muhammad 5.197
Hafiz al-Assad 4.335
hafnium 1.106, 1.149, 6.391
Haganah 4.144
Hagar 9.2, 9.31
Hagerup, Nina 5.193
hagfish 2.149
Haggard, Sir Henry Rider 5.197, 9.26
Haghe, Louis 4.140
Hague Conventions 7.142, 7.328
Hague School 5.197
Hahgwehdiyu 9.15
Hahn, Otto 1.106, 1.149–50, 1.210, 6.218, 6.242
Hahnemann, Samuel 2.159, 6.169
hahnium 1.106
Haig, Douglas, 1st earl 4.144–5, 4.262, 4.319
haiku 5.197
hail 1.150, 1.307

Haile Selassie, emperor ***4.145***, **7.**258
Hainaut ***3.154***
hair ***2.149***, **6.**128
hair-streak ***2.149***
Haiti ***3.154***, ***4.145***, ***7.356***
  leaders **4.**97, **4.**105–6, **4.**347–8, **7.**184
  physical geography and geology **1.**156
  religion and myths **7.**184, **9.**32
    see also Voodoo
  sport and recreation **7.**54
Hajar **9.**18
Haji, Raja **3.**53
ḥajj, see Mecca
hake ***2.149–50***
Hakim, caliph **7.**96
al-Hakīm, Tawfīq **5.**19
Hakluyt, Richard ***1.150***, **5.**460
halberd **6.**165
Haldane, Richard Burdon, viscount Haldane of Cloan ***4.145***
Halder, Franz ***4.145***
Hale, George Ellery ***8.63***, **8.**105
  telescope ***8.63***
Hales, Stephen ***2.150***
Halevi, Yehuda **5.**202
Haley, Bill **5.**358, **5.**384
half-life ***1.150***
half-tone process ***6.164***
halibut ***2.150***
Haliday, Joseph **5.**65
halides ***1.150***, **6.**252
Halidon Hill, battle of ***3.154***
Halifax, Edward Frederick Lindley Wood, 1st earl of ***4.145***, **4.**294
halite ***1.150–1***
  crystal **1.**80, **1.**81
Hall, Asaph **8.**36, **8.**92, **8.**119
Hall, Charles Martin **6.**12, ***6.164***, **6.**213
Hall, James ***1.151***
Hall, Sir Peter ***5.197***
Hall, Sydney Prior **4.**261
al-Hallāj **5.**439
Ḥallaj, al-Asrar ***9.15***
Hallam, Arthur **5.**449
Halley, Edmond **8.**7, ***8.63***, **8.**151, **8.**161, **8.**193
Halley's Comet ***8.63***, **8.**64, **8.**194
  Giotto probe **8.**28, **8.**59
hallmarks ***9.63***
Hallstatt ***3.154***
hallucination ***7.138***
hallucinogens **7.**95
halo **1.**150, ***1.151***
  galactic **8.**54
haloalkane **6.**252
halocarbons ***1.151***
  see also chloro-fluorocarbons; fluorocarbons; PTFE
halogens ***1.151***
  activity series **1.**3
  astatine **1.**22, **1.**106
  halide **1.**150
  tear-gas **6.**345–6
  see also bromine; chlorine; fluorine; iodine
Hals, Frans ***5.197***
Halsey, William Frederick ***4.145***
Halske, Johann **6.**318
Ham ***9.15***, **9.**27
al-Hamadhānī **5.**19
hamadryad **2.**179
Hamersley Range ***1.151***
Hamilcar Barca ***3.154***, **3.**294
Hamilton, Alexander **3.**87, **3.**124, ***3.154–5***, **3.**380, **4.**56, **4.**117, **4.**329
Hamilton, Lady Emma **3.**249, **4.**237
Hamilton, Gavin **5.**315
Hamilton, James, 3rd marquis and 1st duke of ***3.155***, **3.**291
Hamilton, W. D. (1964, zoologist) **7.**294
Hamilton, Sir William (1777, archaeologist) **3.**288
Hamilton, William (1788, painter) **3.**379
Hamilton, Sir William Rowan (1806–65, scientist) **1.**132, ***1.151***
Hammarskjöld, Dag Hjalmar Agne Carl **4.**83, ***4.146***, **4.**359, **7.**322
hammer ***6.164–5***, **6.**227, **6.**357, **6.**358
  pile-driver **6.**270
  steam **6.**236, **6.**331
  trip-hammer **6.**363
hammerhead ***2.150***
hammerkop ***2.150***
Hammerstein, Oscar II **5.**386
Hammett, Dashiel **5.**129
Hammurabi **3.**32, ***3.155***
Hampden, John **3.**127, ***3.155***, **3.**327
Hampton Court Conference ***3.155***
Hampton Roads Conference ***4.146***
hamster ***2.150***
Hamsun, Knut ***5.197***
Han dynasty **3.**72, ***3.155–6***, **5.**95
Han Yu ***5.198***
Hancock, John ***3.156***
Hancock, Thomas **6.**158, ***6.165***, **6.**211
Hancock, Walter **6.**359
hand **2.**118–19, ***2.150–1***, ***9.58***
  nails **2.**225
  see also fingers
hand tools ***6.357***, ***6.358***
  adze **6.**4, **6.**27, **6.**357, **6.**358
  agricultural **6.**6
  axe **6.**27, **6.**165, **6.**357, **6.**358
  chisel **6.**72, **6.**357, **6.**358
  clamp **6.**76
  digging **6.**103
  drilling and boring *see* drilling
  gouge **6.**72
  hammer *see* hammer
  hoe **6.**6, **6.**169
  knife **6.**46, **6.**191, **6.**357
  pick and shovel **6.**103, **6.**227
  plane **6.**211, **6.**212, **6.**270, **6.**357, **6.**358
  pliers, tongs and pincers **6.**273, **6.**358
  power tools **6.**280
  rake **6.**6, **6.**294
  reamer **6.**295
  saw **6.**69, **6.**308, **6.**357, **6.**358, **6.**387
  scissors **6.**310
  screwdriver **6.**311, **6.**358
  scythe **6.**6, **6.**311
  shears **6.**6, **6.**315
  spanner **6.**327, **6.**358
  tap **6.**345
  vice **6.**372, **6.**385
hand weapons ***6.165***
  see also weapons, missiles
Handel, George Frideric **5.**18, **5.**172, ***5.197–8***, **5.**441
handicap *see* disability and handicap
Handke, Peter **5.**175
hands *see* hand
Handsome Lake, Chief **3.**321
Hang Samrin **4.**178
hang-glider ***6.165***, **6.**222
hanging valley **1.**140, ***1.151***
Hanna, Marcus Alonzo ***4.146***
Hannibal **3.**23, **3.**59, **3.**123, ***3.156–7***, **3.**206, **3.**260, **3.**294–5, **3.**314
Hanoum, Latifeh **4.**21
Hanover/House of Hanover **3.**139–40, ***3.157***, **3.**333, ***4.146***
  rulers listed ***9.67***
    see also under rulers *under* Britain
  see also Seven Years War
Hanseatic League ***3.157***
Hansen, Peter **8.**87, **8.**101
Hansom, J. A.: hansom cab **6.**61
Hanukkah ***7.138***
Hanuman ***9.15***
happening **5.**46, ***5.198***, **5.**375
happiness, maximization of **7.**324
Hapsburg *see* Habsburg
ul-Haqa, Zia **4.**256
Hara Takashi ***4.146***
hara-kiri ***7.138***
Harald Fairhair, king of Norway **3.**258
Harald Hardrada, king of Norway **3.**158, **3.**258, **3.**339
Harald V, king of Norway **4.**248
Harappa ***3.157***, **3.**177, **5.**223, **6.**178
harbours *see* ports and harbours
hard disk ***6.165***
hard water **6.**378
hardboard **6.**214
hard-edge painting ***5.198***
Hardenberg, Friedrich Freiherr von **5.**323
Hardenberg, Karl August, prince ***4.146***
Hardie, Keir ***4.146***, **4.**160, **4.**189
Hardin, Lil **5.**238
Harding, Carl **8.**78
Harding, Warren Gamaliel ***4.146***, **4.**339
hardingfele ***5.198***
hardness of water ***1.151–2***
Hardouin-Mansart, Jules ***5.198***
hardpan (and duricrust) **1.**48, **1.**49, ***1.97***, ***1.152***, **1.**189
hardware (computer) ***6.165***
  see also bus; central processing unit; clock; memory; peripheral; printing; visual display units
Hardwick, Philip Charles **5.**474
hardwood **6.** 388
Hardy, Thomas (1752–1832, radical leader) ***3.157***, **3.**209
Hardy, Thomas (1840–1928, writer) ***5.198***, **5.**227, **5.**459
Hare, Robert ***1.152***
hare (animal) ***2.151***, **2.**174
Hare (trickster) ***9.15***
Hare Krishna ***7.138***
harebell **2.**51
Hargobind, Guru **7.**285
Hargreaves, James **6.**92, ***6.165***, **6.**329
Harington, Sir John **5.**21, **6.**377
al-Harīrī, Muhammad ibn al-Qāsim ***5.198–9***, **5.**411
Harishchandra, Bharatendu **5.**206
Hariti **9.**22
Harlem ***4.147***
harlequin bug **2.**307
harlequinade ***5.199***
Harley, Robert, 1st earl of Oxford and Mortimer ***3.157***, **3.**359
Harlington, Lord **4.**357
Harlow, H. F. **7.**207
harmattan ***1.152***
Harmensen, Jakob **3.**23
harmonica ***5.199***
harmonics ***5.199–200***
  harmonic flute ***5.199***
  harmonic mean **1.**209
  harmonic motion, simple **1.**305
harmonium ***5.200***
harmony ***5.200***
  see also chords; harmonics; polyphony; scale; tonality
Harmsworth, Alfred Charles William, 1st viscount Northcliffe ***4.146–7***
Harmsworth, Harold **4.**146
harness and saddlery ***6.165–6***, **6.**300
Harold I, king of Norway ***3.157***
Harold II (Godwinson), king of England ***3.157***, **3.**158, **3.**339, **3.**382
Harpocrates **9.**17
harpoon **6.**133, ***6.166***
harps **5.**56, **5.**138, ***5.200***, **5.**250, **5.**402–3
harpsichord and harpsichordists **5.**28, **5.**115, ***5.200***, **5.**373, **5.**405, **5.**477–8
harpy ***9.15***
harpy eagle ***2.151***, **2.**277
harquebus **6.**130, **6.**131
harrier-eagle 24
harriers (hawks) ***2.151***, **2.**204
Harrington, James ***3.157***
Harrington, Robert **8.**25
Harriot, Thomas (1611, astronomer) **8.**147
Harriott, Thomas (1588, colonialist) **3.**305
Harris, Sir Arthur **4.**44
Harris, Howel **4.**218
Harris, Joel Chandler ***5.200***, **9.**15
Harris, Lawren **5.**73
Harris, Roy ***5.200***
Harrison, Benjamin ***4.147***
Harrison, George **5.**39
Harrison, H. C. **6.**326
Harrison, John (engineer) **6.**297
Harrison, John (1693–1776, horologist) **6.**73, ***6.166***, **6.**303, **8.**142
Harrison, William Henry ***4.147***, **4.**342, **4.**345, **4.**371, **4.**377
harrow **6.**6, ***6.166***
Harsha ***3.157***
Hart, Lorenz **5.**385
Hart, William S. **5.**487
hartebeest ***2.151***
Hartford Convention ***4.147***
Hartman, George **8.**9
hart's tongue ***2.151***
Harun al-Rashid **3.**1, **3.**19, ***3.157–8***
Harut and Marut ***9.15***
Harvard classification of spectral classes **8.**39
harvest mouse ***2.151–2***
harvesting:
  combine harvester **6.**6, **6.**81, **6.**82
  forage **6.**143
  gantry **6.**150
  hand tools **6.**6, **6.**191, **6.**311
  reaper **6.**5, **6.**172, **6.**210, **6.**295
harvestman ***2.151***
harvest-mite ***2.151***
Harvey, William **2.**113, ***2.152***, **3.**319
Harz Mountains ***1.152***
Hasan, Muhammad Abdille Sayyid ***4.147***
Hasan ibn al-Sabbah **3.**25
Hasdrubal ***3.158***
Hašek, Jaroslav ***5.201***
Hasidim ***7.138***
Hassan II, king of Morocco **4.**224
Hastings, Warren **3.**54, ***3.158–9***, **4.**211
Hastings, battle of ***3.158***
hatching (drawing) ***5.201***
Hathor **9.**7, **9.**18
Hatim al Ta'i ***9.15***
Hatshepsut, queen of Egypt ***3.159***, **3.**356
Hauptmann, Gerhart ***5.201***
Hausa **3.**4, ***3.159***, **3.**389, **7.**43, ***7.138–9***
Haushofer, Karl **7.**131
Hausmalerei ***5.201***
havan ***7.139***
Havel, Václav **4.**92, ***4.147***, ***5.201***
Havelock, Sir Henry ***4.147***, **4.**200
Haverty, Joseph **4.**249
Hawaii ***1.152***, ***4.147***
  astronomy **8.**19
  Declaration of (psychiatry) **7.**251
  Omega navigation **6.**250
  peoples **7.**240
  physical geography and geology **1.**152, **1.**153, **1.**358
  religion and myths **9.**25, **9.**35

Hawaii (*cont.*):
surfing **7**.303
Hawaiian goose ***2**.152*
hawfinch **2**.144, ***2**.152*
hawk family ***2**.152*
buzzard **2**.48
caracara **2**.52
fish-hawk **2**.237
harriers **2**.151, **2**.204
kites **2**.160, **2**.179
marsh harrier **2**.204
osprey **2**.237
sparrow-hawk **2**.319
see also eagles; vultures
Hawke, Bob **4**.23, **4**.189
Hawke, Edward, 1st baron ***3**.159*, **3**.298, **3**.324
Hawking, Stephen William ***8**.64*
Hawkins, Sir John **3**.106, **3**.136, ***3**.159*, **3**.337
Hawkins, Sir Richard ***3**.159*
hawkmoths ***2**.152*, **2**.343
Hawksmoor, Nicholas ***5**.201*, **5**.494
Hawkwood, Sir John **3**.84
Hawley-Smoot Tariff Act ***4**.147*
hawthorn ***2**.152–3*
Hawthorne, Nathaniel ***5**.201*, **5**.387
Hawwa **9**.2, **9**.12
Hay, John Milton **4**.148
hay fever ***2**.153*
Hay Treaties ***4**.148*
Haya de la Torre, Victor Raúl **4**.33, ***4**.147–8*
Hayasaka, Shigezo **4**.338
Hayashi, Chushito: Hayashi line ***8**.64*, **8**.65
Hayden, Ferdinand Vandiveer ***1**.152*, **2**.186
Haydn, Joseph **5**.37, **5**.41, ***5**.201–2*, **5**.313, **5**.405, **5**.442
Haydon, Benjamin Robert ***5**.202*
Hayek, Friedrich August von ***7**.139*, **7**.157, **7**.215, **7**.291
Hayes, Rutherford B. **4**.148
haymaking ***6**.166*
see also crop *under* grass
Haymarket Square riot ***4**.148*
haze ***1**.152*
hazel **2**.129, ***2**.152*, **2**.153
Hazlitt, William ***5**.202*
HCFCs (hydrochloro-fluorocarbons) **6**.72
HDTV (high-definition television) **6**.349
head ***2**.153*
migraine **2**.210
see also brain; mouth; nose; skull
heads of state ***7**.139*
see also leaders; monarchy; presidents; prime ministers
heald (heddle) **6**.207–8
Healey, Denis **4**.380
healing **7**.11, **7**.62, ***7**.139*, **7**.282
see also medicine and health care
health care *see* medicine and health care
Healy, G. P. A. **4**.311
Heaney, Seamus **5**.202, **5**.203
Heaphy, T. **4**.374
hearing ***2**.153*
disorders (deafness) ***2**.88*, ***7**.80*
alphabet **9**.48
audiometer **6**.25
hearing aid ***6**.166*
lip-reading **7**.185
otorhinolaryngology **6**.254
see also signing
ears **2**.70, **2**.102
Hearst, William Randolph ***4**.149*
heart ***2**.153–4*, ***9**.60*
disorders ***7**.140*
angina pectoris **2**.10
attack **2**.77
cardiac imaging **6**.243
cardiology and cardiac surgery ***6**.60*, **6**.367
heart (*cont.*):
disorders (*cont.*):
drugs for **6**.60, **6**.266, **6**.387
electrocardiography **6**.112–13, **6**.116, **6**.218
heart-lung machine **6**.156, ***6**.166*, **6**.218
pacemaker **6**.218, **6**.255
transplant surgery **6**.30, **6**.361
valve **6**.370
heart urchin **2**.302
heat ***1**.152*, ***6**.166*
adiabatic change **1**.3
annealing **6**.15
boiler *see* boiler
buildings *see* heating, domestic
capacity **1**.152
central *see* heating, domestic
combined heat and power generation **6**.79, **6**.81, **6**.138, **6**.150
engines *see* engines
exchanger ***6**.166–7*
exothermic reactions **1**.3, **1**.117
see also combustion; neutralization
expansion **1**.117
fuel-oil **6**.147
global warming *see* greenhouse effect
heating *see* heating, domestic
heat-pump ***6**.167*
hypocaust **6**.167, **6**.175
induction **6**.290
latent **1**.188–9
measurement **6**.54–5, **6**.166, **6**.192, **6**.350–1, **6**.354–5, **6**.368
medical treatment (diathermy) **6**.101, **6**.251
natural *see* geothermal
in nuclear reactor **6**.245
power from *see* engines
radio-frequency **6**.290
refraction **1**.278, **1**.279
removed (endothermic reaction) **1**.3, **1**.108
solar *see* solar power
sterilization **6**.334
storage of **6**.121
temperature *see* temperature
thermodynamics *see* thermodynamics
thermography **6**.354, **6**.355
thermostat **6**.89, **6**.185, **6**.193, **6**.355
transfer and exchange:
energy budget **1**.108, **1**.109
thermal radiation **1**.273
see also conduction; convection; exothermic; radiation
treatment **6**.167
waves: reflection **1**.278
waves (weather) ***1**.152*
see also energy; insulation; temperature *and all entries beginning with* therm
Heath, Edward ***4**.149*
heath butterfly ***2**.154*
Heathcoat, John **6**.193
heather family ***2**.154*
bilberry **2**.29
cranberry **2**.80
strawberry tree **2**.327
heating, domestic **6**.166, ***6**.167*
central **6**.66, **6**.67, **6**.104, **6**.167
for cooking **6**.88–9
district **6**.104, **6**.167, **6**.377
gas *see* lighting and heating *under* gas
immersion heater **6**.176
storage heater **6**.167, **6**.335
tumble-drier **6**.364
heatstroke **2**.330
heaven ***7**.140*
Heaviside, Oliver:
Heaviside–Kennelly layer ***1**.153*
heavy chemicals *see* industry *under* chemicals
heavy goods vehicle *see* lorry
heavy industry ***7**.140–1*
see also individual industries
heavy water **1**.360, ***6**.167*, **6**.243, **6**.245
Hebbel, Friedrich ***5**.203*
Hebe (mythology) ***9**.15*
hebe (plant) **2**.319
Hébert, Jacques René **3**.83, **3**.141, **3**.159
Hebrews:
language and alphabet **7**.10, **7**.19, **7**.279, **7**.335, ***9**.46*
literature **5**.9–10, **5**.47, **5**.197, ***5**.202–3*, **5**.216
see also Jews
Hebrides ***1**.153*, ***3**.159*
Hecate ***9**.15*
Hector **9**.2, ***9**.15–16*
Hecuba ***9**.16*
heddle (heald) **6**.207–8
hedge sparrow ***2**.154*
hedgehog ***2**.154*
hedging (investment) ***7**.141*
Hedley, William **6**.293
hedonism ***3**.159–60*, ***7**.141*
Hegel, George Wilhelm Friedrich ***7**.141*
and dialectic **7**.89, **7**.304
and historicism **7**.144
and Kierkegaard **7**.176
and Locke **7**.52
hegemony ***7**.141*
hegira ***3**.160*
Heidegger, Martin **7**.113, ***7**.141*, **7**.223
Heider, Fritz **7**.113, **7**.292, **7**.293–4
Heifetz, Jascha ***5**.203*
height:
above sea-level *see* altitude
histogram **1**.156
Heimdall **9**.13
Heine, Heinrich ***5**.203*
Heinesen, William **5**.404
Heisenberg, Werner Karl ***1**.153*, **1**.349, **7**.42
Heitsi-eibid ***9**.16*
Hejaz ***4**.149*, **4**.159
Hektor (asteroid) ***8**.64*, **8**.193
Hel **9**.2, ***9**.16*, **9**.22
Helen ***9**.16*, **9**.25, **9**.29
Helena, Saint ***3**.160*
Helene (satellite of Saturn) **8**.175
Helgoland ***1**.153*, ***4**.149–50*
heliacal rising and setting ***8**.64*
helicopter **6**.12, **6**.138, ***6**.167*, **6**.190, **6**.225, **6**.319
Heligoland ***1**.153*, ***4**.149–50*
heliocentric system ***8**.64–5*
Copernican **8**.29
heliometer ***8**.65*
heliopause ***8**.65*
Heliopolis ***3**.160*
Helios ***9**.16*
heliosphere ***8**.65*
heliostat ***8**.65*
helium **1**.106, ***1**.153–4*, ***6**.167*, ***8**.65*
in airship **6**.10
alpha particle **1**.10
emission spectrum **1**.316
and hydrogen bomb **6**.246
laser **6**.195
Helix Nebula ***8**.65*
hell ***7**.141*
Hellas Planitia ***8**.65*
hellebore ***2**.154*
Hellenic languages **7**.10–11, **7**.135, ***7**.156*, ***7**.157*, **7**.269
Hellenistic art ***5**.203*
see also arts *under* Greece, ancient
Hellenistic civilization ***3**.160*, **5**.202–3
see also Greece, ancient
Heller, Joseph ***5**.203–4*
Hellespont **3**.2, ***3**.160*
Hellfire Club ***3**.160*
Helmholtz, Hermann Ludwig Ferdinand von ***1**.154*
Helmholtz resonator ***5**.204*
Kelvin–Helmholtz contraction ***8**.79*
helot ***3**.160*
Help Age International **7**.222
Helpmann, Sir Robert ***5**.204*
Helsinki Agreement **7**.272
Helsinki Conference ***4**.150*
Helsinki Final Act **7**.2, **7**.64
Helvetii ***3**.160*
Helvétius, Claude Adrien ***3**.160*
Helwys, Thomas **3**.35
hematite *see* haematite
Hemichordata **2**.2, ***2**.154*
Hemingway, Ernest ***5**.204*
hemiplegia **2**.242
hemlock ***2**.154*
hemlock spruce ***2**.154*
hemp ***2**.154*, **6**.57
cannabis **2**.52, **2**.154
sisal **2**.310
Henare Martua **4**.287
Henderson, Fletcher **5**.47
Henderson, Thomas **8**.4
Heng Samrin **4**.58, **4**.313
Hengist and Horsa ***3**.160*, **3**.193, **3**.375
Henlein, Konrad **4**.332
henna ***2**.154*
Henri, Robert **5**.24
Henrietta Maria ***3**.160*, **3**.161
henry (unit) ***1**.154*, **6**.368
Henry I, king of England (1068–1135) **3**.160–1, **3**.255
Henry I (the Fowler), king of Saxony and Germany (*c*.876–936) ***3**.160*
Henry I de Bourbon (1552–88) **3**.84
Henry II, Holy Roman Emperor (1002–24) **3**.168
Henry II, king of England (1133–89) **3**.13, ***3**.161*, **3**.255, **3**.284
and Becket **3**.38, **3**.39, **3**.161
and Clarendon Constitutions **3**.78
and fyrd **3**.136
and Gascony **3**.137
and Ireland **3**.3, **3**.179
and law **3**.25
and taxes **3**.320
wife (Eleanor of Aquitaine) **3**.18, ***3**.113*, **3**.137, **3**.161, **5**.368
Henry II, king of France (1519–59) **3**.103, **3**.227, **3**.359
Henry II de Bourbon (1588–1646) **3**.84
Henry III, Holy Roman Emperor (1039–56) **3**.168
Henry III, king of England (1207–72) ***3**.161*, **3**.255
and French Wars of Religion **3**.134
resistance to *see* Barons' Wars
and Wardrobe **3**.377
Henry III, king of France (d.1589) **3**.167, **3**.263, **3**.356, **3**.369
Henry IV, Holy Roman Emperor (1056–1106) **3**.148, **3**.167–8, **3**.303
and religion **3**.59, **3**.121, **3**.179
Henry IV, king of England (1366–1413) ***3**.161–2*, **3**.202, **3**.255, **3**.284, **3**.310
and Beauforts **3**.37
and Gascony **3**.137
and Lollards **3**.209
and Wales **3**.142
Henry IV, king of France (Henry of Navarre, 1553–1610) **3**.46, **3**.47, ***3**.162*, **3**.369
and Condé **3**.84

Henry IV, king of France (*cont.*):
and religion **3**.134, **3**.167, **3**.171, **3**.245, **3**.250, **3**.313
and Three Henrys, War of **3**.356
Henry V, Holy Roman Emperor (1105–25) **3**.168, **3**.179, **3**.225
Henry V, king of England (1387–1422) **3**.5, ***3**.162*, **3**.171–2, **3**.202, **3**.255, **3**.284, **3**.310
Henry VI, Holy Roman Emperor (1169–91) **3**.168
Henry VI, king of England (1422–71) ***3**.162–3*, **3**.202, **3**.255, **3**.284, **3**.313
and Hundred Years War **3**.171–2
and Roses, war of **3**.310
wife **3**.222
Henry VII, Holy Roman Emperor (1308–13) **3**.168
Henry VII, king of England (Henry Tudor, 1457–1509) **3**.38, ***3**.163*, **3**.202, **3**.284
battles **3**.47, **3**.128
and exploration **3**.56
and Treaty of Etaples **3**.118
as Tudor **3**.310, **3**.361–2
Henry VIII, king of England (1491–1547) ***3**.163*, **3**.362
battles **3**.126
and Ireland **3**.179
and parliament **3**.83, **3**.270
and religion **3**.13, **3**.236, **3**.282, **3**.290, **3**.346
and sanctuary **3**.316
and Wales **3**.50, **3**.365, **3**.376
wives **3**.16, **3**.44, **3**.64, **3**.90, **3**.91, **3**.137, **3**.170, **3**.270–1, **3**.324, **3**.385
Henry of Blois, Bishop **3**.11
Henry of Huntingdon **3**.60
Henry the Lion **3**.149, ***3**.163*, **3**.318
Henry the Navigator ***1**.154*, **3**.121, ***3**.163*
Henry, Major (d.1898, forger) **4**.104
Henry, Joseph ***1**.154*
Henry, O. ***5**.204*
Henry, Patrick **3**.87, ***3**.163–4*
Henry, William: Henry's Law ***1**.154*
Henry Tudor *see* Henry VII, king of England
Henryson, Robert ***5**.204*
Henze, Hans Werner ***5**.204*, **5**.324
hepatitis ***2**.155*, **6**.87
Hephaestus **9**.6, ***9**.16*
Hepplewhite, George ***5**.204*, **5**.315
Hepworth, Dame Barbara ***5**.205*, **5**.318, **5**.398
Hera **9**.9, ***9**.16*, **9**.29
Heracles *see* Hercules
Heraclius ***3**.164*
heraldry ***3**.164*
herbal medicine ***6**.168*, **6**.387, **7**.46
herbalism *see* drugs *under* plants
Herbert, George ***5**.205*, **5**.342, **5**.481
Herbert, Zbigniew ***5**.205*
herbicides **6**.7–8, ***6**.168*, **6**.178, **6**.271
Herbig–Haro object ***8**.65*
herbivores ***2**.155*
see also in particular cattle; rodents; ruminant mammals
herbs ***2**.155*
see also spices, herbs, and flavourings *and also* drugs *under* plants
Hercegovina *see* Bosnia
Herculaneum ***3**.164*
Herculano, Alexandre **5**.359
Hercules (constellation) ***8**.65*
Hercules beetle **2**.155–6
Hercules (Greek hero) **9**.3, **9**.5, ***9**.16*
Twelve Labours **9**.4, **9**.6, **9**.8, **9**.16, **9**.17
Hercynian episode ***1**.154*
Heredia **5**.340
hereditary rule *see* monarchy
heredity ***2**.156*, **2**.364
behaviour genetics **7**.28–9
see also genetics
Herero Wars **4**.150
heresies *see under* Roman Catholicism
Hereward the Wake ***3**.164*
Hergé **5**.91
Herjolfsson, Bjarni **3**.374
Herman, Woodrow 'Woody' **5**.47, **5**.239
hermaphrodites ***2**.156*, **2**.282, **2**.314
Hermaphroditus ***9**.16*
Hermes (asteroid) ***8**.65*
Hermes (god) ***9**.16*, **9**.29
Hermes Trismegistus ***9**.16*
hermit ***3**.164*
hermit crab **2**.73–4, ***2**.156*
Hernándes de Córdoba, Francisco **3**.253
hernia ***2**.156–7*
Hero ***9**.16*
Hero of Alexandria **6**.84, **6**.92, ***6**.168*, **6**.333
Herod dynasty ***3**.164*
Herod Agrippa, king **3**.164, **9**.19
Herod Antipas, king **3**.164, **9**.20, **9**.32
Herod the Great, king of Palestine **3**.164, **3**.187, **3**.224
Herodotus **3**.40, ***3**.164*, **5**.190, ***5**.205*
heroic couplet **5**.89, **5**.136, ***5**.205*, **5**.357
heroin *see* opium
herons **2**.37, ***2**.157*
Héroult, Paul-Louis-Toussaint **6**.12, ***6**.168*
herpes infections:
Herpes simplex ***2**.157*
Herpes zoster **2**.63, **2**.307
Herrera, Luis Alberto Lacalle **4**.361
Herrick, Robert **5**.148, ***5**.205*
Herrin massacre ***4**.150*
herring family ***2**.157*
gizzard shad **2**.136
pilchard **2**.253–4
shad **2**.305
sprat **2**.323, **2**.366
whitebait **2**.366
herring gull ***2**.157*
Herschel, Caroline Lucretia **8**.46, **8**.66
Herschel, Sir John Frederick William (1792–1871) **5**.73, **8**.14, **8**.60, ***8**.65–6*, **8**.76, **8**.119
Herschel, Sir William Frederick (1738–1822) **8**.24, **8**.65, ***8**.66*
astronomical discoveries by **8**.46, **8**.99, **8**.111, **8**.172, **8**.178, **8**.181
and moon **8**.147
and ultraviolet astronomy **8**.177
Herschel (crater) **8**.99
Hertz, Heinrich Rudolf ***1**.154*, **6**.45, ***6**.168*, **6**.215, **6**.290
hertz ***1**.154*, **6**.146, **6**.368
Hertzog, James Barry Munuik ***4**.150*, **4**.234, **4**.360
Hertzsprung, Ejnar:
Hertzsprung–Russell diagram and gap **8**.64, ***8**.66–7*, **8**.160
Herzen, Alexander Ivanovich ***4**.150*, **4**.275
Herzl, Theodor ***4**.150*, **4**.271, **4**.390, **7**.336
Herzog, August (the Younger) **5**.437
Herzog, Werner ***5**.205*
Hesiod ***5**.205*, **9**.26
Hesperides **9**.16
Hesperus ***8**.67*
Hess, Germain Henri ***1**.154*
Hess, Rudolf ***4**.150*, **4**.248
Hess, Victor Franz (Francis) ***1**.154–5*, **8**.31
Hesse, Hermann ***5**.205*
Hesse (state) ***4**.150*
hessian fly **2**.131
Hestia **9**.9, ***9**.16*
heteroatoms ***1**.155*
heterocyclic compounds **6**.252
heterotroph ***2**.157*
Hevelius, Johannes:
constellations introduced by **8**.20, **8**.29, **8**.82, **8**.84, **8**.88, **8**.146, **8**.148, **8**.185
Hevelius formation ***8**.67*
and moon **8**.67
star atlas **8**.29
Hewish, Antony **8**.105
hexameters **5**.105, ***5**.205*
Latin **5**.121, **5**.147, **5**.212, **5**.335, **5**.477
Hexham, battle of ***3**.164–5*
Heydrich, Reinhard **4**.135, ***4**.150–1*, **4**.327
Heym, Georg **5**.175
Heyward, DuBose **5**.175
Hezekiah ***3**.165*
HFCs *see* hydro-fluorocarbons
HI and HII regions **8**.74
Hiawatha ***9**.16*
hibernation ***2**.157–8*
hibiscus ***2**.158*
hickory ***2**.158*
Hidalgo (asteroid) ***8**.67*
Hidalgo y Costilla, Miguel ***4**.151*, **4**.219, **4**.224
Hidatsa Indians ***3**.92*
Hidāyat, Sadiq **5**.345
Hideyoshi **3**.96, ***3**.165*, **3**.186, **3**.200, **3**.232, **3**.314
hieroglyphs ***3**.165*, **5**.85, **7**.103, **7**.151, **7**.259
high frequency **6**.380
high ground *see* uplands
High Plains ***1**.155*
high (pressure) ***1**.155*, **1**.282, **1**.303, **1**.325
high-definition television **6**.349
higher education **7**.102, ***7**.141–2*
rite of passage **7**.268
see also universities
high-fidelity system **6**.159, ***6**.168*
highland clearances ***4**.151*
high-speed steel **6**.11
high-speed trains ***6**.168–9*, **6**.293, **6**.294
'high-tech' style **5**.162
high-temperature, gas-cooled reactor **6**.4
high-voltage electricity generation **6**.127
highwayman ***3**.165*
hi-jacking ***7**.142*
Hijkata, Tatsumi **5**.141
hijrah ***7**.142*
Hikmet, Nazim ***5**.205*
Hilbert, David ***1**.155*
Hildebrandt, Johann Lucas von ***5**.205–6*
Hill, Christopher **4**.153
Hill, David Octavius ***5**.206*
Hill, George **8**.87
Hill, James Jerome ***4**.151*
Hill, Sir Rowland ***4**.151*
hill ***1**.155*
escarpment **1**.113
hogback **1**.156
kopje **1**.184
see also mountains
Hillary, Sir Edmund **1**.116, **7**.208
Hilliard, Nicholas ***5**.206*, **5**.297
Hillquit, Morris **4**.94
Himalayas **1**.135, **1**.155, ***1**.156*
avalanche **1**.28
Everest **1**.116
Kanchenjunga **1**.183
Himalayas (*cont.*):
Karakoram range **1**.9, **1**.179, **1**.180
Nanga Parbat **1**.224
Himalia ***8**.67*
Himmler, Heinrich ***4**.151*, **4**.327
Hina ***9**.16*, **9**.25
Hind, John Russell **8**.84, **8**.175
Hindemith, Paul ***5**.206*
Hindenburg, Paul von ***4**.151*, **4**.153, **4**.201, **4**.373, **5**.206
Hindi language **7**.157
literature ***5**.206*
Hindu Kush ***1**.156*
Hindu Mahasabha ***4**.151*
Hinduism ***3**.165–6*, ***4**.151–2*, ***7**.142*
Advaita *see* Advaita
afterlife **7**.6
ārtī **7**.19
and arts ***5**.206–7*
architecture *see under* architecture
literature *see below*
Nepal **5**.316
painting and painters **5**.10, **5**.108, **5**.196, **5**.219, **5**.293, **5**.373, **5**.375
rock art **5**.10, **5**.143, **5**.145, **5**.196, **5**.275
asceticism **7**.19
ashram **7**.9, **7**.205
āshrama **7**.19
ātman **7**.6, **7**.21, **7**.295
avatar see avatars
caste system **3**.298, **3**.299, **7**.41, **8**.41
Chōla dynasty **3**.74
demons **9**.31
dharma **7**.89
Dīvālī **7**.93, **7**.142
Durgā-pūjā **7**.87, ***7**.97*, **7**.142
festivals ***7**.142*
and food **7**.123
gods *see under* deities
guru **7**.137
Hare Krishna sect **7**.138
havan **7**.139
heaven **7**.140
hell, multiple **7**.141
heroes and sages **9**.5, **9**.24, **9**.38
Hindu Mahasabha **4**.152
Holi **7**.93, **7**.142, **7**.145
in Java **3**.217
karma **7**.175, **7**.245
language *see* Sanskrit
law **7**.19, **7**.89, ***7**.143*
leaders **4**.230, **4**.369
literature **5**.470, ***7**.144*
Bhagavadgītā **3**.40–1, **5**.46, **5**.276, **7**.31, **9**.5
drama **5**.244, **5**.279, **5**.314
Mahābhārata **3**.40–1, **5**.134, **5**.276, **5**.401, ***7**.189*
Puranas **5**.370
Upanishads **3**.367, **7**.3, **7**.155, **7**.322
see also Sanskrit; *Vedānta*; *Vedas and also* India *under* poetry; novels
mantra **7**.191, **9**.28
marriage **7**.193
māyā **7**.3, **7**.195–6
meditation **7**.196
melā **7**.197
metaphysics **7**.143
millenarianism **7**.202
moksha **7**.3, **7**.6, **7**.203, **7**.205
monasticism **7**.19, **7**.205, **7**.273
movements **4**.20, **4**.48, **4**.344
mysticism **7**.143, **7**.191, **9**.24
nirvana **7**.217
philosophy **7**.155
pilgrimages **3**.281, **3**.370, **7**.236
prasād **7**.244
predestination **7**.245
pūjā **7**.255
Purusha **7**.256
reincarnation **7**.263

Hinduism (*cont.*):
- revivalism 7.143–4
- sacred texts 7.144
- sacrifice 7.272
- saints 7.272
- *samādhi* 7.272
- samsara 7.169, 7.263, 7.273
- *samskāra* 7.273
- *sangha* 7.273
- *sat-cit-ānanda* 7.274
- soul 7.21, 7.273, 7.295
- and suicide 7.301
- tantra 7.306
- temple 3.268, 7.143, 7.144
- texts *see* literature *above*
- yoga *see* yoga

Hine, Lewis Wickes 5.207
Hine-Titama 9.16
hinge 6.169
hip replacement 6.168, 6.169
Hipparchus 8.11, 8.58, 8.62, 8.67–8, 8.90, 8.121, 8.127, 8.161
Hipparcos 8.68
Hippocrates 2.158
hippodrome 3.166
Hippomenes 9.6
hippopotamus 2.158
Hiranyakashipu 9.26, 9.31
Hiranyaksha 9.37
Hirayama families of asteroids 8.9
hire purchase 7.67–8
Hirohito, emperor of Japan 4.152, 4.346
Hiroshige Ando 5.207
Hiroshima 4.152
hirudin 2.186
Hispaniola 1.69, 1.156, 3.166
- *see also* Dominican Republic; Haiti

Hispano-Moresque ware 5.207
hispid bat 2.158
Hiss, Alger: Hiss case 4.152
histogram 1.156
histology 2.29, 2.158
historical linguistics 7.110, 7.144
historical novel 7.207
- *see also under* novels

historicism 7.144
historiography/historians 3.166, 4.153
- annals 3.16
- Arab *see* al-Tabaric; Ibn Khaldun
- Britain 3.141, 3.383, 4.153, 5.43, 5.78, 5.176, 5.408
- China 3.328, 5.95
- Europe *see* Bede; Froissart; Gibbon; Guicciardini; Machiavelli; William of Malmesbury *and also individual countries above and below*
- France 3.83, 3.190, 4.153, 5.184, 5.397
- Germany 4.153, 5.405, 5.490
- Greece, ancient 3.164, 3.287, 3.356, 3.386–7, 5.190, 5.205, 5.454, 5.496
- Iceland 5.423
- Italy 3.49, 5.231
- Jews *see* Josephus
- Latin 5.353, 5.444–5
- Netherlands 5.165
- Rome *see* Livy; Suetonius; Tacitus
- Russia 5.424
- Spain 5.428
- United States 5.5, 5.227
- West Indies 5.59

history *see all references prefixed* 3. *and* 4.
history painting 5.207, 5.381
histosol 1.156, 1.311
Hitchcock, Sir Alfred 5.207–8
Hitchcock, Henry-Russell 5.241
Hitler, Adolf 4.153, 7.90
- anti-Semitism 4.15
- Ardennes Campaign 4.17
- Austria annexed 4.14, 4.24
- and Axis powers 4.26

Hitler (*cont.*):
- bombing offensives 4.44, 4.49–50
- Brownshirts 4.53
- and Czechoslovakia 4.43, 4.92
- death 4.247
- Eastern Front 4.106
- elected 4.373
- and fascism 4.116
- and Franco 4.124
- July plot against 4.145
- and Munich Pact 4.227
- and 'Night of Long Knives' 4.243
- paranoia 7.228
- plots against 4.145, 4.176–7
- and Poland 4.272
- and Rhineland 4.289
- and Third Reich 4.135, 4.342–3
- and Tyrol 4.356
- and von Papen 4.258
- *see also* Nazis

Hitler Youth 4.153
Hittites 3.166, 7.13, 9.18
- art 5.208
- religion and myths 9.18

HIV (Human Immunodeficiency Virus) *see* Aids
hives (disorder) 2.353
Hizbullah 7.166
Hizkil (Ezekiel) 9.12
Hlasko, Marek 5.355
Ho Chi Minh 4.127, 4.154, 4.155, 4.366, 4.367
Hoare, Henry II 5.171
Hoare, William 3.246, 3.283
hoar-frost 1.129
hoatzin 2.158
Hoban, James 5.208
Hobbema, Meindert 5.208
Hobbes, Thomas 3.167, 7.144
- and Bodin 3.43
- and civil society 7.52, 7.294
- and materialism 7.195
- and power politics 7.244
- and social contract 3.331, 7.290

hobbies and pastimes:
- caving 7.42
- coin collecting 7.56
- collecting 7.55, 7.56
- fishing 7.121
- gambling 7.129
- horse-riding 7.147
- kites 7.177
- puzzles 7.256
- stamp collecting 7.298
- *see also* games; leisure; sport

hobby (bird) 2.158
hobgoblin 9.16, 9.30
Hobsbawm, Eric 4.153
Hobson, John Atkinson 7.153
Hoccleve, Thomas 5.294
Hochhuth, Rolf 5.175, 5.351
hockey, field 7.144, 7.145
hockey, ice 7.144–5
Hockney, David 5.208
Hoder 9.6
Hodgkin, Alan Lloyd 2.158
Hodgkin, Dorothy Crowfoot 2.81, 2.158
Hodgkin, Thomas: Hodgkin's disease 2.159
Hodson, Lt. W. S. R. 4.162
Hoe, Richard March 6.169
hoe 6.6, 6.169
Hofer, Andreas 4.356
Hoffa, James R. 4.339
Hoffmann, Ernst Theodor A. 5.208
Hoffmann, Heinrich 5.91
Hoffmann, Josef 5.208
Hofmann, August Wilhelm von 1.156, 1.157, 6.161, 6.260
Hofmannsthal, Hugo von 5.208–9, 5.377
Hogarth, William 5.209, 5.359
hogback 1.156
Hogg, James 5.209
hogs *see* pigs

Hohenstaufen, House of 3.167, 9.65
- *see also under* Holy Roman empire

Hohenzollern and Hohenzollern-Sigmaringen family 3.167, 4.112, 4.154–5, 4.292
- rulers listed 9.67

Hohmann transfer orbit 8.68
Hojo 3.167
Hokan language 7.219
Hokkaido 1.156–7
Hokusai Katsushika 5.209
Holbein, Hans 3.117, 3.241, 5.16, 5.124, 5.209
Holberg, Ludvig, Baron 5.209–10
Hölderlin, Friedrich 5.210
holding company 7.145
Holi 7.93, 7.142, 7.145
Holiday, Billie 5.50
Holinshed, Raphael 5.414
holism 7.21, 7.62–3, 7.214
Holkar, Malhar Rao and Jaswant Rao 3.167
Holl, Elias 5.210
Holland, Henry 5.210
Holland, Sir Sidney George 4.155
Holland *see* Netherlands
Hollar, Wenceslaus 5.210
Hollerith, Herman 6.84, 6.169, 6.189
Holles, Denzil 3.127
Holly, Buddy 5.384
holly 2.159
hollyhock 2.159
Hollywood 5.210
- *see also* United States *under* cinema

holm oak 2.231
Holman Hunt, William 5.214, 5.366
holmium 1.106
Holocaust 4.154, 4.155
- *see also* Nazis *under* Jews

Holocene epoch 1.133, 1.134, 1.157
Holofernes 9.21
holography 5.347, 6.149, 6.169
Holst, Gustav 5.58, 5.210
Holstein-Gottorp-Romanov, House of *see under* tsars *under* Russia
Holt, Harold Edward 4.155
Holub, Miroslav 5.210
Holwards, J. P. 8.100
Holy Alliance 4.155
Holy Ghost 9.14
Holy Land *see* Palestine
Holy Leagues 3.167, 3.206
Holy Roman empire 3.167–8
- arts 5.334
- Cambrai, League of 3.58
- emperors (chronologically)
  - lists of 3.168, 9.65
  - Carolingians (800–911) 3.69, 3.70, 3.137, 3.168, 3.169, 3.210, 3.211
    - *see also* Charlemagne
  - Saxony, House of (919–1024) 3.160, 3.168, 3.265
    - *see also* Otto I (the Great)
  - Salian House (1024–1125) 3.168, 3.179, 3.225, 3.294, 3.341
  - Hohenstaufen, House of (1138–1208, 1212–72) 3.167–8, 3.209, 3.312
    - *see also* Frederick II
  - Habsburgs (1273–91, 1298–1308, 1438–1740) 3.27, 3.124, 3.133, 3.152, 3.153, 3.168
    - *see also* Charles V; Charles VI; Leopold I; Maximilian I
  - Luxemburg, House of (1346–1400, 1410–11) 3.70, 3.168, 3.328, 3.378
  - Habsburg–Lorraine, House of (1745–1806) 3.168
    - *see also* Francis I
- map 3.168

Holy Roman empire (*cont.*):
- politics and government 3.103, 3.113, 3.222

Holyoake, Sir Keith Jacka 4.155
home:
- building 6.171
- country houses 4.87
- heating *see* heating, domestic
- homelessness 7.145–6
- household 7.148
- immigrant (USA) 4.2
- industry *see* crafts
- lake dwelling 3.202
- lighting *see* lighting
- machines in *see* domestic appliances
- manor houses 3.218–20
- violence in 7.94, 7.115
- *see also* buildings; family

Home Guard 4.155
Home Rule, Irish 4.155–6
Homer 3.168–9, 3.244, 3.351, 5.148, 5.210, 5.357, 5.420, 5.484
Homer, Winslow 5.14, 5.211
Homestead Act 4.156
Homestead strike 4.156
homicide 7.146
- *see also* assassin; murder

Hominids/*Homo* spp 2.126, 3.169
- *H. erectus* 2.126, 3.50, 3.120, 3.169, 3.186, 3.200, 3.261, 3.274, 3.346, 3.390
- *H. habilis* 2.216, 3.120, 3.169, 3.199, 3.261
- *H. sapiens* 2.126, 2.200, 3.50, 3.120, 3.169, 3.346
- *H. sapiens neanderthalis* (Neanderthals) 3.120, 3.242, 3.247
- *H. sapiens sapiens* (Cro-Magnons) 3.91, 3.120, 3.367
- *see also* prehistory

homoeopathy 2.7, 2.159, 6.169–70, 6.257
homografting 6.158
homologous series 1.157
homophony 5.356
homosexuality 7.9, 7.146–7
homosphere 8.162
Homusubi 9.19
Honda Kotaro 6.170
Honduras 1.157, 3.169, 4.67, 4.156, 4.251, 7.356
Hone Heke 4.13, 4.294
Honecker, Erich 4.134
Honegger, Arthur 5.211
honey 6.16–17, 6.86
honey badger 2.160, 2.278
honey buzzard 2.160
honeybee 2.159, 2.358
honey-eater 2.160
honeyfungus 2.160
honeyguide 2.160
honeysuckle 2.160–1
Hong Kong 1.157, 4.156, 7.356–7
- and Britain 7.57
- building 6.321
- economy and trade 7.88, 7.309
- education 7.155
- finance 7.155
- languages 7.286
- laws/legal system 7.255
- physical geography and geology 1.157, 1.223
- population density 7.241
- religion/theology 7.47

Hong Wu (Zhu Yuanzhang) 3.232, 3.389
Hong Xiuquan 4.337
honing machine 6.212
Honorius, Roman emperor 3.7, 5.72
Honorius III, pope 3.61
honour 7.147
Honshu 1.157
Honthorst, Gerrit van 5.211
Hooch, Pieter de 5.211

Hood, Samuel, 1st viscount ***3.169***
hooded crow **2.**55
hoodoo **1.**249
hoof **2.**67
hoofed mammals *see* ungulates
Hooke, Robert **2.**59, **6.**369, **8.**61, **8.**147
Hooke's law ***1.157***
Hooker, Joseph (1814–79, American general) **4.**368
Hooker, Sir Joseph Dalton (1817–1911, botanist) ***2.161***
Hooker, Richard ***3.169***
Hooker, Sir William Jackson ***2.161***
hookworm ***2.161***
hoopoe ***2.161***, **2.**370
Hoover, Henry **6.**5
Hoover, Herbert **4.**96, **4.**147, ***4.156–7***
Hoover, J. Edgar **4.**117, ***4.157***
Hoover, William **6.**369
hop **2.**68, ***2.161***
Hopewell cultures ***3.170***
Hopi Indians ***7.147***, **9.**22
Hopkins, Sir Frederick Gowland **6.**141, ***6.170***
Hopkins, Gerard Manley **5.**12–13, ***5.211***
Hopkins, Harry ***4.157***, **4.**381
hoplites ***3.170***, **3.**278
Hopper, Edward ***5.211***
Horace ***5.211–12***
horizon (sky) ***1.157–8***, ***8.68–9***
horizon (soil) ***1.158***
horizontal branch of stars **8.**67
Horkheimer, Max **7.**124
hormones ***2.162***, ***6.170***
and aggression **7.**7
deficiency **6.**29, **6.**170
fertility drug **6.**127
first isolated **6.**344
and *in vitro* fertilization **6.**186
pituitary gland **2.**208, **2.**256–7
plant **6.**271
pregnancy test **6.**280
radioimmunassay **6.**291
steroids from **6.**334
synthetic **6.**87
thyroid gland **2.**140, **2.**340, **6.**243
Horn, Filips van Montmorency, Graaf van ***3.170***
Horn, Cape ***1.158***
Horn of Africa:
famine **7.**117
see also Ethiopia; Somalia
horn (animal) ***2.162***, ***6.170***
hornwork ***5.212***
horn (musical) **5.**59, **5.**113, **5.**163, ***5.212***, **5.**362
hornbeam **2.**30
hornbill **2.**161, ***2.162***
hornblende ***1.158***
Hornblower, Jonathan **6.**330
Horne, General **4.**368
horned toad ***2.162***
Horneman, Christian **5.**41
hornet ***2.162–3***, **2.**212
Horniman, Annie ***5.212***
hornpipe ***5.212***
horntail **2.**294
hornworm ***2.152***, **2.**343
Horologium ***8.69***
Horowitz, Vladimir ***5.212***
horror fiction **5.**208, **5.**354
horse *see* horses
horse latitudes ***1.158–9***
horse fly ***2.163***
horse mackerel ***2.163***
horse-drawn vehicles ***6.61***, **6.**300
carriages and coaches **6.**30, **6.**61–2, **6.**341
carts and wagons **6.**61, **6.**62, **6.**165, **6.**300, **6.**375, **6.**384
chariot **6.**70, **6.**165, **6.**225, **6.**300
harness and saddlery **6.**165–6, **6.**300
horse-drawn vehicles (*cont.*):
steering system **6.**333
tyre **6.**366
see also draught and pack animals; horses
horsehair worm ***2.163***
Horsehead Nebula **8.**68, ***8.69***
horsepower (engines) **6.**368
horseradish ***2.163***
horses ***2.163***
brain **2.**41
cavalry **6.**225
evolution of **2.**109, **2.**208
hybrid (mule) **2.**221
mythical **9.**8, **9.**29, **9.**34
and Plains Indians **7.**236
and sport:
horse-racing ***7.147***
horse-riding ***7.147***
polo **7.**240, **7.**241
zebra **2.**375
see also asses; horse-drawn vehicles; livestock
horseshoe ***6.170***
pitching ***7.147***
horseshoe bat ***2.164***
horseshoe crab **2.**16, **2.**163
horsetails **2.**49–50, ***2.164***
horsts ***1.159***, **1.**288
Horta, Victor, Baron ***5.212***
Horthy de Nagybánya, Nikolaus ***4.157***
horticulture ***6.170***, **6.**371
Horton, Robert Elmer (engineer) ***1.159***
Horton (dancer) **5.**10
horuero **2.**238
Horus **9.**16–17, **9.**28, **9.**32
Hosea ***9.17***, **9.**20
hospital ***7.147***
foundling **3.**130
see also medicine and health care
Hospitallers, Knight **3.**198, **3.**303
hostage-taking **4.**165, **7.**91, ***7.147–8***
hot isostatic pressing **6.**68
hot spring ***1.159***
Hot Springs National Park ***1.159***
hot-melt adhesives **6.**3
Ho-Tsheng-Tien **8.**26
Hotspur *see* Percy, Sir Henry
Houdon, Jean-Antoine **5.**315, **5.**359
Houdry, Eugène: Houdry process **6.**91, ***6.171***
Houghton, Stanley **5.**212
Hounsfield, Geoffrey **6.**85
Houphouët-Boigny, Félix **4.**87, ***4.157***, **4.**170
hour angle **8.**69
houri ***9.17***
house *see* housing
House of Commons ***3.83***, **3.**175, ***4.79***, ***7.148***, **7.**318, **8.**148, **8.**318
House of Lords ***3.210***, ***4.199***, ***7.148***, **7.**318, **8.**148, **8.**318
house martin ***2.164***
House of Representatives ***7.148***, **7.**324
housefly ***2.164***
household ***7.148***
see also home
housemouse ***2.164***
Houses of Parliament **7.**148, **7.**318
see also House of Commons; House of Lords
housing ***6.171***, ***7.148***
government **7.**134
homelessness **7.**145–6
loans **7.**207
slum **7.**288
squatting **7.**298
symbolism **7.**94
see also buildings; home; urban areas
Housman, Alfred Edward ***5.212***
Houston, Samuel ***4.157***, **4.**302
Houtman, Frederick de *see* Keyser
Hovenden, T. **4.**53
hovercraft **6.**79–80, ***6.171***, **6.**172, **6.**175
hoverfly ***2.164***
hover-train **6.**161
Howard, Catherine ***3.170***
Howard, Ebenezer **7.**216
Howard, Lady Frances **3.**265–6
Howard, Henry **5.**49
Howard, John ***3.170***
Howard, Leland Ossian ***2.165***
Howard, Thomas, 4th duke of Norfolk **3.**304
Howe, Elias ***6.171–2***, **6.**321
Howe, Richard, 4th viscount ***3.170***
Howe, William, 5th viscount **3.**49, ***3.170***, **3.**176, **3.**316
Howells, Herbert ***5.212***
Howells, William Dean ***5.212–13***
howitzer **6.**22, ***6.172***
howler monkey ***2.165***
Hoxha, Enver **4.**8
Hoyle, Sir Fred **8.**33, ***8.69***
Hoysala ***3.170***
Hoyte, Desmond **4.**144
Hrabal, Bohumil ***5.213***
Hsia (Xia) **3.**387
Hsiung-nu (Xiongnu) **3.**387
HTGR (high-temperature, gas-cooled reactor) **6.**4
Hu Shi **5.**95
Hua Guofeng **4.**96, ***4.157***
Hua Tuo **6.**14
Huang He **1.**158, ***1.159***
Huáscar, king of Incas **3.**175, **3.**298
Huayana Capac, king of Incas **3.**175–6, **3.**358–9
Hubbard, L. Ron **7.**276
Hubble, Edwin Powell **8.**33, **8.**54, ***8.69***
classification of galaxies ***8.71***
Hubble constant and Hubble's Law ***8.71***
Hubble space telescope **8.**69–70, ***8.71***
Hubble-Sandage variable ***8.71***
Huber, Wolfgang **5.**125
Huchueteotl **9.**38
huckleberry ***2.165***
Hud **9.**2, ***9.17***
Hudson, Henry ***1.159***, **1.**231, **3.**100, **3.**122
Hudson Bay ***1.159***
Hudson River **1.**159
Hudson River School ***5.213***
Hudson's Bay Company ***3.170***
hue and cry ***3.170***
Huerta, Victoriano ***4.157–8***, **4.**219, **4.**368
Huggins, William **8.**120, **8.**157
Hugh Capet *see* Capet, Hugh
Hughes, Charles Evans ***4.157–8***
Hughes, John **5.**486
Hughes, Monica **5.**91
Hughes, Ted **5.**91, ***5.213***, **5.**352
Hughes, Thomas ***5.213***
Hughes, W. H. **4.**189
Hughes, William M. ***4.158***
Hugo, Victor **5.**12, **5.**207, ***5.213–14***, **5.**459, **5.**471
Huguenot ***3.170–1***
huia **2.**362
Huinéng ***9.17***
Huitzilopochtli **3.**31, **9.**9, ***9.17***, **9.**35
Huizong, emperor **5.**98
Hukbalahap movement ***4.158***
Hulagu **3.**238
Hull, Clark **7.**29, **7.**291
Hull, Cordell ***4.158***
Hull, William **4.**371
hull-and-core structure *see* skyscraper
Hulme, T. E. **5.**217
Hülsmeier, Karl **6.**288
Hulst, Hendrik van de **8.**176
human capital ***7.149***
see also labour and employment
human evolution *see* prehistory
Human Immunodeficiency Virus (HIV) *see* Aids
human rights *see* rights, human
human-computer interface **6.**184
human-factors engineering **6.**122
humanism ***3.171***, **3.**252, ***7.149***
see also existentialism; individualism
humanistic psychology ***7.149***, **7.**268
human-powered vehicles:
flight **6.**138, ***6.172***
see also bicycle
Humayun ***3.171***, **3.**235
Humboldt, Friedrich Heinrich Alexander, Baron von ***1.159***
Humboldt Current **1.**255
Hume, David ***3.171***, ***7.149–50***
and atomism **7.**21
and causation **7.**42
and Einstein **7.**234
and Enlightenment **3.**117, **5.**147
and ethics **7.**109
and free will **7.**125
and induction **7.**275
and logical positivism **7.**187
and mercantilism **7.**100, **7.**198
and metaphysics **7.**199
and naturalism **7.**213
and psychology **7.**252
and reason **7.**259
and value judgements **7.**325
Hume, Joseph **4.**79, ***4.158***
humerus **2.**14
Humfrey, Pelham, **5.**50
humidity ***1.159***, **6.**98, **6.**221
Hummel, Johann Nepomuk ***5.214***
hummingbirds **2.**158, ***2.165–6***, **2.**277
humour *see* comedy and humour
humpbacked whale ***2.166***
Humperdinck, Engelbert ***5.214***
Humphrey, Doris ***5.214***, **5.**300
Humphrey, Hubert **4.**175
Humphrey of Gloucester **3.**142–3
humus ***2.166***
Hunac Ceel, king of Incas **3.**227
hundred ***3.171***
'Hundred Days' (Napoleon) ***4.158***, **4.**272
Hundred Days Reform (China) ***4.158***
Hundred Flowers Movement (China) ***4.158***
Hundred Years War **3.**49, ***3.171***, ***3.172***
battles and campaigns **3.**90, **3.**128, **3.**286, **3.**331
Hungary ***1.160***, ***3.172***, ***4.158–9***, ***7.357***
from 1867 to 1919 *see* Austro-Hungarian empire
arts:
dance **5.**252
painting **5.**301
photography **5.**58, **5.**76, **5.**246, **5.**301, **5.**308
sculpture and carving **5.**201
see also literature; music *below*
economy and trade **7.**57, **7.**193
ethnic groups **7.**137
languages **7.**322
literature **4.**150, **5.**5, **5.**28, **5.**76, ***5.214***, **5.**214, **5.**345
music **5.**36–7, **5.**240, **5.**250, **5.**260, **5.**264, **5.**266
physical geography and geology **1.**84, **1.**160
politics and government **4.**12, **4.**94, **4.**178–9, **4.**186–7, **4.**230, **4.**282, **7.**104
religion/theology **3.**341, **4.**220–1, **7.**336, **9.**41, **9.**42
Revolution ***4.158***
rulers **3.**124, **3.**153, **3.**168, **3.**211, **3.**225, **3.**328, **4.**123–4, **4.**157

Hungary (cont.):
technology and science 6.62, 6.259, 6.300, 6.367, 6.384
town 3.236
treaties 4.23, 4.26
wars and defence 3.95, 3.253, 4.159, 4.288
in Warsaw Pact 4.371, 7.328
see also Austro-Hungarian empire; Eastern Europe; Magyars
hunger 2.166
hunger-strike 7.150
see also famine; food
Hungry Forties 4.159
Huns 3.27, 3.64, 3.98, 3.172
Hunt, Henry 'Orator' 4.159, 4.268
Hunt, William Holman 5.214
Hunter, William and John 2.166
hunter-gatherers and hunters (and fishers) 7.150
Africa 7.178, 7.256, 7.285, 7.302
Arctic 7.16, 7.165
Asia and Pacific 7.9, 7.197, 7.211, 7.240
Australia see Aborigines
and nomadism 7.217
North American Indians 7.74, 7.166, 7.178, 7.213, 7.219, 7.236, 7.287, 7.312, 7.316
prehistoric 3.102, 3.247, 3.267, 3.342
South America 7.11
hunting (sport) 6.42, 7.150
hunting dog 2.166
hunting wasp 2.166
Huntington, Collis P. 4.68
Huntsman, Benjamin 6.172, 6.186
hurdy-gurdy 5.214
Hurley, Patrick 4.45
Huron, Lake 1.160
Huron Indians 3.172–3, 7.150
hurricane 1.160
Beaufort scale 9.56
Hurtado, Miguel de la Madrid 4.220
Hus, Jan 5.120
Ḥusain, Imām 7.20, 7.194, 7.284, 9.17
Husák, Gustáv 4.92
al-Husayn ibn Mansur 5.439
Husein, Taha 5.19
Huskisson, William 4.159
Huss, John 3.43, 3.84, 3.86, 3.133, 3.173
hussar 3.173
Hussein, Abdullah ibn, king of Jordan (d.1951) 4.175
Hussein ibn Ali, king of Hejaz (1856–1931) 4.149, 4.159
Hussein, ibn Talal, king of Jordan 4.17, 4.159, 4.175
Hussein, Saddam 4.159, 4.165, 4.166, 7.23
Ba'athism 7.24
invasion of Kuwait and Gulf War 4.143, 7.15
and PLO 7.227
Husserl, Edmund 7.113, 7.199, 7.234
Hussey, Obed 6.172, 6.210
Hussites 3.173
Huston, John 5.214–15, 5.485
Hutchinson, Anne 3.302
hutia 2.166
Hutton, James 1.136, 1.160, 1.260
Huxley, Aldous 5.215, 7.324
Huxley, Sir Julian 2.166
Huxley, Thomas Henry 2.166, 7.7
Huygens, Christiaan 1.127, 1.160, 6.149
in Académie des Sciences 3.2
and catenary 1.54
and pendulum 6.76, 6.260
and Saturn 8.140, 8.145, 8.172
Hwang Ho 1.158, 1.159
hyacinths 2.143, 2.166
Hyacinthus 9.17
Hyades 8.71
Hyatt, Isaac 6.55, 6.172, 6.271
Hyatt, John Wesley 6.172
hybridization 6.36
hybrids 2.166–7, 2.193, 2.221, 2.255
hydatid 2.167
Hyde, Douglas 5.226
Hyde, Edward see Clarendon
Hyder Ali 3.173, 3.244
Hyderabad 4.159
Hydra (astronomy) 8.71
hydra (microorganism) 2.167
Hydra (mythology) 9.17
hydrangea 2.167
hydrant 6.172
hydrate 1.160
hydraulic cement 6.66, 6.86, 6.322
hydraulic power 6.173
brakes 6.35, 6.44, 6.173
cranes 6.92
mining 6.79, 6.172–3
press 6.45, 6.173, 6.259
ram 6.173
hydrocarbons 1.161, 6.58, 6.173
alcohol see alcohol
aliphatic 1.9
benzene see benzene
bitumen see bitumen
cracking see cracking
DDT 1.85, 6.96, 6.180, 6.275
fats see fats, oils, and fatty acids
halocarbon see halocarbons
insecticides 6.96, 6.180
liquefaction see liquefied
naphthalene 1.224
as non-polar solvents 6.325
see also chloro-fluorocarbons; natural gas; petrochemicals; petroleum
hydrochloric acid 1.161, 6.2, 6.70, 6.97, 6.173, 6.308
aqua regia 1.17
hydrochloro-fluorocarbons 6.72
hydrocracking 6.91
hydrodynamics 1.161, 6.138
hydroelectricity 6.173, 6.174, 6.378, 6.379, 7.106
dams 6.95
and locomotives 6.114
percentage used 6.121
pioneer 6.145
pumped storage 6.286
hydrofluoric acid 6.173, 6.177
hydro-fluorocarbons 6.72
hydrofoil 6.173–4
hydrogen 1.106, 1.161, 6.174, 8.71
bonds 1.59, 1.161
from Castner–Kellner cell 6.62
in coal-gas 6.79
compounds see hydrocarbons; hydroxyl; petroleum
emission spectrum 1.316
as fuel 6.174
ions concentration 1.161
see also acids
isotopes 1.88–9, 1.344, 6.369
liquefied 6.150, 6.301
plasma 1.259
production 6.62, 6.70, 6.264
see also cracking and Haber–Bosch under Haber
transport 2.105
see also organic compounds
hydrogen bombs (fusion bombs) 6.240, 6.245, 6.246, 6.350, 6.369
see also nuclear weapons
hydrogen peroxide 1.161, 1.255, 6.40, 6.174
hydrogen sulphide 1.162, 2.111, 6.338
hydrography 1.162, 1.248
hydrograph 1.162
hydrological cycle 6.174, 6.378
hydrology 1.163, 6.174
overland flow 1.242–3
see also runoff; water
hydrolysis 1.163
hydro-mechanical clock 6.340
hydrometer 6.174
hydrophobia 2.274
hydroponics 6.175
hydrosphere 1.163
hydrotreatment 6.175, 6.265
hydroxide 6.202
ion 1.163
see also calcium hydroxide; sodium hydroxide
hydroxyl group 1.163
hydroxyl radical (OH) 8.71
Hydrus 8.71
hyena 2.167
hygrometer 6.221
Hyksos 3.173
Hylas 9.5
Hymen 9.17
hymns 5.215
anthem 5.18
Britain 5.116, 5.486
chorale 5.99
India 5.220
Islam 5.230
Italy 5.46
Pacific Ocean islands 5.325
psalm 5.368
Sikh 5.418
see also sacred music
hyperbola (curve) 1.72, 1.73, 1.163
hyperbole (literature) 5.215
hyperbolic navigation system 6.238, 6.290
Hyperion 8.71–2
hypermarket 7.302
hypersonic flight 6.175, 6.386
hypertext 6.175
hypnosis 2.167, 7.126, 7.150
hypnotic (drug) 6.285, 6.360
hypocaust 6.167, 6.175
hypodermic syringe 6.175
hypophysis 2.256
hypothalamus 2.167
hypothesis 1.163
hyrax 2.168
hyssop 2.168
hysteria 7.126, 7.150

I

IAEA see International Atomic Energy Agency
iambic verse 5.49, 5.205, 5.216, 5.333, 5.343
Iapetus 8.72
IAU (International Astronomical Union) 8.72
Ibarra, Francisco de 3.174
Ibarra, José Maria Velasso 4.108
Ibbit, William 4.163
Iberian Peninsula 1.164
Iberians 3.174
see also Portugal; Spain
Ibert, Jacques 5.216
ibex 2.168
ibis 2.168
Ibiza 1.30
Iblīs 9.10, 9.17
Ibn Abī Rabī'a, 'Umar 5.19
Ibn al-Zarqālah 8.75
Ibn 'Arabi 5.439
Ibn Battuah 1.164
Ibn al-Fāriḍ 5.19, 5.439
Ibn Gabirol, Solomon Ben-Judah 5.216
Ibn Ḥazm 5.19
Ibn Juzayy 1.164
Ibn Khaldun 3.166
Ibn Rushd see Averroës
Ibn Saud, sultan of Najd 4.149
Ibn Saud family see Saud
Ibn Sīnā see Avicenna
Ibn Tulun 3.113
Ibn Tumart 3.10
Ibn al-Zarqalah 8.75
Ibn Zaydūn 5.19
Ibo 7.151, 9.3
Ibrahim Lodi of Delhi 3.32
Ibrahim see Abraham
IBRD see World Bank
Ibsen, Henrik 5.216, 5.218, 5.377, 5.383, 5.459
ibuprofen 6.14
Icarus (asteroid) 8.72
Icarus (mythology) 9.9, 9.17
ICBMs see inter-continental ballistic missiles
ice 1.164–5
ages 1.164
density 1.87
glaciers see glaciation
hail 1.150, 1.307
hockey 7.144–5
iceberg 1.164
ice-breaker 6.175, 6.245
ice-cap 1.164–5
ice-fall 1.140, 1.165
ice-sheet 1.165
ice-wedge 1.165
pack ice 1.244, 1.296
sailing 7.272
sea 1.296
shelf 1.286, 1.287
-skating 6.321, 7.151
see also freezing; snow
ice fish 2.168
ice-cream 6.95
Iceland 1.165, 3.174, 4.160, 7.357–8
arts 5.258, 5.423
in Council of Europe 7.71
in EFTA 7.111
energy 6.104, 6.155, 7.357
infant mortality 7.159
language 7.131, 7.156
literature 5.141, 5.404, 5.421, 5.423
in OECD 7.221

Iceland (*cont.*):
physical geography and geology **1**.165, **1**.327–8
wars and defence **7**.82
Cod War **4**.78
see also NATO
Iceland spar (crystal) **1**.47
Iceni **3**.47, ***3**.174*
ICFTU (International Confederation of Free Trade Unions) **7**.315
ichneumon fly ***2**.168*
ichthyology ***2**.168*
see also fish
ichthyosaur ***2**.168–9*
Ichyosa Kunichika **5**.367
Iconoclastic controversy ***3**.174*
iconoscope **6**.118, ***6**.175*, **6**.349, **6**.391
icons **5**.70, **5**.85–6, ***5**.216–17*, **5**.394
iconography ***5**.217*
icosahedron **9**.51
ICU *see* intensive care unit
ʾĪd ul-Āḍḥā ***7**.152*
IDA (International Development Association) **7**.321, **7**.333
ideal gas **1**.132, ***1**.165*
idealism **7**.151, **7**.175, **7**.324
identikit **6**.175
identity element (set theory) ***1**.165*
identity (personal and social) **7**.233, **7**.290
ideograms **7**.151, **7**.185, **7**.334
ideographs **5**.181
ideology ***7**.151–2*
Marxist views of **7**.11, **7**.151
and one-party state **7**.223
political **7**.118
Idries Shah **9**.26
Idris **9**.11, **9**.16, ***9**.17*
Idris Aloma, king of Kanem-Bornu **3**.193
Idris al-Sanussi, Emir Sayyid of Libya **4**.195
al-Idrisi ***1**.8*
idyll *see* pastoral
Ifa (mythology) ***9**.17*
Ifẽ (mythology) ***9**.17*
Ife (Nigerian city) **3**.4, ***3**.174*
IFOAM (International Federation of Organic Agricultural Movements) **7**.224
ʿifrit ***9**.17–18*
Ignatius of Constantinople **3**.280
igneous intrusion *see* intrusive (plutonic) igneous rocks
igneous rocks ***1**.165*, **1**.166
basalt **1**.32, **1**.33, **1**.70, **1**.138, **1**.301
basement complex **1**.34
batholith **1**.34, **1**.166
columnar jointing **1**.70
diorite **1**.91
dolerite **1**.93
dyke **1**.97, **1**.166
extrusive *see* extrusive (volcanic) igneous rocks
gabbro **1**.130
granite **1**.143, **1**.178, **1**.184, **1**.315
ignimbrite **1**.165–6
intrusive *see* intrusive (plutonic) igneous rocks
laccolith **1**.166, **1**.185
minerals:
amphiboles *see* amphiboles
apatite **1**.16
aquamarine **1**.16–17
asbestos **1**.20–1, **1**.22
augite **1**.26
beryl **1**.38, **1**.108
biotite **1**.39, **1**.143
chalcedony **1**.4–5, **1**.57, **1**.76–7, **1**.238
chlorite **1**.62
feldspar *see* feldspar
fluorite **1**.124
hornblende **1**.158
igneous rocks (*cont.*):
minerals (*cont.*):
kimberlite **1**.182, **1**.183
muscovite **1**.214, **1**.223
olivine **1**.237, **1**.250–1, **1**.256
plagioclase **1**.258, **1**.259
pyroxene **1**.26, **1**.270
quartz *see* quartz
see also ferromagnesian
obsidian **1**.234
pegmatite **1**.250
peridotite **1**.183, **1**.251
plug **1**.357
plutonic rocks *see* intrusive (plutonic) igneous rocks
porphyry **1**.264
pumice **1**.269
pyroclasts **1**.270, **1**.271, **1**.295
rhyolite **1**.282
scoria **1**.295
shields **1**.120, **1**.301
sills **1**.166, **1**.304
stock **1**.320
syenite **1**.329
volcanic *see* extrusive (volcanic) igneous rocks
xenolith **1**.369
see also lava
ignimbrite ***1**.165–6*
ignition system ***6**.175–6*, **6**.327
Igor, prince of Russia **3**.197
IGOs *see* inter-governmental organizations
Iguaçu Falls ***1**.166*
Iguala, Plan of ***4**.160*
iguanas ***2**.169*
anole **2**.11
basilisk **2**.24
sea lizard **2**.300
iguanodon ***2**.169*
Ihara Saikaku ***5**.217*
Ikeda Hayato ***4**.160*
Ikeda Kikunae ***6**.176*
Île de France ***3**.174*
Ilex **2**.159, **2**.231
Iliad **5**.210
Iliescu, Ion **4**.292
Ilium *see* Troy
illegitimacy ***7**.152*
illite ***1**.166*
illiteracy **7**.102, **7**.185
illness *see* diseases and disorders
Illtud, Saint **3**.376
illuminated manuscripts ***5**.217*
Britain **3**.111, **5**.17
Byzantine **5**.70
Celtic **5**.84
France **5**.162, **5**.339
Netherlands and Flemish **5**.265
Ottonian **5**.334
Romanesque **5**.389
illumination *see* lamps; light; lighting
illusionism ***5**.217*, **5**.460
see also perspective
illustration and illustrators *see* books, illustration; drawing; graphic art; illuminated manuscripts; painting *and also* illustrations *under* books
illuviation ***1**.166*
Illuyankas ***9**.18*
Illyria ***3**.174–5*
ilmenite ***1**.166*, **6**.357
ILO *see* International Labour Organization
Iltumish, Mamaluke sultan **3**.218
Ilyas (Elijah) **9**.11
Ilyushin, Sergey Vladimirovich ***6**.176*
image ***7**.152–3*, **7**.299
intensifier ***6**.176*, **6**.375
processing ***6**.176*, **6**.240
imaginary numbers ***1**.166*
imaging photometer **8**.120
Imagists ***5**.217*, **5**.242, **5**.258, **5**.269, **5**.364, **5**.490
imām ***7**.153*, **9**.3, **9**.17
Imbrium Basin ***8**.72*
IMF *see* International Monetary Fund
Imhotep ***3**.175*, **3**.296, ***9**.18*
imitation **7**.64
immersion heater ***6**.176*
immigration **7**.152, ***7**.153*, **7**.319
see also migration *and under individual countries in particular* United States
immobilized cell technology ***6**.176*
immune system/immunity ***2**.169*
allergy **2**.6
antibodies *see* antibodies
disorders **6**.112, **6**.164, **6**.170, ***6**.177*
drugs and **6**.177, **6**.184, **6**.334
immunodeficiency **2**.17
and transplants **6**.361
see also Aids
immune response **6**.164
immunology **2**.170, **6**.177
see also immunization
immunity, diplomatic **7**.91
immunization, inoculation, and vaccination ***2**.169*, ***2**.354*, ***6**.176–7*, ***7**.153*
diphtheria **6**.177, **6**.191
and epidemiology **7**.107
and monoclonal antibodies **6**.232
and Pasteur **2**.243
pioneers **2**.371
rabies **2**.243, **2**.274, **6**.177, **6**.259
smallpox **2**.40, **2**.175, **6**.176–7, **6**.178, **6**.189, **6**.218
see also immune system/immunity
immunodeficiency ***2**.170*
immunoglobin **2**.12, **2**.169
immunology ***2**.170*, ***6**.177*
see also immune system/immunity
impairment *see* disability and handicap
impala **2**.20
impasto ***5**.217–18*, **5**.326
impeachment ***3**.175*, ***7**.153*
impedance **6**.76, ***6**.177*
imperfect competition **7**.222, **7**.233
Imperial College of Science and Technology ***6**.177*
Imperial Preference *see* reform *under* tariff
imperial units **1**.350, **6**.368–9
imperialism ***4**.160*, ***7**.153*
gunboat diplomacy **4**.144
see also colonialism; decolonization *and under individual countries*
imports ***7**.153*, **7**.306
see also trade
Impressed ware cultures ***3**.175*
Impressionism:
in music **5**.39, **5**.126–7, **5**.128, **5**.216, ***5**.218*, **5**.225, **5**.375
in painting ***5**.218*
Britain **5**.318, **5**.417–18, **5**.420, **5**.431
France **5**.56, **5**.87, **5**.286, **5**.351, **5**.380
see also Degas; Manet; Monet
Germany **5**.264
see also Neo-Impressionism; Post-Impressionism
in sculpture **5**.127, **5**.171
imprinting (learning) **2**.185, ***7**.153–4*
imprisonment *see* prisons
imps **9**.16
Imru' al-Qays **5**.19, ***5**.218*
in vitro fertilization ***6**.185–6*, **6**.218, **7**.159–60
incandescent lamp **6**.112, **6**.114, **6**.134, **6**.166, ***6**.177*, **6**.194, **6**.201, **6**.342
incarnation, see *avatars*; reincarnation
Incas ***3**.175–6*
art **5**.425–6
city **3**.298
language **7**.295
religion and myths **9**.24, **9**.37
Incas (*cont.*):
rulers **3**.26, **3**.358–9
technology **6**.219–20, **6**.299
Ince, Thomas **5**.487
incendiary bombs **6**.236
inceptisol ***1**.166*, **1**.311
incest **7**.46, ***7**.154*
inch worm **2**.55, **2**.194
incident of 26 February 1936 ***4**.160*
incidental music ***5**.218–19*
overtures **5**.334, **5**.442
see also music and dance *under* cinema
incineration of waste **6**.376–7
inclined plane ***6**.177–8*
income ***7**.154*
incomes policy **7**.246
support **7**.289
tax ***4**.160*
see also GDP; GNP; national income; pay
incubator ***6**.178*
incunabula ***5**.219*
independence:
nationalism **7**.211–12
secession **7**.277
self-determination **7**.278
sovereignty **7**.296
see also decolonization; freedom
Independence, American War of ***3**.176*
battles and campaigns **3**.40, **3**.45, **3**.46, **3**.49, **3**.53–4, **3**.58, **3**.140, **3**.207, **3**.316, **3**.357, **3**.388
leaders **3**.54, **3**.156, **3**.199, **3**.378
map of **3**.176
minutemen **3**.233
Peace of Paris **3**.269
submarine **6**.377
traitor **3**.34
Valley Forge **3**.369
Independent Churches **7**.50
independent education ***7**.154–5*
Independent Labour Party ***4**.160*
indeterminacy ***5**.219*
index, financial ***7**.155*
indexation ***7**.155*
India ***1**.166–7*, ***3**.177*, ***4**.160–2*, ***7**.358*
Acts ***4**.162*
agriculture **6**.34, **6**.68, **6**.161, **6**.204
aid to **7**.8
arts ***5**.219–21*, **5**.370
and Buddhism **5**.10, **5**.64, **5**.286, **5**.336, **5**.400, **5**.401
cinema **5**.240, **5**.375
metallurgy **5**.99, **5**.196
painting **5**.10, **5**.108, **5**.196, **5**.219, **5**.373
rock **5**.10, **5**.143, **5**.145, **5**.196
see also under architecture; sculpture *and also under* literature; music *below*
Aryans **3**.24
astrology **7**.20
astronomy **6**.248, **8**.144, **8**.146
caste **7**.41
chemistry **6**.10, **6**.275
colonialism in:
British *see* British Raj
French **3**.106, **3**.288
Portuguese **3**.7, **3**.95–6, **3**.107, **4**.136–7
dance *see* music *below*
demography:
census **7**.42
old age **7**.221
population density **7**.241
population growth **7**.242
discrimination, positive **7**.243
dowry **7**.95
economy and trade:
commercial families **4**.39, **4**.339
industrial accident **6**.35, **6**.275

India (*cont.*):
economy and trade (*cont.*):
industrial relations **7.**315
Industrial Revolution **6.**178
tax collector **4.**388
see also East India Company
education **7.**141, **7.**307
energy and fuel **6.**37, **6.**385
ethnic groups **7.**137, **7.**211, **7.**306
exploration **1.**83, **1.**164
famine **7.**116, **7.**118
fishing **6.**146
food **6.**46
government *see* politics and government *below*
household **7.**148
industry *see under* economy and trade *above*
land **4.**37, **4.**298
languages **7.**95, **7.**157, **7.**286
alphabet *see* Devanagari
literature ***5.**220*
Bengali **5.**43
drama **5.**244, **5.**279, **5.**314
epics **5.**134, **5.**276, **5.**373, **5.**401, **5.**447
Hindi **5.**206
Jain **5.**232
Pāli **5.**336
Tamil **5.**445–6
Urdu **5.**466–7
see also Bhagavadgītā; Dravidian; Sanskrit *and under* Hinduism; novels; poetry; stories
maps **1.**167, **3.**151, **3.**177, **3.**235, **3.**323, **4.**161
Marathas **3.**167, **3.**221, **3.**328–9
marriage **7.**193
McMahon Line **4.**204
medicine and health care **6.**176, **6.**178, **6.**340, **6.**342, **7.**24, **7.**139, **7.**293, **7.**333
mining and metallurgy **5.**99, **5.**196, **6.**101, **6.**178
mortality **7.**45
music ***5.**220–1*, **5.**314, **5.**446
dance ***5.**219–20*, **5.**221, **5.**244, **5.**279, **5.**307
instruments ***5.**221*
in Non-Aligned Movement **7.**218
philosophy ***7.**155–6*
see also Advaita; Hinduism; *Upanishads*; *Vedānta*; *Vedas and also* religion/theology *below*
physical geography and geology ***1.**166–7*
climate **1.**65, **1.**220
Deccan **1.**85
desert **1.**334
map **1.**167
mountains *see* Himalayas
regions **1.**185
see also regions *below*
rivers **1.**43, **1.**132
politics and government:
assassination **4.**130, **7.**20
Bill of Rights **7.**32
communist **4.**80
federalism **7.**119
first president **4.**281
hunger-strikes **7.**150
independence **7.**58
see also resistance *under* British Raj
National Congress **4.**83–4
nationalists **4.**37, **4.**46, **4.**97, **4.**137, **4.**230, **4.**262, **4.**276, **4.**282
parties **4.**80, **4.**83–4, **4.**172, **4.**228
prime ministers **4.**97, **4.**130, **4.**172, **4.**236–7
revolutionary **4.**236
separatism **7.**279, **7.**280
Untouchables' leader **4.**10
see also rulers *below*

India (*cont.*):
poverty **7.**244
prehistory **3.**157
prostitution **7.**250
regions and cities **1.**185, ***9.**73*
central *see* Deccan
east **3.**88, **3.**263
megacities **7.**197
new towns **7.**216
north **3.**39, **3.**101, **3.**268, **3.**370
north-west **3.**156, **3.**183–4, **3.**189
slums **7.**288
south **3.**244, **3.**373
west **1.**185, **3.**143, **3.**195, **3.**196
religion/theology:
Buddhism *see under* Buddhism *and also under* arts *above*
caste system **7.**41
Christianity **7.**204, **7.**304, **9.**40, **9.**43
cremation **7.**74, **7.**75
Islam **7.**166, **7.**168, **7.**284
Muslim Brotherhood **7.**166
and politics **4.**182
see also Muslim *under* rulers *below*
Monophysite Christian **7.**304
mystic **3.**193
Zoroastrianism **7.**229
see also Hinduism; Jainism; Sikhism *and* philosophy *above*
rulers and dynasties **3.**173, **3.**326, **3.**357
Buddhist **3.**157
Gupta dynasty **3.**151, **3.**152, **5.**196
Holkar **3.**167
Hoysala **3.**170
Lodi **3.**209
Muslim:
Khalji **3.**195
Mameluke **3.**217
Mauryan empire **3.**25, **3.**225–6, **5.**286–7, **6.**299
Moguls *see* Moguls
Nizam Shahi **3.**254
Sayyid **3.**318
Tughluq **3.**362
Pala **3.**267
Pallava **3.**268, **5.**275
Pandya **3.**268
Rajput **3.**293, **3.**298, **5.**373
Satavahana **3.**316
Sindhia **3.**328
segregation **7.**278
social services and security **7.**293
sport and recreation **7.**75
states and regions **4.**136–7, **4.**159, **4.**313
technology and science **6.**158, ***6.**178*, **6.**346
see also entries above and below prefixed **1.** *and* **6.**
telecommunications **7.**308
textiles **6.**328, **6.**329
thought, contemporary ***7.**156*
tools **6.**195
Union of ***4.**161–2*
urban areas *see* regions and cities *above*
vehicles and transport **6.**64, **6.**70, **6.**196, **6.**299, **6.**301, **6.**332
wars and defence:
Black Hole of Calcutta **3.**42
British resisted *see under* British Raj
Buxar battle **3.**54
with China **4.**163
communal conflict **7.**59
French capture of Madras **3.**30
Indo-Pakistan War ***4.**163*
Maratha **3.**221, **3.**328–9, **4.**211
Mutiny **4.**162, **4.**200–1
nuclear weapons **7.**220
Plassey battle **3.**285
religious **7.**59

India (*cont.*):
wars and defence (*cont.*):
Seven Years **3.**323
technology **6.**40, **6.**42, **6.**343
women's movement **7.**332
Indian bean tree **2.**56
Indian fig **2.**268
Indian Mutiny ***4.**162*, **4.**200–1
Indian National Congress **4.**83–4
Indian Ocean ***1.**167*
islands **4.**208, **4.**214, **4.**310
see also under physical geography *below*
map **1.**168, **3.**107
physical geography and geology ***1.**167*
currents **1.**167, **1.**235
islands **1.**70, **1.**203, **1.**207, **1.**300, **1.**372
see also Madagascar; Zanzibar
map **1.**168
plate tectonics **1.**260
ridges **1.**168
trenches **1.**168
Indians, North American ***3.**11*, ***3.**256*, ***3.**257*
agriculture *see under* subsistence farming
Algonquin **3.**10, **3.**289–90
Anasazi (Pueblo) **3.**12, **3.**79
Apache **3.**17–18, **7.**14, **7.**131, **7.**219
Arapaho **3.**19
arts **5.**14–15, ***5.**321–2*
'basket-makers' **3.**12, **3.**36
Blackfoot **3.**42, **7.**219
Canadian *see under* Canada
Cherokee **3.**73, **4.**71, **5.**322, **7.**44, **7.**219
Chickasaw **3.**73
chiefs **3.**288, **3.**289–90, **4.**77, **4.**88, **4.**135, **4.**176, **4.**252, **4.**315, **4.**339–40, **4.**371
Choctaw **3.**74
Comanche **3.**82
Cree **3.**90, **7.**74, **7.**219
Crow **3.**92, **7.**219
Dakota group **3.**96, **3.**180
Delaware **3.**101
ghost dance **7.**131
Hidatsa **3.**92
Hopi **7.**147
hunters *see under* hunter-gatherers
Huron **3.**172, **3.**173
confederation **7.**150
Iroquois **3.**180, **7.**166
Kwakiutl **7.**178
lacrosse **7.**179
languages ***7.**218–19*, **7.**221, **7.**273
lineage **7.**166, **7.**219, **7.**236, **7.**312, **7.**316
literature ***5.**322*
Mahican (Mohican) ***3.**216*
map of distribution **3.**257
matrilineality *see under* matrilineality
Mississippi cultures **3.**233–4
Mohawk **3.**236
mound-builders **3.**242
music ***5.**14–15*
Natchez **3.**246
Navajo (Navaho) **3.**247, **7.**213, **7.**219
North-west *see* North-west Indians
Ojibwa **7.**219, **7.**221
origins **3.**11
Ottawa **3.**264, **3.**288
patrilinearity **7.**236
Pawnee **3.**272
Plains *see* Plains Indians
Pocahontas **3.**286
Pontiac's confederacy **3.**288
potlatch **7.**219, **7.**243
Pueblo **3.**12, **3.**79, **9.**21
religion and myths **7.**132, **9.**9, **9.**14, **9.**16, **9.**21
reservations ***4.**162*, **4.**250

Indians, North American (*cont.*):
rituals **7.**132
Seneca **3.**321
Sioux **3.**96, **3.**180, **7.**219, **7.**287
superintendent of **3.**190
technology **6.**89
Tlingit **7.**295, **7.**312
totemism **7.**313
Trail of Tears exile **4.**350
Tsimshian **7.**316–17
wars:
with Americans **3.**128, **3.**276, **4.**40, **4.**308, **4.**345, **4.**384
Custer **4.**91, **4.**197
with English **3.**197–8
Fort Stanwix **3.**128
with French **3.**134, **3.**197–8, **3.**202, **3.**284
uprisings **3.**288
writing **7.**221
Indians, South American (and Central American) ***3.**11*
Amazonian **7.**11–12, **7.**335
Arawaks **3.**60
arts, literature, and music **5.**14–15, **5.**425–6
Caribs **3.**61, **7.**295
languages **7.**199–200, ***7.**295*
Olmecs **3.**261, **5.**327
religion and myths **9.**11, **9.**24
deities **9.**9, **9.**15, **9.**17, **9.**28, **9.**30, **9.**35–6, **9.**37, **9.**38
revolt **3.**363
see also in particular Aztecs; Incas; Mayans
indicators in chemistry **1.**197
indigobird **2.**367
indium **1.**106, ***1.**167*
individualism ***7.**156–7*
Indo-Aryan language **7.**156
Indo-China ***1.**167*
refugees **7.**262
wars and defence **7.**136
Indo-Chinese War **4.**98, **4.**99, ***4.**127*, **4.**133, ***4.**163*
Indo-European languages ***7.**156*, ***7.**157*
Albanian **7.**10
Anatolian **7.**13
Armenian **7.**17
Baltic **7.**26
Celtic **7.**42
Germanic **7.**131
Hellenic **7.**10–11, **7.**135, **7.**269
Indo-Iranian **7.**137, **7.**156, **7.**157
Slavic **7.**287–8
see also Italic; Romance
Indo-European people **3.**227, **3.**320
Indo-Greek dynasties ***3.**177–8*
Indo-Iranian languages **7.**137, **7.**156, ***7.**157*
Indonesia ***1.**167*, ***4.**163*, ***7.**358–9*
aid to **7.**8
arts **5.**169, ***5.**221–2*, **5.**222
colonization and decolonization **3.**107, **4.**321
economy and trade **7.**20
see also ASEAN; OPEC
ethnic groups **7.**25–6, **7.**202
Irian **4.**321
Kalimantan **4.**321
Konfratasi with Malaysia **4.**185–6
languages **7.**23, **7.**95
literature ***5.**222*
music ***5.**222*
in OPEC **4.**252
physical geography and geology **1.**42, ***1.**167*
Bali **1.**30–1
Java **1.**177
Krakatoa **1.**184
Moluccas **1.**218
New Guinea **1.**226–7
Sulawesi **1.**326
Sumatra **1.**327
volcanoes **1.**175, **1.**184

Indonesia (cont.):
politics 4.270, 4.302–3, 4.333
population growth 7.242
poverty 7.244
religion/theology 7.302
Revolution 4.163
sport and recreation 7.54
Sulawesi 1.326, 3.214
technology and science 6.7, 6.294, 6.303
see also Bali; Java; Moluccas; Papua New Guinea; Sumatra
Indo-Pakistan War 4.163
Indra 9.18
indri 2.170
induction (electrical):
heating 6.290
motor 6.115, 6.202–3, 6.351
telephony 6.347
see also electromagnetic induction under electromagnetic
induction (logic) 7.157
induction (mathematics) 1.167
inductor/inductance 6.178
inductor compass 6.178
measurement 1.154
indulgences 3.178, 3.269
Indus (constellation) 8.72
Indus Valley 1.167–8, 3.177–8, 6.7
art 5.223
civilisation 3.177–8, 3.236, 5.223
Industrial Revolution 3.178–9, 4.164, 7.158
coal 6.78, 6.79, 6.80
steam power 3.341
technology 6.181, 6.195, 6.221
textiles 6.329
see also industrialization
industrialization/industry/industrial 4.163–4, 7.158
action and relations 7.157–9, 7.235
see also labour and employment; strikes; unions
archaeology 6.178
architecture 5.42, 5.222, 5.408
see also functionalism; Gothic Revival
automation 4.26
democracy 7.158
design 5.42, 5.222–3
see also Arts and Crafts; Bauhaus
diseases 2.241, 2.276
Factory Acts 4.115
firms see companies and business
hazards see chemical and nuclear under accidents
heavy 7.140–1
industrial relations see action above
law 7.158–9
light 7.184
major industries see in particular chemicals; iron and iron and steel; manufacturing; petrochemicals; textiles
mass production 4.213
military-industrial complex 7.202, 7.212
mining see coal; gold
museum 6.87
nationalization 7.212
primary see primary industry
secondary see manufacturing
tertiary see tertiary industry
vertical integration 7.325–6
see also economy; Industrial Revolution and also economy and trade or industry under individual countries
industrialized countries:
North-South relations 7.219
see also Australia; Britain; Canada; Europe; Japan; New Zealand; newly industrializing countries; United States
industry see industrialization
inequality (Moon) 8.101
inert gases see noble gases
inertia 1.168, 8.72–3
measurement see mass
moment of 1.218
inertial navigation 6.1, 6.162, 6.178–9
INF (Intermediate-Range Nuclear Forces) 7.17
infant and child mortality 7.119, 7.140, 7.153, 7.159
infections and communicable diseases 2.74, 2.170, 4.99–100, 7.59
abscesses and boils 2.1
Aids see Aids
causes of see pathogens; vectors
chickenpox 2.63
cholera see cholera
cold 2.72
contagion 2.75
diphtheria 2.93, 6.177, 6.191, 7.153
drugs for see drugs, therapeutic
dysentery 2.101, 7.59
elephantiasis 2.106
epidemics see epidemics and epidemiology
fever see fevers
germs 2.126
hepatitis 2.155, 6.87
herpes 2.157, 2.307
legionnaire's disease 2.186
leprosy 2.188
leptospirosis 2.364
malaria see malaria
measles 2.206, 7.153
meningitis 2.121, 2.207, 6.260
onchocerciasis 2.234, 2.286
peritonitis 2.250
pneumonia 2.260
ringworm 2.285
scabies 2.294–5
scarlet fever 2.296–7
septicaemia 2.303
shingles 2.63, 2.307
smallpox see smallpox
tetanus 2.339, 7.153
tinea 2.341–2
trypanosomiasis 2.17, 2.348, 2.349
tuberculosis see tuberculosis
typhoid see typhoid; paratyphoid
typhus 2.352
venereal see sexually-transmitted diseases
whooping cough 2.367, 7.153
yellow fever 2.373
see also diseases
inferior conjunction of planets 8.122
inferior planet 8.73
infertility 6.127, 6.170, 7.159–60
infiltration 1.168
infinite canon 5.223
inflatable boat 6.179
inflation 7.160
in-flight refuelling 6.179
informal sector 7.160
information:
censorship 7.42
conveying see communication(s); media; speech; telecommunications
freedom of 7.125
retrieval 6.179
systems analysis 6.344
technology 6.179–80, 7.160
see also computers; data
theory 1.168, 6.180
infra-red radiation 1.168, 6.117
astronomy 8.73
detectors 6.375
gun-sight 6.163
missiles 6.162, 6.228
photography 6.268, 6.375
spectroscopy 6.328
infrastructure 7.160
Ingres, Jean-Auguste-Dominique 5.223
inheritance:
biology see genetics
primogeniture 3.292
inhibitor 2.170
initiation 7.6, 7.160–1, 7.188
injunction 7.94, 7.161
injuries 2.242
burn 2.45–6
concussion 2.74
fracture 2.127
pain 2.239
see also accidents and disasters; diseases and disorders
ink 6.180, 6.259
ink-jet printing 6.180, 6.281
Inkerman, battle of 4.164
Inland Sea 1.168–9
inlier 1.169
inner planets 8.73
see also Earth; Mars; Mercury; Venus
Innes, Robert 8.129
Innocent II, pope 3.40, 3.306
Innocent III, pope 3.131, 3.179, 3.203–4, 3.269
Innocent IV, pope 3.133
Innocent VIII, pope 3.384
Inns of Court 7.161
inoculation see immunization
Inönü, Ismet 4.58, 4.164
inorganic chemistry 1.60
input device 6.180
inquest 7.161, 7.174
Inquisition 3.179, 3.336, 3.337
inquisitorial procedure 7.161
inro 5.223–4
insecticides 6.17, 6.180, 7.333
Bhopal accident 6.35
DDT 1.85, 6.96, 6.180, 6.275
pyrethrum 2.272
see also insects
insectivores (insectivorous mammals) 2.171, 2.172
aardvark 2.1
aardwolf 2.1
ant-eaters 2.12, 2.103, 2.277
colugo 2.73
elephant shrew 2.106
golden mole 2.140
hedgehog 2.154
mole 2.204, 2.215
solenodon 2.316, 2.317
tenrec 2.337
tree shrew 2.172
see also bats; shrews
insects 2.170, 6.197, 6.267, 6.339
alder fly 2.6
ant 2.2, 2.11
ant lion 2.12–13
bee see bees
beetle see beetles
bug see bugs
butterfly see butterflies
caddis fly 2.49
chafers 2.288
crickets 2.18, 2.216
damselfly 2.86–7
and disease see under vectors
dragon-flies 2.99, 2.206–7
earwig 2.103
entomology see entomology
fleas 2.120–1, 2.293
flight 2.170–2
grasshoppers see grasshoppers
lacewing 2.181
ladybirds 2.30, 2.181
leaf 2.185
locusts 2.192–3, 2.210
mayfly 2.205
midges 2.131, 2.210
migration 2.211
mite see mites
moth see moths
myriapods 2.59, 2.212, 2.224
insects (cont.):
parasites 2.165, 2.168
pest see under pests
praying mantis 2.267, 2.268
proturan 2.270
pupa 2.272
scorpionfly 2.297
silverfish 2.310
springtail 2.322, 2.323
stick 2.50, 2.325
storefly 2.327
strepsipteran 2.327
termites 2.338
thorax 2.339–40
thrip 2.340
vision 2.308
wasp see wasps
web spinner 2.363
see also insecticides
inselbergs 1.28, 1.169, 1.170
insemination see reproduction
Insha 5.466
insider trading 7.161
insolation 1.169, 8.73
see also Sun/solar
insolvency 7.161
instinct 2.172, 7.109–10, 7.161, 7.169, 7.207, 7.294
institutional care 7.161
community care to replace 7.61
and homosexuality 7.146
and mental handicap 7.198, 7.299
Romanian orphanage 7.378
total institution 7.313
instrument landing system 6.25, 6.290
instrumentalism (philosophy) 7.88
instrumentation and observation 6.180–2
astronomical:
angular measure 8.5, 8.192
astrolabe 8.9
astronomical units 8.10–11, 8.192
bolometer 8.17
CCD camera 8.22
coelostat 8.27
heliostat 8.65
interferometry 8.73, 8.98, 8.156–7
quadrant 8.131
radiometer 8.133
see also observatories; telescopes and astronomy under photography
measurement see measurement
meteorological see under meteorology
see also under laboratories; navigation and also individual instruments
insulation, thermal 6.167, 6.182–3
cavity wall 6.64
double-glazing 6.106
glass fibre 6.156
illustrated 6.171
and solar power 6.324
vacuum as 6.182, 6.370
insulator, electrical 1.32, 6.12, 6.183
insulin 2.22, 2.138, 2.240, 6.29, 6.170
insurance 7.162
actuary 7.2
annuity 7.14
assurance 7.20
Lloyd's of London 7.186
medicine and health care 7.139
national (social) 4.234
see also social security
underwriting 7.319
intaglio 5.224
printing 6.123, 6.134, 6.160, 6.183
see also engraving; etching

integrated circuit **6.**76, **6.**118, **6.**182, ***6.**183*, ***9.**53*
amplifier **6.**13
in clock **6.**77
in computer **6.**85
flip-flop **6.**136
and radio **6.**290
see also microprocessor
Integrated Services Digital Network **6.**346
integration (economics) ***7.**162*
integration (mathematics) ***1.**169–70*
intellectual property ***7.**162*, **7.**230
intelligence ***2.**172*, ***7.**162*
artificial *see* artificial intelligence
tests and quotient (IQ) **7.**32, ***7.**162*
intelligence services ***7.**162*, **7.**278
counter-intelligence **7.**72
see also security
intendant ***3.**179*
intensive care unit **6.**116, ***6.**183*, **6.**298
intentionality **7.**35, ***7.**162*
interactionism *see* symbolic interactionism
interactions (physics) ***1.**170*
see also force (physics)
interactive system (computers) ***6.**183*
Inter-American Commission and Court of Human Rights **7.**12
intercalated month **8.**18
interceptor aircraft **6.**129
inter-continental ballistic missiles **6.**29, **6.**228, **6.**231
intercropping ***6.**183*, **6.**364
interdependence ***7.**163*
interest groups (pressure groups) **7.**135, ***7.**163*, **7.**186
interest and interest rate **7.**26, **7.**74, **7.**120, ***7.**163*
interface, computer ***6.**183–4*
see also bus; central processing unit; disk; memory; peripheral
interference (physics) ***1.**170*, **1.**171, ***6.**382*
interferometry and interferometer ***6.**184*, ***8.**73*, **8.**98
speckle **8.**156–7
very long baseline **8.**184
interferon ***2.**172*, ***6.**184*, **6.**188
intergalactic medium ***8.**73–4*
interglacial ***1.**170*
inter-governmental organizations **7.**165
Interim of Augsburg **3.**27
interior design ***5.**224*
Art Deco **5.**23
art nouveau **5.**23–4
Belgium **5.**212, **5.**470
Biedermeier style **5.**47
Britain **4.**191, **5.**245–6, **5.**304–5, **5.**489
France **5.**146, **5.**259
'Louis Quatorze, Quinze and Seize' styles **5.**56, **5.**259, **5.**261, **5.**269
panelling **5.**338
Restoration **5.**381
United States **5.**455
wallpaper **5.**481
see also carpets and rugs; central heating; clocks; floor; furniture; paint; tapestries; walls
intermediate (theatre) weapons **6.**246
intermediate technology **6.**18
Intermediate-Range Nuclear Forces, Treaty on **7.**17
intermedio ***5.**224*, **5.**328
intermezzo ***5.**224*
intermolecular forces ***1.**170–1*, **1.**354
internal energy **1.**110, **1.**171
internal-combustion engines **6.**122, ***6.**184*
carburettor *see* carburettor
internal-combustions engines (*cont.*):
catalytic converter **6.**9, **6.**62, **6.**64, **6.**124
choke **6.**73
clutch **6.**77–8
compression ratio **6.**83
cooling system **6.**89
diesel *see* diesel
and flying **6.**139
gas-engines **6.**150, **6.**198
gearbox **6.**26, **6.**152
ignition **6.**175–6, **6.**327
knock **6.**191–2, **6.**197, **6.**224, **6.**249
lubricating oil **6.**210
petrol *see* engines *under* petrol
pioneer **6.**94
spark-plug **6.**175, **6.**327
starter motor **6.**330
supercharger **6.**339
warships **6.**357
see also motor vehicles *and under* roads *under individual countries*
international *see entries beginning* global; world
international agreements *see* international organizations and alliances
International Astronomical Union **8.**72
International Atomic Energy Agency ***6.**184*, ***7.**163–4*, **7.**220, **7.**321, **7.**322
International Atomic Time **8.**172
International Bank for Reconstruction and Development *see* World Bank
International Bar Association **7.**255
International Brigades ***4.**164*
International Confederation of Free Trade Unions **7.**315
International Conventions *see* Conventions
International Court of Justice ***4.**164*, **7.**82, ***7.**164*, **7.**321, **7.**322
International Covenants:
on Civil and Political Rights **7.**51–2, **7.**87, **7.**138, **7.**149, ***7.**164*, **7.**322
on Economic, Social and Cultural Rights **7.**149, ***7.**164*, **7.**322
see also Conventions
International Criminal Police Organization (Interpol) **7.**165
international date-line **1.**37, ***1.**171*
International Development Association **7.**321, **7.**333
international distributions *see* world distribution maps and diagrams
International Federation of Organic Agricultural Movements **7.**224
International Finance Corporation **7.**321
International Gothic **5.**16, **5.**173, **5.**175, ***5.**224*, **5.**265, **5.**350, **5.**402, **5.**465
see also Gothic architecture; Gothic Revival; Sienese School
International Labour Organization ***4.**164*, **7.**106, **7.**107, ***7.**164*, **7.**203, **7.**320
international law ***7.**164*
diplomatic immunity **7.**91
extradition **7.**113–14
International Monetary Fund ***4.**164*, ***7.**164–5*, **7.**321, **7.**322
founded **7.**176
and debt crisis **7.**81
and economic adjustment **7.**3
and military expenditure **7.**201
Special Drawing Rights **7.**78, **7.**112, **7.**165
International Monetary Fund (*cont.*):
and World Bank **7.**333
international order *see* international relations
international organizations and alliances ***7.**165*
collective security **7.**56, **7.**82
Arab League **7.**15, **7.**227
Gulf Co-operation **7.**136–7
see also in particular Conference on Security; League of Nations; NATO; OAU; United Nations; Warsaw Pact
Conventions *see* Conventions
economic, social and political:
Group of 7 **7.**136
health *see* World Health Organization
see also ASEAN; CARICOM; COMECON; Commonwealth; EC; EEC; EFTA; GATT; IAEA; OAS; OAU; OECD; OPEC
regional *see* regionalism
see also treaties; United Nations *and entries beginning* International *and* World
International Panel on Climate Change **7.**135
International Phonetic Alphabet **7.**235
international politics **7.**131
International Red Cross **4.**284, **7.**218
international regionalism *see* regionalism
international relations **7.**131
balance of power **7.**25
diplomatic **7.**90–1, **7.**105
see also East–West relations; international law; North-South relations; power (political and social); regionalism; transnational relations; wars
international *see entries beginning* global; world
international society ***7.**165*
International Style *see* Modern Movement
International Telecommunications Union **6.**184, **6.**285, **6.**346, **6.**380, ***7.**258*
international time zones **9.**54
international trade *see* trade
international trusteeship *see* mandate
International Union for Conservation of Nature and Natural Resources **7.**106
Internationals (communist) ***4.**164–5*
interplanetary medium ***8.**74*
Interpol ***7.**165*
interpreter (computers) ***6.**184–5*
interstellar absorption ***8.**74*
interstellar medium ***8.**74*
interstellar reddening ***8.**74*
interstitial compounds ***1.**171*
intertropical convergence zone (ITCZ) ***1.**171*
inter-uterine device **2.**75
interval (music) ***5.**224–5*
intervention ***7.**165*
intestines *see* gastro-intestinal tract
Intimisme **5.**51, ***5.**225*, **5.**480
see also genre
Intolerable Acts *see* Coercive Acts
intra-uterine device (IUD) **6.**87
intrazonal soil ***1.**171*
intrinsic variable **8.**180
introversion ***7.**114*
intrusive (plutonic) igneous rocks ***1.**165*, ***1.**166*, ***1.**260*
batholith **1.**34, **1.**166
dolerite **1.**93
dyke **1.**97, **1.**166
gabbro **1.**130
granite **1.**143, **1.**178, **1.**184, **1.**315
intrusive (plutonic) igneous rocks (*cont.*):
laccolith **1.**166, **1.**185
magma **1.**165, **1.**202
peridotite **1.**183, **1.**251
sills **1.**166, **1.**304
stock **1.**320
volcanic plug **1.**357
see also igneous rocks
Inuit **7.**16, ***7.**165*
language **7.**108
literature ***5.**225*
invar **6.**18, ***6.**185*, **6.**241, **6.**355
invariable plane ***8.**74*
invention
creativity **7.**73–4
see also patent *and also* inventors *and* pioneers
inventory ***7.**165*
Invergordon mutiny ***4.**165*
inverse elements (mathematics) ***1.**171*
inverse function (mathematics) ***1.**171*
inverse square law ***1.**171*
inversions (meteorology) ***1.**171–2*
invertebrates ***2.**173*
rotifer **2.**288, **2.**289
see also cephalopods; echinoderms; molluscs; polyp; worms
Investiture ***3.**179*
investment **7.**120–1, ***7.**165*, **7.**165
cost-benefit analysis **7.**71
trust ***7.**165–6*, **7.**316
see also aid
investment casting, see *cire perdue*
Io (mythology) ***9.**18*
Io (satellite of Jupiter) ***8.**74*
iodine **1.**106, ***1.**172*, ***6.**186*, **6.**243
deficiency **2.**140
Ionesco, Eugène **5.**2, ***5.**225*
Ionian Islands **1.**76, ***1.**172*
ionic *see under* ions
ionization ***1.**172*, **6.**186
ionosphere ***1.**172*, **1.**173, **6.**380, ***8.**74–5*
Appleton layer **1.**16
Heaviside–Kennelly layer **1.**153
ions ***1.**172–3*, **6.**25, ***6.**186*, ***8.**74*
astronomy **8.**3, **8.**74
in battery **6.**32
electroplating **6.**120
engines **6.**326
in gas-discharge lamp **6.**150
hydrogen ion concentration **1.**161
see also acid
hydroxide **1.**163
implantation ***6.**186*, **6.**312
ion-exchange resin ***6.**186*
ionic bond **1.**59, **6.**24
ionic compound ***1.**172*
see also salts
ionizer ***6.**186*
ligand **1.**193
measurement **6.**153
negative **1.**14
neutral **1.**374
in polar solvent **6.**325
positive **1.**55
in xerography **6.**268
zwitterion **1.**374
Iphigenia **9.**2, ***9.**18*
IQ (Intelligence Quotient) **2.**172, **7.**162
Iqbal, Sir Mohammed ***4.**165*, **5.**439, **5.**467, **7.**168
IRA *see* Irish Republican Army
Iram of the Columns **9.**2
Iran (*formerly* Persia) ***1.**173*, ***3.**277*, ***4.**165*, ***7.**359*
abortion illegal **7.**1
arts ***3.**277*, ***5.**344*
architecture **5.**344
ceramics **5.**344
music **5.**487
painting **5.**47
sculpture and carving **3.**97, **3.**276, **5.**344

Iran (*cont.*):
arts (*cont.*):
see also literature *below*
economy and trade *see* OPEC
Hostage Crisis 4.165
Kurds 4.188, 7.178
languages 7.17, 7.231
Indo-Iranian 7.137, 7.156, 7.157
laws/legal system 7.239, 7.284
literature 3.345, 5.220, 5.344–5
see also under poetry
marriage 7.193
in OPEC 4.251
pastoralism 4.188, 7.230
physical geography and geology 1.20, 1.173
desert 1.146
mountains 1.185, 1.372
river 1.300
political prisoner 7.239
politics and government 4.182–3, 4.228, 4.251, 4.289, 7.265, 7.299, 7.310
Qajar tribe 4.279
refugees 7.261
religion/theology 4.27, 4.182–3
āyātollāh 7.24
Bahā'ī 7.25
Islam 3.168, 5.344–5, 7.284
Ismā'īlis 7.168
millenarianism 7.202
myths 9.3, 9.6, 9.10, 9.19, 9.24, 9.32, 9.33, 9.34, 9.39
deities 9.18, 9.25, 9.39
spirits 9.10, 9.20, 9.29
Nestorians 7.214
and politics 7.265
saints 9.40, 9.42, 9.43
Shi'i 7.284
theologian 7.10
see also Zoroastrianism
rulers 3.1, 3.23, 3.181, 3.196, 3.245
ancient 3.2, 3.101, 3.196, 3.245, 3.316, 3.389
see also Abbas; Achaemenid; Artaxerxes; Cyrus; Darius; Safavid; Xerxes
Shahs *see* Reza Shah
technology 6.34, 6.86
urban areas 7.197
wars and defence:
with Greeks 3.101, 3.147, 3.221, 3.314
CENTO 4.68
hostage-taking 4.165, 7.91, 7.147–8
Irangate scandal 4.284
with Iraq 4.165–6, 7.136, 7.231–2, 7.249, 7.277
revolution 7.166, 7.267, 7.331
see also Greece, ancient; Gulf War
writing and script 7.157
Iraq 1.173, 4.166, 7.359
in Arab League 7.15
CENTO 4.68
economy and trade 7.273
see also OPEC
genocide 7.130
Kurds 4.188, 7.44, 7.178
in OPEC 4.251
physical geography and geology 1.173, 1.185
rivers 1.115, 1.300, 1.337
politics and government 4.283, 7.23, 7.24
refugees 7.261
religion/theology 7.214, 7.284
see also Islam
rulers 4.115
technology 6.51, 6.70, 6.263
wars and defence:
hostage-taking 7.148
with Iran 4.165–6, 7.136, 7.231–2, 7.249, 7.277
Iraq (*cont.*):
wars and defence (*cont.*):
special armed services 7.297
weapons 7.220
see also Gulf War
see also Mesopotamia
IRB *see* Irish Republican Brotherhood
Ireland, B. 4.113
Ireland, John 5.225
Ireland 1.173, 3.179–80, 4.166, 7.359
since 1921 *see* Ireland, Republic of; Northern Ireland
adventurer 3.43
arts 5.169
music 5.154, 5.303
painting 5.36, 5.497
see also literature *below*
astronomy 8.6, 8.16, 8.86
economy and trade *see under* Ireland, Republic of
emigrants to United States 4.221–2
Famine 4.167
Land League 4.189
language 5.225–6
laws/legal system (including Republic): 3.290, 4.2, 4.76, 4.344, 7.255
literature 5.225–6, 5.242, 5.326, 5.431, 5.442–3, 7.172
Literary Revival 5.226, 5.442–3, 5.497
see also under drama; novels; poetry
mathematician 1.151
myths and legends 5.225, 9.10, 9.13, 9.23, 9.28
Tuatha Dé Danann 9.9, 9.22, 9.36
philosophy 3.40, 7.30–1
physical geography and geology 1.173, 1.300
physicist 1.151
politics and government (including Republic):
Black-and-Tans 4.40
Catholic emancipation 3.65
Clare election 4.76
and Cromwell 3.180
electoral system 7.250
Fenians 4.118
Home Rule 4.155–6
Irish Republican Brotherhood 4.167
leaders 3.145, 4.64, 4.867
Chartist 4.249–50
list of 9.70
nationalist 4.78, 4.93–4, 4.97–8, 4.112, 4.139, 4.142, 4.261–2, 4.285
prime ministers 4.78, 4.86–7, 4.97–8, 9.70
Land League 4.190
and Lloyd George 4.166
parties 4.119, 4.314
presidents 4.97–8, 4.166, 5.226, 9.70
Prime Ministers 4.78, 4.86–7, 4.97–8, 4.119, 4.166, 9.70
regional policy 7.262
Young Ireland 4.387
prehistory 3.319
regions and provinces 3.85, 9.75
religion/theology:
bishop 3.40
Catholic emancipation 3.65
Celtic Church 3.66
missionary 3.179–80
saints 3.6, 3.49, 3.82, 3.271, 9.40, 9.41, 9.43
theologian 3.368
United Irishmen 3.365–6
see also myths and legends *above*
rulers 3.49–50, 3.179, 3.349–50
see also under Britain
sport and recreation 7.123, 7.147
Ireland (*cont.*):
technology and science 6.89, 6.123, 6.175, 6.258
wars and defence:
with France 3.366
army leaders 3.26
Boyne, battle of 3.48
Curragh incident 4.90
Drogheda, Siege of 3.91
Easter Rising 4.106–7
Irish Republican Army 4.167, 7.150, 7.299, 7.308
nationalist 3.358
no-go areas 7.217
from 1921 to 1937 *see* Irish Free State
see also Ireland, Republic of; Irish Free State; Northern Ireland; Ulster
Ireland, Republic of (from 1937) 4.166
abortion illegal 7.1
aid from 7.8
assassination 4.269
in Council of Europe 7.71
economy and trade 7.107, 7.221, 7.320
see also European Community; European Economic Community
language *see under* Ireland
laws/legal system *see under* Ireland
literature *see under* Ireland
myths *see under* Ireland
neutrality 7.214
politics and government *see under* Ireland
religion *see under* Ireland
sport and recreation 7.123, 7.147
wars and defence *see under* Ireland
Irene 3.180
Ireton, Henry 3.180
Irgun 4.166–7
Irian Jaya 4.321
iridium 1.106, 1.173, 6.186
Irigoyen, Hipólito 4.166
iris family 2.119, 2.136, 2.173, 2.218
Irish elk 2.173
Irish Free State (1921–37) 4.86–7, 4.98, 4.166
see also Ireland, Republic of
Irish Literary Revival 5.226, 5.442–3, 5.497
Irish Republican Army 4.167, 7.150, 7.299, 7.308
Irish Republican Brotherhood 4.167
Irish-language literature 5.225–6
Iron Age 3.154, 3.180, 3.203–4, 3.374
arts 5.149–50, 5.208, 5.226
see also Celtic
iron and iron and steel (industry and uses) 1.106, 1.173, 6.185–8, 6.221, 6.333
alloys 6.11–12, 6.125, 6.220, 6.221
boron 6.43
carbon 6.11
cerium (Auer metals) 6.25
chrome 6.73
chromium 6.11, 6.46, 6.235, 6.330, 6.333
magnetic 6.170, 6.212
manganese 6.214, 6.333
molybdenum 6.333
nickel 6.241, 6.330
pioneer 6.235
silicon 6.333
tantalum 6.345
tungsten 6.11, 6.235, 6.333, 6.365
vanadium 6.333, 6.370
arts (ironwork) 5.226–7
basic oxygen process 6.31, 6.185
Bessemer converter 6.34, 6.185, 6.186, 6.188, 6.190
Iron and iron and steel (industry and uses) (*cont.*):
blast-furnace 6.39–40, 6.186, 6.297
bridges 6.44, 6.47, 6.111, 6.112, 6.187, 6.188, 6.280, 6.293, 6.297
in buildings 5.226–7, 6.50, 6.51, 6.90, 6.112
cable 6.53
casting *see* casting and founding
coke first used in smelting 6.96
corrosion 6.90
corrugated 6.90
crucible process 6.172, 6.186
electric-arc process 6.113, 6.185
electroplating 6.73
extrusion 6.188
ferromagnesian minerals 1.121
forging 6.144, 6.188
foundry *see* casting and founding
frame 6.297
galvanizing 6.149
high-speed 6.11
Indian 6.178
Kelly's process 6.186, 6.190
laboratory equipment 6.193
laminated 6.193
lime used in 6.202
magnetic 6.170, 6.212, 6.213
see also magnet/magnetism
mining 1.174
open-hearth process 6.185, 6.188, 6.250, 6.319
ores 1.149, 1.203, 1.239
pig-iron 6.90, 6.186, 6.285
pioneers 6.214
prefabrication 6.280
prehistoric use 6.280
puddling 6.90, 6.186, 6.285
quenching 6.287, 6.290
rolling mill 6.302
scaffolding 6.308
Schuman Plan 4.305
shell 6.316
sherardizing 6.316
ships and boats 6.49, 6.111, 6.188, 6.385
stainless *see* stainless steel
stove 6.89
tempering 6.351
in thermocouple 6.354
tools 6.4, 6.72, 6.333, 6.357
warship 6.111, 6.188
weapons 6.225
welding 6.188
wheel 6.384
wrought iron 6.90, 6.186, 6.188, 6.285
see also iron compounds *and also* iron and steel *under individual countries in particular* Britain
Iron Bridge 6.187, 6.188, 6.385
iron compounds:
iron aluminium silicate 1.320
iron nickel sulphide 1.250
iron pyrites 1.270
iron silicate 1.141
iron sulphate 6.186
iron sulphide 1.205, 1.250, 1.270
see also iron oxide
Iron Curtain 4.167
'iron lung' 6.298
Iron Mask, Man in 3.225
iron oxide 1.174
and alloys 6.235
ferrite 6.127
limonite 1.195
magnetic tape 6.213
magnetite 1.203, 6.127, 6.186
pigments 6.186, 6.270
rust 1.288
in welding 6.383
Ironclads 4.167, 6.111, 6.188
Ironsides 3.253
irony 5.227

Iroquois Indians 3.180, 7.166, 9.15, 9.16
irradiation (food) 6.188
irrational number 1.174
Irrawaddy 1.174
Irredentism 4.167–8
irrigation 6.7, 6.56, 6.95, 6.161, 6.188, 6.193, 6.378, 7.6
Irving, Sir Henry 5.227
Irving, Washington 5.227
Irwin, James 8.128
Irzykowski, Karol 5.227
'Isa (Jesus) 9.24
Isaac 9.2, 9.18, 9.19
Isaac Newton telescope 8.75
Isaacs, Alick 6.183, 6.188
Isabella I, queen of Castile (1451–1504) 3.180–1
  and Columbus 3.82
  and Granada 3.86, 3.335
  and Inquisition 3.336
  as wife of Ferdinand V 3.19, 3.62, 3.125, 3.335
Isabella II, queen of Spain (1830–1904) 3.46, 4.62, 4.168
Isabella of France (1292–1358) 3.181
Isaiah 3.99, 3.181, 9.8, 9.18
Isandhlwana, battle of 4.168
Ischia 1.174
ISDN (Integrated Services Digital Network) 6.346
Iseult, queen of Ireland 9.36
Iseult of White Hands, princess 9.36
al-Iṣfahānī, Abū al-Faraj 5.249
Ishaq see Isaac
Isherwood, Christopher 5.26, 5.227
Ishmael see Ismā'īl; Ismā'īlis
Ishtar 9.5, 9.14, 9.18
isinglass 6.133
Isis 9.5, 9.18, 9.28
Iskander Shah, prince of Malacca 3.217
Islam/Muslims 3.181, 4.168, 7.166, 7.167
  Africa: map 3.4
    see also under religion under Africa
  afterlife 7.6
  Allah 7.10, 7.133, 7.166, 7.209
  rise of 3.75
  arts 5.227–30
    architecture and carving 5.227–9, 5.420, 5.438
      Abbasid 5.1
      mosques 3.207, 3.208, 7.207, 7.208
      Mughal 5.307, 5.307
      Ottoman 5.333–4
      Safavid 5.397, 5.399
      Seljuq 5.410–11
      Timurid 5.455–6
      Umayyad 5.465–6
    Asia 3.2, 7.166
      see also under individual countries
    ceramics and pottery 5.207, 5.228, 5.229, 6.278
    decorative 5.18, 5.230, 5.344
    Moorish 5.303, 5.304
    music 5.229–30, 5.313
    painting 5.228–9, 5.230
    Persian 5.344
    see also Arabic and also literature below
  asceticism 7.19
  'āshūrā' 7.19–20
  Asia 3.2
    see also under India
  assassins 3.25
  astronomy 6.55, 8.75
  āyātollāh 7.24
  Bábism 4.27
  blasphemy 7.32, 7.33
  calendar 6.54, 7.166, 8.19
  caliphs 7.39
    see also 'Abu Bakr; 'Ali; 'Umar
  caste system 7.41
Islam/Muslim (cont.):
  circumcision 7.50
  cremation 7.75
  dance 5.229–30
  deity see Allah above
  dervish see dervish
  and divorce 7.93
  Druze 4.104, 7.96, 8.96
  education 7.265, 7.275
  Five Pillars 7.121, 7.282
  fundamentalism 7.166, 7.168, 7.331
  God 7.10, 7.133, 7.166, 7.209
  ḥadīth 7.138
  ḥajj, see Mecca
  heaven 7.140
  hell 7.141
  hijrah 7.142
  holy cities 3.187, 3.228, 9.22
    see also Mecca
  holy war 3.188, 7.171
  'Īd ul-Aḍḥā 7.152
  imāms 7.153, 9.3, 9.17
  Ismā'īlis 3.181, 7.168
  jihād 3.188, 7.171
  Ka'ba see Mecca
  Khilafat Movement 4.182
  Kurds 4.188, 7.178
  language see Arabic
  laws/legal system 7.282, 7.302, 7.309
    alcohol abuse 7.10
    punishment 7.40, 7.255
    'ulamā' 7.319
  literature 5.18–19, 5.230–1, 9.17, 9.44
    Koran see Koran
    Persian see under Iran
    Sufi see under Sufism
    Turkish 5.462
    Urdu 5.466–7
  marriage 7.193
  martyrs 7.194
  Mawlid an-Nabi 7.195
  metaphysics 7.10, 7.24, 7.86, 7.301
  Middle East see religion under individual countries
  millenarianism 7.202
  missionaries 7.203
  modernism 7.168
  mosques 5.6, 5.227–8, 7.207, 7.208
  muezzin 7.208
  music 5.229–30, 5.313
  Muslim Brotherhood 7.166
  Muslim League 4.228
  mysticism 3.344, 7.10, 7.24, 7.86, 7.301
  mythical figures and places 9.2–39 passim
  and old age 7.222
  philosophy 3.30, 7.10, 7.23–4
  pilgrimage see Mecca
  politics 4.302–3, 7.265
  predestination 7.245
  prophets 7.250, 9.2, 9.17, 9.18, 9.20, 9.29, 9.32
    see also Muhammad
  Purdah 7.255
  Qur'ān see Koran
  Ramadan 7.258
  resurrection 7.266
  rulers and dynasties 3.369, 9.66
    Aga Khans 4.5
    emirates 3.115, 4.128
    Khalji dynasty 3.195
    Umayyad dynasty 3.365, 5.465–6
    viziers 3.375
    see also Abbasid; caliphates; Fatimid; imām; Mahdi and under rulers under India
  sacred places see holy cities above
  sacrifice 7.271, 7.272
  saints 7.272, 9.3
Islam (cont.):
  Sanusi 4.302
  Sarekat 4.302–3
  Shī'ī see Shī'ī Islam
  sin 7.225, 7.286
  Spain 3.240, 3.350
  spirits 9.20, 9.29
  Sufi see Sufism
  and suicide 7.301
  Sunnī see Sunnī Islam
  technology and science 6.19, 6.54, 6.55, 6.215, 6.278
    see also under Arabs
  texts see literature above
  traders and mercenaries 3.53
  in former USSR 7.282, 7.284
  Wahhābī 4.369–70, 7.328
  wars and defence 3.55, 3.216
    jihād 3.188, 7.171
    see also Crusades
  and women 7.255, 7.332
  see also Arabs; Muhammad
islands 1.174
  Aegean Sea 1.282, 1.317
  arcs 1.174–5, 1.261
  Atlantic see under Atlantic Ocean
  Australia 1.180, 1.230, 1.331–2
  barrier 1.33
  Britain 1.153, 1.241, 1.295, 1.301, 1.318
  Canada 1.30, 1.105, 1.335, 1.353, 1.357
  Channel 1.58
  coral see under coral
  Europe see under individual countries
  France 1.352
  Germany 1.127, 1.153
  Greece 1.82, 1.172, 1.282
  groups of see archipelago
  Indian Ocean see under Indian Ocean
  Indonesia see Indonesia
  Italy 1.174, 1.196
  Japan 1.156–7, 1.185
  Madagascar see Madagascar
  Mediterranean see under Mediterranean
  mythical 3.27, 9.6
  Netherlands 1.127, 1.359
  New Zealand 1.320
  Norway 1.317, 1.328
  Pacific see under Pacific Ocean
  Persian Gulf 1.30
  Tyrrhenian Sea 1.101
  United States 1.7, 1.124, 1.199
Ismā'īl 7.168, 9.2, 9.18, 9.32
  see also Ismā'īlis
Ismail I, king of Persia 3.181, 3.277
Ismail, king of Egypt 4.27, 4.138, 4.332
Ismail Safavi 3.37
Ismā'īlis (Shiite Islam) 3.181, 7.168
isobar 1.175
isolationism 4.168–9
isomerism 1.175
isomorphism 1.175
Isonzo battles 4.169
isorhythm 5.230
isostasy 1.175–6
isothermal change 1.176
isotopes see radioactive elements and isotopes
isotropy 8.75
Isozaki, Arata 5.234
Israel (Jacob) see Jacob
Israel (nation) 1.176, 3.181, 4.169–70, 7.360, 9.18
  aid to 7.8, 7.9
  city see Jerusalem
  economy and trade 7.69
  education 7.245
  family policy 7.116
  immigration 7.152, 7.171
  kibbutzim 4.183
  languages 7.335
  law 7.305
Israel (nation) (cont.):
  map 4.169
  Palestinians in 7.227
  physical geography and geology 1.176
    Dead Sea 1.85
    desert 1.225
    lake 1.130
    river 1.178
    wadi 1.359
  politics and government:
    coalition 7.54
    intelligence services 7.162
    leaders 4.33–4, 4.216, 4.373
    parliamentary 7.229
    and religion 7.265
    see also Zionism
  religion/theology:
    Baha'i 7.25
    Druze 7.96
    Falasha 7.114
    Hasidim 7.138
    Pharisees 3.278
    and politics 7.265
    Sadducees .3.313
    Sephardim 7.280
    see also Christianity; Jews and Judaism
  rulers see in particular David, King of Israel
  technology and science 6.143
  Twelve Tribes of 9.18
  and UN 7.321
  urban areas 3.314
    Jerusalem 3.34, 3.143, 3.187, 3.265, 7.59
  wars and defence:
    armed forces 7.66
    arms trade 7.18
    conventional 7.69
    intelligence services 7.72
    and Jerusalem 7.59
    militia 7.202
    Six-Day 4.315–16
    special armed services 7.297
    Suez 4.332
    weapons 7.220
    Yom Kippur 4.169, 4.386, 6.345, 7.69, 7.224
  see also Judah; Palestine
Israfil 9.18
issuing house 7.168
Istanbul see Constantinople
isthmus 1.176, 1.250
IT see technology under information
Italian Campaign 4.170
Italian Somaliland see Somalia
Italic languages 7.156, 7.157, 7.168
  Latin see Latin
  see also languages under France; Italy; Portugal; Romania; Spain
Italy 1.176, 3.181, 4.170, 7.360
  arts 5.75, 5.83, 7.209
    decorative 5.366
    drawing 5.261
    etching 5.339–40, 5.350, 5.391
    set design 5.345
    see also under architecture; cinema; frescoes; painting; sculpture and also literature; music below
  astronomy 8.31, 8.147
    asteroids 8.9, 8.24
    Bruno 3.51, 5.231, 8.18
    comets 8.38, 8.194
    Schiaparelli 8.19, 8.87, 8.118
    stars 8.3, 8.100, 8.146, 8.177
    see also Galileo
  buildings 3.59, 6.51, 6.308
  chemistry and chemists 1.50, 6.104, 6.236, 6.242, 6.314
  cities and city-states 1.56, 3.139, 3.231, 7.288
    south and central 3.164, 3.245, 3.287–8, 3.309–10

Italy (*cont.*):
city and city-states (*cont.*):
see also Florence; Roman empire; Venice; *and also* regions *below*
clocks **6.**77
colonies **4.**4, **4.**306–7, **7.**316
computers **6.**84
crime **7.**53, **7.**225
economist **7.**104, **7.**228–9
economy and trade:
in Group of Seven **7.**136
industrial relations **7.**304
in OECD **7.**221
see also European Community; European Economic Community
education **7.**206
energy and fuel **6.**78, **6.**95, **6.**155, **6.**374
engines **6.**333
ethnic groups **3.**371, **3.**374
explorers **1.**47, **1.**70, **1.**355
foreign policy:
aid from **7.**8
Council of Europe **7.**71
Munich Pact **4.**227
Rapallo Treaties **4.**283
Stresa Conference **4.**330–1
Triple Alliance **4.**352
Washington Conference **4.**372
Western European Union **4.**375
government *see* politics and government *below*
historiography **3.**149, **5.**231
immigration **7.**105, **7.**153
languages **7.**10, **7.**156, **7.**269
laws/legal system **7.**76, **7.**173
literature:
essays **5.**82
journalism **7.**216
Romantic **5.**261, **5.**281
see also under drama; novels; poetry; Renaissance; stories
map **4.**290
Masonic Order **7.**195
mathematicians **1.**249, **1.**340–1
medicine and health care **6.**116, **6.**248, **6.**328, **7.**116
metallurgy **5.**472
modern food **6.**46
music:
ballet **5.**121
dance **5.**46–7, **5.**49, **5.**83, **5.**121
organ **5.**83, **5.**164, **5.**168, **5.**336
piano **5.**367, **5.**373, **5.**393–4, **5.**409
Romantic **5.**43, **5.**491
violin **5.**113, **5.**172, **5.**335, **5.**447, **5.**478
see also under composers; opera; sacred music; songs
nuclear **6.**127
optics and glass **6.**156
paper **6.**256
philosophy **3.**18, **3.**51, **5.**124, **5.**261, **7.**295
physical geography and geology ***1.**176*
islands **1.**174, **1.**196
lakes **1.**132, ***1.**176*
mountains **1.**10, **1.**16, **1.**207, **1.**220
rivers **1.**260, **3.**312
volcanoes **1.**114, **1.**356, **3.**164, **3.**287–8
physicist **1.**340–1
see also Galileo; Volta
politics and government:
Carbonari **4.**62
Christian Democracy **7.**47
communist **4.**139, **4.**346
electoral system **7.**250
Fascists **4.**41, **4.**76, **4.**78, **4.**228, **7.**118
historian **3.**149
independence **4.**66

Italy (*cont.*):
politics and government (*cont.*):
Irredentism **4.**167–8, **4.**352
Machiavelli **3.**44, **3.**215, **5.**273, **7.**188, **7.**244
Marxist **7.**141
nationalist **3.**374, **7.**212
parliamentary **7.**229
prime ministers **4.**88–9, **4.**94, **4.**136, **4.**212, **4.**224
revolutionary **4.**62, **4.**214, **4.**215, **4.**252
Risorgimento **4.**290–1
socialist **4.**213, **4.**317
theorist **3.**215
women's suffrage **7.**332
Young Italy **4.**387
see also rulers *below*
pottery **6.**278
radio **6.**215
regions:
list of ***9.**75*
north **3.**209, **3.**231, **3.**317–18, **3.**363, **4.**269
south and central **3.**119, **3.**269, **3.**363, **3.**365, **4.**312, **4.**355, **4.**363
see also cities *above*
religion/theology **3.**7, **3.**188, **3.**302
leaders **3.**124, **3.**317, **3.**331–2
see also papacy
saints **9.**40–3
see also individual names
rulers **4.**269, **4.**364–5
listed ***9.**68*
see also Victor Emanuel II *and also* politics *above*
safety **6.**248
sociology **7.**135
sport and recreation **7.**77, **7.**147, **7.**208
technology and science **6.**30, **6.**122, **6.**199, **6.**249, **6.**346
scientific society **6.**1
see also entries above and below prefixed **1.** *and* **6.**
urban areas *see* cities *above*
vehicles and transport:
aircraft **6.**138, **6.**167
bridge **6.**47
railways **6.**169, **6.**294
roads **6.**234, **6.**300
ships **6.**149, **6.**173
tunnels **6.**280, **6.**365
wars and defence:
Austrian Succession **3.**29–30
with Ethiopia **4.**3
leaders **3.**124, **4.**27, **4.**131, **4.**140
mercenary soldiers **3.**84–5, **3.**324–5
Napoleonic **4.**369
Red Brigades **4.**284
Revolutions of 1848 **4.**288–9
terrorism **7.**308
weapons **6.**226
in WEU **7.**330
World War I **4.**62, **4.**169, **4.**369
World War II: Italian Campaign ***4.**170*
see also Axis
see also NATO
see also Roman empire; Sicily
ITCZ (intertropical convergence zone) ***1.**171*
iterative methods ***1.**176*
Ithaca ***3.**181–2*
Ithaca Chasma **8.**171
Ito Hirobumi ***4.**170*, **4.**215
ITU *see* International Telecommunications Union
Iturbide, Agustín de **4.**156, **4.**160, ***4.**170*, **4.**219–20
IUCN (International Union for Conservation of Nature and Natural Resources) **7.**106

IUD (contraception) **2.**75, **6.**87
Ivan I, prince of Russia **3.**243
Ivan III (the Great) of Russia ***3.**182*, **3.**243, **3.**259
Ivan IV (the Terrible), tsar of Russia **3.**48, ***3.**182*, **3.**204, **3.**208, **3.**243, **3.**259, **3.**312
Ives, Charles ***5.**231*
Ives, James M. **5.**120
IVF, see *in vitro* fertilization
Ivory Coast *see* Côte d'Ivoire
ivorywork ***5.**231*
ivy **2.**68, ***2.**173*
Ixchel **9.**18–19
Izanagi and Izanami ***9.**19*
Iznik ware ***5.**231*

J

Jaafar, Dato Onn bin **4.**359
Jabal al-Tariq **3.**141, **3.**350
jacamar ***2.**174*
jacana ***2.**174*
jacaranda ***2.**174*
Jack, K. **6.**318
jackal ***2.**174*
jackdaw ***2.**174*
jackrabbit ***2.**174*
Jackson, Andrew **4.**15, **4.**95, **4.**121, ***4.**171*
and Indians **4.**162, **4.**308
and kitchen cabinet **4.**184
and spoils system **4.**326
and War of 1812 **4.**239
Jackson, A. Y. **5.**73
Jackson, Charles **6.**233
Jackson, Michael **5.**358
Jackson, Robert Houghwout ***4.**171*
Jackson, Thomas Jonathan ('Stonewall') ***4.**171*, **4.**368
Jackson, William **1.**151, **3.**130
Jacob **3.**181, **9.**11, **9.**18, ***9.**19*, **9.**23, **9.**30, **9.**31
Jacobins **3.**129, ***3.**183*
Jacobites and Rebellions ***3.**183*, **3.**197, **3.**214–15
Fifteen **3.**126–7, **3.**326
Forty-Five **3.**123, **3.**128–9, **3.**291
Jacobsen, Arne ***5.**232*
Jacobsen, Jens Peter **5.**404
Jacobsen, Jurgen-Frantz **5.**404
Jacquard, Joseph-Marie: Jacquard loom **6.**84, **6.**169, ***6.**189*, **6.**208
Jacquerie ***3.**183*
jade ***1.**177*
carving ***5.**232*
Jade Emperor **9.**39
jaeger ***2.**174*
Jael ***9.**19*
Jafar, Mir **3.**285
Ja'far Abū Ma'shar **8.**75
Jafar al-Sadia/Ja'far Sādiq, Imām **3.**181, **7.**282
Jagannatha ***9.**19*
Jagiellon dynasty ***3.**183*, **3.**286
jaguar ***2.**174*
Jahan *see* Shah Jahan
Jahangir ***3.**183*, **3.**235, **5.**307
al-Jahiz **5.**19
Jainism ***3.**183*, ***7.**169*
ahimsā **7.**8, **7.**169
art ***5.**232*, **5.**373
caste system **7.**41
deities **9.**26
literature ***5.**232*
monasticism **7.**205
philosophy **7.**155
reincarnation **7.**263
Jaipur ***3.**183–4*
Jaja of Opobo ***4.**171*
Jamaica ***1.**177*, ***3.**184*, ***4.**171*, ***7.**360*
independence **4.**376
infant mortality **7.**159
leaders **4.**57, **4.**131–2
medicine and health care **7.**159
physical geography and geology **1.**177
religion/theology **7.**258
see also Caribbean; CARICOM
Jamal Pasha **4.**387
James I, king of Aragon **3.**19
James I, king of England (and James VI of Scotland) (1566–1625) **3.**50, ***3.**184*, **3.**223, **3.**343
and Divine Right **3.**104

James I (cont.):
Gunpowder plot against 3.150–1
Hampton Court Conference 3.155
and Ireland 3.365
and parliament 3.3
and religion 3.41, 3.89, 3.295
and taxes 3.94
James I, king of Scotland (1394–1437) 3.184, 3.343
and arts 5.294, 5.295, 5.311
James II, king of England and Scotland (1633–1701) 3.152, 3.184, 3.254, 3.343
and America 3.251
battles 3.239, 3.320
and Bill of Rights 3.41
Declarations of Indulgence 3.100
exclusion crisis 3.121
Glorious Revolution 3.14, 3.142
and Ireland 3.209
Monmouth's Rebellion against 3.239
James II, king of Scotland (1430–60) 3.48, 3.184, 3.343
James III, king of Scotland (1452–88) 3.184–5, 3.343
James IV, king of Scotland (1473–1513) 3.128, 3.185, 3.343
James V, king of Scotland (1512–42) 3.185
James VI, king of Scotland see James I, king of England
James (the Just/the Less) 9.5, 9.19
James (the Great), Saint 9.5, 9.19, 9.41
James, Frank 4.171
James, Henry 5.232–3, 5.240
James, Jesse 4.171, 4.378
James, P. D. 5.129
James, William 7.105, 7.169, 7.244, 7.252, 7.278
James River 1.177
Jameson, Leander Storr: Jameson raid 4.171–2
Jamestown 3.185
Jāmī, 'Abd al-Rahmān 5.233, 9.21
Jamīl 5.19
jamming, radar 6.118–19
Jamshid 9.19
Janáček, Leoš 5.233
Janata Party 4.172
Janissaries 3.185
Jansen, Cornelius Otto 3.185
Jansky, Karl Guthe 8.76, 8.133
Janssen, Zacharias 6.223
Jansson, Tove 5.91
Janus (mythology) 9.19
Janus (satellite of Saturn) 8.29, 8.76
Japan 1.177, 3.185–6, 4.172, 7.169, 7.360–1
adoption 7.3
agriculture 6.34, 6.68, 6.76, 6.351
aid from 7.8
arts 5.233–7
architecture 5.233–4, 5.446
and Buddhism 5.236
cinema 5.252
decorative 5.317
painting 5.207, 5.209, 5.236–7, 5.412, 5.465, 5.466, 5.467
print-makers 5.158, 5.207, 5.209, 5.237, 5.467
sculpture and carving 5.237
visual 5.236–7
see also ceramics; literature; music below
astronomy 8.64, 8.147
buildings 6.50, 6.51
ceramics and pottery 5.234–5, 5.237, 5.246, 5.258–9, 5.373, 6.67
chemistry 6.344
city 3.193
colonialism 3.107, 4.321, 7.316
computers 6.129
Japan (cont.):
Constitutions 4.172, 4.215–16
cosmetics 6.91
crime 7.225
divorce 7.93
economy and trade 4.140
fishing 7.296
Greater East Asia Co-Prosperity Sphere 4.140
in Group of Seven 7.136
industrial action 7.157
industrial relations 7.314
Industrial Revolution 6.178
iron and steel industry 6.170
market research 7.192
migrant labour 7.200
multinationals 7.209
in OECD 7.221
paternalism 7.231
post-industrial 7.309
quality control 7.256
robotics 6.301
treaties and treaty ports 4.108, 4.178, 4.267, 4.350
zaibatsu 4.388, 7.336
see also specific industries above and below
education 6.364, 7.20, 7.78, 7.185, 7.275, 7.276
electronics 6.103
energy and fuel 6.381
environment 7.2, 7.260
ethnic group 7.9
feudalism 3.96, 4.92
finance 7.155, 7.299–300
fishing 6.133
food and drink 6.18, 6.46, 6.176
foreign policy:
Anglo-Japanese Alliance 4.13
Anti-Comintern Pact 4.15
China–Japan Peace and Friendship Treaty 4.73
Edo Treaty 4.108
Japan–United States Security Treaty 4.172
Kanagawa Treaty 4.178
Twenty-One Demands 4.356
Washington Conference 4.372
government see politics and government below
immigration 7.200
industry see economy above
infant mortality 7.159
language and alphabet 5.235, 7.169, 9.47
laws/legal system 7.51, 7.78, 7.115, 7.174
literature 5.37, 5.158, 5.235–6, 5.304, 7.216
drama 5.235, 5.236
poetry 5.236
see also under poetry
stories 5.235, 5.445
see also under novels
mathematics 6.1
medicine and health care 6.177, 6.191
monarchy 7.97, 9.66
music 5.236, 5.313, 5.445
occupation 4.172
pacifism 7.227
physical geography and geology 1.177
fumaroles 1.129
islands 1.156–7, 1.185
seas 1.168–9, 1.237
volcano 1.129
physicists 1.258, 1.372
politics and government:
Constitutions 4.172, 4.215–16
corruption 7.71
emperor 4.152
Fujiwara family 3.135
Hojo family 3.167
Liberal Democrats 4.194–5
Meiji Constitution 4.215–16
Japan (cont.):
politics and government (cont.):
Meiji Restoration 4.216
prime ministers see under prime ministers
reformer 4.221
revolutionary 4.184
shogunates see shogunates
religion/theology:
characters and myths 9.20, 9.7, 9.36, 9.39
Buddhism see under Buddhism
Protestantism 7.13
see also Shinto
rulers 7.97, 9.66
emperors 3.24, 3.231, 4.31, 4.152, 4.216, 4.346, 4.367
see also shogunates
science see technology and science below
society 7.169
sport and recreation:
martial arts 7.193, 7.194
softball 7.295
table tennis 7.305
wrestling 7.333, 7.334
technology and science 6.72, 6.217
see also entries above and below prefixed 1. and 6.
telecommunications 6.349, 6.390, 7.258, 7.307
Tokyo Trials 4.346
tools 6.308
trade see economy and trade above
and UN 7.277
urban areas 7.197
vehicles and transport 6.250
bridges 6.341
railways 6.161–2, 6.168, 6.293, 6.294
roads 6.64, 6.234, 6.235
space 6.196, 6.326
tunnels 6.365
wars and defence 4.385
army leaders 4.300, 4.346, 4.385–6
in Axis 4.26
in Manchuria 4.209, 4.227
military coup of 1936 4.160
military technology 6.40, 6.41, 6.165, 6.343
post-war occupation 4.172
Russo-Japanese 4.275, 4.297–8
Sino-Japanese 4.312, 4.314–15
war crimes 7.328
warriors and soldiers 3.165, 3.231–2, 3.260
see also samurai
World War II:
atomic bombs dropped 4.152, 4.230
kamikaze aircraft 4.178
Pacific Campaigns 4.254, 4.255
peace treaty 4.302
Singapore, fall of 4.314
post-war occupation 4.172
see also Axis
whaling 6.384
writing see language above
Japan, Sea of 1.177
Japan Current 1.185
japanning 5.253
Japheth 9.27
Japji 7.169
japonica 2.174
Jaques-Dalcroze, Emile 5.238
jargon 7.169, 7.171
Jarry, Alfred 5.238
Jaruzelski, Wojciech 4.272, 4.318, 4.370
jasmine 2.174–5
Jason 9.5, 9.19
jasper 1.57
Jasper National Park 1.177
Jasperware 5.485
Jātakas 5.238, 9.19
Jatavarman Sundara 3.268
jaundice 2.175
Jaurès, Jean 4.173
Java 1.177, 3.186
agriculture 6.34
decolonization 4.321
Hindu empire in 3.217
Javanese 7.23
physical geography and geology 1.177
port 3.35
prehistory 3.120, 3.186
royalty 4.99
War 4.173
Java man 3.186
Javan, A. 6.214
jaw 2.175, 2.343
see also teeth
Jawara, Sir Dawda Kairaba 4.130
Jawlensky, Alexei von 5.238
jawless fish 2.175
hagfish 2.149
Jaymotius 2.175
lamprey 2.182
pteraspis 2.270
Jay, John 3.87, 3.124, 3.186, 3.269
jay 2.175
Jayadeva 5.401
Jayasi 5.206
Jayavarman II, king of Khmers 3.196
Jayavarman VII, king of Khmers 3.13, 3.196
Jaymotius 2.175
jazz 5.238–9
bandleader 5.144
composers 5.46, 5.112, 5.144, 5.175, 5.339
dance 5.239
Dixieland 5.131
pianist 5.144
saxophonist 5.339
singer 5.22
trumpeters 5.22, 5.126
Jean de Saint Côme 6.340
Jean le Bel 3.135
Jefferies, John 6.138
Jefferies, Richard 5.239
Jefferson, Thomas 3.186, 4.95, 4.173, 4.250, 4.287
and American Philosophical Society 3.11
as architect 7.337
and Declaration of Independence 3.100, 3.331
and Indians 4.250
and Rights of Man 3.304, 3.331
and states' rights 4.329
Jefferys of Wem, George, 1st baron 3.37, 3.43, 3.186
Jeffreys, Alec 6.105
Jehan see Shah Jahan
Jehangir see Jahangir
Jehoiachin of Jerusalem 3.121
Jehoshaphat of Judah 3.6
Jehovah 9.19
Jehovah's Witness 7.171
Jehu 9.11, 9.20
Jellicoe, John Rushworth, 1st earl 4.173, 4.177
jellyfish 2.175
Jena, battle of 4.173
Jenatzy, Camille 6.115
Jencks, Charles 5.363
Jenkins, Nicholas 5.365
Jenkins's Ear, War of 3.186
Jenner, Edward 2.175–6, 4.99, 6.176–7, 6.189, 6.218
Jenney, William Le Baron 5.91, 6.322
Jennings, Sarah (Churchill) 3.223
Jenolan Caves 1.177
Jensen, George 5.24
Jensen, Johannes Vilhelm 5.239
Jephthah 9.20
jerbil 2.135

jerboa 2.175, *2.176*
Jeremiah *3.186*
Jericho *3.187*
Jeroboam I, king of Israel *3.187*, 9.11
Jeroboam II, king of Israel *9.20*
Jerome, Saint 3.41, *3.187*, 9.33, *9.42*
Jerrold, Douglas 5.290
Jersey 1.58
Jerusalem 3.34, 3.143, *3.187*, 3.265, 3.314, 7.59
Jervis, John, 1st earl of St Vincent 3.60, *3.187*
Jesuits 3.58, 3.131, *3.187*, 3.302, *7.171*
Jesus Christ *3.187–8*, *7.171*, 7.265
ascension 7.19
and Bible 7.31
birth 7.16, 9.7, 9.24
Christmas 7.48
cross and crucifixion 3.92, 7.21, *7.76*, 7.134, 7.183, 7.225, 7.271, 9.8
see also Pilate
disciples 3.104
see also Apostles
friends 9.7, 9.22, 9.24, 9.39
Gospels 3.41, 3.57
miracles 7.203, 9.13
mother 9.24
as prophet 7.250
resurrection 7.98, 7.134, 7.266
sacraments 7.271
Second Coming 7.3, 7.47, 7.202
temptation 7.183
trial 9.29
and Trinity 7.316
Turin shroud 3.363
see also Christianity
jet engines (gas-turbine engine) 6.136, 6.150, *6.189–90*, 6.224
airliners 6.8
pioneers and development 6.138, 6.176, 6.249, 6.384–5
ram-jet 6.294
jet (lignite) *1.177*
ornaments *5.239*
Jet Propulsion Laboratory *8.76*
jet stream *1.177–8*
Jevons, William Stanley 7.53, 7.101
Jewel, John 3.66, *3.188*
Jewel Box *8.76*
jewellery *see* ornaments and jewellery
Jewish Revolt 3.35–6, *3.188*
Jew's ear fungus *2.176*
jew's harp *5.240*
Jews and Judaism *3.188*, *3.192*, *4.176*, *7.171–2*
afterlife 7.6
anti-Semitism *4.15*, 4.271, 7.171
and Nazis (Holocaust) 4.154, 4.155, 7.63–4, 7.130, 7.257, 7.336
Ashkenazi 7.19
atonement 7.21
bar mitzvah 7.26
Bible 3.41
blasphemy 7.33
calendar *7.171*, 8.19
cantor 7.39
circumcision 7.50
concentration camps 7.63–4
Conservative 7.66–7
covenant 7.73
cremation 7.75
Dead Sea Scrolls 3.99
diaspora 3.103
eschatology 7.108
ethnic groups 7.170, 7.171
exile 3.121
Exodus 3.121
Falasha 7.114
and food 7.123
ghetto 7.131
God 7.172
Hanukkah 7.138
Hasidim 7.138, 7.139
Jews and Judaism (*cont.*):
heaven 7.140
hell 7.141
historian 3.191
Jerusalem 3.187
Jewish Revolt 3.35–6, *3.188*
Judaism described 3.192, 4.176
Kabbalah 7.175
kosher food 7.178
languages 7.279, 7.335
laws 3.61, 3.316, 7.189, 7.203, 7.257, 7.305–6, 7.309
Liberal 7.183–4
literature 5.197, 5.277, 5.391, 5.420
Hebrew 5.47, 5.197, 5.202–3, 5.216
Maccabees 3.214
martyrs 7.194
messiah 7.202
Midrash 7.200
millenarianism 7.202
Misgnah 7.203
missionaries 7.203
mysticism 7.138, 7.175
mythical figures and places 9.2–39 *passim*
Orthodox 7.225–6, 7.281
in Palestine 4.144, 4.166
Passover 7.229–30
patriarchs *see* patriarchs of Judaism
Pentecost 7.232
Pharisees 3.278
philosophy 3.217, 3.339, 7.189, 7.297, 7.310
physician 3.217–18
politics and government 3.249
priests 3.122
prophets *see under* prophets
Purim 7.256
rabbi 7.257
Reform 7.260
resurrection 7.266
Rosh Hashanah 7.269–70
Rothschilds 4.293
sacrifice 7.272
Sadducees 3.313
saints 3.190, 3.271–2, 3.277
scholarship 3.61
Sephardi 7.279–80
Shabbat 7.281
sin 7.286
suicide 7.301
synagogue 7.303–4
Tabernacles, Feast of 7.305
Talmud 3.348, 7.305–6
Torah 7.312
and women 7.26–7, 7.226
year *7.171*, 7.171
Yom Kippur 7.335
Zionism 7.265, 7.336
see also Israel; Judah; Palestine; Zionism
Jezebel *9.20*, 9.26
Jhabvala, Ruth Prawer *5.240*
Jiang Qing 4.130–1
Jiāngxī (Mazu Dàoyì) 9.25
Jiangxi Soviet *4.173*
Jibril (Gabriel) 9.13
jigger 2.121
jihād *3.188*, *7.171*
'Jim Crow' laws *4.173*
Jiménez, Juan Ramón *5.240*
Jiménez de Quesada, Gonzalo 3.81, 3.252
Jimmu Tenno *9.20*
Jin (Chin) dynasty *3.188*, 3.218
Western 3.380
Jin Shengtan 5.483
jingoism *4.174*
Jinnah, Muhammad Ali 4.161, *4.174*, 4.228, 4.356
jinns 9.10, 9.14, 9.17–18, *9.20*
Jizo *9.20*
Jizo Bosatsu 9.20
Joab 9.2
Joachim, Joseph *5.240*
Joachim of Fiore *3.188*
Joan of Arc, Saint 3.38, 3.70, 3.171–2, *3.189*
Joanna II, queen of Naples 3.9
Joanna the Mad, queen of Castile *3.188–9*
Job *9.20*
job:
evaluation *7.171–2*
satisfaction *7.172*
see also labour and employment
jobbers 7.35
Jodhpur *3.189*
Jodl, Alfred *4.174*
Jodrell Bank and Lovell telescope *8.76*, 8.78, 8.87
Joffre, Joseph Jacques Césaire *4.174*, 4.211, 4.319
John, Saint 3.41, 3.104, 9.12, 9.20
John, king of England *3.189*, 3.202, 3.255, 3.284
and barons 3.36, 3.203
excommunicated 3.121
and Magna Carta 3.216
and Robin Hood 3.306
and taxes 3.108, 3.320
John I (the Great), king of Portugal 3.138, *3.189*
John II (the Perfect), king of Portugal *3.189*
John II, king of France 3.49, 3.111, 3.171–2, 3.286, 3.299
John III (the Pious), king of Portugal (1502–57) *3.189–90*
John III, king of Nicaea 3.253
John III, king of Poland 3.331
John III, king of Sweden 3.127
John IV, king of Ethiopia 4.206
John IV, king of Nicaea 3.231
John IV (the Fortunate), king of Portugal 3.48, *3.190*
John VI, king of Portugal (and Brazil) 4.48, *4.175*, 4.275
John VIII, pope 3.280
John XXII, pope 3.269
John XXIII, pope *4.175*, 4.291
John, Augustus *5.240*
John, Gwen 5.240
John of Austria, Don *3.190*, 3.206
John the Baptist, Saint 2.55, 3.187, *3.190*, 3.198, 9.11, *9.20*, 9.31, 9.33, 9.42
'John Bull' *3.190*
John of the Cross, Saint 3.61, 3.89, *3.190*, 5.311
John Dory *2.176*
John the Evangelist, Saint 9.12, 9.20, *9.42*
John the Fearless, duke 3.54
John Frederick I, king of Saxony 3.318
John of Gaunt *see* Gaunt
John Hunyadi of Hungary 3.172
John Paul II, pope *4.175*, 7.20
John Paul III, pope 4.272, 4.291
John of Salisbury 3.190, *3.190*
Johns, Jasper *5.240–1*
Johnson, Andrew *4.174–5*, 4.284
Johnson, D. W. 2.328
Johnson, Eyvind 5.404
Johnson, James 5.68
Johnson, Lyndon B. 4.95, *4.175*, 4.215, 4.367
Johnson, Philip *5.241*, 5.294
Johnson, Dr Samuel *5.241*, 5.402, 5.460, 7.4, 7.107
Johnson, Sir William (1715–74, fur trader) *3.190*, 3.202
Johnson, William (1966, on sexuality) 7.280
Johnson, Prince Yormie 4.195
Johnston, Edward 5.71
Johnston, Frank 5.73
Johnston, George 4.295
Johnston, Sir H. H. 4.207
Johore *3.190*
joinery *see* carpentry and joinery
joints (anatomy) *2.176*, *9.59*
disorders:
arthritis 2.15–16
orthopaedics 2.236, 6.253
physiotherapy 2.252–3
joints (technology) 6.124, 6.169, 6.369, 6.387–8
joints (geology) *1.178*, 1.301
joint-stock companies 3.71, *7.172*
Joinville, Jean de *3.190*
Jok 9.20
Jókai, Mór 5.214
Joliot-Curie, Jean-Frédéric and Irène 1.81, *1.178*, 1.274
Jolliet, Louis 1.206
Jonah *9.20*
Jonathan *9.20*
Jones, Harold Spencer 8.193
Jones, Inigo 5.140, *5.241*, 5.365, 5.412
Jones, John Gwyllm 5.486
Jones, John Paul *3.191*
Jones, Robert Edmond 5.413
Jones-Griffiths, Philip 5.482
jongleur *5.241*, 5.297
Jonson, Benjamin *5.241–2*, 5.402
Jooss, Kurt *5.242*
Joplin, Scott *5.242*
Jordaens, Jacob *5.242*
Jordan *1.178*, *4.175–6*, *7.361*
aid to 7.8
in Arab League 7.15
cities 3.278, 6.281
monarchy 7.97
Muslim Brotherhood 7.166
Palestinian refugees in 7.227
physical geography and geology 1.85, 1.178
politics and government 7.332
rulers 4.17, 4.159, 4.175–6
Jordan (river) *1.178*
Jörmungand 9.2, 9.23, 9.30, 9.35
Joseph II, Holy Roman Emperor 3.37, 3.133, 3.152, 3.153, 3.168, *3.191*, 3.300
Joseph of Arimathaea (New Testament) 9.6, 9.14, *9.21*
Joseph, Chief *4.176*
Joseph, G. F. 4.281
Joseph of Nazareth, Saint *9.20*, *9.42*
Joseph (New Testament) *see* Joseph of Arimathaea; Joseph of Nazareth
Joseph (Old Testament) 9.7, *9.20–1*, 9.30
Joseph Bonaparte 3.327, 4.265, 4.312
Joseph Ferdinand, duke of Bavaria 3.337, 3.338
Joséphine, Marie Rose Tascher de la Pagerie *4.176*
Josephus, Flavius *3.191*
Joshua 3.121, *9.21*
Josiah *3.191*
Josquin Desprez *5.242*, 5.380
Jostedalsbreen *1.178*
Joubert, Barthélemy 4.288
Joubert, Petrus Jacobus 4.41, 4.276
joule *1.178*, 6.120, 6.166, 6.368
journalism 7.38, *7.172*, *7.215*
Britain 3.381–2, 4.33, 4.125, 4.146–7, 5.127, 5.398, 5.441–2, 8.216
Canada 5.383
France 3.50, 3.102, 5.172, 5.400, 7.216
freedom of press 7.125
Hungary 5.76
Nicaragua 5.125
photo-journalists 5.76, 5.81, 5.207
Reuters 4.288
United States 5.76, 5.117, 5.207, 5.267, 5.276, 5.290, 5.488, 7.216

journeyman *3.191–2*
Jovian *see* Jupiter (planet)
Joyce, James **5**.48, **5**.215, ***5**.242*, **5**.300, **5**.436
Juan Carlos I, king of Spain **3**.46, **4**.62, **4**.124, **4**.322
Juan de Avila **5**.311
Juan de Yepes **5**.311
Juan-Juan (Avars) **3**.30
Juárez, Benito ***4**.176*, **4**.214
Juchen (Manchus) **3**.218
Judah, Rabbi **7**.203, **7**.257
Judah (person and tribe) ***9**.21*
Judah (kingdom) ***3**.192*, ***9**.21*
  fortress **3**.224
  Kings of **3**.165, **3**.191
  see also Israel; Jews and Judaism; Palestine
Judah Ha-Nasi ***3**.192*, **3**.348
Judaism *see* Jews and Judaism
Judas Iscariot **9**.5, ***9**.21*
Judas Maccabeus **3**.214, **7**.138
Jude, Saint **9**.5, ***9**.42*
judges ***7**.172–3*, **7**.279
judicial review **7**.3, ***7**.173*, **7**.245
judiciary ***7**.173*, **7**.279, **7**.318, **7**.324
Judith ***9**.21*
judo **7**.194
Judson, Edward Z.C. **4**.378
Judson, Whitcombe **6**.391
judy (butterfly) **2**.209
Jugnauth, Aneerood **4**.214
Jugurtha ***3**.192*, **3**.260
jujitsu **7**.193
jujube ***2**.176*
Julia, Gaston: Julia sets **1**.126
Julian calendar **8**.18, ***8**.76*
Julian date ***8**.76*
Julian (the Apostate), Roman emperor ***3**.192*
Juliet (Uranian satellite) **8**.178
Julius Caesar *see* Caesar
Julius II, pope **3**.44, ***3**.192*, **3**.269, **3**.301, **5**.293, **5**.378
July Plot (Germany) ***4**.176–7*
July Revolution (France) ***4**.176*
Jumbe, Aboud **4**.387
jumbo jet **6**.385
jumping plant louse ***2**.176*
Jung, Carl Gustav ***7**.173*
  and dreams **7**.95
  and extroversion/introversion **7**.114
  and myth **7**.210
  as neo-Freudian **7**.214
  and psychoanalysis **7**.252
  and symbolism **7**.303
Jungfrau ***1**.179*
junglefowl ***2**.176*
juniper ***2**.176–7*
'Junius' ***3**.192*
junk (vessel) ***6**.190*, **6**.306, **6**.307
junk (waste) *see* waste
junk bond **7**.34
Juno (asteroid) ***8**.78*, **8**.193
Juno (mythology) ***9**.21*
junta ***7**.173*
Juok ***9**.21*
Jupiter (god) ***9**.21*
Jupiter (planet) ***8**.78*
  Great Red Spot **8**.61
  illustrated **8**.77
  orbit **8**.17, **8**.196
  ring **8**.140, **8**.197
  satellites **8**.4, ***8**.76–7*, ***8**.198*
    small **8**.2, **8**.98, **8**.171
    small outer **8**.45, **8**.67, **8**.76–7, **8**.88, **8**.117, **8**.150
    see also Galilean
  statistical details **8**.196
Jura Mountains ***1**.179*
Jurassic period **1**.134, **1**.135, ***1**.179*
jurisdiction ***7**.173–4*
jurisprudence ***7**.174*
jury ***3**.192–3*, **7**.112, ***7**.174*
Jussieu, Antoine Laurent de ***2**.177*
justice ***7**.174*
  natural **7**.3, **7**.213
  social **7**.291
  see also law
Justinian I, Byzantine emperor **3**.38–9, **3**.55, ***3**.192–3*, **6**.95, **7**.269
Justo, Agustín Pedro ***4**.177*
jute ***2**.177*, **6**.16, **6**.53, **6**.364
Jutes (people) **3**.98, **3**.160, ***3**.193*
Jutland ***1**.179*
  battle of ***4**.177*
Juvenal ***5**.242*
juvenile court ***7**.174*
juvenile delinquency ***7**.174*

K

K2 ***1**.179*
Ka'aba ***7**.175*, **9**.21
Kaang ***9**.21*
Ka'ba ***9**.21*
Kabbalah ***7**.175*
Kabil (Cain) **9**.2, **9**.8
Kabīr ***3**.193*, **5**.46, **5**.206
Kabuki **5**.27, **5**.235, **5**.236
Kabwe (Broken Hill) Man **3**.50
Kabylie ***1**.179*
Kachin ***7**.175*
kachina ***9**.21*
Kádár, János **4**.158, ***4**.178*, **4**.230
Kafka, Franz ***5**.243*, **5**.485
kagu ***2**.177*
Kagutsuchi **9**.19
Kahn, Louis **5**.14, ***5**.243*, **5**.294
Kahn, Richard **7**.209
Kaiser, Georg ***5**.243*
Kaiumers **9**.14
kala-azar **2**.293
Kalahari ***1**.179*
kale **2**.49
Kālī **7**.142, ***7**.175*, **7**.281, **7**.285, **9**.10
Kālīdāsa ***5**.243*, **5**.401
Kalila and Dimna ***9**.21–2*
Kalimantan **4**.321
kalimba **5**.400
Kalki **9**.37
Kallinikos **6**.161
Kallman, Chester **5**.436
Kalmar, Union of ***3**.193*
Kalmyk language **7**.11
Kalvos, Andreas **5**.191
Kama **9**.24
Kamakura shogunate ***3**.193*, **3**.231–2
Kamchatka ***1**.180*
  Current ***1**.180*
kame **1**.140, ***1**.180*
Kamenev, Lev Borisovich ***4**.178*
kami ***9**.22*
kamikaze ***4**.178*
al-Kamil, Sultan of Egypt **3**.131
Kampan **5**.446
Kampuchea *see* Cambodia
Kanada **7**.95
Kanagawa, Treaty of ***4**.178*
Kanak ***7**.175*
Kanakas ***4**.178*
Kanchenjunga ***1**.183*
Kandinsky, Wassily **5**.2, ***5**.243–4*
Kane **9**.35
Kane, Paul **5**.73
Kanellopoulos, Kanellos **6**.138, **6**.172
Kanem-Bornu **3**.4, ***3**.193*
al-Kanemi, Muhammad **3**.193
Kang Youwei **4**.158, ***4**.178*
Kangaroo Island ***1**.180*
kangaroo rat (rodent) ***2**.178*
kangaroos ***2**.177*, **2**.278, **2**.347, **2**.359
Kangxi, emperor of China ***3**.193–4*, **3**.250, **3**.296
Kannami Kiyotsugo, Kanze **5**.236
Kannon **9**.6, ***9**.22*
Kano ***3**.194*
Kano Masanobu **5**.237
Kansas–Nebraska Act ***4**.178*
Kant, Hermann **5**.175
Kant, Immanuel ***3**.194*, ***7**.175*
  and aesthetics **7**.4
  and causation **7**.42
  and dialectic **7**.89
  and Enlightenment **3**.116, **5**.147
  and ethics **7**.109
Kant (*cont.*):
  and free will **7**.125
  and Hegel **7**.141
  and idealism **7**.276
  and metaphysics **7**.141, **7**.175, **7**.276
  and ontology **7**.223
  and punishment **7**.255
  and rationalism **7**.259
  and reason **7**.150
  and synthesis **7**.304
  and time **7**.312
  and transcendental argument **7**.315
kantele ***5**.244*
Kanze Kannami Kiyotsugo **5**.236
Kanze Zeami Motokiyo **5**.236
kaolin *see* china clay
Kaplan turbine **6**.378, **6**.379
kapok ***2**.178*
Kapp, Wolfgang: Kapp putsch ***4**.178*
Kaprow, Allan **5**.198
Kapteyn, Jacobus Cornelius ***8**.79*
Kara George, Petrović ***4**.178*, **4**.309
Karacaoglan **5**.462
Karageorgević, Peter **4**.178, **4**.387
Karajan, Herbert von **5**.110
Karakoram range **1**.9, **1**.179, ***1**.180*
Karanta, Sivarama **5**.136
karate **7**.194
Karen ***7**.175*
Karimov, president of Uzbekistan **4**.362
Karinthy, Frigyes **5**.214
Karl Theodore of Bavaria **3**.37
Karle, Guillaume **3**.183
karma ***7**.175*
Karmal, Babrak **4**.3
Karnak ***3**.194*
Károlyi, Mihály, count **4**.159, ***4**.178–9*
Karpeles, Maud **5**.161
karren ***1**.180*
Karsavina, Tamara **5**.30, **5**.340
Karsh, Yousef ***5**.244*
karst ***1**.180*, **1**.181
Karume, Sheikh **4**.389
Kašauskas, Stasys **5**.34
Kasavubu, Joseph ***4**.179*
Kashmir **1**.180, **1**.224, **4**.1, ***4**.179*, **7**.280
Kasprowicz, Jan ***5**.244*
Kassites *see* Babylon
Kästner, Erich **5**.91
katabatic wind ***1**.180*
kathak ***5**.244*
kathakali **5**.134, ***5**.244*
Kato Kimera **9**.22
Kato Komei ***4**.179*
katydid **2**.193
Katyn massacre ***4**.179*
Kauffmann, Angelica **5**.181, ***5**.244*
Kaunda, Kenneth David ***4**.179*, **4**.388
Kaunitz, Wenzel Anton, count von ***3**.194*, **3**.323
kauri pine ***2**.178*
kaval ***5**.244*
Kavir Desert **1**.146
Kawabata Yasunari **5**.236
Kawiti **4**.294
Kay, John **6**.140, ***6**.190*
kayagŭm ***5**.251*
Kaye, Dr Richard **3**.289
Kaygusuz Abdal **5**.462
Kaykobad I, king of Turkey **3**.321
Kazakhstan ***1**.180*, ***4**.180*, **7**.59, **7**.220, ***7**.361*
  physical geography and geology **1**.31, **1**.180, **1**.348
Kazan, Elia ***5**.244*
Kazantzakis, Nikos ***5**.244–5*
Kazi Nazrul Islam **5**.43
kea ***2**.178*
Kean, Edmund **5**.53, ***5**.245*
Kearny, Stephen **4**.218
Keating, Paul **4**.23, **4**.189

Keaton, Buster 5.245
Keats, John 5.245, 5.325, 5.416, 5.424
ked 2.194
keel 6.190
Keeler, James 8.140
Kefauver, Carey Estes 4.180
Keitel, Wilhelm Bodewin Johann Gustav 4.180
Kekulé von Stradonitz, August: Kekulé structure 1.36, 1.37, 1.181
Kell, Joseph 5.66
Keller, Gottfried 5.245
Kellner eyepiece 8.134
Kellogg, Frank B.: Kellogg–Briand Pact 4.180
Kelly, Gene 5.245
Kelly, George 7.233
Kelly, Ned 4.102, 5.318, 5.320
Kelly, Thomas 4.167
Kelly, William 6.186, 6.190
kelp 2.178
Kelvin, Lord William Thomson 1.332, 1.335, 6.351, 8.79
- Kelvin scale of temperature 8.79
- Kelvin time scale 8.79
- Kelvin–Helmholtz contraction 8.79

kelvin (unit) 1.181, 6.321, 6.351, 6.368
Kemal, Mustafa see Atatürk
Kemal, Yasar 5.462
Kemble, John Philip 5.245
Kemény, Zsigmond 5.214
Kempis, Thomas à 3.194, 3.195
kenaf 2.178
kendo 7.194
Kenilworth, siege of 3.194
Kennedy, Edward M. 4.180
Kennedy, Jacqueline 4.175
Kennedy, John Fitzgerald 4.95, 4.180
- assassination 7.20
- and Britain 4.205
- and Cuba 4.32, 4.90
- flexible response 7.121
- and Meredith incident 4.217
- and space exploration 6.17

Kennedy, Robert F. 4.180
Kennelly, Arthur: Heaviside–Kennelly layer 1.153
Kenneth I MacAlpine, king of Scotland 3.194
Kenneth II, king of Scotland 3.194–5
Kent, William 5.245–6, 5.337
Kenward, C. W. 6.301
Kenya 1.181–2, 4.180–1, 7.361–2
- arts 5.318
- colonialism 4.306
- decolonization 4.4
- economy and trade 4.106
- ethnic groups 7.176, 7.188
- famine prevention 7.118
- languages 7.78, 7.217
- Mau Mau 4.213–14
- national parks 7.212
- physical geography and geology 1.181–2
  - see also Great Rift Valley
- political leaders 4.181, 4.215
- population density 7.241
- prehistory 3.199–200
- refugees in 7.117
- religion/theology 7.5
- technology and science 6.34, 6.323

Kenya, Mount 1.182
Kenyatta, Jomo 4.181, 4.257
Kenzaburō, Ōe 5.236
Kenzan, Ogata 5.246, 5.373
Kepler, Johannes 3.301, 3.318, 8.11, 8.79–80
- Kepler's equation 8.80
- Kepler's Laws 8.80
- and solar parallax 8.150
- and supernova 8.106

Keppel, Arnold Joost van, 1st earl of Albemarle 3.195
Kerala 3.195
keratin:
- claws 2.67
- feathers 2.114
- hair 2.149
- horn 2.162
- nails 2.225

Kérékou, Mathieu 4.34
Kerensky, Alexander Feodorovich 4.44, 4.181, 4.297, 4.321
Keret, king of Sidon 9.11
Kern, Jerome 5.246, 5.309, 5.498
kerosene 6.145, 6.264
- see also paraffin

Kerouac, Jack 5.39
Kerr, Sir John 4.378
Kerr, Roy 8.17
Kertész, André 5.246
Keśavdās 5.206
Kesselring, Albrecht 4.170, 4.181
kestrel 2.178
Ketakada 5.43
ketch 6.190, 6.306
ketone 1.182, 6.252
Kett, Robert and William: Kett's Rebellion 3.195
Kettering, Charles 6.330
kettledrums 5.13, 5.246, 5.444, 5.455
kettle-hole 1.182
Ketuvim: books listed 9.44
Keuper 1.182
Kew Gardens 2.38
key (lock) see lock (and key)
key, musical 5.25, 5.48, 5.246–7
keyboard instruments 5.247
- accordion 5.2
- celesta 5.83
- clavichord 5.103–4
- mechanical 5.288
- spinet 5.429
- toccata 5.457
- virginals 5.477–8
- see also harpsichord; organ; pianoforte

Keyes, Roger J. B., 1st Baron 4.389
Keyhole Nebula 8.80
Keynesianism/Keynes, John Maynard 4.181, 7.175–6
- and Bloomsbury Group 5.50
- and budget 7.37, 7.101
- and consumption 7.68
- and deficit financing 7.83
- and depression 7.86
- and discretionary management 7.126
- and fiscal policy 7.120
- and income 7.160
- and interest rates 7.53
- and monetarism 7.101
- and multiplier 7.209
- and reparations 4.287
- and Toynbee Hall 4.348

Keyser, Pieter Dirckszoon and Frederick de Houtman: constellations introduced by 8.7, 8.25, 8.28–9, 8.38, 8.62, 8.71, 8.72, 8.105, 8.117, 8.119, 8.173, 8.175, 8.184
Keystone Kops 5.411
KGB 4.181, 7.162, 7.238
Khachaturian, Aram 5.247
Khadijah 9.22
al-Khadir 9.3
Khajuraho 5.247
Khalid of Saudi Arabia 4.303
Khalifa, Abdallah 4.181
Khalji dynasty 3.195
Khalsa 7.137, 7.176, 7.286
Khama, Sir Seretse 4.46, 4.181–2
khamsin 1.182–3
Khan, Aga see Aga Khan
Khan, Ayub see Ayub Khan
Khan, Fazlur Rahman 5.228
Khan, Ghaffar see Ghaffar Khan
Khan, Liaqat see Liaqat Ali Khan
Khan, Sir Sayyid Ahmad 5.466
khanate 3.195–6
- see also Genghis Khan; Kublai Khan

Khandesh 3.196
Khayyām, 'Omar 5.247
Khephren 9.34
Khepri 9.31
Khilafat Movement 4.182
el-Khizr 9.3
Khlebnikov 5.287
Khmer 3.196
- arts 5.247–8, 5.276, 5.450

Khmer Rouge 4.182, 7.20
Khoisan languages 7.176
Khoisi religion 9.15–16
Kholopov, P. N. 8.22
Khomeini, Ayatollah Ruhollah 4.165, 4.168, 4.182–3, 4.226, 5.394, 7.310
Khonsu 9.4
Khosrau I, king of Persia 3.196, 3.316
Khosrau II, king of Persia 3.196
Khrushchev, Nikita Sergeyevich 4.90, 4.182–3, 4.277, 4.355
Khun Sa 7.225
Khurasan, ruler of 3.216
Khusrau, Amīr of Delhi 5.344
Khwadja Khidr 9.3
al-Khwārizmī, Muhammad ibn-Mūsā 8.75
Khyber Pass 1.183
Ki no Tsurayuki 5.235
Kiakhta, Treaty of 3.196
kiang 2.178
kibbutz 4.183
Kicking Horse Pass 1.183
Kidd, William 3.196–7, 3.282
kidney 2.178, 9.60
- disorders and diseases 2.326, 6.101, 6.164
- urine 2.353

Kiel Canal 4.183
Kielland, Alexander 5.404
Kierkegaard, Søren Aabye 7.14, 7.113, 7.176
Kiev, Grand duke of 3.375
Kiev Rus 3.196, 3.197
Kikuyus 7.176
Kilby, Jack 6.183
Kilij Arslan II, king of Turkey 3.321
Kilimanjaro 1.183
Killiecrankie, battle of 3.197
killifish 2.178
kiln 6.46, 6.190–1, 6.276, 6.278, 6.297
kilogram 1.183, 6.321, 6.368
kilometre 6.368
Kilwa 3.197
Kim Il Sung 4.247, 7.89
kimberlite 1.182, 1.183
Kinby, Rollin 4.238
kinematics, stellar 8.159
kinetic art 5.70–1, 5.248
kinetic energy 1.183, 6.121, 6.189, 6.242, 8.80
kinetic theory 1.183
King, B. B. 5.382
King, Charles Bird 4.40
King, Clarence 1.183
King, John 1.45
King, Mackenzie 4.183, 4.195
King, Martin Luther, Jr 4.97, 4.117, 4.175, 4.183
King, Phillip Parker 4.247
king cobra 2.179
king crab 2.179
King George's War 3.134, 3.197
King Philip's War 3.75, 3.197
King William's War 3.134, 3.197–8
kingbird 2.178
kingcup 2.179
kingdoms (natural history) see fungi; plants; prokaryotes; Protoctista; virus
kingdoms (political) see kingship
kingfisher family 2.179, 2.180
Kingitanga movement 4.183
Kings Canyon National Park 1.183
kingship 7.176–7
- see also Crown; leaders; monarchy and rulers under individual countries

Kingsley, Charles 5.248
Kingsley, Mary Henrietta 1.183
kinkajou 2.179
Kinnock, Neil Gordon 4.183–4, 4.189
Kinsey Reports 7.280
kinship 7.177
- incest 7.46, 7.154
- kin selection 7.11, 7.294
- marriage see marriage
- and nomadism 7.217
- see also family; lineage and ethnic groups under individual countries

Kintu and Nambi 9.22
Kipling, Rudyard 5.151, 5.215, 5.248
Kipping, Frederick Stanley 6.191
Kiprensky, Orest 5.371
Kircher, Athanasius 3.146, 3.357, 5.5
Kirchhoff, Gustav Robert 1.183, 8.157
Kirchner, Ernst Ludwig 5.63, 5.248
Kirghizia 1.183, 1.335, 1.348
Kiribati 1.183, 3.231, 4.184, 7.296, 9.26
Kirk, Sir John 4.388
Kirk, Norman 4.184
Kirkwood, Daniel: Kirkwood gap 8.80
Kirov, Sergei Mironovich 4.184
Kirov Ballet 5.248–9
Kirsch, Sarah 5.175
Kirstein, Lincoln 5.122, 5.318
Kishi Nobusuke 4.184, 4.194
Kishibojin 9.22
Kissinger, Henry 4.97, 4.121, 4.184
Kita Ikki 4.184
Kitāb al-Aghānī 5.249
Kitasoto, Shibasaburo 6.177, 6.191
kitchen cabinet 4.184
Kitchener of Khartoum and Broome, Horatio Herbert, 1st earl 4.184
- and Boer Wars 4.43
- and concentration camps 4.82, 7.63
- and Sudan 4.181, 4.207, 4.211, 4.332
- and World War I 4.126

kitchen-sink drama 5.249, 5.486
kite family (bird) 2.160, 2.179
kites 6.191, 6.221, 7.177
kitsch 5.249
Kitt Peak National Observatory 8.79, 8.80
kitten moth 2.272
kittiwake 2.179
Kitzberg, August 5.34
Kivi, Aleksis 5.155
kiwi (bird) 2.179
kiwi fruit 2.180
Kjellen, R. J. 7.131
Klaus, Václav 4.147
Klee, Paul 5.249
Klein, Felix 1.183–4, 1.193
Klein, Melanie 7.214, 7.252
Kleist, Heinrich von 5.249
Klemperer, Otto 5.110
Klimt, Gustav 5.249
klipspringer 2.180
Klopstock, Friedrich Gottlieb 5.249
klystron 6.354
Kneller, Sir Godfrey 3.16, 5.249–50
Knickerbocker, Diedrich 5.227

Knight, Charles **4**.151
knighthood ***3**.198*
armour **3**.22, **6**.225
chivalry **3**.73–4
outstanding knight **3**.37
scutage **3**.320
squires **3**.339
Teutonic Knights **3**.352
tourney **3**.359
Knights Hospitallers ***3**.198*, **3**.303
Knights of Labor ***4**.185*
Knights of the Red Branch **9**.9
Knights of the Round Table **9**.5, **9**.8, **9**.13–14
see also Arthur, king
Knights Templar ***3**.198*
Knipper, Olga **5**.90, **5**.185
knitting **6**.125, ***6**.191*, **6**.239, **6**.351, **6**.390
knives **6**.46, ***6**.191*, **6**.357
knock (pinking) **1**.15, ***6**.191–2*, **6**.197, **6**.224, **6**.249
Knossos ***3**.198–9*
knot (nautical mile) ***6**.192*
knotweed ***2**.180*
knowledge:
acquisition *see* learning
engineering **6**.124, **6**.179
synthesis **7**.304
three forms of **7**.138
see also cognitive; empiricism; epistemology; science
Know-Nothings ***4**.185*
Knox, Henry ***3**.199*
Knox, John ***3**.199*, **3**.319
Knox, Philander Chase ***4**.185*
koala ***2**.180*
Kōan ***7**.177*
Kōbō Daishi (Kūkai) **9**.9, ***9**.22*
Koch, Heinrich **4**.99
Koch, Helge von: Koch snowflake **1**.126
Koch, Robert ***2**.180*, **6**.191, **6**.192, **6**.218
Kochanowski, Jan ***5**.250*
Kodak camera **5**.347, **6**.111
Kodály, Zoltán ***5**.250*
Koenig **5**.362
Koffka, Kurt **7**.131
Kohl, Helmut **4**.135, ***4**.185*
Köhler, Wolfgang **7**.131
kohlrabi **2**.41, ***2**.180*
Kohoutek, Lubos: Kohoutek's Comet ***8**.80–1*, **8**.196
Koidula, Lydia **5**.34
Koirala, Girija Prasad **4**.237
kokako **2**.362
Kokoschka, Oskar ***5**.250*
Kola Peninsula ***1**.184*
Kolchak, Aleksander Vasileyvich ***4**.185*, **4**.296
Kölhorster, Werner **8**.31
Kolingba, General **4**.67
Kolle, Wilhelm ***6**.192*
Koller, Carl ***6**.192*
Kollwitz, Käthe ***5**.250*
Komenský, Jan (Comenius) **5**.52, **5**.120
Komisarzhevsky, Fyodor ***5**.250*
Komodo dragon ***2**.180*
Komorowski, Tadeusz **4**.371
Konare, Alpha Oumar **4**.208
Konfrontasi ***4**.185–6*
Kongo **3**.4, **3**.199, **7**.217
Koniev, Ivan Stepanovich ***4**.186*, **4**.328
Königgratz *see* Sadowa
Konoe Fumimaro ***4**.186*
Konwicki, Tadeusz **5**.355
Koobi Fora ***3**.199–200*
kookaburra ***2**.180–1*
Kooning, Elaine de **5**.127
kopje ***1**.184*
Köppen, Wladimir **1**.65, **1**.66, ***1**.184*
Köprülü family ***3**.200*, **3**.265
kora ***5**.250*
Koran (Qur'ān) ***3**.200*, **5**.19, ***5**.250*, **7**.166, **7**.177, ***9**.44*
Korda, Sir Alexander ***5**.250*
Korea ***1**.184*, ***3**.200*, ***4**.186*, ***7**.362*
art ***5**.250–1*
economy and trade **7**.88, **7**.113, **7**.157, **7**.277, **7**.319
land reform **7**.180
language ***7**.177*
music ***5**.251*
physical geography and geology **1**.184
politics and government **7**.60, **7**.89
religion/theology **5**.251, **7**.65, **7**.320, **9**.15
sport and recreation **7**.194, **7**.223, **7**.305
technology and science **6**.51, **6**.72, **6**.281
urban areas **7**.197, **7**.323
War **4**.6, ***4**.186*
see also North Korea; South Korea
Korn, Arthur **6**.126
Korngold, Erich Wolfgang ***5**.251*
Koryo ***3**.200*
Kosciusko, Mount ***1**.184*
Kosciuszko, Tadeusz Andrzej Bonawentura ***3**.200–1*, **3**.286, **3**.340
kosher ***7**.178*
Kossovo ***3**.201*
Kossuth, Lajos **4**.158, ***4**.186–7*, **4**.288
Kosygin, Alexei Nikolayevich ***4**.187*
Kosztolányi, Dezsö **5**.214
koto ***5**.251*
Kotzebue, Otto von ***1**.184*
Kowal, Charles **8**.2, **8**.26
Koyo, emperor of Japan **3**.24
Krafft-Ebing, Richard von **7**.146
krait ***2**.181*
Krakatoa ***1**.184*
Král, Janko **5**.422
Krasicki, Ignacy ***5**.251*
Krasiński, Zygmunt ***5**.251–2*
Kravchuk, Leonid **4**.356
Krebs, Hans Adolf: Krebs' cycle ***2**.181*
Kreisler, Fritz ***5**.252*
Kreutzwald, Friedrich Reinhold **5**.34
Krieghoff, Cornelius **5**.73
Kriemhild **9**.8, **9**.33
krill **2**.111
Kripke, Saul ***7**.178*
Krishan, Gopi **5**.244
Krishna **7**.31, **7**.144, ***7**.178*, **7**.326, **9**.19, ***9**.22*
consciousness **7**.138
Krisna Deva Raya **3**.373
Krittibas **5**.43
Kronstadt Mutiny ***4**.187*
Kropotkin, Peter, prince ***4**.187*, **7**.12–13
Krüdener, Baroness Juliane **4**.8
Krüger 60 (star) **8**.80, ***8**.81*
Kruger, Paul **4**.43, ***4**.187*, **4**.276, **4**.350
Krylov, Ivan **5**.151
krypton **1**.106, ***1**.184*
Kshitigarbha **9**.20
Ku Klux Klan ***4**.187*
Kuan-yin **9**.6
Kubiena, Walter L. ***1**.185*
Kubitschek, Juscelino **4**.49, ***4**.187*
Kublai Khan **3**.195, **3**.200, ***3**.201*, **3**.238
and Marco Polo **3**.221
and Song dynasty **3**.333, **3**.388
and Tibet **3**.356
Kubrick, Stanley ***5**.252*
kudu ***2**.181*
Kuiper, Gerard **8**.81, **8**.100, **8**.107, **8**.147, **8**.172
Kuiper Airborne Observatory (KAO) ***8**.81*
Kūkai (Kōbō Daishi) **7**.284, **9**.9, **9**.22
kula ***7**.178*, **7**.197, **7**.316
kulak ***4**.187*
kultarr **2**.266
Kulturkampf ***4**.187*
Kun, Béla **4**.157, **4**.159, ***4**.187*
Kundera, Milan ***5**.252*
kung-fu **7**.194
Kung-San **7**.176, ***7**.178*
Kuomintang **4**.71–2, **4**.73, ***4**.187–8*
Kupka, Frank **5**.332
Kurchatov, Igor: kurchatovium **1**.106
Kurds/Kurdistan ***1**.185*, ***3**.201*, ***4**.188*, ***7**.178*
language **7**.157
leader *see* Saladin
persecuted **7**.44, **7**.130
physical geography and geology **1**.184, **1**.185
Kurile Islands ***1**.185*, **7**.9
Kurma ***9**.22*
Kurosawa Akira ***5**.252*
Kuroshio ***1**.185*
Kursk, battle of ***4**.188*
Kush *see* Nubians
Kut, siege of ***4**.188*
Kutch ***1**.185*
Kutuzov, Mikhail Ilarionovich, prince of Smolensk **4**.45, ***4**.188*
Kuvempu **5**.136
Kuwait ***1**.185*, ***4**.188*, ***7**.362*
agriculture **6**.175
aid from **7**.8
in Arab League **7**.15
economy and trade **7**.224
see also OPEC
in Gulf Co-operation Council **7**.136
invaded *see* Gulf War (1991)
monarchy **7**.97
Muslim Brotherhood **7**.166
in OPEC **4**.251
physical geography and geology **1**.185
politics: franchise **7**.332
refugees **7**.261
al-Kuwarizmi **1**.8
Kvasir ***9**.22*
Kwa languages **7**.217
Kwakiutl Indians ***7**.178*
kwashiorkor ***2**.181*, **7**.190
Kwoth ***9**.22*
Kyd, Thomas **5**.459
Kylian, Jiri **5**.122
kyogen **5**.235
Kyrgyzstan ***1**.185*, ***4**.188*, **7**.59, ***7**.362–3*
Kyushu ***1**.185*

L

La Bruyère, Jean de ***5**.253*
La Chausée **5**.134
La Fayette, Marie-Madeleine, Comtesse de ***5**.253–4*, **5**.323
La Fontaine, Jean de ***5**.254*, **5**.371, **9**.21
La Fontaine, Sir Louis-Hippolyte **4**.28, ***4**.189*
La Palma Observatory ***8**.82*, **8**.83
La Rocca, Nick **5**.131
La Rochefoucauld, François duc de ***5**.256*
La Rochelle ***3**.203*
La Salle, René Robert Cavalier, Sieur de ***1**.188*, ***3**.203*
La Tène ***3**.203–4*
La Tour, Georges de ***5**.258*
La Trobe, Charles Joseph ***4**.190*
La Vérendrye, Pierre Gaultier de Varennes, Sieur de (and sons) ***1**.190*
Laban, Rudolf von **5**.123, ***5**.252*
Labe **1**.101
labelling theory ***7**.179*, **7**.299
Labīd **5**.19
Labor Party (Australia) *see under* politics *under* Australia
laboratories and laboratory equipment ***6**.192–3*
Bunsen burner **6**.25, **6**.52, **6**.193
burette **6**.52
Petri dish **6**.261
pipette **6**.270
test-tube **6**.351, **6**.353
see also instrumentation; microscope
labour camps **7**.64
labour and employment ***7**.105–6*, ***7**.179*
in agriculture **7**.7
child **7**.45, **7**.46
defined **7**.266
division **7**.93
ergonomics **7**.108
forced **7**.64
see also slavery
job *see* job
manpower planning **7**.191
migrant *see under* migration
pay *see* pay
percentage in different fields **7**.106
personnel management **7**.234
professional occupation **7**.249
theory of value **4**.290, ***7**.179*, **7**.303
time-and-motion study **7**.312
unemployment *see* unemployment
unions *see* unions
see also International Labour Organization; labour movements
labour movements **4**.143
International Labour Organization **4**.164
see also Labour Parties; socialism; unions *and also* action *under* industrialization
Labour Parties **7**.179
Britain *see under* politics *under* Britain
Independent (Britain) **4**.160
New Zealand *see under* New Zealand
Labourers, Statute of ***3**.201*
Labrador ***1**.185*
Labrador Current **1**.185
labradorite **1**.259
Labrouste, Henri ***5**.252–3*
laburnum ***2**.181*

labyrinth fish ***2.181***
fighting fish **2.**117
gourami **2.**142
snake-head **2.**314
Lacaille, Nicolas Louis de:
constellations introduced by **8.**6, **8.**18, **8.**20, **8.**26, **8.**29, **8.**52, **8.**69, **8.**96, **8.**98, **8.**109, **8.**112, **8.**121, **8.**130, **8.**138, **8.**146
laccolith **1.**166, ***1.185***
lace **5.**252, ***5.253***, **6.**125, ***6.193***, **6.**351
lacebug ***2.181***
Lacerta ***8.82***
lacewing ***2.181***
lackey moth **2.**337–8
Laclos, Pierre Choderlos de **5.**148, ***5.253***
lacquer **5.**469, ***6.370–1***
arts (lacquerwork) **5.**223–4, **5.**237, ***5.253***
lacrosse ***7.179***
lactic acid **2.**115, **6.**127
lactose **6.**337
Ladino language **7.**156, **7.**280
Ladislas I, king of Poland **3.**183, **3.**286, **3.**352
Ladislas II, king of Bohemia **3.**183
Ladislas III, king of Poland **3.**183
Ladislas V, king of Austria, Hungary, and Bohemia **3.**153
Ladoga, Lake ***1.185***
lady fern **2.**115
Lady of the Lake (Nineve) **9.**12, **9.**25
ladybird **2.**30, ***2.181***
Laënnec, René-Théophile Hyacinthe ***6.193***, **6.**218, **6.**335
Laetoli ***3.201***
Lafayette, Marie Joseph, marquis de ***3.201–2***
lager *see* brewing
Lagerkvist, Pär ***5.254***
Lagerlöf, Selma ***5.254***
lagoon ***1.185–6***
Lagoon Nebula ***8.82***
Lagrange, Joseph Louis, Comte de ***1.186***, **1.**261, **8.**140, **8.**174–5
Lagrangian points ***8.82***
lai **5.**258
Laing, Alexander **3.**357
Laing, R. D. **7.**116, **7.**149
laissez-faire **3.**202, **7.**100, **7.**139, ***7.179***
Lake District ***1.186***
lake dwelling ***3.202***
Lake Erie, battle of ***4.189–90***
Lake George, battle of ***3.202***
Lake of the Woods ***1.186***
lakes ***1.186***
Africa **1.**57, **1.**203, **1.**331, **1.**356
Asia **1.**30, **1.**31, **1.**54, **1.**185, **1.**339
Australia **1.**118, **1.**130, **1.**340
Bolivia **1.**338, **1.**339
Britain **1.**186, **1.**331
Canada **1.**23, **1.**57–8, **1.**145, **1.**146, **1.**186, **1.**367–8
Ethiopia **1.**331
Europe **1.**54, **1.**73, **1.**132, **1.**176, **1.**185, **1.**301
eutrophication **2.**111
evaporites **1.**116
finger **1.**121
and glaciation **1.**121, **1.**331
hydrology **1.**163
Israel **1.**130
Kampuchea **1.**339
lagoon **1.**185–6
New Zealand **1.**332, **1.**359
ox-bow **1.**242, **1.**243
Peru **1.**338, **1.**339
playas **1.**259–60
Russia and Siberia **1.**30, **1.**185
shott **1.**302
South America **1.**338, **1.**339
lakes (*cont.*):
thermocline **1.**334
United States **1.**41, **1.**48, **1.**57–8, **1.**79, **1.**145, **1.**146, **1.**186, **1.**237
see also salt lakes
Lakshmi **7.**93, **7.**281, **7.**326, ***9.22***
avatars (incarnations) **9.**22, **9.**34
Lakshmi Bai **4.**162
Lakshmisha **5.**134
Lāl, Bihārī **5.**206
Lāl, Lallūjī **5.**206
Lalande, Joseph Jerome le Français de **8.**9, **8.**21, **8.**51
Lalande 21185 (star) ***8.82***
Lalique, René ***5.254***
Lally, Thomas Arthur, comte de, baron de Tollendal ***3.202***, **3.**377
Lalo, Edouard ***5.254***
Lamaism ***3.202***, **3.**356
Lamarck, J.-B. P. A. de Monet, Chevalier de **2.**111, ***2.181***
Lamartine, Alphonse de ***5.254***
Lamb, Charles ***5.254***, **5.**284
Lambert (1731, artist) **3.**109
Lambert, John (1619–83, soldier) ***3.202***
Lambton, John George, earl of Durham (1792–1840) **4.**105
lamellaphone **5.**8
laminar flow **6.**138
laminate **6.**82, **6.**156, ***6.193***, **6.**271
lammergeyer ***2.181***, **2.**358
Lamming, George ***5.254***
lamp shell ***2.182***
lampoon ***5.265***
lamprey ***2.182***
lamps:
arc **6.**19–20
Argand **6.**20, **6.**201
Davy's safety **1.**85
fluorescent **6.**138–9, **6.**177, **6.**202
gas-discharge **6.**76, **6.**150, **6.**202
mercury vapour **6.**194
oil **6.**20, **6.**30, **6.**201, **6.**249
sodium **6.**323
see also incandescent lamp; lighting
Land, Edwin **6.**274
Lanby buoy ***6.193***
Lancastrians **3.**138, ***3.202***, **3.**284, **3.**310
see also under rulers *under* Britain
lance **6.**165
lancelet ***2.182***
land:
common **3.**83, **3.**115, **7.**57–8
defined as resource **7**:266
drainage *see* reclamation *below*
enclosure **3.**115
feudal fief **3.**126
franklin **3.**132
freehold **3.**388
glebe **3.**142
landforms ***1.187***
see also geology; geomorphology
manorialism **3.**220
marches/marche/mark **3.**221, **3.**222
nationalization **3.**339
ownership **7.**56, **7.**69
reclamation and drainage **1.**201, **1.**263, **1.**374, **6.**79, ***6.193***, **6.**376
see also irrigation
reform ***7.179–80***
relief *see* landforms *above*
socage **3.**331
squatting **7.**298
tenancy *see* sharecropping
tenure **7.**179
use ***7.180***
vassal **3.**370
see also agriculture *and also under individual countries*
land (economics) ***7.179***
Land League ***4.190***
landau **6.**61
Lander, Harold **5.**393
Lander, Richard Lemon ***1.186–7***
Landine, Francesco **5.**379
landing aircraft **6.**9, **6.**10, **6.**25, **6.**290, **6.**330
landing craft ***6.193***
Ländler **5.**482
landlord, absentee **3.**1
Landor, Walter Savage ***5.254***
Landsat *see* Earth resources satellite
landscape formation *see* geology; geomorphology
landscape gardening and design *see* garden
landscape painting and painters ***5.254–5***
Australia **5.**26–7
Britain:
18th century **5.**116–17, **5.**118, **5.**168, **5.**179, **5.**269, **5.**348–9, **5.**490, **5.**495
19th century **5.**110–11, **5.**114, **5.**118, **5.**130, **5.**165, **5.**322, **5.**338, **5.**348–9, **5.**462–3, **5.**468–9
20th century **5.**431, **5.**441
Canada **5.**73
China **5.**96
Danube School **5.**125
France **5.**34–5, **5.**56, **5.**88, **5.**103, **5.**113, **5.**472
see also Barbizon School
Germany **5.**13, **5.**117, **5.**165, **5.**248, **5.**264, **5.**494
Italy **5.**42, **5.**74, **5.**268, **5.**383
Netherlands and Flemish **5.**27, **5.**55, **5.**63, **5.**120
Hague School **5.**197
17th century **5.**27, **5.**49, **5.**55, **5.**120, **5.**208, **5.**333, **5.**394, **5.**396, **5.**410
Sublime **5.**438
United States **5.**13, **5.**14, **5.**211, **5.**213
urban **5.**74
see also painting and painters
landscape photography **5.**1, **5.**5, **5.**347, **5.**431, **5.**441
Landseer, Sir Edwin, John, and Thomas ***5.255***
landslide ***1.187***
Landsteiner, Karl **6.**40, ***6.194***
Lane-Fox, St George **6.**114
Lanfranc, archbishop ***3.202***, **3.**256
Lanfranco, Giovanni ***5.255***
Lang, Fritz ***5.255***
Lange, C. G. **7.**105
Lange, David **4.**189
Lange, Oskar **7.**192–3
Langhans **5.**315
Langland, William **3.**306, ***5.255***
Langley, Samuel **6.**139
Langmuir, Irving ***1.187***, **6.**20, ***6.194***
Langren, Michael van **8.**147
Langstruth, L. L. **6.**16
Langtoft, Peter **3.**375
Langton, Stephen ***3.202–3***, **3.**216
language ***7.180***
bilingualism **7.**31
classification and major groups ***7.180–1***
African *see under* Africa
Afro-Asiatic **7.**30, **7.**43, **7.**78, **7.**109, **7.**278–9
Altaic **7.**11
Australian Aborigine **7.**22
Austro-Asiatic **7.**22–3
Austronesian **7.**23, **7.**192, **7.**197, **7.**228, **7.**240
Baltic **7.**26
Basque **7.**28
Berber **7.**30
Causcasian **7.**41–2
Celtic **7.**42
language (*cont.*):
classification and major groups (*cont.*):
Chadic ***7.43***
Chinese *see under* China
Cushitic ***7.78***
Dravidian ***7.95***
Egyptian ***7.103***
English *see* English language
Eskimo–Aleut ***7.108***
European *see under* Europe
Germanic ***7.131***
Greek *see under* Greece, modern
Indo-European *see* Indo-European languages
Indo-Iranian **7.**137, **7.**156, ***7.157***
Italic ***7.168***
Japanese ***7.169***
Khoisan ***7.176***
Korean ***7.177***
Kwa ***7.178***
Niger–Congo ***7.217***
Nilo-Saharan ***7.217***
North American Indian ***7.218–19***, **7.**221, **7.**273
Palaeosiberian **7.**227
Papuan ***7.228***
Romance **7.**108, ***7.269***
Semitic **7.**109, ***7.278–9***
Sino-Tibetan ***7.286–7***
South and Central American Indians ***7.199–200***, ***7.295***
Tai ***7.305***
Uralic ***7.322–3***
as code *see* codes; semiotics
and communal conflict **7.**59
and computers **7.**63
deconstruction **7.**81–2
development **7.**35, **7.**180
dialect *see* dialect
dictionaries **7.**107, **7.**196
diglossia **7.**90
Esperanto **7.**108
figures of speech *see* figures of speech
grammar *see* grammar
jargon **7.**169–70
lingua franca **7.**185, **7.**217, **7.**279, **7.**295
linguistics *see* linguistics
machine translation **6.**211
malapropism **5.**277
metaphor **5.**292
palindrome **5.**337
pidgin and creole **7.**235–6
private **7.**247
proverb **5.**368
and religion/theology **7.**19
semantics **7.**196, **7.**278
semiotics *see* semiotics
sign **7.**218, **7.**285
simile **5.**420
slang **7.**287
spoken *see* speech
universals ***7.181***
written *see* writing
see also under individual countries
Languedoc ***3.203***
lanner ***2.182***
Lansbury, George ***4.190***
Lansdowne, Henry Charles Keith Petty-Fitzmaurice, 5th marquis of ***4.190***
Lansdowne, William Petty Fitzmaurice, 1st marquis and 2nd earl of Shelburne **3.**269, ***3.326***
L'Anse aux Meadows ***3.203***, **3.**373
Lanston, Tolbert **6.**232
lantern fish ***2.182***
lantern fly ***2.182***, **2.**183
lanthanides (rare earths) ***1.187***, ***6.295***
scandium **1.**294
yttrium **1.**372
see also cerium; germanium
lanthanum **1.**106

Lao She *5.256*
Laos *1.187*, *3.203*, *4.190*, *7.363*
agriculture **7**.363
and ASEAN **7**.20
crime **7**.225
decolonized **4**.4
languages **7**.305
literature *5.255–6*
physical geography and geology **1**.187, **1**.210
politics and government **4**.262–3, **4**.321, **7**.94
refugees **7**.261
religion/theology **7**.36
Lǎozi (Lao-tzu) *3.203*, **3**.349, **7**.79–80, **9**.13
laparoscopy **6**.248, **6**.340
lapiés **1**.180
lapis lazuli *1.187–8*
Laplace, Pierre-Simon, marquis de **8**.17, **8**.33, *8.82*
Laplace nebular hypothesis *8.82–3*
Lapland *1.188*
lappet moth **2**.104
lapping machine **6**.212
Lapps **7**.16, *7.181*, **7**.322
lapse rate *1.188*
lapwing *2.182–3*
larch *2.183*
Lares and Penates *9.22*
Large Magellanic Cloud **8**.88–9
Largo, Caballero Francisco *4.190*
lark family *2.183*, **2**.198, **2**.313
Larkin, Philip **5**.256
Larkin, William **3**.51
larkspur **2**.90
Larra, Mariano José de **5**.428
Larrea, Osvaldo Hurtado **4**.108
Lartigue, J. H. **5**.347
larvae *2.183*
axolotl **2**.19, **2**.226
maggot **2**.165, **2**.198
neoteny **2**.226
rat-tailed maggot **2**.165
salamander **2**.292
sawfly **2**.294
tadpole **2**.333
wireworm **2**.67–8
woodworm **2**.130
see also caterpillars
laryngoscope **6**.120, **6**.218
larynx *2.183*, **2**.357
Las Casas, Bartolomé de **3**.86, *3.203*
Las Casas, Francisco de **3**.169
Lasalle, Ferdinand **4**.317
Lascaux Cave *1.188*, **3**.342
laser *6.194–5*
argon **6**.240
bent **6**.251
cutting **6**.211
dating **6**.96
dentistry **6**.99
engraving *6.195*
holography **6**.149, **6**.169
interferometer **6**.184
in medicine **6**.99, **6**.251, **6**.340
monochromatic light **1**.219
pioneer **6**.214
printing *6.195*
surgery **6**.340
surveying **6**.340
weapons *6.195*, **6**.336
Lassell, William **8**.8, **8**.71, **8**.174, **8**.177
Lassen Volcanic National Park *1.188*
Lassus, Orlande de *5.256*
lateen rig **6**.101, **6**.143, *6.195*, **6**.306, **6**.307
latent heat *1.188–9*
lateral line *2.183*
Lateran Council *3.204*
Lateran Treaties *4.190*
laterite (rock) **1**.195
laterite (soil) *1.189*, **1**.311
late-type star *8.83*
latex **2**.87, *2.183–4*, **2**.241, **2**.289–90, **2**.294
lath and plaster **6**.146, *6.195*
lathe **6**.42–3, *6.195*, **6**.211, **6**.212, **6**.217, **6**.294
Latimer, Hugh *3.204*
Latin **7**.156, *7.168*, **7**.185, **7**.288
literature **3**.43, **5**.189, *5.257–8*, *5.387–9*, **7**.168
comedy *5.387*
elegy *5.388*
epics **5**.121, **5**.147, **5**.270, **5**.335, **5**.477
hexameters **5**.121, **5**.147, **5**.212, **5**.335, **5**.477
historians **3**.208, **3**.314, **3**.344, **3**.347, **5**.353, **5**.444–5
philosophy **5**.101
satire *5.389*
see also under drama; poetry
Latin America:
dance *5.257*
literature *5.256*
music *5.256–7*
see also Central America; South America
latitude and longitude *1.189*, *6.195*, *8.128*
celestial **8**.23
degrees **1**.86
eastings **1**.99
galactic **8**.54
horse latitudes **1**.158–9
international date line **1**.37, **1**.171
libration in **8**.85
measurement **6**.73, **6**.77, **6**.237, **6**.238, **6**.303
parallels of **1**.247
variation of **8**.180
see also maps and charts; meridians
latosol *1.189*
Latrobe, Benjamin *5.258*
lattice **6**.337
Latvia *1.189*, *3.204*, *4.190*, *7.363*
in Baltic Entente **4**.29–30
chemist **1**.359–60
language **7**.26, **7**.156
Livonia **3**.208
physical geography and geology **1**.189
Laud, William **3**.115, **3**.130, *3.204*, **3**.295
laudanum **2**.235
Laue, Max Theodore Felix von *1.189*
laughing gas **6**.13
Launay, Bernard de **3**.36
Launcelot **9**.5
launch site *6.196*
launch vehicle **6**.64, *6.196*, **6**.326
Laurasia **1**.135, *1.189*
laurel **2**.25, *2.184*
Laurentian Shield **1**.50
Laurentides *1.190*
Laurier, Sir Wilfrid *4.190*, **4**.195
Lausanne, Treaty of **4**.364
lava *1.190*
erosion **1**.96
glass, volcanic **1**.141
intrusion **1**.89
and plate tectonics **1**.261
pyroclasts **1**.270, **1**.271, **1**.295
scoria **1**.295
volcanic bomb **1**.357
see also basalt; extrusive (volcanic) igneous rocks; igneous rocks; volcanoes
Laval, Bishop of Quebec **3**.135
Laval, Carl Gustaf Patrik de **6**.52, *6.196*
Laval, Pierre *4.191*, **4**.364
lavender *2.184*
laver *2.184*
Lavoisier, Antoine Laurent **1**.60, *1.190*, **1**.267
Lavrentev, M. A. **6**.242
Lavrovsky, Leonid **5**.51
Law, Bonar *4.191*
Law, John *3.204*
law and justice *7.181*
act *see* act
administrative **7**.3, **7**.213, **7**.319
adversarial procedure **7**.3
advocate **7**.3, **7**.4
amnesty **7**.12
appeal **7**.15, **7**.318
arrest **7**.18–19
bail **7**.25
barrister **7**.27
Buddhist **7**.89
canon **7**.39
capital punishment **3**.150
case **7**.41, **7**.245
charter **3**.71
Christian **3**.59
citizenship **7**.50
civil *see* civil law
class action **7**.52
commercial **7**.57, **7**.198, **7**.282
common *see* common law
community penalty **7**.61
company **7**.19, **7**.62, **7**.197, **7**.319
compensation **3**.379, **7**.62, **7**.75, **7**.79
competition **7**.62
conspiracy **7**.67
consumer **7**.68
contempt of court **7**.68
contingency fee to lawyer **7**.68
contract **7**.69
conveyancing **7**.69
copyright **7**.70
court-martial **7**.73
courts *see* courts of law
covenant **7**.73
criminal **7**.76, **7**.318, **7**.319
customary **7**.78–9
damages **7**.79
detention without trial **7**.87
discrimination **7**.91
due process in **7**.97
equity **7**.58, **7**.108
euthanasia **7**.111
evidence **7**.112
family **7**.55, **7**.115
forensic medicine **6**.143, **6**.217
frankpledge **3**.132
habeas corpus **7**.138
Hindu **7**.19, **7**.89, **7**.143
homicide *see* homicide
illegality *see* crime
impeachment **7**.153
industrial relations **7**.158–9
injunction **7**.161
Inns of Court **7**.161
inquest **7**.161
inquisitorial procedure **7**.161
intellectual property **7**.162
international *see* International Court of Justice; international law
Islamic *see under* Islam
Judaic *see under* Jews and Judaism
judge **7**.172–3, **7**.279
judicial review **7**.3, **7**.173, **7**.245
judiciary **7**.173, **7**.279, **7**.318, **7**.324
jurisprudence **7**.174
jury **3**.192–3, **7**.112, **7**.174
lawyer **7**.181
legal aid **7**.181–2
legislation *see* act; legislation *and* laws/legal system *under individual countries*
legitimacy **7**.182
liability **7**.301, **7**.326
magistrate **7**.189, **7**.318
maritime **7**.272
marriage **7**.55
martial **7**.194
mercantile **7**.198
military **7**.209
minimum wage **7**.203
law and justice (*cont.*):
Napoleonic, see *Code Napoléon*
natural **7**.213
negligence **7**.214
notary **7**.220
nuisance **7**.220
obligations **7**.221
and old age **7**.222
ordeal **3**.262–3
and order *see* police
perjury **7**.233
philosophy of **7**.174
plaintiff **7**.236
plea bargaining **7**.237
police *see* police
political trial **7**.239
Pragmatic Sanction **3**.290
precedent **7**.245
primogeniture **3**.292
prison *see* prisons
privacy **7**.247
private *see* private law
property **7**.250, **7**.331
prosecution **7**.250
public *see* public law
punishment *see* punishment
Roman *see* Roman law
rule of **7**.270
Salic **3**.314
salvage **7**.272
sanctuary **3**.315–16
Scots **7**.4, **7**.276, **7**.318, **7**.319
segregation **7**.278
sentences *see* sentences (court)
settlement **7**.280
solicitor **7**.295
statute *see* statute law
strict liability **7**.301
and suicide **7**.301
tort **7**.312–13
tribunal **7**.315–16
trust **7**.316
ultra vires **7**.319
vicarious liability **7**.326
violent crime **7**.326
wills **7**.248, **7**.331
see also justice; legislature *and under* laws/legal system *under individual countries*
Lawes, Henry *5.258*
Lawes, Sir John Bennet *2.184*, **6**.6, **6**.196
lawn-mower *6.196*
Lawrence, Saint *9.42*
Lawrence, David Herbert **4**.94, *5.258*, **7**.243
Lawrence, Ernest Orlando *6.196–7*, **6**.258
lawrencium **1**.106
Lawrence, Sir Henry **4**.191
Lawrence, John Laird Mair, 1st Baron *4.191*
Lawrence, T. E. **4**.115, **4**.142, *4.191*, **4**.218
Lawrence, Sir Thomas **4**.218, *5.258*
Laws of Manu **9**.24
Lawson, H. J. **6**.35
lawyer *7.181*
Laxness, Halldór *5.258*
lay (lai) *5.258*, **5**.359
see also stories
Laye, Camara **5**.7
Lazarus *9.22*
Lazlo I and II, kings of Hungary **3**.183
LD process **6**.188
Le Bon, Gustave **7**.77
Le Châtelier, Henri-Louis *1.191*
Le Corbusier **5**.64, **5**.219, *5.259–60*, **5**.370, **6**.51, **7**.216
Le Fanu, Joseph Sheridan *5.260*
Le Luu **5**.475
Le Nain, Antoine, Louis, and Mathieu *5.261*
Le Nôtre, André **3**.372, *5.261*
Le Pen, Jean-Marie **4**.116

Le Roy, Pierre **5**.105
Le Vau, Louis *5.262*
Le Verrier, Urbain Jean-Joseph **8**.2, **8**.11, **8**.17, ***8.84***, **8**.117, **8**.185
Leach, Bernard *5.258*, **5**.398
Leach, Edmund **7**.175
leaching (soil) **1**.108, **1**.190
Leacock, Stephen Butler *5.259*
lead **1**.106, ***1.190***, **1**.191, ***6.197***
- in accumulator **6**.1
- and alchemy **6**.10
- alloys **6**.265
- azide **6**.100
- chamber process ***6.197***, **6**.339
- extraction **6**.70, **6**.197, **6**.221, **6**.253
- lead/acid battery **6**.31
- ores **1**.130, **1**.308
- pollutant **6**.377
- roof **6**.274
- and rubber vulcanization **6**.158
- white **6**.384

lead carbonate **6**.197, **6**.384
lead chromate **6**.197, **6**.270
lead dioxide **6**.1
lead oxide ***1.190–1***, **6**.156, **6**.157, **6**.197, **6**.296
lead sulphide **1**.130, ***1.191***, **6**.197, **6**.253
lead tetraethyl **6**.147, ***6.197***, **6**.224, **6**.261, **6**.262, **6**.297, **6**.323
leaders and rulers:
- assassination **7**.20
- authority **7**.23
- chiefs **7**.44
- church **7**.48
  - *see also* papacy
- dictatorship **7**.89–90
- hereditary *see* monarchy
- kings and queens *see* monarchy
- state as (hegemony) **7**.141
- *see also* heads of state; power (political and social); monarchy *and also* rulers *under individual countries*

lead-line ***6.197***
leaf ***2.184***, **2**.185, **2**.355
leaf beetle ***2.184***
leaf insect ***2.185***
leaf miner **2**.253
leaf-chinned bat ***2.184***
leaf-cutter bee ***2.184***
leafhopper ***2.185***
leaf-nosed bat ***2.185***
League of Nations ***4.191***, **7**.56, **7**.262
- *see also* mandate

Leah **9**.7, **9**.19
Leakey, Mary **1**.237
Lean, David *5.259*
Leander **9**.16
Leandro **5**.37
leap seconds **8**.62
leap year **8**.18
Lear, Edward *5.259*
learning ***2.185–6***
- animals **7**.110
- conditioning *see* conditioning
- and expectation **7**.113
- imprinting **7**.153–4
- learned helplessness ***7.181***
- memory **2**.207
- motivation **7**.207
- psychology of **2**.340
- social **7**.7, **7**.126, **7**.207, **7**.233, **7**.291
- *see also* education; knowledge; memory; reading; scholasticism; science

Leary of Ireland **3**.350
least squares method ***1.191***
leather ***6.197***, **6**.281, **6**.345
- leathercraft *5.259*

leaves ***2.184***, **2**.185, **2**.355
Leavis, Frank Raymond *5.259*
Leavis, Queenie Dorothy **5**.259
Leavitt, Henrietta **8**.23, **8**.118
Leballo, Potlako **4**.257
Lebanon ***1.191***, ***3.205***, ***4.191***, ***7.363–4***
- in Arab League **7**.15
- civil war **7**.96, **7**.193
- environment **7**.83
- Palestinian refugees in **7**.227
- physical geography and geology **1**.191
- politics and government **7**.67
- religion/theology **7**.59, **7**.96, **7**.193, **7**.284
- wars and defence:
  - civil **7**.52, **7**.96, **7**.193, **7**.364
  - communal conflict **7**.59
  - hostage-taking **7**.148
  - peacekeeping forces **7**.277

Lebensraum ***4.191***
Lebesgue, Henri Léon ***1.191***
Leblanc, Nicolas: Leblanc process **6**.70, **6**.96, ***6.197***
Lebon, Philippe **6**.150, ***6.197***, **6**.386
Lebrun, Albert ***4.191–2***
Lebrun, Charles *5.259*
Lechfeld, battle of ***3.205***
lecithin **6**.141
Leclanché, Georges: Leclanché cell **6**.31–2, ***6.197***
Leclerc, Philippe, vicomte **4**.125, **4**.245
Leconte de Lisle, Charles-Marie-René *5.259*
LED ***6.201***, **6**.204, **6**.314
Leda **9**.8, ***9.22***
Ledbetter, Huddie **5**.50
Ledoux, Claude-Nicolas *5.260*
Lee, Ann **3**.325
Lee, Arthur **3**.34
Lee, Robert Edward **4**.10, **4**.15, **4**.16, **4**.136, ***4.192***, **4**.368
Lee, William **6**.191
Lee Kuan Yew ***4.192***, **4**.314
leech ***2.186***
leek ***2.186***
Leeuwenhoek, Anton van ***2.186***, **6**.197, ***6.198***
Leeward Islands **1**.52
Left (politics) ***7.181***
- New Left **7**.192, **7**.215
- Popular Front **4**.273
- *see also* communism; Labour Parties; socialism

leg ***2.186***
- foot **2**.124

legal aid ***7.181–2***
Legazpi, Miguel de **3**.280
Legba ***9.22***
Legendre, Adrien Marie ***1.191***
Léger, Fernand *5.260*, **5**.395
Legge, Walter **5**.407
legionnaires' disease ***2.186***
legions, Roman **3**.308
legislation ***7.182***
- bills **3**.41, **7**.31, **7**.32
- *see also* act; judicial review; law and justice; statute law *and also* laws/legal system *under individual countries*

legislature ***7.182***
- bicameral assembly **7**.31, **7**.182
- impeachment **7**.153
- lobbying **7**.163, **7**.186
- parliamentary **7**.148
  - *see also* Parliament *under* Britain
- presidential **7**.182
  - *see also* presidents
- separation of powers **7**.279
- *see also* legislation *and also* laws/legal system *under individual countries*

legitimacy **7**.66, ***7.182***
Leguía, Augusto Bernardino ***4.192***, **4**.267
legumes ***2.186***
- alfalfa **2**.6
- beans **2**.25–6, **2**.42

legumes (*cont.*):
- carob **2**.54–5
- catalpa **2**.56
- chick-pea **2**.63
- clover **2**.69
- lentil **2**.187
- liquorice **2**.191
- mesquite **2**.208
- peanut **2**.246
- peas *see* pea family
- sainfoin **2**.292
- soya bean **2**.318
- vetch **2**.355

Lehár, Franz *5.260*
Leibniz, Gottfried Wilhelm, Baron von ***1.191***, **3**.51, ***7.182***
- and calculator **6**.54, **6**.84
- and calculus **1**.48, **1**.90, **1**.191, **1**.227
- and functions **1**.129
- and monads **1**.191
- and rationalism **7**.259
- and time **7**.312

Leicester, Robert Dudley, earl **3**.114, ***3.205***
Leicester, Thomas William Coke, earl **2**.72, **3**.5, **3**.271, **3**.278, ***6.80***
Leichhardt, Friedrich Wilhelm Ludwig ***1.191***, **4**.122
Leichner, Ludwig **5**.277
Leidy, Joseph ***2.186***
Leif Ericsson ***1.112***, **3**.118, **3**.374
Leigh, Augusta **5**.69
Leighton, Frederic *5.260*
Leila **9**.24
Leipzig, battle of ***4.192–3***
Leisler, Jacob ***3.205***
leisure ***7.182–3***
- *see also* games; hobbies and pastimes; sport and recreation; tourism

leitmotif *5.260*
lek **2**.305
Lekes, Harold **4**.381
Lely, Sir Peter *5.260–1*
Lem, Stanisław *5.261*
Lemaitre model **8**.34
Lemass, Sean **4**.166
Lemercier, Jacques *5.261*
lemma ***1.191***
lemming ***2.186–7***
lemon ***2.187***
lemon sole ***2.187***
lemurs **2**.170, ***2.187***
Lemus, José M. **4**.111
Lena River ***1.191–2***
Lenclos, Anne ***3.205***
Lend–Lease Act ***4.193***
L'Enfant, Pierre-Charles *5.261*
length measurement ***6.198***
- angström **1**.14
- kilometre **6**.368
- metre ***1.213***, *5.292*, **6**.321, **6**.368
- odometer **6**.198, **6**.249
- and surveying **6**.340
- units ***9.49***
- Vernier scale **6**.198, **6**.371
- *see also* interferometer; micrometer

Lenin, Vladimir Ilyich and Leninism ***4.193***, **4**.357
- and Bolsheviks **4**.44
- and Brest-Litovsk **4**.49
- and CHEKA **4**.71
- and Civil War **4**.297
- and Comintern **4**.79, **4**.165
- and communism **7**.60–1
- and imperialism **7**.153
- and Kronstadt mutiny **4**.187
- and Marxism **7**.194
- and prison camps **4**.277, **7**.64
- and Revolution **4**.297
- tomb of **5**.287

Leningrad, siege of ***4.193***
Lennon, John **5**.39
Leno, Dan **5**.310
Lenoir, Jean-Joseph Étienne **6**.150, ***6.198***
lenses and lens systems ***6.198–9***
- achromatic **6**.2, **6**.310
- apochromatic **6**.1
- camera obscura **6**.56
- contact **6**.87
- Fresnel **6**.146, **6**.201
- instruments with *see* camera; microscope; telescope
- lighthouse **6**.201
- photochromic **6**.267
- photographic ***6.199***, **6**.268
- spectacles **6**.327
- *see also* optics/optical

Lent ***7.183***
lenticular galaxy ***8.84***
lentils ***2.187***
Lenya, Lotte **5**.485
Lenz, Sigfried **5**.175
Leo I, St, pope **3**.27, ***3.205***
Leo III, Byzantine emperor **3**.174
Leo III, pope **3**.61, **3**.68, ***3.205***
Leo V, Byzantine emperor **3**.174, **3**.280
Leo IX, St ***3.205–6***
Leo X, pope **1**.192, **3**.228, **3**.301, **5**.378
Leo XIII, pope **3**.119
Leo, Johannes (Leo Africanus) ***1.192***
LEO (computer) **6**.84
Leo (constellation) ***8.84***
Leo Minor (constellation) ***8.84***
León **3**.9, **3**.124–5, ***3.206***
Leonardo da Vinci *5.261*, ***6.199***
- and arts **3**.131, **5**.5, **5**.16, **5**.78, **5**.87, **5**.261, **5**.262–3, **5**.359
- and technology **6**.122, **6**.167, **6**.199, **6**.234, **6**.310, **6**.360

Leoni, Raul **4**.363
Leonidas of Sparta ***3.206***, **3**.354
Leonids ***8.84***, **8**.194, **8**.195
Leoninus **5**.289
Leonov, Alexei **4**.323
leopard **2**.69, ***2.187–8***, **2**.316
leopard moth ***2.188***
leopard snake ***2.188***
Leopardi, Giacomo *5.261*
Leopold I, king of Belgium **4**.33, ***4.193***
Leopold I, Holy Roman Emperor (1640–1705) **3**.27, **3**.132, **3**.153, **3**.168, ***3.206***
Leopold II, king of Belgium **4**.33, ***4.193***, **4**.388
Leopold II, Holy Roman Emperor (1747–92) **3**.153, **3**.168
Leopold II, king of Tuscany **4**.355
Leopold III, king of Belgium **4**.33
Leopold of Hohenzollern-Sigmaringen, prince **4**.112, **4**.155
Lepanto, battle of ***3.206***
Lepidodendron ***2.188***
Lepidus **3**.28, **3**.307
leprosy ***2.188***
leptons **1**.105, ***1.192***
- flavour **1**.123
- muons **1**.222–3
- neutrinos **1**.123, **1**.226
- *see also* electrons

leptospirosis **2**.364
Lepus ***8.84***
Lêr ***9.23***
Lerebours **8**.120
Lermontov, Mikhail *5.261–2*
Lerner, Abba P. **7**.193
Lesage, Georges **6**.346
lesbians **7**.146, **7**.332
Lesches of Mytilene **6**.201
Leśmian, Boleslaw *5.262*
Lesotho (Basutoland) ***4.193–4***, ***7.364***
- colonization and decolonization **4**.4, **4**.306
- contraception **7**.116
- economy and trade **7**.296, **7**.364

Lesotho (Basutoland) (*cont.*):
monarchy **7**.97
physical geography and geology **1**.95, **1**.192
less developed countries *see* developing countries
Lesseps, Ferdinand Marie, vicomte de **4**.257, ***6**.199*, **6**.256, **6**.337
lesser panda **2**.280
Lessing, Doris May ***5**.262*
Lessing, Gotthold Ephraim ***5**.262*
Lesueur brothers **5**.313
Leszczynski *see* Stanislaus
Lethbridge, John **6**.97
Lethe ***9**.23*
Leticia dispute ***4**.193–4*
Letsie III, king of Lesotho **4**.193
letter of credit ***7**.183*
letterpress printing **6**.123, ***6**.199*, **6**.282, **6**.334, **6**.366
lettuce ***2**.188*
leucocytes ***2**.188–9*, **2**.272, **6**.164
leukaemias **1**.82, ***2**.189*
Levant ***3**.206*, **3**.247, **3**.280
Levassor, Emile **6**.94
levee ***1**.192*
Levellers ***3**.207–8*
levelling: spirit-level **6**.329
Lever, William Hesketh **7**.216
lever **6**.92, ***6**.199*, **6**.200, **6**.217
Levers, John **6**.193
Levi ***9**.23*
Levi, Primo **5**.231
Leviathan ***9**.23*
Levison, Swift Nick **3**.165
Lévi-Strauss, Claude **7**.132, ***7**.183*
and incest **7**.154
and kinship **7**.14, **7**.154
and linguistics **7**.301
and myth **7**.11, **7**.210
and symbolism **7**.303
and totem **7**.313
Lewes, G. H. **5**.143
Lewin, Kurt **7**.113
Lewis, Clive Staples ***5**.262*
Lewis, David **4**.238
Lewis, Gilbert Newton ***1**.192–3*
Lewis, John Llewellyn **4**.10, ***4**.194*
Lewis, Matthew **9**.7
Lewis, Meriwether (and William Clark) **1**.63, ***1**.64*, ***1**.193*, **3**.122, ***4**.194*
Lewis, M. G. **5**.186
Lewis, Sir Samuel ***4**.194*
Lewis, Saunders **5**.486
Lewis, Sinclair ***5**.262*
Lewis, Thomas **6**.113
Lewis, Wyndham *see* Wyndham Lewis
Lexell's Comet ***8**.84–5*, **8**.194
Lexington and Concord, battle of ***3**.207*
Leyster, Judith ***5**.262*
Leyte Gulf, battle of ***4**.194*
li ***7**.183*
Li Bai ***5**.262*, ***5**.264*
Li Dazhao **4**.74
Li He **5**.98
Li Hongzhang ***4**.196*, **4**.308
Li Qingzhao ***5**.265–6*
Li Yuan **3**.344, **3**.349
Li Yue **6**.159
Li Zicheng **3**.232, **3**.296
liability **7**.301, **7**.326
lianas **2**.39–40, **2**.173, ***2**.189*
Liang K'ai **5**.264
Liang Lingzam **6**.76
Liang Qichao **4**.158
Liao dynasty **3**.72, ***3**.207*
Liaqat Ali Khan, Nawabzada ***4**.194*, **4**.256
Libby, Willard Frank **6**.20, **6**.48, **6**.111, ***6**.199–200*, **6**.290, **8**.32
libel **7**.82
liberal democracy:
Liberal Democratic Party (Japan) ***4**.194*
see also constitutional government *under* constitutions
Liberal Judaism ***7**.183–4*, **7**.304
Liberal Parties:
Australia **4**.94, ***4**.195*
Britain **4**.145, ***4**.195*, **4**.290
Canada ***4**.195*
Japan **4**.194
New Zealand **4**.29, **4**.120, **4**.285, **4**.308, **4**.370–1
liberalism ***7**.183*
and economics *see* free market; *laissez-faire*
and politics *see* Liberal Parties; Locke; Mill, John Stuart; privacy; rights; social contract *and* constitutional government *under* constitutions
toleration **7**.312
liberalization (economics) ***7**.183*
see also adjustment; deregulation; free market; libertarianism; New Right; privatization
liberation theology ***7**.184*
Liberia ***1**.193*, ***4**.195*, ***7**.364–5*
colonized and decolonized **4**.4, **4**.306–7
physical geography and geology **1**.193
president **4**.354
refugees from **7**.261
technology **6**.250
libertarianism **7**.125, ***7**.184*, **7**.220
see also liberalism
liberty ***3**.207*
see also freedom *and entries beginning* free
Libra ***8**.85*
library ***7**.184*
libration ***8**.85*
Libya ***1**.193*, ***3**.207*, ***4**.195*, ***7**.365*
aid from **7**.8
in Arab League **7**.15
decolonization **4**.4
economy and trade *see* OPEC
languages **7**.30
laws/legal system **7**.284
leader (Qaddafi) **4**.195, ***4**.279*
physical geography and geology **1**.193
politics and government **7**.299
region **3**.361
religion/theology **7**.284
see also Numidia
lice ***2**.194*
biting (bird) **2**.32
book **2**.38, **2**.88
and disease **2**.352
jumping plant **2**.176
sucking **2**.329
lichens **2**.188, ***2**.189*, **2**.282
Lichtenstein, Roy ***5**.264*
Licinius **3**.86
Lick, James: Lick Observatory **8**.85
Lie, Sophus ***1**.193*
Lie, Trygve Halvdan ***4**.195–6*, **7**.322
lie detector **6**.275
Liebermann, Carl **6**.260
Liebermann, Max **5**.264
Liebig, Justus, Baron von ***1**.193*, **6**.72
Liebknecht, Karl **4**.202, **4**.324
Liebknecht, Wilhelm ***4**.196*, **4**.317
Liechtenstein ***1**.193*, **7**.319, ***7**.365*
in Council of Europe **7**.71
physical geography and geology **1**.193, **1**.282
royalty **7**.97
Lied **5**.142, **5**.203, ***5**.264*
Lienas, Don Juan de **5**.257
Lieutaud, Balthazar **5**.105
Lievens, Jan ***5**.264*
Lifar, Serge **5**.122
life ***2**.189*
cycle ***2**.189*
see also growth of living organisms
expectancy **7**.6, ***7**.184*
life-crisis rituals **7**.268
see also rites of passage *under* rituals
see also living organisms
lifeboat **6**.20, ***6**.200*, **6**.317
life-jacket and lifebuoy **6**.52, ***6**.200–1*
life-support machine **6**.201
lift ***6**.201*, **6**.322
lift-to-drag ratio **6**.135
ligand ***1**.193*
Ligeti, György ***5**.264*
light ***1**.193–4*
and astronomy ***8**.85–6*
ashen **8**.8
diffraction **8**.37
light-year ***8**.86*
photon **8**.121
reflection **8**.3
refraction **8**.137
velocity of **8**.1, **8**.181
zodiacal **8**.190
see also colour
bulb *see* incandescent lamp
colour *see* colour
detection *see* vision
diffraction *see* diffraction
-emitting diode (LED) ***6**.201*, **6**.204, **6**.314
energy changed into heat energy **6**.166
fluorescence *see* fluorescence
interference **6**.184
lighting *see* lighting
light-time effect **8**.174
measurement **6**.327–8
meter **6**.124
microscope **6**.223
mirage **1**.215, **1**.216, **1**.279
monochromatic **1**.219
phosphorescence **1**.256, **1**.257, **6**.267
photomultiplier **6**.269
and plants *see* photosynthesis
polarized *see* polarization
pollution **7**.240
prism *see* prism
reflection *see* reflection
refraction *see* refraction
scattered **1**.275
seeing *see* vision
spectrum *see* spectra
speed **1**.123, **1**.317, **8**.1, **8**.181
stroboscope **6**.336
Very **6**.371
waves **1**.31, **1**.275, **1**.278, **1**.363
see also laser; lenses; optics/optical; photography *and also entries beginning* photo
Light Brigade, Charge of **4**.27–8
light industry ***7**.184*
see also tertiary industry
lighter (barge) **6**.30
lighter (cigarette) **6**.73
lighthouse ***6**.201*, **6**.202, **6**.322
lighting ***6**.201–2*
candle **6**.57
gas *see under* gas
lamps *see* lamps
navigational **6**.239
searchlight **6**.311
stage ***5**.264–5*
traffic lights **6**.360
lightning ***1**.194*
conductor **6**.146, ***6**.202*
lightship ***6**.202*
light-year ***8**.86*
lignite **1**.67, **1**.177, ***1**.194*, **6**.78, **6**.91
lignum vitae ***2**.190*
lilac ***2**.190*
Lilburne, John ***3**.207–8*
Lilienthal, Otto **6**.138, **6**.139, **6**.157, ***6**.202*, **6**.389
Lilith ***9**.23*
lily family ***2**.190*
aloe **2**.7
arum **2**.16
aspidistra **2**.17
autumn crocus **2**.19
butcher's broom **2**.47
chive **2**.64
fritillary **2**.127
garlic **2**.133
hyacinths **2**.143, **2**.166
Madonna **2**.198
nerine **2**.226
onions and leeks **2**.186, **2**.234, **2**.306
scilla **2**.297
Star of Bethlehem **2**.324
tulip **2**.349–50
waterlilies **2**.194, **2**.289, **2**.361–2
limb (astronomy) ***8**.86*
brightening and darkening ***8**.86*
limb (body):
arm ***2**.14*, **2**.29
artificial **6**.21
leg **2**.124, ***2**.186*
limbo ***9**.23*
limbo dancing **5**.78
Limburg, Herman, Jean, and Paul ***5**.265*
lime (*Citrus*) ***2**.190*
lime (chemical) ***1**.48*, ***1**.194*, **6**.53, **6**.86, **6**.127, **6**.161, ***6**.202*, **6**.378
see also calcium carbonate
lime tree (*Tilia*) ***2**.190*
lime water ***1**.195*
limestone ***1**.194*, ***6**.202*
calcium carbonate **1**.47
calcium in **6**.53
in cement **6**.66
and chemical industry **6**.70
dolomite **1**.94
erosion:
caves **1**.53, **1**.180, **1**.181, **1**.318, **1**.319
doline **1**.93, **1**.180, **1**.181
gorge **1**.53
karren **1**.180
karst **1**.180, **1**.181
polje **1**.180, **1**.263
pothole **1**.265
sink-hole **1**.305, **1**.306
see also dry valleys
escarpment **1**.113
metamorphosed (marble) **1**.47, **1**.205, **5**.282
oolitic and oolith **1**.238
Purbeck marble **1**.269
quarrying **6**.287
tufa **1**.346
wold **1**.368
see also chalk
limited liability ***7**.184*
limonite ***1**.195*
limpet ***2**.189*, **2**.190
limpkin ***2**.190*
Limpopo ***1**.195*
Lin Biao **4**.74, **4**.90, ***4**.196–7*, **4**.210
Lin Zexu ***4**.196*
Linacre, Thomas ***2**.190*
Lincoln, Abraham **4**.10, **4**.45, ***4**.196*, **4**.287
Emancipation Proclamations **4**.15, **4**.316
Lind, Jenny ***5**.265*
Lindbergh, Charles **6**.138
Linde, Carl Paul Gottfried von ***6**.202*, **6**.297
Linder, Leslie **5**.363
Lindgren, Astrid **5**.91
Lindsey, Theophilus **7**.320
line (astronomy):
of apsides ***8**.86*
of nodes ***8**.86*
see also spectral lines

lineage *7.184–5*
  see also kinship; matrilineality; patrilineality
Linear A and B **3**.233, **7**.135
linear accelerator **6**.196, **6**.258
linear momentum **1**.218–19
linear motor **6**.201, *6.202–3*
linearity *1.195–6*
linen **2**.120, **6**.16, *6.203*, **6**.390
liner *6.203*, **6**.333
lines of force **1**.195, *1.196*
ling (animal) *2.190*
ling (plant) **2**.154
lingua franca *7.185*, **7**.217, **7**.279, **7**.295
linguistics *7.185*
  clinical **7**.53
  comparative **7**.144
  computational **7**.63
  ethnolinguistics **7**.109
  grammar *see* grammar
  historical **7**.110, **7**.144
  meaning *see* meaning
  philosophy **7**.47, *7.185*
  phonology **7**.235
  psycholinguistics **7**.252, **7**.273–4
  relativity **7**.273–4
  semantics **7**.196, **7**.278
  sociolinguistics **7**.294
  structuralism **7**.301
  see also language
Linna, Väinö **5**.155
Linnaeus, Carl **2**.2, *2.190–1*, **2**.229
linnet **2**.191
linoleum *6.203*
Linotype *6.203*, **6**.219, **6**.366
linseed **2**.120
  oil **6**.109, **6**.255, **6**.286
Linz–Donawitz (LD) process **6**.188
lion roar (drum) **5**.164–5
lions **2**.58, *2.191*
Lipari Islands *1.196*
Lipchitz, Jacques *5.265*
lipids **2**.114, *2.191*
  see also fats, oils, and fatty acids; steroids
Lippershey, Hans **6**.347
Lippi, Fra Filippo *5.265*
lip-reading *7.185*
liquefied coal **6**.343
liquefied gas **1**.203, **6**.76, **6**.150, **6**.237
liquefied hydrogen **6**.301
liquefied natural gas *1.196*, *6.203*, **6**.237
liquefied oxygen **6**.150, **6**.301
liquefied petroleum gas *see* liquid petroleum gas
liquid *1.196*, *6.23–4*
  azeotrope **1**.28
  boiling-point **1**.41
  cavitation **6**.64
  crystal *see* liquid-crystal
  diffusion **1**.91
  distillation **6**.11, **6**.104
  electrorheological fluid **6**.120
  elements *see* bromine; mercury
  evaporation **1**.116, **1**.162, **2**.345
  expansion **1**.117
  flow *see* flow
  fluidics **6**.137
  heat exchanger **6**.166–7
  immiscible, mixture of *see* emulsion
  solids in *see* supsension
  solution *see* solvents and solutions
  specific gravity measurement **6**.174
  surface tension and surfactants **1**.327, **6**.11, **6**.339, **6**.340
  suspension *see* suspension
  ultrasound **6**.367
  venturi tube **6**.371
  volatile **6**.34
  see also elements; fluid; hydraulic; liquefied; vapour; water
liquid petroleum gas *1.196*, *6.204*, **6**.237, **6**.264
  see also butane; propane
liquidation (finance) **7**.161
liquid-crystal *1.196*
  display (LCD) *6.204*
  polymer *6.204*, **6**.247
  television **6**.350
liquidity (finance) *7.185*
liquidizer **6**.142
liquorice *2.191*
lira (musical instrument) *5.266*
Lisle, Rouget de **5**.313
Lismer, Arthur **5**.73
LISP **6**.86
Lissitzky, El *5.266*
List, Guido von **4**.334
list sytem of proportional representation **7**.250
Lister, Joseph, Baron *2.191*, **6**.16, *6.204*, **6**.259, **6**.340
Liszt, Franz **5**.10, *5.266*, **5**.367, **5**.382, **5**.442
litchi *2.191*
literacy *7.185*
literature:
  children's books *see under* children
  copyright law **7**.70
  libraries **7**.184
  PEN **7**.232
  pornographic **7**.242–3
  see also biography; books; comedy; critics, literary; drama; essays; figures of speech; genre; myths and legends; novels; poetry; romances; Romanticism; satire; stories; tragedy *and under individual countries*
lithification **1**.82
lithium **1**.106, *1.196*, *6.204*
  in aluminium alloys **6**.11, **6**.12
  in glass **6**.156, **6**.310
lithography **5**.152, *5.266–7*, **5**.359, *6.205*, **6**.249, **6**.282
lithosol *1.196*
lithosphere **1**.98
lithotripsy *6.205*, **6**.340, **6**.367
Lithuania *1.197*, *4.197*, *7.365*
  in Baltic Entente **4**.29–30
  language **7**.26, **7**.31, **7**.156
  nationalism **7**.211
  physical geography and geology **1**.197
litmus *1.197*
litter (vehicle) *6.205*
litter (rubbish) *see* waste management
Little Big Horn, battle of *4.197*
Little Dipper **8**.178–9
Little Entente *4.197*
little owl *2.191*
Little Rock *4.197*
Littlewood, Joan **5**.452
liturgical drama *5.267*
liturgy *7.185–6*
  see also Baptism; Mass
Litvinov, Maxim Maximovich *4.197*
Liu Bang **3**.155
Liu E **5**.96
Liu Shaoqi **4**.90, *4.197*
Liu Xiu **3**.156
Live Aid **7**.308
live broadcasting *7.186*
liver *2.191–2*
  bile **2**.29
  disorders **2**.66, **2**.155, **2**.175, **2**.192, **6**.87
  urea **2**.280
liver fluke *2.192*
Liverpool, Robert Banks Jenkinson, 2nd earl of *4.197–8*, **4**.315
liverwort *2.192*
livery company *3.208*
livestock **6**.7, **6**.204, *6.205–6*
  breeding **6**.7, **6**.14–15, **6**.28, **6**.125, **6**.155, **6**.283
livestock (*cont.*):
  cattle *see* cattle
  dairy *see* dairy farming
  draught *see under* horse-drawn
  factory farming **6**.125, **6**.205, **6**.279
  and fall-out **6**.72
  food *see* feedstuffs, animal
  fur-farming **6**.148, **6**.205
  genetic engineering **6**.153
  genetic resources **6**.155
  horn **6**.170
  leather **6**.197, **6**.281, **6**.345
  machinery **6**.6, **6**.225, **6**.227
  meat *see* meat
  organic farming **6**.253
  pigs *see* pigs
  poultry **6**.279
  progeny testing **6**.14, **6**.21, **6**.283
  sheep *see* sheep
  veterinary medicine **6**.371
  see also animals; domesticated animals; wool
living organisms **2**.30
  animals *see* animals
  biochemistry **2**.29
  biodiversity **2**.29
  biology **2**.30
    see also botany; ecology; zoology
  biomass **2**.30
  biophysics **2**.30
  body parts *see* physiology
  cell structure **6**.65
  changed form *see* metamorphosis
  electricity in **2**.105
  and environment *see* ecology
  extraterrestrial **8**.49, **8**.148
  growth *see* growth of living organisms
  and machinery *see* biotechnology
  metabolism *see* metabolism
  migration **2**.210–11
  mimicry **2**.212–13
  morphology **2**.218–19
  movement **2**.210–11, **2**.220
  organic compounds *see* organic compounds
  pest control by **6**.37
  plants *see* plants
  protection *see* protective coloration
  reproduction *see* reproduction
  see also biogeochemical cycles; life; micro-organisms
living standard **7**.4–5, **7**.299
Living Theatre *5.267*
Livingston, M. S. **6**.196
Livingstone, David *1.197*, **1**.320, **1**.356, *4.198*, **4**.207, **4**.221, **4**.329, **4**.388
Livingstone daisy **2**.208
Livonia *3.208*
Livy **3**.77, *3.208*
lizards *2.192*
  agamid **2**.4, **2**.124
  anole **2**.11
  chameleon **2**.11, **2**.50, **2**.60, **2**.61
  flying **2**.124
  frilled **2**.127
  gecko **2**.134
  gila monster **2**.136
  glass snake **2**.137
  iguanas **2**.11, **2**.24, **2**.169, **2**.300
  Komodo dragon **2**.180
  monitor **2**.217
  pygopodid **2**.272
  sand **2**.293
  skink **2**.311, **2**.312
  slow worm **2**.313
llamas **2**.146, *2.192*, **2**.356
Llanos *1.197*
Lloyd, Marie **5**.310
Lloyd George of Dwyfor, David, 1st earl **4**.21, *4.198*, **4**.364
  and convoy system **4**.173
  and foreign affairs **4**.90–1
  and Ireland **4**.166
Lloyd George of Dwfor (*cont.*):
  and 'People's Budget' **4**.266
  and social security **4**.318
Lloyd Webber, Andrew *5.267*
Lloyd's of London *7.186*, **7**.319
Llywelyn II, prince of Wales **3**.376
Llywelyn the Great, prince of Wales *3.208*, **3**.376
Llywelyn ap Gruffydd *3.208*
loach *2.192*
load (rivers) **1**.35, **1**.92, *1.197*
loadstone **1**.203
loam *1.197*
loans *7.186*
  collateral **7**.56
  debenture **7**.80
  mortgage **7**.207
  see also debt
Lobachevsky, Nikolai Ivanovich **1**.136, *1.197*
lobbying **7**.163, *7.186*
lobefins **2**.38, **2**.71, **2**.195–6
lobelia *2.192*
Lobengula of Ndebele *4.198*, **4**.236, **4**.390
loblolly pine *2.192*
lobster *2.192*
lobster moth *2.192*
local area network (computing) **6**.240
local government *7.186*, **7**.320
Local Group of galaxies *8.86*
  Andromeda **8**.5
  Galaxy of our solar system **8**.56
  Magellanic Clouds **8**.88–90
Local Supercluster *8.86*
Local System **8**.59–60
Locarno, Treaties of *4.198*
location of firms *7.186*
lock (and key) *5.267*, **6**.45, **6**.205, *6.206*, **6**.304
lock (canal) **6**.56
Locke, John *3.208–9*, *7.186–7*
  and civil society **7**.52
  and consent **7**.66
  and kingship **3**.104
  and metaphysics **7**.199
  and personal identity **7**.233
  and primary and secondary qualities **7**.31
  and psychology **7**.252
  and social contract **3**.331, **7**.221, **7**.290
  and substance **7**.301
  Transcendentalism as reaction against **5**.459
lockjaw **2**.338–9, **7**.153
Locle, Camille du **5**.471
locomotives:
  articulated **6**.20
  bogie **6**.41
  diesel **6**.102, **6**.150, **6**.294
  electric **6**.114–15, **6**.293–4
  fireless **6**.132
  steam *see* locomotives *under* steam power
  tank **6**.345
  see also engines; railways; rolling stock
locus of control *7.187*
locust *2.192–3*, **2**.210
locust tree **2**.287
lodestone **1**.203
Lodge, Henry Cabot *4.198*
Lodi family *3.209*
loess *1.197–8*
Loesser, Frank **5**.309
Loest, Erich **5**.175
Loewy, Raymond **5**.223
Lofting, H. **5**.91
log (navigation) *6.206*
log drum **5**.422
Logan, James *2.193*
Logan, Sir William Edmond *1.198*
Logan, Mount *1.198*
loganberry *2.193*
logarithms *1.198*, **1**.224, **6**.53, **6**.322

logic 1.198, 7.187
  Aristotelian 7.16
  formal 7.187
  induction 7.157
  logical positivism 7.24, 7.187, 7.187, 7.315, 7.325
  and mathematics *see under* mathematics
  modal 7.203
  'modern' 7.221
  modern *see* Frege
  paradox 7.228
  philosophical 7.234
  and reason 7.259
  symbolic 1.87
  validity 7.325
logic gate (computers) 6.206–7
logicism 7.187
Lohengrin 9.23
Loire 1.198
lokapalas 9.23
Loki 9.2, 9.6, 9.13, 9.23, 9.30
Lollard 3.209
Lomax, John 5.50
Lombard, Peter 7.1
Lombardy 3.98, 3.168, 3.209, 3.328
Lombroso, Cesare 7.76
Lomé Conventions 4.198, 7.103
Lomonosov, Mikhail Vasilyevich 1.198, 5.267
Lon Nol 4.178
London, Jack 5.267
London:
  Contemporary Dance Theatre 5.267
  Corresponding Society 3.209
  Dockers' strike 4.198–9
  Fire of 3.127
  livery companies 3.208
Londonderry 3.209
lone pair 1.199
Long, Huey 4.199
long (dragon) 9.23
Long Island 1.199
Long Island Sound 1.199
Long March 4.199
Long Parliament 3.144, 3.209–10
long run 7.187
long wave *see* long-wave radiation
Longfellow, Henry Wadsworth 5.267
Longhena, Baldassare 5.267–8
Longhi, Alessandro 5.268
Longhi, Pietro 5.268
longhorn beetle 2.193
longhorn grasshopper 2.193, 2.194
longhorn moth 2.193
Longinus 5.268, 5.438
longitude *see* latitude and longitude
longitudinal waves 1.198, 1.199, 6.381
long-line fishing 6.133, 6.207
Longshan 3.210
longship 3.210
longshore drift 1.199
long-tailed duck 2.193–4
long-wall mining 6.79, 6.207, 6.228
Lóngwáng 9.23
long-wave radiation 1.103, 6.380, 9.54
Lönnrot, Elias 5.155
loofah 2.142, 2.194
loom 6.62, 6.140, 6.169, 6.189, 6.190, 6.207–8
  see also weaving
loon 2.95
Loop Nebula *see* Cygnus Loop
looper 2.194, 2.198
Loos, Adolf 5.268
López, Carlos Antonio and Narciso 4.199
López, Francisco Solano 4.199, 4.259–60
López, Marciso 4.199
Lopez, Vicente 4.168
Lopokova, Lydia 7.176
loquat 2.194
Loran (navigation) 6.208–9
lorandite 6.353
Lorca, Federico García 5.169, 5.170
Lord Appellant 3.210
Lord Lieutenant 3.210–11
Lord Ordainer 3.210
Lords, House of *see* House of Lords
Lorelei 9.23
Lorentz, Hendrik Antoon 1.199, 8.86
  Fitzgerald–Lorentz contraction 1.122–3, 8.86
  Lorentz transformation 8.86–7
Lorenz, Konrad Zacharias 2.28, 2.194, 2.359, 7.109–10, 7.154
Lorenzetti, Pietro and Ambrogio 5.268, 5.418
Lorenzo the Magnificent 3.228
Loris, Guillaume de 3.74
loris 2.194
Lorrain, Claude 5.103
Lorraine 3.150, 3.211
  dukes of *see* Guise
Lorris, Guillaume de 5.387
lorry 6.8, 6.44, 6.209, 6.341
loss (bereavement) 7.30
loss (finance) 7.187
lost generation 5.268–9
  see also cummings; Fitzgerald; Hemingway; Pound; Stein
lost-wax process 6.184, 6.185
Lot 9.23
Lothair I, king of West Franks 3.69, 3.210
Lothair II, Holy Roman Emperor 3.168
Lotto, Lorenzo 5.269
lotus 2.194
Lotus Sutra 7.187
Loudoun, Lord John 3.134
loudspeaker 6.13, 6.209, 6.289
Louis I (the Great), king of Hungary (778–840) 3.211
Louis I (the Pious), king of Franks (1326–82) 3.137, 3.211
Louis I, duc de Bourbon (1279–1341) 3.46, 3.47
Louis I, king of Spain 3.46
Louis I de Bourbon (1530–69) 3.84
Louis II, king of Hungary 3.172, 3.236
Louis II de Bourbon (the Great Condé) 3.84, 3.135, 3.144
Louis IV, king of Hesse 4.150
Louis IV, Holy Roman Emperor (1314–46) 3.168, 3.384
Louis VI, king of France 3.60, 3.161
Louis VII, king of France 3.18, 3.113
Louis IX, king of France (St Louis) 3.93, 3.211, 3.300, 3.363, 5.339
Louis XI, king of France 3.54, 3.211, 3.340
Louis XII, king of France 3.50, 3.263, 3.369
Louis XIII, king of France 3.16, 3.46, 3.47, 3.304
  and arts 5.365, 5.480
Louis XIV, king of France 3.46, 3.47, 3.211–12
  and arts 5.30, 5.181
    'Louis Quatorze' style 5.56, 5.259, 5.261, 5.269
  and astronomy 8.116
  and Austrian Succession, War of 3.29
  and Divine Right 3.104
  and French Revolution 3.134, 3.306
  Grand Siècle 3.144
  and law 3.179, 3.270
  mistress 3.239
  politics and government 3.84, 3.128, 3.130, 3.302, 3.304
  queen 3.222
Louis XIV (*cont.*):
  and religion 3.49, 3.245
  resistance to 3.28, 3.58, 3.135
  and science 3.2, 6.1
  and Siam 3.327
  tutor 3.128
  and Versailles 3.372
  and Versailles 3.372
  wars and conquests:
    Anglo-Dutch 3.14, 3.70
    Flanders 3.128, 3.154, 3.337
    Franche-Comté and War of Devolution 3.54, 3.84, 3.102, 3.130
    Nine Years War 3.254, 3.312
    Palatinate 3.267
    Spanish Succession 3.217
  zoo 2.375
Louis XV, king of France 3.46, 3.47, 3.212
  mistresses 3.106, 3.287
  style ('Louis Quinze') 5.269
Louis XVI, king of France 3.46, 3.47, 3.212, 4.69
  and America 3.132
  and States-General 3.340
  style ('Louis Seize') 5.269
  and Tennis Court Oath 3.246
  Varennes, Flight to 3.370
  and Versailles 3.372
Louis XVII, king of France 3.46, 3.47
Louis XVIII, king of France 3.46, 3.47, 4.69, 4.155, 4.199, 4.366
Louis, prince de Condé 3.46, 3.47
Louis the German 3.69, 3.211
Louis, Morris 5.269
Louis Napoleon *see* Napoleon III
Louis Philippe *see* Louis-Philippe
Louis, Pierre-Charles-Alexandre 6.218
Louisburg 3.212
Louisiana 3.213
  Purchase 4.199–200
Louis-Philippe I, king of French (1773–1850) 4.123, 4.177, 4.199, 4.200
Louis-Philippe I, duc d'Orléans (1725–85) 3.46, 3.263
Louis-Philippe II, duc d'Orléans (1747–93) 3.46, 3.263
louse *see* lice
louse fly 2.194
lousewort 2.195
Loutherbourg, Philippe-Jacques de 3.178, 5.269, 5.430
love, courtly 5.116
love-bird 2.195
Lovelace, Augusta Ada King, Countess of 6.209
Lovelace, Richard 5.269
love-lies-bleeding 2.195
Lovell, Sir (Alfred Charles) Bernard:
  and telescope 6.291, 8.76, 8.78, 8.87
Lovelock, James E. 7.129
Lovett, William 4.70, 4.200
Löw, Rabbi Juda 9.14
Low Countries *see* Belgium; Flanders; Luxemburg; Netherlands
low frequency 1.103, 6.380
low (pressure) 1.199
Lowell, Amy 5.269
Lowell, Francis Cabot 4.200
Lowell, James Russell 5.269
Lowell, Percival 8.19, 8.87
  Observatory 8.87
Lowell, Robert 5.270
Lower Canada *see* Quebec
low-ground-pressure vehicle 6.209–10
Lowry, Laurence Stephen 5.270
Lowry, Malcolm 5.270
Loyola, Saint Ignatius 3.89, 3.187, 3.213, 7.171
LPG *see* liquid petroleum gas
LSE (liquid solvent extraction) 6.343
Lu Chi 6.159
Lu Xun 5.272
Luba kingdom 3.4, 3.213
Lubetkin, Berthold 5.300, 5.448
Lubitsch, Ernst 5.270
lubricant 6.145, 6.160, 6.210
  oil 6.32, 6.145, 6.204, 6.210, 6.264
Lucan 5.270
Lucan, Lord (1854) 4.27
Lucas van Leyden 5.270
lucerne 2.6
Lucheni (anarchist) 4.124
Lucian 5.190
Lucifer *see under* Devil
Lucilius 5.389
Lucknow, siege of 4.200–1
Lucretius 5.270
Luddite 4.201
Ludendorff, Erich 4.178, 4.201, 4.213, 4.227, 4.262, 4.376
Ludwig II and III, kings of Bavaria 4.32
Ludwig, Christa 5.270
Lug of the Long Arm 9.9
Lugard, Frederick Dealtry, 1st Baron 4.128, 4.201, 4.242
Lugh 9.23
lugworm 2.195
Luis, Agustina Bessa 5.359
Luke the Evangelist, Saint 3.41, 5.216, 9.12, 9.23, 9.42, 9.42
Lully, Jean-Baptiste 5.270, 5.328, 5.359
lumbar puncture 6.210
Lumière, Auguste and Louis 4.81, 5.270–1, 6.75, 6.210, 6.268
luminosity:
  and astronomy (brightness) 8.87
    Bayer letters 8.14
    classes of stars 8.67, 8.87
    list of brightest stars 8.199
    mass-luminosity relation 8.93
    period–luminosity law 8.118
  luminous paint 6.210, 6.249
  measurement 1.50, 6.328, 6.368
luminous beetle 2.119
lumpsucker 2.195
Lumumba, Patrice 4.83, 4.179, 4.201, 4.388
lunar *see* Moon/lunar
lunation (synodic month) 8.100, 8.168
lunette 1.199
lungfish 2.195–6
lungless salamander 2.196
lungs 2.195, 2.196, 9.60
lungworm 2.196
luni-solar precession 8.127
Luo Guanzhong 5.387, 5.483
lupin 2.196
Lupus 8.88
Luqman 9.23
Lusaka Declaration 7.296
Lusinchi, Jaime 4.363
Lusitania 4.201
Lustig, Arnošt 5.120
lustreware 5.207, 5.229, 5.271
Lut 9.23
lutes 5.271
  lutenists 5.134, 5.175
  mandora 5.278
  opharion 5.332
  pipa and *biwa* 5.349
  saöng-gauk 5.402–3, 5.427
  saz 5.403
  shamisen 5.414
  tār 5.446–7
lutetium 1.106
Luther, Martin 3.213, 7.187
  Bible translated 3.41
  and Council of Trent 3.360
  and Diet of Worms 3.103, 3.385–6

Luther (*cont.*):
and indulgences 3.269, 3.352
and peasants 3.274
and printing 3.293
prose style 5.174
and Reformation 3.300, 3.301, 7.310
Lutheran Church 7.50, 7.112, *7.187*
Luthuli, Albert John 4.5, *4.201*, 4.391
Lutoslawski, Witold *5.271*
Lutyens, Sir Edwin 5.219, *5.271*
Lützen, battle of *3.214*, 3.355
Luxemburg, Rosa *4.201–2*, 4.324
Luxemburg *1.199–200*, *3.214*, *7.365*
in Council of Europe 7.71
economy and trade 7.221
see also EC; EEC
finance 7.307
physical geography and geology 1.199–200
rulers 3.168, 3.328, 7.97
see also under Holy Roman empire
wars and defence 7.330
see also NATO
in WEU 4.375
Lvov, prince 4.297
lychee 2.191
Lycurgus 3.338
lyddite 6.124
Lydgate, John 5.294
Lydia 3.91, *3.214*
Lyell, Sir Charles 1.133, *1.200*
lymph glands *2.196*
adenoids 2.4
elephantiasis 2.106
tonsils 2.343
Lynch, Jack 4.166
lynx (animal) *2.196*
Lynx (constellation) *8.88*
Lyons, Joseph Aloysius 4.189, *4.202*
Lyr 9.23
Lyra *8.88*
lyre 5.119, 5.191, *5.272*
lyre-bird *2.196*
lyric poetry and odes 5.211, *5.272*
Bengali 5.445
Britain 5.107, 5.189, 5.212, 5.242, 5.269, 5.296, 5.342, 5.416, 5.428–9, 5.449, 5.496
elegy 5.142–3
France 5.100, 5.214, 5.310–11, 5.476
Germany 5.142, 5.173, 5.181–2, 5.205, 5.210, 5.481
Greece, ancient 5.191, 5.349, 5.402
haiku 5.197
hymn 5.215
India 5.401
Italy 5.162, 5.261, 5.345–6, 5.355
Latin 5.82, 5.211–12, 5.477
Persia 5.25, 5.233
Poland 5.250, 5.262, 5.294
Provençal 5.368
Romanticism 5.390
Russia and USSR 5.10, 5.43, 5.50, 5.261, 5.326, 5.341, 5.396, 5.461
Slovak 5.422
Sufi 5.197, 5.233, 5.394, 5.397, 5.439
United States 5.17, 5.130–1, 5.386
Wales 5.486
see also sonnets
Lyrids *8.88*, 8.195
Lysander 3.3, *3.214*, 3.275
Lysenko, T. D. 6.371
Lysimachus 3.321
Lysippus *5.272*
Lysithea *8.88*
Lytton, Lord 4.105

M

al-Ma'arrī 5.19
Maasai 7.6, *7.188*
Maastricht, Treaty of 4.114, *4.202*, 4.375, 7.99, 7.103, 7.111, 7.331
Ma'at *9.24*
Mabinogion *5.273*
Mabuse *see* Gossaert
McAdam, John Loudon 4.350, *6.210*, 6.299, 6.300
Macao *see* Macau
macaque monkeys 2.22, *2.197*
McArthur, Douglas *4.202*, 4.203
Macarthur, John 4.295
Macartney, Lord 3.297
Macassar *3.214*
Macau *1.200*, *3.214*, 7.241, *7.365–6*
Macaulay, Thomas Babington, 1st Baron 4.153, *4.203*, 5.258, *5.273*
macaw *2.197*
Macbeth, earl of Moray *3.106*
McBride, Sean 4.167
Maccabees *3.214*
McCarran Act *4.202*
McCarran–Walter Act *4.202*
McCarthy, Joseph Raymond *4.202–3*, 7.255, 7.278
McCartney, Paul 5.39
McClellan, George Brinton *4.203*
McClintock, Barbara *2.205*
McClure, Robert 4.17
McCormick, Cyrus Hall 6.172, *6.210*, 6.295
MacCready, Paul 6.138, 6.172
McCullers, Carson *5.273*
McCullin, Don 5.482
MacCumail 9.13
MacDiarmid, Hugh 5.215, *5.273*
Macdonald, Alexander 5.409
Macdonald, Flora *3.214–15*
MacDonald, J. E. H. 5.73
Macdonald, Sir John Alexander 4.60, 4.85, *4.204*, 4.255, 4.354
Macdonald, Margaret 5.274
MacDonald, Ramsay 4.28, *4.203*, 4.234, 4.339
Macdonnell Ranges *1.200*
McDougall, William 7.207
mace 2.230, 2.231, 6.165
mace gas 6.346
Macedonia (*formerly* Yugoslavian republic) *1.200*, *4.204*
Macedonia (Greece) *1.200*, *3.215*, *7.366*
language 7.288
physical geography and geology 1.200
rulers and dynasties 3.278, 3.321
of Egypt *see* Ptolemy
see also Alexander III (the Great)
wars *3.215*, 3.279
McEwen, John 4.234
McGonagall, William 5.132
McGough, Roger 5.91
McGready, James 4.288
MacGregor, Chris 5.239
McGuire, Edward 5.203
Mach, Ernst *1.200*, *7.188*
Mach number 6.140, 6.175, *6.211*, 6.339, 7.188
Mach's Principle *8.88*
Mácha, Karel Hynek *5.273*
Machado, Manuel 5.172–3
Machado y Morales, Gerardo 4.32, 4.89
Machaut, Guillaume de 5.30, *5.273*
Machel, Samora Moises 4.126, *4.204*, 4.225
Machiavelli, Niccolò di Bernardo dei 3.44, *3.215*, *5.273*, *7.188*, 7.244
machine code *6.210*
machine translation *6.211*
machine-gun 6.131, *6.210*, 6.217, 6.226, 6.337
machinery:
agricultural *see under* agriculture
automated *see* automation; mass production; robotics
engines *see* engines
ergonomics 7.108
in home *see* domestic appliances
and living creatures *see* biotechnology
medical *see* medicine
motors *see* motors
office *see* office equipment
textile *see under* textiles
tools *see* machine-tools
vehicles *see* motor car; motor vehicles; vehicles
see also engineering; instrumentation
machine-tools *6.210–11*, 6.308, 6.357
brakes 6.44
controlled *see* numerically controlled machine-tools
gear production 6.151
grinding machine 6.161
and hydraulic power 6.173
lathe 6.195
mass-production 6.216
pioneers 6.45, 6.217, 6.236, 6.385
steam-hammer 6.236, 6.332
transfer machine 6.361
trip-hammer 6.363
Machonaland *see* Zimbabwe
Macintosh, Charles 6.158, *6.211*, 6.304
MacIver, David 4.90
Mackay, Andrew 3.197
McKellan, Ian 7.146
Mackenzie, Sir Alexander (1755–1820, explorer) *1.200–1*, 3.122
Mackenzie, Alexander (1822–92, Canadian prime minister) 4.195, *4.204*
Mackenzie, Sir Compton *5.273*
Mackenzie, Sir James 6.275
Mackenzie, William Lyon: Mackenzie's Rebellion *4.204*, 4.361
Mackenzie river *1.200*
mackerel 2.37, 2.163, *2.197–8*, 2.318
McKim, Mead, and White *5.273*, 5.383
McKinlay, John *1.201*, 1.324
McKinley, William 4.53, 4.146, *4.204*
McKinley, Mount *1.201*
Mackintosh, Charles Rennie *5.273–4*
Mackmurdo, Arthur H. *5.274*
Maclaurin, Colin 1.201
McLean, John 1.63
MacLean, Sorley 5.409
McLellan, George 4.15, 4.192, 4.269, 4.368
Macleod, John James Rickard 6.29
Macleod, Mary 5.409
McLuhan, Marshall *7.196*, 7.197, 7.308
Maclure, William 1.201
MacMahon, Marie Edme Patrice Maurice, comte de 4.124, *4.204*, 4.307
McMahon, Sir Henry: McMahon Line *4.204*
McMath solar telescope *8.96*
MacMillan, Sir Kenneth *5.274*
Macmillan, Kirkpatrick 6.35
Macmillan, Maurice Harold, 1st earl of Stockton *4.205*, 4.380
MacNeice, Louis *5.274*
Macpherson, James 5.367
Macquarie, Lachlan 4.38, 4.111, *4.205*, 4.295
Macready, William Charles *5.274*
macro-economics *7.188*
balance of payments 7.25
classical *see under* classical economics
consumption 7.68
demand *see* demand
employment *see* labour and employment
exchange rate 7.111, 7.112–13, 7.136
fiscal policy 7.120
interest rate *see* interest
investment *see* investment
national income *see* national income
see also GNP; Keynesianism; monetarism
macromolecule *1.201*
'Mad Mullah' 4.147
Mada, Gajah 3.217
Madách, Imre 5.214
Madagascar *1.201*, *3.215*, *4.205*
colonized and decolonized 4.4, 4.306
ethnic group 7.199
physical geography and geology 1.201
madder 6.260
Maddox, Richard 6.268
made land *1.201*
Madeira (island) *1.201*, *3.215*
Madeira (river) 1.201
Maderna, Carlo 5.35
Madero, Francisco Indalécio *4.205*, 4.219, 4.368
Madhva 7.325
Mādhyamika 7.155
Madison, James 3.86, 3.87, 3.124, 3.129–30, *3.215–16*, 4.205
madness 7.123
see also mental illness
Madonna 5.358
Madonna lily 2.198
Madrid Conference on Security and Co-operation in Europe 7.64
madrigals *5.274–5*, 5.283
Maelstrom *1.201*
Maenads *9.24*
maestà *5.275*
Maeterlinck, Maurice 5.73, *5.275*
Maeve 9.9, 9.25
Maeztu, Ramiro de 5.173
Maffei, Paulo: Maffei galaxies *8.88*
Mafia *4.206*, 7.53, *7.188*
Magadha *3.216*
maganeura 2.206–7
Magdala *4.206*
Magdalena *1.202*
Magdalenian 3.367
Magellan, Ferdinand *1.202*, 3.121, 3.122, *3.216*, 3.301, 6.58
Magellan, Strait of 1.201, *1.202*
Magellanic Clouds *8.88–90*
Magenta, battle of *4.206*
maggots 2.165, *2.198*
Māgha 5.401
al-Maghili 3.25, 3.159
Magi *9.24*
magic 7.11, 7.112, *7.188–9*, 7.195
see also tricksters; witchcraft
magic lantern 6.74
magic numbers *1.202*
Maginot Line *4.206–7*
magistrate *7.189*, 7.318
maglev 6.162, 6.293
magma 1.165, *1.202*
see also intrusive (plutonic) igneous rocks
Magna Carta *3.216*
magnesia *1.202*
magnesite *1.202*, 6.297

magnesium **1**.106, ***1**.202*, ***6**.212*
alloys **6**.11–12, **6**.212–13, **6**.213
discovered **6**.96
ferromagnesian minerals **1**.121
ore **1**.202, **6**.297
steatite **1**.320
magnesium carbonate **1**.93–4, **1**.202
magnesium oxide **1**.202
Magnox reactor **6**.3–4, **6**.54, **6**.213
magnesium silicate **1**.210, **1**.299, **1**.330–1
magnesium sulphate **6**.378
magnet/magnetism/magnetic ***1**.202–3*, ***6**.113–14*, ***6**.212*
alloy ***6**.212*
anomalies ***1**.202*
ceramic **6**.116
compass **6**.72, ***6**.212–13*, **6**.215
dipole **1**.92, **6**.15, **6**.104
field ***1**.202*, ***8**.90*
and lines of force **1**.195, **1**.196
measurement **6**.104
flux ***1**.202*
geomagnetism **1**.136
measurement **1**.334, **6**.104
and North Pole **1**.139
ore separation ***6**.213*
palaeomagnetism **1**.244
polarity ***1**.202*
pole **6**.104
resonance imaging ***6**.213*
solenoid **6**.325
superconducting **6**.241
tape **6**.105, ***6**.213*, **6**.345
see also electromagnet *under* electromagnetic radiation
magnetite ***1**.203*, **6**.127, **6**.186
magnetograph ***8**.90*
magnetohydrodynamics ***8**.90*
generator ***6**.213*
magnetometer ***8**.90*
magnetopause **8**.90
magnetosheath **8**.90
magnetosphere ***8**.90*
magnetron **6**.288
magnifying glass **6**.223
magnifying power ***8**.90*
magnitude of galaxies **8**.38
magnitude of stars ***8**.90*
bolometric scale **8**.17
colour index **8**.27
distance modulus **8**.38
magnolias ***2**.198*, **2**.350
Magnox reactor **6**.3–4, **6**.54, ***6**.213*
Magnum Photos **5**.81
Magog ***9**.14*
magpie lark ***2**.198*
magpie moth ***2**.198*
magpies **2**.28, ***2**.198*
Magritte, René ***5**.275*
Magyars **3**.205, ***3**.216*
Mahabalipuram ***5**.275*
Mahābhārata **3**.40–1, **5**.134, ***5**.276*, **5**.401, ***7**.189*
Maharaj, Shambhu **5**.244
Mahathir bin Mohamed **4**.208
Mahavira, Vardhamana **3**.41, **3**.183, **7**.169
Mahāyāna Buddhism **7**.36, ***7**.189*, **7**.211, **9**.5, **9**.20, **9**.24
bodhisattva **7**.16, **7**.33
buildings **7**.37
Lotus Sutra **7**.187
Pure Land **7**.255, **9**.4
saints **7**.272
sangha **7**.273
shunyata **7**.285
Tibetan *see* Buddhism *under* religion *under* Tibet
Trikāya **7**.316
Mahdi/*mahdī* ***4**.207*, **7**.202
see also Muhammad Ahmad bin Abdallah
Mahendra of Nepal **4**.237
Mahfouz, Neguib **5**.19, **5**.276
Mahican Indians ***3**.216*
mahjong ***7**.189*
Mahler, Gustav ***5**.276*
Mahmud II, Ottoman Sultan **4**.207, **4**.215, **4**.253
Mahmud of Ghazna **3**.3, **3**.141, ***3**.216*, **3**.345
Mahmud of Malacca **3**.191
mahogany ***2**.198*
Mahoney, William Henry **8**.193
Mahuike **9**.25
maidenhair fern ***2**.198*
maidenhair tree **2**.136
mail (armour) *see* armour *under* body
mail (post) **4**.273, **6**.278
mail cheek fish ***2**.198*
Mailer, Norman ***5**.276*
Maillart, Robert ***6**.213–14*
Maillol, Aristide ***5**.276*
Maiman, Theodore Harold **6**.194, ***6**.214*
Maimonides, Moses ***3**.216–17*, **3**.348, **5**.203, ***7**.189*, **7**.310
main sequence stars **8**.67
see also dwarf stars
Maine (US state) ***3**.217*
Maine (battleship) ***4**.207*
mainframe computer ***6**.214*
maintenance after divorce ***7**.189*
maiolica ***5**.276*
Maire, Quett **4**.46
Maistre, Joseph de **7**.66
Maitreya ***7**.190*
maize ***2**.198*, **2**.332
Majapahit **3**.186, ***3**.217*
Maji-Maji ***4**.207*
Majnūn ***9**.24*
Major, John ***4**.207*, **4**.342
major axis of ellipse **8**.45
major planet ***8**.90*
see also Earth; Jupiter; Mars; Mercury; Neptune; Pluto; Saturn; Uranus; Venus
Majorca **1**.30
majoritarian representation **7**.104, ***7**.190*
Mājūj **9**.38
Majumdar, Charu **4**.236
Makarios III, Archbishop **4**.91, **4**.112, ***4**.207*
Makarova, Natalia **5**.30, **5**.248
Makeda **9**.33
make-up, theatrical ***5**.276–7*
see also arts *under* body
Maksutov telescope **8**.135, **8**.136
Malacca ***3**.217*
Malagasy *see* Madagascar
Malamud, Bernard ***5**.277*
Malan, Daniel François ***4**.207*, **4**.234, **4**.360
malapropism ***5**.277*
malaria ***2**.198–9*, **6**.72, **6**.96, **6**.169, **6**.377, **7**.59, **7**.107, **7**.333
Malaviya, Pandit Mohan **4**.152
Malawi (*earlier* Nyasaland) ***1**.203*, ***4**.207*, ***7**.366*
colonized **4**.306
crops **6**.34
decolonized **4**.4
leaders **4**.30, **4**.73, **4**.207
physical geography and geology **1**.203
in SADCC **7**.296
Malawi, Lake ***1**.203*
Malaya (*now* West Malaysia) ***3**.217*, ***4**.207–8*
arts and literature **3**.67, **5**.277
Chams people **3**.67–8
colonialism in **3**.7
decolonization **4**.321
Johore sultanate **3**.190
Malay language **7**.23
Malay literature ***5**.277*
Malayan Campaign ***4**.208*
Malayan Emergency ***4**.208*, **7**.72
port **3**.217
see also Malaysia
Malayalam **7**.95
Malaysia ***1**.203*, ***4**.208*, ***7**.366–7*
advertising **7**.4
decolonization **4**.321
economy and trade **6**.303, **7**.20, **7**.215
Konfrontasi with Indonesia **4**.185–6
land reform **7**.180
languages **7**.23, **7**.95
old age **7**.221
physical geography and geology **1**.42, **1**.203
politics **4**.281, **4**.359
poverty **7**.244
religion/theology **7**.302
states of **4**.299, **4**.302
West *see* Malaya
Malcolm III (Canmore), king of Scotland **3**.161, ***3**.217*, **3**.222, **3**.319
Malcolm IV, king of Scotland **3**.99
Malcolm X (black leader) **4**.40, **4**.175, ***4**.208*
Maldives ***1**.203*, ***4**.208*, ***7**.367*
Maldon, battle of ***3**.217*
male fern ***2**.199*
male mammals:
penis **2**.248
sperms **2**.303, **2**.319, **2**.338
testicle **2**.338
Malevich, Kasimir ***5**.277*, **5**.395
Malherbe, François de ***5**.277*
Mali ***1**.203*, **3**.4, **3**.141, ***3**.217*, ***4**.208–9*, ***7**.367*
decolonization **4**.4
ethnic groups **7**.94
finance **7**.124
languages **7**.30, **7**.217
physical geography and geology **1**.203
religion and myths **9**.27
rulers **3**.220, **3**.345–6, **3**.357
al-Malīk, ʿAbd **5**.228
Malik Ahmad of Ahmadnagar **3**.254
al-Malik al Muizz **3**.218
al-Malik an-Nasir **3**.218
Malik Shah of Turkey **3**.321
Malinowski, Bronislaw Kaspar **7**.14, ***7**.190*
and fieldwork **7**.120
and functionalism **7**.127
and gifts **7**.132, **7**.178
and magic **7**.189
and myths **7**.73, **7**.210
and personality **7**.252
and social evolution **7**.112
and society **7**.294
and totem **7**.313
and Trobriand Islanders **7**.316
mall **7**.302
mallard ***2**.199*
Mallarmé, Stéphane ***5**.277–8*
Malle, Louis **5**.323
mallee fowl ***2**.199–200*
mallow ***2**.200*
malnutrition ***7**.190*
Malory, Sir Thomas ***5**.278*, **9**.13
Malpighi, Marcello ***2**.200*
Malplaquet, battle of ***3**.217*
Malraux, André ***5**.278*
Malta ***1**.203*, ***3**.217*, ***7**.367*
in Council of Europe **7**.71
and EC **7**.99
and Knights Hospitaller **3**.198
language **7**.279
physical geography and geology **1**.203
politics and government **7**.250
religion **9**.43
Malter, L. **6**.391
Malthus, Thomas Robert ***4**.209*, **4**.273, **7**.52, **7**.53, **7**.88, **7**.101
maltose **6**.337
Maluku *see* Moluccas
Malvinas *see* Falkland Islands
Mama Occlo **9**.24
Mamaloni, Solomon **4**.318
mamba ***2**.200*
Mameluke dynasties (Egypt) **3**.4, **3**.217–18
art ***5**.278*
Mameluke sultanate (Delhi) ***3**.217*
mammals ***2**.200*
aquatic *see* mammals *under* water
armoured *see* edentates
badgers **2**.20, **2**.278
bat *see* bats
bear *see* bears
brain **2**.41
camel-like *see* camelids
egg-laying **2**.103, **2**.259
elephants **2**.105–6, **2**.200, **2**.205
endangered species **2**.108
female *see* female mammals
giraffe **2**.136
hair **2**.149
hares **2**.151, **2**.174
heart **2**.153
hibernation **2**.157–8
hippopotamus **2**.158
hoofed *see* ungulates
humans *see* man
hyena **2**.167
insectivorous *see* insectivores
Linnaeus and **2**.191
male *see* male mammals
marsupials *see* marsupials
Mustelidae *see* weasel family
otters **2**.237, **2**.301
pig *see* pigs
Pleistocene megafauna **2**.259
primates *see* primates
rabbit **2**.121, **2**.274
racoons **2**.70, **2**.179, **2**.240–1, **2**.275, **2**.285
rhinoceros **2**.284, **2**.371
rodents *see* rodents
ruminant *see* ruminant mammals
seals **2**.299, **2**.359
sleep **2**.313
sweat gland **2**.250
vision **2**.308
weasel family *see* weasel
mammary glands ***2**.200*
disorders **2**.205, **6**.214
mammography ***6**.214*
see also milk
mammoth ***2**.200*, **2**.201
Mammoth Cave National Park ***1**.204*
al-Mamun **3**.19
man ***2**.200*
brain **2**.41
see also H. sapiens under Hominids
management:
accounting ***7**.190–1*
buy-out ***7**.191*
production **7**.248
risk analysis **7**.268
science ***7**.191*
manakin ***2**.200–1*
manatee ***2**.201*, **9**.25
Manchester School ***4**.209*
see also Bright; Cobden; Smith, Adam; Ricardo
Manchu dynasty **3**.72, **3**.218, **7**.11
Manchuko ***4**.209*, **7**.255
Manchuria ***1**.204*, ***3**.218*, **7**.255
Mukden incident **4**.227
physical geography and geology **1**.204
puppet state **4**.209, **7**.255
rulers **3**.188, **3**.207
Manco Capac **3**.175, ***9**.24*
mandala **7**.173, ***9**.24*
mandarin (Chinese) ***3**.218*
mandarin (fruit) ***2**.201*
mandate **4**.169, ***4**.209*, ***7**.191*, **7**.316
Mandela, Nelson Rolihlahla **4**.5, ***4**.209*, **7**.238
Mandela, Winnie **4**.209

Mandelbrot, Benoit *1.204*
Mandelshtam, Osip *5.278*
Mandeville, Sir John 5.294
mandolin *5.278*
mandora *5.278*
mandrake *2.201*
mandrill *2.201*, 2.202
Manen, Hans van 5.122
Manes 3.218
Manet, Édouard *5.278–9*, 5.299, 5.304
Manfred, king of Sicily 3.69, 3.167, *3.218*, 3.327
manganese 1.106, *1.204*, *6.214*
  in alloys 6.45, 6.185, 6.235, 6.333
  ore 1.239
mangel-wurzel (mangold) 2.28
mango *2.201–2*
mangosteen *2.202*
mangrove *2.202*
Manhattan Project 1.121, 6.127, 6.200, *6.214*, 6.243, 6.245, 6.251, 6.369
mania 6.360
Maniapoto, Rewi 4.183
manic depression 7.96, *7.191*
Manichaeism *3.218*, 7.19
manifest destiny *4.209*
Manila Bay, battle of *4.209*
Manilow, Barry 5.358
Manipuri *5.279*
Manitoba *1.204*
Manjusri *9.24*
Manley, Michael 4.171
man-made fibre *see* synthetic fibres
Mann, Heinrich *5.279*
Mann, Thomas *5.279*, 5.478
manna 2.334
Manna-heim 9.25
manned space programme *6.214*
  see also space-flight and space exploration
Mannerheim, Carl Gustav Emil, Baron von 4.120, *4.209–10*
Mannerism 5.61, 5.83, 5.161, *5.279–80*, 5.339, 5.366, 5.469, 5.475, 5.499
Mannesmann, Max and Reinhard *6.214*, 6.364
Mannheim School *5.280*, 5.430
Manning, Henry Edward *4.210*
manometer *6.214*
manor house *3.218–20*
manorial system *3.220*
manpower planning *7.191*
  see also labour and employment
Mansa Musa 3.217, *3.220*
Mansart, François 5.280
Mansfield, Katherine *5.280*
manslaughter 7.146
manta ray *2.202*
Mantegna, Andrea *5.280–1*
mantis shrimp 2.202
mantissa 1.198
mantle of Earth 1.98–9, *8.90–1*
Manto 5.467
mantra *7.191*, 9.28
Manu *9.24*, 9.25
Manu Vaivasvata 9.24
Manueline style *5.281*
manufactured board *6.214*, 6.387
manufacturing (secondary industry) *7.277*
  aircraft 6.3, 6.98
  assembly line 6.23, 6.143, 6.216, 6.278
  automated *see* automation
  carpets 6.60–1
  clothes 6.77
  and computer 6.26, 6.85, 6.179
  conveyor 6.88
  employment percentage 7.106
  flexible system 6.134, 6.211
  footwear 6.142
  home *see* crafts
  light industry 7.184
manufacturing (secondary industry) (*cont.*):
  mass production *see* mass production
  motor car 6.23, 6.26, 6.129, 6.216
  nationalized 7.212
  paper *see* paper
  production *see* production
  recycling 7.260
  research and development 7.266
  robots *see* robotics
  textiles *see* textiles
  see also goods; industrialization
manure *see* compost and manure
manuscript illumination *see* illuminated manuscripts
Manx language 7.42, 7.156
Man'yōshū *5.281*
Manzikert, battle of *3.221*
Manzoni, Alessandro *5.281*
Mao Dun 5.95
Mao Zedong (Mao Tse-tung) 4.71, *4.210*
  and Civil War 4.73
  collectivization 4.78
  and Confucianism 7.65
  and Cultural revolution 4.90, 4.285, 7.191, 7.255, 7.267
  and Great Leap Forward 4.140, 7.191
  and guerrilla warfare 4.143, 7.136
  Hundred Flowers Movement 4.158
  and Jiangxi Soviet 4.173
  Long March 4.74, 4.199
  Maoism *7.191–2*
  and Peasants' Bureau 4.265
  and Red Guards 4.285
  and USSR 4.183
Mao Zonggang 5.387
Maoris *4.210*, *7.192*
  Anglo-Maori wars 4.13–14, 4.294, 4.338
  arts *5.281*
  land 4.287
  language 7.23
  leaders 4.275, 4.283, 4.339, 4.340, 4.341
  movements 4.183, 4.255–6, 4.287
  music *5.282*
  Waitangi Treaty 4.370
maple 2.89, *2.202*
mappings (mathematics) 1.129, *1.204*
maps and charts *1.204*, 1.205, *6.214–15*, 6.219
  cartography 1.53–4, 1.204, 1.210–11, 1.241–2
  contour 1.74
  eastings 1.99
  globe 1.141
  grid system 6.391
  Moon (selenography) 8.147
  oceans 1.208
  projections *see* projections
  scale 1.294
  stars 8.158–9
  weather 1.175
  see also latitude and longitude; surveying; world distribution *and under individual countries*
maqām 5.487
Maquis *4.211*
Mar, John Erskine, earl of 3.126, 3.127, 3.326
Mara *9.24*
Marabai 5.206
marabou *2.202*, 2.203
Maracaibo, Lake *1.204–5*
maracas *5.282*
Marat, Jean Paul 3.87, *3.221*
Maratha people and Wars 3.167, *3.221*, 3.328–9, *4.211*
Marathon (place) 3.221
marathon (race) *7.192*
Maratta, Carlo *5.282*
marble 1.47, *1.205*, *5.282*
Marbut, Curtis Fletcher *1.205*
Marc, Franz *5.282*
marcasite *1.205*
Marceau, Marcel *5.282*, 5.296
march *5.282–3*
March, earls of 3.110, 3.181, 3.241
Marchand, Jean-Baptiste *4.211*
marcher lords *3.221*, 3.241, 3.276
Marco Polo *1.205*, 3.121, 3.136, 3.201, 3.215, *3.221*, 3.328, 3.389
Marconi, Guglielmo *6.215*, 6.290, 6.291
Marcos, Ferdinand Edralin *4.211*, 4.268–9
Marcos, Imelda 4.211
Marcus Aurelius, Roman emperor 3.17, *3.221*, 3.342
Marcuse, Herbert 7.124, *7.192*, 7.215
Marcy, William 4.326
Marduk *9.24*
mare (on Moon) *8.91*
Marenzio, Luca 5.275, *5.283*
Marey, Etienne-Jules 6.74
Margai, Sir Milton 4.312
Margaret I, queen of Denmark 3.102
Margaret of Anjou *3.222*, 3.241, 3.313, 3.352, 3.360
Margaret of Antioch, Saint *9.42*
Margaret, Maid of Norway *3.221–2*
Margaret, Saint, queen of Scotland *3.222*, 3.319
margarine *6.215*, 6.384
marginal analysis *7.192*
marginal sea *1.205*
margrave *3.222*
Mari Jata 5.7
maria (craters) *see* mare
Maria Theresa, empress of Austria 3.152, 3.153, *3.222*, 3.290
  and Austrian Succession, War of 3.6, 3.29, 3.30, 3.133
  and Seven Years War 3.323–4
Marianas Islands 1.147, 3.231
Marianas Trench *1.206*
Marie Antoinette *3.222*, 3.372
marigold *2.202–3*
marijuana *see* cannabis
marimba 5.496
Marín, Muñoz 4.278
marine areas *see* oceans and seas
marine fish *see* fish *under* oceans and seas
marine iguana 2.300
marine mammals *see* mammals *under* water
marine navigation 6.237–9
  see also navigation
marine painting *5.283*
marine resource management *6.215*
marine vehicles *see* ships and boats
Mariner *8.91*
Marinetti, Filippo Tommaso 5.166
Marino, Giambattista *5.283*
Maris, Jacob, Matthijs, and Willem 5.197
maritime law 7.272
Maritime strike *4.211*
Marius, Gaius *3.222*, 3.308, 3.314, 3.352
Marivaux, Pierre Carlet de Chamblain de *5.283*
marjoram *2.203*
Mark, Saint 3.41, 9.12, 9.42, *9.42*
Mark Antony *3.222–3*
  and Actium 3.3, 3.6
  and Cicero 3.76
  and Cleopatra 3.79
  and Octavian 3.51, 3.307
  in Triumvirate 3.28
Markab *8.91*
Markarian, Beniamin 8.55, 8.91
  Markarian galaxy *8.91–2*
market (economics) *7.192*
  commodity 7.57
  common *see* common markets
market (economics) (*cont.*):
  free 7.125, 7.126, 7.139
  futures (forward) 7.128
  Pareto utility 7.229
  see also capitalism; trade
market research *7.192*
market socialism *7.192–3*
marketing *7.192*
markhor *2.203*
Markievicz, Constance 4.381
Markova, Alicia 5.30
marl *1.206*
Marlborough, John Churchill, 1st duke of *3.223*
  Blenheim and Blenheim Palace 3.89
  and Sedgemoor, battle of 3.320–1
  and Spanish Succession, War of 3.43, 3.89, 3.265, 3.338
  and Sweden 3.257
Marley, Bob 7.258
marlin *2.203*
Marlowe, Christopher 5.49, *5.283*, 5.304, 5.341, 9.13
Marmara, Sea of 1.84
Marmion, Simon 3.133
marmoset *2.203*
marmots *2.203–4*, 2.370
Marne, battles of *4.211*
Maronites *4.211*, *7.193*
Marot, Clément *5.283*
Marprelate tracts *3.223*
Marquesas Islands *1.206*, *4.212*
marquetry 5.168, *5.283*, 5.436
Marquette, Jacques *1.206*
Márquez, Gabriel García 5.169–70
marram *2.204*
marriage *7.193*, 7.193, 7.268
  common law 7.55
  divorce 7.93–4
  dowry 7.95
  outside kin group 7.154
  as sacrament 7.271
  see also family
marrow *2.204*
Mars (god) *9.24*
Mars (planet) *8.92*
  canals of 8.19–20
  illustrated 8.94–5
  impact basins 8.8, 8.26, 8.65
  orbit 8.17, 8.196
  plain 8.46
  satellites 8.36, 8.119, 8.198
  statistical details 8.196
  volcano 8.117
Marsh, Othniel Charles *2.204*
marsh *see* swamps and marshes
marsh deer *2.204*
marsh gas *see* methane
marsh harrier *2.204*
Marshal, William and Richard *see* Pembroke
Marshall, Alfred 7.187, *7.193*, 7.249, 7.265
Marshall, George Catlett: Marshall Plan 4.74, *4.212*
Marshall, John *4.212*
Marshall, Thurgood 4.53
Marshall Islands *1.206*, 3.231, 4.38, *4.212*, 7.296
marshalling yard *6.215*, 6.216
marshes *see* swamps and marshes
Marston, John 5.127
Marston Moor, battle of *3.223*
marsupials *2.204*
  bandicoots 2.21, 2.274
  flying phalanger 2.124
  kangaroos 2.177, 2.278, 2.347, 2.359
  koala 2.180
  marsupial mole *2.204*
  opossum 2.235, 2.278
  phalanger 2.251
  possum 2.266
  pouched mouse 2.266
  rabbit-eared bandicoot 2.274

marsupials (*cont.*):
rat kangaroo **2**.278
rat opossum **2**.278
Tasmanian devil **2**.336
Tasmanian wolf **2**.336
wallaby **2**.359
wombat **2**.369–70
martagon lily **2**.190
martens ***2**.204*, **2**.291
Martha ***9**.24*
Martial **5**.148, ***5**.283*
martial arts ***7**.193–4*
martial law ***7**.194*
Martin V, pope **3**.146
Martin, E. Hall **4**.225
Martin, Emile and Pierre **6**.188
Martin, Jean-Baptiste **3**.372
Martin, John ***5**.284*
Martin du Gard, Roger ***5**.284*
Martin of Tours, Saint ***9**.42*
martin (bird) **2**.164, ***2**.204*, **2**.293
Martinique **1**.221, **1**.233, **7**.367
Martins, Peter **5**.393
Martinson, Harry **5**.404
Martinů, Bohuslav ***5**.284*
Martov, L. **4**.44, ***4**.212*
Martyn, Edward **5**.226
martyrs ***3**.223*, **7**.194, **7**.272
Marut **9**.15
Marvell, Andrew ***5**.284*, **5**.325
Marx, Karl/Marxism ***4**.212–13*, ***7**.194*
and alienation **7**.10
and atheism **7**.21
and capitalism **7**.39
and class struggle **7**.53, **7**.289
and communism **7**.60, **7**.61
and Communist manifesto **4**.80
and critical theory **7**.76
and dialectical materialism **7**.89, **7**.194, **7**.195
and Engels **4**.112
failure of **7**.194
and Frankfurt School **7**.124
and historiography **4**.153
as humanist **7**.149
and ideology **7**.11, **7**.151
and Internationals **4**.164
and labour theory of value **4**.290, **7**.179
Marxists **4**.193, **4**.271, **7**.11, **7**.53, **7**.135, **7**.141, **7**.194, **7**.195, **7**.291
see also Engels
and millenarianism **7**.202
and socialism **4**.317, **4**.318, **7**.291
and society **7**.294
and surplus value **7**.249, **7**.303
Marx, Roberto Burle **5**.170
Mary I, queen of England and Ireland (1516–58) ***3**.224*, **3**.286, **3**.362, **3**.365, **3**.386
Mary II, queen of England, Scotland, and Ireland (1662–94) ***3**.224*, **3**.343, **5**.489
Mary, queen of Scots (1542–87) ***3**.223–4*
adviser **3**.304–5
and Elizabeth I **3**.114
execution **3**.66
husbands **3**.47, **3**.97
plots to put on throne **3**.32, **3**.256, **3**.304, **3**.356
Mary Magdalene, Saint ***9**.24*, ***9**.42*
Mary of Guise **3**.40
Mary the Virgin **7**.49, ***9**.24*
Maryland ***3**.224*
Masaccio ***5**.284*
Masada ***3**.224*
Masaoka Shiki **5**.197
Masaryk, Jan ***4**.213*
Masaryk, Tomáš Garrigue **4**.25, **4**.34, **4**.92, ***4**.213*, **7**.35
Mascagni, Pietro ***5**.284*, **5**.471
mascon ***8**.92*
Masefield, John ***5**.284*
maser ***6**.215*, ***8**.92*
Mashonaland **4**.390
Masina, Giulietta **5**.153
Maskelyne, J. N. **6**.25
Maskelyne, Nevil (Astronomer Royal) **8**.193
masks **5**.3, **5**.7, ***5**.284–5*, **5**.289, **7**.151
Masolino da Panicale ***5**.285*
Mason, Charles: Mason–Dixon line ***3**.224*
Mason, George **3**.129
Mason, John **3**.252
mason bee ***2**.204*
mason wasp ***2**.204*
Masonic Order ***3**.133–4*, **4**.15, ***7**.195*
masonry ***5**.285*, **6**.51, ***6**.216*
bridge **6**.47, **6**.48, **6**.372
dam **6**.95
skyscraper **6**.322
masons *see* Masonic Order
Masowe, John **7**.5
masque **5**.50, **5**.258, ***5**.285*
masquerades **5**.7
Mass (Eucharist) ***5**.285*, **5**.380–1, **7**.123, **7**.185–6, **7**.271, **7**.272
France **5**.89, **5**.137, **5**.256, **5**.325
see also sacred music
mass (matter) ***1**.206*, ***8**.92–3*
and astronomy **8**.92–3, **8**.192
centre of **8**.61
mass–luminosity relation ***8**.93*
missing mass problem **8**.100
biomass **2**.30
conservation **1**.73
density **1**.87
measurement **1**.28, **1**.183, **6**.368
moment of inertia **1**.218
relative atomic **1**.280
relative molecular **1**.280
mass defect (nuclear physics) ***1**.206*
mass flow (geomorphology) ***1**.206*, **1**.341–2
avalanches **1**.27, **1**.28
cambering **1**.49
creep **1**.49, **1**.79
landslides **1**.187
mudflows **1**.222, **1**.223
overland ***1**.242–3*
rock glaciers **1**.284–5
rockfalls **1**.284
mass media *see* mass *under* media
mass production ***4**.213*, **6**.143, ***6**.216*, **6**.326, **6**.384, **7**.10, **7**.248
see also automation
mass spectroscopy **6**.186, ***6**.216–17*
Massachusetts ***3**.225*, **3**.383
Institute of Technology ***6**.216*
Masséna, André **4**.265
Massenet, Jules ***5**.285*, **5**.366
Massey, William Ferguson ***4**.213*
Massif Central ***1**.206–7*
Massine, Léonide ***5**.285*
Massinger, Philip ***5**.285–6*
Massys, Quentin, Jan, and Cornelis ***5**.286*
Master of the Queen's and King's Music ***5**.286*
Masters, Virginia **7**.280
mastic **6**.66
mastication ***2**.205*
mastitis ***2**.205*
mastodon ***2**.205*
al-Masudi ***1**.10*
Masurian Lakes ***4**.213*
Matabeleland *see* Zimbabwe
matamata ***2**.205*
Match Girls' strike ***4**.213*
matches **6**.132
matchlock **6**.131
materialism ***7**.195*
dialectical **7**.89, **7**.194, **7**.195
and mind **7**.203
maternity leave **7**.228
see also pregnancy
mathematics ***1**.207*
Abelian group **1**.1
algebra *see* algebra
algorithm **1**.8, **6**.11
analysis **1**.13
applied *see* applied mathematics
arithmetic *see* arithmetic
associativity **1**.22
averages **1**.27, **1**.209–10, **1**.217
axiom **1**.28
binary **1**.39
Boolean algebra **1**.41, **1**.198, **6**.207
calculators **6**.1, **6**.53–4, **6**.84, **6**.259, **6**.322
calculus *see* calculus
catastrophe theory **1**.54
catenary **1**.54
chaos **1**.58
closure **1**.66
commutativity **1**.70
complex numbers **1**.18, **1**.70
convergent series **1**.74
coordinates *see* coordinates
De Moivre's theorem **1**.86
derivative **1**.88
determinant **1**.88
differential equations **1**.90
differentiation **1**.90, **1**.91, **1**.98
dimensions **1**.91
divergent series **1**.93
ergodic theory **1**.112
existence theorems **1**.116
exponential function **1**.118
factorial **1**.119
Fibonacci numbers **1**.121
Fourier analysis **1**.125
functions **1**.129
game theory **1**.131, **6**.240, **7**.259
geometric series **1**.136
geometry *see* geometry
graphs *see* graphs
group theory **1**.147
hyperbola **1**.72, **1**.73, **1**.163
identity element **1**.165
imaginary numbers **1**.166
induction **1**.167
information theory **1**.168, **6**.180
inverse elements **1**.171
inverse function **1**.171
irrational numbers **1**.174
isomorphism **1**.175
iterative methods **1**.176
lemma **1**.191
linearity **1**.195–6
logarithms **1**.198, **1**.224, **6**.53, **6**.322
mappings **1**.129, **1**.204
mathematicians *see under individual countries*
matrices **1**.207
mechanics **1**.209
Möbius strip **1**.217
modelling **6**.17, **6**.110, **6**.126–7, **6**.320
modulus **1**.217
Monte Carlo method **1**.220
numbers *see* numbers
operations **1**.239
paradox **1**.247
parameter **1**.247–8
philosophy **7**.34, **7**.86, **7**.182, **7**.331
logicism **7**.187
see also logic
postulate **1**.247, **1**.264
power **1**.266
power series **1**.266
probability **1**.268
pure **6**.385
Pythagorean *see* Pythagoras
quantum numbers **1**.271
random numbers **1**.276
real numbers **1**.277
reciprocal **1**.277
regression analysis **1**.280
root-mean-square ***1**.286*
mathematics (*cont.*):
round-off errors **1**.286
saddle points **1**.289
scalars **1**.294
sequences **1**.121, **1**.299
series *see* series (mathematics)
sets *see* set theory
shapes and forms
illustrated ***9**.51*
see also in particular angle; circle; conic section; triangle
significant figures **1**.304
simulation **6**.17, **6**.320
slide-rule **6**.322
statistical *see* statistics
symbols ***9**.50*
Taylor expansion **1**.332
tensor **1**.333
transcendental numbers **1**.341
trigonometry *see* trigonometry
truth tables **1**.345
uniqueness theorems **1**.350
vectors *see* vector (mathematics)
Venn diagram **1**.355
volume **1**.358
see also quantum mechanics *and under individual countries*
Mather, Cotton **3**.131, ***3**.225*, **3**.314
Mather, Increase ***3**.225*
Mathura art ***5**.286*
Matilda ***3**.225*
Matisse, Henri ***5**.286*
Mato Grosso ***1**.207*
matriarchy ***7**.195*
matrices ***1**.207*
matrilineality **7**.86, ***7**.185*
Africa **7**.18, **7**.19, **7**.30
Asia and Pacific **7**.9, **7**.197, **7**.202
see also Trobriand Islanders
North American Indians **7**.166, **7**.213, **7**.219, **7**.236, **7**.312, **7**.316
Matsuo Bashō **5**.37
Matsya ***9**.24–5*
Matteotti, Giacomo ***4**.213*
matter ***1**.207*
annihilation of **1**.14
contraction of **1**.74
expansion of **1**.117
phases of **1**.256
states of **1**.320
see also gas; liquid; liquid crystal; plasma; solid
waves of **1**.85
Matterhorn ***1**.207*
Matthew, Saint **3**.41, **3**.104, **9**.12, ***9**.42*
Matthias I (Corvinus), king of Hungary (1443–90) **3**.172, **3**.173, ***3**.225*, **3**.226
Matthias, Holy Roman Emperor (1557–1619) **3**.153, **3**.168
Mattioli, Ercolo Antonio ***3**.225*
Matua, Henare **4**.287
Mau Mau ***4**.213–14*, **7**.176
Mauchly, John William **6**.85, ***6**.217*
Maud (Matilda) **3**.225
Maude, Sir Frederick **4**.217
Maudling, Reginald **4**.103
Maudslay, Henry **6**.45, ***6**.217*, **6**.236, **6**.247, **6**.385
Maugham, William Somerset ***5**.286*
Maui-tiki-tiki ***9**.25*
Mauna Kea **8**.73, ***8**.93*, ***8**.96*
Maunder minimum ***8**.96*
Maupassant, Guy de ***5**.286*, **5**.330
Mauriac, François ***5**.286*
Maurice, duke of Saxony **3**.318
Maurice, Michael Denison **4**.63
Maurice of Nassau, prince **3**.106, **3**.108, **3**.262
Mauritania ***1**.207*, **3**.141, ***4**.214*, ***7**.367–8*
in Arab League **7**.15
colonization and decolonization **4**.4, **4**.306

Mauritania *(cont.)*:
languages **7**.30
physical geography and geology **1**.207
refugees **7**.261
urban areas **7**.282
Mauritius ***1**.207*, ***4**.214*, ***7**.368*
Maurolyko, Francesco **8**.177
Maurras, Charles **4**.2
Maury, Matthew Fontaine ***1**.207–8*
Mauryan empire ***3**.225–6*
art ***5**.286–7*
technology **6**.299
see also Ashoka
mausoleum ***5**.287*, **5**.307, **5**.456
see also tombs
Mauss, Marcel **7**.132
mauveine **6**.109, **6**.260
al-Mawdudi, Syed Abu ***4**.168*
Mawlid an-Nabi ***7**.195*
Mawu-Lisa ***9**.25*
Max Planck Society for the Promotion of Sciences ***6**.217*
Maxim, Sir Hiram Stevens **6**.210, ***6**.217*
Maximilian I, king of Bavaria (1818) **4**.32
Maximilian I, Holy Roman Emperor and king of Romans (1459–1519) **3**.152, **3**.153, **3**.168, ***3**.226*
and Belgium **3**.38
and Diets **3**.103
and Fuggers **3**.135
Maximilian II, Holy Roman Emperor and king of Romans (1527–76) **3**.153, **3**.168, **5**.13
Maximilian, Archduke and emperor of Mexico (1832–67) ***4**.214*, **4**.220
Maximilian Joseph, king of Bavaria **3**.37
Maxwell, James Clerk **1**.143, **1**.154, ***1**.208*, **6**.168, **6**.290, **8**.145
Maxwell–Boltzmann distribution ***1**.208*
Maxwell Davies, Sir Peter **5**.126
Maxwell Montes **8**.182
may blossom **2**.152–3
May Fourth Movement ***4**.214–15*
māyā (Hinduism) **7**.3, ***7**.195–6*
Mayakovsky, Vladimir ***5**.287*
Mayans ***3**.226–7*
arts ***5**.287–8*
astronomy **8**.18
calendar **3**.227, **6**.54
languages **7**.199, **7**.219
religion/theology **9**.18–19
technology **6**.54, **6**.287, **6**.346
Mayapán ***3**.227*
Maybach, Wilhelm **6**.94, **6**.234
Mayflower **3**.282
mayfly ***2**.205*
Mayhew, Henry ***4**.215*
Mayr, Ernst Walter **2**.205
Mazama, Mount **1**.79
Mazarin, Jules **3**.84, **3**.135, ***3**.227*
Mazaryk *see* Masaryk
Mazeppa **3**.88, **3**.257, **3**.287
Mazowiecki, Tadeuz **4**.370
Mazrui family **4**.299
Măzŭ Dàoyì ***9**.25*
Măzŭpō ***9**.25*
mazurka ***5**.288*
Mazzini, Giuseppe **4**.131, **4**.214, ***4**.215*, **4**.290, **4**.387
Mboya, Tom ***4**.215*
Mead, George Herbert **7**.278, **7**.292, **7**.303
Mead, Margaret ***7**.196*, **7**.273, **7**.280
Meade, General **4**.136
meadow pipit ***2**.205*
meadow saffron **2**.19
meadow soil ***1**.208–9*, **1**.311
meadowsweet ***2**.206*
mealybug **2**.295
mean ***1**.209*
mean anomaly **8**.6
mean sea-level ***1**.209*
mean solar day **8**.36, **8**.109
mean solar time **8**.172
mean solar year **8**.189
mean sun **8**.36, **8**.109, **8**.172
meandering ***1**.209*, **1**.261
meaning ***7**.196*
image **7**.152–3
semantics **7**.278
see also propaganda
Meany, George **4**.11
Meares, John **3**.254
measles ***2**.206*, **7**.153
measurement ***6**.180–2*, ***9**.49–50*
acceleration **6**.1
acidity **6**.267
air and wind speed **6**.14
altitude *see* altitude
amount of substance **1**.218
angle **6**.157
capacitance **1**.119
capacity **9**.49
climate *see* meterology
cubic **9**.49
depth **6**.111, **6**.197
dimensions **1**.91
displacement **6**.184
distance **6**.249, **6**.259
electricity **1**.11, **1**.78, **1**.237, **6**.114, **6**.120
current **6**.13, **6**.149, **6**.368
voltage **6**.368, **6**.374
electromagnetic radiation **6**.291
electronic signal **6**.97
energy **1**.104–5, **1**.178
see also calories
force **1**.124, **1**.227
frequency **1**.154
hearing **6**.25
heat *see* temperature *below*
humidity **6**.221
inductance **1**.154
inertia *see* mass
ions **6**.153
latitude *see* latitude and longitude
length *see* length
light **6**.327–8
polarized **6**.274, **8**.126
longitude *see* latitude and longitude
luminosity **1**.50, **6**.328, **6**.368
luminous intensity **1**.50
magnetism **1**.334, **6**.104
mass **1**.28, **1**.183, **6**.368
micrometer **6**.199, **6**.222, **6**.294
nuclear radiation **6**.106, **6**.152–3, **6**.186
potential difference **1**.358
power **1**.362
precipitation **6**.221
pressure *see under* pressure
radioactivity **1**.35
remote **6**.347
resistance *see* resistance
sensor **6**.312
sound **6**.97, **6**.184
specific gravity **6**.174
speed *see* velocity *below*
square ***9**.49*
strain **6**.335
surveying *see* surveying
temperature *see under* temperature
thermodynamic **1**.181, **6**.368
time **1**.298
see also clocks
units *see* units
velocity and speed **6**.14, **6**.192, **6**.206, **6**.328, **6**.368, **6**.386
Vernier calliper **6**.198, **6**.371
volume **6**.368
water consumption **6**.377
water depth **6**.111, **6**.197
weight **1**.28, **9**.49
see also instrumentation
measuring worm **2**.194
meat ***6**.217*
abbatoir **6**.1
beef **2**.57
cane rat **2**.52
chicken **2**.63
dormouse **2**.97
frozen **6**.297
guinea pig **2**.148
llama **2**.186
manatee **2**.201
pork **6**.270
reindeer **2**.282
walrus **2**.359
see also livestock
Mecca ***3**.227*, **3**.281
ḥajj and Ka'aba **7**.121, **7**.166–7, ***7**.175*, **7**.209, **7**.236, **7**.271, **7**.281
Méchain, Pierre **8**.46, **8**.97
mechanical advantage **6**.151–2, **6**.199, ***6**.217*, **6**.285
mechanical energy **6**.121, ***6**.151–2*, **6**.153, **6**.166
mechanical engineering **6**.122, ***6**.217*
crane **6**.91–2, **6**.158
dynamometer **6**.110
lift **6**.201, **6**.322
lock **6**.45, **6**.205, **6**.206, **6**.304
pioneers **6**.49
robotics **6**.26, **6**.259, **6**.301, **6**.326, **6**.383
servo-mechanisms **6**.312, **6**.314
soil mechanics **6**.324
see also civil engineering; engines; locomotives; machinery; machine-tools; vehicles
mechanical musical instruments ***5**.288*
mechanical transmission *see* transmission in vehicles
mechanics (mathematics) ***1**.209*
mechanics (physics):
celestial **8**.22
fluid **6**.138
see also quantum mechanics
mechanisms of chemical reactions ***1**.209*
mechanization *see* automation
mechatronics **6**.179, ***6**.217*
Mečiar, Vladimir **4**.92
Meck, Nadezhda von **5**.448
medals ***5**.288–9*, **5**.350
Medbh **9**.9, ***9**.25*
Medea **9**.19, **9**.38
Medes ***3**.227*
media ***7**.196–7*
art ***5**.289*
see also paint
astronomy:
intergalactic **8**.73–4
interplanetary **8**.74
interstellar **8**.74
mass ***7**.195*
advertising **7**.3–4
medium as message **7**.196, **7**.197
see also journalism; news; persuasion; radio; television
median ***1**.209–10*
mediation **7**.92
Medicare ***4**.215*
Medici family **3**.363
and arts **3**.301, **5**.158, **5**.230, **5**.294, **5**.337, **5**.369, **5**.378, **5**.379
Catherine de Medici **3**.81, ***3**.227*, **3**.258, **3**.313, **3**.369, **5**.337
Cosimo de Medici ***3**.227–8*
Giuliano de Medici **3**.273
Lorenzo de Medici ***3**.228*, **3**.273, **5**.230, **5**.369, **5**.379
and Machiavelli **3**.215
Marie de Medici **3**.340
Pazzi Conspiracy against **3**.272–3
medicine and health care and medical technology ***2**.206*, ***4**.148–9*, ***6**.217–18*, **6**.346, ***7**.139–40*
acupuncture **6**.3
alternative *see* complementary medicine
ambulance **6**.12
anaesthetics *see* anaesthesia
anticoagulant **6**.16
antisepsis *see* antisepsis
artificial limb **6**.21
asepsis **6**.22
Ayurvedic **2**.19, **7**.24
biopsy **6**.37
blood *see* blood
CAT scan **6**.85, **6**.213, **6**.218, **6**.292, **6**.356, **6**.357
catheter **6**.64
Christian Science **7**.48
chronology **6**.218
cohort study **7**.56
in community **7**.61
complementary *see* complementary medicine
diathermy **6**.101, **6**.251
diseases *see* diseases and disorders of humans
doctors *see* physicians
education **2**.9, **2**.37–8, **2**.190, **2**.237, **2**.241, **2**.364
electrocardiography **6**.112–13, **6**.116, **6**.218
electroconvulsive therapy **6**.116
electroencephalograph **6**.34, **6**.116, **6**.218, **6**.240–1
endoscopy **6**.37, **6**.120, **6**.248, **6**.251, **6**.340
ethics ***7**.196*
abortion **7**.1, **7**.115
euthanasia **7**.111–12, **7**.301
finance ***7**.139–40*
flying doctor **7**.340
forensic **6**.143, **6**.217
grafting **6**.158
haematology **6**.112, **6**.164, **6**.218
haemodialysis **6**.101, **6**.164
healing **7**.11, **7**.62, **7**.139, **7**.282
hospital **3**.130, **7**.147
hypnosis **2**.167, **7**.126, **7**.150
and infant mortality **7**.159
intensive care **6**.116, **6**.183, **6**.298
laser **6**.99, **6**.251, **6**.340
leech **2**.186
lithotripsy **6**.205, **6**.340, **6**.367
lumbar puncture **6**.210
magnetic resonance imaging **6**.213
medical profession *see* physicians
Medicare **4**.215
nuclear *see* nuclear medicine
occupational **2**.276
and old age **7**.222
operations *see* surgery
ophthalmoscope **6**.120, **6**.218, **6**.251
optical fibre **6**.251
orthopaedics **2**.236, **6**.218, **6**.253
otorhinolaryngology **6**.254
pacemaker **6**.218, **6**.255
paramedical workers **7**.228
pathology *see* pathology
pharmacology *see* drugs, therapeutic
physiotherapy **2**.253
positron emission tomography **6**.243, **6**.259, **6**.277–8, **6**.357
preventive medicine **7**.246
primary **7**.159, **7**.246–7
prosthetics *see* prosthesis
public health **7**.254
radiation therapy **6**.57, **6**.218, **6**.292
radiology **6**.30, **6**.210, **6**.291
see also tomography
rehabilitation **7**.263
respirator **6**.298

medicine and health care and medical technology (*cont.*):
screening **6**.68, **6**.214, ***7**.140*
speculum **6**.68, **6**.328
sphygmomanometer **6**.218, **6**.328
sterilization of equipment **6**.22, **6**.334
stethoscope **6**.193, **6**.218, **6**.335
surgery *see* surgery
Susruta's treatise **6**.342
teaching *see* education *above*
ultrasound **6**.60, **6**.218, **6**.340, **6**.357, **6**.367
veterinary **6**.371
see also diseases; drugs, therapeutic; obstetrics and gynaecology; surgery; World Health Organization *and under individual countries*
medieval music **5**.274–5, ***5**.289*, **5**.297, **5**.306, **5**.332, **5**.351, **5**.368
see also polyphony
Medina ***3**.228*, **7**.142, **7**.207, **7**.209
Medina Sidonia, Alonso Pérez de Guzmán, duke of ***3**.228*, **3**.335
meditation ***7**.196*, **7**.205
Mediterranean:
physical geography and geology ***1**.210*
climate **1**.65, **1**.66
islands:
Balearics **1**.30
Corfu **1**.76
Corsica **1**.77
Cyclades **1**.82
Gozo **1**.309
Ionian **1**.172
Sardinia **1**.293
see also Crete; Cyprus; Malta; Sicily
seas **1**.3
strait **1**.211–12
Sea Peoples **3**.320
technology and science **6**.195, **6**.209, **6**.357
agriculture **6**.7, **6**.76, **6**.351
ships and boats **3**.248, **6**.143, **6**.149, **6**.195, **6**.276, **6**.307, **6**.316, **6**.333, **6**.337, **6**.375
see also individual countries and islands
medium (art) ***5**.289*
see also paint
medium (spiritual) **7**.297
medium (communication) *see* mass *under* media
medium-wave (frequency) **1**.103, **6**.380, **9**.54
medlar ***2**.206*
Medusa **9**.14
meerkat ***2**.206*
meerschaum ***1**.210*
megacity ***7**.197*
megaliths ***3**.228*, **3**.229, **5**.434, ***6**.218*, **6**.335
megalosaurus ***2**.206*
Meganeura ***2**.206*
megapode **2**.199–200
Megasthenes **3**.226
megatherium **2**.135
Mège-Mouriès, H. **6**.215
Megiddo, Mount of **9**.5
Mehemet Ali of Egypt (Mohammad Ali) **3**.218, **4**.141, **4**.207, ***4**.215*, **4**.332
Mehmed II, Ottoman sultan ***3**.228*, **3**.265
Mehmed VI, Ottoman sultan **4**.254
Mehta, Narsimha **5**.46
Meighen, Arthur **4**.85
Meiji Constitution ***4**.215–16*
Meiji Restoration ***4**.216*
Meikle, Andrew **6**.355
Meinhof, Ulrike **4**.27
Meinong, Alexius ***7**.197*
meiosis ***2**.207*, **2**.214
Meir, Golda ***4**.216*, **4**.381
Meistersinger **5**.297, **5**.397
Meitner, Lise **1**.150, ***1**.210*, **6**.127, ***6**.218*, **6**.242
Mekeren, Jan van **5**.283
Mekong ***1**.210*
melā ***7**.197*
melamine-formaldehyde **6**.273
Melanchthon, Philip **3**.27, ***3**.228*
Melanesia ***3**.228*, ***7**.368*
art ***5**.289*
cargo cults **7**.40
languages **7**.23
physical geography and geology **1**.315
religion and myths **7**.14, **9**.25
society **3**.228, **7**.175, ***7**.197*
see also Papua New Guinea; Trobriand Islanders
Melba, Dame Nellie ***5**.289*
Melbourne, Bruce, viscount **4**.53
Melbourne, William Lamb, 2nd viscount ***4**.216*, **4**.217
Meleager ***9**.25*
Méliès, Georges ***5**.290*
Mellon, Andrew William ***4**.216*
melodeon **5**.2
melodrama ***5**.290*, **5**.390
melody ***5**.290*
melon ***2**.207*
melting-point ***1**.210*
Melville, Herman ***5**.290*, **5**.459, **5**.485
membrane ***2**.207*
Memlinc, Hans ***5**.290*, **5**.369
memoirs *see* autobiography
Memorandum of Association ***7**.197*
memory (electronics) **6**.183, ***6**.218–19*
memory (mind) ***2**.207*, ***7**.197*
collective *see* myths and legends; stereotype
disorder ***7**.197–8*
and hypnosis **7**.150
see also mind
memory metal ***6**.219*
Memphis ***3**.228–9*
men *see* patriarchy; patrilineality
Mena, Juan de **5**.428
Menander of Athens (342–290 BC) ***5**.290*
Menander of Bactria (155–30 BC) **3**.33
menarche **2**.208
Mencius **3**.73, ***3**.229*, **3**.390, **7**.265
Mencken, Henry Louis ***5**.290*
Mendel, Gregor Johann **2**.134, ***2**.207*, **6**.36, **7**.80
Mendeleyev, Dimitri Ivanovich ***1**.210*, **1**.251, **6**.25
mendelevium **1**.106
Mendelsohn, Erich ***5**.290*
Mendelssohn, Moses **4**.176
Mendelssohn-Bartholdy, Felix **5**.88, ***5**.291*
Mendès-France, Pierre ***4**.216*
Mendez Montenegro, Julio Cesar **4**.142
Mendoza, Antonio de ***3**.229*
Mendoza, López **5**.428
Menelaus, king of Sparta **9**.16, ***9**.25*
Menelik I, emperor of Ethiopia **9**.33
Menelik II, emperor of Ethiopia **4**.3, **4**.113, ***4**.216–17*
Menem, president of Argentina **7**.248
Meneses, Jorge de **4**.259
Menger, Carl **7**.53
Mengistu, Colonel **4**.113
Mengs, Anton Raffael ***5**.291*, **5**.490
menhir ***3**.229*
see also megaliths
meningitis **2**.121, ***2**.207*, **6**.260
Menno Simons **7**.198
Mennonites ***3**.229*, ***7**.198*
Menon, Krishna ***4**.217*
Menon, V. P. **4**.262
menopause ***2**.208*
Menotti, Gian Carlo ***5**.291*
Mensa (constellation) ***8**.96*
Menshevik **4**.44
Menshikov, Alexander Danilovich **3**.64, **3**.257
Menshikov, Alexander Sergeievich **4**.88, **4**.164
menstruation ***2**.208*
mental faculty *see* brain; memory; mind
mental handicap **7**.161, ***7**.198*, **7**.313
mental illness ***7**.198*
autism **7**.23
dementia **7**.83, **7**.222
depression *see* depression and anxiety
drug treatment *see* psychotropic drugs
institutional care **7**.161
obsession **7**.221
personality disorder *see under* personality
phobias **7**.14, **7**.30, **7**.214, **7**.235
stress **7**.300–1
and technology **6**.116, **6**.285, **6**.360
schizophrenia **6**.116, **6**.241, **6**.285, **6**.360
stereotaxic surgery **6**.334
see also neuroses; psychiatry; psychoanalysis; psychoses; psychotherapy
menu (computer) ***6**.219*
Menuhin, Sir Yehudi ***5**.291*
Menzies, Sir Robert Gordon **4**.195, ***4**.217*
Mephistopheles **9**.13
mercantile law ***7**.198*
mercantilism ***3**.229*, **7**.100, ***7**.198*
Mercator, Gerardus: and Mercator projection **1**.204, ***1**.210–11*, **1**.241, **1**.268, **1**.270, **6**.215, ***6**.219*, **6**.284, **8**.27
mercenary ***7**.198*
merchandizing ***7**.198–9*
Merchant Adventurer ***3**.229*
merchant banks **7**.1, **7**.168
merchant shipping *see under* ships and boats
Merchant Stapler ***3**.229*
Mercia **3**.15, **3**.109, ***3**.229*, **3**.260, **3**.275
Mercier, Honoré ***4**.217*
Mercier, Philippe **5**.198
Mercury (god) ***9**.25*
mercury (metal) **1**.106, ***1**.211*, ***6**.219*
amalgam **1**.11, **6**.12
in barometer **6**.30
in Castner–Kellner cell **6**.62
in manometer **6**.214
mercury chloride **6**.147
mercury fulminate **6**.100, **6**.124
mercury sulphide **1**.63
ore **1**.63, **6**.219
as poison and pollutant **6**.248, **6**.377, **7**.2
in pressure gauge **6**.281
vapour lamp **6**.194
Mercury (planet) ***8**.96*
Caloris Basin **8**.19
orbit **8**.17, **8**.196
statistical details **8**.196
Meredith, George ***5**.291*
Meredith, James: Meredith incident ***4**.217*
mergansers ***2**.208*, **2**.314
Mergenthaler, Ottmar **6**.203, ***6**.219*
merger **7**.65, ***7**.199*
Merian, Matthaeus (the Elder) **3**.290
meridians **1**.211, ***8**.97*
celestial **8**.97
see also Greenwich Mean Time
Mérimée, Prosper ***5**.291*, **5**.477
Merina ***7**.199*
meritocracy ***7**.199*
Merleau-Ponty, Maurice ***7**.199*
MERLIN (multi-element radio linked interferometer network) ***8**.105*
merlin (bird) ***2**.208*
Merlin (mythology) **9**.5, ***9**.25*
mermaid ***9**.25*
Meroë ***3**.229*
Mérovée, king of Franks **3**.229
Merovingians **3**.51, ***3**.229–30*
Merrill, John **5**.421
Merrimack ***1**.211*
Mersey ***1**.211*
Merton, Robert K. **7**.14
mesa ***1**.211*
mescalin **2**.251, **7**.306
mesembryanthemum ***2**.208*
Meseta **1**.164
mesite ***2**.208*
Mesoamerica *see* Central America
Mesohippus ***2**.208*
Mesolithic ***3**.230*, **5**.434
meson ***1**.211*
mesopause ***8**.97*
Mesopotamia (and Tigris and Euphrates) ***1**.211*, ***3**.230*
under Saladin **3**.314
civilization in **3**.77
Fertile Crescent **3**.125
languages and writing **7**.13, **7**.278
Mesopotamia Campaign ***4**.217–18*
myths and legends **9**.36
peoples *see* Sumerians
physical geography and geology **1**.211
regions and cities **3**.6, **3**.26, **3**.32, **3**.254
technology and science ***6**.112*, **6**.158, **6**.346
buildings **6**.46, **6**.49, **6**.51, **6**.82, **6**.287
canal **6**.54
food and drink **6**.37
nails **6**.236
oil lamps **6**.201
pottery **6**.278
vehicles and transport **6**.165, **6**.299, **6**.384
writing and scripts **7**.13
see also Iraq
mesosphere ***1**.211*, ***8**.97*
Mesozoic era **1**.133, **1**.134, ***1**.211*
periods *see* Cretaceous; Jurassic; Triassic
mesquite ***2**.208*
Messerschmitt, Willy ***6**.220*
Messiaen, Olivier ***5**.291–2*, **5**.382
Messier, Charles: Messier catalogue ***8**.97*
Messina, Antonello da **5**.42
Messina, Strait of ***1**.211–12*
metabolism ***2**.208–9*
see also fermentation; liver; respiration
metacarpals **2**.150
Metacomet, chief of Wampanoag **3**.197
metal *see* metals
metal mark (butterflies) **2**.127, ***2**.209*
metallic glass ***6**.220*
metalloids ***1**.212*, ***6**.24*
antimony **1**.16, **1**.106, **6**.16, **6**.265
arsenic **1**.20, **1**.106, **6**.20, **6**.74, **6**.312
bismuth **6**.12, **6**.38
germanium **1**.106, **1**.137, **6**.155–6, **6**.312, **6**.391
silicon *see* silicon
tellurium **1**.107, **1**.332, **6**.350

metals and metallurgy 1.212, 6.24, 6.220–1
actinide *see* actinide metals
alkali *see* alkali metals
alkaline *see* alkaline earth metals
alloying *see* alloys
aluminium *see* aluminium
amalgam 1.11, 6.12
annealing 6.15, 6.272, 6.304, 6.388
arts:
Britain 5.79, 5.176, 5.177
Celtic 5.84
China 5.92–3, 5.96–7
cire-perdue 5.101, 6.184, 6.185
coins 5.106
damascening 5.121
filigree 5.154
Greece 5.190
Hittite 5.208
India 5.99, 5.196
Islam 5.230
Italy 5.83
medals 5.288–9
niello 5.318
Russia 5.151
sculpture 5.79, 5.176, 5.177
United States 5.423
Viking 5.475
see also arts *under* bronze; copper; enamel; engraving; gold; iron; lacquer
assaying 6.5, 6.23, 6.357, 6.359
barium 6.30, 6.96
beryllium 1.38, 1.106, 6.34
bimetallic strip 6.36, 6.350, 6.355
bismuth 1.39, 1.106, 6.12, 6.38
blacksmith *see* forging
brazing 6.45
cadmium *see* cadmium
caesium *see* caesium
calcium *see* calcium
casting *see* casting and founding
cerium 6.11, 6.25
cermet 6.68, 6.82
chlorides 1.61
chromium *see* chromium
coating *see* enamel
cobalt *see* cobalt
copper *see* copper
corrosion *see* corrosion
creep 6.221
detector 6.220
die-casting 6.102
electric-arc furnace 6.113, 6.185, 6.188, 6.214
electrochemical series 1.102
electrolysis *see* electrolysis
electrons in 6.220
electroplating 6.73, 6.89, 6.119–20, 6.319
as ER fluids 6.120
extrusion 6.124, 6.188, 6.221
fatigue 6.220, 6.221
forging *see* forging
foundry *see* casting and founding
gallium *see* gallium
galvanizing 6.149
germanium 6.155–6, 6.312, 6.391
glass, metallic 6.220
gold *see* gold
grinding machine 6.161, 6.211, 6.212
hafnium 1.106, 1.149
indium 1.106, 1.167
intermediate *see* metalloids
investment casting, see *cire perdue*
iridium 1.106, 1.173, 6.186
iron *see* iron and iron and steel
lanthanides *see* lanthanides
lathe 6.195, 6.217
lead *see* lead
lightning conductor 6.202
lithium 6.156, 6.204, 6.310
magnesium *see* magnesium
manganese *see* manganese

metals and metallurgy (*cont.*):
memory 6.219
mercury *see* mercury
metal-matrix composite 6.82
molybdenum *see* molybdenum
monel 6.11, 6.241
nickel *see* nickel
niobium 1.106, 1.229, 6.241
noble *see* noble metals
ores *see* ores
organometallic compounds 1.240
osmium 1.106, 1.242, 6.177
oxides *see* oxides
palladium 1.107, 1.246, 6.256
panning 6.253
platinum *see* platinum
plutonium 6.126, 6.214, 6.245, 6.246, 6.274, 6.290
powder metallurgy 6.221, 6.279, 6.387
precious *see* noble metals
prehistoric use 6.280
production *see* electrolysis; mining; ores; smelting
quenching 6.287, 6.290
radioactive *see* radioactive elements
radium *see* radium
rare earths *see* lanthanides
rhenium 1.107, 1.282, 6.241, 6.298
rhodium 1.107, 1.282, 6.298, 6.354
riveting 6.299
rolling mill 6.301, 6.302
rubidium 6.96
ruthenium 1.107, 1.288, 6.304
screw 6.310–11
silver *see* silver
soldering 6.324–5, 6.356
speculum 8.158
strontium 6.96, 6.125, 6.290, 6.336
tantalum *see* tantalum
technetium 1.107, 1.332
tempering 6.351
thallium 1.334, 6.353
thorium 6.25, 6.243, 6.285, 6.355
tin *see* tin
titanium *see* titanium
transition *see* transition metals
transmutation 1.7, 6.10
tungsten *see* tungsten
type 6.356, 6.366
uranium *see* uranium
vanadium *see* vanadium
welding 6.188, 6.194, 6.321, 6.383
wire 6.298, 6.341, 6.386–7
work-hardening 6.388
zinc *see* zinc
zirconium *see* zirconium
see also alloys; mining; ores; smelting *and under individual countries*
metamorphic rocks 1.212
anthracite *see* anthracite
basement complex 1.34
foliation 1.124
gneiss 1.141
granulites 1.143
marble 1.47, 1.205, 5.282
minerals:
amphiboles 1.11–12
asbestos 1.20–1, 1.22
augite 1.26
biotite 1.39, 1.143
chlorite 1.62
garnet 1.132
graphite *see* graphite
hornblende 1.158
lapis lazuli 1.187–8
muscovite 1.214, 1.223
pyroxene 1.26, 1.270
staurolite 1.320
mylonite 1.223
phyllite 1.257
quartzite 1.272
schist 1.124, 1.294

metamorphic rocks (*cont.*):
shields 1.120, 1.301
slate 1.307
metamorphosis (animals) 2.49, 2.170, 2.209, 2.272
metamorphosis of themes 5.442
metaphor 5.108, 5.292, 5.339
Metaphysical painting 5.80, 5.98, 5.292
Metaphysical poets 5.133, 5.205, 5.284, 5.292, 5.384, 5.469
metaphysics 7.199
of Aquinas 7.15
in arts *see* Metaphysical
atomism 7.21
causation 7.42
Duns Scotus 7.97
Hindu 7.143, 7.191
Islamic 7.10, 7.24, 7.86, 7.301
of Kant 7.141, 7.175, 7.276
ontology 7.223
stoicism 7.300
substance 7.199, 7.301
time 7.296, 7.313
metasequoia 2.209
Metastasio, Pietro 5.329
metatarsals 2.124
Metaxas, Ioannis 4.218
Metchnikoff, Elie 6.112
meteor 1.212, 8.97
fireball 8.50
meteoric dust 1.212
showers 8.5, 8.7, 8.58, 8.88, 8.97–8
list of names 8.195
stream 8.98
swarm or storm 8.84, 8.98
meteorites 1.212, 1.213, 8.97
chondrite 8.26
craters 1.79
meteoroid 8.97
stream 8.98, 8.194
meteorology 1.212, 1.364
air *see entries prefixed 1. under* air and atmosphere
and altitude (orographic effect) 1.241, 1.274–5
anticyclone 1.15
clouds *see* clouds
condensation nuclei 1.71
cyclone 1.83, 1.344, 1.345
depressions 1.39–40, 1.88, 1.128, 1.262
dew 1.89–90
doldrums 1.93, 1.171
energy budget 1.108, 1.109
fog and smog 1.124, 1.307, 1.308
fronts 1.128, 1.262
frost 1.128, 1.129, 1.254
heat waves 1.152
humidity 1.159
hurricane 1.160
instruments 6.221–2
anemometer 6.14
balloons 6.221, 6.291–2
barometer 6.30, 6.294
satellite 6.221, 6.222
thermometer *see* thermometer
intertropical convergence zone 1.171
inversions 1.171–2
isobar 1.175
jet stream 1.177–8
lightning 1.194
meteorologists *see under individual countries*
monsoons 1.220
precipitation *see* precipitation (meteorology)
pressure *see* pressure, atmospheric
rainbow 1.274, 1.275
shower 1.302
smog 1.307, 1.308
snow *see* snow
storms *see* storm
subsidence 1.325

meteorology (*cont.*):
symbols 9.56
trough 1.345
visibility 1.357
waterspout 1.361
wind *see* winds
World Meteorological Organization 6.389
see also climate and climatology
meters *see* measurement
methadone 2.235
methanal 1.212–13, 6.297
methane 1.213, 6.222
in coal-gas 6.79
as greenhouse gas 7.135
in landfill 6.194
in natural gas 1.132, 6.237
production 6.36–7, 6.222, 6.264
methanoic acid 1.213, 6.222
methanol 1.213, 6.222, 6.261
method acting *see* Stanislavsky
Methodists 3.218, 3.230, 3.379, 3.380, 4.218, 7.50, 7.199
Church established 3.14, 3.230, 4.218
Methodius, Saint 3.95, 7.288
Methuselah 9.25
methyl alcohol *see* methanol
methyl chloride 6.261
methyl isocyanate 6.35
methyl methacrylate 6.276
methylated spirits 1.213, 6.222
methylbenzene 1.339
Metis 8.98
Meton 8.98
Metonic cycle 8.98
metric units 1.350, 6.368–9, 9.49
in astronomy 8.99, 8.176
kilometre 6.368
metre 1.213, 5.292, 6.321, 6.368
prefixes 9.52
metronome 5.292
Metsu, Gabriel 5.102, 5.292, 5.387
Metternich Winneburg, Clemens Wenzel Nepomuk Lothar, prince 4.25, 4.134, 4.218, 4.278, 4.288, 4.366
Meusnier de la Place, J.-B.-M.-C. 6.10
Mexican burrowing toad 2.209
Mexico 1.213, 3.230–1, 4.219–20, 7.368–9
arts 5.59, 5.85, 5.257, 5.327, 5.384, 5.457, 5.498
architecture 5.74, 5.84–5, 5.101, 5.449, 5.458
literature 5.85–6
sculpture and carving 3.31, 3.261, 3.298
conferences in 4.69
conquered 3.88
economy and trade 7.215
emigration 4.72, 7.200
ethnic groups 7.306
see also Aztecs; Mayans; Mixtecs; Toltecs
industrial relations 7.54
languages 7.199–200, 7.269
literature 5.85–6
maps 3.31, 3.226, 4.67
physical geography and geology 1.213
Gulf 1.213–14
mountains 1.285, 1.303
region 1.332
river 1.283
volcanoes 1.248, 1.264
politics and government:
constitutional plan 4.160
presidents 4.57, 4.62, 4.64, 4.98, 4.157–8, 4.176, 4.205, 4.249, 4.302
population growth 7.242
prehistory 3.102, 3.261

Mexico (*cont.*):
regions and urban areas **1**.322, **3**.227, **3**.351, **3**.389, **7**.197, **7**.298
religion and myths *see under* Aztecs; Toltecs
sport and recreation **7**.44, **7**.54, **7**.223
technology and science **6**.45, **6**.161, **6**.220, **6**.294, **6**.351
wars and defence:
leaders **4**.58, **4**.151, **4**.157–8, **4**.170, **4**.223–4, **4**.249, **4**.368, **4**.389
Mexican Revolution **4**.58, ***4**.219*, **4**.249, **4**.388–9
Mexican–American Wars **4**.7, **4**.129, ***4**.218–19*, **6**.81
and Zimmermann letter **4**.390
Mexico, Gulf of ***1**.213–14*
Meyerbeer, Giacomo **5**.82, **5**.265, ***5**.292–3*
Meyerhold, Vsevolod ***5**.293*
Meyrink, Gustav **9**.14
mezzanine finance **7**.34
mezzotint ***5**.293*
MHD (magnetohydrodynamic generator) ***6**.213*
MI5 and MI6 ***4**.220*
mica ***1**.214*
biotite **1**.39, **1**.143
muscovite **1**.214, **1**.223
Micah ***9**.25*
mice *see* mouse group
Michael (archangel) ***9**.25*
Michael, Saint ***9**.42*
Michael II, Byzantine emperor **3**.174
Michael VIII Palaeologus, Byzantine emperor **3**.93, ***3**.231*, **3**.253
Michael Romanov, emperor of Russia **3**.309, **3**.312
Michaelmas daisy **2**.86
Michaux, Pierre and Ernest **6**.35
Michelangelo Buonarroti **3**.301, **5**.133, ***5**.293–4*, **5**.410
dome **5**.132
film about **5**.376
sculpture **5**.349, **5**.379, **5**.410
Michelino, Domenico di **5**.125
Michelozzo di Bartolommeo ***5**.294*
Michels, Robert **7**.104
Michelson, Albert Abraham ***1**.214*, **1**.221, **8**.98
Michelson interferometer ***8**.98*
Michelson–Morley experiment **8**.48
Michigan, Lake **1**.145
Mickiewicz, Adam ***5**.294*
microbiology ***2**.209*
microchip *see* integrated circuit
microclimate ***1**.214*
microcomputer ***6**.222*, **6**.223
microdensitometer ***8**.98*
micro-economics **7**.120, **7**.193, ***7**.200*
see also companies and businesses; demand; marginal analysis; Marshall, Alfred; Pareto; supply; Walras
microelectronics *see* electronics *and* technology *under* information
microfiche ***6**.222*
microlight ***6**.222*
micrometer **6**.198, ***6**.222*, **6**.294, **8**.50
micron ***1**.214*
Micronesia ***3**.231*, ***7**.369*
languages **7**.23
physical geography and geology **1**.315
in South Pacific Forum **7**.296
see also Kiribati
micro-organisms:
diatoms **2**.232–3
microbiology **2**.209
one-celled *see* unicellular
see also bacteria; virus
microphone **6**.13, **6**.128, ***6**.222*, **6**.289, **6**.326, **6**.347
microphotography **6**.268
microprocessor **6**.85, **6**.179, ***6**.222–3*
chips *see* integrated circuit
and numerical control **6**.246–7
in pacemaker **6**.255
microscope **6**.192, ***6**.223*
electron **6**.65, **6**.119, **6**.192, **6**.223, **6**.268
microtome **6**.224
photography **6**.268
pioneer **6**.1, **6**.197–8
Microscopium ***8**.98*
microsurgery ***6**.224*, **6**.254
microtome ***6**.224*
microwaves ***1**.214*, **6**.117, **6**.215, **6**.380, **9**.54
background radiation ***8**.98–9*
cooker ***6**.224*
waveguide **6**.380
Midas ***9**.25*
Mid-Atlantic Ridge ***1**.214–15*
Middle Ages ***3**.231*
see also feudalism; medieval music
Middle East:
and aid **7**.8
Arabs *see* Arabs
arts **5**.230, **5**.344, **5**.397, **5**.486–7
see also under Arabs
Bedou **7**.28, **7**.29
demography **7**.116, **7**.159
dowry **7**.95
economy and trade **7**.200, **7**.203
education **7**.265
exploration **1**.30
government *see* politics and government *below*
Gulf Co-operation Council **7**.136–7
immigration **7**.200
land reform **7**.180
languages **7**.278–9
law, Islamic *see under* Islam
mandates **7**.191
monarchy **7**.97
music **5**.486–7
Muslim Brotherhood **7**.166
peace agreement **4**.59
physical geography and geology:
climate **1**.220
deserts **1**.146, **1**.225
lake **1**.130
mountains **1**.141–2, **1**.185, **1**.372
rivers **1**.114–15, **1**.178, **1**.300, **1**.337
seas **1**.85, **1**.277–8, **1**.326
wadi **1**.359
see also under individual countries
politics and government:
coup d'état **7**.73
electoral system **7**.123
franchise **7**.332
military **7**.201
state terrorism **7**.299
poverty **7**.244
refugees **7**.261
religion/theology **7**.96
Eastern Orthodox **7**.98
Maronites **7**.193
Monophysite Christian **7**.304
Sephardim Jews **7**.280
see also Islam
Sassanian empire **3**.316
Semites **3**.321
technology and science **6**.15, **6**.34, **6**.51, **6**.167, **6**.263, **6**.368
agriculture **6**.7, **6**.64, **6**.106, **6**.204, **6**.273
desalination **6**.99
energy and fuel **6**.270
mathematics **6**.1
optics and glass **6**.156
paper **6**.256
pottery **6**.278
Middle East (*cont.*):
technology and science (*cont.*):
textiles **6**.191, **6**.208
tools **6**.195, **6**.280
vehicles and transport **6**.70, **6**.101, **6**.299, **6**.300, **6**.301, **6**.337
urban areas **7**.288
see also Jerusalem
wars and defence:
civil **7**.52
communal conflict **7**.59
hostage-taking **7**.147–8
see also Arabs; Islam *and individual countries*
Middle English literature **5**.175, **5**.255, ***5**.294*
Middle Way **7**.217
Middleton, Thomas ***5**.295*
Middleton Murry, John **5**.280
Midgard ***9**.25*, **9**.26
midges **2**.131, ***2**.209*
Midgley, Thomas, Jr. **6**.197, ***6**.224*
Midlothian Campaigns ***4**.220*
mid-oceanic ridges ***1**.214–15*, **1**.236
Midrash ***7**.200*
midwife toad ***2**.210*
midwifery ***7**.200*
Mies van der Rohe, Ludwig **5**.14, **5**.167, **5**.294, ***5**.295*
Miezko I, king of Poland **3**.286
Mighty Handful *see* Five, The
migraine ***2**.210*
migration (animals) ***2**.210*, **2**.236
migration (people) ***4**.274–5*, ***7**.200–1*
depopulation **7**.85
emigration **4**.274–5, **7**.105, **7**.171
immigration **7**.152, **7**.153, **7**.319
migrant labour **7**.191, ***7**.200*, **7**.201, **7**.334
nomadic *see* nomadism
and old age **7**.222
refugees *see* refugees
tourism **7**.212, **7**.313–14, **7**.370
transhumance **7**.217, **7**.230
urbanization **7**.323
world ***4**.274*, **4**.275
Mihailovich, Draza ***4**.220*, **4**.387
Mikoyan, Anastas Ivanovich ***4**.220*
Mikszáth, Kálmán **5**.214
Milan **3**.11, ***3**.231*
Milbanke, Annabella **5**.69
mildew ***2**.210*
Miletus ***3**.231*
Milgram, Stanley **7**.221
Milhaud, Darius ***5**.295*, **5**.421
military activities:
court-martial **7**.73
expenditure ***7**.201*
military government ***7**.201–2*
junta **7**.173
martial law **7**.194
military law **7**.209
military-industrial complex ***7**.202*, **7**.212
see also military technology; wars
military technology ***6**.224–6*, **6**.249
aircraft **6**.41–2, **6**.129, **6**.139, **6**.167, **6**.176, **6**.220, ***6**.224–5*, **6**.296, **6**.365, **7**.17
see also missiles
amphibious vehicle **6**.13
armoured vehicles *see* armoured vehicles
balloon barrage **6**.29
body armour **6**.39, **6**.40–1, **6**.69, **6**.225, **6**.226
bridges **6**.260–1
camouflage **6**.56
chemicals *see* warfare *under* chemicals
electronic **6**.118–19, **6**.375
low-ground-pressure vehicle **6**.209–10
satellite **6**.224, ***6**.225*
military technology (*cont.*):
seas *see in particular* submarines; warships
shelter and bunker **6**.10
siege warfare **6**.225, **6**.318, **6**.371
tanks *see* tank
war chariots **6**.165
see also armed forces; artillery; defence; explosives; firearms; wars; weapons *and also* weapons *under* individual countries
militia ***3**.231*, **3**.360, ***7**.202*
milk **2**.57, ***2**.210–1*, **2**.282
milking machine **6**.6, ***6**.225*, **6**.227
pasteurization **6**.259
milkfish ***2**.211*
milkweed ***2**.211–12*
milkweed butterfly **2**.216
Milky Way **8**.56
Mill, John Stuart ***4**.220*, **5**.27, **7**.52, **7**.124, **7**.141, ***7**.202*, **7**.324
mill:
gig **6**.156
rolling **6**.156
treadmill **6**.362
water-operated **6**.355–6, **6**.377–8, **6**.379
wind **6**.285
Millais, Sir John Everett ***5**.295*, **5**.366
Millay, Edna St Vincent ***5**.295*
millenarianism (millenialism) ***7**.202*
Adventists **7**.3, **7**.202
cargo cults **7**.40
Christadelphians **7**.47
Druze **7**.96
eschatology **7**.108
fundamentalism **7**.127
ghost dance **7**.131
imām **7**.153
Jehovah's Witnesses **7**.171
see also Mahdi; Mormons; Zionism
Miller, Arthur **5**.215, ***5**.295–6*, **5**.459
Miller, Glenn **5**.47
Miller, Henry ***5**.295–6*
Miller, Neal **7**.7, **7**.291
Miller, Oskar von **6**.114
miller's thumb **2**.45
Millet, Jean-François ***5**.296*
millet ***2**.211*
Millett, Kate **7**.231, **7**.332
Millikan, Robert Andrews ***1**.215*, **8**.31
millimetre astronomy ***8**.99*
milling (cereal) **6**.142, ***6**.225–6*
milling (machine-tool) **6**.211, **6**.212
millipede ***2**.212*
millisecond pulsars **8**.130
Mills, Charles Wright **7**.104, **7**.244
millstone grit ***1**.215*
Milne, Alan Alexander ***5**.296*
Milne, David **5**.73
Milner, Alfred, viscount **4**.43, **4**.51, **4**.82, ***4**.220*
Milner, Thomas **6**.304
Milosevic, Slobodan **4**.309
Milosz, Czeslaw ***5**.296*
Milstein, César ***2**.212*
Milton, John ***5**.296*
Blake and **5**.49
masque **5**.285
poetry **5**.142, **5**.335, **5**.340, **5**.424
Paradise Lost **5**.249, **5**.284
polemic **5**.355, **7**.125
Mimas ***8**.99*
mime **5**.36, **5**.282, ***5**.296*, **5**.338
mimicry in nature ***2**.212*
mimosa **2**.2
Minamoto Yoritomo **3**.186, **3**.193, ***3**.231–2*, **3**.327
Minangkabau ***7**.202*
mind **7**.16, **7**.319
disorders *see* mental illness
handicap *see* mental handicap

mind (*cont.*):
philosophy of **7.**29, **7.**127, **7.**162, **7.**199, ***7.**202–3*
see also dualism *under* Descartes
unconscious *see* unconscious
see also brain; memory
Minden, battle of ***3.**232*
Mindszenty, József ***4.**220–1*
mine (weapon) **6.**134, ***6.**226–7*
mine-detector ***6.**227*
mine-layer ***6.**227*
minesweeper ***6.**227*
mine (mineral extraction) *see* mining
mineral water ***1.**215*
minerals ***1.**215*
carbonates *see* carbonates
chalcedony *see* chalcedony
classified **1.**83
cleavage **1.**64
in food **6.**140, **6.**141
gypsum *see* gypsum
halides **1.**150
mineralogy ***1.**215*
nitrates *see* nitrates
oxides *see* oxides
phosphates *see* phosphates
in rocks *see under* igneous rocks; metamorphic rocks; sedimentary rocks
silica *see* silica
silicates *see* silicates
spar **1.**316
sulphates *see* sulphates
sulphides *see* sulphides
see also elements; metals; mining
miner's cat **2.**285
miner's friend ***6.**227*
Minerva ***9.**25*
minesweeper ***6.**227*
Ming dynasty **3.**72, ***3.**232*, **3.**349, **3.**350, **3.**388
Minghuang, emperor of China ***3.**232*, **3.**349
miniatures and miniaturists ***5.**297*
Europe **5.**5, **5.**112, **5.**162, **5.**206, **5.**327
Middle East **5.**230, **5.**344, **5.**397
minicomputer ***6.**227*
Minié, Claude-Etienne **6.**52
minimal art **5.**240–1, ***5.**297*, **5.**397
minimal music **5.**179, ***5.**297*, **5.**377
minimum tillage farming **6.**19, ***6.**227*
minimum wage ***7.**203*
mining **6.**5, ***6.**227–8*, **7.**343
adit **6.**3, **6.**79
bord-and-pillar **6.**42, **6.**79, **6.**228
coal **6.**78, **6.**79, **6.**173, **6.**207, **6.**228, **6.**270
and commodity market **7.**58
diamond **1.**90, **1.**182, **1.**183
drag-line **6.**106
drift **6.**106–7
employment percentage **7.**106
explosives **6.**228
gold **6.**157, **6.**173
hydraulic **6.**79, **6.**172–3
iron **1.**174
long-wall **6.**79, **6.**207, **6.**228
open-cast **6.**78, **6.**250
outcrop **1.**242
pit-prop **6.**79, **6.**270
rock-salt **6.**308
soda ash **1.**8
tunnel **6.**228
uranium **7.**58
winding gear **6.**385
see also metals; minerals; ores
minivet **2.**83
mink ***2.**213*
Minkhelers **6.**150
Minkowski, Hermann ***1.**215*, **8.**156
Minnesinger **5.**297
Minoan civilization **3.**198–9, ***3.**232–3*
art ***5.**297*
king **3.**232, **9.**9
myths (*cont.*):
myths and legends **9.**9, **9.**25
see also Crete
minor axis of ellipse **8.**45
minor planet ***8.**99*
see also asteroids
Minorca **1.**30
Minos, king of Crete **3.**232, **9.**9
see also Minoan
Minotaur **9.**9, ***9.**25*
Minsky, Marvin **7.**19
minstrels ***5.**297*, **5.**368
mint family ***2.**213*
lavender **2.**184
marjoram **2.**203
patchouli **2.**244
pennyroyal **2.**248
peppermint **2.**249
sage **2.**292, **2.**293
salvia **2.**292–3
Minto, Gilbert John, 4th earl **4.**224
minuet ***5.**297–8*
minuteman ***3.**233*
Miocene epoch **1.**133, **1.**134, ***1.**216*
Mīr (poet) **5.**466
Mir (spacecraft) **6.**327, ***8.**99*
Mir Dard **5.**439
Mira Ceti ***8.**99–100*
Mīrābāī **5.**46
Mirabeau, Honoré Riqueti, vicomte de ***3.**233*
miracle ***7.**203*, **7.**272, **9.**13
miracle plays **5.**294, **5.**311
mirage **1.**215, ***1.**216*, **1.**279
Miranda, Francisco de (1750–1816) ***3.**233*
Miranda, Francisco de Sá de (1485–1558) **5.**359
Miranda (satellite of Uranus) ***8.**100*
mirid bug ***2.**213*
Miró, Joan ***5.**298*, **5.**441
mirror ***6.**228*, **6.**229
coelostat **8.**27
Foucault's knife-edge test **8.**52
see also reflecting *under* telescopes
MIRV **6.**235, ***6.**357*
miscarriage **7.**1
misericord ***5.**298*
Mishima Yukio ***5.**298*
Mishnah ***7.**203*
missel-thrush ***2.**213*
missiles ***6.**228–31*
anti-ballistic **6.**15, **6.**231, **6.**335–6
ballistic *see* ballistic missiles
cruise **6.**93, **6.**162, **6.**228, **6.**246
developed **6.**301
guided **6.**162, **6.**228, **6.**231, **6.**375
hand-held *see under* weapons
launch site **6.**196
warheads **6.**246, **6.**375
see also rocketry
missing mass problem ***8.**100*
missionaries **3.**131, **3.**271–2, ***4.**221*, ***7.**203*
see also Paul, Saint
Mississippi cultures ***3.**233–4*
Mississippi River ***1.**216*
see also under United States *under* rivers
Mississippian period **1.**134, ***1.**216*
Missouri Compromise ***4.**221*
Missouri River **1.**192, ***1.**216*
see also under United States *under* rivers
mist ***1.**216*
mistletoe ***2.**213*
Mistral, Frédéric ***5.**298*, **5.**368
mistral ***1.**216*
Mitchel, John (Irish, 1815–75) **4.**387
Mitchell, John (manufacturer, 1828) **6.**259
Mitchell, Margaret **5.**207
Mitchell, Sir Thomas Livingstone ***1.**216*
mites ***2.**214*
cheese-mite **2.**61
mites (*cont.*):
harvest-mite **2.**151
red spider **2.**280
scabies **2.**294–5
Mithra ***9.**25*
Mithras ***3.**234*
Mithridates II, king of Pontus **3.**27, **3.**67
Mithridates IV and V, kings of Pontus **3.**288
Mithridates VI, king of Pontus (120–63 BC) **3.**91, ***3.**234*, **3.**282, **3.**288
mithuna ***5.**298*
mitochondrion ***2.**214*
mitosis ***2.**214–15*
Mitra, Dinabandhu **5.**43
Mitterand, François **4.**123, ***4.**221*
Mixcoatl **9.**9
mixed economy ***7.**203*
mixing (air) ***1.**216*
Mixtecs ***3.**234*
mixtures (chemistry) **1.**68, ***1.**216–17*, **1.**273
Miyamoto, Shotaro **8.**147
Mizar ***8.**100*
Mizuno Tadakuni ***4.**221*
Mnemosyne **9.**25
moa ***2.**215*
mobiles **5.**70–1, ***5.**298*
mobility:
geographical *see* migration
social **7.**291
Möbius, August Ferdinand: Möbius strip ***1.**217*
Mobutu, Sese Seko **4.**83, **4.**179, ***4.**221*, **4.**388
Mochica culture ***3.**234*, **5.**426
mock orange ***2.**215*
mockingbirds **2.**56, ***2.**215*, **2.**340
modal logic ***7.**203*
mode (music) ***5.**299*
mode (statistics) ***1.**217*
Model, Walter **4.**188
model, animated **6.**25
Model Parliament ***3.**234*
modelling, mathematical **6.**17, **6.**110, **6.**126–7, **6.**320
modem **6.**179, **6.**180, ***6.**231*
modern art ***5.**299*
see also abstract art; Abstract Expressionism; conceptual art; Cubism; Expressionism; Fauvism; functionalism; Impressionism; Modern Movement; photography; Post-Modernism; Surrealism
modern dance **5.**39, **5.**70, **5.**139, ***5.**299–300*, **5.**397, **5.**489
choreographers **5.**120, **5.**139, **5.**187–8, **5.**214, **5.**397, **5.**415–16, **5.**447, **5.**450
companies **5.**13, **5.**39, **5.**68, **5.**120, **5.**122, **5.**126, **5.**187–8, **5.**214, **5.**267, **5.**299, **5.**397, **5.**489
indeterminacy **5.**219
see also dance *under* Expressionism *and under* Post-Modernism
modern literature ***5.**300*
drama **5.**2, **5.**59, **5.**225, **5.**255, ***5.**300*, **5.**428
novels:
Britain **5.**110, **5.**242, **5.**493–4
Europe **5.**243, **5.**323, **5.**368, **5.**428
Russia **5.**43
United States **5.**152–3
see also science fiction
poetry:
Europe **5.**340, **5.**371, **5.**428
United States and Britain **5.**119–20, **5.**143–4, **5.**217, **5.**242, **5.**284, **5.**303, **5.**364, **5.**480
modern literature (*cont.*):
stream of consciousness **5.**242, **5.**436, **5.**493
Modern Movement in architecture **5.**299, ***5.**300*
Austria **5.**208, **5.**268, **5.**480
Belgium **5.**212
Britain **5.**448, **5.**480
France *see* Le Corbusier
Italy **5.**356
United States **5.**241, **5.**243, **5.**290, **5.**295, **5.**317–18
Gropius **5.**38, **5.**194–5
Mies van der Rohe **5.**14, **5.**167, **5.**293, **5.**294
Sullivan and Chicago School **5.**6, **5.**14, **5.**91, **5.**439, **5.**494–5
modern music *see* twentieth-century music
modernism ***5.**300*
see also modern art; modern dance; modern literature; Modern Movement; Post-Modernism; twentieth-century music
Modersohn-Becker, Paula **5.**494
Modigliani, Amedeo ***5.**300*
Modred **9.**5, **9.**14
modulation **5.**247, **6.**128, ***6.**231*, **6.**231, **6.**285, **6.**289, **6.**290, **6.**291, **6.**346
see also AM; FM
modulus ***1.**217*
Mogadishu ***3.**234–5*
Moguls (Mughal) ***3.**235–6*
arts and architecture ***5.**307*, **5.**456
emperors ***9.**66*
Durbar **4.**105
15th and 16th centuries **3.**171, **3.**183, **3.**235
see also Akbar; Babur
17th and 18th centuries **3.**183, **3.**235, **5.**307
see also Akbar; Aurangzeb
Mohács ***3.**236*
mohair **6.**128, ***6.**231*
Mohammed *see* Muhammad, prophet
Mohawk Indians ***3.**236*, **7.**219
Mohenjo-daro **3.**177, ***3.**236*, **5.**223
Mohican Indians ***3.**216*
Moholy-Nagy, László and Lucia ***5.**301*
Mohorovičić, Andrija: Mohorovičić discontinuity (Moho) **1.**98, ***1.**217*
Moi, Daniel Arap **4.**181, **4.**225
moisture *see* humidity
moksha **7.**3, **7.**6, ***7.**203*, ***7.**205*
molarity ***1.**217*
molasses **6.**338
Moldova ***1.**217–18*, ***4.**221*, **7.**59, **7.**269, ***7.**369*
mole (animal) **2.**204, ***2.**215*
mole (unit) ***1.**218*, **6.**321, **6.**368
mole cricket ***2.**215*
mole plough ***6.**231*
mole rat ***2.**215–16*
mole salamanders **2.**19, ***2.**216*
molecules/molecular ***1.**218*, ***6.**23–5*
benzene ring **1.**36, **1.**37, **1.**181, **1.**339
diatomic **1.**90
dissociation **1.**20, **1.**92
distillation ***6.**231*
formula ***1.**218*
intermolecular forces **1.**170–1, **1.**354
ligand **1.**193
macromolecule **1.**201
mass **1.**280
models ***1.**218*
monomer **1.**219–20
orbitals **1.**239
polar **1.**262–3
relative molecular mass **1.**280
resonance effects **1.**281
stereochemistry **1.**320

molecules/molecular (*cont.*):
  weight **1**.280
  see also atoms
Molenaer, Jan Miense ***5.262***
Molière **3**.299, ***5.301***, **5**.309, **5**.371, **5**.398, **5**.402, **5**.496
Molina, Tirso de **5**.428
Mollet, Guy ***4.221***
mollisol ***1.218***, **1**.311
molluscs ***2.216***
  abalone **2**.1
  drill **2**.99
  ramshorn snail **2**.276
  sea slug **2**.300, **2**.301
  sea snail **2**.301
  shellfish **2**.307
  slug **2**.313
  winkle **2**.368
  see also bivalves; cephalopods; gastropods; snails
Molly Maguires ***4.221–2***
Molnár, Ferenc **5**.214
Moloch ***9.25***
Molotov, Vyacheslav Mikhailovich ***4.222***, **6**.232
  Molotov cocktail ***6.232***
Moltke, Helmuth, Graf von **4**.124, ***4.222***, **4**.299
Moluccas ***1.218***, ***3.236***
  exploration and colonization **1**.201, **1**.202, **3**.11
  physical geography and geology **1**.201, **1**.202, **1**.218
molybdenum **1**.106, ***1.218***, **6**.11, ***6.232***, **6**.241, **6**.333
Mombasa ***3.236***
moment (physics) ***1.218***
  moment of force **6**.357
  moment of inertia ***1.218***
momentum ***1.218–19***, ***6.232***, ***8.100***
  angular **1**.14, **8**.5–6
  uncertainty principle **1**.349
Momin **5**.466
Mon people **5**.66–7, **5**.450
Monaco ***1.219***, **7**.97, **7**.319, ***7.369***
monadnock ***1.219***
monads **7**.182
monarch butterfly **2**.209, ***2.216***
monarchy **7**.139, **7**.188, ***7.205***
  Divine Right of Kings **3**.104
  historical *see* rulers *under individual countries*
  kingship **7**.176–7
  major present-day countries with:
    in Africa *see* Lesotho; Swaziland
    in Asia *see* Japan; Thailand
    in Europe *see* Belgium; Britain; Denmark; Liechtenstein; Luxemburg; Monaco; Netherlands; Norway; Spain; Sweden
    in Middle East *see* Jordan; Saudi Arabia
    in Pacific *see* Tonga
  royal progress **3**.293
  see also Crown
monasticism ***3.236–7***, **3**.258, **7**.204, ***7.205***
  arts **5**.65, **5**.454–5
  asceticism **7**.19
  Benedictine **3**.39, **3**.80, **7**.205
  Buddhist *see under* Buddhism
  Carmelite **3**.61
  Carthusian **3**.51, **3**.62
  Cistercian **3**.40, **3**.77, **3**.237, **7**.205
  Dissolution of Monasteries ***3.236***
  Gildas **3**.141
  Hindu **7**.19, **7**.205, **7**.273
  nunnery **3**.260
  scholarship *see* Bede
  Welsh chronicler **3**.250
  see also friars
Moncayo, José Pablo **5**.257
Monck, George, 1st duke of Albemarle **3**.70, **3**.210, ***3.237***, **3**.302
Mond, Ludwig ***6.232***
Mondale, Walter **4**.284
Mondragon co-operative **7**.69
Mondrian, Piet ***5.301***, **5**.316
monel metal **6**.11, **6**.241
Monet, Claude **5**.56, **5**.218, ***5.301***
monetarism **7**.101, **7**.126–7, ***7.205***
money and monetary systems ***3.237–8***, ***7.205***
  compensation **7**.62
  European **7**.103, **7**.111
  gold bullion **7**.133–4
  gold standard **4**.138, **7**.134
  laundering ***7.205–6***
  liquidity **7**.185
  quantity theory of **7**.120, **7**.126, **7**.256–7
  supply ***7.206***
  see also banks; currency; exchange rate; finance
Monge, Gaspard **6**.122
Monge, Luis Alberto **4**.87
Mongkut, king of Siam ***4.222***, **4**.341
Mongol empire ***3.238***
  archers **3**.22
  founder of **3**.139
  Golden Horde **3**.143
  people **3**.369
  religion **3**.202
  rulers **3**.348–9, **3**.388–9
  see also Mongolia
Mongolia ***1.219***, ***3.238–9***, ***4.222***, ***7.369***
  economy and trade **7**.57
  language **7**.11
  literature ***5.301–2***
  physical geography and geology **1**.141, **1**.219
  rulers **3**.188
  Xiongnu **3**.387
  see also Mongol empire
mongolism **2**.98–9
Mongoloids ***3.239***
mongoose family ***2.217***
  civet **2**.66–7, **2**.240
  genet **2**.134
  meerkat **2**.206
Monier, Joseph **6**.51
monism **7**.141, **7**.203, ***7.206***, **7**.297
  see also materialism
monitor lizard **2**.180, ***2.217***
Monitor v. Merrimack ***4.222***, **4**.223
Monk, Meredith **5**.363
Monk, Thelonious **5**.239
monkey *see* monkeys
Monkey (Journey to the West) ***5.302***
monkey puzzle tree ***2.217***
monkeynut **2**.246
monkeys ***2.217***
  baboons **2**.20, **2**.201
  capuchin **2**.52, **2**.277
  Diana **2**.91–2
  howler **2**.165
  macaque **2**.22, **2**.197
  mythical **9**.15
  night **2**.228
  proboscis **2**.269
  spider **2**.277, **2**.320
  see also apes
monkfish ***2.217–18***
Mon-Khmer languages **7**.22
monks *see* monasticism
monkshood **2**.2
Monmouth, James, duke of: Monmouth's rebellion **3**.43, ***3.239***, **3**.320–1
Monnet, Jean **4**.114, ***4.222***, **4**.305
Monoceros ***8.100***
monochromatic light ***1.219***
monocline ***1.219***
monoclonal antibodies **6**.57, **6**.153, ***6.232***
Monocotyledons ***2.218***
  bamboo **2**.21
  iris family **2**.119, **2**.136, **2**.173, **2**.218
Monocotyledons (*cont.*):
  lilies *see* lily family
  orchids **2**.123, **2**.236, **2**.354
  yam **2**.373
  see also cereals and grains; grass; palms; sedge
monody ***5.302***
monoglyceride **6**.141
monologue ***5.302***, **5**.424
Monomatapa **3**.244
monomer ***1.219–20***
Monophysite Christian Churches **7**.47
  Armenian Apostolic **7**.16–17
  Coptic **3**.87, **7**.69–70, **7**.217
  Ethiopian Orthodox **7**.109
  Syrian Orthodox **7**.304
monoplane **6**.5, **6**.40, **6**.232
monopoly ***7.206***, **7**.266
  natural **7**.212
  oligopoly **7**.222
monopsony ***7.206***
monosaccharides (sugars) **2**.52–3
  see also glucose
monosodium glutamate **6**.141, **6**.176
monotremes **2**.10, **2**.259
Monotype ***6.232***, **6**.366
monoxide *see* carbon monoxide
Monroe, James: and Monroe Doctrine ***4.222–3***
Monroe, Marilyn **7**.281
Mons ***3.239***
Monsengwo, Mgr. Laurent **4**.388
monsoon ***1.220***
  forest **2**.125
monsters **9**.9, **9**.14–15, **9**.17, **9**.22, **9**.25, **9**.32
  see also mythical *under* animals
Mont Blanc **1**.79, ***1.220***
Mont Pelée ***1.221***, **1**.233
Mont Tremblant ***1.221***
montage ***5.302***
Montagu, Edwin: Montagu-Chelmsford Proposals ***4.223***
Montagu, John Neville, earl of **3**.164–5
Montagu, Mary **6**.176
Montaigne, Jacques **8**.15
Montaigne, Michel de ***5.302***
Montale, Eugenio ***5.302***, **5**.441
Montanari, Geminiano **8**.3
montbretia ***2.218***
Montcalm-Gozon, Louis-Joseph, marquis de ***3.239***, **3**.284
Monte Carlo method ***1.220***
Monte Carlo Rally **7**.208
Montemayor, Jorge de **5**.428
Montenegro ***1.220***, **1**.301, ***4.223***
Montespan, Francoise Athénais, marquise de ***3.239***
Montesquieu, Charles de Secondat, baron de la Brède et de **3**.115, ***3.239***, ***5.302***, ***7.279***
Montessori, Maria ***7.206***, **7**.246
Monteverdi, Claudio **5**.275, ***5.302–3***, **5**.328, **5**.334, **5**.405
Montezuma of Mexico **3**.31, **3**.88, ***3.239***, **9**.30
Montfort, Simon de, earl of Leicester ***3.239–40***
  and Albigensian Crusade **3**.7
  and barons **3**.36, **3**.161
  and Evesham defeat **3**.110, **3**.120
  and Gascony **3**.137
  and parliament **3**.45, **3**.83, **3**.270
Montgolfier, Joseph-Michel and Jacques-Étienne **6**.29, **6**.138, ***6.232***
Montgomery, Lucy Maud ***5.303***
Montgomery of Alamein, Bernard Law **4**.7, **4**.170, ***4.223***, **4**.246, **4**.247
month ***8.100–1***, **9**.55
  sidereal **8**.100, **9**.55
  synodic **8**.168, **9**.55
Montherlant, Henri de ***5.303***
montmorillonite ***1.220–1***
Montreal Convention on hi-jacking **7**.142
Montreal Protocol on ozone depletion **6**.254, **7**.206, **7**.226
Montrose, James Graham, 5th earl and 1st marquis of ***3.240***
Montserrat ***1.221***
Monument Valley **1**.46
Moon, Sun Myung **7**.320
moonfish **2**.234–5
Moonies **7**.320
Moon/lunar ***8.101–4***, ***8.198***
  calendar ***8.87***
  corona **1**.77
  craters **1**.79, **8**.57, **8**.114–15
  eclipse **8**.42
  ejecta blanket **8**.67
  exploration ***8.87***
  illustrated **8**.102–3
  landing and exploration **6**.17, **8**.87, **8**.123, **8**.128
  libration of **8**.85
  luni-solar precession **8**.127
  mapping **8**.147
  mare **8**.63, **8**.72, **8**.91, **8**.112, **8**.148
  module and lunar roving vehicle, see *Apollo*
  moonquakes **8**.101
  motion **8**.48
  mountains **8**.53, **8**.140
  orbit **8**.101, ***8.196***
  phases **8**.118–19
  rille **8**.139–40
  ring plains **8**.140
  study of **8**.147–8
  sublunar point **8**.163
  temperature **1**.221, **1**.273
  theory ***8.87–8***
  and tides **1**.336–7
  transient lunar phenomena **8**.172–3
  year **8**.189
moonstone ***1.221***
moor (land) ***2.218***
Moor (Saracen) *see* Moors
Moore, G. E. **7**.270
Moore, George ***5.303***
Moore, Harriet **6**.125
Moore, Henry ***5.303***
Moore, Hyram **6**.81
Moore, Sir John ***4.223***, **4**.265
Moore, Julia A. **5**.313
Moore, Marianne ***5.303***
Moore, Thomas ***5.303***
moorhen **2**.131
Moors ***3.240***, **3**.350
  arts ***5.303***, **5**.304
moose ***2.218***
Mor, Anthonis ***5.303***
moraines **1**.140, ***1.221***
moral right (copyright) **7**.70
morality:
  casuistry **7**.41
  emotivism **7**.105
  honour **7**.147
  justice **7**.174
  moral philosophy **7**.108–9
  moral right and copyright **7**.70
  and science **7**.125
  value judgement **7**.325
  see also ethics; virtues
morality plays **5**.294, ***5.303–4***, **5**.421
Moravia, Alberto ***5.304***
Moravia (country) **1**.325, ***3.240***, ***4.43***
Moravian Church ***7.207***
moray eel ***2.218***
morbidity rate ***7.207***
Morcar, earl of Northumbria **3**.164, **3**.339
mordants **6**.12, **6**.89, **6**.110, ***6.232***
More, Sir Thomas **2**.190, **3**.117, ***3.240–1***, **5**.258, **7**.60, **7**.324

Moreau, Gustave *5.304*
Moreau, Jean Victor **4**.288
Morel, Edmund **4**.193
morel *2.218*
Morelos y Pavón, José María **4**.219, *4.223–4*
Moreno, Mariano *4.224*
Moreschi, Alessandro **5**.82
Morgan, Sir Henry **3**.51, *3.241*, **3**.282
Morgan, John Pierpont **4**.77, *4.224*
Morgan, William **4**.15
Morgenstern, Oskar **1**.131, **1**.226
Mori Ōgai *5.304*
Móricz, Zsigmond **5**.214
Morier, David **3**.93
Morínigo, Higinio *4.224*
Morison, Stanley **5**.464
Morisot, Berthe *5.304*
Morley, Edward Williams **1**.214, *1.221*
Morley, John: Morley–Minto reforms *4.224*
Morley, Thomas *5.304*
Mormons *4.224*, **4**.317, **4**.386, *7.207*
morning star *see* Venus (planet)
Moro, Aldo *4.224*, **4**.284
Moro, Antonio **5**.303
Morocco *1.221*, *3.241*, *4.224–5*, *7.369–70*
in Arab League **7**.15
colonization and decolonization **4**.4, **4**.306–7
explorer **1**.164
languages **7**.30
leaders **3**.124, **3**.325, **4**.1
meetings **4**.64
physical geography and geology **1**.221, **1**.234
tourism **7**.370
wars **3**.8
morphine *see* opium
morpho butterfly *2.218*, **2**.277
morphology (anatomy) **2**.9, *2.218–19*
morphology (grammar) **7**.134
Morrelli, Joe **4**.299
Morrice, Norman **5**.373
Morris, May **5**.305
Morris, William (1834–96) **4**.213, **4**.318, *5.304–5*, **5**.367
Morris, William Richard (Lord Nuffield, 1877–1963) **4**.248
Morris dance **5**.159–60, **5**.162
Morrison, Philip **8**.115
Morrison of Lambeth, Herbert Stanley Morrison, Baron *4.225*
Morse, Samuel Finley Breese: and Morse code **4**.80–1, **6**.93, *6.232*, **6**.290, **6**.346, **7**.257, *9.48*
Morsztyn, Jan Andrzej and Zbigniew **5**.355
mortality:
infant and child **7**.119, **7**.140, **7**.153, **7**.159
rate *7.207*
see also death
mortar (building) **6**.5, **6**.66, *6.232*
mortar (pestle and) **6**.261
mortar (weapon) **6**.22, **6**.232–3
mortgage *7.207*
Mortimer, Roger, 8th Baron of Wigmore and 1st earl of March **3**.110, **3**.181, *3.241*
Mortimer's Cross, battle of *3.241*
Morton, Ferdinand ('Jelly Roll') **5**.238
Morton, G. A. **6**.391
Morton, John *3.241*
Morton, William Thomas Green *6.233*, **6**.320
MOS (metal oxide semiconductor) **6**.70
mosaic *5.305*, **5**.387, **5**.440
Mosby, John S. **4**.142
Mosca, Gaetano **7**.104
Moscovici, Serge **7**.64, **7**.292
Moscow *see* Muscovy
Moseley, Henry Gwyn Jeffreys *1.221*
Moses **3**.121, **3**.192, *3.241*, **9**.5, *9.25–6*
Moses, Grandma *5.305*
Moshoeshoe I, king of Basutoland **4**.193
Mosley, Sir Oswald Ernald **4**.116, *4.225*
mosques **5**.6, **5**.227–8, *7.207*, **7**.208
mosquitoes **1**.85, **2**.198–9, *2.219*, **2**.281, **2**.289, **2**.354, **2**.373
moss *2.219*, **2**.282, **2**.319–20
moss animal *2.219*
Mossadeq, Muhammad *see* Mussadegh
Mossi **3**.4, *3.241–2*
motet **5**.274–5, **5**.289, *5.306*
mother-of-pearl **1**.17
ornamentation *5.306*
Motherwell, Robert *5.306*
moths *2.219–20*
army worm **2**.14
bagworm **2**.20
burnet **2**.46
camouflage **2**.50
carpet **2**.55
case-bearer **2**.56
caterpillars *see* caterpillars
china-mark **2**.63–4
clearwing **2**.67
clothes **2**.68–9
codlin **2**.71
cutworm **2**.84
eggar **2**.104
emperor **2**.107, **2**.310
footman **2**.124–5
ghost **2**.332
goat **2**.139
gold **2**.141
grass **2**.144
gypsy **2**.148
hawkmoth **2**.152, **2**.343
leopard **2**.188
lobster **2**.192
longhorn **2**.193
magpie **2**.198
owlet *see* owlet moths
peppered **2**.225, **2**.249
pigmy **2**.253
plume **2**.260
primitive **2**.269
prominent **2**.269
pug **2**.271
puss **2**.192, **2**.269, **2**.272, **2**.273
silk **2**.17–18, **2**.310
silver-Y **2**.310
swift **2**.332
tiger **2**.341
tobacco hornworm **2**.343
tortrix **2**.344–5
tussock **2**.148, **2**.351–2, **2**.354
underwing **2**.352–3
vapourer **2**.354
wainscot **2**.358
yucca **2**.374
Motilal, B. K. **7**.156
motion *1.221*, **1**.222
equations of **1**.110
force *see* force
momentum *see* momentum
perpetual **6**.260
pictures *see* cinema
prime mover **6**.281
of satellite **8**.48
of star **8**.128, **8**.161
see also engine; locomotives; motor; vehicles
motivation **7**.113, *7.207*
motmot *2.220*
motor:
armature **6**.20
distinct from engine **6**.121–2
motor (*cont.*):
electric *see* linear motor *and also* motor *under* electricity
induction **6**.202–3, **6**.351
linear **6**.201, **6**.202–3
rocket **6**.157, **6**.301
starter **6**.330
see also engines; machinery; motor car; motor vehicles; petrol
motor car *6.233–4*
amphibious **6**.13
braking *see* braking system
cruise control **6**.92–3
diesel *see* diesel
electric **6**.147, **6**.234
engines *see* internal combustion engine *and* engines *under* petrol
exhaust **6**.62, **6**.64, **6**.124
ferry **6**.60, **6**.127
gears **6**.26, **6**.103, **6**.152
history and pioneers **4**.225, **4**.350, **6**.34, **6**.94, **6**.143, *6.234*
manufacture **6**.23, **6**.26, **6**.129, **6**.216
momentum **1**.219
motor-racing *6.288*, *7.207–8*
pollution:
cause *see* lead tetraethyl
prevention *see* catalytic converter
safety or laminated glass **6**.156
safety-belt **6**.304
steering **6**.333–4
suspension **6**.341
transmission *see* transmission in vehicles
tyre **6**.366
universal joint **6**.369
wheel **6**.384
see also motor vehicles; roads
motor cycles **6**.77, **6**.94, *6.234–5*
racing *7.207*, **7**.297
motor skills, development of *7.208*
motor vehicles *4.225*
all-terrain **6**.209–10
ambulance **6**.12
amphibious **6**.13
buses and coaches **6**.292, **6**.363
car *see* motor car
caterpillar tracks **6**.64, **6**.65
engines *see* internal-combustion engine *and* engines *under* petrol
fire-engine **6**.131
'gasohol' **6**.37
gear *see* gear
lorry **6**.209
low-ground-pressure *6.209–10*
military *see* military technology
steering system **6**.334
tractor **6**.103, **6**.359–60
transmission *see* transmission in vehicles
wheel **6**.384
see also motor car; motor cycles; roads
motorbikes *see* motor cycles
motor-drive *6.235*
motorway **6**.92–3, **6**.300, **6**.360
Mott, Lucretia **4**.117
mottling in soils *1.221*
mouflon *2.220*
mould *2.220*
mould-board ploughing **6**.6
moulding **6**.62, **6**.66, *6.235*, **6**.272, **6**.303
metal *see* casting
moulin *1.221–2*
mound-builders *3.242*
Mount Badon *3.242*
Mount Palomar Observatory *8.104*
Mount Rainier National Park **1**.247, **1**.274
Mount Wilson Observatory *8.104–5*
mountain ash **2**.289
mountain beaver *2.220*
Mountain Men *4.225*
mountaineering **1**.366, *7.208*
mountains *1.222*
Afghanistan **1**.156, **1**.183, **1**.246
Africa **1**.23–4, **1**.89, **1**.95, **1**.135, **1**.179, **1**.182, **1**.183
Alaska **1**.201, **1**.285, **1**.289
Antarctica **1**.111, **1**.112
Asia **1**.14, **1**.135, **1**.156, **1**.180, **1**.246, **1**.335
see also Himalayas
Australia **1**.27, **1**.123, **1**.145, **1**.151, **1**.184, **1**.200, **1**.211, **1**.309
Austria **1**.10, **1**.16
block **1**.40
Britain:
England **1**.250
Scotland **1**.37, **1**.47, **1**.135, **1**.142, **1**.241
Wales **1**.49, **1**.62, **1**.309
building **1**.135, **1**.154, *1.222*, **1**.241
Canada **1**.16, **1**.54, **1**.67, **1**.183, **1**.198, **1**.221, **1**.285, **1**.289
Central America **1**.285, **1**.303
chain *1.222*
China **1**.246, **1**.335, **1**.369
Europe *see under individual countries*
fold *see under* folds
France **1**.10, **1**.79, **1**.179, **1**.206–7, **1**.220, **1**.269–70, **1**.359
Germany **1**.152
glaciated *see* erosion *and* deposition *under* glaciation
Greenland **1**.135
Iran **1**.185, **1**.372
Iraq **1**.185
Italy **1**.10, **1**.16, **1**.207, **1**.220
Kashmir **1**.180, **1**.224
Kenya **1**.182
Mexico **1**.285, **1**.303
Middle East **1**.141–2, **1**.185, **1**.372
Nepal **1**.14, **1**.116, **1**.155, **1**.183
New Zealand **1**.75, **1**.100, **1**.287, **1**.314, **1**.339
North Africa **1**.89, **1**.179
Pakistan **1**.156, **1**.179, **1**.183
Papua New Guinea **1**.246
piedmonts **1**.258
and rainfall (orographic effect) **1**.241, **1**.274–5
ring **8**.140
Russia **1**.135, **1**.351
Scandinavia **1**.135
Siberia **1**.351
sierra **1**.303
Sikkim **1**.183
South America **1**.13, **1**.23–4, **1**.78, **1**.135
Spain **1**.269–70, **1**.303–4
Switzerland **1**.10, **1**.19, **1**.179, **1**.207
Syria **1**.141–2
Tanzania **1**.183
Tibet **1**.116
Turkey **1**.17, **1**.185
United States:
central **1**.17, **1**.40
east **1**.3, **1**.9, **1**.16, **1**.41, **1**.55, **1**.135, **1**.146, **1**.360, **1**.365
west **1**.54, **1**.67, **1**.237–8, **1**.285, **1**.301, **1**.303, **1**.365, **1**.366, **1**.371
see also hill; volcanoes
Mountbatten, Louis Francis A. V. N., 1st earl **4**.55, *4.225*
mounting (telescope) *8.104*
mourning **7**.30
mouse (computer) **6**.180, *6.235*
mouse deer **2**.62–3
mouse group *2.220*
field mouse **2**.370–1
hamster **2**.150
harvest mouse **2**.151–2
jerboa **2**.175, **2**.176

mouse group (*cont.*):
lemming **2**.186–7
marsh rat **2**.362
pocket mice **2**.178, **2**.260
rats *see* rats
voles *see* voles
water rat **2**.362
wood mouse **2**.370
mousebird ***2**.220*
mouse-tailed bat ***2**.220*
Moussorgsky *see* Musorgsky
Mousterian ***3**.242*
mouth ***9**.59*
jaw **2**.175, **2**.343
tongue **2**.343, **9**.61
see also teeth
mouth music ***5**.310*
mouth organ **5**.199, ***5**.306*, **5**.427
movement (living organisms) **2**.210–11, ***2**.220*
movement (population) *see* migration
movement (physics and technology) *see* motion
moving cluster **8**.113
Mowbray, Thomas, 3rd earl of Nottingham **3**.210
mowing grass **6**.6, **6**.196, **6**.235
mowing machine ***6**.235*
Moyne, Lord **4**.329
Mozambique ***1**.222*, ***3**.242*, ***4**.225*, ***7**.370*
decolonization **4**.4
economy and trade **7**.296
finance **7**.201
land reform **7**.180
language **7**.269
leader **4**.204
physical geography and geology **1**.195, **1**.203, **1**.222, **1**.372
refugees **7**.261
wars and defence **4**.126, **7**.52, **7**.201
Mozart, Anna Maria **5**.306
Mozart, Nannerl **5**.306
Mozart, Wolfgang Amadeus ***3**.306*
and musical instruments **5**.338, **5**.347, **5**.362,
operas **5**.40, **5**.48, **5**.194, **5**.329, **5**.338
and Salieri **5**.398
Mpande **4**.99
mṛdaṃga ***5**.306*
MRI (magnetic resonance imaging) **6**.213
Mrożek, Slawomir **5**.355
MS-DOS operating system **6**.250
Mswati III, king of Swaziland **4**.335
Mubarak, president **4**.109, **4**.309
Mucha, Alphonse ***5**.307*, **5**.362
mud ***1**.222*
building **6**.49, ***6**.235*
mud eels **2**.75, **2**.310
mud puppy ***2**.220–1*
mudflow ***1**.222*, **1**.223
mudrā ***5**.307*
mudskipper ***2**.221*
muezzin ***7**.208*
mufti **7**.282
Mugabe, Robert Gabriel **4**.225, ***4**.226*, **4**.390
Mughal *see* Moguls
Muhammad, Prophet **3**.115, ***3**.242*, **4**.168, **7**.153, **7**.166, ***7**.208–9*, **7**.250
birthday **7**.195
blasphemy against **7**.32
daughter **9**.12
Five Pillars of Islam **7**.121
hadīth **7**.138
hegira **3**.16
hijrah **7**.142
holy cities **3**.187, **3**.227–8, **7**.166
see also Mecca; Medina
and Koran **7**.177
and miracles **7**.203
as prophet **7**.250
Muhammad, Prophet (*cont.*):
teachings **3**.181, **3**.200, **3**.330, **3**.346
wife **9**.3
see also Islam
Muhammad II (Mehmed) **3**.228
Muhammad V, king of Morocco **4**.224
Muhammad Ahmad bin Abdallah (Mahdi) (1843–85) **4**.138, **4**.207, **4**.332
Muhammad Ali of Egypt *see* Mehemet Ali
Muhammad Ali of India (1878–1931) **4**.182
Muhammad Ali of Persia (1907–9) **4**.279
Muhammad of Ghor **3**.218
Muhammad Reza Shah Pahlavi, Shah of Iran **4**.165, **4**.182–3, ***4**.226*, **4**.289, **7**.148, **7**.309, **7**.331
Muhammad Rumfa of Kano **3**.159
Muir, Edwin ***5**.307*
Muir, Thomas **3**.319
al-Muizz, al-Malik **3**.218
Mujaji ***9**.26*
Mujibar Rahman, Sheikh ***4**.227*
Mukden incident ***4**.227*
Mukundaram **5**.43
mulberry ***2**.221*
Muldoon, Sir Robert ***4**.227*
mule ***2**.221*
mulgara **2**.266
ul-Mulk, Asaf Jah Nizam **4**.159
al-Mulk, Nizam of Turkey **3**.321
Mullard Radio Astronomy Observatory ***8**.105*
mullein ***2**.221*
Müller, Paul **1**.85, **6**.96
Müller, W. **6**.152
Mullerian mimicry **2**.212
mullet ***2**.221*
Mulligan, Gerry **5**.239
Mulroney, Brian **4**.85
multi-element radio linked interferometer network ***8**.105*
multilateral aid **7**.9
multinational corporations **7**.83, ***7**.209*, **7**.315, **7**.326
multiparty system ***7**.209*
multiple independently targetable re-entry vehicle ***6**.235*, **6**.375
multiple sclerosis ***2**.221–2*, **6**.241
multiple stars ***8**.105*
first-magnitude **8**.4
multiple-mirror telescope **8**.93, ***8**.105*
multiple-unit train ***6**.235*
multiplexing **6**.69, **6**.112, ***6**.235*, **6**.346, **6**.347
multiplier ***7**.209*
multi-wire chamber **6**.259
Mulungu ***9**.26*
al-Mumin, Abd **3**.10
mummers' plays ***5**.307*
mummy ***3**.243*
Munch, Edvard ***5**.307–8*
Muni, Bharata **5**.220
Munich 'beer-hall' putsch ***4**.227*
Munich Pact ***4**.227*
Munkacsi, Martin ***5**.308*
Munkar ***9**.26*
Munro, Thomas **4**.298
Munshi Abdullah **5**.277
muntjac ***2**.222*
muon ***1**.222–3*
Muqi (Fachang) ***5**.308*
Murad Bey of Egypt **3**.296
Murad I and II, Ottoman sultans **3**.201, ***3**.243*
mural quadrant ***8**.105*
murals *see* frescoes and murals
Murasaki Shikibu **3**.135, **5**.445
Murat, Joachim ***4**.227–8*, **4**.312
Murchison, Sir Roderick Impey ***1**.223*, **1**.298
murder ***7**.146*
capital punishment **3**.150, **7**.40, **7**.146
compensation for **3**.379
extra-judicial execution **7**.114
genocide **7**.130
see also assassin; suicide
Murdoch, Iris ***5**.308*
Murdoch, Rupert **7**.216
Murdock, William **6**.359
Murillo, Bartolomé Esteban ***5**.308–9*
Murmean Sea **1**.33
Murnau, Friedrich Wilhelm ***5**.309*
Murray, Jerome **6**.303
Murray River ***1**.223*
Murry, John Middleton **5**.280
Mursilis I, king of Hittites **3**.166
Musa al-Kazim, imām **3**.181
Musa ibn Nusayr **3**.241
Musca ***8**.105*
muscles **2**.29, ***2**.222*, **2**.252–3, ***9**.59*
muscovite **1**.214, ***1**.223*
Muscovy **3**.182, ***3**.243*
Company ***3**.243*
musculo-skeletal system **6**.104, **6**.266
Muses **9**.8, **9**.9, ***9**.26*
museums ***7**.209*, **7**.210
of technology and science **6**.87, **6**.255, **6**.310, **6**.322–3
Museveni, Yoweri Kaguta III **4**.356
Mushet, Robert Forester ***6**.235*
mushroom rock **1**.249
mushrooms ***2**.222*, **6**.148
see also fungi
music:
abstract **5**.367, **5**.496
aleatory **5**.12
arabesque **5**.18
Ars Antiqua and Ars Nova **5**.23
atonality **5**.25, **5**.439, **5**.484, **5**.485
bagatelle **5**.29
Baroque *see under* Baroque
binary form **5**.48
bitonality **5**.48
Buddhist **5**.65
cadence **5**.70
capriccio **5**.77
chamber **5**.88
chorale **5**.99
chord **5**.70, **5**.99
choreography *see* choreographers
church *see* sacred music
coda **5**.106
composers *see* composers
computer **5**.12, **5**.108
concerto *see* concerto
concerto grosso **5**.109
conductors *see* conductors
consort **5**.110
continuo **5**.111
copyright law **7**.70
counterpoint *see* counterpoint
country **5**.115
crescendo **5**.118
dance *see* dance
divertimento **5**.131
drone **5**.13
electronic *see* music *under* electronics
ethnomusicology **5**.415
Expressionist *see under* Expressionism
fantasia **5**.152
film *see under* cinema
folk *see under* folk music
harmonics **5**.199–200
harmony **5**.200
incidental *see* incidental music
indeterminacy **5**.219
instruments *see* musical instruments
interval **5**.224–5
jazz *see* jazz
key **5**.25, **5**.48, **5**.246–7
leitmotif **5**.260
music (*cont.*):
march **5**.282–3
Master of the Queen's (King's) Music **5**.286
medieval *see* medieval music
melody **5**.290
metronome **5**.292
minimal **5**.179, **5**.297, **5**.377
mode **5**.299
mouth **5**.306
musicals *see* musicals
musicology ***5**.310*
musique concrète ***5**.310*
Neoclassical *see under* Neoclassicism
notation **5**.322–3, **5**.444
opera *see* opera
orchestra *see* orchestra
orchestration **5**.330
overture **5**.334
partita **5**.340
passion **5**.341
pitch **5**.351, **5**.352
pizzicato **5**.351
polyphony *see* polyphony
pop *see* popular music
prelude **5**.365–6
programme **5**.367
recording *see* recording *under* sound
Renaissance *see under* arts *under* Renaissance
rhapsody **5**.382
rhythm *see* rhythm
Romantic *see under* Romanticism
rondo **5**.390–1
scale *see* scale
scherzo **5**.405
score **5**.407
serenade **5**.411
serialism *see* serialism
singing *see* songs
solmization **5**.424
sonata and sonata form **5**.103, **5**.424, **5**.442
staccato **5**.429
standing waves **1**.319
stave **5**.322–3
suite **5**.439
synthesizer **5**.443, **6**.343
tablature **5**.444
temperament **5**.448–9
ternary form **5**.449–50
time and tempo **5**.455
toccata **5**.457
tonality **5**.458
tremolo **5**.460
trio **5**.460
variation and theme **5**.468
see also under individual countries
musical box **5**.288
musical instruments:
Africa **5**.8, **5**.321
American Indian **5**.15
bullroarer **5**.65
China **5**.96
folk *see under* folk
fret **5**.164
güiro **5**.196
India **5**.220, **5**.221
Indonesia **5**.222
Japan **5**.236
Korea **5**.251
mechanical **5**.288
musical glasses **5**.179
North Africa **5**.321
Pacific Ocean islands **5**.325
Papua New Guinea **5**.339
South and Central America **5**.257, **5**.282
South-East Asia **5**.427
stamping tube **5**.430
West Asia **5**.487
see also brass; keyboard; music; percussion; reed; stringed; woodwind

musicals (films or plays) 5.309
composers and writers 5.45, 5.174, 5.175, 5.246, 5.359, 5.385–6, 5.424, 5.491
music-hall 5.309–10, 5.357, 5.469
musicology 5.310
Musil, Robert 5.310
musique concrète 5.310
musk deer 2.222–3
musk ox 2.223
musk rat 2.223
muskeg 1.223
muskellunge 2.223
musket 6.130, 6.131, 6.384
musketeer 3.22, 3.243
Muslim see Islam/Muslims
Muslim Brotherhood 7.166
Muslim League 4.228
Musorgsky, Modest Petrovich 5.87, 5.310, 5.370
Muspelheim 9.6, 9.25, 9.26
Mussadegh, Muhammad 4.228
mussels 2.223, 2.246, 2.331
Musset, Alfred de 5.310–11
Mussolini, Benito 4.26, 4.76, 4.228, 6.127, 7.90
Abyssinian Campaign 4.2, 4.27, 4.113, 4.170
decline 4.170
and fascism 4.116, 7.118
and Lateran Treaty 4.190, 4.363
and Munich Pact 4.227
and Tyrol 4.356
Mustafa IV, Ottoman sultan (d. 1808) 4.253
Mustafa Resid Pasha, Ottoman sultan (1800–58) 4.338
mustard family 2.223, 2.299
see also brassica
mustard gas 6.72, 6.235
al-Musta'ṣimī, Yāqūt 5.71
Mustel, Auguste 5.83
Mustelidae see weasel family
Mut 9.4
al-Mu'tadid 9.16
al-Mutanabbī, Abū al-Tayyib 5.311
mutation 2.223–4, 6.104
see also evolution; genetics
mute swan 2.224
Mutesa I, Kabaka of Buganda 4.229, 4.356
Mutesa II, Sir Edward Frederick, Kabaka of Buganda 4.229, 4.356
mutilid wasps 2.355
mutinies 7.209–10
Bounty 4.47
Invergordon 4.165
Kronstadt 4.187
Mutiny Act 3.243–4
Nore 3.255
Spithead 3.339
Mutsuhito, emperor of Japan 4.216
mutton bird 2.224
mutual fund 7.166
mutualism see symbiosis
Muvarak 4.188
Muybridge, Eadweard 5.311, 6.74
Muzaffar al-Din 4.165
Muzorewa, Bishop 4.390
MVD 4.229
Mwene Mutapa 3.244, 3.312
Mwinyi, Ali Hassan 4.388
Mwinyi, Ndugu Ali Hassan 4.338
Myall Creek massacre 4.229
Myanmar (Burma) 1.223, 3.54, 4.229, 7.370–1
art 5.66–7
city 3.267
crime 7.225
decolonization 4.321
dynasty 3.359
ethnic groups 7.175
languages 7.22, 7.286–7, 7.305
law 7.143
literacy 7.185
Myanmar (Burma) (cont.):
literature 5.67
people 3.239, 3.325
physical geography and geology 1.174, 1.223, 1.292
politics and government 7.58, 7.84
leaders 4.22–3, 4.238, 4.361–2
refugees 7.261
region 3.19
religion/theology 7.36, 7.143
wars and defence 4.13
Burma Campaigns 4.55–6
Mycenaean civilization 3.244
art 5.311
mycorrhiza 2.129, 2.224
mylonite 1.223
mynah bird 2.224, 2.324
myocardial infarction 2.77
Myrdal, Gunnar 7.210
myriapods 2.16, 2.224
centipede 2.59
millipede 2.212
Myrica, Gaspar à 6.219
Myron 5.190
myrtle family 2.69, 2.224, 2.254
Myśliwski, Wiesław 5.355
Mysore 3.173, 3.244, 3.357
mystery plays 5.230, 5.294, 5.311, 5.341
mysticism 7.210
Buddhist 7.191
dervish 7.86
Hindu 7.143, 7.191, 9.24
Islam 3.344, 7.10, 7.24, 7.86, 7.301
Judaism 7.138, 7.175
mantra 7.191
Mystics 5.311, 5.359, 5.428, 5.438–9
Mytens, Daniel (Elder and Younger) 5.311
myths and legends 7.210, 9.1–39
anthropology 7.11, 7.210
and arts:
Aboriginal 5.1
Africa 5.7
Arabic 5.19
Czech 5.120
fables 5.6, 5.151
Greece 5.6
Iceland 5.141
India 5.370
Ireland 5.225
Middle English 5.294
Norse sagas 5.397
Slovak 5.422
Tibet 5.455
creation see creation myths
meaning as 7.196
and psychology 7.210
see also stories and under individual countries
Myxophyceae 2.84
Mzilikazi 4.229, 4.236, 4.390

N

Na Atibu 9.26
nabi rasūl 9.29
Nabiyev, Rakhmon 4.337
nabob 3.245
Nabokov, Vladimir 5.252, 5.312
Nabonidus, king of Chaldeans 3.95
Nabopolassar, king of Assyria 3.26, 3.32, 3.67
Naboth 9.20, 9.26
Nachtigal, Gustav 1.224
Nadar 5.312
nadir 8.106
Nadir Shah, king of Persia 3.101, 3.245
naga 9.26
Nāgārjuna 7.211
Nagas 7.211
Nagasaki 4.230
naginata 6.165
Nagy, Imre 4.158, 4.178, 4.230
Nagy, László 5.214
al-Nahayan, Sheikh Zayed bin Sultan 4.358
Naiads 9.26, 9.27
nail (body) 2.67, 2.225
nail (technology) 6.236
Naipaul, Shiva 5.312
Naipaul, Vidiadhar Surajprasad 5.312
naïve art 5.305, 5.312, 5.359, 5.392
Nakasone Yasuhiro 4.194, 4.230, 4.338
naker 5.246
Nakir and Munkar 9.26
NAM (Non-Aligned Movement) 4.244, 7.218, 7.311
Namaliu, Rabbie 4.259
Nambi 9.22
Nāmdev, Marathis 5.46
Namibia 1.224, 4.230, 7.371
decolonization 4.4
economy and trade 1.90, 7.296, 7.319
languages 7.176
laws/legal system 7.40
as mandate 7.191
physical geography and geology 1.224, 1.237, 1.372
religion and myths 9.15–16
SWAPO 4.334
and UN 7.321
wars and defence 7.82
Nana Sahib 4.230
Nanak, Guru 3.245, 3.328, 5.46, 5.418, 7.137, 7.169, 7.211, 7.285
Nandi 9.33
Nanga Parbat 1.224
Nanjing, Treaty of 4.230
Nannaya 5.134
Nansen, Fridtjof 1.224
Nantes, Edict of 3.171, 3.245
Naoise 9.10
Naoroji, Dadabhai 4.230
nap shearer 6.236
napalm 6.225, 6.236
naphtha and naphthalene 1.224, 6.79, 6.91, 6.145, 6.161, 6.264, 6.296
Napier, John 1.224, 6.53
Napier, Robert Cornelis, 1st Baron of Magdala 4.230–1
Naples 3.245
rulers 3.9, 3.69, 3.70, 3.125, 3.153, 3.279, 4.227–8
Napoleon I, emperor of France 3.245–6, 4.123, 4.231
and army 3.23
and Concordat 4.82, 4.291
and Confederation of Rhine 4.82–3
and Directory 3.104
and Egypt 3.113, 3.254
and 'Hundred Days' 4.158, 4.272
and Italy 3.181, 4.104
and law, see Code Napoléon
and Malta 3.198, 3.217
and Russia 4.120, 4.125
and Sicily 4.312
and Sweden 4.70
and Tilsit treaties 4.345
and Trafalgar 4.349
and Turkey 3.218
and Tyrol 4.356
and Waterloo 4.372
wife 4.176
see also Napoleonic Wars
Napoleon II, emperor of France 4.123
Napoleon III, emperor of France 4.232–3
and arts 5.146
assassination plot 4.252
and Cavour 4.206
deposed 4.41
and Franco-Prussian War 4.124, 4.307–8
and Piedmont 4.66, 4.291
and Villafranca di Verona treaty 4.368
wife 4.113
Napoleonic Wars 4.231–2
battles and campaigns 3.60, 3.296, 4.23, 4.42, 4.45, 4.192–3, 4.232, 4.288, 4.349, 4.369
Congress of Vienna 4.365–6
leaders 4.70, 4.94, 4.227–8, 4.240, 4.319
map 4.232
Quadruple Alliance 4.279
see also Peninsular War
nappe 1.224
Naqi'a 9.32
naqqāra 5.246
Narasinha 9.26
Narayan, Rasipuram Krishnaswami 5.312
narcissus (flower) 2.225
Narcissus (mythology) 9.11, 9.26
narcotic 2.225, 6.14, 6.236
see also drugs
Nareau 9.26
narrative verse 5.269, 5.312, 5.333
narrowboat see barge under canal
narrowcasting 7.211
Narva, battle of 3.246
Narváez, Pánfilo de 3.56, 3.88, 3.246
narwhal 2.225
NASA (National Aeronautics and Space Administration) 8.106
see also space-flight
Naseby, battle of 3.246
Nash, John 4.286, 5.312–13, 5.377
Nash, Ogden 5.313
Nash, Paul 4.50, 5.313, 5.362
Nash, Richard 3.246
Nash, Sir Walter 4.233
an-Nasir, al-Malik 3.218
Nasmyth, James 6.45, 6.217, 6.236, 6.331
Nasreddin Hoja 9.26
Nasreddin Mahmud 9.26
Nassau-Dillenberg, count of 3.382
Nasser, Gamal Abdul 4.109, 4.115–16, 4.233, 4.332, 7.218
Naströnd 9.16
nasturtium 2.225
Natal 1.95, 1.224, 4.233, 4.306
Naṭarāja 9.33

Natchez Indians ***3**.246*
Nathan **9**.7, ***9**.26*
Nation, Carry ***4**.233*
National Academy of Sciences ***6**.236*
National Aeronautics and Space Administration *see* NASA
national anthem ***5**.313*
National Assembly ***3**.246*, **7**.182
national debt ***7**.211*
National Federation of Labour **4**.285
National Government ***4**.233–4*
national income **7**.68, **7**.209, ***7**.211*, **7**.219
see also GNP
national insurance ***4**.234*
national parks ***1**.224*, **2**.225, ***7**.212*
Canada (and Provincial Parks):
Alberta **1**.177, **1**.362, **1**.371
British Columbia **1**.132, **1**.140, **1**.322, **1**.371
Ontario **1**.8
Quebec **1**.190, **1**.221
New Zealand **1**.100, **1**.122
United States (and National Monuments):
Arches **1**.18, **1**.225
Arkansas **1**.159
Badlands **1**.29–30
Big Bend **1**.39
Bryce **1**.45
California **1**.93, **1**.183, **1**.188, **1**.299, **1**.371
Devil's Tower **1**.89
Glacier **1**.140
Hot Springs **1**.159
Kentucky **1**.204
Kings Canyon **1**.183
Lassen **1**.188
Mammoth Cave **1**.204
Montana **1**.140, **1**.362
Mount Rainier **1**.247, **1**.274
Natural Bridges **1**.224
Sequoia **1**.299
South Dakota **1**.29–30
Texas **1**.39
Utah **1**.18, **1**.45, **1**.224, **1**.225, **1**.293, **1**.373
Washington (state) **1**.247, **1**.274
Waterton–Glacier International Peace Park ***1**.362*
Wyoming **1**.89, **1**.138, **1**.151, **1**.370
Yellowstone **1**.138, **1**.151, **1**.370
Yosemite **1**.93, **1**.371
Zion **1**.293, **1**.373
National Party (South Africa) ***4**.234*
National Party of Australia ***4**.234*
see also Nationalists *under* politics *under* Australia
National Physical Laboratory ***6**.236*
National Radio Astronomy Observatory ***8**.106*
national security state ***7**.212*
National Socialist German Workers' Party *see* Nazis
national theatre ***5**.313–14*
nationalism ***7**.211–12*
see also communal conflict; ethnic groups; National Party *and* self-determination *under* self *and also under* politics *under individual countries in particular* Australia; China; India; Ireland; Italy; Scotland
nationality ***7**.212*
naturalization **7**.213
nationalization ***7**.212*
Nations, Battle of (Leipzig) **4**.192–3
nation-state ***7**.212*
see also state
nativism ***7**.213*
see also in particular Chomsky; Descartes; Kant; Plato
NATO (North Atlantic Treaty Organization) ***4**.234*, **7**.10, **7**.82, ***7**.213*
air force **4**.6
alphabet **9**.47
early-wraning system **6**.111
flexible response **7**.121
Natsagorj, Dashdorijiya **5**.302
Natsume Sōseki ***5**.314*
Natta, Giulio ***6**.236*, **6**.391
natterjack ***2**.225*
Natufian culture ***3**.247*
natural bridge ***1**.224*, **1**.225
Natural Bridges National Monument ***1**.224*
natural gas ***1**.132*, ***6**.236–7*
exploration and recovery **6**.107–9, **6**.263
in gas-engines **6**.150
for heating **6**.167
for lighting **6**.150
liquefied ***1**.196*, ***6**.203*, **6**.237
pipeline **6**.270
products from:
carbon-black **6**.58, **6**.261
helium **6**.167
hydrogen **6**.164, **6**.174
methane **6**.36–7, **6**.222
see also petrochemicals; plastics
replacing coal gas **6**.79
resources **6**.121
see also gas
natural justice **7**.3, ***7**.213*
natural law ***7**.213*
natural selection **2**.194, ***2**.225*, **7**.11
sexual selection **2**.304–5
see also Darwin; evolution
naturalism and art **5**.26, **5**.342–3, **5**.488
naturalism in philosophy ***7**.213*
naturalism and realism in literature ***5**.314*
Britain **5**.127, **5**.143, **5**.179, **5**.303, **5**.415
France **5**.34, **5**.125, **5**.140, **5**.157, **5**.184, **5**.286, **5**.314, **5**.330, **5**.499
Germany **5**.201
Italy **5**.471
in novels *see* social life and realism *under* novels
Russia and USSR **5**.3, **5**.184, **5**.185, **5**.430–1, **5**.457–8
Scandinavia **5**.216, **5**.218, **5**.377, **5**.383, **5**.436, **5**.459
Slovak **5**.422
Spain **5**.10, **5**.343, **5**.428
United States **5**.117, **5**.136, **5**.213
see also realism
naturalists:
botanists *see under* botany
Buffon **2**.44, **7**.336
evolutionists *see under* evolution
Lamarck **2**.111, **2**.181
Leeuwenhoek **2**.186, **6**.197, **6**.198
Pliny as **2**.259
Sloane **2**.313
Wallace **2**.30, **2**.111, **2**.359
White **2**.366
zoologists *see under* zoology
see also Darwin
naturalization ***7**.213*
nature:
and art *see* landscape; naturalism and art; Romanticism; Sublime
in literature *see* naturalism and realism in literature
natural history *see* animals; ecology; naturalists; plants *and entries prefixed* **2.**
and nurture argument **7**.28–9, **7**.146
reserve ***2**.225*
see also national parks
Nāṭyaśāstra ***5**.314*
Naunet **9**.27
Nauru ***1**.225*, ***4**.234*, **7**.296
nausea ***2**.225*
Nautical Almanac **8**.142
nautical mile (knot) **6**.192
nautiluses **1**.121, **2**.241, **2**.245, **2**.246
Navajo (Navaho) Indians ***3**.247*, ***7**.213*, **7**.219
Navarre **3**.162
navel ***2**.225*
navies ***3**.247*, ***4**.234–6*
naval architecture ***6**.237*, **6**.316
see also warships
navigation ***6**.237–9*
air *see under* aircraft
artificial horizon **6**.21
astrolabe **6**.23, **8**.9
automation **6**.237
azimuth **1**.29, **8**.22
buoys **6**.52, **6**.193
charts **6**.215
chronometer **6**.73, **6**.77, **6**.166, **6**.303
compass *see* compass
cross-staff **6**.92
Decca Navigator **6**.97
dip-circle **6**.104
gyro-compass **6**.163, **6**.328
hyperbolic system **6**.238, **6**.290
inertial **6**.1, **6**.162, **6**.178–9
latitude *see* latitude and longitude
lead-line **6**.197
light ***6**.239*, **9**.end papers
lighthouse **6**.201, **6**.202, **6**.322
lightship **6**.202
log **6**.206
long range (Loran) **6**.208–9
maps *see* maps and charts
Navigation Acts **3**.247
quadrant **6**.287
radar *see* radar
radio **6**.208–9, **6**.237, **6**.238, **6**.249–50, **6**.290
river **6**.30, **6**.106, **6**.299
satellite ***6**.239*
sextant **6**.294, **6**.315
sonar *see* sonar
see also exploration
Navigation Acts ***3**.247*
Navstar **6**.239
Navy *see* navies
Naxalite Movement ***4**.236*
nay ***5**.314*
Nazarenes ***5**.314*
Nazca ***3**.247*
Nazis ***4**.236*, ***7**.213–14*
and academic freedom **7**.1
Brownshirts **4**.53
concentration camps **4**.82, **4**.154, **4**.155, **7**.63–4
and eugenics **7**.110
geopolitics **7**.131
Hitler Youth **4**.153, **4**.304
and Jews **4**.82, **4**.109, **4**.154–5, **7**.63–4, **7**.130, **7**.257, **7**.336
July Plot **4**.176–7
leaders:
Borman **4**.45
Eichmann **4**.109
Halder **4**.145
Jodl **4**.174
Ribbentrop **4**.290
Schirach **4**.304
Seyss-Inquart ***4**.310*
Speer **4**.324
Streicher **4**.330
see also Goebbels; Goering; Heydrich; Hitler
Lebensraum **4**.191
Munich 'beer-hall' putsch **4**.227
Nazi–Soviet Pact **4**.179, ***4**.236*
Nuremberg Trials **4**.248
organizations **4**.53, **4**.135, **4**.153
propaganda and doctrine **4**.137, **4**.192, **4**.330
Nazis (*cont.*):
racism *see* Jews *above*
Reichstag burnt **4**.286
SS **4**.327
state terrorism **7**.299
symbol **4**.334
Third Reich **4**.342, **4**.343, **9**.71
as totalitarianism **7**.313
N-body problem ***8**.106*
Ndebele **4**.198, **4**.229, ***4**.236*, ***7**.214*
Ndembu **7**.6, **7**.268
NDT (non-destructive testing) **6**.367
Ne Win **4**.229, ***4**.238*
Neanderthals **3**.120, **3**.242, ***3**.247*
neap tide ***1**.225*, **1**.337, **8**.106
Nebuchadnezzar I, king of Babylon **3**.32
Nebuchadnezzar II, king of Babylon **3**.32, **3**.67, ***3**.247*
and Daniel **9**.9
and Hanging Gardens **3**.322, **5**.28
and Judah **3**.187, **3**.192, **3**.267
and Tyre **3**.364
nebulae ***8**.106*
Crab **8**.34
Cygnus Loop **8**.35–6
Dumb-bell **8**.39
extragalactic **8**.49
Great Leopard **8**.35
Great Looped **8**.35–6
Great Nebulae in Orion **8**.61
Gum **8**.62
Helix **8**.65
Horsehead **8**.69
Keyhole **8**.80
Lagoon **8**.82
Laplace nebular hypothesis **8**.82–3
Omega **8**.112
Owl **8**.115
planetary **8**.123, **8**.124
reflection **8**.137
Ring **8**.140
Tarantula **8**.169
Trifid **8**.173
Veil **8**.180
Necker, Jacques ***3**.249*, **3**.340
nectarine ***2**.225*
Nedīm **5**.462
needle ***6**.239*
needlefish **2**.133
needlework *see* embroidery; textiles
Nefertiti ***3**.249*
negative charge:
beta particle **1**.38
cation **1**.14
electrostatics **1**.105
see also cathode; electrons
negative (photography) **6**.101
Negev Desert ***1**.225*
negligence ***7**.214*
Nègre, Charles ***5**.314*
négritude movement **5**.7, **5**.86, **5**.411
Negroids ***3**.249*
Neguib, Muhammad **4**.233
Neguib Mahfouz **5**.19
Nehemiah ***3**.249*
Nehru, Jawaharlal **4**.83, **4**.161, **4**.174, ***4**.236–7*, **7**.156, **7**.218
Nehru, Motilal ***4**.237*
Nei Teukez **9**.26
Neilson, James **6**.186
Nekrasov, Nikolay ***5**.314*
Nelson, Horatio, viscount ***3**.249–50*, ***4**.237*
and battles:
Aboukir Bay **4**.232
Cape St Vincent **3**.60
Copenhagen **4**.86
Nile **3**.254
Trafalgar **4**.349
and Emma Hamilton **3**.249, **4**.237
flagship **4**.235
and Sicily **4**.312
nematodes *see* roundworms
Němcová, Božena **5**.120

nemertine **2**.284
nemesia ***2**.226*
Nemesis ***9**.26*
ne-ne **2**.152
Nennius **3**.23, **3**.139, **3**.242, ***3**.250*
neo-classical economics **7**.53, **7**.101
see also classical economics *and also in particular* Marshall, Alfred; Pareto; Walras
Neoclassicism ***5**.314–16*
architecture **5**.56, **5**.258, **5**.260, **5**.425, **5**.490
see also under Britain *below*
Britain:
architecture **5**.3, **5**.4, **5**.72–3, **5**.105, **5**.106, **5**.121–2, **5**.312–13, **5**.351, **5**.377, **5**.423
ceramics **5**.105, **5**.485
furniture **5**.3, **5**.5, **5**.105, **5**.204, **5**.315, **5**.351, **5**.416–17
sculpture **5**.157
Sheffield Plate **5**.416
ceramics **5**.105, **5**.315, **5**.485
Denmark **5**.454
drama **5**.5
France **5**.125–6, **5**.131, **5**.146, **5**.170–1, **5**.223, **5**.269, **5**.375
architecture **5**.56, **5**.260, **5**.425
Germany **5**.204, **5**.206, **5**.291, **5**.490
interior design and furniture **5**.131, **5**.146, **5**.170–1
see also under Britain *above*
Italy **5**.75
literature **5**.316
metalware **5**.315, **5**.416
music **5**.204, **5**.206, **5**.375
see also Stravinsky
painting **5**.13, **5**.112, **5**.125–6, **5**.207, **5**.291, **5**.486
sculpture **5**.75, **5**.157, **5**.454
United States **5**.13, **5**.112, **5**.207, **5**.486
see also classicism; Greece, ancient; Roman empire
Neo-Confucianism **7**.46, **7**.65, **7**.305
neodymium **1**.106
neo-Freudians ***7**.214*, **7**.252, **7**.253
see also Jung
Neogene period **1**.134, ***1**.225*
see also Miocene; Pliocene
neo-imperialism **7**.153
Neo-Impressionism **5**.171–2, **5**.182–3, **5**.286, ***5**.316*, **5**.354, **5**.362, **5**.413, **5**.418, **5**.458
see also Impressionism
Neolithic ***3**.250*
art and artefacts **3**.35, **3**.175, **5**.434
Beaker cultures **3**.37
grave-mounds **3**.36
Impressed ware culture **3**.175
Longshan site **3**.210
Megalithic *see* megaliths
settlement **3**.64
Stonehenge **3**.342
Western **3**.380
Yangshao **3**.387
see also prehistory
neon **1**.106, ***1**.225*, **6**.195, ***6**.239*
Neo-Plasticism **5**.301, ***5**.316*
Neoplatonism **3**.192, ***3**.250*, **3**.285, **7**.237, **7**.310
see also Augustine of Hippo; Plato
neoprene (polychloroprene) **6**.60, **6**.304
Neo-Romanticism **5**.312, ***5**.316*, **5**.350, **5**.362, **5**.441
see also Romanticism
neoteny ***2**.226*
Nepal ***1**.225*, ***3**.250*, ***4**.237*, ***7**.371*
aid to **7**.8
arts ***5**.316*
languages **7**.157, **7**.286
Nepal (*cont.*):
physical geography and geology ***1**.225*
mountains **1**.14, **1**.183
see also Himalayas
rivers **1**.317, **1**.333
nephrite **1**.177
Neptune (mythology) ***9**.26*
Neptune (planet) ***8**.106–7*
orbit **8**.196
rings **8**.197
satellites **8**.174, **8**.198
inner ***8**.107*
statistical details **8**.196
neptunium **1**.106
Nerchinsk, Treaty of ***3**.250*
Nereid ***8**.107*
Neri, Philip, Saint ***3**.250*
nerine ***2**.226*
Nernst, Walther Hermann ***1**.226*
Nero, Claudius Caesar Drusus Germanicus **3**.57, **3**.101, ***3**.250–1*, **3**.309
tutor *see* Seneca
Nersenne, Marin **6**.1, **6**.53
Neruda, Pablo ***5**.316–17*, **5**.371
Nerval, Gérard de ***5**.317*
nerve ***2**.226*
see also nervous system
nerve gas **6**.72, **6**.239–40
Nervi, Pier Luigi **5**.317
nervous system **2**.158, ***2**.226*, ***2**.227*
autonomic **2**.18, **2**.222
brain *see* brain
disorders:
decompression **6**.98
and dioxin **6**.104
multiple sclerosis **2**.221–2, **6**.241
neurology and neurosurgery **2**.227, **6**.240–1
shingles **2**.307
see also poliomyelitis; rabies
drugs acting on **6**.34, **6**.266
see also anaesthesia
and muscles **2**.222
reflex **2**.281
nest ***2**.226–7*
Nestor (mythology) ***9**.26*
Nestorians and Nestorius ***3**.251*, **7**.47, ***7**.214*
Nestroy, Johann Nepomuk ***5**.317*
net *see* nets
netball ***7**.214*
Netherlands ***1**.226*, ***3**.251*, ***4**.237–8*, ***7**.371*
adoption **7**.3
aid from **7**.8
arts:
architecture **5**.468
drawing **5**.140
engraving and engravers **5**.49, **5**.270
etching **5**.378, **5**.410
graphic art and design **5**.148, **5**.264
literature **5**.165
print **5**.148
sculpture **5**.396
see also under landscape; painting; portraiture *and also* music *below*
astronomy **1**.199, **6**.347, **8**.19, **8**.27, **8**.29, **8**.112, **8**.159, **8**.176
see also Huygens; Keyser; Lorentz
chambers of commerce **7**.43
chemist **1**.354
city **3**.49
clocks **6**.76
demography **7**.221, **7**.241
economy and trade **7**.209, **7**.221
see also EC; EEC
education **7**.155
empire/colonialism **3**.11, ***3**.107–8*, **3**.344, **3**.350, **4**.384, **7**.371–2
Netherlands (*cont.*):
empire/colonialism (*cont.*):
decolonization **4**.321
explorers **1**.33, **1**.331
see also Dutch East India Company; Dutch East Indies; Dutch Guiana; Indonesia
energy and fuel **6**.339, **6**.385
explorers **1**.33, **1**.331
foreign policy **4**.372, **4**.375
Council of Europe **7**.71
government *see* politics and government *below*
historiography **5**.165
industry **6**.208, **6**.256, **6**.278
land reclamation **6**.193
languages **7**.131, **7**.136
laws/legal system **3**.149, **7**.111, **7**.164, **7**.165
literature **5**.165
mathematician **1**.160
medicine and health care **6**.112–13, **6**.198, **7**.196
music and dance **5**.122, **5**.242, **5**.256, **5**.441
optics **6**.198, **6**.223
philosophy **3**.339, **7**.149
physical geography and geology ***1**.226*
floods **1**.321
islands **1**.127, **1**.359
meteorology **1**.46
rivers **1**.282, **1**.294
Zuiderzee **1**.374
physicists **1**.160, **1**.199, **1**.353
politics and government:
Christian Democracy **7**.47
coalition **7**.54
consociationalism **7**.67
parliamentary **7**.229
statesmen **3**.261, **3**.384
statholder **3**.340–1
provinces **1**.374, **3**.337
listed ***9**.75*
religion/theology **3**.23, **3**.185, **3**.229, **3**.339, **7**.198
saints **9**.40, **9**.41
rulers **4**.238, **4**.378, **7**.97, **7**.139
listed ***9**.68*
scholar **3**.117
social services and security **7**.221, **7**.330
sport and recreation **7**.151
technology and science *see entries above and below prefixed* **1**. *and* **6**.
tools **6**.358
United Provinces of **3**.366
vehicles and transport **6**.31, **6**.56, **6**.143, **6**.277
wars and defence:
against Britain **3**.14, **3**.58, **6**.375
Acheh War **4**.2
army leader **3**.195
Dutch Revolts (against Spain) ***3**.108*, **3**.141
Franco-Dutch **3**.254
Java War **4**.173
navy leaders **3**.312, **3**.361
Revolutionary Wars **4**.288
in WEU **7**.269
World War II **4**.124
Xhosa Wars **4**.384
see also NATO
see also Flanders
Netherlands Antilles ***7**.371–2*
Netherlands Dance Theatre **5**.122
nets **6**.133, **6**.239, **6**.240
netsuke ***5**.317*
nettle ***2**.227*, **2**.247
nettle rash **2**.353
Netungati, Appu **5**.136
network (technology) **6**.179, ***6**.240*
neural **6**.21, **6**.180, **6**.240
Neue Sachlichkeit **5**.131, **5**.195
Neuilly, Treaty of **4**.364
Neumann, Balthasar ***5**.317*
Neumann, John von **1**.131, ***1**.226*, **6**.84–5, ***6**.240*
neumes **5**.323
neural network **6**.21, **6**.180, ***6**.240*
neuroleptics **6**.285, **7**.96
neurology and neurosurgery ***2**.227*, ***6**.240–1*
neurophysiology **2**.158
neuroses ***7**.214*
hysteria **7**.126, **7**.150
obsession **7**.221
see also depression and anxiety; phobias
Neutra, Richard ***5**.317–18*
neutral equilibrium **8**.47
neutrality **7**.82, ***7**.214*
Neutrality Acts ***4**.238*
neutralization (chemistry) **1**.59, ***1**.226*
neutrinos **1**.123, ***1**.226*, ***8**.107*
neutron stars ***8**.107*
see also pulsars
neutrons ***1**.226*, **6**.23–4, **6**.245
different number of *see* radioactive elements and isotopes
'neutron bomb' **6**.246
névé **1**.122
Nevi'im: books listed ***9**.44*
Neville, John, earl of Montagu **3**.164–5
Neville's Cross, battle of ***3**.251*
Nevis *see* Saint Kitts and Nevis
Nevis, Ben ***1**.37*
New Age movement ***7**.214*
New Amsterdam *see* New York (city)
New Australians ***4**.238*
New Brunswick ***1**.226*, ***3**.251*
New Caledonia **3**.228, ***4**.238*, **7**.175
New Deal ***4**.238*, **4**.318, **4**.374
New Democratic Party ***4**.238*
New Economic Policy ***4**.238*
New England ***1**.226*, ***3**.251*
colonized **3**.86, **3**.217, **3**.225, **3**.251, **3**.252, **3**.282, **3**.286, **3**.302, **3**.383–4
Glass Company ***5**.318*
New English Art Club ***5**.318*
New France **3**.135, ***3**.251–2*
New Granada ***3**.252*, **4**.43
New Guinea:
Highlands ***7**.214–15*
see also Papua New Guinea
New Hampshire ***3**.252*
New Hebrides *see* Vanuatu
New International Economic Order ***7**.215*, **7**.218, **7**.219, **7**.224, **7**.311, **7**.333
New Jersey ***3**.252*
New Learning ***3**.252*
New Left **7**.192, ***7**.215*
New Model Army ***3**.253*
new moon **8**.118, **8**.119
New Orleans, battle of ***4**.238–9*
New Plymouth **3**.286
New Right **7**.69, **7**.86, ***7**.215*
New South Wales ***1**.227*, ***4**.239–40*
New Spain ***3**.253*
new technology *see* computers
new town ***7**.216*
New World Information Order **7**.195, **7**.308, **7**.320
New York (city) **3**.253
Ellis Island **4**.111
New York City Ballet ***5**.318*
School ***5**.318*
New York State ***3**.253*
New Zealand ***1**.228*, ***4**.240*, ***7**.372–3*
agriculture **7**.192
arts **5**.80, **5**.280
astronomy **8**.17
discovered and colonized **1**.75, **1**.331, **3**.350, **4**.240
administration **4**.13–14, **4**.93, **4**.142, **4**.190, **4**.294, **4**.338, **4**.370

New Zealand (cont.):
  economy and trade 7.203, 7.221
  environment 7.135, 7.296
  ethnic groups 7.240
    see also Maoris
  explored see discovered above
  foreign policy:
    aid from 7.8
    in ANZUS 7.14
    in South Pacific Forum 7.296
  government see politics and government below
  immigration 7.105
  industrial relations 4.285, 7.159
  land: Repudiation Movement 4.287
  languages 7.106
  laws/legal system:
    compensation 7.62, 7.75
    habeas corpus 7.138
    industrial relations 7.159
    International Court of Justice 7.164
  literature 5.146–7, 5.280
  Maori people see Maoris
  music 5.80
  National Federation of Labour 4.285
  physical geography and geology 1.228
    cliffs 1.70, 1.322
    coasts 1.70, 1.122, 1.278, 1.283, 1.301, 1.322
    fiords 1.122
    geysers and hot springs 1.137, 1.286, 1.326
    island 1.320
    lakes 1.332, 1.359
    landslide 1.187
    marsh 1.206
    mountains 1.75, 1.100, 1.287, 1.314, 1.339
    national parks 1.100, 1.122
    peninsula 1.251
    reef 1.278
    rias 1.283
    rivers 1.39, 1.43, 1.86, 1.353, 1.359
    seas 1.75, 1.332
    shore platform 1.301
    sink-holes 1.305
    volcanoes 1.100, 1.287, 1.339
  politics and government:
    cabinet 7.38
    as Dominion 7.58
    electoral system 7.123, 7.260, 7.332
    Green 7.135
    Labour 4.124, 4.184, 4.189, 4.233, 4.303
    Liberal 4.29, 4.120, 4.285, 4.308, 4.370–1
    National Party 4.155, 4.227
    ombudsman 7.223
    prime ministers: listed 9.71
    Reform Party 4.77, 4.213
    women's suffrage 4.381, 7.332
  Red Feds 4.285
  regions listed 9.73
  religion/theology 4.255–6, 7.187, 9.25, 9.35
  social services and security 4.317, 4.318, 7.45, 7.330
  sport and recreation 7.75, 7.272, 7.297, 7.303
  technology and science 6.94, 6.104, 6.143
  telecommunications 7.307
  unions 4.285
  wars and defence 4.16, 4.127, 8.14
    Anglo-Maori 4.13–14, 4.294, 4.338
  welfare state 4.240, 4.374, 7.330
New Zealand Company 4.240
Newcastle, Thomas Pelham-Holles, 1st duke of 3.251
Newcomen, Thomas: Newcomen engines 3.341, 6.32, 6.227, 6.241, 6.330, 6.380
Newfoundland 1.226, 3.203, 3.251
Newlands, John Alexander Reina 1.227
newly industrializing countries 7.87, 7.88, 7.113, 7.215, 7.311
  major see Hong Kong; Singapore; South Korea; Taiwan
  putative see Brazil; Malaysia; Mexico; Thailand
Newman, John Henry 4.12, 4.238, 4.239, 5.412
Newport Rising 4.239
news 7.215
  agency 7.215–16
  freedom of press 7.125
  newspaper 7.216
  see also journalism; media; press
newt 2.79, 2.227
newton (unit) 6.368
Newton, Sir Isaac 1.227, 6.149, 8.107
  and alchemy 1.7
  and calculus 1.48, 1.90, 1.120, 1.191, 1.201
  and elasticity 1.157
  and geodesy 1.133
  Laws 1.227–8
    of Gravitation and astronomy 8.11, 8.60, 8.61, 8.107, 8.108
      and celestial mechanics 8.22
      and cosmology 8.33
      and galactic rotation 8.54
      and Halley's comet 8.63
      and Laplace 8.82
      and Le Verrier 8.84
      and mass 8.93
      and relativity 8.159
    of Motion 1.57, 6.143
      and astronomy 8.11, 8.22, 8.29, 8.51, 8.92, 8.108, 8.172
  Newtonian telescope see reflecting under telescopes
  and prism 6.282
  and satellites 8.144
  and scientific revolution 3.319
  and spectrum 8.85
  and thought experiments 7.311
  and time 7.312
Newton, John 5.116
newton 1.227
Ney, Michel, duc d'Elchingen 4.240, 4.372
Nezval, Vítězslav 5.318
Ngewo 9.26
Ngo Dinh Diem 4.240, 4.241, 4.366, 4.367
NGOs see non-governmental organizations
Nguema, Macias and Obiang 4.113
Ngugi wa Thiong'o 5.318
Nguni people 4.240–1, 4.384
  see also Zulus
Nguyen Anh (later emperor Gia-Long) 4.14, 4.367
Nguyen Cao Ky 4.241
Nguyen Du 5.475
Nguyen Tat Thanh see Ho Chi Minh
Nguyen Van Thieu 4.241
niacin 2.357
Niagara River and Falls 1.228
Niall of Nine Hostages 3.350
Nibelung 9.27
Nibelungenlied 5.318
Nicaea, Councils of 3.253, 7.74, 7.309
Nicaean empire 3.253
Nicaragua 1.228, 3.253, 4.67, 4.241, 7.373
  arts 5.125
  economy 4.251
  education 7.93
Nicaragua (cont.):
  exploration 1.47
  law 7.164
  physical geography and geology 1.228
  politics and government 4.302, 4.319
  refugees 7.261
  Sandinistas 7.73
  wars and defence 7.52, 7.299
Nicephorus, emperor of Bulgaria 3.53
niche 2.3, 2.227
Nicholas, Saint 9.43
Nicholas I, emperor of Russia 4.241–2
Nicholas II, emperor of Russia 4.104, 4.242, 4.283, 4.296, 4.297, 4.365
Nicholas V, pope 3.133
Nicholson, Ben 5.2, 5.318, 5.398
Nicholson, Sir William 5.318
nichrome 6.241
nickel 1.106, 1.228, 6.241
  alloys 6.11, 6.45, 6.185, 6.212, 6.241, 6.330, 6.345
  in battery 6.32
  catalysts 6.164, 6.241
  electroplating 6.120
  magnets 6.212
  ore 1.250
  plating 6.340
  refining 6.232
  steel 6.11
nickel oxide 6.156
nickel sulphide 1.250, 6.241
Nicobar Islands and language 7.22
Nicodemus 9.21, 9.27
Nicolas of Verdun 5.146
Nicolet, Jean 1.228
Nicollet, Joseph Nicolas 1.228
Nicopolis, battle of 3.253
Nicotiana 2.227–8
nicotine 2.228, 2.343, 6.357
nicotinic acid 2.357
NICs see newly industrializing countries
Nidhogg 9.16, 9.38
Nidud 9.38
Niebuhr, Barthold Georg 4.153
Niebuhr, Reinhold 7.184
niello 5.318
Nielsen, Carl 5.318, 5.320
Niemcewicz, Julian 5.320
Niemeyer, Oscar 5.320, 5.425
Niemöller, Martin 4.242
NIEO see New International Economic Order
Niepce, Joseph Nicéphore 5.121, 6.241, 6.268, 8.119
Nietzsche, Friedrich 4.236, 7.21, 7.113, 7.216–17
Niflheim 9.6, 9.16, 9.25, 9.26, 9.38
Nigari 3.35
Niger 1.228, 4.242, 7.373
  colonization and decolonization 4.4, 4.306
  ethnic groups see Hausa
  finance 7.124
  languages 7.30, 7.43, 7.217
  physical geography and geology 1.57, 1.86, 1.228, 1.229
Niger River 1.229
Niger–Congo languages 7.30, 7.217
Nigeria 1.229, 4.242–3, 7.374
  aid to 7.8
  arts 5.427
  Civil War and Biafra 4.38, 7.130, 7.151, 7.277
  colonization and decolonization 4.4, 4.306
  economy and trade 7.373
    see also OPEC
  ethnic groups and kingdoms 3.39–40, 3.159, 3.260, 7.151, 7.335
Nigeria (cont.):
  ethnic groups and kingdoms (cont.):
    see also Hausa
  genocide 7.130
  languages 7.43, 7.217
  leaders 4.26, 4.28, 4.33–4, 4.139, 4.171, 4.362
  literature 5.427
  in OPEC 4.252
  physical geography and geology 1.57, 1.229
  politics 4.362
  population growth 7.242
  prehistory 3.254
  religion and myths 7.5, 9.17
    deities 9.3, 9.28, 9.32
    tricksters 9.11, 9.15
  states 4.34–5, 4.38
  urban areas 3.174, 3.194, 3.266, 7.323
  wars and defence 7.52
night 1.229
night monkey 2.228
'Night of the Long Knives' 4.243
Nightingale, Florence 4.88, 4.243
nightingale 2.228
nightjar 2.228
nightshade 2.228
nihilism 4.243
Nijinska, Bronislava 5.320
Nijinsky, Vaslav 5.30, 5.320, 5.340
Nijmegen, Treaty of 3.254
Nikisch, Artur 5.110
Nikolayev, Leonid 4.184
Nile 1.229, 1.276
  battle of 3.254
  languages 7.217
  map 3.259
  see also Egypt; Nubians; Sudan
nilgai 2.228
Nilo-Saharan languages 7.217
Nilsson, Birgit 5.320
nimbus and nimbostratus clouds 1.128, 1.229
al-Nimeiri, Gaafar Muhammad 4.243, 4.332
Nimitz, Chester William 4.243, 4.254
nimonic 6.11, 6.241
Nimrod 9.27
Nine Years War 3.254, 3.312
Nineve (Lady of the Lake) 9.12, 9.25
Nineveh 3.254
Ning Chen Yang 1.329
Ninhursaga 9.11
Ninon de Lenclos 3.205
Niobe 9.27
niobium 1.106, 1.229, 6.241
nipa palm 2.228
Nipkow, Paul: Nipkow disc 6.241
Nirala 5.206
nirvana 7.217, 9.5
nitrates 1.229, 6.241
  explosives 6.147, 6.162, 6.241
  fertilizers 6.13, 6.124, 6.128, 6.241
  pollutant 6.377
  silver 1.305
nitre see potassium nitrate
nitric acid 1.229–30, 6.241
  aqua regia 1.17
  explosives 6.242
  fertilizer 6.71
  manufacture 6.2, 6.13
nitrile 1.230
nitrobenzene 6.14
nitro-cellulose and cellulose trinitrate 6.66, 6.153, 6.241–2, 6.268, 6.310, 6.387
nitrogen 1.106, 1.230, 6.242
  assimilated by plants 2.18
  carbon–nitrogen cycle 8.20
  cycle 2.228, 2.229
  and decompression disease 6.97–8
  and fertilizer 6.127–8

nitrogen (*cont.*):
fixation **2**.228, **6**.38, **6**.43, **6**.71, **6**.128, **6**.164
see also legumes
liquefaction **6**.150
silicon nitride production **6**.320
see also nitrates; nitric acid *and entries beginning* nitro
nitrogen dioxide ***1**.230*, **7**.2
nitrogen oxides **6**.2
nitrogen trihydride *see* ammonia
nitro-glycerine **1**.118, **6**.110, **6**.153, ***6**.242*
nitrous oxide **6**.14, **7**.135
Niue **7**.296
nival plant ***2**.229*
Nix Olympica ***8**.108*
Nixon, Richard Milhous **4**.95, ***4**.243–4*, **4**.287
and détente **4**.97
impeachment **7**.153
and Kissinger **4**.184
and Watergate scandal **4**.372
Niyazov, Suparmuradi **4**.354
Nizam Shahi ***3**.254*
Niẓāmī **5**.344, **9**.3, **9**.6, **9**.24
Njegos, Nicholas Petrovic **4**.223
Njörd **9**.2, **9**.13
Nkomo, Joshua Mqabuko Nyongolo **4**.226, ***4**.244*, **4**.390
Nkrumah, Kwame **4**.136, **4**.181, ***4**.244*, **4**.257
NKVD ***4**.244*
No theatre **5**.235, **5**.236
Noah **7**.73, **9**.5, **9**.15, ***9**.27*
Nobel, Alfred **6**.99, **6**.110, **6**.124, **6**.153, ***6**.242*, **6**.242
nobelium **1**.106
Nobile, Umberto **1**.105
Noble Eightfold Path **7**.97, **7**.123, ***7**.217*
noble gases ***1**.230*
argon **1**.20, **1**.106
helium *see* helium
krypton **1**.106, **1**.184
neon **1**.106, **1**.225
radon **1**.107, **1**.274
xenon **1**.107, **1**.369
noble metals:
electrolysis **6**.157, **6**.253
iridium **1**.106, **1**.173, **6**.186
osmium **1**.106, **1**.242, **6**.177
palladium **1**.107, **1**.246, **6**.256
rhodium **1**.107, **1**.282, **6**.298, **6**.354
ruthenium **1**.107, **1**.288, **6**.304
see also gold; platinum; silver
nocturne ***5**.320*
noddy ***2**.229*
nodes (astronomy) ***8**.108*
line of **8**.86
nodical month **8**.100
nodes (anatomy) *see* lymph glands
nodes (physics): antinode **1**.16
nodules in rocks **1**.71
no-go areas ***7**.217*
noise pollution **7**.240
Nok culture ***3**.254*
Nolan, Sir Sidney ***5**.318*, **5**.320
Nolde, Emil ***5**.320*
Nollekens, Joseph ***5**.320*, **5**.359
nomadism ***7**.217*
Bedou (Bedouin) **7**.28, **7**.29
Berbers *see* Berbers
gypsies **7**.137
Huns **3**.172–3
Numidia **3**.260
Saracens **3**.316
transhumance **7**.217, **7**.230
Turks **3**.363
Xiongnu **3**.387
see also hunter-gatherers; migration; pastoralism
nomenclature and classification of living creatures ***2**.67*, ***2**.229*, ***2**.258*
family **2**.113
genus **2**.134
Linnean **2**.2, **2**.190–1
order **2**.236
phylum **2**.252
race **2**.275
species **2**.319
nomenclature in science *see* symbols
nomenklatura ***7**.218*
nominalism **7**.1, ***7**.218*, **7**.221
Nommo ***9**.27*, **9**.28
Non-Aligned Movement ***4**.244*, ***7**.218*, **7**.311
Nonconformists ***3**.254*
see also Protestants
non-co-operation ***4**.244–5*
non-destructive testing **6**.367
non-electrolytes ***1**.230*
non-governmental organizations **7**.66, **7**.135, **7**.165, **7**.203, ***7**.218*, **7**.256, **7**.327
non-metallic elements ***1**.230*, ***6**.24*
arsenic **6**.20, **6**.72, **6**.312
boron **1**.42, **1**.106, **6**.43, **6**.212, **6**.245, **6**.312
carbon *see* carbon
phosphorus *see* phosphorus
selenium **6**.268, **6**.312, **6**.349
sulphur *see* sulphur
tellurium **6**.350
see also gas; halogens
Nono, Luigi ***5**.320*
Non-Proliferation Treaty **7**.220
non-saturation and gravity **8**.60
nonsense poetry **5**.81, **5**.121, **5**.259, ***5**.320*
non-verbal communication **7**.114, ***7**.218*, **7**.219, **7**.281
non-violence **4**.125
see also peace *and under* Gandhi, Mohandas
nonwoven fabric **6**.125, **6**.127, ***6**.242*, **6**.351
noon ***8**.108–9*
Nootka Sound ***3**.254*
Nordenskiöld, Baron Nils Adolf Erik ***1**.230*, **1**.231, **4**.17
Nordmeyer, A. H. **4**.184
Nordraak, Rikard **5**.193
Nore mutiny ***3**.255*
Norfolk, Thomas Howard, 4th duke of **3**.304
Norfolk four-course system **6**.359
Norfolk Island ***1**.230*
Norgay, Tenzing **1**.116
Norie, Alexander **4**.14
Norie, Orlando **4**.381
Noriega, General Manuel **4**.257
norm ***7**.218*
Norma ***8**.109*
normalization (mental handicap) **7**.198
Norman, Jessye ***5**.321*
Normand, Benjamin **6**.363
Normand, Mabel **5**.411
Norman architecture *see* Romanesque
Normandy ***3**.256*
Campaign and Landings ***4**.245–6*
see also Normans
Normans **3**.150, ***3**.255–6*
Conquest **3**.158, ***3**.256*, **3**.339
rulers *see under* rulers *under* Britain
Norris, Frank **5**.213
Norseman *see* Viking
North, Frederick, Lord **3**.80, **3**.158, ***3**.256*, **4**.122
North, Robert **5**.373
North Africa:
and aid **7**.8
in Arab League **7**.15
emigration **7**.105
North Africa (*cont.*):
ethnic groups **7**.137
Bedou **7**.28, **7**.29
see also Berbers; Moors
languages **7**.30, **7**.78, **7**.269, **7**.278
laws/legal system **7**.221, **7**.284
maps **3**.266, **3**.307, **4**.246
music ***5**.321*
Numidia **3**.260
old age **7**.221
physical geography and geology **1**.8, **1**.101, **1**.193, **1**.221
climate **1**.65
desert **1**.89
mountains **1**.89, **1**.179
oasis **1**.234
ravine **1**.277
poverty **7**.244
religion/theology **3**.28, **3**.105, **7**.166, **7**.280, **7**.284, **7**.302
saints **3**.28
wars and defence **4**.7
North African Campaigns ***4**.246*
see also Algeria; Egypt; Libya; Morocco; Tunisia
North America: physical geography and geology ***1**.230*, **1**.231
map **1**.231
plate tectonics **1**.260
see also Canada; United States of America
North America Nebula ***8**.109*
North American Indians *see* Indians, North American
North Atlantic Drift ***1**.230*
North Atlantic Treaty **7**.315
see also NATO
North Carolina ***3**.256*
north celestial pole **8**.109
north galactic pole **8**.54
north galactic spur ***8**.109*
North Korea ***4**.247*, ***7**.362*
economy and trade **7**.319
languages **7**.177
politics and government **7**.60, **7**.89
sport and recreation **7**.305
see also Korea
north polar distance **8**.22, ***8**.109*
North Pole ***1**.231*
see also Arctic; pole/polar
North Sea **1**.93, ***1**.231*
oil **6**.108, **6**.236
North Vietnam *see* Vietnam
Northcliffe, Alfred viscount **4**.147
North-East Passage ***1**.231*
northern constellations *see* constellations
Northern Expedition ***4**.246*
Northern Hemisphere Observatory **8**.82
Northern Ireland ***1**.231*, ***3**.365*, ***4**.246–7*
city **3**.209
law **7**.114, **7**.239, **7**.319
physical geography and geology **1**.136, **1**.231
politics, government and leaders **3**.364, **4**.117, **4**.330
devolution **7**.88
hunger-strikes **7**.150
Ulster Unionist Party **4**.357
regions listed ***9**.73*
religion/theology **7**.59
wars and defence:
IRA *see* Irish Republican Army
no-go areas **7**.217
sectarian **7**.59
Ulster Volunteers **4**.357
see also Britain; Ireland
Northern Lights **1**.173, **8**.12–13
Northern Renaissance **5**.73
Northern Rhodesia *see* Zambia
Northern Rising ***3**.256*
Northern Territory ***1**.231*, ***4**.247*
Northern War **3**.246, ***3**.256–8*, **3**.260, **3**.287
Northern Wei ***3**.258*
North–South relations ***7**.219*
see also developing countries; industrialized countries
Northumberland, Sir Hugh Smithson, duke of **3**.276
Northumberland, John Dudley, duke of **3**.110, **3**.195, ***3**.258*
Northumberland: Percy family **3**.265, **3**.276
Northumbria **3**.15, **3**.109, ***3**.258*, **3**.264
North-West Europe Campaign ***4**.247*
North-West Frontier ***4**.247–8*
North-west Indians ***7**.219*
Kwakiutl ***7**.178*
potlatch **7**.219, ***7**.243*
Tlingit **7**.295, ***7**.312*
totemism **7**.313
Tsimshian ***7**.316–17*
North-West Passage ***1**.231*
Northwest Territories (Canada) ***1**.231–2*
Northwest Territory (United States) ***3**.258*
Norton, Arthur P. **8**.159
Norway ***1**.232*, ***3**.258*, ***4**.247–8*, ***7**.374*
aid from **7**.8
arts:
museums **7**.209
music **5**.156, **5**.192–3
painting **5**.192–3, **5**.307–8
see also literature *below*
in Council of Europe **7**.71
economy and trade **7**.99, **7**.107, **7**.111, **7**.221
environment **7**.260
ethnic groups **7**.181
explorers **1**.12, **1**.224, **3**.118
government *see* politics and government *below*
languages **7**.131, **7**.156
literature **5**.48, **5**.197, **5**.209–10, **5**.216
see also under poetry
mathematician **1**.193
music **5**.156, **5**.192–3
physical geography and geology **1**.232
Fennoscandian Shield **1**.120
fiords **1**.122, **1**.310
ice-sheet **1**.178
islands **1**.317, **1**.328
Lapland **1**.188
Maelstrom **1**.201
meteorology **1**.39–40
Skagerrak **1**.306
politics and government **4**.195–6, **4**.280, **7**.250
religion/theology **3**.261, **7**.112
rulers **7**.97, **7**.139
9th to 11th centuries **3**.59–60, **3**.157, **3**.158, **3**.258, **3**.261, **3**.339
13th to 15th centuries **3**.8, **3**.74, **3**.152, **3**.193, **3**.258
19th to 20th centuries **4**.69–70, **4**.144, **4**.248, **4**.335
Social Democrats **4**.317
sport and recreation **7**.287
technology and science **6**.38, **6**.96, **6**.250, **6**.301, **6**.322, **6**.384
wars and defence *see* NATO
Norway lobster **2**.100
Norwegian Sea **6**.209
Norwich School ***5**.322*
Norwid, Cyprian ***5**.322*
nose ***2**.229*, ***9**.59*
disorders **6**.254, **6**.342
no-see-um **2**.210
Nostoc ***2**.229–30*
Nostradamus ***3**.258–9*
notary ***7**.220*
notation ***5**.322–3*, **5**.444
notch flute ***5**.323*, **5**.414

notornis *2.230*
noun **7**.229
nouveau roman 5.323
Nouvelle Vague 5.323, **5**.460–1
nova *8.109–10*
dwarf **8**.21
see also supernova
Nova Scotia *1.232*, *3.259*
nova-like variables **8**.21
'Novalis' *5.323*
Novaya Zemlya *1.232*
novella **5**.324
novels and novelists *5.323–4*
Africa **5**.2, **5**.7, **5**.184, **5**.318, **5**.334, **5**.402, **5**.406
Arabic **5**.19
Australia **5**.57, **5**.103, **5**.383, **5**.431, **5**.488
Austria **5**.60, **5**.243, **5**.310, **5**.405, **5**.486
Belgium **5**.420
Bengali **5**.43, **5**.445
Bildungsroman **5**.47–8
Britain:
adventure **5**.197
Angry Young Men **5**.15, **5**.17, **5**.333
autobiographical **5**.188, **5**.270, **5**.286, **5**.382, **5**.490
Bloomsbury Group *5.50*, **5**.162, **5**.259, **5**.435, **5**.436, **5**.493–4, **5**.550
'cabbage patch' **5**.36
children and childhood **5**.81, **5**.188, **5**.213, **5**.239, **5**.248, **5**.296
classical themes **5**.188–9
crime and detective **5**.134
epistolary **5**.383
expatriate **5**.227, **5**.258, **5**.270
family life **5**.143, **5**.168–9, **5**.490
feminist **5**.491
Gothic *see* Gothic novels
historical **5**.40, **5**.141, **5**.162, **5**.408, **5**.457
India **5**.248, **5**.394
journalistic **5**.127
modernist **5**.110
moralistic and philosophical **5**.41, **5**.192, **5**.248, **5**.262, **5**.308, **5**.383, **5**.428
parables and fables **5**.183
picaresque **5**.423
Picturesque **5**.348–9
political and civil service **5**.90, **5**.423
Romantic **5**.408, **5**.432
rural life **5**.48, **5**.198, **5**.239
satire and humour **5**.68, **5**.154, **5**.183, **5**.291, **5**.342, **5**.423, **5**.432, **5**.483
science fiction **5**.416, **5**.485
social life and realism **5**.67
19th century **5**.26, **5**.43, **5**.60, **5**.130, **5**.131, **5**.171, **5**.179, **5**.198, **5**.303, **5**.450, **5**.460
20th century **5**.43, **5**.168–9, **5**.198, **5**.258, **5**.286, **5**.303, **5**.333, **5**.366
thrillers and mystery **5**.100, **5**.106–7, **5**.134, **5**.162
Wales **5**.486
wildlife **5**.490
Bulgaria **5**.74
Burma **5**.67
Cambodia **5**.72
Canada **5**.26, **5**.126, **5**.303, **5**.422
children's *see* literature *under* children
China **5**.13, **5**.95, *5.96*, **5**.256, **5**.302, **5**.483
Colombia **5**.169–70
Czechoslovak **5**.76, **5**.120, **5**.140, **5**.201, **5**.210, **5**.213, **5**.252, **5**.318, **5**.422, **5**.468
Denmark **5**.49, **5**.239
novels and novelists (*cont.*):
Dravidian **5**.136
epilogue **5**.148
epistolary **5**.148
Finland **5**.155
France **3**.103
aircraft **5**.397
artistic **5**.172
autobiography and childhood **5**.10, **5**.163, **5**.368
detective **5**.420
existentialist **5**.40, **5**.73, **5**.173, **5**.278, **5**.402
historical **5**.139, **5**.475
humour **5**.125
moralistic and religious **5**.45, **5**.286, **5**.366, **5**.397
nouveau roman **5**.177, **5**.323
pastoral **5**.467
psychological **5**.111
Romanticism **5**.89, **5**.139, **5**.291, **5**.310–11, **5**.400, **5**.475
satire **5**.163
science fiction **5**.472
sentiment **5**.283
social life and realism **5**.253–4, **5**.284, **5**.392
19th century **5**.34, **5**.157, **5**.165, **5**.184, **5**.214, **5**.286, **5**.386, **5**.431–2, **5**.499
Surrealism **5**.106
Symbolism **5**.368, **5**.499
on women **5**.303
Germany **5**.59, **5**.142, *5.175*, **5**.181–2
existentialism **5**.188
horror **5**.208
individuality **5**.243
Romantic and lyrical **5**.205, **5**.210, **5**.323
satire **5**.51, **5**.194
social life and realism **5**.161–2, **5**.201, **5**.205, **5**.279
17th century **5**.194
war **5**.377–8
Gothic *see* Gothic novels
Greece, ancient **5**.189
Greece, modern **5**.191, **5**.244–5
Guatemala **5**.25
Hebrew **5**.9–10
historical:
Europe **5**.139, **5**.148, **5**.281, **5**.418, **5**.475
Russia **5**.326, **5**.341, **5**.370, **5**.457
United States **5**.112, **5**.117, **5**.133, **5**.474
see also under Britain *above*
Hungary **5**.28, **5**.214
Iceland **5**.258
India **5**.16, **5**.129, **5**.240, **5**.248, **5**.312, **5**.374, **5**.394, **5**.467
Ireland **5**.40–1, *5.226*
historical **5**.141
innovative **5**.242
satire and humour **5**.183, **5**.324
social life and realism **5**.141, **5**.303, **5**.326
Italy **5**.50–1, *5.231*, **5**.350, **5**.441
autobiographical **5**.124
fantasy **5**.72
historical **5**.281
political **5**.342, **5**.478
religious and moralistic **5**.159
rural life **5**.342
social life and realism **5**.162, **5**.304, **5**.342, **5**.471
Japan *5.235–6*, **5**.445
autobiographical **5**.304
colloquial **5**.314
erotic **5**.446
realist **5**.217
right-wing **5**.298
Roman Catholic **5**.146
Kenya **5**.318
Latin **5**.189
novels and novelists (*cont.*):
Mongolia **5**.302
naturalism *see* social life *below*
Nigeria **5**.427
Norway **5**.48, **5**.197
Pakistan **5**.467
Persia **5**.345
Peru **5**.468
picaresque **5**.348, **5**.398, **5**.423, **5**.428
Picturesque **5**.348–9
Poland **5**.320, **5**.345, *5.355*
didactic **5**.333
experimental **5**.44
historical **5**.418
immaturity **5**.183–4
moralistic **5**.16
psychological **5**.227
science fiction **5**.261
social life and realism **5**.333, **5**.381, **5**.491, **5**.498
Portugal **5**.35, **5**.359
realism *see* naturalism and realism
Romantic *see under* Romanticism
Russia and USSR *5.396*
dissident and out of favour **4**.277, **5**.10, **5**.65, **5**.133, **5**.341, **5**.424, **5**.497, **5**.498
historical **5**.326, **5**.341, **5**.370, **5**.457
idealist **5**.462
modernist **5**.43
pacifist **5**.458
psychological and social **5**.133–4, **5**.182–3, **5**.184, **5**.261–2
Romanticism **5**.370
rural life **5**.457, **5**.462
Socialist Realist **5**.184, **5**.417, **5**.424
war **5**.65, **5**.457
Scotland **5**.408, **5**.432
Senegal **5**.334
Serbo-Croat **5**.16
social life and realism *see under countries above and below*
South Africa **5**.184, **5**.406
South and Central America **5**.114, **5**.256
Spain **5**.10, **5**.86, **5**.172–3, **5**.323, **5**.333, **5**.343, **5**.372, **5**.387, **5**.428, **5**.466, **5**.467, **5**.470
stream of consciousness **5**.436
Sweden **5**.254, **5**.436
Switzerland **5**.111, **5**.165
Turkey **5**.462
United States **5**.5, **5**.201, **5**.267, **5**.326
allegorical **5**.359
animals **5**.267
autobiographical **5**.133, **5**.295–6, **5**.463, **5**.491
Beat Movement **5**.39, **5**.68
black **5**.29, **5**.495
for children **5**.11, **5**.200
detective and horror **5**.353–4
fantasy **5**.371, **5**.486
historical **5**.112, **5**.117, **5**.133, **5**.474
humour **5**.491
'international' **5**.212–13, **5**.232–3
Jazz Age **5**.156
Jewish **5**.277, **5**.391, **5**.420
journalistic **5**.76, **5**.276
Middle America **5**.82, **5**.262
mythology **5**.466
naturalistic **5**.117
New York **5**.90, **5**.112, **5**.488
picaresque **5**.398
psychological **5**.43, **5**.99, **5**.156, **5**.352
religion **5**.325
rural life **5**.431
satire and humour **5**.90, **5**.203–4, **5**.276, **5**.474, **5**.486, **5**.491
novels and novelists (*cont.*):
United States (*cont.*):
science fiction **5**.479
sea adventures **5**.290
southern states **5**.71, **5**.76, **5**.99, **5**.152–3, **5**.325, **5**.435
travel **5**.296
war **5**.117, **5**.276
Vietnam **5**.475
Wales **5**.486
West Indies **5**.254, **5**.312
see also science fiction; stories
Noverre, Jean-Georges *5.324*
Novgorod *3.259*
Novosibirsk State University *6.242*
Nozick, Robert **7**.215, *7.220*, **7**.291
nozzle *6.242*
N-type galaxies **8**.55–6
Nubecula Major and Minor **8**.88–9
Nubian sandstone *1.232*
Nubians *3.259–60*
nuclear energy and power *6.242–3*, *6.245*
EURATOM *see* European Atomic Energy Community
fall-out *see under* radioactivity
IAEA *see* International Atomic Energy Agency
nuclear-powered ships and submarines **4**.235, **6**.175, **6**.243, *6.245*, **6**.337
see also nuclear reactors and power stations; radioactivity
nuclear explosions **1**.118
see also nuclear weapons
nuclear fission *see* fission, nuclear
nuclear fuels *6.243*, **6**.245
see also nuclear reactors
nuclear fusion *see* fusion, nuclear
nuclear magnetic resonance **6**.213, **6**.218
nuclear medicine *6.243*, **6**.292
see also tomography
nuclear physics:
gluons **1**.69, **1**.141
graviton **1**.144
magic numbers **1**.202
mass defect **1**.206
neutron *see* neutrons
pioneers **1**.57, **1**.67, **1**.120, **1**.149–50, **1**.178, **1**.226
Curie family **1**.81–2, **1**.106, **1**.178
see also Manhattan Project
see also atomic physics; electrons; elementary particles; physics
nuclear power *see* nuclear energy and power
nuclear proliferation *7.220*
nuclear radiation *6.288*, **7**.240
fall-out *see under* radioactivity
measurement and detection **6**.106, **6**.152–3, **6**.288
medical *see* nuclear medicine
nuclear winter **6**.245, *6.246*
radioactivity *see* radioactivity
waste **6**.243, **6**.245
see also radioactive elements and isotopes; radioactivity
nuclear reaction **1**.73, **6**.69, **6**.242, **6**.245
nuclear reactors and power stations **6**.243, **6**.244, *6.245*
accidents *see under* accidents *and also* fall-out *under* radioactivity
advanced gas-cooled **6**.3–4, **6**.243, **6**.244
atomic pile **6**.23
Calder Hall **6**.54
CANDU **6**.57, **6**.167, **6**.243, **6**.244, **6**.369
fast breeder **6**.126, **6**.243, **6**.244, **6**.245, **6**.274
fuel **6**.243, **6**.245
graphite in **6**.3, **6**.160, **6**.213, **6**.245

nuclear reactors and power stations (cont.):
hafnium in 6.391
heavy water 6.167, 6.243, 6.245
IAEA 6.184
Magnox 6.3–4, 6.54, 6.213
percentage of energy from 6.121
pioneer 6.127
pressurized-water reactor 6.243, 6.244, 6.281
sodium as coolant 6.323
thermal 6.57, 6.244
zirconium 6.391
nuclear strategy 7.220
nuclear synthesis 8.110
Nuclear Test-Ban Treaty 4.18, 4.248
nuclear time-scale 8.79
nuclear wastes 6.243, 6.245
nuclear weapons 6.42, 6.245–6, 7.220
arms control treaties 7.17–18
atomic bombs dropped 4.152, 4.230
Conference on Disarmament 7.64
control 4.18, 4.99
campaigns for 4.58, 4.59
Nuclear Test-Ban Treaty 4.18, 4.248
SALT talks 4.330
development see Manhattan Project
fall-out see nuclear radiation
flexible response 7.121
fusion 6.148
nuclear proliferation 7.220
nuclear strategy 7.220
nuclear-free zone 7.296
strategy 7.220
nuclear winter 6.245, 6.246
nucleation 1.233
nuclei see nucleus
nucleic acids 2.230
denaturation 2.90
see also DNA; RNA
nucleons 1.233
protons 1.26, 1.269
see also neutrons
nucleophile 1.233
nucleotides 2.134, 2.230
nucleus (astronomy) 8.110
nucleus (cell) 2.230
nucleus (meteorology) 1.71
nucleus (physics) 6.23–4
atomic 1.26, 1.233
see also electrons
atomic number 1.26
see also nucleons
nudibranch 2.301
nuée ardente 1.233
Nuer 7.185, 7.217, 7.220, 9.22
Nuffield, William Richard Morris, viscount 4.248
Nuffield Radio Astronomy Laboratories see Jodrell Bank
Nugent, Elliot 5.454
Nūh (Noah) 9.27
nuisance 7.220
Nujoma, Samuel 4.230
nullah 1.233
nullification 4.248
Num 9.27
numbers 1.233
complex 1.18, 1.70
equivalents listed 9.52
Fibonacci 1.121
imaginary 1.166
irrational 1.174
logarithms 1.198, 1.224
magic 1.202
modulus 1.217
prime 1.267–8
quantum 1.271
random 1.276
real 1.277
round-off errors 1.286
significant figures 1.304
transcendental 1.341
numbers (cont.):
see also mathematics
numerically controlled machine-tools 6.134, 6.179, 6.211, 6.246–7
Jacquard loom 6.84, 6.169, 6.189, 6.208
Numidia 3.105, 3.192, 3.260
see also Libya
Nun 9.27
nunnery 3.260
see also monasticism
Nupe 3.4, 3.260
Nuremberg Trials 4.248, 7.328
Nureyev, Rudolf 5.324
Nurhachi of Manchus 3.218
nurse shark 2.55
nursery education 7.220, 7.245
nursery rhymes 5.91, 5.324
nurses 7.200
Nut (goddess) 9.27
nut (plant) see nuts and nut trees
nutation 8.110
nuthatch 2.230–1
nutmeg 2.230, 2.231, 6.271
nutria 2.80, 2.362
nutrient 2.231
nutrition 2.231
see also drinks; food
nuts and bolts 6.247
nuts and nut trees 2.129, 2.230
almond 2.7
brazil nut 2.41
butternut 2.48
cashew 2.56
chestnut 2.62
hazel 2.129, 2.152, 2.153
peanut 2.245, 2.246
pecan 2.246–7
pistachio 2.256
walnut 2.359
Nyahuma of Rozvi empire 3.68
Nyasaland see Malawi
Nyāya 7.155
nyckelharpa 5.324
Nye, Gerald P.: Committee 4.248
Nyerere, Julius Kambarage 4.248, 4.338
Nyikang 9.21
nylon 6.204, 6.247, 6.272, 6.273
carpet 6.61
pioneer 6.60, 6.384
rope 6.302
nymphs 9.8, 9.11, 9.16, 9.27
Nystad, Treaty of 3.260

O

oak 2.231
oak apple gall 2.132, 2.231
OAPEC (Organization of Arab Petroleum Exporting Countries) 7.224
oarfish 2.231, 2.232
oars 6.333
OAS see Organization of American States; Organization de l'Armée secrète
oasis 1.234
Oastler, Richard 4.249
Oates, Lawrence Edward Grace 1.295
Oates, Titus 3.289
oats 2.231–2
see also cereals and grains
OAU see Organization of African Unity
Ob 1.234
obedience 7.64, 7.221
Oberon (mythology) 9.27
Oberon (satellite of Uranus) 8.111
Oberth, Hermann 4.323, 6.45, 8.144
objective (in telescope) 8.137
objective prism 8.111
objects and objectives (philosophy) 7.197
objet trouvé 5.324
oblateness 8.111
of Earth 1.234
obligation, political 7.221
obligations, law of 7.221
obliquity of ecliptic 8.43, 8.127
oboe 5.113, 5.324, 5.352, 5.376, 5.415
Obote, Milton 4.249, 4.356
Obregón, Alvaro 4.219, 4.249
Obrenović dynasty 4.178, 4.249, 4.309
O'Brien, Flann 5.324
O'Brien, Smith 4.387
Obrisset, John 5.212
observation device see instrumentation; measurement
observatories, astronomical 6.248, 8.111
Arecibo 8.7, 8.8
Cerro Tololo Inter-American 8.24
European Southern 8.48
Jodrell Bank 8.76
Kitt Peak National 8.80
Kuiper Airborne 8.81
La Palma 8.82, 8.83
Lick 8.85
Lowell 8.87
Mount Palomar 8.104
Mount Wilson 8.104–5
Mullard Radio Astronomy 8.105
National Radio Astronomy 8.106
Paris 8.116–17
Roque do los Muchachos 8.82, 8.83
Royal (Edinburgh) 8.142
Royal Greenwich 6.303, 8.142
Siding Spring 8.150
Steward 8.161–2
Uraniborg 8.178
Yerkes 8.189
see also telescopes
observer's meridian 8.22
obsession 7.30, 7.119, 7.221
obsidian 1.234, 6.136, 6.156
obstetrics and gynaecology 2.166, 6.248
amniocentesis 2.8, 6.13, 6.218
anaesthesia 6.320
cervical screening 6.68
ultrasound 6.367
in vitro fertilization 6.185–6, 6.218, 7.159–60
see also birth; pregnancy; sexual reproduction
Obstfelder, Sigbjørn 5.404
ocarina 5.324
OCAS see Organization of Central American States
O'Casey, Sean 5.324
Occam see William of Occam
occasional verse 5.325
occlusion (meteorology) 1.234
occult see supernatural
occultation 8.111–12
occupation:
occupational disease 6.22, 6.248–9
profession 7.249
see also labour and employment
ocean see oceans and seas
ocean basin 1.235
Oceania 1.315
music 5.325
see also Pacific Ocean
Oceanids 9.27
oceanography 1.208, 1.235–6
oceans and seas 1.234–6
animals see living organisms below
Arctic see Arctic Ocean
Atlantic see Atlantic Ocean
Baltic see Baltic Sea under Baltic
bathing 3.335
coasts see coasts
creatures see fish; living organisms below and living organisms under water
currents see currents (ocean)
elements from sea-water see halogens
energy and power see under tides/tidal; waves (ocean)
exploration 2.27
fish:
anchovy 2.10
angelfish 2.10
anglerfish 2.10
barracuda 2.23
barrel-fish 2.23
bass 2.24, 2.146
beard-fish 2.26
blenny 2.34
Bombay duck 2.38
bonito fish 2.38
bream 2.41, 2.265, 2.280, 2.298
bullhead 2.45
butterfly fish 2.48
cardinal fish 2.53, 2.54
catfish 2.57, 2.105
charr 2.61
clingfish 2.68
coalfish 2.69–70
cod see cod family
coelacanth 2.71
conger eel 2.74
dab 2.86
deep-sea 2.89–90
dogfish 2.96–7
dragonfish 2.99
eel 2.104
electric 2.105
flatfish see flatfish
flying fish 2.123
flying guinard 2.124
garfish 2.133
ghost pipefish 2.135
goby 2.139
grouper 2.146
guitarfish 2.148
gurnard 2.148
haddock 2.149
hagfish 2.149

oceans and seas (*cont.*):
fish (*cont.*):
hake **2.**149–50
halibut **2.**150
herring *see* herring
horse mackerel **2.**163
ice fish **2.**168
John Dory **2.**176
lamprey **2.**182
lantern fish **2.**182
ling **2.**190
lumpsucker **2.**195
mackerel **2.**37, **2.**163, **2.**197–8, **2.**318
manta ray **2.**202
marlin **2.**203
milkfish **2.**211
monkfish **2.**217–18
mudskipper **2.**221
mullet **2.**221
oarfish **2.**231
opah **2.**234–5
parrotfish **2.**242–3
pilchard **2.**253–4
pilot fish **2.**254
pipefish **2.**255
pollack **2.**262–3
pompano **2.**263
pufferfish **2.**271, **2.**272
rabbitfish **2.**274
ragfish **2.**275
rat-tail **2.**278
red sea bream **2.**280
redfish **2.**280
remora **2.**282
ribbon-fish **2.**284
sailfish **2.**292
salmon **2.**61, **2.**292
scorpionfish **2.**297
sea bream **2.**265, **2.**298
sea horse **2.**299
sea moth **2.**300
sea perch **2.**301
shad **2.**305
shark *see* sharks
shrimp fish **2.**309
silverside **2.**310
skate **2.**105, **2.**311
snapper **2.**315
snipe-eel **2.**315
snipefish **2.**315
snoek **2.**315–16
Spanish mackerel **2.**318
spearfish **2.**319
squirrelfish **2.**323–4
stickleback **2.**325
stone fish **2.**327
sturgeon **2.**328
sunfish **2.**329
swordfish **2.**332
tarpon **2.**335
thread-fin **2.**34
toadfish **2.**342
triggerfish **2.**347–8
trumpetfish **2.**348
trunkfish **2.**348
tube-mouth fish **2.**349
tuna **2.**38, **2.**318, **2.**350
unicorn fish **2.**353
weeverfish **2.**363–4
whitebait **2.**366
yellowtail **2.**373
see also fishing
floor *see* physical geography *below*
Indian *see* Indian Ocean
level *see* sea-level
living organisms in or near:
animals *see* animals *under* water
birds *see* wading birds *and also* birds *under* water
coral *see* coral
fish *see* fish *above*
graptolite **2.**143
Hemichordata **2.**154
mammals *see* mammals *under* water

oceans and seas (*cont.*):
living organisms in or near (*cont.*):
plankton *see* plankton
see also living organisms *under* water
maps **1.**168, **1.**208, **1.**235, **1.**245
maritime law **7.**272
navigation *see* navigation
Pacific *see* Pacific Ocean
physical geography and geology **1.**234–6
abyssal plain **1.**1, **1.**236
basin **1.**235
boundary layer **1.**42
brine **1.**44
circulation and movement **1.**63–4
see also currents, ocean; tides/tidal
coastline *see* coasts
continental rise **1.**73, **1.**236
continental shelf **1.**73–4, **1.**236
continental slope **1.**74, **1.**236
convection **1.**74
convergence **1.**74
currents *see* currents, ocean
doldrums **1.**93
evaporites **1.**116
geomagnetic dating **1.**136
hydrography **1.**162
hydrological cycle **1.**162
iceberg **1.**164
islands *see* islands
isthmus **1.**176, **1.**250
map **1.**208
marginal seas **1.**205
oceanography **1.**208, *1.235–6*
pack ice **1.**244, **1.**296
plate tectonics **1.**260
ridges **1.**24, **1.**168, **1.**245, **1.**261, **1.**295
mid-oceanic **1.**214–15, **1.**236
salinity **1.**290
sea ice **1.**296
sea-level *see* sea-level
seamounts **1.**236, **1.**245, **1.**297
seawater **1.**290, **1.**297–8
sediments **1.**236, **1.**238
straits *see* straits
submarine canyons **1.**325
Tethys Sea **1.**135, **1.**334
thermocline **1.**334
tides *see* tides/tidal
trenches *see* trench, submarine
turbidity current **1.**347
upwelling **1.**351
water mass **1.**360–1
waves *see* waves (ocean)
whirlpool **1.**365
wind waves **1.**363
plants *see* plants *under* water
seaweed **6.**18, **6.**186
underwater activities *see* diving; submarines
vessels *see* ships and boats; submarines
Oceanus (mythology) *9.27*
Oceanus Procellarum (astronomy) *8.112*
ocelot **2.**277
Ockeghem, Johannes *5.325*
Ockham *see* William of Occam
O'Connell, Daniel **3.**65, **4.**65, **4.**76, **4.**249, **4.**387
O'Connor, Feargus Edward **4.**70, *4.249–50*
O'Connor, Flannery *5.325*
O'Connors of Tara **3.**85
OCR (optical character recognition) **6.**251
octahedron *9.51*
octane number and rating **6.**224, *6.249*, **6.**296
Octanus *8.112*
Octavian *see* Augustus, Caesar

October Revolution (1917) *see under* Revolutions *under* Russia
October War *see* Yom Kippur War
octopus **2.**232, **2.**241, **2.**308
ODA (Overseas Development Assistance) **7.**9
Oda Nobunaga **3.**24, **3.**186, *3.260*
Odaenathus of Palmyra **3.**268
ODECA *see* Central American Common Market
Oder *1.236*
Oder–Neisse Line *4.250*
odes **5.**211–12, *5.325*, **5.**349
see also lyric poetry
Odilo, Abbot **3.**80
Odin **9.**2, **9.**5, **9.**22, *9.27*, **9.**30, **9.**38–9
Odoacer *3.260*
odometer **6.**198, *6.249*
Odum, E. P. **2.**328
Odysseus (crater) **8.**171
Odysseus (mythology) **9.**8, **9.**9, *9.27*, **9.**29
Odyssey **5.**210
OECD (Organization for Economic Co-operation and Development) *4.251*, **7.**221
Oedipus **9.**4, *9.27*, **9.**34
Oedipus complex (psychology) **7.**253
Oehlenschläger, Adam Gottlob *5.326*
Oersted, Hans Christian **1.**119, *1.236–7*, **6.**125
oesophagus **2.**232, **2.**331
O'Faolain, Sean *5.326*
Offa, king of Mercia **3.**15, **3.**229, *3.260*, **3.**380
Offenbach, Jacques *5.326*, **5.**365, **5.**441
office equipment:
calculator **6.**54
duplicating **6.**109, **6.**267–8
fax **6.**126, **6.**278, **6.**347
paper **6.**59
photocopier **6.**60, **6.**101, **6.**267–8
typewriter **6.**318, **6.**366
word processor **6.**179, **6.**388
workstations **6.**179, **6.**389
see also computers
off-road vehicle *6.249*
offset lithography **6.**134, *6.249*, **6.**282
Ogaden *1.237*
Ogata Korin and Ogata Kenzan **5.**246
Ogden, Peter Skene *1.237*
Oglethorpe, James Edward **3.**140, *3.260–1*
Ogo **9.**27, **9.**28
OGPU *4.250*
Ohain, Hans Pabst von **6.**189, *6.249*, **6.**385
O'Hara, John *5.326*
O'Higgins, Ambrosio *4.250*
O'Higgins, Bernardo **4.**72, *4.250*
Ohio River *1.237*
Ohm, George **1.**236
Ohm's Law **6.**75–6
ohm *1.237*, **6.**368
see also impedance; reactance; resistor *under* resistance
oil, animal **2.**144
see also fats, oils, and fatty acids
oil, mineral:
diesel *see* diesel
drying **6.**109
exploration fossil *2.232–3*
gasification *6.249*
industry *see under* petroleum
lamp **6.**20, **6.**30, **6.**201, *6.249*
lubricating **6.**32, **6.**145, **6.**204, **6.**210, **6.**264
rig *see* drilling rig
shale **6.**263, **6.**315
see also petroleum
oil, vegetable **2.**233
almond **2.**7

oil, vegetable (*cont.*):
anise **2.**11
bay rum tree **2.**25
birch **2.**30
butternut **2.**48
cinnamon **2.**66
citronella **2.**66
coconut (copra) **2.**75–6
corn **2.**198
cotton seed **2.**78
essential **6.**123, **6.**260
eucalyptus **2.**110
fir (turpentine) **2.**119
flax **2.**120
geranium **2.**134–5, **2.**247
gingko **2.**136
groundnut **2.**246
jasmine **2.**175
kenaf **2.**178
lavender **2.**184
linseed **6.**109, **6.**255, **6.**286
maize (corn) **2.**198
oil-palm **6.**271
olive **2.**234
palm **2.**233, **2.**240
patchouli **2.**244
peanut **2.**246
peppermint **2.**249
pimento **2.**254
rape **2.**278
rosemary **2.**288
safflower **2.**291
St John's wort **2.**292
sesame **2.**304
soya bean **2.**318
sunflower **2.**330
turpentine **6.**366
wintergreen **2.**368
witch-hazel **2.**369
ylang ylang **2.**374
oil beetle **2.**232
oil painting **5.**157, *5.326*, **5.**336
oil plant *see* oil, vegetable
oilbird **2.**232, **2.**233
oilskin *6.249*
Oisin **9.**28
Oistrakh, David *5.326*
Ojibwa Indians **7.**219, *7.221*
Ojukwu, Colonel **4.**38, **4.**139
Okada Keisuke **4.**160
okapi *2.233*
Okara, Gabriel **5.**7
Okavanga *1.237*
Okeechobe, Lake *1.237*
O'Keeffe, Georgia *5.326*
Okhotsk, Sea of *1.237*
Okigbo, Christopher **5.**7
Okinawa *4.250*
Oklahoma Indian Territory *4.250*
okra **2.**233–4
Okri, Ben **5.**7
Okudzhava, Bulat *5.326*
Olaf I and II, kings of Norway **3.**258, *3.261*
Olav V, king of Norway **4.**248
Olbers, Heinrich **8.**112, **8.**116, **8.**183
Olbers' paradox *8.112*
old age **7.**221, *7.222–3*
ageing *see* ageing
benefit **7.**222
diseases *see under* ageing
family support in *7.222*
institutional care **7.**161
Old English literature **5.**44, *5.326*
Old Vic **5.**327
Oldcastle, Sir John **3.**209
Oldenbarneveldt, Johan van **3.**108, **3.**149, *3.261*
Oldowan *3.261*
Olduvai Gorge *1.237*, **3.**201, *3.261*, **3.**391
oleander **2.**234
oleum *1.237*, **6.**87
olfactory sense *see* smell
Oligocene epoch **1.**133, **1.**134, *1.237*
oligopoly **7.**222

Oliphants River **1**.195
olive family **2**.17, **2**.125, **2**.190, ***2**.234*
Oliver, Bernard **8**.148
Oliver, Isaac ***5**.327*
Oliver, Joseph 'King' **5**.238
Oliver, Peter **5**.327
Olivier, Laurence **5**.24, ***5**.327*, **5**.481
olivine ***1**.237*, **1**.250–1, **1**.256
Ollier, Claude **5**.323
olm ***2**.234*
Olmec ***3**.261*
  art **3**.261, ***5**.327*
Olorun **9**.11, ***9**.28*
Olson, Harry F. **6**.343
Olympia ***3**.261–2*
Olympic Games ***7**.222–3*
  athletics **7**.21, **7**.232
  boycott **7**.34
  first Olympiad **3**.262
  gymnastics **7**.137
  hockey **7**.145
  ice-skating **7**.151
  marathon **7**.192
  martial arts **7**.194
  pentathlon and decathlon **7**.232
  revived **4**.326
  wrestling **7**.333
Olympic Mountains (USA) ***1**.237–8*
Olympus, Mount ***1**.238*, **9**.28
Olympus Mons (Mars) **8**.108
Om ***9**.28*
O'Mahony, John **4**.118
Oman (*formerly* Muscat and Oman) ***1**.238*, ***3**.262*, ***4**.250*, **7**.136, ***7**.374*
  in Arab League **7**.15
  in Gulf Co-operation Council **7**.136
  physical geography and geology **1**.186, **1**.238
  ruler **4**.299–300
Omar *see* Umar ibn al-Khattab
ombudsman ***7**.223*
Omega (Swan) Nebula ***8**.112*, **8**.167
Omega (radio navigation) ***6**.249–50*
Omega Workshops ***5**.327*
Ometecuhtli ***9**.28*
O-mi-t'o **9**.4
Omri dynasty **3**.6, ***3**.262*
onchocerciasis ***2**.234*, **2**.286, **6**.377
O'Neill, Eugene ***5**.327–8*
one-party state ***7**.223*
oni ***9**.28*
onion family **2**.186, ***2**.234*, **2**.305–6
Onnes, Heika Kamerlingh **6**.339
Ono, Kazuo **5**.141
onomatopoeia ***5**.328*
Onsager, Lars ***1**.238*
Ontario ***1**.238*
ontological argument ***7**.223*
ontology ***7**.223*
onyx ***1**.238*
oolite and oolith ***1**.238*
  oolitic **1**.238
Oort, Jan H. **8**.112, **8**.159
  Oort–Öpik Cloud ***8**.112*
  Oort's constants ***8**.112*
ooze ***1**.238*
Op art ***5**.328*, **5**.383, **5**.469
opacity ***8**.113*
opah ***2**.234–5*
opal ***1**.238*, **1**.239
OPEC **4**.251–2, **7**.57, **7**.215, **7**.219, ***7**.223–4*, **7**.326
open cluster ***8**.113*
open universe ***8**.113*
open-cast mining **6**.78, ***6**.250*
open-hearth furnace **6**.185, **6**.188, ***6**.250*
open-loop system **6**.88
opera ***5**.328–9*
  aria **5**.21
  Austria **5**.44, **5**.306
  bel canto **5**.42
opera (*cont.*):
  Britain:
    17th and 18th centuries **5**.5, **5**.22, **5**.172, **5**.258, **5**.370
    19th and 20th centuries **5**.48, **5**.210, **5**.430, **5**.456–7
    Covent Garden **5**.116
  China **5**.94, **5**.96
  classical music **5**.103
  Czechoslovak **5**.140, **5**.233, **5**.422
  France:
    17th and 18th centuries **5**.270, **5**.373–4
    19th century **5**.48, **5**.87, **5**.126, **5**.128, **5**.187, **5**.254, **5**.326, **5**.398, **8**.285
    20th century **5**.126, **5**.138, **5**.153, **5**.285, **5**.291–2, **5**.392
  Germany:
    18th century **5**.181, **5**.197
    19th century **5**.214, **5**.292–3, **5**.429, **5**.435–6, **5**.480–1, **5**.484
    20th century **5**.204, **5**.206, **5**.214, **5**.435–6
  intermezzo **5**.224
  Italy:
    16th and 17th centuries **5**.83, **5**.302–3
    18th century **5**.90, **5**.101, **5**.398, **5**.404–5
    19th century **5**.43, **5**.90, **5**.133, **5**.284, **5**.368, **5**.391, **5**.398, **5**.471
    20th century **5**.121, **5**.320, **5**.368–9, **5**.381, **5**.491
  Russia and USSR:
    19th century **5**.53, **5**.180, **5**.310, **5**.383, **5**.448
    20th century **5**.367, **5**.384, **5**.417, **5**.436
  South and Central America **5**.257
  Spain **5**.10
  United States **5**.174, **5**.175, **5**.251
  see also operetta; songs and singers
opéra-ballet ***5**.329*
operating system (computers) **6**.219, ***6**.250*
operation (mathematics) ***1**.239*
operation (medical) *see* surgery
operations/operational research ***6**.250*, **6**.253, **6**.344, ***7**.224*
operetta **5**.260, **5**.326, ***5**.329*, **5**.435, **5**.439
  see also musicals; opera; songs and singers
Ophelia (Uranian satellite) **8**.178
ophicléïde **5**.329
Ophiuchids ***8**.113*
Ophiuchus ***8**.113*
ophthalmic surgery ***6**.250–1*
ophthalmoscope **6**.120, **6**.218, ***6**.251*
Ophüls (Oppenheimer), Max ***5**.329–30*
Öpik, Ernst: Oort–Öpik Cloud ***8**.112*
opinion poll ***7**.224*
opioid ***2**.235*
opium and derivatives (morphine and heroin)
  addiction **6**.192, **7**.95, **7**.96
  as analgesics **6**.14
  as narcotics **6**.236
  Opium Wars and trade **4**.230, ***4**.250–1*
  poppy ***2**.235*
opossum ***2**.235*, **2**.278
Oppenheimer, Sir Ernest ***4**.251*
Oppenheimer, Julius Robert **6**.214, ***6**.251*, **6**.350
Oppenheimer, Max ***5**.329–30*
opposition, planets in **8**.122
optics/optical:
  activity (crystals) ***1**.239*
  and astronomy *see* telescopes
optics/optical (*cont.*):
  character recognition ***6**.251*, **6**.259, **6**.278
  depth ***8**.113*
  disk *see* compact disc
  double ***8**.113*
  fibre **6**.156, **6**.218, ***6**.251*
    arthroscope **6**.253
    cable **6**.53
    catheter **6**.64
    endoscopy **6**.120
    laser **6**.195
    light-emitting diode **6**.201
    telephony **6**.347
  glass **6**.156, ***6**.251*
    binoculars **6**.36
    fibre *see* fibre *above*
    instruments *see* microscope; telescopes; theodolite
    lens *see* lenses
    periscope **6**.260
    pioneers **6**.125, **6**.294, **6**.391
    see also theodolite
  isomerism **1**.175
  mirage **1**.216
  optic sense *see* vision
  parallax **1**.247
  pulsar ***8**.113–14*
  Ritchey–Chrétien ***8**.140*
  sound reproduction **6**.325–6
  see also light
optimization ***6**.251*, ***6**.253*
opuntia **2**.268
Opus Dei ***7**.224*
ora *see* komodo dragon
orache ***2**.235*
oracles ***3**.262*, **9**.10
oral traditions **5**.7, **5**.15, **5**.141, **5**.160
  see also folk; myths and legends; speech
Orange (French principality) ***3**.262*
  rulers of **3**.382, ***9**.68*
Orange Free State (South Africa) **1**.95, ***1**.239–40*, ***4**.251*, **4**.306
Orange River ***1**.240*
orange (fruit) ***2**.235*
orang-utan ***2**.235–6*
oratorio **5**.21, **5**.143, **5**.197–8, **5**.201–2, **5**.291, ***5**.330*, **5**.405
  see also choirs; Handel
oratory **5**.190–1
orbital model of benzene rings **1**.36
orbitals (atomic physics) **1**.239, ***1**.240*, **8**.131
orbits ***8**.114*
  co-orbital satellites **8**.29, **8**.47, **8**.76
  determination ***8**.114*
  Earth **8**.17, **8**.196
  elements of **8**.45, ***8**.114*
  ellipse **8**.45–6
  geostationary **8**.58
  geosynchronous **8**.58
  Hohmann transfer **8**.68
  Moon **8**.101
  perturbations in **8**.22
  see also satellites; space-flight
Orcagna, Andrea ***5**.330*
orchestra ***5**.330*, **5**.331
  gamelan **5**.169, **5**.222
  Mannheim School **5**.280
  percussion **5**.169
  South-East Asia **5**.427
  symphony **5**.280, **5**.298, **5**.442
  see also classical music; composers; conductors
orchestration ***5**.330*
orchids **2**.123, ***2**.236*, **2**.354
orchis ***2**.236*
ordeal ***3**.262–3*
order (animal and plant classification) ***2**.236*
order, world **7**.333
Orders of architecture **5**.330
Ordination **7**.271
Ordovician period **1**.133, **1**.134, ***1**.240*
oregano **2**.203
Oregon Boundary dispute ***4**.251*
Oregon Trail ***4**.251*
ores **1**.239, ***1**.240*, ***6**.253*
  aluminium *see* bauxite
  antimony **6**.16
  assaying *see* assaying
  barium **1**.33
  boron **6**.43
  chromium **6**.73
  copper **1**.239, **6**.89, **6**.221, **6**.253
  cut-off grade **1**.82
  and electricity *see* electrolysis
  flotation **6**.89, **6**.241, ***6**.253*
  grade **1**.82, **1**.142
  iron **1**.149, **1**.203, **1**.239
    see also iron and iron and steel
  lead **1**.130, **1**.308, **6**.197, **6**.253
  magnesium **1**.202, **6**.297
  magnetic ore separation **6**.213
  manganese **1**.239, **6**.214
  mercury **1**.63, **6**.219
  mining *see* mining
  nickel **1**.250, **6**.241
  niobium **6**.241
  prospecting **6**.228, **6**.285
  radium **6**.292
  silver **1**.19
  smelting *see* smelting
  strontium **6**.336
  tin **1**.54, **6**.356
  titanium **6**.357
  tungsten **1**.368, **6**.365
  uranium **1**.351–2
  waste (gangue) **1**.132
  zinc **1**.373, **6**.221, **6**.253, **6**.391
  zirconium **6**.391
Orestes **9**.2, **9**.11, ***9**.28*
orfe, golden **2**.140
Orff, Carl **5**.78, ***5**.330*
organ (body) *see* physiology of living organisms; transplant surgery
organ and organists ***5**.331–2*
  Austria **5**.63
  Britain **5**.50, **5**.57, **5**.65, **5**.69, **5**.103, **5**.176, **5**.370, **5**.485, **5**.486
  France **5**.115, **5**.153, **5**.163, **5**.187
  Germany **5**.28, **5**.68, **5**.377
  harmonium **5**.200
  hydraulic **6**.93
  Italy **5**.83, **5**.164, **5**.168, **5**.336
  mechanical **5**.288
  Netherlands **5**.441
  portative **5**.138, **5**.359
organelles **2**.285
organic chemistry ***1**.240*, ***6**.253*
  see also hormones; hydrocarbons; organic compounds
organic compounds ***6**.252*, **6**.253
  acid anhydride **1**.2
  alcohol *see* alcohol
  aldehyde **1**.7, **1**.212–13
  alicyclic **1**.8
  aliphatic **1**.8, **1**.9
  alkenes **6**.252, **6**.261
  amine **1**.11
  amino acids **6**.140, **6**.252
  aromatic *see* aromatic compounds
  ether *see* ether
  homologous series **1**.157
  ketone **1**.182
  lipids **2**.114, **2**.191
  nitrile **1**.230
  production **6**.70, **6**.264
  urea **2**.353
  see also carbohydrates; hormones; hydrocarbons; steroids
organic farming **6**.18, **6**.183, ***6**.253*, ***7**.224*
organic fertilizer **6**.127–8
organic fuel *see* biofuel
organic growth (economics) ***7**.224*

organic growth (living organisms) *see* growth of living organisms
organic life *see* living organisms
organic pollutants **6**.377
organic rocks *see* coal
Organization de l'Armée secrète ***4**.251*
Organization for Economic Co-operation and Development ***4**.251*, ***7**.221*
Organization of African Unity ***4**.251*, **7**.5, ***7**.224–5*
Organization of American States ***4**.251*, **7**.12, **7**.225
Organization of Arab Petroleum Exporting Countries **7**.224
Organization of Central American States ***4**.251*
Organization of Petroleum Exporting Countries ***4**.251–2*, **7**.57, **7**.215, **7**.219, ***7**.223–4*, **7**.326
organizations:
business *see* companies and business
international *see* international organizations
organizational theory ***7**.224*
voluntary **7**.327
organized crime ***7**.225*
organochlorine **6**.180
organometallic ***1**.240*
organophosphate **6**.180
organophosphorus compounds **6**.240
organ-pipe coral ***2**.236*
organs of body *see* physiology of living organisms
organum ***5**.332*
Orient *see* Asia
Oriental Orthodox Churches **7**.47
Armenian **7**.16–17
Coptic **7**.69–70, **7**.217
Syrian **7**.394
see also Eastern Orthodox
Orientale Basin ***8**.114–15*
Orientalism ***7**.225*
Origen **5**.411
Origin of Species **2**.88
original sin ***7**.225*
Orinoco River ***1**.240–1*
oriole family ***2**.236*
blackbird **2**.33
bobolink **2**.37
cuckoo-shrike **2**.83
grackle **2**.142
see also thrush family
Orion (constellation) ***8**.115*
Great Nebula in **8**.61
Orion association ***8**.115*
Orion's Belt and Sword ***8**.115*
Orion (mythology) ***9**.28*
Orion arm (spiral arm of Galaxy) ***8**.115*
Orion variables (stars) **8**.175
Orissa ***3**.263*
Orkney Islands ***1**.241*, **1**.318, ***3**.263*
Orlando, Vittorio Emanuele ***4**.252*
Orléans, ducs de **3**.50, ***3**.263*, **3**.369
Ormazd **9**.3
ormolu **5**.332
Ormonde, James Butler, 2nd duke of **3**.262, ***3**.263*
Ormonde, James Butler, 12th earl and 1st duke of ***3**.263*
ornamental fish *see* aquarium and ornamental fish
ornamental motif ***5**.332*
ornaments and jewellery ***5**.239–40*
Africa **3**.24
alabaster **5**.10
amber **5**.13
amulets **5**.15–16
Britain **3**.10, **3**.180, **3**.346
Byzantine **5**.70
Celtic **5**.84
China **5**.92
coral **5**.112–13
ornaments and jewellery (*cont.*):
crown jewels **5**.118–19
cutlery **5**.416, **5**.444
Egypt **5**.142
Etruscan **5**.149
France **3**.203, **5**.254
gold *see* arts *under* gold
Greece, ancient **5**.190
hornwork **5**.212
investment casting, see *cire perdue*
jade **5**.232
jet **5**.239
mother-of-pearl **5**.306
Phoenician **5**.346
Russia **5**.151
Scythian **5**.410
Sheffield plate **5**.416
Stone Age **5**.434
Sumerian **5**.305
Vietnam **5**.475
Viking **5**.475
West Indies **5**.488
see also decorative arts; gemstones
ornithology and ornithologists *see under* birds
orogeny and orogenesis **1**.154, ***1**.241*
see also building *under* mountains
orographic effects ***1**.241*, **1**.274–5
oropendola **2**.236
Orpen, Sir William **4**.364
orpharion ***5**.332*
Orpheus **9**.12, ***9**.28*
Orphic mysteries ***3**.263*
Orphism **5**.87, **5**.128, ***5**.332*
orrery ***8**.115*
orris root **2**.173
Orsini, Felice ***4**.252*
Ortega y Gasset, José ***5**.333*
Ortelius, Abraham ***1**.241*
orthoboric acid **6**.43
orthoclase feldspar **1**.143
orthodontics **6**.99
Orthodox Church *see* Eastern Orthodox
Orthodox Judaism **7**.138, ***7**.225–6*, **7**.281, **7**.304, **7**.305
orthopaedics ***2**.236*, **6**.218, **6**.253
Ortiz, Roberto **4**.177
ortolan ***2**.236*
Orton, Joe **5**.152, ***5**.333*
Òrundmìlà **9**.17
Orwell, George **4**.164, **5**.12, **5**.151, ***5**.333*, **7**.324
oryx **2**.108, ***2**.236–7*
Orzeszkowa, Eliza ***5**.333*
OS/2 operating system **6**.250
Osborne, John James **5**.327, ***5**.333*
Osborne, Thomas Danby, 1st earl of ***3**.96*
Osborne, W. V.: Osborne judgment ***4**.252*
Oscar II, king of Sweden **8**.125
Osceola ***4**.252*
oscillation:
in clocks and watches **6**.77, **6**.377
damping **6**.317
free **1**.127
frequency **1**.127
oscillating universe ***8**.115*
oscillators **6**.215, ***6**.253*
resonance **1**.281
oscilloscope **6**.45, **6**.64, ***6**.253–4*
Osei Bonsu of Asante **4**.20
Osei Tutu of Asante **3**.24
osiers **2**.368
Osiris **9**.5, **9**.16, ***9**.28*, **9**.32
Osler, Sir William ***2**.237*
Osman ('Uthmān), third caliph **3**.58, ***3**.369*, **7**.39, **7**.177
Osman I, Ottoman emperor **3**.263, ***3**.264*, **3**.265
osmium **1**.106, ***1**.242*, **6**.177
osmosis ***2**.237*
Osorio, Oscar **4**.111
osprey ***2**.237*
Ossian **9**.28
Ostade, Adriaen and Isaak van ***5**.333*
Ostpolitik ***4**.252*, **7**.87
ostracism ***3**.264*
ostracod **2**.232
ostrich ***2**.237*
Ostrogoths **3**.98, ***3**.264*, **3**.354
philosophy **7**.33
see also Goths
Ostrovsky, Aleksandr ***5**.333*
Ostwald, Wilhelm ***1**.242*
O'Sullivan, John **4**.209
Oswald, Saint, king of Northumbria **3**.15, **3**.56, **3**.109, ***3**.264*, **3**.264, **3**.275, **3**.380
Oswald, king of Ireland **3**.6
Oswald, Lee Harvey **4**.180, ***4**.252*
Oswy, king of Northumbria **3**.15, ***3**.264*, **3**.380
Otis, James **3**.2, ***3**.264*
otorhinolaryngology ***6**.254*
otoscope ***6**.254*
Ottakar I, king of Bohemia **3**.43
ottava rima **5**.21, ***5**.333*, **5**.355, **5**.369
Ottawa Agreements (1932) ***4**.252–3*
Ottawa Conference on Security and Co-operation in Europe (1985) **7**.64
Ottawa Indians ***3**.264*, **3**.288
otter ***2**.237*, **2**.301
otter shrew **2**.171
Otterburn, battle of ***3**.265*
Otto I (the Great), Holy Roman Emperor **3**.141, **3**.167, **3**.168, **3**.264, ***3**.265*
and Magyars **3**.205, **3**.216
Ottonian art **5**.334
Otto I, king of Greece **4**.141
Otto III, Holy Roman Emperor, king of Germany **3**.168, ***3**.265*
Otto IV, Holy Roman Emperor, king of Brunswick **3**.168
Otto, Nikolaus August **4**.225, **6**.94, ***6**.254*
Ottoman empire ***3**.265–6*, ***4**.253–4*
arts **5**.231, ***5**.333–4*
laws/legal system **7**.282
leaders **3**.35, **3**.185, **3**.271, **4**.10, **4**.112, **4**.200
sultans **3**.37, **3**.228, **3**.243, **3**.264, **3**.321, **3**.344, **3**.345, **4**.207
maps **3**.266, **4**.253
politics **4**.10, **4**.112, **4**.200
religion/theology **7**.280, **7**.282
Tanzimat reforms **4**.338
wars and defence **3**.35, **3**.185, **3**.201, **3**.206, **4**.3, **4**.217–18, **4**.310
Young Turks **4**.387
end of **4**.309–10
see also Turkey
Ottonian arts ***5**.334*
Otway **5**.418
Oudenarde, battle of ***3**.265*
oued **1**.20
Ouegbadja, king of Abomey **3**.96
Ousmane, Sembène ***5**.334*
Outback ***4**.254*
outcrop ***1**.242*
outer planets ***8**.115*
see also Jupiter; Neptune; Pluto; Saturn; Uranus; Vulcan
outlier ***1**.242*
output device ***6**.254*
Outremer ***3**.265*
outwash, glacial ***1**.242*
fan **1**.140
oval *see* ellipse
ovary ***2**.237*
oven **6**.193
temperatures ***9**.62*
see also kiln
ovenbird **2**.237, ***2**.238*
Overbeck, Friedrich **5**.314
overburden ***1**.242*
Overbury, Sir Thomas ***3**.265–6*
overheads **7**.71
overland flow ***1**.242*
Overland Stage ***4**.254*
overpopulation ***7**.226*
Overseas Development Assistance **7**.9
overshot wheel **6**.367
overthrust fault ***1**.243*
overture ***5**.334*, **5**.442
Ovid ***5**.334–5*, **9**.27, **9**.30
oviduct **2**.113
ovum **2**.104, ***2**.238*
Owen, Goronwy **5**.486
Owen, Robert **4**.86, ***4**.254*, **4**.317
Owen, Wilfred ***5**.335*
Owings, Nathaniel **5**.421
Owl Nebula ***8**.115*
owlet moths ***2**.238*
army worm **2**.14
cutworm **2**.84
silver-Y **2**.310
underwing moth **2**.352–3
wainscot moth **2**.358
owls ***2**.238*
barn **2**.23
eagle **2**.101–2
little **2**.191
short-eared **2**.308
snowy **2**.316
tawny **2**.336
ownership *see* property
ox ***2**.238*
ox-drawn vehicle **6**.165
see also cattle family
ox-bow lakes **1**.242, ***1**.243*
Oxenstierna, Axel Gustafsson, count **3**.75, ***3**.266*, **3**.318, **3**.347, **3**.355
Oxfam **7**.8, **7**.218, **7**.256
Oxford, earls of *see* Asquith, Herbert; Harley, Robert
Oxford Movement **4**.238
oxhorn **5**.65
oxidation and reduction ***1**.243*, ***1**.278*, **6**.253, ***6**.254*
antimony **6**.16
anti-oxidant **6**.16
barium **6**.30
copper **6**.89
hydrogen peroxide **6**.174
manganese **6**.214
mercury **6**.219
molybdenum **6**.232
nitric acid as agent **6**.241
oxidizing agent **1**.243
prevention (antioxidant) **1**.16
silver **6**.93
see also oxygen
oxides ***1**.243*
aluminium *see* aluminium oxides
anodizing **6**.15
copper *see* copper oxide
glazing **6**.156, **6**.157
iron *see* iron oxide
lead *see* lead oxide
magnesium *see* magnesium oxide
monoxide *see* carbon monoxide
in paint **6**.255
refractory **6**.297
semiconductor (MOS) **6**.70
superoxides **1**.327
zinc **1**.373, **6**.391
see also corrosion; dioxides; hydroxide; Magnox
oxidizing agent ***1**.243*
oxisol ***1**.243*
oxy-acetylene **6**.123, **6**.383
oxygen **1**.107, ***1**.243*, ***6**.254*
from barium oxide **6**.43
in blood **6**.40
cycle ***2**.238*
from hydrogen peroxide **6**.174
hydroxyl group **1**.163
liquefaction **6**.150, **6**.301
ozone *see* ozone
reaction prevented **6**.16
removal (reduction) **1**.278

oxygen (*cont.*):
see also oxidation and reduction; oxides
oxymoron *5.335*
Oyo, Yoruba empire of **3**.4, *3.266*, **4**.386
oyster *2.238*, **2**.246
oystercatcher *2.238*
Oz, Amos **5**.203
Özal, Turgut **4**.354
Ozark Plateau *1.243*
Ozenfant, Amédée **5**.370
Ozma Project *8.115*
ozone layer and ozone depletion *1.243*, **6**.72, *6.254*, *7.226*, *8.115*
Montreal Protocol on **7**.206, **7**.226
ozonosphere **8**.115

P

paca *2.239*
pacemaker **6**.218, *6.255*
Pachacutec (Pachacuti), king of Incas **3**.175–6, **5**.426
Pachomius, Saint **3**.236
Pacific, War of *4.254*
Pacific Campaigns *4.254–5*
Pacific Ocean area *1.244*
ACP countries **7**.103
arts **5**.325
Colombo Plan **4**.79, **7**.57
education **7**.93
ethnic groups **4**.178
see also Trobriand Islanders
exploration **1**.42, **1**.95, **1**.97, **1**.366, **3**.96, **3**.106
islands:
Caroline **3**.231, **4**.63–4
Guam **1**.147, **4**.142
Kurile **1**.185, **7**.9
Marianas **1**.147, **3**.231
Marquesas **1**.206, **4**.212
Nauru **1**.225, **4**.234, **7**.296
New Caledonia **3**.228, **4**.238, **7**.175
Norfolk **1**.230
Society **1**.309, **1**.330, **4**.336–7
Solomon **1**.312, **4**.318, **7**.296
Tahiti **1**.330, **4**.336–7
Tuvalu **1**.348, **4**.355, **7**.296
Vanuatu **1**.354, **4**.362, **7**.296
see also Fiji; Hawaii; Marshall Islands; Melanesia; Micronesia; Philippines; Polynesia; Samoan; Taiwan; Tonga
languages **7**.23
Lomé Convention **4**.198
map **1**.245
music *5.325*
patrilineality **7**.197, **7**.273, **7**.312
physical geography and geology *1.244*, *1.245*
Antipodes Islands **1**.16
climate **1**.108
currents **1**.49, **1**.180, **1**.185, **1**.235, **1**.255
earthquakes and volcanoes **1**.99
Galápagos **1**.130
gulfs **1**.49
islands **1**.245
see also islands *above*
Krakatoa **1**.184
map **1**.245
plate tectonics **1**.260
ridges **1**.245
seamounts **1**.297
seas **1**.17, **1**.38, **1**.61, **1**.76, **1**.168–9, **1**.177, **1**.332, **1**.370
trenches **1**.206, **1**.245, **1**.342
religion and myths **9**.16, **9**.25, **9**.31, **9**.35
South Pacific Forum **7**.296
technology and science **6**.57, **6**.209, **6**.292
wars and defence **4**.41–2, **4**.254–5, **7**.14
Pacific scandal (Canada) *4.254–5*
Pacific Security Treaty **7**.14
Pacifico, Don **4**.102
pacifism *see* peace
pack animals *see under* draught and pack animals
pack ice *1.244*, **1**.296
packaging technology *6.255*
paddle worm **2**.42
paddlefish *2.239*
paddle-wheel *6.255*
Padei-tha-ya-za **5**.67
Paderewski, Ignacy Jan *5.335*
Padova, Marchetto da **5**.379
paean *5.335*
Paets, Konstantin **4**.113
Páez, José Antonio *4.255*, **4**.324, **4**.363
Pagan (city) *3.267*, **5**.66–7
Paganini, Niccolò *5.335*
Page, Sir Earle (Christopher Grafton) **4**.234
Page, R. M. **6**.288
pageants *5.335*, **5**.474
Paget, Sir Arthur **4**.90
paging system *6.255*
Pagneaux, Monica **5**.296
Pagnol, Marcel *5.335*
pagoda tree **2**.127
Pahlavi *see* Reza Shah Pahlavi
Pai Marire **4**.14, *4.255–6*
pain *2.239*
pain-killers *see* analgesia
torture **7**.114, **7**.313
Paine, Thomas **3**.117, **3**.209, *3.267*, **5**.491
paint **5**.289, *5.335–6*, *6.255*, **6**.261
acrylic **5**.2–3
emulsion **6**.120, **6**.255
luminous **6**.210, **6**.249
oil **5**.157, **5**.326
palette **5**.336
primary colours **5**.366
red lead **6**.296
scumble **5**.409
tempera **5**.201, **5**.448
see also painting; pigment; water-colour
painted snipe *2.239*
painted-lady butterfly *2.239*
painting and painters **5**.207
Abbasid **5**.1
Aboriginal **5**.1
abstract *see* abstract art; Abstract Expressionism
action **5**.3
Africa **5**.6
air brush **5**.10
anamorphosis **5**.16
arabesque **5**.18
Art Deco **5**.23
Australia **5**.1, **5**.26–7, **5**.319–20
Austria **5**.249–50, **5**.405
automatism **5**.27
Baroque *see under* Baroque
Belgium **5**.147, **5**.275, **5**.470
bistre **5**.48
Britain:
abstract **5**.318, **5**.350
allegorical **5**.483
animals **5**.255, **5**.437–8
Baroque **5**.454
Camden Town Group **5**.72, **5**.417–18
candlelit **5**.495
contemporary **5**.29
conversation pieces **5**.129
educationalist **5**.140
Fauvist **5**.423
genre **5**.489
horses **5**.437–8
Impressionist **5**.318, **5**.417–18, **5**.420, **5**.431
landscape *see under* landscape
Op art **5**.383
Picturesque **5**.348–9
Pop art **5**.208
portraits *see under* portraiture
Pre-Raphaelite Brotherhood **5**.62, **5**.214, **5**.295, **5**.366, **5**.391, **5**.442
religion **5**.428
Romantic **5**.48, **5**.67, **5**.110–11, **5**.284, **5**.390, **5**.462–3
painting and painters (*cont.*):
Britain (*cont.*):
royalty **5**.499
Scotland **5**.179
St Ives School **5**.398
theatrical **5**.499
theorist **5**.381–2, **5**.394
urban and industrial life **5**.270
Vorticism **5**.418, **5**.479–80, **5**.496
war **5**.428
water-colour **5**.179, **5**.394, **5**.400, **5**.462–3, **5**.468–9
Byzantine **5**.70
Canada **5**.73
canvas **5**.75
capriccio **5**.77
chiaroscuro **5**.90, **5**.382
China **5**.91, *5.96*, *5.98*, **5**.196, **5**.308, **5**.438, **5**.486, **5**.495, **5**.496, **5**.498
concrete **5**.109
continuous representation **5**.111
contrapposto **5**.111–12
conversation pieces **5**.112, **5**.129, **5**.168, **5**.499
craquelure **5**.117
crayon **5**.118
Cubist *see* Cubism
Czechoslovak **5**.210, **5**.250, **5**.307
Dada *see* Dada
diptych **5**.131
encaustic **5**.146
Etruscan **5**.150
Fatimid **5**.152
France **5**.258, **5**.391–2, **5**.457, **5**.467, **5**.480
abstract **5**.128, **5**.244, **5**.260
Barbizon School **5**.34–5, **5**.296, **5**.392
Baroque **5**.88, **5**.259, **5**.480
Burgundian school **5**.66
classical **5**.364–5
Cubist *see under* Cubism
Expressionist *see* Expressionism in painting
Fauvist *see* Fauvism
genre **5**.89, **5**.192
illuminators **5**.162, **5**.339
Impressionist *see* Impressionism; Neo-Impressionism
Intimisme **5**.51, *5.225*, **5**.480
landscape *see under* landscape painting
miniaturist **5**.162
modernism **5**.137
Nabis group **5**.51–2, **5**.129
Neoclassical *see* France *under* Neoclassicism
Op art **5**.469
Orphism **5**.87, **5**.128, **5**.332
portraiture **5**.105, **5**.152, **5**.261
Post-Impressionist *see* Post-Impressionism
Purism **5**.260, **5**.370
Realist **5**.116
Renaissance **5**.162–3, **5**.366
Rococo **5**.55–6, **5**.89, **5**.162, **5**.483
Romantic **5**.127–8, **5**.165, **5**.173, **5**.174
Salon **5**.398
still life **5**.58, **5**.87, **5**.89, **5**.370
Surrealism **5**.23, **5**.87, **5**.148
Symbolism **5**.376
Synthetism **5**.443
Tachisme **5**.444
theatrical **5**.483
genre *see* genre
Germany:
abstract **5**.244
Blaue Reiter **5**.49, **5**.243–4, **5**.249, **5**.282

painting and painters (*cont.*):
Germany (*cont.*):
Brücke, Die **5**.63, **5**.248, **5**.320
collage **5**.407
Dada **5**.195, **5**.407
Expressionism **5**.41, **5**.63, **5**.248, **5**.282
frescoes and murals **5**.498
Impressionism **5**.264
landscape **5**.13, **5**.117, **5**.165, **5**.248, **5**.264, **5**.494
miniatures and illustrations **5**.209
Neoclassical **5**.291
Neue Sachlichkeit **5**.131, **5**.195
portraiture **5**.117, **5**.131, **5**.209, **5**.264, **5**.491
Renaissance **5**.139
Romantic **5**.165, **5**.195, **5**.394, **5**.405
Sezession **5**.413
Worpswede School **5**.494
gesso **5**.175
golden section **5**.183, **5**.368
Greece, ancient **5**.18, **5**.190, **5**.192, **5**.498
grisaille **5**.194
hard-edge **5**.198
Hindu *see* India *below*
Hungary **5**.301
impasto **5**.217–18, **5**.326
Impressionist *see* Impressionism
India **5**.10, **5**.108, **5**.196, **5**.219, **5**.293, **5**.373, **5**.375
Indians, North American **5**.321
Ireland **5**.36, **5**.497
Islam **5**.228–9, **5**.230
Italy **5**.177–8
Baroque **5**.45–6, **5**.114, **5**.195, **5**.255, **5**.282
Classical **5**.132
decorative **5**.382–3
Futurist **5**.29, **5**.80
genre **5**.268
International Gothic **5**.173
landscape **5**.42, **5**.74, **5**.268, **5**.383
Mannerist **5**.61, **5**.179, **5**.279–80
Metaphysical **5**.80, **5**.98, **5**.292
oil **5**.18
portraiture *see under* portraiture
Renaissance *see under* arts *under* Renaissance
Rococo **5**.455
Romantic **5**.391
symbolic **5**.21
see also Florentine; Futurism; Sienese School; Venetian *and under* arts *under* Renaissance
Japan **5**.207, **5**.209, **5**.237, **5**.412, **5**.465, **5**.466, **5**.467
Korea **5**.251
landscape *see* landscape painting
Mannerism **5**.279–80
marine **5**.283
medium **5**.289
Melanesia **5**.289
Metaphysical **5**.80, **5**.98, **5**.292
Mexico **5**.384
minimal art **5**.297
Minoan **5**.297
naïve **5**.305, **5**.312, **5**.359, **5**.392
Neoclassical *see* Neoclassicism
Nepal **5**.316
Netherlands and Flemish **5**.73, ***5**.157–8*, **5**.184, **5**.265, **5**.270, **5**.286, **5**.407
abstract **5**.301
Baroque **5**.393
Belgium **5**.147, **5**.275, **5**.470
bizarre and fantastic **5**.54–5
genre **5**.13, **5**.62, **5**.211, **5**.262, **5**.292, **5**.431, **5**.449
Hague School **5**.197
interiors **5**.472, **5**.473

painting and painters (*cont.*):
Netherlands and Flemish (*cont.*):
landscape *see under* landscape painting
marine **5**.283
Northern Renaissance **5**.73
portraits *see under* portraiture
Post-Impressionist **5**.182
Stijl, De **5**.301, ***5**.433*
Norway **5**.192–3, **5**.307–8
oil **5**.326
Op art **5**.328, **5**.383, **5**.469
panel **5**.338, **5**.356
pentimento **5**.343
Persia **5**.47
perspective *see* perspective
Picturesque **5**.348–9
pietà **5**.349
pointilism **5**.354
Poland **5**.496
polyptych **5**.356
portraits *see* portraiture
post-painterly abstraction **5**.363, **5**.431
predella **5**.365
Pre-Raphaelite *see* Pre-Raphaelite
putto **5**.371
realism *see under* realism
Rococo *see* Rococo
Romanesque **5**.389
Romantic *see under* Romanticism
Russia and USSR ***5**.395–6*
Constructivism **5**.266, **5**.346, **5**.447
Fauvism **5**.238
icons **5**.238, **5**.394
realism **5**.380, **5**.447
Suprematism **5**.266, **5**.277, **5**.440
Safayid **5**.397
Sezession **5**.413
Spain **5**.189, **5**.499
Baroque **5**.74
bizarre and fantastic **5**.187
chiaroscuro **5**.382
Cubism **5**.194, **5**.348
genre **5**.309
naturalism **5**.308–9
portraiture **5**.189, **5**.470
still life **5**.470
Surrealism **5**.121, **5**.298, **5**.348
Sri Lanka **5**.429
Suprematism **5**.277, **5**.440
Switzerland **5**.166, **5**.175–6, **5**.244, **5**.249
Tachisme **5**.444
Tibet **5**.454–5
Timurid **5**.456
triptych ***5**.460*
trompe-l'oeil ***5**.460*
United States (and America): ***5**.13–14*, **5**.402, **5**.496
Abstract Expressionism **5**.2, **5**.14, **5**.127, **5**.184, **5**.306, **5**.355, **5**.375, **5**.391
art nouveau **5**.455
colour field **5**.269
genre **5**.141
Impressionist **5**.82
landscape **5**.13, **5**.14, **5**.211, **5**.213
Minimalist **5**.240–1
modernist **5**.326
naïve **5**.305
naturalists **5**.26, **5**.342–3
Pop art **5**.240–1, **5**.264, **5**.482
portraiture **5**.13–14, **5**.112, **5**.141, **5**.342, **5**.486
post-painterly abstraction **5**.363, **5**.431
still life **5**.342
Surrealism **5**.184, **5**.391
see also Ashcan
varnish **5**.469
war **5**.428, **5**.441

painting and painters (*cont.*):
see also drawing; frescoes and murals; illuminated manuscripts; illustration; landscape; miniatures; paint; portraiture; water-colour *and* caves *under* rock art
pair-bonding ***2**.239–40*
Paisley, Ian **4**.247, **4**.357
Pakenham, Sir Edward **4**.239
Pakhet **9**.7
Pakistan ***1**.244*, ***4**.256*, ***7**.374*
agriculture **6**.267
aid to **7**.8
boundaries in 1948 **4**.161
city **3**.350
economy and trade:
child labour **7**.45
employment **7**.106
fishing **7**.45
migrant labour **7**.200
emigration **7**.200
ethnic groups **4**.262, **7**.231
languages **7**.95, **7**.157
laws/legal system **7**.284
leaders **4**.26
literacy **7**.185
literature **5**.466–7
Muslim Brotherhood **7**.166
North-West Frontier **4**.247–8
physical geography and geology **1**.244
desert **1**.334
mountains **1**.156, **1**.179, **1**.183
river *see* Indus
politics and government **7**.58, **7**.119
leaders **4**.37–8, **4**.174, **4**.194, **4**.390
secession *see* Bangladesh
population **7**.200, **7**.242
poverty **7**.244
prehistory **3**.157, **3**.177
see also Indus
province **3**.328
refugees **7**.262
religion/theology **7**.271, **7**.284, **7**.302
sport and recreation **7**.75
urban areas **7**.197
wars and defence **4**.68, **4**.163, **7**.52, **7**.220
see also Bangladesh; East Pakistan; Indus
Pal, Bipin Chandra **7**.143
Pala dynasty ***3**.267*
Palacký, Frantisek **4**.43, ***4**.256*
Palade, George Emil ***2**.240*
Palaeocene epoch **1**.134, **1**.135, ***1**.244*
Palaeogene period **1**.133, **1**.134, ***1**.244*
Palaeolithic ***3**.267*
art **5**.434
cave-dwellers **3**.66
Mousterian culture **3**.242
Oldowan **3**.261
Upper **3**.367
palaeomagnetism ***1**.244*
palaeontology and palaeontologists **1**.4, **2**.84, **2**.186, **2**.204, ***2**.240*
see also fossils
Palaeosiberian languages ***7**.227*
palaeosol ***1**.244–5*
Palaeozoic era **1**.133, **1**.134, ***1**.245*
periods *see* Cambrian; Carboniferous; Devonian; Ordovician; Permian; Silurian
Palais de la Découverte ***6**.255*
Palamas, Kostis **5**.191
Palatinates ***3**.267*
Elector **3**.133
pale ***3**.267*
Pale Fox **9**.27, ***9**.28*

Paleocene epoch **1**.133, **1**.134
Palestine ***1**.246*, ***3**.267–8*, ***4**.256*, ***7**.227*
in Arab League **7**.15
Balfour Declaration **4**.29
British Mandate **4**.169
Canaan **3**.58–9
languages **7**.279
Muslim Brotherhood **7**.166
peoples **3**.280, **7**.227
physical geography and geology **1**.246
politics and government **7**.191
see also Zionism
refugees **7**.227, **7**.261
religion/theology **7**.193, **9**.40–3
revolt **3**.389
rulers **3**.164, **3**.281
Talmud **7**.305
terrorism and militant organizations **4**.329
Black September **4**.41
al-Fatah **4**.117
Haganah **4**.144
Irgun group **4**.166–7
PLO **4**.169, ***4**.256*, **7**.130, **7**.227, **7**.239, **7**.308
and UN **7**.321
urban areas *see* Jerusalem
Zealots **3**.188
see also Canaan; Israel; Jews; Judah; Zionism
Palestine Liberation Organization **4**.169, ***4**.256*, **7**.130, **7**.227, **7**.239, **7**.308
Palestrina, Giovanni Pierluigi da ***5**.336*
palette ***5**.336*
Pāli language **7**.28, **7**.157, **7**.305, **7**.316
literature ***5**.336*
palindrome ***5**.337*
Palissy, Bernard ***5**.337*
Palladio, Andrea and Palladianism **3**.301, ***5**.337–8*, **5**.359, **5**.402
Britain **5**.66, **5**.73, **5**.80, **5**.241, **5**.245–6, **5**.365, **5**.447–8
palladium **1**.107, ***1**.246*, ***6**.256*
Pallas (asteroid) ***8**.116*, **8**.193
Pallava ***3**.268*, **5**.275
palm *see* palms
palm civet ***2**.240*
Palma, Ricardo **5**.256
palm-chat ***2**.240*
Palme, Olof **4**.335, **7**.58
Palmer, A. Mitchell: Palmer raids ***4**.256*
Palmer, Samuel **5**.149, ***5**.338*
palmer ***3**.268*
Palmerston, Henry John Temple, 3rd viscount **4**.79, **4**.217, ***4**.257*
first liberal government **4**.195
gunboat diplomacy **4**.144
and Opium Wars **4**.250
palmetto ***2**.240*
palms ***2**.240*
coconut **2**.71, **2**.75–6
date **2**.88
dwarf fan **2**.240
nipa **2**.228
palm oil **2**.233, **6**.364
raffia **2**.275
rattan **2**.278
talipot **2**.333–4
Palmyra ***3**.268*
Palomar, Mount (observatory) **8**.104
palynology ***2**.240*
see also pollen
Pamirs ***1**.246*, **1**.369
pampas ***1**.246*, ***2**.240*, **2**.241
pampas grass ***2**.240*
pamphlets *see* essays and essayists
Pan ***9**.28*
Pan gu ***9**.28*

Pan-African Congress (South Africa) *4.257*
Pan-Africanism *4.257*
Panama *3.268*, *4.257–8*, *7.374–5*
  Canal 4.67, 4.148, *4.257–8*, 6.199, *6.256*
  economy and trade 7.309
  independence 4.43
  physical geography and geology 1.246
Pan-Americanism *4.258*
Pañcatantra *5.338*, 9.21
Pancho Villa 4.368
pancreas *2.240*
  disorder *see* diabetes
pandas *2.240–1*, 2.280
Pāṇḍava 9.22
Pandora *9.28*
Pandya *3.268*
panegyrics 5.7, *5.338*
panel *5.338*, 5.356
panelling *5.338*
Pangaea 1.135, *1.246*
pangolin *2.241*
Panhard, René 6.94
Pānini 5.328
Panipat *3.268*
Pankhurst, Christabel 4.258, 7.332
Pankhurst, Emmeline *4.258*, 4.333, 7.332
Pankhurst, Sylvia 4.258
panning for metals 6.253
Panoan language 7.295
panpipes 5.15, *5.338*
Pan-Slavism *4.258*
pansy 2.356
Pant 5.206
panther *see* leopard
pantomime *5.338*
pantomimus 5.296, *5.338*
Paoli, Pasquale de 3.88
Papa *9.31*
papacy/popes *3.268–9*, 7.48, *7.240–1*, *9.71–2*
  list of *9.71–2*
  5th century 3.27, 3.117, 3.205
  6th to 9th centuries 3.61, 3.68, 3.204, 3.280
    see also Gregory I
  10th and 11th centuries 3.205–6, 3.216
    see also Gregory VII; Urban II
  12th and 13th centuries 3.3, 3.40, 3.61, 3.69, 3.121, 3.132–3, 3.161, 3.179, 3.306, 3.327
  14th century 1.29, 3.44, 3.146, 3.211, 3.269
  15th century 3.146, 3.283, 3.317, 3.359, 3.383
    see also Sixtus IV
  16th century 3.70, 3.89, 3.131, 3.283, 3.329
    see also Gregory XIII; Julius II; Leo X; Paul III
  17th to 19th centuries 3.119, 3.187, 3.367–8, 5.46
    see also Pius VII; Pius IX
  20th century 4.175, 4.190, 4.270, 4.272, 4.291, 4.363, 7.20
  anti-papal legislation 3.290
  antipope *see* antipopes
  Cambrai, League of 3.58
  Lateran Council and Treaties 3.204, 4.190
  Papal States *3.269*, 4.363
  Peter's pence (tax) 3.277
  see also Vatican
Papadopoulos, Christos 8.22
Papagos, Alexandros 4.140
Papandreou, Andreas 4.140
Papanicolaou, George: 'pap' smear 6.68
papaya *2.241*
Papen, Franz von *4.258*
paper *6.256–7*
  bleach 6.40
paper (*cont.*):
  carbonless transfer 6.59
  carbon-paper 6.59
  cardboard 6.59–60, 6.255
  China 6.72, 6.256
  chromatography 1.61
  innovators 5.95, 6.72, 6.105
  manufacture 6.145, *6.256–7*
  papier mâché *5.338*
  papyrus 3.269, 6.257
  parchment/vellum 6.258
  photographic 6.27, 6.256
  recycling 6.296, 7.260
  size 6.321
  watermark 6.377
  wood-pulp for 6.143, 6.387
paper nautilus *2.241*
Papin, Denis 6.281, 6.304
Papineau, Louis-Joseph: Papineau's rebellion 4.258–9, 4.279, 4.361
Papua New Guinea (New Guinea) *1.226–7*, *1.246*, *4.259*, *7.375*
  ethnic groups 3.228, 7.197
  exploration 1.75
  highlands 7.214–15
  independence 4.321
  languages 7.23, *7.228*, 7.236
  music *5.338–9*
  physical geography and geology 1.26, 1.226–7, 1.246
    New Guinea highlands 7.214–15
  rituals 7.132
  in South Pacific Forum 7.296
  see also Trobriand Islanders
papyrus *3.269*, *6.257*
parables *5.339*
  see also stories and tales
parabola 1.72, 1.73, *1.246–7*
parabolic dune 1.97
Paracas culture 5.426
Paracelsus, Philippus Aureolus 1.7, *2.241*, *6.257*
paracetamol 6.14
parachute 6.8, 6.113, 6.138, *6.257*
Paradise Cave *1.247*
paradox *1.247*, *7.228*
paraffin 6.79, *6.257*, 6.264, 6.323
  oil lamp 6.20, 6.30, 6.201, 6.249
  wax 6.57, *6.257*, 6.382
Paraguay *1.247*, *3.269*, *4.259*, *7.375*
  economy and trade 4.320
  independence 4.43
  physical geography and geology 1.142, 1.247, 1.248
  politics and government 7.23
    leaders 4.123, 4.199, 4.224, 4.331
  population 4.320
  technology 6.91, 6.95
  wars and defence 4.68, 4.259–60
parakeet *2.241–2*
Parakramabahu I, king of Sri Lanka 3.339
parallax *1.247*, *8.116*
parallel connection 6.74, 6.76
parallel postulate *1.247*
parallel processing (computing) 6.180, *6.257*
parallel reliability 6.297
parallelogram 9.51
parallels of latitude *1.247*
  see also latitude and longitude
paralysis *2.242*
Paramecium *2.242*
paramedical worker *7.228*
parameter (mathematics) *1.247–8*
Paraná River *1.248*
paranoia *7.228*
paranormal *see* parapsychology
paraplegia *2.242*
parapsychology *7.228*
paraquat 6.168
parasites and semi-parasites *2.242*, 7.107
  control 7.107
  ichneumons 2.168
  lousewort 2.195
  mistletoe 2.213
  mould 2.220, 2.313
  nematodes *see* roundworm
  spiny-headed worms 2.322
  tapeworms 2.167, 2.334
  tongue-worm 2.343
  toothwort 2.343–4
  see also flukes; pathogens; pests; platyhelminth; protozoa; roundworm
paratyphoid fever *2.352*
  see also typhoid
parchment *6.258*
pardolotes 2.122
pardoner *3.269*
Paré, Ambroise 6.21
parental leave *7.228*
Pareto, Vilfredo 7.104, *7.228–9*
Parícutin *1.248*
Parini, Giuseppe *5.339*
Paris Basin *1.248*
Paris (city):
  Bastille 3.36
  Commune *4.260*
  Conference on Security and Co-operation in Europe (1990) 7.64
  Congress of (1856) *4.260*
  Fronde 3.135
  Île de France 3.174
  Observatory *8.116–17*
  Opera Ballet 5.122
  Pact of (1928) 4.180
  Peace of (1785) *3.269*
  Peace Settlement (1919) 4.363–4, 7.315
  School of *5.339*
  Treaty (1763) *3.269–70*
  Treaty (1952) 7.99
  Tuileries 3.362
Paris (Greek hero) *9.28–9*
parish *3.270*
parity 1.329
Park, Mungo 1.186, *1.248*, *4.260*
Park, Robert E. 7.44
Park Chung Hee 4.320
Parker, Charles *5.339*
Parker, Dorothy *5.339*
Parker, Sir Hyde 3.249, 4.86, 4.237
Parker, Richard 3.255
Parkes, Sir Henry *4.260*
parking, car 6.360
Parkinson, James: Parkinson's disease *2.242*, 6.241, 6.334
Parkinson, Sydney 5.325
parks *see* garden; national parks
Parlement *3.270*
Parliament (Britain) *3.270*, *4.260–1*
  see also under Britain
parliamentary government 3.270, 4.260–1, *7.229*
  civil service 7.52
  prime-ministerial 7.247
  see also legislature; Parliament
Parma, Alessandro Farnese, duke of 3.124
Parmigianino *5.339–40*
Parnassians 5.38, 5.172, 5.259, *5.340*, 5.471–2
Parnassus *9.29*
Parnell, Charles Stewart 4.48, 4.155, 4.189, *4.261–2*, 4.269
parody 5.66, 5.81, 5.242, *5.340*
parole *7.229*
Parr, Catherine *3.270–1*
parrot family *2.242*, 2.277
  budgerigar 2.44
  cockatoo 2.70
  kea 2.178
  love-bird 2.195
  macaw 2.197
parrot family (*cont.*):
  parakeet 2.241–2
parrotfish *2.242–3*
Parry, Sir Hubert *5.340*
Parry, Sir William Edward and Sir John Franklin *1.248*
parsec *8.117*
Parsi 7.23, *7.229*
  see also Zoroastrianism
parsley *2.243*
parsnip *2.243*
parson *3.271*
Parsons, Sir Charles Algernon *6.258*, 6.333
Parsons, Talcott 7.127, *7.229*
parthenogenesis *2.243*
Parthia 3.93, *3.271*
Partial Test Ban Treaty 7.220
particle(s):
  accelerator 6.186, 6.196, *6.258*, 6.292, 6.370
  detector *6.258–9*
  elementary *see* elementary particles
  in fluids *see* suspension
  see also dust
partita *5.340*
partnership *7.229*
Parton, Dolly 5.115
partridge *2.243*
parts of speech 7.22, 7.181, *7.229*
part-song *5.340*
Pārvatī 7.285, 9.10
pas de deux *5.341*
Pascal, Blaise 1.120, 1.248, *3.271*, *5.340*, 6.1, 6.53, *6.259*
PASCAL (computers) 6.86
pascal (unit) *1.248*, 6.368
Pascoli, Giovanni *5.340*
Pasek, Jan Chryzostom *5.341*
pasha *3.271*
Pašić, Nikola *4.262*, 4.387
Pasiphae *8.117*
Pasolini, Pier Paolo *5.341*
pasque-flower *2.243*
Pass, J. 2.96
passacaglia 5.87
Passchendaele, battle of *4.262*
passenger pigeon *2.243*
passion fruit 2.142, 2.243
passion music *5.341*
passion play *5.341*, 9.17
passion vine *2.243*
passive resistance *4.262*
  see also under Gandhi, Mohandas
Passover *7.229–30*
pastel 5.336, *5.341*
Pasternak, Boris *5.341*
Pasteur, Louis *2.243*, 4.99, 6.16, 6.37, 6.140, 6.177, 6.204, 6.218, *6.259*
pasteurization *6.259*
pastiche *5.341*
pastimes *see* hobbies and pastimes; sport and recreation
pastoral and idyll (arts) 5.182, *5.217*, *5.341*, 5.428
pastoralism *7.230*
  in Africa *7.5*
    Bedouin 7.28, 7.29
    Maasai 7.6, 7.188
    Ndebele 7.214
    Nuer 7.185, 7.217, 7.220
    Tuareg 7.217, 7.317
    Xhosa 7.334
  in Arctic 7.16, 7.181
  in Europe 7.217
  in Middle East 7.28, 7.29, 7.188
    see also Kurds
  see also nomadism; transhumance
pasture 2.69, *2.243–4*
Patagonia *1.248–9*, *4.262*
Patañjali 7.144, 7.335
patchouli *2.244*
patchwork *5.341*
Patel, Sardar Vallabhbhai *4.262*
patella (kneecap) 2.186

patent *6.259*, **6.**260, *7.230*
Pater, Walter *5.341–2*
paternalism *7.230–1*
path of totality *8.117*
Pathan *4.262*, **7.**231
Pathet Lao *4.262–3*
pathogens *2.244*
asbestos **6.**22
bacteria *see under* bacteria
dioxin **6.**104
radiation **6.**125
see also diseases and disorders; parasites; poison; vectors of disease
pathology and pathologists **2.**122, *2.244*, **2.**354, **2.**364
patience (card-game) **7.**40
Patiño, Simon Iture *4.263*
patriarchs of Judaism **3.**192, **9.**11, **9.**25
see also Abraham; Isaac; Jacob; Moses
patriarchy *7.231*
patrician *3.271*
Patrick, Saint **3.**66, *3.271*, **3.**350, **9.**43
patrilineality **7.**86, *7.185*
Africa **7.**24, **7.**28, **7.**151, **7.**176, **7.**220, **7.**336
Asia **7.**9, **7.**175, **7.**215
Australia **7.**1
North American Indians **7.**236
Pacific **7.**197, **7.**273, **7.**312
South America **7.**11, **7.**335
patriotism **4.**174
patronage **4.**236, *7.231*
pattern poetry *5.342*
pattern recognition (computer) *6.259*
patterned ground **1.**248, *1.249*
pattern-making **6.**62
Patton, George Smith **4.**110, **4.**170, **4.**245, *4.263*
Paul, Saint *3.271–2*, *9.43*
as Apostle **3.**104
Epistles **3.**41, **3.**59
and homosexuality **7.**146
and Jerusalem **3.**187
travels **3.**74, **3.**75
Paul I, emperor of Russia **3.**198, *3.271*
Paul III, pope **3.**124, **3.**187, **3.**213, **7.**171
Paul VI, pope *4.263*, **4.**291
Paul, Leslie **5.**17
Pauli, Wolfgang Ernst **1.**116, *1.249*, **8.**107
Pauling, Linus Carl *2.244–5*
Paulus, Friedrich von *4.263*, **4.**328
Pausanias **5.**460
pavan *5.342*
Pavarotti, Luciano *5.342*
Pavelič, Ante **4.**89
Pavese, Cesare *5.342*
Pavia, battle of *3.272*
Pavlov, Ivan Petrovich **2.**244, *2.245*, **7.**29, *7.231*
Pavlova, Anna **5.**30
Pavo (constellation) *8.117*
Pavo Cluster *8.117*
Pavonis Mons *8.117*
pawl **6.**295
Pawlak, Waldemar **4.**272
Pawnee Indians *3.272*
pawpaw **2.**241
Paxton, Sir Joseph **4.**140, *5.342*, **5.**362, **5.**363, **6.**161
Paxton, Steve *5.342*
Paxton Boys *3.272*, **3.**273
pay **7.**98
equal **7.**107
fringe benefits **7.**127
indexation **7.**155
wages **7.**203, **7.**328
see also income
Payne, Henry **4.**200
Paz, Octavio **5.**256
Paz Estenssoro, Victor **4.**44, *4.263*
Pázmány, Péter **5.**214
Pazzi Conspiracy *3.272–3*
p'Bitek, Okot **5.**7
PCR (polymerase chain reaction) **6.**153
pea family *2.245*
fenugreek **2.**115
flamboyant tree **2.**119
fruit **2.**129
gorse **2.**43, **2.**142
laburnum **2.**181
mesquite **2.**208
pod **2.**260
Robinia **2.**287
sensitive plant **2.**303
sweet pea **2.**68, **2.**332
see also legumes
Peabody, George *4.263*
peace:
movement *7.232*
pacifism **7.**65–6, *7.227*
peacekeeping *7.231–2*, **7.**277
see also non-violence
Peace of Augsburg **3.**27
Peace River *1.249*
peach **2.**225, *2.245*
Peacock, Thomas Love *5.342*, **5.**348
peacock *2.245–6*
peacock butterfly *2.245*
peacock moth **2.**107
peacock worm **2.**113–14
Peake, Robert **3.**292
Peale, Charles Wilson **3.**191, *5.342–3*
Peale, Raphaelle **5.**342
Peale, Rembrandt **5.**343
Peale, Titian Ramsay II **5.**343
Peano, Giuseppe *1.249*
peanut **2.**245, *2.246*
peanut worm **2.**246
pear *2.246*
Pearl **5.**172
Pearl Harbor *4.263*
pearl mussel *2.246*
pearl oyster *2.246*
pearly nautilus **2.**245, *2.246*
Pears, Peter **5.**60
Pearse, Pádraic **4.**107, **4.**167
Pearson, Karl **1.**280, *2.246*
Pearson, Lester Bowles *4.263–4*
Peary, Robert Edwin *1.249*
peas *see* pea family
peasants *7.232*
Peasants' Revolt *3.273*
Peasants' War *3.273–4*
see also agriculture; feudalism; serfs
peat **1.**67, **1.**120, *1.249*, **1.**311, **6.**89, **6.**167, *6.259*
pebble *1.249*
pecan *2.246–7*
peccary **2.**246, *2.247*
pectin **6.**141
ped *1.249*
pedalfer *1.249*
Peddana, Allasani **5.**134, **5.**483
pedestal rock *1.249*
pediment *1.249*, *5.343*
pedipalp *2.247*
pedocal *1.249–50*
pedology *1.250*
pedometer *6.259*
Pedro I, emperor of Brazil **4.**12, **4.**48, **4.**175, *4.264*
Pedro II, emperor of Brazil **4.**12, **4.**48, *4.264*
Pedro III, king of Aragon **3.**327
Pedro the Cruel, king of Castile and León **3.**7–8, **3.**111
PEEK (polyether etherketone) **6.**129
Peel, Sir Robert **4.**15, **4.**79, **4.**217, *4.264*
and Corn Laws repealed **4.**101
and income tax **4.**160
and police **4.**272
Peel, Sir Robert (*cont.*):
and Tamworth Manifesto **4.**84, **4.**337
peewit **2.**183
Pegasus (mythology) **9.**7, **9.**14, *9.29*
Pegasus (constellation) *8.117*
Pegasus Cluster *8.117*
pegmatite **1.**250
Péguy, Charles *5.343*
Peirce, Charles **7.**244, **7.**285
pekin robin **2.**20
Peking Man *3.274*
Pelagius, Saint (of Jerusalem): and Pelagianism *3.274*
pelargonium *2.247*
Pelée, Mont *1.221*, **1.**233
Pelham, Henry *3.274*
pelican **2.**247
pellagra **2.**90
Pellegrini, Carlos **4.**31
pellet bells **5.**375
pellitory **2.**247
Peloponnesian area **3.**19, **3.**21, *3.117*
Peloponnesian War **3.**3, **3.**8, *3.274–5*
Pelops *9.29*
pelota *7.232*
Pelouze, T.-J. **6.**242
Peltier pump and effect **6.**354
Pelton, L. A.: Pelton wheel **6.**379
pelvis **2.**247–8
pelycosaur **2.**333
Pembroke, Richard Marshal, 3rd earl of **3.**161
Pembroke, William Marshal, 1st earl of **3.**102, **3.**161
pen *6.259*
PEN (literary association) *7.232*
penal system:
penal settlements, Australia *4.264*
see also under penal colony *under* Australia
see also convicts; prisons; punishment
Penn family **3.**131
Penance **7.**271
Penates *9.22*
Penaud, Alphonse **6.**311
pencil *5.343*
Penck, Albrecht and Walther *1.250*
Penda, king of Mercia **3.**56, **3.**229, *3.275*, **3.**380
Penderecki, Krzysztof *5.343*
Pendleton, George H.: Pendleton Act *4.264*
pendulum **6.**76, **6.**149, **6.**166, *6.259–60*
Foucault's **1.**125, **8.**52–3
Penelope *9.29*
peneplain *1.250*
Peng Pai **4.**74, *4.264–5*
penguin **2.**247, *2.248*, **2.**298
penicillin **6.**72, *6.260*
discovered and developed **2.**121, **2.**122, **6.**16, **6.**69, **6.**134, **6.**137
penillion *5.343*
Peninsular War **4.**19, *4.265*, **4.**368–9
map **4.**265
peninsulas *1.250*, **1.**251
Jutland **1.**179
Kamchatka **1.**180
Kola **1.**184
Sinai **1.**304, **1.**306, **1.**325
Ungava **1.**350
Yucatán **1.**372
see also Scandinavia
penis *2.248*
penis worm *2.248*
Penn, William *3.275*, **3.**297
Pennines *1.250*
Penning gauge **6.**281
Pennington, Isaac **3.**310
Pennsylvania *3.275*
Pennsylvanian period **1.**134, *1.250*
penny post **4.**151
penny whistle **5.**156
pennyroyal *2.248*
pension **7.**222, **7.**290
penstock *6.260*
pentagon (mathematical figure) **9.**51
Pentagon Papers *4.265*
pentagram **9.**51
pentameter **5.**121, **5.**205, **5.**259, *5.343*
pentane **6.**354
Pentateuch **7.**312
pentathlon *7.232*
Pentecost *7.232*
Pentecostal Movement *7.232*, **7.**441
pentimento *5.343*
pentlandite *1.250*, **6.**241
Pentrich Rising *4.266*
pentstemon *2.248*
penumbra *8.40*
Penzias, Arno **8.**15, **8.**33, **8.**98
peon *4.266*
peony *2.249*
'People's Budget' *4.266*
People's Party *see* Populist Party *under* populism
People's Republic of China *see* China
Pepin I, king of Franks **3.**132, **3.**230, *3.275*
Pepin II, king of Franks **3.**230, **3.**275
Pepin III (the Short), king of Franks **3.**61, *3.275*
pepper **2.**52, **2.**248, *2.249*
peppered moth **2.**225, *2.249*
Pepperell, William **3.**197
peppermint *2.249*
pepper-shrike *2.249*
peptic ulcer **2.**352
Pepys, Samuel **3.**127, **3.**247, *3.275–6*, *5.343*
Pequot War *3.276*
perception (philosophy) *7.232*
perception (psychology) *7.232–3*, **7.**234, **7.**279
Perceval, Spencer *4.266*
perch *2.249–50*
climbing **2.**68
darter **2.**87
sea **2.**301
trout **2.**348
perched rock **1.**249
perched water-table *1.250*
Percier, Charles **5.**146
Percival, A. E. **4.**314
percussion instruments *5.343*
bells **5.**43
castanets **5.**82
clappers **5.**102
cymbals **5.**120
dulcimers **5.**138, **5.**402, **5.**497
gamelan **5.**169
glass harmonica **5.**179
glockenspiel **5.**180–1
gong **5.**184
güiro **5.**343
North Africa **5.**321
rattles **5.**282, **5.**375, **5.**420–1
sansa **5.**400
sanṭūr **5.**402
steel band **5.**431
tabor **5.**349–50, **5.**444
xylophone **5.**8, **5.**343, **5.**496
yangqin **5.**497
see also bells; drums; plucked instruments
Percy, Henry de, 1st earl of Northumberland *3.276*
Percy, Sir Henry 'Hotspur' **3.**265, *3.276*
Perdiccas I, king of Macedonia **3.**215
peregrine **2.**250
perennial plants *2.250*
acanthus **2.**2
antirrhinum **2.**12
aquilegia **2.**13
arrowroot **2.**15
asphodel **2.**17

perennial plants (*cont.*):
aubergine **2**.18
autumn crocus **2**.19
banana **2**.21
broom **2**.43
broomrape **2**.43
bryony **2**.43
bugle **2**.45
bulbs *see* bulbs
campion **2**.51
cardamom **2**.53
chilli **2**.63
chrysanthemum **2**.65
clover **2**.69
comfrey **2**.73
coneflower **2**.74
corms *see* corms
cranesbill **2**.80
crocus **2**.19, **2**.81–2, **2**.291–2
daffodil **2**.86, **2**.316
dahlia **2**.86
delphinium **2**.90
dock **2**.96
duckweed **2**.100
fennel **2**.115
fleabane **2**.121
forget-me-not **2**.125
foxglove family **2**.29, **2**.50, **2**.126, **2**.319, **2**.369
galingale **2**.130
garlic **2**.133
geranium **2**.134–5, **2**.247
geum **2**.135
ginger **2**.136
gladiolus **2**.136
gourd **2**.142
grape hyacinth **2**.143
ground elder **2**.145
gunnera **2**.148
hellebore **2**.154
hemlock **2**.154
hollyhock **2**.159
horseradish **2**.163
iris **2**.119, **2**.136, **2**.173, **2**.218
knotweed **2**.180
lily *see* lily family
liquorice **2**.191
lobelia **2**.192
lupin **2**.196
mallow **2**.200
mandrake **2**.201
meadowsweet **2**.206
milkweed **2**.211–12
montbretia **2**.218
narcissus **2**.225
nemesia **2**.226
Nicotiana **2**.227–8
onions **2**.186, **2**.234, **2**.306
pasque flower **2**.243
pelargonium **2**.247
pellitory **2**.247
pentstemon **2**.248
peony **2**.249
pepper **2**.249
phlox **2**.252
pink **2**.255
plantains **2**.21, **2**.258
potato **2**.266
primrose *see* primrose family
pyrethrum **2**.272
rhizomes *see* rhizome
rhubarb **2**.284
rock rose **2**.287
rush **2**.290–1
safflower **2**.291
saffron **2**.291–2
sainfoin **2**.292
St John's wort **2**.292
salvia **2**.292–3
scabious **2**.295
scorzonera **2**.297
sedges **2**.130, **2**.302
sensitive plant **2**.303
snowdrop **2**.316
soapwort **2**.316
Solomon's seal **2**.318

perennial plants (*cont.*):
sorrel **2**.317
spiderwort **2**.320–1
spiraea **2**.322
spurges **2**.56, **2**.183, **2**.289–90, **2**.323
sunflower *see* sunflower family
taro **2**.335
thistle **2**.339
tobacco **2**.342–3
toothwort **2**.343–4
trees *see* forests and trees
valerian **2**.354
violet **2**.356
wallflower **2**.359
watercress **2**.361
water-lily **2**.361–2
willowherb **2**.368
woad **2**.369
yucca **2**.374
see also deciduous; evergreen; fruit plants
perestroika ***7**.132*
Peretz, Y. L. **5**.497
Pérez, Carlos Andrés **4**.363
Perez, Juan **3**.254
Pérez de Cuéllar, Javier **4**.111, **4**.166, ***4**.266*, **7**.322
Pérez Galdós, Benito ***5**.343*
perfect competition ***7**.233*
Performance Art **5**.198
performing arts *see* ballet; cinema; dance; drama; music; opera; set design; television
performing rights ***7**.233*
perfume ***6**.260*
see also aromatic
Pergamum ***3**.276*
Pergolesi, Giovanni Battista ***5**.343–4*
peri ***9**.29*
Peri, Jacopo **5**.328
periapse **8**.7
periastron ***8**.118*
Pericles **3**.2, **3**.26, **3**.101, **3**.275, ***3**.276*, **9**.6
peridot ***1**.250*
peridotite **1**.183, ***1**.251*
perigalacticon ***8**.118*
perigee ***8**.118*
Perignon, Dom **3**.68
perihelion ***8**.118*
argument of **8**.114
period (astronomy) ***8**.118*
period-luminosity law ***8**.118*
synodic **8**.168
period (chemistry) ***1**.254*
period (geology) ***1**.254*
see also under Cainozoic; Mesozoic; Palaeozoic
period (menstruation) **2**.208
periodic groups ***1**.251*
periodic law ***1**.251*
periodic table ***1**.251–3*, **6**.24, **6**.25
peripheral, computer **6**.180, ***6**.260*
see also terminal *and* computer *under* printing
periscope ***6**.260*
peritonitis ***2**.250*
periwinkle (animal) **2**.368
periwinkle (plant) ***2**.250*
perjury ***7**.233*
Perkin, Sir William Henry ***6**.260*
Perls, Fritz **7**.131
permafrost ***1**.254*
permalloy **6**.11, **6**.241
permeability of rock ***1**.254*
Permian period **1**.133, **1**.134, ***1**.254*
Permo-Trias period ***1**.255*
Perón, Eva ***4**.266*
Perón, Juan Domingo **4**.18, ***4**.266–7*
Péron, María Estela Martínez de **4**.267
Perotinus **5**.289
peroxide ***1**.255*
Perpendicular style ***5**.344*, **5**.461
perpetual motion ***6**.260*

Perrault, Charles **5**.91, **5**.371, **9**.9
Perrault, Claude **6**.51
Perret, Auguste **5**.260
Perrin, Jean-Baptiste ***1**.255*
Perronet, Jean-Rodolphe ***6**.260–1*
Perrotin, Henri **8**.19
Perry, Matthew Calbraith: and Perry Treaty **4**.172, **4**.178, ***4**.267*
Perry, Oliver Hazard **4**.190
Perseids ***8**.118*, **8**.194, **8**.195
Persephone ***9**.29*
Persepolis **3**.97, ***3**.276*
Perseus (Greek hero) **9**.4, **9**.6, **9**.14, ***9**.29*
Perseus (constellation) **8**.118
Perseus, king of Macedonia (168 AD) **3**.175
Perseus, king of Macedonia (from 179 AD) **3**.215
Perseus arm (spiral arm of Galaxy) ***8**.118*
Perseus Cluster (Perseus A) ***8**.118*
Pershing, John J. ***4**.267*, **4**.368, **4**.376
Persia *see* Iran
Persian Gulf *see* Gulf
persimmon ***2**.250*
person perception ***7**.234*
personal computer **6**.222, **6**.223
personal identity ***7**.233*
personality ***7**.233*
aggression **7**.7
conditioning **7**.64, **7**.231
disorder ***7**.233*
psychopathic personality **7**.14, **7**.252–3
see also psychoanalysis; psychoses
extroversion/introversion **7**.114, **7**.173
individuation **7**.173
locus of control **7**.187
psychosexual development **7**.253
social learning **7**.233
see also behaviour; emotion; motivation; self
personnel management ***7**.234*
perspective **5**.5, **5**.16, **5**.64, **5**.284, ***5**.345*, **5**.465
Perspex **6**.276
perspiration ***2**.250*
persuasion ***7**.234*
see also advertising; marketing; propaganda; public relations
perturbations in orbits **8**.22
Peru ***1**.255*, ***3**.277*, ***4**.267*, ***7**.375–6*
agriculture **6**.7, **6**.267
Andean Group **7**.13
arts **3**.71, **5**.467–8
city **3**.94
diplomat *see* Pérez de Cuéllar
diseases and disorders **7**.59
economy and trade **4**.12, **4**.320
education **7**.321
ethnic groups and prehistory **7**.375
Chavín culture **3**.71
Indian revolt **3**.363
Mochica culture **3**.234
Nazca culture **3**.247
see also Incas; Indians, South American
exploration **1**.88
fishing **6**.133
myths and legends *see under* Incas
national park **7**.212
physical geography and geology **1**.112, **1**.255, **1**.338, **1**.339
mountains *see* Andes
politics and government **4**.33, **4**.43, **4**.65, **4**.147–8, **4**.192, **4**.267
population **4**.320
prehistory *see* ethnic groups *above*
state **3**.73
wars and defence **4**.193–4, **4**.254, **4**.336
Peru Current ***1**.255*
Perugino, Pietro ***5**.345*

Peruzzi, Baldassare ***5**.345*, **5**.412
Pessoa, Fernando ***5**.345*
pest *see* pests
pesticides and pest control **6**.7–8, **6**.147, ***6**.261*
biological **2**.30, **2**.250, **6**.37, **6**.267
fungicide **6**.147, **6**.311
green revolution **6**.161
herbicides **6**.7–8, **6**.168, **6**.178, **6**.271
pollution **6**.9, **6**.377, **7**.240
see also insecticides; pests
pestle and mortar ***6**.261*
pests ***2**.250*
animals as:
flying fox **2**.123
from fur farms **2**.213, **2**.223
housemouse **2**.164
marmot **2**.203
toad **2**.342
see also parasites
insects as ***2**.172*
and animals:
bot fly **2**.39
ked **2**.194
stable fly **2**.324
tsetse fly **2**.348, **2**.349
see also lice
biological control **2**.30, **2**.250
and humans:
bedbug **2**.26
clothes moth **2**.68–9
cockroach **2**.70–1
firebrat **2**.119
midges **2**.131, **2**.210
silverfish **2**.119, **2**.310
see also mosquitoes; wasps
and plants:
aphid **2**.13
cabbage white **2**.49
caterpillar **2**.198
chafer **2**.60
clearwing moth **2**.67
click-beetles **2**.67–8, **2**.83
Colorado beetle **2**.72
eelworm **2**.104
locust **2**.192–3
mirid bug **2**.213
scale insect **2**.30
shieldbug **2**.307
weevil **2**.364
whitefly **2**.30, **2**.366–7
plants as *see* weeds
see also pathogens; pesticides; vectors
pet (animal) *see* pets
PET (polyethene terephthalate) **6**.271, **6**.273
PET (positron emission tomography) **6**.243, **6**.259, ***6**.277–8*, **6**.357
Pétain, Henri-Philippe ***4**.268*, **4**.364
pétanque (boule) ***7**.234*, **7**.287
Peter, Saint **3**.104, **3**.268–9, ***3**.277*, **7**.241, **7**.269, **9**.5, ***9**.43*
Peter I (the Great), tsar of Russia ***3**.277*, **3**.309, **3**.312
and boyars **3**.48
and Northern War **3**.246, **3**.256, **3**.260, **3**.287
and science **6**.1
Peter I Karageorgević, king of Serbia **4**.178, **4**.387
Peter II, king of Yugoslavia **4**.220
Peter III, tsar of Russia **3**.64, **3**.324
Peter the Hermit ***3**.277–8*
Peterkiewicz, Jerzy ***5**.345*
Peterloo massacre ***4**.268*
Peters, Karl **4**.135
Peters, Lenrie **5**.7
Peters map projection **6**.284
Peter's pence ***3**.277*
Petilius Cerealis **3**.50
Petipa, Marius ***5**.345*
Petit, Roland ***5**.345*
Petition of Right ***3**.278*

Petöfi, Sándor ***5**.345*
Petra ***3**.278*
Petrarch **3**.301, **5**.141, ***5**.345–6*
petrels ***2**.250*
mutton bird **2**.224
shearwater **2**.305, **2**.306
storm **2**.327
Petri, J.: Petri dish ***6**.261*
petrified wood ***1**.255*
petrochemicals ***6**.261–4*
see also refining and petrochemicals *under* petroleum
petrol ***6**.262*, **6**.264
engines **6**.233–4, ***6**.262–3*
pioneers **6**.34, **6**.254
see also under motor car
see also internal-combustion engines; knock; lead tetraethyl; motor vehicles
octane rating **6**.224, **6**.249, **6**.296
see also petroleum
petroleum ***1**.255*, ***6**.263–5*
'capping' fire **6**.132
consumption **6**.121
countries producing *see* OPEC
exploration and recovery **6**.107–9, ***6**.263–4*
for heating **6**.167
oil slick treatment **6**.27, **6**.37
pipeline **6**.265, **6**.269, **6**.270
price rises **7**.57
refining and petrochemicals ***6**.261–5*
bitumen **6**.263
butadiene **6**.261, **6**.303–4
butene **6**.261
catalytic cracking **6**.91, **6**.171, **6**.175
cracking **6**.265
crystallization **6**.265
detergents **6**.100
diesel oil (gas oil) **6**.91, **6**.102–3, **6**.145, **6**.249, **6**.264
ethene *see* ethylene
fuel-oil **6**.147, **6**.265
gasoline *see* petrol
hydrotreatment **6**.175, **6**.265
lubricating oil **6**.32, **6**.145, **6**.204, **6**.210, **6**.264
phenol-formaldehyde **6**.27, ***6**.267*, **6**.271, **6**.273
propene **6**.261, **6**.276
reforming **6**.296–7
sodium carbonate **6**.323
solvent extraction **6**.264
sulphuric acid **6**.338–9
toluene **1**.339, **6**.261, **6**.262, **6**.357
vacuum distillation **6**.210, **6**.264, **6**.370
xylene **6**.262
see also in particular ammonia; aromatic compounds; aviation fuel; butane; fractional distillation; hydrocarbons; lead tetraethyl; liquid petroleum gas; naphtha; paraffin; petrol; plastics; polymers; sulphur; synthetic fibres
resources **6**.121
shale oil **6**.263, **6**.315
storage **6**.265
tanker **6**.345
see also oil, mineral; petrol
petrology ***1**.255*, **1**.256
see also geology
Petronius **5**.389
pets:
birds *see* cage-birds
cat **2**.56
dog **2**.96
fish *see* aquarium and ornamental fish
guinea pig **2**.147–8
hamster **2**.150
mouse **2**.164
pets (*cont.*):
see also domesticated animals
petty sessions ***3**.278*
petunia ***2**.250*
Pevsner, Antoine **5**.346
Pevsner, Naum ***5**.168*
pewter/pewterware ***5**.346*, ***6**.265*
peyote ***2**.250*
Pfeifer, August **6**.192
pH **1**.161, **6**.2
meter ***6**.267*
Phaethon ***9**.29*
phaeton **6**.61
phalange **2**.124
phalanger **2**.124, ***2**.251*
phalanx ***3**.278*
phalarope ***2**.251*
phanerogam ***2**.251*
see also angiosperms; gymnosperms
Phanerozoic Age **1**.133, **1**.134, ***1**.255*
Cambrian period **1**.49, **1**.133, **1**.134
Pharaohs of Egypt ***3**.112*, ***3**.278*
Pyramids/Valley of Kings **3**.296, **3**.369
queens **3**.3, **3**.78, **3**.79, **3**.159, **3**.249
Seti I **3**.2, **3**.243
see also Akhenaten; Ramesses; Thutmose; Tutankhamun
Pharisee ***3**.278*
pharmaceutics and pharmacology *see* drugs, therapeutic
Pharnaces I, king of Pontus **3**.288
Pharsalus, battle of ***3**.278*
phase (astronomy) ***8**.118–19*
phases of matter ***1**.256*
pheasant family ***2**.251*
junglefowl **2**.176
partridge **2**.243
peacock **2**.245
quail **2**.48, **2**.273
turkey **2**.351
Pheidippedes **3**.221
Pheidon of Argos **3**.21
phenocryst ***1**.256*
phenol (carbolic acid) ***1**.256*, **6**.16, **6**.204, ***6**.266–7*
phenomenology **7**.68, **7**.199, ***7**.234*
phenotype **2**.223–4, ***2**.252*
phenylalanine **6**.342
pheromone ***2**.252*, **6**.170, ***6**.267*
Phidias **3**.322, ***5**.346*
philately **7**.298
Philemon and Baucis ***9**.29*
Philip (Apostle) **9**.5
Philip I (the Handsome), king of Castile (1478–1506) **3**.152, **3**.153, ***3**.278*
Philip I, king of Portugal *see* Philip II, king of Spain
Philip II (Augustus), king of France (1165–1223) **3**.13, **3**.60, ***3**.278–9*
Philip II, king of Macedonia (382–336 BC) ***3**.278*
wars with Greece **3**.27, **3**.67, **3**.101, **3**.354
wars with Rome **3**.215, **3**.356
Philip II, king of Portugal *see* Philip III, king of Spain
Philip II, king of Spain, and Philip I of Portugal (1527–98) **3**.152, **3**.153, ***3**.279*, **3**.335
and arts **5**.457
and Counter-Reformation **3**.89
and England **3**.114, **3**.386
and Frisians **3**.135
and Inquisition **3**.336
and Netherlands **3**.49, **3**.108, **3**.337
and Portugal **3**.289
Philip III, king of Portugal *see* Philip IV, king of Spain
Philip III, king of Spain, Naples, and Sicily and Philip II, king of Portugal (1578–1621) **3**.153, **3**.279
Philip IV (the Fair), king of France (1268–1314) **3**.44, **3**.60, **3**.68, **3**.89, **3**.198, ***3**.279*
Philip IV, king of Spain, Naples, and Sicily and Philip III of Portugal (1605–65) **3**.128, **3**.153
Philip V, king of Macedonia (3rd century) **3**.90, **3**.215
Philip V, king of Spain (1683–1746) **3**.29, **3**.46, **3**.47, ***3**.279*, **3**.369
Bourbon and Habsburg ancestors of **3**.338
Philip VI, king of France (1293–1350) **3**.90, **3**.171–2, ***3**.279*, **3**.369
Philip the Bold (1342–1404), duke of Burgundy **3**.54, ***3**.280*
Philip the Good (1396–1467), duke of Burgundy **3**.154, ***3**.280*
Philip of Hesse, prince (1504–67) **3**.274, ***3**.279*, **3**.318
Philip Johnson, Kahn and Jacobs **5**.294
Philip Mountbatten, duke of Edinburgh **4**.110–11
Philip of Swabia, king of Romans (1198–1208) **3**.168
Philippa of Lancaster **3**.138
Philippi, battle of ***3**.279*
Philippines ***1**.256*, ***3**.279–80*, ***4**.268–9*, ***7**.376*
agriculture **6**.161, **6**.351
decolonization **4**.321
economy and trade **7**.20, **7**.200
emigration **7**.200
environment **7**.240
Hukbalahap movement **4**.158
industrial relations **7**.54
languages **7**.23
laws/legal system **7**.40
old age **7**.221
physical geography and geology **1**.256
politics and government **4**.5, **4**.211, **4**.280, **4**.291, **7**.72
sport and recreation **7**.295
urban areas **7**.197
wars and defence **4**.194, **4**.209
Philippines Campaign ***4**.269*
Philistines ***3**.280*, **9**.14
Phillip, Captain Arthur **4**.46
Phillips, J. **2**.21
Phillips, T. **6**.346
Philo of Byzantium **6**.168
Philoctetes ***9**.29*
philology **7**.144
philosophical logic ***7**.234*
see also logic
philosophy ***7**.234–5*
aesthetics **7**.4
analytic **7**.12, **7**.68, **7**.173, **7**.185
atomism **7**.21
authority *see* authority
behaviourism **7**.29, **7**.203
British *see under* Britain
Buddhist **7**.155, **7**.211
causation **7**.42
China ***7**.46*, ***7**.64–5*, ***7**.78–80*
continental *see* continental philosophy
deduction **7**.82
dialectic **7**.89, **7**.141, **7**.294
dualism **7**.96–7, **7**.202–3, **7**.233, **7**.301
emotivism **7**.105, **7**.109
empirical *see* empiricism
Enlightenment **5**.147
environmentalism **7**.107
ethical *see* ethics
existentialist *see* existentialism
free will **7**.109, **7**.125, **7**.199
philosophy (*cont.*):
humanism **7**.149
idealism **7**.151, **7**.175, **7**.324
India ***7**.155*, ***7**.156*
instrumentalism **7**.88
intentionality **7**.35, **7**.162
Islam **3**.30, **7**.10, **7**.23–4
Judaism **3**.217, **3**.339, **7**.189, **7**.297, **7**.310
of knowledge *see* epistemology
of law **7**.174
linguistic **7**.47, **7**.185
logic *see* logic
mathematical *see under* mathematics
meaning *see* meaning
metaphysical *see* metaphysics
of mind *see under* mind
monism *see* monism
moral **7**.108–9
see also ethics; morality
nativism *see* nativism
naturalism **7**.213
nominalism **7**.1, **7**.218, **7**.221
objects and objectives **7**.197
Ostrogoth **7**.33
perception **7**.232
phenomenology **7**.68, **7**.199, **7**.234
post-structuralism **7**.68
pragmatism **7**.88–9, **7**.244
a priori **7**.15, **7**.178, **7**.259, **7**.312
private language **7**.247
rationalism **7**.86, **7**.259
realism **7**.259
reason **7**.259–60
reductionism **7**.223, **7**.260
scepticism **7**.275
scholasticism *see* scholasticism
of science **7**.188, **7**.241, **7**.276
social contract **7**.221, **7**.259, **7**.289–90
solipsism **7**.295
sophists **7**.295
soul *see* soul
space **7**.296
stoicism **7**.300
substance **7**.199, **7**.301
Taiji **7**.305
teleology **7**.307
thought experiment **7**.311
time **7**.296, **7**.312
Transcendentalism **5**.145, **5**.459, **7**.315
truth **7**.306, **7**.316
universals **7**.1, **7**.181, **7**.221, **7**.322
utilitarianism **7**.109, **7**.202, **7**.259, **7**.324
Utopianism **7**.324
value judgement **7**.325
verificationism **7**.187, **7**.325
see also under individual countries
Phipps, Sir William **3**.59, **3**.197, **3**.314
phloem ***2**.252*
phlox ***2**.252*
phobias **7**.14, **7**.30, **7**.214, ***7**.235*
Phobos ***8**.119*
Phoebe ***8**.119*
Phoenicians ***3**.280*
alphabet **7**.10
art ***5**.346–7*
religion and myths **9**.5, **9**.6, **9**.10, **9**.11, **9**.20, **9**.35
technology **6**.237
Tyre **3**.364
Phoenicids ***8**.119*, **8**.195
Phoenix (constellation) ***8**.119*
phoenix (mythology) ***9**.29*
Phoenix Park murders ***4**.269*
Phomvihane, Kaysone **4**.190
phonemes **7**.235, **7**.275
phonetics **7**.11, ***7**.235*, **7**.259
phonograph *see* gramophone
phonology ***7**.235*
phosgene **6**.72, ***6**.267*
phosphates ***1**.256*, ***6**.267*
calcium (apatite) **1**.16

phosphates (*cont.*):
copper aluminium (turquoise) **1**.348
in detergents **6**.376
in fertilizers **6**.2, **6**.128, **6**.338
phospholipids **2**.191
phosphor **6**.138, ***6**.267*, **6**.268, **6**.295
in television **6**.64, **6**.81, **6**.295, **6**.349–50
phosphorescence ***1**.256*, **1**.257, **6**.267
phosphoric acid **6**.2, **6**.71, ***6**.267*
phosphorus **1**.107, ***1**.256*, ***6**.267*
alloys **6**.45, **6**.49
and fertilizer **6**.127–8
on matches **6**.132
poisoning **6**.248
in semiconductor **6**.149, **6**.312
Phosphorus (Venus) ***8**.119*
Photius ***3**.280*
photocell ***6**.267*, **6**.322
photochemical reaction ***1**.256–7*
photochromic lens ***6**.267*
photoconductive cell **6**.267
photocopier **6**.60, **6**.101, ***6**.267–8*
photodiode **6**.267
photoelectric cell **6**.267, **6**.322
photoelectric effects ***1**.257*
photoengraving **6**.283
photography ***5**.347*, ***6**.268*, ***8**.119–20*
aerial **6**.215, **6**.268
and astronomy **8**.11, **8**.98, ***8**.119–20*
astrograph **8**.9
photographic zenith tube (PZT) ***8**.119*
Britain:
atmospheric and abstract **5**.105–6, **5**.350
documentary **5**.58, **5**.347
experimental **5**.445
landscape **5**.441
naturalistic **5**.145
Pop art **5**.208
portraiture **5**.40, **5**.73, **5**.81, **5**.206, **5**.377, **5**.384
Scotland **5**.206
technology **6**.73, **6**.74, **6**.268, **6**.344
see also under cinema
camera *see* camera
Canada **5**.244
chromotography **6**.74
cinema *see* cinema; cinematography
colour **6**.75, **6**.81, **6**.109, **6**.130, **6**.210, **6**.268
daguerrotype **6**.94
developing **6**.100, **6**.101, **6**.130, **6**.268
emulsion **6**.111, **6**.129–30, **6**.268
experimental and innovative **5**.103, **5**.121, **5**.312, **5**.375, **5**.445
exposure and exposure meter **6**.16, **6**.25, **6**.124, **6**.318
fashion **5**.19, **5**.308
fax **6**.126, **6**.278, **6**.347
film *see* film, photographic
flash **6**.134
France:
architecture **5**.25
experimental and pioneer **5**.103, **5**.121, **5**.312, **5**.375
photo-journalist **5**.81, **5**.347
pictorial **5**.128
see also under cinema
Germany **5**.347
half-tone process **6**.164
history of ***5**.347*
holography **5**.347
Hungary **5**.58, **5**.76, **5**.246, **5**.301, **5**.308
infra-red **6**.268, **6**.375
landscape **5**.1, **5**.5, **5**.431, **5**.441
lens **6**.199, **6**.268
Mexico **5**.59
microfiche **6**.222
microphotography **6**.268

photography (*cont.*):
motor-drive **6**.235
paper **6**.27, **6**.256
photo-journalists **5**.76, **5**.81, **5**.207
pioneers **5**.347, **6**.210, **6**.241, **6**.342, **6**.344
portraiture **5**.1, **5**.27, **5**.40, **5**.73, **5**.81, **5**.206, **5**.244, **5**.347, **5**.377, **5**.431
printing **6**.101, **6**.268, **6**.269
projector **6**.75, **6**.284
satellite **6**.310
as sign **7**.285
stereophotography **6**.215, **6**.334, **6**.340
thermography **6**.354, **6**.355
tripod **6**.363
United States **5**.347
animals **5**.311
art **5**.432–3, **5**.435
cityscapes **5**.1
experimental **5**.375
fashion **5**.19, **5**.431
Group f.64 **5**.5, ***5**.195*, **5**.488
landscapes **5**.5, **5**.431
portraiture **5**.1, **5**.27, **5**.431
technology **6**.75, **6**.111, **6**.268, **6**.274
war **5**.431
see also under cinema
war **5**.76, **5**.431, **5**.482
warning and detection system **6**.375
see also cinema *and under individual photographers*
photogravure **6**.123, **6**.160, **6**.282
photometer and photometry ***8**.120–1*
colour index **8**.27
spectrophotometry **6**.40, **6**.328, **8**.157
photomicrography **6**.268
photomultiplier ***6**.269*, **6**.391
photon ***1**.257*, ***8**.121*
see also light
photopolymer ***6**.269*, **6**.272
photosphere ***8**.121*
photosynthesis **2**.53, ***2**.252*
chlorophyll **2**.64
and oxygen **2**.238
plants without *see* fungi
sap **2**.294
see also cyanobacteria; leaf
phototypesetting **6**.366
photovoltaic cell **6**.267
Phraortus of Medes **3**.227
Phrygia **3**.117, ***3**.280*
Phyfe, Duncan **5**.153
phyllite ***1**.257*
phylloquinone **2**.357
phylum ***2**.252*, **2**.298, **2**.299
acanthocephalan **2**.322
brachiopods **2**.182
bryozoan **2**.219
cephalochordate **2**.182
chaetognathan **2**.15
ctenophoran **2**.298, **2**.299
hemichordate **2**.2, **2**.154
nematodes *see* roundworms
poriferan **2**.322
rotifer **2**.288, **2**.289
sipunculid **2**.310
urochordate **2**.301–2
see also annelids; arthropods; cnidarian; echinoderms; molluscs; platyhelminth; vertebrates
physical geography *see* geology; geomorphology; maps and charts; *and also* physical geography *under individual countries*
physical vapour deposition **6**.370
physicians, major ***7**.196*
Boerhaave **2**.37–8
Boylston **2**.40
Fracastero **2**.126–7

physicians (*cont.*):
Galen **1**.332, **2**.130, **2**.190, **6**.350
Hippocrates **2**.158
Linnaeus **2**.2, **2**.190–1, **2**.229
Paracelsus **1**.7, **2**.241, **6**.257
Ramazzini **2**.276
Sloane **2**.313
Sydenham **2**.332–3
Weil **2**.364
Withering **2**.369, **6**.387
Woodville **2**.371
see also physiologists *under* physiology *and* surgeons *under* surgery
physics ***1**.257*
action and reaction **1**.2
air resistance **1**.5–6
amplitude **1**.12
angular momentum **1**.14
angular velocity **1**.14
antinode **1**.16
atomic *see* atomic physics
biophysics **2**.30
capillary action **1**.51
catenary **1**.54
centrifugal force **1**.56–7
centripetal force **1**.57
colour used metaphorically **1**.69
complementarity **1**.70, **1**.349, **1**.363
conservation laws **1**.73
corona *see* corona
cosmic rays **1**.77–8, **8**.31–2, **8**.98–9
diffraction *see* diffraction
diffusion **1**.91
elastic collision **1**.101
elasticity **1**.101, **1**.157, **6**.113, **6**.330
electricity *see* electricity
electromagnetism *see* electromagnetic radiation
electrons *see* electrons
elementary particles *see* elementary particles
energy *see* energy
equilibrium **1**.111, **8**.47
excitation **1**.116
exclusion principle **1**.116
expansion **1**.117
fermion *see* fermions
field *see* field
Fitzgerald–Lorentz contraction **1**.122–3, **8**.86
flotation **1**.124
Fourier analysis **1**.125
free oscillation **1**.127
frequency *see* frequency
friction **1**.5–6, **1**.127, **6**.146, **8**.171
geophysics **8**.58
gluons **1**.69, **1**.141
grand unified theory **1**.143
graviton **1**.144
gravity *see* gravitation/gravity
heat *see* heat
horizon **1**.157–8
ideal gas **1**.132, **1**.165
inertia *see* inertia
interactions **1**.170
inverse square law **1**.171
light *see* light
magic numbers **1**.202
mass defect **1**.206
moment **1**.218
moment of inertia **1**.218
motion *see* motion
neutron *see* neutrons
nuclear *see* nuclear physics
optical *see* optics/optical
orbitals **1**.239, **1**.240, **8**.131
parameter **1**.248
particle *see* elementary particles
physicists *see under individual countries*
pion **1**.222–3, **1**.258
plasma **1**.258, **1**.259
power **1**.265–6

physics (*cont.*):
quantum *see* quantum mechanics and quantum theory
radiation *see* electromagnetic radiation; radiation
relativity *see* relativity
resonance **1**.281, **1**.356
root-mean-square ***1**.286*
spectrum *see* spectra
strangeness **1**.322
string theory **1**.323–4
strong force **1**.324
temperature *see* heat; temperature
tunnelling **1**.346
uncertainty principle **1**.70, **1**.349
wavelengths *see* wavelengths
waves *see* waves (physics)
work **1**.368
physiocrats **7**.100
physiology of living organisms ***9**.58–61*
abdomen **2**.1
arms **2**.14, **2**.29
bladder **2**.34
blood *see* blood
body fluid **2**.37
bones *see* bone
brain *see* brain
bursa **2**.46
cartilage **2**.55–6
coelom **2**.72
diaphragm **2**.1, **2**.195
digestion *see* gastro-intestinal tract
diseases *see* diseases and disorders
ear *see* ears; hearing
fingers *see* fingers
fistula **2**.119
foot **2**.124
gall bladder **2**.130
glands *see* glands
hand *see* hands
head **2**.153
heart *see* heart
joints *see* joints
kidney *see* kidney
legs **2**.124, **2**.186
liver *see* liver
lungs **2**.195, **2**.196
lymphoid tissue **2**.4
muscles **2**.29, **2**.222, **2**.252–3, **9**.59
nails **2**.225
nervous system *see* nervous system
nose **2**.229
pancreas **2**.240
physiologists and anatomists, major:
Banting **2**.22, **6**.29
Beaumont **2**.26
Bell **2**.28
Bernard **2**.28
Bichat **2**.29
Fabricius **2**.113
Falloppio **2**.113
Graaf **2**.142
Hales **2**.150
Hodgkin **2**.158
Leidy **2**.186
Malpighi **2**.200
Milstein **2**.212
Vesalius **2**.355, **3**.301
see also physicians
respiration *see* respiration and respiratory tract
senses *see* senses
skin *see* skin
spleen **2**.322
sweat gland **2**.250
thorax **2**.339
tongue **2**.343
urinary tract **2**.34, **2**.326
vestigial organ **2**.355
wastes *see* body *under* wastes
see also cells of living organisms; medicine and health care
physiotherapy ***2**.252–3*, **6**.367
phytogeography **2**.30

pi *1.257–8*
Piaget, Jean 7.32, 7.54, *7.235*, 7.246, 7.252, 7.273, 7.294
Piano, Renzo 5.386
pianoforte and pianists *5.347–8*
  Australia 5.188
  Austria 5.214
  Britain 5.60, 5.104, 5.154, 5.225, 5.393
  chamber music 5.88
  France 5.87, 5.153, 5.398
  Germany 5.57, 5.68, 5.291
  Hungary 5.36–7, 5.266
  Italy 5.367, 5.373, 5.393–4, 5.409
  piano mechanism 6.200
  Poland 5.99, 5.335
  Russia and USSR 5.367, 5.373, 5.393–4, 5.409
  sonata 5.424
  Spain 5.10, 5.82, 5.188
  street (mechanical) 5.288
  United States 5.46, 5.112, 5.175, 5.188, 5.212, 5.242, 5.391
pianola 5.288
Piazzi, Giuseppe 8.9, 8.24
pibcorn 5.212
Pibul Songgram *4.269*, 4.341
picaresque novels *5.348*, 5.398, 5.423, 5.428
Picasso, Pablo 4.325, *5.348*, 5.403
piccolo *5.348*, 5.352
pickerel *2.253*
Pickering, Edward 8.100
Pickering, William 8.119
picketing 7.158, *7.235*
Pickett, General 4.136
pickling 6.141
picks (tools) 6.103, 6.227
Pico della Mirandola, Giovanni 5.231
Picot, François Georges 4.335
Picquart, Colonel 4.104
picric acid *see* TNT
pictograms 7.123, *7.235*, 7.334
Pictor *8.121*
Picts *3.281*
Picturesque *5.348–9*
  see also landscape
piddock *2.253*
pidgin and creole languages *7.235–6*
Pied Piper of Hamelin 3.73
Piedmont 4.88, *4.269*
piedmonts *1.258*
pier 6.337
Pierce, Franklin *4.269*
Piero della Francesca *5.349*
Piero di Cosimo *5.349*
Pierrot 5.282, *5.349*
pietà *5.349*
piezo-electric effect *1.258*, 6.1, 6.77, 6.116, 6.175, *6.269*, 6.383
pig *see* pigs
Pigalle, Jean-Baptiste 5.315
pigeon family 2.98, 2.243, *2.253*, 2.371
  pigeon-racing *7.236*
pig-iron 6.90, 6.186, 6.285
pigment 5.326, 5.341, 5.346, *5.349*, 5.411, 6.80, 6.109, 6.255, *6.270*, 6.384
  see also paint
pigmy moth *2.253*
Pigou, Arthur Cecil 7.101
pigs *2.253*
  boar 2.37
  bush 2.26
  farming *6.269–70*
    see also livestock
  peccary 2.247
  warthog 2.360
  wild boar 2.367
pika *2.253*
Pike, Zebulon Montgomery *1.258*
pike family 2.223, *2.253*
pike-perch *2.253*
Pilate, Pontius 3.188, *3.281*, 9.6, *9.29*
pilchard *2.253–4*
pile-driver *6.270*
Pilgrim Fathers 3.48, 3.62, *3.282*, 3.286
pilgrimage *3.281*, *7.236*, 9.24
  Buddhist 3.281
  Islam *see* Mecca
  palmers 3.268
  Pilgrimage of Grace 3.281–2
  Shinto 7.284
piling *6.270*
Pilkington, Alistair 6.136
pill, contraceptive 2.75
  see also drugs, therapeutic
pill beetle *2.254*
pill bug 2.370
Pillai, Takazi Sivasankara 5.136
Pillai, Vetanayakam 5.446
pillory and stocks *3.282*
Pillow Book of Sei Shōnagon *5.349*
pilot fish *2.254*
Pilsudski, Joseph Klemens *4.269*, 4.271
Piltdown Man *3.282*
pimento *2.254*
pimpernel *2.254*
pin oak 2.231
pincers *6.273*
pinchbeck *see* brass instruments
Pinchot, Gifford *2.254*
Pinckney, Charles Cotesworth *4.269*
Pindar 5.325, 5.335, *5.349*
Pindaris *4.269*
pine marten 2.204
pineal gland *2.254*
pineapples 2.42–3, *2.254–5*
Pinero, Sir Arthur Wing 5.73, *5.349*
pines *2.254*
  kauri 2.178
  loblolly (yellow) 2.192
  Scots 2.89, 2.297
  spruce 2.254, 2.323
  in succession 2.328
Pinget, Robert 5.323
pink *2.255*
Pinkerton, Allan *4.269*
pink-footed goose 2.254, *2.255*
Pinkie, battle of *3.282*
pinking *see* knock
Pinochet, Augusto 4.10, 4.73, *4.269*, 7.23
Pinsker, Leo 4.390
pintail *2.255*
Pinter, Harold *5.349*, 5.383
pinworm *2.255*
Pinyin 7.11, 7.286–7
piobaireachd music *5.349*
pion *1.258*
  muon 1.222–3
Pioneer (spacecraft) *8.121*
pipa *5.349*
pipe snake 2.85
pipe and tabor 5.138, *5.349–50*
pipefish *2.255*
pipeline *6.270*
Piper, John *5.350*
pipe-roll *3.282*
pipes *see* tubes and pipes
pipette *6.270*
pipistrelle *2.255*
pipits *see* wagtails and pipits
piracy (copyright) 7.70, 7.162
piracy (sea) *see* pirates
Piraeus *3.282*
Pirandello, Luigi *5.350*
Piranesi, Giovanni Battista *5.350*
piranha *2.255*
Pirani gauge 6.281
pirarucu 2.14
pirates *3.282*
  buccaneers 3.51
  corsairs 3.35, 3.88
  filibusters 4.119
  privateers 3.196–7, 3.293
Pisanello 5.288–9, *5.350*
Pisano, Andrea and Nino *5.350*
Pisano, Antonio *see* Pisanello
Pisano, Nicola and Giovanni *5.350*
Piscator, Erwin Friedrich Max *5.351*
Pisces *8.121*
Piscis Austrinus (or Australis) *8.121*
Pisistratus 3.27, *3.282*
Pissarro, Camille and Lucien *5.351*
pistachio *2.256*
pistol and revolver 6.81, 6.130, 6.131, *6.270*
Pistrucci, Benedetto 5.106
pit viper *2.257*, 2.278
Pitcairn, Major John 3.207
pitch (aircraft flight) 6.136
pitch (from coal-tar) 6.79
pitch (music) *5.351*, 5.352
pitchblende 1.351–2, 6.292
pitcher plant *2.256*
pitchstone 1.234
Pitman, Isaac 6.318
pit-props 6.79, *6.270*
Pitt, William (the Younger) *3.283*, 4.35, 4.79, 4.122, *4.270*
  and income tax 4.160
  and Ireland and Catholics 3.140, 4.2, 4.65
  and Reform Acts 4.285
  and taxes 3.94, 3.383
  and war 3.58, 3.134
  as Whig 3.359
Pitt, William, 1st earl of Chatham (the Elder) *3.283*, 3.324
pitta *2.256*
pituitary gland 2.208, *2.256–7*
Pius II, pope *3.283*
Pius V, St, pope 3.89, *3.283*
Pius VII, pope 3.187, 4.82, 4.231, *4.270*, 4.291
Pius IX, pope 4.131, 4.215, *4.270*, 4.289, 4.291, 6.1
Pius XI, pope 4.190, 4.363
Pius XII, pope *4.270*, 4.291
pivot *see* lever
pixels 8.22
Pizarro, Francisco 3.33–4, 3.44, 3.86, *3.283–4*, 5.426
  and Incas 3.26, 3.94, 3.109, 3.175–6, 3.277
Pizarro, Gonzalo 3.109, *3.284*
Pizarro, Hernando 3.283
Pizzaro, Francisco 1.88
pizzicato *5.351*
PKI (Indonesian Communist Party) *4.270*
Place, Francis 3.275, 4.70, 4.79, 4.158, *4.270*
placenta *2.257*, 2.352
placoderm *2.257*
plages 8.121
plagioclase *1.258*, 1.259
plague 3.42, 3.145–6, *3.284*, 6.191, 7.74
plaice 2.121, *2.257*
Plaid Cymru *4.270*
plains *1.258*
  abyssal 1.1, 1.236
  floodplain 1.123–4
  peneplain 1.250
  ring (on Moon) 8.140
  South America 1.142, 1.197, 1.246
  United States 1.145–6, 1.155
Plains Indians *7.236*
  Apache 3.17–18, 7.14, 7.131, 7.219
  Cherokee 3.73, 4.71, 5.322, 7.44, 7.219
  Cree 3.90, 7.74, 7.219
  ghost dance 7.131
  Iroquois 3.180, 7.166
  Sioux 3.96, 3.180, 7.219, 7.287
Plains of Abraham, battle of *3.284*
plainsong 5.332, *5.351*
plaintiff *7.236*
planaria 2.120
Planchon, Roger 5.238
Plancius, Petrus 8.19, 8.27, 8.29
Planck, Max Karl Ernst Ludwig 1.41, *1.258*
  Max Planck Society *6.217*
  and quantum mechanics and quantum theory 1.258, 1.271, 8.121, 8.131
    Planck's Law *8.121*
  and uncertainty principle 1.349
plane (aeroplane) *see* aircraft
plane (astronomy):
  invariable 8.74
  reference 8.29, 8.86
plane (inclined) 6.177–8
plane (tool) 6.211, 6.212, *6.270*, 6.357
plane figures *9.51*
plane tree *2.257–8*
planet *see* planets
planetarium *8.122*
planetary nebula *8.123*, 8.124
planetesimal *8.123*
planets *8.121–2*
  orbits 8.17, 8.196
  periods and motions 8.196
  physical data 8.196
  planetary configurations *8.122*
  planetary exploration *8.122–3*
  planetology *8.123*
  proto-planet 8.129
  rings 8.140, 8.197
  symbols of *9.62*
  see also Earth; giant planets; inner planets; outer planets
planisphere *8.123*
plankton *1.258*, 2.15, 2.257, *2.258*
planned economy *7.236–7*
Plantagenets *3.255*, *3.284*, *9.67*
  see also under rulers *under* Britain
plantains 2.21, *2.258*
  plantain-eater 2.350
plantation *3.284–5*, *6.270–1*
  beverage crops 6.7, 6.34, 6.271, 6.364
  intercropping 6.183, 6.364
  propagation 6.271
  rubber 6.7, 6.271, 6.303
  sugar 6.7, 6.271
  tobacco 6.357
  workers *7.237*
plantcutter *2.258*
Plante, Gaston 6.1
plants 2.6, 2.7, 2.52, *2.258*
  agricultural *see* crops
  algae *see* algae
  alpine *see* alpine plants
  angiosperms *see* angiosperms
  annual *see* annual plants
  biogeography 2.30
  botany *see* botany
  breeding 6.36, 6.155, *6.271*
  bryophytes *see* bryophytes
  buttercup *see* buttercup
  carbon cycle 2.53
  carnivorous *see* carnivorous plants
  cells 2.59, 6.65
  cellulose 2.59
  cereals *see* cereals and grains
  chlorophyll 2.64
  classification *2.258*
  cloning 6.77
  commensalism 2.73
  coniferous *see* conifers
  conservation 2.75
  Convention on endangered species 7.106
  cotyledon 2.78
  cultivars 6.93
  cuticle 2.84
  deciduous *see* deciduous trees and shrubs
  defence *2.258*
  Dicotyledons *see* Dicotyledons
  diseases 6.147

plants (*cont.*):
drugs from:
alkaloid 2.6
cannabis 2.52
chamomile 2.60
cinchona (quinine) 2.667
comfrey 2.73
Culpepper's work 2.83
digitalis 2.369
ergot 2.110
fleabane 2.121
fumitory 2.129
fungi 2.129
galingale 2.130
geum 2.135
heroin 2.235
hyssop 2.168
lignum vitae 2.190
male fern 2.199
mallow 2.200
mandrake 2.201
medical botanist 2.371
pokeweed 2.261
quassia 2.274
St John's wort 2.292
sarsaparilla 2.294
sassafras 2.294
sea holly 2.299
tamarind 2.334
tormentil 2.344.
valerian 2.354
wintergreen 2.368
yam 2.373
see also opium
dye *see under* dyes and dyeing
ecology *see* ecology
endangered species 2.107, 2.108, 7.106
essential oil from 6.123
evergreen *see* evergreen plants
evolution *see* evolution
fern *see* ferns
fibre *see* plants *under* fibre
flavourings from *see* spices, herbs, and flavourings
forest *see* forests and trees
fossil *see under* fossils
fungus *see* fungi
genetic engineering 6.153, 6.155
grafting 6.147, 6.158
grains *see* cereals and grains
greenhouse 6.161, 6.170, 6.175
gum from 6.162
gymnosperms *see* gymnosperms
heather *see* heather
herbicides 6.7–8, 6.168, 6.178, 6.271
herbs *see* spices, herbs, and flavourings
hormones *6.271*
leaf 2.184, 2.185, 2.355
Monocotyledons *see* Monocotyledons
moss 2.219, 2.282, 2.319–20
nitrogen 2.228
oil from *see* oil, vegetable
ovary 2.237
oxygen cycle 2.235
perennial *see* perennial plants
perfume from *see* aromatic and scented
phloem 2.253
photosynthesis *see* photosynthesis
placenta 2.257
plankton 1.258, 2.15, 2.257, 2.258
pollen *see* pollen
pollination 2.193, 2.263
primrose *see* primrose
propagation *6.271*
protected species 2.270
protective coloration 2.270
pteridophytes *see* pteridophyte
root 2.288
sap 2.294
scented *see* aromatic and scented
seaweed 6.18, 6.186
plants (*cont.*):
shore 2.308
spices from *see* spices, herbs, and flavourings
starch 2.56, 2.292, 2.324, 2.335, 6.330
stem 2.325
study of *see* botany
succulent *see* succulent plants
sunflower *see* sunflower
symbiosis 2.188, 2.189, 2.224, 2.282
transpiration 1.341, 2.345
trees *see* forests and trees
weeds *see* weeds
see also botany
plasma (blood) 2.116, *2.258–9*
membrane 2.207
plasma (physics) *1.258*, 1.259, *8.123*
engine 6.326
plasma-arc torch 6.221
plasmapheresis *6.271*
plasmids 6.153
Plasmodium 2.198
Plassey, battle of *3.285*
plaster 6.195, *6.271*
of Paris 6.163
plasterwork *5.351*
see also stucco
plasticizer *6.272*
plastics *6.271–2*
bakelite 6.27, 6.271
bullet *6.272*
cellophane 6.65
celluloid 6.66, 6.172, 6.271
cellulose acetate 6.66, 6.92, 6.129, 6.153, 6.272
elastomer 6.113, 6.271
epoxy resin 6.3, 6.122, 6.129, 6.272
explosive *6.272*
extrusion 6.124, 6.272
fibre *see* synthetic fibres
fibre-reinforced 6.5, 6.82, 6.129, 6.156, 6.157, 6.204, 6.247
fireproofing 6.132
as insulator 6.183
laminated 6.271
moulding 6.235, 6.272
nylon *see* nylon
packaging 6.255
phenol-formaldehyde 6.27, 6.267, 6.271, 6.273
pioneer 6.172
plasticizer 6.272
polymers *see* polymers; thermoplastics; thermosetting plastics
record 6.296
Plataea *3.285*
Plate, River *1.258*
battle of 4.270, *4.271*
plate (geology) *see* plates of Earth's crust
plate (sundial) 8.165
plate, tin 6.356
plate girder *6.272*, 6.336
plate glass 6.136, *6.272*
plateaux *1.258–9*
Canada 1.145–6
Ethiopia 1.237
France 1.206–7
horst 1.159, 1.288
South America 1.207, 1.248–9
and tectonic plates 1.261
United States 1.69, 1.145–6, 1.243
see also Deccan
platelets 6.164
platemaking *6.272*, 6.366
Plateresque *5.352*
plates of Earth's crust *1.259*
plate margins 1.262
plate tectonics *1.259*, *1.260*, *1.261*
and earthquakes 1.99
geosyncline 1.137
sea-floor spreading 1.296
plates of Earth's crust (*cont.*):
plate tectonics (*cont.*):
subduction 1.324–5
platforms, shore 1.301
Plath, Sylvia *5.352*
platinum 1.107, *1.259*, *6.272–3*
alloys 6.186, 6.272
catalyst 6.273, 6.296
electroplating 6.120
powder metallurgy 6.387
in thermocouple 6.354
Plato *3.285*, *5.353*, 7.234, *7.237*
and Academy 3.2
and aesthetics 7.4
and Aristotle 3.21
and astronomy 8.61
and Atlantis 3.27, 9.6
and dialectic 7.89
and Forms (universals) 7.237, 7.322
and justice 7.174
and Orphic Mysteries 3.263
and Plotinus 7.237
and rhetoric 5.382
and society 7.294
and Socrates 3.332
and sophists 7.295
and Sparta 3.338
Platt, Thomas Collier *4.271*
Platte River *1.259*
platy *2.259*
platyhelminth *2.259*
flatworms 2.120, 2.259
flukes 2.35, 2.122, 2.192
tapeworms 2.167, 2.334
platypus *2.259*
Plautus 5.257, 5.387
play *7.237*, 7.286, 7.300
see also games
play (theatre) *see* drama
playa *1.259–60*
Playfair, James 5.353
Playfair, John 1.136, *1.260*
Playfair, William Henry *5.353*
Playford, John 5.160
playing cards *5.353*
plays *see* drama
plea bargaining *7.237*
plebiscite (referendum) 7.260
plebs *3.285*
plectrum *5.353*
Pléiade 5.136, 5.391
Pleiades *8.123–5*
Pleione *8.125*
Pleistocene epoch 1.133, 1.134, 1.170, *1.260*
megafauna *2.259*
Plekhanov, Georgi Valentinovich 4.44, *4.271*, 4.275
Plenzdorf, Ulrich 5.175
plesiosaurus *2.259*
pliers *6.273*, 6.358
Plimsoll, Samuel: plimsoll line *6.273*
Pliny, Gaius Plinius Secundus (the Elder) *2.259*, *3.285*, 3.288, *5.353*
Pliny the Younger *3.285*, 3.288, *5.353*
Pliocene epoch 1.133, 1.134, *1.260*
PLO *see* Palestine Liberation Organization
Plotinus 3.250, *3.285*, *7.237*
plough (agriculture) 6.6, 6.7, 6.231, *6.273–4*
Plough (astronomy) *8.125*
plover family 2.182–3, *2.259*
plucked instruments:
bandora *5.34*
banjo *5.34*
bouzouki *5.56*
cittern *5.101*, 5.102
jew's harp *5.240*
orpharion *5.332*
pizzicato *5.351*
plectrum *5.353*
plucked drums *5.353*
plucked instruments (*cont.*):
psaltery *5.368*, 5.369
sansa *5.400*
sarod 5.376
sitar 5.414, *5.421*
tambūrā *5.445*
tromba marina *5.460*
vihuela *5.475*
zither *5.498–9*
see also guitar; harp; lutes; lyre; zither
'Plug' strikes *4.271*
plum 2.87, 2.129, 2.144, *2.259*
plumbago *2.259–60*
plumbing *6.274*
plumb-line *6.274*
plume moth *2.260*
Plummer, Henry 8.159
pluralism *7.237*
Plutarch *3.285–6*, *5.353*
Pluto (planet) *8.125*, 8.126
orbit 8.196
satellite 8.25, 8.126, 8.198
statistical details 8.196
plutonic rocks *see* intrusive (plutonic) igneous rocks
plutonium 1.107, *1.260*, *6.274*
first produced 6.214
from nuclear fission 6.126, 6.214, 6.245, 6.290
and nuclear weapons 6.246
plutonium dioxide 6.243
pluvials 1.260
Plymouth Brethren *7.237*
Plymouth Colony *3.286*
plywood 6.193
pneumatics 6.83, 6.109, *6.274*
pneumonia *2.260*
Po *1.260*
poa grass *2.260*
poaching *3.286*
Pocahontas *3.286*, 3.290
pochards *2.260*, 2.297, 2.349
pocket borough *4.271*
pocket gopher 2.141
pocket mice 2.178, *2.260*
pod 2.129, *2.260*
podzol *1.260–1*, 1.269, 1.311
podzolic *1.261*
Poe, Edgar Allan *5.353–4*
Poet Laureate *see* Poets Laureate
poetry and poets *5.354*
for children 5.90, 5.259, 5.296, 5.324, 5.432
Aboriginal 5.1
acrostic 5.2
Africa 5.7, 5.86, 5.427
alexandrine 5.12
alliteration 5.12–13, 5.294, 5.326, 5.486
anthology 5.18
Arab 5.2, 5.18–19, 5.198–9, 5.218, 5.311
Austria 5.175, 5.194, 5.208–9, 5.318, 5.383, 5.486
Baltic 5.34
bathos 5.38
Belgium 5.275
Bengali 5.43, 5.445
blank verse 5.49, 5.136, 5.165, 5.262, 5.269, 5.296, 5.441, 5.453
Britain 5.15, 5.22, 5.89, 5.90, 5.105, 5.172, 5.254, 5.422, 5.466, 5.481
for children 5.259, 5.296, 5.432
allegories 5.187, 5.255, 5.391, 5.428
Augustan 5.5, 5.26, 5.117, 5.183, 5.357, 5.384, 5.441–2
blank verse 5.136, 5.296, 5.441, 5.453
Cavalier 5.82, 5.148, 5.205, 5.269, 5.284
Chaucerian 5.204

poetry and poets (*cont.*):
Britain (*cont.*):
classical **5**.241–2
dialect and colloquial **5**.67–8, **5**.204, **5**.255
eclogues and pastorals **5**.204, **5**.296, **5**.418, **5**.428
epics **5**.44, **5**.254, **5**.296, **5**.357, **5**.428
experimental **5**.291
Georgian **5**.61–2, **5**.128, **5**.174, **5**.188–9, **5**.258, **5**.284, **5**.402
heroic couplets **5**.89, **5**.136, **5**.357
humour **5**.43, **5**.259
hymns **5**.116
Imagists **5**.217, **5**.242, **5**.258, **5**.364
India **5**.248
Irish *see* Ireland *below*
'Lake Poets' **5**.13, **5**.27, **5**.30, **5**.106, **5**.325, **5**.424, **5**.427, **5**.494, **5**.495
lyrics and odes *see under* lyric poetry
Medieval **5**.187
Metaphysical **5**.133, **5**.205, **5**.284, **5**.292, **5**.384, **5**.469
Middle English **5**.172, **5**.255
modernist **5**.143–4, **5**.421
mystic **5**.117–18
nature **5**.102, **5**.213
see also 'Lake Poets' *above*
'new writing' **5**.428
nonsense verse **5**.81, **5**.259, **5**.320
nursery rhymes **5**.91, **5**.324
Old English **5**.44, **5**.326
parody **5**.242
pre-Raphaelite **5**.391, **5**.442
provincial and suburban life **5**.467
realistic **5**.284
Roman Catholic **5**.211
romances **5**.187, **5**.408, **5**.418, **5**.449
Romantics (and precursors) **5**.48–9, **5**.69–70, **5**.89, **5**.107, **5**.189, **5**.209, **5**.245, **5**.254, **5**.390, **5**.416, **5**.449, **5**.453, **5**.494
rural life **5**.67–8, **5**.116, **5**.198, **5**.256
satire **5**.136, **5**.357, **5**.384, **5**.402, **5**.421, **5**.423, **5**.441–2, **5**.496
Scotland **5**.67–8, **5**.105, **5**.139, **5**.204, **5**.209, **5**.307, **5**.408, **5**.453
Shakespearean times **5**.241–2, **5**.283, **5**.418, **5**.428–9
sonnets **5**.296, **5**.418, **5**.429, **5**.441, **5**.496
Wales **5**.35, **5**.273, **5**.452, **5**.486
war **5**.61–2, **5**.335
women **5**.41, **5**.391
see also Poets Laureate
Burma **5**.67
Cambodia **5**.72
Canada **5**.25–6
Celtic **5**.35
Central America *see* South and Central America *below*
Chile **5**.316–17
China ***5**.98*
first known **5**.372
epics **5**.483
Shijing **5**.417
Six dynasties **5**.446
Song dynasty **5**.265–6, **5**.438
Tang dynasty **5**.29, **5**.137–8, **5**.198, **5**.262, **5**.264
woman **5**.265–6
clerihew **5**.104

poetry and poets (*cont.*):
couplet **5**.115
Czechoslovak **5**.120, **5**.140, **5**.210, **5**.252, **5**.318, **5**.410
dactyl **5**.121
Dada **5**.121
Denmark **5**.209–10, **5**.326, **5**.404
doggerel **5**.132
dramatic monologue **5**.134
Dravidian **5**.136
eclogue **5**.141
elegy **5**.142–3
epics *see* epics
epigram **5**.148
epithalamion **5**.148
figures of speech *see* figures of speech
Finland **5**.155, **5**.404
France **9**.31
allegory **5**.387–8
ballade **5**.30, **5**.100, **5**.283, **5**.476
black **5**.86
classical **5**.90, **5**.254, **5**.277, **5**.467
fables **5**.254
innovative **5**.106
lyric **5**.100, **5**.214, **5**.310–11, **5**.476
modernist **5**.18
Parnassians **5**.38, **5**.172, **5**.259, **5**.340, **5**.471–2
Pléiade **5**.136, **5**.391
Provençal **5**.298, **5**.368
Romantic **5**.38, **5**.172, **5**.213–14, **5**.254, **5**.310–11, **5**.317, **5**.475
satire **5**.51
Surrealist **5**.18, **5**.59, **5**.86, **5**.145, **5**.317
Symbolism **5**.38, **5**.43, **5**.103, **5**.238, **5**.277–8, **5**.317, **5**.383, **5**.471–2
troubador **3**.361
woman **5**.100
free verse **5**.163, **5**.244, **5**.400, **5**.486, **5**.488
Gaelic **5**.225–6, **5**.409
Germany **5**.186, **5**.188, **5**.202
ballad **5**.181
blank verse **5**.262
classical **5**.249
courtly love **5**.491
Enlightenment **5**.489
epic **5**.59, **5**.181, **5**.249, **5**.318, **5**.351
idylls **5**.182
Lied **5**.142, **5**.203
lyric **5**.142, **5**.173, **5**.181–2, **5**.205, **5**.210, **5**.481
Meistersinger **5**.397
psychological **5**.499
Romantic **5**.203, **5**.205, **5**.210, **5**.323
satire **5**.203
Sturm und Drang **5**.405
Symbolism **5**.173
Greece, ancient **5**.71–2, **5**.121, ***5**.191*, **5**.325
comic **5**.21
epic **5**.210
lyric **5**.191, **5**.349, **5**.402
paean **5**.335
see also Homer
Greece, modern **5**.82, **5**.191, **5**.244–5, **5**.410
haiku **5**.197
Hebrew **5**.47
heroic couplet **5**.89, **5**.136, **5**.205, **5**.357
hexameter *see* hexameters
Hungary **5**.5, **5**.28, **5**.214, **5**.345
Iceland **5**.141, **5**.404, **5**.421, **5**.423
idyll *see* pastoral and idyll

poetry and poets (*cont.*):
India **5**.129, **5**.134, **5**.220, **5**.243, **5**.276, **5**.373, ***5**.401*, **5**.447, **5**.466, **7**.258
Ireland **5**.41, **5**.183, **5**.202, **5**.225–6, **5**.242, **5**.303, **5**.489, **5**.497
Italy **4**.92, **5**.82, **5**.281, **5**.294, **5**.302, **5**.341, **5**.342
epics **5**.21, **5**.51, **5**.369, **5**.447
flamboyant **5**.283
lyrics and odes **5**.162, **5**.261, **5**.345–6, **5**.355
modernist **5**.340, **5**.371
pagan **5**.124
religious and moralistic **5**.124–5
revolutionary **5**.77–8
Romantic **5**.261
satire **5**.12, **5**.339
sonnets **5**.12, **5**.162, **5**.345–6
Symbolism **5**.466
see also Dante
Japan **5**.37, **5**.197, **5**.217, **5**.235, ***5**.236*, **5**.281, **5**.380
Korea **5**.251
Latin ***5**.257–8*
epics **5**.121, **5**.147, **5**.270, **5**.335, **5**.477
epigrams **5**.148, **5**.283
hexameters **5**.121, **5**.147, **5**.212, **5**.335, **5**.477
love **5**.334–5, **5**.388
lyrics and odes **5**.82, **5**.211–12, **5**.477
satire **5**.212, **5**.242, **5**.389
lay (lai) **5**.258
lyric *see* lyric poetry
Malay **5**.277
metre (*or* meter) **5**.292
Middle English **5**.294
narrative verse **5**.269, **5**.312, **5**.333
Netherlands and Flemish **5**.165
Nicaragua **5**.125
Nigeria **5**.427
nonsense **5**.81, **5**.121, **5**.259, **5**.320
Norway **5**.48, **5**.141, **5**.209–10, **5**.216, **5**.404, **5**.421
nursery rhymes **5**.90, **5**.324
occasional verse **5**.325
ode *see* odes
onomatopoeia **5**.328
ottava rima **5**.333, **5**.369
Pakistan **5**.466
pattern **5**.342
pentameter **5**.121, **5**.205, **5**.259, **5**.343
Persia **5**.247, ***5**.344–5*, **5**.439
allegory **5**.25
epic **5**.155, **5**.345
lyric **5**.25, **5**.233
sonnets **5**.197, **5**.394, **5**.397
Peru **5**.467
Poland:
colloquial **5**.322
free verse **5**.244
lyric **5**.250, **5**.262, **5**.294
Romantic **5**.294, **5**.422
satire **5**.377
sonnets **5**.294
symbolism **5**.262, **5**.422
Portugal **5**.73, **5**.345, ***5**.359*, **5**.359, **5**.474
Provençal *see under* France
quatrain **5**.371
rhyme **5**.382
Romantic *see under* Romanticism
Russia and USSR **5**.65, **5**.129, **5**.267, ***5**.396*, **5**.462, **5**.497
epics **5**.10, **5**.396, **5**.461, **5**.463
Futurist **5**.287
intellectual **5**.60
lyrics **5**.10, **5**.43, **5**.50, **5**.261, **5**.326, **5**.341, **5**.461
Romantic **5**.261–2, **5**.370
Symbolism **5**.43, **5**.49–50, **5**.278, **5**.396

poetry and poets (*cont.*):
Sanskrit **5**.134, **5**.243, **5**.276, **5**.373, **5**.401
scansion **5**.404
Scotland **5**.67–8, **5**.105, **5**.139, **5**.204, **5**.209, **5**.307, **5**.408, **5**.453
Senegal **5**.411
Slovak **5**.422
sonnet *see* sonnets
South and Central America **5**.53, **5**.256, **5**.316–17
Spain **5**.86, **5**.125, **5**.170, **5**.184, **5**.216, **5**.372, **5**.427–8, **5**.470
Generation of '27 **5**.11, **5**.169, **5**.170, **5**.173, **5**.240
stanza **5**.431
Sufi **5**.19, **5**.197, **5**.233, **5**.394, **5**.397, ***5**.439*
Sweden **5**.165, **5**.254, **5**.404, **5**.436
Switzerland **5**.245
Thai **5**.450, **5**.452
Tibet **5**.455
Turkey **5**.205, **5**.462
United States ***5**.13–14*, **5**.290, **5**.384, **5**.431, **5**.432
Beat Movement **5**.39–40
blank verse **5**.165, **5**.269
colloquial **5**.165, **5**.400
'confessional' **5**.270, **5**.352
free verse **5**.400, **5**.488
humour **5**.313, **5**.339
Imagists **5**.217, **5**.269, **5**.490
lyric **5**.17, **5**.130–1, **5**.386
modernist **5**.119–20, **5**.143–4, **5**.303, **5**.364
narrative **5**.269
rural life **5**.165
satire **5**.269
sonnets **5**.295
Symbolism **5**.354
Transcendentalist **5**.145, **5**.453–4, **5**.488
women **5**.130–1, **5**.295, **5**.303, **5**.339, **5**.352
Urdu **5**.466
Vietnam **5**.475
Wales **5**.35, **5**.273, **5**.452, **5**.486
West Indies **5**.59, **5**.481
Yiddish **5**.497
see also songs
Poets Laureate ***5**.354*
listed **9**.71
17th and 18th centuries **5**.101
see also Dryden; Jonson
19th century **5**.449, **5**.494
see also Tennyson; Wordsworth
20th century **5**.46, **5**.60, **5**.91, **5**.213, **5**.284, **5**.352
pogrom ***4**.271*
Pogson, Norman **8**.90
Poincaré, Jules Henri ***1**.261*, **8**.25, **8**.114, ***8**.125*
Poincaré, Raymond ***4**.271*
poinsettia **2**.323
point bar ***1**.261*
Point Four Program ***4**.271*
pointers **8**.178
pointillism **5**.316, ***5**.354*
point-of-sale terminal ***6**.274*
poison and poison gases **2**.260, ***3**.345*
ammonia **6**.13
arsenic **6**.20, **6**.72, **6**.312
Bhopal accident **6**.35, **6**.275
chemical warfare **6**.71–2, **6**.98, **6**.164, **6**.235, **6**.239–40, **6**.267
chlorine **1**.61–2, **1**.106
curare **6**.94
cyanide **1**.82
dioxin **6**.91–2, **6**.98, **6**.104, **6**.314
food **2**.39
hydrogen sulphide **1**.162
lead **6**.197
mercury **6**.248

poison and poison gases (*cont.*):
  nitrogen dioxide **1**.230
  phenol **6**.266–7
  phosphorus **6**.248
  selective toxicity **6**.266
  sulphur dioxide **6**.338
  thallium **6**.353
  toxicology **2**.345
  in water **6**.377
  wood preservative **6**.387
  see also poisonous; pollution
poisonous animals *2.260–1*
  arrow-poison frog **2**.14–15
  fire salamander **2**.360
  jellyfish **2**.175
  puss moth **2**.272, **2**.273
  scorpion **2**.297
  sea anemone **2**.297, **2**.298
  spiders **2**.369
  see also bees; poisonous fish; poisonous snakes; wasps
poisonous fish *2.261*
  pufferfish **2**.271
  sting-rays **2**.261–2, **2**.326
  stone fish **2**.327
  toadfish **2**.342
  trunk-fish **2**.348
  weever **2**.363–4
poisonous plants *2.258*
  algal blooms **2**.6
  alkaloid **2**.6
  arrow-poison tree **2**.353
  autumn crocus **2**.19
  cassava **2**.56
  chantarelle **2**.61
  death cap **2**.88
  ergot **2**.110
  fritillary **2**.127
  fungi **2**.61
  hellebore **2**.154
  hemlock **2**.154
  laburnum **2**.181
  milkweed **2**.211–12
  nettle **2**.227
  Nicotiana **2**.227–8
  nightshade **2**.228
  poison ivy *2.260*
  rhubarb **2**.284
  spurges **2**.183, **2**.323
  thorn apple **2**.340
  upas tree **2**.353
poisonous snakes:
  adder **2**.3
  boomslang **2**.73
  cobras **2**.70, **2**.179, **2**.285–6
  colubrid **2**.73
  copperhead **2**.75
  coral snake **2**.76–7
  gaboon viper **2**.130, **2**.131
  king cobra **2**.179
  krait **2**.181
  mamba **2**.200
  pit vipers **2**.257
  puff adder **2**.271
  rattlesnake **2**.278
  rinkal **2**.285–6
  sea snake **2**.301
  taipan **2**.333
  tiger snake **2**.341
  venom vaccine **6**.177
  vipers *see* viper family
Poisson, Siméon Denis *1.261*
Poitiers, battle of *3.286*
poker (card-game) **7**.40
pokeweed *2.261*
Pol Pot **4**.58, **4**.178, **4**.182, *4.273*, **7**.130
Poland *1.261–2*, *3.286*, *4.271–2*, *7.376*
  anthropology *see* Malinowski
  arts: painting **5**.496
    see also literature; music *below*
  astronomy **8**.20, **8**.29
    see also Copernicus; Hevelius
  in COMECON **7**.57
Poland (*cont.*):
  environment **7**.2
  frontier **4**.250
  language **7**.156, **7**.288
  literature **5**.227, **5**.294, *5.355*, **5**.406, **5**.422
    see also under drama; novels; poetry
  mathematics **1**.204, **7**.306
  music **5**.99, **5**.271, **5**.335, **5**.343, **5**.443
  physical geography and geology **1**.65, **1**.261–2, **1**.304, **1**.357
  physicist **1**.81
  Polish Corridor *4.272*
  politics and government **4**.318, **7**.99, **7**.223
    politicians and statesmen **3**.200–1, **4**.91, **4**.138, **4**.201–2, **4**.269, **4**.313, **4**.318
    Solidarity **4**.318, **7**.332
    women's suffrage **7**.332
  regions **3**.136, **4**.272
  religion/theology **7**.98, **7**.138
  rulers **3**.29, **3**.62, **3**.183, **3**.211, **3**.331, **3**.340
  sport and recreation **7**.297
  wars and defence:
    Katyn massacre **4**.179
    leaders **3**.200–1, **4**.269, **4**.313
    Polish Succession *3.287*
    Warsaw Pact **4**.371, **7**.328
    see also Northern War
  see also Eastern Europe
polar *see* pole/polar
polar bear **2**.261, *2.262*
Polaris (Pole Star) *8.126*
polarity, magnetic **1**.202
polarization *8.126*
  binary star **8**.15
  polarimetry *6.274*, *8.126*
  polarized light **1**.17, **1**.239, *1.262*, **6**.223, **6**.274
polarography **6**.274
Polaroid Land camera *6.274–5*
polars (stars) **8**.4
Pole, Reginald *3.286*
polder *1.263*, **1**.374
pole (geometry) *1.263*
Pole Star **8**.126
polecat **2**.115, *2.262*
polemic *5.354–5*
pole/polar *8.126*
  aurora **8**.12–13
  axis *8.125*
  caps *8.125–6*
  celestial **8**.22
  climate **1**.65
  coordinates *1.262*, **1**.263
  distance **8**.22, **8**.109
  front *1.262*
  galactic **8**.54
  low *1.262*
  magnetic **6**.104
  molecules *1.262–3*
police *4.272*, *7.237–8*, **7**.318
  arrest **7**.18–19
  Bow Street Runners **3**.47
  fingerprints **7**.120
  forensic science **6**.217
  'lie detector' **6**.275
  security services **7**.278
  state *7.238*
  tear-gas **6**.346
  water cannon **6**.377
  see also law and justice; secret services
poliomyelitis *2.262*, **6**.253, **6**.298, **7**.153
polis *3.286–7*
  see also regions and city-states *under* Greece, ancient
polish *6.275*
Politburo *4.272*
politics/political *7.239*
  Arabs **7**.24
  asylum *7.238*
  Centre and Centre-Right **7**.43, **7**.47
  citizenship **7**.50
  clientelism **7**.53
  confederation **7**.64
  conservative *see* conservatism
  convention *see* Conventions
  corporatism **7**.70
  corruption **7**.71–2
  domino theory **7**.94
  elections *see* electoral system; voting
  executive **7**.113, **7**.279, **7**.318
    see also government
  fascist *see* fascism
  federalism *see* federalism
  freedom **7**.124
  functionalism **7**.127
  geopolitics **7**.131
  Green *see* Green (politics)
  individualism **7**.156–7
  interdependence **7**.163
  and legislation *see* legislature
  liberal *see* liberalism
  obligation **7**.221
  organization *7.238*
  parties *7.238*
    coalition government **7**.54
    Left *see* Left
    multiparty system **7**.209
    one-party state **7**.223
    two-party system **7**.317
    see also Christian Democrats; communism; Conservative; fascism; Labour; Left; Liberal; social democracy; socialism; Right *and also under individual countries*
  patronage **7**.231
  pluralism **7**.237
  political asylum **7**.238
  populism **7**.242
  power (realism) **7**.244
    see also power (political and social)
  prisoner *7.238–9*
  protest *7.239*
  purge *see* purges
  and religion **7**.263, **7**.264, **7**.265, **7**.309
  Right *see* Right (politics)
  social contract *see* social contract
  theatre *5.355*
  totalitarianism **7**.16, **7**.125, **7**.313
  trial *7.239*
  trusteeship **7**.316, **7**.321, **7**.322
  violence *7.239*
    see also terrorism
  see also government; revolution; state *and also* politics and government *under individual countries*
Poliziano, Angelo *5.355*
polje **1**.180, *1.263*
Polk, James Knox **4**.251, *4.273*
polka *5.355*
poll tax **3**.273, *3.287*, **7**.306
pollack *2.262*
Pollaiuolo, Antonio and Piero *5.355*
pollen *2.262*
  allergy to **2**.153
  palynology **2**.240
  pollen-feeding moths **2**.141
  stamen **2**.324
pollination **2**.193, *2.263*
Pollock, Jackson **5**.2, **5**.27, *5.355*
pollution *1.5*, *2.263*, *6.2*, *6.9*, *6.275*, *6.377*, *7.239–40*
  acid rain **6**.2, **7**.2, **7**.240
  agrochemical **6**.7–9, **6**.377
  boiler **6**.41
  CFCs *see* chloro-fluorocarbons
  fall-out *see* nuclear radiation
  and fishing **6**.146
pollution (*cont.*):
  fossil fuels **6**.2, **6**.9, **6**.62–3, **6**.78, **6**.144, **6**.161, **6**.263, **6**.275, **6**.300–1
    see also lead tetraethyl
  greenhouse effect *see* greenhouse effect and global warming
  and health hazards **7**.140
  industrial accidents **6**.35, **6**.275, **6**.314
  nuclear accidents **6**.9, **6**.72, **6**.245, **6**.281, **6**.355
  oil slick treatment **6**.27, **6**.37
  ozone *see* ozone layer
  power stations **6**.280
  prevention **6**.62–3
  road traffic *see* fossil fuel *above*
  sewage **6**.32, **6**.314, **6**.377, **7**.240
  thermal **6**.377
  water **6**.146, **6**.314, *6.377*, **6**.378
  see also poison *and under* sulphur dioxide
Pollux (mythology) *9.8*
Pollux (star) *8.126*
Polo, Marco **5**.460
polo *7.240*, **7**.241
polonaise *5.355–6*
polonium **1**.107, *1.263*
Poltava, battle of *3.287*
polyamide *see* nylon
polyanthus **2**.269
Polybius *3.287*
polycarbonate **6**.273, *6.275*
polychloroprene (neoprene) **6**.60, **6**.304
Polyclitus of Argos *5.356*, **5**.368
polyester *6.275*, **6**.302, **6**.384
polyethene *see* polythene
polyether etherketone (PEEK) **6**.129
polyethylene *see* polythene
polygon *1.263*
polygraph *6.275*
polyhedron *1.263*
polymerase chain reaction **6**.153
polymers **1**.201, **6**.272, **6**.273, *6.275–6*
  acrylic fibre **6**.2–3, **6**.58, **6**.61
  listed **6**.273
  natural inorganic *see* diamond; graphite; quartz
  natural organic *see* DNA; proteins; rubber
  polymerization *1.263*, **6**.265
  synthetic:
    acrylic **6**.2–3, **6**.58, **6**.61
    ion-exchange resin **6**.186
    liquid crystal **6**.204, **6**.247
    polymerization **6**.265
    PTFE **6**.273, **6**.285
    silicone **6**.191, **6**.320, **6**.378
    texturing **6**.390
    thermoplastics *see* thermoplastics
    thermosetting *see* thermosetting plastics
    urea-formaldehyde resin **6**.272, **6**.273, **6**.369
  see also plastics; synthetic rubber
polymethyl methacrylate **6**.271, **6**.273, *6.276*
Polynesia *7.376*
  arts *5.356*, **5**.357
  ethnic groups:
    Polynesians *3.287*, *7.240*, **7**.312
    see also Maoris; Samoan Islands
  languages **7**.23
  Marquesas Islands **1**.206, *4.212*
  physical geography and geology **1**.315
  religion and myths **7**.305, **9**.25, **9**.35
  technology **6**.64, **6**.237
Polynices **9**.4
polynomial *1.264*
polyp *2.263*
  hydra **2**.167

polyp (*cont.*):
  Portuguese man-of-war 2.265, 2.266
  sea anemone 2.297, 2.298
Polyphemus 9.9
polyphony *5.356*
  anthem 5.18
  Britain 5.69, 5.139, 5.176, 5.393
  Europe 5.89, 5.256, 5.336, 5.365, 5.439
  motet 5.306
  Oceania 5.525
  organum 5.332
  Renaissance 5.379–80
  see also harmony
polypody (fern) *2.263*
polypoidy (chromosomes) 2.65
polypropylene 6.61, 6.236, *6.273*, *6.276*, 6.302, 6.390, 6.391
polyptych *5.356*
polysaccharides 2.53
  glycogen 2.138
  see also starch
polystyrene 6.236, 6.261, 6.272, 6.273, 6.275, *6.276*
polytetrafluoroethene (PTFE) 1.151, 6.273, 6.285
polytheism 7.133
  see also deities *and in particular* deities *under* Hinduism
polythene 6.236, 6.261, 6.272, 6.273, 6.275, *6.276*, 6.391
  polyethene terephthalate 6.271, 6.273
polytonality *5.356*
polyunsaturated fats 6.215
polyurethane 1.218, 6.273, 6.275, *6.276*
polyvinyl acetate 6.273
polyvinyl chloride *see* PVC
Poma de Ayala, Guaman 3.175
Pomare II, king of Tahiti 4.337
Pombal, Sebastião José de Carvalho e Mello, marquis of *3.287*, 3.289
pome *2.263*
pomegranate *2.263*
pomelo 2.143
Pomerania *3.287*
Pompadour, Jeanne Antoinette Poisson, marquise de *3.287*
pompano *2.263*
Pompeii *3.287–8*, 5.315
Pompey 3.57, 3.60, 3.278, *3.288*
  conquests 3.17, 3.76, 3.268, 3.307–8
Pompidou, Georges Jean Raymond 4.123, *4.273*
Ponce de León, Juan 3.128, *3.288*, 3.294
Pond, James 8.142, 8.193
pond skater *2.263*
pond snail *2.263–4*
Pondicherry *3.288*
ponds *see* lakes
pondweed *2.264*
Pons, Louis 8.15
Pons, Stanley 6.148
Pont-Aven School *5.356*
Ponte, Lorenzo da 5.306
Ponti, Gio *5.356*
Pontiac *3.288*
Pontius Pilate 3.188, *3.281*, 9.6, *9.29*
pontoon bridge 6.47
Pontoppidan, Henrik 5.404
Pontus 3.234, *3.288*
Pony Express *4.273*
pool (game) 7.289
Poor Laws *3.288–9*, 3.344, 3.385, 4.12, *4.273*
Pop art *5.356–7*
  Britain 5.208
  United States 5.240–1, 5.264, 5.482
  see also Hockney
pop music *see* popular music
Pope, Alexander 5.66, *5.357*
popes *see* papacy
Popish Plot *3.289*
poplar *2.264*
Popocatepetl *1.264*
Popov, Oleg 5.105
Popper, Karl Raimund *2.264*, *7.241*
poppies 2.110, 2.235, *2.264*
Popular Front *4.273*
popular music *5.357–8*
  Beatles 5.39
  jazz *see* jazz
  musicals *see* musicals
  ragtime 5.242, 5.373
  rhythm and blues 5.358, 5.382, 5.425
  rock 5.39, 5.358, 5.384–5
  soul 5.358, 5.425
population (species) *2.264–5*
  regulation *2.265*
population and demography *7.84*, *7.241–2*
  age *7.6*, *7.241*
    ageing *see* old age
    dependency ratio 7.85
    structure 7.241
  birth rate 7.32
  census 7.42–3
  cohort 7.56
  demographic transition 7.84
  density *7.241*
  depopulation 7.85
  eugenics 7.110
  growth *7.241–2*, 7.336
    see also urban areas
  life expectancy 7.6, 7.184
  movement *see* migration
  overpopulation 7.226
  sex ratio *7.241*
  and water supply 7.328
  world *see under* world
  see also ethnic groups; race *and also* demography *under individual countries*
Populations I and II (astronomy) 8.161
populism *7.242*
  Populist Movement (Russia) 4.27, *4.275*, 7.242
  Populist Party (USA) *4.275*
porbeagle *2.265*
porcelain *5.358–9*, *5.360–1*, *6.276*
  bone china 5.51, 6.276
  famille rose, *famille verte* *5.152*
  firing 6.68
  innovator 6.43
  insulators 6.183
  Japanese 5.234
porcupine 2.264, *2.265*
pores (astronomy) *8.127*
porgy *2.265*
poriferan 2.322
pornography 7.46, *7.242–3*
porosity *1.264*
Porphyry (philosopher) 7.237
porphyry (rock) *1.264*
porpoise *2.265*
Porsche, Ferdinand 6.234
Port Jackson shark *2.265*
port *see* ports
Portal, Charles Frederick Algernon, viscount *4.275*
portal frame *6.276*
Portales, Diego *4.275*
portative organ 5.138, *5.359*
Porteous, John: Porteous Riots *3.289*
Porter, Cole *5.359*
Porter, Katherine Anne *5.359*
Portia (Uranian satellite) 8.178
portico *5.359*
Portland, dukes of (Bentinck) 4.35
Portland cement 6.66, 6.86, 6.271, *6.276*
Portland stone *1.264*
portraiture *5.359*
  America 5.13–14
  Britain:
    18th century 5.129–30, 5.168, 5.209, 5.249–50, 5.258, 5.373, 5.374, 5.381–2, 5.390
    19th century 5.258, 5.373
    20th century 5.148, 5.164, 5.240, 5.441
    see also under Britain *under* photography
  busts 5.68
  caricature 5.78
  China 5.96
  conversation pieces 5.112, 5.129, 5.168, 5.499
  Coptic 5.112
  France 5.105, 5.152, 5.261
  Germany 5.117, 5.131, 5.209, 5.264, 5.491
  Greece, ancient 5.190
  Italy 5.42, 5.261, 5.300, 5.374, 5.457
  Netherlands and Flemish 5.303
    15th century 5.100, 5.151, 5.290, 5.488
    17th century 5.120, 5.140, 5.197, 5.242, 5.262, 5.264, 5.311, 5.378, 5.393, 5.449
  Norway 5.192–3
  photographic 5.1, 5.27, 5.40, 5.73, 5.81, 5.206, 5.244, 5.347
  Roman empire 5.387
  sculpture 5.12, 5.88, 5.148, 5.392
  Spain 5.189, 5.470
  United States 5.13–14, 5.112, 5.141, 5.342, 5.486
  see also miniatures
ports and harbours *6.276–7*
  cranes 6.91–2
  docks 6.105, 6.109, 6.297
  dredging 6.106, 6.277
  free port 7.125–6
  ice-breaker 6.175
  tug 6.364
Portsmouth, Treaty of *4.275*
Portugal *1.264*, *3.289*, *4.275*, *7.377*
  arts 5.281
    see also literature *below*
  colonies *see* empire *below*
  in Council of Europe 7.71
  economy and trade 7.221, 7.315
    see also European Community; European Economic Community
  empire and colonialism *3.289*, 4.58
    Africa 3.7, 3.102, 3.103, 3.107, 4.306–7
    Atlantic 3.30, 3.215
    decolonization 4.4, 4.126, 4.321
    explorers and navigators 1.154, 3.122
    India 3.7, 3.95–6, 3.107, 4.136–7
    map 3.122
    South America 3.49, 3.334
    South-East Asia 3.7, 3.11, 3.107, 3.216
    see also Portuguese *and* exploration *below*
  exploration 1.47, 1.154, 3.122
    see also Da Gama; Diaz de Novaes; Magellan *and* empire *above*
  languages 7.156, 7.235, 7.269
  literature 5.35, 5.73, 5.345, *5.359*, 5.359, 5.474
  physical geography and geology 1.264, 1.330
  politics and government 3.287, 4.58, 4.300, 7.73, 7.201
  religion/theology 7.280, 9.43
  rulers 3.48, 3.153, 3.163, 3.189–90, 4.175
  sport and recreation 7.37–8
Portugal (*cont.*):
  technology 6.47, 6.201
    ships 6.23, 6.29, 6.58, 6.149, 6.306
  wars and defence 3.8, 3.359
    in WEU 7.331
    see also NATO; Peninsular War
  Washington Conference 4.372
Portuguese East Africa *see* Mozambique
Portuguese Guinea *see* Guinea-Bissau
Portuguese man-of-war *2.265–6*
Portuguese South West Africa *see* Angola
POS (point-of-sale terminal) *6.274*
Poseidon 9.9, 9.27, *9.29*
positive charge:
  anion 1.14
  electrostatics 1.105
  see also anode
positive discrimination *7.243*
positive economics *7.243*
positivism 7.63, 7.292
positron *1.264*
positron emission tomography 6.243, 6.259, *6.277–8*, 6.357
possum *2.266*
postal services 4.273, *6.278*
poster art *5.359*, *5.362*
posthorn *5.362*
Post-Impressionism 5.87, 5.171–2, 5.182–3, 5.351, *5.362*, 5.465
  see also Impressionism
post-industrialism *7.243*, 7.309
post-mill 6.385
Post-Modernism *5.363*, *7.243*
  in architecture 5.241, 5.299, 5.363, 5.471, 7.243
  in dance 5.62, 5.70, 5.120, 5.342, *5.362–3*, 5.373, 5.397
    see also modern dance
  in literature 5.363, 5.394, 7.243
  see also modernism
post-painterly abstraction *5.363*, 5.431
post-structuralism 7.68
postulates (mathematics) 1.247, *1.264*
Potana 5.134
potash *1.264*
  caustic 1.264–5
potassium 1.107, *1.264*, 6.96, 6.127–8, *6.278*
potassium aluminium sulphate *see* alum
potassium carbonate (potash) 1.264
potassium chlorate 1.265
potassium chloride 6.100, 6.141
potassium hydroxide *1.264–5*
potassium iron silicate 1.141
potassium nitrate *1.265*, 6.141, 6.162, 6.241
potassium permanganate *1.265*
potassium sorbitate 6.141
potassium sulphate 1.326
Potatau Te Wherowhero 4.183, *4.275*
potato family *2.266*, 6.302
  Cape gooseberry 2.141
  mandrake 2.201
  Nicotiana 2.227–8
  nightshade 2.228
Potemkin, Gregory 3.65, *3.289*
Potemkin (battleship) *4.275–6*
potential (astronomy) *8.127*
potential (electrical):
  measurement 1.358
  potential difference *1.265*, 1.358, 6.117
potential energy *1.265*
  gravitational 1.133, 6.121
Potentilla 2.91
potentiometer *6.278*
Potgieter, Andries Hendrik 4.140
potholes *1.265*
  moulin 1.221–2

potlatch **7**.178, **7**.219, ***7**.243*
Potock, Waclaw **5**.355
Potomac ***1**.265*
potoo ***2**.266*
Potsdam Conference ***4**.276*, **4**.385
Pott, Percivall ***2**.266*
Potter, Beatrice **4**.372
Potter, Beatrix ***5**.363*
potter wasp **2**.204–5
potter's wheel ***6**.278*
pottery *see* ceramics and pottery
potto ***2**.266*
pouched mice ***2**.266*
pouches, animals with *see* marsupials
Poulenc, Francis ***5**.364*, **5**.421
Poulsen, Valdemar **6**.345
poultry and poultry farming **2**.63, ***2**.266*, ***6**.279*
see also livestock
Pound, Ezra **5**.217, ***5**.364*, **5**.480
Poussin, Nicolas ***5**.364–5*
Pouvanaa a Oopa of Tahiti **4**.337
poverty **7**.222, ***7**.243–4*, **7**.250–1
powder metallurgy **6**.221, ***6**.279*, **6**.387
Powderly, Terence V. **4**.185
Powell, Anthony ***5**.365*
Powell, John Wesley ***1**.265*
Powell, Michael ***5**.365*
power (energy) ***1**.265–6*, **1**.362, **6**.120
  loom *see* machinery *under* textiles
  tools **6**.69, ***6**.280*, **6**.308
    see also machine-tools
  see also energy; power stations
power (mathematics) ***1**.266*
power (political and social) **7**.123, ***7**.244*
  authoritarian state **7**.23, **7**.223, **7**.296
  centralization **7**.43
  corruption **7**.70–1
  dictatorship **7**.89–90
  great power **7**.135, **7**.302
  international order **7**.25
  separation of powers **7**.279
  superpower **7**.302
  will to power **7**.216
  see also leaders
power politics ***7**.244*
  see also power (political and social)
power series (mathematics) ***1**.266*
power stations ***6**.114*, ***6**.279–80*
  boiler **6**.41, **6**.280
  coal-fired *see* thermal *below*
  combined-cycle technology **6**.81, **6**.280
  gas-turbine **6**.150
  generator **6**.153
  geothermal **6**.155
  hydroelectric *see* hydroelectricity
  magnetohydrodynamic generator **6**.213
  nuclear *see* nuclear reactors and power stations
  pollution **6**.275, **6**.377
  thermal **6**.57, **6**.78, **6**.244, **6**.275, **6**.279
  tidal *see under* tides/tidal
  see also electricity generation; steam-turbine
power tools **6**.69, ***6**.280*, **6**.308
Powers, Hiram **5**.315
Powhatan ***3**.289–90*
Powys, John Cowper ***5**.365*
Poynings, Sir Edward: Poynings' Law ***3**.290*
Pozzo, Andrea **5**.35
PR *see* proportional representation; public relations
Prado, Manuel **4**.267
Praemunire, Statute of ***3**.290*
Praesepe ***8**.127*
praetorian ***3**.290*
Praetorius, Michael ***5**.365*
Pragmatic Sanction ***3**.290*
pragmatism **7**.88–9, ***7**.244*
Prague, Defenestration of ***3**.290*
Prague, Peace of **3**.290
prairie ***2**.266*
  soil ***1**.266*, **1**.311
prairie chicken ***2**.266–7*
prairie dog ***2**.267–8*
Prasad, Jayashankar **5**.206
Prasad, Rajendra ***4**.276*
Prasad, Sunder **5**.244
prasād ***7**.244–5*
praseodymium **1**.107
Prasutagus of Iceni **3**.47, **3**.174
pratincole ***2**.267*
Pratt, Sir Roger ***5**.365*
prawn ***2**.267–8*
Praxiteles **5**.189, ***5**.365*
praying mantis **2**.267, ***2**.268*
Pre-Cambrian era **1**.18, **1**.133, **1**.134, ***1**.266*
  see also Archaean; Proterozoic
pre-cast concrete **6**.44, **6**.76, ***6**.280*, **6**.343
precedent ***7**.245*
precession ***8**.127*
  Chandler wobble **8**.25
  of equinoxes ***1**.266*
  of gyroscope **6**.163
precious metals *see* noble metals
precious stones *see* gemstones
precipitation (chemistry) **1**.59, **1**.116, ***1**.266*, **1**.342
precipitation (meteorology) ***1**.266*
  and altitude **1**.241, **1**.274–5
  bar chart **1**.144
  drizzle **1**.96
  and erosion *see* weathering *and also* erosion *under* limestone
  fronts **1**.126, **1**.252
    see also depression (meteorology)
  hail **1**.150, **1**.307
  hydrological cycle **1**.162
  lack of *see* drought
  and latitude **1**.274–5
  measurement **6**.221
  monsoons **1**.220
  polluted *see* acid rain
  rain-shadow **1**.274–5
  runoff *see* runoff
  shower **1**.302
  sleet **1**.307
  world map **1**.267
  see also rain; snow
precipitator, electrostatic **6**.120
precognition **7**.228
Pre-Columbian:
  art and architecture **5**.84–5, **5**.425–6
  technology **6**.219–20
  see also Indians, South American
predella ***5**.365*
predestination ***7**.245*, **7**.260
prefabrication **6**.44, **6**.146, ***6**.280*, **6**.317, **6**.343
  see also Crystal Palace
Pregl, Fritz ***1**.266*
pregnancy ***2**.268*, ***6**.248*
  amniocentesis **2**.8, **6**.13, **6**.218
  maternity leave **7**.228
  obstetricians **2**.166
  placenta **2**.257
  prevention *see* contraception
  termination *see* abortion
  test ***6**.280*
  umbilical cord **2**.352
  see also birth; obstetrics and gynaecology; sexual reproduction
prehistory and human evolution ***3**.120–1*
  African sites:
    Acheulian **3**.2, **3**.50
    Ethiopia **3**.152
    Kenya **3**.199–200
    Nigeria **3**.254
    Olduvai area **3**.201, **3**.261
prehistory and human evolution (*cont.*):
  African sites (*cont.*):
    South Africa **3**.341
  archaeology **3**.19–20
  Asian sites **3**.120, **3**.157, **3**.177, **3**.186, **3**.210, **3**.239, **3**.274, **3**.390
  astronomy ***8**.127–8*
  European sites **3**.2, **3**.120, **3**.154, **3**.174, **3**.203–4, **3**.368
    Britain **3**.342, **3**.345
    megaliths **3**.228, **3**.229
  hunter-gatherers **3**.102
  North American sites **3**.80, **3**.170, **3**.203, **3**.233–4, **3**.242, **3**.335
  Piltdown Man hoax **3**.282
  South and Central American sites **3**.234, **3**.247, **3**.261
  technology **6**.136, **6**.219–20, ***6**.280–1*
    dating *see* dating
    medicine **6**.340
    stone monuments *see* megaliths
  tells **3**.350–1
  see also australopithecines; Bronze Age; Copper Age; evolution; Iron Age; Hominids; Neanderthals; Stone Ages
prejudice *see* discrimination and prejudice
prelude ***5**.365–6*
Premadasa, Ranasinghe **4**.327
premature babies **6**.178
Premchand **5**.206
Prempeh I, king of Asante **4**.20, ***4**.276*
Prempeh II, king of Asante **4**.276
Prepertius **5**.388
preposition **7**.229
Pre-Raphaelite Brotherhood **5**.62, **5**.214, **5**.295, ***5**.366*, **5**.391, **5**.442
Presbyterians **3**.223, ***3**.291*, **7**.50, ***7**.245*
  see also United Reformed Church
pre-school education **7**.206, **7**.220, ***7**.245*
Prescott, William **3**.54
preservation:
  food *see under* food
  wood **6**.387
presidents **7**.113, **7**.139, ***7**.245–6*
  cabinet **7**.38
  civil service **7**.52
  of France **4**.191–2, **4**.271, **4**.273, **7**.246, **9**.70
  of Germany **9**.70
  of Ireland **4**.97–8, **4**.166, **5**.226, **9**.70
  legislature **7**.182
  of Mexico **4**.57, **4**.62, **4**.64, **4**.98, **4**.157–8, **4**.176, **4**.205, **4**.249, **4**.302
  of Soviet Union: listed **9**.70
  of United States **7**.246
    list of ***9**.69*
    1789–1850:
      Democrat **4**.171, **4**.273, **4**.362
      Federalist **3**.3, **3**.377–8, **4**.2
      Republican **4**.173, **4**.205, **4**.222–3
      Whig **4**.119, **4**.147, **4**.339, **4**.356
    1850–1901:
      Democrat **4**.54, **4**.77, **4**.269
      Republican **3**.100, **3**.186, **3**.215–16, **4**.18, **4**.131, **4**.139, **4**.148, **4**.174–5, **4**.196, **4**.204
      Whig **4**.119
    1901–45:
      Democrat **4**.292–3, **4**.379–80
      Republican **4**.85–6, **4**.146, **4**.156–7, **4**.293, **4**.336
presidents (*cont.*):
  of United States (*cont.*):
    since 1945:
      Democrat **4**.64, **4**.175, **4**.180, **4**.353
      Republican **4**.109–10, **4**.121, **4**.243–4, **4**.283–4
Presley, Elvis **5**.384
press, private **5**.305, **5**.367
  see also journalism; news
Pressburger, Emeric ***5**.365*
pressgang **3**.291
pressure ***1**.266–7*
  atmospheric **1**.25, ***1**.266–7*
    airlock **6**.9
    baroclinicity **1**.33
    high **1**.155, **1**.282, **1**.303, **1**.325
    isobar **1**.175
    lows **1**.199, **1**.262
    symbols **9**.56
    vapour pressure **1**.354
    and winds **1**.367
  blood **2**.35
  defined **8**.52
  measurement **1**.248, **6**.294, **6**.368
    altimeter **6**.12, **6**.290
    gauge **6**.214, ***6**.281*
  pressure-cooker ***6**.281*
  radiation **8**.133
  and temperature:
    standard **1**.320
    triple point **1**.344
  transducer, electrical **6**.181
  welding **6**.383
pressure groups *see* interest groups
pressurized-water reactor **6**.243, **6**.244, ***6**.281*
Prester John **3**.119, ***3**.291*
Preston, battle of ***3**.291*
Prestonpans, battle of ***3**.291*
pre-stressed concrete **6**.146, ***6**.281*, **6**.372
pretenders ***3**.291–2*
  to English throne **3**.291–2, **3**.310, **3**.328, **3**.377
    see also Stuart, Charles Edward; Stuart, James Francis Edward
  to Russian throne **3**.292, **3**.294
Pretorius, Andries **4**.41, **4**.140, ***4**.276*, **4**.350
Pretorius, Martinus Wessel ***4**.276*
preventive medicine ***7**.246*
Prévert, Jacques **5**.79
Prévost d'Exiles, Antoine-François, abbé ***5**.366*
Priam ***9**.29*
prices ***7**.246*
  adjustment **7**.2
  commodity **7**.57
  and demand **7**.83
  discount **7**.92
  equilibrium **7**.107
  hedging **7**.141
  and incomes policy ***7**.246*
  inflation **7**.160
  resale price maintenance **7**.266
  retail price index **7**.266
  system ***7**.246*
  and taxation **7**.306
  see also under petroleum
prickleback **2**.3
prickly pear ***2**.268*
Pride, Thomas **3**.210
  Pride's Purge ***3**.292*
Priestley, John Boynton ***5**.366*
Priestley, Joseph **1**.190, ***1**.267*, **1**.268
Priestman brothers **6**.102
primary colours ***5**.366*
primary education **7**.102, **7**.180, **7**.206, ***7**.246*
primary elections ***4**.276*, ***7**.246*
primary health care **7**.159, ***7**.246–7*

primary industry 7.247
  primary commodities 7.57
  see also in particular agriculture; fishing; mining
primary medicine and health care see medicine and health care
primates 2.268
  aye-aye 2.19
  bushbaby 2.46, 2.47
  hominids 2.126
  lemur 2.187
  loris 2.194
  man see man
  marmoset 2.203
  nails 2.67
  potto 2.266
  tarsier 2.335
  tree shrew 2.172
  see also apes; monkeys
Primaticcio, Francesco 5.366
prime meridian see Greenwich Mean Time; latitude and longitude
prime ministers 7.247, 9.69–71
  of Australia:
    listed 9.69
    1901–15 4.23, 4.32, 4.94, 4.120, 4.286, 4.377
    1915–39 4.53, 4.158, 4.189, 4.202, 4.307
    since 1939 4.23, 4.72, 4.90, 4.155, 4.195, 4.217, 4.234, 4.378
  of Britain:
    listed 9.69
    1707–1801 3.56, 3.144, 3.283, 3.306, 3.376–7, 4.270
    1801–30 4.2–3, 4.35, 4.61, 4.137, 4.142, 4.197–8, 4.266, 4.270, 4.374
    1830–68 4.1, 4.97, 4.142, 4.216, 4.217, 4.257, 4.264, 4.295
    1868–1902 4.100–1, 4.136, 4.293, 4.300
    1902–40 4.21, 4.28–9, 4.59, 4.69, 4.191, 4.198, 4.203
    since 1940 4.22, 4.58, 4.75–6, 4.103, 4.108, 4.149, 4.205, 4.380
    prime minister's cabinet 7.38
  of Canada:
    listed 9.70
    19th century 4.190, 4.195, 4.204, 4.354
      see also Macdonald, Sir John
    1911–48 4.45, 4.85, 4.183, 4.190, 4.195
    since 1948 4.85, 4.98, 4.195, 4.198, 4.263–4, 4.300, 4.353
  of India 4.97, 4.130, 4.172, 4.236–7
  of Ireland 4.78, 4.86–7, 4.97–8, 4.119, 4.166
    listed 9.70
  of Italy 4.88–9, 4.94, 4.136, 4.212, 4.224
  of Japan:
    19th century 4.170, 4.385–6
    20th century 4.146, 4.160, 4.179, 4.184, 4.186, 4.230, 4.303, 4.337, 4.346, 4.386
  of New Zealand:
    listed 9.71
    19th century 2.142, 4.14, 4.21, 4.29, 4.184, 4.308, 4.369
    1900–39 4.77, 4.120, 4.189, 4.213, 4.303, 4.308, 4.370
    since 1940 4.124, 4.155, 4.184, 4.227, 4.233
  prime ministerial government 7.247
  see also cabinet; heads of state; presidents; and also Chancellors under politics under Germany
prime mover 6.281
prime number 1.267–8
prime vertical 8.128
primitive art 5.366
primitive moth 2.268–9
primitivism 5.367
Primo de Rivera, António 4.115
Primo de Rivera, Miguel 4.276, 4.322, 4.324
primogeniture 3.292
primordial fireball 8.34, 8.50
primrose family 2.85, 2.111, 2.254, 2.268, 2.269
Primrose League 4.276
prince 5.358
Prince Edward Island 1.53, 1.268
Princely states 4.276–7
Pringle, John 6.16
print and print-makers (arts) 5.367
  Belgium 5.147
  Britain 5.208
  cotton 5.115
  France 5.133, 5.171–2
  Germany 5.131
  Japan 5.158, 5.207, 5.209, 5.237, 5.467
  Netherlands and Flemish 5.148
  silk-screen 5.419
  technology see printing
  transfer 5.459
  United States 5.82, 5.120, 5.240–1
  woodcut 5.492
  see also engraving; etching; intaglio; lithography; relief; typography
printed circuit 6.281
printing 3.292–3, 6.281–2
  braille 6.44
  China 6.72
  collotype 6.80
  colour 6.81, 6.164
  computer 6.180, 6.195, 6.281
  diazo 6.101–2
  electronic page make-up 6.118
  film 6.101
  flexography 6.134
  graphic image manipulation 6.159
  gravure 6.123, 6.160, 6.282
  half-tone process 6.164
  ink 6.180
  ink-jet 6.180, 6.281
  innovation 3.66, 6.42, 6.105
  intaglio see printing under intaglio
  invention of 3.171
  laser 6.195
  letterpress 6.123, 6.199, 6.282, 6.334, 6.366
  Linotype 6.203, 6.219, 6.366
  lithography 6.205, 6.249, 6.282
  Monotype 6.232, 6.366
  offset 6.134, 6.249, 6.282
  photography 6.101, 6.268, 6.269
  pioneers 6.163, 6.169
  platemaking 6.272, 6.366
  screen 6.310, 6.353
  silk-screen 5.419
  stereotype 6.334
  teleprinter 6.347
  textiles 6.353, 6.387
  typesetting 6.203, 6.219, 6.232, 6.281–2
  wood-block 6.387
  see also etching; print and print-makers (arts); publishing; writing
Pripet Marshes 1.268
Priscus of Rome 3.27, 3.350
prism 1.68, 1.279, 6.64, 6.282, 6.328
  objective 8.111
prisons 7.123, 7.154
  Amnesty International 7.12, 7.218, 7.239, 7.313
  Bastille 3.36
  camps 4.12, 4.277
    concentration 4.82, 4.154, 7.63–4
prisons (cont.):
  life sentence 7.146
  parole 7.229
  political prisoners 7.238–9
  prisoners of conscience 7.12
  prisoners of war 7.328
  ransom for release 3.299
  for rape 7.258
  reform 3.170, 3.298, 4.101, 4.128
  see also convicts; law and justice; sentences
Prithviraj, king of Delhi 3.101, 3.293
privacy, right of 7.247
private education 7.154–5
private language 7.247
private law 7.247
  contract 7.69
  family 7.55, 7.115
  property 7.250, 7.331
  tort 7.312–13
  see also civil law
private press movement 5.305, 5.367
private property see property
private sector 7.203, 7.247
privateers 3.196–7, 3.293
  see also pirates
privatization 7.69, 7.86, 7.247–8
privet 2.269
probability 1.268
probate 7.248
probation 7.248
problem solving, optimization 6.251, 6.253
proboscis monkey 2.268, 2.269
proboscis worm 2.284
Probstein, Ronald F. 8.8
procedure 6.282–3
Procellarum Oceanus 8.112
process engraving 6.283
process manufacture 7.248
processor (computer) see central processing unit; microprocessor
Proclamation Line 3.293, 3.354
Procyon 8.128
production 7.248
  factors of 7.114, 7.249
    see also capital; labour; land
  management 7.248
  method 7.10, 7.248–9
  quality control 7.256
  research and development 7.266
  see also industrialization; manufacturing; mass production
productivity 7.249
profession 7.249
profile (geology) 1.268, 1.269
profit 7.249
Profumo, John 4.205
progeny testing 6.14, 6.21, 6.283
program, computer 6.282–3
  applications 6.17, 6.179, 6.329–30
  flow chart 6.137, 6.283
  see also software
programme music 5.367
programming language see language under computers
progress, royal 3.292, 3.293
Progressive Movement 4.277
Progressive Parties 4.277
Prohibition era 4.277, 4.369, 7.10
Project Apollo see Apollo
projectiles see missiles and under weapons
projection (psychology) 7.82
projections, map 1.268, 6.283–4
  see also Mercator
projector, film 6.75, 6.284
prokaryotes 2.269
  see also bacteria; cyanobacteria
Prokhorov, A. M. 6.214
Prokofiev, Sergey 5.367
proletariat see working class
PROLOG 6.86
prologue 5.367–8
Prometheus (Titan) 9.6, 9.29
promethium 1.107
prominences 8.128, 8.129
prominent moth 2.269
pronghorn 2.269
prontosil 6.338
proof-correction marks 9.64
proofing, digital 6.104
propaganda 7.4, 7.35, 7.196, 7.249, 7.249, 7.299
Propaganda (Roman Catholic missionaries) 7.203
propagation, plant 6.271
propane 1.268, 6.91, 6.150, 6.204, 6.237, 6.264, 6.284
propanone 1.268, 6.284, 6.295
propellant see explosives; rocketry
propeller 6.284
  -driven aircraft 6.25, 6.136, 6.284
propene 6.261, 6.276
proper motion 8.128
property/ownership:
  collective 7.56
  crime 7.250
  intellectual 7.162, 7.230
  land 7.56, 7.69
  law 7.250, 7.331
  public 7.212
  social 7.212
Prophet Messengers 9.2, 9.18, 9.29, 9.32
prophets 7.250
  Christianity 7.250
    see also John the Baptist
  Islam 7.250, 9.2, 9.17, 9.18, 9.20, 9.29, 9.32
    see also Muhammad
  Judaism:
    Amos 9.4
    Daniel 9.10
    Elijah 3.59, 3.114, 9.11
    Elisha 3.59, 3.114, 9.11
    Ezekiel 3.121, 3.122, 9.12
    Isaiah 3.99, 3.181, 9.18
    Jeremiah .3.186
    Nathan 9.26
    Samuel 3.314, 9.317
    see also John the Baptist
proportion 5.368
proportional representation 7.54, 7.104, 7.250
propylene 6.261
prose see biography and biographers; essays; fiction; novels; philosophy; stories
prosecution 7.3, 7.250
  see also courts of law
prospecting 6.228, 6.285
prosthesis 6.36, 6.285
  artificial limb 6.21
  haemodialysis 6.101, 6.164
  hearing aid 6.166
  hip replacement 6.169
  pacemaker 6.255
prostitution 7.46, 7.250–1
protactinium 1.107
Protagoras 7.295
protea 2.269–70
protected species 2.270
protectionism 7.251, 7.257
  tariffs 4.338–9, 7.306
protective coloration 2.270
  camouflage (cryptic) 2.50
  displays 2.94–5
  flash 2.119–20
  mimicry 2.212
  startle 2.324–5
  warning 2.360
Protectorates 3.293, 7.57
proteinases 6.122
proteins 2.270
  denaturation 2.90
  fibrinogen 2.116
  in food 2.92, 6.140, 6.141
  interferon 6.184
  single-cell (SCP) 6.320, 6.321

proteins (*cont.*):
see also amino acids; antibodies; enzymes
Proterozoic aeon **1**.133, **1**.134, ***1**.269*
Protestant ethic ***7**.251*, **7**.330
Protestants ***3**.293*, ***7**.251*
Adventists **7**.3, **7**.202
see also eschatology; millenarianism
Anabaptists **3**.12
Anglican *see* Anglicans
Baptists **3**.35, **4**.31, **4**.183, **7**.26, **7**.50
Black Churches **7**.32, **7**.44, **7**.50
'Brownists' **3**.51
Camisards **3**.58
charismatic **7**.44, **7**.232, **7**.250
Church of England *see* Anglicans
Churches of Christ **7**.48
Congregationalists **3**.85, **7**.65, **7**.245
Covenanters **3**.89–90
Cranmer **3**.90
Denefenestration of Prague **3**.290
Diet of Worms **3**.385–6
evangelicalism **7**.65, **7**.112
Fundamentalist *see* fundamentalism
Huguenots **3**.170–1
Jewel **3**.188
liturgy **7**.186
Lutherans *see* Luther; Lutheran Church
Melanchthon **3**.228
Mennonites **3**.229, **7**.198
Methodists *see* Methodists
missionaries **7**.203
Moravian Church **7**.207
as Nonconformists **3**.254
Pentecostal **7**.232, **7**.441
and philosophy **7**.176
Plymouth Brethren **7**.237
predestination **7**.245, **7**.260
Presbyterians *see* Presbyterians
Puritans *see* Puritans
Quakers *see* Quakers
Reformed *see* Reformed Churches
reformers **3**.384, **3**.391
sacraments **7**.271
Salvation Army **7**.272
Schmalkaldic War **3**.318
Shakers **4**.325
theology **7**.310
see also Calvin; Luther
Toleration Act **3**.358
Unitarians **4**.358, **7**.50, **7**.320
United Church of Christ **7**.65
United Reformed *see* United Reformed Church
wars *see* Thirty Years
see also Reformation
Proteus ***9**.29*
prothrombin **2**.357
protocol (computer) ***6**.285*
Protoctista **2**.270, **2**.313
see also algae; protozoa
protons **1**.26, ***1**.269*, **6**.23–4
accelerator **6**.258
microscope **6**.223
proton–proton reaction ***8**.128–9*
see also accelerator *under* particle(s)
proto-planet ***8**.129*
protoplasm **2**.86
proto-star ***8**.129*, **8**.160
protozoa ***2**.270*
amoeba **2**.8
and disease *see* infections
flagellate **2**.119
foraminiferan **2**.125
Paramecium **2**.242
radiolarian **2**.275
protractor (goniometer) **6**.157
proturan ***2**.270*
Proudhon, Pierre Joseph **4**.11, ***4**.278*
Proust, Marcel ***5**.368*, **5**.436
Prout, William **1**.280
Provence ***3**.293–4*
language **7**.156
literature **5**.297, **5**.298, ***5**.368*
proverbs **5**.7, **5**.148, ***5**.368*
Provisional IRA **4**.167
Proxima Centauri ***8**.129*
Prus, Boleslaw ***5**.368*
Prussia ***3**.294*, ***4**.278*
astronomy **8**.22, **8**.159
in Concert of Europe **7**.135
language **7**.26, **7**.156
political leaders **4**.62, **4**.146, **4**.329
provinces **4**.154–5, **4**.278, **4**.289
rulers:
listed ***9**.67*
17th and 18th centuries **3**.133, **3**.294
see also Frederick I; Frederick II (the Great)
19th and 20th centuries *see* Frederick William IV; William I, emperor; William II, king of Prussia
Three Emperors' League **4**.344
Tilsit Treaties **4**.345
wars and defence **7**.328
army leaders **4**.42, **4**.62, **4**.77, **4**.222, **4**.303–4
Austrian Succession **3**.29–30
Austro-Prussian **4**.26, **4**.299
Bavarian Succession **3**.37
with France *see* Franco-Prussian War
Napoleonic **4**.279, **4**.372
Revolutionary **4**.288
see also Seven Years
see also Germany
prussic acid **2**.56
Prynne, William ***3**.294*
Przybyszewski, Stanislaw **5**.355
psalm ***5**.368*
psaltery ***5**.368*, **5**.369
PSBR (public sector borrowing requirement) ***7**.254*
psephology **7**.224
pseudoscorpion ***2**.270*
Psyche ***9**.29*
psyche *see* soul
psychiatry ***7**.251*
analytical psychology **7**.173
disorders *see* mental illness *and under* personality
family therapy **7**.116
psychopathic personality **7**.14, ***7**.252–3*
see also psychoanalysis; psychology; psychoses
psychic phenomena *see* supernatural
psychoanalysis ***7**.252*
defence mechanisms **7**.82–3
dream interpretation **7**.95
hysteria **7**.150
motivation **7**.207
neo-Freudian *see* neo-Freudians
see also Freud, Sigmund; personality
psychokinesis **7**.228
psycholinguistics ***7**.252*, **7**.273–4
psychology ***7**.252*
aggression **7**.7
analytical **7**.173
anxiety *see* depression and anxiety
attachment *see* attachment
behaviour *see* behaviour; behaviourism (psychology)
child **7**.35, **7**.55
clinical **7**.53
cognitive *see* cognitive psychology
conditioning *see* conditioning
conformity **7**.64
consciousness **7**.150, **7**.169
counselling **7**.30, **7**.71–2, **7**.116
cross-cultural **7**.77
crowd **7**.77
dream interpretation **7**.95
education **7**.35
expectation **7**.113
psychology (*cont.*):
experimental **7**.334
fantasy **7**.118
feelings *see* emotion
and genetics **7**.129
Gestalt **7**.131
hedonism **7**.141
and homosexuality **7**.146
humanistic **7**.149, **7**.268
intelligence **7**.32, **7**.162
of learning **2**.340
locus of control **7**.187
meaning **7**.152–3, **7**.196, **7**.278
memory **7**.150, **7**.197–8
mental illness *see* mental illness
motivation **7**.113, **7**.207
myths **7**.210
obedience studies **7**.221
parapsychology **7**.228
perception **7**.232–3, **7**.234, **7**.279
person perception **7**.234
personality *see* personality
rationalism **7**.86, **7**.259
rationalization **7**.82
reaction-formation **7**.82
regression **7**.82
relationships *see* relationships
repression **7**.82
self *see* self
sex differences **7**.129, **7**.280
social *see* social psychology
test ***7**.252*
see also psychiatry *and entries beginning* psycho
psychopathic personality **7**.14, ***7**.252–3*
psychoses ***7**.253*
obsession **7**.221
see also hysteria; manic depression; mental illness; neuroses; obsession; paranoia; psychoses; schizophrenia
psychosexual development **7**.126, ***7**.253*
psychosomatic disorders ***7**.253*
psychotherapy ***7**.253*, **7**.268
behaviour therapy **7**.30
counselling **7**.30, **7**.71–2, **7**.116
family **7**.116
Gestalt therapy **7**.131
group therapy **7**.136
psychotropic drugs ***6**.285*, **6**.360, **7**.14, **7**.86, ***7**.96*, **7**.233, **7**.275
Ptah **9**.5, ***9**.29–30*
ptarmigan ***2**.270*, **2**.280
pteraspis ***2**.270*
pteridophyte ***2**.270–1*
clubmoss **2**.69
horsetails **2**.49–50, **2**.164
lepidodendron **2**.188
Rhynea **2**.284
see also ferns
pterodactyl ***2**.271*
PTFE (polytetrafluoroethene) **1**.151, **6**.273, ***6**.285*
Ptolemy (Claudius Ptolemaeus) ***1**.269*, **1**.270, ***6**.285*
Almagest and Ptolemaic system **6**.285, ***8**.3–4*, **8**.11, **8**.62, ***8**.129–30*
constellations **8**.27, **8**.28, **8**.47, **8**.158
and Eratosthenes **8**.48
geocentric system **8**.58
Hipparchus **8**.68
and Islam **8**.75
stellar brightness **8**.121
geography **1**.204, **1**.205
Ptolemy dynasty (Egypt) **3**.3, **3**.78, **3**.79, ***3**.294*, **3**.323, **6**.215
ptyalin **2**.292
Pu Songling ***5**.370*
Pu Yi **4**.278
puberty ***2**.271*
rites ***7**.253*
public choice theory ***7**.253–4*, **7**.259
public debt **7**.211
public finance ***7**.254*
public health ***7**.254*
see also medicine and health care
public law ***7**.254*
see also administrative law
public limited companies **7**.172
public opinion **7**.224
see also propaganda
public ownership **7**.212
public relations ***7**.254*
public sector **7**.203, ***7**.254*
borrowing requirement ***7**.254*
see also government
public service broadcasting ***7**.254*
public transport:
tram **6**.360
trolley-bus **6**.363
see also buses; railways
publishing ***7**.254–5*
desk-top **6**.85, **6**.99, **6**.118
see also printing
Puccini, Giacomo **5**.71, **5**.81, **5**.290, **5**.366, ***5**.368–9*
Puck (mythology) ***9**.30*
Puck (Uranian satellite) **8**.178
puddling pig-iron **6**.90, **6**.186, ***6**.285*
Puea, Te **4**.184
Pueblo Indians **3**.12, **3**.79, **9**.21
Puerto Rico ***1**.269*, ***3**.294*, ***4**.278*, **7**.54, ***7**.377*
puff adder ***2**.271*
puffball ***2**.271*
puffbird ***2**.271*
pufferfish **2**.271, **2**.272
puffin ***2**.271*
pug moth ***2**.271*
Pugachev, Emelian Ivanovich **3**.64, **3**.88, **3**.292, ***3**.294*
Pugin, Augustus Welby Northmore ***5**.369*
pūjā ***7**.255*
Pulci, Luigi **5**.231, ***5**.369*
pulley **6**.217, ***6**.285*
Pullman, George **6**.292
Pullman strike ***4**.278*
pulsars ***8**.130*
binary **8**.15
optical **8**.113–14
Vela **8**.180
pulsating universe **8**.115
pulsating variable *see under* variable stars
pulse code **6**.285, **6**.346, **6**.347
modulation (PCM) ***6**.285*
pulsometer **6**.227
puma ***2**.271–2*
pumice ***1**.269*, **6**.1
pump **6**.93, ***6**.285–6*, **6**.383
Peltier **6**.167, **6**.354
vacuum **6**.227, **6**.370
water **6**.285–6, **6**.385
pumped storage ***6**.286*
pumpkin family **2**.194, ***2**.272*, **2**.323
punched-card machines *see* Hollerith; Jacquard; numerically controlled
punctuation marks ***9**.64*
Punic wars ***3**.294–5*, **3**.319
punishment ***7**.255*
capital **3**.150, **7**.40, **7**.146
see also execution; prisons; sentences
Punjab ***4**.278*, **4**.282–3, **7**.279
language **7**.157
religion/theology *see* Sikhism
punkies **2**.210
pupae **2**.65, ***2**.272*
puppet state ***7**.255*
puppetry **5**.108, **5**.188, **5**.235, **5**.236, ***5**.369–70*, **5**.413
Puppis (constellation) ***8**.130*
Puppis Ā (supernova remnant) ***8**.130*
Purāṇas ***5**.370*
Purandaradasa **5**.134
Purbeck marble ***1**.269*

Purcell, Edward **8.**176
Purcell, Henry **5.**18, **5.**50, ***5.**370*
purdah ***7.**255*
Pure Land Buddhism **7.**140, ***7.**255*, **9.**4
purgatory ***9.**30*
purges ***7.**255*
see also under Stalin
purification:
air **6.**120
water **6.**378–9
Purim ***7.**256*
Purism **5.**260, ***5.**370*
Puritans ***3.**295*
Diggers **3.**103–4
Fifth-rulers Men **3.**127
general **3.**123
Levellers and leader **3.**207–8
ministers **3.**37, **3.**112, **3.**225
pamphleteer **3.**294
radical **3.**383
see also Roundheads
purple emperor butterfly ***2.**272*
purse seine **6.**239
Purusha **7.**41, ***7.**256*
Pūrva Mīmāmsa **7.**155
pus **2.**1, ***2.**272*
Pushkin, Aleksandr ***5.**370*, **5.**371
puss moths **2.**192, **2.**269, ***2.**272*, **2.**273
Putnam, Israel ***3.**295*
putrefaction ***2.**272*
Puttana, M. S. **5.**136
putto ***5.**371*
putty ***6.**268*
Puvis de Chavannes, Pierre ***5.**371*
Puyi (Pu Yi), emperor of China **4.**209, **4.**279
puzzles ***7.**256*
PVC (polyvinyl chloride) **1.**218, **6.**71, **6.**106, **6.**271, **6.**274, ***6.**287*
PVD (physical vapour deposition) **6.**370
PWR (pressurized-water reactor) **6.**243, **6.**244, ***6.**281*
Pygmalion **9.**10, ***9.**30*
pygmy ***7.**256*
pygmy shrew **2.**200
pygopodid ***2.**272*
Pym, John **3.**127, **3.**144, **3.**209, **3.**270, ***3.**295–6*
Pynchon, Thomas Ruggles ***5.**371*
pyramidal peak **1.**140
Pyramids ***3.**296*, ***6.**287*
battle of ***3.**296*
Pyramus and Thisbe ***9.**30*
Pyrenees ***1.**269–70*
pyrethrum ***2.**272*, **6.**180
pyridoxine **2.**357
pyrite ***1.**270*
pyrochlore **6.**241
pyroclasts ***1.**270*, **1.**271, **1.**295
pyrolusite **6.**214
pyrometer **6.**351
pyroxene **1.**26, ***1.**270*
Pyrrhus **3.**117, ***3.**296*
pyruvic acid **2.**138
Pythagoras of Samos and Pythagoreans ***1.**270–1*, **1.**343, **5.**403, ***7.**256*, **8.**61
Pytheas ***1.**271*
Python (mythology) ***9.**30*
python (snake) ***2.**272*
Pyxis ***8.**130*

Q

Qaboos bin Said **4.**250
al-Qaddafi, Muammar **4.**195, ***4.**279*
Qajar tribe ***4.**279*, **5.**344
Qat **9.**30
Qatar ***1.**271*, **4.**252, ***4.**279*, **7.**15, **7.**136, ***7.**377*
economy and trade *see* OPEC
Qattara Depression ***1.**271*
Qianlong, emperor of China ***3.**296*, **3.**297
qin (zither) **5.**251
Qin dynasty **3.**72, ***3.**296*, **3.**326
see also Shi Huangdi
Qing dynasty ***3.**296*, ***4.**279*
Cixi **4.**48, **4.**76, **4.**178, **4.**279
Kangxi **3.**193–4, **3.**250, **3.**296
and Manchus **3.**218
Puyi **4.**209, **4.**278
Qianlong **3.**296, **3.**297
Taiping rebellion **4.**337
see also Cixi
Qu Yuan ***5.**372*
quadrant ***6.**287*, ***8.**131*
Quadrantids ***8.**131*, **8.**195
quadrature **8.**122
quadrilaterals **9.**51
quadriplegia **2.**242
Quadruple Alliance ***4.**279*
quagga ***2.**273*
quahog **2.**67
quail **2.**48, ***2.**273*
Quakers **3.**130, **3.**148, **3.**275, ***3.**296–7*, ***7.**256*
quality assurance and control **6.**216, ***7.**256*
quango ***7.**256*
quantity theory of money **7.**120, **7.**126, ***7.**256–7*
quantum mechanics and quantum theory ***1.**271–2*, **6.**17, ***8.**131–2*
and atom **1.**24, **1.**26
and Born **1.**41–2
complementarity **1.**70
de Broglie wavelengths **1.**85, **1.**295
and Dirac **1.**92
and Einstein **1.**258, **1.**271, **8.**45, **8.**131
Heisenberg *see* uncertainty *below*
and Planck **1.**258, **1.**271, **8.**121, **8.**131
quantum number ***1.**271*
Schrödinger equation **1.**294–5
Stern–Gerlach experiment ***1.**320*
uncertainty principle and Heisenberg **1.**153, **1.**349, **7.**42
zero-point energy **1.**373
see also complementarity; elementary particles
Quarenghi, Giacomo ***5.**371*
quarks **1.**69, **1.**123, ***1.**272*, **1.**322
Quarrel of Ancients and Moderns ***5.**371*
quarrying ***6.**287*
quartz **1.**143, ***1.**272*, **6.**275
amethyst **1.**11
clock and watch **6.**77, **6.**377
crystal **1.**80, **1.**81
onyx **1.**238
rock-crystal **1.**284
see also chalcedony; sand
quartzite ***1.**272*
quasar ***8.**132*
Quasimodo, Salvatore ***5.**371*
quassia ***2.**274*
Quaternary period **1.**133, **1.**134, ***1.**272*
see also Holocene; Pleistocene
quatrain **5.**247, ***5.**371*
Quatremère de Quincy, Antoine **5.**315
quattrocento **5.**371
Quebec ***1.**273*, **3.**284, ***3.**297*, ***4.**279–80*
Liberation Front ***4.**280*
Québécois ***7.**257*
Quechua language **5.**426, **7.**295
Queen Anne style ***5.**371–2*, **5.**415
Queen Anne's War **3.**134, ***3.**297*, **3.**338
Queen's shilling ***3.**297*
Queensberry, Marquess of: boxing rules **7.**34
Queensland ***1.**273*, ***4.**280*
quenching ***6.**287*, **6.**290
Quercia, Jacopo della ***5.**372*
Quesnay, François ***3.**297*, **7.**100
quetzal ***2.**274*
Quetzalcóatl **3.**31, **3.**88, ***3.**297–8*, **3.**358, ***9.**30*
Quevedo y Villegas, Francisco Gómez de ***5.**372*
Quezon, Manuel Luis ***4.**280*
Quiberon Bay, battle of ***3.**298*
quicklime *see* lime
quicksand ***1.**273*
Quilley, Dennis **5.**327
quilting ***5.**372*
quince ***2.**274*
quinine **2.**66, **6.**72, **6.**169–70
Quintilian ***5.**372*
Quinton, Anthony **7.**192
Quirius ***9.**30*
Quisling, Vidkun Abraham Lauritz Jonsson **4.**247, ***4.**280*
Quito **3.**175, ***3.**298*
Quito Protocol **7.**13
quoits **7.**147
quotas ***7.**257*
Qur'ān *see* Koran
Qutb ud-Din Aibak, sultan of Delhi **3.**101, **3.**218

R

R Coronae Borealis star ***8.**134*
Ra **9.**5
Raabe, Wilhelm **5.**175
rabbi ***7.**257*
rabbit **2.**121, ***2.**274*
rabbit-eared bandicoot ***2.**274*
rabbitfish ***2.**274*
Rabearivelo, Jean-Joseph **5.**7
Rabeh, az-Zubayr ***4.**280*
Rabelais, François ***5.**372*, **5.**402
Rābī'a of Syria **9.**30
Rabi'ah al-'Adawiya of Basra **5.**439, ***9.**30*
rabies **2.**243, ***2.**274*, **2.**354, **6.**177, **6.**259
Rabin, Itzak **4.**170
Rabinowitz, Sh. **5.**497
Rabuka, Sitiveni **4.**119
race ***2.**275*, ***7.**257*
genocide **7.**130
Race Relations Act ***4.**280*
racial types **3.**65, **3.**239, **3.**249
see also black people; ethnic groups; racism
racemic mixture ***1.**273*
Rachel (1820–58, actress) ***5.**372*
Rachel (Jewish matriarch) **9.**19, ***9.**30*
Rachmaninov *see* Rakhmaninov
Racine, Jean ***5.**372*
racing:
car ***6.**288*, ***7.**207–8*
horse **7.**147
running **7.**192, **7.**221
see also sport and recreation
racism **4.**137, ***7.**257*, **7.**278
see also civil rights; race
rack-and-pinion railway ***6.**288*, **6.**334
racket (musical instrument) ***5.**372*
racketball **7.**298
Rackham, Arthur **5.**53
racoon family ***2.**275*
coati **2.**70
kinkajou **2.**179
panda **2.**240–1
Radama II, king of Madagascar **4.**205
radar ***6.**288*
and aircraft **6.**9, **6.**10, **6.**26–7, **6.**139, **6.**224, **6.**379–80
astronomy ***8.**132*
AWACS **6.**26–7
homing *see* guided missiles
jamming **6.**118–19
Lanby buoy **6.**193
length measurement **6.**198
pioneer **6.**379–80
waves **6.**382
Radcliffe, Ann **5.**348, ***5.**372–3*
Radek, Karl ***4.**280*
Radetzky, Josef, count ***4.**280–1*
Rādhā **9.**22
Radhakrishnan, Sir Sarvepalli ***4.**281*, **7.**156
radial velocity ***8.**132*
radian (geometry) ***1.**273*
radiant (astronomy) ***8.**132*
radiation:
background **1.**29
belts ***8.**133*
black-body **1.**40
Cerenkov **8.**24
cosmic rays **1.**77–8, **8.**31–2, **8.**98–9
diffusion **1.**91
extinction **8.**49
long-wave **1.**103, **6.**380, **9.**54

radiation (*cont.*):
medical *see* radiology; radiotherapy
microwave background 8.98–9
nuclear *see* nuclear radiation
pressure 8.133
short-wave 1.103, 1.301, 6.380, 9.54
thermal 1.273–4
Van Allen radiation belt 1.353, 8.179–80
see also electromagnetic radiation; luminosity *and also under* Sun/solar
radiator:
in car 6.89
in house *see* central heating
radicals, free 1.127
radio 6.117, 6.288–9, 7.257–8
altimeter 6.12, 6.290
amplifier 6.14
and astronomy *see* radio astronomy; radio galaxies; radio telescopes
British *see under* telecommunications *under* Britain
broadcasting *see* broadcasting
CB 6.65
Decca Navigator 6.97
and education 7.93
frequency 9.54
modulation *see* modulation
navigation 6.208–9, 6.237, 6.238, 6.249–50, 6.290
paging system 6.255
pioneers 1.154, 6.20, 6.128, 6.168, 6.215
propaganda 7.35, 7.249
receiver 6.290, 6.291, 6.361
source count 8.33
-telephone 6.65, 6.292, 6.347
telescopes *see* radio telescopes
transponder 6.361
waves 1.274, 6.382
waveband 6.380, 9.54
see also long-wave; medium-wave; short-wave; ultra-high frequency; very high frequency
wireless telegraphy 6.45
see also AM; FM; radar
radio astronomy 8.11, 8.33, 8.133
aperture synthesis 8.6
radio source 8.133
radio source counts 8.33
see also radio galaxies; radio telescopes
radio galaxies 8.23, 8.35, 8.133
radio telescopes 6.292, 6.347–8, 8.134
dish 8.38
Effelsburg 8.43
Lovell (Jodrell Bank) 6.291, 8.76, 8.78, 8.87
RATAN 600 8.134
see also telescopes
radioactive elements and isotopes 1.176, 6.188, 6.290
dating 1.274, 6.96, 6.199–200, 6.290, 8.133
deuterium (hydrogen) 1.88–9
in heavy water 6.167
hydrogen 6.369
iodine 6.186
in medicine *see* nuclear medicine
as nuclear fuel 6.243, 6.245
radioimmunassay 6.290–1
strontium-90 6.336
thorium 6.25, 6.243, 6.285, 6.355
see also nuclear radiation; plutonium; radioactivity; radium; uranium
radioactivity 1.274
fall-out and accidents 6.125, 6.246, 6.249, 6.281, 6.355
Chernobyl 6.9, 6.72, 6.243, 6.245, 6.275, 7.91
radioactivity (*cont.*):
half-life 1.150
pioneers 1.81–2
see also under nuclear physics
polonium 1.107, 1.263
unit (becquerel) 1.35
waste disposal 6.243, 6.245
see also gamma-radiation; radioactive elements and isotopes *and entries beginning* nuclear
radio-carbon dating *see* dating *under* radioactive elements
radio-frequency heating 6.101, 6.224, 6.251, 6.290
radioimmunoassay 6.290–1
radiolarian 2.275
radiology 6.30, 6.210, 6.291
see also tomography
radiometer 6.222, 6.291, 8.133
radiometric dating 1.133, 1.274
radiosonde 6.291–2
radio-telephone 6.65, 6.292, 6.347
radiotherapy 6.57, 6.218, 6.292
radish 2.275
radium 1.81–2, 1.107, 1.274, 6.218, 6.292
radius (bone) 2.14
radius vector 8.80
radon 1.107, 1.274
Raeburn, Sir Henry 5.359, 5.373
Raeder, Erich 4.281
raffia 2.275
Raffles, Sir Thomas Stamford 4.281, 4.313
Rafsanjani, Hashemi 4.165
raft 6.292
rāga 5.220, 5.221
ragfish 2.275
Raglan, FitzRoy James Henry Somerset, 1st Baron 4.27, 4.88, 4.281
Ragnarök 9.23, 9.25, 9.30, 9.39
ragtime 5.242, 5.373
ragworm 2.42
ragwort 2.275
Rahman, Sheikh Mujibur 4.26, 4.30, 4.256
Rahman, Tungku (Prince) Abdul 4.208, 4.281
Rai, Lala Rajpat 4.20
Raikes, Robert 3.298, 4.333
rail (bird) 2.275
coot 2.75
gallinule 2.131
moorhen 2.131
notornis 2.230
weka 2.364
rail-bus 6.292
railways 4.281–2, 6.292–4
automatic train control 6.25–6
brakes 6.44
brake-van 6.44
bridges 6.49, 6.111, 6.293
buffers 6.49
carriage 6.292
Channel Tunnel 6.69, 6.365
construction 6.292–3
coupling 6.91
engines *see* locomotives
funicular 6.148
gauge 6.150–1
goods wagon 6.157
high-speed train 6.168–9, 6.294
history 6.292, 6.293–4
marshalling yard 6.215, 6.216
multiple-unit train 6.235
pioneers 6.334
rack-and-pinion 6.288, 6.334
rolling stock 6.44, 6.257, 6.292
signal system 6.294, 6.312, 6.359
suspension 6.341
track 6.294
track circuit 6.359
tunnel 6.69, 6.293, 6.365, 7.67
railways (*cont.*):
underground 6.114, 6.132, 6.161, 6.294, 6.367
viaduct 6.293, 6.371–2
see also under individual countries
rain 1.274
acid *see* acid rain
and altitude 1.241, 1.274–5
bar chart 1.144
and depression 1.88
drizzle 1.96
and erosion *see* weathering *and also* erosion *under* limestone
fronts 1.126, 1.262
see also depression (meteorology)
lack of *see* drought
and latitude 1.274–5
measurement 6.221
monsoons 1.220
rain-shadow 1.274–5
runoff *see* runoff
shower 1.302
sleet 1.307
rain forest *see* tropical rain-forest
rain frog 2.276
rainbow 1.274, 1.275
rainbow trout 2.275–6
Rainer, Yvonne 5.362, 5.373
Rainey, P. M. 6.285
Rainier, Mount 1.274
Rainis, Jānis 5.34
raised beach 1.275
raisins 2.143
Rajagopalachariar, Chakravarti 4.282
Rajaraja I, king of Chōla 3.74
Rajendra I, king of Chōla 3.74
Rajpat Rai, Lala 4.152
Rajput 3.283, 3.298
painting 5.373
rake 6.6, 6.294
Rakhmaninov, Sergey 5.366, 5.373
Rakhsh 9.33, 9.34
Rakosi, Matyas 4.282
Rakshasa 9.31
raku 5.373
Raku Jizo 9.20
Raleigh, Sir Walter 1.84, 1.139, 1.275, 3.298–9, 3.361
RAM (random access memory) 6.83, 6.218–19
Ram Das 5.418
Rāma (goddess) 7.326, 9.15, 9.30–1
Rama I, king of Siam 3.327, 4.54
Rama III and IV, king of Siam 4.222
Rama V, king of Siam 4.75
Ramadan 7.258
Ramaena, Colonel Elias 4.193
Ramakrishna, Sri 4.152
Raman, Chandrasekhara: Raman effect 1.275
Rāmānuja 3.299, 7.325
Rāmāyaṇa 5.373, 5.401, 7.258
Ramazzini, Bernadino 2.276
Rambert Dance Company (Ballet Rambert) and Marie Rambert 5.13, 5.24, 5.62, 5.299, 5.373, 5.461
Rambouillet, Catherine de Vivonne, marquise de 3.299
rambutan 2.191
Rame, Franca 5.159
Rameau, Jean-Philippe 5.373–4
Ramesses II (the Great), pharaoh 3.2, 3.243, 3.278, 3.299
Ramesses III, pharaoh 3.299
Ramillies, battle of 3.299
ram-jet 6.294
Ramos, Fidel 4.267
ramps 6.177–8
Ramsay, Allan 5.359, 5.374
Ramsay, Sir Andrew Crombie 1.275
Ramsay, Sir William 1.149, 1.275–6, 1.309
Ramsden, Jesse 6.294
Ramsden eyepiece 8.134
ramshorn snail 2.276
Ranade, Mahadev Govind 4.152
Ranariddh, Norodom 4.58
Rance barrage 6.294–5
Rand 1.368
Randolph, Asa Philip 4.282
random access memory (RAM) 6.83, 6.218–19
random number 1.276
Ranggawarsita 5.222
Rangi and Papa 9.31
Ranjit Singh 3.328, 4.282–3, 7.9
Ranke, Leopold von 4.153
ransom 3.299
Ransome, J. R. and A. 6.359
ranunculus *see* buttercup
Rao, Narasimha 4.161
Rao, Raja 5.374
Rao Jodha, Raja of Jodhpur 3.189
Raoult, François Marie 1.276
Rapallo, Treaties of 4.283
rape (crime) 7.258
punishment for 7.40, 7.255
rape (plant) 2.276, 2.278
Raphael, Saint 9.43
Raphael (painter) 3.301, 5.42, 5.64, 5.146, 5.359, 5.366, 5.374
Rapid Eye Movement sleep 7.95
rapids 1.276
rare earths *see* lanthanides
rare gases *see* noble gases
rash 2.353
Rashid Ali al-Ghailani 4.283
Rasmussen, Knud Johan Victor 1.276
raspberry 2.129, 2.278
rasps 6.129
Rasputin, Grigori Yefimovich 4.283
Rastafarianism 7.258
Rastrelli, Bartolomeo Francesco 5.375
rat *see* rats
rat kangaroo 2.278
rat opossum 2.278
RATAN 600 8.134
Ratana, Tahupotiki Wiremu 4.283
ratchet and pawl 6.295
rate of reaction (chemistry) 1.60
ratel 2.278
ratfish 2.63
rath 5.375
Rathenau, Walther 4.283
rational choice theory 7.253, 7.259
rational communication 7.138
rationalism 7.86, 7.259
rationalization (psychology) 7.82
rats 2.278
diseases from 2.278, 2.364
learning 2.186
marsh 2.362
rat flea 2.121
water 2.362
see also rodents
Ratsiraka, Didies 4.205
rat-tail (fish) 2.278
rat-tailed maggot 2.165
rattan 2.278
rattles 5.282, 5.375, 5.420–1
rattlesnake 2.278
Rauschenberg, Robert 5.375
Ravaillac, François 3.162
Ravana 9.15, 9.31
Ravel, Maurice 5.10, 5.375
raven 2.278
ravine 1.276–7
Rawdon, Lord 3.88
Rawlings, Flight-Lieutenant Jerry 4.136
Rawls, John 7.259, 7.290, 7.291
Ray, Bharatachandra 5.43
Ray, Man 5.347, 5.375
Ray, Satyajit 5.375
ray *see* rays
Rayet, Georges: Wolf-Rayet star 8.187
Rayleigh, Lord *see* Strutt

rayon **6**.66, **6**.92, **6**.128, ***6**.295*
Rayonnant style ***5**.375*
rays (fish) ***2**.278–9*
manta **2**.202
sawfish **2**.294
sting-ray **2**.261–2, **2**.326
torpedo fish **2**.105, **2**.344
rays (physics) *see* radiation
Razin, Stenka **3**.88
razorbill ***2**.279*
razor-shell ***2**.279*
RDF (refuse-derived fuel) **6**.377
RDX (explosive) **6**.124, **6**.272
Re **9**.4, ***9**.31*
reactance **6**.76, ***6**.295*
reaction-formation (psychology) **7**.82
Read, Richard (Jr) **4**.205
Read, W. G. **4**.294
reading ***7**.259*
dyslexia **7**.97
see also literacy; literature
Reagan, Ronald ***4**.283–4*, **4**.287
assassination attempt **7**.20
and Central America **4**.87, **4**.241
and New Right **7**.215
and SDI **4**.330, **6**.335, **7**.300
reagent ***1**.277*
real numbers ***1**.277*
real time ***6**.295*
realism (arts) ***5**.375–6*
film **5**.101, **5**.391
literature *see* naturalism and realism
paintings **5**.116, **5**.308–9, **5**.375–6, **5**.380, **5**.447, **5**.472
realism (philosophy) ***7**.259*
image ***7**.152*
māyā **7**.195
see also utilitarianism
realism (power politics) **7**.244
Socialist **5**.380, **5**.394, **5**.422, **5**.424, **5**.447
reality and science ***1**.277*
see also realism
reamer ***6**.295*
reaper **6**.5, **6**.172, **6**.210, ***6**.295*
reason (philosophy) ***7**.259–60*
Reason, Age of ***3**.299–300*
reasoning **7**.89
see also logic
rebab ***5**.376*
rebec ***5**.376*
Rebecca (Jewish matriarch) ***9**.31*
Rebecca riots (19th century) ***4**.284*
receiver:
radio **6**.290, **6**.291, **6**.361
telephone **6**.347
television **6**.349–50
Recent epoch *see* Holocene
receptor and drugs **6**.265–6
recession (economics) ***7**.260*
recession of galaxies ***8**.134*
recharge ***1**.277*
reciprocal ***1**.277*
reciprocating engines **6**.92, ***6**.295–6*
beam-engine **6**.32, **6**.241, **6**.330
diesel engine **6**.33, **6**.102, **6**.263, **6**.339
gas-engine **6**.150, **6**.198
Stirling engine **6**.335
see also engines *under* petrol
reciprocating pumps **6**.286
recitative **5**.328
Reck, Hans **1**.237
reclamation *see under* land
recombinant DNA **6**.153
reconnaissance aircraft ***6**.296*
Reconstruction Acts ***4**.284*
record **6**.159, ***6**.296*, **6**.326, **6**.334
player **6**.33, **6**.112, **6**.158–9
recorder (musical instrument) ***5**.376*
recording:
sound *see under* sound
visual image *see* video
recreation *see* games; leisure; sport and recreation; tourism
recrystallization ***1**.277*
rectifier and rectification ***6**.296*, **6**.323
diode **6**.104
half-wave **6**.74
in radio receiver **6**.289, **6**.291
rectum ***2**.279*
recurrent novae **8**.109
recusant ***3**.300*
recycling ***6**.296*, **6**.377, ***7**.260*
red admiral butterfly ***2**.279*
Red Army Faction (Germany) **7**.308
Red Army (USSR) **4**.188, ***4**.284*
Red Brigades (Italy) ***4**.284*, **7**.308
red bug ***2**.279*
red clover **2**.160
red coral ***2**.279*
Red Cross ***4**.284*, **7**.218
red currant ***2**.279*
red deer ***2**.279–80*
red dwarfs (stars) ***8**.39*
Red Feds (New Zealand) ***4**.285*
red giants **8**.3, **8**.7, ***8**.134*
red grouse ***2**.280*
Red Guards (China) ***4**.285*
red lead ***6**.296*
red oak **2**.231
red panda ***2**.280*
Red River ***1**.277*
Settlement ***4**.285*
Red Sea ***1**.277–8*, **1**.326
red sea bream ***2**.280*
red shift ***8**.134–5*
Red Shirts (India) ***4**.285*
red spider mite ***2**.280*
red supergiants **8**.6
red-backed vole **2**.21–2
reddening, interstellar **8**.74
Reddi, Duvurri Rami **5**.136
redfish ***2**.280*
Redl, Alfred ***4**.285*
Redmond, John Edward ***4**.285*
Redon, Odilon ***5**.376*
redshank ***2**.280*
redstart ***2**.280–1*
reduction (chemistry) ***1**.278*, **6**.254
see also oxidation and reduction
reductionism (philosophy) **7**.223, ***7**.260*
redundancy **6**.297
redwing ***2**.281*
redwood ***2**.281*
Reed, Sir Carol ***5**.376*, **5**.383, **5**.485
Reed, Walter ***2**.281*
reed ***2**.281*
reed instruments ***5**.376–7*
bagpipes **5**.13, **5**.29, **5**.349, **5**.376
bassoon **5**.36, **5**.38, **5**.352, **5**.376
crumhorn **5**.119
hornpipe **5**.212
oboes **5**.324, **5**.352, **5**.376
racket **5**.372
sansa **5**.400
saxophone **5**.352, **5**.376, **5**.403
shahnāī **5**.413
zummāra **5**.499
see also clarinets; free-reed; shawms
reed warbler ***2**.281*
reedfish **2**.29
reefs ***1**.278*
coral **1**.24, **1**.33, **1**.145, ***1**.278*
reeve ***3**.300*
Reeves, A. H. **6**.285
Reeves, Jim **5**.115
Reeves, William Pember ***4**.285*
reference plane **8**.29, **8**.86
reference tables ***9**.45–75*
referendum ***7**.260*
reflation ***7**.260*
reflecting telescopes *see under* telescopes
reflection ***1**.278*, ***1**.279*, **6**.228, **6**.251, **6**.382, **8**.3
nebula ***8**.137*
reflex **2**.245, ***2**.281*
Reform Acts ***4**.285*
Reform Judaism ***7**.260*, **7**.304
Reformation **3**.273–4, ***3**.300*, **3**.301
see also Protestants
Reformed Churches **7**.65, ***7**.260*
United **7**.50, **7**.65, **7**.112, **7**.245, **7**.322
see also Calvin; Church of Scotland
reforming process (chemistry) ***6**.296–7*
catalytic **6**.34, **6**.262
hydrotreatment **6**.175
refracting telescope (or refractor) **8**.65, ***8**.137*
refraction ***1**.278*, ***1**.279*, **6**.382, ***8**.137*
and lenses **6**.2, **6**.199, **6**.251
and prism **6**.251, **6**.282
refractive index ***1**.278*, **1**.280
Snell's law **1**.308
telescope *see* refracting telescope
wave **1**.363–4
refractory **6**.47, ***6**.297*
refrigeration **6**.141, **6**.147, ***6**.297*
chloro-fluorocarbons **6**.72
heat-pump **6**.167
and nitrogen **6**.242
pioneer **6**.202
see also freezing
refugees **4**.161, **7**.227, **7**.238, ***7**.260–2*
UN Commisssion on **7**.262, **7**.321, **7**.322
see also migration (people)
refuse-derived fuel **6**.377
reg ***1**.280*
Regency ***4**.285–6*
style **5**.312–13, ***5**.377*
Reger, Maximilian ***5**.377*
reggae **5**.358
regional policy ***7**.262*
regionalism (international) ***7**.262*
see also in particular European Community; Organization of African Unity; Organization of American States *and* Arab League *under* Arabs
regionalization **7**.88, ***7**.262*
register (computing) **6**.67
regolith ***1**.280*, ***8**.137*
regosols ***1**.280*
regression (psychology) **7**.82
regression analysis (mathematics) ***1**.280*
regulation, government **7**.86, **7**.120, ***7**.262–3*
see also act; law
Regulators ***3**.300*
Regulus, Marcus Atilius ***3**.300*
Regulus (astronomy) ***8**.137*
rehabilitation **7**.262, ***7**.263*
Rehoboam, king of Israel ***3**.300*, ***9**.31*
Re-Horakhty **9**.17
Reich, Steve **5**.297, ***5**.377*
Reich, Wilhelm **7**.214
Reichenbach, H. M. **6**.111
Reichstag ***4**.286*
Reid, Sir George Houstoun ***4**.286*
reincarnation **7**.6, ***7**.263*, **7**.276
see also Dalai Lama
reindeer ***2**.282*
reindeer moss ***2**.282*
reinforced composite *see under* composite
reinforced concrete **6**.51, **6**.82, ***6**.297*
bridge **6**.213–14
caisson **6**.53
dams **6**.95
foundation **6**.144
frame **6**.146
glass-fibre **6**.86, **6**.156
pioneer **6**.213–14
roof **6**.302
skyscrapers **6**.322
reinforced plastics *see* fibre-reinforced plastics
reinforcement of learning **7**.64
Reinhardt, Max ***5**.377*
Reiniger, Lotte **5**.17, **5**.413
Reinmuth, Karl **8**.7, **8**.65
Reiss, William **1**.78
Reisz, Karel **5**.163
Reith, John Charles Walsham, 1st Baron ***4**.286*, **7**.28, **7**.254
Rej, Mikolaj ***5**.377*
Rejlander, Oscar Gustave ***5**.377*
rejuvenation of rivers ***1**.280*, **1**.333
relationships ***7**.263*
see also attachment; children; family; groups; marriage; sexuality; society
relative atomic mass ***1**.280*
relative molecular mass ***1**.280*
relativism ***7**.263*, **7**.296
relativity and relativity theory **1**.258, ***1**.280–1*, **6**.17
dimensions **1**.91
general **6**.17, **8**.17, **8**.33–4, **8**.44, **8**.60, **8**.69, **8**.96, ***8**.137*, **8**.181
special **1**.281, ***8**.137–8*
speed of light **1**.123, **1**.317
as thought experiment **7**.311
see also space–time
relay ***6**.297*
reliability ***6**.297*
relic ***3**.300*
relief (art) **5**.344, ***5**.377*, **5**.387
relief (land) *see* geomorphology; topography
relief printing **6**.199
Religion, French Wars of ***3**.134*, **3**.245, **3**.313, **3**.356
religion and theology ***7**.263*, ***7**.265*, ***7**.309–10*
afterlife **7**.5–6, **7**.140, **7**.141, **7**.217
ancestors *see* ancestor worship
asceticism **7**.9, **7**.19, **7**.86, **7**.210
Baha'ism **4**.27, ***7**.25*
blasphemy **7**.32, **7**.33
canon law **7**.39
cathedrals, mosques, and churches *see* architecture; stained glass
Christian Science **4**.75, **7**.48
and communal conflict **7**.59
conversions *see* missionaries
covenant **7**.73
deities *see* deities
Druids **3**.106
education ***7**.265*
fetish **7**.119
flagellants **3**.127
and food **7**.123
see also totemism
free will **7**.125
gods *see* deities
and healing **7**.139
and law **7**.39
see also under individual religions in particular Islam
liberation theology **7**.184
Manichaeism **3**.218, **7**.19
monasteries *see* monasticism
and music *see* sacred music
mystic *see* mysticism
mythical figures and places ***9**.1–39*
oracle **3**.262
and politics **7**.263, **7**.264, ***7**.265*, **7**.309
Rastafarianism **7**.258
ritual *see* rituals
scholasticism **3**.318, **3**.319, **7**.1, **7**.275, **7**.309–10
Scientology **7**.276
sermon **5**.411–12
shamanism **7**.11, **7**.16, **7**.181, ***7**.282*, **7**.335
soul *see* soul
spirits *see* spirits
theocracy **7**.309, **7**.310
theosophy **7**.299, **7**.310
toleration **3**.100, **7**.312
totems *see* totemism
Unification Church **7**.320

religion and theology (*cont.*):
see also Buddhism; Christianity; Daoism; Hinduism; Islam; Jews; Mormons; Protestants; Roman Catholicism; Salvation Army; Shinto; Sikhism; Voodoo; Zoroastrianism; *and under individual countries*
REM sleep **7**.95
Remarque, Erich Maria ***5**.377–8*
Rembrandt Harmensz van Rijn **5**.90, **5**.359, ***5**.378*, **5**.410
Reményi, Eduard **5**.57
remora ***2**.282*
Remsen, Ida **6**.342
Remus **9**.31
ren **7**.65, ***7**.265*
Renaissance ***3**.301–2*, **6**.19
and Arab technology **6**.19
arts ***5**.378–80*
architecture, Italian ***5**.378*
Florence *see under* Florentine art
in France **5**.411
historian **5**.469
Milan **5**.57
Palladian *see* Palladio
Perugia **5**.345
Rome **5**.57–8, **5**.293–4, **5**.345, **5**.374, **5**.400, **5**.475
theorist **5**.10–11
Venice **5**.400, **5**.401–2
Verona **5**.400
dance ***5**.379*
France **5**.162–3, **5**.366, **5**.411
Germany **5**.139
grotesque **5**.195
Italy *see* architecture *above and* painting *below*
literature ***5**.379*
Britain **5**.28–9, **5**.418
see also Shakespeare; Spenser
France **5**.302, **5**.372
Italy **5**.50–1, **5**.148, **5**.345–6
see also Dante; Petrarch
Spain **5**.86, **5**.170, **5**.323, **5**.348, **5**.387, **5**.470
music **5**.137, **5**.139, **5**.242, **5**.273, **5**.332, **5**.336, **5**.376, ***5**.379–80*, **5**.474
Northern **5**.73, **5**.139, **5**.407
painting in France **5**.162–3, **5**.366
painting in Italy **5**.35, **5**.77, **5**.139, **5**.158, **5**.349, **5**.407, **5**.470–1
Bassano **5**.37
Bologna **5**.80–1, **5**.380
in England **5**.173
Florence *see* Florentine art
in France **5**.366
historian of **5**.469
Mantua **5**.179, **5**.280–1
Milan **5**.261
Orvieto **5**.418
Padua **5**.280–1
Parma **5**.113–14, **5**.339
Perugia **5**.345
Rome **5**.261, **5**.265, **5**.293–4, **5**.339, **5**.345, **5**.350, **5**.410
see also Michelangelo
Sansepolcro **5**.349
Siena *see* Sienese School
Venice *see* Venetian art
Verona **5**.350
see also Giotto; International Gothic; Leonardo; Mannerism; Michelangelo; perspective; Raphael
sculpture in Italy **5**.111–12, **5**.132–3, **5**.175, **5**.293–4, **5**.350, **5**.372, **5**.384, **5**.401–2
humanism **7**.149
man *see* Alberti; Leonardo
Renaud, Madeleine **5**.36
Renaudel, Pierre **4**.173
Rendell, Ruth **5**.129
rendzina ***1**.281*, **1**.311
René, France Albert **4**.310
René of Aix **3**.294
renewable resource **1**.109
see also alternative energy
renga ***5**.380*
Reni, Guido ***5**.380*
rennet **6**.70
Rennie, John ***6**.297*
Renoir, Auguste ***5**.380*
Renoir, Jean ***5**.380*, **5**.437
Renold, Hans **6**.34
rent ***7**.265*
Renwick, Thomas **6**.323
reparations ***4**.286–7*
repeater (telephony) ***6**.297*, **6**.347
repertory theatre ***5**.380*
Repin, Ilya **5**.310, ***5**.380*
Reppe, J. W.: Reppe synthesis ***6**.297*
representative government *see* democracy; pluralism
Representatives, US House of ***4**.287*
repression (psychology) **7**.82
reprocessing, nuclear *see* nuclear fuels
reproduction (procreation) ***2**.282*
asexual **2**.68, **2**.282
see also clones
hermaphrodites **2**.156, **2**.282
parthenogenesis **2**.243
pollination **2**.263
vegetative **2**.282
see also sexual reproduction
reproduction (sound) *see under* sound
reptiles ***2**.282–3*
alligator **2**.7
amphisbaenid **2**.9
cayman **2**.58
crocodiles **2**.7, **2**.58, **2**.81, **2**.135
endangered species **2**.108
extinct **2**.168–9, **2**.259, **2**.271, **2**.333
see also dinosaurs
gharial **2**.135
heart **2**.153
lizard *see* lizards
mythical **9**.23
scales **2**.295
snake *see* snakes
tuatara **2**.349
turtles *see* turtles
Repton, George **5**.380
Repton, Humphry ***5**.380*
Repton, John Adey **5**.380
republic ***7**.265–6*
see also presidents
Republic of Ireland *see* Ireland
Republic of South Africa *see* South Africa
Republican Party *see under* politics *under* United States of America
Repudiation Movement ***4**.287*
requiem ***5**.380–1*
resale price maintenance ***7**.266*
research:
agricultural **6**.6–7
and development ***7**.266*
operational/operations **6**.250, **6**.253, **6**.344, **7**.224
scientific **6**.67
social **7**.273
reservoir *see* dams and reservoirs
residential care *see* institutional care
resin **2**.254, **2**.295
epoxy **6**.3, **6**.122, **6**.129, **6**.272
fossilized (amber) **1**.11
ion-exchange **6**.186
laminate **6**.193
in paint **6**.255
phenol-formaldehyde **6**.267
on photographic paper **6**.256
urea-formaldehyde **6**.369
resistance (electrical) ***1**.281*, **6**.75, **6**.76, ***6**.298*
potentiometer **6**.278
rectifier *see* rectifier
resistor ***6**.298*, **9**.53
zero **6**.339
resistance (living organisms):
to antibiotics **6**.16
to disease *see* immune system
to drugs **6**.260, **6**.338
to pesticides **6**.96, **6**.261
resistance (social):
movements ***4**.287–8*
see also revolution (political)
resistance (technology):
air **1**.5–6
streamlining to reduce **6**.336
welding **6**.383
Resnais, Alain **5**.323
resolving power ***8**.138*
resonance effects (chemistry) ***1**.281*
resonance (physics) ***1**.281*
vibrations **1**.356
resources (economics) ***7**.266*
scarcity **7**.275
underemployed **7**.260
see also assets; capital; labour; land
Respighi, Ottorino ***5**.381*
respiration and respiratory tract ***2**.283*, **6**.140, **6**.266, ***9**.60*
disorders **6**.22, **6**.248, **6**.254, **6**.266, ***9**.60*
legionnaire's disease **2**.186
and nutrition **6**.140
occupational **6**.22, **6**.248
otorhinolaryngology **6**.254
pneumonia **2**.260
whooping cough **2**.367, **7**.153
electron transport chain **2**.105
gills **2**.136
glycolysis **2**.138
Kreb's cycle **2**.181
lungs **2**.195, **2**.196, **9**.60
and nitrogen **2**.228
nose, mouth and throat **2**.229, **9**.59
respirator ***6**.298*
throat **9**.59
trachea **2**.345
responsibility, criminal **7**.76, **7**.146
Restoration **3**.65, ***3**.302*
comedy **5**.101, **5**.110, **5**.136, **5**.152, ***5**.381*, **5**.467–8, **5**.496
style ***5**.381*
see also Charles II, king of England
restrictive practices ***7**.266*
see also in particular cartels; closed shop; monopoly
resurrection ***7**.266*
retail price index ***7**.266*
retail trade **7**.93, ***7**.266–7*, **7**.314
auction **7**.22
bar code **6**.30, **6**.179
cash register **6**.62
chain stores **7**.43
co-operatives **7**.69
employment percentage **7**.106
point-of-sale terminal **6**.179, **6**.274
supermarket **7**.302
see also prices
retaining wall ***6**.298*
rete (in astrolabe) **8**.9
Reticulum ***8**.138*
Retief, Piet **4**.99, **4**.140
retinol **2**.357
retinopathy ***2**.283*
retrograde motion **8**.144
Retz, Paul de Gondi **3**.135, ***3**.302*
Reuben ***9**.31*
Réunion **6**.250, ***7**.377*
Reuter, Admiral von **4**.303
Reuter, Paul Julius, Baron ***4**.288*
Reuther, Walter Philip **4**.11, ***4**.288*, **4**.339
Revere, Paul **3**.45, **3**.207, ***3**.302*, **5**.153
reverse fault **1**.120, ***1**.281*
reversed winds **1**.108, **1**.200
reversing layer of Sun's atmosphere **8**.26
revisionism ***7**.267*
revivalism ***4**.288*
revolution (political) ***7**.267*
America *see* Independence, American War of
counter-revolution **7**.73
Glorious (England) **3**.142
Iran **7**.166, **7**.267, **7**.331
Revolutionary Wars ***4**.288*
Revolutions of 1848 ***4**.288–9*
see also civil wars *and* revolution(s) *under* China; France; Hungary; Russia
revolution of Earth (astronomy) ***1**.281–2*, ***8**.139*
revolver *see* pistol and revolver
revue **5**.309, ***5**.381*, **5**.498
reward and learning **2**.185
Reymont, Wladyslaw Stanislaw ***5**.381*
Reynaud, Paul ***4**.289*
Reynolds, Sir Joshua **5**.117, **5**.359, ***5**.381–2*, **5**.393
Reza Shah Pahlavi (1878–1944), Shah of Iran **4**.165, **4**.279, ***4**.289*
Reza Shah Pahlavi, Muhammad (1941–79), Shah of Iran **4**.165, **4**.182–3, ***4**.226*, **4**.289, **7**.148, **7**.309, **7**.331
RGO *see* Royal Greenwich Observatory
rhapsody ***5**.382*
rhea (bird) ***2**.283–4*, **6**.41
Rhea (moon of saturn) **8**.138, ***8**.139*
Rhea (mythology) **9**.9, ***9**.31*
Rhee, Syngman ***4**.289*, **4**.320
rhenium **1**.107, ***1**.282*, **6**.241, ***6**.298*
rheostat ***6**.298*
rhetoric **5**.338, ***5**.382*, **5**.732
rheumatism **6**.101
Rhine, Joseph B. **7**.228
Rhine River ***1**.282*
Rhineland ***4**.289*
rhinoceros ***2**.284*, **2**.371
rhizome, plants with ***2**.284*
begonia **2**.28
iris **2**.119, **2**.136, **2**.173, **2**.218
Solomon's seal **2**.318
see also corms; tubers
Rhode Island ***3**.302*, **3**.383
Rhodes, Cecil **4**.51, **4**.61, **4**.198, ***4**.289*, **4**.350, **4**.388, **4**.390
Rhodes (island) ***1**.282*, **3**.302, **3**.303
Rhodesia **4**.67, ***4**.289–90*, **4**.373, **7**.273
Northern *see* Zambia
Southern *see* Zimbabwe
Rhodesia and Nyasaland, Federation of **4**.67
rhodium **1**.107, ***1**.282*, ***6**.298*, **6**.354
rhodochrosite **1**.239
rhododendron ***2**.284*
rhodonite **1**.80
rhombus **9**.51
Rhône ***1**.282*
rhubarb family ***2**.284*, **2**.317
rhyme ***5**.382*
Rhynea ***2**.284*
rhyolite ***1**.282*
Rhys, Jean ***5**.382*
rhythm ***5**.382*
and blues **5**.358, ***5**.382*, **5**.425
isorhythm **5**.230
and melody **5**.290
rhythm method of contraception **2**.75
ria ***1**.282*, **1**.283
rib (bone) ***2**.284*
Ribbentrop, Joachim von ***4**.290*
ribbon-fish ***2**.284*
ribbon-worm ***2**.284*

Ribeiro, Bernadim **5**.359
Ribera, José de ***5**.382*
riboflavin **2**.357
ribonucleic acid *see* RNA
ribosomes **2**.240, ***2**.285*
ribs ***2**.284*
Ricardo, David ***4**.290*, ***7**.267–8*
as classical economist **7**.52
and division of labour **7**.93
and labour theory of value **7**.179
in Manchester School **4**.209
and poverty **7**.88
and rent **7**.53, **7**.265
and trade **7**.267, **7**.314
Ricardou, Jean **5**.323
Ricci, Marcò **5**.383
Ricci, Matteo **3**.186, ***3**.302*
Ricci, Sebastiano ***5**.382–3*
Riccioli, Giambattista **8**.100, **8**.147
Rice, Tim **5**.267
rice ***2**.285*, ***6**.298*, ***7**.268*
Rich, Adrienne **7**.332
Richard I (the Lionheart), king of England **3**.255, **3**.284, ***3**.302–3*
and Crusades **3**.93, **3**.314
and Cyprus **3**.95
ransom for **3**.299
and scutage **3**.320
Richard II, king of England **3**.161, **3**.255, **3**.284, ***3**.303*
and barons **3**.36
and Commons **3**.83
and Lords Apellant **3**.210
and Peasants' revolt **3**.273, **3**.364
and religion **3**.209
and war **3**.171–2
Richard III, king of England **3**.110, **3**.255, **3**.284, ***3**.303*
and Bosworth, battle of **3**.47, **3**.310
Richard of York *see* York, Richard Plantagenet
Richardson, Dorothy **5**.436
Richardson, F. **5**.91, **5**.148
Richardson, Henry Handel ***5**.383*
Richardson, Henry Hobson ***5**.383*
Richardson, Sir Ralph ***5**.383*
Richardson, Samuel ***5**.383*
Richardson, Tony **5**.163
Richelieu, Armand Jean du Plessis, duc de **3**.144, **3**.179, **3**.245, ***3**.304*, **5**.2
richest field telescope ***8**.139*
Richler, Mordecai ***5**.383*
Richmond, George **4**.300
Richter, Charles F.: Richter scale ***1**.282*, **6**.312
Richter, Hans **5**.110
rickets **2**.90, **2**.357
Rickettsia **2**.287, **2**.353
Rickman, Thomas **5**.127
Rickover, Hyman **6**.245
rickshaw ***6**.298*
Rida, Rashid **7**.168
riddles **5**.7, **5**.326
ridge (meteorology) ***1**.282*
ridge, topographical ***1**.282*
ocean **1**.24, **1**.168, ***1**.214–15*, **1**.236, **1**.245, **1**.261, **1**.295
Ridgway, Matthew **4**.172, **4**.186
Ridley, Nicholas (1553) **3**.204, **3**.355
Ridolfi, Roberto di: Ridolfi Plot **3**.151, ***3**.304*
Ridpath, I. **8**.22
Rie, Lucie **5**.438
Riebeeck, Jan van ***3**.304*
Riefenstahl, Leni **5**.132
Riel, Louis: Riel Rebellions ***4**.290*
Riemann, Bernhard **1**.136, ***1**.282–3*
Rienzo, Cola di ***3**.304*
Riet, Richard van der **8**.193
rifle **6**.130, **6**.131, **6**.162–3, **6**.226, ***6**.298*
rift valleys ***1**.283*
graben **1**.142
and tectonic plates **1**.261
rift valleys (*cont.*):
see also Great Rift Valley
rig, oil **6**.107–9, **6**.263
rig, sailing ***6**.298*, **6**.302, **6**.304
Bermuda **6**.143, **6**.305, **6**.306
gaff **6**.143, **6**.306
lateen **6**.101, **6**.143, **6**.195, **6**.306, **6**.307
yawl **6**.390
see also fore-and-aft; sailing ships; square-rigged ships
Rigel ***8**.139*
Right (politics) ***7**.268*
New Right **7**.69, **7**.86, **7**.215
see also conservatism
right ascension of star **8**.139
right whale ***2**.285*
rights, animal **7**.13
rights, human ***7**.149*
African Charter on Human and People's Rights **7**.5
Bills of Rights **3**.41, **7**.32
Charter 77 **4**.70
of children **7**.319
citizenship **7**.50
civil **7**.51–2
Conventions on *see* American Convention *and under* European Conventions
copyright **7**.70
European Court of Human Rights **7**.110
freedom of expression **7**.124–5
Helsinki Conference **4**.150
International Covenant on Economic, Social, and Cultural Rights **7**.149, ***7**.164*, **7**.322
Rights of Man and Citizen ***3**.304*
Universal *see* Universal Declaration of Human Rights
see also civil rights
rights issue (shares) **7**.282
Rigil Kentaurus **8**.4
Riley, Bridget ***5**.383*
Riley, Terry **5**.297
Rilke, Rainer Maria **5**.143, ***5**.383*
rille (lunar valleys) ***8**.139–40*
rills (channels) ***1**.283*, **1**.284, **1**.301
Rim Kin **5**.72
Rimbaud, Arthur ***5**.383*
Rimsky-Korsakov, Nikolay **5**.87, **5**.310, ***5**.383–4*
ring, planetary ***8**.140*, **8**.145, **8**.197
ring dove **2**.371
ring dykes **1**.166
ring mains **6**.387
ring mountains ***8**.140*
Ring Nebula ***8**.140*
ring plains ***8**.140*
ring-frame spinning **6**.329
ringlet butterfly ***2**.285*
ring-ouzel ***2**.285*
ring-tailed cat ***2**.285*
ringworm ***2**.285*, **2**.341–2
rinkal ***2**.285–6*
Rinuccini, Ottavio **5**.328
Rio de Oro **4**.306
Rio Grande ***1**.283*
Riopelle, Jean-Paul **5**.73
Riot Act ***3**.304*
rip current ***1**.283–4*
Ripon, Frederick Goderich, earl of ***4**.137*
Ripon, George Frederick Samuel Robinson, 1st marquess of ***4**.290*
Riquet de Bonrepos, Baron Pierre-Paul **6**.56, ***6**.298–9*
RISC **6**.84, ***6**.299*
rising (astronomy) ***8**.140*
risk analysis ***7**.268*
Risorgimento ***4**.290*
Ritchey, George: Ritchey–Chrétien optics ***8**.140*
rites *see* rituals
Rittenhouse, David **3**.11
Ritter, Johann **8**.177
Ritter, Karl **7**.130
rituals **7**.112, ***7**.268*
gifts **7**.132
magic *see* magic
rites of passage **7**.5, **7**.132, ***7**.268*
bar mitzvah **7**.26
circumcision **7**.50, **7**.51
funerals **7**.74–5, **7**.127 8
initiation **7**.6, **7**.160–1, **7**.188
marriage *see* marriage
puberty **7**.253
ritualization in animal behaviour ***2**.286*
sacrifice **7**.271–2
and symbolism **7**.303
taboo *see* taboos
totems *see* totemism
see also magic; myths and legends; religion and theology
Riva-Rocci, Scipione **6**.328
river blindness **2**.234, ***2**.286*
river dolphin ***2**.286*
Rivera, Diego **5**.164, ***5**.384*
rivers ***1**.284*
Africa **1**.72, **1**.195, **1**.237, **1**.246, **1**.346, **1**.352, **1**.356, **1**.358, **1**.372, **2**.240
see also Nile
Angola **1**.237, **1**.372
Argentina **1**.19, **1**.258
Asia **1**.114–15
see also China; India; Nepal; Siberia; Tibet *below*
Australia **1**.75, **1**.84, **1**.123, **1**.223
Austria **1**.84, **1**.282
bayou **1**.35
Belgium **1**.294
birds *see* birds *under* water
Botswana **1**.195, **1**.237
Brazil **1**.166, **1**.201, **1**.248
see also Amazon
Britain **1**.67, **1**.211, **1**.300, **1**.334, **1**.336, **1**.342
Canada:
central **1**.277, **1**.293–4, **1**.334
east **1**.228, **1**.290, **1**.300, **1**.335
north **1**.63, **1**.200, **1**.372
west **1**.69–70, **1**.127, **1**.249
China **1**.12, **1**.158, **1**.159, **1**.210, **1**.277, **1**.369, **1**.370
Colombia **1**.202, **1**.241
Czech Republic **1**.101, **1**.236
deposition **1**.114
alluvium *see* alluvium
bars **1**.33, **1**.261
braiding **1**.43, **1**.140
conglomerates **1**.72
deltas **1**.39, **1**.86
floodplain **1**.123–4
gravel **1**.144
levee **1**.192
meandering **1**.209, **1**.261
slip-off **1**.307
divide **1**.93, **1**.361
drainage basin **1**.94–5
drainage pattern **1**.95, **1**.342
Ecuador **1**.100
Egypt *see* Nile
ephemeral (arroyo) **1**.20
erosion:
base level **1**.34
canyon **1**.50
cave **1**.55
channel **1**.58
cliffs **1**.64
coomb **1**.75
crevass **1**.79
dry valleys **1**.75, **1**.96–7, **1**.233, **1**.367
escarpment **1**.113
estuary **1**.114
floodplain **1**.123–4
gorge *see* gorges
meandering **1**.209, **1**.261
rivers (*cont.*):
erosion (*cont.*):
natural bridge **1**.224, **1**.225
ox-bow lake **1**.243
rias **1**.282, **1**.283
saltation **1**.291
spur **1**.317, **1**.318
terraces **1**.333
undercut bank **1**.350
wadi **1**.20, **1**.359
water-gap **1**.360
wind-gap **1**.367
see also valleys
Ethiopia **1**.229
Europe *see under individual countries above and below*
eutrophication **2**.111
fish *see* freshwater fish
flow:
baseflow **1**.34
discharge **1**.92
hydrograph **1**.162
stage-discharge **1**.319
France **1**.198, **1**.282, **1**.294, **1**.298
Germany **1**.84, **1**.282, **1**.364
Ghana **1**.358
hydrography **1**.92, **1**.162
hydrological cycle **1**.162
hydrology **1**.163
India **1**.43, **1**.132
Iran **1**.300
Iraq **1**.115, **1**.300, **1**.337
Ireland **1**.300
Israel **1**.178
Italy **1**.260, **3**.312
Jordan **1**.178
Kampuchea **1**.210
Laos **1**.210
load **1**.35, **1**.92, **1**.197
Mexico **1**.283
Middle East **1**.114–15, **1**.178, **1**.300, **1**.337
Mozambique **1**.195, **1**.372
Myanmar (Burma) **1**.174, **1**.292
Namibia **1**.237, **1**.372
navigation **6**.30, **6**.106, ***6**.299*
Nepal **1**.317, **1**.333
Netherlands **1**.282, **1**.294
New Zealand **1**.39, **1**.43, **1**.86, **1**.353, **1**.359
Pakistan *see* Indus
Paraguay **1**.248
Peru **1**.112
Poland **1**.357
Portugal **1**.330
rapids **1**.276
rejuvenation **1**.280, **1**.333
river blindess *see* onchocerciasis
Russia **1**.357–8
Siberia **1**.12, **1**.191–2, **1**.234, **1**.371
South Africa **1**.195, **1**.240, **1**.246, **1**.352
South America **1**.19, **1**.112, **1**.166, **1**.201, **1**.202, **1**.240–1, **1**.248, **1**.258
see also Amazon
Spain **1**.330
Sudan **1**.229
Switzerland **1**.282
Syria **1**.114, **1**.178
Tibet **1**.167, **1**.210, **1**.291–2
tidal bore **1**.336
tidal stream **1**.336
transportation **1**.35, **1**.92, **1**.197
Turkey **1**.114, **1**.337
Ukraine **1**.84, **1**.93
United States:
central and south **1**.63, **1**.209, **1**.242, **1**.277, **1**.283
Mississippi and Missouri and tributaries **1**.177, **1**.188, **1**.192, **1**.216, **1**.237, **1**.259, **1**.370

rivers (*cont.*):
United States (*cont.*):
east 1.60, 1.73, 1.86, 1.159, 1.177, 1.211, 1.265, 1.290, 1.301, 1.328, 1.334, 1.335
north 1.228, 1.283
west 1.50, 1.68, 1.69–70, 1.142–3, 1.289, 1.308
Uruguay 1.248, 1.258
waterfalls *see* waterfalls
watershed 1.93, 1.361
weir 6.383
whirlpool 1.365
Zambia 1.356, 1.372
Zimbabwe 1.356, 1.372
riveting *6.299*
rivulin 2.178
Rizal, José 4.268, *4.291*
Rizzio, David 3.97, *3.304–5*
RNA 2.134, 2.230, *2.286*, 6.8, 6.153, 6.154, 6.373
roach (cockroach) 2.70–1
roach (fish) *2.286*
road-runner *2.286*
roads *4.350*, *6.299–301*
building *6.299–300*
see also Telford
Cat's-eye 6.64
grader 6.158
of Incas 3.175
limestone ballast 6.202
military 6.260
motorways 6.92–3, 6.300, 6.360
pioneers 6.64, 6.363
see also Telford
road roller *6.300*
Roman *see under* Roman empire
stone for 6.287
Tarmac(adam) 6.299
traffic control 6.301, 6.360
transport *6.300–1*
see also bicycle; buses; carriages; carts; horse-drawn; lorry; motor car; motor vehicles
wheel 6.384
see also roads *or* vehicles and transport *under individual countries*
Roanoke Island *3.305*
roaring forties *1.284*
roasting *see* oxidation
Rob Roy *3.305*
Robbe-Grillet, Alain 5.323
'robber baron' *4.291*
robber crab 2.71
robber fly *2.286*
robbery 7.326
Robbia, Luca della *5.384*
Robbins, Jerome *5.384*
Robbins, Lionel 7.275
Robert I (the Bruce), king of Scotland 3.35, 3.110, *3.305*, 3.319
Robert II, king of Scotland *3.305*, 3.343
Robert III, king of Scotland *3.305*, 3.343
Robert of Clermont 3.46, 3.47
Robert of Molesme, Saint 3.77
Robert of Normandy (Curthose) 3.109, 3.255
Robert, Nicholas 6.145
Robertian House of France: list of rulers *9.68*
Roberts, Joseph Jenkins 4.195
Roberts, Richard 6.329
Roberts, Tom 5.26
Roberts, William 5.479
Robertson, Sir William Robert *4.291*
Robeson, Paul *5.384*
Robespierre, Maximilien François Marie Isidore de 3.12, 3.32, 3.97, 3.183, *3.305*
Robey, George 5.310
robin 2.287
Robin Goodfellow 9.30
Robin Hood *3.306*, *9.31*
Robinia *2.287*
Robinson, Edwin Arlington *5.384*
Robinson, Henry Peach 5.347, *5.384*
Robinson, Mary 4.166
Robinson, Ronald 7.153
robotics 6.26, 6.259, *6.301*, 6.326, 6.383, 7.19
rocaille *5.384*
Roche, Edouard 8.140
Roche limit *8.140*
Roche lobe 8.138, *8.140*, *8.142*
roche moutonnée *1.284*
Roches, Peter des 3.161
Rochester, John Wilmot, 2nd earl of *5.384*
rock (geology) *see* geology and rocks
rock art:
Africa 5.8
caves 5.1, 5.8, 5.83, 5.434
India 5.10, 5.143, 5.145, 5.196
marble 5.282
rock-crystal carving *5.384*
rock-cut temples 5.143, 5.145, 5.275–6
rock flour *1.284*
rock glacier *1.284–5*
rock lobster 2.80
rock music 5.39, 5.358, *5.384–5*
rock rose *2.287*
rock-crystal *1.284*
Rockefeller, John D. and family *4.291*
rocketry:
fuel 6.174, 6.301
history of 6.45, 6.139–40, 6.157, 6.228, *6.301*
launch site and launch vehicle 6.64, 6.196, 6.326
motor 6.157, *6.301*
see also ballistic missiles; missiles; space-flight
rockfall *1.284*
rockfish 2.280
Rockingham, Charles Watson-Wentworth, 2nd marquess of 3.54, *3.306*, 3.339, 3.380
rockling 2.287
rocks *see* geology and rocks
rock-salt 1.150–1
Rocky Mountain School 5.213
Rocky Mountain spotted fever *2.287*, 2.352
Rocky Mountains 1.135, *1.285*
Rococo art *5.385*
architecture 5.100–1, 5.168, 5.317, 5.498
painting 5.55–6, 5.89, 5.162, 5.385, 5.455, 5.483
rodents *2.287*
agouti 2.5
beaver 2.26, 2.220
cane rat 2.52
capybara 2.52
cavy *see* cavy and cavy-like
chinchilla 2.64
chinchilla rat 2.64
chipmunk 2.64
coypu 2.80, 2.362
dormouse 2.98
gopher 2.141
ground-squirrel 2.141, 2.145–6
guinea pig 2.147, 2.148
gundi 2.148
hamster 2.150
harvest mouse 2.151
as herbivores 2.155
housemouse 2.164
hutia 2.166
jerboa 2.175, 2.176
kangaroo rat 2.178
lemming 2.186–7
marmots 2.203–4, 2.370
marsh rat 2.362
mole rat 2.215–16
mountain beaver 2.220
rodents (*cont.*):
mouse *see* mouse group
musk rat 2.223
paca 2.239
pika 2.253
porcupines 2.264, 2.265
prairie dog 2.267–8
scaly-tailed squirrel 2.295–6
spiny rat 2.322
spring haas (hare) 2.323
squirrel *see* squirrel group
tuco-tuco 2.349
viscacha 2.356
vole *see* voles
water rat 2.362
woodchuck 2.370
woodmouse 2.370
Rodgers, Richard (composer) *5.385–6*
Rodin, Auguste *5.386*
Rodney, George Brydges, 1st Baron *3.306*, 3.313
Rodrigo, Joaquín *5.386*
Rodríguez, General Andrés 4.259
roe deer *2.287*
Roebling, John 6.341
Roebuck, John 6.197
Roelas, Juan de 3.212
Roemer, Ole Christensen 8.43, 8.117, *8.142*, 8.173, 8.181
Roentgen *see* Röntgen
Roethke, Theodore *5.386*
Roger II, king of Sicily *3.306*
Roger III, king of Sicily 3.327
Rogers, Carl Ranson 7.149, *7.268–9*
Rogers, Ginger 5.25, 5.309
Rogers, Richard (architect) 5.162, *5.386*
Rogers, Robert *3.306*
Roget, P. M. 6.73
Roggeveen, Jakob 3.109, 4.301
Roh Tae Woo 4.320
Röhm, Ernst 4.53, 4.243
Rokossovsky, Konstantin Konstantinovich *4.291*, 4.371
Roland *9.31*
Roland, Jean-Marie 3.306
Roland de la Platière, Marie-Jeanne Philipon 3.141, *3.306*
role *7.269*
Rolfe, John 3.286
roll (aircraft flight) 6.136
Rolland, Romain *5.386*
Rolle, Richard 5.294
roller (bird) *2.287–8*
roller-printing 6.353
roller-skating 6.321, *7.269*
rolling bearing 6.32, 6.33, *6.301*
rolling friction 6.146
rolling mill *6.301*, 6.302
rolling stock 6.44, 6.157, 6.292
see also locomotives
Rollo 3.255, 3.256, *3.306*, 3.374
roll-on, roll-off ferry 6.60, 6.127
ROM (read-only memory) 6.83, 6.218–19
Romains, Jules *5.386*
Roman alphabet and script *see under* writing and scripts
Roman Britain 3.17, *3.306*, 3.309
Hadrian's Wall 3.154
roads 3.118, 3.129, 3.378
rulers 3.5, 3.375
tribes 3.47, 3.49–50, 3.174
see also Roman empire
Roman Catholicism *3.306–7*, 4.65, *4.291*, 7.40, *7.269*
and abortion 7.1
afterlife 7.6
anticlericalism 3.17
arts 5.311
see also mass; sacred music *and also* literature *below*
canon law 7.39
Catholic Leagues 3.167, 3.206
Roman Catholicism (*cont.*):
charismatics 7.44
Concordats 4.82
conspiracies 3.32, 3.150, 3.289, 3.304, 3.356
Counter-Reformation 3.89
and cremation 7.75
and divorce 7.94
Donatist breakaway 3.105
ecumenical councils 3.67, 3.84
education 7.265
emancipation 3.65, 4.65
excommunication 3.121
expansion to 1500 *3.75*
and freemasons 7.195
French Gallicanism 3.136–7
Gordon riots against 3.143
heresies *3.164*
Albigensians 3.7, 3.64, 3.105
Arianism 3.21
Cathars 3.7, 3.64
council to combat 3.86
Hussites 3.173
and Inquisition 3.179
Manichaeism 3.218, 7.19
Pelagianism 3.274, 9.30
Waldenses 3.376
hermits 3.164
Holy Alliance 4.155
and homosexuality 7.146
indulgences and pardoners 3.178, 3.269
investiture 3.179
Jansenism 3.185
Jerusalem 3.187
Jesuits 3.58, 3.131, 3.187, 3.302, 7.171
Lateran Council 3.204
law 3.59
leaders 7.48
see also papacy
liberation theology 7.184
literature 5.146, 5.211
see also Bible
liturgy 7.185–6, 7.271
see also Mass (Eucharist)
Lollards 3.209
map 3.75
Maronites 4.211, 7.193
martyrs 3.223, 7.194
military orders 3.198
missionaries 7.203
monasteries *see* monasticism
Nicaea, Councils of 3.253
Opus Dei 7.224
pope *see* papacy
Propaganda 7.203
rebellion against 3.290
relics 3.300
religious orders 3.28
sacraments 7.271
saints *see under* saints
Schism 3.109, 3.146
see also Eastern Orthodox
Schmalkaldic War 3.318
in Scotland 3.37
simony 3.328
Sonderbund 4.319
Trent, Council of 3.360
Virgin Mary 7.49, 9.24
wars and defence *see* Crusades; Thirty Years
see also Bible; Christianity; friars; Jesus; missionaries; Orthodox; papacy; religion
Roman de la Rose *5.387–8*
Roman empire *3.307–8*
agriculture 6.295
annals 3.16
arts 3.11–12, 5.15, 5.305, *5.386–8*
architecture 5.330, 5.386–7, 5.478, 6.373
literature *see* literature *under* Latin
see also under sculpture and carving
astronomy 8.18
Britain *see* Roman Britain

Roman empire (*cont.*):
buildings **6**.51, **6**.66
calendar **3**.57
chemistry **6**.110
emperors *see* emperors of Rome
energy and fuel **6**.79, **6**.167, **6**.377, **6**.379, **6**.385
heating **6**.67, **6**.175
entertainment **3**.77, **3**.82, **3**.141–2, **3**.338–9
fasces **3**.124
forum **3**.129
hippodrome **3**.166
language *see* Latin *and under* writing
laws/legal system *see* Roman law
literature:
novel ***5**.189*
see also literature *under* Latin
map **3**.307
medicine and health care **6**.14
mining **6**.227
peoples **3**.313
philosopher **3**.321
plebs **3**.285
politics and government:
comitia **3**.82
consuls **3**.300, **3**.347
Cursus Honorum **3**.94
dictators **3**.57, **3**.76–7
governors **3**.5, **3**.281
Kings **3**.350
patricians **3**.271
republic ***3**.308*
Senate and senators **3**.285, ***3**.308–9*
statesmen **3**.6, **3**.65, **3**.76, **3**.77, **3**.79, **3**.345
tribunes **3**.144, **3**.360
Triumvirates **3**.90
see also emperors of Rome
pottery **3**.11–12
provinces **3**.138
recreation *see* entertainment *above*
religion and myths ***3**.308*
augurs **3**.28
Christians:
and catacombs **3**.64
see also under Constantine I
goddesses **9**.8, **9**.10, **9**.13, **9**.21, **9**.25, **9**.27, **9**.37
gods **9**.6, **9**.9, **9**.12, **9**.17, **9**.19, **9**.21–6 *passim*, **9**.29, **9**.30, **9**.32, **9**.33, **9**.37
humans **9**.2, **9**.29, **9**.31, **9**.33
Mithras cult **3**.234
saints **3**.11, **3**.160
Vestal virgins **3**.372–3
roads **3**.18, **3**.278, ***3**.308*, **6**.62, **6**.299–300, **6**.375
in Britain **3**.118, **3**.129, **3**.378
rulers *see* politics *above and also* emperors of Rome
scholasticism **3**.318, **3**.319
script *see under* writing and scripts
sport and recreation **3**.77, **3**.82, **3**.141–2, **3**.338–9
technology and science **6**.57, ***6**.158*, **6**.178, **6**.368
see also entries above and below prefixed **1**. *and* **6**.
tools **6**.358, **6**.383
vehicles and transport **6**.56, **6**.170, **6**.201
bridges **6**.47, **6**.372
ships **6**.143, **6**.219, **6**.307
water **6**.56, **6**.62, **6**.143, **6**.219, **6**.307
see also roads *above*
villa ***3**.309*
wars and defence:
army:
centurion **3**.22, **3**.67
leaders **3**.38–9, **3**.51, **3**.57, **3**.90, **3**.123, **3**.222–3, **3**.288, **3**.344
legion ***3**.308*

Roman empire (*cont.*):
wars and defence (*cont.*):
army (*cont.*):
praetorian guard **3**.290
religion **3**.234
barbarian attacks **3**.23, **3**.98, **3**.173
see also in particular Goths; Huns; Vandals
Cannae battle **3**.59
with Carthage **3**.59, **3**.294–5, **3**.319
civil ***3**.307*
Dacian **3**.95
Jewish rebellion **3**.35–6
legion **3**.308
Macedonian **3**.215
navy **3**.348
Samnite **3**.314
technology **6**.40, **6**.343
see also Punic wars
water **6**.18, **6**.19, **6**.146
writing *see under* writing
Roman law **3**.285, **7**.4, ***7**.269*, **7**.276
see also civil law
Roman numerals ***9**.52*
Romance languages **7**.108, **7**.269
Romance of the Three Kingdoms ***5**.387*
romances **5**.278, **5**.294, ***5**.387*, **5**.428, **5**.497
in British poetry **5**.187, **5**.408, **5**.418, **5**.449
Romanenko, Yuri **4**.323
Romanesque **5**.179, **5**.383, ***5**.388–9*
Romania ***1**.286*, ***4**.291–2*, ***7**.377–8*
arts **5**.58
in COMECON **7**.57
family policy **7**.116
institutions **7**.313
languages **7**.156, **7**.269
leaders **4**.16, **4**.63, **4**.292
physical geography and geology **1**.53, **1**.84, **1**.286, **1**.342
politics and government **7**.238, **7**.239
regions **1**.342, **3**.360, **4**.350
religion/theology **7**.98
wars and defence **3**.95, **4**.26, **4**.197, **4**.371
in Warsaw Pact **7**.328
see also Balkan States; Eastern Europe
Romanos Diogenes IV, Byzantine emperor **3**.221
Romanov, house of ***3**.309*
rulers listed ***9**.68*
see also under tsars *under* Russia *and also in particular* Catherine II; Peter I
Romanticism ***5**.389–90*
literature ***5**.390*
Britain **5**.26, **5**.78, **5**.207, **5**.408, **5**.432
see also Gothic novels *and* Britain *under* poetry
France **5**.27, **5**.148, **5**.291, **5**.367, **5**.400, **5**.429, **5**.431–2
drama **5**.139, **5**.214, **5**.310–11, **5**.475
novels **5**.89, **5**.139, **5**.291, **5**.310–11, **5**.400, **5**.475
poetry **5**.38, **5**.172, **5**.213–14, **5**.254, **5**.310–11, **5**.317, **5**.475
Germany **5**.203, **5**.205, **5**.208, **5**.210, **5**.323, **5**.405, **5**.438
Italy **5**.261, **5**.281
Poland **5**.294, **5**.422
Portugal **5**.359
Russia **5**.261–2, **5**.370, **5**.371, **5**.396
Scandinavia **5**.326, **5**.404
Spain **5**.428
United States **5**.112, **5**.145, **5**.353–4, **5**.453–4, **5**.459

Romanticism (*cont.*)
music **5**.143, ***5**.390*
Austria **5**.44, **5**.276, **5**.406
Czech **5**.140
France **5**.45, **5**.153, **5**.266
Germany **5**.57, **5**.291, **5**.406, **5**.407, **5**.435–6, **5**.480–1, **5**.484
Italy **5**.43, **5**.491
Poland **5**.99
Russia **5**.156, **5**.247, **5**.373, **5**.448
United States **5**.34, **5**.251
painting ***5**.390*
Britain **5**.48, **5**.67, **5**.110–11, **5**.284, **5**.390, **5**.462–3
France **5**.127–8, **5**.165, **5**.173, **5**.174
Germany **5**.165, **5**.195, **5**.394, **5**.405
Italy **5**.391
Switzerland **5**.166
United States **5**.13
see also Pre-Raphaelite; Gothic Revival
see also Picturesque; romances; Sublime
Romany language **7**.137, **7**.157
Rome ***3**.309–10*
painters **5**.261, **5**.265, **5**.293–4, **5**.339, **5**.345, **5**.350, **5**.410
Treaties of (EEC) ***4**.292*, **7**.62, **7**.99, **7**.103, **7**.315
see also Roman empire
Romero, Oscar **7**.194
Rommel, Erwin **4**.7, **4**.9, **4**.177, **4**.245, **4**.246, ***4**.292*, **4**.345
Romney, George **5**.359, ***5**.390*
Romulus **9**.30, ***9**.31*
Romulus Augustus, Roman emperor **3**.308
rondo ***5**.390–1*, **5**.442
Ronsard, Pierre de **5**.12, ***5**.391*
Röntgen, Wilhelm Konrad von **1**.80, ***1**.286*, **1**.370, **2**.372, **6**.218, **6**.389
roof **6**.51, **6**.145, **6**.146, **6**.274, ***6**.301–2*, **6**.337
rook ***2**.288*
Rooke, Sir George **3**.141
Roosevelt, Franklin Delano **4**.95, **4**.96, ***4**.292–3*
and Atlantic Charter **4**.22
and Cairo Conference **4**.58
and Casablanca Conference **4**.64
and conservation **2**.254
and currency **4**.87
and Destroyer–Bases Deal **4**.97
and Good Neighbor Policy **4**.138, **4**.258
and Hay Treaties **4**.148
and isolationism **4**.168
and legal system **4**.158
and Lend–Lease Act **4**.193
and Neutrality Acts **4**.238
as New York governor **4**.157
and nuclear bomb **1**.101, **6**.127
and Philippines **4**.202
and Teheran Conference **4**.340
and Yalta Conference **4**.385
Roosevelt, Theodore **4**.271, **4**.287, ***4**.293–4*
and Panama **4**.258
and Progressive Movement **4**.277
and Rough Riders **4**.293–4, **4**.324
and Russo-Japanese War **4**.275
and strike **4**.15
Root, Elihu ***4**.293*
root ***2**.288*
crops ***6**.7*, ***6**.302*
breeding **6**.271, **6**.371
organic **6**.18
rotation **6**.303
swede and turnip **2**.331, **2**.351
see also potato family
Root-and-Branch Petition ***3**.310*

root-mean-square ***1**.286*
rope ***6**.302–3*
see also cable; rig, sailing
Roque do los Muchachos Observatory **8**.82, **8**.83
ro-ro ferry **6**.60, **6**.127
rorqual **2**.119, **2**.288
Rosa, Salvator ***5**.391*
Rose, Sir Hugh **4**.162
Rose, John **5**.34
Rosalind (Uranian satellite) **8**.178
Rosas, Juan Manuel de ***4**.293*
rose ***2**.288*
rose of Jericho **2**.*288–9*
rosebay willowherb **2**.368
Rosebery, Archibald Philip Primrose, 5th earl of ***4**.293*
rosechafer ***2**.288*
rosefinch ***2**.288*
rosemary ***2**.288*
Rosen, Jelka **5**.128
Rosen, Michael **5**.91
Rosenberg, Isaac **5**.174
Rosenberg, Julius and Ethel: Rosenberg case ***4**.293*
Rosencrans, W. **4**.71
Rosenman, Samuel **4**.238
Roses, Wars of ***3**.310*, **3**.377
battles and campaigns **3**.36, **3**.164–5, **3**.241, **3**.313, **3**.352, **3**.360, **3**.376
Rosetta stone ***3**.310*, **3**.311
Rosette Nebula **8**.139, ***8**.142*
Rosh Hashanah ***7**.269–70*
Rosicrucian ***3**.310–11*
Roskilde, Treaty of ***3**.311*
Ross, James Clark **1**.111, ***1**.286*, **4**.14
Ross, Sir John ***1**.286*
Ross, Sir Ronald ***2**.289*
Ross, Tony **5**.91
Ross Ice Shelf ***1**.286*, **1**.287
Rossby, Carl-Gustav Arvid ***1**.286*
Rosse, Lord **8**.16
Rossellini, Roberto ***5**.391*
Rossetti, Christina ***5**.391*
Rossetti, Dante Gabriel **5**.366, ***5**.391*
Rossi, Bruno **8**.31
Rossini, Gioachino **5**.40, ***5**.391*
Rosskopf, G. F. **6**.377
Rostand, Edmond **5**.120
Rostropovich, Mstislav ***5**.391*
rotary baler **6**.6
rotary engine ***6**.303*
rotation (astronomy):
Earth **1**.229, ***1**.286*, **8**.141, ***8**.142*
galactic **8**.54
rotation, crop **3**.91–2, **6**.7, **6**.253, ***6**.303*
rotation, torque **6**.357
rotational time **1**.337–8, **6**.356
Roth, Philip ***5**.391*
Rothko, Mark ***5**.391*
Rothschild family **4**.29, ***4**.293*
rotifer **2**.288, ***2**.289*
Rotorua ***1**.286*
rotten borough ***4**.293*
Rotter, Julian **7**.187
rotwari system **4**.298
Rouault, Georges ***5**.391–2*
Roubiliac, Louis-François ***5**.392*
Rough Riders ***4**.293–4*
roughage in diet **6**.141
round (canon) **5**.75
Round Table Conferences ***4**.294*
rounders ***7**.270*
Roundheads **3**.118, **3**.180, **3**.253, ***3**.311*
round-off errors ***1**.286*
roundworms ***2**.289*
ascarid **2**.16–17, **2**.242
eelworm **2**.104
and elephantiasis **2**.106
hookworm **2**.161
lungworm **2**.196
and onchocerciasis **2**.234, **2**.286
pinworm **2**.255

roundworms (*cont.*):
Trichinella spiralis **2**.347
Rousseau, Henri **5**.312, **5**.359, ***5**.392*
Rousseau, Jean-Jacques ***3**.311*, ***5**.392*, ***7**.270*
autobiography **5**.27
and democracy **7**.83
as Encyclopedist **3**.115
and Enlightenment **3**.117
novel by **5**.148
and psychology **7**.149
Rousseau, Pierre-Étienne-Théodore ***5**.392*
Roussel, Albert ***5**.392*
rove beetle ***2**.289*
row cultivation **6**.170, ***6**.303*
rowan ***2**.289*
rowing ***7**.270*
boat *see* dinghy; galley and trireme
Rowland, Henry Augustus ***1**.286–7*
Rowlands, Daniel **4**.218
Rowlandson, Thomas ***5**.392–3*
Rowlatt Act (India) ***4**.294*
Rowley, John **8**.115
Rowley, Thomas **5**.40, **5**.89
Rowley, William **5**.162, **5**.295
Rowntree, Joseph ***4**.294*
Rowntree, Seebohm **4**.294
Roxas, Manuel **4**.268
Roy, Jamini **5**.219
Roy, Raja Ram Mohun **7**.143
Roy, Ram Mohan **4**.151, ***4**.294*
Royal Academy of Arts, London **5**.381–2, ***5**.393*
Royal African Company ***3**.311*
Royal Ballet **5**.274, ***5**.393*
Royal Danish Ballet ***5**.393*
Royal Greenwich Observatory ***6**.303*, ***8**.142*
Royal Institution ***6**.303*
Royal Observatory, Edinburgh ***8**.142*
Royal Society ***3**.311–12*, ***6**.303*
Royal Society for the Encouragement of Arts, Manufactures, and Commerce ***6**.303*
royal waterlily ***2**.289*
royalty (fee) ***7**.270*
see also copyright; franchise
royalty (monarchy) *see* Crown; leaders; monarchy
Rozvi empire **3**.4, **3**.244, ***3**.312*
RPI *see* retail price index
RPM *see* resale price maintenance
RR Lyrae star ***8**.142*
Ru **9**.16
Ruanda-Urundi *see* Burundi; Rwanda
Ruapehu, Mount ***1**.287*
Ruapekapeka ***4**.294*
rubber (natural) ***2**.289–90*, ***6**.303*
cables **6**.53
extrusion **6**.303
growing **6**.7, **6**.271, **6**.303
as insulator **6**.183
latex **2**.183–4
rubberized fabric ***6**.304*
see also vulcanization; waterproofing
tyre **6**.366
rubber (synthetic) **6**.60, **6**.236, **6**.261, **6**.271, ***6**.303–4*
Rubbra, Edmund ***5**.393*
Rubens, Sir Peter Paul **5**.64, **5**.359, ***5**.393*
Rubicon ***3**.312*
rubidium **1**.107, ***1**.287*, **6**.96
Rubinstein, Anton ***5**.393–4*, **5**.448
Rublyov, Andrey ***5**.394*
ruby ***1**.287*
Rudaki **5**.344
rudbeckia **2**.74
rudd ***2**.290*
rudder ***6**.304*
Rudolf I, king of Romans (1218–91) **3**.152, **3**.153, **3**.168
Rudolf II, Holy Roman Emperor (1552–1612) **3**.153, **3**.168
Rudolf III, duke of Austria (1282–1307) **3**.153
Rudolph, Paul ***5**.394*
ruff **2**.94, ***2**.290*
rugby football **4**.326, **7**.122, **7**.123
Ruggles, Carl Sprague ***5**.394*
rugs *see* carpets and rugs
Ruhr ***4**.294*, **4**.295
Ruisdael, Jacob van ***5**.394*
Ruiz, Juan **5**.428
Ruiz, Martínez **5**.172–3
rule of law ***7**.270*
see also law and justice
rulers *see* leaders; monarchy
Rum Rebellion ***4**.294–5*
Rumania *see* Romania
rumba ***5**.394*
Rūmī, Jalāl al-Dīn **3**.344, ***5**.394*
ruminant mammals ***2**.290*
camels/camel-like *see* camelids
giraffe ***2**.136*, **2**.137, **2**.233
as herbivores **2**.155
stomach **2**.290, **2**.326
see also cattle; deer; goat and goat-antelope; sheep
Rump Parliament **3**.210, ***3**.312*
Rundstedt, Fieldmarshal von ***4**.295*
Runeberg, Johan Ludvig **5**.155, **5**.404
Runge, Philipp Otto ***5**.394*
running **7**.76–7, **7**.192
runoff ***1**.287*
gully **1**.148–9
hydrological cycle **1**.162
ravines **1**.276–7
rill **1**.283, **1**.284
sheet erosion **1**.300
Rupert (15th century, Wittelsbach) **3**.168
Rupert, prince (1619–82) **3**.110, **3**.246, ***3**.312*
Rurik, prince of Novgorod **3**.259
rush ***2**.290–1*
Rushdie, Salman **5**.363, ***5**.394*, **7**.32, **7**.243
Rushmore, Mount **1**.40
Rusimatov, Faroukh **5**.248
Ruskin, John **4**.213, ***5**.394*
Russell, Lord (1549) **3**.380
Russell, Lord (1683) **3**.312
Russell, Bertrand ***7**.270*
and analytic philosophy **7**.12
and descriptions **7**.197
and logicism **7**.187
and mathematics **1**.365
and millenarianism **7**.202
and ontology **7**.223
as pacifist **4**.59, **4**.94
and pragmatism **7**.244
Russell, Charles **7**.171
Russell, George **5**.226
Russell, Henry Norris ***8**.142–3*
Hertzsprung–Russell diagram ***8**.66–7*, **8**.160
Russell, John, 1st earl (1792–1878) **4**.79, **4**.217, ***4**.295*
Russell, John Scott (1857, shipbuilder) **6**.49
Russell, W. H. (1850s, journalist) **4**.88
Russia ***1**.287–8*, ***3**.312*, ***4**.295–6*, ***7**.378*
for 1922–91 *see* Union of Soviet Socialist Republics
since 1991 *see* Commonwealth of Independent States; Russian Federation
arts (including USSR) **5**.151, ***5**.394–6*
architecture **5**.371, **5**.375, **5**.394–5
cinema **5**.132, **5**.142, **5**.447
critics **5**.278
Enlightenment **5**.396
graphic art and design **5**.266, **5**.298
Russia (*cont.*):
arts (*cont.*):
metallurgy **5**.151
painting *see under* painting
sculpture and carving **5**.21, **5**.111, **5**.346
set and stage design **5**.29, **5**.293, **5**.430
see also literature; music *below*
astronomy *see under* Union of Soviet Socialist Republics
Boyars **3**.47–8
chemists **1**.154, **1**.198, **1**.210
Civil War ***4**.296*, **4**.297, **4**.378
Concert of Europe **7**.135
Cossacks **3**.88, **3**.89
economy and trade **3**.243
energy *see under* technology and science *below*
ethnic groups **3**.6–7, **7**.109
exploration **1**.36, **1**.58, **1**.184
foreign policy:
Congress of Paris **4**.260
Three Emperors' League **4**.344
Tilsit Treaties **4**.345
see also under Union of Soviet Socialist Republics
government *see* politics and government *below*
kulaks **4**.187
land reform **7**.179
languages **7**.156, **7**.227, **7**.288, **7**.322
alphabet *see* Cyrillic
literature (including USSR) **4**.150, **5**.60, **5**.65, **5**.278, ***5**.396*, **5**.424, **7**.216, **7**.272
myths **9**.27
naturalism and realism **5**.3, **5**.184, **5**.185, **5**.430–1, **5**.457–8
Romanticism **5**.261–2, **5**.370, **5**.371, **5**.396
symbolism **5**.43, **5**.49–50, **5**.278, **5**.396
see also under drama; novels; poetry; stories
map **4**.296
mathematician **1**.197
music (including USSR):
concertos **5**.247, **5**.373, **5**.436, **5**.448
dance **5**.285
see also under ballet; ballet and dance companies; choreographers
piano **5**.367, **5**.373, **5**.393–4, **5**.409
Romanticism **5**.156, **5**.247, **5**.373, **5**.448
songs and singers **5**.87, **5**.310, **5**.313, **5**.370, **5**.448
violin **5**.326
see also under composers; opera; symphonies
physical geography and geology **1**.287–8
lake **1**.185
Lapland **1**.188
mountains **1**.135, **1**.351
Novaya Zemlya **1**.232
peninsulas **1**.180, **1**.184
regions **1**.290
river **1**.357–8
soil scientist *see* Dokuchaiev
physicist **1**.290
pogroms **4**.271
politics and government **4**.44, **4**.94, **4**.104
anarchists **4**.27, **4**.187, **7**.12
leaders **4**.241–2, **4**.326, **4**.330, **4**.352, **4**.380
see also Yeltsin
Populists **4**.27, **4**.275, **7**.242
Social Democrats **4**.317
socialism **4**.275, **4**.317
Russia (*cont.*):
politics and government (*cont.*):
see also Revolutions; tsars; wars *below and under* Union of Soviet Socialist Republics
port **4**.309
regions and cities **1**.290, **3**.196, **3**.208, **3**.259, **4**.197, **9**.74
see also Georgia; Muscovy; Russian Federation; Turkistan; Ukraine
religion **3**.375, **4**.283, **7**.98
Revolutions:
in 19th century **4**.27, **4**.94, **4**.150, **4**.243, **4**.271
of 1905 **4**.275–6, ***4**.297*, **4**.352
of 1917 ***4**.297*, **7**.53, **7**.267
leaders **4**.54, **4**.181, **4**.193, **4**.212, **4**.327, **4**.352
see also under politics *under* Union of Soviet Socialist Republics
rulers *see* tsars *below*
space exploration *see under* Union of Soviet Socialist Republics
technology and science *see under* Union of Soviet Socialist Republics
tsars (chronologically):
listed ***9**.68*
before 1598 **3**.197, **3**.312
see also Ivan III; Ivan IV
Godunov, House of (1598–1606) **3**.45, **3**.312
Shuiski, House of (1606–10) **9**.68
Romanov, House of (1613–1762) **3**.64, **3**.300, **3**.309, **3**.312
see also Elizabeth Petrovna; Peter I
Holstein-Gottorp-Romanov, House of (1762–1917) **3**.64, **3**.198, **3**.271, **3**.324, **4**.9, **4**.241–2
see also Alexander I; Alexander II; Catherine II; Nicholas II
urban areas *see* regions and cities *above*
wars and defence:
army leaders **3**.8, **3**.289, **3**.346, **4**.53, **4**.188, **4**.384
Caucasus occupied **4**.310
China, treaties with **3**.196, **3**.250
Civil ***4**.296*, **4**.297, **4**.378
Cossack uprising **3**.294
Crimean **4**.27–8, **4**.88, **4**.164, **4**.260
navy leader **4**.185
Polish Succession **3**.287
Quadruple Alliance **4**.279
Revolution *see* Revolutions *above*
Russo-Japanese War **4**.275, ***4**.297–8*
Russo-Turkish Wars **4**.3, ***4**.298*
World War I **4**.213
see also Napoleonic Wars; Northern War; Seven Years *and under* Union of Soviet Socialist Republics
see also (*under* Union of Soviet Socialist Republics) economy; education; foreign policy; politics; religion; wars; weapons
Russian Federation **4**.296
republics listed ***9**.74*
rust ***1**.288*
rust fungus ***2**.291*
Rustam **9**.33, ***9**.34*
rutabaga **2**.331
Rutan, Dick **6**.138
Ruth ***9**.31*
ruthenium **1**.107, ***1**.288*, ***6**.304*
Rutherford, Ernest, 1st Baron Rutherford of Nelson **1**.41, **1**.150, **1**.221, ***1**.288*, **1**.309
rutherfordium **1**.107

rutile **6**.357
Ruwenzori ***1**.288*
Ruysdael, Salomon van ***5**.396*
Ruyter, Michiel Adrianszoon de **3**.14, ***3**.312*
RV Tauri star ***8**.143*
Rwanda ***1**.288*, **4**.4, ***4**.298*, **7**.9, **7**.261, ***7**.378*
Rydberg, Johannes Robert ***1**.288*
rye **2**.110, ***2**.291*
Rye House Plot ***3**.312*
rye-grass **2**.244
Ryle, Gilbert **7**.185
Ryle, Martin **8**.105, **8**.193
Rynd, Francis **6**.175
ryotwari system ***4**.298*
Rysbrack, John Michael ***5**.396*
Ryswick, Treaty of ***3**.312*

S

SA (Brownshirts) **4**.53
Saar ***1**.289*, ***4**.299*
Saarinen, Eero **5**.14, ***5**.397*
Saarinen, Eliel **5**.397
Sabah ***4**.299*, **4**.321
Sabatier, Paul **1**.147
Sabbato, Giovanni **5**.416
Sabines ***3**.313*
sable ***2**.291*
sabretoothed tiger ***2**.291*
saccharin **6**.342
Sacco, Nicola **4**.299
Sacco–Vanzetti Case ***4**.299*
Sacheverell, Henry ***3**.313*
Sachs, Hans ***5**.397*
Sachs, Nelly **5**.10
sackbut ***5**.397*
Sacramento River ***1**.289*
sacraments ***7**.271*
  see also Mass (Eucharist)
sacred music **5**.351, **5**.380–1
  Britain **5**.50, **5**.139, **5**.176, **5**.212, **5**.370, **5**.430, **5**.486
  France **5**.89, **5**.137, **5**.256, **5**.325
  Germany **5**.365, **5**.406–7
  Italy **5**.90, **5**.303, **5**.336, **5**.343–4, **5**.398
  see also hymns; Mass (Eucharist)
Sacred Thread **7**.273
sacrifice ***7**.271–2*
Sadat, Anwar **4**.59, **4**.109, ***4**.299*, **7**.15, **7**.20
SADCC (Southern African Development Co-ordination Conference) ***7**.296*
Saddam Hussein *see* Hussein, Saddam
saddle (col) **1**.68
saddle points (mathematics) ***1**.289*
saddleback (bird) **2**.362
saddlery **6**.165–6, **6**.300
Sadducee ***3**.313*
Saʿdī, Muṣliḥ al-Dīn ***5**.397*
Sadiq, al-Mahdi **4**.332
Sadler, William **3**.366
Sadler's Wells Companies **5**.274
Sadowa, battle of ***4**.299*
Safavid dynasty **3**.181
  art **5**.344, ***5**.397*, **5**.399
safe ***6**.304*
safety:
  -belt ***6**.304*
  -critical system **6**.297
  glass **6**.156, **6**.304
  laboratory **6**.192, **6**.193
  life-jacket and lifebuoy **6**.52, **6**.200–1
  in nuclear reactor **6**.245
  valve ***6**.304*, **6**.370
  see also accidents and disasters
safflower ***2**.291*
saffron ***2**.291–2*
Safi ud-Din **3**.181
sagas ***5**.397*, **5**.404, **5**.423
sage ***2**.292*, **2**.293
Sagitta ***8**.143*
Sagittarius (constellation) ***8**.143*
Sagittarius A (radio source) ***8**.143*
Sagittarius arm (spiral arm of Galaxy) ***8**.143*
Sagittarius B2 (radio source) ***8**.143–4*
sago ***2**.292*
Saha, Meghnad: Saha's equation ***8**.144*
Sahara ***1**.289*, **7**.261
Sahel **7**.86
Saibou, Ali **4**.242
Said, Edward **7**.225
Said, Khedive Mohammed **6**.199
Said ibn Sultan of Oman and Zanzibar **4**.250, ***4**.299–300*, **4**.388
Saidu Momoh, Joseph **4**.312
Saigo Takamori ***4**.300*
sail/sailing ***6**.304–5*
  principles ***6**.305*
  solar **6**.326
  as sport **6**.104, **6**.308, **6**.390, ***7**.272*, **7**.331
  see also rig, sailing; sailing ships and boats
sailfish ***2**.292*
sailing ships and boats ***6**.305–8*
  barque **6**.30, **6**.298, **6**.306
  brig **6**.47, **6**.298, **6**.408
  clipper **6**.76, **6**.219
  corvette **6**.90
  cutter **6**.94
  frigate **6**.146, **6**.188
  galleon **6**.149, **6**.306, **6**.307
  junk **6**.190, **6**.306, **6**.307
  ketch **6**.190, **6**.306
  sails *see* rig, sailing
  schooner **6**.76, **6**.298, **6**.310
  sloop **6**.90, **6**.322
  warships **6**.375
  yachts *see* yachts
  see also sail/sailing; ships and boats
sailors *see* navies
sainfoin **2**.292
saint *see* saints
St Albans ***3**.313*
St Anthony's fire **2**.110
St Bartholomew's Day Massacre ***3**.313*
St Bartholomew's Fair **3**.123
Saint Christopher and Nevis *see* Saint Kitts and Nevis
St Denis, Ruth ***5**.397*, **5**.415
Saint Elias Mountains ***1**.289*
Saint Elmo's fire ***1**.289*
St Germain-en-Laye, Treaty of **4**.363–4
Saint Helena ***1**.289*
Saint Helens, Mount ***1**.289–90*
St Ives School ***5**.398*, **5**.438
St John, Order of **3**.198, **3**.303
Saint John River ***1**.290*
Saint John's wort ***2**.292*
Saint Kitts and Nevis ***1**.290*, ***7**.378–9*
  see also Caribbean; CARICOM
St Laurent, Louis Stephen **4**.195, ***4**.300*
Saint Lawrence River ***1**.290*, **1**.335
Saint Leon **5**.122
Saint Lucia ***1**.290*, ***7**.379*
  see also Caribbean; CARICOM
St Petersburg (*previously* Leningrad) **4**.193, **4**.297
Saint Vincent and Grenadines ***1**.290*, ***7**.379*
  see also Caribbean; CARICOM
Sainte-Beuve, Charles-Augustin ***5**.397*
Saintes, Les, battle of ***3**.313*
Saint-Exupéry, Antoine de ***5**.397*
Saint-Gaudens, Augustus ***5**.398*
Saint-John Perse ***5**.398*
Saint-Just, Louis Antoine Léon florelle de ***3**.313–14*
saints ***7**.272*, ***9**.31*, ***9**.40–3*
  Buddhist **7**.272
  Christian ***7**.272*, ***9**.40–3*
    list of ***9**.40–3*
  Islamic **7**.272, **9**.3
  list of ***9**.40–3*
  see also individual names
Saint-Saëns, Camille **5**.124, **5**.155, **5**.181, ***5**.398*
Saint-Simon, Claude-Henri de Rouvroy, comte de ***4**.300*, **4**.317, **7**.63, **7**.307
saithe pollock **2**.69–70
Saiwi Jai Singh II, Raja of Jaipur **3**.183–4
Sakaida Kakiemon **5**.235
saker falcon ***2**.292*
Sakhalin ***1**.290*
Sakharov, Andrey Dmitriyevich ***1**.290*
Saki ***5**.398*
Śakti **9**.32
salad *see* vegetables
Saladin **3**.92, **3**.198, **3**.201, **3**.303, ***3**.314*
Salamanca, Daniel **4**.44
salamander family ***2**.292*
  axolotl **2**.19, **2**.226
  congo eel **2**.75
  fire **2**.260
  giant **2**.135
  lungless **2**.196
  mole **2**.19, **2**.216
  mud puppy **2**.220–1
  newt **2**.227
  olm **2**.234
  siren **2**.310
Salamis, battle of ***3**.314*
salaries *see* pay
Salazar, Antonio de Oliveira **4**.275, ***4**.300*
Saleh, Ali Abdullah **4**.386
Salem witch trials ***3**.314*
sales:
  auction **7**.22
  merchandizing **7**.198–9
  see also retail trade; trade
Salian House *see under* Holy Roman empire
Salic law ***3**.314*
Salieri, Antonio ***5**.398*
al-Salih, sultan of Turkey **3**.218
Salinas de Gortari, Carlos **4**.220
Salinger, Jerome David ***5**.398*
salinity *see under* salts
Salisbury, Robert Arthur Talbot Gascoigne-Cecil, 3rd Marquess of ***4**.300*
saliva ***2**.292*
al-Sallal, Abdullah **4**.386
Sallé, Marie **5**.30
Sallust ***3**.314*
Salman al-Farisi ***9**.31*
salmon **2**.61, ***2**.292*
  see also trout
Salmonella **2**.352
Salò, Gasparo da **5**.477
Salome **9**.20, ***9**.31*
Salon **5**.299, ***5**.398*
Sālote Tupou III, queen of Tonga **4**.347
salp **2**.301–2
salpiglossis ***2**.292*
salsify ***2**.292*
salt *see* salts
SALT *see* Strategic Arms Limitation Talks
salt lakes:
  Asia **1**.31, **1**.54
  Australia **1**.118, **1**.130, **1**.340
  Europe **1**.54, **1**.130
  evaporites **1**.116
  Israel **1**.130
  lagoon **1**.185–6
  salt-flats **1**.291
  United States **1**.41, **1**.146
Salt March ***4**.301*
saltation ***1**.291*
saltpetre (nitre) *see* gunpowder; potassium nitrate
salts ***6**.2*, ***6**.308*
  common *see* sodium chloride
  dome ***1**.291*
  as ionic compounds **1**.172
  and ions **6**.186
  lakes *see* salt lakes
  in paints **6**.255
  from plant **2**.106, **2**.137

salts (*cont.*):
  plug ***1.291***
  salinity ***1.290***
  salinization (soil) ***1.290–1***, **6.**188
  salt-flats ***1.291***
  salt-marshes ***1.291***
  salt-pan ***1.291***
  sodium *see entries beginning* sodium
  sulphites **1.**326
  see also nitrates; phosphates; sulphates
salvage ***7.272***
Salvation Army **4.**45, **4.**238, ***4.301***, **7.**218, ***7.272***
salvias ***2.292–3***
  sage **2.**292, **2.**293
Salween River ***1.291–2***
Salyut **6.**327
samādhi ***7.272***
Samaria ***3.314***
  Samaritans **7.**302
samarium **1.**107
samba ***5.398***
samizdat ***7.272***
Sāmkhya **7.**155
Sammartini, Giuseppe and Giovanni Battista ***5.398***
Sammuramat **9.**32
Samnite wars **3.**314
Samoan Islands ***1.292***, ***4.301***, **7.**240
  anthropology **7.**196
  physical geography and geology **1.**292, **1.**365
  Samoans ***7.273***
  in South Pacific Forum **7.**296
Samori Touré ***4.301–2***
sample ***7.273***
sampler ***5.398***, **5.**400
Samrin, Heng **4.**58, **4.**313
SAMs *see* surface-to-air missiles
samsara **7.**169, **7.**263, ***7.273***
samskāra ***7.273***
Samson ***9.31***
Samudra Gupta of India **3.**151
Samuel ***3.314***, **3.**317
Samuel, Herbert **4.**234
samurai ***3.314–15***, **4.**300, ***4.302***, **7.**38, **7.**194
  body armour **6.**41
San Andreas Fault ***1.292***
San Francisco Conference ***4.302***
San Jacinto, battle of ***4.302***
San Marino ***1.293***, **7.**71, **7.**319, ***7.379***
San Martín, José de **4.**43, **4.**72, **4.**250, **4.**267, ***4.302***, **4.**319
Sanchi ***3.315***, ***5.400***
Sancho II, king of Castile **3.**113
Sancho IV, king of Castile and León **3.**9
Sancroft, William **3.**100
sanctions **7.**34, ***7.273***
  against South Africa **7.**34, **7.**59, **7.**225
sanctuary ***3.315–16***
Sand, George ***5.400***, **5.**401
sand ***1.292***
  animals **2.**293
  -bar ***1.292***
  blasting **6.**1, ***6.308***
  dunes **1.**33, **1.**97, **1.**298
  erg **1.**112
  quicksand **1.**272
  rock formed from *see* sandstone
  sand-spit ***1.292–3***, **1.**339
  sea ***1.292***
  see also deposition *under* coasts
sand boa ***2.293***
sand eel ***2.293***
sand fly ***2.293***
sand grouse ***2.293***
sand lizard ***2.293***
sand martin ***2.293***
sand paintings **5.**321–2
sand wasp ***2.294***
Sandage, Alan R.: Hubble–Sandage variable ***8.71***
sandblasting **6.**1, ***6.308***
Sandburg, Carl ***5.400***
Sandby, Paul and Thomas ***5.400***
sand-dollar ***2.293***
sanderling ***2.293***
sandhopper ***2.293***
Sandinistas and Sandinista Liberation Front **4.**241, ***4.302***, **7.**73
Sandino, Augusto César ***4.302***
sandpiper family ***2.293–4***
  greenshank **2.**144
  redshank **2.**280
  ruff **2.**290
  sanderling **2.**293
sandstone **1.**146, **1.**147, **1.**215, ***1.293***
  metamorphosed (quartzite) **1.**272
Sandwich Islands *see* Hawaii
Sanford, M. M. **3.**378
Sanges, Louis de **4.**56
sangha **7.**205, ***7.273***
Sanguinetta, Julio **4.**361
Sanhedrin ***3.316***
sanitation ***7.273***
  sanitary engineering **6.**218
  water supply **6.**378–9
  see also sewage
Sanjurjo, José **4.**324
Sankaracharya ***3.316***
Sanmicheli, Michele ***5.400***
Sann, Son **4.**58, **6.**313
sansa ***5.400***
sans-culotte ***3.316***
Sanskrit language **7.**11, **7.**156, **7.**157, **7.**305
  literature **5.**314, **5.**338, ***5.400–1***, **9.**21–2
    poetry and epics **5.**134, **5.**243, **5.**276, **5.**373, ***5.401***
  see also literature *under* Hinduism
Sansovino, Jacopo ***5.401–2***
Santa Anna, Antonio López de **4.**7, **4.**170, **4.**219–20, ***4.302***, **4.**341
Santa Maria, president of Chile **4.**29
Santander, Francisco **4.**78
Santillana, Marqués de **5.**428
santūr ***5.402***
Sanusi ***4.302***
al-Sanusi al-Idrisi, Sidi Muhammad **4.**302
São Tomé and Príncipe ***1.293***, **4.**4, ***7.379***
sap ***2.294***
Sapir, Edward: Sapir–Whorf hypothesis ***7.273–4***
sapodilla ***2.294***
sapphire ***1.293***
Sappho ***5.402***, **7.**146
saprolite ***1.293***
saprophytes **2.**200, ***2.294***
  see also fungi; mould
sapsucker **2.**371
saraband **5.**439
Saracens **3.**240, ***3.316***
Saragossa ***3.316***
Sarah (Abraham's wife) **9.**2, **9.**18, ***9.31–2***
Sarah (Tobias' wife) **9.**36
Sarajevo assassination **4.**32, ***4.123***, **7.**20
  as precursor of World War I **4.**29, **4.**46, **4.**135
  and Serbia and Black Hand **4.**40, **4.**309
sārangī ***5.402***
Sarasvati ***9.32***
Saraswati, Swami Dayananda **4.**20, **4.**152, **7.**34, **7.**143
Saratoga campaign ***3.316***
Sarawak **4.**52, ***4.302***, **4.**321
sardine **2.**254
Sardinia ***1.293***, **4.**139, **4.**206, **4.**260, **4.**364
sardonyx **1.**238
Sarekat Islam ***4.302–3***
Sargasso Sea ***1.293***
Sargent, John Singer ***5.402***
Sargon of Akkad **3.**6, **3.**230
sarin **6.**239
Sarne, José **4.**49
Saros ***8.144***
Sarraute, Nathalie **5.**323
sarsaparilla ***2.294***
Sarto, Andrea del **3.**131
Sartre, Jean-Paul **5.**323, ***5.402***, **7.**113, ***7.274***
Saskatchewan ***1.293***
Saskatchewan River ***1.293–4***
sassafras tree ***2.294***
Sassanian empire **3.**93, ***3.316***
  art **5.**344
Sassetta ***5.402***
Sassoon, Siegfried ***5.402***
Sassou-Nquesso, Denis **4.**83
Sastre, Alfonso **5.**428
Satakarni, Gautamiputra **3.**316
Satan *see under* Devil
Satavahana ***3.316***
sat-cit-ānanda ***7.274***
satellites, artificial **6.**307, ***6.308***, ***7.274***, **8.**68, ***8.144***
  broadcasting ***7.274***
  communications **6.**81, **6.**308, **6.**347, **6.**349, **7.**39, **7.**274
  dish ***6.308***
  Earth resources **6.**111, **6.**215, **6.**285
  geostationary orbit **8.**58
  geosynchronous orbit **8.**58
  ground station **6.**161–2, **6.**222, **6.**308
  launch vehicle **6.**94, **6.**196, **6.**326
  meteorological **6.**221, **6.**222
  military **6.**224, **6.**225
  navigation **6.**239
  photography from **6.**268
  and rocketry **6.**301
  scientific **6.**310
  Sputnik **6.**330
  transmission ***7.274***
  X-ray **8.**188
  see also space-flight
satellites, natural ***8.144–5***, ***8.198***
  co-orbital **8.**29, **8.**47, **8.**76
  McMath solar **8.**96
  mountains **8.**140
  ring plains **8.**140
  shepherding **8.**12, **8.**149
  Trojan **8.**175
  see also Moon *and under individual planets*
Satie, Erik ***5.402***, **5.**403
satire and satirists ***5.402***
  Britain **5.**68, **5.**154, **5.**284, **5.**342, **5.**483
    poetry **5.**136, **5.**357, **5.**384, **5.**402, **5.**421, **5.**423, **5.**441–2, **5.**496
  France **5.**51, **5.**120, **5.**125, **5.**253, **5.**372, **5.**479
  Germany **5.**51, **5.**194, **5.**203
  Ireland **5.**324
  Italy **5.**12, **5.**339
  lampoon **5.**254
  Latin **5.**212, **5.**242, **5.**389
  Netherlands **5.**148
  Poland **5.**377
  Russia and USSR **5.**396, **5.**498
  Spain **5.**372, **5.**428
  United States **4.**136, **5.**269, **5.**474, **5.**486
  Yiddish **5.**497
  see also caricature; comedy
Sato, Eisaku **4.**194, ***4.303***
satrap ***3.316–17***
satsuma **2.**201
saturated aliphatic hydrocarbons **1.**9
saturated fats **6.**126
Saturn (god) ***9.32***
Saturn (planet) ***8.145***
Saturn (*cont.*):
  belts **8.**145
  orbit **8.**17, **8.**196
  rings ***8.145***, **8.**197
  satellites ***8.198***
    Calypso **8.**175
    co-orbital **8.**29, **8.**47
    Dione **8.**37
    Enceladus **8.**46
    Epimetheus **8.**29, **8.**47
    Helene **8.**175
    Hyperion **8.**71–2
    Janus **8.**29, **8.**76
    largest **8.**172
    Mimas **8.**99
    Phoebe **8.**119
    Rhea **8.**139
    Telesto **8.**170, **8.**175
    Tethys **8.**171
    Titan **8.**172
    Trojan **8.**170, **8.**175
  statistical details **8.**196
  zones **8.**145
Saturn (rocket) **6.**17, **6.**196, **6.**301
satyagraha ***4.303***
Satyanarayana, Visvanatha **5.**136
Satyr ***9.32***
Saud, Abd al-Aziz ibn **4.**303, **4.**370
Saud family ***4.303***, **4.**370
Saudā **5.**466
Saudi Arabia ***1.294***, ***4.303***, ***7.379–80***
  aid from **7.**8
  in Arab League **7.**15
  economy and trade *see* OPEC
  in Gulf Co-operation Council **7.**136
  laws/legal system **7.**284
  Mecca **3.**227
  physical geography and geology **1.**294
  politics and government **7.**332
  religion/theology **7.**284, **7.**302, **7.**328
  rulers **4.**115, **4.**303, **7.**97
  segregation **7.**278
  technology and science **6.**51, **6.**209
  see also Arabia
Saukara, Thomas **4.**55
Saul, king of Israel **3.**314, ***3.317***
saùng-gauk **5.**402–3, **5.**427
Saussure, Ferdinand de **7.**180, ***7.274–5***, **7.**278, **7.**285, **7.**301
Savage, Michael Joseph **4.**189, ***4.303***
savannah ***1.294***, ***2.294***
Savart, Félix **8.**178
  Biot–Savart law (1820) **1.**39
Save the Children Fund **7.**327
Savery, Thomas **6.**227
Savimbi, Jonas **4.**14
saving:
  friendly societies **7.**127
  savings-and-loan associations **7.**37, **7.**207
Savonarola, Girolamo ***3.317***
savory ***2.294***
Savoy **3.**119, ***3.317–18***, ***9.68***
saw **6.**69, ***6.308***, **6.**357, **6.**358, **6.**387
Saw Maung, General **4.**229
sawfish ***2.294***
sawflies **2.**57, ***2.294***, **2.**371
Sawyer, Amos **4.**195
Saxe, Maurice, Comte de **3.**128, ***3.318***
Saxe-Coburg-Gotha, House of **4.**108, **4.**365
  see also Windsor
saxhorn ***5.403***
saxifrage ***2.294***, **2.**295
Saxons ***3.318***
  see also Anglo-Saxons *and under* rulers *under* Britain
Saxony **3.**160, **3.**163, ***3.318***, **4.**303
  House of *see under* Holy Roman empire
Saxony spinning wheels **6.**329

saxophone 5.352, 5.376, *5.403*
Say, Jean-Baptiste: Say's Law 7.52, 7.53
Sayers, Dorothy L. 5.129
Sayyid dynasty *3.318*
saz (lute) *5.403*
scabies 2.294–5
scabious *2.295*
scaffolding *6.308–10*
scalars 1.87, *1.294*
scalawag *4.303*
scale (animal) *2.295*
  see also edentates; fish; reptiles
scale (music) 5.299, *5.403–4*
  see also harmony
scale, geometrical *1.294*
scale worm 2.295
scale-insects 2.30, 2.142, *2.295*
scallop *2.295*, 2.296
scaly-tailed squirrel *2.295–6*
scampi 2.100
Scandinavia *1.294*, *3.318*
  aid from 7.8
  arts *see under individual countries*
  astronomy 8.3
    see also under Denmark
  cohabitation 7.55
  in Council of Europe 7.71
  economy and trade:
    and EC 7.99
    and EFTA 7.111
    in OECD 7.221
    pay 7.107
  environment 7.2, 7.140, 7.260
  ethnic groups 7.181
  explorers 1.12, 1.38, 1.224, 1.230, 1.231, 1.276, 3.118
  languages 7.131, 7.156, 7.322–3
  laws/legal system 7.62, 7.75, 7.255
  literature *4.404*
    see also under individual countries
      medicine and health care 7.140, 7.196, 7.198
  monarchy 7.97, 7.139
    see also rulers *under individual countries*
      physical geography and geology *1.122*, *1.232*, *1.294*, *1.328–9*
    Fennoscandian Shield 1.120
    fiords 1.122, 1.310
    ice-sheet 1.178
    islands 1.317, 1.328
    Lapland 1.188
    mountains 1.135
    seas 1.201, 1.306
  physicists 1.41, 1.236, 1.288
  politics and government 7.54, 7.250, 7.290
    see also under individual countries
  religion/theology 3.261, 7.112, 7.187
  rulers *see under individual countries*
  social services and security 7.45, 7.91
  sport and recreation 7.145, 7.151, 7.287, 7.297
  statistics 7.326
  technology and science 6.62, 6.68, 6.171, 6.384
    see also under individual countries
  wars and defence 7.202
    see also under Denmark; Sweden
  see also Denmark; Finland; Iceland; Norway; Sweden; Vikings
scandium 1.107, *1.294*
Scanian War *3.318*
scanning:
  body *see* tomography
  electronic 6.118, 6.164
  thyroid 6.243
  tunnelling microscope 6.223
scansion *5.404*
Scapa flow *4.303*
scapula 2.14
scarab *2.296*
scarcity *7.275*
Scarfe, Gerald 5.78
Scarlatti, Alessandro and Domenico *5.404–5*
scarlet fever *2.296–7*
scarp *see* escarpment
scat singing 5.239, 5.306
scatter diagram 1.280
scattering (astronomy) *8.145*
scaup *2.297*
scenery, natural *see* nature
scenery, theatrical 5.412, 5.429–30
scent *see* aromatic; perfume
scepticism *7.275*
Schacht, Hjalmar *4.303*
Schachter, Stanley 7.105
Schaeffer, Pierre 5.310
Schaerer, Edvard 4.259
Scharnhorst, Gerhard Johann David von *4.303–4*
Schaufuss, Peter 5.122, 5.393
Schawlow, Arthur Leonard 6.214
Scheele, Karl Wilhelm *1.294*
scheelite 6.365
Scheer, Reinhard 4.177, *4.304*
Scheldt *1.294*
Schent, P. 3.195
Scherk, Jöel 1.324
scherzo *5.405*
Schiaparelli, Giovanni 8.19, 8.87, 8.118
Schickard, Wilhelm 6.53
Schiele, Egon *5.405*
Schikaneder, Emmanuel 5.306
Schiller, Friedrich von *5.405*, 5.471
Schilling, Baron 6.88
Schinkel, Karl Friedrich *5.405*
Schirach, Baldur von *4.304*
schist 1.124, *1.294*
schistosomiasis/bilharzia 2.29, 6.377, 7.59
schizanthus *2.297*
schizophrenia *7.275*
  and criminal responsibility 7.76
  drug treatment 6.285, 6.360, 7.96
  and ECT 6.116
  and neurology 6.241
  and paranoia 7.228
Schlegel, August Wilhelm and Friedrich von *5.405*
Schleswig-Holstein *4.304*
Schlieffen, Alfred, Graf von: Schlieffen Plan *4.304*, 4.375
Schliemann, Heinrich 3.244, 3.361
Schmalkaldic War *3.318*
Schmidt, Bernhard: Schmidt telescope 8.135, 8.136
Schmidt, Helmut *4.304*
Schnitzler, Arthur 5.329, *5.405*
Schoenberg, Arnold 5.25, 5.44, *5.405–6*, 5.411
scholasticism 3.318–9, 3.319, 7.1, *7.275*, 7.309–10
Schönbein, Christian Friedrich 6.242, *6.310*
Schönberg, Erich M.: Schönberg–Chandrasekhar limit *8.146*
school system *see* education
schooner 6.76, 6.298, *6.310*
Schopenhauer, Arthur *7.276*
Schott, Otto *6.310*, 6.391
Schracher, Gerhard 6.239
Schreiner, Olive *5.406*
Schrödinger, Erwin: Schrödinger equation *1.294–5*
Schrötner, Anton von 6.132
Schubert, Franz 5.194, *5.406*
Schulz, Bruno *5.406*
Schumacher, E. F. 6.18
Schuman, Robert (1950, politician): Schuman Plan *4.305*
Schumann, Clara 5.406, 5.407
Schumann, Robert (1810–56, composer) 5.57, *5.406*, 5.407
Schumpeter, Joseph *7.276*
Schuschnigg, Kurt von 4.14, *4.305*
Schutz, A. 6.375
Schütz, Heinrich 5.341, *5.406–7*
Schwartz, John 1.324
Schwarz, Hans 5.289
Schwarzenberg, Felix, prince of 4.25, *4.305*
Schwarzenberg, Karl Philipp, prince of 4.192, 4.280, *4.305*
Schwarzkopf, Elisabeth *5.407*
Schwarzkopf, General Norman 4.143
Schwarzschild, Karl 8.17, 8.146
  Schwarzschild radius *8.146*
Schweitzer, Albert 4.221
Schwind, Moritz von 5.406
Schwitters, Kurt *5.407*
sciatica 2.94
science:
  hypothesis 1.163
  and knowledge:
    Age of Reason 3.299–300
    Renaissance 3.301–2
    Royal Society 3.311, 6.303
    Scientific Revolution *3.318–19*
    see also education
  management 7.191
  methodology 2.264
  and morality 7.125
  philosophy of 7.188, 7.241, 7.276
  and reality 1.277
  societies *see* Societies and Academies
  see also biology; chemistry; geology; physics; technology; UNESCO *and entries beginning* 1., 2. *and* 6. *and* technology and science *under individual countries*
science fiction *5.407*
  in cinema 5.252, 5.255, 5.407, 5.429
  in literature 5.186, 5.261, 5.376, 5.407, 5.416, 5.472, 5.479, 5.485, 5.498
Science Museum *6.310*
Scientific Revolution *3.318–19*
scientific satellite *6.310*
Scientology *7.276*
scilla *2.297*
Scilly Isles *1.295*
scintillation 8.145
Scipio, Aemilianus Africanus Numantinus 3.260, 3.295, *3.319*
Scipio, Publius Cornelius 3.132, 3.308, *3.319*
scissors *6.310*
Sclater, P. L. 2.30
Scopas of Paros *5.407*
Scopes, John T.: Scopes case 4.128, *4.305*
score, musical 5.330, *5.407*
Scorel, Jan van *5.407*
scoria *1.295*
scorpion *2.297*
scorpionfish 2.280, *2.297*
scorpionfly *2.297*
Scorpius (constellation) *8.146*
Scorpius X-1 (X-ray source) *8.146*
scorzonera *2.297*
scoter *2.297*
Scotia Ridge *1.295*
Scotland *1.295*, *3.319*, *4.305*
  agriculture 6.295
  arts 5.73, 5.147, 5.179, 5.206, 5.353
    see also literature *below*
  astronomy 8.142
  buildings 6.51
  chemistry and chemists 1.40, 1.142, 1.275–6, 6.211
  clans 3.78, 7.52
  economists 3.331, 4.220
  engines 6.380
Scotland (*cont.*):
  explorers 1.44–5, 1.200–1, 1.216, 1.248, 1.286, 1.324, 1.335
  geologist 1.275
  highland clearances 4.151
  industry 4.62–3
  islands 1.153, 1.241, 1.301, 1.318, 3.159, 3.263, 3.326
  language 7.42, 7.59
  law 7.4, *7.276*, 7.318, 7.319
  literature 4.62, 5.432
    novelist 5.26, 5.207, 5.408
    Scottish-Gaelic *5.408–9*
    see also under poetry
  mathematicians 1.201, 1.224, 1.260
  medicine and health care 6.204
  myths and legends 9.13
  New Age Movement 7.214
  nobility 3.47, 3.97, 3.155
  nuclear technology 6.243
  philosophy 3.106, 3.171, 3.331, 4.220, 7.149–50
    see also Hume; Smith, Adam
  physical geography and geology *1.295*
    cliffs 1.32
    geologists 1.133, 1.200, 1.223
    islands *see* islands *above*
    mountains 1.37, 1.47, 1.135, 1.142, 1.241
    river 1.67
    schist 1.294
    turbidite 1.347
  politics and government:
    devolution 7.88
    nationalism 7.212
    radical 3.157
    and religion 7.265
    Scottish Martyrs *3.319–20*
    Scottish Nationalist Party *4.305–6*
    statesman 3.54
  prehistory 3.281, 3.319
  privateer 3.196–7
  regions 3.342, *9.73*
  religion/theology:
    Catholics 3.37
    Covenanters 3.89–90
    leaders 3.54
    and politics 7.265
    Protestants 3.90, 3.199, 3.384, 7.13, 7.50, 7.245
    saints 3.222, 9.41, 9.43
  Romans in 3.57
  rulers:
    listed *9.67*
    9th to 11th centuries 3.106, 3.194–5, 3.342
      see also Malcolm III
    12th and 13th centuries 3.8, 3.98–9, 3.221–2, 3.319, 3.382
    14th and 15th centuries 3.40, 3.251, 3.305, 3.343
      see also James II; James III; Robert I
    16th and 17th centuries 3.185
      see also James I, king of England; James IV; Mary, queen of Scots
    after 1625 *see* rulers *under* Britain
  Scapa flow 4.303
  Scots *3.319*
  Scottish Martyrs *3.319–20*
  tools 6.236
  urban areas 7.323
  vehicles and transport 6.47, 6.210, 6.332
  wars 3.305
    with England:
      Bannockburn 3.35
      Berwick treaties 3.40
      Bishops' wars 3.41–2, 3.90
      Culloden 3.93
      Dunbar 3.106

Scotland (*cont.*):
wars (*cont.*):
with England (*cont.*):
Falkirk **3**.123
Flodden Field **3**.128
Glencoe massacre **3**.142
Halidon Hill **3**.154
leader **3**.376
Montrose and **3**.240
Neville's Cross **3**.251
Otterburn **3**.265
Pinkie **3**.282
Prestonpans **3**.291
see also Jacobites
Porteous riots **3**.289
Scots pine **2**.89, *2.297*
Scott, Sir George Gilbert and sons (George Gilbert, John Oldrid, Adrian Gilbert, and Sir Giles Gilbert) **3**.109, *5.407–8*
Scott, Léon **6**.158
Scott, Robert Falcon *1.295*, **1**.300, **4**.15
Scott, Thomas A. **4**.62
Scott, Sir Walter **5**.26, **5**.207, *5.408*
Scott, Winfield **4**.218, *4.305*
Scottsboro case *4.306*
SCP (single-cell protein) **6**.320, *6.321*
'Scramble for Africa' *4.306–7*
see also colonies *under* Africa
scrambler *6.310*
scraper *6.310*
screamer *2.297*
scree *1.295–6*
screen (arts) *5.409*
printing *6.310*, **6**.353
screening (health) **6**.68, **6**.214, **7**.140
screw *6.310–11*
-cutting lathe **6**.294
propulsion **6**.123, **6**.160, **6**.284
screwdriver *6.311*, **6**.358
Scriabin, Aleksandr *5.409*
scrimshaw *5.409*
scripts *see* writing and scripts
scrub warbler and scrub-wren **2**.360
Scruton, Roger **7**.215
Scud **2**.163
Scullin, James Henry **4**.189, *4.307*
sculpin **2**.45
Sculptor (constellation) *8.146*
Sculptor Group (galaxies) *8.146*
sculpture and carving *5.409*
Africa **3**.40, **3**.174, **3**.254, **5**.8–9
Assyrian **3**.26, **3**.280, **5**.24–5
Aztec **5**.27
Babylon **3**.155
bone carving **5**.51
Britain **5**.260, **5**.320
abstract **5**.205, **5**.303
allegories **5**.483
Anglo-Saxon **5**.17
animals **5**.165, **5**.255
art nouveau **5**.177
medievalist **5**.177
metal **5**.79, **5**.176, **5**.177
modernist **5**.148
Neoclassical **5**.157
portrait busts and figures **5**.88, **5**.148, **5**.303, **5**.392
Vorticism **5**.148, **5**.480
Buddhist **5**.169
Burma **5**.67
busts **5**.68
cameo **5**.72
Carolingian **5**.79
caryatid **5**.81
casting **5**.82
Celtic **5**.84
Central America and Mexico **5**.85, **5**.327, **5**.457, **5**.498
China **3**.155, **5**.91–2, **5**.94
cire-perdue **5**.101
Coptic **5**.112
coral **5**.112–13
Crete **5**.189
sculpture and carving (*cont.*):
Cyprus **3**.50
Denmark **5**.454
Easter Island **3**.108
Egypt **3**.78, **3**.111, **3**.243, **3**.249, **3**.299
Etruscan **5**.149–50
Fatimid **5**.153
France:
abstract **5**.23, **5**.286
Burgundian school **5**.66
Cubism **5**.265, **5**.498
Impressionist **5**.127, **5**.171
junk materials **5**.137
nudes **5**.276, **5**.386
Romanesque **5**.179
gargoyle **5**.170
Germany **5**.46, **5**.248, **5**.250
Greece, ancient *5.190*
Mycenaean **3**.244, **5**.311
Phidias **3**.322, **5**.346
Polyclitus **5**.356, **5**.368
Praxiteles **5**.189, **5**.365
Scopas **5**.407
statues **3**.21, **3**.332, **3**.386, **5**.356
Hittite **3**.166, **5**.208
Hungary **5**.201
India *5.219*
Ajaṇṭā **5**.10, **5**.196
Amarāvatī **5**.13
Chōla **5**.99
Gupta **5**.196
Mathura **5**.286
Mauryan **3**.25, **3**.215, **5**.286–7
rock-cut caves and shrines **3**.268, **5**.10, **5**.143, **5**.145, **5**.196, **5**.275–6
Indonesia **5**.221
Italy:
Baroque **5**.45–6
Futurist **5**.29, **5**.51
International Gothic **5**.175
Mannerist **5**.83
Medieval **5**.32
modernist **5**.300
Neoclassical **5**.75
portrait busts **5**.12
terracotta **5**.384
see also under Florentine art; Renaissance arts *and also* Roman *below*
jade **5**.232
Jain **5**.232
Japan **5**.237
Khmer **5**.247–8
kinetic art **5**.70–1, **5**.248
Korea **5**.251
Malaya **3**.67
Maori **5**.281
marble **5**.282
mausoleum **5**.287
Mayan **5**.287–8
Melanesia **5**.289
Mesopotamia **3**.230
Mexico **3**.31, **3**.261, **3**.298
minimal art **5**.297
Minoan **5**.298
misericords **5**.298
mobiles **5**.298
Moorish **5**.303
Neoclassical **5**.75, **5**.157, **5**.454
Nepal **5**.316
Netherlands and Flemish **5**.396
North American Indians **5**.321
Olmec **5**.327
Ottonian **5**.334
Persia **3**.97, **3**.276, **5**.344
Peru **3**.71
Polynesia **5**.356, **5**.357
proportion **5**.368
relief **5**.344, **5**.377, **5**.387
rock *see* rock art
Roman **3**.21, **3**.28, **3**.57, **3**.77, **3**.95, **3**.124, **3**.250, **3**.330, **3**.373, **3**.387
Romanesque **5**.389
sculpture and carving (*cont.*):
Romania **5**.58
Russia and USSR **5**.21, **5**.111, **5**.346
scrimshaw **5**.409
Scythian **5**.410
South American Indians **5**.426
Spain **5**.74, **5**.189, **5**.348
Sri Lanka **5**.429
Stone Age **5**.434
stone-carving **5**.25, **5**.141, **5**.275, **5**.434
Sumerian **5**.440
Switzerland **5**.175–6
Thai **5**.450–1
Tibet **5**.455
United States **5**.14, **5**.168, **5**.240–1, **5**.264, **5**.398, **5**.423
Vietnam **5**.475
Viking **5**.476
waxwork **5**.483–4
West indies **5**.488
woodcarving **5**.491–2
see also bronze; Gothic architecture
Sculthorpe, Peter *5.409*
scumble *5.409*
scutage *3.320*
Scutum *8.146*
Scylla *9.32*
scythe **6**.6, *6.311*
Scythians *3.320*
art *5.409–10*
SDI *see* Strategic Defense Initiative
SDRs *see* Special Drawing Rights
sea *1.296*
creatures *see* living organisms *under* oceans and seas; water
Sea Peoples **3**.320
see also exploration; oceans and seas; pirates; ships and boats *and* routes *under* trade
sea anemone *2.297*, **2**.298
sea bass **2**.24, **2**.146
sea birds *see* birds *under* water
sea bream *2.298*
sea butterfly **2**.301
sea cow *2.298*
sea cucumber *2.298*
sea dragon **2**.299
sea duck *2.298*
see also birds *under* water
sea eagle *2.298*
sea fan *2.298–9*
sea gooseberry **2**.298, *2.299*
sea holly *2.299*
sea horse *2.299*
sea ice *1.296*
sea kale *2.299*
sea lily *2.299*
sea lion *2.299–300*
sea lizard *2.300*
sea mammals *see* mammals *under* water
sea mat **2**.219
sea moth *2.300*
sea mouse *2.300–1*
sea otter **2**.237, **2**.300, *2.301*
Sea Peoples *3.320*
sea perch *2.301*
sea scorpion *2.301*
sea slug **2**.300, *2.301*
sea snail *2.301*
sea snake *2.301*
sea spider *2.301*
sea squirt **2**.300, *2.301–2*
seabirds *see* birds *under* water
sea-dragon **2**.50
sea-floor spreading **1**.296
Seaga, Edward **4**.171
seal *2.299*, **2**.359
sea-level *1.296–7*
eustatic change **1**.115–16
mean **1**.209
storm surge **1**.321
sea-lily **2**.299
seamounts **1**.236, **1**.245, *1.297*
seaplane **6**.138, **6**.139, **6**.140, *6.311*
searchlight *6.311*
Searhaddon of Assyria **3**.26
seasons *1.297*, *8.146*, **8**.147, **9**.62
see also equinoxes; spring (season); summer; winter
seat-belt **6**.304
SEATO (South-East Asia Treaty Organization) *4.320*
sea-urchin **2**.293, **2**.302
seawater **1**.290, *1.297–8*, **1**.309
seaweeds **2**.184, *2.302*, **6**.18, **6**.186
Sebastian, Saint *9.43*
Sebastian II, king of Portugal **3**.8
Sebastiano del Piombo *5.410*
Sebuktegin **3**.141
Secchi, (Pietro) Angelo
Secchi classification **8**.39
Secchi, Angelo *8.146*, **8**.157
secession **7**.212, *7.277*
see also separatism
second (time) *1.298*, **6**.321, **6**.356, **6**.368, *9.55*
leap **8**.62
Second Coming of Jesus Christ **7**.3, **7**.47, **7**.202
Second Front *4.307*
Second Reich **4**.134
secondary colour **5**.366
secondary education **7**.102, *7.277*
secondary industry *see* manufacturing
secret services and societies **7**.5
Britain **4**.220
Mafia **4**.206, **7**.188
Masonic order **7**.195
Nazi **4**.327
Serbia **4**.40
USSR **4**.71, **4**.181, **4**.229
Secretaries-General of UN *see under* United Nations
secretary bird *2.302*, **2**.303
sectarianism **7**.59
secular (astronomy) *8.146–7*
secularism **7**.265
Securitate **7**.238
securities (finance) *7.277*
bonds **7**.34
debenture **7**.80
futures markets **7**.128
gilt-edged **7**.132
indexation **7**.155
regulation **7**.120
speculator **7**.297
stock and Stock Exchange **4**.330, **7**.299–300
see also shares
Security Council (UN) *4.307*, **7**.273, *7.277*, **7**.321, **7**.322, **7**.326
security services *7.277–8*, **7**.303
see also defence; secret services; wars
Sedan, battle of *4.307–8*
sedan chair **6**.205
sedatives **6**.285, **6**.360
Seddon, Richard John *4.308*
sedge **2**.130, *2.302*
sedge warbler *2.302–3*
Sedgemoor, battle of *3.320–1*
Sedgwick, Adam *1.298*
sediment *see* sediments
sedimentary rocks *1.298*
arenaceous **1**.18, **1**.20
argillaceous **1**.19–20
bedding *see* bedding
caliche **1**.48, **1**.49
chalk *see* chalk
clay *see* clay
coal *see* coal
concretion **1**.70, **1**.71
conformity **1**.72
greensand **1**.147
grits **1**.147, **1**.215
lignite *see* lignite
limestone *see* limestone
meerschaum **1**.210
millstone grit **1**.215

sedimentary rocks (cont.):
minerals:
alabaster 1.6
aragonite 1.17
borax 1.41
chalcedony 1.4–5, 1.57, 1.76–7, 1.238
chert 1.60
flint see flint
garnet 1.132
glauconite 1.141
illite 1.166
marcasite 1.205
montmorillonite 1.220–1
muscovite 1.214, 1.223
mudstone 1.222
sandstone see sandstone
shale 1.300
siltstone 1.222, 1.305
slate 6.287
strata 1.322
strike 1.323
stripped land surface 1.324
tillite 1.337
travertine 1.180, 1.342
turbidite 1.347
unconformity 1.349–50
see also deposition; fossils; sediments
sedimentation tank (sewage) 6.314, 6.379
sediments/sedimentation (geology) 1.298
diagenesis 1.90
load see load
see also deposition; sedimentary rocks
sedition 7.278
Sedum 2.327
Seebeck effect 6.354
Seeckt, Hans von 4.308
seeds 2.303, 6.227
dressing 6.311
drill 6.6, 6.311, 6.364
seeing (astronomy) 8.147
seeing (sight) see vision
seer (fish) 2.318
Seferis, George 5.410
Seghers, Hercules 5.410
segmented mirror telescope 8.105
Segovia, Andrés 5.410
segregation 7.278
Séguin, Camille 6.311
Séguin, Marc 6.311, 6.341
Sei Shōnagon 5.349
seif dune 1.298
Seifert, Jaroslav 5.410
seigneurialism 3.220
seine net 6.239
Seine River 1.298
seismology 1.298, 1.299, 8.147
seismograph 6.311–12, 6.382, 6.391
see also earthquakes
Sekhmet 9.7
Seku Ahmadu Lobbo 4.308
Seku Hamadu of Fulani 3.136
Selassie, Haile 4.113
Selden, John 3.321
selection see natural selection
selectors (Australian farming) 4.308
Selene 9.11, 9.32
selenium 1.107, 1.298–9, 6.268, 6.312, 6.349
selenography 8.147
selenology 8.147–8
Seleucid 3.321
Seleucus I, king of Macedonia 3.321
self:
egoism 7.103
James on 7.169
Rogers on 7.268
self-actualization 7.149
self-concept 7.278
self-determination 7.278
self-help 4.131, 4.316
self-presentation 7.278
Self-Denying Ordinance 3.321
self-propelled gun 6.312
Self-Strengthening Movement 4.308
Seligman, Martin 7.181
Selim I, Ottoman sultan 3.37, 3.181, 3.321
Selim III, Ottoman Sultan 4.207
Seljuq 3.321
art and architecture 5.344, 5.410–11
Selkirk, Alexander 1.83, 3.96
Selkirk, Thomas Douglas, 5th earl of 4.285
selling see sales
Selye, Hans 7.300
semantics 7.196, 7.278
semaphore 6.69, 6.312, 9.48
Semele 9.10
semen 2.303
semi-automatic rifle 6.131
semiconductor 1.299, 6.118, 6.312, 6.313
band theory 1.32
electricity conduction through see electronics
elements in 6.20, 6.149, 6.156, 6.312, 6.319
ion implantation 6.186, 6.312
laser 6.194
in solar cell 6.324
solid-state device 6.325
see also diode; integrated circuit; transistor
semi-diameter (astronomy) 8.148
semi-major axis (astronomy) 8.148
Seminole Wars 4.308
semiotics 7.81–2, 7.274, 7.278, 7.285
see also signs
semi-precious stones see gemstones
Semiramis 9.32
Semites 3.321
Semitic languages 7.109, 7.278–9
Sempervivum 2.327
'Semtex' 6.272
Sen, Amartya 7.116
Sen, Keshab Chandra 4.152, 7.143
Senanayake, Don Stephen and Dudley 4.308, 4.327
Senate, Roman 3.285, 3.308–9
Senate, US 4.308–9, 7.279, 7.324
Sendero Luminoso 4.267
Sendic, Raúl 4.354
Seneca, Lucius Annaeus 3.79, 3.321, 3.342, 5.411, 5.459
Seneca Indians 3.321
senecio 2.303
Senefelder, Alois 6.205
Senegal 1.299, 3.141, 4.308–9, 7.380
aid to 7.8
arts 5.334, 5.411
colonization and decolonization 4.4, 4.306
exploration 1.248
finance 7.124
physical geography and geology 1.299
senescence 2.303
see also ageing
Senghor, Léopold Sédar 4.309, 5.411
Sennacherib, king of Assyria 3.26, 3.32, 3.254, 3.321
Sennett, Mack 5.411
senses:
sense data 7.279
sight see vision
smell (olfactory) 2.41, 2.229, 2.313–14
sound see hearing
taste 2.336
touch 2.345
sensitive plant 2.303
sensor 6.37, 6.312
sentences (court) 7.279
capital punishment 3.150, 7.40, 7.146
community penalty 7.61
compensation 7.62
parole 7.229
probation 7.258
for rape 7.258
see also prisons
separation of powers 7.279
separatism 7.279, 7.280
Sephardi 7.279–80
sepia 5.411
septic tank 6.312
septicaemia 2.303
Septimius Severus see Severus
sequence (mathematics) 1.121, 1.299
Sequoia (redwood) 2.281
Sequoia National Park 1.299
Sequoyah (Cherokee Indian) 5.322
seraphim 9.8–9
Serapis 9.32
Serbia 1.299, 3.321–2, 4.309
language 7.156, 7.288
literature 5.16
physical geography and geology 1.299
religion/theology 7.98
rulers and leaders 3.341, 4.178, 4.220, 4.249, 4.262, 4.309, 4.387, 9.67
secret societies 4.40
Serbo-Croat language 7.156, 7.288
literature 5.16
serenade 5.411
Serenitatis Basin 8.148
serfs 3.160, 3.322, 3.374
see also peasants
serial processing 6.312, 6.314
serialism 5.320, 5.411, 5.417, 5.484
Austria 5.405, 5.484
France 5.291
Germany 5.204, 5.434
Italy 5.44–5, 5.121, 5.320
Russia 5.436
seriema 2.303
series connection (electricity) 6.74, 6.76
series (mathematics) 1.299
convergent 1.74
divergent 1.93
expansion 1.332
geometric 1.136
power 1.266
Serlio, Sebastiano 5.411
sermon 5.19, 5.411–12
serow 2.303–4
Serpens 8.148
serpent (brass instrument) 5.59, 5.412
serpent star see brittle star
serpentine 1.299
serum 6.177, 6.194
service tree 2.289
services see goods and services; tertiary industry
Servius Tullus of Rome 3.82
servo-mechanism 6.312, 6.314
sesame 2.304
Sesshu 5.412
Sessions, Roger 5.412
set design (and stage design) 5.412–13
Britain 5.117, 5.208, 5.241, 5.269
France 5.58, 5.87, 5.348
Germany 5.351, 5.405
Italy 5.345
lighting, stage 5.264–5
machinery 5.351, 5.429–30
Russia and USSR 5.29, 5.293, 5.430
Spain 5.194
Set/Seth (Egypt) 9.16, 9.32
set theory in mathematics 1.50, 1.299–300
closure 1.66
identity element 1.165
set theory in mathematics (cont.):
inverse elements 1.171
paradox 1.247
Venn diagram 1.355
Seth (Bible) 9.32
Seti I, pharaoh 3.2, 3.243
SETI (search for extraterrestrial intelligence) 8.148
Seto-Naikai 1.168–9
setting (astronomy) 8.148
settlement (engineering) 6.314
settlement (law) 7.280
Settlement, Act of 3.322
Seurat, Georges 5.316, 5.354, 5.413
Sevastopol 4.309
Seven Sisters (stars) 8.123–5
Seven Sleepers of Ephesus 9.32
Seven Weeks War see Austro-Prussian War
Seven Wonders of the World 3.322–3, 5.413
Seven Years War 3.134, 3.323, 3.323–4
battles and campaigns 3.232, 3.298
Treaty of Paris 3.269–70
Seven Years War of the North 3.324
seven-segment display 6.314
Seventh Approximation 1.311
Seventh-day Adventist 7.3, 7.202
Severn 1.300
Severus, Lucius Septimius Roman emperor 3.17, 3.308, 3.324
Seveso accident 1.91–2, 6.314
Sévigné, Madame de 3.299
Seville, Alcázar of 3.7
Sèvres, Treaty of 4.310, 4.364, 7.178
sewage and sewage treatment 6.314, 7.273
biofuel from 6.37
pioneer 6.32
pollution 6.32, 6.314, 6.377, 7.240
septic tank 6.312
and storm-water 6.335
see also wastes
Seward, William Henry 4.209, 4.310, 4.352
sewellel 2.220
sewing:
sewing-machine 6.77, 6.171–2, 6.239, 6.314–15
inventor 6.321
stitch 6.335
see also clothing; textiles
sex:
determination 2.304
differences 7.280
see also gender
hermaphrodite 2.156
ratio in population 7.241
see also sexual reproduction; sexuality and sexual activity
sexism 7.280
Sextans 8.148
sextant 6.294, 6.315
sexton beetle 2.46
sexual behaviour see sexuality
sexual reproduction 2.282
artificial insemination 6.15, 6.21
cell differentiation 2.92
cell division 2.207, 2.215
courtship in animals 2.78
embryo 2.107
fallopian tubes 2.113
fertilization 2.115–16
gametes 2.132
gonads 2.141
of hermaphrodites 2.314
hybrid 2.166
life cycle 2.189
ovary 2.237
ovum 2.238
penis 2.248
pheromones 2.252
placenta 2.257
pregnancy 2.268, 2.352
prevention see contraception

sexual reproduction (*cont.*):
puberty **2**.271
reproductive system illustrated ***9**.59*
semen **2**.303
spawn **2**.319
sperms **2**.303, **2**.319
sterilization **6**.87, **6**.334
of toads **2**.331
umbilical cord **2**.352
in vitro fertilization **6**.185–6, **6**.218
see also birth; breeding; female mammals; male mammals; obstetrics and gynaecology; pregnancy; sexuality
sexual selection ***2**.304–5*
sexuality and sexual activity ***7**.280–1*
abuse **7**.46
and Aids **7**.9
and children **7**.46, **7**.126
and disease *see* sexually-transmitted diseases
harassment ***7**.280–1*
homosexuality **7**.146–7
incest **7**.46, **7**.154
offences **7**.40, **7**.46, **7**.255, **7**.258, ***7**.281*
pheromone **6**.267
pornography **7**.46, ***7**.242–3*
prostitution **7**.46, **7**.250–1
psychosexual development **7**.253
reproduction *see* sexual reproduction
sexually-transmitted diseases ***2**.304*, ***7**.281*
gonorrhoea **2**.141
herpes simplex **2**.157
syphilis *see* syphilis
test for **6**.177, **6**.192
treatment **6**.72, **6**.112, **6**.257
see also Aids
Seychelles ***1**.300*, ***4**.310*, **7**.380
Seychelles frog ***2**.305*
Seyfawa family **3**.193
Seyfert, Carl **8**.56, **8**.149
Seyfert galaxy ***8**.148–9*
Seymour, Edward **3**.324
Seymour, Jane ***3**.324*
Seyss-Inquart, Arthur ***4**.310*
Sezession ***5**.413*
Sforim, Mendele Moykher **5**.497
Sforza family ***3**.324–5*
Shaba (Congo Crisis) **4**.83
Shabaka of Kush and Egypt **3**.259
Shabbat ***7**.281*
Shackleton, Sir Ernest Henry ***1**.300*, **4**.15
shad **2**.136, ***2**.305*
Shaddad **9**.2
shaddock **2**.66, ***2**.305*
shadow bands ***8**.149*
shadow plays **5**.19, ***5**.413*
shaduf ***6**.315*
Shaftesbury, Anthony Ashley Cooper, 1st earl of (1621–83) **3**.70, ***3**.325*
Shaftesbury, Anthony Ashley Cooper, 7th earl of (1801–85) **4**.115, ***4**.310*
shag **2**.77, ***2**.305*
shaggy ink cap **2**.304, ***2**.305*
Shah Jahan I, emperor of India **3**.101, ***3**.325*
and arts **5**.219, **5**.287, **5**.307
shahnāī ***5**.413*
Shahrazad ***9**.32*
Shahrbanu **9**.17
Shahs of Iran *see* Reza Shah
Shaitan *see* Satan *under* Devil
Shaka, chief **4**.99, **4**.236, **4**.240, ***4**.310*, **4**.391
Shaker ***3**.325*
furniture ***5**.413*
Shakespeare, William ***5**.413–14*
burlesque **5**.66
cinema **5**.327
history play **3**.303
Shakespeare, William (*cont.*):
malapropisms **5**.277
masques **5**.285
morality plays **5**.304
music for **5**.218
and myths **9**.30
Old Vic **5**.327
pastoral **5**.341
Pope's amendments **5**.357
productions **5**.61, **5**.101, **5**.197, **5**.250, **5**.370, **5**.384
Shakespearean actors and actresses **5**.24, **5**.66, **5**.177, **5**.227, **5**.245, **5**.327, **5**.383, **5**.450, **5**.454
sonnets **5**.424
theatre **5**.453
tragedies **5**.459
translated **5**.371, **5**.489
Shakti **7**.142, ***7**.281*, **7**.285, ***9**.32*
shakuhachi ***5**.414*
shale ***1**.300*
shale oil **6**.263, ***6**.315*
shallot ***2**.305–6*
Shalmaneser II, king of Assyria **3**.67
shamanism **7**.11, **7**.16, **7**.181, ***7**.282*, **7**.335
Shamba ***9**.32*
shame **7**.147
Shamil ***4**.310*
Shamir, Yitzak **4**.170
shamisen **5**.235, ***5**.414*
Shamshi-Adad I, king of Assyria **3**.26
Shang dynasty **3**.72, ***3**.325*
Shango ***9**.32–3*
Shankar, Ravi ***5**.414*
Shankar, Uday **5**.220
Shankara **7**.3, **7**.325
Shankaracharya **3**.316
Shannon, Claude **6**.180
Shannon River ***1**.300*
Shans ***3**.325*
shanty town ***7**.282*, **7**.283
shaping machines **6**.212
Shapley, Harlow **8**.79, ***8**.149*
sharecropping ***4**.310*, **7**.179, **7**.232, **7**.268, ***7**.282*
shares ***7**.282*
blue chip **7**.33
dividend **7**.93
flotation **7**.121
insider trading **7**.161
investment trust **7**.165–6
joint-stock company **7**.172
limited liability **7**.184
privatization **7**.247
stock and Stock Exchange **4**.330, **7**.299–30
take-over **7**.305
see also securities
sharī'ah ***7**.282*, ***7**.284*
Sharif, Mian Mohammad Nawaz **4**.256
Sharifian ***3**.325*
sharks ***2**.305*
basking **2**.24
carpet **2**.55
dogfish **2**.96–7
hammerhead **2**.150
porbeagle **2**.265
Port Jackson **2**.265
whale **2**.365
Sharp, Cecil **5**.160, **5**.161, ***5**.415*
sharpbill ***2**.306*
Sharpeville massacre ***4**.310–11*
Shatt al-Arab Waterway ***1**.300*
Shaukat Ali **4**.182
Shaw, Artie **5**.239
Shaw, George Bernard **4**.115, **5**.73, ***5**.415*, **5**.450, **5**.454
Shaw, Sir Napier ***1**.300*
Shaw, Percy **6**.64
Shaw, Richard Norman **5**.372, ***5**.415*
shawms **5**.138, **5**.347, **5**.376, **5**.413, ***5**.415*
Shawn, Ted **5**.397, ***5**.415–16*
Shays, Daniel: Shays's Rebellion ***3**.325–6*
Shayṭān *see* Satan *under* Devil
shearing **6**.6
shearers' strikes ***4**.311*
shears ***6**.315*
shearwater ***2**.305*
sheathbill ***2**.306*
sheath-tailed bat ***2**.306*
Sheba, Queen of **3**.119, ***9**.33*
sheep **2**.14, **2**.29, **2**.220, ***2**.306*, **6**.15, ***6**.315*
parasites **2**.194, **2**.306
see also livestock; shearing; wool
sheep-ked **2**.194
sheep-tick ***2**.306*
sheet erosion ***1**.300*
sheet flood (geomorphology) ***1**.301*
sheeting (geology) **1**.301
Sheffield plate ***5**.416*, ***6**.315*
Sheherazade **9**.32
Shelburne, William Petty Fitzmaurice Lansdowne, 1st marquis and 2nd earl of **3**.269, ***3**.326*
shelduck ***2**.306*
shell (building) **6**.51, **6**.316
shell (living organism) *see* shellfish; shells
shell (military) ***6**.315–16*, **6**.318
shell star ***8**.149*
shellac **2**.295, **6**.296
Shelley, Mary Wollstonecraft **5**.186, ***5**.416*
Shelley, Percy Bysshe **5**.143, **5**.215, **5**.342, ***5**.416*
shellfish ***2**.307*
farming **6**.18, **6**.133, ***6**.316*
see also shells
shells ***2**.306–7*
fibonacci numbers **1**.121
see also arthropods; bivalves; molluscs; shellfish; snails
shellwork ***5**.416*
Shelton, Ian **8**.166
Shem **9**.27
Shén ***9**.33*
Shen Fu ***5**.416*
Shenandoah River ***1**.301*
Shénrén **9**.33
she-oak **2**.56
Shephard, Alan **8**.154
shepherding satellites **8**.12, ***8**.149*
Sheppard, Jack **3**.165
Shepstone, Sir Theophilus **4**.68
Sher Shah Suri ***3**.326*
sherardizing ***6**.316*
Sheraton, Thomas **5**.315, ***5**.416–17*
Shere Ali, amir of Afghanistan ***4**.311*
Sheridan, Philip Henry ***4**.311*
Sheridan, Richard Brinsley **5**.177, **5**.383, **5**.413, ***5**.417*
sheriff (England) ***3**.326*
sheriff (Scotland) **7**.276
Sheriffmuir, battle of ***3**.326*
Sherman, Roger **3**.129
Sherman, William Tecumseh **4**.71, ***4**.311*, **4**.364, **4**.368
Shermarke (of Somalia) **4**.318
Shetland Islands ***1**.301*, ***3**.326*
Shevardnadze, Eduard **4**.133, **4**.138
Shi Huangdi, emperor of China ***3**.326*
and Great Wall of China **3**.146, **6**.161
and measurement **1**.350, **6**.368
and writing **3**.73
Shi Naian **5**.483
Shidehara Kijuro **4**.179
shield excavator **6**.49, ***6**.316*, **6**.365, **6**.367
shield (geology) **1**.120, ***1**.301*
shield volcano ***1**.301*
shieldbug ***2**.307*
shield-tail snake ***2**.307*
shifting cultivation ***6**.316*, ***7**.284*
Asia and Pacific **7**.175, **7**.197, **7**.211
see also Trobriand Islanders
South America **7**.335
see also subsistence farming
Shī'ī Islam ***3**.326*, **7**.166, ***7**.284*
'āshūrā' ***7**.19–20*
āyātollāh **7**.24
Bábism **4**.27
and caliphs **7**.10
see also 'Ali
Druze **4**.104, **7**.96, **8**.96
imāms **7**.153, **9**.3, **9**.17
Ismaī'īlis **3**.181, ***7**.168*
Kurds **7**.178
law **7**.282
martyr **7**.194
pilgrimage **7**.236
saints **7**.272
see also Islam/Muslims
Shijing ***5**.417*
Shikoku ***1**.301*
Shils, Edward **7**.152
Shimabara ***3**.326*
Shimonoseki, Treaty of ***4**.312*
Shin Rat-hta-tha-ra **5**.67
Shin Thi-la-wun-tha **5**.67
shingle (deposit) ***1**.301*
shingles (disorder) **2**.63, ***2**.307*
Shingon Buddhism ***7**.284*, **9**.9, **9**.22
Shinto ***3**.326*, **7**.169, ***7**.284*
ancestor worship **7**.13
deities **9**.3–4, **9**.22, **9**.34
ship money ***3**.327*
Ship Rock ***1**.301*
Shipley, William **6**.303
ships and boats ***3**.247–8*, ***3**.316–17*, ***4**.234–6*
anchor **6**.14
Anglo-Saxon burial **3**.346
automatic pilot **6**.25
barge **6**.30, **6**.56, **6**.306
bearing **6**.33
building ***6**.316–17*
cable-ship **6**.53
canoe **6**.57, **6**.375
capstan **6**.58
caravel **3**.60
caulking **6**.64, **6**.316
cavitation **6**.64, **6**.367
centreboard **6**.67
convoy system **4**.85
coracle **6**.89
design **6**.237, **6**.316
dinghy **6**.67, **6**.104, **6**.308, **6**.317
dredger **6**.106, **6**.277
ferry **6**.60, **6**.127, **6**.172, **6**.277
fishing **6**.133, **6**.239, **6**.362
galleon **3**.136, **6**.149, **6**.306, **6**.307
galley *see* galley *and* trireme
gas-turbine **6**.150
hovercraft **6**.79–80, **6**.171, **6**.172, **6**.175
hulkhead **6**.51
hydrofoil **6**.173–4
ice-breaker **6**.175, **6**.245
inflatable **6**.179
iron **6**.49, **6**.111, **6**.188
junk **6**.190, **6**.306, **6**.307
keel **6**.190
landing-craft **6**.193
lifeboat **6**.20, **6**.200, **6**.317
lightship **6**.202
liner **6**.203, **6**.333
merchant and trading ***6**.219*, **6**.306, **6**.307
bulk carrier **6**.51, **6**.219
caravel **6**.58
container **6**.87, **6**.219, **6**.277
dhow **6**.101
Plimsoll line **6**.273
mine-layer **6**.227
minesweeper **6**.227
missiles in **6**.230–1
navies *see* navies; warships

ships and boats (*cont.*):
navigation *see* navigation
nuclear-powered **6**.175, **6**.243, **6**.245, **6**.337
paddle-wheel **6**.255
petrol engines **6**.262
propellor **6**.284
raft **6**.292
and roaring forties **1**.284
rudder **6**.304
sailing *see* sail; sailing ships
screw propulsion **6**.123, **6**.160, **6**.284
ship of the line **6**.306, ***6**.317*
ship's boat ***6**.317*, **6**.390
speed measurement **6**.192, **6**.206
spirit **9**.13
stabilizer **6**.330
steam *see* steamships *under* steam power
steering-oar **6**.333
submarines *see* submarines
and superstition **7**.302
tanker **6**.51, **6**.219, **6**.265, **6**.345
trawler **6**.239, **6**.362
trireme *see* galley and trireme
tug **6**.364
Viking longships **3**.210, **3**.248, **5**.475, **6**.306, **6**.373
war *see* warships
see also mutinies; oceans; ports and harbours
ship-to-air missiles **6**.231
ship-to-ship missiles **6**.230
shipworm ***2**.307*
shire ***3**.327*
Shishupala **9**.31
Shiva **7**.133, **7**.142, ***7**.284–5*, ***9**.33*
Shivaji (Sivaji) **3**.329
Shkodër, Lake ***1**.301*
Shnitke, Alfred ***5**.417*
shoals **1**.93
shock (medical) ***2**.307–8*
shock absorber ***6**.317*, **6**.341
Shockley, William Bradford **6**.291, ***6**.317–18*, **6**.361
shoddy **6**.388
shoebill **2**.365
shoestring rill ***1**.301*
shogunates ***3**.327*, ***4**.312*, ***9**.66*
Ashikaga **3**.24
Kamakura **3**.193, **3**.231–2
see also Tokugawa
Shoji Hamada **5**.258
Sholes, Christopher Latham ***6**.318*
Sholokhov, Mikhail ***5**.417*
Shona ***7**.285*
shooting *see* weapons
shooting star **8**.97
see also meteor
shooting star (plant) **2**.269
shopping *see* retail trade
shore:
birds *see* wading birds
plants ***2**.308*
see also plants *under* water
platform ***1**.301*
shoreline *see* coasts
short stories *see* stories and tales
Short Take-Off and Landing **6**.25, **6**.225, **6**.335
short wave *see* short-wave
short-eared owl ***2**.308*
shorthand ***6**.318*, ***7**.285*
shorthorn grasshoppers **2**.192–3, ***2**.308*
short-range weapons **6**.246
short-tailed bat ***2**.308*
short-wave radiation **1**.103, ***1**.301*, **6**.380, **9**.54
Shostakovich, Dmitry **5**.326, **5**.417
shotgun ***6**.318*
Shotoku Taishi **3**.135, **3**.185, ***3**.327*, **3**.387
shott ***1**.302*
shovel **6**.103, **6**.227
shoveller (bird) **2**.307, ***2**.308*
showboat ***5**.417*
shower ***1**.302*
Shrapnel, Henry: shrapnel ***6**.318*
shrew opossum **2**.278
shrews ***2**.308–9*
otter **2**.171
pygmy **2**.200
tree **2**.172
water **2**.362
shrikes **2**.249, ***2**.309*
shrimp fish ***2**.309*
shrimps ***2**.309*
euphausid **2**.110, **2**.111
mantis **2**.202
shrink-resistant fabrics ***6**.318*
shrubs ***2**.309*, **2**.328
deciduous *see* deciduous trees and shrubs
evergreen *see* evergreen plants
Shu'ayb ***9**.33*
Shuiski, House of **9**.68
Shuja al-Mulk, Shah **4**.13
shūnyāta ***7**.285*
shutter (photography) ***6**.318*
shuttering (building) **6**.144
shuttle **6**.140, **6**.190, **6**.208, ***6**.318*
Sì dà Tiānwáng ***9**.33*
Si Kiang **1**.369
SI units ***1**.302*, ***1**.306*, **1**.350, ***6**.321*, **6**.369, ***9**.50*
amount **1**.218
basic seven listed **1**.306
capacitance **1**.119
electric charge **1**.78
electrical resistance **1**.237
electricity **1**.11
force **1**.124, **1**.227
frequency **1**.154
inductance **1**.154
joule **1**.178
length *see* length
list of **1**.302
luminous intensity **1**.50
magnetic flux or induction **1**.334
mass **1**.183
potential difference **1**.358
power **1**.362
pressure **1**.248
radioactivity **1**.35
tables of **6**.368, **9**.50
thermodynamic temperature **1**.181
time **1**.298, **6**.356
see also measurement; metric units; units
sial ***1**.302*
sialon ***6**.318*
Siam *see* Thailand
Siamese fighting fish **2**.117
Sibelius, Jean ***5**.417*
Siberia ***1**.302–3*
ethnic group **3**.387
languages **7**.11, **7**.108, **7**.227, **7**.322
physical geography and geology ***1**.302–3*
climate **1**.303
lake **1**.30
mountains **1**.351
rivers **1**.12, **1**.191–2, **1**.234, **1**.371
sea and islands **1**.185, **1**.237
Trans-Siberian Railway **4**.350
Siberian high ***1**.303*
Sibirtsev, Nikolai **1**.93, ***1**.303*
Sibley, Antoinette **5**.393
Sibyl ***9**.33*
Sichuan ***1**.303*
Sicilies, Kingdom of the Two ***4**.312*
Sicily ***1**.303*, ***3**.327*
Mafia **4**.206, **7**.188
philosophy **7**.295
physical geography and geology **1**.114, **1**.196, **1**.303
religion **9**.41
Sicily (*cont.*):
rulers **3**.69, **3**.70, **3**.124–5, **3**.153, **3**.218, **3**.279, **3**.306
Sicilian Vespers ***3**.327*
sick building syndrome **6**.8
Sickert, Walter Richard ***5**.417–18*
sickle **6**.6, **6**.191
Siddārtha Gautama *see* Buddha
Siddons, Sarah ***5**.418*
side-necked turtle ***2**.309*
sidereal time **1**.286, ***8**.149–50*, ***9**.55*
day **8**.36, **9**.55
month **8**.100, **9**.55
year **8**.189, **9**.55
see also astronomy *under* time
siderolites **8**.97
Siding Spring Observatory ***8**.150*
Sidney, Algernon **3**.312
Sidney, Sir Philip **5**.341, **5**.387, ***5**.418*, **5**.424
Siebe, Augustus **6**.97
siege warfare **6**.225, ***6**.318*, **6**.371
Siegfried **9**.8, **9**.26, ***9**.33*
Siegfried Line ***4**.312*
Siemens, Sir Charles William **6**.188, ***6**.319*
Siemens–Martin furnace *see* open-hearth
Siemens, Ernst Werner von **6**.114, **6**.293, ***6**.319*
Siemens, Frederick **6**.319
Sienese School **5**.268, **5**.350, **5**.372, **5**.402, ***5**.418*, **5**.420
see also Duccio; International Gothic
Sienkiewicz, Henryk ***5**.418*
sierozem ***1**.303*, **1**.311
sierra ***1**.303*
Sierra Leone ***1**.303*, ***3**.327–8*, ***4**.312*, ***7**.380*
colonization and decolonization **4**.4, **4**.306
education **7**.103
physical geography and geology **1**.303
religion and myths **9**.26
Sierra Madre ***1**.303*
Sierra Nevada ***1**.303–4*
Sieyès, Emmanuel Joseph ***3**.328*
Sigeberht, king of East Anglia **3**.108
sight *see* optics/optical; vision
Sigismund, Holy Roman Emperor **3**.168, **3**.172, **3**.253, ***3**.328*
Sigismund II Augustus, king of Poland **3**.183
sign *see* signs
Signac, Paul **5**.418
signals ***9**.48*
electrical **6**.319
-to-noise ratio ***6**.319*
processing (computer) **6**.319
railway **6**.294, **6**.312, **6**.359
traffic light **6**.360
Very light **6**.371
see also Morse; semaphore; signs
significant figures ***1**.304*
signing and sign language **7**.218, ***7**.285*, ***9**.48*
Signorelli, Luca ***5**.418*
signs ***7**.285*
ideograms **7**.151
language *see* signing
pictograms **7**.123, **7**.235, **7**.334
semiotics **7**.81–2, **7**.274, **7**.278, **7**.285
of Zodiac **8**.190, **9**.62
see also signals; symbols; symbolism; writing and scripts
Sigyn **9**.23
Sihanouk, King Norodom **4**.58, **4**.178, ***4**.312–13*
Sikelianos, Angelos **5**.191
síkhara ***5**.417*
Sikhism ***3**.328*, ***7**.285–6*
Akali sect **7**.9
cremation **7**.75
Sikhism (*cont.*):
festivals ***7**.285*
Gurdwara **7**.137
gurus *see* gurus
Khalsa **7**.137, **7**.176, **7**.286
literature ***5**.418*
pilgrimage **7**.236
texts **5**.418, **7**.2, **7**.80, **7**.169
Wars ***4**.313*
Sikkim **1**.183, ***1**.304*, ***4**.313*
Sikorski, Vladislav **4**.179, ***4**.313*
Sikorsky, Igor Ivan ***6**.319*
silage and silo **6**.142, **6**.143, **6**.166, **6**.235, ***6**.319*, ***6**.320*
silencer ***6**.319*
Silenus ***9**.33*
Silesia ***1**.304*, ***4**.313*
silhouette ***5**.418–19*
silica (silicon dioxide) ***1**.304*
chert **1**.60, **1**.123
glass **6**.68, **6**.156, **6**.319
glazes **6**.157
insulator **6**.183
microchip **6**.182
opal **1**.239
see also chalcedony; flint; quartz
silicates ***1**.304*
aluminium *see* aluminium silicates
beryl **1**.38, **1**.108
and ceramics **6**.156
chalcedony **1**.4–5, **1**.57, **1**.76–7, **1**.238
chlorite **1**.62
garnet **1**.132
glauconite **1**.141
hornblende **1**.158
illite **1**.166
iron **1**.141
jade **1**.177
lapis lazuli **1**.187
magnesium **1**.210, **1**.299, **1**.330–1
montmorillonite **1**.220–1
pyroxene **1**.26, **1**.270
serpentine **1**.299
sodium *see* sodium silicates
staurolite **1**.320
talc **1**.320, **1**.330–1
topaz **1**.339, **1**.341
tourmaline **1**.341
zeolite **1**.373
zirconium **1**.373–4
see also china clay; feldspar *and also* minerals *under* igneous rocks *and* metamorphic rocks
silicon **1**.107, ***1**.304*, ***6**.319*, **6**.391
alloys **6**.11, **6**.185, **6**.212, **6**.333
chip *see* integrated circuit
in magnets **6**.212
pioneer **6**.191
p-type **6**.182
in semi-conductor **6**.312
see also silicates
silicon carbide (carborundum) **6**.1, **6**.68, **6**.82, **6**.297, ***6**.319–20*
silicon chip *see* integrated circuit
silicon dioxide *see* silica
silicon nitride **6**.318, ***6**.320*
silicon oxide *see* silica
silicone **6**.191, ***6**.320*, **6**.378
silicosis **6**.248
Siline, Ignazio **5**.231
silk ***5**.419*, **6**.128, ***6**.320*
moths **2**.17–18, ***2**.310*
Ottoman **5**.334
Silk Route and China **3**.328, ***3**.329*, **5**.94
silk-screen printing ***5**.419*
silkworm **2**.243, **2**.309, **2**.310, **6**.259
spinning **6**.328, **6**.329
silky anteater **2**.277
silky flycatcher ***2**.310*
sill (rock) **1**.166, ***1**.304*
Sillanpää, Frans Eemil **5**.155
Sillitoe, Alan **5**.17
silo *see* silage

silt 1.3, 1.197–8, 1.304–5
siltstone 1.222, 1.305
Silurian period 1.133, 1.134, 1.305
Silva, Anibal Cavaço 4.275
Silvanus 9.33
silver 1.107, 1.305, 6.320
- alloy 6.12, 6.320
- in amalgam 6.12
- arts (silverware) 5.416, 5.419–20, 5.426
- assaying 1.341, 6.357, 6.359
- electroplating 6.120
- extraction by cupellation 6.93–4
- hallmarks 9.63
- nickel 6.241
- ore 1.19
- salts (halide) 6.129–30, 6.256, 6.268
- Sheffield plate 5.416, 6.315
- silvering mirror 8.150
- silverpoint 5.419
- steel 6.333
- vein of 1.303

silver bromide 1.305
silver nitrate 1.305
silver sulphide 1.19
silverfish 2.119, 2.310
silvering 8.150
silverpoint 5.419
silverside (fish) 2.310
silverware see arts under silver
silver-Y moth 2.310
sima (geology) 1.305
Sima Guang 5.95
Sima Qian 3.328, 5.95
Sima Yen (Yan) 3.356, 3.380
Simenon, Georges 5.420
simile 5.108, 5.420
Simla Conference 4.204
Simnel, Lambert 3.291, 3.310, 3.328
Simon, Claude 5.323
Simon, Théodore 7.32
Simon the Canaanite 9.5
Simon the Leper 9.7
Simon Maccabeus 3.214
Simon Peter see Peter, Saint
Simon 'the Zealot' 3.104, 3.389
Simone Martini 5.420
Simons, Menno 3.229
simony 3.328
simple harmonic motion 1.305–6
Simpson, Sir James Young 6.73, 6.320
Simpson, Mrs Wallis 4.1
Simpson Desert 1.306
simulation (mathematics) 1.220, 6.17, 6.320
Simurgh 9.33
sin 7.225, 7.286
Sinai, Mount 9.33
Sinai peninsula 1.304, 1.306, 1.325
Sinan 5.132, 5.334, 5.420
Sinatra, Frank 5.118, 5.358
Sinclair, Upton 5.420
Sind 3.328
Sindbad the Sailor 9.33
Sindhia 3.328
sine 1.306
sinfonia 5.334, 5.442
Singapore 1.306, 4.313–14, 7.380–1
- astrology 7.20
- decolonization 4.321
- economy and trade 7.20, 7.88, 7.126, 7.309, 7.372
- fall of 4.314
- languages 7.286
- physical geography and geology 1.306
- population density 7.241
- statesman 4.192
- technology and science 6.79, 6.193
- telecommunications 7.211
- and UN 7.320
- and war 4.134

Singer, Isaac Bashevis (b.1904, writer) 5.420
Singer, Isaac Merrit (1811–75, inventor) 6.321
Singer, Israel Joshua 5.420
Singh, Atomba and Amubi 5.279
Singh, Maharaja Sawai Pratap 3.183
Singh, Ranjit 4.278, 4.313
Singh, V. P. 4.172
singing games 7.286
- see also songs and singers

Single European Act 7.99, 7.111
single transferable vote 7.250
single-cell protein (SCP) 6.320, 6.321
single-celled organisms see unicellular
single-channel photometer 8.120
Singspiel 5.329
singularity 8.150
sink-hole 1.305, 1.306
Sinkiang 1.369
Sinn Fein 4.107, 4.314
Sino-French war 4.314
Sino-Japanese war 4.312, 4.314–15
Sinope 8.150
Sino-Soviet border dispute 4.315
Sino-Tibetan languages 7.286–7
sintering 6.68, 6.320, 6.321, 6.373
sinusoidal wave 6.381
Sioux (Dakota) Indians 3.96, 3.180, 7.219, 7.287
siphon 6.321
Siprat 5.452
sipunculid 2.310
Siraj ud-Daula, nawab of Bengal 3.42, 3.79, 3.285
siren (salamander) 2.310
Sirens (mythology) 9.33–4
Sirius 8.150
sirocco 1.306
sisal 2.4, 2.310, 6.271
Sisera 9.19
siskin 2.310–11
Sisley, Alfred 5.420
sistrum 5.420–1
Sisyphus 9.34
Sītā 9.22, 9.30–1, 9.34
sitar 5.414, 5.421
sitatunga 2.311
Sitha, Saint 9.43
sit-in 7.157–8
Sitting Bull 4.315, 4.378, 4.384
Sitwell, Dame Edith, Sir Osbert, and Sir Sacheverell 5.421, 5.481
Síva see Shiva
Sivaji, king of Maratha 3.221, 3.329
Six, Les 5.211, 5.295, 5.364, 5.421
Six Acts 4.315
Six-Day War 4.169, 4.315–16
Sixtus IV, pope 3.272, 3.329, 3.336
Sixtus V, pope 3.329
Siyad Barrah, Muhammad 4.319
sizing textiles 6.321, 6.353
Skagerrak 1.306
skaldic poetry 5.421
Skallagrimsson, Egil 5.423
skandha 7.287
skateboards 1.56
skates (fish) 2.105, 2.311
skates (sport) 6.321
- roller-skating 6.321, 7.269

skeleton 9.58
- endoskeleton 2.64, 2.109
- see also bone; exoskeleton

Skelton, John 5.132, 5.304, 5.421
Skidmore, Louis 5.421
Skidmore, Owings & Merrill (SOM) 5.14, 5.229, 5.421–2
skiing and skis 6.322, 7.287
skills:
- loss of (deskilling) 7.86–7
- social 7.293

skimmer 2.311
skin 2.311–12, 9.61
- animal 6.258
- disorders 2.90, 2.352, 6.104
- grafting 6.158
- sweat gland 2.331

skink 2.311, 2.312
Skinner, Burrhus Frederic 7.29–30, 7.64, 7.180, 7.252, 7.287
skipjack 2.67–8
skipper 2.312
Skippon, Philip 3.253
skis see skiing
skittles 7.34, 7.287
Skokov, Yuri 4.386
Skryabin, Alexander see Scriabin
skua 2.312
- jaeger 2.174

skull 2.54, 2.312, 9.58
skunk 2.312–13
Škvorecký, Josef 5.422
sky 1.306–7
Skylab 6.214, 6.325
skylark 2.313
skyscraper 5.14, 6.51, 6.201, 6.321, 6.322
slack water 1.307
Sládkovič, Andrej 5.422
slaked lime see calcium hydroxide
slander 7.82
slang 7.287
slapstick 5.422
slash-and-burn 7.284
slate 1.306, 1.307, 6.287
slater see woodlouse
slavery/slave trade 3.329–30, 7.287
- abolition and emancipation 4.1, 4.52–3, 4.103, 4.128, 4.354
  - US Proclamation for Emancipation of 4.316, 7.237
  - see also under United States
- African 3.330
- American see under United States
- Arab 3.330
- British 3.25
- Caribbean 4.36
- Mamelukes 3.217–18
- and plantations 7.237
- Spanish 3.25
- voodoo 7.327

Slavs 3.330–1, 4.258
- Slavic languages 7.156, 7.157, 7.287–8

sledge 6.300, 6.322
Sleeman, William 4.344
sleep 2.313, 7.95, 7.126
sleeping sickness 2.348, 2.349
sleet 1.307
slide-rule 6.322
sliding friction 6.146
Slim, William Joseph, 1st viscount 4.316
slime mould 2.313
sling 6.64, 6.322
slip 5.422
slip-off 1.307
slit drum 5.422
slit-faced bat 2.158
Sloane, Sir Hans 2.313
sloe 2.33–4
sloop 6.90, 6.322
slope, ground 1.113, 1.307–8
sloths 2.135, 2.312, 2.313
Slovakia (now Slovak Republic) 1.308, 4.92, 7.381
- language 5.422, 7.156, 7.288
- literature 5.422
- physical geography and geology 1.53, 1.84, 1.308
- see also Czechoslovakia

Slovenia 1.308, 4.316, 7.381
- Slovene language 7.156, 7.288

slow nova 8.110
slow worm 2.313
Slowacki, Juliusz 5.422
sludging 1.312
slug 2.313
sluice 6.322
slum 7.288, 7.334
slurry explosive 6.124
Sluter, Claus 5.66
Sluys, battle of 3.331
Small Magellanic Cloud 8.88
smallpox 2.313
- eradicated 7.153
- vaccination and inoculation 2.40, 2.175, 6.176–7, 6.178, 6.189, 6.218

Smart, Christopher 5.422
'smart' fluid 6.120
Smeaton, John 6.51, 6.66, 6.86, 6.201, 6.322, 6.330
smell:
- sense of (olfactory) 2.41, 2.229, 2.313–14
- see also aromatic; perfume

smelting 6.96, 6.253, 6.322, 6.356, 6.391
- see also iron and steel industry

Smetana, Bedřich 5.422
smew 2.314
Smiles, Samuel 4.316
Smirke, Sir Robert and Sydney 5.422
Smith, Adam 3.331, 7.100–1, 7.288
- and civil society 7.52
- and Enlightenment 3.117, 5.147
- and free trade 4.125
- and labour theory of value 7.179
- and laissez-faire 3.202, 7.179
- in Manchester School 4.209

Smith, Alfred E. 4.157, 4.316
Smith, Bessie 5.50, 5.238
Smith, Charles 6.138
Smith, David 5.14, 5.423
Smith, Francis Pettit 6.284
Smith, Sir Harry 4.276
Smith, Ian 4.226, 4.390, 7.273
Smith, Sir James Edward 2.314
Smith, Jedediah Strong 4.225, 4.316–17
Smith, John (1580–1631) 3.290, 3.331
Smith, Joseph 4.224, 4.317, 7.207
Smith, Sir Matthew 5.423
Smith, Oliver 5.14
Smith, Stevie 5.423
Smith, 'Whispering' Jack 5.118, 5.358
Smith, William 1.133, 1.308
Smith, Willoughby 6.349
Smithson, Sir Hugh 3.276, 5.5
Smithson, James 6.322
Smithsonian Institution 6.322, 6.323
smithsonite 1.308
smog 1.307, 1.308, 7.239–40
smoke detector 6.322–3
smokeless fuel 6.323
smokeless powder see nitro-cellulose
smoking 2.51, 2.343, 7.288–9
- see also tobacco

smoking fish 6.132–3
smoky bat 2.314
Smollett, Tobias 5.348, 5.423, 5.460
smooth snake 2.314
smoothing circuit 6.323
Smrek, Jan 5.422
smuggler 3.331
smut fungus 2.314
Smuts, Jan Christian 4.150, 4.317, 4.360
Smyth, Charles Piazzi 8.82
Smyth, John 3.35
Smythson, John 5.423
Smythson, Robert 5.423
snails 2.314
- pond 2.263–4
- ramshorn 2.276
- sea 2.301
- smooth snail 2.314
- as vectors 2.29
- whelk 2.365

snake see snakes
Snake River 1.308
snake venom vaccine 6.177
snakebirds (darter and wryneck) 2.87, 2.371, 2.372
snakefly 2.314
snakehead fish 2.314

snake-necked turtles 2.205, 2.309, 2.314–15
snakes 2.314
  anaconda 2.9
  blind 2.34–5
  boas 2.36–7, 2.293, 2.346
  colubrid 2.72–3, 2.188
  cylinder 2.85
  grass 2.144
  leopard 2.188
  mythical 9.30
  python 2.272
  shield-tail 2.307
  sunbeam 2.329
  thread 2.340
  venom vaccine 6.177
  venomous *see* poisonous snakes
  vine 2.277
  wart 2.361
snake's head (plant) 2.127
snapdragon 2.12
snapper 2.315
snapping turtle 2.315
Snell, Willebrord: Snell's law 1.308
Snider-Pellegrini, Antonio 1.73
snipe 2.239, 2.315
snipe-eel 2.315
snipefish 2.315
snoek 2.315–16
snooker 7.289
snorkel 6.323
Snorri Sturluson 5.423
Snow, Charles Percy 5.423
Snow, Helen 8.104
Snow, John 4.99
Snow, Michael 5.73
snow 1.308–9
  blizzard 1.40
  deposition 1.33
  and depression 1.88
  firn 1.122
  nival plant 2.229
  sleet 1.307
  snow-line 1.309
  see also glaciation; ice
snow flea 2.297, 2.316
snow goose 2.316
snow leopard 2.316
snowberry 2.316
Snowden, Philip, 1st viscount 4.317
Snowdonia 1.309
snowdrop 2.316
Snowy Mountains 1.309
snowy owl 2.315, 2.316
snuff-boxes and bottles 5.94, 5.212
Soane, Sir John 5.423
soap 6.99–100, 6.267, 6.323, 6.367, 6.384
soapstone 1.320
soapwort 2.316
Soares, Mario 4.275
Sobhuza II, king of Swazi 4.335
Sobieski, John 3.286, 3.331
Sobrero, Ascanio 6.242
Sobukwe, Robert 4.257
socage 3.331
soccer 7.122, 7.123
social animals 2.316–17
social assistance 7.289
  see also social security
social Buddhism 7.36
social care *see* social services
social class 7.289
  and accent 7.1
  aristocracy 7.16
  caste 7.41
  class struggle 7.53
  Marxist views of 7.194
  social classification 7.289
  social mobility 7.291
  status 7.299
  stratification 7.293
  underclass 7.319
  working class 7.194
  see also élite
social classification 7.289
  see also social class
social contract 3.331, 7.221, 7.259, 7.289–90
social control 7.290
  statism 7.299
  see also law and justice; police; power (political and social) *and also* laws/legal system *under individual countries*
social credit 4.317
social dance 5.423–4
  see also ballroom dancing
Social Darwinism 4.317
social democracy and Social Democrats 4.317, 7.290
social identity theory 7.290
social institutions 7.112
social insurance 7.290
  see also social security
social justice 7.259, 7.291
social learning 7.7, 7.126, 7.207, 7.233, 7.291
social mobility 7.291
social ownership 7.212
social policy 7.291
  family policy 7.116
  see also social security; social services
social psychology 7.35, 7.291–2
  aggression 7.7
  altruism 7.11
  conformity 7.64
  person perception 7.234
  prejudice *see* discrimination and prejudice
  social understanding 7.293–4
  see also communication(s); persuasion; psychology; relationships; self; sexuality
Social Realism 5.424
social research 7.273
social sciences 7.292–3
  human geography 7.130–1
  see also anthropology; demography; economics; politics; psychology; sociology and society
social security and welfare 4.318, 7.293
  child benefit 7.45
  insurance 7.290
  old age pension 7.222
  social assistance 7.289
  unemployment benefit 4.357, 7.320
  welfare state 4.374, 7.330
  see also Beveridge
social services 7.289, 7.292, 7.293
  for disabled 7.91
  employment percentage 7.106
  expenditure cuts 7.3
  housing 7.134
  and old age 7.222
  social work 7.294
  welfare state 4.374, 7.330
  see also education; medicine and health care; social security
social skills 7.293
social stratification 7.293
  status 7.299
  see also social class
social structure 7.293
  see also sociology and society
social understanding 7.293–4
social work 7.294
  see also social services
socialism 4.317–18, 7.290–1
  Christian 4.75
  Fabians *see* Fabians
  failure of 7.194
  Internationals 4.164–5
  market 7.192–3
  revisionism 7.267
  Social Democrats 4.317, 7.290
  Socialist League 4.318
  Socialist Realism 5.380, 5.394, 5.422, 5.424, 5.447
socialism (*cont.*):
  see also communism; labour movements; Labour Party; Marx; social democracy *and under* politics *under individual countries*
socialization 7.291
Societies and Academies, scientific:
  Britain 6.47, 6.217, 6.236, 6.303
  China 6.72
  France 6.1, 6.87
  Germany 6.217
  Italy 6.1
  United States 6.12–13, 6.236
  USSR 6.1
society 7.294
  see also sociology and society
Society Islands 1.309, 1.330, 4.336–7
Society of Friends *see* Quakers
Socinus, Faustus 3.332
Socinus, Laelius 3.331–2
sociobiology 7.294
sociolinguistics 7.294
sociology and society 7.97, 7.294
  beginning of 7.63
  age *see under* population and demography
  alienation 7.10, 7.194
  and anarchism 7.12–13
  anomie 7.14, 7.97
  anthropology *see* anthropology
  arts *see* art
  authority 7.23
  bureaucracy 7.38, 7.52, 7.330
  Chicago School *see* symbolic interactionism
  citizenship 7.50
  civil 7.52
  class *see* social class
  clientelism 7.53
  cohort 7.56
  collectivism 7.56–7
  community *see* community (society)
  critical theory 7.76
  culture 7.11, 7.78
  daily life 7.109, 7.133
  demography *see* population and demography
  deviance 7.88, 7.179
  economics *see* economics
  ethnic groups *see* ethnic groups *and under individual countries*
  family *see* family
  Frankfurt School *see* Frankfurt School
  functionalism 7.14, 7.112, 7.127, 7.190, 7.229
  gender 7.129
  gifts *see* gifts
  groups *see* ethnic groups; groups
  health *see* medicine and health care
  individualism 7.156–7
  international 7.165
  kinship *see* kinship
  labelling theory 7.179
  law *see* law and justice
  legitimacy 7.182
  meaning 7.196
  medicine *see* medicine and health care
  meritocracy 7.199
  norms 7.218
  organization *see* politics
  philosophy of *see in particular* Rousseau
  politics *see* politics
  population *see* population and demography
  post-industrial 7.243
  power *see* power (political and social)
  Protestant ethic 7.251, 7.330
  psychology *see* psychology
  relationships *see* family; home; marriage; relationships
sociology and society (*cont.*):
  role 7.269
  sex *see* sex; sexism; sexuality
  social evolution 7.112
  social problems *see* crime; domestic violence; drugs, addictive; police
  sociobiology 7.294
  sociolinguistics 7.294
  status 7.299
  stigma 7.299
  see also social sciences *and other entries beginning* social *and under individual countries*
Socrates 3.285, 3.332, 5.21, 5.353, 7.294
soda/soda ash *see* sodium carbonate; sodium hydroxide
Soddy, Frederick 1.309
Södergran, Edith 5.404
sodium 1.107, 1.309–10, 6.323
  in battery 6.32
  from Castner–Kellner cell 6.62, 6.71
  discovered 6.96
  ionic bonding 6.24
  lamp 6.323
sodium benzoate 6.142
sodium bicarbonate 1.310
sodium borate (borax) 1.41, 6.42, 6.43, 6.156
sodium carbonate (soda ash) 1.310, 6.323
  mining 1.8
  from plant 2.137
  production and chemical industry 6.53, 6.70, 6.96–7, 6.197, 6.308, 6.325
sodium chloride 1.61, 1.310, 6.24
  chemical industry and production 6.2, 6.70, 6.72, 6.308
  crystal 1.81
  desalination 6.99
  food preservation 6.132, 6.140, 6.141
  glazing 6.157
  in petroleum 6.263
  seawater 1.290, 1.297–8, 1.309
  and soap 6.323
  sodium from 6.323
  see also salts
sodium cyanide 6.323
sodium hydrogencarbonate 1.310
sodium hydroxide 1.86, 1.310, 6.2, 6.62, 6.71, 6.295, 6.308, 6.323–4
sodium hypochlorite 1.310, 6.40, 6.353
sodium nitrate 6.128, 6.241
sodium peroxide 6.323
sodium propionate 6.141
sodium silicates 1.59, 1.177, 1.310
  zeolites 1.373, 6.120, 6.378
sodium sorbitate 6.141
Sodom and Gomorrah 9.34
soft coral 2.317
soft drink 6.324
soft water 6.378
softball 7.294–5
soft-shelled turtle 2.317
software, computer 6.282–3, 6.324
  applications 6.17, 6.179
  computer graphics 6.159
  interpreter 6.184–5
  mechatronics 6.179, 6.217
  operating system 6.219, 6.250
  see also program, computer
softwood 6.387, 6.388
Sōgi 5.380
Soglo, Nicéphore 4.34
Sogne Fiord 1.310
Sohrab and Rustam 9.34
soils 1.310, 1.311, 2.317
  alfisol 1.7, 1.311
  alkali 1.9
  andosol 1.14

soils (*cont.*):
aridisol **1**.20, **1**.311
azonal **1**.29, **1**.311
bog **1**.311
brown **1**.44, **1**.311
brown earth **1**.44
catena **1**.54
chelation **1**.59
chernozem **1**.60, **1**.269, **1**.311
chestnut **1**.60, **1**.311
classification ***1**.311*
consistence **1**.73
creep **1**.49, ***1**.79*
duricrust *see* hardpan *below*
eluviation **1**.108
entisol **1**.110, **1**.311
erosion **1**.310, **1**.311
creep **1**.79
dustbowl **1**.97
solifluction **1**.312
ferrisol **1**.121
gleying **1**.141
hardpan and duricrust **1**.48, **1**.49, **1**.97, **1**.152, **1**.189
histosol **1**.156, **1**.311
horizon **1**.158
ice-wedge **1**.165
inceptisol ***1**.166*, **1**.311
intrazonal **1**.171
landslide **1**.187
laterite **1**.189, **1**.311
latosol **1**.189
leaching **1**.108, **1**.190
list of **1**.311
lithosol **1**.196
loam **1**.197
meadow **1**.208–9, **1**.311
mechanics ***6**.324*
mollisol **1**.218, **1**.311
mottling **1**.221
oxisol **1**.243
palaeosol **1**.244–5
peat *see* peat
ped **1**.249
pedalfer **1**.249
pedocal **1**.249–50
pedology **1**.250
permafrost **1**.254
podzol **1**.260–1, **1**.269, **1**.311
podzolic **1**.261
prairie **1**.266
profiles **1**.268, **1**.269
regosols **1**.280
rendzina **1**.281, **1**.311
salinization **1**.290–1
sierozem **1**.303, **1**.311
solonchak **1**.311, **1**.312
solonetz **1**.312
solum **1**.313
spodosol **1**.311, **1**.317
taxonomy ***1**.311–12*
terra rossa **1**.333–4
topsoil **1**.340
tundra **1**.269, **1**.346
ultisol **1**.311, **1**.349
vertisol **1**.355
zonal **1**.311, **1**.374
see also compost and manure; fertilizers
Sokolow, Anna **5**.122
Sokoto **4**.128
solar *see* Sun/solar; solar power
solar power ***1**.312*, ***6**.324*, **7**.106, ***8**.150*
cell **6**.115, ***6**.324*
furnace ***6**.324*
heating **6**.167
panel ***6**.324*
satellite **6**.308
solar sail **6**.326
solar system ***8**.151*, **8**.152–3
age of **8**.2
geocentric system **8**.58
heliocentric system **8**.64–5
models of **8**.8, **8**.58, **8**.115
solar system (*cont.*):
symbols **8**.192
see also Moon; planets; Sun
soldering ***6**.324–5*, **6**.356
soldiers *see* armies
sole (fish) ***2**.317*
Solemn League and Covenant ***3**.332*
solenodon ***2**.316*, ***2**.317*
solenoid **6**.20, ***6**.325*
Solferino, battle of ***4**.318*
solicitor ***7**.295*
solid ***1**.312*, ***6**.23–4*
in liquids *see* suspension
see also elements
Solidarity ***4**.318*, **7**.332
solid-state device ***6**.325*
camera ***8**.151*
see also semiconductor; transistor
solifluction ***1**.312*
soliloquy ***5**.424*
solipsism ***7**.295*
solitaire (game) **7**.40
solitaire bird ***2**.317*
solmization ***5**.424*
Solomon, king of Israel ***3**.332*, ***9**.34*
and Ethiopia **3**.119
and Jerusalem **3**.187
and myths **9**.7, **9**.26, **9**.33, **9**.34
Solomon Islands ***1**.312*, ***4**.318*, **7**.296
Solomon's seal ***2**.317*
Solomos, Dionysios **5**.191
Solon **3**.27, ***3**.332*, **5**.190
solonchak **1**.311, ***1**.312*
solonetz ***1**.312*
solstice **1**.297, ***1**.312*, ***8**.151*, **8**.163
solum ***1**.313*
solutes ***1**.313*
colligative properties **1**.68
see also solvent and solution
solution *see* solvent and solution
Solutrian **3**.367
Solvay, Ernest: Solvay process **6**.70, **6**.71, **6**.232, ***6**.325*
solvent and solution ***1**.313*, ***6**.325*, **6**.338
caustic *see* alkali; alkaline
dialysis **6**.101
dry-cleaning **6**.109
extraction **6**.264, **6**.343
molarity **1**.217
propanone **6**.284
solvent abuse **7**.95
Solzhenitsyn, Aleksandr **4**.277, ***5**.424*
Somadeva **5**.401
Somalia (*earlier* Somaliland and Somali Republic) ***1**.313*, ***4**.318–19*, ***7**.381*
aid to **7**.121
in Arab League **7**.15
city **3**.234–5
colonization and decolonization **4**.4, **4**.306
economy and trade **7**.87
famine **7**.117
language **7**.78
leader **4**.147
literacy **7**.185
physical geography and geology **1**.313
refugees **7**.261
technology **6**.379
soman **6**.240
Sombrero Galaxy **8**.151
Somers, John Somers, Baron ***3**.332–3*
Somerset, duke and earls of *see* Beaufort
Somerset, Edward Seymour, 1st earl of Hertford and duke of **3**.110, **3**.282, ***3**.333*
Somme, battle of ***4**.319*, **4**.352
Somoza, Anastasio, Luis, and Anastasio Debayle **4**.241, ***4**.319*
Son Sann **4**.58, **6**.313
sonar **6**.111, **6**.133, **6**.159, ***6**.325*, **6**.375, **6**.382
sonata and sonata form **5**.103, ***5**.424*, **5**.442
Sonderbund ***4**.319*
Sondheim, Stephen ***5**.424*
Song (Sung) dynasty **3**.72, ***3**.333*
Song Jiaoren **4**.187
song thrush ***2**.317*
Songgram, Pibul **4**.208
Songhay **3**.4, **3**.25, ***3**.333*
songs and singers:
Africa **5**.7–8, **5**.321
aria **5**.21
Australia **5**.1, **5**.289, **5**.313, **5**.441
Austria **5**.44, **5**.276, **5**.306, **5**.313, **5**.491
birds **2**.31–2
Britain **5**.22, **5**.69–70, **5**.134, **5**.212, **5**.225–6, **5**.258, **5**.313, **5**.343
hymns **5**.116, **5**.486
see also under opera
canon **5**.75, **5**.223
cantata **5**.75
canticle **5**.75
canzona **5**.75
cappella **5**.76
carole or carol **5**.79
castrato **5**.82
chanson **5**.88
China **5**.29, **5**.94, **5**.96
consort **5**.110
Czech **5**.140, **5**.210, **5**.233, **5**.422
dirge **5**.131
epithalamion **5**.148
folk **5**.29–30, **5**.160–1
France **5**.66, **5**.103, **5**.137, **5**.153, **5**.187, **5**.242, **5**.256, **5**.270, **5**.313
see also under opera
Gaelic **5**.409
Germany **5**.68, **5**.78, **5**.155–6, **5**.210, **5**.264, **5**.270, **5**.313, **5**.397, **5**.407
see also under opera
India **5**.446
Ireland **5**.303
Islam **5**.230
Italy **5**.42, **5**.46, **5**.81, **5**.83, **5**.168, **5**.175, **5**.283, **5**.302, **5**.342
see also under opera
Japan **5**.313
jazz **5**.22, **5**.50
Lied **5**.142, **5**.203, **5**.264
madrigal **5**.274–5, **5**.283
Maori **5**.282
medieval **5**.289
minstrels **5**.297, **5**.368
monody **5**.302
motet **5**.274–5, **5**.289, **5**.306
mouth music **5**.306
music imitating (barcarole) **5**.35
national anthems **5**.313
North Africa **5**.321
Norway **5**.156
Pacific Ocean islands **5**.325
paean **5**.335
Papua New Guinea **5**.339
part-song **5**.340
penillion **5**.343
pitch **5**.352
plainsong **5**.289, **5**.351
polyphony *see* polyphony
pop *see* popular music
Provençal **5**.368
psalm **5**.368
Renaissance **5**.379–80
Russia and USSR **5**.87, **5**.310, **5**.313, **5**.370, **5**.448
see also under opera
South and Central America **5**.14–15, **5**.257
Spain **5**.10, **5**.132, **5**.313, **5**.440
spiritual **5**.429
Sweden **5**.48, **5**.265, **5**.320
songs and singers (*cont.*):
United States **5**.18, **5**.71, **5**.175, **5**.246, **5**.313, **5**.321, **5**.359, **5**.384, **5**.385–6, **5**.424, **5**.429
opera **5**.174, **5**.175, **5**.251
voice **5**.478–9
Wales **5**.343
West Indies **5**.72, **5**.78
see also choirs; hymns; Mass (Eucharist); musicals; opera; operetta; oratorio
sonnets ***5**.424–5*
Britain **5**.296, **5**.418, **5**.429, **5**.441, **5**.496
Italy **5**.12, **5**.162, **5**.345–6
Persia **5**.197, **5**.394, **5**.397
Poland **5**.294
United States **5**.295
see also lyric poetry
Sonni Ali ***3**.333*
Sonoran Desert ***1**.313*
Sons of Liberty ***3**.333*
Soong Ching-Ling **4**.334
Soong Mei-ling **4**.71
Sophia, Electress of Hanover ***3**.333*
sophists ***3**.334*, ***7**.295*
Sophocles **5**.191, **5**.377, ***5**.425*, **5**.454, **5**.459
sorbitol **6**.141
sorcery *see* witchcraft
sorghum ***2**.317*
sorrel ***2**.317*
Soufflot, Jacques-Germain ***5**.425*
soul ***7**.295*
afterlife **7**.5–6
Aristotelian view of **7**.16
in Hinduism **7**.21, **7**.273, **7**.295
and resurrection **7**.266
see also reincarnation
soul music **5**.358, ***5**.425*
Soult, Nicholas Jean de Dieu **4**.199
Soumangourou, king of Sosso **3**.346
sound and sound waves ***1**.313–14*, ***6**.325*, ***6**.381*
acoustics **6**.2, **6**.223
amplification **6**.14
barrier *see* supersonic flight
communications *see* radio; telecommunications; telephone
diffraction **1**.90–1
Doppler effect **1**.94, **6**.105–6
echo **1**.100
energy changed into heat energy **6**.166
film **6**.75
frequency **6**.325
as longitudinal waves **1**.198, **1**.199
loudspeaker **6**.209
measurement **1**.85–6, **6**.97, **6**.184
recording and reproduction **6**.33, ***6**.325–6*
acoustics **6**.2
compact disc **6**.81–2, **6**.179, **6**.180, **6**.372
Dolby system **6**.105, **6**.345
gramophone and record-player **6**.33, **6**.112, **6**.158–9
high-fidelity **6**.159, **6**.168
microphone **6**.13, **6**.128, **6**.222, **6**.289, **6**.326, **6**.347
record **6**.159, **6**.296, **6**.326, **6**.334
stereophonic **6**.40, **6**.159, **6**.334
tape and tape recorder **6**.103, **6**.105, **6**.213, **6**.326, **6**.345
reflection **1**.278, **1**.279
refraction **1**.278, **1**.279
speech *see* speech
speed **1**.317, **6**.140, **6**.175, **6**.211, **6**.339
synthesizer **6**.343
ultrasound *see* ultrasound
vibrations **1**.356
voice *see* voice

sound and sound waves (*cont.*):
see also hearing; music
sounding lead 6.197
Souphanouvong, prince 4.190, 4.263, 4.321
soursop *2.317–18*
Sousa, John Philip 5.283
Sousa, Martim Alfonso de 3.49, *3.334*
sousaphone 5.461
South *see* developing countries; Third World
South Africa *1.314*, *4.319*, *7.381–2*
Afrikaners 4.5
agriculture 6.34
apartheid *see under* black people *below*
arts 5.117, 5.165–6, 5.184, 5.406
black people:
apartheid 4.16, 7.334
sanctions against 7.34, 7.59, 7.225
Bantustans 4.30–1
leaders 4.38, 4.39, 4.201, 4.209
Pan-African Congress 4.257
Sharpeville massacre 4.310–11
Soweto 4.322
see also apartheid *above and* ethnic groups *and* ANC *under* politics *below*
Boers *see* Boers
decolonization 4.4
economy and trade 4.33, 4.251
mining 1.182
SADCC 7.296
sanctions 7.34, 7.59, 7.225
energy and fuel 6.385
ethnic groups 7.334, 7.336
Uitlanders 4.356
see also black people *above*; *and also* Boers; Nguni; Zulus
exploration and discovery 1.197, 3.122
government *see* politics and government *below*
languages 7.131, 7.136, 7.176
laws/legal system 7.87
literature 5.165–6, 5.184, 5.406, *5.426*
medicine and health care 6.30
migration 7.241
mining 6.101, 7.58
music 5.117
old age 7.221
physical geography and geology *1.314*
Cape of Good Hope 1.142
mountains 1.95, 1.330
regions *see* provinces *below*
rivers 1.195, 1.240, 1.346, 1.352
politics and government:
ANC (African National Congress) 4.4–5, 4.201, 4.209, 7.238
consociationalism 7.67
as Dominion 7.58
Dutch governor 3.304
emergency powers 7.105
fascism 7.118
National Party 4.234
Pan-African Congress 4.257
political prisoners 7.238
president 4.187
prime ministers 3.304, 4.46, 4.150, 4.207, 4.317, 4.364
United Democratic Front 4.359
United Party 4.360
see also black people *above*
prehistory 3.120, 3.341
provinces and regions 1.355
see also Cape Province; Natal; Orange Free State; Transvaal
refugees 7.261
religion/theology:
Coptic Church 7.70
Dutch Reformed Church 7.260
South Africa (*cont.*):
religion (*cont.*):
Jews and Judaism 7.171, 7.335
myths 9.16, 9.26, 9.36
new 7.5
Sikhism 7.286
segregation *see* apartheid *above*
sport and recreation 7.34, 7.75, 7.303
technology and science *see entries above and below prefixed* 1. *and* 6.
townships 4.311, 4.322
Uitlanders 4.356
vehicles and transport 6.293, 6.332
wars and defence 7.66, 7.82, 7.220
Boer Wars 4.42–3, 4.82, 4.363, 7.63
Jameson raid 4.171–2
leaders 4.46, 4.150, 4.317
Xhosa 4.384
Zulu wars 4.42, 4.168, 4.391
Zulus *see* Zulus and Zululand
South America *3.334*, *4.319–20*
adoption 7.3
agriculture 4.132, 6.267, 7.232, 7.237
technology 6.7, 6.34, 6.68, 6.204, 6.351
aid to 4.320
Andean group 7.13
arts *5.257*, *5.425–6*
architecture 5.100–1, 5.320, *5.425–6*
see also literature; music *below*
astronomy 8.24
colonization 3.49, 3.122, 3.252, 3.334, 4.43, 4.127
demography 4.320
ageing 7.6
birth rate 7.32
contraception 7.115
fertility rate 7.119
infant mortality 7.159
infertility 7.159
life expectancy 7.184
mortality 7.207
population growth 7.226, 7.241–2
diplomat *see* Pérez de Cuéllar
discovery, colonization, and exploration 1.88, 1.159, 1.275, 1.355–6, 3.106, 3.159
decolonization 4.43, 4.324
see also *conquistadors*, *under* Spain
diseases and disorders 7.9, 7.59
economy and trade 7.88
agriculture *see* agriculture *above*
Andean Common Market 4.12
Andean Group 7.13
debt 7.2
dependency theory 7.85
drugs 7.96
energy and fuel 6.37, 6.95
foreign investment 4.320
mining 6.16, 6.101
NICs 7.215
pay 7.203, 7.328
unemployment 7.320
see also under individual countries
education 7.93, 7.101, 7.102, 7.141, 7.185, 7.277, 7.321
energy and fuel 6.37, 6.95, 7.106
environment *see* Amazon
ethnic groups 3.71, 7.11, 7.137, 7.335, 7.375
see also Indians, South American
exploration *see* discovery *above*
fishing 6.133
food and drink 2.41, 6.324, 7.190
forests 6.111, 7.83
see also Amazon
government *see* politics and government *below*
hacienda 4.144
South America (*cont.*):
independence 4.43
Indians *see* Indians, South American
industrial relations 7.53
land reform 7.180
languages 7.269, *7.295*, 7.335
laws/legal system 7.40, 7.51, 7.54, 7.174
literature 5.53, 5.114, 5.256, 7.216
novels 5.169–70, 5.468
malnutrition 7.190
maps 1.313, 4.43
medicine and health care 7.9, 7.59, 7.140, 7.159
mining 6.101, 7.343
music 5.9, 5.14–15, 5.256–7, 5.257, 5.282, 5.476
see also ballroom dancing
peon 4.266
physical geography and geology 1.313, *1.314*
climate 1.65
continental drift 1.135
desert 1.23
earthquakes and volcanoes 1.99
Horn, Cape 1.158
lake 1.338, 1.339
Magellan, Strait of 1.201, 1.202
map 1.313
mountains 1.13, 1.23–4, 1.78, 1.135
see also Andes
plains 1.142, 1.197, 1.246
plateaux 1.207, 1.248–9
rivers *see under* rivers
volcanoes 1.78, 1.99
waterfalls 1.14
see also under individual countries
poison 6.94
politics and government:
anti-government movements 7.217
caudillo 4.66
Christian Democracy 7.47
corruption 7.71
coup d'état 7.73
independence 4.43–4
leaders 3.233, 4.143
military 4.66, 7.173, 7.201
Pan-Americanism 4.258
see also under individual countries
population *see* demography *above*
poverty 7.244
refugees 7.261–2
in regional organizations *see* Organization of American States
religion/theology:
charismatic 7.44
Jews and Judaism 7.171, 7.335
liberation theology 7.184
myths 9.11, 9.33
pentecostal 7.232
shamanism *7.282*
voodoo *see* Voodoo
see also under Indians, South American
rulers *see under individual countries*
slavery 7.287
social services and security 7.293
sport and recreation 7.37, 7.217, 7.232, 7.295, 7.303
technology and science:
vehicles and transport 6.234, 6.250, 6.292, 6.299
see also entries above and below prefixed 1. *and* 6.
telecommunications 6.250, 7.35
urban areas 7.217, 7.283, 7.288, 7.298
wars and defence:
Chaco War 4.68
guerrilla 4.143
leaders *see under* Argentina
South America (*cont.*):
wars and defence (*cont.*):
Leticia dispute 4.193–4
Paraguayan 4.259–60
Spanish-South American Wars of Independence 4.43–4, 4.324
Tacna-Arica conflict 4.336
War of Pacific 4.254
weapons 6.41, 7.220
see also under individual countries
see also Organization of American States; *and also individual countries*
South Asia:
Colombo Plan 7.57
societies *7.295–6*, 7.315
gypsies 7.137
Nagas 7.211
Pathans 4.262, 7.231
see also Tamils
see also in particular Bangladesh; India; Nepal; Pakistan; Sri Lanka
South Australia *1.314*, *4.320*
South Carolina *3.334–5*
south celestial pole 8.22
South China Sea 1.61, 6.209
south galactic pole 8.54
South Georgia *1.314–15*
South Korea *4.320*, *7.362*
economy and trade 7.88, 7.113, 7.157, 7.277, 7.319
land reform 7.180
languages 7.177
President 4.289
religion/theology 7.65, 7.320
sport and recreation 7.194, 7.223, 7.305
urban areas 7.197, 7.323
see also Korea
South Orkney Islands *1.315*
South Pacific Forum *7.296*
South Pole *1.315*
see also Antarctica; pole/polar
South Sandwich Islands *1.315*
South Seas *1.315*
South Sea Bubble 3.334, *3.335*
see also Pacific Ocean area
South Shetland Islands *1.315*
South Vietnam *see* Vietnam
South West Africa *see* Namibia
South West Africa People's Organization *4.334*, 7.130
South Yemen *4.321*
see also Yemen
South-East Asia:
agriculture 7.268
aid to 7.8
ASEAN 4.21, 4.321, 7.20
Bandung Conference 4.30
Colombo Plan 4.79
colonialism in 3.7, 3.11, 3.106–7, 3.216
decolonization *4.320*, 4.321
ethnic groups *see* societies *below*
languages 7.22–3, 7.95
literature 5.427
map 4.321
music and dance *5.426–7*
physical geography and geology 1.210, 1.277
see also under individual countries
politics and government 7.73, 7.94
see also under individual countries
poverty 7.244
religion/theology 7.36, 7.143, 7.302
see also Theravāda
social services and security 7.293
societies 3.196, 7.25–6, 7.175, 7.202, *7.296*
see also Trobriand Islanders *and under individual countries*
sport and recreation 7.194
technology and science 6.53, 6.132, 6.178, 6.307

South-East Asia (*cont.*):
Treaty Organization **4**.320
urban areas **7**.288
wars and defence:
French Indo-China War **4**.98, **4**.99, ***4**.127*, **4**.133
see also under individual countries in particular Vietnam
see also individual countries
Southern African Development Co-ordination Conference ***7**.296*
Southern Alps ***1**.314*
Southern Coalsack ***8**.151*
southern constellations *see under* constellations
Southern Cross **8**.35
Southern Crown **8**.29–3
Southern cult ***3**.335*
Southern fish **8**.121
Southern Ocean ***1**.314*
Southern Rhodesia *see* Zimbabwe
Southern Triangle **8**.173
Southey, Robert ***5**.427*
Soutine, Chaïm ***5**.427*
Soutpansberge **1**.374
Souvanna Phouma, prince **4**.190, ***4**.321*
souvenirs *see* collecting
sovereign *see* Crown; monarchy
sovereignty ***7**.296*
Soviet ***4**.321–2*
Soviet Union *see* Commonwealth of Independent States; Russia; Union of Soviet Socialist Republics
sow bug *see* woodlouse
Sowerby, James **2**.314
Soweto ***4**.322*
sowing seed *see* seeds
soya bean ***2**.318*
Soyinka, Wole ***5**.427*
Soyuz see *Apollo*
spa town ***3**.335*
Spaak, Paul-Henri **4**.114, ***4**.322*
space (astronomy) ***1**.315–16*, ***8**.154*
ether **8**.48
exploration *see* space–flight
ionized particles **8**.3
space telescope **6**.310, **8**.68, **8**.69–70, **8**.71
and time *see* space–time
see also astronomy
space (philosophy) ***7**.296*
space frame (building) **6**.51, ***6**.326*
space-flight and space exploration ***4**.322*, **4**.323, ***8**.122*, ***8**.156*
astronautics **8**.10
ESA **8**.48, **8**.68
ground station **6**.161–2, **6**.222, **6**.308
interferometer **6**.184
launch site **6**.196
manned **6**.214
moon landing **6**.17
NASA **8**.106
principles of ***6**.326*
race **6**.301
robot ***6**.326*
spacecraft **8**.59, **8**.68, **8**.74, **8**.91, **8**.99, **8**.121, ***8**.122*, ***8**.154–6*
airlock **6**.9
American *see under* space *under* United States
launch vehicle **6**.64, **6**.196, **6**.326
probe ***6**.326*
Russian *see under* space *under* Russia
Space Shuttle (Spacelab) **6**.301, ***6**.326–7*
Spacelab ***8**.156*
space-plane ***6**.326*
station ***6**.327*
see also *Apollo*
spacesuit ***6**.327*
space-flight and space exploration (*cont.*):
see also astronomy; rocketry; satellites *and under* China; Union of Soviet Socialist Republics; United States
space–time ***8**.156*
continuum ***1**.316*
curvature **8**.35
dimension **7**.296
gravitational waves **8**.60
spade **6**.103
spadefoot toad ***2**.318*
Spain ***1**.316*, ***3**.335*, ***4**.322*, ***7**.382*
arts:
architecture **5**.74, **5**.100–1, **5**.171, **5**.189, **5**.352
ceramics and pottery **5**.348
cinema **5**.65–6
graphic art and design **5**.187, **5**.194, **5**.348
painting *see under* painting
sculpture and carving **5**.74, **5**.189, **5**.348
see also literature; music *below*
conquistadors ***1**.47*, **1**.88, **3**.85, ***3**.86*
Balboa **3**.33–4, **3**.268
Narváez **3**.56, **3**.88, **3**.246
see also Cortés; de Soto; Pizarro, Francisco *and also* empire *below*
economy and trade:
co-operatives **7**.69
industrial relations **7**.304, **7**.315
in OECD **7**.221
slaves **3**.25
unemployment **7**.320
see also European Community; European Economic Community
education **7**.155
empire, colonialism and exploration **3**.107, **3**.122, **3**.131, **3**.174, **3**.203, **3**.252, **3**.253, **3**.288, ***3**.336*, **4**.4, **4**.306–7
see also *conquistadors* above
ethnic group *see* Basques
foreign policy **4**.9
Council of Europe **7**.71
see also wars *below*
government *see* politics and government *below*
historiography **5**.428
language **7**.156, **7**.269
literature **5**.333, ***5**.427–8*
see also under drama; novels; poetry
maps **4**.265, **4**.325
music **5**.10, **5**.82, **5**.132, **5**.188, **5**.313, **5**.348, **5**.440
see also under composers
philosophy **7**.23, **7**.189
physical geography and geology ***1**.316*
mountains **1**.269–70, **1**.303–4
regions **1**.13, **1**.139
river **1**.330
politics and government **3**.143, **7**.118, **7**.201
Carlists **4**.62
Falange **4**.115–16
fascism **7**.118
leaders **3**.7, **4**.124, **4**.190, **4**.276
Socialists **4**.317
see also rulers *below*
pollution **7**.2
prehistory **3**.174
regions:
listed ***9**.75*
northern **3**.19, **3**.62, **3**.64, **3**.136, **3**.206, **3**.316
southern **1**.13, **3**.12, **3**.141, **3**.144
see also Gibraltar
Spain (*cont.*):
religion/theology **3**.190, **3**.359, **7**.280
Islam **3**.240, **3**.250
Jews **3**.103, **3**.218
missionaries **3**.131, **3**.203
saints **9**.41, **9**.42, **9**.43
see also individual names
Spanish Inquisition **3**.336, **3**.337
rulers:
listed ***9**.68*
Bourbons ***3**.46*, **3**.70, **3**.74, **3**.338, **4**.62, **4**.124, **4**.168, **7**.97
see also Ferdinand VII; Juan Carlos; Philip V
Habsburgs **3**.128, **3**.153, **3**.279, **3**.337
see also Charles V; Philip II
Savoy **4**.168
see also politics *above*
sport and recreation **7**.37–8, **7**.223
technology and science **6**.73, **6**.75, **6**.178, **6**.256
vehicles and transport **6**.25, **6**.47, **6**.58, **6**.149, **6**.169
see also entries above and below prefixed **1**. *and* **6**.
wars and defence:
army leaders **3**.7, **3**.56, **3**.190
see also *conquistadors* above
Austrian Succession **3**.29–30
with Britain **3**.32, **3**.186, **3**.335–6
Cambrai, League of **3**.58
Civil War **4**.6, **4**.82, ***4**.324*, **4**.325, **7**.13
Devolution (with France) **3**.102–3
Dutch Revolts against **3**.108, **3**.141
fortresses **3**.7–8
leader **4**.276
Moors, invasions of **3**.240, **3**.350
navy **3**.190, **3**.335–6
Revolutionary Wars **4**.288
Spanish Armada ***3**.335–6*
Spanish–American War **4**.207, **4**.209, ***4**.324*
Spanish-South American Wars of Independence **4**.43–4, ***4**.324*
Tordesillas Treaty with Portugal **3**.359
Trafalgar **4**.349–50
in WEU **7**.331
see also NATO; Peninsular War; Seven Years War; Spanish Succession
spalling **1**.315, ***1**.316*
Spanish Armada ***3**.335–6*
Spanish Civil War **4**.6, **4**.82, ***4**.324*, **4**.325, **7**.13
Spanish fly ***2**.318*
Spanish Guinea *see* Equatorial Guinea
Spanish Inquisition ***3**.336*, **3**.337
Spanish mackerel ***2**.318*
Spanish Main ***3**.336–7*
Spanish Morocco **4**.4, **4**.306
Spanish Netherlands ***3**.337*
Spanish Succession, War of **3**.297, ***3**.337–8*, **3**.369
battles and campaigns **3**.43, **3**.217, **3**.265, **3**.299
Spanish–American War **4**.207, **4**.209, ***4**.324*
Spanish-South American Wars of Independence **4**.43–4, ***4**.324*
spanner ***6**.327*, **6**.358
spanworm **2**.194
spar ***1**.316*
Spark, Muriel ***5**.428*
spark:
chamber **6**.258
erosion **6**.211
spark (*cont.*):
-plug **6**.175, ***6**.327*
transmitter **6**.168
sparrow family **2**.44, **2**.263, ***2**.318–19*, **2**.347, **2**.363
sparrowhawk **2**.152, **2**.318, ***2**.319*
Sparta **3**.206, **3**.214, ***3**.338*
Peloponnesian War **3**.3, **3**.8, **3**.274–5
Spartacus **3**.90, **3**.92, ***3**.338–9*
Spartakist Movement ***4**.324*
spawn ***2**.319*
spear **6**.226, ***6**.327*
spearfish **2**.203, ***2**.319*
special armed services ***7**.296–7*
Special Drawing Rights **7**.78, **7**.112, **7**.165
special education ***7**.297*
special relativity **1**.281, **8**.137–8
species **2**.229, ***2**.319*
endangered **2**.107, **2**.108
extinct *see* extinct species; fossils
population **2**.264–5
specific gravity: measurement **6**.174
specific heat capacity **1**.152
speckle interferometry ***8**.156–7*
spectacles **6**.146, **6**.198, ***6**.327*, **6**.328
spectra ***1**.316–17*
analysis *see* spectroscopy
and astronomy ***8**.157*
Draper classification **8**.39
electromagnetic **8**.45
emission **8**.157
flash **8**.51
spectral classes **8**.39, ***8**.157*
see also spectral lines
electromagnetic **1**.103, **8**.45
neon **1**.225
spectrometer **8**.157
visible **6**.117
see also spectroscopy
spectral lines **8**.51, **8**.157
forbidden line **8**.51
Fraunhofer **8**.53
red shift **8**.134–5
spectroheliograph ***8**.157*
spectrometer ***8**.157*
spectrophotometry **6**.40, **6**.328, ***8**.157*
spectroscopic binary star **8**.15, **8**.16
spectroscopy ***1**.316*, **6**.192, ***6**.327–8*
diffraction grating **6**.103
helium discovered **6**.167
mass **6**.186, **6**.216–17
spectrophotometry **6**.40, **6**.328, **8**.157
spectroscope **8**.11
spectroscopic parallax **8**.116
X-ray **1**.43
spectrum *see* spectra
speculator ***7**.297*
speculum (medicine) **6**.68, ***6**.328*
speculum metal ***8**.158*
speech ***2**.319*
disorder **7**.15
free **7**.82, **7**.124–5
lip-reading **7**.185
monologue **5**.302
phonetics **7**.235
phonology **7**.235
recognition **6**.240, ***6**.328*
rhetoric **5**.338, **5**.382, **5**.732
scrambler **6**.310
sign language and signing **7**.218, **7**.285, **9**.48
vocoder **6**.373–4
see also communication; conversation; figures of speech; language; oral traditions; telephone
speed *see* velocity and speed
speedway racing ***7**.297*
speedwell ***2**.319*
Speer, Albert ***4**.324*
Speke, John Hanning **1**.46, ***1**.317*, **4**.56, ***4**.324*, ***4**.326*, **4**.356

Spence, Sir Basil 5.428
Spence, Thomas 3.339
Spence, William Guthrie 4.326
Spencer, Herbert 4.317, 4.326, 7.157
Spencer, Sir Stanley 5.428
Spencer Jones, Harold 8.193
Spender, Sir Stephen 5.26, 5.428
Spenser, Edmund 5.12, 5.141, 5.387, 5.418, 5.424, 5.428–9
Spenta Mainyu 9.2
Speransky, Mikhail Mikhailovich, count 4.326
sperm whale 2.319
spermaceti 6.328, 6.382, 6.384
sperms 2.303, 2.319, 2.338
  spermicide 6.87
Sperry, Elmer Ambrose 6.163, 6.328
sphagnum 2.319–20
sphalerite 1.373
sphere:
  armillary 8.8
  celestial 8.22–3
spherical astronomy 8.158
spherically symmetric gravitation 8.60
spheroidal free oscillation 1.127
spheroidal weathering 1.317
sphinx 9.34
sphinx moths 2.152, 2.343
sphygmomanometer 6.218, 6.328
Spica 8.158
Spice Islands *see* Moluccas
spices, herbs, and flavourings 2.155, 2.320, 6.140
  anise 2.11
  caper 2.52
  cardamom 2.53
  chilli 2.63
  chive 2.64
  cinnamon 2.66
  clove 2.69
  dill 2.92–3
  fennel 2.115
  fenugreek 2.115
  geum 2.135
  ginger 2.136
  horseradish 2.163
  hyssop 2.168
  liquorice 2.191
  marjoram 2.203
  nutmeg 2.231
  oregano 2.203
  parlsey 2.243
  pepper 2.249
  pimento 2.254
  rosemary 2.288
  saffron 2.291–2
  sage 2.292, 2.293
  savory 2.294
  tamarind 2.334
  tarragon 2.335
  thyme 2.340
  turmeric 2.351
  vanilla 2.354
  see also aromatic and scented plants *and* food *under* additive
spicules 8.158
spider *see* spiders
Spider, Mr 9.4
spider crab 2.320
spider monkey 2.277, 2.320
spiders 2.320
  black widow 2.34
  mythical 9.38
  sun 2.330
  tarantula 2.277, 2.335
  trapdoor 2.345
  wolf 2.369
spiderwort 2.320–1
Spielberg, Steven 5.429
spies 7.278, 7.303
spin (physics) 1.317
  electron 6.119
spina bifida 2.321, 6.13
spinach 2.321
spinach beet 2.321
spindle 6.329
spindle tree 2.321
spine 2.321, 9.58
  animals with *see* vertebrates
  animals without *see* invertebrates
  disc 2.93–4
  disorders 2.93–4, 2.321
  vertebrae 2.56, 2.70, 2.93, 2.355, 9.58
spinet 5.429
Spink, Ian 5.122, 5.126
spinning 6.128, 6.328–9, 6.390
  cotton and wool 6.20, 6.92, 6.165, 6.328–9, 6.388
  machines 6.20, 6.92, 6.165, 6.329
  silk 6.320
  spindle 6.329
  spinning-jenny 6.329
  synthetic fibre 6.2, 6.295, 6.329
  see also textiles
Spinola, Ambrose 3.49
Spinoza, Baruch de 3.51, 3.339, 7.206, 7.259, 7.297
spiny anteater 2.103
spiny dormouse 2.322
spiny rat 2.322
spiny rock lobster 2.80
spiny-headed worm 2.322
Spiraea 2.322
spiral arms 8.158
spiral galaxy 8.14, 8.55, 8.158
spirit-level 6.329
spirits 7.335
  Celtic 9.6
  demons and imps 9.14, 9.16
  exorcism 7.139
  fairies 9.12
  ghosts 9.14
  Islam 9.20, 9.29
  jinns 9.10, 9.14, 9.17–18, 9.20
  Persian 9.10, 9.20, 9.29
  possession 7.282, 7.297
  ship 9.13
  spiritualism 7.297–8
  see also ancestor worship; angels; animism; supernatural; tricksters; Voodoo
spiritual (music) 5.429
Spirogyra 2.321, 2.322
spit, sand 1.292–3
Spithead mutiny 3.339
Spitsbergen 1.317
spittlebug 2.127
spleen 2.322
Spode, Josiah 6.276
spodosol 1.311, 1.317
Spohr, Louis 5.110, 5.429
spoils system 4.326
spokeshave 6.358
sponge 2.322
spoonbill 2.322
Sporades 1.317
sporadic meteor 8.97
Spörer's Law 8.158
spores 2.322–3
  see also cryptogram; fungi
sport and recreation 4.326, 7.182–3
  and animals *see under* animals
  archery 7.15–16
  athletics 7.21, 7.232
  badminton 7.25
  ballooning 7.26
  baseball 7.27
  basketball 7.27–8
  bowling 7.34
  bowls 7.34
  boxing 7.34, 7.305
  bull-fighting 7.37–8
  canoeing 7.39
  chariot racing 6.70
  cock-fighting 7.54
  cricket 7.75
  croquet 7.76
  cross-country running 7.76–7
sport and recreation (*cont.*):
  cycling 6.35, 7.79
  darts 7.80
  diving 7.93
  fencing 7.119
  fives 7.121
  flying 7.26, 7.132
  game-shooting 7.129
  gliding 6.202, 7.132
  golf 7.134
  greyhound racing 7.136
  gymnastics 7.137
  hang-gliding 6.165
  hobbies *see* hobbies and pastimes
  hockey 7.144–5
  horse-racing 7.147
  horseshoe-pitching 7.147
  ice-skating 7.151
  lacrosse 7.179
  marathon 3.221, 7.192
  martial arts 7.193–4
  microlighting 6.222
  motorbike-racing 7.207
  motor-racing 7.207–8
  mountaineering 1.366, 7.208
  netball 7.214
  pelota 7.232
  pétanque 7.234
  pigeon-racing 7.236
  polo 7.240, 7.241
  roller-skating 7.269
  rounders 7.270
  rowing 7.270
  running 3.221, 7.76–7, 7.192
  sailing 6.104, 6.308, 6.390, 7.272
  skating 6.321
  skiing 6.322, 7.287
  skittles 7.287
  softball 7.294–5
  speedway racing 7.297
  squash 7.298
  surfing 7.303
  swimming 7.303
  table tennis 7.305
  Taiji Quan 7.305
  tennis 7.308
  textiles for 6.353
  volleyball 7.326–7
  water 7.39, 7.93, 7.270, 7.272, 7.303, 7.331
  windsurfing 7.331
  wrestling 7.333
  see also games; leisure; Olympic Games *and under individual countries*
sporting birds *see* game birds
Sprague, Frank 6.114
sprats 2.323, 2.366
spreadsheet 6.329–30
spring (season):
  equinox 1.111, 1.297, 8.47
  tide 1.318, 1.337, 8.171
spring (technology) 6.330
spring (water) 1.318
  artesian 1.20
  geysers and hot springs 1.137, 1.138, 1.159
springbok 2.323
springhaas 2.323
Springsteen, Bruce 5.358
springtail 2.322, 2.323
sprinkler systems 6.188, 6.330
sprout, Brussels 2.43
spruces 2.154, 2.323
SPS (super proton accelerator) 6.258
spurges 2.56, 2.183, 2.289–90, 2.323
Spurs, Battle of Golden 3.89
spurs (geomorphology) 1.317, 1.318
  truncated 1.140, 1.345, 3.318
Sputnik 6.301, 6.330
sputtering 6.370
spying 7.278, 7.303
squall 1.318
square 9.51
  measurement 9.49
Square of Pegasus 8.117
square root: root-mean-square 1.286
square-rigged ships 6.304, 6.306, 6.330
  barque 6.30, 6.298, 6.306
  clipper 6.76, 6.219
  cog 6.80, 6.307
  frigate 6.146, 6.188
  see also brig
squash (plant) 2.204, 2.323
squash (sport) 7.298
squatter 4.326, 7.298
squid 2.323
squire 3.339
squirrel group 2.323
  beavers 2.26, 2.220
  gopher 2.141
  marmots 2.203–4, 2.370
  prairie dog 2.267–8
  spring hare 2.323
  woodchuck 2.370
  see also ground squirrels
squirrelfish 2.323–4
Sri Lanka 1.318, 3.339, 4.326–7, 7.382
  agriculture 6.34, 6.191
  arts 5.429
  ethnic group *see* Tamils
  infant mortality 7.159
  languages 7.95
  literature:
    Tamil 5.445–6
    see also under Dravidian
  medicine and health care 7.24, 7.140, 7.159
  physical geography and geology 1.318
  politics 4.30, 4.308
  refugees 7.261
  religion/theology: Buddhism 3.339, 7.36, 7.37
    see also Theravāda
  sport and recreation 7.75
  technology and science 5.429, 6.34, 6.57, 6.191
  wars and defence 7.59
Srikanthayya, B. M. 5.136
Srivijaya 3.339
SS (*Schutzstaffel*) 4.327
SSMs *see* surface-to-surface missiles
stability (atmosphere) 1.318–19
stabilization policies 7.298
stabilizer (food) 6.140
stabilizer (ship) 6.330
stable equilibrium 8.47
stable fly 2.324
staccato 5.429
stack (geomorphology) 1.318, 1.319
stacking system (aircraft) 6.330
Staël, Madame de 5.429
Stafford Smith, John 5.313
stag beetle 2.324
stage design *see* set design
stage machinery 5.429–30
stage-coach 6.61
stage-discharge 1.319
Stagnelius, Erik Johan 5.404
stag's horn fern 2.324
stained glass 3.18, 5.67, 5.87, 5.350, 5.389, 5.430
Stainer, Jacob 5.477
stainless steel 6.188, 6.330
  chromium in 6.73
  cutting 6.221
  developed 6.11, 6.46
  instruments 6.193, 6.340
  niobium in 6.241
stairs: escalator 6.123
stalactites and stalagmites 1.180, 1.181, 1.318, 1.319
Stalin, Josef 4.327
  and collectivization 4.78
  and COMECON 4.79
  and Comintern 4.79
  dictatorship 7.90
  and East Germany 4.287
  and labour camps 7.64

Stalin (*cont.*):
paranoia **7**.228
and Poland **4**.271
and Potsdam Conference **4**.276
purges **4**.54, **4**.178, **4**.184, **4**.244, **4**.277, **4**.284, **4**.355, **7**.255
and Stalingrad **4**.328
and Teheran Conference **4**.340
totalitarianism **7**.313
and Yalta Conference **4**.385
Stalingrad, battle of ***4**.327–8*
Stalwarts ***4**.328*
Stamboliyski, Alexander **4**.55
stamen ***2**.324*
Stamford Bridge, battle of ***3**.339*
Stamitz, Johann and Carl **5**.28, ***5**.430*
Stamp Act ***3**.339*
stamp collecting ***7**.298*
stamping tube ***5**.430*
standard deviation ***1**.319*
standard of living **7**.4–5, ***7**.299*
see also poverty; wealth
standard temperature and pressure ***1**.320*
standing waves **1**.319, ***1**.320*
Stanford, Sir Charles Villiers ***5**.430*
Stanford, Leland **4**.68
Stanhope, James, 1st earl ***3**.339–40*
Stanislaus II, king of Poland ***3**.340*
Stanislaus Leszczynski **3**.29, **3**.211, **3**.257, **3**.287
Stanislavsky, Konstantin: and method acting **5**.3, **5**.185, ***5**.430–1*
Stanley, Lord (1485) **3**.47
Stanley, Edward *see* Derby, Edward
Stanley, Sir Henry Morton ***1**.320*, ***4**.328–9*
and Congo **4**.33
and Livingstone **4**.198
as MP **1**.319
and Sudan **4**.112
as travel writer **5**.460
and Uganda **4**.356
and Zaïre **4**.388
Stanley crane (bird) ***2**.324*
Stanton, Edwin **4**.175
Stanton, Elizabeth Cady **4**.117
Stanwix, John **3**.128
Stanwix, Fort ***3**.128*
stanza ***5**.431*
star *see* stars
Star Chamber ***3**.340*
star of Bethlehem ***2**.324*
'Star Wars' *see* Strategic Defense Initiative
starch ***2**.324*, ***6**.330*
cassava **2**.56
sago **2**.292
taro **2**.335
starfish **2**.82, **2**.324, **2**.325
stargazer (fish) **2**.105
Stark, Freya **5**.460
Stark, John **3**.40
Starley, James **6**.36, **6**.384
Starley, John **6**.35
starling family **2**.215, **2**.224, ***2**.324*
Starr, Ellen Grates **4**.2
Starr, J. W. **6**.342
Starr, Ringo **5**.39
stars/stellar ***8**.158*
age of **8**.2
AM Herculis **8**.4
association **8**.159
astrology **7**.20–1
atlas **8**.22
black hole as final state **8**.17
brightness *see under* luminosity
catalogues **8**.21–2
chart ***8**.158–9*
circumpolar **8**.26
clock **8**.26
clusters *see* clusters of stars
degenerate **8**.36
dynamics and kinematics ***8**.159*
evolution ***8**.159–61*
stars/stellar (*cont.*):
flamsteed numbers **8**.51
flare **8**.51
giant *see* giant stars; supergiants
Gould's Belt **8**.59–60
heliacal rising and setting **8**.64
Hertzsprung–Russell **8**.64, ***8**.66–7*, **8**.160
late-type **8**.83
lists of brightest and nearest **8**.199
Local System **8**.59–60
magnitude *see* magnitude
mass ***8**.161*
motion **8**.128, ***8**.161*
neutron **8**.107
nova *see* nova; supernova
parallax **8**.116
paths **1**.222
populations ***8**.161*
proto-star **8**.129
shell **8**.149
spectrum *see* astronomy *under* spectra
structure ***8**.161*
substellar point **8**.163
symbiotic **8**.167
T Tauri **8**.65, **8**.175
temperature **8**.43, ***8**.161*
triple **8**.173
twinkle **8**.145
Wolf–Rayet **8**.187
see also astronomy; binary stars; dwarf stars; variable stars
starter motor ***6**.330*
startle coloration ***2**.324–5*
starvation *see* famine
Stasi **7**.238
state ***7**.299*
and anarchism **7**.12–13
authoritarian **7**.23
confederation **7**.64
see also federalism
employees *see* civil service
great power **7**.135
head of *see* heads of state
hegemony **7**.141
jurisdiction **7**.173–4
national security **7**.212
nationalism *see* nationalism
nationality **7**.212, **7**.213
nation-state **7**.212
neutral **7**.82, **7**.214
one-party **7**.223
police **7**.238
power *see* power (political and social)
republic **7**.265–6
sovereignty **7**.296
states' rights doctrine ***4**.329*
statism ***7**.299*
terrorism ***7**.299*
theory **7**.188
totalitarianism **7**.16, **7**.125, **7**.313
treason **7**.315
unitary **7**.320
see also communal conflict; constitutions; government; monarchy; politics *and* politics and government *under individual countries*
state of matter ***1**.320*
see also gas; liquid; liquid-crystal; plasma; solid
States-General ***3**.340*
statholder ***3**.340–1*
static electricity **1**.102, **1**.194
static equilibrium **8**.47
static universe **8**.34, **8**.159
statics **6**.138
stationary point ***8**.159*
stationary waves **1**.319, **1**.320
stationery *see* paper
statistics ***1**.320*
actuary **7**.2
correlation **1**.77, **7**.70
econometrics **7**.100
statistics (*cont.*):
hypothesis **1**.163
median **1**.209–10
mode **1**.217
and psychology **7**.129
representation *see* graphs
standard deviation **1**.319
statistician **2**.246
stochastic processes **1**.220, **1**.320
variables **1**.354
variance **1**.354
vital **7**.326
Statius **3**.18
statues *see* sculpture and carving
status ***7**.299*
statute law **7**.41, ***7**.299*
see also legislation
Staufer, Johann Georg **5**.23
Stauffenberg, Count Berthold von **4**.177
Stauffenberg, Claus Graf von **4**.177
staurolite ***1**.320*
stave **5**.322–3
Stead, Christina ***5**.431*
Steady State theory ***8**.159*
Stealth Bomber **6**.224
steam power ***3**.341*, ***4**.328*, ***4**.329*
boiler *see* boiler
cranes **6**.91
distillation ***6**.330*
locomotives **6**.20, **6**.294, **6**.311, **6**.330, ***6**.331–2*, **6**.334, **6**.345
see also railways
pump **6**.227
steam-engines ***6**.330–1*
beam engines **6**.32, **6**.241, **6**.330
Gramme dynamo **6**.158
pioneers **6**.43, **6**.93, **6**.123, **6**.241, **6**.363, **6**.380
triple-expansion **6**.33, **6**.263, **6**.330, **6**.363
steam-hammer **6**.236, ***6**.331*
steam-powered vehicles **6**.234, **6**.300, ***6**.332*, **6**.359
see also locomotives; steam-engines *above and* steamships *below*
steamroller **6**.300
steamships **4**.350, **6**.308, **6**.330, ***6**.332–3*
battleship **6**.32
first steam-turbine **6**.258
liner **6**.203
merchant ships **6**.219
see also under shipping *under* Britain
steam-turbine **6**.25, **6**.114, **6**.168, **6**.196, **6**.280, ***6**.333*, **6**.333
see also steamships *above*
steatite ***1**.320*
steel band ***5**.431*
Steele, Sir Richard ***5**.431*
Steen, Jan ***5**.431*
Steer, Philip Wilson ***5**.431*
steering system ***6**.333–4*
steering-oar ***6**.333*
Stefan, Joseph: and Stefan's Law ***1**.320*, **8**.159
Steichen, Edward Jean **5**.347, ***5**.431*
Stein, Andreas **5**.348
Stein, Gertrude ***5**.431*
Stein, Heinrich Friedrich Karl, Baron von ***4**.329*
Steinbeck, John **1**.97, ***5**.431*, **5**.432
Steiner, George **7**.78
Steiner, Rudolf **5**.43, ***7**.299*
Stella, Frank ***5**.431*
stellar *see* stars
stem ***2**.325*
see also rhizome
stencil **6**.310, ***6**.334*
Stendhal ***5**.431–2*
stenotypy **6**.318
Stepanov, Vladimir **5**.123
Stéphan, Edouard **8**.161
Stephan Dushan, king of Serbia **3**.322, ***3**.341*
Stéphan's Quintet ***8**.161*, **8**.162
Stephen, Saint (d.35) **3**.223, **7**.194, ***9**.43*
Stephen I, Saint, king of Hungary (975–1036) **3**.216, ***3**.341*
Stephen, king of England **3**.255, ***9**.43*, ***3**.341*
Stephens, Alexander H. **4**.82
Stephens, James (1824–1901, Irish patriot) **4**.118, **4**.167
Stephens, James (1882–1950, poet) **5**.226
Stephenson, George **4**.281, **6**.293, **6**.332, ***6**.334*
Stephenson, Robert **6**.293, **6**.331–2, **6**.334
stepper motor ***6**.334*
steppes ***2**.325*
art **5**.409–10
stereochemistry ***1**.320*
stereophonic reproduction **6**.40, **6**.159, ***6**.334*
stereophotography **6**.215, ***6**.334*, **6**.340
stereoscopic microscope **6**.223
stereotaxic surgery ***6**.334*
stereotype (printing) ***6**.334*
stereotype (psychology) ***7**.299*
sterile environment **6**.193
sterilization (medical equipment) **6**.22, ***6**.334*
sterilization (reproductive) **6**.87, ***6**.334*, **7**.115
Sterkfontein ***3**.341*
Stern, Otto: Stern–Gerlach experiment ***1**.320*
Stern, William **7**.32
Stern Abraham: Stern Gang ***4**.329*
Sternberg, Josef von ***5**.432*
Sterne, Laurence ***5**.432*, **5**.460
sternum **2**.284, ***2**.325*
steroids ***2**.325*, ***6**.334*
anabolic **6**.170
bile acids **2**.29
cholesterol **2**.64, **2**.325
drugs **6**.177
vitamin D **2**.357
see also hormones
stethoscope **6**.193, **6**.218, ***6**.335*
Stevens, Siaka **4**.312
Stevens, Thaddeus ***4**.329*
Stevens, Wallace ***5**.432*
Stevenson, Adlai ***4**.329*
Stevenson, Sir John **5**.303
Stevenson, Robert Louis ***5**.432*
Steward Observatory ***8**.161–2*
Stewart, Balfour **8**.74
Stewart *see* Stuart
Stewart Island ***1**.320*
stibnite **6**.16
stick insects **2**.50, ***2**.325*
stickleback ***2**.325*
Stickney ***8**.162*
Stieglitz, Alfred **5**.326, **5**.347, ***5**.432–3*, **5**.435, **5**.488
stifftail ***2**.325–6*
Stifter, Adalbert **5**.175
stigma ***7**.299*
Stijl, De **5**.301, ***5**.433*
Stilicho **3**.7
still life ***5**.433*
China **5**.96
France **5**.58, **5**.87, **5**.89, **5**.370
Spain **5**.470
United States **5**.342
see also Cubism; Fauvism
stilt ***2**.326*
Stilwell, Joseph Warren ***4**.329*
Stimson, Henry Lewis ***4**.329*
stimulus, conditioned **2**.185
stingrays **2**.261–2, ***2**.326*
stink bug **2**.307
stinkhorn ***2**.326*
Stirling, James **5**.363, ***5**.433*

Stirling, Robert: Stirling engines *6.335*
Stirner, Max 7.103
stitch *6.335*
STM (scanning tunnelling microscope) 6.223
stoat *2.326*
stochastic processes 1.220, *1.320*
stock (geology) *1.320*
stock (security) *7.299*
stock exchange *7.299–300*
Stock Market Crash (1929) *4.330*
Stockhausen, Karlheinz 5.382, *5.434*
stocking frame 6.191
Stockmar, Baron 4.8
stocks and pillory 3.282
stoichiometry *1.320–1*
Stoics/stoicism 3.117, *3.341–2*, *7.300*
  see also Brutus, Marcus; Cato; Cicero; Marcus Aurelius; Seneca; Zeno of Citium
Stoker, Bram 5.309, 9.10
Stokes, Sir George Gabriel *1.321*
Stokowski, Leopold 5.110
STOL (Short Take-Off and Landing) aircraft 6.25, 6.225, *6.335*
Stolypin, Piort Arkadevich 4.295, *4.330*
stomach 2.290, *2.326*
Stone Ages *3.342*
  art *5.434*
  technology 6.357
  see also Mesolithic; Neolithic; Palaeolithic
stone crab *2.327*
stone fish *2.327*
stone (geology) *1.321*
  as building material *see* masonry
  prehistoric monuments 6.218, 6.335
  quarry 6.287
  road 6.300
  stone-carving 5.25, 5.141, 5.275, *5.434*
    see also sculpture and carving
  tracery 5.459
  see also geology; masonry; rocaille; sculpture and carving
stone (in body organs) 2.131–2, *2.326*
stonechat *2.326–7*
stonecrop *2.327*
stone-curlew *2.327*
stonefly *2.327*
Stonehenge *3.342*, *6.335*
stones *see* stone
stoneware *5.434*
stonework *see* masonry
Stopes, Marie Charlotte Carmichael *4.330*
Stoppard, Tom 5.317, *5.434–5*
storage:
  data (computer) 6.137, 6.183
  energy 6.121, 6.140, 6.286
  grain 6.158
  heater 6.167, *6.335*
  petroleum 6.265
  pumped 6.286
  rings 6.258
  water 6.96, 6.297, 6.379
Storasta, Vilius (Vydūnas) 5.34
Storey, David 5.383
stories and tales *5.417*
  Arabic 5.19, 5.454
  Argentina 5.53, 5.114
  Austria 5.243, 5.310
  Baltic 5.34
  Britain 5.41, 5.128, 5.248, 5.286, 5.398, 5.486, 5.490
  Buddhist 5.238, 9.19
  Burma 5.67
  for children *see* literature *under* children
  China 5.95, 5.270, 5.370
stories and tales (*cont.*):
  Colombia 5.169–70
  Czechoslovak 5.120, 5.201, 5.213
  Denmark 5.16, 5.49
  detective 5.129, 5.134, 5.353–4, 5.420
  Dravidian 5.134
  Finland 5.155
  France 5.34, 5.286, 5.291, 5.372
  Gaelic 5.226, 5.409
  Germany 5.51, 5.205, 5.279
    for children 5.194
    private myth 5.243
    psychological 5.499
    Romantic 5.208
    tragic 5.249
  Hebrew 5.9–10
  India 5.129, 5.220, 5.312, 5.338, 5.374, 5.401, 5.466, 9.21–2
  Ireland 5.225, 5.326
  Italy *5.231*, 5.342, 5.350
    bourgeoisie 5.34
    narrative 5.50–1
    pagan 5.124
    psychological 5.72
    rural life 5.471
  Japan 5.235, 5.445
  New Zealand 5.280
  Norse sagas 5.397
  Pakistan 5.466
  parables 5.339
  Persia 5.345
  Poland 5.406
  Romanticism 5.390
  Russia and USSR 5.90, 5.184, *5.396*
    fantasy 5.65
    Romantic 5.370
    rural life 5.182, 5.457
    urban life 5.182–3
    war 5.28, 5.65
  South Africa 5.184
  South and Central America 5.256
  South-East Asia 5.427
  Sufi 5.25, 5.439
  Switzerland 5.139–40
  United States 5.43, 5.76, 5.117, 5.201, 5.267, 5.326, 5.371, 5.398
    black 5.495
    burlesque 5.227
    children 5.200
    expatriate 5.204, 5.232–3
    horror and grotesque 5.325, 5.353–4
    Jazz Age 5.156
    Jewish and Yiddish 5.277, 5.391, 5.420
    maritime 5.290
    middle America 5.82, 5.262, 5.466
    New York 5.90
    psychological 5.359
    rural life 5.431
    satire and humour 5.90, 5.227, 5.262, 5.339
    science fiction 5.479
    southern states 5.71, 5.99, 5.152–3, 5.325, 5.486
  Wales 5.486
  see also fables; lays; myths and legends; novels; parables; science fiction
stork family 2.150, 2.202, *2.327*, 2.365
Storm, Theodor 5.175
storm *1.321*
  Beaufort scale 9.56
  blizzard 1.40
  cyclone 1.83, 1.344, 1.345
  hurricane 1.160
  storm-water drainage 6.314, *6.335*
  thunderstorm 1.289, 1.335
  tornado 1.340, 1.361
  typhoon 1.348
  waterspout 1.361
storm (*cont.*):
  see also tropical revolving storm
storm petrel *2.326*
storm surge *1.321*
Stormont *4.330*
stove *see* cookers and stoves
Stowe, Harriet Beecher 4.1, 4.33, *5.435*
Strabo 8.48
Strachey, Lytton *5.435*
Stradivari, Antonio 5.84, 5.477
Strafford, Thomas Wentworth, 1st earl of *3.342*
strain gauge *6.335*, 6.383
straits *1.322*
  Cook 1.75
  Magellan 1.201, 1.202
  Messina 1.211–12
  Skagerrak 1.306
Straits Settlements *4.330*
Stramm, August 5.175
Strand, Paul *5.435*
strand line *1.322*
strangeness *1.322*
Strasser, Captain Valentino 4.312
Strassmann, Fritz 1.149–50, 1.210, 6.242
strata 1.322
Strategic Arms Limitation Talks *4.330*, 6.15, 7.17, 7.87
Strategic Defense Initiative *4.330*, 6.15, *6.335–6*, *7.300*
strategic weapons 6.246, 7.220
Strathclyde and Cumbria (kingdom) *3.15*, *3.342*
Strathcona Provincial Park *1.322*
stratification (geology) 1.35
stratigraphy 1.136, *1.322*
Stratius 5.257
stratopause *1.322*, *8.162*
stratosphere 1.109, *1.322*, *8.162*
stratus and stratocumulus clouds 1.66, 1.128, *1.323*
Strauss, Botho 5.175
Strauss, Eduard 5.435
Strauss, Franz Josef 4.32
Strauss, Johann (father and son) 5.270, 5.355, *5.435*
Strauss, Josef 5.435
Strauss, Richard 5.209, 5.329, 5.367, 5.407, *5.435–6*
Stravinsky, Igor 5.13, 5.315–16, 5.407, *5.436*, 5.439
  influence 5.112, 5.211, 5.284, 5.316, 5.364, 5.421, 5.443
strawberry *2.326*
strawberry tree *2.326*
straw-work *5.436*
stream of consciousness 5.242, *5.436*, 5.493, 7.169
streamline (meteorology) *1.323*
streamlining 6.5, 6.135, *6.336*
streams *see* rivers
street theatre 5.108
street-games *7.300*
Streicher, Julius *4.330*
Streisand, Barbra 5.358
Strelitzia 2.31
strepsipteran *2.327*
streptomycin 6.16, 6.375
Stresa Conference *4.330–1*
Stresemann, Gustav *4.331*
stress (mental) *7.300–1*
striations *1.323*
strict liability *7.301*
strike (geology) 1.323
strikes 7.158
  hunger-strike 7.150
  see also unions *and also* unions *under individual countries*
strike-slip (tear) fault 1.120, *1.323*
Strindberg, August *5.436*
string theory *1.323–4*
stringed instruments *5.436–7*
  aeolian harp 5.5
  arpeggione 5.23
stringed instruments (*cont.*):
  balalaika 5.29
  baryton 5.37
  bow 5.56
  bow, musical 5.56
  bow harp 5.56
  bumbass 5.65
  chamber music 5.88
  crwth 5.119
  double bass 5.134
  hurdy gurdy 5.214, 5.215
  lira 5.266
  mandolin 5.278
  pitch 5.352
  plucked *see* plucked instruments
  rebab 5.376
  rebec 5.376
  tromba marina 5.460
  ukulele 5.465
  vīṇā 5.476
  viol 5.437, 5.476
  viola 5.352, 5.476–7
  see also cello; fiddle; harps; lutes; pianoforte; violin
stripped land surface *1.324*
strips (land) *3.342–3*
stroboscope *6.336*, 6.381
Strode, William 3.127
Stroessner, Alfredo 4.259, *4.331*, 7.23
Stroheim, Erich von *5.437*
stroke 2.242, *2.327*
stromatolite *2.327–8*
Strömgren, Bengt: Strömgren sphere *8.162*
Strong, John 3.123
strong force *1.324*
strontianite 6.336
strontium 1.107, *1.324*, 6.96, 6.125, 6.290, *6.336*
Strowger, Almon 6.347
structural engineering *6.336–7*
  see also bridges; building; crane; conveyor; frame
structural isomerism 1.175
structuralism 7.14, 7.183, *7.301*
structure, geological *1.324*
Strutt, John William (Lord Rayleigh) 1.275, *1.324*, 6.303
Strutt, Robert John (Lord Rayleigh) *1.324*
Struve, Friedrich, Georg, Hermann, Ludwig, Otto, and Otto Wilhelm *8.162*
Stuart, Charles Edward (Bonnie Prince Charlie) 3.93, 3.129, 3.291–2, 3.343
Stuart, Herbert Ackroyd 6.102
Stuart, James (1713–88, architect) *5.437*
Stuart, James Ewell Brown (1833–64, general) *4.331*
Stuart, James Francis Edward (1688–1766, Old Pretender) 3.126, 3.128, 3.292, 3.343
Stuart, John McDouall *1.324*, 4.247
Stuarts *3.343*
  monarchs listed *9.67*
  Restoration 3.302
  see also Stuart, Charles; Stuart, James Edward *and under* rulers *under* Britain
Stubbs, George *5.437–8*
stucco 5.344, 5.351, 5.366, 5.387, *5.438*
student revolts *4.331–2*, 7.215
studio potter 5.258, *5.438*
stupa *5.438*
Štúr, Ľudovít 5.422
sturdy beggars *3.344*
sturgeon *2.328*
Sturm und Drang 5.181–2, 5.405, *5.438*
Sturt, Charles *1.324*
Stuttgart Ballet 5.122
Stuyvesant, Peter *3.344*

styrene 6.271, 6.304
  see also polystyrene
Styx 9.34
Su Dongpo 5.438
Su Shi see Su Dongpo
Su Song 6.76, 6.340
Subandhu 5.401
subatomic particles see electrons; elementary particles; neutrons; protons
subduction 1.324–5
subdwarf 8.163
subgiants 8.2, 8.3, 8.59
sublimation (chemistry) 1.325, 6.186
sublimation (psychology) 7.82
Sublime 5.438
sublunar point 8.163
sub-machine-gun 6.130, 6.131, 6.337
submarine bomb 6.99
submarine cable 6.53, 6.347
submarine canyon 1.325
submarine exploration see diving
submarine landforms see physical geography under oceans and seas
submarines 6.323, 6.337
  nuclear-powered 4.235, 6.243, 6.245, 6.337
  periscope 6.260
  snorkel 6.323
  sonar 6.325
  torpedo 6.337, 6.357
  see also warships
submersible 6.31, 6.33, 6.97, 6.337
subsidence (engineering) 6.314
subsidence (geology and meteorology) 1.325
subsidy 7.301
subsistence farming 7.232, 7.284
  Africa 7.19, 7.94, 7.151, 7.176, 7.199, 7.285
  Asia and Pacific 7.25, 7.202, 7.215, 7.231, 7.240, 7.312
  Central America 7.306
  New Zealand 7.192
  North American Indians 7.14, 7.74, 7.147, 7.150, 7.166, 7.221, 7.236
  rice 7.268
  see also shifting cultivation
subsolar point 8.163
substance (philosophy) 7.199, 7.301
substellar point 8.163
substitution reaction 1.60
subtropical high 1.325
succession 2.328
succulent plants 2.328–9
  aloe 2.7
  baobab 2.22
  cacti 2.49, 2.251
  cotyledon 2.78
  mesembryanthemum 2.208
  spurge see spurges
  stonecrop (Sempervivum) 2.327
Suchinda Kraprayoon, General 4.342
sucker-footed bat 2.329
sucking lice 2.329
Suckling, Sir John 5.82
Sucre, Antonio José de 4.44, 4.108, 4.267, 4.332
sucrose 6.337
Sudan 1.325, 4.332, 7.382
  aid to 7.121
  anthropology 7.112
  in Arab League 7.15
  colonialism 4.4, 4.306
  ethnic groups 3.259–60, 7.24, 7.185, 7.220
  famine 7.117
  kingdom of 3.136
  languages 7.78, 7.217
  laws/legal system 7.284
Sudan (cont.):
  Muslim Brotherhood 7.166
  physical geography and geology 1.229, 1.325
  political leaders 4.181, 4.243, 4.280
  refugees 7.260, 7.261
  religion/theology 7.70, 7.284
    myths 9.21, 9.22
  wars and defence 4.280
  see also under Kitchener
Sudermann, Hermann 5.175
Sudetenland 4.332
Sudety 1.325–6
Suetonius, Gaius Suetonius Tranquillus 3.344, 5.48
Suetonius Paulinus 3.47, 3.106
Sueves 3.98
Suez:
  Canal 4.332, 6.76, 6.199, 6.337, 6.337, 6.342, 6.345
  Gulf of 1.325, 1.326
  War 4.332, 6.193, 6.345
suffrage 7.123–4, 7.332
  movements 3.209, 4.159
    for women (suffragettes) 4.258, 4.333, 4.381, 7.123, 7.332
  see also electoral system; voting
Sufism 3.236, 3.344, 7.166, 7.301
  literature 5.19, 5.25, 5.197, 5.233, 5.394, 5.397, 5.438–9
  monasticism 7.205
  saints 7.272
sugarbird 2.329
sugars 2.329, 6.140, 6.141, 6.338
  confectionery 6.86–7
  crystal 1.81
  and DNA 2.95
  glucose 2.138
  growing 6.7, 6.271
  nucleotide 2.230
  polymer 6.66
  processing 6.259, 6.338
  sugar cane 2.329
  sugar-beet 2.329, 6.302
  types of 6.337
  see also monosaccharides
Suger of St Denis, Abbot 5.177, 5.185
Sugriva 9.15
Suharto 4.333
Suhrawardy, H. S. 4.26
Sui dynasty 3.344
suicide 7.97, 7.138, 7.275, 7.301–2
suite (music) 5.439, 5.442
Suits, Gustav 5.34
Suk, Josef 5.439
Sukarno, Achmad 4.163, 4.208, 4.333
Sulawesi (Celebes) 1.326, 3.214
Sulayman (Solomon) 9.34
Suleiman I, Ottoman sultan 3.236, 3.302, 3.344, 3.345
Suleiman II, Ottoman emperor 3.236
sulfur see sulphur
Sulla, Lucius Cornelius 3.27, 3.90, 3.282, 3.308, 3.314, 3.344
Sullivan, Sir Arthur 5.439
Sullivan, Harry Stack 7.214
Sullivan, Louis 5.6, 5.14, 5.91, 5.439, 5.494–5
Sully, Maximilien de Béthune, duc de 3.345
Sully-Prudhomme, René 5.340
sulphates 1.326
  ammonium 1.326
  calcium see calcium sulphate
  copper 1.59, 1.76, 6.89, 6.147
  fertilizer 6.2, 6.13, 6.338
  in medicine 6.30
  in paint 6.255
  potassium 1.326
  potassium aluminium 1.10
  in water 6.378
  zinc 1.373, 6.391
sulphides 1.326
  copper 6.253
  iron 1.205, 1.250, 1.270
  lead 1.130, 1.191, 6.197, 6.283
  mercury 1.63
  nickel 1.250
  silver 1.19
  zinc 1.373, 6.253, 6.391
sulphites 1.326
sulphonamide 6.105, 6.338
sulphur 1.107, 1.326, 6.338
  in battery 6.32
  crystal 1.80
  extraction 6.146, 6.232
  as fungicide 6.147
  in gunpowder 6.162
  in petroleum 6.262, 6.263, 6.265
  and rubber see vulcanization
sulphur dioxide 1.326–7
  bleach 6.40, 6.338
  and contact process 6.87
  pollution 6.2, 6.275, 6.338, 7.2, 7.239
sulphur trioxide 6.87
sulphuric acid 1.327, 6.338–9
  in accumulator 6.1
  fuming (oleum) 1.237
  in nitro-glycerine 6.241
  production 6.2, 6.70, 6.71, 6.87, 6.197, 6.262, 6.267, 6.338–9
sultanas 2.33, 2.84, 2.143, 2.279
sultanate 3.345
Sulzer weaving machine 6.208
sumacs 2.261, 2.329
Sumatra 1.327, 3.345
  Acheh and Acheh Wars 3.2, 4.2
  decolonization 4.321
  ethnic group 7.202
  port 3.339
Sumerians 3.141, 3.345
  arts 5.305, 5.439–40
  cities 3.367, 3.368
  religion and myths 9.11, 9.14, 9.35
  technology 6.62, 6.311, 6.327
summer 1.297
  solstice 8.163
  triangle (astronomy) 8.163
summoner 3.345
Sumner, Charles 4.333
Sumner, William Graham 7.157
sumo wrestling 7.333
Sumter, Fort 4.122
Sumuabum 3.32
Sun/solar 8.163–5
  activity 8.163
  apex 8.6
  apex of Sun's way 8.6
  atmosphere 8.26, 8.29, 8.51
  constant 8.150
  corona 1.77
  cycle 8.165
  day 8.26
    mean 8.36, 8.109
  eclipse 8.29, 8.40, 8.42
  energy see solar power
  equinox see equinoxes
  faculae 8.50
  filament 8.50
  flare 8.51
  granulations 8.60
  halo 1.150, 1.151
  heliocentric system 8.64–5
  heliosphere 8.65
  hydrogen plasma 1.259
  illustrated 8.164
  insolation 8.73
  luni-solar precession 8.127
  motion 8.150
  parallax 8.150–1
  power see solar power
  prominences 8.128, 8.129
  radiation 1.169, 1.312, 6.166, 8.73, 8.150
    albedo 1.6
    energy budget 1.108, 1.109
Sun/solar (cont.):
  radiation (cont.):
    insolation 1.169
    see also solar power
  solstice 1.297, 1.312
  spectrum 8.151
  subsolar point 8.163
  sunspots see sunspots
  sunstroke 2.330
  symbol 9.62
  system see solar system
  telescope 8.96
  time 1.286, 8.172
  wind 8.151
  year, mean 8.189
sun bittern 2.329
Sun Houzu 9.15
sun spider 2.330
Sun Yat-sen 4.74, 4.187, 4.334, 4.352
sunbeam snake 2.329
sunbird 2.329
Sunday School 3.298, 4.333
Sundbach, Gideon 6.391
Sunderland, Robert Spencer, 2nd earl of 3.345
sundew 2.329, 2.330
sundial 6.76, 8.165
Sundjata Keita 3.141, 3.217, 3.345–6
sunfish 2.24, 2.329–30
sunflower family 2.330
  artemisia 2.15
  artichoke 2.16
  chrysanthemum 2.65
  cornflower 2.77
  daisy 2.86
  dandelion 2.87
  elecampane 2.105
  marigold 2.202–3
  ragwort 2.275
  salsify 2.292
  scorzonera 2.297
  senecio 2.303
  thistle 2.339
  zinnia 2.375
Sung (Song) dynasty 3.333
  arts 5.92–6
Sunnī Islam/Sunnites 3.346, 7.166, 7.302
  'āshūrā' 7.19–20
  āyātollāh 7.24
  Berbers 7.30
  butterfly diagram 8.18, 8.164
  and caliphs 7.10
  cycle 8.165
  imām 7.153
  Kurds 7.178
  law 7.282
  Wahhābī 4.369–70, 7.328
  see also Islam/Muslims
sunspots 8.165
  Evershed effect 8.48
  Maunder minimum 8.96
sunstroke 2.330
Suomi see Finland
super proton accelerator 6.258
Superbus of Rome 3.350
supercharger 6.339
Supercluster, Local 8.86
supercomputer 6.85, 6.339, 7.19
superconductivity 1.327, 6.114, 6.116, 6.241, 6.339
supergiants 8.14, 8.50, 8.165
  in binary stars 8.6
  first-magnitude 8.20, 8.37
  Hubble–Sandage 8.71
superheterodyne reception 6.291
Superior, Lake 1.145
superior conjunction of planets 8.122
superior planet 8.165
supermarket 7.302
supernatural 7.297–8, 7.302
  see also spirits
supernova 8.165–6
  remnants 8.35–6, 8.166, 8.167

supernova (*cont.*):
  Supernova 1987A ***8.166***
superoxide ***1.327***
superphosphates **6.**344
superplasticity ***6.339***
superpower ***7.302***
  see also Union of Soviet Socialist Republics; United States
supersonic flight **6.**5, **6.**138, **6.**139–40, ***6.339–40***
superstition ***7.302***
Supervia, Conchita ***5.440***
Suppiluliumas, king of Hittites **3.**166
supply (economics) **7.**193, ***7.302***
  and demand, law of ***7.302–3***
  marginal analysis **7.**192
Supremacy, Acts of ***3.346***
Suprematism **5.**266, **5.**277, ***5.440***
Supreme Court (Federal) ***4.334***
Surdas **5.**206
surface gravity ***8.166***
surface temperature ***8.166–7***
surface tension and surfactants ***1.327***, **6.**11, **6.**339, ***6.340***
surface-to-air missiles **6.**229, **6.**230
surface-to-surface missiles **6.**230, **6.**231
surfactant *see* surface tension
surfing ***7.303***
surgeonfish **2.**353
surgery ***6.340***
  anaesthetics *see* anaesthesia
  asepsis **6.**22
  bone-marrow transplants **6.**164
  cardiac **6.**30, **6.**60, **6.**361
  dentistry *see* dentistry
  diathermy **6.**101, **6.**251
  endoscopy **6.**37, **6.**120, **6.**248, **6.**251, **6.**340
  grafting **6.**158
  heart transplant **6.**30, **6.**361
  heart-lung machine **6.**156, **6.**166, **6.**218
  hip replacement **6.**168, **6.**169
  history of **6.**178, **6.**218
  instrument **6.**340
  laparoscopy **6.**248, **6.**340
  laser **6.**99, **6.**251, **6.**340
  lithotripsy **6.**205, **6.**340, **6.**367
  microsurgery **6.**224, **6.**254
  neurosurgery **6.**240–1
  operating theatre **6.**250
  ophthalmic **6.**250–1
  orthopaedics **6.**253
  otorhinolaryngology **6.**254
  stereotaxic **6.**334
  sterilization (reproductive) **6.**87, **6.**334
  sterilization of equipment **6.**22, **6.**334
  stitches (sutures) **6.**335
  surgeons **2.**28, **2.**144, **2.**191, **2.**266
  Susruta's treatise **6.**342
  transplant *see* transplant surgery
  trephining **6.**340, **6.**362–3
  ultrasound **6.**60, **6.**340
  see also diseases and disorders; medicine
surges ***8.167***
Suri dynasty ***9.66***
suricate **2.**206
Surinam toad ***2.330***, **2.**331
Surinam(e) ***1.327***, ***4.334***, ***7.382–3***
surplus value **7.**249, ***7.303***
Surrealism **5.**339, ***5.440–1***
  in cinema **5.**65–6
  in literature **5.**106, **5.**318
    French poetry **5.**18, **5.**59, **5.**86, **5.**145, **5.**317
  in painting **5.**275, **5.**292, **5.**313
    France **5.**87, **5.**148
    Spain **5.**121, **5.**298, **5.**348
    United States **5.**184, **5.**391
  in photography **5.**375
Surrey, Henry Howard, earl of ***5.441***
Surrey, Thomas Howard, earl of **3.**128
Surt **9.**13
Surtsey ***1.327–8***
surveillance ***7.303***
  freedom from **7.**247
surveying **6.**215, ***6.340***, **6.**341
  aerial **6.**268
  baseline **1.**34
  bench-mark **1.**36
  geology **1.**35
  theodolite **6.**168, **6.**294, **6.**340, **6.**341, **6.**353
  see also maps and charts
survival **7.**11
Surya ***9.34***
Suryavarman II, Khmer ruler **3.**13
Susan-o-o **9.**3, **9.**19, ***9.34***
Susenyos, emperor of Ethiopia **3.**119
suspended load ***1.328***
suspended structure ***6.340***, **6.**351
suspension (of solids in liquids):
  Brownian motion **1.**44
  centrifuge **6.**67
  filtration **6.**130
  fog **1.**124
  gel **6.**153
  haze **1.**152
  mist **1.**216
  suspended load ***1.328***
  see also aerosol
suspension bridges **6.**47, ***6.48***, **6.**311, ***6.341***, **6.**350
suspension system (vehicle) **6.**120, **6.**317, ***6.341***
Susquehanna ***1.328***
Susruta **6.**178, ***6.342***
Sutcliffe, Frank Meadow ***5.441***
Sutcliffe, Rosemary **5.**91
Sutherland, Edwin **7.**331
Sutherland, Graham **5.**359, **5.**362, ***5.441***
Sutherland, Dame Joan **5.**441
suttee **7.**75, **7.**301
Sutter, J. A. **4.**137
Sutton Hoo ***3.346***
Suvarov, Alexander Vasilevich, count Rimnikskу, prince Italysky **3.**65, ***3.346***
Svalbard ***1.328***
Svevo, Italo ***5.441***
Swabia ***3.346***
Swahili language **7.**185, **7.**217
swallow family ***2.330–1***
  house-martin **2.**164
  martin **2.**204
  sand martin **2.**293
  swallow **2.**330
swallow hole **1.**181
swallowing ***2.331***
swallowtail butterfly ***2.331***
swamps and marshes ***1.206***, ***1.328***
  Belarus **1.**268
  birds *see* birds *under* water
  bog *see* bogs
  Everglades **1.**116
  fen **1.**120
  marsh buck **2.**311
  marsh deer **2.**204, ***2.204***
  marsh harrier **2.**204, ***2.204***
  marsh marigold **2.**179
  muskeg **1.**223
  salt-marshes **1.**291
  shott **1.**302
  see also wetlands
Swan, Sir Joseph Wilson **6.**177, ***6.342***
Swan (Omega) Nebula **8.**112, **8.**167
swan (bird) *see* swans
swan mussel ***2.331***
Swanland ***1.328***
swans ***2.331***, ***2.332***
  Bewick's **2.**28
  mute **2.**224
  trumpeter **2.**348
swans (*cont.*):
  whistling **2.**366
  whooper **2.**367
Swanscombe Man ***3.346***
SWAPO (South West Africa People's Organization) ***4.334***, **7.**130
swash ***1.328***
swastika ***4.334***
Swaziland ***1.328***, ***4.334–5***, ***7.383***
  colonization and decolonization **4.**4, **4.**306
  in SADCC **7.**296
sweat gland ***2.331***
swede ***2.331***
Sweden ***1.328–9***, ***3.346–7***, ***4.335***, ***7.383***
  adoption **7.**3
  aid from **7.**8
  arts **5.**24, **5.**44, **5.**165, **5.**254, **5.**404, **5.**436
    see also literature; music *below*
  astronomy **8.**3
  chemists **1.**20, **1.**38, **1.**294
  in Council of Europe **7.**71
  diplomacy **4.**36, **4.**146
  economy and trade **7.**210
    and EC **7.**99
    in EFTA **7.**111
    industrial relations **7.**315
    in OECD **7.**221
    pay **7.**107
  education **7.**277
  energy, renewable **7.**106
  environment **7.**2, **7.**140
  ethnic groups **7.**181
  explorers **1.**230, **1.**231
  government *see* politics and government *below*
  languages **7.**131, **7.**156
  laws/legal system **7.**62, **7.**75
  literature **5.**44, **5.**254, **5.**436
    poetry **5.**165, **5.**254, **5.**404, **5.**436
  medicine and health care **7.**140
  music **5.**48, **5.**49, **5.**265, **5.**320
  neutrality **7.**214
  parental leave **7.**228
  physical geography and geology **1.**328–9
    Fennoscandian Shield **1.**120
    Lapland **1.**188
    meteorology **1.**37–8
  physicist **1.**288
  politics and government **3.**266, **4.**317, **7.**223, **7.**262, **7.**332
  pollution **7.**2, **7.**140
  rulers (chronologically):
    listed ***9.67***
    before 16th century **3.**59–60, **3.**127, **4.**69–70
    16th to 17th centuries **3.**74, **3.**127, **3.**318, **3.**324
      see also Charles X; Charles XII; Christina; Gustavus I; Gustavus II
    18th and 19th centuries **3.**347, **4.**69–70, **4.**335
      see also Charles XII
    20th century **4.**335, **8.**125
  social services and security **4.**357, **7.**330
  sport and recreation **7.**145, **7.**297
  technology and science **6.**196, **6.**351
    chemistry **6.**242
    computers **6.**84
    energy and fuel **6.**89, **6.**385
    nuclear **6.**242, **6.**245
    pollution **6.**2
    textiles **6.**391
    vehicles and transport **6.**169, **6.**335
    see also Nobel *and also entries above prefixed* **1.** *and* **6.**
  wars and defence **3.**324, **7.**202
    with Denmark *see under* Denmark
Sweden (*cont.*):
  wars and defence (*cont.*):
    neutrality **7.**82, **7.**214
    see also Northern War; Seven Years War
Sweelinck, Jan Pieterszoon ***5.441***
sweeping **6.**61
sweet bay **2.**25
sweet corn ***2.332***
sweet pea **2.**68, ***2.332***
sweet pepper ***2.332***
sweet potato ***2.332***, **6.**302
sweetener **6.**140, ***6.342***
  see also sugars
sweets **6.**86–7
Sweyn Forkbeard, king of Denmark **3.**60, **3.**119, ***3.347***
swidden **7.**284
Swiddle, Willem **3.**76
Swift, Jonathan ***5.441–2***
swift **2.**80, ***2.332***
swift moth ***2.332***
swim bladder ***2.332***
swimming ***7.303***
Swinburne, Algernon Charles ***5.442***
swine *see* pigs
'Swing, Captain' ***4.335***
swing *see* jazz
swing bridge ***6.342***
swing-wing aeroplane ***6.342***
switch **6.**75, **6.**136, **6.**297, ***6.342***, ***9.53***
  switchgear ***6.343***
Switzerland ***1.329***, ***3.347***, ***4.335***, ***7.383–4***
  aid from **7.**8
  arts **5.**249
    painting **5.**166, **5.**175–6, **5.**244, **5.**249
    see also literature; music *below*
  astronomy **8.**87
  building **6.**213–14
  ceramics **6.**116
  chemistry **1.**364, **6.**96
  computers **6.**86
  in Council of Europe **7.**71
  economy and trade **7.**99, **7.**107, **7.**111, **7.**221, **7.**314, **7.**319
  education **7.**275, **7.**276
  energy and fuel **6.**95, **6.**249
  language **7.**131, **7.**269
    semiotics *see* Saussure
  literature **5.**111, **5.**139–40, **5.**165, **5.**245
  mathematicians **1.**18, **1.**38, **1.**101, **1.**114
  medicine and health care **3.**118, **6.**257
  music **5.**211, **5.**238
  myths **9.**35
  philosophy *see* Piaget; Rousseau
  physical geography and geology **1.**329
    lakes **1.**73, ***1.329***
    mountains **1.**10, **1.**19, **1.**179, **1.**207
    rivers **1.**282
  physicists **1.**31
  politics and government **7.**119, **7.**250, **7.**260, **7.**332
  prehistory **3.**203–4
  psychiatry *see* Jung
  regions (cantons) listed ***9.75***
  religion **3.**118, **3.**300, **3.**391, **9.**43
  Sonderbund **4.**319
  technology and science **6.**25
    see also entries above and below prefixed **1.** *and* **6.**
  telecommunications **6.**346
  urban areas **3.**139
  vehicles and transport **6.**20, **6.**209, **6.**365
  wars and defence **3.**22, **7.**50, **7.**66, **7.**202
    neutrality **7.**82, **7.**214
  watches **6.**377

swordfish *2.332*
swords **6**.165, **6**.226, *6.343*
  mythical **9**.12
swordtail *2.332*
sycamore *2.332*
Sydenham, Thomas *2.332–3*
syenite *1.329*
Sykes, Sir Mark: Sykes–Picot Agreement *4.335*
syllogism *7.303*
Sylvester I, pope **3**.86
Sylvester II, pope **3**.216
symbiosis *2.333*
  fern and cyanobacteria **2**.361
  fungi **2**.129
  lichen **2**.188, **2**.189, **2**.282
symbiotic star *8.167*
symbolic interactionism (Chicago School) **7**.44, **7**.278, *7.303*, **7**.393
  see also Goffman
symbolism **7**.77, **7**.268, *7.303*
  house **7**.94
  see also signs; symbolic interactionism; Symbolism in arts
Symbolism in arts *5.442*
  drama **5**.103, **5**.178, **5**.238, **5**.275
  novels **5**.368, **5**.499
  painting **5**.67, **5**.304, **5**.307–8, **5**.371, **5**.376
  poetry **5**.173, **5**.275, **5**.345, **5**.354, **5**.466
    France **5**.38, **5**.43, **5**.103, **5**.238, **5**.277–8, **5**.317, **5**.383, **5**.471–2
    Poland **5**.262, **5**.422
    Russia **5**.43, **5**.49–50, **5**.278, **5**.396
symbols:
  chemical **1**.38, **1**.60
  mathematics *9.50*
  meteorology **9**.56
  see also signs
Syme, James **6**.204
symmetry *1.329*
  symmetric gravitation **8**.60
sympathetic nervous system **2**.18
  see also nervous system
symphonic poem *5.442*
symphonies *5.442*
  Britain **5**.340, **5**.393, **5**.457
  France **5**.138, **5**.392, **5**.398
  Germany **5**.204, **5**.280, **5**.291, **5**.429, **5**.430, **5**.435
  Russia and USSR **5**.53, **5**.180, **5**.367, **5**.373, **5**.384, **5**.409, **5**.417, **5**.448
  see also classical music
synagogue *7.303–4*
synapse **2**.226
synapsid *2.333*
synchromesh **6**.152
synchronous (astronomy) *8.167*
  geosynchronous orbit *8.58*
synchronous motor **6**.115
synchrotron **6**.258
  radiation *8.168*
syncline **1**.125, **1**.137, *1.329*
syncopation **5**.382
syndicalism *4.335*, *7.304*
Synge, John Millington *5.442–3*, **9**.10
Synnott, S. P. **8**.98, **8**.171
synod **7**.48
synodic month (lunation) **8**.100, *8.168*
synodic period **8**.168
syntax **7**.134
synthesis (astronomy):
  aperture **8**.6
  nuclear **8**.110
  telescopes **8**.73
synthesis (chemistry) *1.329*
  reaction **1**.59
synthesis (knowledge) *7.304*
synthesizer *5.443*
  music *6.343*
synthetic adhesives **6**.4
synthetic (artificial) *see entries beginning* artificial
synthetic diamonds **6**.101
synthetic drugs *see* antibiotics
synthetic dyes **6**.14, **6**.79, **6**.109–10, **6**.112, **6**.260
synthetic fertilizer *see under* fertilizers
synthetic fibres *6.128*, **6**.261, *6.343*
  acrylic **6**.2–3, **6**.58, **6**.61
  canvas **6**.57
  carpet **6**.61
  nonwoven **6**.242
  nylon *see* nylon
  polyester **6**.275, **6**.302, **6**.384
  polymer *see* polymers
  rayon (viscose and acetate) **6**.66, **6**.92, **6**.128, **6**.295
  rope **6**.302
  spinning **6**.2, **6**.295, **6**.329
  yarn **6**.390
  see also carbon fibre; fibre-reinforced plastics; plastics
synthetic fluorescence and phosphorescence **6**.267
synthetic fuel *6.343*
synthetic hormones **6**.87
synthetic non-metal *see* plastics
synthetic polymers *see* polymers
synthetic resins *see* resin
synthetic rubber **6**.60, **6**.236, **6**.261, **6**.271, *6.303–4*
synthetic sweetener **6**.342
Synthetism *5.443*
syphilis **2**.126–7, *2.333*
  identified **2**.126–7
  test for **6**.177, **6**.192
  treatment **6**.72, **6**.112, **6**.257
Syria *1.329*, *3.347*, *4.335*, *7.384*
  aid to **7**.8
  in Arab League **7**.15
  Ba'athism *7.24*, **7**.223
  castle **3**.63
  city **3**.268
  ethnic groups **3**.316, **4**.188, **7**.178
  and Gulf Co-operation Council **7**.137
  Kurds **7**.178
  languages **7**.13
  law **7**.221, **7**.284
  Muslim Brotherhood **7**.166
  old age **7**.221
  Palestinian refugees in **7**.227
  physical geography and geology **1**.114, **1**.141–2, **1**.178, **1**.329
  religion/theology:
    Christianity **7**.214, **7**.304, **7**.394, **9**.42, **9**.43
    Islam **7**.96, **7**.168, **7**.284
    Syrian Orthodox Church *7.394*
  under Saladin **3**.314
  technology and science **6**.156, **6**.161, **6**.365
  in United Arab Republic **4**.358–9
  wars and defence **7**.148
Syringa **2**.190
Syrtis Major *8.168*
syrup from plants **2**.198, **2**.202, **2**.294
system building **6**.280, *6.343–4*
  see also prefabrication
system identification **6**.110
systems analysis *6.344*
systems engineering **6**.253, *6.344*
systems theory **7**.224
syzygy **1**.336–7, *8.168*
Szasz, Thomas **7**.251
Szechwan *see* Sichuan
Szymanowski, Karol *5.443*

T

T associations **8**.175
T Tauri stars **8**.65, *8.175*
al-Ta'a'ishi, Muhammad **4**.181
al-Tabaric **3**.166
Tabei, Junko **1**.116
Tabernacles, Feast of *7.305*
Tabinshweti of Burma **3**.359
tabla *5.444*
tablature *5.444*
table games *see under* games
Table Mountain *1.330*
table tennis *7.305*
tablelands *see* plateaux
tableware *5.444*
taboos **7**.14, *7.305*
  excreta disposal **7**.273
  incest **7**.46, **7**.154
  and tantra **7**.306
  see also totemism
tabor **5**.138, *5.444*
tabulating machine **6**.169
tabun **6**.239
tachinid fly **2**.41
Tachisme *5.444*
Tacitus, Publius Cornelius **3**.5, **3**.23, **3**.106, *3.347*, *5.444–5*
Tacna–Arica Conflict *4.336*
Tadhzikistan **7**.59, **7**.157
tadpole *2.333*
Taff Vale case *4.336*
Taft, William Howard **4**.277, **4**.336
  Taft–Hartley Act *4.336*
Tagalog language **7**.23
Taglioni, Marie **5**.30
Tagore, Debendranath (1817–1905) **4**.151, **7**.143
Tagore, Rabindranath (1861–1941) *5.445*
Tagus *1.330*
Taha Husein **5**.19
Taharka of Kush and Egypt **3**.259
Ṭāhir, Bābā **5**.344
Tahiti *1.330*, *4.336–7*
Tahuantinsuyu **5**.426
tai kwan do **7**.194
Tai languages *7.305*
taiga *2.333*
Taiji *7.305*
Taiji Quan *7.305*
tail (comet) *8.169*
tailed frog *2.333*
taille (tallage) **3**.347
Tailleferre, Germaine **5**.421
taipan *2.333*
Taiping rebellion **4**.336, *4.337*
Taira Kiyomori **3**.231
Taiwan *1.330*, *3.347*, *4.337*, *7.384*
  decolonization **4**.321
  economy and trade **7**.88, **7**.113, **7**.277, **7**.296, **7**.314
  land reform **7**.180
  languages **7**.286
  physical geography and geology **1**.330
  religion/theology **7**.47
  sport and recreation **7**.194
Taizong, emperor of China *3.347*, **3**.348, **3**.349, **3**.387
Taj Mahal **5**.307
Tajfel, Henri **7**.290
Tajikistan **1**.246, *1.330*, **1**.348, *4.337*, *7.384*
takahe **2**.230
Takamine, Jokichi *6.344*
Takemitsu, Tōru *5.445*
take-over **7**.65, **7**.224, *7.305*
Takeshitu **7**.71
takin *2.333*
Taklamakan Shamo *1.330*
Taksin, General **3**.327
tāla **5**.221
Talat Pasha **4**.387
Talbot, William Henry Fox *5.445*, **6**.268, *6.344*
talc **1**.320, *1.330–1*
Tale of Genji *5.445*
Tale of Heike **5**.235
tales *see* stories and tales
talipot palm *2.333–4*
talismans **9**.13
talking drums *5.445*
tallage *3.347*
Tallard, Camille de **3**.43
Talleyrand-Périgord, Charles Maurice de *3.348*, **3**.386, **4**.199, *4.337*
Tallis, Thomas *5.445*
Talman, William *5.445*
Talmud *3.348*, *7.305–6*
talus **1**.295–6
tamandua **2**.334
tamarind *2.334*
tamarisk *2.334*
tambourin (tabor) **5**.444
tambourine *5.445*
tambura *5.445*
Tamerlane (Timur) **3**.140, *3.348–9*, **3**.362, **5**.455–6
  Timurid art **5**.344, *5.455–6*
Tamihana, Wiremu **4**.183
Tamils **3**.74, **3**.268, **7**.95, **7**.109, *7.306*
  literature *5.445–6*
Tammany Hall *4.337*
Tammuz *9.35*
Tampuran, Kottyatu **5**.134
tam-tam (gong) **5**.184
Tamworth Manifesto *4.337*
Tana, Lake *1.331*
Tanach: books listed *9.44*
tanager *2.334*
Tanaka Chojiro **5**.373
Tanaka Kakuei **4**.194, *4.337*, **4**.338
Tane **9**.16, **9**.31, *9.35*
Tang dynasty **3**.72, **3**.232, **3**.347, **3**.348, *3.349*, **3**.387
Tang Xianzu *5.446*
Tang Yin **5**.98
Tanganyika *see* Tanzania
Tanganyika, Lake *1.331*
Tange, Kenzo **5**.234, *5.446*
tangent (geometry) *1.331*
tangent (trigonometry) *1.331*
tangerine **2**.201
tango *5.446*
Tanit *9.35*
Tanizaki Jun'ichirō *5.446*
tank (military vehicle) **6**.13, **6**.20, **6**.134, **6**.226, *6.344–5*
  anti-tank gun **6**.16, **6**.131
tank locomotive *6.345*
tanka *5.446*
tanker (ship) **6**.51, **6**.219, **6**.265, *6.345*
tanning **2**.344, **6**.197, *6.345*, **6**.387
Tansen *5.446*
tantalite **6**.345
tantalum **1**.107, *1.331*, **6**.177, **6**.241, *6.345*
Tantalus **9**.29, *9.35*
Tantia Topi **4**.162
tantra *7.306*
Tantric yoga **7**.173
al-Tanukhī **5**.19
Tan-Zam Railway *4.337–8*
Tanzania (*earlier* Tanganyika) *1.331*, *4.338*, *7.384–5*
  Arusha Declaration **4**.20
  decolonization **4**.4
  diseases and disorders **7**.9
  economy and trade **4**.106, **7**.296
  ethnic groups **7**.188

Tanzania *(cont.)*:
islands **1**.372, **3**.197
languages **7**.78, **7**.217
literacy **7**.185
physical geography and geology ***1**.331*
gorge *see* Olduvai
islands **1**.372
lakes **1**.203, **1**.331
mountains **1**.183
political leader **4**.248
prehistory **3**.201, **3**.261, **3**.391
railway **4**.337–8
wars: rebellion **4**.207
see also Zanzibar
Tanzimat reforms **4**.338
Tao Qian ***5**.446*
Taoism *see* Daoism
tap ***6**.345*
tapaculos ***2**.334*
tape recorder and audio tape **6**.103, **6**.105, **6**.213, **6**.326, ***6**.345*
tapestries **5**.39, **5**.64, **5**.94, **5**.181, ***5**.446*
see also embroidery
tapeworm **2**.167, ***2**.334*
tapioca **2**.56
tapir **2**.277, ***2**.334–5*
tar **6**.79, **6**.261
tār (lute) ***5**.446–7*
Tara ***3**.349–50*
Tarahumara ***7**.306*
Taraki, Nur Mohammad **4**.3
Taranaki Wars ***4**.338*
tarantula **2**.277, ***2**.335*
Tarantula Nebula ***8**.169*
tardigrade **2**.361
tare **2**.355
tariff ***7**.306*
reform ***4**.338–9*
Tariq ibn Zaid **3**.240, ***3**.350*
Tarkaratna, Ramnarayan **5**.43
Tarkovsky, Andrei ***5**.447*
Tarmac(adam) **6**.299
tarn **1**.331
taro ***2**.335*, **6**.302
Tarot cards **5**.353, **7**.40
tarpon ***2**.335*
Tarquin, king of Rome **3**.119, ***3**.350*
tarragon ***2**.335*
Tárrega, Francisco **5**.196
tarsals **2**.124
tarsier ***2**.335–6*
Tarski, Alfred ***7**.306*
Tartars ***3**.143*, ***3**.350*
Tartarus ***9**.35*
Tartini, Giuseppe ***5**.447*
Taruc, Luis **4**.158
Tasman, Abel Janszoon ***1**.331*, ***3**.350*
Tasman Sea ***1**.332*
Tasmania ***1**.331–2*, **3**.29, **3**.236, ***4**.339*
Tasmanian devil ***2**.336*
Tasmanian wolf ***2**.336*
Tasso, Torquato **5**.148, ***5**.447*
taste ***2**.336*
Tata ***4**.339*
Tatchell, Peter **7**.146
Tati, Jacques ***5**.447*
Tatlin, Vladimir ***5**.447*
tattooing ***5**.447*
Tau Ceti ***8**.169*
Taupo ***1**.332*
Taurids ***8**.169*, **8**.194, **8**.195
Taurus (constellation) ***8**.169*
Taurus A (radio source) ***8**.169*
Taurus X-1 (X-ray source) ***8**.169*
Taurus–Littrow (Moon valley) ***8**.169*
Taverner, John ***5**.447*
Tawfīq al-Hakīm **5**.19
Tawhiao, king of Maoris **4**.183, ***4**.339*
tawny owl ***2**.336*
taxation ***7**.306*
Britain *see under* Britain
customs and excise **7**.79
taxation *(cont.)*:
and education **7**.102
fiscal policy **7**.120
haven ***7**.307*
income **4**.160
medicine and health care **7**.139
Peter's pence **3**.277
poll **3**.287, **7**.306
scutage **3**.320
ship money **3**.327
and social security **7**.290
Stamp Act **3**.339
subsidy **7**.301
tallage/taille **3**.347
tariff **4**.338–9, **7**.306
tithe **3**.357
tunnage and poundage **3**.363
VAT **7**.306
whisky **3**.380
window **3**.383
taxidermist **2**.142
Taxila ***3**.350*
Taya, Colonel Moaouia Ould Sidi **4**.214
tayberry ***2**.336*
Taylor, A. J. P. **4**.153
Taylor, Brook: Taylor expansion ***1**.332*
Taylor, Cecil **5**.239
Taylor, Charles **4**.195
Taylor, F. B. **1**.73
Taylor, Frederick W. **6**.26, **6**.356
Taylor, Hannah **5**.400
Taylor, Jeremy ***3**.350*
Taylor, Paul ***5**.447*
Taylor, Sir Robert ***5**.447–8*
Taylor, Zachary **4**.218, ***4**.339*, **4**.377
ta'ziya (plays) **9**.17
Tchaikovsky, Pyotr **5**.370, ***5**.448*
TCP (trichlorophenol) **6**.267
Te Kooti Rikirangi Te Turuki **4**.14, ***4**.340*
Te Puea Herangi **4**.183, ***4**.341*
Te Rauparaha **4**.13
Te Whiti ***4**.341*
tea ***2**.336*, **6**.34, **6**.271
clippers **6**.76
teaching ***7**.307*
art **5**.140
dance *see* choreographers
music **5**.56, **5**.112, **5**.121
see also under education
teak ***2**.336–7*
teal ***2**.337*
team games *see* sport and recreation
Teamsters, International Brotherhood of ***4**.339*
Teannaki, Teatao **4**.184
Teapot Dome scandal ***4**.339*
tear fault **1**.120, **1**.323
tear-gas ***6**.345–6*
tears ***2**.337*
'tears of St Lawrence' **8**.118
teasel ***2**.337*
technetium **1**.107, ***1**.332*
technicolor **6**.7
technocracy ***7**.307*
technology ***6**.346*
Technological Revolution and high technology ***4**.339*
see also science; *and all references prefixed* **6**.
Tecton ***5**.448*
tectonics ***1**.332*, ***8**.169–70*
see also under plates of Earth's crust
Tecumseh **4**.147, ***4**.339–40*, **4**.342, **4**.345, **4**.371
teeth **2**.54, **2**.205, ***2**.343*, ***9**.58*
tusk **2**.351
see also dentistry
teflon **1**.151, **6**.273, **6**.285
Tegh Bahadur, Guru **7**.194, **7**.285
Tegnér, Esaias **5**.404
Teheran Conference ***4**.340*
Tehuantepec ***1**.332*
Tejada, Lydia Guelier **4**.44
tektite ***8**.170*
telecommunications ***4**.80*, ***4**.81*, **6**.179–80, ***6**.346*, ***7**.307*
channel **6**.69
electronic mail **6**.118, **6**.180
modem **6**.179, **6**.180, **6**.231
multiplexing **6**.69, **6**.112, **6**.235, **6**.346, **6**.347
narrowcasting **7**.211
network **6**.240
news *see* news
optical fibre **6**.251, **6**.347
radio *see* radio
repeater **6**.297, **6**.347
satellite **6**.81, **6**.308, **6**.347, **6**.349
telegraphy *see* telegraphy
telephone *see* telephone
teletext **6**.180, **6**.348
television *see* television
telex **6**.347, **6**.350
waveband **6**.288–9, **6**.380
see also broadcasting; International Telecommunications Union; satellites, artificial
telegraphy ***6**.346–7*
cable **6**.53, **6**.319
fax **6**.278, **6**.347
modulation **6**.231
pioneers **6**.69, **6**.88, **6**.112, **6**.232, **6**.319, **6**.384
radio wireless **6**.45
semaphore **6**.69
teleprinter **6**.347
telex **6**.347, **6**.350
Tel-el-Kebir, battle of ***4**.340*
Telemann, Georg Philipp ***5**.448*
telemetry ***6**.347*
teleology ***7**.307*
telepathy **7**.228
telephone and telephony ***6**.347*
answering machine **6**.15
bandwidth **6**.29
cable **6**.53, **6**.347
coding **6**.93
cordless **6**.65, **6**.292, **6**.347
exchange ***6**.347*
fax **6**.126, **6**.278, **6**.347
optical fibre **6**.251, **6**.347
pioneers **6**.33, **6**.112
privatization **7**.248
radio **6**.65, **6**.292, **6**.347
receiver ***6**.347*
repeater **6**.297, **6**.347
teleprinter ***6**.347*
telescopes ***6**.347–8*, ***8**.170*
and aberration **8**.1
alt-azimuth system **8**.22, **8**.104
astrograph **8**.9
Australia Telescope National Facility **8**.13
Dawes limit **8**.36
focus **8**.51
Galileo and **8**.11, **8**.137
innovator **6**.149
Kellner eyepiece **8**.134
Lovell *see under* radio telescopes
mounting **8**.104
multiple-mirror **8**.93, **8**.105
observatories *see* observatories
radio *see* radio telescopes
Ramsden eyepiece **8**.134
reflecting (Newtonian) ***8**.135–6*, **8**.137
Anglo-Australian **8**.150
Canada–France–Hawaii **8**.19
Cassegrain **8**.135
coudé **8**.135, **8**.136
Hale **8**.63
Hubble space **8**.69–70, **8**.71
Isaac Newton **8**.75
McMath **8**.96
Maksutov **8**.135
Schmidt **8**.135
Zelenchukskaya **8**.189
refracting **8**.65, **8**.137
richest field **8**.139
telescopes *(cont.)*:
segmented mirror *(cont.)*:
space **6**.310, **8**.68, **8**.69–70, **8**.81
synthesis **8**.73
on theodolite **6**.353
X-ray **8**.188
Telescopium ***8**.170*
Telesto ***8**.170*, **8**.175
teletex(t) **6**.180, ***6**.348*, **7**.35
television **6**.118, ***6**.348–9*, ***7**.307–8*
acting **5**.3
animation **5**.17
bandwidth **6**.29
BBC **7**.28
broadcasting **6**.47–8
cable **6**.53, **7**.38–9
camera **6**.175, **6**.241, **6**.348, ***6**.349*
closed-circuit **6**.77
colour **6**.28, **6**.81, **6**.349, **6**.350
copyright **7**.70
eidophor **6**.112, **6**.113
405-line system **6**.40
high-definition **6**.349
iconoscope **6**.118, **6**.175, **6**.349, **6**.391
phosphors in **6**.64, **6**.81, **6**.295, **6**.349–50
pioneers **6**.28, **6**.391
propaganda **7**.249
receiver ***6**.349–50*
satellite dish **6**.308
on telephone circuits **6**.347
teletex and teletext **6**.180, **6**.348, ***7**.35*
video *see* video
videotex **6**.180, **6**.373
wavebands **6**.288–9
telex **6**.347, ***6**.350*
Telford, Thomas **4**.350, **6**.47, **6**.210, **6**.299, **6**.300, ***6**.350*
Tell, William ***9**.35*
tell (archaeological) ***3**.350–1*
Teller, Edward ***6**.350*
tellurium **1**.107, ***1**.332*, ***6**.350*
Telstar **6**.81
Telugu **7**.95
Tempel, William (1864, German astronomer): Tempel's Comet ***8**.170*, **8**.194
tempera **5**.201, ***5**.448*
temperament (music) ***5**.448–9*
temperament (personal) *see* personality
Temperance Movements ***4**.340*
temperate climates **1**.65, **1**.66
temperature ***1**.332–3*, ***8**.170*, ***9**.54*
absolute zero **1**.1
advection **1**.3
and altitude (lapse rate) **1**.188
astronomy ***8**.170*
effective **8**.43
Kelvin scale of **8**.79
stellar **8**.43, **8**.161
surface **8**.166–7
of atmosphere **1**.25
body **2**.330, **2**.337
see also fevers
boiling-point **1**.41
global *see* greenhouse effect
inversions **1**.171–2
isothermal change **1**.176
measurement and scales **6**.54–5, **6**.192, **6**.350–1, **6**.354–5, **6**.368, **9**.54
Celsius **1**.55–6, **1**.332, **6**.166, **6**.351, **9**.54
Fahrenheit ***1**.119*, **1**.332, **6**.350
melting-point **1**.210
Moon **1**.221, **1**.273
oven **9**.62
and pressure:
standard **1**.320
triple point **1**.344
symbols **9**.56
thermodynamic **1**.181

temperature (*cont.*):
thermostat **6**.89, **6**.185, **6**.193, **6**.355
see also cooling; freezing; heat *and entries beginning* therm
tempering ***6**.351*
Temple, Sir William (1628–99) ***3**.351*
Temple, William (1881–1944) ***4**.340*
Temple of Solomon **3**.187
Templer, Sir Gerald ***4**.340–1*
temples **5**.64, **5**.247–8, **5**.251, **7**.137, **7**.143, **7**.144
Indian *see* India *under* architecture
tempo **5**.455
Temur of China **3**.389
tench ***2**.337*
tendons **2**.186
Tenerife **1**.50
Teniers, David (the Elder and Younger) ***5**.449*
Tennant, Charles **6**.40, ***6**.351*
Tennent, Gilbert **3**.145
Tennessee Valley Authority ***4**.341*
Tenniel, Sir John **5**.53, **5**.80
tennis **4**.326, **7**.305, ***7**.308*
Tennis Court Oath ***3**.351*
Tennyson, Alfred, Lord **5**.12, **5**.134, **5**.143, **5**.217, **5**.371, **5**.430, ***5**.449*
Tenochtitlán **3**.31, ***3**.351*, **6**.220
tenrec ***2**.337*
tension structure **6**.51, ***6**.351*
tensor ***1**.333*
tent caterpillar ***2**.337–8*
tenth planet ***8**.170–1*
Tenzin Gyatso **7**.79
Tenzing Norgay **7**.208
Teotihuacán ***3**.351*, ***5**.449*
tephra ***1**.333*
terbium **1**.107
Terborch, Gerard ***5**.449*
Terbrugghen, Hendrick ***5**.449*
Terence **5**.257, **5**.379, **5**.387
Terentius Varro **3**.59
Teresa, Mother **4**.221
Teresa of Ávila, Saint **3**.61, **3**.89, ***3**.351*, **3**.352, **5**.311, ***9**.43*
Tereshkova, Valentina **4**.323, **8**.154
terminal (computer) ***6**.351*
terminal velocity ***1**.333*
terminator ***8**.171*
termites **2**.129, ***2**.338*
ternary form ***5**.449–50*
terns **2**.229, **2**.311, ***2**.338*
terra rossa ***1**.333–4*
terrace cultivation **6**.351, ***6**.352*
terrace (geomorphology) ***1**.333*
terracotta **5**.288, **5**.384, ***5**.450*
terrapin ***2**.338*
Terreblanche, Eugene **7**.118
terrestrial planets *see* inner planets
terrestrial radiation **1**.108, **1**.109
territory (ecology) ***2**.338*
Terror, Reign of **3**.316, ***3**.351–2*
terrorism and anarchism ***3**.341*, ***4**.11–12*, ***7**.12–13*, ***7**.308–9*
Armenia **7**.17
assassination **7**.20
Germany **4**.27
hi-jacking **7**.142
hostage-taking **4**.165, **7**.91, **7**.147–8
Serbia **4**.40
state **7**.299
see also under Palestine
Terry, Alfred **4**.91
Terry, Dame Ellen ***5**.450*
Terry, Eli **5**.105
tertiary education **7**.102
see also higher education
tertiary industry ***7**.184*, ***7**.309*
financial *see* banks; insurance *and also* finance *under individual countries*
services *see* goods and services
social *see* social services
tertiary industry (*cont.*):
tourism *see* tourism
see also post-industrialism
Tertiary period **1**.133, **1**.134, ***1**.334*
Eocene **1**.110, **1**.133, **1**.134
Holocene **1**.133, **1**.134, **1**.157
Miocene **1**.133, **1**.134, **1**.216
Neogene **1**.133, **1**.134, **1**.225
Oligocene **1**.133, **1**.134, **1**.237
Palaeocene **1**.133, **1**.134, **1**.135, **1**.244
Palaeogene **1**.133, **1**.134, **1**.244
Pliocene **1**.133, **1**.134, **1**.260
Tertullian, Quintus Septimius Florens ***3**.352*
Terylene **6**.275, **6**.384
Tesla, Nikola **1**.334
Tesla coil ***6**.351*, **6**.383
tesla (unit) ***1**.344*, **6**.368
Test Acts ***3**.352*
Testelin, Henri **3**.211
testicle ***2**.338*
testosterone **6**.334, **7**.7
tests:
assessment in education **7**.20
chemistry: reagent **1**.277
intelligence **7**.32, **7**.162
psychological **7**.252
test-tube ***6**.351*, **6**.353
baby **6**.185–6, **6**.218
Tet Offensive ***4**.341*
tetanus ***2**.339*, **7**.153
Tethys (satellite of Saturn) ***8**.171*
Tethys Sea **1**.135, ***1**.334*
Tetley, Glen **5**.122, **5**.373, ***5**.450*
tetra ***2**.339*
tetrachloroethane **6**.253
tetrachloromethane **1**.52
tetracyclines **6**.16, **6**.72
tetrafluoroethene **6**.139, **6**.285
tetrahedron **9**.51
Tetricus **3**.29
Tetzel, Johann ***3**.352*
Teutonic Knight **3**.198, ***3**.352*
Teutons ***3**.352*
Tewkesbury, battle of ***3**.352*
Tewodros II, emperor of Ethiopia **3**.119, **4**.206
tex (unit) **6**.390
Texas **4**.302, ***4**.341*
Texas Rangers **4**.341
text:
image **7**.152
textual analysis **5**.150, **7**.81–2
see also critics, literary
textiles ***6**.125*, **6**.128, ***6**.353*
Africa **5**.7
Batik **5**.38
bleaching **6**.40, **6**.174, **6**.351
canvas **6**.57
carding **6**.60, **6**.165, **6**.329, **6**.390
carpets *see* carpets and rugs
China **5**.91, **5**.94
clothes *see* clothing
Coptic **5**.112
costume **5**.114
cotton *see* cotton
dyes *see* dyes and dyeing
elastic **6**.113
felt **6**.125, **6**.127, **6**.242
fibre *see* fibre
finishing ***6**.353*
fireproofing **6**.132
France **5**.138
fulling **6**.147
glass-fibre **6**.86, **6**.129, **6**.156
see also fibre-reinforced
ironing **6**.186
Islam **5**.230
Japan **5**.237
knitting *see* knitting
lace **5**.252, **5**.253, **6**.125, **6**.193, **6**.351
leathercraft **5**.259
linen **6**.16, **6**.203, **6**.390
textiles (*cont.*):
machinery and industry ***6**.353*
bleaching **6**.40, **6**. 351
bush roller chain **6**.34
carding **6**.60, **6**.165
gig mill **6**.156
hackling **6**.164
knitting **6**.191
lace-making **6**.193
loom *see* weaving
nap shearer **6**.236
sewing *see* sewing-machine
spinning **6**.20, **6**.92, **6**.165, **6**.329
weaving **6**.62, **6**.140, **6**.169, **6**.189, **6**.190, **6**.207–8, **6**.318
wool-comber **6**.388
mohair **6**.231
needle **6**.239
nonwoven **6**.125, **6**.127, **6**.242, **6**.351
North American Indians **5**.321
oilskin **6**.249
Ottoman **5**.334
patchwork **5**.341
printing ***6**.353*, **6**.387
quilting **5**.372
rags for paper **6**.256
rubberized **6**.304
Safayid **5**.397
sewing **6**.335
shrink-resistant **6**.318
silk *see* silk
size **6**.321, **6**.353
South American Indians **5**.426
spinning *see* spinning
synthetic *see* synthetic fibres
texturing **6**.390
Timurid **5**.456
waterproof **6**.211, **6**.249, **6**.378
weaving *see* weaving
wool *see* wool
yarn **6**.390
zip **6**.391
see also clothing; embroidery; sewing; tapestries
textual *see* text
texturing, yarn **6**.390
texturized vegetable protein **6**.142
Tezcatlipoca ***9**.35*
Thābit ibn Qurrah **8**.75
Thackeray, William Makepeace **5**.252, ***5**.450*
Thailand (Siam) ***1**.334*, ***3**.327*, ***4**.341–2*, ***7**.385*
adoption **7**.3
art ***5**.450*
contraception **7**.115–16
crime **7**.225
economy and trade **7**.20, **7**.200, **7**.215
education **7**.93
emigration **7**.200
languages **7**.305
laws/legal system **7**.40, **7**.143
leaders **4**.75, **4**.222, **4**.269, **7**.97, **7**.139
literature ***5**.450–1*
physical geography and geology **1**.334
prostitution **7**.250–1
religion/theology **7**.36, **7**.143
social services and security **7**.293
sport and recreation **7**.194
urban areas **7**.197
Thakurs *see* Rajput
thalassaemia ***2**.339*
Thales of Miletus **8**.61
thalidomide **6**.265
thallium **1**.107, ***1**.334*, ***6**.353*
Thames, battle of (Canada) ***4**.342*
Thames, River (England) ***1**.334*
Than Shwe, General **4**.229
thane ***3**.352*
al-Thani, Shaikh Khalifa bin Hamad **4**.279
Thar Desert ***1**.334*
Tharp, Twyla ***5**.452*
Tharsis Ridge **8**.171
Thatcher, Margaret **4**.85, ***4**.342*, **4**.381, **7**.20, **7**.215
theatre (operating) **6**.250
theatre (plays) *see* drama
theatre (intermediate) weapons **6**.246
Theatre Workshop ***5**.452*
Thebe (satellite) ***8**.171*
Thebes (Egypt) ***3**.352*, **3**.353, **3**.369
Thebes (Greece) ***3**.352–3*
theft **7**.250
thematic maps **6**.215
Themis ***9**.35*
Themistocles **3**.147, **3**.264, **3**.282, **3**.314, ***3**.353*
theocracy ***7**.309*, **7**.310
Theocritus **5**.190, **5**.217
theodolite **6**.168, **6**.294, **6**.340, **6**.341, ***6**.353*
Theodora, Byzantine Empress ***3**.353–4*
Theodore I, tsar of Russia **3**.312
Theodore I and II, emperors of Nicaea **3**.253
Theodore, archbishop **3**.380
Theodoric I, king of Visigoths (418–51) ***3**.43*, **3**.354
Theodoric the Great, king of Ostrogoths (455–526) **3**.97, **3**.264, ***3**.354*
Theodosius I, Flavius (the Great), Roman emperor (349–95) **3**.55, **3**.74, **3**.101, **3**.142, **3**.262, **3**.307, ***3**.354*
Theodosius II, pope **3**.117
theology ***7**.309–10*
see also religion and theology
Theophilus, Byzantine emperor **3**.174
Theophrastus of Eresos **1**.215, **2**.38, ***2**.339*
theosophy **7**.299, ***7**.310*
therapy *see* complementary medicine; psychotherapy
Theravāda Buddhism **7**.36, ***7**.310*
arhat **7**.16
Jataka Tales **9**.19
monasticism **7**.205
nirvana **9**.5
saints **7**.272
sangha **7**.273
Tripitaka and Tipitaka **7**.316
Wesak **7**.330
Theresa, Mother **7**.204–5
thermal (air current) ***1**.334*
thermal *see* heat
thermal conductivity **1**.71
thermal cracking **6**.91
thermal efficiency **6**.166, **6**.332
thermal insulation *see* insulation, thermal
thermal nuclear reactor **6**.57, **6**.244
thermal pollution **6**.377
thermal power stations **6**.57, **6**.78, **6**.244, **6**.279
thermal radiation **1**.273
see also radiation
thermionic converter ***6**.353–4*
thermionic valve ***6**.354*
in computer **6**.84
in hearing aid **6**.166
pioneers **6**.98, **6**.134
and radar **6**.288
radio **6**.118, **6**.291
radioactive isotopes **6**.290
thermite **6**.383
thermocline ***1**.334*
thermocouple **6**.350, **6**.351, ***6**.354*
thermodynamics ***1**.334*
enthalpy **1**.110
entropy **1**.110

thermodynamics (*cont.*):
measurement **6**.368
temperature **1**.181
thermoelectric device **6**.166, **6**.290, ***6**.354*
thermography ***6**.354*, **6**.355
thermoluminescence **6**.96
thermometer **1**.196, **6**.166, **6**.218, **6**.350, ***6**.354–5*, **6**.382
thermonuclear bomb *see* hydrogen bombs
thermoplastics **6**.271, **6**.272, ***6**.273*
acrylic fibre **6**.2–3, **6**.58, **6**.61
cellulose acetate **6**.66
polycarbonate **6**.275
polymethyl methacrylate **6**.271, **6**.273, **6**.276
polytetrafluoroethene (PTFE) **6**.273, **6**.285
recycling **6**.296
welding **6**.383
see also polypropylene; polystyrene; polythene; polyvinyl chloride; thermosetting plastics
Thermopylae ***3**.354*
thermos flask **6**.370
thermosetting plastics **6**.271, ***6**.273*, **6**.369
epoxy resins **6**.122, **6**.129, **6**.272
melamine-formaldehyde **6**.273
nylon *see* nylon
phenol-formaldehyde **6**.27, **6**.267, **6**.271, **6**.273
polycarbonate **6**.273, **6**.275
polyester **6**.275, **6**.302, **6**.384
polyethene terephthalate **6**.271, **6**.273
polyurethane **6**.273, **6**.275, **6**.276
urea-formaldehyde **6**.272, **6**.273
see also plastics; thermoplastics
thermosphere ***1**.334*
thermostat **6**.89, **6**.185, **6**.193, ***6**.355*
Theseus **9**.5, **9**.25, ***9**.35*
Thesiger, Wilfred **5**.460
Thespis **5**.191
Thessaly ***3**.354*
Thetis **9**.2
thiamine **2**.357
Thibaud, Jacques **5**.82
Thibaw, king of Burma **4**.13, **4**.229
thickhead ***2**.339*
thick-knee **2**.327
Thiers, Louis Adolphe **4**.260, ***4**.342*
Thing ***3**.354*
thinking *see* thought
third quarter of moon **8**.119
Third Reich ***4**.342–3*, **9**.70
see also Nazis
Third World ***7**.310–11*
see also developing countries
thirst ***2**.339*
Thirteen Colonies ***3**.354–5*
see also states *under* America
Thirty Years War **3**.49, **3**.214, ***3**.355*, **3**.380, **3**.381
Thirty-Nine Articles ***3**.355*
Thisbe **9**.30
thistle ***2**.339*
Thistlewood, Arthur **4**.65–6
Thollon, Louis **8**.19
Thom, Alexander **8**.127
Thom, René **1**.54, ***1**.334–5*, **8**.22
Thomas, Brandon **5**.152
Thomas, Dylan ***5**.452*
Thomas, Edward ***5**.452–3*
Thomas, John **7**.47
Thomas, Sidney **6**.188
Thomas, William **6**.172
Thomas à Kempis **5**.258
Thomas (Didymus), Saint **9**.5, ***9**.43*
Thomas of Lancaster, duke of Clarence **3**.78
Thomas of Woodstock **3**.210, **3**.303, ***3**.355*
Thomé de Jesus, Frei **5**.359
Thompsen, Bill **4**.61
Thompson, Alexander 'Greek' **5**.315
Thompson, Benjamin, count von Rumford ***1**.335*, **6**.303
Thompson, E. P. **4**.153
Thomson, Elihu ***6**.355*
Thomson, George (1757–1851, musicologist) **5**.68
Thomson, Sir George Paget (1892–1975, physicist) ***1**.335*
Thomson, James ***5**.453*
Thomson, Joseph (1858–95, explorer) **1**.80, ***1**.335*
Thomson, Sir Joseph John (1856–1940, physicist) ***1**.335*, **6**.118
Thomson, R. W. **6**.109
Thomson, Tom **5**.73
Thomson, Virgil ***5**.453*
Thomson, William *see* Kelvin, Lord
Thor **9**.2, **9**.10, **9**.30, ***9**.35*
thorax ***2**.339–40*
Thoreau, Henry David **5**.355, ***5**.453–4*
Thorez, Maurice ***4**.343*
Thorfinn Karlsefni **3**.374
thorium **1**.107, **6**.25, **6**.243, **6**.285, **6**.355
thorn apple ***2**.340*
thornbill **2**.360
Thorndike, Edward Lee ***2**.340*, **7**.64
Thorndike, Dame Sybil ***5**.454*
Thorneycroft, John **6**.171
Thornhill, Sir James ***5**.454*
Thornthwaite, Charles Warren **1**.65, **1**.66, ***1**.335*, **1**.360
Thorvald **3**.374
Thorvaldsen, Bertel ***5**.454*
Thoth **9**.16, ***9**.35*
thought:
experiment ***7**.311*
psychology of *see* cognitive psychology
Thousand Islands ***1**.335*
Thousand and One Nights ***5**.454*
characters **9**.3, **9**.17–18, **9**.33
Thrace ***3**.355–6*
thrasher (bird) ***2**.340*
thread snake ***2**.340*
thread-fin ***2**.340*
threadworm **2**.250
Three Emperors' League ***4**.344*
'Three Fs' ***4**.344*
Three Henrys, War of ***3**.356*
Three Kingdoms **3**.72, **3**.356
Three Kings, battle of **3**.8
Three Marks of Existence **7**.97, **7**.263, ***7**.311–12*
Three Mile Island accident **6**.243, **6**.245, **6**.281, ***6**.355*
three-body problem **8**.106
Threefold Refuge (Three Jewels) ***7**.311–12*
threshing machine ***6**.355*
thrip ***2**.340*
throat ***9**.59*
disease of **6**.254
Throckmorton, Francis: Plot **3**.151, ***3**.356*
thrombosis **2**.77, ***2**.340*
throstle (spinning machine) **6**.329
throwing weapons *see* missiles *under* weapons
thrush family ***2**.340*
blackbirds **2**.33, **2**.94
bluebird **2**.36
chats **2**.61, **2**.280–1
fieldfare **2**.116
missel-thrush (mistle-thrush) **2**.213
nightingale **2**.226
redwing **2**.281
ring-ouzel **2**.285
robin **2**.287
song thrush **2**.317
stonechat **2**.326–7
wheatear **2**.365
see also oriole family
Thuban ***8**.171*
Thucydides **3**.264, **3**.274–5, **3**.284, ***3**.356*, **5**.190, ***5**.454*
thug **4**.344
thulium **1**.107
Thulstrup, Thore de **4**.148
thumbless bat **2**.314
thunder ***1**.335*
thunder fly **2**.340
thunderclouds **1**.81
thunderstorms **1**.289, ***1**.335*
see also storm
Thurber, James **5**.151, ***5**.454*
Thutmose I, pharaoh **3**.279, ***3**.356*, **3**.369
Thutmose III, pharaoh **3**.278, ***3**.356*
thyme ***2**.340*
thymine **2**.95
thymus gland ***2**.340*
thyroid gland **2**.140, ***2**.340*
scans **6**.243
Ti Zang **9**.20
Tiahuanaco ***3**.356*
Tiamat **9**.24, ***9**.35*
Tian Han (Chinese playwright) **5**.94
Tian Shan ***1**.335*
Tiananmen Square massacre **4**.262, **4**.332, **7**.239
Tiānfēi **9**.25
Tianjin, Treaty of **4**.250–1
Tiberius, Claudius Nero **3**.175, ***3**.356*
Tibet ***1**.335–6*, ***3**.356*, **4**.204, ***4**.344*, ***7**.385*
art **5**.64, ***5**.454–5*
building **3**.357
languages **7**.286–7
literature ***5**.455*
McMahon Line **4**.204
pastoralism **7**.230
physical geography and geology **1**.335–6
mountain **1**.116
rivers **1**.167, **1**.210, **1**.291–2
politics and government **7**.265, **7**.309
refugees **7**.261
religion/theology **3**.202
Buddhism **3**.202, **5**.64, **5**.454–5, **7**.36, **7**.37, **7**.273, **7**.306, **7**.311, ***7**.312*, **9**.6, **9**.7
see also Dalai Lama
and politics **7**.265, **7**.309
Tibeto-Burman languages **7**.286
tibia **2**.186
Tibullus **5**.388
ticker-tape machine **6**.112
ticket-of-leave ***4**.344*
ticks **2**.306, ***2**.341*
and disease **2**.287, **2**.352
Ticonderoga **3**.356–7
tides/tidal ***1**.336–7*, ***8**.171*
bore ***1**.336*
bulge **8**.171
ebb **1**.99–100
energy and power ***1**.336*, ***6**.355*, **7**.106
barrage **6**.294–5
tide-mill ***6**.355–6*, **6**.378
turbine **6**.379
friction **8**.171
neap **1**.225, **1**.337, **8**.106
slack water **1**.307
spring **1**.318, **1**.337
stream **1**.336
wave **1**.123, **1**.345–6
Tieck, Ludwig **5**.174
Tiepolo, Giambattista ***5**.455*
Tierra del Fuego ***1**.337*
Tiffany, Louis Comfort ***5**.455*
tiger **2**.291, ***2**.341*, **2**.342
tiger beetle ***2**.341*
tiger lily **2**.190
tiger moths **2**.124–5, ***2**.341*
tiger snake ***2**.341*
tigerfish ***2**.341*
Tiglath-Pileser I, king of Assyria **3**.26
Tiglath-Pileser III, king of Assyria **3**.26, **3**.364
Tigranes the Great, king of Armenia **3**.23
Tigre languages **7**.279
Tigris ***1**.337*
and Euphrates *see* Mesopotamia
Tijou, Jean **5**.227
Tikhana **5**.134
Tikopia **7**.240, ***7**.312*
Tilak, Bel Gangadhar ***4**.344*, **7**.143
tilapia ***2**.341*
tiles **6**.51, **6**.76, **6**.146, **6**.302, ***6**.356*
till **1**.42, ***1**.337*
Till Eulenspiegel ***9**.12*
tillage **6**.227, **6**.273–4
Tillemans, Peter **3**.270
Tillett, Benjamin ***4**.344*
Tilley, Sir Samuel Leonard ***4**.345*
Tilley, Vesta **5**.310
tillite ***1**.337*
Tilly, Johannes Tserklaes, count of **3**.49, ***3**.357*
Tilsit, Treaties of ***4**.345*
timber:
framing ***5**.455*
see also forests and trees; wood
Timbuktu **3**.357
time ***1**.337*, ***7**.313*, ***9**.55*
-and-motion studies **6**.216, ***6**.356*, **7**.312
in astronomy ***8**.171–2*, ***8**.192*
equation of **8**.47
month **8**.100–1, **8**.168
second **8**.62
time-scale **8**.79
year **8**.189
see also calendar; sidereal; space–time
calendar *see* calendar
dating systems *see* dating
dilation ***1**.337*
fossil history ***9**.57*
geological *see* timescale *under* geology
GMT *see* Greenwich Mean Time and Meridian
history *see* historiography
international time zones **1**.338, **9**.54
measurement **1**.298, ***6**.76–7*, **6**.356
see also chronometer; clocks; watch
in music (and tempo) ***5**.455*
-scale ***1**.337–8*, **1**.356
see also under geology
sidereal **1**.286
solar **1**.286
zones ***1**.338*, ***9**.54*
see also space–time
Timor **4**.321, ***4**.345*
Timothy ***9**.35*
timothy grass ***2**.341*
timpani ***5**.455*
TIMs (transparent insulation materials) **6**.324
Timur *see* Tamerlane
Timurid art **5**.344, ***5**.455–6*
tin **1**.107, ***1**.338*, **6**.221, ***6**.356–7*
alloys **6**.11–12, **6**.49, **6**.265, **6**.356
canning **6**.57
electroplating **6**.120
fluoride **6**.357
ore (cassiterite) **1**.54
Tin Pan Alley **5**.357–8
tin whistle **5**.156
tinamou ***2**.341*
Tinbergen, Nikolaas **2**.28, **2**.194, ***2**.341*, **2**.358, **7**.109–10
Tindemans, Leo **4**.33
tinder-box **6**.132
tinea infection ***2**.341–2*
Tiniran ***9**.35*
Tinith **9**.35
tinstone **6**.213

Tintoretto, Jacopo **3**.104, ***5**.456*
Tipitaka **7**.316
Tippecanoe, battle of **4**.345
Tippett, Sir Michael **5**.441, ***5**.456–7*
Tipu Sultan **3**.88, **3**.244, ***3**.357*
Tir na nÓg **9**.27, **9**.36
Tirion, Wil **8**.159
Tiro, Marcus Tullius **6**.318
Tirpitz, Alfred von **4**.135, ***4**.345*
Tissot, James ***5**.457*
tissues ***2**.342*
tit (bird) ***2**.342*
Titan (mythology) *see* giants and Titans
Titan (satellite of Uranus) ***8**.172*
Titania (mythology) ***9**.27*
Titania (satellite of Uranus) ***8**.172*
titanium **1**.107, ***1**.338*, **6**.220, ***6**.357*
alloys **6**.5, **6**.11, **6**.82, **6**.220, **6**.235, **6**.241, **6**.339, **6**.357
anodized **6**.15
production **6**.323
surgical instruments **6**.340
titanium dioxide **6**.357
titanium oxide **1**.166
tithe ***3**.357*
Titian **3**.104, **3**.301, **5**.159, **5**.178, **5**.359, ***5**.457*
Titicaca, Lake ***1**.338*, **1**.339
Tito, Santi de (artist) **3**.215
Tito, Marshal **4**.223, **4**.309, ***4**.345*, **4**.387, **7**.218
Titokowaru **4**.14
Titov, Gherman **4**.323
titrations **1**.2, ***1**.338*
tits (birds) ***2**.342*
Titus, Roman emperor **3**.188
Tiunit **9**.35
tjurunga ***9**.35*
Tlaloc ***9**.35–6*
Tlatilco ***5**.457*
Tlingit Indians **7**.295, ***7**.312*
TNT **1**.118, **6**.124, ***6**.357*
toadfish ***2**.342*
toads ***2**.342*
horned **2**.162
Mexican burrowing **2**.209
midwife **2**.210
natterjack **2**.225
reproduction **2**.331
spadefoot **2**.318
tadpoles **2**.333
toadstool **2**.342
Toba Wei *see* Northern Wei
tobacco ***2**.342–3*, **6**.7, **7**.288–9
pests **2**.343
plant **2**.227
processing ***6**.357*
see also smoking
tobacco hornworm **2**.152, ***2**.343*
tobacco mosaic virus ***2**.343*
Tobago *see* Trinidad and Tobago
Tobias ***9**.36*
Tobit and Tobias ***9**.36*
Tobruk, siege of ***4**.345–6*
toccata **5**.457
Tocharian languages **7**.156, **7**.157
tocopherol **2**.357
Tocqueville, Alexis, comte de ***4**.346*
Todi, Jacopone da **5**.230
tody ***2**.343*
toenails **2**.225
Toer, Pramoedya Ananta **5**.222
Togliatti, Palmiro ***4**.346*
Togo/Togoland ***1**.338*, ***4**.346*, ***7**.385*
colonization **4**.306
diet **7**.190
in Franc Zone **7**.124
physical geography and geology **1**.338–9
religion and myths **9**.25, **9**.37
Togodumnus of Catuvellauni **3**.94
Tojo Hideki ***4**.346*
tokamak **6**.148
Tokugawa era and shogunate ***3**.357–8*, **4**.312
Tokugawa Ieyasu **3**.186, **3**.357, ***3**.358*
Tokugawa Yoshimune ***3**.358*
Tokyo declaration on terrorism **7**.309
Tokyo Trials ***4**.346*, **7**.328
Toland, Gregg **5**.72
toleration ***7**.312*
Toleration Act ***3**.358*
toleware **5**.253
Tolkien, John Ronald Reuel **5**.91, ***5**.457*
Toller, Ernst ***5**.457*
Tolman, Edward **7**.29, **7**.113, **7**.207
Tolpuddle Martyrs ***4**.346–7*
Tolstoy, Count Lev **5**.207, ***5**.457–8*, **7**.227
Toltecs ***3**.358*
art ***5**.458*
religion and myths **3**.297–8, **9**.30, **9**.35–6
technology **6**.287, **6**.346
toluene ***1**.339*, **6**.261, **6**.262, **6**.357
Tom Thumb ***9**.36*
tomato **2**.129, ***2**.343*
Tombalbaye, François **4**.68
Tombaugh, Clyde **8**.87, **8**.125
tombolo ***1**.339*
tombs **5**.410, **5**.475
mausoleum **5**.287, **5**.307, **5**.456
pyramids **3**.296, **6**.287
Tomislav, Kneze of Croatia **3**.91
Tomlinson, Charles **6**.346
tomography **6**.356, ***6**.357*
positron emission **6**.243, **6**.259, ***6**.277–8*, **6**.357
see also computerized axial tomography; radiology
Tomoyuki, Yamashita **4**.208
Tompion, Thomas **5**.104
Tonacatecutli **3**.31
tonality **5**.442, ***5**.458*
polytonality **5**.356
see also harmony
Tonatiuk **9**.36
Tone, Theobald Wolfe ***3**.358*
tone languages **7**.235
tone-poem **5**.442
Tonga ***1**.339*, ***4**.347*, **7**.97, **7**.296, **7**.319
Tongariro, Mount ***1**.339*
tongs **6**.273
tongue ***2**.343*, ***9**.61*
tongue-worm ***2**.343*
Tongzhi, emperor of China **4**.76
Tonkin Gulf Resolution ***4**.347*
Tonlé Sap ***1**.339*
Tönnies, Ferdinand **7**.61
tonsils ***2**.343*
Tonypandy ***4**.347*
Tooke, John Horne ***3**.358*
tools:
digging **6**.103
file and rasps **6**.129
history of **6**.280–1, ***6**.357–8*
pile-driver **6**.270
pneumatic **6**.274
power *see* power tools
steel **6**.333
vice **6**.372
see also cutting tools; hand tools; machine-tools
tooth *see* teeth
tooth carp (fish) **2**.178
toothed whales ***2**.343*
beaked whale **2**.25
dolphin **2**.97
narhwal **2**.225
river dolphin **2**.286
sperm whale **2**.319
toothwort ***2**.343–4*
Topa Inca, emperor of Incas **3**.175–6, ***3**.358–9*
topaz ***1**.339*, **1**.341
top-minnow *see* killifish
topography ***1**.339*
topographic survey and maps **1**.2–4, **6**.215, **6**.340
topographical ridge **1**.282
see also geomorphology
topology ***1**.339–40*
topsoil ***1**.340*
tor ***1**.340*
Torah ***7**.312*, ***9**.44*
Tordesillas, Treaty of ***3**.359*
Torelli, Giacomo **5**.430
tormentil ***2**.344*
tornado ***1**.340*, **1**.361
torpedo **6**.337, ***6**.357*
torpedo fish **2**.105, ***2**.344*
torpedo-boat ***6**.357*
torpor **2**.158
torque **6**.103, **6**.110, **6**.152, ***6**.357*
Torquemada, Tomás de **3**.336, ***3**.359*
Torre, Haya de la **4**.267
Torrens, Lake ***1**.340*
Torres, Juan José **4**.44
Torres Jurado, Antonio de **5**.196
Torrey, John **2**.144, ***2**.344*
Torricelli, Evangelista ***1**.340–1*, **6**.30
Torrijos, Omar **4**.257
torsional free oscillation **1**.127
tort ***7**.312–13*
tortoise ***2**.344*
tortoiseshell butterfly **2**.43
tortrix moth ***2**.344–5*
torture **7**.114, ***7**.313*
Tory ***3**.359*
see also Conservative Party
Tosa Mitsunobu **5**.237
Toscanini, Arturo **5**.110
Tōson **5**.236
Tostig **3**.339
total eclipse of Sun **8**.42
total institution ***7**.313*
total internal reflection **6**.251
totalitarianism **7**.16, **7**.125, ***7**.313*
totality, path and zone of **8**.117
totemism **7**.1, **7**.11, **7**.24, **7**.197, **7**.236, ***7**.313*
totem-poles **5**.321, **7**.219
see also
toucan **2**.277, ***2**.345*
touch ***2**.345*
touchstone ***1**.341*, ***6**.357*, **6**.359
Toulouse-Lautrec, Henri de **5**.362, ***5**.458*
Toungoo ***3**.359*
touracos **2**.350
Touré, Lieut. Colonel Amadou Toumani **4**.208
Touré, Ahmed Sékou **4**.143, ***4**.347*
Touré, Samori **4**.143
tourism **7**.212, ***7**.313–14*, **7**.370
tourmaline ***1**.341*
tourney ***3**.359*
Tours, Council of **7**.3
Tourte, François **5**.56
Toussaint-l'Ouverture, François Dominique **4**.145, ***4**.347–8*
town gas ***1**.341*, ***6**.359*
town planning **5**.425
Townes, C. H. **6**.214
towns *see* urban areas
Townsend, Francis Everett ***4**.348*
Townshend, Charles, 2nd viscount (1674–1738) **3**.5, ***3**.360*, ***6**.359*
Townshend Acts ***3**.359*
Townshend, Sir Charles Vere Ferrers (20th century) **4**.188
Towton, battle of ***3**.360*
toxicity/toxicology ***2**.345*
see also poison; poisonous; pollution
Toynbee Hall ***4**.348*
Toyotomi Hideyoshi **5**.373
toys ***5**.459*
automata **6**.25
optical **6**.73–4
trace elements ***2**.345*
trace fossils ***2**.345*
tracery ***5**.459*
trachea ***2**.345*
track (for vehicles) **6**.294, **6**.375
tracked vehicle **6**.64, **6**.65
track circuit (electrical) ***6**.359*
traction-engines **6**.332, ***6**.359*
tractor **6**.51, **6**.103, ***6**.359–60*
trade ***7**.314*
adjustment **7**.2
agreements and treaties **4**.252–3
see also economic *under* international organizations and alliances
arms **7**.18
balance of **7**.25
bill of exchange **7**.31
bill of lading **7**.31
boycott **7**.34
Bugis traders **3**.53
chartered company **3**.71
commodity **7**.57
comparative advantage **7**.267, **7**.314
and contract law **7**.69
credit **7**.74
cycle ***7**.314*
in endangered species **7**.106
exchange rate *see* exchange rate
exhibition **6**.160–1
exports **7**.113, **7**.314
free **4**.49, **4**.77, ***4**.125*, **4**.159, **4**.286, **7**.314
see also common markets
fur **3**.136
guilds **3**.149–150
imports **7**.153, **7**.306
mercantilism **3**.229, **7**.198
Merchant Adventurers and Staplers **3**.229
opium *see under* opium
primary commodities **7**.57
primary industry **7**.247
protectionism **7**.251
retail *see* retail trade
routes **3**.42, **3**.122, **3**.329
caravanserai **3**.60
Silk Route **3**.328, **3**.329
and spread of Black Death **3**.42
see also sea; ships
Royal African Company **3**.311
sanctions **7**.34
by ship *see* merchant *under* ships
slaves *see* slavery
smuggling **3**.331
tariffs **4**.338–9, **7**.306
tax on **3**.359
wholesale **7**.93, **7**.331
winds *see* trade winds
wool **3**.229, **3**.385
see also East India Company; market *and* economy and trade *under individual countries*
trade mark **7**.123, ***7**.314*
trade unions *see* unions, trade
trade winds ***1**.341*, **1**.367
Trafalgar, battle of ***4**.349–50*
traffic control ***6**.360*
air **6**.9, **6**.10, **6**.330
roads **6**.301, **6**.360
traffic lights ***6**.360*
tragedy ***5**.459*
Britain **5**.162, **5**.282, **5**.295, **5**.459
catharsis **5**.82
France **5**.113, **5**.372, **5**.459
Germany **5**.202, **5**.249, **5**.262
Greece, ancient **5**.5–6, **5**.150, **5**.425, **5**.459
Italy **5**.12, **5**.21
Roman **5**.411
United States **5**.53
Traherne, Thomas ***5**.459*
Trail of Tears ***4**.350*
trained band ***3**.360*
trains *see* locomotives; railways

Trajan, Roman emperor **3**.18, **3**.95, **3**.308, ***3**.360*
Trajan's column **3**.95
trajectory ***8**.172*
Trakl, Georg **5**.175
tram ***6**.360*
tramway **6**.375
trance, hypnotic **2**.167
Tranquillitatis Mare ***8**.172*
tranquillizers **6**.285, ***6**.360*, **7**.95
transcendence **7**.263
see also deities
transcendental argument ***7**.315*
Transcendental Meditation **7**.196
transcendental numbers ***1**.341*
Transcendentalism **5**.145, **5**.453–4, ***5**.459*, **5**.488
transducer **6**.13, **6**.181, **6**.267, ***6**.360–1*
transfer machine **6**.211, ***6**.361*
transfer orbit **8**.68
transfer printing ***5**.459*
transformer **6**.114, ***6**.361*, **9**.53
transgenic organism ***2**.345*, **6**.153
transhumance **7**.217, **7**.230
transient lunar phenomena ***8**.172–3*
transistor **6**.182, **6**.312, **6**.313, ***6**.361*, ***9**.53*
and amplifier **6**.13–14
charge-coupled device **6**.70
computer **6**.85
pioneer **6**.317–18
and radio **6**.290, **6**.291
in switch **6**.342
transit (astronomy) ***8**.173*
transit system (satellite) **6**.239
transition metals ***1**.341*
cadmium *see* cadmium
chromium *see* chromium
cobalt *see* cobalt
copper *see* copper
gold *see* gold
hafnium **1**.106, **1**.149, **6**.391
iridium **1**.106, **1**.173, **6**.186
iron *see* iron
manganese *see* manganese
molybdenum *see* molybdenum
nickel *see* nickel
niobium **1**.106, **1**.229, **6**.241
organometallic compounds **1**.240
osmium **1**.106, **1**.242, **6**.177
palladium **1**.107, **1**.246, **6**.256
platinum *see* platinum
rhenium **1**.107, **1**.282, **6**.241, **6**.298
ruthenium **1**.107, **1**.288, **6**.304
silver *see* silver
tantalum **1**.331
technetium **1**.332
titanium *see* titanium
tungsten *see* tungsten
vanadium **1**.353
zinc *see* zinc
zirconium **1**.374
Transjordan *see* Jordan
Transkei ***1**.341*
translation, machine **6**.211
translators **3**.41, **5**.156
Bible **3**.41, **3**.187, **3**.364, **3**.386
translocation ***2**.345*
transmission in vehicles:
automatic **6**.26, **6**.152
belt and chain drives **6**.33–4
clutch **6**.77–8
differential **6**.108
gears *see* gear
hydraulic power **6**.35, **6**.44, **6**.173
transmission of electricity **6**.114
transmission of information *see* telecommunications
transmutation of metals **1**.7, **6**.10
transnational relations **7**.209, ***7**.315*
multinational corporations **7**.83, **7**.209, **7**.315, **7**.326
transnational relations (*cont.*):
see also international organizations and alliances; non-governmental organizations
transnationals **7**.209
transparent insulation materials **6**.324
transpiration ***2**.345*
atmospheric ***1**.341*
hydrological cycle **1**.162
transplant surgery **6**.218, **6**.340, ***6**.361*, ***7**.225*
bone-marrow **6**.164
grafting **6**.158
heart **6**.30
and immunology **6**.177, **6**.334
transponder ***6**.361*
transport revolution ***4**.350*
see also vehicles and transport
transportation (convicts) ***4**.85*
see also penal colony *under* Australia
transportation (geomorphology) **1**.82, ***1**.341–2*
by rivers **1**.35, **1**.92, **1**.197
see also deposition
transporter bridge ***6**.361*, **6**.362
transputer ***6**.361*
Trans-Siberian Railway ***4**.350*
transuranic elements **1**.106–7, ***1**.342*
see also in particular plutonium
Transvaal **1**.95, ***1**.342*, **1**.368, **1**.374, ***4**.350*
transverse wave ***1**.342*, **6**.381
Transylvania ***1**.342*, ***3**.360*, ***4**.350*
trapdoor spider ***2**.345*
Trapezium (astronomy) ***8**.173*
trapezium (geometry) ***1**.342*, **9**.51
Trapezium-like object (astronomy) **8**.173
Trappists **7**.205
traps **6**.133, ***6**.362*
Travancore *see* Kerala
travel:
travel-writing ***5**.460*
Britain **5**.41, **5**.54, **5**.68, **5**.421, **5**.432
Japan **5**.37
see also exploration; vehicles and transport
travelator (escalator) **6**.123
travelling waves ***1**.342*
Travers, Morris **1**.275
Travers, P. L. **5**.91
travertine **1**.180, ***1**.342*
Travis, William B. **4**.7
trawler and trawling **6**.239, ***6**.362*
treadmill ***6**.362*
treason ***7**.315*
treaties ***7**.315*
arms control **7**.17–18
significant *see* Central Treaty Organization; North Atlantic Treaty; Versailles; Warsaw Pact *and* Treaties *under* Rome
see also international organizations and alliances *and* foreign policy *under individual countries*
treaty ports ***4**.350–1*
tree *see* forests and trees
tree boa **2**.277, ***2**.346*
tree fern ***2**.346*
tree frog **2**.123–4, **2**.277, ***2**.346*
tree kangaroo **2**.346, ***2**.347*
tree night-hawk **2**.266
tree of heaven ***2**.347*
tree shrew **2**.172
tree sparrow ***2**.347*
tree squirrels **2**.123, **2**.323
tree tomato ***2**.347*
treecreeper ***2**.346*
treehopper ***2**.346*
trees *see* forests and trees
tree-shrew ***2**.347*
Trefousel, J. **6**.105
Treitschke, Heinrich von **4**.236
trellis drainage ***1**.342*
Tremblant, Mont **1**.221
tremolo ***5**.460*
trench, submarine **1**.24, **1**.168, **1**.206, **1**.236, **1**.245, **1**.261, ***1**.342*
trench warfare ***4**.351–2*
Trenchard, Hugh Montague, 1st viscount ***4**.351*
Trent, Council of ***3**.360*, **7**.94
Trent affair ***4**.352*
Trent River ***1**.342*
trepang **2**.298
trephining (trepanning) **6**.340, ***6**.362–3*
Trésaguet, Pierre Marie Jérôme **6**.210, **6**.299, ***6**.363*
Trevithick, Richard ***6**.363*
first steam locomotive **4**.281, **4**.350, **6**.293, **6**.331, **6**.332
steam road vehicle **4**.225
steam-engine **6**.42, **6**.330
Triad Societies ***4**.352*
trials:
by ordeal **3**.262–3
see also courts of law
triangle ***1**.343*, ***9**.51*
see also trigonometry
triangulation **6**.341
Triangulum ***8**.173*
Triangulum Australe ***8**.173*
Trianon, Treaty of **4**.364
Triassic period **1**.134, **1**.135, **1**.182, ***1**.343–4*
tribe (biology) ***2**.133*
tribe (society) ***7**.315*
Tribhuvan, king of Nepal **4**.237
tribology **6**.146
tribunals **7**.181–2, ***7**.315–16*
tribunes **3**.144, ***3**.360*
tricarboxylic cycle **2**.181
triceratops ***2**.347*
Trichinella spiralis ***2**.347*
trickle-feed irrigation **6**.188
tricksters ***7**.316*
Africa **9**.11, **9**.15, **9**.22
Europe **9**.12
in Judaism *see* Jacob
in Norse and Germanic myths **9**.12, **9**.13, **9**.23
North American Indians **9**.9
Oceanic **9**.25
West Indies **9**.4
tricycle ***6**.363*
Trident **6**.230
Trieste ***4**.352*
Trifid Nebula ***8**.173*
triggerfish ***2**.347–8*
trigonometry ***1**.344*, ***9**.51*
cosine **1**.77
hyperbola **1**.163
sine **1**.306
tangent **1**.331
trigonometric parallax **8**.116
see also geometry
Trikāya ***7**.316*
trilobite ***2**.348*
Trinidad and Tobago ***1**.344*, ***3**.360–1*, ***7**.385–6*
independence **4**.376
physical geography and geology **1**.344
religion and myths **7**.41, **9**.33
statesman **4**.379
see also Caribbean; CARICOM
Trinity ***7**.316*
trio ***5**.460*
triode **6**.134
see also thermionic valve
trip-hammer ***6**.363*
Tripitaka ***7**.316*
Triple Alliance (1882, Germany, Austria, and Italy) ***4**.352*
Triple Alliance, War of (1864–70, Paraguay, Uruguay, Brazil, and Argentina) **4**.259–60
triple point ***1**.344*
triple star ***8**.173–4*
triple-alpha process ***8**.173*
triple-expansion engines **6**.263, **6**.330, **6**.333, ***6**.363*
tripod ***6**.363*
Tripolitania **3**.361, ***4**.306–7*
Tripolitan War **4**.352
triptych ***5**.460*
trireme *see* galley and trireme
Tristan de Cunha ***1**.344*
Tristram **9**.5, ***9**.36*
tritium ***1**.344*, **6**.204, **6**.246, **6**.369
Triton (god) ***9**.36*
Triton (satellite of Neptune) ***8**.174*
Trobriand Islanders **7**.73, **7**.132, **7**.178, **7**.190, **7**.197, ***7**.316*
trogon ***2**.348*
Trojan group of asteroids **8**.2, **8**.64, ***8**.174–5*
Trojan satellites **8**.170, ***8**.175*
Trojans *see* Troy
troll ***9**.36*
trolleybus ***6**.363*
Trollope, Anthony ***5**.460*
Trollope, Frances **5**.460
tromba marina ***5**.460*
trombone **5**.59, **5**.352, **5**.397, ***5**.460*
Tromp, Cornelius **3**.14, **3**.312, **3**.361
Tromp, Maarten Harpertszoon **3**.14, **3**.312, ***3**.361*
trompe-l'oeil ***5**.460*
trophic level **2**.124
Tropic of Cancer ***1**.344*
Tropic of Capricorn ***1**.344*
tropical agriculture ***6**.363–4*
see also plantation
tropical month **8**.100
tropical rain-forests ***2**.276*, **2**.277, **6**.111, **7**.83, **7**.349
deforestation **6**.143, **7**.83
see also deforestation *under* forests and trees
tropical revolving storm **1**.160, **1**.340, ***1**.344*, **1**.345, **1**.348, **1**.361
tropical year **8**.189
tropical zones ***1**.344*
tropical climate **1**.65, **1**.66
tropic-bird ***2**.348*
tropopause ***1**.345*, ***8**.175*
troposphere **1**.93, **1**.109, ***1**.345*, ***8**.175*
Trotsky, Leon **4**.178, **4**.193, ***4**.352*
and Brest-Litovsk Treaty **4**.49
and Internationals **4**.165
and Red Army **4**.284, **4**.296
troubadours ***3**.361*, **5**.297, **5**.368
trough (meteorology) ***1**.345*
trout **2**.275–6, ***2**.348*
trout perch ***2**.348*
trouvère **5**.297
Troy and Trojan War ***3**.361*
characters **9**.2, **9**.3, **9**.16, **9**.25, **9**.29
Trucial States ***4**.352–3*
truck *see* lorry
Truck Acts ***4**.353*
Trudeau, Pierre Elliott **4**.195, ***4**.353*
true anomaly **8**.6
Truffaut, François **5**.323, ***5**.460–1*
truffle ***2**.348*
Trujillo, Rafael **4**.102, ***4**.353*
Truman, Harry S. **4**.202, ***4**.353*
and desegregation **4**.282
Doctrine ***4**.353*
Point Four Program **4**.271
and Potsdam Conference **4**.276
Trumbull, John **3**.100
trumpet and trumpeters **5**.22, **5**.59, ***5**.461*
alphorn **5**.13
Britain **5**.22

trumpet and trumpeters (*cont.*):
conch **5**.109
didjeridu **5**.131
pitch **5**.352
trumpeter swan ***2**.348*
trumpetfish ***2**.348*
Trumpler, Robert: Trumpler's classification ***8**.175*
truncated spur **1**.140, **1**.318, ***1**.345*
trunkfish ***2**.348*
truss (building) **6**.48, **6**.336, **6**.337
trust (conglomerate) **7**.316
trust (finance) **7**.165–6, **7**.316
trust (law) ***7**.316*
trusteeship (politics) ***7**.316*, **7**.321, **7**.322
truth **7**.306, ***7**.316*
truth tables (mathematics) ***1**.345*
trypanosomiasis ***2**.348*
Ts'ao Ts'ao (Cao Cao) **3**.60
tsars of Russia *see under* Russia
Tschirnhausen, E. W. von **6**.43
Tseng Kuo-fan (Zeng Guofan) **4**.389
tsetse fly **2**.348, ***2**.349*
Tshombé, Moise **4**.83, ***4**.353*
Tsimshian Indians ***7**.316–17*
Tsiolkovsky, Konstantin **4**.322, **4**.323, **6**.301, **8**.10, **8**.144
Tso Tsung-tang **4**.391
Tsoede of Nupe **3**.260
Tsukuba Science City ***6**.364*
Tsukumi **9**.19
tsunamis **1**.123, ***1**.345–6*
Tsung Dao Lee **1**.329
Tsvetayeva, Marina ***5**.461*
Tuareg **7**.217, ***7**.317*
tuatara ***2**.349*
Tuatha Dé Danann **9**.9, **9**.22, ***9**.36*
tuba **5**.59, **5**.352, ***5**.461*
tube-mouth fish ***2**.349*
tuberculosis ***2**.349*, **6**.253, **6**.375, **7**.59, **7**.153
tubers, plants with ***2**.349*
begonia **2**.28
dahlia **2**.86
gloxinia **2**.138
peony **2**.249
yam **2**.373
see also corm; rhizome
tubes and pipes ***6**.364*
copper **6**.274
drainage **6**.193
gas **6**.150, **6**.236, **6**.270
manometer **6**.214
petroleum **6**.265, **6**.269, **6**.270
PVC **6**.106, **6**.274
sewage **6**.314
siphon **6**.321
steel **6**.214
test-tube **6**.351, **6**.353
tube tunnel **6**.365
tube-railway *see* underground railways
valve **6**.370
venturi **6**.371
waveguide **6**.380
Tubman, Harriet ***4**.353*
Tubman, William Vacanarat Shadrach **4**.195, ***4**.354*
tubular bell ***5**.461*
TUC (Trades Union Congress) **4**.348–9
Tucana ***8**.175*
Tucanoan languages **7**.295
Tucker, Albert **5**.27
Tucker, Benjamin **7**.12
tuco-tuco ***2**.349*
Tudor, Antony ***5**.461*
Tudor, Owen **3**.361
Tudors ***3**.361–2*, ***9**.67*
architecture **5**.423, ***5**.461–2*
family trees **3**.310, **3**.362
see also under rulers *under* Britain
tufa ***1**.346*
tufted duck ***2**.349*
tug ***6**.364*
Tugela ***1**.346*
Tughluq dynasty ***3**.362*
Tughrul Beg **3**.321
Tuileries ***3**.362*
Tukulti-Ninurta I, king of Assyria **3**.26
tulip ***2**.349–50*
tulip tree ***2**.350*
Tull, Jethro ***6**.364*
Tullock, Gordon **7**.253
Tulsīdās **5**.206, **5**.373
tumble-drier ***6**.364*
tumbril **6**.61
tumours **2**.51, ***2**.350*
see also cancer
Tumulus culture ***3**.362*
tuna family **2**.38, **2**.318, ***2**.350*
tundra ***2**.350*
soil **1**.269, ***1**.346*
tuner **6**.291, ***6**.364–5*
tungsten **1**.107, ***1**.346*, **1**.368, **6**.177, ***6**.365*
in steel **6**.11, **6**.235, **6**.333, **6**.365
tungsten carbide **6**.68, **6**.365
Tunguska Event ***8**.175*
Tunisia ***1**.346*, ***3**.362–3*, ***4**.354*, ***7**.386*
in Arab League **7**.15
colonialism **4**.4, **4**.306, **4**.354, **7**.191
law **7**.221
old age **7**.221
physical geography and geology **1**.277, **1**.346
statesman **4**.47
see also Carthage
tunnage and poundage ***3**.363*
tunnel ***6**.365*
Channel Tunnel **6**.69, **6**.365, **7**.67
mining **6**.3, **6**.228
railway **6**.69, **6**.293, **6**.365
tunnelling and shield excavator **6**.49, **6**.223, **6**.316, **6**.365, **6**.367
wind **6**.4, **6**.386
tunnelling (physics) ***1**.346–7*
tunny *see* tuna
Tupac (Topa) **5**.426
Tupac Amarú ***3**.363*
Tupamaro ***4**.354*
tupelo ***2**.350*
Tupolev, Andrey Nikolayevich ***6**.365*
Tupper, Sir Charles ***4**.354*
turaco ***2**.350*
turbidite ***1**.347*
turbidity current ***1**.347*
turbine ***6**.365–6*
gas *see* gas-turbine
Kaplan **6**.378, **6**.379
steam *see* steam-turbine *under* steam power
water **6**.145, **6**.173, **6**.379
water wheels **6**.260, **6**.367, **6**.379
windmills **6**.385
turbocharger **6**.339
turboprop **6**.5, **6**.189–90
turbot ***2**.350*
turbulence ***1**.347*
Turenne, Henri de la Tour d'Auvergne, vicomte de **3**.102, **3**.135, **3**.144, ***3**.363*
Turgenev, Ivan Sergeyevich ***5**.462*
Turgot **3**.117
Turin Shroud ***3**.363*
Turing, Alan Mathison **6**.84, ***6**.366*
Turkey ***1**.347–8*, ***4**.354*, ***7**.386*
arts **5**.462
architecture **5**.333–4, **5**.410–11, **5**.420
see also literature; music *below*
in Council of Europe **7**.71
economy and trade **7**.200, **7**.221
education **7**.265
emigration **7**.31, **7**.200
ethnic groups **4**.188, **7**.17, **7**.178
games **7**.24
Turkey (*cont.*):
Hittites **3**.166, **5**.208
languages **7**.11, **7**.13, **7**.17
laws/legal system **7**.282
literature **5**.205, ***5**.462*
music **5**.486–7
physical geography and geology **1**.347–8
Dardanelles **1**.84
mountains **1**.17, **1**.185
region **1**.20
rivers **1**.114, **1**.337
politics and government **7**.201, **7**.265
leaders **4**.21, **4**.164
and religion **7**.265
prehistory **3**.64
regions **3**.76, **3**.253, **4**.18
religion/theology **7**.16, **7**.265, **7**.282, **7**.304, **9**.40–3
saints **9**.40–3
see also individual names
rulers and dynasties **3**.141, **3**.195, **3**.321
technology and science **6**.34, **6**.176, **6**.201, **6**.225, **6**.258, **6**.341
Turks ***3**.363*
urban areas **3**.86, **3**.276, **7**.298
wars and defence:
with Byzantium **3**.221
CENTO **4**.68
Congress of Paris **4**.260
Crimean **4**.27–8, **4**.88, **4**.260
leaders **4**.164
with Russia **4**.3, **4**.298
World War I **4**.129, **4**.188
see also NATO
see also Balkan States; Hittites; Ottoman empire
turkey (bird) ***2**.351*
turkey oak **2**.231
turkey-vulture ***2**.351*
Turkic languages **7**.11
Turkistan ***1**.348*, ***3**.363*, ***4**.354*
Turkmenistan ***1**.348*, ***4**.354*, **7**.59, ***7**.386*
Turks ***3**.363*
Turk's cap lily **2**.190
turmeric ***2**.351*
Turner, F. J. C. **7**.282
Turner, Joseph Mallord William **5**.5, **5**.179, **5**.462–3, **5**.482
Turner, Nat ***4**.355*
Turner, Ted **7**.38
turnip ***2**.351*
turn-off point ***8**.175–6*
turnover ***7**.317*
turpentine **2**.119, ***6**.366*, **6**.387
Turpin, Dick **3**.165
turquoise ***1**.348*
turtles ***2**.351*
greaved **2**.144
matamata **2**.205
side-necked **2**.309
snake-necked **2**.205, **2**.309, **2**.314–15
snapping **2**.315
soft-shelled **2**.317
terrapin **2**.338
tortoise **2**.344
Tuscany **3**.119, ***3**.363*, ***4**.355*
al-Tūsī, Nāsir al-Dīn **8**.75
tusk ***2**.351*
Tussaud, Marie **5**.483
tussock moths **2**.148, ***2**.351–2*, **2**.354
tussock-grass ***2**.351*
Tutankhamun, Pharaoh of Egypt **3**.243, ***3**.363–4*
Tuthmosis *see* Thutmose
Tutuola, Amos **5**.7
Tuvalu ***1**.348*, ***4**.355*, **7**.296
Tuwim, Julian ***5**.463*
Tvardovsky, Aleksandr ***5**.463*
Twain, Mark **4**.136, ***5**.463–4*
Twardowski, Samuel **5**.355
Tweed, William Marcy 'Boss': and Tweed Ring **4**.337, ***4**.355*
Twenhofel, William Henry ***1**.348*
Twentieth Congress ***4**.355–6*
twentieth-century music ***5**.464*
twenty-one centimetre line ***8**.176*
Twenty-One Demands ***4**.356*
twilight ***8**.176*
twill weave **6**.208, **6**.382
twinkling stars **8**.145
twins, studies of **7**.29, **7**.137, **7**.146
twister *see* tornado
twite ***2**.352*
two-body problem **8**.106
two-party system ***7**.317*
two-stroke engines **6**.34, **6**.262–3
Tycho (Moon crater) ***8**.176*
Tycho Brahe **8**.11, ***8**.176*
and Kepler's Laws **8**.80
observatory **8**.178
solar parallax **8**.150
and supernova **8**.21, **8**.106
Tychonic system **8**.176–7
Tycho's Star ***8**.177*
Tyler, John ***4**.356*
Tyler, Wat **3**.273, ***3**.364*
Tyndale, William **3**.41, **3**.90, ***3**.364*
Tyndall, John **1**.68, **6**.303
type metal **6**.356, ***6**.366*
typesetting and type-casting **6**.281–2, ***6**.366*
Linotype **6**.203, **6**.219, **6**.366
Monotype **6**.232, **6**.366
see also letterpress
typewriter **6**.318, ***6**.366*
typhoid fever **2**.281, ***2**.352*, **6**.192, **6**.377
typhoon ***1**.348*
typhus fever ***2**.352*
typography **5**.177, **5**.219, ***5**.464–5*
Tyr **9**.2, ***9**.36*
tyrannosaurus ***2**.352*
Tyre (place) ***3**.364*
tyre (vehicle) **6**.5, **6**.35, **6**.109, ***6**.366*, **6**.384
Tyrian purple **6**.110
Tyrol ***4**.356*
Tyrone, Hugh O'Neill, 2nd earl of **3**.179, ***3**.364*
Tyrrhenian Sea: islands **1**.101, **1**.196
Tz'u hsi *see* Cixi

U

U Kyin U 5.67
U Nu 4.229
U Pon-nya 5.67
U San Yu 4.229
U Thant 4.361–2, 7.322
Ubaidallah (Mahdi) 3.124
Uccello, Paolo 5.465
Udall, Nicholas 3.252
UDI (unilateral declaration of independence) 7.273
UFO (Unidentified Flying Object) 8.177
Uganda 1.349, 4.356, 7.387
  colonialism 4.4, 4.306
  diseases and disorders 7.9
  economy and trade 4.106
  ethnic groups 7.25
  genocide 7.130
  languages 7.217
  leaders 4.11, 4.229, 4.249
  physical geography and geology 1.65, 1.276, 1.288, 1.349
  refugees from 7.260
  religion and myths 9.20, 9.22
UHF *see* ultra-high frequency
Uhlenbeck, George 1.142
Uitlanders 4.356
Ujigami 9.36
Ujung Pandang *see* Macassar
UK *see* Britain (and United Kingdom)
ukiyo-e 5.207, 5.209, 5.465, 5.466, 5.467
Ukraine 1.349, 3.365, 4.356–7, 7.59, 7.386–7
  languages 7.156, 7.288
  physical geography and geology 1.29, 1.80, 1.349
    rivers 1.84, 1.93
  religion/theology 7.98
  wars and defence 7.220
  see also Crimea
ukulele 5.465
Ulaids 9.27
'ulamā' 7.319
Ulanova, Galina 5.30, 5.51
Ulate, Otilio 4.87
Ulbricht, Walter 4.134, 4.357
ulcer 2.352
ulcerative colitis 2.72
Ulfsson, Gunnbjörn 1.112
Ullur 5.136
Ulm, C. T. P. 6.138
Ulmanis, Karlis 4.190
ulna 2.14
Ulster 3.365
  Ulster Unionist Party 4.357
  Ulster Volunteers 4.357
  see also Northern Ireland
ultisol 1.311, 1.349
ultra vires 7.319
ultracentrifuge 6.67
ultrafiltration 6.130
ultra-high frequency 1.103, 6.15, 6.380, 9.54
ultra-microscopy 6.223
ultrasound 1.349, 6.367
  drilling 6.211
  in medicine 6.60, 6.218, 6.340, 6.357, 6.367
ultraviolet radiation 1.349, 6.117, 6.268, 6.328, 7.226, 8.177
Ulysses *see* Odysseus (mythology)
'Umar, second caliph 7.39
Umar ibn al-Khattab, caliph 3.58, 3.160, 3.365
Umar ibn Said Tal 4.357
Umayyad dynasty 3.365
  art and architecture 5.465–6
umbilical cord 2.352
umbra 8.40, 8.42
umbrella 6.367
umbrella bird 2.352, 2.353
Umbria 3.365
Umbriel 8.177
UN *see* United Nations
Unamuno, Manuel Machado 5.172
Unamuno, Miguel de 5.466
uncertainty principle 1.70, 1.349
unconformity (geology) 1.349–50
unconscious mind 7.77, 7.126, 7.173, 7.319
UNCTAD 4.359–60, 7.129, 7.219, 7.319, 7.321, 7.322
underclass 7.319
undercut bank 1.350
under-developed countries *see* developing countries
underground cables 6.53, 6.347
underground passage *see* tunnel
Underground Railroad (escaping slaves) 4.357
underground railways 6.114, 6.132, 6.161, 6.294, 6.367
underpinning 6.367
undershot wheel 6.367
underwater *see* submarines
underwing moth 2.352–3
underwriting 7.319
UNDP (United Nations Development Programme) 7.201, 7.321
unemployment 7.320
  assistance/benefit 4.357, 7.320
UNEP (United Nations Environment Programme) 7.66, 7.86, 7.321
Unequal Treaty 4.357
UNESCO 7.320, 7.321, 7.322
  and literacy 7.185
  and mass media 7.195
UNGA (United Nations General Assembly) 4.132, 7.320–1, 7.322
Ungaretti, Giuseppe 5.466
Ungava Peninsula 1.350
ungulates (hoofed mammals):
  chevrotain 2.62–3
  horn 2.162
  hyrax 2.168
  pronghorn 2.269
  tapir 2.277, 2.334–5
  see also antelopes; camelids; cattle; goat and goat-antelope family; horses; pigs; sheep
unicameral assembly 7.182
UNICEF 7.321
  aid 7.8
  and child labour 7.46
  and economic adjustment 7.3
  and education 7.102, 7.297
  and medicine and health care 7.140
unicellular organisms 6.65
  amoeba 2.8
  plankton 2.15, 2.247, 2.258
  see also algae; protozoa
Unicode 6.22
unicorn 9.36
unicorn fish 2.353
Unidentified Flying Object *see* UFO
Unification Church 7.320
unified field theory 1.143
Uniformity, Acts of 3.365
Unilateral Declaration of Independence 4.390
Union, Acts of 3.365
Union of Soviet Socialist Republics (1922–91) 4.357–8
  before 1922 *see* Russia
  since 1991 *see* Commonwealth of Independent States; Russian Federation
  abortion 7.1
Union of Soviet Socialist Republics (*cont.*):
  agriculture 4.78, 7.232
  aid from 7.8
  alcohol abuse 7.10
  arts *see under* Russia
  astronomy (including Russia) 8.4, 8.53, 8.55, 8.112, 8.135, 8.156, 8.162, 8.189
    see also under space exploration *below*
  crime 7.225
  and developing countries 7.153
  discrimination, positive 7.243
  and Eastern Europe 4.92, 4.157, 4.159, 7.6, 7.99
    see also Warsaw Pact
  economy and trade:
    adjustment 7.3
    COMECON 4.79, 7.57
    and Group of Seven 7.136
    industrial relations 7.158, 7.315
    military–industrial complex 7.202
    nationalization 7.212
    New Economic Policy 4.238
    planned 7.236
    retail 7.266–7
  education 7.245–6, 7.265, 7.275–6, 7.316
  emigration 7.152
  ethnic groups 7.109
  finance 7.56, 7.201
  foreign policy:
    Cold War 4.78
    Helsinki Conference 4.150
    Nazi–Soviet Pact 4.179, 4.236
    United Nations 4.104
    see also Warsaw Pact *and under* Russia *and also* wars and defence *below*
  genetics (1922–91) 7.1
  glasnost and *perestroika* 5.396, 7.59, 7.132, 7.218
  government *see* politics and government *below*
  languages 7.269, 7.288, 7.335
    see also under Russia
  laws/legal system:
    capital punishment 7.40
    Gulags 7.64
    Islamic 7.282, 7.284
    property 7.250
    show trial 7.239
  literature *see under* Russia
  map 4.296
  music *see under* Russia
  philosophy 7.89
  physical geography and geology *see under* Russia
  politics and government 4.79
    anarchist 4.187
    communist 7.60
    confederation 7.64
    coup d'état attempted 7.73
    élite 7.104
    federalism 7.119
    as great power 7.135, 7.302
    intelligence services 7.162
    leaders 4.49, 4.138, 4.183, 4.193
      listed 9.70
      see also in particular Gorbachev; Khruschev; Lenin; Stalin
    Marxism 7.194
      see also Lenin
    nationalism 7.212
    nomenklatura 7.218
    one-party state 7.223
    police state 7.238
    Politburo 4.272
    prison camps 4.277
    protest 7.239
    purges 7.255
    and religion 7.264–5
Union of Soviet Socialist Republics (*cont.*):
  politics and government (*cont.*):
    revolutionaries 4.178, 4.184, 4.197, 4.220, 4.390
      see also Lenin *and also under* politics *under* Russia
    security and secret police 4.54, 4.71, 4.181, 4.229, 4.244, 4.250
    sedition against 7.278
    Soviets 4.321–2
    state terrorism 7.299
    Twentieth Congress 4.355–6
    women's suffrage 7.332
    see also under Russia *and also* foreign policy *above*
  population growth 7.242
  psychology 7.54, 7.231
  regions 4.354, 4.356
    see also under Russia
  religion/theology:
    Islam 7.282, 7.284
    Jews and Judaism 7.171, 7.335
    and politics 7.264–5
    see also under Russia
  republics listed 9.74
  revolutions *see under* Russia *and also* revolutionaries *under* politics *above*
  social services and security 7.45
  space exploration 4.322, 4.323
    vehicles 6.162, 6.301
      launch 6.196
      manned 6.214
      military 6.225
      Mir 6.327, 8.99
      navigation 6.239
      robots 6.326
      Sputnik 6.301, 6.330
      see also Apollo and *Apollo-Soyuz*
  sport and recreation 7.145, 7.151, 7.223
  as superpower 7.135, 7.302
  technology and science (*including* Russia)
    Academy 6.1
    agriculture 6.34, 6.68
    buildings 6.343
    chemistry 6.25, 6.101
    energy and fuel 6.78, 6.95, 6.213, 6.236, 6.355
    fishing 6.133
    genetics 6.371
    mathematics 6.1, 6.53
    nuclear 6.9, 6.72, 6.243, 6.245, 6.246, 6.275
    telecommunications 6.319, 6.349, 7.216
    universities 6.242
    vehicles and transport 6.42, 6.245, 6.301
      aircraft 6.129, 6.138, 6.140, 6.167, 6.176, 6.319, 6.365
      bridges 6.293
      railways 4.350, 6.150, 6.292–3, 6.367
      roads 6.363
      ships 6.175, 6.376
      space *see under* space exploration *above*
    weapons *see* weapons *below*
    see also entries above and below prefixed 1. *and* 6.
  and UN 7.277, 7.326
  urban areas 7.197
  vehicles *see under* technology and science *above*
  wars and defence:
    Afghanistan invaded 7.6, 7.34, 7.255
    armed forces 4.6, 7.66
    arms *see* weapons *below*
    and China 4.315
    civil 7.50
    communal conflict 7.59

Union of Soviet Socialist Republics (cont.):
wars and defence (cont.):
Cuban Missile Crisis 4.90
Czechoslovakia invaded 7.6
expenditure 7.201
Finnish–Russian 4.120
Hungarian Revolution 4.159
leaders 4.96, 4.186, 4.291, 4.389
'Prague spring' 4.92
Red Army 4.284
Revolution see under Russia and also revolutionaries under politics above
special armed services 7.297
terrorism 7.308
in Warsaw Pact 7.328
weapons see weapons below
World War I 4.49, 4.85
World War II 4.85, 4.106, 4.107, 4.188, 4.193
see also foreign policy above and also Warsaw Pact and under Russia
weapons:
arms control 4.248
CFE 4.17, 7.64
SALT talks 4.330, 6.15, 7.17, 7.87
flexible response 7.121
mines 6.226
missiles 6.15, 6.29, 6.228, 6.229–30, 6.235
Molotov cocktail 6.232
nuclear 6.245, 6.246, 7.220
Nuclear Test-Ban Treaty 4.248
SDI, response to 7.300
trade 7.18
see also Commonwealth of Independent States; East–West relations; Russia; Russian Federation
unions, trade 4.349, 7.314–15
closed shop 7.53–4
collective bargaining 7.56
industrial action 7.157–8, 7.235
Solidarity 4.318
syndicalism 7.304
Trades Union Congress 4.348–9, 7.314
see also under individual countries
uniqueness theorems 1.350
unit see units
unit trust 7.166
Unitarianism 4.358, 7.50, 7.320
unitary state 7.320
United Arab Emirates 1.350, 4.252, 4.352–3, 4.358, 7.15, 7.387
economy and trade see OPEC
United Arab Republic 4.358–9
United Australia Party 4.53, 4.202, 4.217
United Church of Christ 7.65
United Democratic Front 4.359
United Empire Loyalists 3.365
United Irishmen 3.365–6
United Kingdom 1.350, 7.387–8
Union, Acts of 3.365
see also Britain (and United Kingdom); England; Northern Ireland; Scotland; Wales
United Malays National Organization 4.359
United Nations 4.359, 7.321–2
and ageing 7.6, 7.221
and aggression 7.6
aid 7.121
and capital punishment 7.40
Children's Fund see UNICEF
and children's rights 7.45, 7.46, 7.149, 7.319
and collective defence 7.56
Conference on Disarmament 7.64
United Nations (cont.):
Conference on Trade and Development see UNCTAD
Conventions see Conventions
court see International Court of Justice
Covenants see International Covenants
and crime 7.96
demilitarized zone 4.169
and demography see population below
Development Programme 7.201, 7.321
and diplomacy 7.90–1
and drug offences 7.96
and Dumbarton Oaks 4.104
Economic and Social Council 7.321
and economy 4.49, 7.321
and education 7.246
Educational, Scientific, and Cultural Organization see UNESCO
Environment Programme 7.66, 7.86, 7.321
Food and Agriculture Organization 4.120, 7.180, 7.321
GATT see General Agreement on Tariffs and Trade
General Assembly 4.132, 7.320–1, 7.322
and Gulf War (1991) 7.164
Habitat 7.197, 7.282, 7.321
High Commission for Refugees, Office of 7.262, 7.321, 7.322
and homelessness 7.145
and housing 7.197, 7.282
IAEA see International Atomic Energy Agency
ILO see International Labour Organization
IMF see International Monetary Fund
International Bank for Reconstruction and Development see World Bank
International Development Association 7.321, 7.333
International Finance Corporation 7.321
International Telecommunications Union 7.258
and justice see International Court of Justice
and labour see International Labour Organization
and life expectancy 7.184
NIEO see New International Economic Order
and old age 7.6, 7.221
and pacifists 7.227
and pay 7.107
peacekeeping 7.231–2
and PLO 7.109
and population 7.32, 7.42, 7.221, 7.226, 7.241, 7.321
and refugees 7.238, 7.261, 7.262, 7.321, 7.322
Secretaries-General 4.195–6, 4.361–2, 7.321, 7.322, 9.71
Hammarskjöld 4.83, 4.146, 4.359
listed 9.71
Security Council see Security Council
and slavery 7.287
and social services 7.293
and space 7.274
technology 6.184, 6.346, 6.389
Trust territories 4.321
Trusteeship Council 7.316, 7.321, 7.322
UNCTAD 4.359–60, 7.129, 7.219, 7.319, 7.321, 7.322
United Nations (cont.):
WHO see World Health Organization
World Bank see World Bank
World Intellectual Property Organization 7.162, 7.230, 7.321
United Party 4.360
United Provinces of Central America 4.67
United Provinces of Netherlands 3.366
United Reformed Church 7.50, 7.65, 7.112, 7.245, 7.322
United States of America 1.350–1, 3.366, 4.360
before 1783 see America
agriculture 6.7
agrochemicals 6.128, 6.180
apiculture 6.16
crops 6.68, 6.76, 6.351
economics 7.7
Granger Movement 4.139
land reclamation 6.193
machinery 6.5, 6.81, 6.172, 6.210, 6.295, 6.355, 6.359
Populist Party 4.275
reformer 4.370
sharecropping 4.130
sheep 6.315
aid from 4.74, 4.212, 4.271, 7.8, 7.9, 7.57, 7.121
aircraft:
airliner 6.8
airship 6.10
biplane 6.38
early warning 6.27
glider 6.157, 6.165, 6.389
helicopter 6.12
history 6.138, 6.139
military 4.6, 6.42, 6.129, 6.224, 6.296, 6.319, 6.342, 6.374
seaplanes 6.140, 6.311
supersonic 6.339–40
wide-bodied 6.385
anthropology 7.14, 7.273–4
see also Mead, Margaret
arts:
art nouveau 5.455
artists' society 5.318
copyright 7.70
furniture and cabinet-making 5.153, 5.295, 5.413
graphic art and design 5.264, 5.488
interior design 5.455
literature see literature below
museums 7.209
music see music below
Neoclassical 5.13, 5.112, 5.207, 5.486
Pop art 5.240–1, 5.264, 5.482
print-makers 5.82, 5.120, 5.240–1
see also under Abstract Expressionism; architecture; cinema; painting; photography; sculpture
astronomy 8.11
asteroids 8.26, 8.72, 8.80
astronautics see space exploration below
comets 8.186
constellations 8.28
galaxies 8.56, 8.149
measurement 8.14
Moon 8.87
observatories 8.24, 8.60, 8.63, 8.80, 8.81, 8.87, 8.104–5
optics 8.140
photography 8.120
planets and their satellites 8.4, 8.19, 8.36, 8.71, 8.92, 8.98, 8.119, 8.171
radio 8.33, 8.76, 8.115, 8.133
SETI 8.148
United States of America (cont.):
astronomy (cont.):
stars 8.14, 8.23, 8.59–60, 8.100, 8.118, 8.142–3, 8.175
evolution 8.25, 8.146
heat and radiation 8.57, 8.98
spectra 8.51, 8.176
see also Einstein; Hubble; Kuiper
black people see civil rights below
buildings 6.50, 6.51, 6.171, 6.201, 6.299, 6.322
see also under architecture
census 7.42–3
chemistry and chemists:
electrostatics 6.120
liquids 1.152, 1.187
measurement 1.121
physical 1.138–9
plastics 6.27, 6.60, 6.66, 6.172
radio-carbon dating 6.199–200
thermodynamics 1.192–3, 1.238
civil rights/black people 7.51–2
Bill of Rights 3.41, 7.32
Black Power 4.40–1
campaign and desegregation 4.55, 4.57, 4.97, 4.101, 4.125, 4.141, 4.197
discrimination, negative 4.173, 4.187
discrimination, positive 7.243
leaders 4.104, 4.183–4, 4.208, 4.282, 4.353, 4.355, 4.371, 4.378
legislation 3.41, 4.76, 4.173, 7.32
see also slavery below
climate 1.60, 1.65, 1.194, 7.136
computers 6.84, 6.85, 6.86, 6.129, 6.180, 6.211, 6.217, 6.240, 7.19
Constitution see under politics below
cowboys 4.87–8
criminals and outlaws 4.38, 4.39, 4.61–2, 4.88, 4.171, 7.188, 7.331
demography:
abortion 7.1
dependency ratio 7.85
divorce 7.94
illegitimacy 7.152
life expectancy 7.6
population growth 7.242
see also immigration below
disease see medicine below
divorce 7.94
domestic appliances 6.61, 6.146, 6.171, 6.172, 6.297, 6.314, 6.321, 6.369
economics 7.120–1, 7.149
economy and trade:
'affluent society' 7.4
agreements 4.108, 4.178
agriculture see agriculture above
business leaders 4.146
cars see under roads below
cattle trails 4.66
coinage 4.125
division of labour 7.93
finance see finance below
fur trade 4.20–1
'Gilded Age' 4.136
imports 4.147
industry see industry below
iron and steel see under metallurgy below
and law 4.147
merchandizing 7.199
migrant labour 7.200
military–industrial complex 7.202
mining see under metallurgy below
multinationals 7.209
New Deal 4.238, 4.318, 4.374
Nye Committee 4.248
in OECD 7.221
oil see under energy below

United States of America (*cont.*):
economy and trade (*cont.*):
paternalism **7.**231
post-industrial **7.**243, **7.**309
public relations **7.**254
retail trade **7.**266–7
'robber barons' **4.**291
Teapot Dome fraud **4.**339
Tennessee Valley Authority **4.**341
trusts **4.**15–16, **7.**314, **7.**331
and World Bank **7.**333
see also finance; unions and labour; vehicles and transport *below and also individual industries above and below*
education **7.**276
desegregated **4.**57
educators **4.**371–2, **4.**379
expenditure **7.**102
pre-school **7.**245
progressive **7.**88–9
secondary **7.**63, **7.**277
segregated **4.**53
teaching **7.**307
for women **4.**379
electricity:
generation and supply **6.**114, **6.**173, **6.**213, **6.**351, **6.**355, **6.**370
hydro-electricity **6.**173, **6.**379
wind power **6.**386
lamps **6.**53, **6.**89, **6.**116, **6.**177, **6.**194, **6.**383
see also energy *below and under* nuclear *below*
electronics and recording ***6.**118*, **6.**183, **6.**317–18, **6.**325–6, **6.**343, **6.**372
energy and fuel:
alternative and renewable **4.**386, **6.**18, **6.**36, **6.**385
coal **6.**78
natural gas **6.**236
petroleum **4.**291, **6.**263, **6.**269, **6.**270
see also electricity *above*
engines **6.**41, **6.**123, **6.**330
environment:
pollution **7.**2
recycling **7.**260
see also under national parks
ethnic groups **7.**109
gypsies **7.**137
see also immigration *below*
see also Indians, North American
eugenics **7.**110
explored *see under* exploration
explorers from **1.**46, **1.**75, **1.**105, **1.**146, **1.**249, **1.**366
finance:
alimony **7.**189
banks **7.**26, **7.**43
bonds and stocks **7.**34, **7.**172
credit cards **6.**92
currency **4.**87
debt crisis **7.**81
dollar diplomacy **4.**102
financiers **4.**216, **4.**224, **4.**263
gold standard **4.**53–4, **4.**138
index **7.**155
military **7.**201
regulation **7.**120
savings and loan associations **7.**37, **7.**207
securities **7.**120, **7.**299
stock exchange **7.**299
Stock Market Crash (1929) **4.**330
tax **3.**380
trust **7.**316
see also aid *above*
fishing **6.**133
food and drink **6.**38, **6.**134, **6.**342
foreign policy:
in ANZUS **7.**14

United States of America (*cont.*):
foreign policy (*cont.*):
colonialism **4.**321
containment **4.**85
dollar diplomacy **4.**102
Edo Treaty **4.**108
Eisenhower Doctrine **4.**109–10
Fourteen Points peace programme **4.**122
Hay Treaties **4.**148
isolationism **4.**168–9
Latin America **4.**67, **4.**138
Monroe Doctrine **4.**223
neutrality **4.**238
OAS **4.**251, **7.**12, **7.**225
Truman Doctrine **4.**353
Washington Conference **4.**372
Webster–Ashburton Treaty **4.**373
see also aid; economy *above and* United Nations; wars and defence; weapons *below*
genetics **7.**29
glass *see* optics and glass *below*
government *see* politics and government *below*
illustrations **5.**53
immigration **4.**72, **4.**202, **4.**221–2, **7.**105, **7.**153, **7.**200
housing **4.**2
see also demography *above*
industry **7.**277
automation and mass production **6.**23, **6.**26, **6.**216, **6.**301, **6.**356, **6.**384
industrial relations *see* unions and labour *below*
Industrial Revolution **6.**178, **7.**158
see also economy and trade *above and individual industries and entries prefixed* **6.** *above and below*
Ku Klux Klan **4.**187
land:
cattle trails **4.**66
exploration *see under* exploration
Louisiana Purchase **4.**199–200
squatters **4.**326
territorial expansion **4.**156, **4.**209, **4.**376, **4.**377
see also under exploration
west **4.**368, **4.**376, **4.**377, **4.**378
see also agriculture *above*
languages **7.**28, **7.**106, **7.**335
laws/legal system ***7.**323–4*
advocate **7.**4
anti-communist **4.**202
Anti-Trust **4.**15–16, **7.**316, **7.**331
case **7.**41
civil **7.**52, **7.**323
civil rights **4.**76, **4.**173
common **7.**323
Congress **4.**84
contingency fee **7.**68
court-martial **7.**73
courts **4.**55, **4.**93, **4.**152–3, **4.**158, **7.**15, **7.**73, **7.**174, **7.**250
crime *see* criminals *above*
defamation **7.**82
detective **4.**269
discrimation **7.**92
discriminatory **4.**173
due process in **7.**97
and economy **4.**147
euthanasia **7.**112
Factory **4.**115
FBI **4.**117
finance **7.**120
Force **4.**121
habeas corpus **7.**138
homosexuality **7.**147
illegitimacy **7.**152
immigration **4.**202

United States of America (*cont.*):
laws/legal system (*cont.*):
impeachment **7.**153
imports **4.**147
injunction **7.**161
International Court of Justice **7.**164
judges **7.**173
judicial review **7.**173
jurisdiction **7.**174
jurisprudence **7.**174
labour and industrial relations **4.**369, **7.**159
leaders **4.**55, **4.**171, **4.**212, **4.**371
legal aid **7.**182
Lend Lease **4.**193
migration west **4.**156
murder case **4.**299
neutrality **4.**238
nullification **4.**248
plea bargaining **7.**237
police **4.**341, **7.**238
and politics **4.**55, **4.**158, **4.**171
precedent **7.**245
prisons **4.**101
privacy **7.**247
Prohibition era **4.**277, **4.**369, **7.**10
property **7.**250
punishment **7.**255, **7.**279
rape case **4.**306
Reconstruction **4.**284
relief **4.**381
Scopes case **4.**305
segregation **7.**278
slavery **4.**128, **4.**221, **4.**316
Supreme Court **4.**334
surveillance **7.**303
Taft–Hartley Act **4.**336
literature:
biography **5.**400
children's **5.**11, **5.**200
historians **5.**5, **5.**227
libraries **7.**184
naturalism and realism **5.**117, **5.**136, **5.**213
see also under autobiography; critics; drama; essays; journalism; novels; poetry; Romanticism; satire; stories
mail service **4.**273
maps **1.**231, **4.**360, **4.**377
mathematicians **1.**101, **1.**138–9, **1.**141, **1.**226, **6.**385
medicine and health care:
Aids **7.**9
American Red Cross **4.**31
finance **7.**139–40
medical profession **7.**196
Medicare **4.**215
mental handicap **7.**198
psychiatric **4.**101, **7.**116, **7.**251
technology **6.**14, **6.**69, **6.**98, **6.**113, **6.**192, **6.**194, **6.**233, **6.**320, **6.**375
women in **4.**31, **4.**41, **4.**101
metallurgy and mining **1.**8, **6.**146, **6.**164
arts **5.**423
borax mining **6.**42, **6.**43
iron and steel **4.**62–3, **6.**34, **6.**39, **6.**186, **6.**190
museums **6.**322, **6.**323
music ***5.**14–15*, **5.**22
Afro-American **5.**9
cinema **5.**45, **5.**128, **5.**175, **5.**245, **5.**251
conductors **5.**46, **5.**391
jazz *see* jazz
opera **5.**174, **5.**175, **5.**251
Romanticism **5.**34, **5.**251
violin **5.**203, **5.**252
see also under ballet and dance companies; choreographers; composers; pianoforte; songs

United States of America (*cont.*):
national parks *see under* national parks
New Age Movement **7.**214
newspapers **4.**149
nuclear technology:
heavy water **6.**369
reactors and power **6.**23, **6.**127, **6.**242–3, **6.**281, **6.**355
see also under physicists *and* weapons *below*
office equipment **6.**60, **6.**169, **6.**259, **6.**318, **6.**366
oil industry *see* petroleum *under* energy *above*
optics and glass **1.**214, **1.**221, **6.**156, **6.**220, **6.**229, **6.**310, **6.**328
outlaws *see* criminals *above*
paper **6.**59
petroleum *see under* energy *above*
philosophy **7.**178, **7.**259
analytic **7.**12
Dewey **7.**88–9, **7.**105, **7.**244, **7.**246
James **7.**105, **7.**169, **7.**244, **7.**252, **7.**278
libertarian **7.**220
logical positivism **7.**187
see also symbolic interactionism
physical geography and geology ***1.**350–1*
caves **1.**53, **1.**247, **1.**367, **1.**368
climate and weather **1.**60, **1.**65, **1.**194, **7.**136
coast **1.**60
continental drift **1.**135
deserts **1.**85, **1.**145, **1.**313
Driftless Area **1.**96
earthquakes and volcanoes **1.**99
ethologist **1.**265
Everglades **1.**116
geologists **1.**57, **1.**83, **1.**97, **1.**139, **1.**151, **1.**152, **1.**183, **1.**201, **1.**265, **1.**348
geomorphologists **1.**84, **1.**139, **1.**159
Grand Canyon **1.**142–3
Gulf of Mexico **1.**213–14
hydrographer **1.**207–8
islands **1.**7, **1.**124, **1.**199
lakes *see under* lakes
map **1.**231
meteorite crater **1.**213
meteorologists and climatologists **1.**113, **1.**120–1, **1.**286, **1.**335
mountains *see under* mountains
national parks *see under* national parks
petrified wood **1.**255
plains **1.**145–6, **1.**155
plateaux **1.**69, **1.**145–6, **1.**243
regions *see* states *below*
rivers *see under* rivers
soils **1.**205, **1.**311
tornado **1.**340
volcanoes **1.**99, **1.**274, **1.**289–90, **1.**301
weathering **1.**284
physicists:
atomic and nuclear **1.**84, **1.**120, **1.**142, **1.**215, **1.**226, **1.**249, **6.**148, **6.**196–7, **6.**350
see also Manhattan Project *and also* nuclear *above*
electricity/electromagnetics **1.**71, **1.**153, **1.**154–5, **1.**286–7
optics **1.**214, **1.**221
quantum mechanics **1.**138–9, **1.**142
plastics **6.**172
politics and government ***7.**323*, **7.**324
anarchism **7.**12
assassination **7.**20

United States of America (*cont.*):
politics and government (*cont.*):
bicameral **7.**31
Bill of Rights **7.**32
black people *see* civil rights *above*
Bonus Army **4.**45
cabinet **7.**38
carpetbaggers and scalawags **4.**64, **4.**303
CIA **4.**68
civil service **4.**264, **7.**52
communism **4.**122
attitudes to **4.**38–9, **4.**202, **4.**203
Compromise of 1850 **4.**80, **4.**82
Confederacy **4.**82
confederation **7.**64
Congress **4.**84, **4.**287, **4.**308–9, **7.**324
conservative *see* New Right
Constitution ***3.****86–7*, **4.**122, **7.**67, **7.**97, **7.**250, **7.**265
separation of powers **7.**279
Continental Congress **3.**87
Declaration of Independence **3.**100
defence consultant **4.**184
Democrats **4.**49, **4.**53–4, **4.**56, **4.**95, **4.**180, **4.**199, **4.**316, **4.**329, **4.**370
see also under presidents
deregulation **7.**86
diplomats and administrators **4.**55, **4.**184
domino theory **7.**94
economist **4.**185
electoral system and suffrage **4.**276, **7.**104, **7.**123, **7.**124, **7.**260, **7.**332
élite **7.**104
ethnologist **4.**129
executive **7.**113
Fair Deal **4.**115
Federalist **3.**3, **3.**124, **3.**186, **3.**377–8, **4.**2, **4.**117, **4.**147, **4.**269, **7.**119
foreign policy *see* foreign policy *above*
fraud **4.**355
as great power **7.**135
House of Representatives **7.**148, **7.**324
impeachment **7.**153
intelligence services **7.**162
interest groups **7.**163, **7.**186
labour *see* unions *below*
law **4.**55, **4.**158, **4.**171
legislature **7.**182
Molly Maguires **4.**221–2
New Deal ***4.****328*
New Right **7.**69
ombudsman **7.**223
Pan-Americanism **4.**258
parties:
major *see* Democrats *above* and Republicans *below*
minor **4.**15, **4.**117, **4.**125, **4.**141, **4.**185, **4.**275, **7.**242
patriotic **3.**3, **4.**93
pluralism **7.**237
Populist Party and populism **4.**275, **7.**242
presidents: listed **9.**69
primaries **7.**246
protest **7.**239
purge **7.**255
and religion **7.**265
Republicans ***4.****287*
conservative **4.**328
early **4.**126, **4.**310, **4.**333, **4.**369
and foreign policy **4.**198, **4.**329, **4.**362
'Half Breed' **4.**41
military **4.**126, **4.**293
Party **4.**287

United States of America (*cont.*):
politics and government (*cont.*):
Republicans (*cont.*):
Secretaries of State **4.**41, **4.**157–8
'Stalwarts' **4.**271, **4.**328
see also under presidents
secret revolutionary **4.**118
security services **7.**278
sedition against **7.**278
self-help **3.**131–2
Senate **4.**308–9, **7.**279, **7.**324
spies **4.**293
spoils system **4.**326
'Stalwarts' **4.**328
state terrorism **7.**299
states' rights doctrine **4.**329
Tammany Hall **4.**316, **4.**337
two-party system **7.**317
veto **7.**326
vice-presidents **4.**56, **4.**58
Watergate scandal **4.**372
Whigs **4.**77, **4.**147, **4.**329, **4.**369, **4.**373, **4.**377
women politicians **4.**266
women's suffrage **7.**332
XYZ affair **3.**386
see also civil rights; foreign policy; laws *above and also* politics *under* America
pollution **6.**2
population *see* demography *above*
printing **6.**169, **6.**219, **6.**232
professions **7.**249
Prohibition **4.**277, **4.**369, **7.**10
psychology **7.**252
humanistic **7.**268–9
learning and education **7.**33, **7.**35, **7.**113, **7.**291
psycholinguistics **7.**273–4
sexuality **7.**280
social **7.**35, **7.**181, **7.**292
see also behaviourism; Skinner
radar **6.**288
railways **4.**68, **4.**151, **4.**362, **6.**114, **6.**292–4, **6.**367
in regional organizations *see* Organization of American States
religion/theology:
Adventists **7.**3
Amish Church **7.**198
Anglican **7.**13
Baptist **4.**183
Black Churches **7.**32
carnival **7.**40
Christadelphians **7.**47
Christian Science **4.**75
Churches of Christ **7.**48
Evangelical Church and evangelism **4.**101, **7.**65
evangelist **4.**101
fundamentalism **7.**127
Jehovah's Witnesses **7.**171
Jews and Judaism **7.**19, **7.**66–7, **7.**138, **7.**171, **7.**172, **7.**260, **7.**335
Lutherans **7.**187
Maronites **7.**193
Methodists **7.**50, **7.**199
millenarianism **7.**47, **7.**202
Mormons **4.**224, **4.**386, **7.**207
Nestorians **7.**214
and politics **7.**265
Presbyterian **4.**33
Quakers **7.**256
Reformed Church **7.**65
revivalist **4.**288
Scientology **7.**276
Sikhism **7.**286
spiritualism **7.**297
Temperance **4.**233, **4.**340
theosophy **7.**310
Unitarians **7.**320
United Church of Christ **7.**65
voodoo **7.**327

United States of America (*cont.*):
religion/theology (*cont.*):
witchcraft **7.**331
retail trade **6.**62, **7.**43
'Stalwarts' **4.**271, **4.**328
riots and uprisings **3.**380, **4.**148, **4.**150
see also under slavery *below*
roads:
automobile industry **4.**121, **6.**64, **6.**116, **6.**143, **6.**209, **6.**216
engines **6.**210, **6.**224, **6.**303, **6.**330, **6.**383
highways **6.**300
traffic-lights **6.**360
rubber **6.**157
safety **6.**249
science *see* technology and science *below*
segregation **7.**278
sewage **6.**314
shipping **6.**32, **6.**89, **6.**105, **6.**203, **6.**316
navigation **6.**163, **6.**249–50, **6.**315, **6.**328
sailing **6.**76, **6.**94, **6.**310
screw-propelled **6.**123, **6.**284
steam **6.**147, **6.**332
submarines **6.**245, **6.**337
slavery **7.**287
abolition **4.**1, **4.**52–3, **4.**103, **4.**316, **4.**354
laws **4.**128, **4.**221, **4.**316
Proclamation of Emancipation **4.**316
revolts and escapes **4.**353, **4.**355, **4.**357, **4.**364
see also civil rights *above*
social class **7.**319–20
social services and security **4.**318, **4.**348, **4.**381–2, **7.**91, **7.**289, **7.**293, **7.**294
sociology **7.**14, **7.**126, **7.**133, **7.**179, **7.**229, **7.**331
see also Chicago School; Goffman; symbolic interactionism
space exploration ***4.****322*, **6.**196
rocketry **6.**45, **6.**157, **6.**301
satellites **6.**81, **6.**111, **6.**162, **6.**225, **6.**239, **6.**307, **6.**308
spacecraft **6.**17, **6.**214, **6.**326, **6.**327, **8.**76, **8.**91, **8.**121
see also *Apollo*
sport and recreation:
baseball **7.**27
bowling **7.**34
football **7.**122, **7.**123
horse-racing **7.**147
horseshoe-pitching **7.**147
ice hockey **7.**145
ice-skating **7.**151
lacrosse **7.**179
motor-racing **7.**208
Olympic Games **7.**223
pool **7.**289
sailing **7.**272
softball **7.**295
squash **7.**298
surfing **7.**303
tennis **7.**308
states:
central **4.**250, **4.**275
listed ***9.****74*
north-east **3.**86, **3.**217, **3.**225, **3.**252, **3.**253, **3.**302, **3.**363, **3.**372, **3.**383
see also New England *and also* colonies *under* America
south **3.**40, **3.**100–1, **3.**128, **3.**140, **3.**213, **3.**256, **3.**334–5, **3.**374, **4.**302, **4.**341
west **4.**251
see also Alaska; Hawaii
as superpower **7.**302

United States of America (*cont.*):
technology and science:
computers *see* computers *above*
explosives **6.**124
lightning conductor **6.**202
locks **6.**206
measurement **1.**221, **1.**350, **6.**369
scientific societies and academies **6.**12–13, **6.**236
tools **6.**98
see also references above and below prefixed **1.** *and* **6.**
telecommunications:
broadcast database **7.**35
multiplexing **6.**235
news agency **7.**216
radio **6.**20, **6.**48, **6.**98, **6.**208, **6.**249–50, **6.**290, **7.**258
satellite **6.**81, **6.**225, **7.**274
telegraphy **6.**232, **6.**346
telephony **6.**33, **6.**65, **6.**112, **6.**347
television **6.**349, **6.**391, **7.**38–9, **7.**53, **7.**254, **7.**274, **7.**307
textiles **4.**200, **6.**60, **6.**171–2, **6.**247, **6.**314, **6.**384, **6.**391
unions and labour:
American Federation of Labor and Knights of Labor **4.**185, **7.**314
industrial democracy **7.**158
law **4.**369, **7.**159
leaders **4.**94, **4.**194, **4.**288
membership **7.**315
strikes **4.**15, **4.**94, **4.**122, **4.**150, **4.**156, **4.**278
Teamsters **4.**339
and United Nations **4.**104, **7.**277, **7.**320, **7.**321, **7.**326
university **6.**216
urban areas **1.**307, **7.**197, **7.**216, **7.**323
see also New York
vehicles and transport:
air *see* aircraft *above*
amphibious **6.**13
automobiles *see* roads *above*
bridges **6.**111, **6.**293, **6.**341
bulldozer **6.**51
canals **6.**30
diving **6.**31, **6.**33
off-road **6.**249
railway *see* railways *above*
road *see* roads *above*
ships *see* shipping *above*
space *see* space exploration *above*
tunnels **6.**365
Wells, Fargo and Company **4.**374
wars and defence:
airforce **4.**6
see also military *under* aircraft *above*
arms *see* weapons *below*
army **4.**19
19th century **4.**32–3, **4.**93, **4.**126, **4.**157, **4.**267, **4.**305, **4.**378; Civil War **4.**139, **4.**171, **4.**192, **4.**203, **4.**311; Confederacy **4.**121, **4.**171, **4.**331
20th century **4.**48, **4.**202, **4.**212, **4.**263, **4.**329
Bay of Pigs **4.**32
cavalry **4.**293–4
CENTO **4.**68
coalition leaders *see* Gulf War (1991)
conscription **7.**66
Cuban Missile Crisis **4.**90
1812 with Britain **4.**136, **4.**189–90, **4.**238–9, **4.**342, **4.**371
expenditure **7.**201
Gulf War **4.**143–4

United States of America (*cont.*):
wars and defence (*cont.*):
intelligence services **7**.72
Iran Hostage Crisis **4**.165
Iran–Iraq **7**.249
Japan–United States Security Treaty **4**.172
and Mexico **4**.7, **4**.129, **4**.218–19
militia **7**.202
national security state **7**.212
navy **4**.222, **4**.223
leaders **4**.90, **4**.94, **4**.98, **4**.145, **4**.243, **4**.267
Pentagon Papers **4**.265
Spanish–American **4**.207, **4**.209, **4**.324
Tonkin Gulf Resolution **4**.347
treaties **4**.16, **4**.68, **4**.172
Tripolitan War **4**.352
war crimes **7**.328
World War I **4**.390
World War II **4**.22, **4**.85, **4**.194, **4**.254, **4**.255
see also NATO *and also* weapons *below and under* Indians, North American
water **6**.19, **6**.95, **6**.111, **6**.377, **6**.379
weapons and military technology:
chemical **4**.71
control **4**.248, **7**.64
see also CFE; SALT
early-warning system **6**.111
firearms **6**.81, **6**.162, **6**.210, **6**.318
flexible response **7**.121
missiles **6**.15, **6**.228–30
napalm **6**.236
nuclear **6**.245–6, **7**.220
see also Manhattan Project
SDI *see* Strategic Defense Initiative
trade **7**.18
Very light ***6.371***
see also wars and defence *above*
welfare *see* social services *above*
west *see under* land *above*
women and women's movement **4**.93, **4**.379, **7**.332
see also America; American; East–West relations; Indians, North American; presidents
units ***1.350***, ***6.367–9***
astronomical **8**.10–11, **8**.192
imperial **1**.350, **6**.368–9
metric **1**.350, **6**.368–9
see also measurement; SI units
UNIVAC **6**.84, **6**.85
Universal Declaration of Human Rights **7**.5, **7**.110, **7**.149, **7**.164, ***7.322***
freedom of expression **7**.125
as natural law **7**.213
prisoners **7**.12
property **7**.250
torture **7**.313
universal joint ***6.369***
Universal Time **8**.62, **8**.172, ***8.177–8***
see also Greenwich Mean Time
universalism (Hindu) **7**.143
Universalist Church **7**.320
universals (philosophy) **7**.1, **7**.181, **7**.221, ***7.322***
universe ***8.178***
age of **8**.2
end (Big Crunch) **8**.15
expanding **8**.49
grand unified theory **8**.60
models of **8**.37, **8**.60
open **8**.113
origin:
Big Bang theory **8**.15
cosmogony **8**.32–3
oscillating **8**.115
primordial fireball **8**.34, **8**.50
universe (*cont.*):
pulsating **8**.115
static **8**.34, **8**.159
see also entries prefixed **8.**
universities ***3.366–7***, ***4.361***, **7**.141–2
distance education **7**.92–3
revolts *see* students
rite of passage **7**.268
of technology and science **6**.216, **6**.242, **6**.364
UNIX **6**.84, **6**.86, **6**.250
Unkulunkulu ***9.36***
unloading (geology) ***1.351***
unstable equilibrium **8**.47
Upanishads ***3.367***, **7**.3, **7**.155, ***7.322***
see also Vedānta
upas tree ***2.353***
Updike, John ***5.466***
uplands:
slope **1**.113, **1**.307–8
wold **1**.368
see also mountains; plateaux; ridge, topographical
uplift (geology) **1**.82, ***1.351***
see also faults *and* building *under* mountains
Upper and Lower Canada ***4.361***
Upper Palaeolithic ***3.367***, **6**.280
Upper Volta *see* Burkina Faso
upthrow ***1.351***
upwelling **1**.351
Ur ***3.367***
Ural Mountains **1**.135, ***1.351***
Uralic languages ***7.322–3***
Uraniborg ***8.178***
uraninite ***1.351–2***
uranium **1**.107, ***1.352***, **6**.285, ***6.369***
centrifuged **6**.67
mining **7**.58
ore **1**.351–2
uranium-235 **6**.67, **6**.126, **6**.134, **6**.214, **6**.243, **6**.246
uranium-238 **6**.67, **6**.243
see also nuclear energy and power; nuclear reactors
uranium dioxide **6**.4, **6**.243, **6**.369
Uranus (god) ***9.36***
Uranus (planet) ***8.178***
orbit **8**.196
rings **8**.140, **8**.197
satellites **8**.8, **8**.100, **8**.111, **8**.149, **8**.172, **8**.177, ***8.198***
inner satellites of ***8.178***
statistical details **8**.196
Urartu **9**.5
urban areas/urbanization ***4.361***, ***7.323***
borough **3**.45
commune **3**.83
developing countries **7**.336
megacity **7**.197
new town **7**.216
renewal ***7.323***
shanty town **7**.282, **7**.283
slum **7**.288, **7**.334
squatting **7**.298
transport *see* public transport
see also housing *and under individual countries*
Urban II, pope **3**.51, **3**.83, **3**.92, ***3.367***
Urban IV, pope **3**.69, **3**.327
Urban VI, pope **3**.146
Urban VIII, pope ***3.367–8***, **5**.46
al-'Urdī **8**.75
Urdu language **7**.323
literature ***5.466–7***
urea ***2.353***
urea-formaldehyde resin **6**.272, **6**.273, ***6.369***
Urey, Harold Clayton ***6.369***
Urfé, Honoré d' **5**.314, ***5.467***
Uriah **9**.7
urinary tract **2**.34
disorders **2**.326
urine **2**.34, ***2.353***
Urnfield cultures ***3.368***
urochordate **2**.301–2
Ursa Major and Minor and Cluster and System (constellations) ***8.178–9***
Ursids ***8.179***, **8**.194, **8**.195
Ursula, Saint ***9.43***
urticaria ***2.353***
Uruguay ***1.352***, ***4.361***, ***7.388***
economy and trade **4**.320
independence **4**.43
leaders **4**.20, **4**.32
National Liberation Movement **4**.354
physical geography and geology **1**.248, **1**.258, **1**.352
population **4**.320
wars **4**.259–60, **4**.270, **4**.271
Uruguay Round of GATT talks **4**.132, **7**.129
Uruk ***3.368***
USA *see* United States of America
Ushant ***1.352***
U-shaped valley **1**.140
Ussher, James ***3.368***
USSR *see* Union of Soviet Socialist Republics
UT *see* Universal Time
Utamaro Kitagawa **5**.466, ***5.467***
uterus **2**.113, ***2.353***
'Uthmān, caliph *see* Osman, caliph
Uthman dan Fodio **3**.136, **3**.188, **4**.128, ***4.362***
utilitarianism (ethics) ***4.362***, **7**.109, **7**.202, **7**.259, ***7.324***
utility (economics) **7**.229, ***7.324***
Utnapishtim ***9.36***
Utopia Planitia ***8.179***
Utopianism ***7.324***
Utrecht, Peace of **3**.368, ***3.369***
Utrillo, Maurice ***5.467***
Uttacchan, Tuncattu **5**.134
U-2 Incident **4**.110
Utzon, Jørn ***5.467***
Uzbekistan ***1.352***, ***4.362***, **7**.59, ***7.388–9***
physical geography and geology **1**.348, **1**.352
Uzbeks ***3.369***

V

V1 and V2 rockets **6**.29, **6**.93
Vaal river ***1.352***
vaccination *see* immunization
vacuum ***1.352***
cleaner ***6.369***
distillation **6**.210, **6**.264, ***6.370***
electricity conducted through *see* electronics
flask ***6.370***
as insulator **6**.182, **6**.370
pump **6**.227, ***6.370***
vagabonds **3**.344
vagina ***2.354***
Vainamoinen **9**.27, **9**.37
Vairocana **9**.24
Vaisheshika **7**.155
Valadon, Suzanne ***5.467***
Valdés, Alfonso **5**.428
Valdés, Juan **5**.428
valency ***1.352***
Valens, Roman emperor **3**.3, **3**.374
valerian ***2.354***
Valéry, Paul-Ambroise ***5.467***
Valhalla (astronomy) **8**.19
Valhalla (mythology) **9**.5
validity ***7.325***
Valkyries **9**.5, ***9.37***
Vallathol **5**.136
Vallee, Rudy **5**.118
Valle-Inclán, Ramón María del **5**.173, ***5.467***
Vallejo, César Abraham ***5.467***
Valles Marineris ***8.179***
Valley Forge ***3.369***
Valley of the Kings ***3.369***
valley train ***1.353***
valleys ***1.352–3***
dry *see* dry valleys
glaciers **1**.140
gorge *see* gorges
graben **1**.142
hanging **1**.140, **1**.151
rift *see* rift valleys
wadi **1**.359
see also rivers
Valmiki **5**.373
Valmy, battle of ***3.369***
Valois, House of **3**.70, **3**.89, **3**.171–2, ***3.369***
listed ***9.68***
see also Charles VII; Charles VIII; John II; Louis XI; Philip VI
value (economics) ***7.325***
labour theory of **7**.179
value added tax **7**.306
value judgement ***7.325***
valve **5**.59, **6**.304, ***6.370***
Vāmana ***9.37***
vampire (legendary) ***9.37***
vampire bat ***2.354***
Van Allen, James: Van Allen radiation belts ***1.353***, ***8.179–80***
van Alsloot **5**.336
Van Blarenberghe **3**.30, **3**.388
van Breen, Adam **3**.282
Van Buren, Martin **4**.71, **4**.184, ***4.362***
van Campen, Jacob ***5.468***
van de Graaff, Robert: generator **6**.258, ***6.370***
van de Passe, Simon **3**.286
van de Velde *see* Velde
van der Lubbe (arsonist) **4**.286
van der Waals, Johannes Diderik ***1.353–4***
van der Waals' forces ***1.354***

Van Diemen's Land *see* Tasmania
van Druten, John **5**.227
van Dyck *see* Dyck
van Dyke, Willard **5**.488
van Eyck *see* Eyck
van Gogh *see* Gogh
van Loo, Louis Michel **3**.103
van Steenwyck the Younger, Hendrik **3**.161
vanadium **1**.107, ***1**.353*, **6**.11, **6**.87, **6**.333, ***6**.370*
Vanbrugh, Sir John **5**.445, ***5**.467–8*
Vancouver, George ***1**.353*
Vancouver Island ***1**.353*
Vančura, Vladislav ***5**.468*
vandalism **7**.250
Vandals **3**.7, **3**.98, **3**.139, ***3**.369*
Vanderbilt, Cornelius **4**.291, ***4**.362*
Vanderbilt, William Henry **4**.291
Vanderlyn, John **5**.315
Vandernberg, Arthur Hendrick ***4**.362*
Vane, Sir Henry ***3**.369–70*
vanilla ***2**.354*
Vanir **9**.2
van't Hoff, Jacobus Hendricus ***1**.354*
Vanuatu ***1**.354*, ***4**.362*, **7**.296
Vanzetti, Bertolomeo **4**.299
vapour:
  condensation *see* condensation
  deposition ***6**.370*
  evaporation *see* evaporation
  pressure ***1**.354*
  vaporization ***1**.354*
  water **1**.362
  see also gas
vapourer moth ***2**.354*
Varāha ***9**.37*
Varanaṣi ***3**.370*
Varennes, flight to ***3**.370*
Varèse, Edgard ***5**.468*
Vargas, Getúlio **4**.48, ***4**.362–3*
Vargas Llosa, Mario ***5**.468*
variable stars ***8**.180*
  Betelgeuse **8**.14
  cataclysmic **8**.21, **8**.180
    symbiotic **8**.167
    see also nova; supernova
  dwarf **8**.37
  eruptive **8**.180
    Hubble–Sandage **8**.71
    Proxima Centauri **8**.134
    R Corona Borealis **8**.134
    shell stars **8**.149
    T Tauri **8**.65, **8**.175
    see also supergiants
  extrinsic **8**.180
    Cor Caroli **8**.29
    eclipsing binaries **8**.3, **8**.14, **8**.43
    see also pulsars
  first-magnitude **8**.14
  intrinsic *see* cataclysmic; eruptive; pulsating; X-ray *above and below*
  pulsating **8**.180
    Cepheid **8**.23–4
    Mira Ceti **8**.99–100
    red giants **8**.3, **8**.7, **8**.134
    RR Lyrae **8**.142
    RV Tauri **8**.143
  rotating *see* Sun/solar
  X-ray binaries **8**.36, **8**.188
variable-jet carburettor **6**.59
variables (astronomy) *see* variable stars
variables (statistics) ***1**.354*
variance (statistics) ***1**.354*
variation (in Moon's motion) ***8**.180*
variation, theme and (music) ***5**.468*
variation of latitude ***8**.180*
variolation **6**.176–7, **6**.178
Varley, Cornelius and William **5**.469
Varley, F. H. **5**.73
Varley, John ***5**.468–9*
Varman, Mahadevi **5**.206
varnish ***5**.469*, ***6**.370–1*
Varuna ***9**.37*
Vasa, House of ***9**.67*
Vasarely, Victor ***5**.469*
Vasari, Giorgio **5**.16, **5**.18, **5**.23, **5**.55, ***5**.469*
Vasco da Gama *see* Da Gama
Vashti **9**.11, ***9**.37*
Vasquez de Coronado, Francisco **3**.12, **3**.88
vassal ***3**.370*
VAT (value added tax) **7**.306
Vatican ***4**.363*
  Councils:
    First **7**.241
    Second **4**.175, **7**.48, **7**.171, **7**.186, **7**.203, **7**.241
  Lateran Treaties **4**.190
  see also papacy/popes
Vauban, Sébastien le Prestre de ***3**.370*, ***5**.469*, **6**.225, ***6**.371*
Vaucanson, J. de **6**.87
Vaucouleurs, Gérard de **8**.184
vaudeville **5**.154, ***5**.469*
Vaughan, Henry and Thomas ***5**.469*
Vaughan, Keith **5**.316
Vaughan Williams, Ralph **5**.382, **5**.457, **5**.464, ***5**.469*
vault **5**.344, ***5**.469–70*, **6**.49, **6**.51, **6**.302, **6**.337, ***6**.371*
Vavilov, Nikolai Ivanovich ***6**.371*
VDU (visual display unit) **6**.64, **6**.84, ***6**.373*
vector (genetic engineering) ***2**.354*
vector (mathematics) ***1**.354–5*
  components of **1**.70–1
  quantity and force **1**.124
  tensor **1**.333
vectors (disease) **2**.186, ***2**.354*
  amoeba **2**.101
  bacilli **2**.101, **2**.364
  bacteria *see* infections
  black fly **2**.234
  Cape buffalo **2**.44
  fleas **2**.121, **2**.352
  flukes *see* flukes
  fungi **2**.129, **2**.211, **2**.260, **2**.285, **2**.291, **2**.314, **2**.341–2
  hookworm **2**.161
  housefly **2**.164
  housemouse **2**.164
  lice **2**.352
  mites **2**.214, **2**.292–3
  pathogens **2**.244
  protozoa *see* infections
  rats **2**.278, **2**.364
  Rickettsia **2**.287, **2**.353
  roundworm *see* roundworms
  Salmonella **2**.352
  sand fly **2**.293
  snails **2**.29
  streptococci **2**.296
  ticks **2**.287, **2**.352
  tsetse flies **2**.348, **2**.349
  vampire bat **2**.354
  virus *see* infections
  see also mosquitoes; parasites; pathogens
Vedānta **7**.155, ***7**.325*
  Advaita **7**.3, **7**.155, **7**.156
  see also *Upanishads*
Vedas ***3**.370–1*, **5**.220, ***5**.470*, **7**.142, **7**.155, ***7**.325*
  deities **9**.18, **9**.32, **9**.34, **9**.37
Vega, Suzanne **5**.358
Vega (star) ***8**.180*
Vega Carpio, Lope de ***5**.470*
vegetables:
  adhesives **6**.3
  artichoke **2**.16
  asparagus **2**.17
  aubergine **2**.18
  beans **2**.25–6, **2**.42
  beetroot **2**.28
  beets **2**.321, **2**.329
  brassica *see* brassica
  brussels sprout **2**.43
  cabbage *see* cabbage
  capsicum (pepper) **2**.52, **2**.63
vegetables (*cont.*):
  carrots *see* carrot family
  cassava **2**.56
  cauliflower **2**.58
  celery **2**.59
  chick-pea **2**.63
  chicory **2**.63
  chive **2**.64
  corn salad **2**.77
  cress **2**.80
  cucumber **2**.83
  cultivation ***6**.371*
  endive **2**.107–8
  fennel **2**.115
  forage crops **6**.142
  garlic **2**.133
  gourd **2**.142
  horticulture **6**.170, **6**.371
  kohlrabi **2**.180
  leek **2**.186
  lettuce **2**.188
  mallow **2**.200
  marrow **2**.204
  nasturtium **2**.225
  onions **2**.186, **2**.234, **2**.306
  parsnip **2**.243
  pepper **2**.249
  plantain **2**.21
  potato **2**.266
  radish **2**.275
  rape **2**.278
  rhubarb **2**.284
  root **6**.7, **6**.142, **6**.271, **6**.302, **6**.303, **6**.371
  salsify **2**.292
  scorzonera **2**.297
  sea kale **2**.299
  shallot **2**.305–6
  shore plants **2**.308
  sorrel **2**.317
  spinach **2**.321
  swede **2**.331
  sweet pepper **2**.332
  sweet potato **2**.321
  taro **2**.335
  textured vegetable protein **6**.142
  watercress **2**.361
  yam **2**.373
  see also fruit; legumes
vegetarianism **7**.8, ***7**.325*
vegetation:
  chaparral **2**.61
  forest *see* forests and trees
  grasslands *see* grasslands
  moor **2**.218
  pampa **2**.240, **2**.241
  prairie **2**.266
  savannah **1**.294, **2**.294
  shrub *see* shrubs
  steppe **2**.325
  succession **2**.328
  taiga **2**.333
  tundra **2**.350
  see also crops; plants; vegetables
vegetative reproduction **2**.282
vehicles and transport ***4**.350*
  air *see* aircraft; space-flight
  all-terrain **6**.209–10
  automobiles *see* motor car
  bicycle *see* bicycle
  boats *see* ships and boats
  brakes *see* braking system
  cable-car **6**.53
  caravan **6**.58
  caterpillar tracks **6**.64, **6**.65
  electric **6**.115–16, **6**.147, **6**.363
  employment percentage **7**.106
  ferries **6**.60, **6**.127, **6**.172, **6**.277
  friction **6**.146
  gear *see* gear; transmission
  hi-jacking **7**.142
  hovercraft **6**.79–80, **6**.171, **6**.172, **6**.175
  lift **6**.201
  litter **6**.205
vehicles and transport (*cont.*):
  motor *see* motor vehicles
  off-road **6**.249
  public *see* public transport
  rail *see* railways
  road *see* roads
  ships *see* ships and boats
  skates **6**.321
  skis **6**.322
  sledge **6**.300, **6**.322
  space *see* space-flight
  steering system **6**.333–4
  suspension system **6**.120, **6**.317, **6**.341
  transport revolution **4**.30
  tricycle **6**.363
  tyre **6**.5, **6**.35, **6**.109, **6**.366, **6**.384
  wheel *see* wheel
  see also under individual countries
Veil Nebula ***8**.180*, **8**.181
vein **2**.340, ***2**.355*
Vela (constellation) ***8**.180*
Vela pulsar ***8**.180*
Velasco, Torrejón y **5**.257
Velázquez, Diego Rodríguez de Silva **3**.49, **3**.93, **5**.359, ***5**.470*
veld ***1**.355*
Velde, Henri van de ***5**.470*
Velde, Willem van de, the Younger **5**.283
vellum **6**.258
velocipede **6**.35, **6**.234
velocity and speed ***1**.355*, ***8**.180–1*
  acceleration *see* acceleration; accelerator
  angular **1**.14, **8**.6
  cruise control **6**.92–3
  escape **8**.48
  and force **8**.51–2
  and gear **6**.152
  governor **6**.158
  hypersonic **6**.175, **6**.386
  of light **1**.123, ***1**.317*, **8**.1, ***8**.181*
  measurement **6**.14, **6**.192, **6**.206, **6**.328, **6**.368, **6**.386
  radial **8**.132
  of sound ***1**.317*, **6**.140, **6**.175, **6**.211, **6**.339
  speedometer **6**.249, ***6**.328*
  terminal **1**.333
  triangle **6**.238
  velocity ratio ***6**.371*
  velocity–distance relation **8**.71
  winds **1**.35, **6**.14, **9**.55
velvet ant ***2**.355*
Vendée ***3**.371*
Vendelinus **8**.150
Vendôme, Louis Joseph, duc de ***3**.371*
veneer ***5**.470*
Venera ***8**.181*
venereal disease *see* sexually transmitted diseases
Veneti ***3**.371*
Venetiaan, Ronald **4**.334
Venetian art ***5**.470–1*
  architecture **5**.400, **5**.401–2
  painting **5**.43, **5**.79, **5**.177–8, **5**.195, **5**.269, **5**.382–3, **5**.456, **5**.457
  sculpture **5**.472
  see also Venice
Veneto, Bartolomeo **3**.45
Venezuela ***1**.355*, ***4**.363*, ***7**.389*
  Andean Group **7**.13
  economy and trade **4**.12, **4**.251, **4**.320
    see also OPEC
  ethnic groups **7**.335
  exploration **1**.275
  independence **4**.43
  physical geography and geology **1**.355
    lake **1**.204–5
    rivers **1**.14, **1**.241
    waterfalls **1**.14

Venezuela (*cont.*):
politics and government **4**.36–7, **4**.138, **4**.144, **4**.255, **7**.47
population **4**.320
Venice ***3**.371*
adventurers and travellers **1**.205, **3**.56, **3**.62, **3**.221
Cambrai, League of **3**.58
politics and government **3**.96, **3**.104, **3**.105
see also Venetian art
Venizélos, Eleuthérios ***4**.363*
Venn, John: Venn diagram ***1**.355*
venomous snakes *see* poisonous snakes
Venturi, Robert ***5**.471*
venturi tube ***6**.371*
Venus (goddess) **9**.29, ***9**.37*
Venus (planet) ***8**.104*, ***8**.182*
ashen light **8**.8
Greek name for **8**.67
orbit **8**.17, **8**.196
statistical details **8**.196
Venus's fly-trap ***2**.355*
verb **7**.22, **7**.181, **7**.229
Vercingetorix **3**.137, **3**.138, **3**.306, ***3**.371*
Verdi, Giuseppe ***5**.471*
opéra-ballet **5**.329
singers of his works **5**.71, **5**.132, **5**.270, **5**.289, **5**.321, **5**.342
verdin **2**.342
Verdon, Pierre **6**.142
Verdun, Treaty of **3**.141, ***3**.372*
Vereeniging, Treaty of ***4**.363*
Verga, Giovanni **5**.140, ***5**.471*
verification (arms control) **7**.17–18
verificationism (philosophy) **7**.187, ***7**.325*
Verlaine, Paul ***5**.471–2*
Vermeer, Jan **5**.196, ***5**.472*, **5**.473
Vermont ***3**.372*
vernacular architecture ***5**.472*
vernal (spring) equinox **1**.111, **1**.297, **8**.47
Verne, Jules ***5**.472*
Vernet family ***5**.472*
Vernier, Pierre: Vernier calliper and scale **6**.198, ***6**.371*, ***8**.182*
Vernon, Edward **3**.186, ***3**.372*
Veronese ***5**.472*
Veronica **2**.319
Verrazzano, Giovanni da ***1**.355*, ***3**.372*
Verrocchio, Andrea del ***5**.472*
Versailles ***3**.372*
building and interior **5**.56, **5**.259, **5**.262
park **5**.261
Peace Settlement ***4**.363–4*, **7**.315
Treaty (Frankfurt) **4**.124
verse *see* poetry and poets
vertebrae **2**.56, **2**.70, **2**.93, ***2**.355*, ***9**.58*
vertebrates ***2**.355*
see also amphibians; birds; fish; mammals; reptiles; spine
vertical, prime **8**.128
vertical integration ***7**.325–6*
vertical take-off and landing *see* VTOL aircraft
vertisol ***1**.355*
Vertov, Dziga **5**.132
Verwoerd, Hendrik Frensch ***4**.364*
Very, E. W.: Very light ***6**.371*
very high frequency **1**.103, **6**.15, **6**.380, **6**.390, **9**.54
Very Large Array (VLA) ***8**.182–3*
very long baseline interferometry ***8**.184*
Vesalius, Andreas ***2**.355*, **3**.301
Vesey, Telemaque ***4**.364*
Vespasian, Roman emperor **3**.50, **3**.82, **3**.188, **3**.307, ***3**.372*
Vespucci, Amerigo ***1**.355–6*, ***3**.372*
vessel flute **5**.324, ***5**.472*, **5**.474
vessels *see* ships and boats
Vesta (asteroid) ***8**.183*, **8**.193
Vesta (mythology) ***9**.37*
Vestal virgin ***3**.372–3*
vestigial organ **2**.13, ***2**.355*
Vesuvius ***1**.356*, **3**.164, **3**.287–8
vetch ***2**.355*
veterinary medicine ***6**.371*
see also diseases and disorders of animals
veto ***7**.326*
VHF *see* very high frequency
VHS (Video Home System) **6**.373
viaduct **6**.293, ***6**.371–2*
vibrations ***1**.356*
Viburnum **2**.147
vicarious liability ***7**.326*
vice (tool) **6**.358, ***6**.372*
Vicente, Gil ***5**.474*
Vichy government ***4**.364*
Vicksburg Campaign ***4**.364*
Vico, Giovanni Battista **7**.144
Victor Emanuel I, king of Italy **4**.269
Victor Emanuel II, king of Italy **4**.170, **4**.269, **4**.290–1, ***4**.364*, **4**.365, **4**.369
and Mafia **4**.206
Victor Emanuel III, king of Italy ***4**.364–5*
Victoria, queen of Great Britain **4**.146, ***4**.365*
and anaesthesia **6**.320
and arts **5**.491
as Empress of India **4**.50, **4**.52, **4**.101, **4**.105, **4**.277
and Melbourne **4**.217
Prince Consort **4**.8, **4**.140, **4**.352, **4**.365
see also Victorian architecture
Victoria, Guadalupe **4**.170
Victoria, Tomás Luis de ***5**.474*
Victoria (Australian state) ***1**.356*, ***4**.365*
Victoria (carriage) **6**.61
Victoria Falls ***1**.356*
Victoria Island ***1**.357*
Victoria, Lake ***1**.356*
Victoria waterlily **2**.289
Victorian architecture **5**.66, **5**.68, **5**.342, **5**.369, ***5**.474*
see also Gothic Revival
vicuña ***2**.356*
Vidal, Gore **5**.207, ***5**.474*
video:
certificates ***9**.62*
recording ***6**.372–3*
camera ***6**.372*
compact disc **6**.81–2
films **7**.39
imaging **6**.104
recorder **6**.372–3, **7**.308
scrambler **6**.310
videotex **6**.180, ***6**.373*
Vidyasagar, Isvara Chandra **7**.143
Vieille, P.-M.-E. **6**.242
Vieira, João **4**.143
Vienna, Congress of ***4**.365–6*
Vienna Circle **7**.187, **7**.188
Vienna Conventions:
on Diplomatic Relations **7**.90
against Illicit Traffic in Narcotic Drugs **7**.96
Vierzehnheiligen **5**.385
Vietcong ***4**.366*
Vietminh ***4**.366–7*
Vietnam ***1**.357*, ***4**.367*, ***7**.389–90*
agriculture **7**.232
arts ***5**.474–5*
and ASEAN **7**.20
Champa kingdom **3**.67–8
in COMECON **7**.57
decolonization **4**.321
guerilla movements **4**.366–7
land reform **7**.179
languages **7**.22–3, **7**.305
literature ***5**.475*
Vietnam (*cont.*):
physical geography and geology **1**.357
politics, government, and leaders **4**.154, **4**.240, **4**.241, **7**.60, **7**.94
refugees **7**.261
regions **3**.16, **3**.80, **4**.14
religion/theology **7**.36
wars and defence **7**.255
see also Vietnam War
Vietnam War ***4**.367*
conscription **7**.66
counter-insurgency **7**.72
map **4**.367
objections to **7**.215, **7**.232
technology **4**.6, **6**.10, **6**.12, **6**.98, **6**.167
war crimes **7**.328
viewdata **6**.373
viewfinder **6**.65
vigilante ***4**.368*
Vignola, Giacomo Barozzi da **5**.475
Vigny, Alfred de ***5**.475*
Vigo, Jean **5**.154
vihuela ***5**.475*
Vijayanagar, emperor of India **3**.244, ***3**.373*
Viking (spacecraft) ***8**.183*
Vikings/Norsemen **3**.318, ***3**.373–4*
arts ***5**.475–6*
explorers and leaders **1**.112, **3**.203, **3**.306, **3**.373, **3**.374
language **7**.107, **7**.156
longships **3**.210, **3**.248, **5**.475, **6**.306, ***6**.373*
parliament **3**.354
religion and myths **9**.30, **9**.37, **9**.38–9
deities **9**.2, **9**.6, **9**.13, **9**.22, **9**.23, **9**.27, **9**.35, **9**.36
dwarfs and elves **9**.10, **9**.36, **9**.38
giants **9**.6, **9**.14, **9**.36
places **9**.5, **9**.16, **9**.25, **9**.26
tricksters **9**.13, **9**.23
Villa, Francisco 'Pancho' **4**.64, **4**.219, ***4**.368*
Villafranca di Verona, Treaty of ***4**.368*
Villa-Lobos, Heitor ***5**.476*
Villanovans ***3**.374*
Villard d'Honnecourt ***5**.476*
Villeda Morales, Ramón **4**.156
Villehardouin, Geoffroi de ***3**.374*
villein ***3**.374*
Villeneuve, Pierre C. J. B. S. de **4**.349
Villiers, George, 1st duke of Buckingham **3**.51–2
Villon, François **5**.30, ***5**.476*
Vimy Ridge, battle of ***4**.368*
vīṇā ***5**.476*
Vincent, Henry **4**.239
vine snake **2**.277
vinegar fly *see* fruit fly
vines *see* grape
Vinland **3**.373, ***3**.374*
Vinogradov, Oleg **5**.248
vinyl chloride *see* PVC
viol **5**.154, **5**.437, ***5**.476*
viola **5**.352, ***5**.476–7*
violence:
domestic **7**.94, **7**.115
political **7**.239
violent crime ***7**.326*
see also terrorism and anarchism; wars; weapons *and also* wars; weapons *under individual countries*
violets **2**.97, ***2**.356*
violin and violinists ***5**.477*
American Indian **5**.15
Austria **5**.435
Bohemia **5**.47
Britain **5**.291
Czech **5**.439
violin and violinists (*cont.*):
Denmark **5**.318, **5**.320
France **5**.254, **5**.270
Germany **5**.62, **5**.280, **5**.291, **5**.429, **5**.430
Hungary **5**.240
Italy **5**.113, **5**.172, **5**.335, **5**.447, **5**.478
pitch **5**.352
Russia **5**.326
United States **5**.203, **5**.252
see also fiddle
Viollet-le-Duc, Eugène-Emmanuel ***5**.477*
violoncello *see* cello and cellists
viper family ***2**.356*
adder **2**.3
gaboon viper **2**.130, **2**.131
puff adder **2**.271
rattlesnake **2**.278
Viracocha ***9**.37*
Virchow, Rudolf Carl ***2**.356*
vireo ***2**.356*
Viresalingam **5**.136
Virgil **5**.45, **5**.136, **5**.141, **5**.147, **5**.441, ***5**.477*
Virgin Islands ***1**.357*, ***4**.368*
virginals ***5**.477–8*
Virginia **3**.40, ***3**.374*
Campaigns ***4**.368*
Virginia creeper **2**.68, ***2**.356*
Virginius incident ***4**.368*
Virgo (constellation) ***8**.183–4*
Virgo A (radio source) ***8**.184*
Virgo Cluster (of galaxies) ***8**.184*
virions **2**.356
virtues **7**.109, **7**.237
see also ethics; morality
virus ***2**.356*, ***6**.373*
DNA and RNA **2**.286, **2**.356
HIV *see* Aids
infections *see* infections
and interferons **2**.172, **6**.184, **6**.188
virus, computer **7**.63
Viśākhadatta **5**.401
viscacha ***2**.356*
Visconti, Filippo Maria and Gian Galeazzo ***3**.374*
Visconti, Luchino **5**.279, ***5**.478*
viscose *see* rayon
viscosity ***1**.357*, **6**.138
Vishnevskaya, Galina **5**.391
Vishnu **7**.34, **7**.133, **7**.142, ***7**.326*, **9**.26, ***9**.37*
avatars (incarnations) **7**.23, **9**.22, **9**.24–5, **9**.26, **9**.30–1, **9**.37
see also Krishna
Vishnusharman **5**.338
Vishtaspa, king of Persia **3**.389
visibility (meteorology) ***1**.357*
Visigoths **3**.3, **3**.7, **3**.51, **3**.98, **3**.354, ***3**.374*
see also Goths
vision and eyes ***2**.112*, ***2**.308*, ***2**.309–10*, ***9**.61*
and brain **2**.41
disorders/blindness ***2**.34*, **6**.87, **6**.266, **6**.377, ***7**.33*
braille **6**.44, **7**.35, ***9**.48*
cataract **2**.56
colour blindness **2**.72
conjunctivitis **2**.75
contact lens **6**.87
drugs for **6**.266
glaucoma **2**.137
ophthalmic surgery **6**.250–1
and pollution **6**.377
retinopathy **2**.283
river blindness **2**.234, **2**.286, **3**.77
spectacles **6**.327
and vitamin deficiency **2**.90
insects **6**.197
mirage **1**.216

vision and eyes (*cont.*):
  persistence 6.73
  tears 2.337
  see also light
Visnuvardhana of India 3.170
Vistula *1.357*
visual arts *see* architecture; ceramics; decorative; engraving; painting; sculpture
visual binary 8.15
visual display unit 6.64, 6.84, *6.373*
vital statistics 7.326
vitamins 2.92, *2.357*, 6.140, 6.141
  A 2.55, *2.357*
  B complex *2.357*
  C 2.308, *2.357*
    see also citrus
  D *2.357*
  deficiency 2.90, 6.373, 7.198
  discovery 6.170
  E *2.357*
  K *2.357*
  treatment *6.373*
Vitellius, Roman emperor 3.290, 3.307
Vitoria, battle of *4.368–9*
vitrification *6.373*
Vitruvius 5.379, *5.478*, 6.175, *6.373*, 6.379
Vitry, Philippe de 5.379
Vittorini, Elio *5.478*
Vittorio Veneto, battle of *4.369*
Vivaldi, Antonio *5.478*
Vivekananda, Swami 4.152, *4.369*, 7.3, 7.143
vizier *3.375*
Vladimir I, St *3.375*
Vladimir, grand duke 3.8–9
Vlaminck, Maurice de *5.478*
VLBI (very long baseline interferometry) *8.184*
Vo Nguyen Giap 7.136
vocational education *7.326*
vocoder *6.373–4*
Vodun *9.37*
  see also Voodoo
Vogel, Sir Julius *4.369*
voice and voice box 2.183, *2.357*, 2.358, 5.478–9
  see also speech
Volans *8.184*
volatile liquid 6.34
volcanic rocks *see* extrusive (volcanic) igneous rocks
volcanoes *1.357*, 1.358
  Africa 1.99
  Antarctica 1.112
  Asia 1.99, 1.129, 1.184
  caldera *see* caldera
  Caribbean 1.221, 1.233
  craters 1.49, 1.79
  deposits and projectiles:
    agglomerate 1.5
    ash 1.21
    cinder cone 1.63
    fumarole 1.129
    ignimbrite 1.165–6
    nuée ardente 1.233
    pyroclasts 1.270, 1.271, 1.295
    scoria 1.295
    tephra 1.333
    volcanic bomb *1.357*
  Indonesia 1.175, 1.184
  islands *see* islands *under* Pacific Ocean area
  Italy 1.114, 1.356, 3.164, 3.287–8
  Japan 1.129
  Kenya 1.182
  Mexico 1.248, 1.264
  New Zealand 1.100, 1.287, 1.339
  Pacific 1.99
  and plate tectonics 1.261
  rocks associated with *see* extrusive (volcanic) igneous rocks
  shield 1.301
  South America 1.78, 1.99
volcanoes (*cont.*):
  Surtsey island 1.327–8
  United States 1.99, 1.274, 1.289–90, 1.301
  volcanic plug *1.357*
  see also igneous rocks
voles *2.357–8*
  bank 2.21–2
  field 2.117
  lemming 2.186–7
  water 2.194, 2.289, 2.357, 2.362
Volga *1.357–8*
volleyball *7.326–7*
Vollmöller 5.377
Volstead, Andrew: Volstead Act *4.369*
volt *1.358*, 6.368
Volta, Alessandro Giuseppe Antonio, Conte *1.358*, 6.32, *6.374*
Volta River *1.358*
voltage 6.75
  bias 6.35
  electrometer 6.118, 6.267
  high-voltage electricity generation 6.127
  and locomotives 6.115
  measurement 6.368, 6.374
  multiplier 6.258
  signal, electrical 6.319
  transformer 6.361
  and transmission 6.114
  voltmeter 6.374
voltaic cell 6.267
  see also battery
voltaic pile 6.374
Voltaire, François-Marie Arouet *3.375*, *5.479*
  as Encyclopedist 3.115, 3.300
  on English 3.55
  and Enlightenment 3.117, 3.300
  novel and satire 5.402, 7.182
  and Russian monarchs 3.64, 3.133, 3.300
voltmeter *6.374*
volume *1.358*
  density 1.87
  measurement 6.368
voluntary organization *7.327*
Von Frisch, Karl 2.194, *2.358*
von Neumann, John 6.84–5, 6.240
Vonnegut, Kurt, Jr *5.479*
Voodoo *7.327*, 9.22, 9.37, 9.39
Voortrekkers 4.140
Vörösmarty, Mihály 5.214
Vorticism (art) 5.148, 5.418, *5.479–80*, 5.496
vorticity (fluids) *1.359*
Vortigern 3.160, *3.375*
Vosges *1.359*
voting:
  behaviour *7.327*
  franchise *see* suffrage
  rights 7.123, 7.332
  veto 7.326
  for women 4.258, 4.333, 4.381, 7.123, 7.332
  see also electoral system
Vouet, Simon *5.480*
Voyager 8.184, *8.185*
voyages of discovery *see* exploration
Voysey, Charles Francis Annesley *5.480*
VTOL (Vertical Take-Off and Landing) aircraft 6.138, 6.225, *6.374*
  see also helicopter
Vuillard, Edouard *5.480*
Vulcan (god) *9.37*
Vulcan (hypothetical planet) *8.185*
vulcanite 6.296
vulcanization of rubber 6.303, 6.304, 6.338, 6.350, *6.374*
  pioneers 6.157–8, 6.165
  and tellurium 6.350
  see also waterproofing
vulcanoid 8.185
Vulpecula *8.185*
vultures *2.358*
  condor 2.74, 2.358
  griffon 2.144, 2.145, 2.358
  lammergeyer 2.181, 2.358
  turkey-vulture 2.351
VX 6.240
Vydūnas (Vilius Storasta) 5.34
Vygotsky, Lev 7.54, 7.273

W

Wackenroder, Wilhelm Heinrich 5.174
Wade, Benjamin Franklin *4.369*
Wade, General 3.129
wadi 1.20, *1.359*
wading birds *2.358*
  avocet 2.19, 2.316
  curlew 2.84, 2.327
  dunlin 2.100
  flamingo 2.119
  godwit 2.140
  greenshank 2.144
  herons 2.37, 2.157
  ibis 2.168
  jacana 2.174
  oystercatcher 2.238
  phalarope 2.251
  plovers 2.182–3, 2.259
  pratincole 2.267
  redshank 2.280
  ruff 2.290
  sanderling 2.293
  sandpiper *see* sandpiper
  snipe 2.239, 2.315
  stilt 2.326
  stone-curlew 2.327
  storks *see* stork
  woodcock 2.370
  see also birds *under* water
Wafd *4.369*
wages *7.328*
  see also pay
Wagner, Otto *5.480*
Wagner, Richard *5.480–1*
  atonality 5.25
  opéra-ballet 5.329
  operas based on myths 9.13, 9.23
  singers of operas 5.132, 5.156, 5.321
  and trombone 5.460
Wagner, Robert: Wagner–Connery Act *4.369*
wagon *see* carts and wagons; goods wagon; lorry
wagon-way *6.375*
Wagram, battle of *4.369*
wagtails and pipits 2.113, 2.205, *2.255*, *2.358*
Wahhābī/Wahhābīsm *4.369–70*, *7.328*
Waikato *1.359*
Wain, John 5.17
wainscot moth *2.358*
waisted drum *5.481*
wait *see* minstrel
Waitangi, Treaty of *4.370*
Wakatipu, Lake *1.359*
Wakefield, Arthur 4.240
Wakefield, Edward Gibbon 4.240, *4.370*
Wakefield, battle of *3.376*
Waki, Idris Abdul 4.388
Waksman, Selman Abraham 6.16, *6.375*
Walcheren *1.359*
Walcott, Derek *5.481*
Waldemar IV of Denmark 3.102
Walden, Paul *1.359–60*
Waldenses *3.376*
Waldheim, Kurt 7.322
Wales *1.360*, *3.376*, *4.370*
  arts 5.35, 5.273, 5.343, 5.486
  buccaneer 3.241
  castle 3.63
  education 7.180
  language *4.485–6*, 7.42, 7.59, 7.156, 7.180

Wales (*cont.*):
literature ***4**.485–6*
marches **3**.221
physical geography and geology **1**.34, **1**.306, **1**.323, **1**.331, ***1**.360*
mountains **1**.49, **1**.62, **1**.309
see also physical geography *under* Britain
politics and government **3**.142, **4**.270
devolution **7**.88
regions listed ***9**.73*
religion/theology **3**.99, **3**.139, **3**.250, **7**.13, **9**.41
myths **9**.7–8, **9**.10, **9**.14, **9**.25
see also myths *and* religion *under* Britain
riots **4**.239, **4**.284
rulers **3**.56, **3**.208
princes **3**.110, **3**.111, **3**.208
technology and science **6**.89, **6**.293, **6**.341
Tonypandy dispute **4**.347
see also Britain
Walesa, Lech **4**.272, **4**.318, ***4**.370*
Walī **5**.466
Walker, Alice **5**.429
Walker, George **6**.191
Walker, William **4**.68, **4**.241
walking fern ***2**.358*
walking stick (insect) **2**.325
walking worm ***2**.358–9*
Wall Street **7**.299
wallaby ***2**.359*
Wallace, Alfred Russel **2**.30, **2**.111, ***2**.359*
Wallace, George Corley **4**.370
Wallace, Henry Agard **4**.277, ***4**.370*
Wallace, Sir William **3**.123, ***3**.376*
Wallenstein, Albrecht Wenzel Eusebius von **3**.214, ***3**.376*
Waller, Edmund **5**.82
wallflower ***2**.359*
wallpaper ***5**.481*
Wallraff, Günter **5**.175
walls **6**.145, **6**.146, **6**.337
cavity **6**.64
cob **6**.79
curtain **6**.8, **6**.50, **6**.51, **6**.94, **6**.146, **6**.322
damp-proof course **6**.96
Great Wall of China **6**.161
retaining **6**.298
stone **6**.216
walnut ***2**.359*
Walpole, Horace ***5**.481*
Walpole, Sir Robert, 1st earl of Orford ***3**.376–7*
and cabinet **3**.56
and South Sea Bubble **3**.335
and taxes **3**.94
and war **3**.186
and Whigs **3**.380
Walras, Marie-Esprit Léon **7**.228, ***7**.328*
walrus ***2**.359*
Walsh, Thomas J. **4**.339
Walsingham, Sir Francis **3**.32, **3**.356, ***3**.377*
Walter, Hubert ***3**.377*
Walther von der Vogelweide ***5**.481*
Waltman, William **8**.115
Walton, Ernest **1**.67, **6**.258
Walton, Izaak ***5**.481*
Walton, Sir William **5**.421, ***5**.481*
waltz **5**.435, ***5**.482*
Walwyn, William **3**.207
Wandewash, battle of ***3**.377*
Wang Anshi ***3**.377*
Wang Hongwen **4**.130–1
Wang Mang **3**.156
Wang Shifu **5**.94
Wang Wei **5**.98
Wang Xizhi **5**.71
Wankel, Felix: Wankel engines **6**.303
wapiti ***2**.360*
war *see* wars and defence
War of 1812 **4**.136, **4**.189–90, **4**.238–9, ***4**.371*
Warbeck, Perkin **3**.291, ***3**.377*
warble fly **2**.39
warblers ***2**.360*
chats **2**.61, **2**.280–1
chiffchaff **2**.360
redstart **2**.280–1
reed **2**.281
scrub **2**.360
sedge **2**.302–3
willow **2**.368
wood **2**.371
Ward, F. T. **4**.337
Ward, Sir Joseph George ***4**.370–1*
Wardrobe ***3**.377*
warhead **6**.246, ***6**.375*
Warhol, Andy ***5**.482*
Warin, Jean **5**.289
Warlock, Peter ***5**.482*
warlord ***4**.371*
warning coloration ***2**.360*
warning and detection system ***6**.375*
see also in particular radar
warp **6**.125, **6**.189, **6**.207, **6**.320, ***6**.375*
Warren, Earl **4**.371
Warren, Mercy Otis **4**.117
wars and defence ***7**.82*, ***7**.328*, **7**.329
aggression **7**.6
ANZUS **7**.14
armed forces *see* armed forces
arts:
novels **5**.65, **5**.377–8, **5**.457
painting **5**.428, **5**.441
photography **5**.76, **5**.431, ***5**.482*
poetry **5**.61–2, **5**.335
chemical and biological **7**.44
civil defence **7**.50
collective security **7**.56
see also under international organizations and alliances
conventional **7**.69
crimes ***7**.328*
Geneva Conventions **4**.133
guerrilla **4**.143, **4**.366–7, **7**.136
militia **7**.202
pacifist attitude to **7**.227, **7**.232
prisoners of war **7**.328
technology of *see* military technology
see also World War *and under individual countries and wars*
Warsaw Pact ***4**.371*, **6**.246, **7**.6, **7**.213, ***7**.328*
Warsaw Rising ***4**.371*
warships ***3**.247*, ***4**.234–6*, **6**.307, ***6**.375–6*
aircraft carrier **6**.8, **6**.64
battleships **4**.167, **6**.32, **6**.188
corvette **6**.90
cruiser **6**.93
cutter **6**.94
destroyer **6**.99, **6**.357
frigate **6**.146, **6**.188
galley *see* galley and trireme
iron **6**.188
ironclads **6**.111, **6**.188
merchant ships **6**.219
nuclear-powered ships and submarines **6**.175, **6**.243, **6**.245, **6**.337
ship of the line **6**.306, **6**.317
sonar **6**.325
Spanish Armada **3**.335–6
torpedo-boat **6**.357
trireme *see* galley amd trireme
Viking *see under* Vikings
see also submarines
Warsitz, Erich **6**.138
wart snake ***2**.361*
warthog ***2**.360*
Warwick, Richard Neville, earl of **3**.36, **3**.310, **3**.313, ***3**.377*
Warwick, Thomas de Beauchamp, earl of **3**.210
wash (arroyo) **1**.20
Washakie ***4**.371*
washing:
dishwasher **6**.104
-machine ***6**.376*
powder ***6**.376*
bleach **6**.40
phosphate in **6**.267
see also detergent
Washington, Booker T. ***4**.371–2*
Washington, George ***3**.377–8*
and armed forces and war **3**.233, **3**.369, **3**.388
and Federalist Party **3**.124, **4**.117
as Founding Father **3**.129–30
and New York **3**.253
Washington (DC) designed **5**.261
Washington, Mount ***1**.360*
Washington Conference ***4**.372*
wasps ***2**.361*
as biological control **2**.30
chalcid **2**.60
cuckoo **2**.83
fig **2**.118
gall **2**.132, **2**.231
hornet **2**.162–3, **2**.212–13
hunting **2**.166
ichneumon fly **2**.168
mason **2**.204
mimicry **2**.213
sand **2**.294
velvet ant **2**.355
Wasserman, August von: test **6**.177, **6**.192
wastes and waste management ***6**.376–7*
body:
excreta **2**.111
faeces **2**.113
urine **2**.353
see also sewage
ores (gangue) **1**.132
pollution **6**.377, **7**.240
radioactive **6**.243, **6**.245
recycling **6**.296, **6**.377, **7**.260
water **7**.330
see also sanitation
watch **6**.76, ***6**.377*
watch and ward ***3**.378*
water ***1**.360*, ***6**.377–9*
animals:
anaconda **2**.9
capybara **2**.52
lancelet **2**.182
sea anemone **2**.297, **2**.298
sea butterfly **2**.301
sea gooseberry **2**.298, **2**.299
sea scorpion **2**.301
sea snake **2**.301
sea spider **2**.301
sea squirt **2**.300, **2**.301–2
sipunculids **2**.310
sponge **2**.322
water bear ***2**.361*
see also cephalopods; crustaceans; echinoderms; molluscs; *and also* birds, insects *and* mammals *below*
aquaculture **6**.18, **6**.133, **6**.316
aqueduct **6**.18–19, **6**.379
aquifer **1**.17, **1**.20
balance (meteorology) ***1**.360*
birds ***2**.297–8*
albatross **2**.5
auks **2**.18, **2**.271, **2**.279
bay ducks **2**.24–5, **2**.260, **2**.297, **2**.298, **2**.349
Bewick's swan **2**.28–9
bittern **2**.32, **2**.329
booby **2**.38
coot **2**.75
cormorant **2**.77, **2**.305
darter **2**.87
dipper **2**.93
water (*cont.*):
birds (*cont.*):
diver **2**.95
duck *see* ducks
eider **2**.104
finfoot **2**.118
flamingo **2**.119
frigate bird **2**.127
fulmar **2**.129
gannet **2**.132
garganey **2**.133
goose *see* geese
grebes **2**.86, **2**.144
greylag **2**.144
guillemot **2**.147
gulls **2**.148, **2**.157, **2**.179
herons **2**.37, **2**.157
herring gull **2**.157
hoatzin **2**.158
kingfisher **2**.179, **2**.180
kittiwake **2**.179
limpkin **2**.190
long-tailed duck **2**.193–4
mallard **2**.199
mergansers **2**.208, **2**.314
mute swan **2**.224
pelican **2**.247
penguin **2**.247, **2**.248, **2**.298
petrels *see* petrels
pintail **2**.255
pochard **2**.260, **2**.297, **2**.349
puffin **2**.271
scoter **2**.297
sea ducks **2**.104, **2**.193, **2**.297, **2**.298
shelduck **2**.306
shoveller **2**.307, **2**.308
skua **2**.312
smew **2**.314
spoonbill **2**.322
stifftail **2**.325–6
swan *see* swans
teal **2**.337
terns **2**.229, **2**.311, **2**.338
tropic-bird **2**.348
trumpeter swan **2**.348
tufted duck **2**.349
whistling swan **2**.366
whooper swan **2**.367
wigeon **2**.36
see also wading birds
brine **1**.44
canals *see* canals
cannon ***6**.377*
chlorine **1**.62
clock **6**.76
composition of **1**.55
compounds without (anhydrous) **1**.14
condensation *see* condensation
cooling system **6**.89
Coriolis force **1**.76
of crystallization **1**.100, **1**.160, ***1**.361*
dams and flood control **6**.95, **6**.96, **6**.136, **6**.294–5, **6**.379
deposition **1**.43, **1**.64, **1**.72
see also under coasts; rivers
depth measurement **6**.111, **6**.197
dew **1**.89–90
diving *see* diving
drowning prevention **6**.52, **6**.200–1
energy *see* power *below*
engineer **6**.146–7
erosion **1**.77, **1**.93
see also under coasts; river
evaporation *see* evaporation
falls *see* waterfalls
flood *see* floods
fluoridation **6**.139
fog **1**.124
in food **6**.140, **6**.141
groundwater **1**.147, **1**.250
hardness **1**.151–2
heating **6**.176
in heating systems **6**.167

water (*cont.*):
heavy 6.167, 6.245, 6.369
humidity 1.159
hydrant 6.172
hydraulics *see* hydraulic cement; hydraulic power
hydrography 1.148, 1.162
hydrological cycle 1.162, 6.378
hydrology 1.162, 1.163, 1.241–3, 6.174
hydrolysis 1.163
hydrometer 6.174
hydroponics 6.175
hydrosphere 1.163
hydrotreatment 6.175, 6.265
ice *see* glaciation; ice
infiltration 1.168
inland waterways *see* canals; rivers
insects:
damselfly 2.86–7
pond skater 2.264
water beetle *2.361*
water boatman *2.361*
water flea 2.87
water scorpion *2.362*
water-bug 2.135
lifting 6.19
living organisms in and near:
amphibious *see* amphibians
animals *see* animals *above*
birds *see* birds *above*
flatworm *see* flatworms
insects *see* insects *above*
mammals *see* mammals *below*
plants *see* plants *below*
see also living organisms *under* oceans and seas
mammals:
beaver 2.26, 2.220
dolphins 2.97, 2.286
dugong 2.100
manatee 2.201
musk rat 2.223
porpoise 2.265
sea cow 2.298
sea lion 2.299–300
sea otter 2.237, 2.300, 2.301
seal 2.299, 2.359
water rat *2.362*
water shrew *2.362*
water voles 2.194, 2.289, 2.357, *2.362*
whale *see* whales
mass (oceans) *1.360–1*
meter *6.377*
mill 6.355–6, 6.377–8, 6.379
mineral 1.215
oceans *see* coasts; oceans and seas
osmosis 2.237
permeability of rock 1.254
plants *2.362*
bladderwort 2.33, 2.34
duckweed 2.100
eelgrass 2.104
kingcup 2.179
mangrove 2.202
nipa palm 2.228
sea holly 2.299
sea kale 2.299
seeweeds 2.184, 2.302, 6.18, 6.186
water fern *2.361*
water hyacinth 2.363
watercress *2.361*
waterlilies 2.194, 2.289, 2.361–2
see also algae; moss
pollution 6.314, *6.377*, 6.378
porosity 1.264
power 6.18, *6.377–9*
turbine 6.145, 6.173, *6.379*
water-wheel 6.260, 6.367, *6.379*
waves 1.363, 6.380–1, 7.106
see also dams; hydroelectricity *and also under* tides/tidal
precipitation *see* precipitation (meteorology)
water (*cont.*):
pump 6.285–6, 6.385
purification 6.378–9
rain *see* rain
refraction 1.278, 1.279–80
removal 6.193
see also drying
resources management *6.378*
rice grown in 6.298
rivers *see* rivers; waterfalls
seas *see* coasts; oceans and seas
shaduf 6.315
sluice 6.322
snow *see* snow
softening *6.378*
solid particles in *see* suspension
solids dissolved in *see* solvent and solution
sports 7.39, 7.93, 7.270, 7.272, 7.303, 7.331
sprinkler 6.188, 6.330
storage 6.96, 6.297, 6.379
storm 6.314, 6.335
supply and treatment 6.139, *6.378–9*, *7.328*, *7.330*
thermocline 1.334
thirst 2.339
as transport channel *see* canals; oceans and seas; *and* navigation *under* rivers
turbine 6.145, 6.173, *6.379*
vapour *1.362*
waste *see* sewage
water-table *1.361–2*
drawdown 1.95
perched 1.250
recharge 1.277
springs and geysers 1.20, 1.137, 1.138, 1.139, 1.318
waves:
interference 1.171
transverse 1.342, 6.381
see also waves (ocean)
well 6.379, 6.383
see also coasts; glaciation; irrigation; liquid; rivers; steam *and entries beginning* aqua *and other entries beginning* water
water drum 5.430
Water Margin *5.483*
water melon *2.362*
Water Snake (hydra) 8.71
Waterbearer (Aquarius) 8.7
water-buffalo 6.106
water-closet *6.377*
water-colour *5.482*
Britain 5.179, 5.394, 5.400, 5.462–3, 5.468–9
Czechoslovakia 5.210
gouache 5.186–7
waterfalls *1.360*
Africa 1.356
Brazil 1.166
and glaciation 1.140
North America 1.63, 1.228
rapids 1.276
Venezuela 1.14
see also rivers
water-frame 6.329
water-gap *1.360*
Watergate scandal *4.372*, 7.153
Waterhouse, Alfred *5.482–3*
Waterloo, battle of *4.372*
Waterman, Lewis 6.259
watermark *6.377*
waterproofing 6.64, 6.211, 6.249, 6.303, *6.378*
see also vulcanization
Waters, Muddy 5.382
watershed 1.93, *1.361*
waterspout *1.361*
water-table *see under* water
Waterton–Glacier International Peace Park *1.362*
water-vapour *1.362*
water-wheel 6.260, 6.367, *6.379*
waterworks 6.378–9
Watling Street *3.378*
Watson, James Dewey *2.80–1*, 2.95, 6.105
Watson, John Broadus: behaviourism 7.29, 7.252, *7.330*
and conditioning 7.30, 7.64, 7.233
and emotion 7.105
Watson, Thomas 6.33
Watson-Watt, Sir Robert Alexander 6.288, *6.379–80*
Watt, James *1.362*, 6.297, *6.380*
and mass production 4.213
patent 6.363
and steam engines 3.341, 4.328, 4.329, 6.385
beam-engine 6.32
governor 6.158
watt (unit) *1.362*, 1.362, 6.368
Watteau, Jean-Antoine *5.483*
wattle (plant) 2.2
wattle and daub 6.146, 6.235, *6.380*
wattlebird *2.362*
wattled crow 2.362
Watts, George Frederick 4.57, *5.483*
Waugh, Evelyn 5.402, *5.483*
wave mechanics *see* quantum mechanics
wave power *see* energy *under* waves (ocean)
wave *see* waves
waveband 6.288–9, *6.380*
wavefunction (physics) 1.363
waveguide (tube) *6.380*
wavelengths (physics) *1.363*, 6.117, 6.146, 6.290, *6.380*, 6.381, *8.186*
Balmer series 1.31
Doppler effect 8.38
Raman effect 1.275
see also waves
Wavell, Archibald Percival, 1st earl 4.246, 4.345, *4.372*
wave-particle duality 1.70, *1.363*
waves (ocean):
backwash 1.29
energy and power *1.363*, *6.380–1*, 7.106
erosion *1.363*
see also erosion *under* coasts
fetch 1.121
refraction *1.363–4*
swash 1.328
tidal bore 1.336
tsunami 1.123, 1.345–6
wave-cut platform 1.301
wind *1.363*
see also tides/tidal
waves (physics) *1.362*, *6.381–2*
De Broglie 1.85
diffraction 1.90–1
Doppler effect 6.105–6
electromagnetic *see* electromagnetic radiation
frequency 1.127
gravitational 8.60
interference 1.170, 6.184
light 1.31, 1.275, 1.278, 1.363
reflection *see* reflection
refraction *see* refraction
sound *see* sound
transverse 1.342
travelling 1.342
water 1.171
see also waves (ocean)
wave mechanics *see* quantum mechanics
wavefunction 1.363
wavelengths *see* wavelengths
wave-particle duality 1.70, *1.363*
see also frequency
waves (meteorology), atmospheric *1.362–3*
see also winds
wax 2.191, *6.382*
beeswax 6.33
wax (*cont.*):
candle 6.57
paraffin 6.57, 6.257, 6.382
spermaceti 6.328
waxwork *5.483–4*
see also *cire perdue*
Wax Girl 9.4
waxbill *2.363*
waxwing *2.363*
Wayland the Smith *9.38*
Wayne, Anthony *3.378*
WCC *see* World Council of Churches
weak force *1.364*
wealth 7.288, *7.330*
weapons/arms *5.22*, *6.225–6*
banning *see* disarmament
bombs *see* bomb
chemical warfare *6.71–2*, 6.98, 6.164, 6.235, 6.239–40, 6.267
control *7.17–18*
see also disarmament
flamethrower 6.134
grenade 6.64, 6.161, 6.226
guns *see* firearms
laser 6.195, 6.336
limitation talks 6.15
mine 6.134, 6.226–7
missiles and projectiles 6.228
ballista 6.22, 6.29
bolas 6.41
boomerang 6.42
bow, arrows, and crossbow 6.29, 6.44, 6.92, 6.225, 6.226
catapult 6.22, 6.64, 6.225, 6.318
harpoon 6.133, 6.166
sling 6.64, 6.322
spear 6.226, 6.327
see also ballistic missiles; missiles
nuclear *see* nuclear weapons
prehistoric 6.280
race *7.18*, 7.201, 7.202
SDI 4.330, 6.15, 6.335–6, 7.300
on ships *see* warships
siege warfare 6.225, 6.318, 6.371
strategic 6.246
see also missiles
sword 6.165, 6.226, 6.343
tanks *see* tank
theatre (intermediate) 6.246
torpedo 6.337
trade *7.18*
see also artillery; explosives; firearms; military technology *and also* weapons *under individual countries*
weasel family *2.363*
badgers 2.20, 2.278
ermine 2.326
ferret 2.115
martens 2.204, 2.291
mink 2.213
otters 2.237, 2.301
polecat 2.115, 2.262
sable 2.291
skunk 2.312–13
stoat 2.326
wolverine 2.369
weather *1.364*
and geomorphology *see* weathering
lightning conductor 6.202
satellite 6.221, 6.222
symbols 9.56
see also climate; meteorology
weather-boarding 6.76
weathering *1.364*
chelation 1.59
colluvium 1.68
denudation 1.87
exfoliation 1.28, 1.94, 1.116
felsenmeer 1.120
gully 1.148–9
joints 1.178
landslide 1.136
mudflow 1.222
patterned ground 1.249

weathering (*cont.*):
pedestal rock **1**.249
ravines **1**.276–7
rills **1**.283, **1**.284, **1**.301
rock glacier **1**.284–5
rockfall **1**.284
runoff *see* runoff
saprolite **1**.293
scree **1**.295–6
solifluction **1**.312
spalling **1**.315, **1**.316
spheroidal **1**.317
see also erosion; mass flow
Weaver, James B. **4**.275
Weaver, John *5.484*
weaverbird (weavers) **2**.44, *2.363*
weaving *5.484*, *6.207–8*, **6**.281, *6.351*, *6.382*
Africa **5**.7
carpets and rugs *5.79–80*
China **5**.94
loom **6**.62, **6**.140, **6**.169, **6**.189, **6**.190, **6**.207–8
Ottoman **5**.334
shuttle **6**.140, **6**.190, **6**.208, **6**.318
silk **5**.94, **5**.334, **5**.419, **6**.320
South American Indians **5**.426
warp and weft **6**.125, **6**.189, **6**.207, **6**.208, **6**.320, **6**.375, **6**.382
weave *6.382*
yarn **6**.390
see also embroidery; tapestries; textiles
web offset printing **6**.249
web spinner *2.363*
Webb, Beatrice **4**.115, **4**.372
Webb, Philip *5.484*
Webb, Sidney James, Baron Passfield **4**.115, **4**.189, *4.372–3*
Weber, Carl Maria von **5**.334, *5.484*
Weber, Joseph **8**.60
Weber, Max *7.330*
and authority **7**.23
and location theory **7**.186
and magic **7**.189
and Protestant ethic **7**.251
and social class **7**.289
and social sciences **7**.292, **7**.294
Webern, Anton *5.484*
Webster, Daniel *4.373*, **4**.377
Webster–Ashburton Treaty *4.373*
Webster, John **5**.459, *5.484*
webworm **2**.144
Wedekind, Frank **5**.377, *5.484*
wedge **6**.177–8
Wedgwood, Josiah **5**.105, *5.485*, **6**.278, **6**.382
weeds *2.363*
bindweed **2**.29
couch grass **2**.78
dock **2**.96
dodder **2**.96
ground elder **2**.145
hoe **6**.169
knotweed **2**.180
pondweed **2**.264
ragwort **2**.275
Robinia **2**.287
water hyacinth **2**.363
weedkiller *see* herbicides
see also pests
Weelkes, Thomas **5**.18, *5.485*
weeverfish *2.363–4*
weevil *2.364*
weft **6**.125, **6**.207, **6**.208, **6**.320, *6.382*
Wegener, Alfred **1**.73, **1**.142, **1**.189, *1.364*
Weierstrass, Karl **1**.13
Weigel, Helene **5**.45, **5**.59
Weighing of the Hearts ceremony **9**.38
weight *1.364*, *8.186*
avoirdupois **1**.28, **9**.49
raising and moving **6**.91–2
weight (*cont*):
weighing machine **6**.192, **6**.382, *6.383*
weightlessness *8.186*
Weil, Adolf *2.364*
Weill, Kurt **5**.172, *5.485*
Weimar Republic *4.373*
leaders listed **9**.70
weir *6.383*
Weismann, August *2.364*
Weiss, Peter **5**.23, **5**.61
Weizsacker, Carl Friedrich von **8**.11, **8**.33
Weizsacker's theory *8.186*
Weizmann, Chaim *4.373*, **4**.390
weka *2.364*
Welch, William Henry *2.364*
welding **6**.188, **6**.194, **6**.321, *6.383*
Welensky, Sir Roy *4.373*
Welf family *see* Guelph
welfare economics **7**.243
welfare state *4.374*, *7.330*
see also social security; social services *and* social services and security *under individual countries*
well **6**.379, *6.383*
natural *1.364*
Welles, Orson **5**.72, **5**.485, **5**.495
Wellesley *see* Wellington
Wellington, Arthur Wellesley, 1st duke of *4.374*
and Catholic emancipation **4**.65
and Maratha wars **4**.211
and Peninsular war **4**.265
and Privy Council **4**.217
and Waterloo **4**.372
Wellingtonia **2**.281
Wells, Fargo and Company *4.374*
Wells, Herbert George **5**.376, *5.485*
Wells, Horace **6**.233
Welsbach, Carl Auer, Freiherr von **6**.25, **6**.150
Welsh, Jane **5**.78
Welsh Nationalist Party **4**.270
Welsh-language literature *5.485–6*
see also Wales
Welty, Eudora *5.486*
Wen Yiduo **5**.95, **5**.98
Wen Zhengming *5.486*
Wenceslas, Saint, duke of Bohemia (907–29) *3.378–9*, *9.43*
Wenceslas II, king of Bohemia (1361–1419) **3**.168, *3.378*
Wendi **5**.95
Wentworth, Thomas *see* Strafford
werewolf *9.38*
Werfel, Franz *5.486*
wergild *3.379*
Werner, Abraham Gottlob *1.364*
Werner, Alfred *1.364*
Wernicke, Carl **7**.15
Wertheimer, Max **7**.131
Wesak *7.330*
Weser *1.364*
Wesker, Arnold *5.486*
Wesley, Charles **3**.379, **3**.380, **5**.486
Wesley, John *3.379*, **3**.380, **7**.199, **7**.207
Methodist Church established **3**.14, **3**.230, **4**.218
Wesley, Samuel and Samuel Sebastian *5.486*
Wessex **3**.15, *3.379–80*
rulers **3**.10, **3**.26, **3**.110, **3**.111, **3**.112, **3**.118, **3**.143
West, Benjamin **5**.13, **5**.112, **5**.207, *5.486*
West, Nathanael *5.486*
West Africa:
chiefdoms and kingdoms **3**.39–40, **3**.96, **3**.241–2
see also Asante; Fulani; Hausa; Songhay; Yoruba *and individual countries*
city **3**.357
West Africa (*cont.*):
colonies **4**.306
Economic Community **4**.108
exploration **1**.33, **1**.64, **1**.90, **1**.154, **1**.183, **1**.186–7, **1**.248, **4**.260
famine **7**.117–18
physical geography and geology **1**.138, **1**.228, **1**.358
drought **1**.86
lake **1**.57
river **1**.229
religion and myths **4**.308, **9**.4, **9**.26, **9**.32
deities **9**.25, **9**.27, **9**.28, **9**.37
tricksters **9**.11, **9**.15, **9**.22
see also individual countries
West Asia:
music *5.486–7*
see also Middle East
West Germany *see* German Federal Republic
West Indies *see* Caribbean and West Indies
westerlies *1.364–5*, **1**.367
Western Australia *1.365*, *4.374–5*
Western countries *see* industrialized countries
western elongation **8**.46
Western European Union *4.375*, *7.330–1*
Western films **5**.128, **5**.162, *5.487*
Western Front *4.375–6*
Western Jin **3**.72, *3.380*
Western Neolithic *3.380*
western prime vertical **8**.128
Western Rising *3.380*
Western Samoa *1.365*
see also Samoan Islands
westernization *7.331*
Westinghouse, George **6**.351, *6.383*
Westmacott, Mary **5**.100
Westminster, Statute of *4.376*
Weston, Edward **5**.347, **5**.488
Westphalia, Treaty of *3.380*, **3**.381
westward expansion, American **4**.156, **4**.368, **4**.376, *4.377*, **4**.378
wetlands *2.364*
convention **7**.66
see also bogs; swamps and marshes; living organisms *under* water
'wetting agents' **6**.340
WEU (Western European Union) **4**.375, **7**.330–1
Weyden, Rogier van der **5**.290, *5.488*
Weygand, Maxime *4.377*
WFP (World Food Programme) **7**.321
WFTU (World Federation of Trade Unions) **7**.315
Whale (astronomy) **8**.24
whale shark *2.365*
whalebone whales **2**.36, **2**.119, **2**.285, **2**.288, *2.364–5*
whale-headed stork *2.365*
whales *2.364*
beaked **2**.25
blue **2**.36, **2**.200
blue whale **2**.36
dolphin **2**.97
fin whale **2**.119
fishing for *see* whaling
grey **2**.144
humpbacked **2**.166
narhwal **2**.225
right whale **2**.285
river dolphin **2**.286
rorqual **2**.119, **2**.288
sperm whale **2**.319
toothed *see* toothed whales
whalebone *see* whalebone whales
whaling **6**.133, **6**.166, **6**.328, *6.383–4*
Wharton, Edith *5.488*
wheat *2.365*
see also cereals and grains
wheatear *2.365*
Wheatley, Phyllis **5**.7
Wheatstone, Sir Charles *1.365*, **4**.80, **4**.81, **6**.88, **6**.232, *6.384*
wheel *6.384*
aircraft **6**.5
axle **6**.27, **6**.300, **6**.384
cam **6**.55
flywheel **6**.140, **6**.163
paddle-wheel **6**.255
potter's **6**.278
pulley **6**.217, **6**.285
spinning **6**.329
treadmill **6**.362
tyre **6**.5, **6**.35, **6**.109, **6**.366, **6**.384
undershot **6**.367
water **6**.260, **6**.367, **6**.379
wheel animalcule **2**.288, **2**.289
Wheel of Life **7**.85
wheelchair *6.384*
wheel-lock (firearm) **6**.131
whelk *2.365*
whey **6**.95
Whigs:
in Britain **3**.251, **3**.256, **3**.306, **3**.332–3, **3**.376–7, *3.380*, **4**.377
exclusion crisis **3**.121
Rye house Plot **3**.312
in United States **4**.77, **4**.147, **4**.329, **4**.369, **4**.373, *4.377*
Whinfield, John Rex **6**.275, *6.384*
Whipple, Fred L. **8**.38, **8**.186
Whipple's Comet *8.186*
Whipple, George **8**.120
whip-poor-Will **2**.228
whirligig **2**.361
whirlpool **1**.201, *1.365*
Whirlpool Galaxy *8.186*
whirlwind *1.365*
see also tornado; waterspout
whiskey, corn **2**.198
Whisky Rebellion *3.380*
whist **7**.40
Whistler, James Abbott McNeill *5.488*
whistlers **2**.339, *2.366*
whistling swan *2.366*
Whit Sunday **7**.232
Whitbread, Samuel *4.377*
Whitby, Synod of *3.380*
White, E. B. (writer) **5**.454
White, Edwin (painter) **4**.122
White, Gilbert (1720–93, naturalist) *2.366*
White, Gilbert (b.1912, novelist) *5.488*
White, John **3**.305
White, Patrick *5.488*
white (butterfly) *2.366*
white admiral butterfly *2.366*
white ant **2**.338
White Australia policy *4.377*
white dwarfs *8.186–7*
white hole *8.187*
white lead *6.384*
White Lotus Society *3.380–1*
White Mountains *1.365*
white oak **2**.231
White Paper **7**.31
White Russians *4.378*
see also Belarus
White Sea *1.365–6*
whitebait *2.366*
whitebeam **2**.289
white-collar crime *7.331*
white-eye *2.366*
Whitefield, George **3**.145, *3.380*
whitefly **2**.30, *2.366*
Whitehead, Alfred North *1.365*
Whitehead, Robert **6**.357
white-out **1**.40
whiting *2.366–7*
Whitlam, Gough: crisis *4.378*

Whitman, Walt 5.143, 5.325, 5.488, 5.489
Whitney, Eli 4.213, 6.217, 6.384
Whitney, Mount 1.366
Whittier, John Greenleaf 5.488
Whittle, Sir Frank 6.189, 6.384–5
Whitworth, Sir Joseph 6.217, 6.385
WHO *see* World Health Organization
wholesale trading 7.93, 7.331
whooper swan 2.367
whooping cough 2.367, 7.153
Whorf, Benjamin Lee 7.273
whydah bird 2.367
Whymper, Edward 1.366, 7.208
wide-area network 6.240
wide-bodied aircraft 6.385
wide-screen cinema 5.488–9
Wieck, Friedrich 5.406
Wieland, Christoph Martin 5.489
Wien, Wilhelm: Wien's Displacement Law 8.187
Wiener, Norbert 6.385
wigeon 2.367
Wigman, Mary 5.489
Wilans, P. W. 6.331
see also Westwood expansion
Wilberforce, William 4.316, 4.378
Wilbrand, J. 6.357
Wild, Sam 4.164
wild boar 2.367
wild cat 2.367
wild ox 2.18
'Wild West' 4.378
see also Westward expansion
Wilde, Oscar 5.177, 5.250, 5.489
wildebeest 2.138–9
Wilder, Thornton 5.244, 5.489
Wilderness Road 3.381
Wildman, John 3.207
Wilfrid, Saint 3.135, 3.258, 3.380
Wilhelmina, queen of Netherlands 4.238, 4.378
Wilkes, Charles 1.83, 1.366
Wilkes, John 3.381–2
Wilkie, Sir David 4.217, 5.408, 5.489
Wilkins, Sir Hubert 1.366
Wilkins, Maurice 2.81
Wilkins, Roy 4.378
Wilkinson, A. B. 6.51
Wilkinson, James 4.378–9
Wilkinson, John 6.43, 6.385
will 7.248, 7.331
Willard, Emma 4.117, 4.379
William I (the Conqueror), king of England (1028–87) 3.202, 3.255, 3.382
and Domesday Book 3.104
Norman Conquest 3.256
rebellion against 3.109
and tax 3.277
William I, king of Holland 4.193
William I, emperor of Prussia (and German empire) 4.39, 4.112, 4.154, 4.278, 4.379
William I (the Lion), king of Scotland (1143–1214) 3.382
William I (the Silent), prince of Orange (1533–84) 3.108, 3.262, 3.382
William II, emperor of Prussia and German Kaiser (1859–1941) 4.37, 4.55, 4.154, 4.172, 4.378, 4.379
William II (Rufus), king of England (1056–1100) 3.106, 3.202, 3.255, 3.383
William III, king of England, Ireland, and Scotland (William of Orange, 1650–1702) 3.28, 3.43, 3.54, 3.96, 3.262, 3.383
finances 3.35
Glorious Revolution 3.142
and Ireland 3.365
and parliament 3.270, 3.380
William III, king of England (*cont.*):
and religion 3.14
and Scots 3.78
and war 3.254, 3.312, 3.337–8
wife *see* Mary II
William-and-Mary style 5.489
William IV, king of Great Britain and Ireland 4.379
William V, king of Orange 3.366
William IX, count of Anjou 3.361
William of Aquitaine, duke (909) 3.80
William of Malmesbury 3.383
William of Occam (Ockham) 3.318, 3.383
Occam's (Ockham's) razor 4.383, 7.221, 7.223
William of Orange *see* William III
William of Wykeham 3.383
William-and-Mary style 5.489
Williams, Emlyn 5.454
Williams, Eric 4.379
Williams, Frederick 6.85
Williams, Raymond 7.78, 7.215
Williams, Roger 3.35, 3.302, 3.383
Williams, Tennessee 5.459, 5.489–90
Williams, William Carlos 5.490
Williamson, Henry 5.490
Willoughby, Hugh 1.58
willow 2.367–8
willow grouse 2.280
willow warbler 2.368
willowherb 2.368
Wills, William John 1.45, 1.201
Willstatter, Richard 1.366
Wilmot, David 4.125
Wilson, Alexander 2.368
Wilson, Sir Angus 5.490
Wilson, Charles Thomson Rees 1.366
Wilson, Colin 5.17
Wilson, Edmund 5.490
Wilson, Edward (b.1929, sociologist) 7.294
Wilson, Edward Adrian (1872–1912, polar explorer) 1.295
Wilson, F. N. 6.113
Wilson, Harold 4.380
Wilson, James 3.129
Wilson, John Burgess 5.66
Wilson, Richard 5.490
Wilson, Robert 8.15, 8.33, 8.98
Wilson, Woodrow 4.54, 4.95, 4.287, 4.379–80
Fourteen Points 4.122, 4.363
and isolationism 4.168
and Mexico 4.158
and Nye Committee 4.248
and Progressive Movement 4.277
and Versailles 4.364
Wilson, Mount (observatory) 8.104–5
Wimshurst, James: Wimshurst machine 6.385
Winchester disk 6.165
Winckelmann, Johann Joachim 5.23, 5.490
wind *see* winds
Wind Cave 1.367
wind instruments 5.490
see also brass; woodwind
'Wind of Change' 4.380
wind power *see under* winds
wind scorpion 2.330
wind-gap 1.367
winding gear 6.385
Windischgrätz, prince Alfred 4.25
windmill 6.385
windows (building) 6.106
window tax 3.383
windows (computer) 6.385–6
wind-pipe 2.345
winds 1.366, 1.367
anabatic 1.12
Beaufort scale 1.35
blizzard 1.40
bora 1.41
breeze 1.43–4
winds (*cont.*):
chinook 1.61
circumpolar vortex 1.64
deposition:
clastic sediments 1.43, 1.64, 1.72
dunes 1.33, 1.97, 1.298
loess 1.197–8
lunette 1.199
easterlies 1.99
erosion:
deflation 1.86
dustbowl 1.97
pedestal rock 1.249
saltation 1.291
Föhn 1.61, 1.124
gale 1.130
harmattan 1.152
horse latitudes 1.158–9
intertropical convergence zone 1.171
jet stream 1.177–8
katabatic 1.180
khamsin 1.182–3
map 1.367
measurement 6.14, 9.56
mistral 1.216
monsoon 1.220
power 1.366, 6.385, 6.386, 7.106
reversed 1.108, 1.200
roaring forties 1.284
sirocco 1.306
speed 1.35, 6.14, 9.55
squall 1.318
storm force 1.321
symbols 9.56
trades 1.341, 1.367
tunnel 6.4, 6.386
westerlies 1.284, 1.364–5, 1.367
Windsor, House of 4.380, 9.67
see also under rulers *under* Britain
windsurfing 7.331
Windward Islands 7.350, 7.367, 7.379
physical geography and geology 1.33, 1.94, 1.147, 1.221, 1.233, 1.290, 1.379
see also Barbados; Caribbean; Grenada
wine 6.10–11, 6.386
bottle sizes 9.62
Wingate, Orde Charles 4.55, 4.380
winkle 2.368
Winnipeg, Lake 1.367–8
winnowing 6.386
Winsor, Frederick Albert 6.150, 6.386
Winstanley, Gerrard 3.103, 3.383
winter 1.297
nuclear 6.246
'Winter War' (Finnish-Russian) 4.120
wintergreen 2.368
Winterhalter, Franz Xavier 5.491
Wintersteen, Rachel Boone 5.342
Winthrop, John (1588–1649, Massachusetts) 3.225, 3.383
Winthrop, John (1606–76, Connecticut) 3.86, 3.383–4
WIPO (World Intellectual Property Organization) 7.162, 7.230, 7.321
wire 6.298, 6.341
wire-drawing 6.386–7
wiring, electrical 6.387
wireless telegraphy 6.45
see also radio
Wiremu Kingi 4.338
wireworm 2.67–8
Wirz, Henry 4.12
Wishart, George 3.37, 3.384
wisteria 2.368
wit *see* comedy and humour
witan 3.384
witchcraft 3.384, 7.331, 9.38
Aborigines 7.15
Azande 7.24, 7.92, 7.112, 7.263
familiar 9.12
witchcraft (*cont.*):
Salem trials 3.314
sorceress 9.9
witchetty grub 2.139
witch-hazel 2.368–9
Withering, William 2.369, 6.387
Witkiewicz, Stanislaw Ignacy 5.491
Witt, Georg 8.48
Witt, Jan de 3.14, 3.384
Witte, Sergei Yulyevich, Count 4.380
Wittelsbach 3.384–5
Wittenberg, Johann Titius von 8.17
Wittgenstein, Ludwig 7.331–2
and analytic philosophy 7.12
and logical positivism 7.187
and mind 7.203
and naturalism 7.213
and private language 7.203, 7.247, 7.274
Witwatersrand 1.368
Wladyslaw I, king of Poland 3.62
woad 2.369, 6.110
Wodehouse, Sir Pelham Grenville 5.491
Woden *see* Odin
Wöhler, Friedrich 1.193, 1.368
Wolaston, William Hyde 6.387
wold 1.368
Wolf, Charles: Wolf–Rayet star 8.187
Wolf, Christa 5.175
Wolf, Sir Drummond 4.276
Wolf, Hugo 5.491
Wolf, Max 8.2
wolf 2.369
mythical 9.38
wolf spider 2.369
Wolfe, James 3.134, 3.251, 3.284, 3.324, 3.385
Wolfe, Thomas 5.491
Wolfendale, Arnold W. 8.193
Wolff, Caspar Friedrich 2.369
Wolff, Robert Paul 7.221
Wolf-Ferrari, Ermanno 5.491
wolfram 1.368
Wolfram von Eschenbach 5.491
wolfram and wolframite 6.213, 6.365
Wollaston, William (1802, astronomer) 8.53
Wollaston, William Hyde (1766–1828, metallurgist) 6.387
Wollstonecraft, Mary 4.117, 5.491, 7.332
Wolpe, Joseph 7.30
Wolseley, Garnet Joseph, 1st viscount 4.340, 4.381
Wolsey, Thomas 3.385
wolverine 2.369
womb *see* uterus
wombat 2.369–70
women:
authority *see* matriarchy
and descent *see* matrilineality
discrimination against 7.280–1
and domestic violence 7.94
education 4.32
equal pay 7.107
feminism *see* women's movement
franchise *see* suffrage *below*
gypsies 7.137
and health 4.57, 7.9
see also birth; obstetrics and gynaecology; pregnancy *and* breast *under* cancer
illiteracy 7.102, 7.185
initiation 7.188
and Islam 7.255, 7.332
and Judaism 7.26–7, 7.226, 7.257
liberation *see* women's movement
longevity 7.6
in myths 9.4
goddesses *see* deities
see also witchcraft
old 7.222
Opus Dei 7.224
ordination 7.50

women (*cont.*):
and patriarchy **7.**231
and pornography **7.**243
and reproductive system
abortion **7.**1, **7.**115
fertility rate **7.**119
maternity leave **7.**195
physiology *see* female mammals
see also pregnancy
segregated **7.**278
sexual behaviour **7.**280
sexual harassment **7.**280–1
sexual offences against (including rape) **7.**40, **7.**255, **7.**258, **7.**281
shame **7.**147
smoking **7.**288
and sport **7.**214
suffrage **4.**258, **4.**333, ***4.**381*, **7.**123, ***7.**332*
and trade **7.**70
as underclass **7.**319
see also under individual countries
women's movement ***4.**117–18*, ***4.**381*, ***7.**332*
developing world ***7.**332*
and pornography **7.**243
and sexism **7.**280
suffrage *see under* women
Western world ***7.**332*
Wonder, Stevie **5.**358
Wongar **9.**3
Wood, Henry **5.**110
Wood, John (Elder and Younger) ***5.**491*
Wood, Leonard **4.**293
wood and woodwork ***5.**492–3*, **6.**4, ***6.**387–8*
in bridges **6.**47
Britain **5.**46, **5.**176
see also under furniture
in buildings **5.**455, **6.**51, **6.**145–6, **6.**301–2, **6.**387
carpentry and joinery **6.**60
see also furniture
carving *see* woodcarving *below*
charcoal from **6.**69
Egypt **5.**142
fibre from *see* cellulose acetate; rayon
fireproofing **6.**132
as fuel **6.**36, **6.**89, **6.**167
joints ***6.**387–8*
laminate **6.**193
lathe **6.**195
nail **6.**236
naphtha and wood spirit *see* methanol
petrified **1.**255
plane *see* plane
preservative ***6.**387*
processing and utilization ***6.**387*
production **6.**143
pulp for paper **6.**143, **6.**387
scaffolding **6.**308
screw **6.**310
timber framing **4.**455
veneer **5.**470
Viking **5.**475–6
wood-block printing ***6.**387*
woodcarving **5.**176, ***5.**491–2*
woodcut ***5.**492*
wood-engraving **5.**367, ***5.**492*, **5.**493
Zapotec **5.**498
see also forests and trees; furniture; *ukiyo-e*; *and under* lacquer
wood hoopoe ***2.**370*
wood mouse ***2.**370–1*
wood swallow ***2.**371*
wood warbler ***2.**371*
woodbine **2.**161
woodchuck ***2.**370*
woodcock ***2.**370*
woodcreeper ***2.**370*
woodlouse ***2.**370*
woodpecker **2.**370, ***2.**371*
woodpigeon ***2.**371*
woodruff ***2.**371*
Woodville, William ***2.**371*
woodwasp ***2.**371*
wood-white butterfly ***2.**371*
woodwind **5.**490, ***5.**492*
basset horn **5.**37
bassoon **5.**36
chamber music **5.**88
cornett **5.**113
flageolet **5.**156
oboe **5.**324
panpipes **5.**338
pitch **5.**352
recorder **5.**376
serpent **5.**412
see also brass; flute; reed instruments
woodwork *see* wood
woodworm **2.**130
woody plants:
bamboo **2.**21
begonia **2.**28
peony **2.**249
pepper **2.**249
shrubs **2.**309
see also bark; deciduous trees and shrubs; evergreen plants; lianas; forests and trees
wool **6.**128, ***6.**388*
carpet **6.**61
fulling **6.**147
guanaco **2.**146
knitting *see* knitting
llama **2.**192
spinning **6.**20, **6.**92, **6.**328–9, **6.**388
staple ***3.**385*
trade **3.**229, **3.**385
wool-comber ***6.**388*
worsted **6.**388, **6.**390
yarn **6.**390
see also sheep; weaving
Woolf, Leonard **5.**50, **5.**493
Woolf, Virginia **5.**436, ***5.**493–4*, **5.**550
woolly rhinoceros ***2.**371*
Wootton, Frank **4.**307
Worcester, Thomas, earl of **3.**276
word processor **6.**179, ***6.**388*
Worde, Wynkyn de **3.**269
words *see* language
Wordsworth, William **5.**27, **5.**325, **5.**424, **5.**491, ***5.**494*
work (physics) ***1.**368*
work-bench ***6.**388*
workforce *see* labour; unions
work-hardening ***6.**388*
workhouse ***3.**385*
working class **7.**194
Works Project Administration ***4.**381–2*
workstation **6.**179, ***6.**389*
world:
line ***8.**187*
maps *see* world distribution maps and diagrams
models (cosmology) **8.**34
order ***7.**333*
organizations *see* international organizations
population ***4.**273–4*, ***9.**75*
migrations **4.**274, **4.**275
see also entries beginning global; international; World
World Bank ***4.**382*, **7.**321, ***7.**333*
founded **7.**176
and commodity market **7.**57
and developing countries **7.**87
and economic adjustment **7.**3
and education **7.**93, **7.**102, **7.**307
and gross national product **7.**133
and housing **7.**282
and malnutrition **7.**190
and military expenditure **7.**201
and poverty **7.**244
World Bank (*cont.*):
and sanitation **7.**273
and water supply **7.**330
World Council of Churches ***4.**382–3*, **7.**13, **7.**48, **7.**269, ***7.**333*
world distribution maps and diagrams:
aid **7.**8
British empire **4.**51
continental drift **1.**73, **1.**135
debt **7.**81
earthquakes and volcanoes **1.**99
education **7.**102
energy resources **6.**121
exploration **3.**122
family planning **7.**115
fertility rate **7.**119
gross national product **7.**133
growth **7.**242
infant mortality **7.**159
navigation **6.**238
ocean currents **1.**208, **1.**235
plate tectonics **1.**260
population migrations **4.**274
precipitation **1.**267
projections **1.**270, **6.**284
Ptolemy's **1.**205
refugees **7.**261
winds **1.**208
world fairs ***4.**383*
World Federation for Mental Health **7.**198
World Federation of Trade Unions **7.**315
World Food Council **7.**321
World Food Programme **7.**321
World Health Organization ***4.**383*, ***6.**389*, **7.**321, **7.**322, ***7.**333*
and abortion **7.**1
and Aids **7.**9
and blindness **7.**33
and cancer **7.**39
and children **7.**21, **7.**140
and communicable diseases **7.**59
and contraception **7.**115
and dentistry **7.**84
and disability **7.**91
and epidemiology **7.**107
and finance **7.**140
and immunization **7.**153
and malnutrition **7.**190
and maternal death **7.**45
and mental illness **7.**198
and preventive medicine **7.**246
and primary medicine and health care **7.**246–7
and sanitation **7.**273
and schizophrenia **7.**275
and smoking **7.**28
and water supply **7.**328
World Intellectual Property Organization **7.**162, **7.**230, **7.**321
World Meteorological Organization ***6.**389*
World War I ***4.**382*, ***4.**383*
airforce **4.**6–7, **4.**178
see also under vehicles *below*
army **4.**19
British Expeditionary Force **4.**51–2
see also land *under* battles *below and* vehicles *below*
battles and campaigns:
land:
Caporetto **4.**62
Gallipoli **4.**129
Kut **4.**188
Marne **4.**211
Masurian Lakes **4.**213
Mesopotamia **4.**217–18
Passchendaele **4.**262
Somme **4.**319
trench warfare **4.**351–2
Vimy Ridge **4.**368
Vittorio Veneto **4.**369
World War I (*cont.*):
battles and campaigns (*cont.*):
land (*cont.*):
Western Front ***4.**375*, **4.**376
sea **4.**129, **4.**177, **4.**389
chemical warfare **4.**71, **7.**44
defensive lines **4.**206–7, **4.**312
iron and steel industry **6.**170
leaders *see under individual countries*
maps **4.**375, **4.**382
navy **4.**234
battles **4.**129, **4.**177, **4.**389
convoy system **4.**85
weapons **6.**99, **6.**357
see also under vehicles *below*
plastics and rubber **6.**66, **6.**304
technology **6.**41, **6.**56, **6.**170, **6.**311
see also entries above and below prefixed **6.**
vehicles and transport **6.**13
aircraft **6.**8, **6.**66, **6.**129, **6.**220, **6.**257
armoured **6.**20, **6.**344
motor **6.**12, **6.**20, **6.**235
ships and boats **6.**8, **6.**32, **6.**99, **6.**227, **6.**322, **6.**357
submarines **6.**337
weapons **7.**44
depth-charges **6.**99
flame-thrower **6.**134
gas **6.**72, **6.**235, **6.**267
grenade launchers **6.**64
grenades **6.**161
guns **6.**312, **6.**318
shells **6.**316
torpedo **6.**357
Versailles Peace Settlement **4.**363–4
reparations **4.**286–7, **4.**387
see also under individual countries
World War II ***4.**383*
airforce **4.**6–7
see also air *under* battles *below and under* vehicles *below*
army **4.**19
Home Guard **4.**155
see also land *under* battles *below and under* vehicles *below*
Axis Powers **4.**22, **4.**26
battles and campaigns:
air:
Arnhem **4.**18
Battle of Britain **4.**49–50
Blitzkrieg **4.**42
bombing offensives **4.**44, **4.**45
Dresden **4.**103
Pearl Harbor **4.**263
land:
Ardennes **4.**17
Arnhem **4.**18
Blitzkrieg **4.**42
Burma **4.**55
Malaya **4.**208
Meuse crossed at Sedan **4.**308
Normandy **4.**245
North Africa **4.**7, **4.**246
North-West Europe **4.**247
Singapore **4.**314
Soviet Union **4.**188, **4.**193
Stalingrad **4.**327–8
Tobruk siege **4.**345–6
maps **4.**245, **4.**246, **4.**255
sea:
Atlantic **4.**21–2
Dieppe **4.**98
Dunkirk evacuation **4.**104–5
Leyte Gulf **4.**194
Normandy Landings **4.**245–6
Pacific **4.**255, **4.**256
Philippines Campaign **4.**269
Plate River **4.**270, **4.**271
camouflage **6.**56
civil defence **7.**50
electronic warfare **6.**118
fascism **7.**118

World War II (cont.):
fascism (cont.):
see also Nazis
leaders see under individual countries
Lend-Lease Act 4.193
maps 4.245, 4.246, 4.255, 6.215
medicine and health care 6.40, 6.137, 6.260, 6.338
meetings and pacts:
Cairo Conference 4.58
Casablanca Conference 4.64
Destroyer-Bases Deal 4.97
Potsdam Conference 4.276, 4.385
San Francisco Conference 4.302
Teheran Conference 4.340
Yalta 4.385
navy 4.234–5
convoy system 4.85
see also sea under battles above and under vehicles below
postwar recovery 4.212
see also under Japan
radio 6.208, 7.35, 7.258
resistance 4.211, 4.287–8
Second Front 4.307
technology:
helmet 6.41
life-jacket 6.200
metal-detector 6.220
operational research 6.250
radio 7.35, 7.258
searchlight 6.311
sonar 6.325
see also entries above and below prefixed 6.
vehicles and transport:
aircraft 6.42, 6.91, 6.139, 6.140, 6.176, 6.220, 6.224, 6.385
amphibian 6.13
balloons 6.29
bridges 6.47
diving 6.97, 6.323
motor 6.102, 6.235, 6.249
ships and boats 6.32, 6.87, 6.90, 6.93, 6.97, 6.99, 6.146, 6.179, 6.322, 6.357
submarines 6.325, 6.337
tank 6.16, 6.226
weapons:
bombs 6.42, 6.99
see also Manhattan Project
guns 6.16, 6.131, 6.226, 6.312
missiles 6.29, 6.93, 6.228
Molotov cocktails 6.232
napalm 6.236
shells 6.316
see also under individual countries
World Wide Fund for Nature 7.66, 7.135, 7.327
worm-lizard 2.9
Worms, Diet of 3.385–6
worms 2.371
acorn 2.2
annelid see annelids
ascarid 2.16–17
bristle 2.42
earthworm 2.102–3
eelworm 2.104
fanworm 2.113–14
hookworm 2.161
horsehair 2.163
hyatid 2.167
leech 2.186
lugworm 2.195
lungworm 2.196
peanut 2.246
penis 2.248
pinworm 2.255
ribbon-worm 2.284
scale 2.295
sea mouse 2.300–1
spiny-headed 2.322
tapeworm 2.334
walking 2.358–9, 2.371
see also roundworm
wormwood 2.15
Worpswede School 5.494
worship see religion and theology
worsted 6.388, 6.390
Worth, William H. 4.308
Wounded Knee, battle of 4.384
Wrangel, Piotr Nikolayevich 4.296–7, 4.384
wrasse 2.371
Wren, Sir Christopher 3.127, 5.132, 5.494, 8.142
wren 2.371–2
wren thrush 2.372
wrench 6.258, 6.327
wrestling 7.333, 7.334
Wright, Almroth 6.134
Wright, Frank Lloyd 5.14, 5.234, 5.494–5
Wright, Joseph 1.7, 5.495
Wright, Richard 5.495
Wright, Wilbur and Orville 4.350, 6.38, 6.139, 6.157, 6.389
wrinkle ridge 8.187
writers:
PEN 7.232
see also literature
writing and scripts 3.386, 7.333–4
Arabic see under Arabic
Armenian 7.17
Austro-Asiatic 7.22
Baltic languages 7.26
Berber 7.30
braille 7.35
calligraphy 5.71, 5.91, 5.344
Chinese 7.151, 7.177
cuneiform see cuneiform script
Cyrillic 7.11, 7.28, 7.323
Devanagari 7.11, 7.157, 7.305
dyslexia 7.97
hieroglyphs see hieroglyphs
ideograms 7.151, 7.185, 7.334
Indo-Iranian 7.157
Indus Valley 3.178
ink 6.180, 6.259
Japan 5.235, 7.169, 9.47
Korean 7.177
Linear A and B 3.233, 7.135
Linear B 7.135
literacy 7.185
logographic 7.286, 7.334
Malay 7.23
Nilo-Saharan 7.217
North American Indians 7.221
parchment 6.258
pen 6.259
Phoenician 5.346
pictogram 7.235
pictograms 7.123, 7.235, 7.334
reading see reading
Roman 6.318, 7.10, 7.11, 7.23, 7.43, 7.108, 7.269, 7.288, 7.323
shorthand 6.318, 7.285
Sumerian 5.305
systems 3.386
typewriter 6.318
Vietnam 5.475
see also alphabets; historiography; language; literacy; literature; paper; printing; publishing
wrought iron 6.90, 6.186, 6.188, 6.285
wryneck 2.371, 2.372
Wu Cheng'en 5.302
Wu Daozi 3.84, 5.495
Wŭ Dì, emperor 9.38
Wu Jingzi 5.96
Wu Sangui 3.296
Wu the Martial 3.390
Wuchang uprising 4.384
Wudi, emperor of China 3.155, 5.95
Wundt, Wilhelm Max 7.252, 7.334
Wurtz, Charles-Adolphe 1.368
Wuruk, Hayam of Java 3.217
Wyandotte Cave 1.368
Wyatt, James 5.495
Wyatt, Sir Thomas (1503–42, poet) 5.496
Wyatt, Sir Thomas: Wyatt's Rebellion (1554) 3.386
Wyattville, Sir Jeffrey 5.495
Wycherley, William 5.496
Wyck, Jan 3.48
Wyclif, John 3.34, 3.41, 3.84, 3.130, 3.209, 3.386
Wyeth, Andrew Newell 5.448, 5.496
Wyndham Lewis, Percy 5.144, 5.327, 5.479, 5.496
Wyspiański, Stanislaw 5.496

X

Xangô 9.33
Xenakis, Iannis 5.496
Xenocrates 3.386
xenolith 1.369
xenon 1.107, 1.369
Xenophon 3.95, 3.274, 3.332, 3.386–7, 5.496
Xenopus 2.372
xerography 6.60, 6.101, 6.267–8
Xerxes I, king of Persia 3.2, 3.97, 3.387
and Greek-Persian Wars 3.147, 3.160, 3.314
and Persepolis 3.276
wives 9.11–12, 9.37
Xhosa 7.334
Wars 4.384
Xi Jiang (river) 1.369
Xia dynasty 3.72, 3.387
Xia Gui 5.496
xīān (in Daoism) 9.38
Xi'an incident 4.384
xīāng shēng xīāng kè 9.38
Xinjiang (region) 1.369, 3.387, 4.384
Xiongnu 3.387
Xipe Totec 3.31
Xiuhtecuhtli 9.38
Xīwángmŭ 9.38
X-rays 1.369–70, 2.372, 6.117, 6.389
and astronomy:
sources 8.18
X-ray binary 8.36, 8.188
X-ray burster 8.18
X-ray satellite 8.188
X-ray telescope 8.188
crystallography 6.389
diffraction 1.43, 1.90, 2.158
discoverer see Röntgen
microscopy 6.223
and orthopaedics 6.253
particle accelerator 6.258
pregnancy test 6.280
silicosis 6.248
spectroscopy 1.43
tomography 6.357
see also radiology
Xu Zhimo 5.95
Xuanzang 9.15
XUV astronomy 8.188
xylene 6.261, 6.262
xylophone 5.8, 5.343, 5.496
XYZ Affair 3.386

Y

yachts **6**.308, ***6**.390*
  catamaran **6**.64
  centreboard **6**.67
  rig and sail **6**.143, **6**.304
  sailing principles **6**.305
  sloop **6**.322
  see also sailing ships
yaffle **2**.371
Yagi, Hidetsugu: Yagi antenna **6**.15, ***6**.390*
Yahya **9**.20
  see also John the Baptist
Yahya, Imām of Yemen **4**.386
Yājūj and Mājūj ***9**.38*
yak ***2**.373*
Yakovlev, I. **6**.176
Yakub Beg **4**.384
Yale, Linus (Junior): Yale lock **6**.205, **6**.206
Yalta Conference ***4**.385*
yam ***2**.373*, **6**.302
Yama **9**.37, ***9**.38*
Yamagata Aritomo **4**.170, **4**.215, ***4**.385–6*
al-Yaman **3**.333
Yamashita Tomoyuki **4**.314, ***4**.386*
Yamato ***3**.387*
Yanauluha **9**.38
Yang Qien **3**.344
yangqin ***5**.497*
Yangshao ***3**.387*
Yangtze Kiang ***1**.370*
Yánluó **9**.38
Yanomamo ***7**.335*
Yao Wenyuan **4**.130–1
Yaoundé Convention ***4**.386*
Ya'qub (Jacob) **9**.19
yarn **6**.191, ***6**.390*
  texturing ***6**.390*
yaw (aircraft flight) **6**.136
yawl ***6**.390*
Yazdagird **9**.17
Yazid, caliph **9**.17
Yazoo stream ***1**.370*
Yeager, Charles **6**.138
Yeager, Jeana **6**.138
year ***8**.189*
  sidereal **8**.189, **9**.55
yeast ***2**.373*, **6**.127, **6**.147
Yeats, John Butler **5**.442, ***5**.497*
Yeats, William Butler **5**.325, ***5**.497*
Yefanov, J. **4**.327
Yehoshua, Abraham B. **5**.203
yellow (butterfly) ***2**.373*
yellow fever **2**.281, ***2**.373*
Yellow River **1**.158, **1**.159
Yellow Sea ***1**.370*
yellowhammer ***2**.373*
Yellowstone National Park **1**.138, **1**.151, ***1**.370*
yellowtail ***2**.373–4*
Yeltsin, Boris **4**.138, **4**.296, ***4**.386*, **7**.300
Yemen ***1**.370–1*, ***3**.387–8*, ***4**.321*, ***7**.390*
  in Arab League **7**.15
  physical geography and geology **1**.370–1
  religion and myths **7**.168, **7**.284, **9**.33
  Republic of **4**.386
  wars and defence **7**.52
Yenisei ***1**.371*
yeoman ***3**.388*
Yerkes, Charles Tyson: Yerkes Observatory ***8**.189*
Yesenin, Sergey ***5**.497*
Yevtushenko, Yevgeny ***5**.497*
yew ***2**.374*
Yggdrasil ***9**.38–9*
Yi Song-gye of Korea **3**.200
Yi Xing **6**.76
Yiddish language **7**.19, **7**.131, **7**.156, ***7**.335*
  literature ***5**.497*
Yijing ***7**.335*
Yima **9**.19, ***9**.39*
yin and yang **7**.46, ***7**.335*, ***9**.39*
ylang ylang ***2**.374*
Ymir **9**.6
yoga **7**.155, **7**.173, **7**.196, **7**.205, **7**.272, ***7**.335*
yoghurt **6**.94
Yoho National Park ***1**.371*
yoke **6**.165
Yom Kippur ***7**.335*
Yom Kippur War **4**.169, ***4**.386*, **6**.345, 7.69, **7**.224
Yongle, emperor of China **3**.350, ***3**.388*
York, Richard Plantagenet, 3rd duke of **3**.310, **3**.313, **3**.376, ***3**.388*
Yorkists **3**.284, **3**.310, ***3**.388*
  see also under rulers *under* Britain
Yorktown ***3**.388*
Yoruba ***4**.386*, **7**.316, ***7**.335*
  religion and myths **9**.11, **9**.15, **9**.17, **9**.28, **9**.32
Yorugu **9**.28
Yosemite National Park **1**.93, ***1**.371*
Yoshida Shigeru ***4**.386*
Yoshitsune **3**.231–2
Young, Brigham **4**.224, **4**.317, ***4**.386*
Young, Charles **8**.51
Young, Michael **7**.199
Young, Owen D.: Young Plan ***4**.387*
Young, Thomas (1773–1829, physicist) **1**.127, ***1**.371–2*
Young, Thomas (1820s, archaeologist) **3**.310
Young England ***4**.386–7*
Young Ireland ***4**.387*
Young Italy ***4**.387*
Young Turks ***4**.387*
Ypsilanti, Alexander **4**.140–1, ***4**.387*
ytterbium **1**.107
yttrium **1**.107, ***1**.372*, **6**.295
Yu Dafu **5**.95
Yuan dynasty **3**.72, **3**.201, ***3**.388–9*
Yuan Mei **5**.98
Yuan Shikai **4**.74, **4**.187, ***4**.387*
Yucatán ***1**.372*, **3**.227, ***3**.389*
yucca ***2**.374*
yucca moth ***2**.374*
yueqin ***5**.497*
Yugoslavia ***4**.387*, ***7**.390*
  economy and trade **7**.158, **7**.193, **7**.221
  languages **7**.10, **7**.269, **7**.288
  leaders **4**.8, **4**.262, **4**.345
  Little Entente **4**.197
  Non-Aligned Movement **7**.218
  partisans **4**.220
  politics and government **4**.262, **4**.345, **7**.64, **7**.99
  Rapallo Treaties **4**.283
  regions and republics *see* Bosnia-Herzegovina; Croatia; Macedonia (Yugoslavia); Montenegro; Serbia; Slovenia
  wars and defence **3**.201, **7**.52, **7**.202
  see also Balkan States
Yu-Hsi **8**.26
Yùhuáng ***9**.39*
Yukawa, Hideki **1**.258, ***1**.372*
Yukon River ***1**.372*
Yukon Territory ***1**.372*
Yunnan ***3**.389*
Yunus (Jonah) **9**.20
Yusuf **9**.19, **9**.21
  see also Joseph
Yusuf ibn Tashfin **3**.11

Z

Zacchaeus ***9**.39*
Zach, Nathan **5**.203
Zacharias ***9**.39*
Zadkine, Osip ***5**.498*
Zaghlul Pasha **4**.369
Zagros Mountains ***1**.372*
Zagwe ***3**.389*
Zaharoff, Sir Basil ***4**.388*
Zahhak **9**.19
zaibatsu ***4**.388*, ***7**.336*
Zaïre ***1**.372*, ***4**.388*
  decolonization **4**.4
  ethnic groups **7**.24, **7**.256
  kingdom **3**.4, **3**.213
  physical geography and geology **1**.331, **1**.372
  political leader **4**.221
  refugees **7**.261
  religion/theology **7**.5, **9**.20
  technology **6**.285
  see also Congo
Zaïre River (Congo River) **1**.72
Zakariya (Zacharias) **9**.39
Zakharov, Andreyan **5**.315
Zal ***9**.39*
Zambezi ***1**.372*
Zambia (*earlier* Northern Rhodesia) ***1**.372*, ***4**.388*, ***7**.391*
  aid to **7**.8
  colonization and decolonization **4**.4, **4**.306
  economy and trade **7**.296
  ethnic groups **7**.30
  physical geography and geology **1**.331, **1**.372
  political leader **4**.179, **4**.388
  prehistory **3**.50
  Tan–Zam railway **4**.337–8
  technology **6**.285
Zamenhof, Ludwig Lazarus **7**.108
zamindar ***4**.388*
Zamorro, Jaime Paz **4**.44, **4**.263
Zamyatin, Yevgeny ***5**.498*
Zamzam ***9**.39*
Zanahary **9**.39
zander **2**.253
Zanzibar ***1**.372*, **3**.236, ***3**.389*, **4**.299–300, ***4**.388–9*
  decolonization *see* Tanzania
Zàojūn ***9**.39*
Zapata, Emiliano **4**.219, **4**.368, ***4**.389*
Zapotec art ***5**.498*
Zarathustra ***3**.389*, **3**.391
  see also Zoroastrianism
Zaria ***3**.389*
Zarnov, Andrej **5**.422
al-Zarqalah, Ibn **8**.75
zarzuela **5**.329
Zauq **5**.466
Zealots ***3**.389*
Zeami Motokiyo, Kanze **5**.236
zebras **2**.273, ***2**.375*
Zeebrugge raid ***4**.389*
Zeeman, Pieter: Zeeman effect ***1**.373*, ***8**.189*
Zeiss, Carl **6**.1, **6**.310, ***6**.391*
Zelaya, José Santos **4**.241
Zel'dovich, Yakov Borisovich ***8**.189*
Zelenchukskaya telescope ***8**.189*
Zen Buddhism ***3**.389–90*, ***7**.336*, **9**.17, **9**.25
  Koan **7**.177
  monasticism **7**.205
Zend-Avesta **9**.10
Zeng Guofan **4**.308, **4**.337, ***4**.389*
zenith ***1**.373*, ***8**.189–90*

zenith (*cont.*):
distance **8**.22
tube **8**.119, ***8**.190*
Zeno of Citium **1**.48, **1**.247, **3**.5, **3**.341, ***3**.390*, **7**.300
Zeno of Elea ***3**.390*
Zenobia of Palmyra **3**.268
zeolite ***1**.373*, **6**.120, **6**.378
Zephyrus **9**.17
Zeppelin, count Ferdinand von: Zeppelins **6**.10, **6**.138, **6**.139, **6**.224, ***6**.391*
zero, absolute **1**.1
zero population growth ***7**.336*
Żeromski, Stefan ***5**.498*
zero-point energy ***1**.373*
Zeus ***9**.39*
offspring **9**.4, **9**.5, **9**.6, **9**.10, **9**.12, **9**.15, **9**.25, **9**.29
relationships and activities **9**.8, **9**.9, **9**.11, **9**.12, **9**.13, **9**.18, **9**.22, **9**.28, **9**.29, **9**.39
Zeuxis ***5**.498*
Zhang Chungqiao **4**.130–1
Zhang Heng **6**.311, **6**.312, ***6**.391*
Zhao Shuli **5**.96
Zheng, prince *see* Shi Huangdi
zheng ***5**.251*
Zheng He **3**.121, **3**.122, **3**.217, **3**.235, ***3**.390*
Zhivkov, Todor **4**.55
Zhou dynasty **3**.79, ***3**.390*
Zhou Enlai **4**.384, ***4**.389*
Zhoukoudian ***3**.390*
Zhu Da ***5**.498*
Zhu De **4**.173, **4**.210, ***4**.389*
Zhu Xi (Chu Hsi) **7**.65, **7**.305
Zhu Yuanzhang *see* Hong Wu
Zhu Zaiyu, prince **5**.96
Zhuangzi (Zhuang Zhou and *Zhuangzi* text) ***3**.390–1*, ***5**.498*, **7**.80, ***7**.336*
Zhukov, Georgi Konstantinovich **4**.188, **4**.328, ***4**.389*
Zia ul-Haq, General Mohammed **4**.30, ***4**.390*
zidovudine **6**.8
Ziegfeld, Florenz ***5**.498*
Ziegler, Karl **6**.45, **6**.236, ***6**.391*
ziggurat **6**.51, **6**.287
Zimbabwe (*earlier* Southern Rhodesia) ***1**.372*, ***3**.391*, ***4**.391*, ***7**.391*
colonization and decolonization **4**.4, **4**.306
cricket **7**.75
economy and trade **7**.70, **7**.296
ethnic groups **7**.214, **7**.285
famine prevention **7**.118
physical geography and geology **1**.356, **1**.372, **1**.373
politics and government:
independence and UDI **7**.59, **7**.273
leaders **4**.158, **4**.226, **4**.244
religion, new **7**.5
technology **6**.21
wars and defence **7**.136
Zimmermann, Alfred: Zimmermann note ***4**.390*
Zimmermann, Dominikus **5**.385, ***5**.498*
Zimmermann, Johann Baptist **5**.498
zinc **1**.107, ***1**.373*, ***6**.391*
alloys **6**.11, **6**.45, **6**.49, **6**.391
in battery **6**.32, **6**.374
electroplating **6**.120
galvanizing **6**.149
ore **6**.221, **6**.253, **6**.391
oxides **1**.373, **6**.391
production **6**.221, **6**.253
sherardizing **6**.316
zinc chloride **6**.391
zinc oxide ***1**.373*, **6**.391
zinc sulphate ***1**.373*, **6**.391
zinc sulphide ***1**.373*, **6**.253, **6**.391
'Zinjanthropus' ***3**.391*
zinnia ***2**.375*
Zinoviev, Grigori Yevseyevich **4**.178, ***4**.390*
Zion ***9**.39*
Zion National Park **1**.29, ***1**.373*
Zionism ***4**.390*, **7**.265, ***7**.336*
groups **4**.166–7, **4**.329
leader **4**.150, **4**.373
see also Jews
zip **6**.77, ***6**.391*
zircon **1**.80, ***1**.373–4*
zirconia **6**.297, ***6**.391*
zirconium **1**.107, ***1**.374*, **6**.11, **6**.12, **6**.323, ***6**.391*
oxides **6**.297, **6**.391
zirconium silicate **1**.373–4
Zita, Saint ***9**.43*
zither **5**.244, **5**.251, ***5**.498*
Žižka, Jan **3**.173
zodiac ***8**.190*
constellations of **8**.7, **8**.8, **8**.20, **8**.58, **8**.85, **8**.121, **8**.146
signs of ***8**.190*, ***9**.62*
zodiacal light ***8**.190*
Zoffany, Johann **3**.140, ***5**.499*
Zog, king of Albania **4**.8, ***4**.391*
Zola, Émile **4**.104, **5**.140, **5**.314, ***5**.499*
Zollverein ***4**.391*
zombi ***9**.39*
zonal soils ***1**.374*
listed **1**.311
zone (astronomy) **8**.78, **8**.145
of avoidance ***8**.190*
of totality **8**.117
zone refining **6**.155, ***6**.391*
zoo ***2**.375*, ***7**.336*
zoogeography **2**.30
zoology ***2**.375*
classification *see* nomenclature
zoology (*cont.*):
zoologists **2**.44, **2**.194, **2**.205
see also naturalists
see also animals; sociobiology
Zorilla, José **5**.428
zorilla **2**.262
Zoroastrianism **3**.391
Avesta **7**.23
first man **9**.14
founder **3**.389, **3**.391
god **9**.25
Parsis **7**.23, **7**.229
resurrection **7**.266
sacred book **9**.3
and Zurvan **9**.39
Zoser **9**.18
Zoutpansberg ***1**.374*
Zrínyi, Miklós **5**.214
Zu **9**.24
Zuazo, Herman **4**.263
Zuccaro, Taddeo and Federico ***5**.499*
Zuiderzee ***1**.374*
zukra **5**.212
Zulaikha **9**.21
Zulus and Zululand ***4**.391*, ***7**.336*
Inkatha Movement **4**.31, ***4**.391*, **7**.336
kings **4**.68, **4**.98, **4**.310
religion and myths **9**.36
wars **4**.42, **4**.168, ***4**.391*
zummāra **5**.376, ***5**.499*
Zuo Zongtang **4**.308, ***4**.391*
Zurbarán, Francisco ***5**.499*
Zurich relative sunspot number **8**.165
Zurvan ***9**.39*
Zuse, Konrad **6**.84, **6**.86
Zweig, Stefan ***5**.499*
Zwide **4**.99
Zwingli, Ulrich **3**.293, **3**.300, **3**.360, ***3**.391*
zwitterion ***1**.374*
Zworykin, Vladimir Kosma **6**.175, **6**.349, ***6**.391*

Countries of the world

Senegal, Somalia, Sweden, Togo, Turks and Caicos Islands, Uruguay, Wales

Seychelles, South Africa, Switzerland, Tonga, Tuvalu, Uzbekistan, Western Samoa

Sierra Leone, Spain, Syria, Transkei, Uganda, Vanuatu, Yemen

Singapore, Sri Lanka, Tajikistan, Trinidad/Tobago, Ukraine, Vatican City, Yugoslavia

Slovakia, Sudan, Taiwan, Tunisia, UAE, Venda, Zaire

Slovenia, Suriname, Tanzania, Turkey, United Kingdom, Venezuela, Zambia

Solomon Islands, Swaziland, Thailand, Turkmenistan, USA, Vietnam, Zimbabwe

International organizations

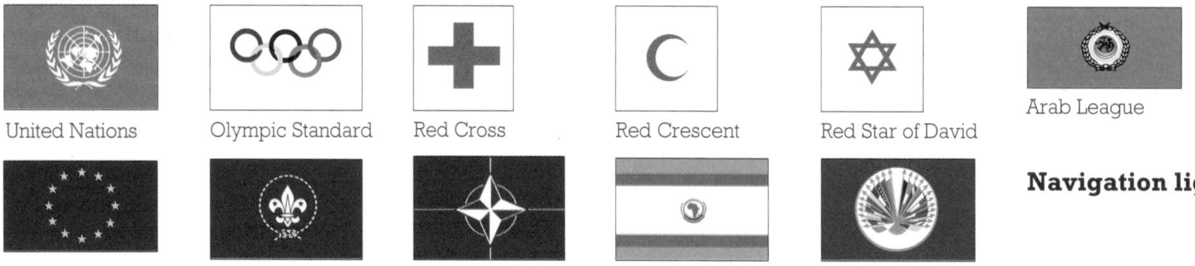

United Nations, Olympic Standard, Red Cross, Red Crescent, Red Star of David, Arab League

European Community, Scouts and Guides, NATO, Organization of African Unity, Organization of American States

Flags produced by Lovell Johns Ltd., Oxford, UK, and authenticated by the Flag Research Center, Winchester, MA 01890, USA

Navigation lights

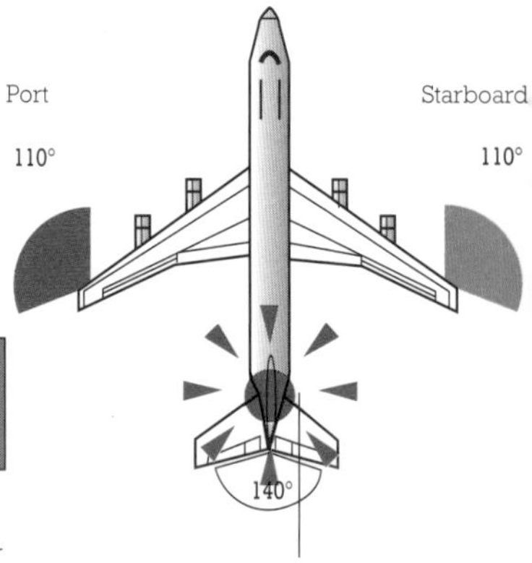

Runway markings

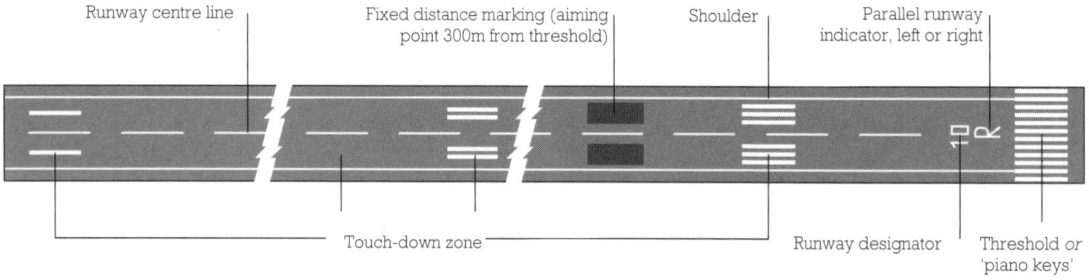